PTC C ametric 3.0

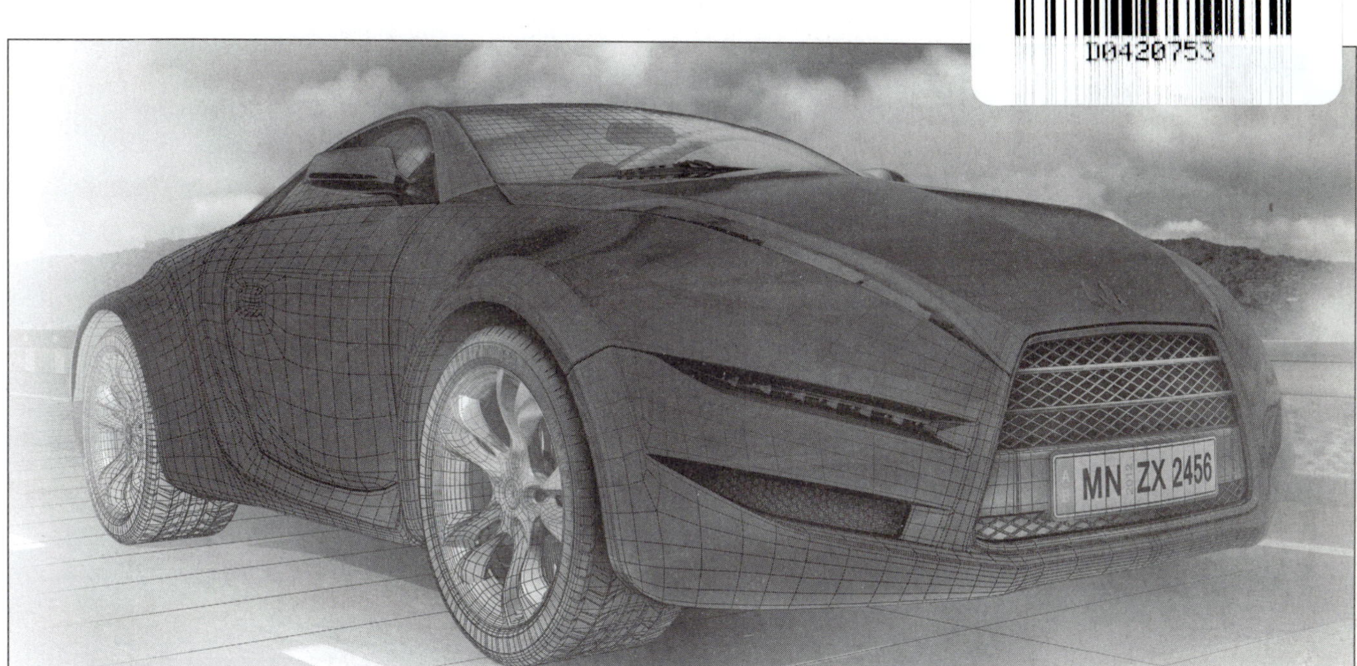

Louis Gary Lamit

De Anza College

Australia • Brazil • Japan • Korea • Mexico • Singapore • Spain • United Kingdom • United States

PTC Creo™ Parametric 3.0
Louis Gary Lamit

Publisher, Global Engineering: Timothy L. Anderson

Senior Development Editor: Mona Zeftel

Senior Content Project Manager: Jennifer Ziegler

Senior Art Director: Michelle Kunkler

Media Assistant: Ashley Kaupert

Production Management and Composition: S4Carlisle

IP Analyst: Christine Myaskovsky

IP Project Manager: Sarah Shainwald

Senior Manufacturing Planner: Doug Wilke

Marketing Manager: Kristin Stine

Team Assistant: Sam Roth

Cover Designer: c miller design

Cover Images: © Mikhail/Shutterstock; © Vitalii Culibacinii/iStockphoto; © penfold/iStockphoto

ISBN-13: 978-1-305-25318-6

Cengage Learning
20 Channel Center Street
Boston, MA 02210
USA

Cengage Learning is a leading provider of customized learning solutions with office locations around the globe, including Singapore, the United Kingdom, Australia, Mexico, Brazil, and Japan. Locate your local office at: **www.cengage.com/global**

Cengage Learning products are represented in Canada by Nelson Education, Ltd.

To learn more about Cengage Learning Solutions, visit **www.cengage.com**

Purchase any of our products at your local college store or at our preferred online store **www.cengagebrain.com**

Printed in the United States of America
Print Number: 01 Print Year: 2014

Table of Contents

Downloadable Lessons

Lesson 13 Patterns
- *Weldment*
- *Feature pattern*
- *Hole pattern*
- *Direction pattern*
- *Dimension pattern*
- *Component Pattern*
- *Fill Pattern*
- *3DModelSpace > Pro/Library*
- *Weldment drawings (assembly and part detail)*

Lesson 14 Blends
- *Blend Sections*
- *Blend Options*
- *Parallel Blends*
- *Blend Tool*
- *Sketch Options*
- *Polar Grid*
- *Axial Pattern*
- *Shell Tool*
- *Analysis Measure*
- *Section*
- *Enhanced Realism*
- *Scenes*
- *Drawing and Model Annotations*

Lesson 15 Sweeps
- *Sweep Forms*
- *Sweep Options*
- *Sweep Tool*
- *Trajectory*
- *Sweep Section*
- *Polar Grid*
- *Enhanced Realism*
- *Rendering Scene*
- *Spot Lights*
- *Drawings and Model Annotations*

Lesson 16 Helical Sweeps and Annotations
- *Helical Sweeps*
- *Annotations*
- *Helical Compression Spring*
- *Helical Sweep Tool*
- *Notes/Annotations*
- *URL Links*
- *Springs*
- *Annotation Features*
- *Digital Product Definition Data Practices ASME Y14.41*
- *Driving Dimensions*
- *Annotation Orientation*
- *Geometric Tolerance*
- *Annotation Feature*
- *Surface Finish*
- *Active Annotation Orientation*

Lesson 17 Shell, Reorder, and Insert Mode
- *Creating Shells*
- *Reordering Features*
- *Inserting Features*
- *Draft Tool*
- *Context Sensitive Help*
- *Shell Tool*
- *Pattern*
- *Pattern Table*
- *Reorder*
- *Insert Mode*
- *Model Setup*
- *Scenes*
- *Color Editor*
- *Lights*
- *Spot Light*
- *Distance Light*
- *Sky Light*
- *3D PDF*
- *Adobe Reader for PDF U3D*
- *Perspective View on PDF*
- *Drawing Views*
- *Model Annotations*
- *Drawing Tree*
- *ECO*

Lesson 18 Drafts, Suppress, and Text Extrusions
- *Drafts*
- *Suppressing and Resuming Features*
- *Text Extrusions*
- *Draft Tool*
- *Shell Tool (non-default thickness)*
- *Group*
- *Mirror*
- *Surface Round*
- *Section*
- *Suppress*
- *Extrude Tool (Text)*
- *Resume*
- *ModelCHECK Geometry Check*
- *Render Setup*
- *Spotlight*
- *Scenes*
- *Color Editor*
- *Moving Lights*
- *Focus*
- *Distance Light*
- *Rendering Rooms*
- *Drawing Views*
- *Flexible modeling*

Lesson 19 Gear Assembly *(Video Lesson)*
Lesson 20 Valve Assembly *(Video Lesson)*
Lesson 21 Pulley Assembly *(Video Lesson)*
Lesson 22 Coupling Assembly *(Video Lesson)*
- *Shaft*
- *Taper Coupling*
- *Straight Coupling*
- *Coupling Assembly*
- *Coupling Assembly Drawing*

About the Author

Louis Gary Lamit is currently retired but still teaches part time (Creo Parametric) at De Anza College (since 1984) in Cupertino, CA. He is the founder of Scholarships for Veterans at www.scholarshipsforveterans.org. Mr. Lamit has worked as a drafter (1966), designer, numerical control (NC) programmer, technical illustrator, and engineer in the automotive, aircraft, and piping industries. A majority of his work experience is in the area of mechanical and piping design. He started as a drafter in Detroit (as a job shopper) in the automobile industry, doing tooling, dies, jigs and fixture layout, and detailing at Koltanbar Engineering, Tool Engineering, Time Engineering, and Premier Engineering for Chrysler, Ford, AMC, and Fisher Body. Mr. Lamit has worked at Remington Arms and Pratt & Whitney Aircraft as a designer, and at Boeing Aircraft and Kollmorgan Optics as an NC programmer and aircraft engineer. He also owns and operates his own consulting firm (CAD-Resources.com-Lamit and Associates), and has been involved with advertising, and patent illustration. He is the author of over 40 books, journals, textbooks, workbooks, tutorials, and handbooks, including children's journals and books (www.walkingfishbooks.com). Mr. Lamit received a BS degree from Western Michigan University in 1970 and did Masters' work at Wayne State University and Michigan State University. He has also done graduate work at the University of California at Berkeley and holds an NC programming certificate from Boeing Aircraft. Since leaving industry, Mr. Lamit has taught at all levels (Melby Junior High School, Warren, Mi.; Carroll County Vocational Technical School, Carrollton, Ga.; Heald Engineering College, San Francisco, Ca.; Cogswell Polytechnical College, San Francisco and Cupertino, Ca.; Mission College, Santa Clara, Ca.; Santa Rosa Junior College, Santa Rosa, Ca.; Northern Kentucky University, Highland Heights, Ky.; and De Anza College, Cupertino, Ca.). He also continues to provide week-long industry trainings for Creo Parametric and Creo Updates. His textbooks include:

- *Industrial Model Building*, with Engineering Model Associates, Inc. (1981),
- *Piping Drafting and Design* (1981),
- *Piping Drafting and Design Workbook* (1981),
- *Descriptive Geometry* (1983),
- *Descriptive Geometry Workbook* (1983), and
- *Pipe Fitting and Piping Handbook* (1984), Prentice-Hall.
- *Drafting for Electronics* (3rd edition, 1998),
- *Drafting for Electronics Workbook* (2nd edition 1992), and
- *CADD* (1987), Charles Merrill (Macmillan-Prentice-Hall Publishing).
- *Technical Drawing and Design* (1994),
- *Technical Drawing and Design Worksheets and Problem Sheets* (1994),
- *Principles of Engineering Drawing* (1994),
- *Fundamentals of Engineering Graphics and Design* (1997),
- *Engineering Graphics and Design with Graphical Analysis* (1997), and
- *Engineering Graphics and Design Worksheets and Problem Sheets* (1997), West Publishing (ITP/Delmar).
- *Basic Pro/ENGINEER in 20 Lessons* (1998) (Revision 18) and
- *Basic Pro/ENGINEER (with references to PT/Modeler)* (1999), PWS.
- *Pro/ENGINEER 2000i* (1999), and
- *Pro/ENGINEER 2000i² (Pro/NC and Pro/SHEETMETAL)* (2000), Brooks/Cole Publishing (ITP).
- *Pro/ENGINEER Wildfire* (2003), Brooks/Cole Publishing (ITP).
- *Introduction to Pro/ENGINEER Wildfire 3.0* (2004), SDC.
- *Moving from 2D to 3D CAD for Engineering Design* (2007), BookSurge, eBook by MobiPocket.
- *Pro/ENGINEER Wildfire 3.0 Tutorial* (2007), BookSurge, eBook by MobiPocket.
- *Pro/ENGINEER Wildfire 3.0* (2007), Cengage.
- *Pro/ENGINEER Wildfire 4.0 Tutorial* (2008), BookSurge eBook by MobiPocket.
- *Pro/ENGINEER Wildfire 4.0* (2008), Cengage.
- *Pro/ENGINEER Wildfire 5.0* (2010), Cengage.
- *Creo Parametric (2012),* Cengage.
- *PTC Creo Parametric 3.0 (2015),* Cengage.

Mr. Lamit also owns and writes for WalkingFish Books. See **www.walkingfishbooks.com**:

- *Fishing Journal, My Fishing Journal, Resources for Fishing and Boating Santa Clara County.*
- *Golfing Journal, My Golfing Journal, Golf Journal and Sketch Pad, Golfing Scorecard and Journal.*
- *Wally the WalkingFish Coloring Book.*
- *Wally the WalkingFish meets Madison and Cooper, Wally the WalkingFish meets Aspen and Cooper*
- *The Essential Home Bar*
- *Garden Journal*
- *My Bird Watching Journal*

Dedication

This book is dedicated to my granddaughter: **Juniper "Suvi" Pace Lamit**

gate gate pāragate pārasagate bodhi svāhā

Preface

PTC Creo Parametric 3.0 is one of the most widely used CAD/CAM software programs in the world today. Any aspiring engineer will greatly benefit from the knowledge contained herein, while in school or upon graduation as a newly employed engineer.

Significant changes, upgrades, and new capabilities including have made PTC Creo Parametric 3.0 a unique product. *This is not a revised textbook* but a new book covering all the necessary subjects needed to master this high-level CAD software. There are few if any comprehensive texts on this subject so we hope this text will fill the needs of both schools and professionals alike.

The text involves creating a new part, an assembly, or a drawing, using a set of commands that walk you through the process systematically. Lessons and Projects all come from industry and have been tested for accuracy and correctness as per engineering standards. Projects are downloadable as a PDF with live links and 3D embedded models.

Resources

Lessons 13-18 (downloadable PDF)**, Video Lessons 19-22,** and **Lesson Projects** (downloadable PDF) are not included in the printed version to keep the length and cost to the user down- lessons and projects not in the printed portion can be downloaded at: **www.cengage.com and navigate to the Authors Resource page for PTC Creo Parametric 3.0.**

A complete set of **Recorded Lectures** for this book is available for *free* at **The Cengage Author Resource page.** The Lectures are in WMV format and run between 15 and 40 minutes for each lesson/project. The Lectures are recorded using a commercial version of PTC Creo Parametric 3.0 and are the exact content presented in the classroom by the author. Lectures can be downloaded at: **www.cengage.com and navigate to the Free author's resource page for PTC Creo Parametric 3.0.**

PTC Creo Parametric 3.0 Free Schools Edition software is available at www.ptc.com.

Contact

If you wish to contact the author concerning orders, questions, changes, additions, suggestions, comments, or to get on our email list, please send an email to one of the following:

Web Site: **www.cad-resources.com**
Email: **lgl@cad-resources.com**

Introduction

This text guides you through parametric design using PTC Creo Parametric 3.0™. While using this text, you will create individual parts, assemblies, and drawings.

Parametric can be defined as *any set of physical properties whose values determine the characteristics or behavior of an object.* **Parametric design** enables you to generate a variety of information about your design: its mass properties, a drawing, or a base model. To get this information, you must first model your part design.

Parametric modeling philosophies used in PTC Creo Parametric 3.0 include the following (Fig. 1):

Feature-Based Modeling Parametric design represents solid models as combinations of engineering features.

Creation of Assemblies Just as features are combined into parts, parts may be combined into assemblies.

Capturing Design Intent The ability to incorporate engineering knowledge successfully into the solid model is an essential aspect of parametric modeling.

Figure 1 Parts and Assembly Design

Parametric Design

Parametric design models are not drawn so much as they are *sculpted* from solid volumes of materials. To begin the design process, analyze your design. Before any work is started, take the time to ***tap*** into your own knowledge bank and others that are available. **T**hink, **A**nalyze, and **P**lan. These three steps are essential to any well-formulated engineering design process.

Break down your overall design into its basic components, building blocks, or primary features. Identify the most fundamental feature of the object to sketch as the first, or base, feature. Varieties of **base features** can be modeled using extrude, revolve, sweep, and blend tools.

Sketched features (*extrusions, sweeps, etc.)* and pick-and-place features called **referenced features** (*holes, rounds, chamfers, etc.*) are normally required to complete the design. With the SKETCHER, you use familiar 2D entities (points, lines, rectangles, circles, arcs, splines, and conics) (Fig. 2). There is no need to be concerned with the accuracy of the sketch. Lines can be at differing angles, arcs and circles can have unequal radii, and features can be sketched with no regard for the actual objects' dimensions. In fact, exaggerating the difference between entities that are similar but not exactly the same is actually a far better practice when using the SKETCHER.

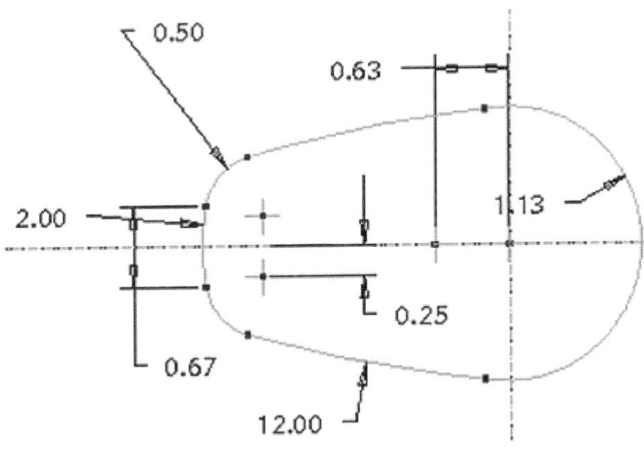

Figure 2 Sketching

Geometry assumptions and constraints will close ends of connected lines, align parallel lines, and snap sketched lines to horizontal and vertical (orthogonal) orientations. Additional constraints are added by means of **parametric dimensions** to control the size and shape of the sketch.

Figure 3 Feature Design

Features are the basic building blocks you use to create an object (Fig. 3). Features "understand" their fit and function as though "smarts" were built into the features themselves. For example, a hole or cut feature "knows" its shape and location and the fact that it has a negative volume. As you modify a feature, the entire object automatically updates after regeneration. The idea behind feature-based modeling is that the designer constructs an object so that it is composed of individual features that describe the way the geometry is supposed to behave if its dimensions change. This happens quite often in industry, as in the case of a design change. Feature-based modeling is diagramed in Figure 4.

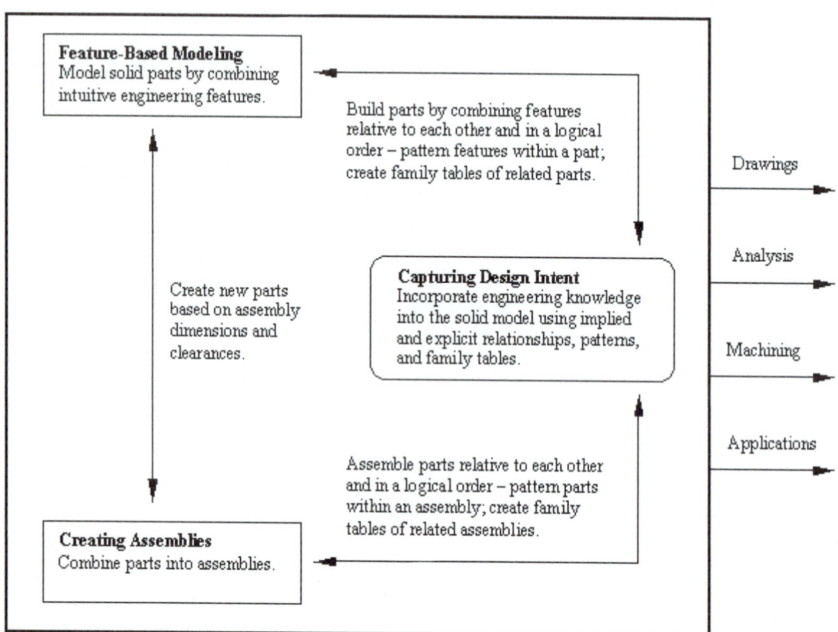

Figure 4 Feature-based Modeling

Parametric modeling is the term used to describe the capturing of design operations as they take place, as well as future modifications and editing of the design. The order of the design operations is significant. Suppose a designer specifies that two surfaces be parallel, such that surface two is parallel to surface one. Therefore, if surface one moves, surface two moves along with surface one to maintain the specified design relationship. The surface two is a **child** of surface one in this example. Parametric modeling software allows the designer to **reorder** the steps in the object's creation.

Various types of features are used as building blocks in the progressive creation of solid objects. Figures 5(a-c) illustrates base features, datum features, sketched features, and referenced features. The "chunks" of solid material from which parametric design models are constructed are called **features**.

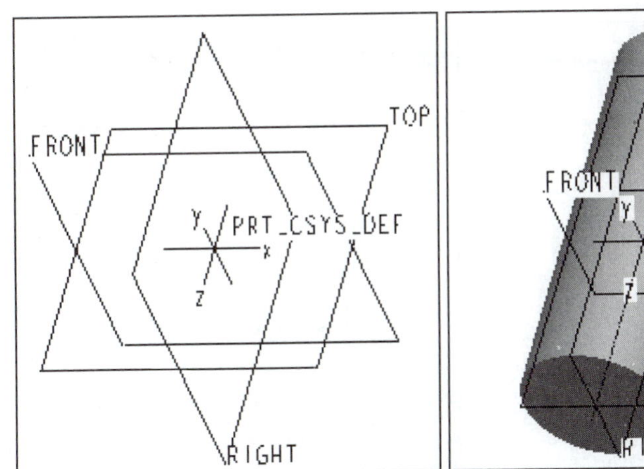
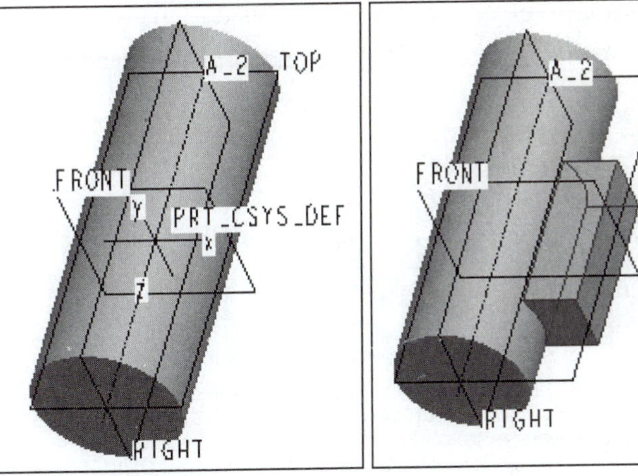

Figures 5(a-c) Features

Features generally fall into one of the following categories:

Base Feature The base feature is normally a set of datum planes referencing the default coordinate system. The base feature is important because all future model geometry will reference this feature directly or indirectly; it becomes the root feature. Changes to the base feature will affect the geometry of the entire model [Figs. 6(a-c)].

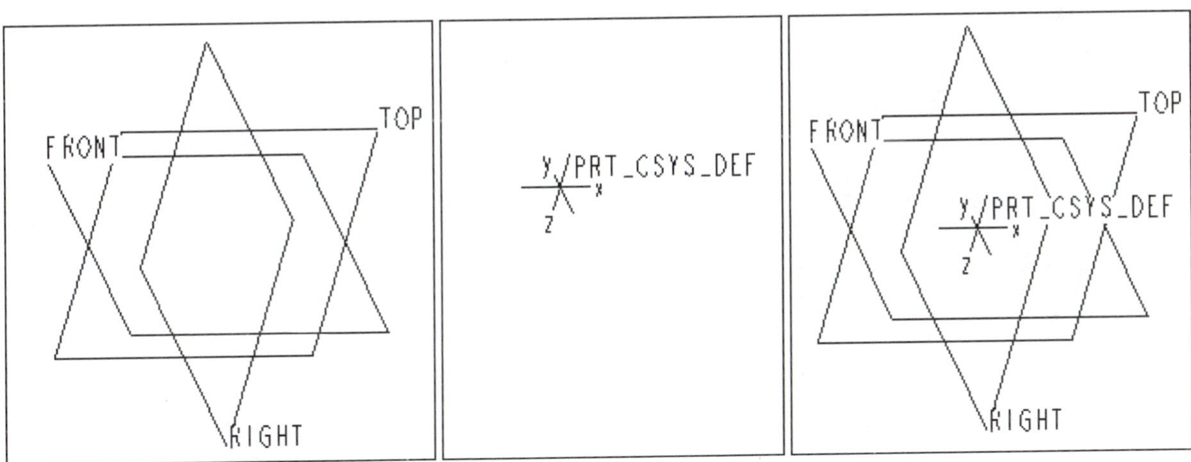

Figures 6(a-c) Base Features

Datum Features Datum features (lines, axes, curves, and points) are generally used to provide sketching planes and contour references for sketched and referenced features. Datum features do not have volume or mass and may be visually hidden without affecting solid geometry (Fig. 7).

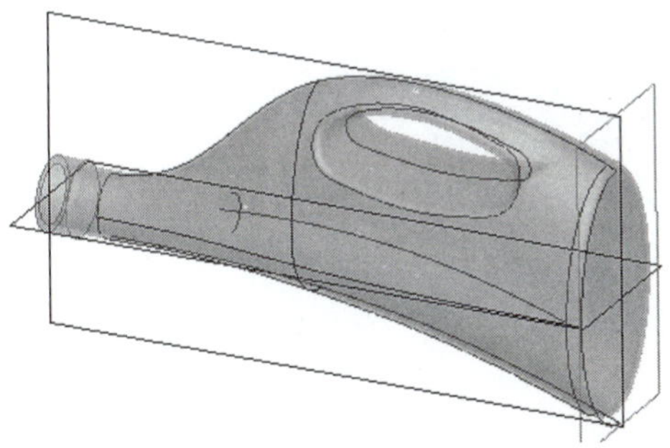
Figure 7 Datum Curves

Sketched Features Sketched features are created by extruding, revolving, blending, or sweeping a sketched cross section. Material may be added or removed by protruding or cutting the feature from the existing model (Fig. 8).

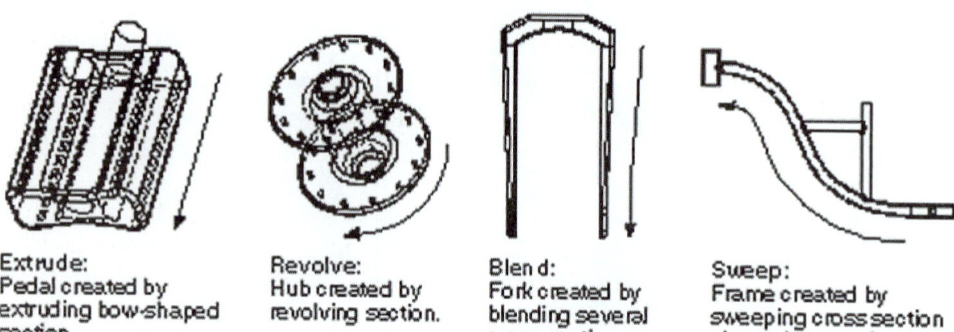

Extrude:
Pedal created by extruding bow-shaped section.
Revolve:
Hub created by revolving section.
Blend:
Fork created by blending several cross sections.
Sweep:
Frame created by sweeping cross section along shown trajectory.

Figure 8 Sketched Features

Referenced Features Referenced features (rounds, holes, shells, and so on) utilize existing geometry for positioning and employ an inherent form; they do not need to be sketched [Figs. 9(a-b)].

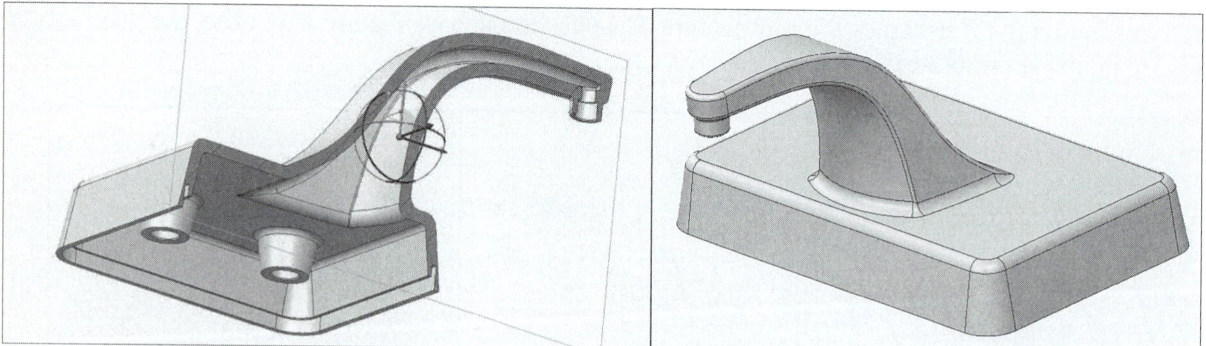

Figures 9(a-b) Referenced Features- Shell and Round *(Spout is a Swept Blend Feature)*

A wide variety of features are available. These tools enable the designer to make far fewer changes by capturing the engineer's design intent early in the development stage (Fig. 10).

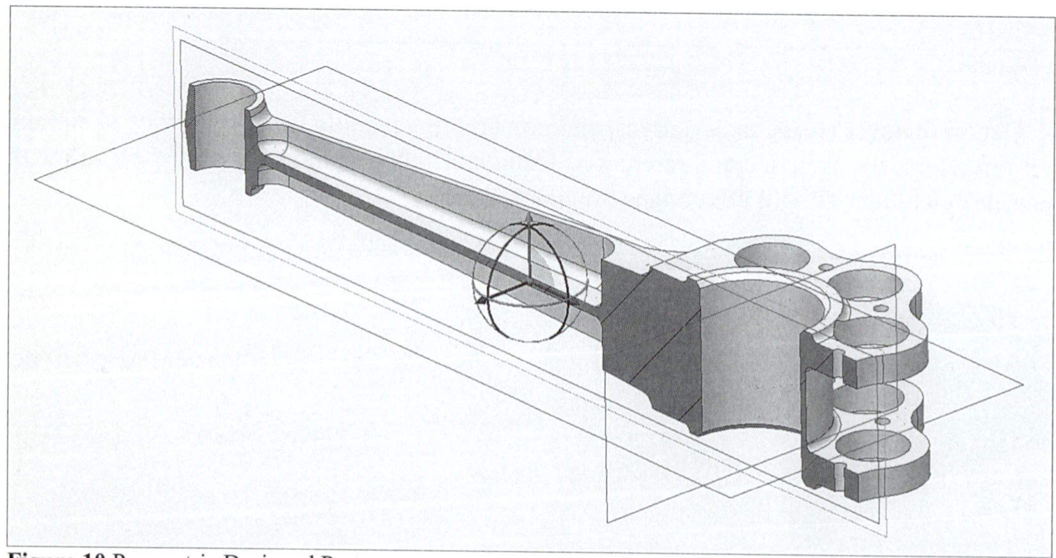

Figure 10 Parametric Designed Part

Fundamentals

The design of parts and assemblies, and the creation of related drawings, forms the foundation of engineering graphics. When designing with Creo Parametric, many of the previous steps in the design process have been eliminated, streamlined, altered, refined, or expanded. The model you create as a part forms the basis for all engineering and design functions.

The part model contains the geometric data describing the part's features, but it also includes non-graphical information embedded in the design itself. The part, its associated assembly, and the graphical documentation (drawings) are parametric. The physical properties described in the part drive (determine) the characteristics and behavior of the assembly and drawing. Any data established in the assembly mode, in turn, determines that aspect of the part and, subsequently, the drawings of the part and the assembly. In other words, all the information contained in the part, the assembly, and the drawing is interrelated, interconnected, and parametric (Fig. 11).

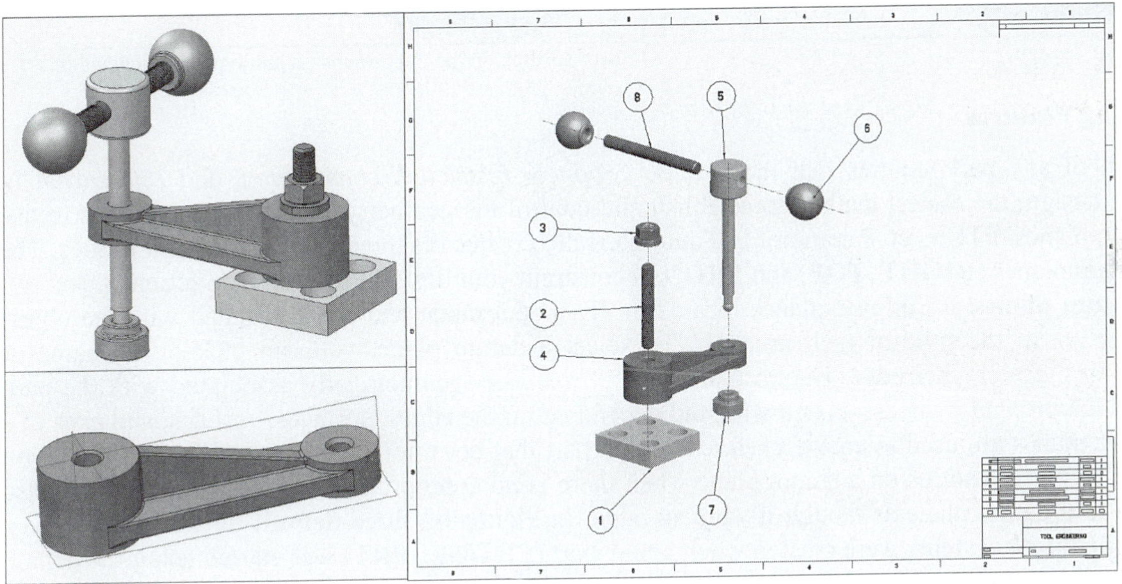

Figure 11 Assembly Drawing, Part, and Assembly

Part Design

In many cases, the part will be the first component of this interconnected process. The *part* function in PTC Creo Parametric 3.0 is used to design components.

During part design (Fig. 12), you can accomplish the following:

- Define the base feature
- Define and redefine construction features to the base feature
- Modify the dimensional values of part features (Fig. 13)
- Embed design intent into the model using tolerance specifications and dimensioning schemes
- Create pictorial and shaded views of the component
- Create part families (family tables)
- Perform mass properties analysis and clearance checks
- List part, feature, layer, and other model information
- Measure and calculate model features
- Create detail drawings of the part

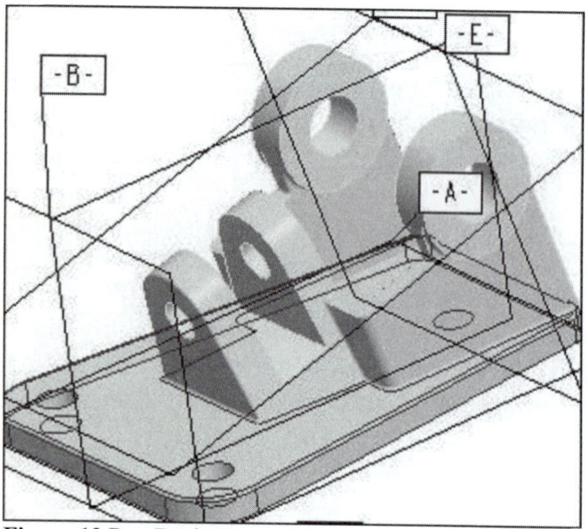

Figure 12 Part Design

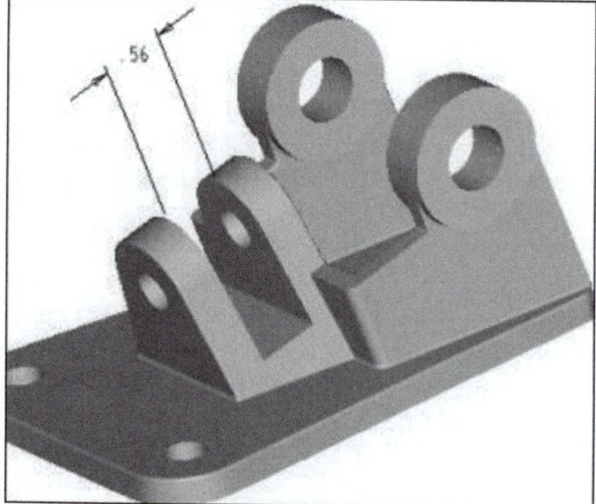

Figure 13 Pick on the **.56** dimension and type in a new value

Establishing Features

The design of any part requires that the part be *confined*, *restricted*, *constrained*, and *referenced.* In parametric design, the easiest method to establish and control the geometry of your part design is to use three datum planes. PTC Creo Parametric 3.0 automatically creates the three **primary datum planes**. The default datum planes (**RIGHT**, **TOP**, and **FRONT**) constrain your design in all three directions.

Datum planes are infinite planes located in 3D model mode and are associated with the object that was active at the time of their creation. To select a datum plane, you can pick on its name or anywhere on the perimeter edge. Datum planes are *parametric*--geometrically associated with the part. Parametric datum planes are associated with and dependent on the edges, surfaces, vertices, and axes of a part. Datum planes are used to create a reference on a part that does not already exist. For example, you can sketch or place features on a datum plane when there is no appropriate planar surface. You can also dimension to a datum plane as though it were an edge. In Figure 14, three **default datum planes** and a **default coordinate systems** were created when a new part (**PRT0001.PRT**) was started using the default template. Note that in the **Model Tree** window, they are the first four features of the part, which means that they will be the *parents* of the features that follow.

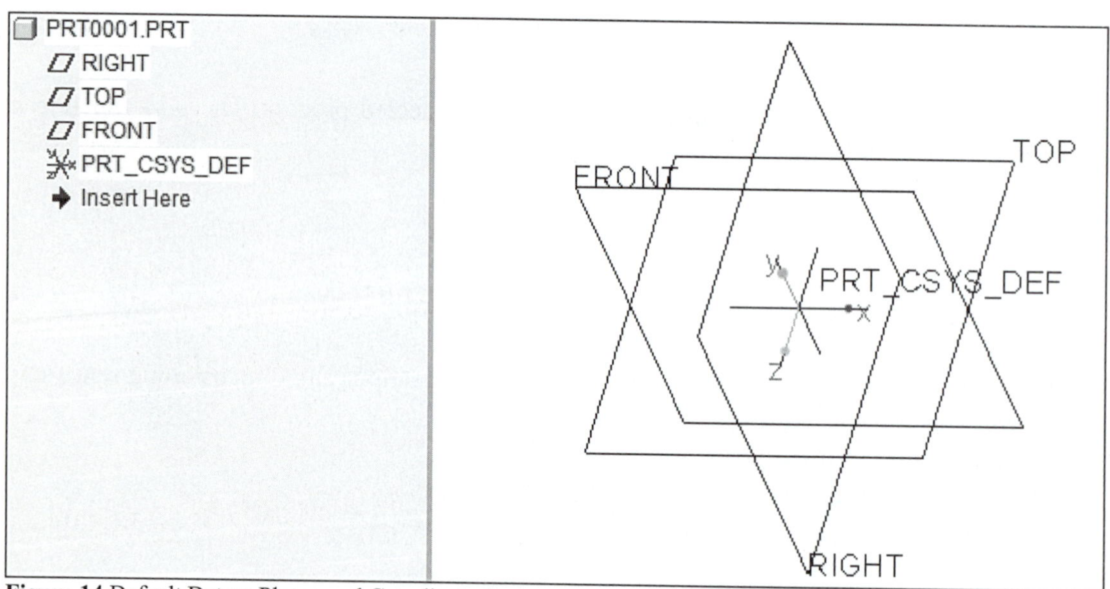

Figure 14 Default Datum Planes and Coordinate System

Datum Features

Datum features are planes, axes, and points you use to place geometric features on the active part. Datums other than defaults can be created at any time during the design process.

As we have discussed, there are three (primary) types of datum features: **datum planes** (Fig. 14), **datum axes**, and **datum points** (there are also ***datum curves*** and ***datum coordinate systems***). You can display all types of datum features, but they do not define the surfaces or edges of the part or add to its mass properties. In Figure 15, a variety of datum planes are used in the creation of the casting.

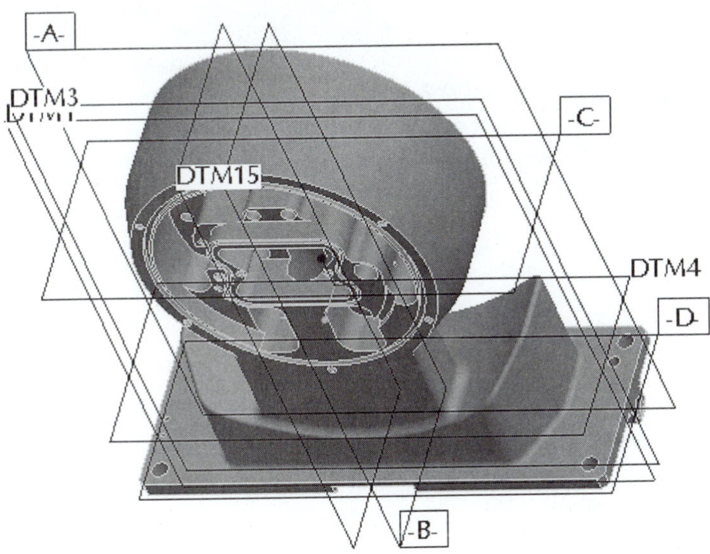

Figure 15 Datums in Part Design

Specifying constraints that locate it with respect to existing geometry creates a datum. For example, a datum plane might be made to pass through the axis of a hole and parallel to a planar surface. Chosen constraints must locate the datum plane relative to the model without ambiguity. You can also use and create datums in assembly mode.

Besides datum planes, datum axes and datum points can be created to assist in the design process. You can also automatically create datum axes through cylindrical features such as holes and solid round features by setting this as a default in your Creo Parametric configuration file. The part in Figure 16 shows the default axes for a variety of holes.

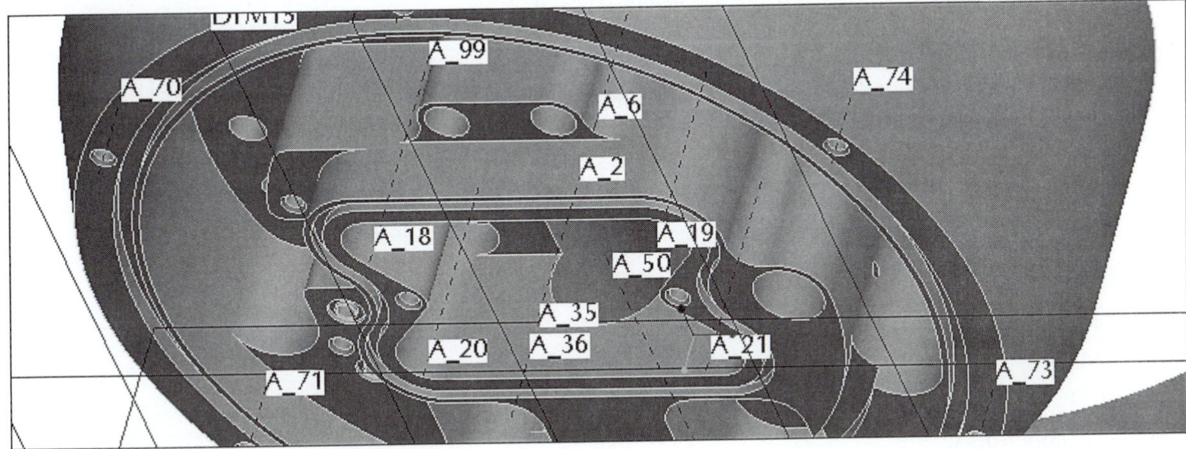

Figure 16 Feature Default Datum Axes

Parent-Child Relationships

Because solid modeling is a cumulative process, certain features must, by necessity precede others. Those that follow must rely on previously defined features for dimensional and geometric references. The relationships between features and those that reference them are termed ***parent-child relationships***. Because children reference parents, parent features can exist without children, but children cannot exist without their parents. This type of CAD modeler is called a history-based system. Using Creo Parametric's information command will list a model's parent-child references and dependencies (Fig. 17).

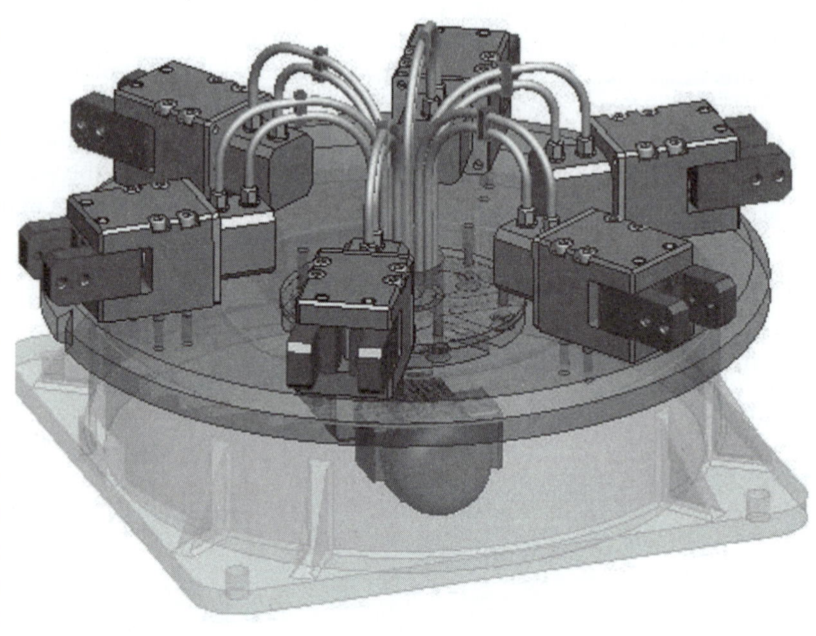

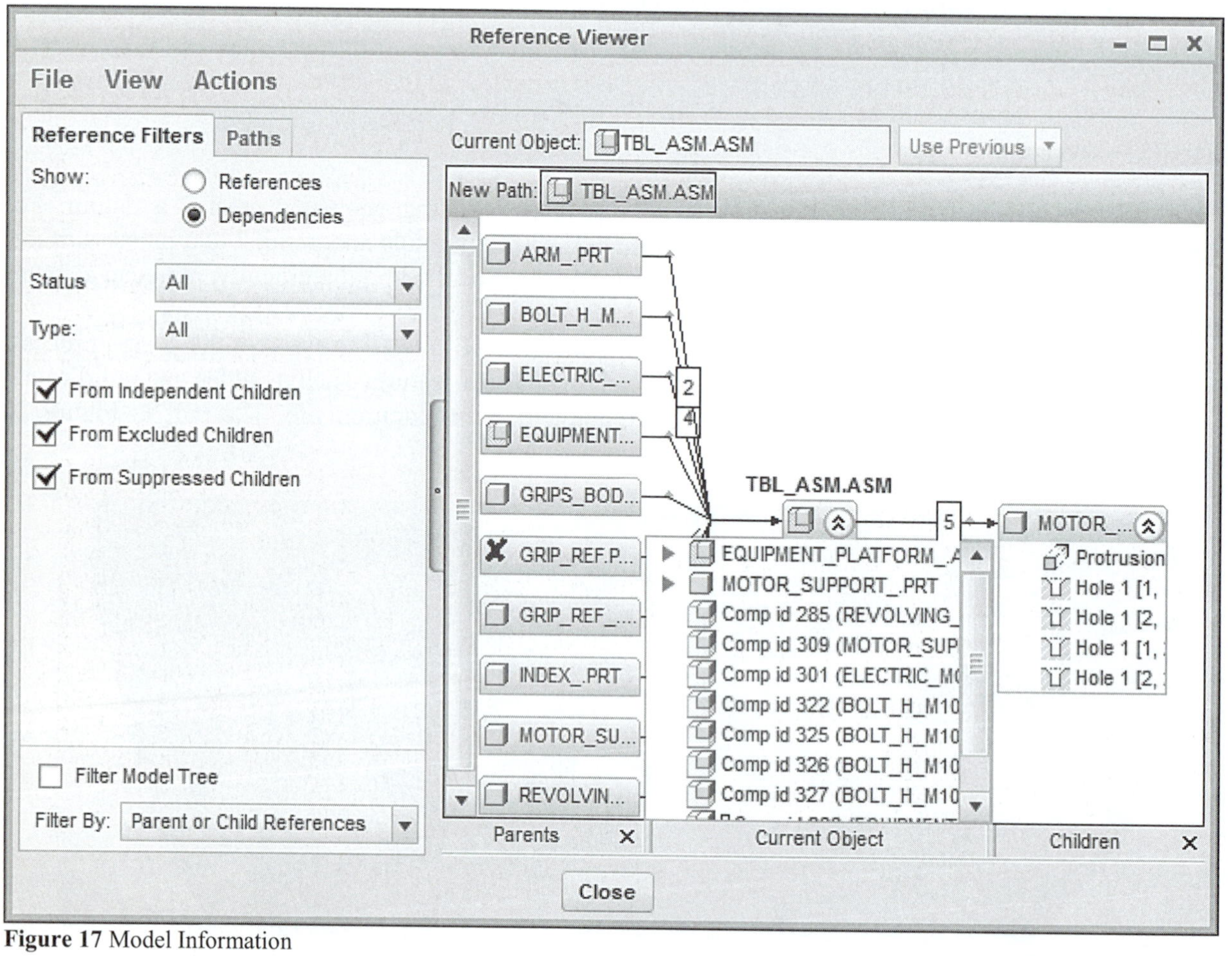

Figure 17 Model Information

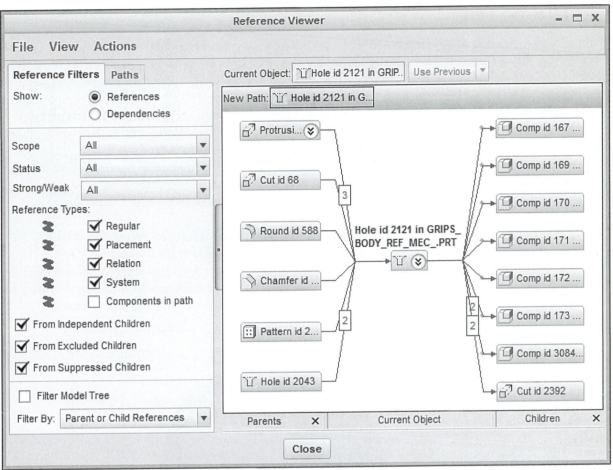

Figure 18 Parent-Child Information

The parent-child relationship (Fig. 18) is one of the most powerful aspects of parametric design. When a parent feature is modified, its children are automatically recreated to reflect the changes in the parent feature's geometry. It is essential to reference feature dimensions so that design modifications are correctly propagated through the model/part. Any modification to the part is automatically propagated throughout the model (Fig. 19) and will affect all children of the modified feature.

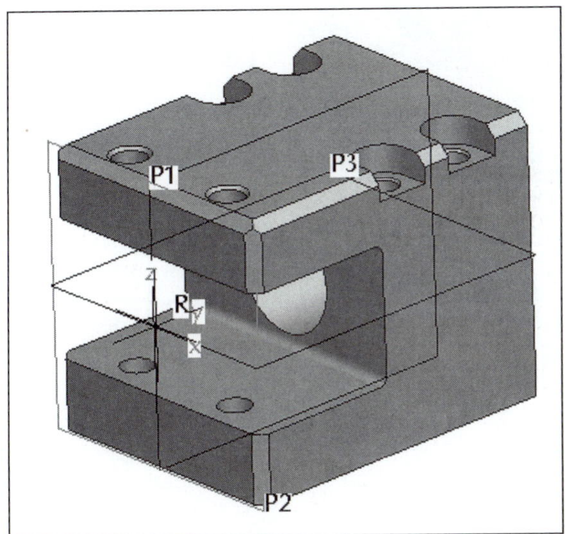

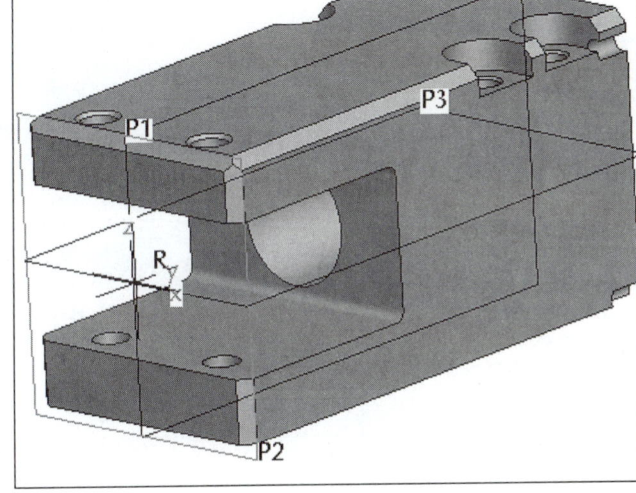

Figure 19 Original Design and Modification

Capturing Design Intent

A valuable characteristic of any design tool is its ability to *render* the design and at the same time capture its *intent* (Fig. 20). Parametric methods depend on the sequence of operations used to construct the design. The software maintains a *history of changes* the designer makes to specific parameters. The point of capturing this history is to keep track of operations that depend on each other. Whenever Creo Parametric is told to change a specific dimension, it can update all operations that are referenced to that dimension.

For example, a circle representing a bolt hole circle may be constructed so that it is always concentric to a circular slot. If the slot moves, so does the bolt circle. Parameters are usually displayed in terms of dimensions or labels and serve as the mechanism by which geometry is changed. The designer can change parameters manually by changing a dimension or can reference them to a variable in an equation (**relation**) that is solved either by the modeling program itself or by external programs such as spreadsheets.

Features can also store non-graphical information. This information can be used in activities such as drafting, numerical control (NC), finite-element analysis (FEA), and kinematics analysis.

Capturing design intent is based on incorporating engineering knowledge into a model by establishing and preserving certain geometric relationships. The wall thickness of a pressure vessel, for example, should be proportional to its surface area and should remain so, even as its size changes.

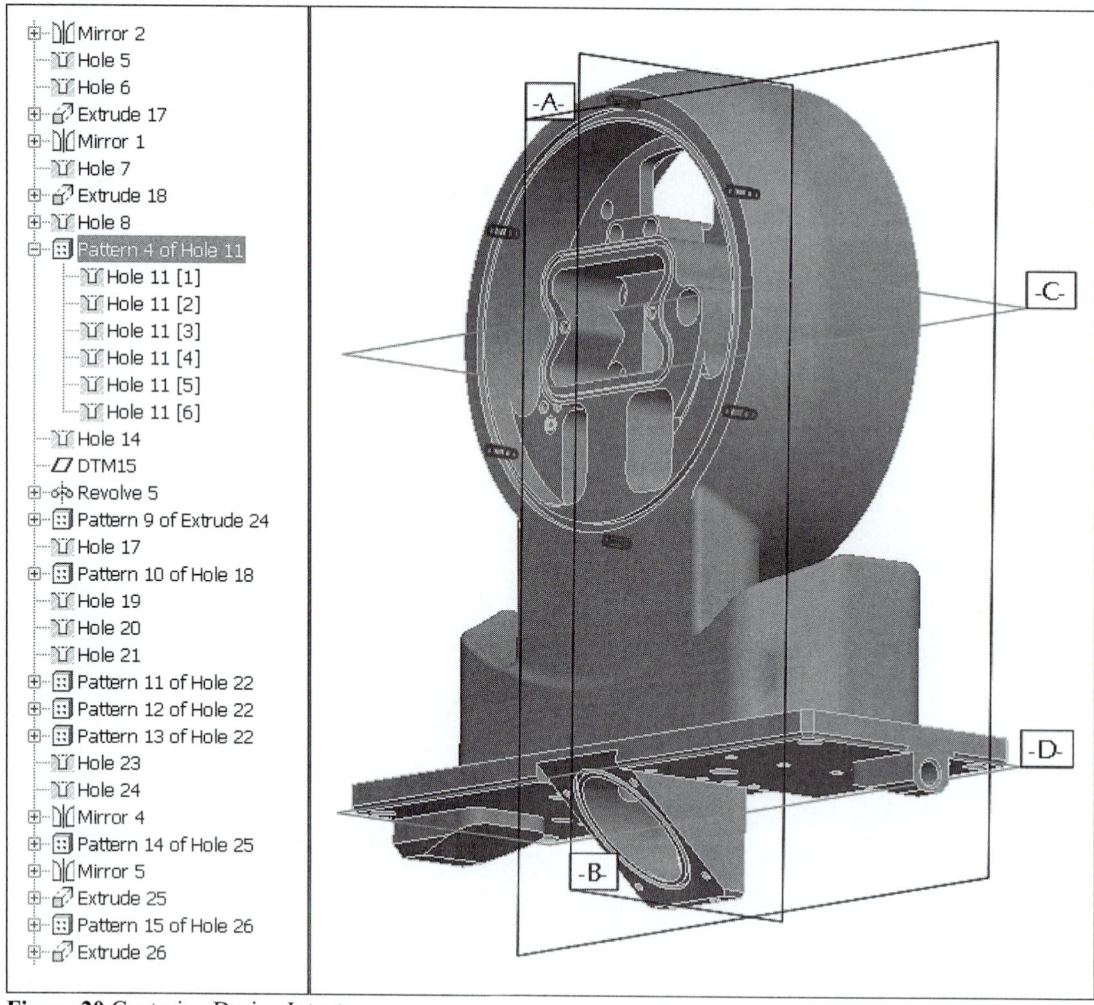

Figure 20 Capturing Design Intent

Parametric designs capture relationships in several ways:

Implicit Relationships Implicit relationships occur when new model geometry is sketched and dimensioned relative to existing features and parts. An implicit relationship is established, for instance, when the section sketch of a tire (Fig. 21) uses rim edges as a reference.

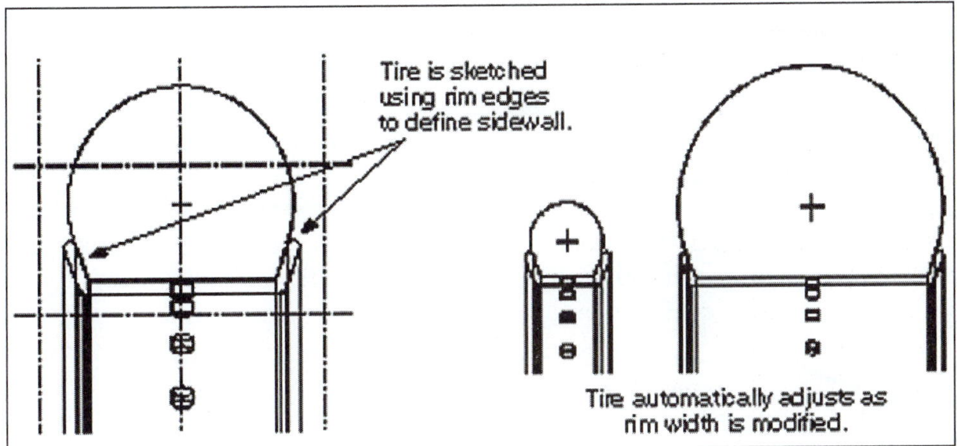

Figure 21 Tire and Rim

Patterns Design features often follow a geometrically predictable pattern. Features and parts are patterned in parametric design by referencing either construction dimensions or existing patterns. One example of patterning is a wheel hub with spokes (Fig. 22). First, the spoke holes are radially patterned. The spokes can then be strung by referencing this pattern.

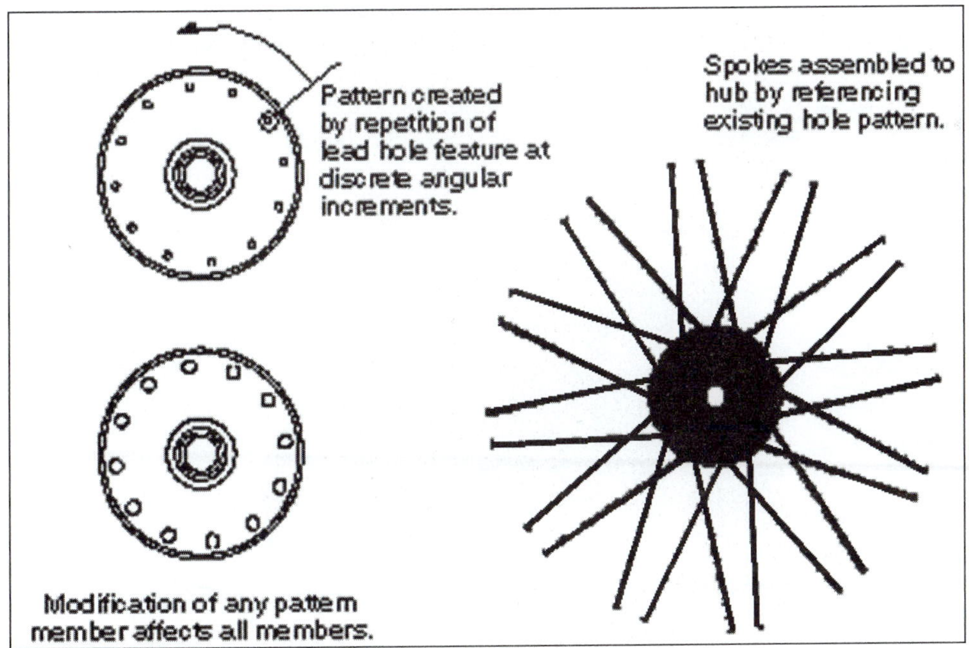

Figure 22 Patterns

Modification to a pattern member affects all members of that pattern. This helps capture design intent by preserving the duplicate geometry of pattern members.

The modeling task is to incorporate the features and parts of a complex design while properly capturing design intent to provide flexibility in modification. Parametric design modeling is a synthesis of physical and intellectual design (Fig. 23).

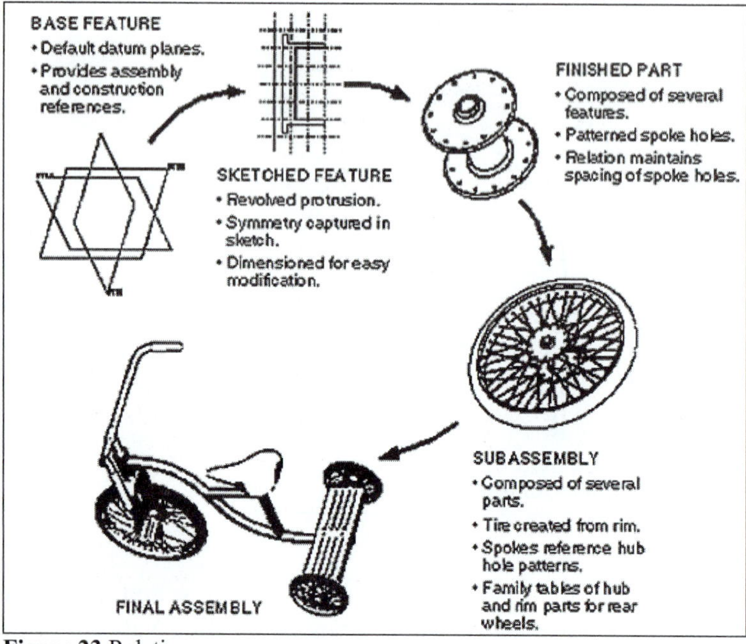

Figure 23 Relations

Explicit Relations Whereas implicit relationships are implied by the feature creation method, the user mathematically enters an explicit relation. This equation is used to relate feature and part dimensions in the desired manner. An explicit relation (Fig. 24) might be used, for example, to control sizes on a model.

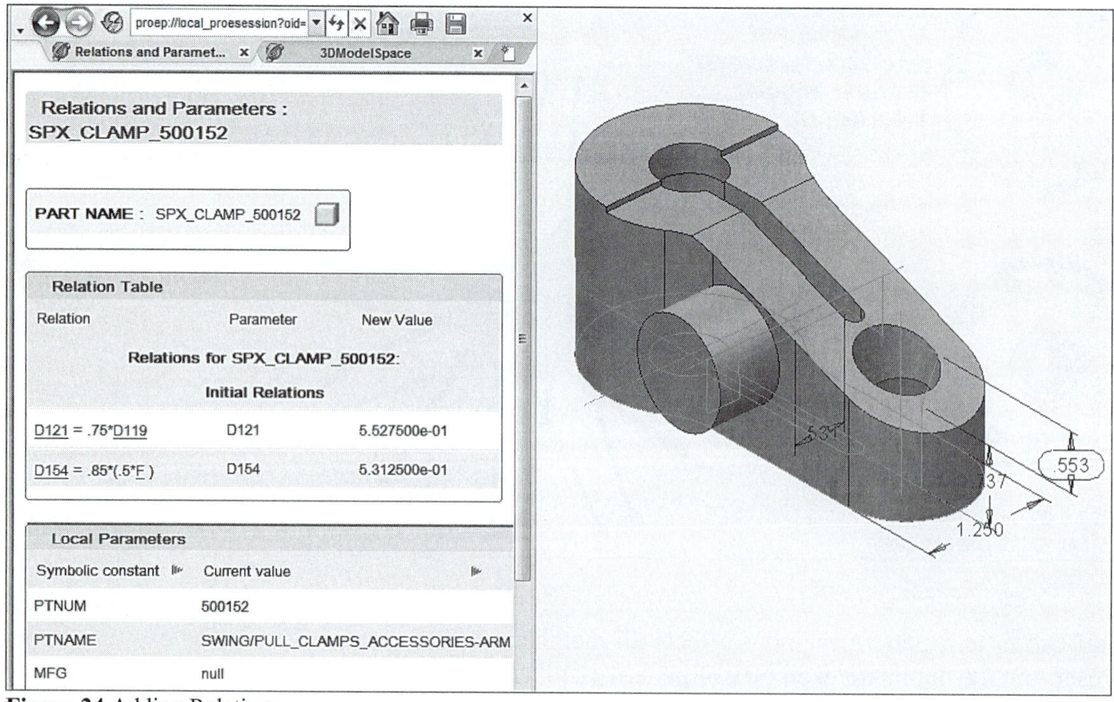

Figure 24 Adding Relations

Family Tables Family tables are used to create part families [Figs. 25(a-c)] from generic models by tabulating dimensions or the presence of certain features or parts. A family table might be used, for example, to catalog a series of couplings with varying width and diameter as shown in Figure 26.

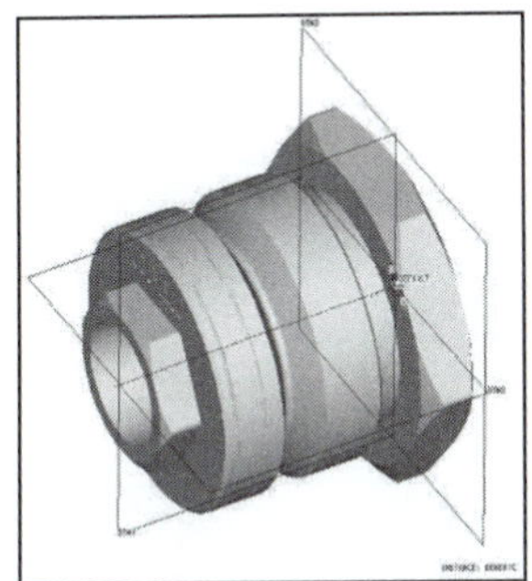
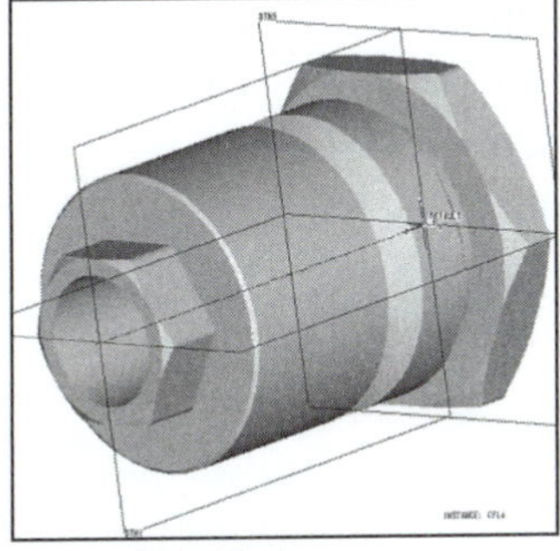
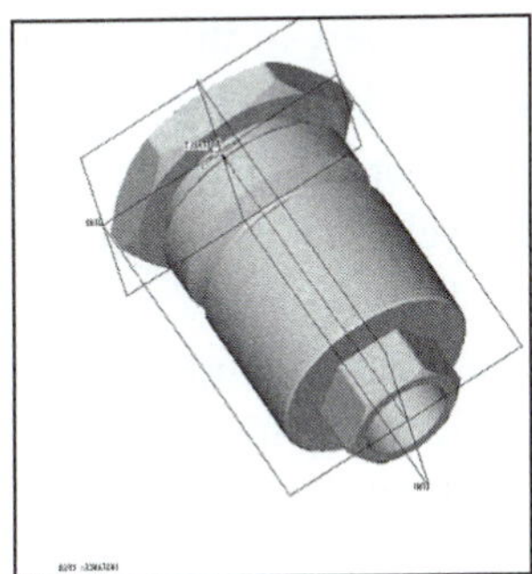

Figures 25(a-c) Family of Parts- Coupling-Fitting

Type	Instance Name	Common Name	d30	d46	F1068 [CUT]	F829 [SLOT]	F872 [SLOT]	F2620 COPIED_G...
	COUPLING-FITTI...		2.06	0.3130	Y	Y	Y	Y
	CPLA		3.00	0.5000	N	N	N	N
	CPLB		3.25	0.6250	N	Y	N	N
	CPLC		3.50	0.7500	Y	N	Y	Y

Figure 26 Family Table for Coupling-Fitting

Assemblies

Just as parts are created from related features, **assemblies** are created from related parts. The progressive combination of subassemblies, parts, and features into an assembly creates parent-child relationships based on the references used to assemble each component (Fig. 27).

The *Assembly* functionality is used to assemble existing parts and subassemblies.

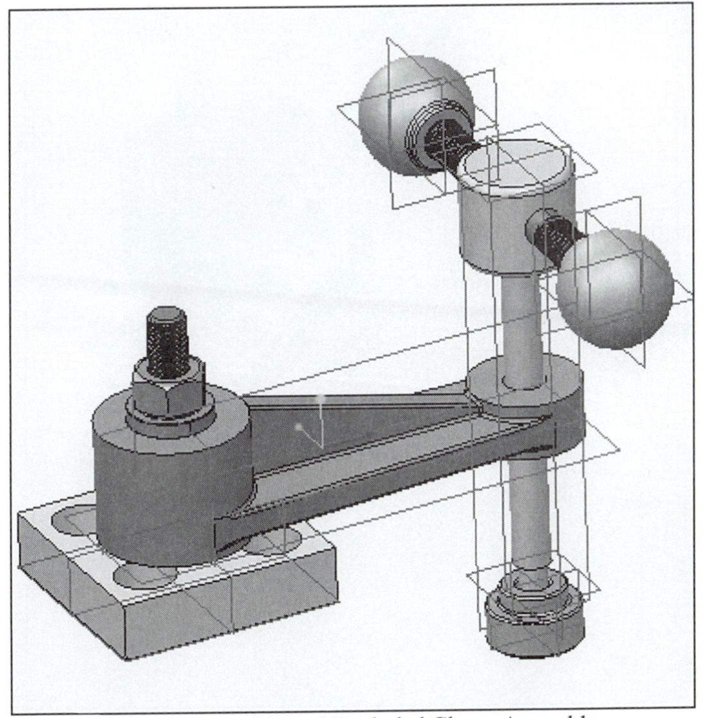
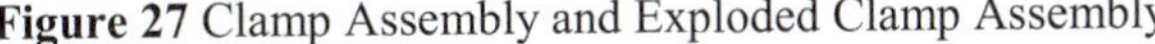
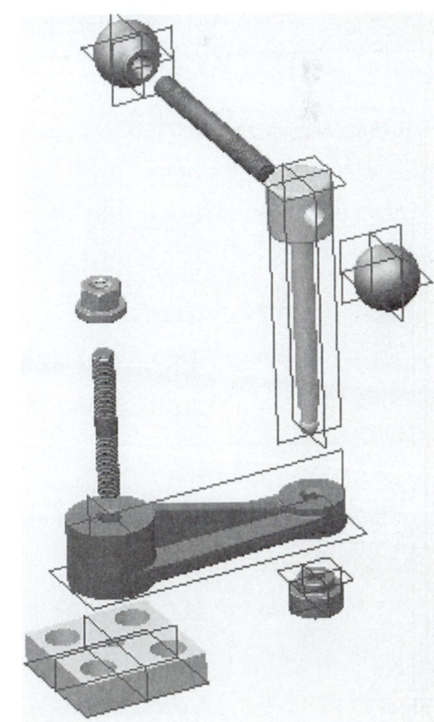

Figure 27 Clamp Assembly and Exploded Clamp Assembly

During assembly creation, you can:

- Simplify a view of a large assembly by creating a simplified representation
- Perform automatic or manual placement of component parts
- Create an exploded view of the component parts
- Perform analysis, such as mass properties and clearance checks
- Modify the dimensional values of component parts
- Define assembly relations between component parts
- Create assembly features
- Perform automatic interchange of component parts
- Create parts in Assembly mode
- Create documentation drawings of the assembly

Just as features can reference part geometry, parametric design also permits the creation of parts referencing assembly geometry. **Assembly mode** allows the designer both to fit parts together and to design parts based on how they should fit together. In Figure 28, an assembly *Bill of Materials* report is generated.

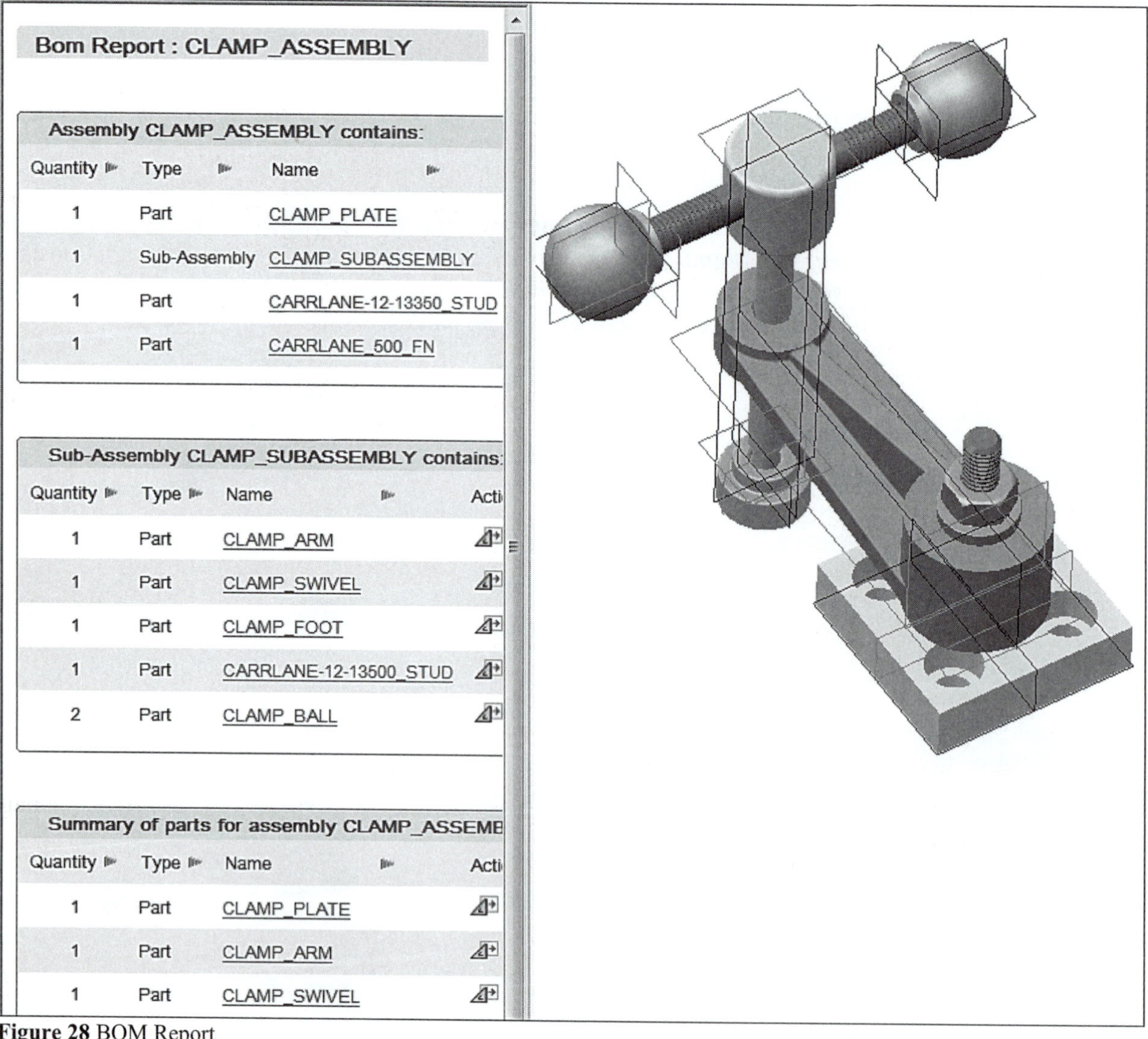

Bom Report : CLAMP_ASSEMBLY

Assembly CLAMP_ASSEMBLY contains:

Quantity	Type	Name
1	Part	CLAMP_PLATE
1	Sub-Assembly	CLAMP_SUBASSEMBLY
1	Part	CARRLANE-12-13350_STUD
1	Part	CARRLANE_500_FN

Sub-Assembly CLAMP_SUBASSEMBLY contains:

Quantity	Type	Name	Acti
1	Part	CLAMP_ARM	
1	Part	CLAMP_SWIVEL	
1	Part	CLAMP_FOOT	
1	Part	CARRLANE-12-13500_STUD	
2	Part	CLAMP_BALL	

Summary of parts for assembly CLAMP_ASSEMB

Quantity	Type	Name	Acti
1	Part	CLAMP_PLATE	
1	Part	CLAMP_ARM	
1	Part	CLAMP_SWIVEL	

Figure 28 BOM Report

Drawings

You can create drawings of all parametric design models (Fig. 29). All model views in the drawing are *associative:* if you change a dimensional value in one view, other drawing views update accordingly. Moreover, drawings are associated with their parent models. Any dimensional changes made to a drawing are automatically reflected in the model. Any changes made to the model (e.g., addition of features, deletion of features, dimensional changes, and so on) in Part, Sheet Metal, Assembly, or Manufacturing modes are also automatically reflected in their corresponding drawings.

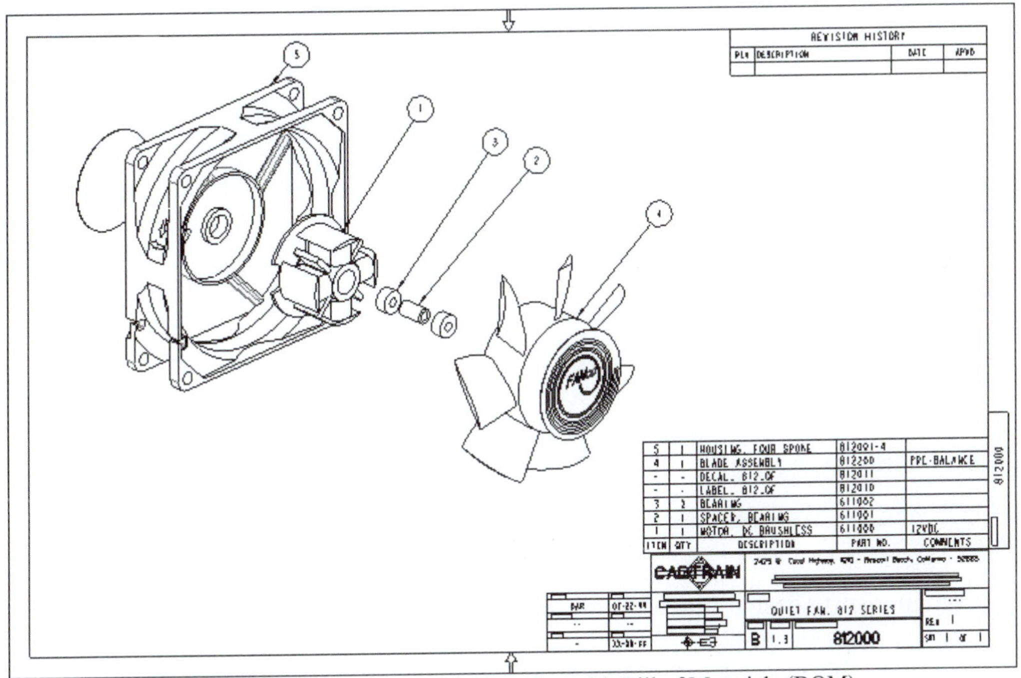

Figure 29 Ballooned Exploded View Assembly Drawing with Bill of Materials (BOM)

The **Drawing** functionality is used to create annotated drawings of parts and assemblies. During drawing creation, you can:

- Add views of the part or assembly
- Show existing dimensions
- Incorporate additional driven or reference dimensions
- Create notes on the drawing
- Display views of additional parts or assemblies
- Add sheets to the drawing
- Create draft entities on the drawing
- Balloon components on an assembly drawing (Fig. 30)
- Create an associative BOM

You can annotate the drawing with notes, manipulate the dimensions, and use layers to manage the display of different items on the drawing.

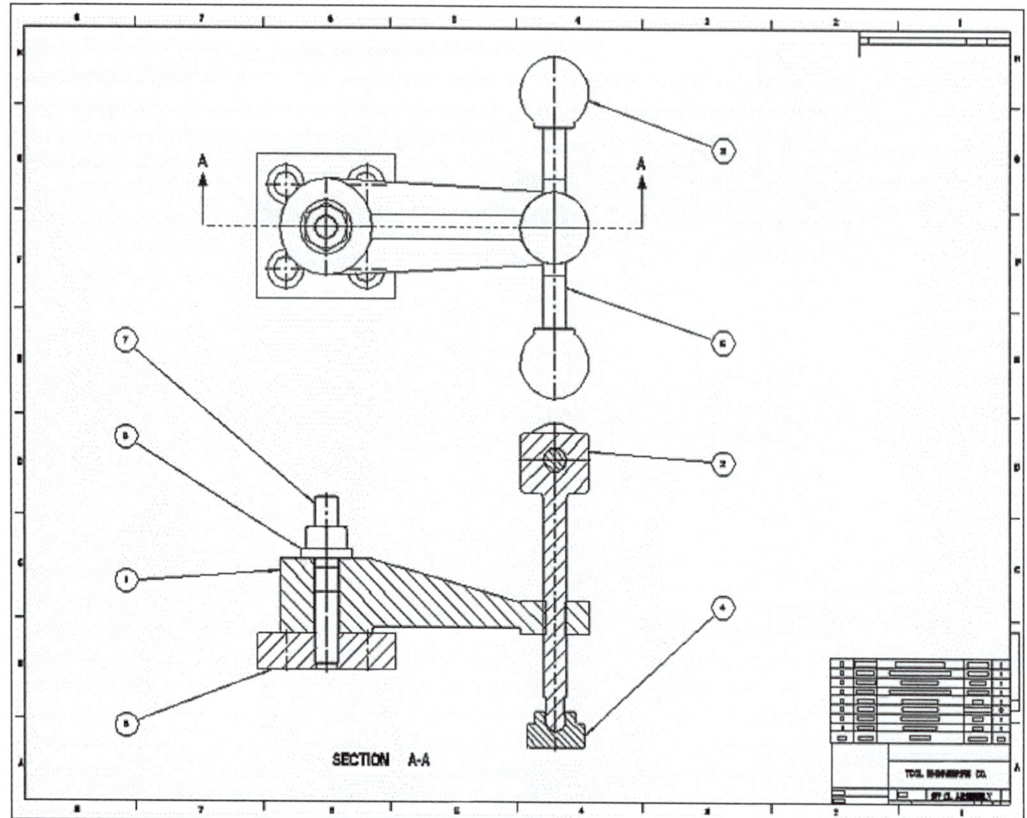

Figure 30 Ballooned Assembly Drawing

Drawing mode in parametric design provides you with the basic ability to document solid models in drawings that share a two-way associativity.

Changes that are made to the model in Part mode or Assembly mode will cause the drawing to update automatically and reflect the changes. Any changes made to the model in Drawing mode will be immediately visible on the model in Part and Assembly modes. The model shown in Figure 31 has been detailed in Figure 32.

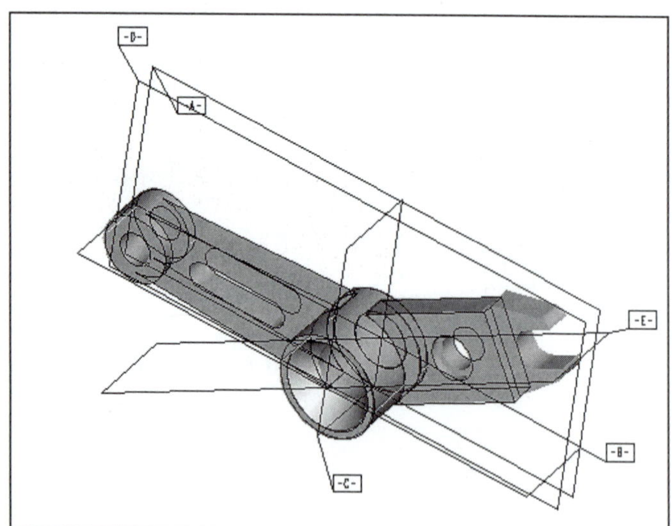

Figure 31 Angle Frame Model

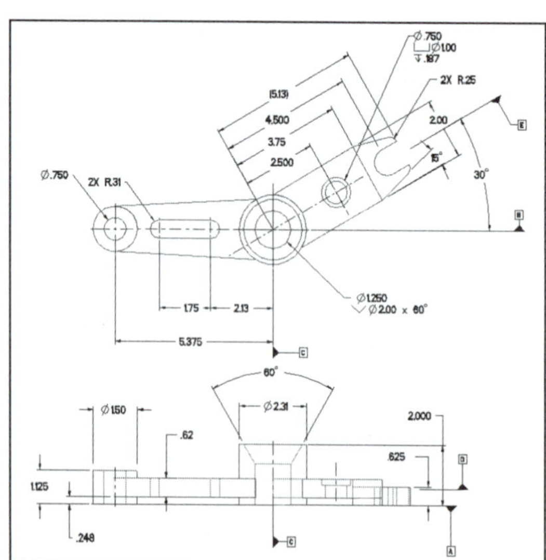

Figure 32 Angle Frame Drawing

Flexible Modeling

Flexible modeling allows you to experiment with your design without being committed to changes. You can make explicit modifications to selected geometry while ignoring pre-existing relationships. With this new capability within Creo Parametric you can utilize some basic Creo Direct capabilities. The ribbon in Figure 33 shows the new interface. Design changes are done directly on the model without parent-child relationships being imbedded (Fig. 34).

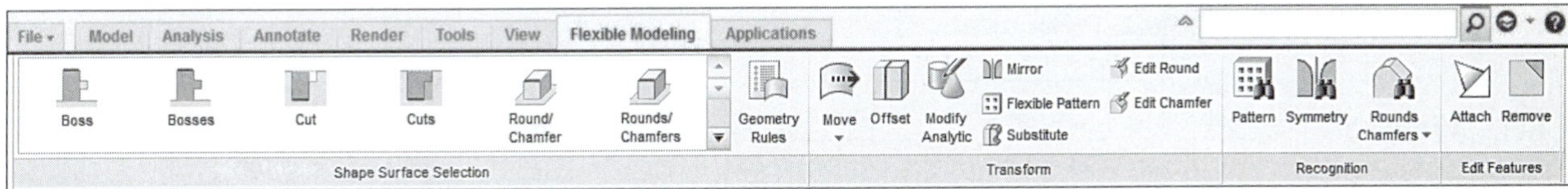

Figure 33 Flexible Modeling Ribbon

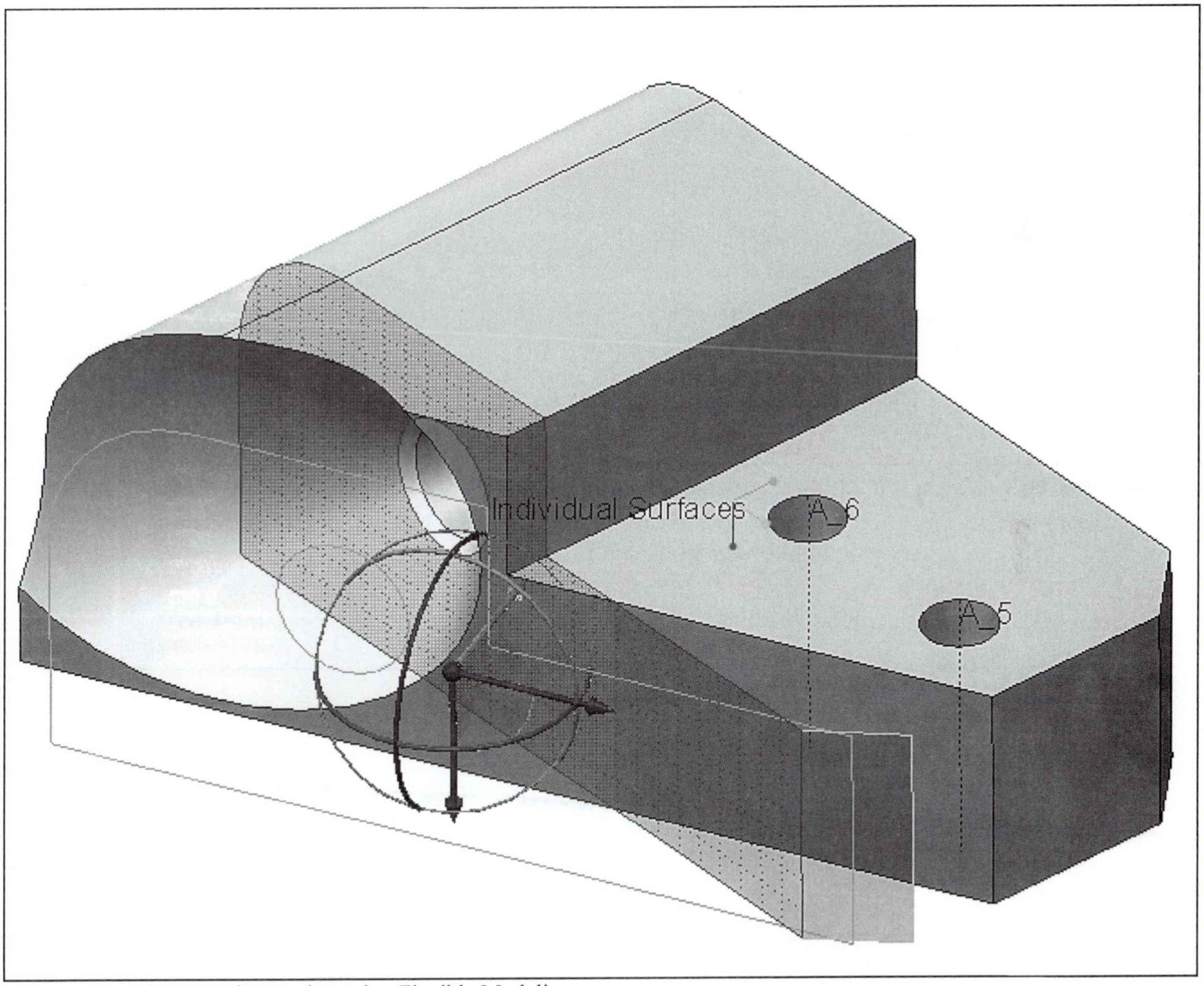

Figure 34 Design experimentation using Flexible Modeling

Using the Text

The text utilizes a variety of command boxes and descriptions to lead you through construction sequences. Also, see the downloadable Creo Parametric Quick Reference Cards at www.cad-resources.com. The following icons, symbols, shortcut keys, and conventions will be used *(command sequences are always in a box)*:

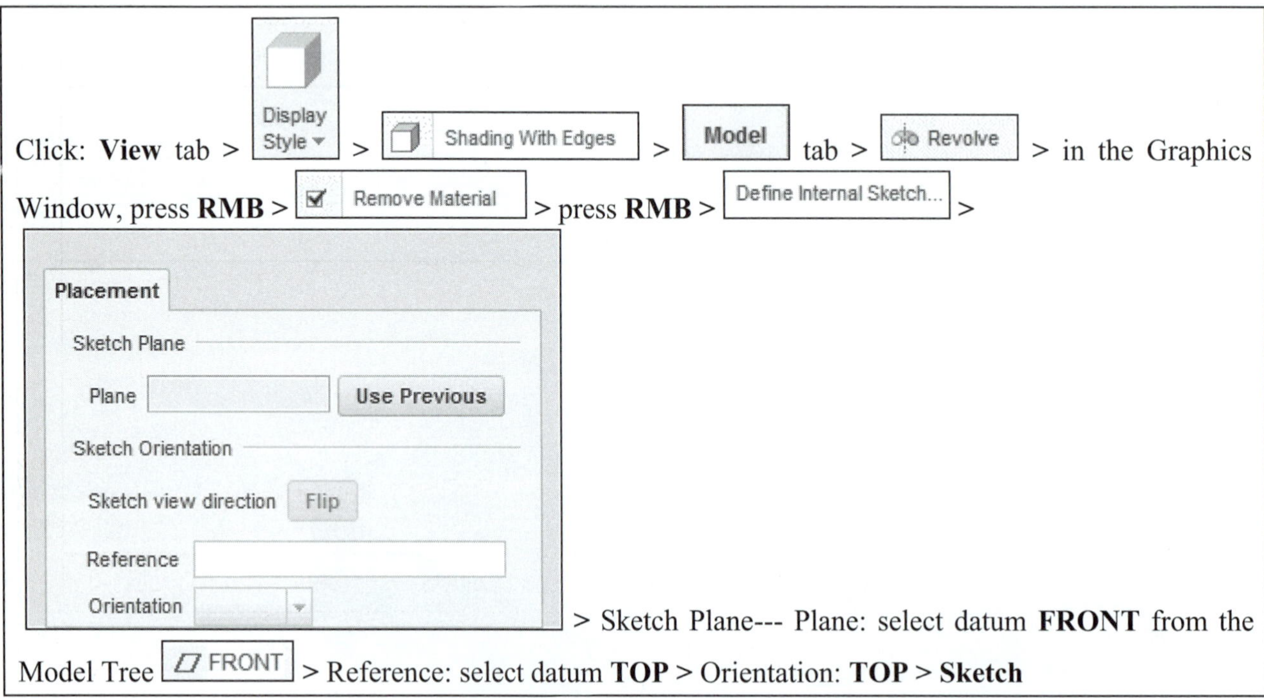

Click: **View** tab > Display Style > Shading With Edges > **Model** tab > Revolve > in the Graphics Window, press **RMB** > Remove Material > press **RMB** > Define Internal Sketch... > Sketch Plane--- Plane: select datum **FRONT** from the Model Tree FRONT > Reference: select datum **TOP** > Orientation: **TOP** > **Sketch**

Commands:

- \> Continue with command sequence or screen picks using **LMB**
- Line Chain — Icon (with description) indicates command to pick using **LMB**

Mouse or keyboard terms used in this text:

- **LMB** — **L**eft **M**ouse **B**utton
  - or "**Pick**" — term used to direct an action (i.e., "Pick the surface")
  - or "**Click**" — term used to direct an action (i.e., "Click on the icon")
  - or "**Select**" — term used to direct an action (i.e., "Select the feature")
- **MMB** — **M**iddle **M**ouse **B**utton (accept the current selection or value)
  - or **Enter** — press **Enter** key to accept entry
  - or ✓ — Click on this icon to accept entry
- **RMB** — **R**ight **M**ouse **B**utton
  - *Click* (toggles to next selection)
  - *Press and hold* (provides a list of available commands)

Shortcut Keys:

Keystroke	Action
ALT	•Disables filters temporarily during selection when the pointer is in the graphics window. Hold down the ALT key during selection. •Activates KeyTips on the Ribbon. With your pointer anywhere in the Creo window (Ribbon, Graphics Window, Model Tree, etc.), press the ALT key.
CTRL	Temporarily hides the Live Toolbar in 3D part and assembly modes.
CTRL+A	Activates a window.
CTRL+C	Copies the selection to the clipboard.
CTRL+D	Displays a model in standard view.
CTRL+N	Opens the **New** dialog box.
CTRL+O	Opens the **File Open** dialog box.
CTRL+R	Repaints a window.
CTRL+S	Opens the **Save Object** dialog box.
CTRL+V	Pastes the clipboard contents to the selected location.
CTRL+Y	Performs a single step redo operation.
CTRL+Z	Performs a single step undo operation.
ESC	Performs one of the following operations depending on the modeling context: • Cancels an operation • Clears the selection in the graphics window • Cancels a dragger operation •Closes the Live Toolbar • Cancels the currently active tool
ENTER	Accepts the changes and closes the Live Toolbar in 3D and assembly modes. Also clears the selection in the graphics window.
F1	Opens **Help** for the current context.
G	Hides the current guide for the geometry in 2D mode.
L	Locks or unlocks the current guide in 2D mode.
P	Shows or hides precision panels in 2D mode temporarily.
S	Switches between the Line and Arc Creation modes in 2D mode.
TAB	Pre-highlights a set of surfaces for selection based on shape selection rules. Refer to the PTC Creo Parametric 3.0 Flexible Modeling Help for more information on shape selection rules.

Text Organization

- **Lesson 1** has you create two simple parts, an assembly, and two drawings using default settings.
- **Lesson 2** introduces Creo Parametric's interface and embedded Browser.
- **Lesson 3** provides uncomplicated instructions to model a variety of simple-shaped parts.
- **Lessons 4** and **5** involve part modeling using a variety of commands and tools.
- **Lessons 6** through **12** involve modeling the parts, creating the assembly [Fig. 35(a)], and documenting a design with detail and assembly drawings [Fig. 35(b)].
- **Lessons 13-18, Video Lessons 19-22 (Download)** Instructions to model and document parts using more advanced features and techniques. Available as a download to minimize the size and cost of the text.
- **Projects (Download)** Provides over 100 pages of project drawings as a PDF download so as to minimize the size and cost of the text.

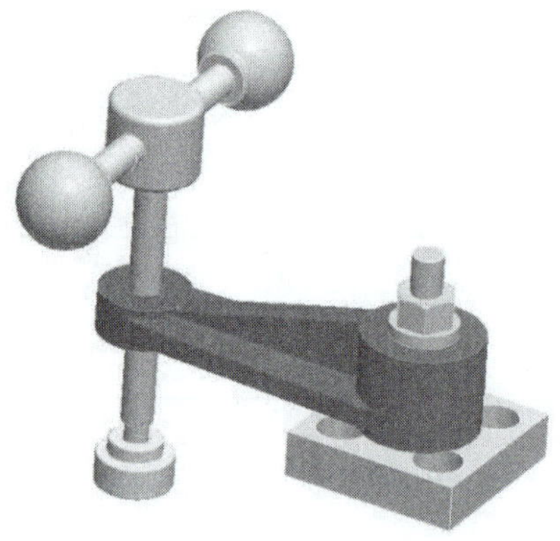

Figure 35(a) Clamp Assembly

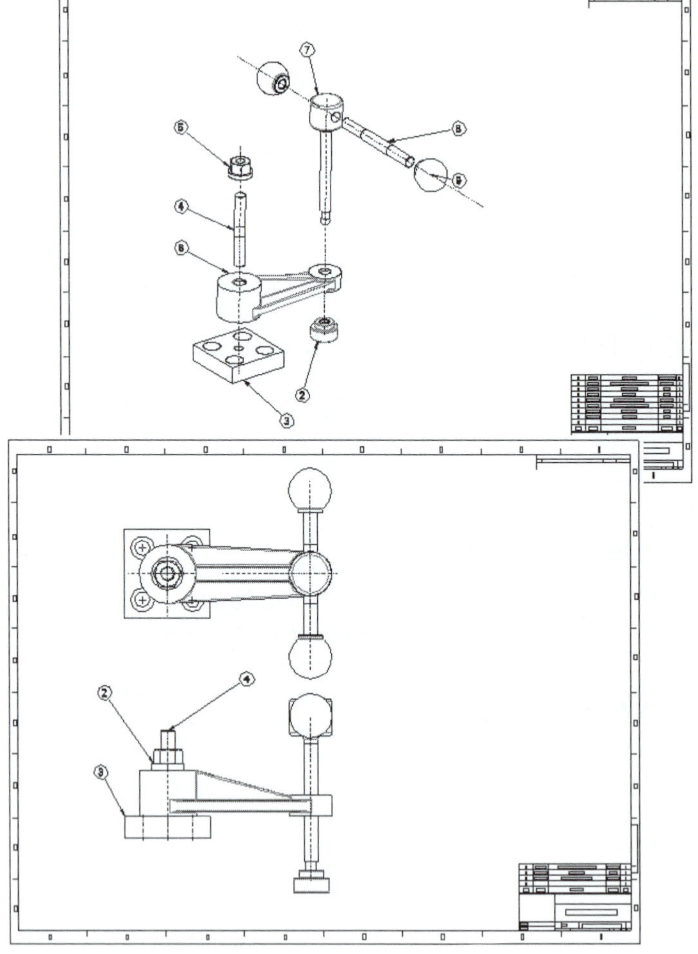

Figure 35(b) Clamp Assembly Drawing

Lesson 1 PTC Creo Parametric 3.0 Overview

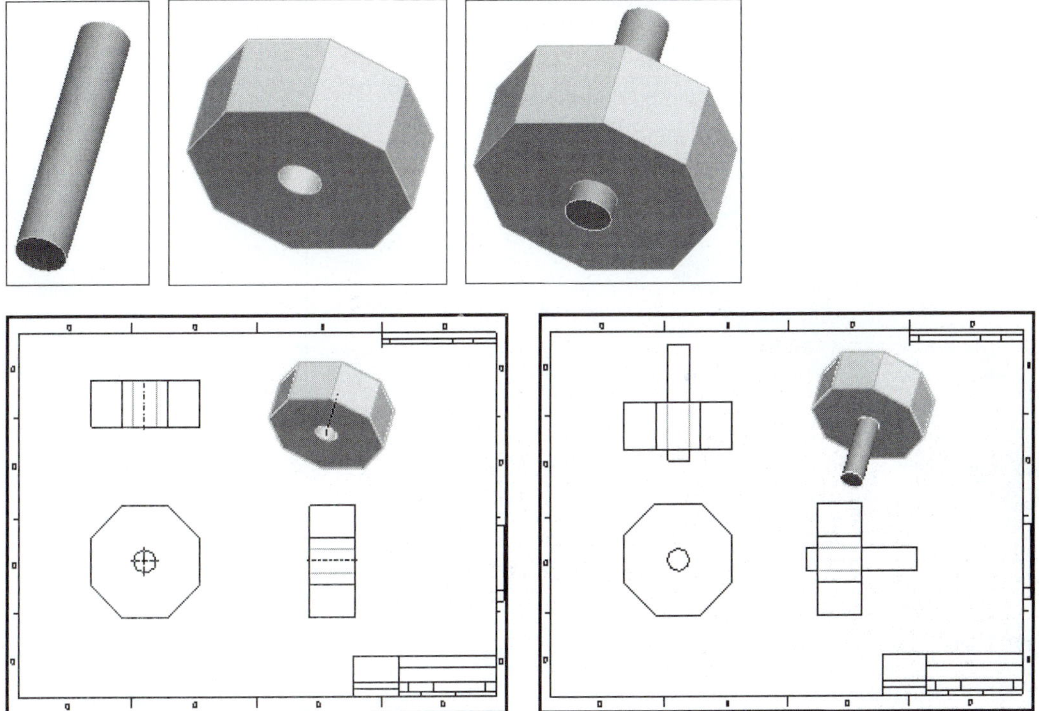

Figure 1.1 Pin Model, Plate Model, Assembly Model, Plate Drawing, and Assembly Drawing

OBJECTIVES

- Model two parts
- Assemble parts to create an assembly
- Construct a part drawing
- Make an assembly drawing

REFERENCES AND RESOURCES

For **Resources** go to **www.cad-resources.com** > click on the image of your Book cover

- Lesson 1 Lecture
- Lesson 1 3D models embedded in a PDF
- Quick Reference Card
- Configuration Options

PTC Creo Parametric 3.0 Overview

This lesson (Fig. 1.1) will allow you to experience the part (Fig. 1.2), assembly, and drawing modes of PTC Creo Parametric 3.0 by creating two simple parts, assembling them, and creating drawings. A minimum of explanation and instruction is provided here. This lesson will quickly get you up and running.

Lesson 1 STEPS

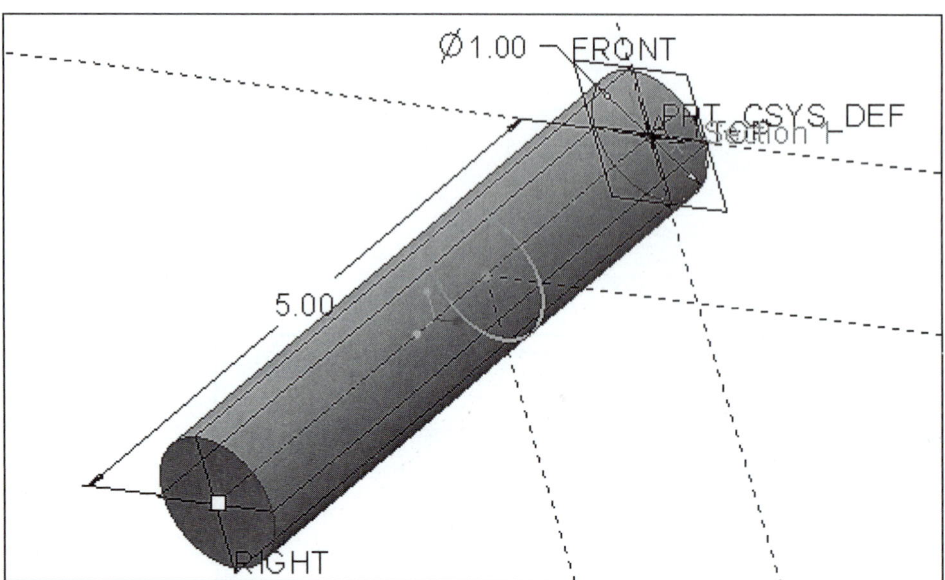

Figure 1.2 Pin

Creating the Pin Part

In this lesson most commands and picks will be accompanied by the window or dialog box that will open as the command is initiated. After this lesson, the step-by-step boxes will *not* show every window and dialog box, as this would make the text extremely long. Appropriate illustrations will be provided.

Throughout the text, a box surrounds all commands and menu selections.

Open **PTC Creo Parametric 3.0** using a shortcut icon on your Desktop *(or with WINDOWS, click: **Start** > **All Programs** >* **PTC Creo Parametric 3.0***)*. Creo Parametric will open on your computer.
Click: **Select Working Directory**

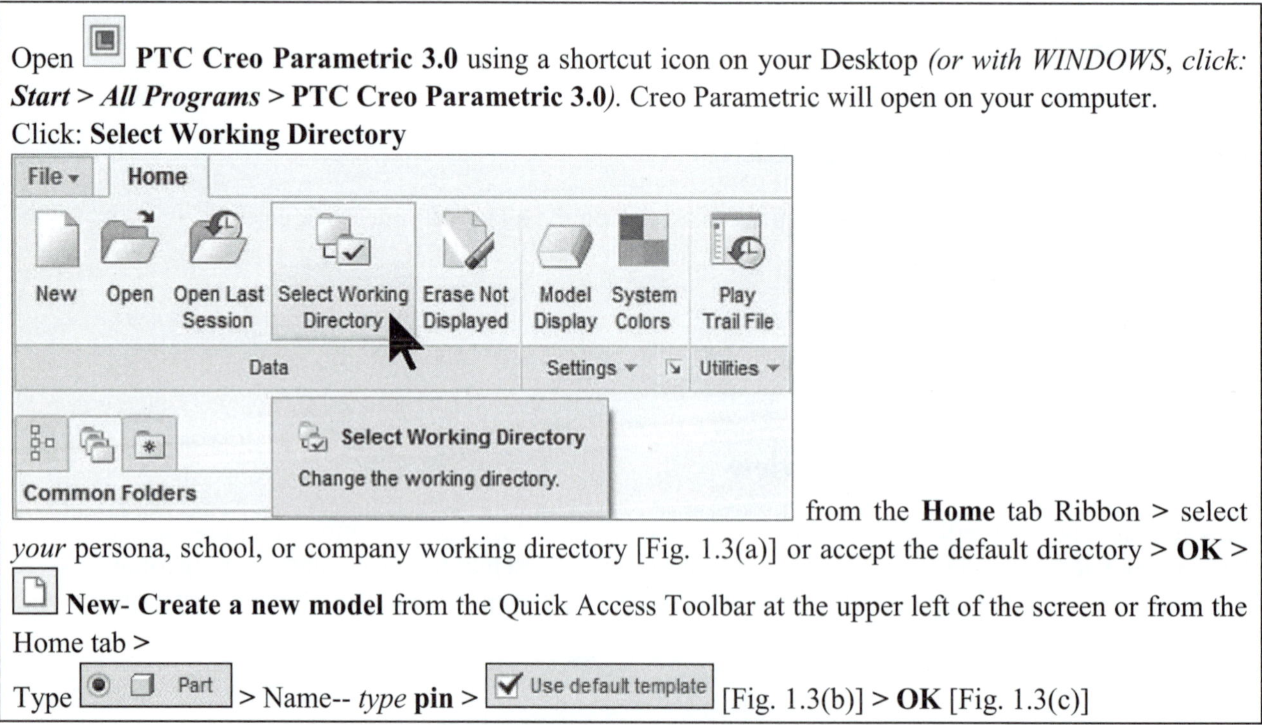

from the **Home** tab Ribbon > select *your* persona, school, or company working directory [Fig. 1.3(a)] or accept the default directory > **OK** > **New- Create a new model** from the Quick Access Toolbar at the upper left of the screen or from the Home tab >
Type Part > Name-- *type* **pin** > Use default template [Fig. 1.3(b)] > **OK** [Fig. 1.3(c)]

Select Working Directory

« CAD-Resources ▸ CREO ▸ Creo_3-0 ▸ PTC_Creo-3 ▸ 3-0

Search...

Organize Views Tools

Common Folders

Desktop

My Documents

cad-resources_7

Working Directory

Favorites

Creo_3-0

Folder Tree

File name: . Date modified: 19-Jul-14 11:00:36 AM

File name: .

Type Directories

OK Cancel

Figure 1.3(a) Select Working Directory Dialog Box ***(your directory will be different)***

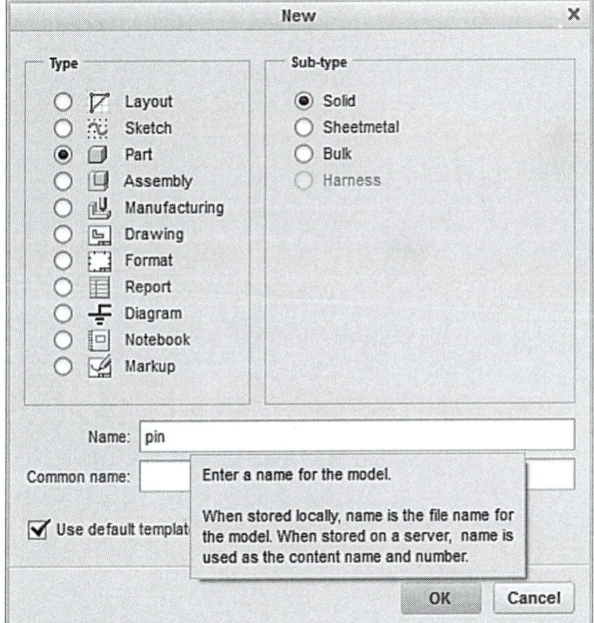

Figure 1.3(b) New Dialog Box

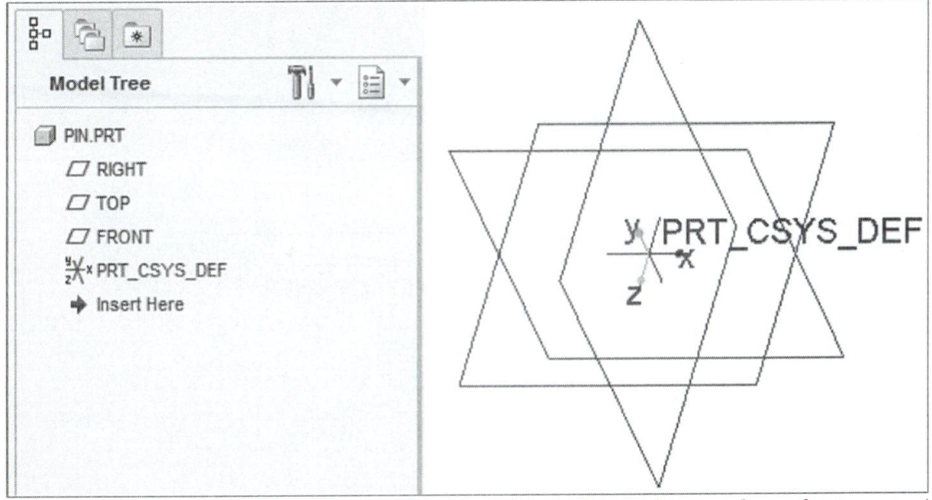

Figure 1.3(c) Model Tree and Graphics Window *(default datum planes and coordinate system)*

From the keyboard, **Ctrl+S** (saves the object) > **OK** > Extrude **Extrude** from Model tab Ribbon > **Placement** tab in the Dashboard [Fig. 1.4(a)] > **Define** [Fig. 1.4(b)]

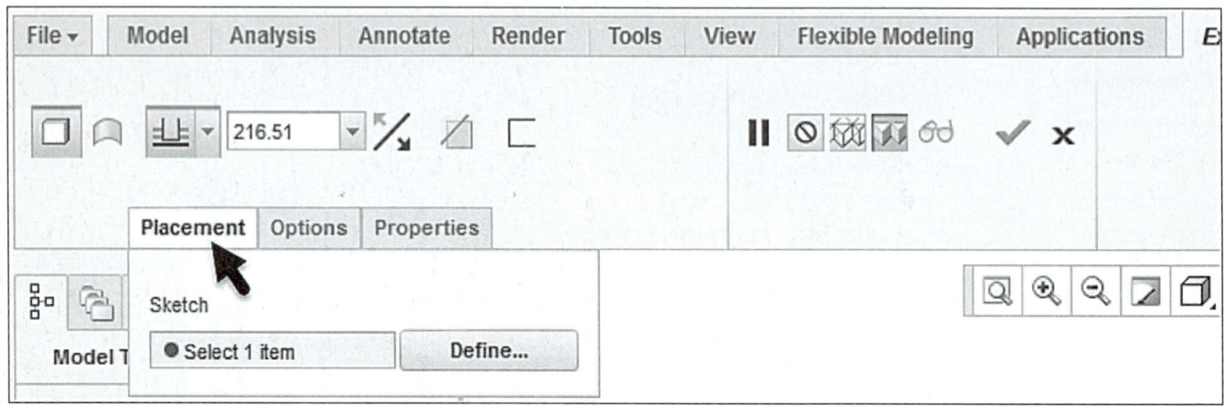

Figure 1.4(a) Extrude Placement Tab

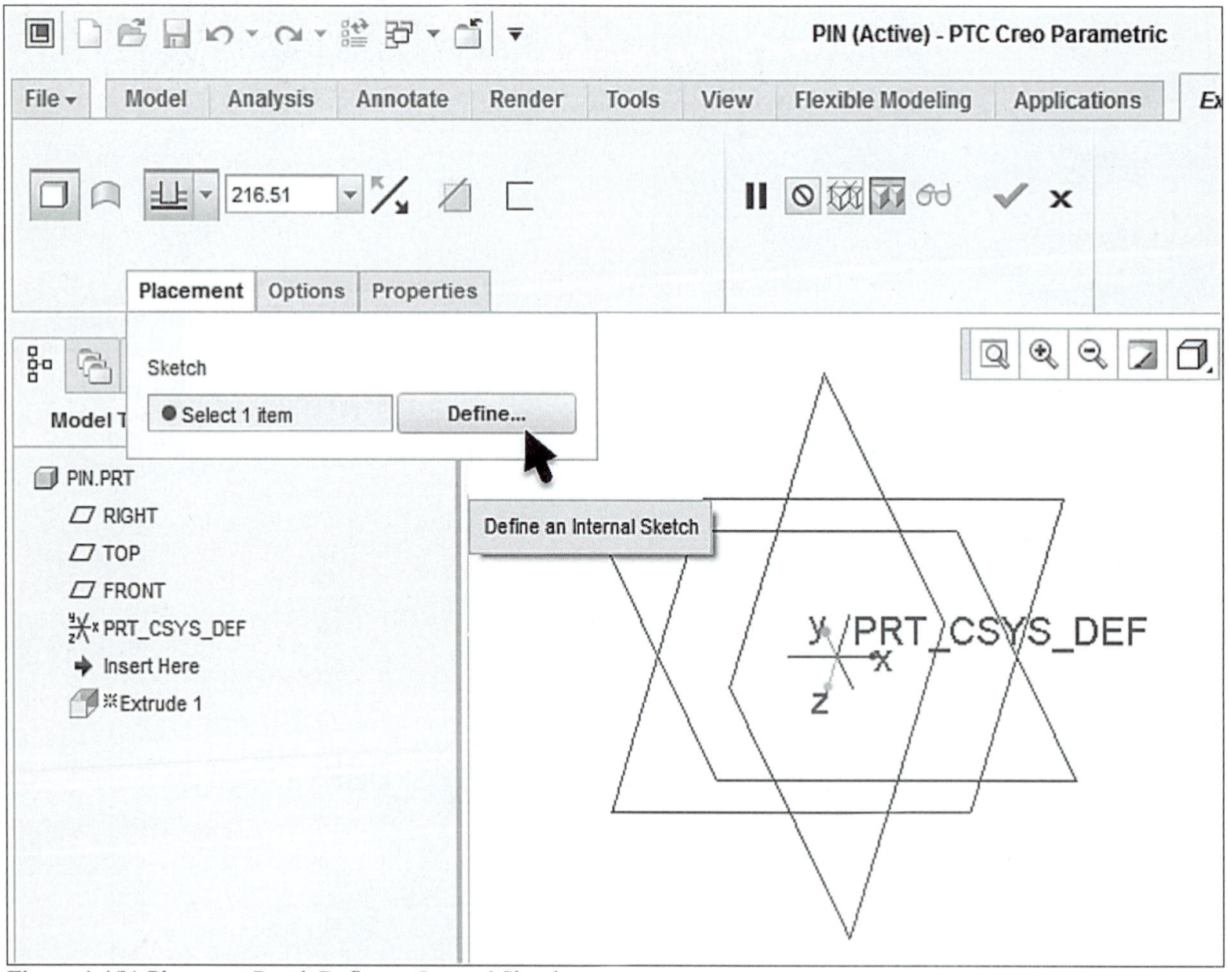

Figure 1.4(b) Placement Panel, Define an Internal Sketch

Select the **FRONT** datum plane from the Graphics Window (or Model Tree) [Fig. 1.4(c)]. The FRONT datum plane is the Sketch Plane and the RIGHT datum plane is automatically selected as the Sketch Orientation Reference with an Orientation of Right [Fig. 1.4(d)].

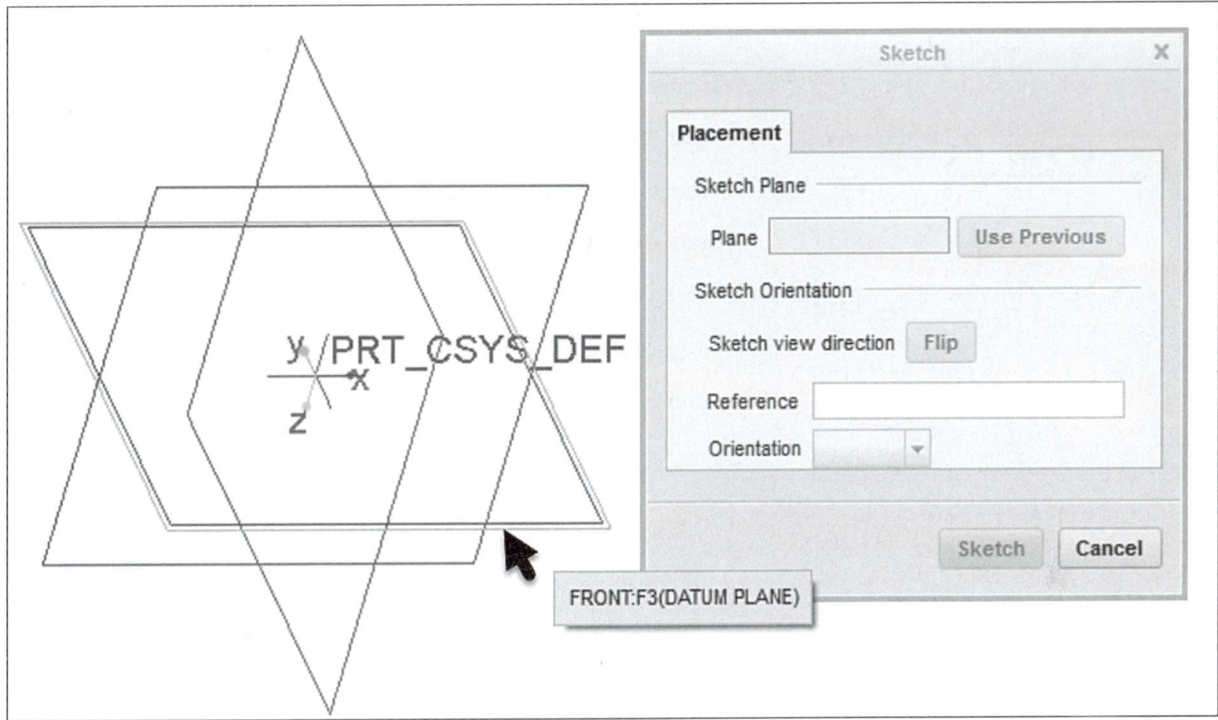

Figure 1.4(c) Select the FRONT Datum Plane

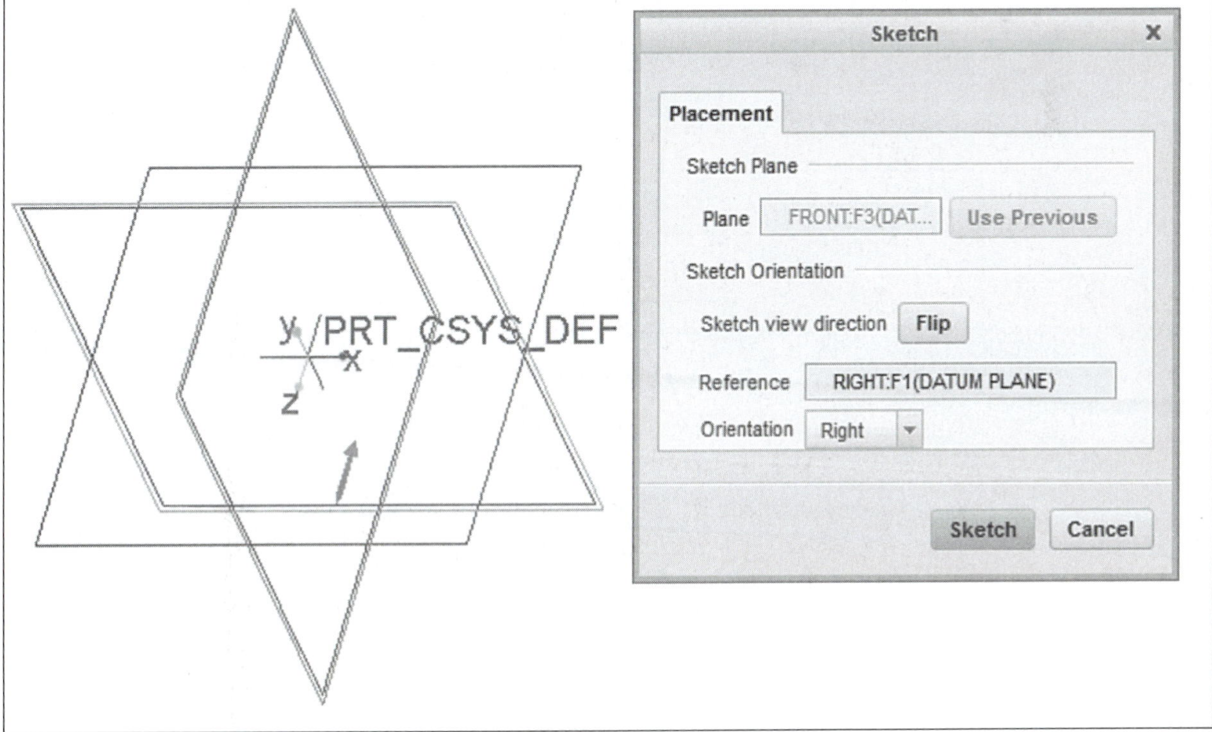

Figure 1.4(d) Sketch Dialog Box

Click: **Sketch** from the Sketch dialog box [Fig. 1.4(e)] > **Center and Point- Create circle by picking the center and a point on the circle** from the Sketch tab Ribbon [Fig. 1.4(f)]

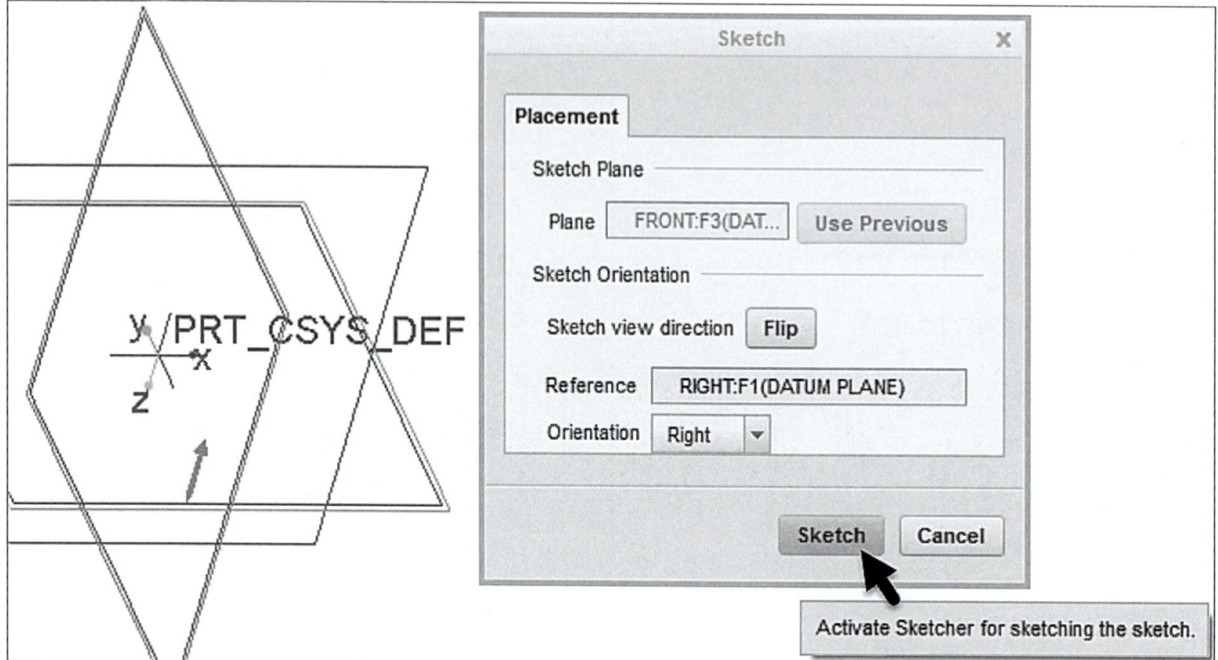

Figure 1.4(e) Activate the Sketcher

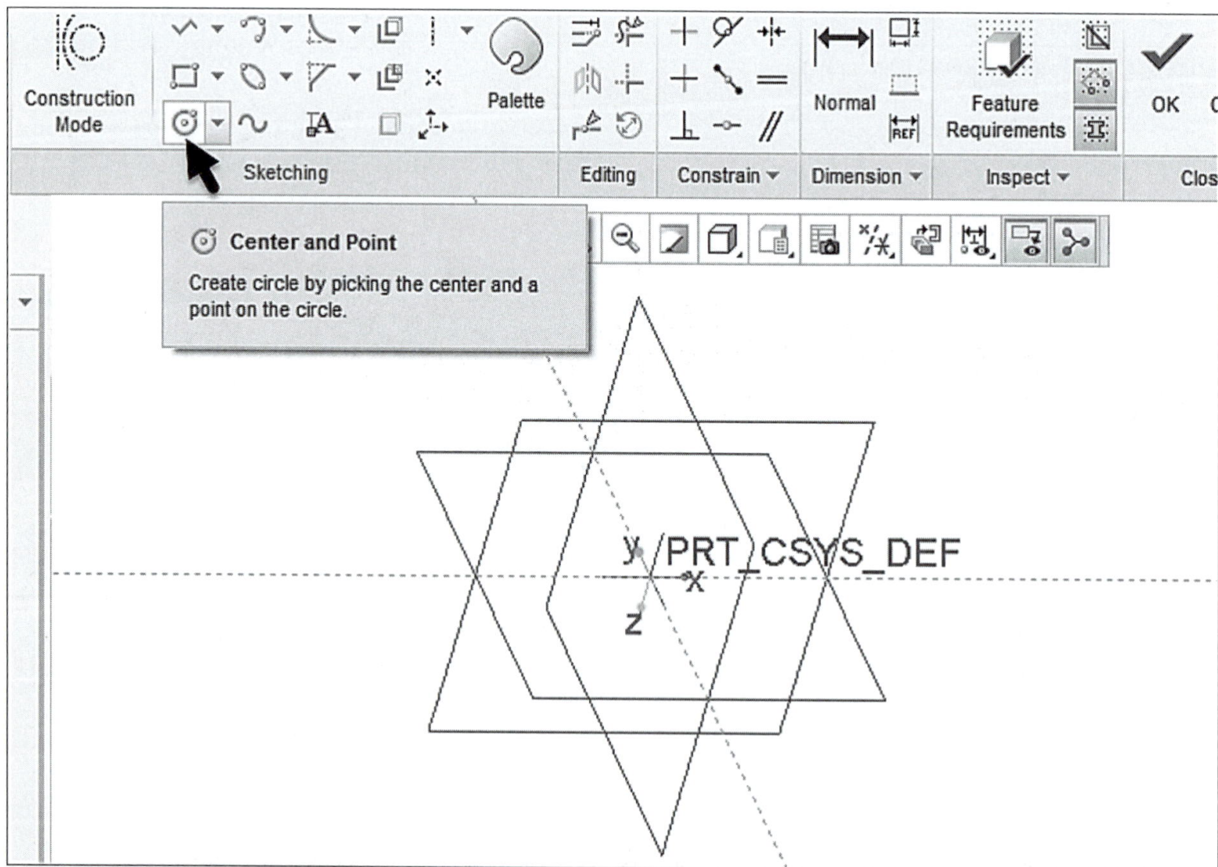

Figure 1.4(f) Select the Circle Tool

Pick the origin for the circle's center [Fig. 1.4(g)] > drag the pointer away from the center point and pick a point on the circle's edge [Fig. 1.4(h)] > move the pointer off of the circle's edge > **MMB** (**M**iddle **M**ouse **B**utton) [Figs. 1.4(i-l)] > double-click on the dimension > *type* **1.00** > **Enter**

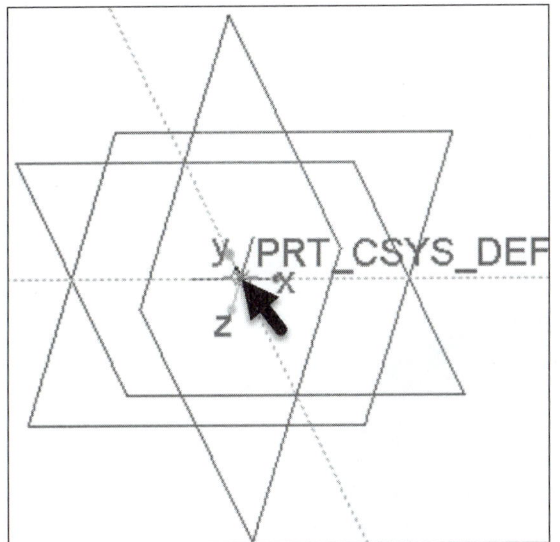

Figure 1.4(g) Pick the Circle's Center

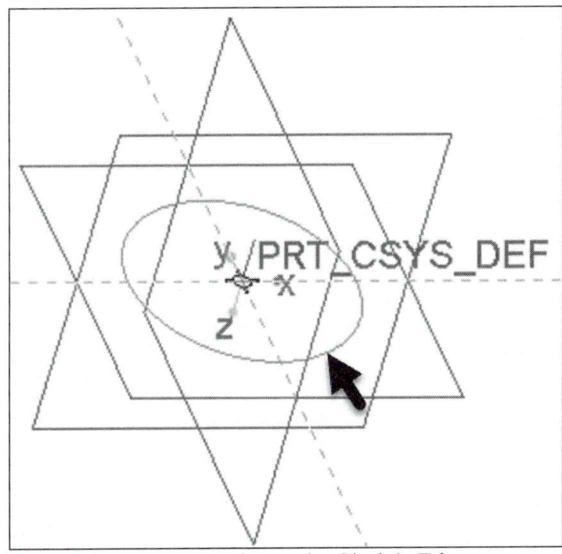

Figure 1.4(h) Pick a Point on the Circle's Edge

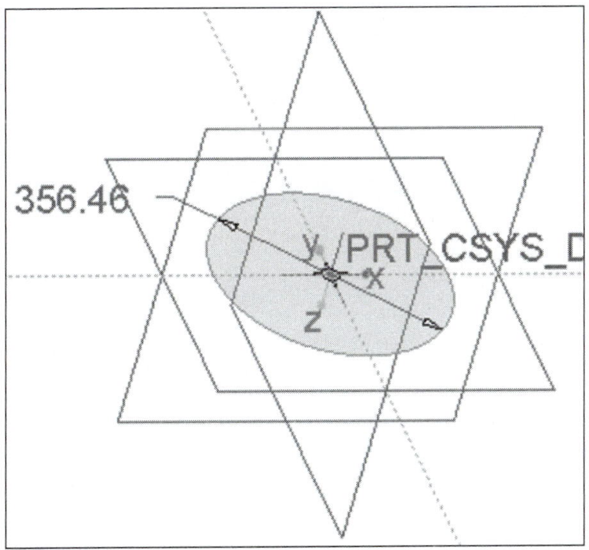

Figure 1.4(i) Circle and Dimension

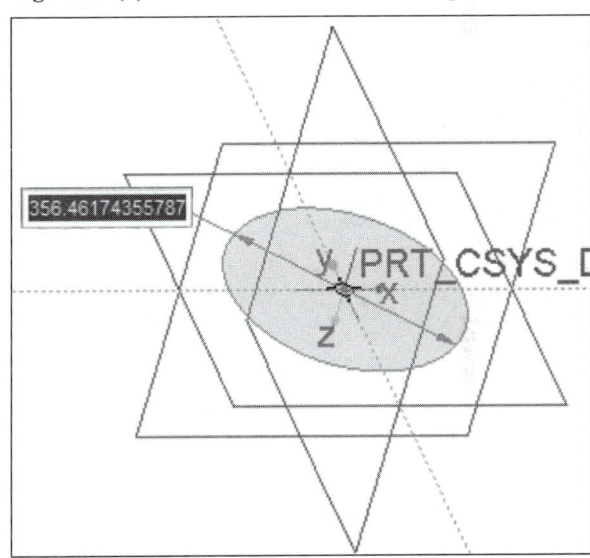

Figure 1.4(j) Double-click on the Dimension

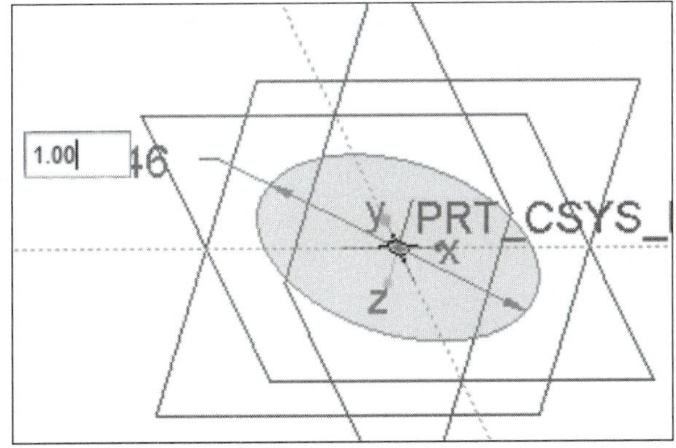

Figure 1.4(k) Type **1.00**

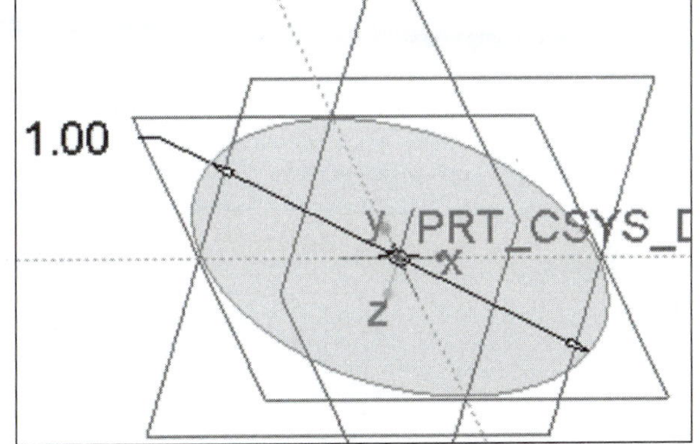

Figure 1.4(l) Completed Sketch

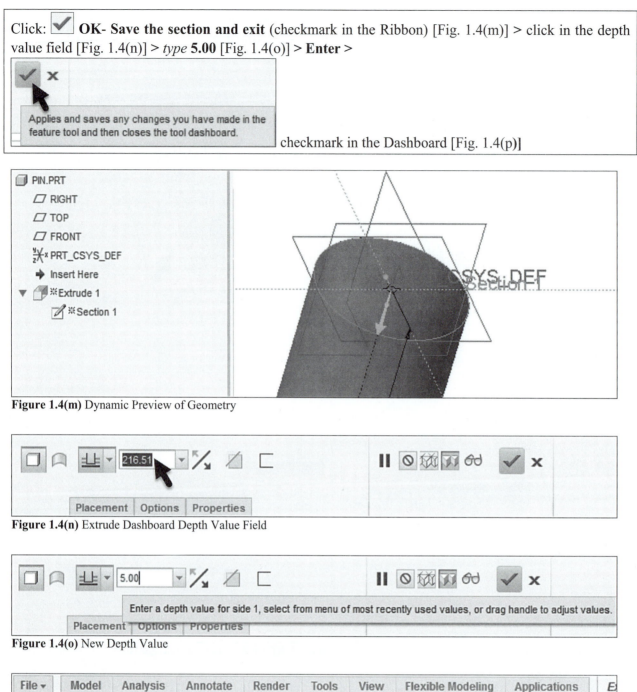

Click: ✓ **OK- Save the section and exit** (checkmark in the Ribbon) [Fig. 1.4(m)] > click in the depth value field [Fig. 1.4(n)] > *type* **5.00** [Fig. 1.4(o)] > **Enter** >

Applies and saves any changes you have made in the feature tool and then closes the tool dashboard.

checkmark in the Dashboard [Fig. 1.4(p)]

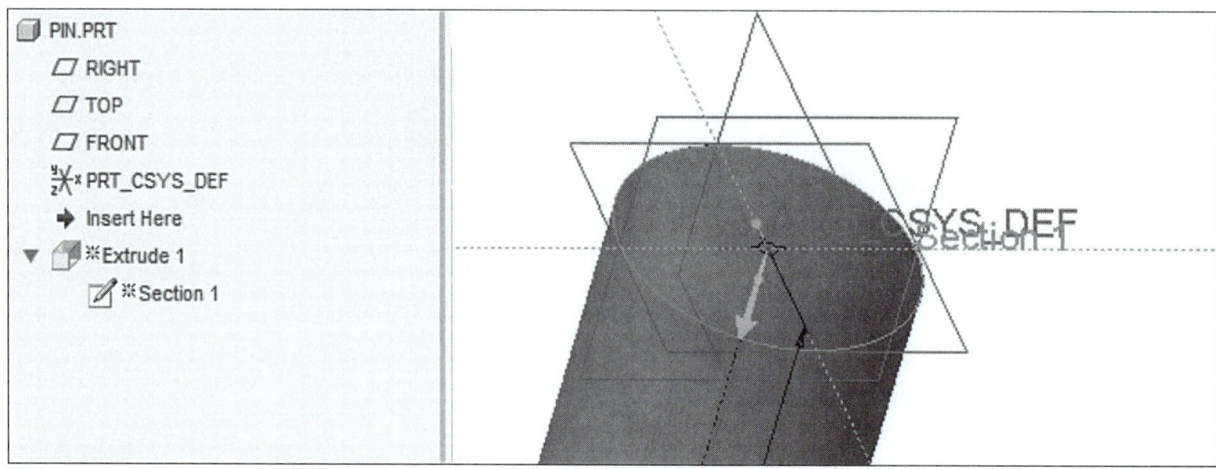

Figure 1.4(m) Dynamic Preview of Geometry

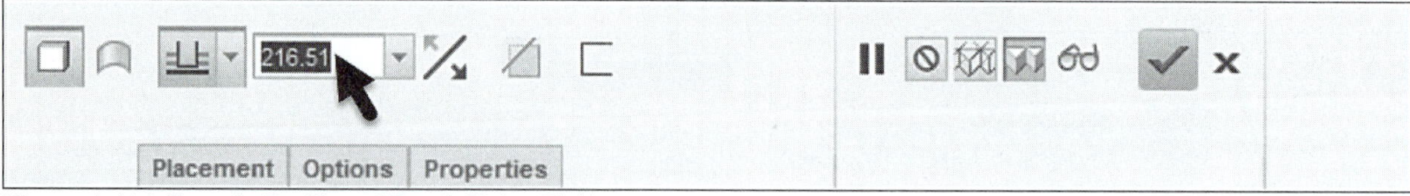

Figure 1.4(n) Extrude Dashboard Depth Value Field

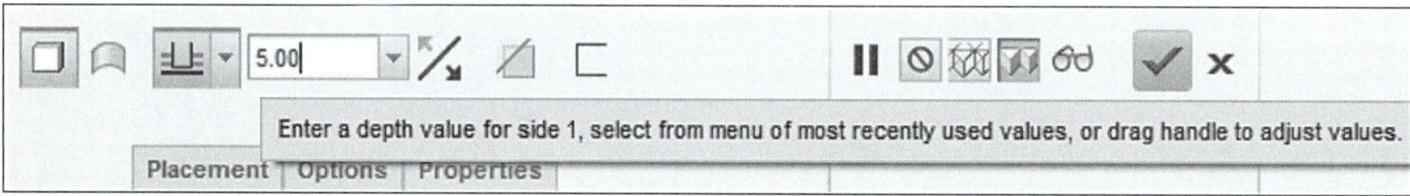

Figure 1.4(o) New Depth Value

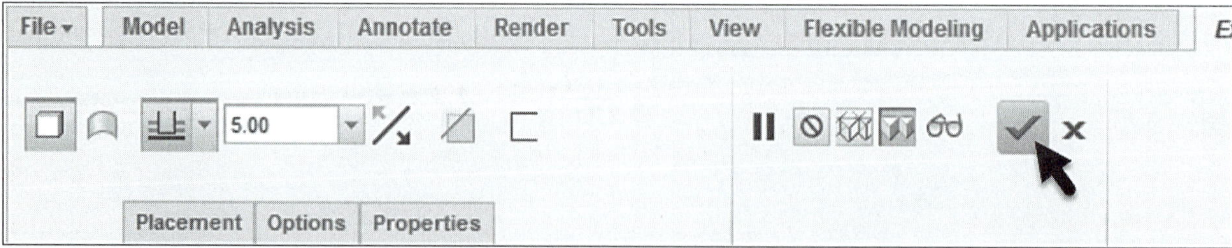

Figure 1.4(p) Complete the Extrusion

Click: Refit from the Graphics Toolbar [Figs. 1.4(q-r)] > **File** > **Save** [Fig. 1.4(s)] > **Close** from the Quick Access Toolbar *(or File > Close)* [Fig. 1.4(t)]

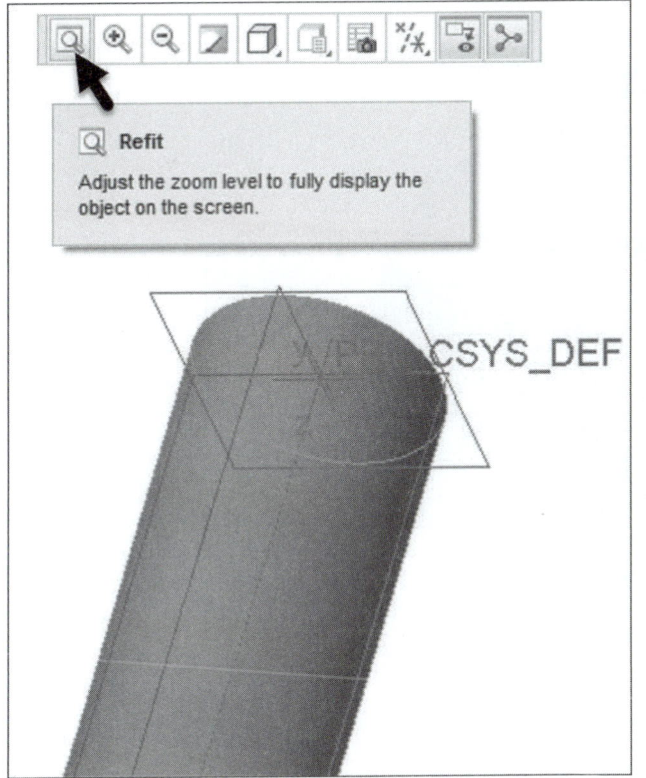

Figure 1.4(q) Refit

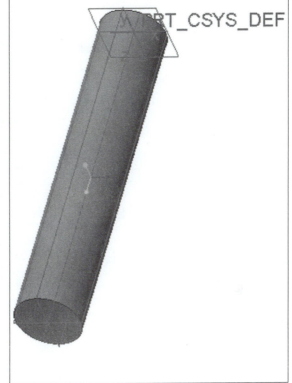

Figure 1.4(r) Completed Pin

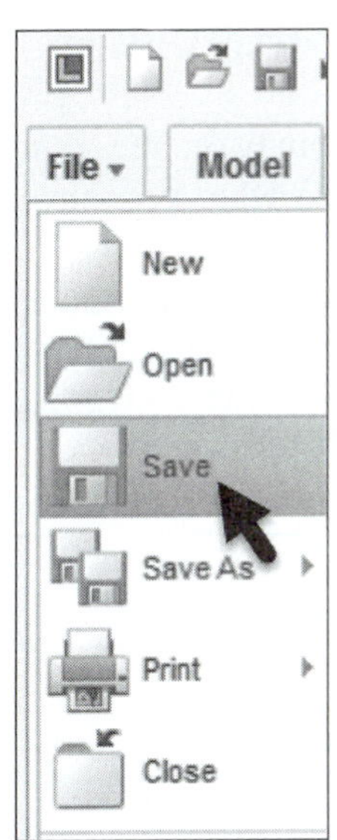

Figure 1.4(s) Save

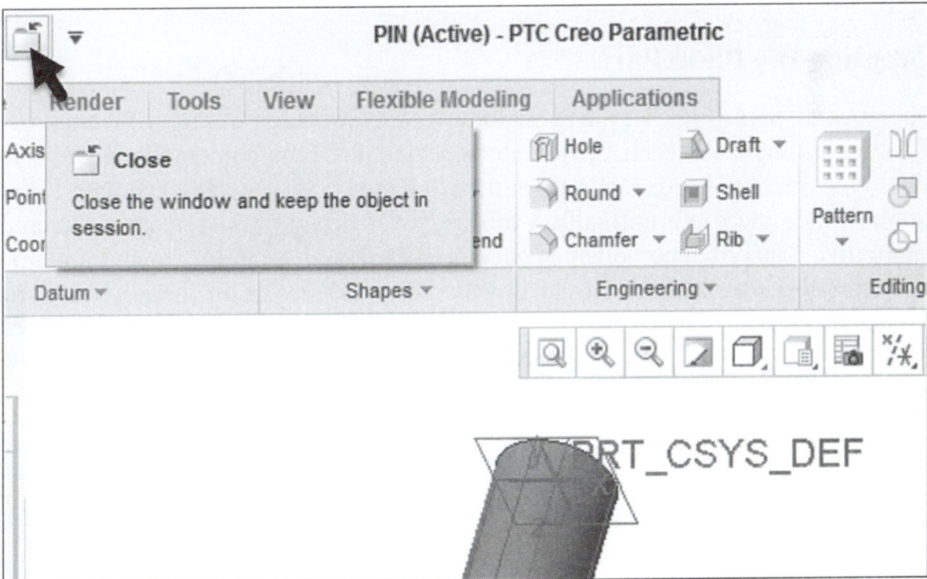

Figure 1.4(t) Close the window and keep the object in session.

You have completed your first Creo Parametric component.

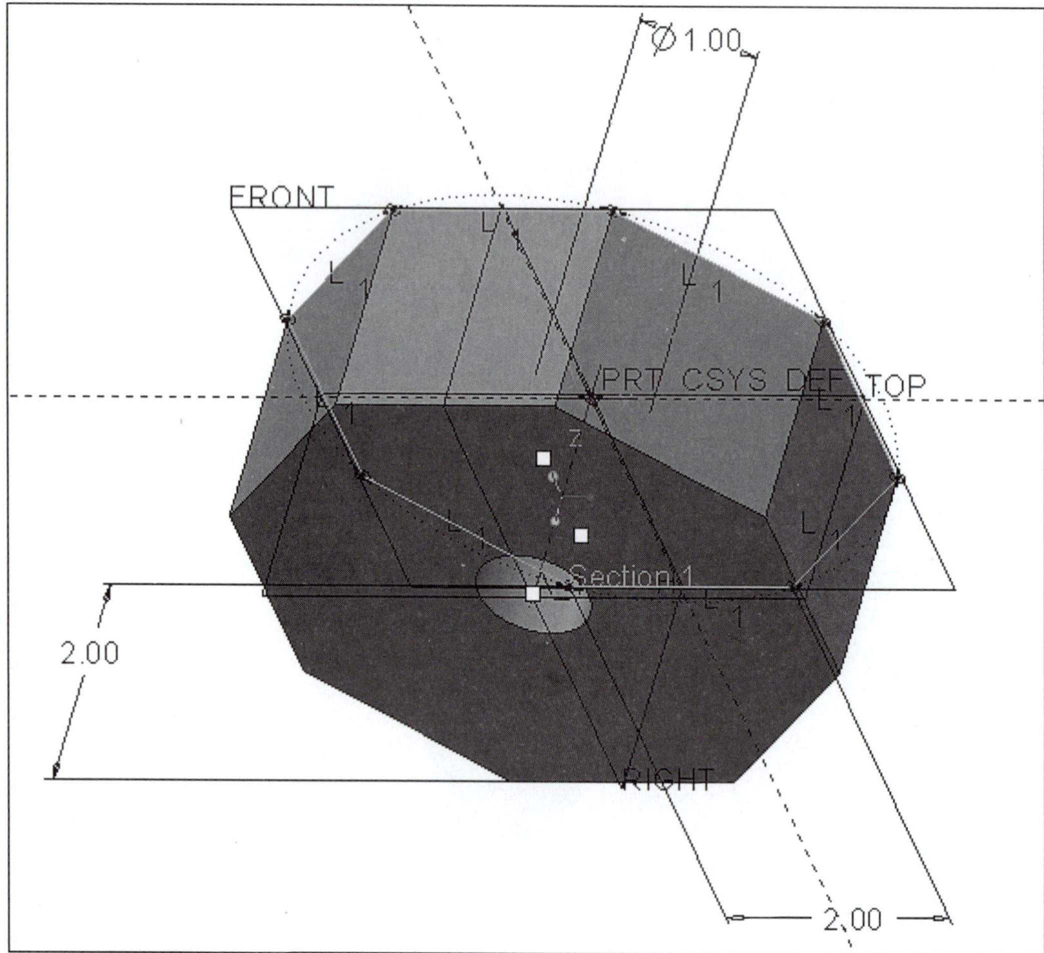

Figure 1.5 Plate

Creating the Plate Part

The second part (Fig. 1.5) will include new commands along with ones used previously in the Pin. Throughout the book each Lesson has repeated and new commands in order to reinforce what you have already learned and adds new capabilities to increase your experience with the software.

Since there are normally 3 to 4 ways of doing almost every command, the book will introduce alternative ways of accomplishing each activity. As an example, to end a command such as Extrude, you can click the ***check mark*** on the Dashboard, press ***Enter*** on the keyboard, or click the ***MMB (Middle Mouse Button)***.

Click: Select Working Directory from the Home tab Ribbon > select *your* working directory [Fig. 1.3(a)] or accept the default directory > **OK**

(Do not change directories if you have one assigned as a default by your school or company.)

Click: Create a new model > Type Part [Fig. 1.6(a)] > Sub-type **Solid** [Fig. 1.6(b)] > Name-- *type* **plate** > Use default template > click the **MMB** (**M**iddle **M**ouse **B**utton) *(or press **Enter** or pick **OK**)* [Fig. 1.6(c)] > **Save** from the Quick Access Toolbar *(PTC Creo Parametric 3.0 does NOT automatically save for you. So manually save after the completion of each feature.)* > **OK**

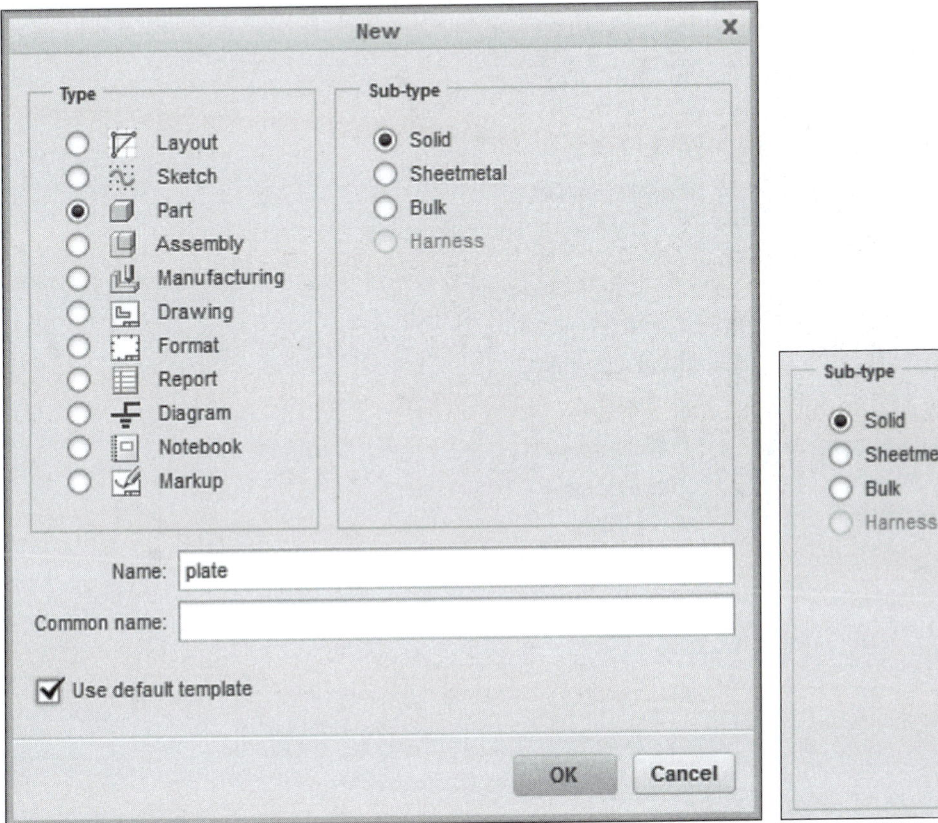

Figure 1.6(a) New Dialog Box, Part is the Default

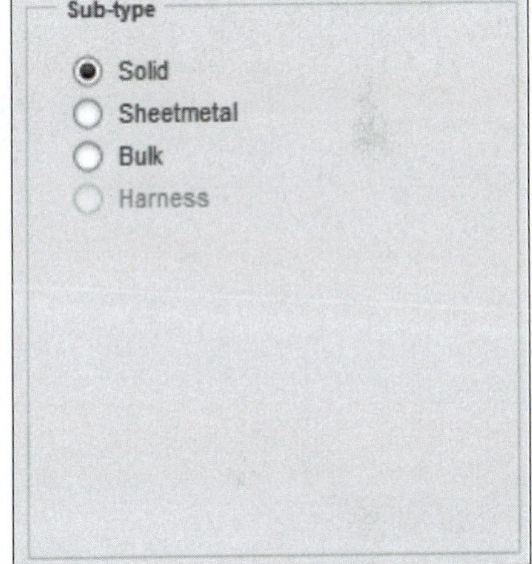

Figure 1.6(b) Sub-type Solid is the Default

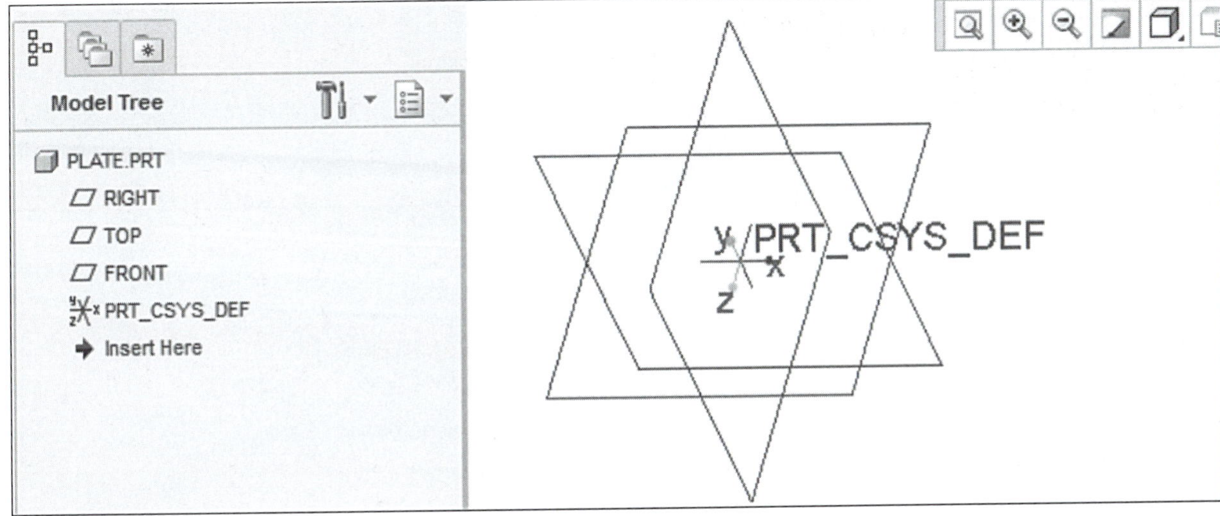

Figure 1.6(c) Plate

Click: **File** tab > **Options** [Fig. 1.6(d)] > **Sketcher** > scroll down if necessary > ☑ Show the grid [Fig. 1.6(e)] > ☑ Snap to grid > **OK** > **No**

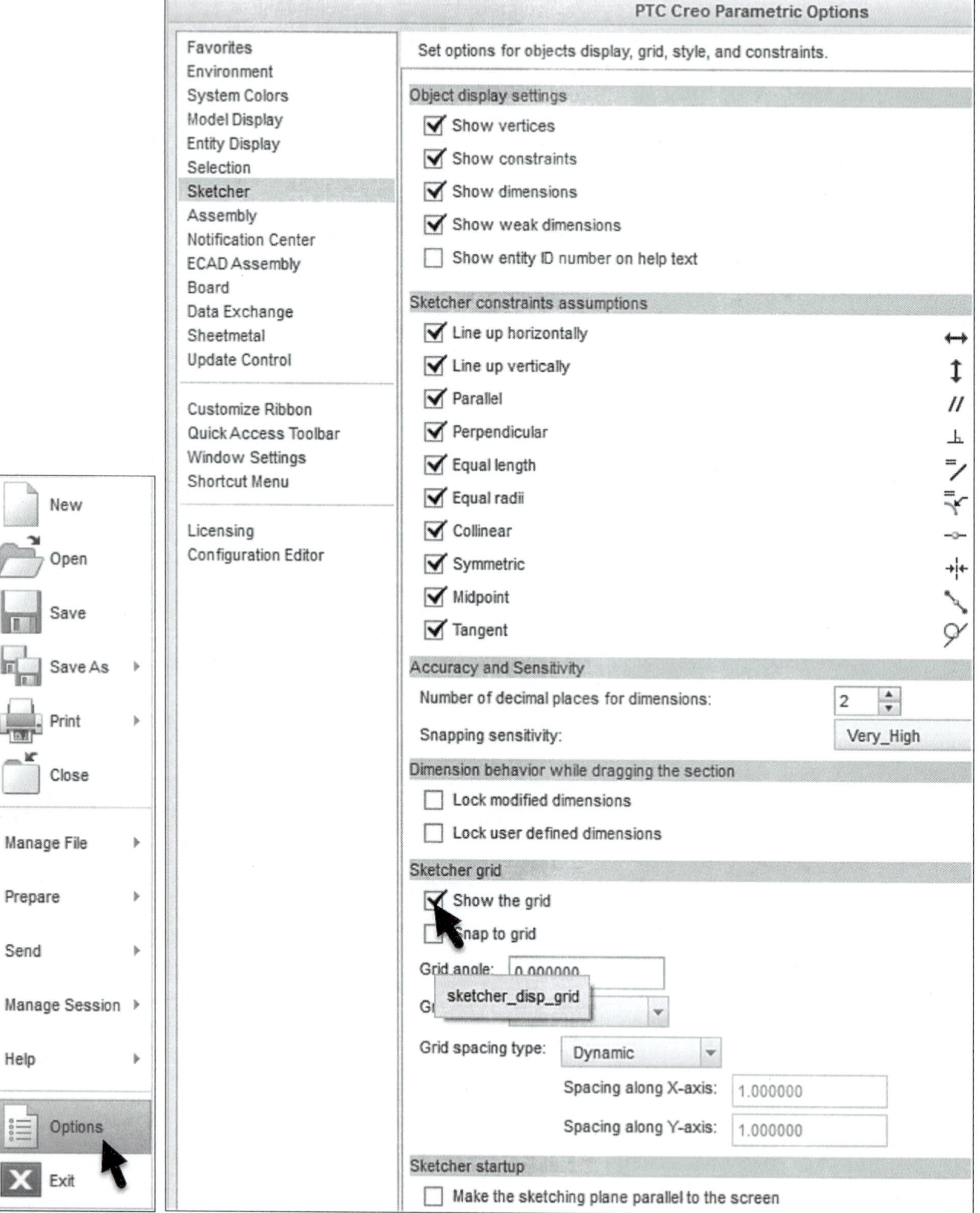

Figure 1.6(d) Options

Figure 1.6(e) Sketcher Options

Pick the **FRONT** datum plane [Fig. 1.7(a)] > Extrude from the Model tab Ribbon > from the Graphics Toolbar [Fig. 1.7(b)] or Sketch tab Ribbon Setup Group [Fig. 1.7(c)], click: **Sketch View- Orient the sketching plane parallel to the screen**

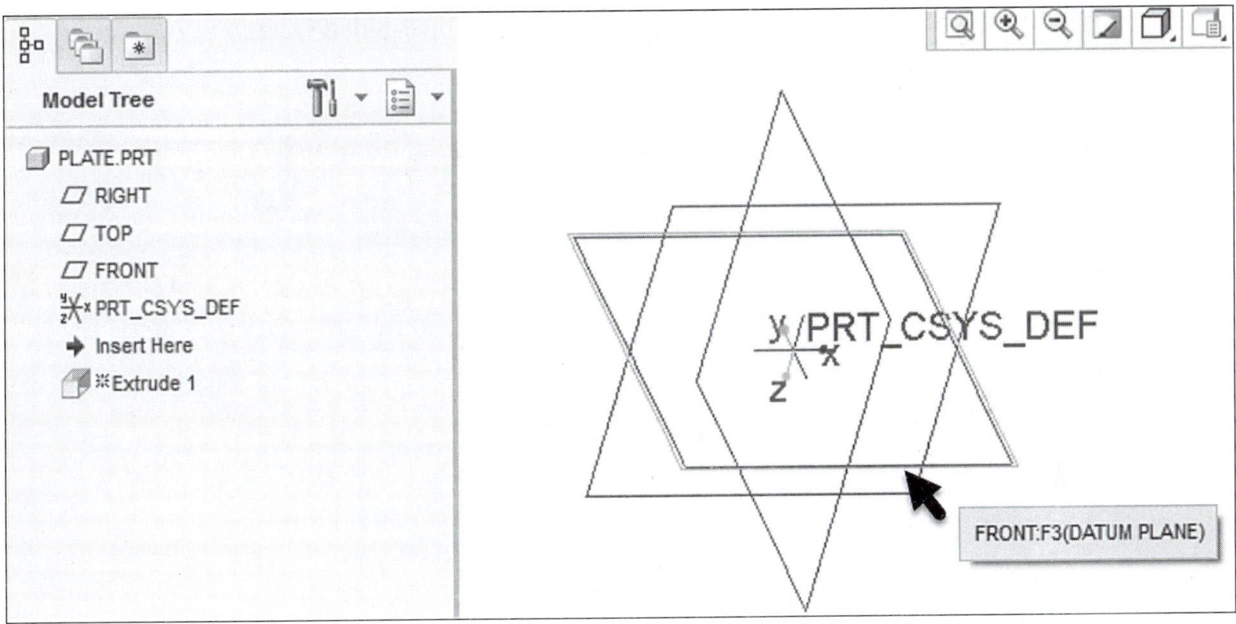

Figure 1.7(a) Select the Front Datum Plane

Figure 1.7(b) Sketch View from the Graphics Toolbar

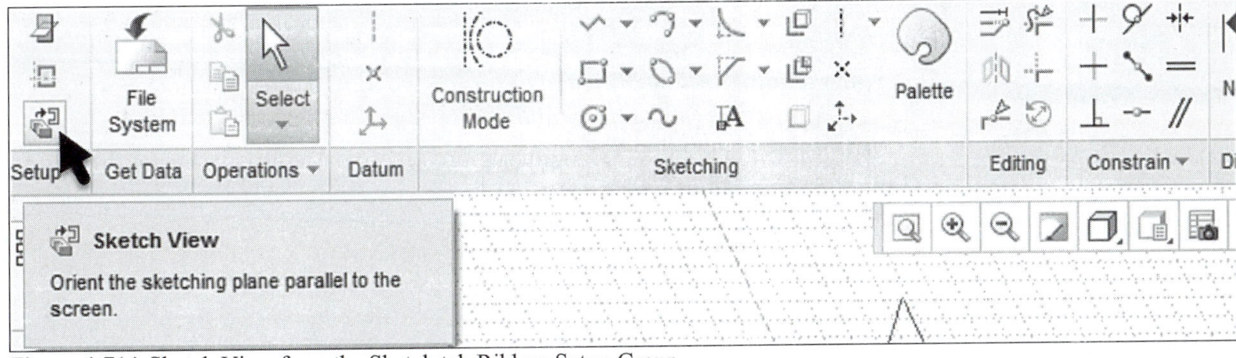

Figure 1.7(c) Sketch View from the Sketch tab Ribbon Setup Group

Click: **View** tab > from the Ribbon, Display Style toggle all *on* > **Refit** from the Graphics Toolbar > **Sketch** tab [Fig. 1.7(d)] > Palette **Import data from a palette into the active object** [Fig. 1.7(e)] > within the Polygons tab (default selection), scroll down to see the octagon 8-Sided Octagon > double-click LMB (or Drag & Drop) to select on **8-Sided Octagon** [Fig. 1.7(f)]

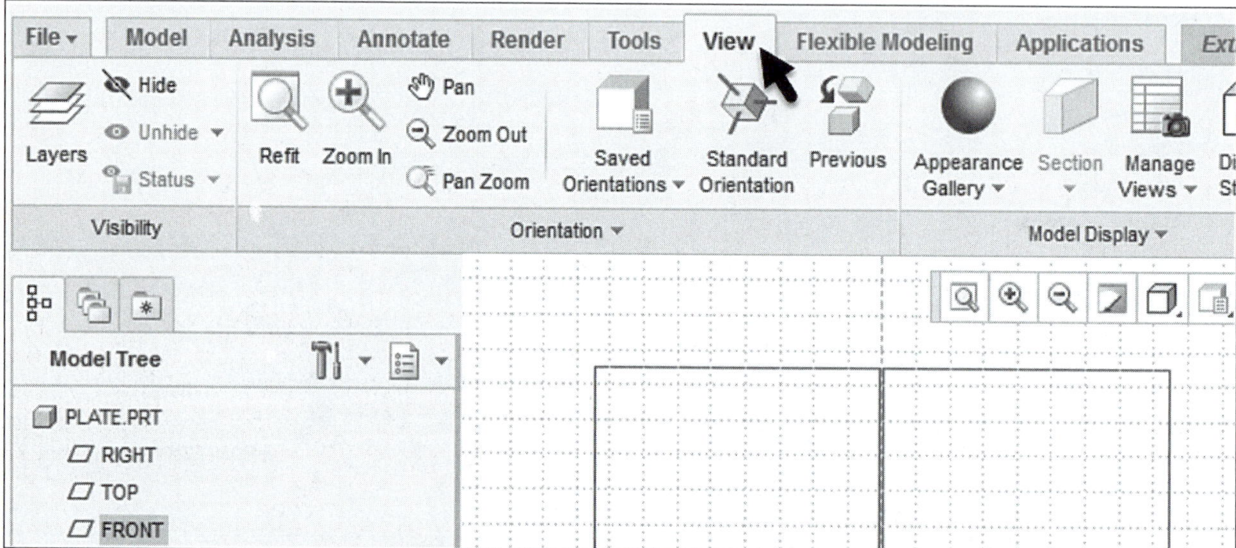

Figure 1.7(d) View Tab. Sketch Plane parallel to the screen

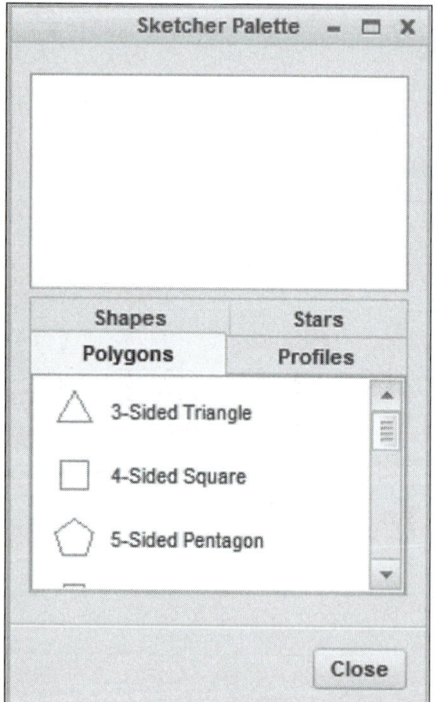

Figure 1.7(e) Sketcher Palette

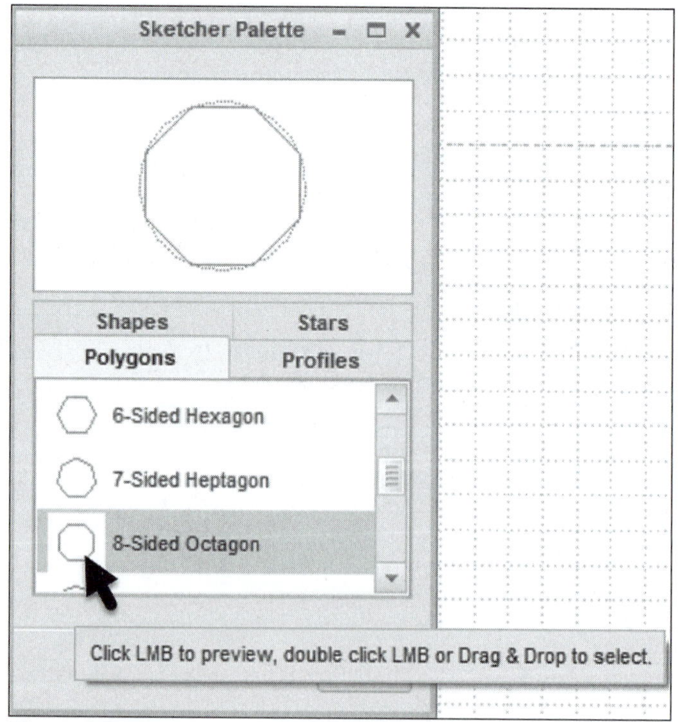

Figure 1.7(f) Polygons tab: 8-Sided Octagon

Click: **LMB** to place the octagon on the sketch (pick a position **not** on the CYSY) [Fig. 1.7(g)] > place the pointer at the center of the octagon > press and hold down the **LMB** (**L**eft **M**ouse **B**utton) to drag (move) the octagon > drop (release the **LMB** at) the center of the octagon at the origin > modify the Scaling Factor value to **2** [Fig. 1.7(h)] > **Enter** > ✓ in the Ribbon > **Close** from the Sketcher Palette

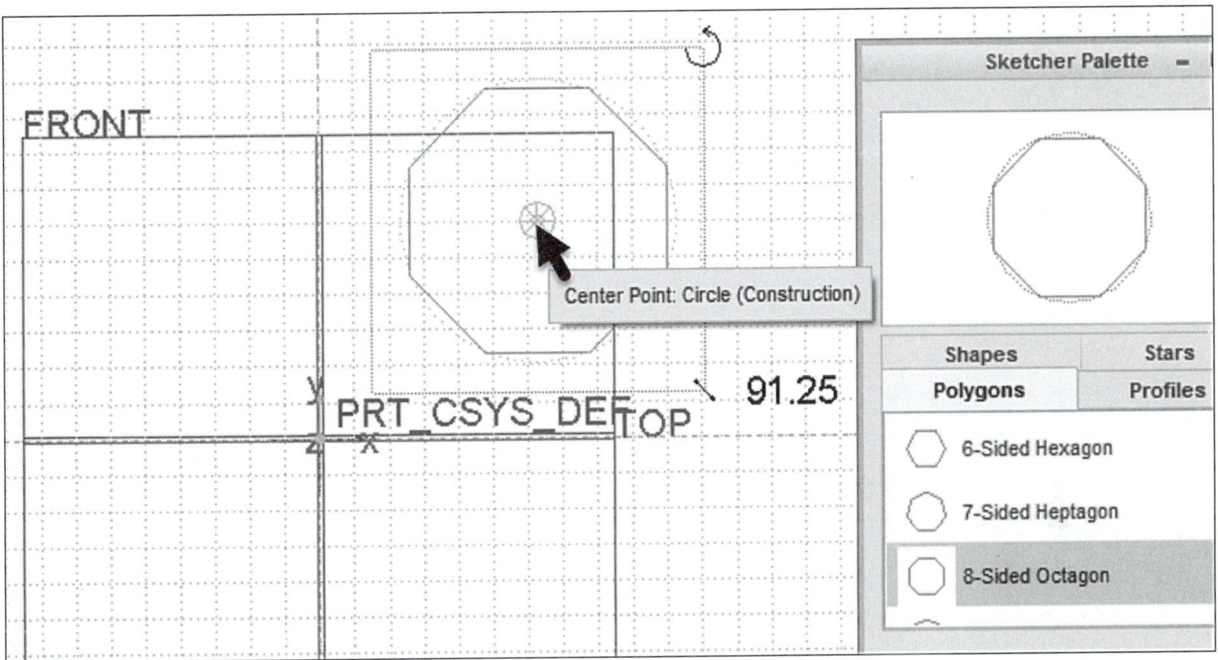

Figure 1.7(g) Place Octagon anywhere on the Sketch by picking a Position with the **LMB**

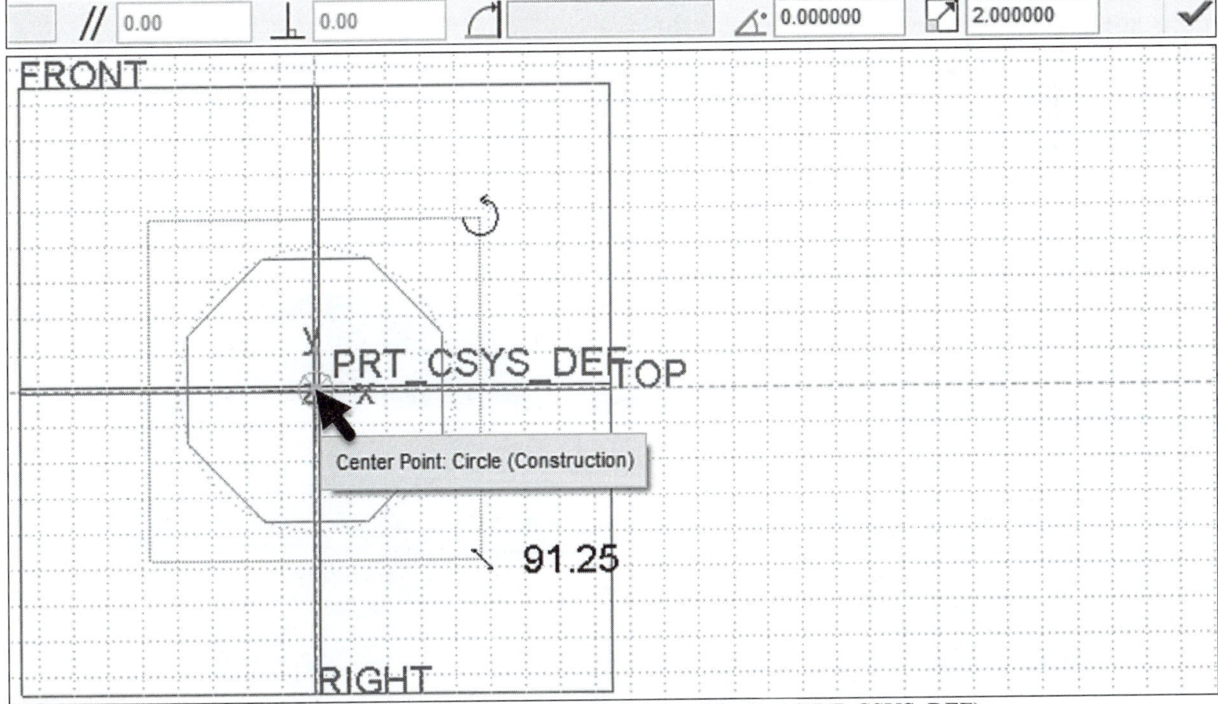

Figure 1.7(h) Drag and Drop the Center of the Octagon at the Origin of the Part (PRT_CSYS_DEF)

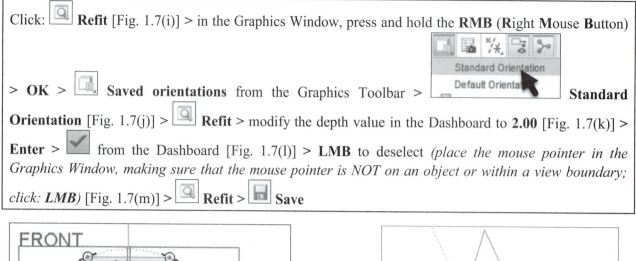

Click: **Refit** [Fig. 1.7(i)] > in the Graphics Window, press and hold the **RMB** (**R**ight **M**ouse **B**utton) > **OK** > **Saved orientations** from the Graphics Toolbar > Standard Orientation > **Standard Orientation** [Fig. 1.7(j)] > **Refit** > modify the depth value in the Dashboard to **2.00** [Fig. 1.7(k)] > **Enter** > from the Dashboard [Fig. 1.7(l)] > **LMB** to deselect *(place the mouse pointer in the Graphics Window, making sure that the mouse pointer is NOT on an object or within a view boundary; click: **LMB**)* [Fig. 1.7(m)] > **Refit** > **Save**

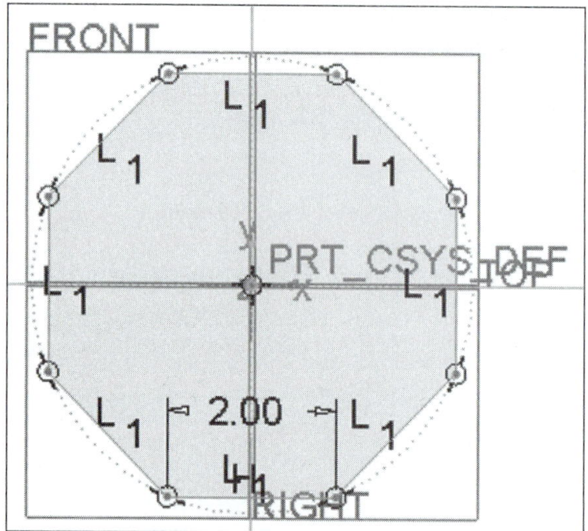

Figure 1.7(i) Octagon Sketch

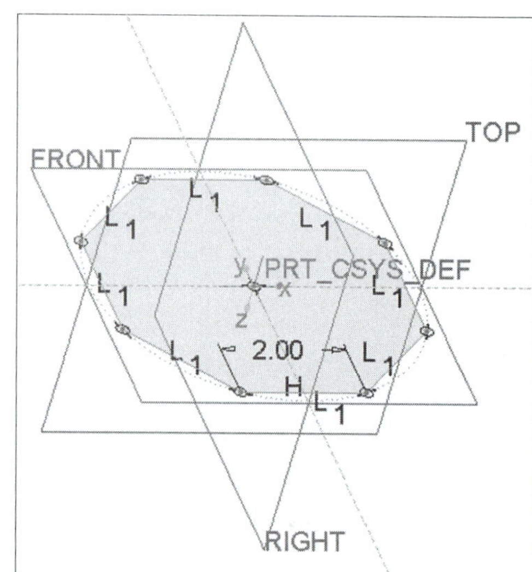

Figure 1.7(j) Standard Orientation

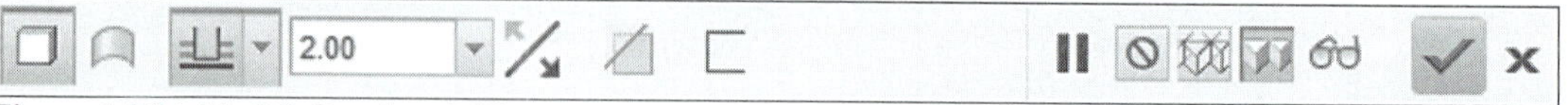

Figure 1.7(k) Modify Depth to **2.00**

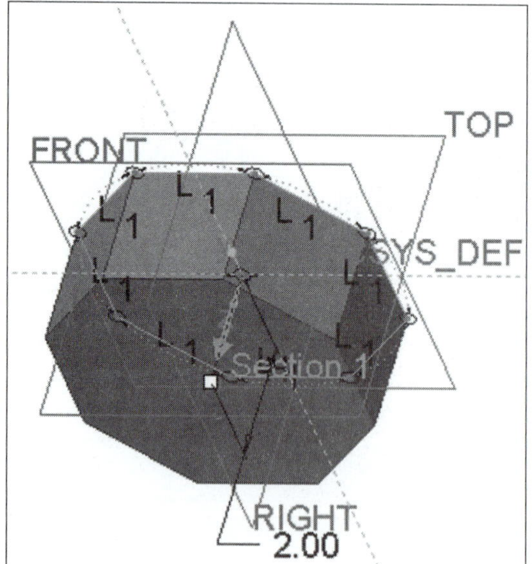

Figure 1.7(l) Depth Preview

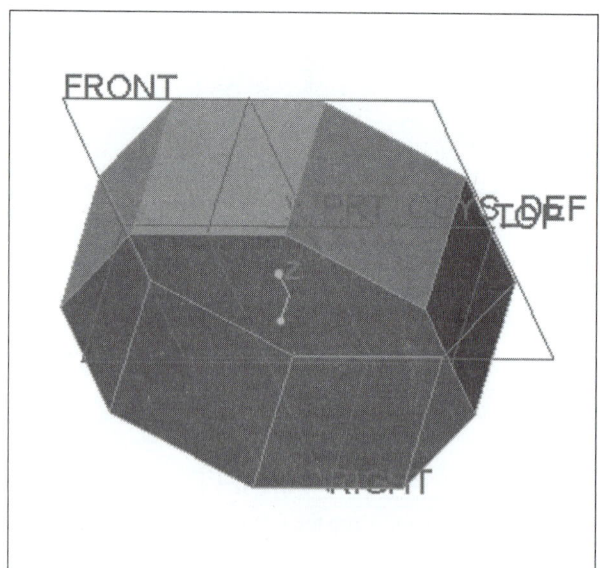

Figure 1.7(m) Extruded Octagon

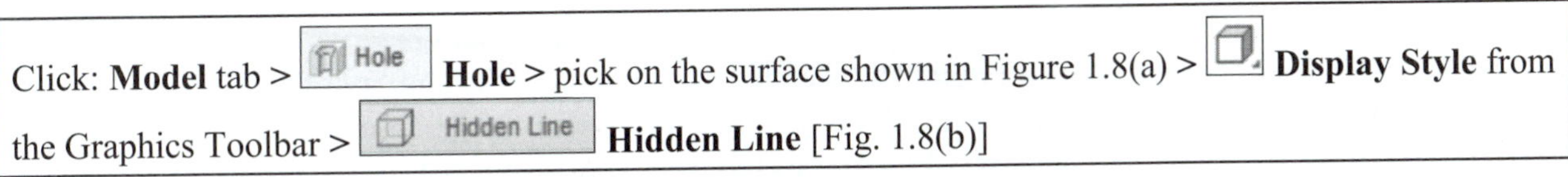

Click: **Model** tab > Hole **Hole** > pick on the surface shown in Figure 1.8(a) > **Display Style** from the Graphics Toolbar > Hidden Line **Hidden Line** [Fig. 1.8(b)]

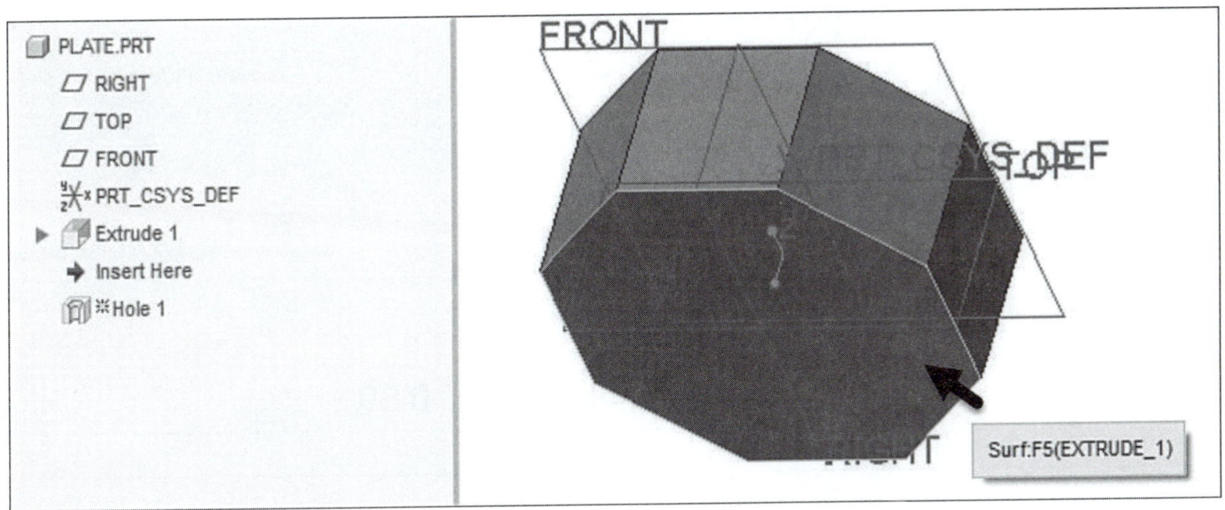

Figure 1.8(a) Select the Surface

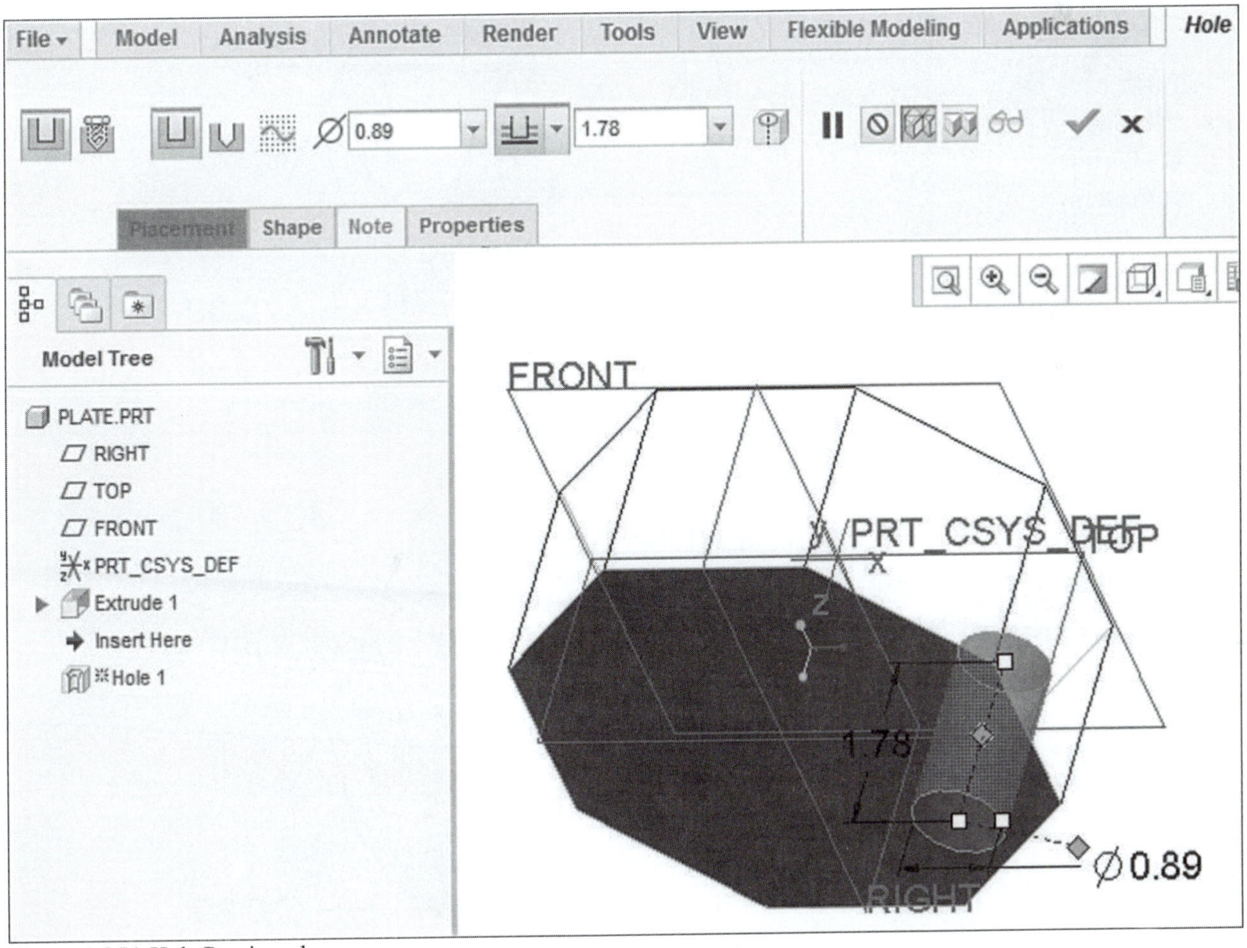

Figure 1.8(b) Hole Previewed

In the Graphics Window *(making sure that the pointer is NOT on an object or within a view boundary)*, press and hold: **RMB** [Fig. 1.8(c)] > **Offset References Collector** > press and hold the **Ctrl** key > select datum **RIGHT** from the Model Tree > with the **Ctrl** key still pressed, select datum **TOP** from the Model Tree [Fig. 1.8(d)] > release the **Ctrl** key > **Placement** tab from the Dashboard [Fig. 1.8(e)]

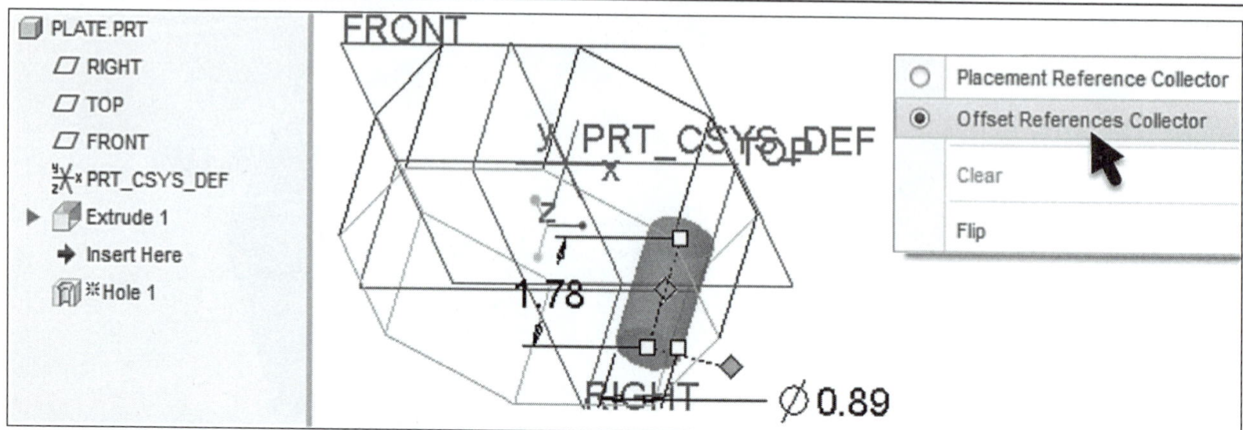

Figure 1.8(c) RMB > Offset References Collector

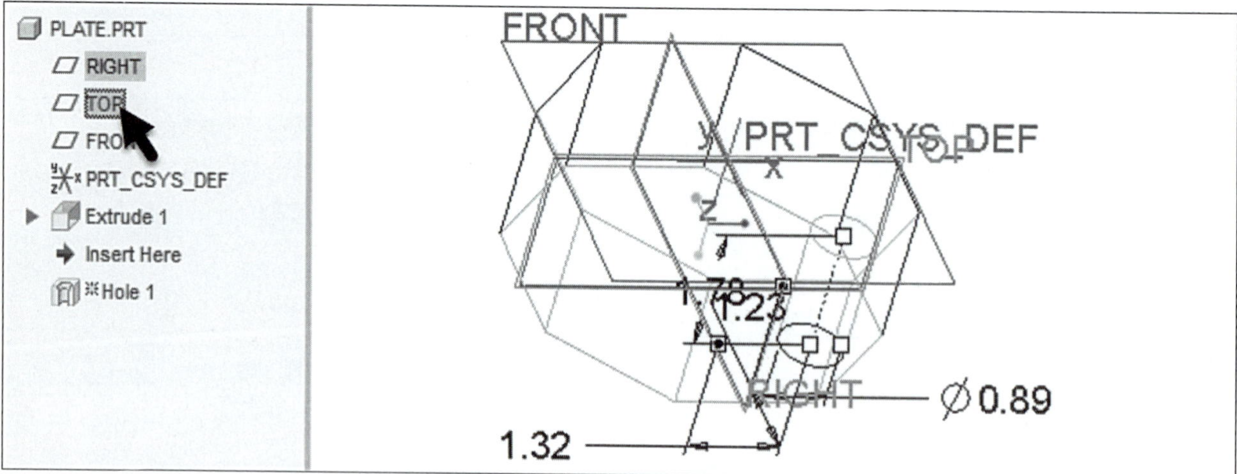

Figure 1.8(d) Select the RIGHT and the TOP Datum Planes (from the Model Tree) as Offset References

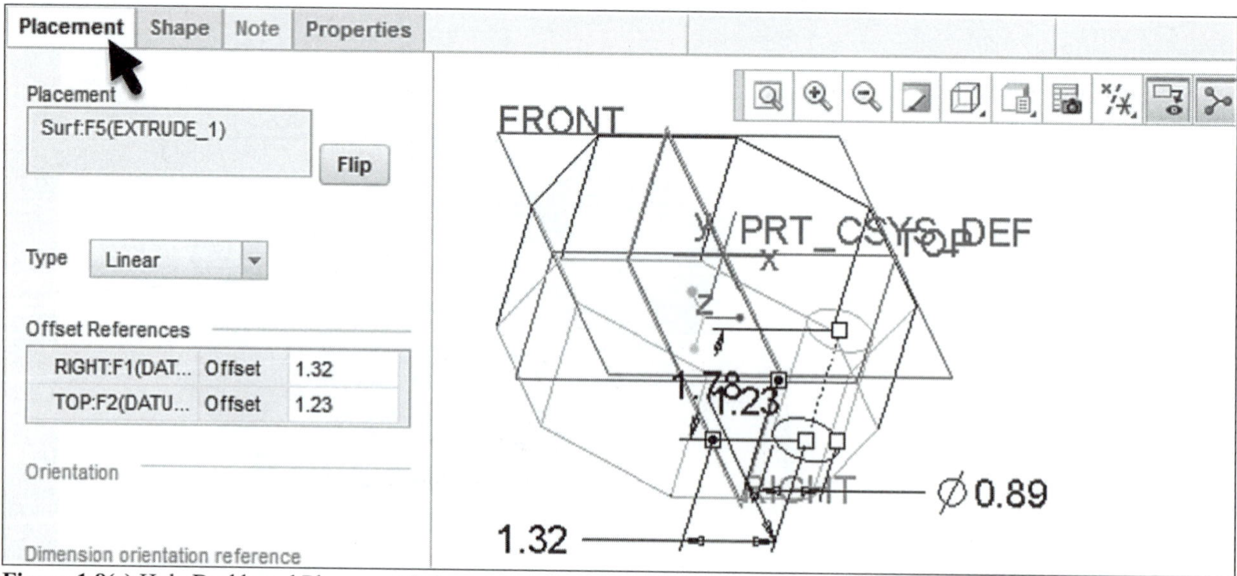

Figure 1.8(e) Hole Dashboard Placement Tab

In the Offset References collector, click on **Offset**, next to RIGHT [Fig. 1.8(f)] > click on **Offset** > **Align** [Fig. 1.8(g)] > click on **Offset**, next to TOP [Fig. 1.8(h)] > click on **Offset** > **Align** [Figs. 1.8(i-j)]

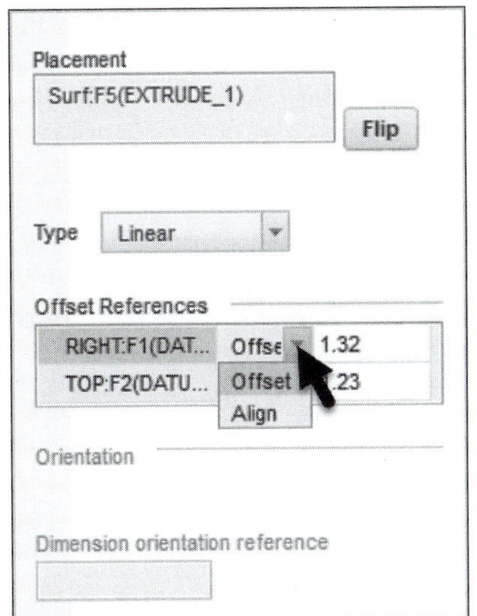

Figure 1.8(f) Click on Offset, next to RIGHT

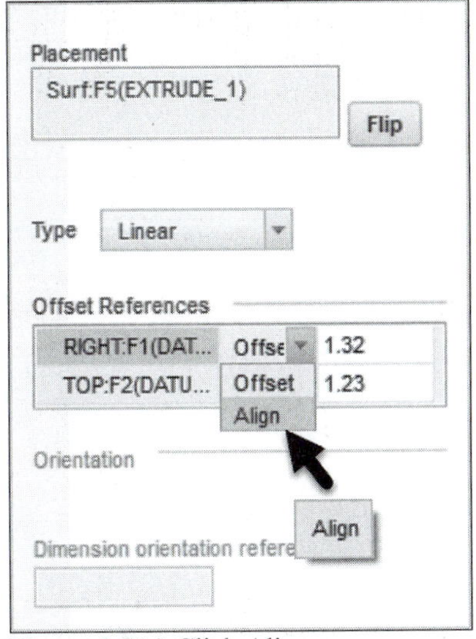

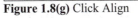

Figure 1.8(g) Click Align

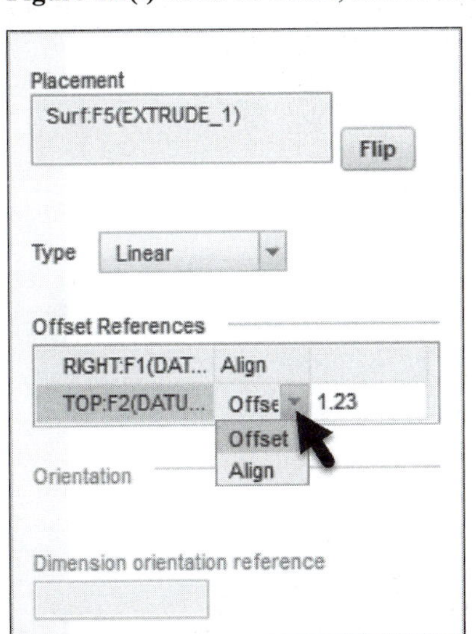

Figure 1.8(h) Click on Offset, next to TOP

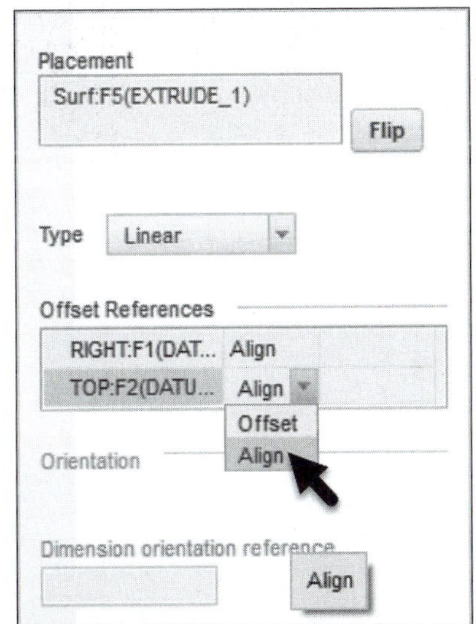

Figure 1.8(i) Click Align

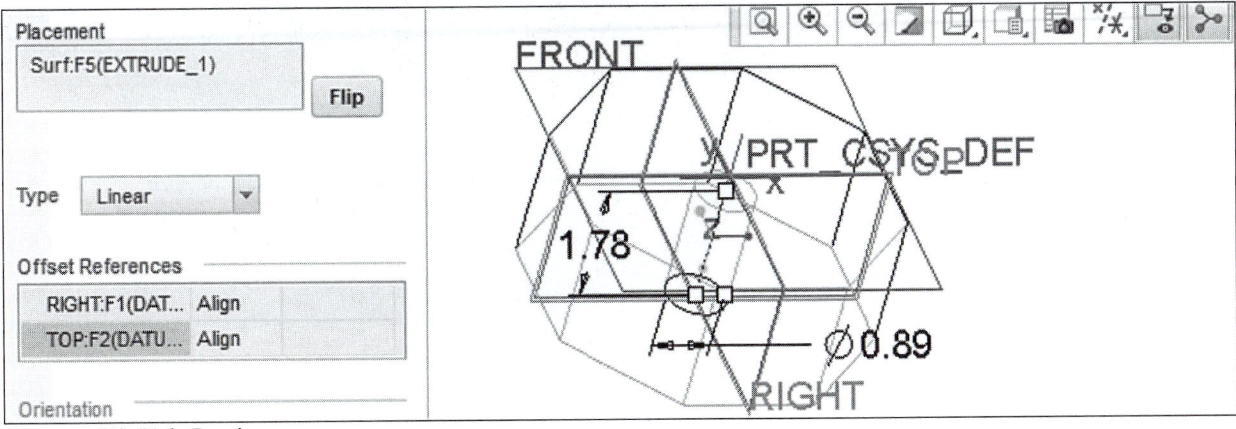

Figure 1.8(j) Hole Preview

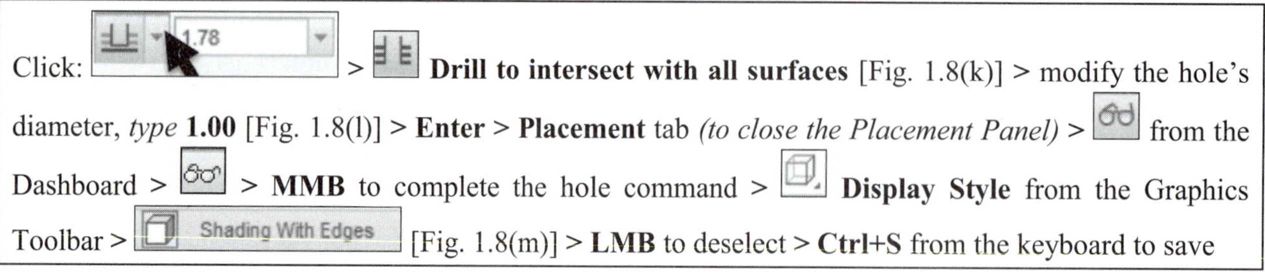

Click: 1.78 > **Drill to intersect with all surfaces** [Fig. 1.8(k)] > modify the hole's diameter, *type* **1.00** [Fig. 1.8(l)] > **Enter** > **Placement** tab *(to close the Placement Panel)* > from the Dashboard > > **MMB** to complete the hole command > **Display Style** from the Graphics Toolbar > Shading With Edges [Fig. 1.8(m)] > **LMB** to deselect > **Ctrl+S** from the keyboard to save

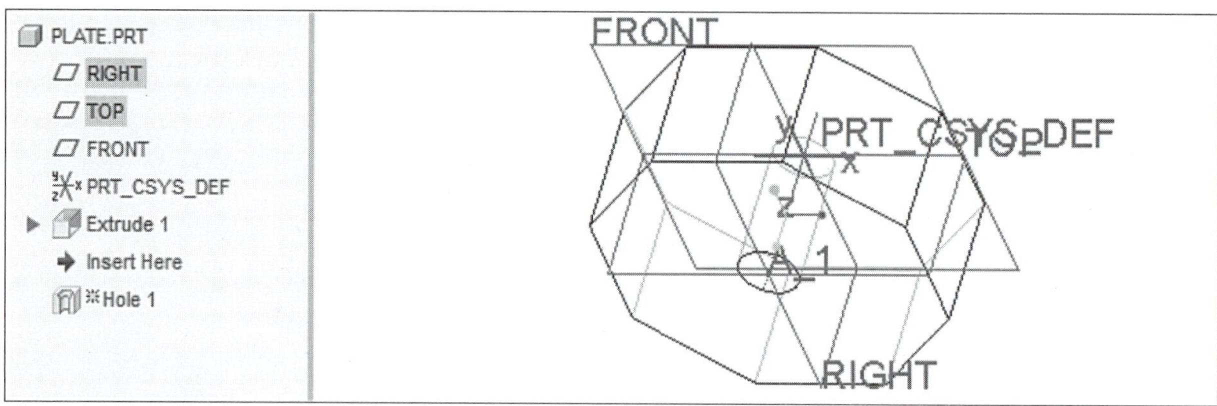

Figure 1.8(k) Drill to intersect with all surfaces

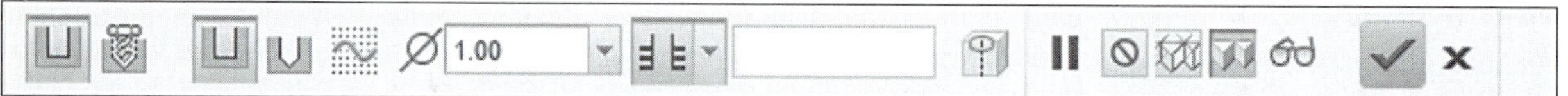

Figure 1.8(l) Diameter **1.00**

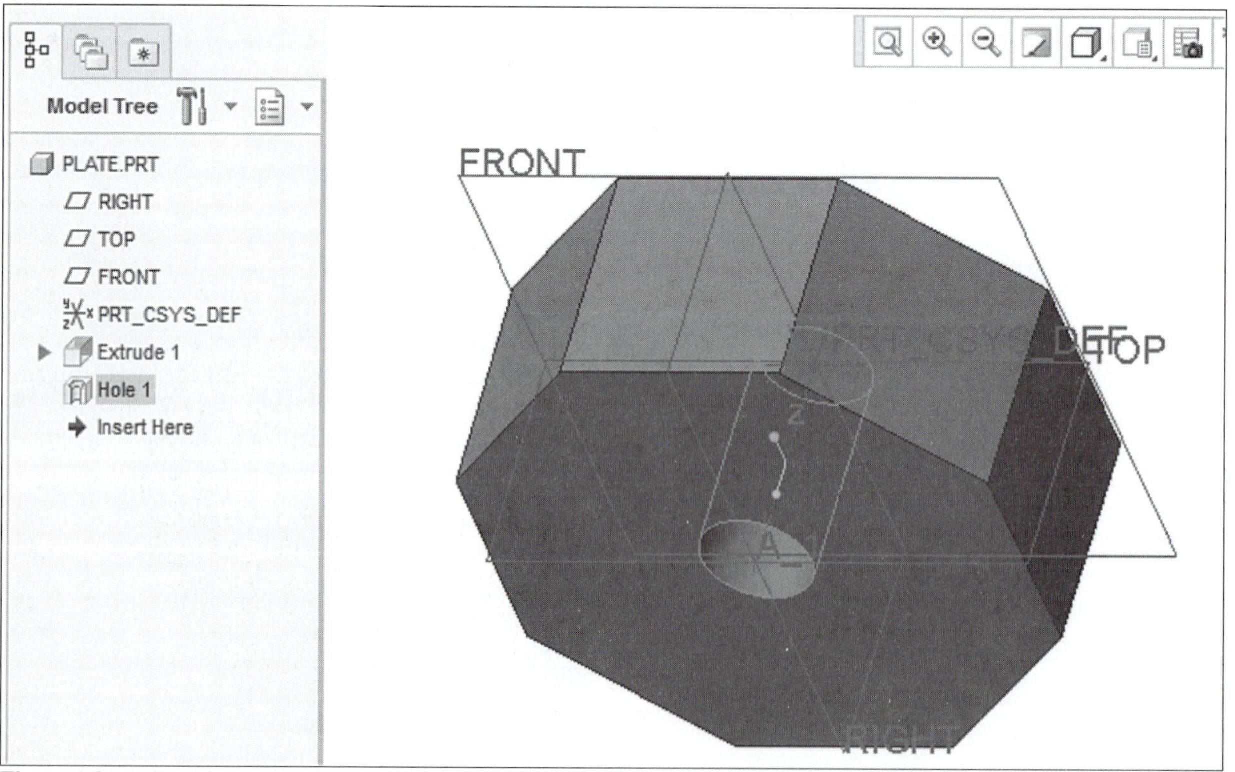

Figure 1.8(m) Completed Hole

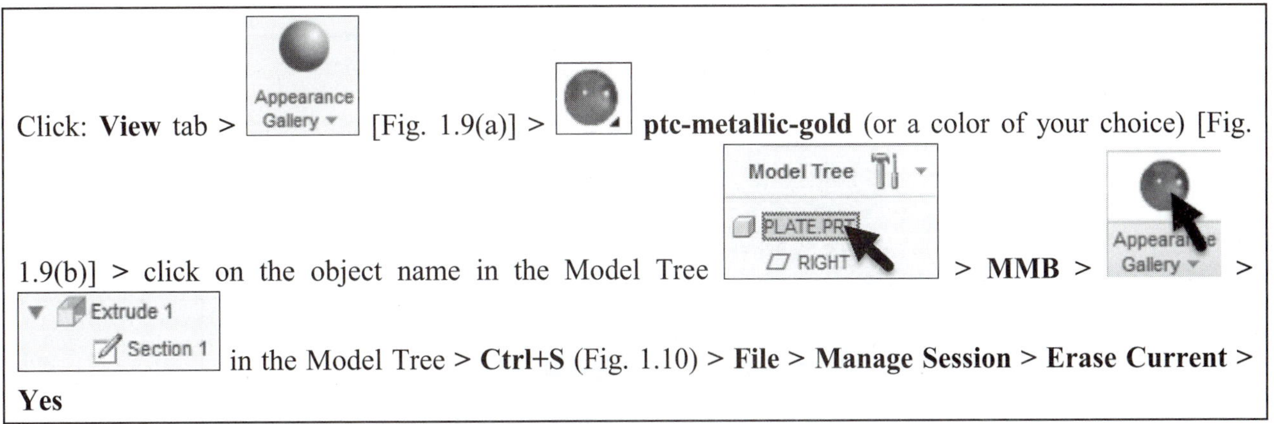

Click: **View** tab > Appearance Gallery [Fig. 1.9(a)] > **ptc-metallic-gold** (or a color of your choice) [Fig. 1.9(b)] > click on the object name in the Model Tree > **MMB** > Appearance Gallery > Extrude 1 in the Model Tree > **Ctrl+S** (Fig. 1.10) > **File** > **Manage Session** > **Erase Current** > **Yes**

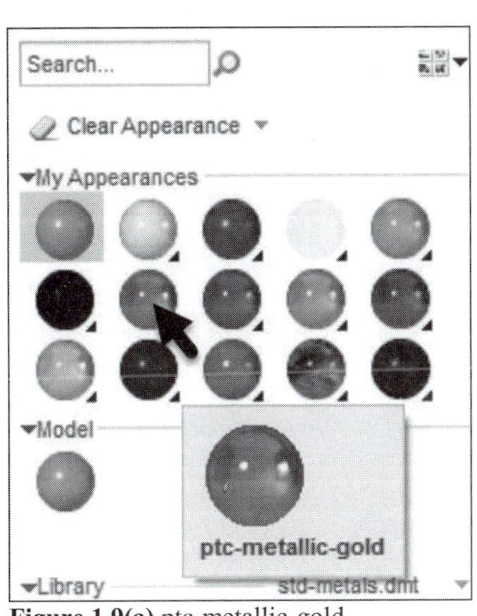

Figure 1.9(a) ptc-metallic-gold

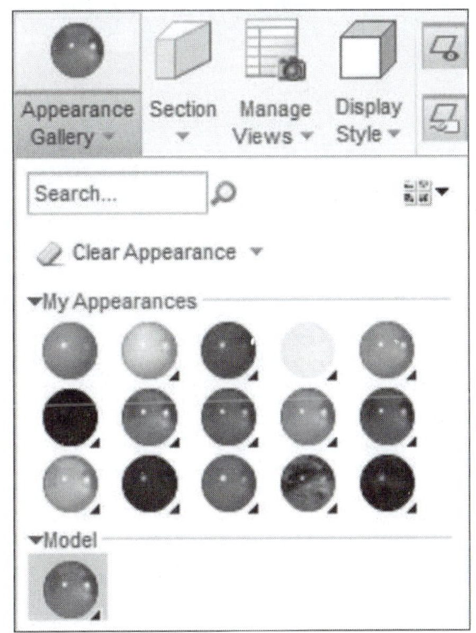

Figure 1.9(b) Appearance Gallery

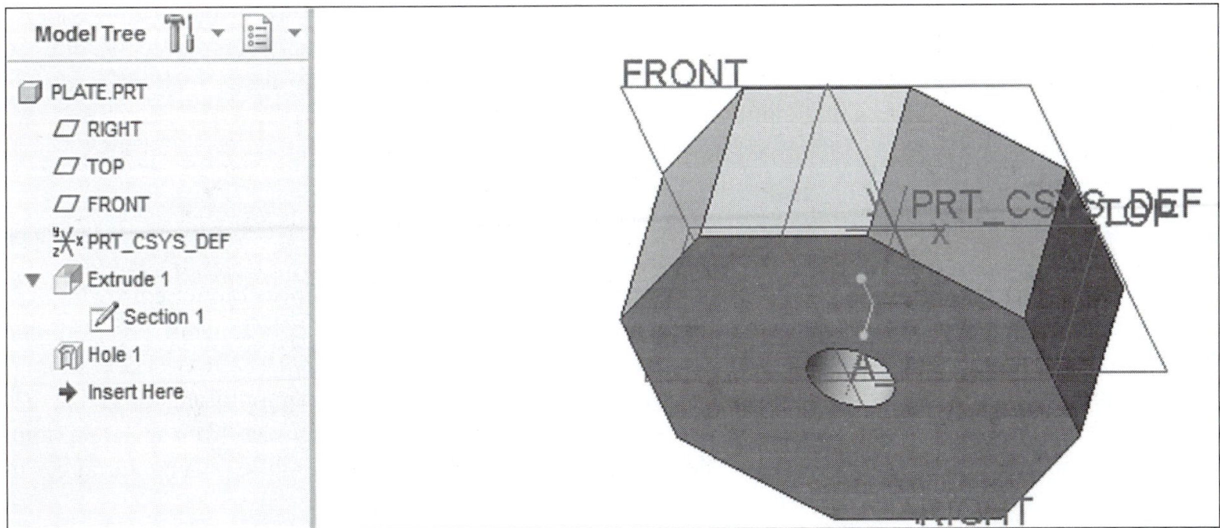

Figure 1.10 Completed Plate

You have completed your second Creo Parametric component. Next, you will put the parts together to create an assembly.

Figure 1.11 Assembly

Creating the Assembly

Using the two parts just modeled you will now create an assembly (Fig. 1.11). Just as you can combine features into parts, you can also combine parts into assemblies. The Assembly mode in PTC Creo Parametric 3.0 enables you to place component parts and subassemblies together to form assemblies.

As with a part, an assembly starts with default datum planes and a default coordinate system. To create a subassembly or an assembly, you must first create datum features. You then create or assemble additional components to the existing component(s) and datum features.

The 3D Dragger is a graphical tool for precision handling of assembly components. The *axes* are called **draggers**. Pull or rotate the draggers to make changes to the components position and orientation. The draggers enable movement in one or more *degrees of freedom* (DOFs). They support linear, planar, free, translation, and angular movements. You can pull a dragger along a 2D or 3D trajectory. When you move a geometric entity using a dragger, the relative distances from the start point of the entity appear in the graphics window. You can hold down the SHIFT key while pulling a dragger to snap the dragger to a geometric reference.

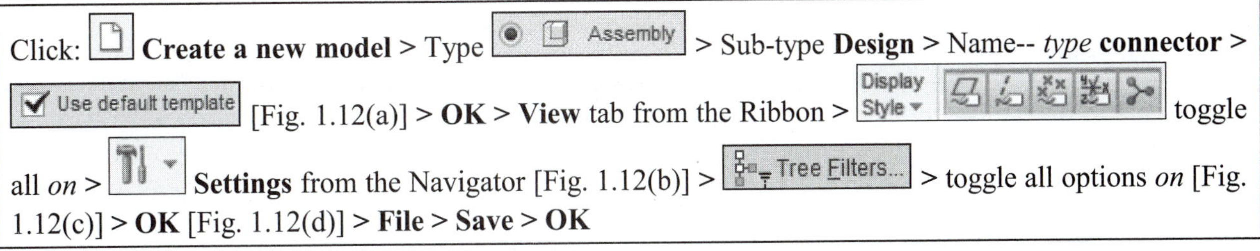

Click: **Create a new model** > Type Assembly > Sub-type **Design** > Name-- *type* **connector** > Use default template [Fig. 1.12(a)] > **OK** > **View** tab from the Ribbon > Display Style toggle all *on* > **Settings** from the Navigator [Fig. 1.12(b)] > Tree Filters... > toggle all options *on* [Fig. 1.12(c)] > **OK** [Fig. 1.12(d)] > **File** > **Save** > **OK**

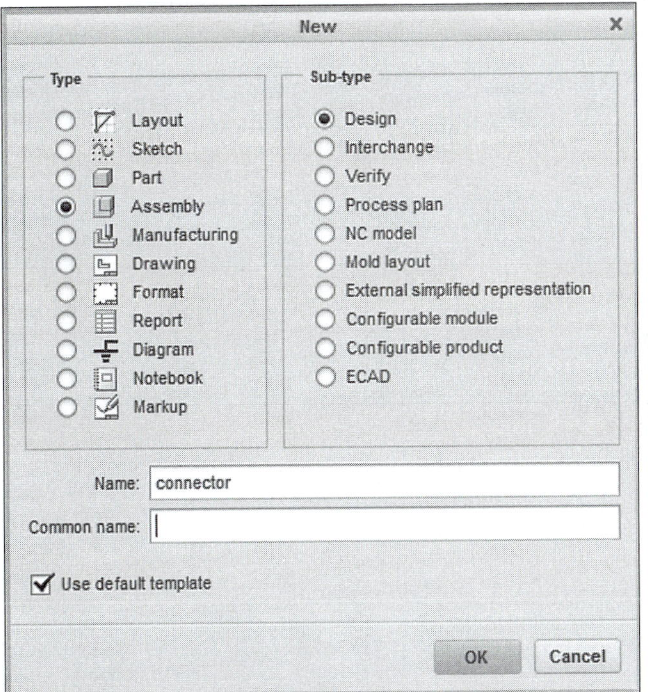

Figure 1.12(a) New Dialog Box, Assembly

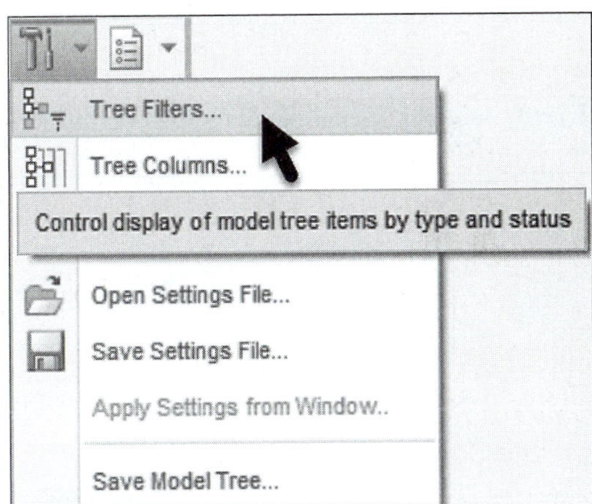

Figure 1.12(b) Tree Filters

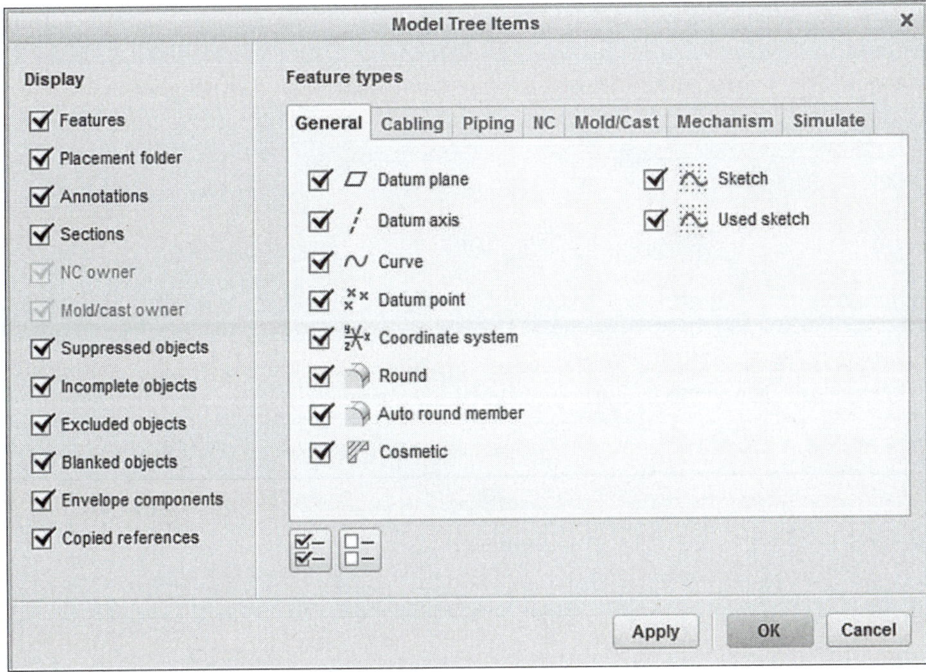

Figure 1.12(c) Model Tree Items Dialog Box

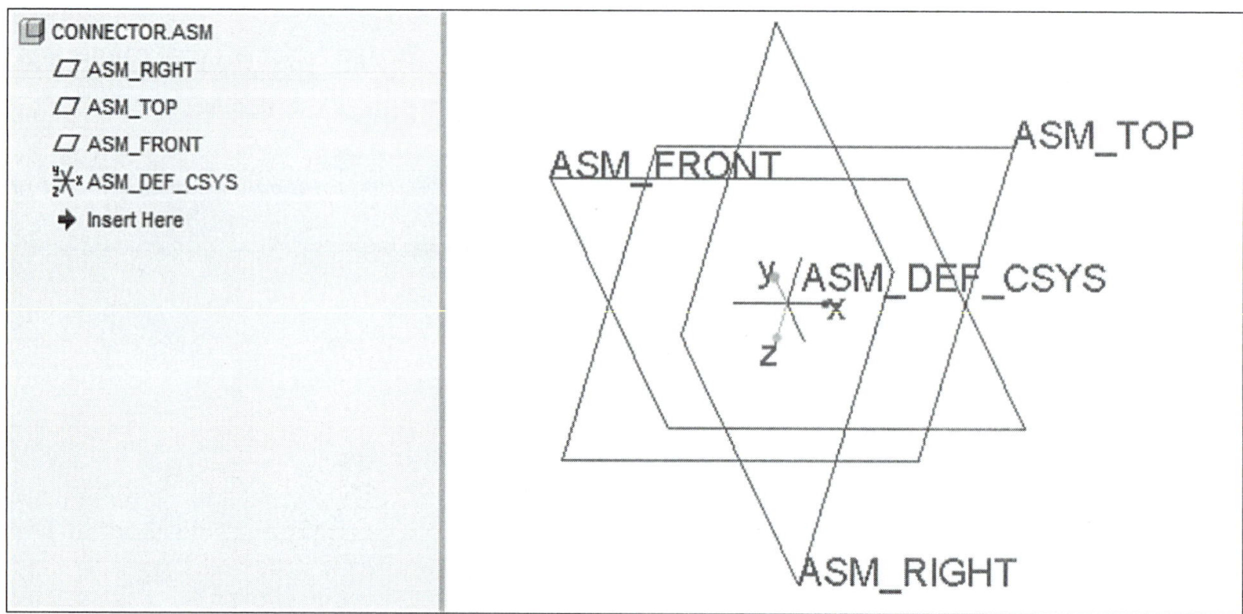

Figure 1.12(d) Assembly Default Datums and Coordinate System

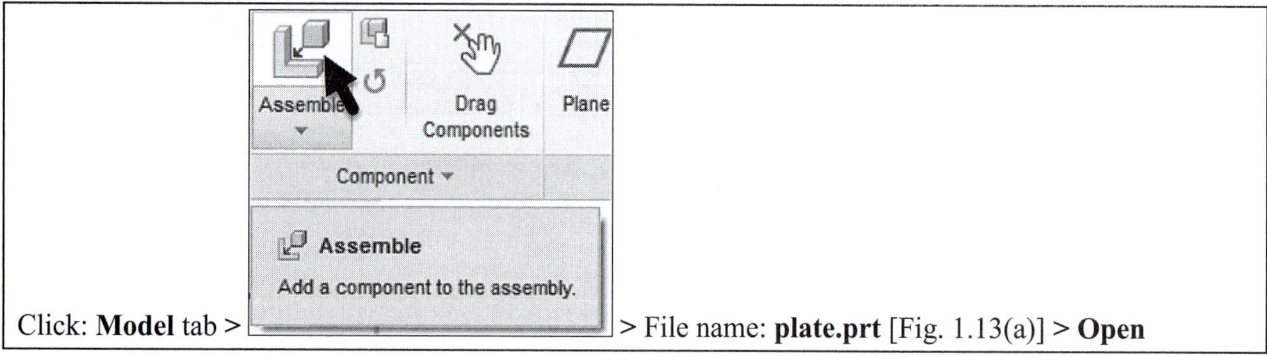

Click: **Model** tab > > File name: **plate.prt** [Fig. 1.13(a)] > **Open**

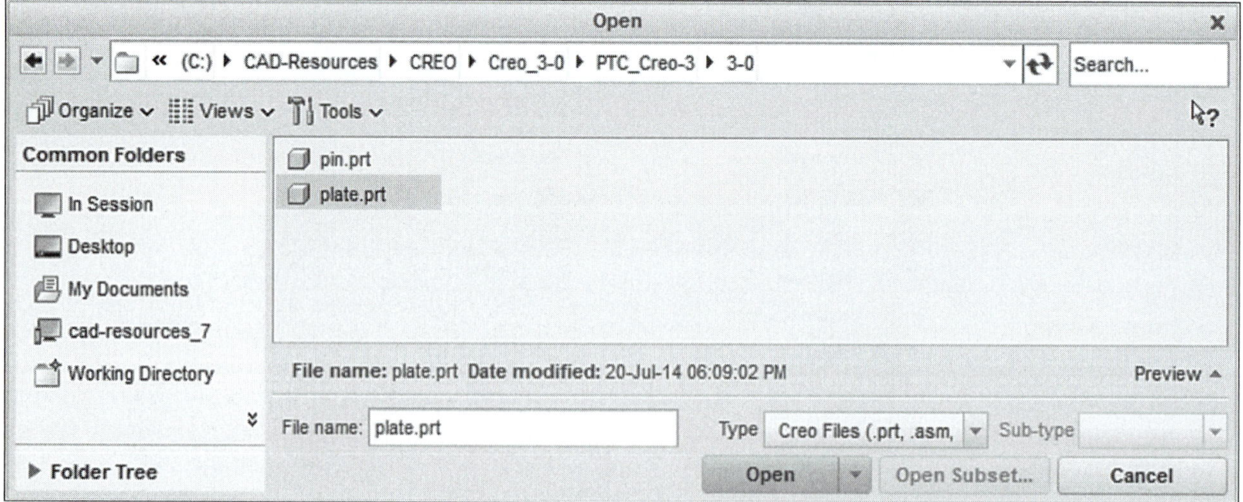

Figure 1.13(a) Open Dialog Box, Type Creo Files

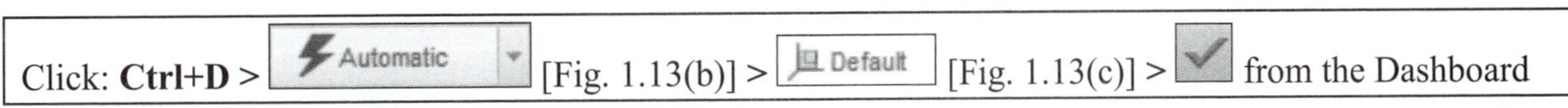

Click: **Ctrl+D** > Automatic [Fig. 1.13(b)] > Default [Fig. 1.13(c)] > ✓ from the Dashboard

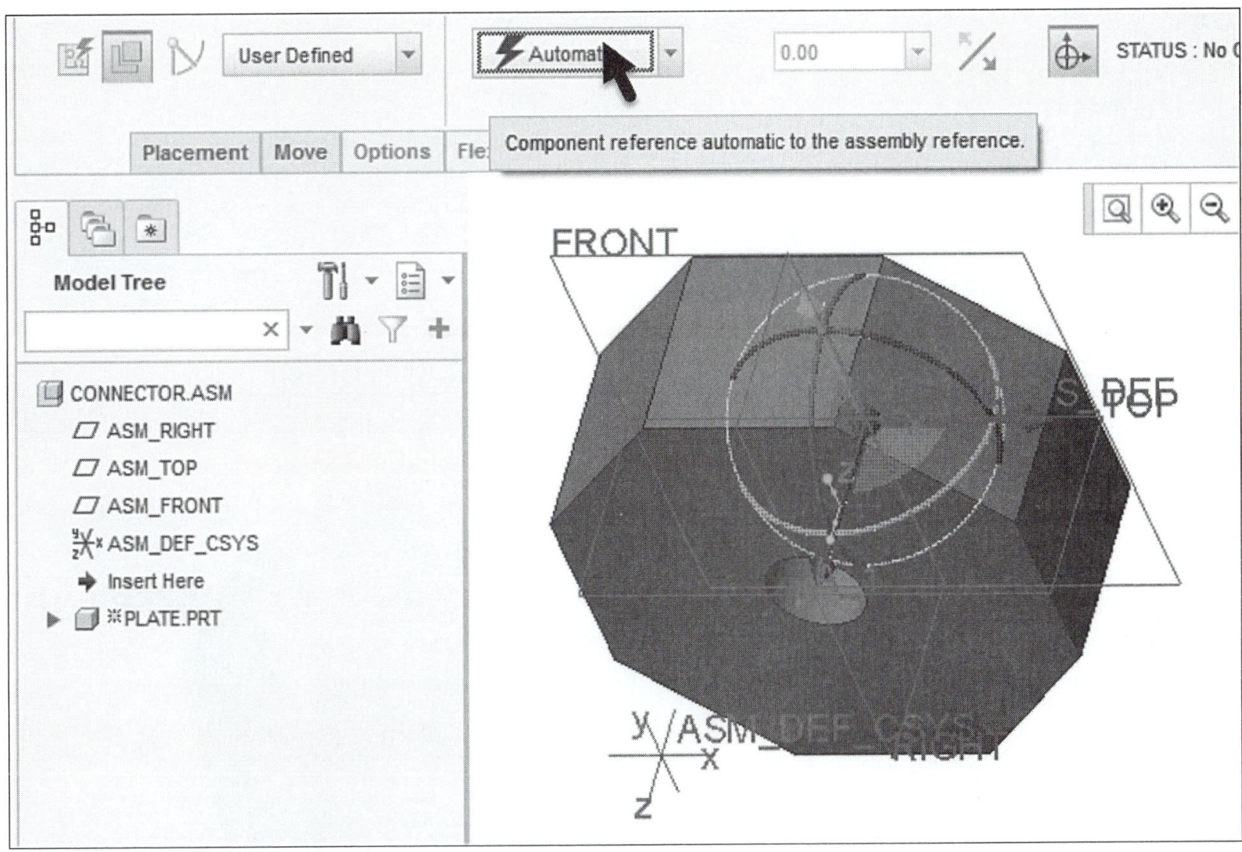

Figure 1.13(b) Assembly Dashboard

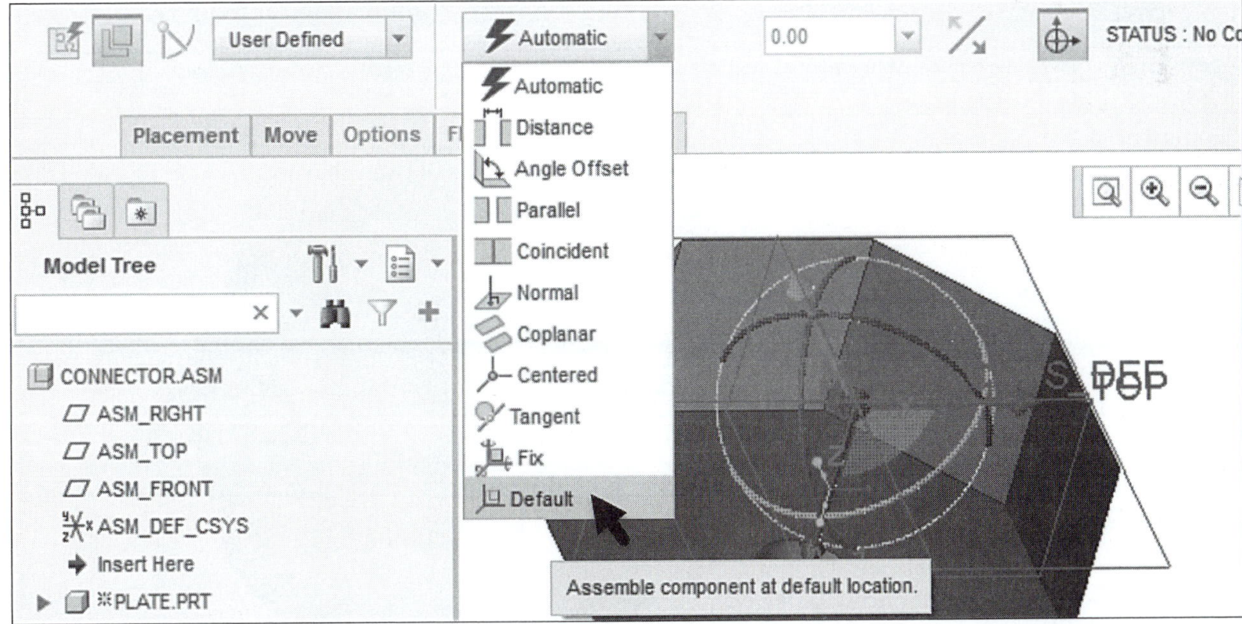

Figure 1.13(c) Default Constraint

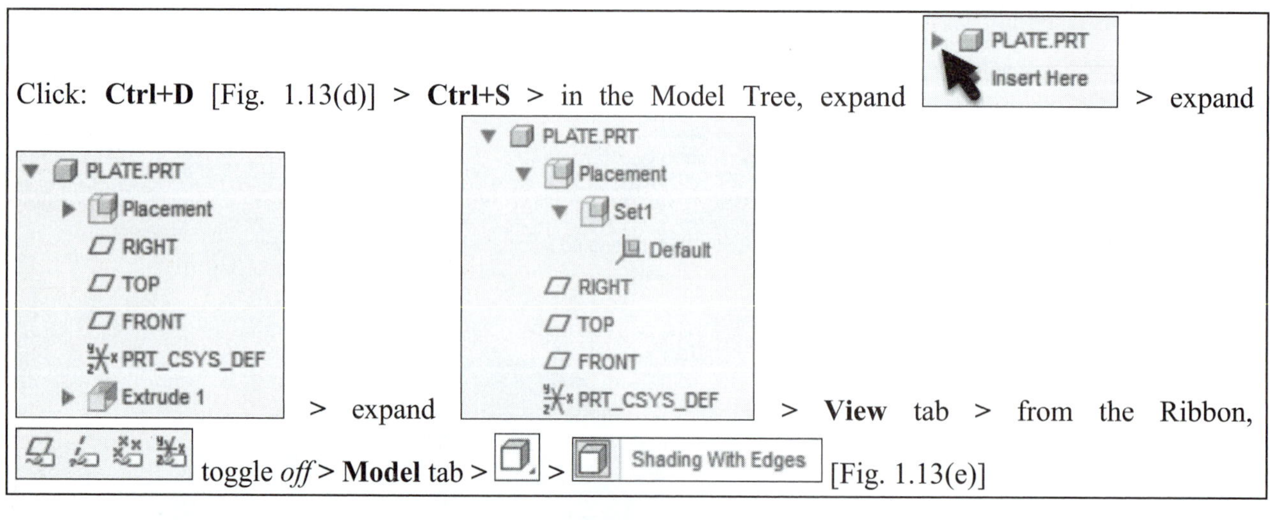

Click: **Ctrl+D** [Fig. 1.13(d)] > **Ctrl+S** > in the Model Tree, expand PLATE.PRT (Insert Here) > expand Placement > expand Set1 > **View** tab > from the Ribbon, toggle *off* > **Model** tab > > Shading With Edges [Fig. 1.13(e)]

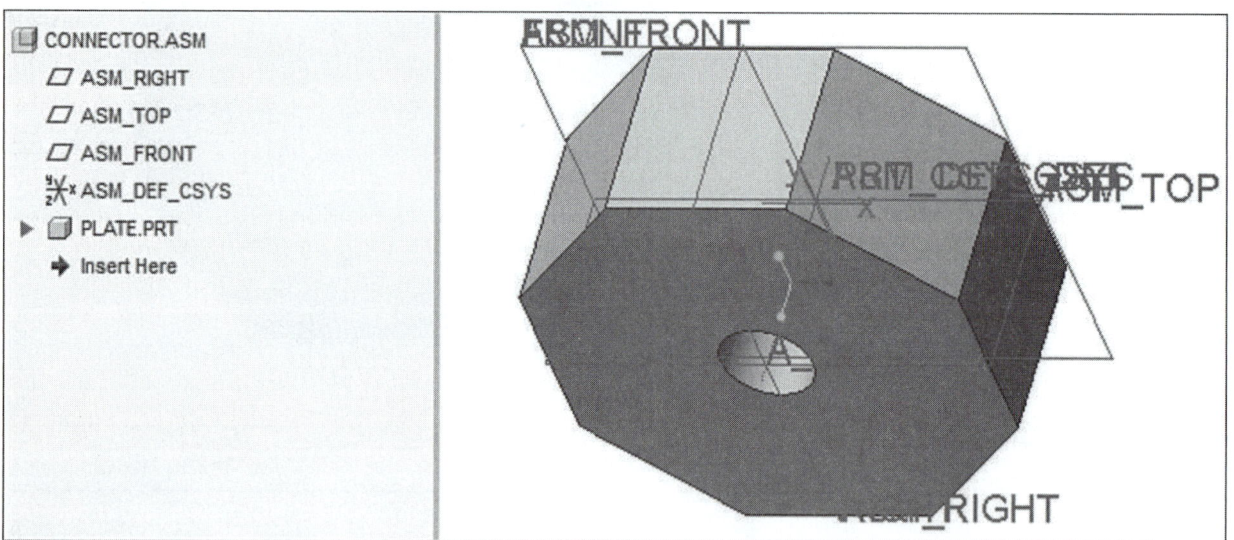

Figure 1.13(d) Fully Constrained Component at the Default Position *(coordinate system to coordinate system)*

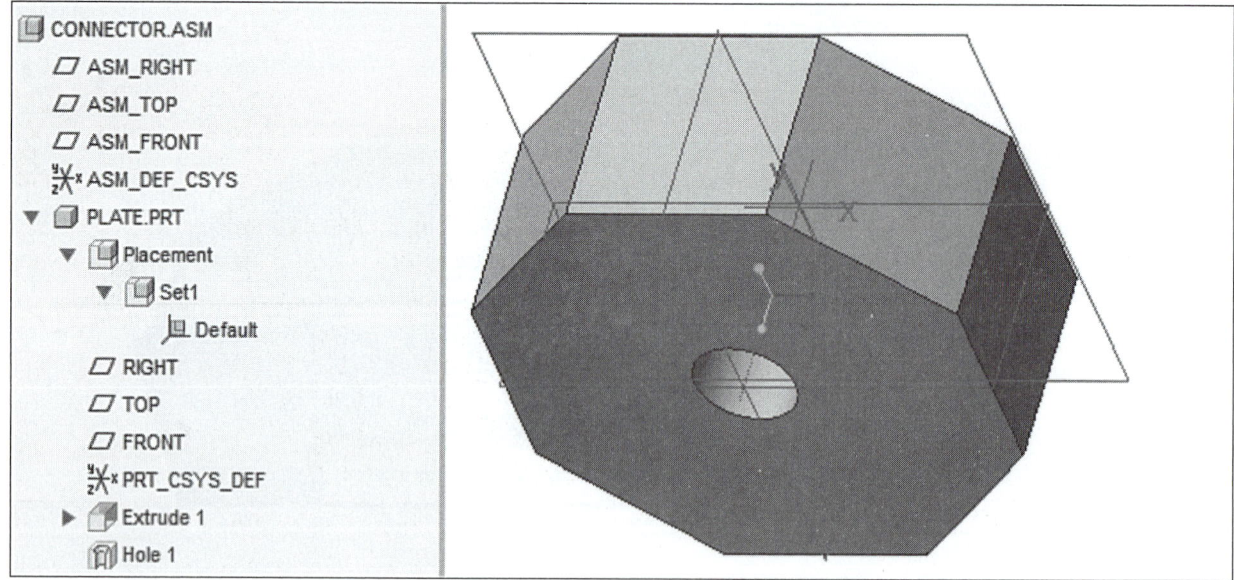

Figure 1.13(e) CONNECTOR Assembly

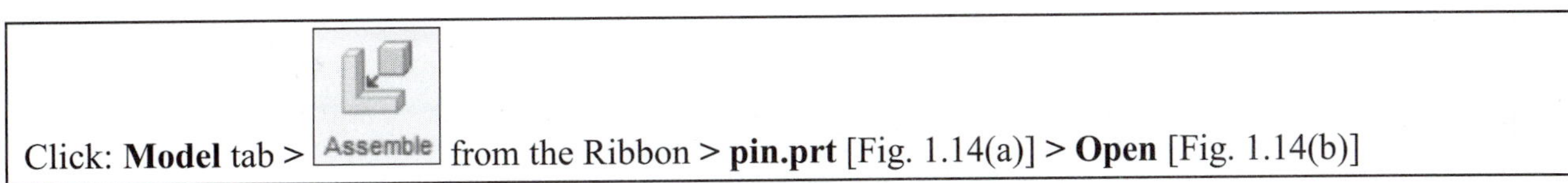

Click: **Model** tab > Assemble from the Ribbon > **pin.prt** [Fig. 1.14(a)] > **Open** [Fig. 1.14(b)]

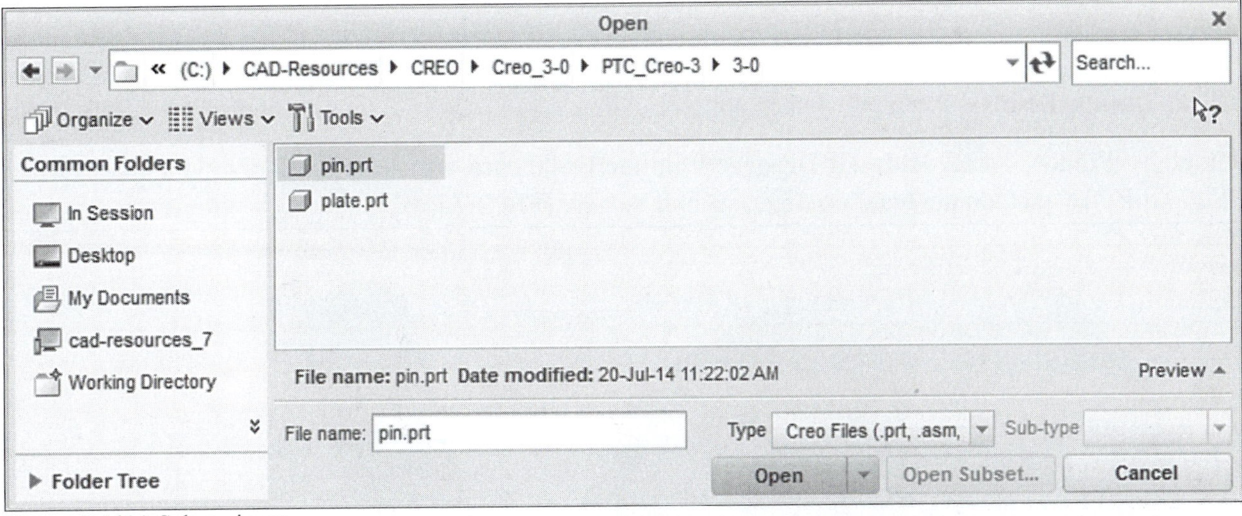

Figure 1.14(a) Select pin.prt

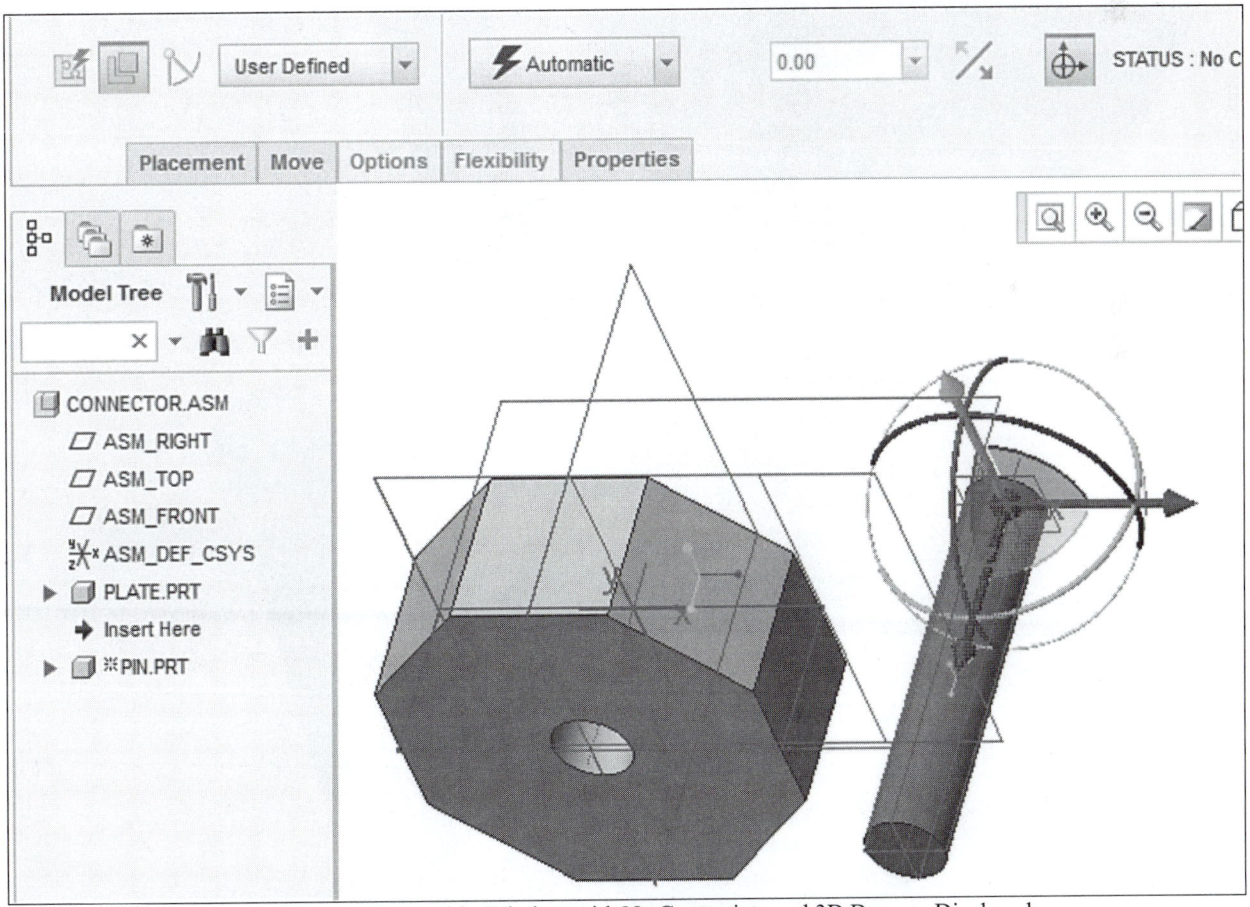

Figure 1.14(b) Component Shown in Assembly Window with No Constraints and 3D Dragger Displayed

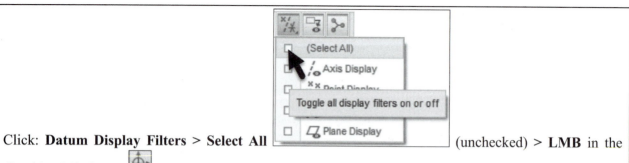

Click: **Datum Display Filters > Select All** 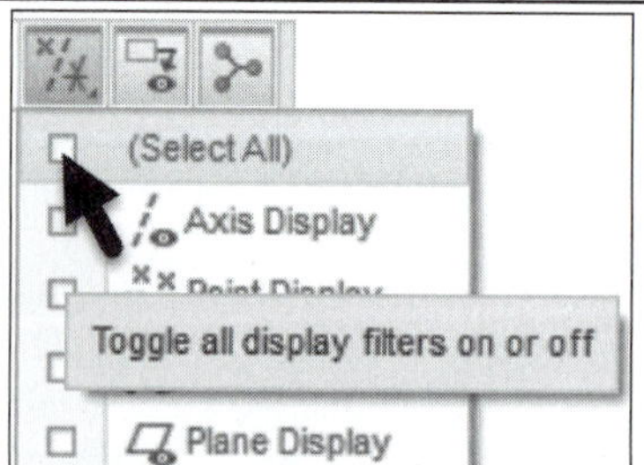

(unchecked) > **LMB** in the Graphics Window > **Hide 3D Dragger** from the Dashboard > pick on the Pin's cylindrical surface [Fig. 1.14(c)] > pick on the Plate's cylindrical hole surface [Fig. 1.14(d)]

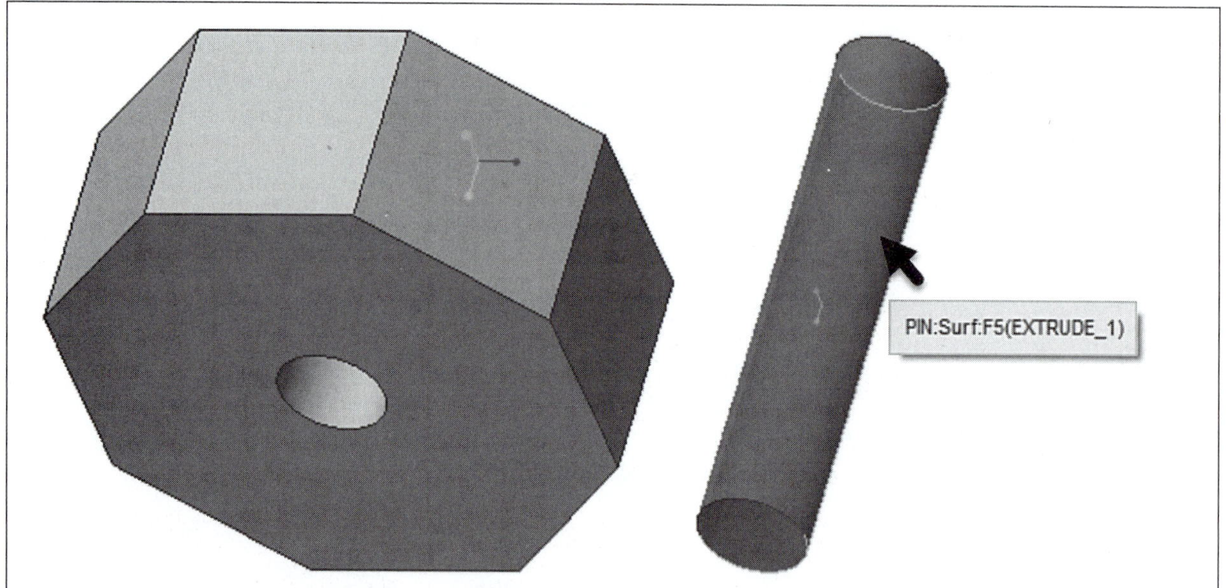

Figure 1.14(c) Select the Pin's Cylindrical Surface

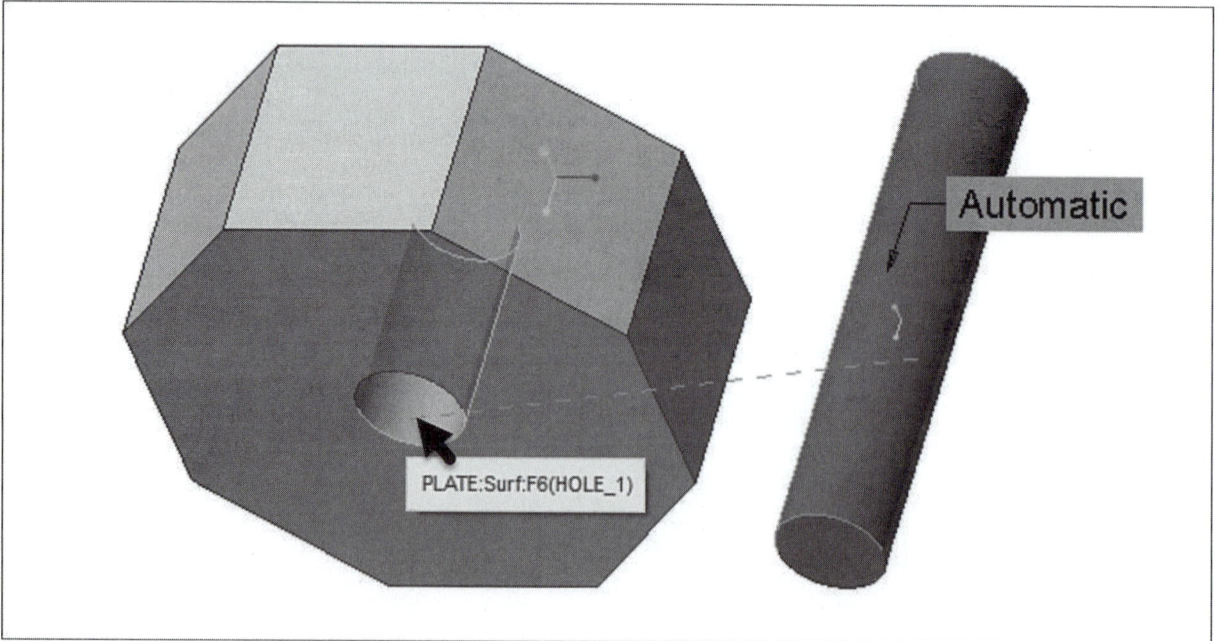

Figure 1.14(d) Select the Plate's Cylindrical Hole Surface

Click: **Placement** tab > **New Constraint** [Fig. 1.14(e)]

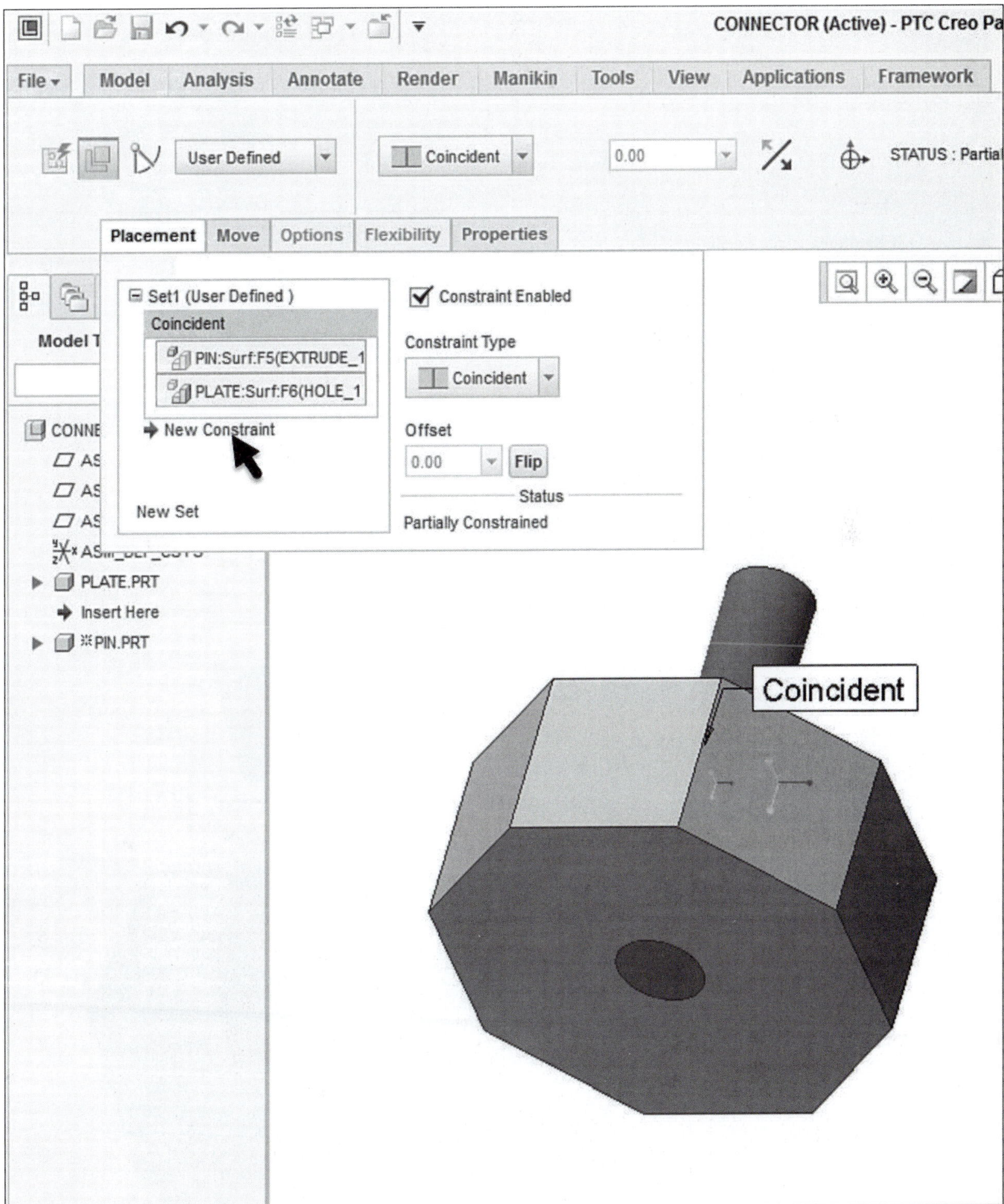

Figure 1.14(e) New Constraint

Pick on the Plate's surface [Fig. 1.14(f)]

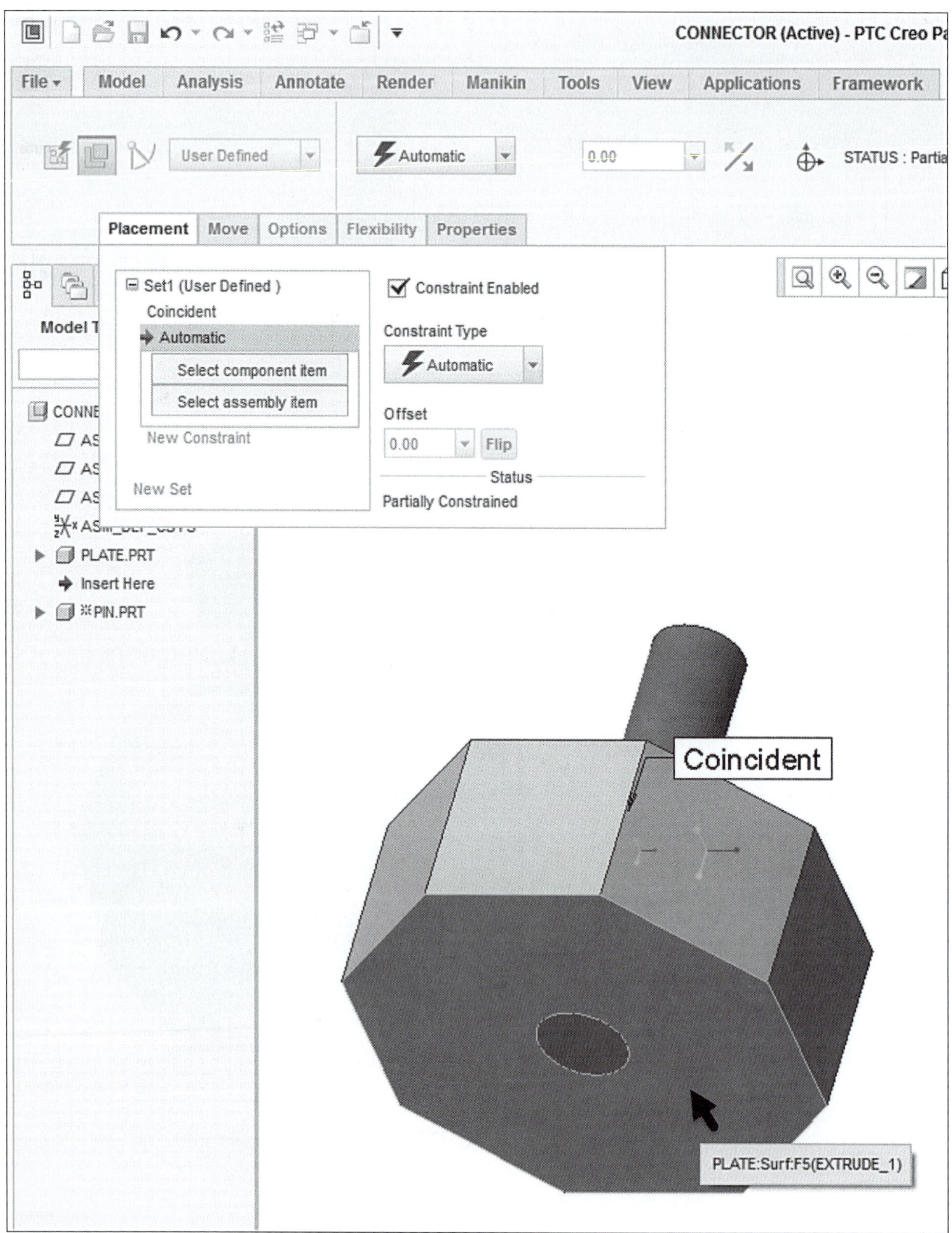

Figure 1.14(f) Select the Plate's Surface

Pick on the Pin's end face surface [Fig. 1.14(g)]

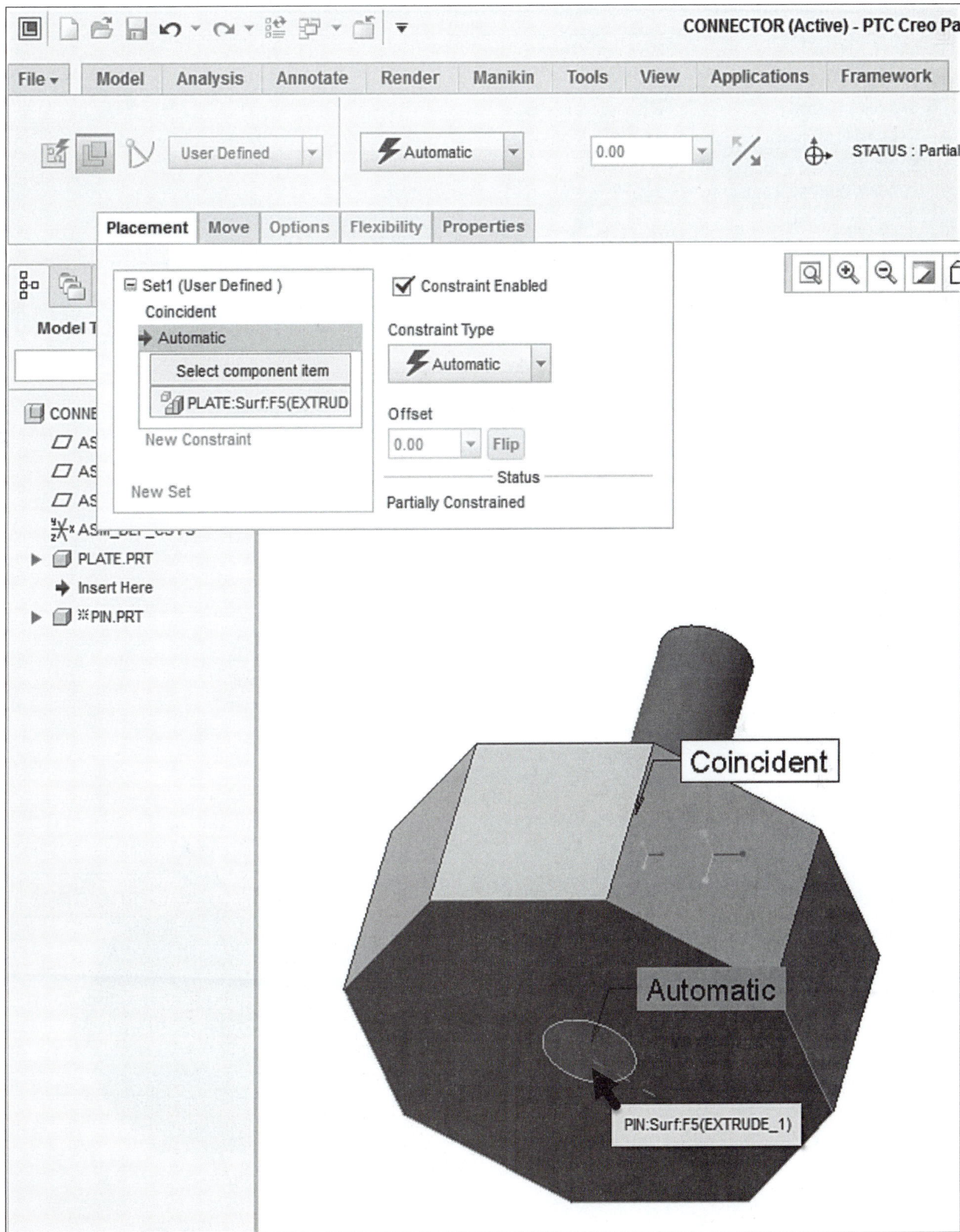

Figure 1.14(g) Select the Pin's End Face Surface

Click: **Coincident** [Fig. 1.14(h)] > **Distance** [Fig. 1.14(i)] > click in the Offset value field, *type* **.50** [Fig. 1.14(j)] > **Enter**

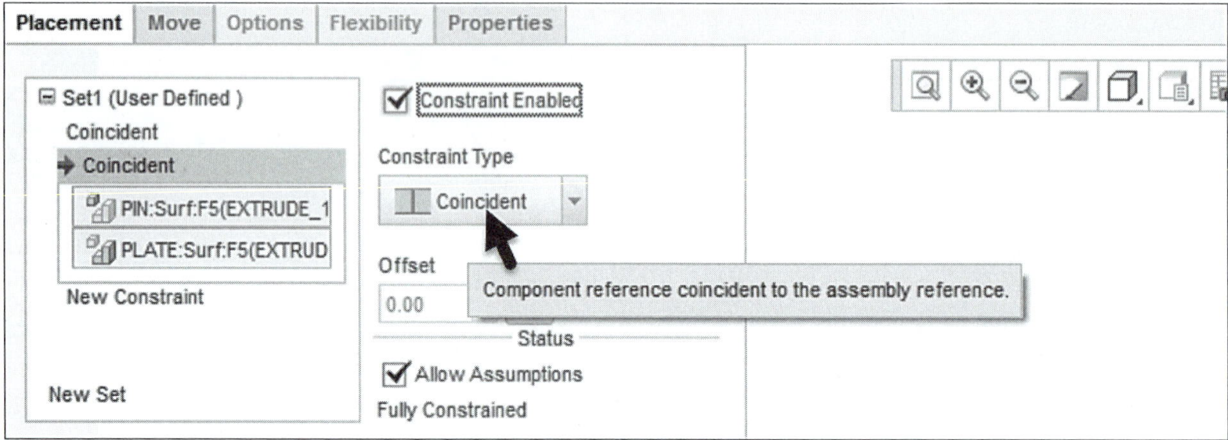

Figure 1.14(h) Select Coincident

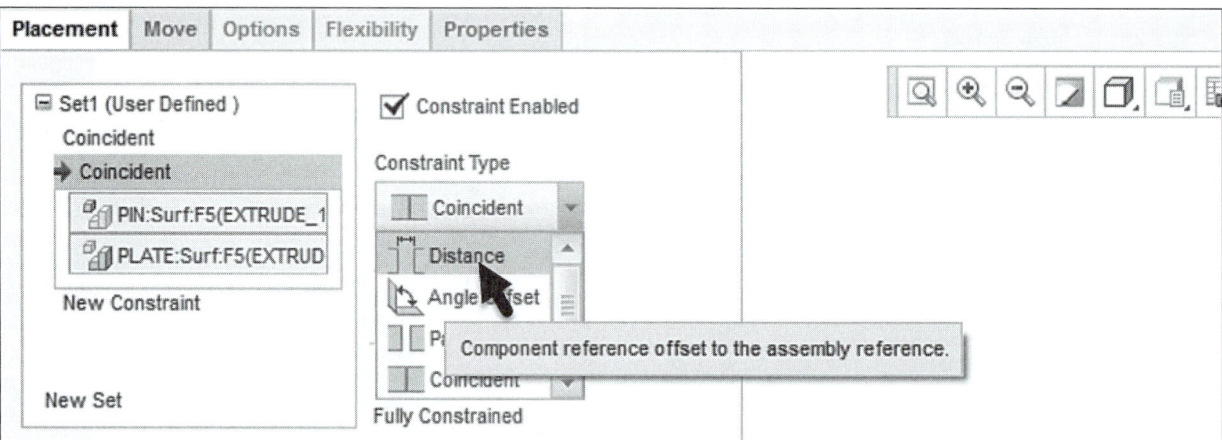

Figure 1.14(i) Changing Constraint Type to Distance

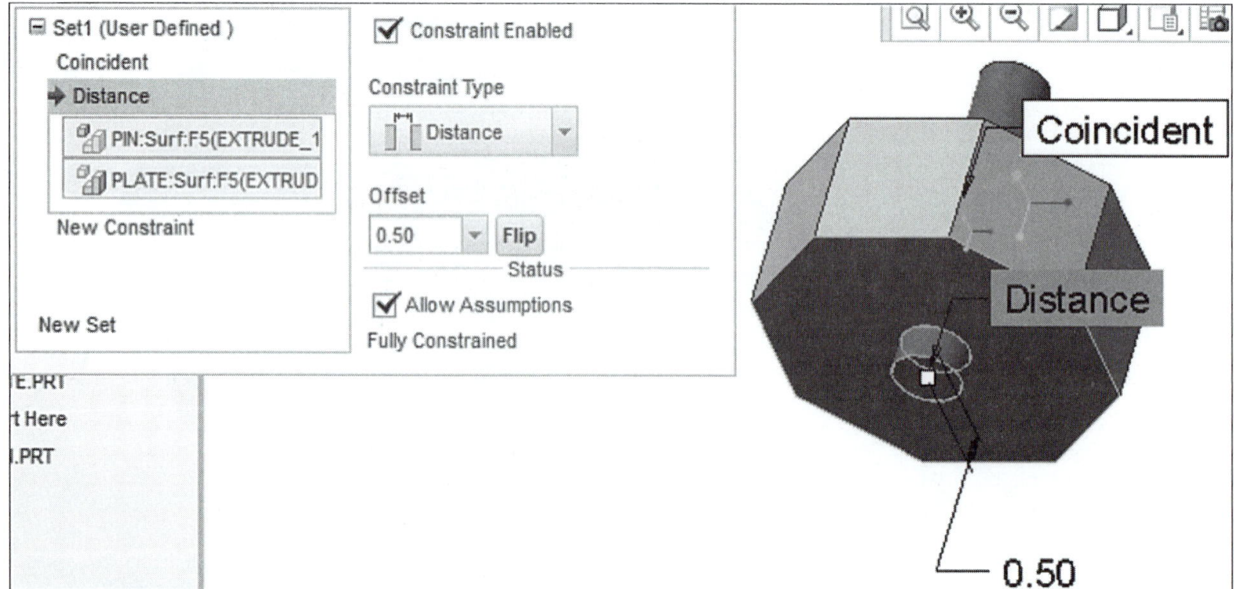

Figure 1.14(j) Offset **.50** *(if your offset is in the opposite direction, enter a value of -.50)*

Click: from the Dashboard [Fig. 1.14(k)] > expand all in Model Tree > **Datum Display Filters** from the Graphics Toolbar > (Select All) [Fig. 1.14(l)] > **LMB** in the Graphics Window > **Refit** > **Ctrl+S**

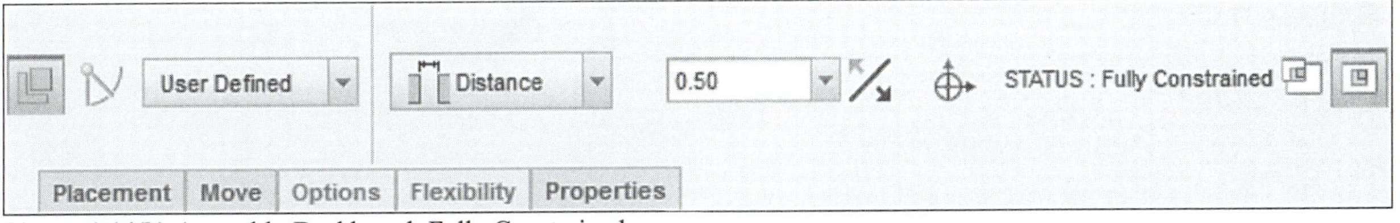

Figure 1.14(k) Assembly Dashboard: Fully Constrained

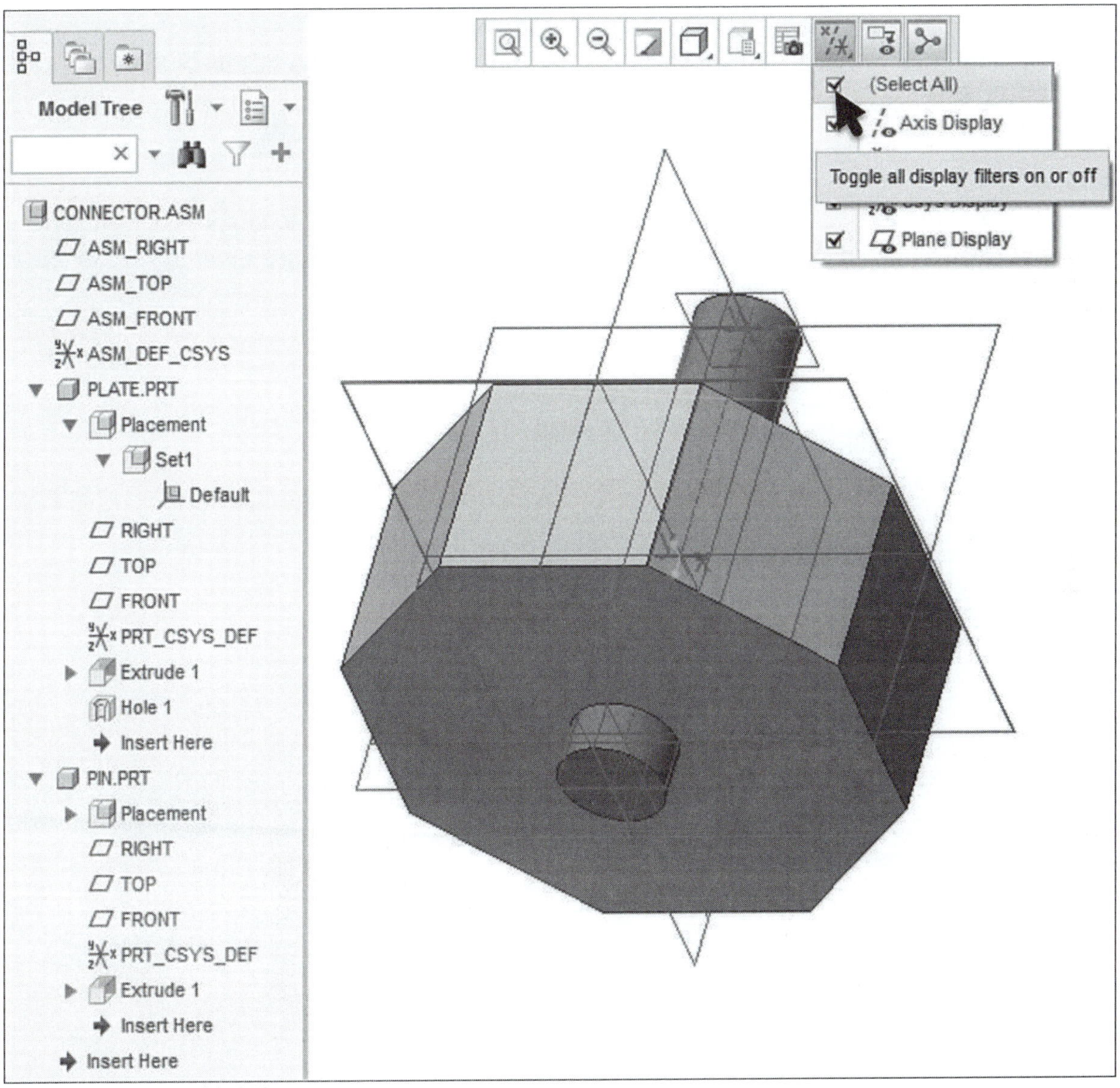

Figure 1.14(l) Assembled Pin

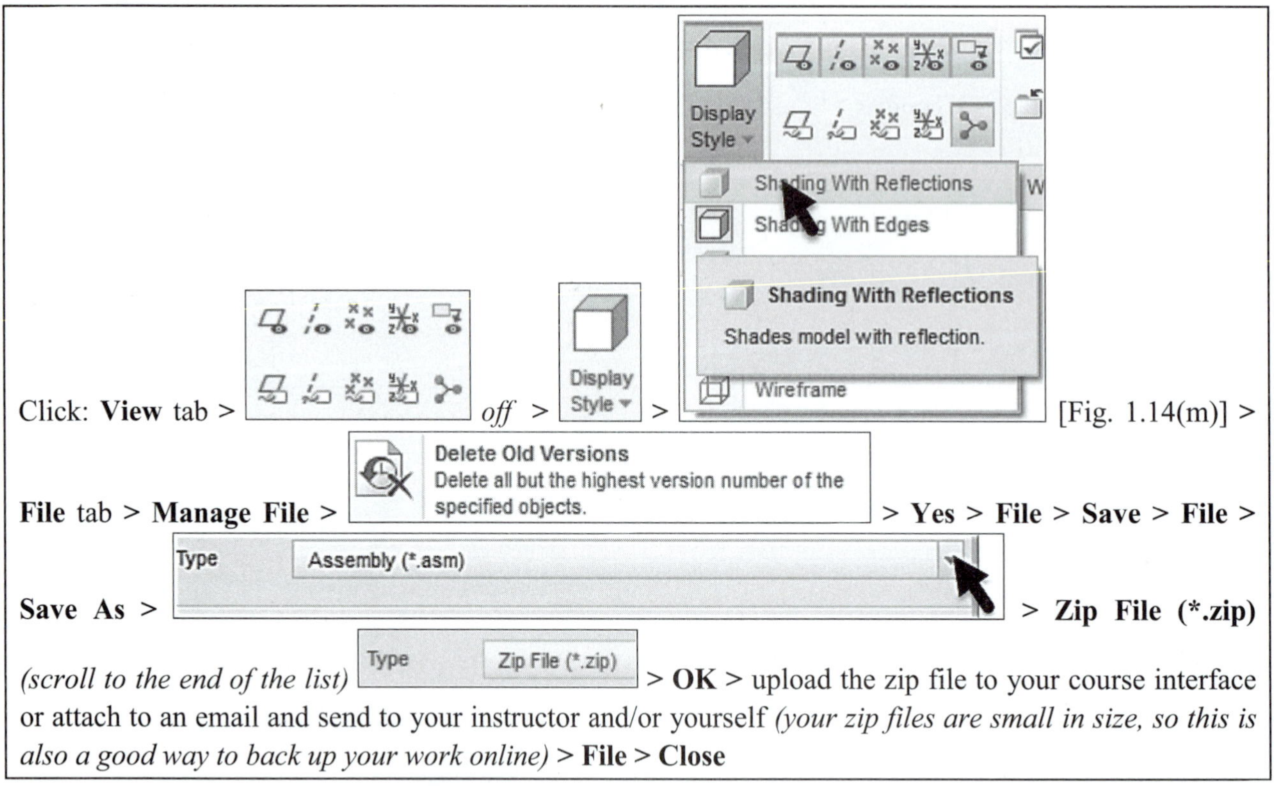

Click: **View** tab > *off* > > [Fig. 1.14(m)] > **File** tab > **Manage File** > > **Yes** > **File** > **Save** > **File** > **Save As** > > **Zip File (*.zip)** *(scroll to the end of the list)* > **OK** > upload the zip file to your course interface or attach to an email and send to your instructor and/or yourself *(your zip files are small in size, so this is also a good way to back up your work online)* > **File** > **Close**

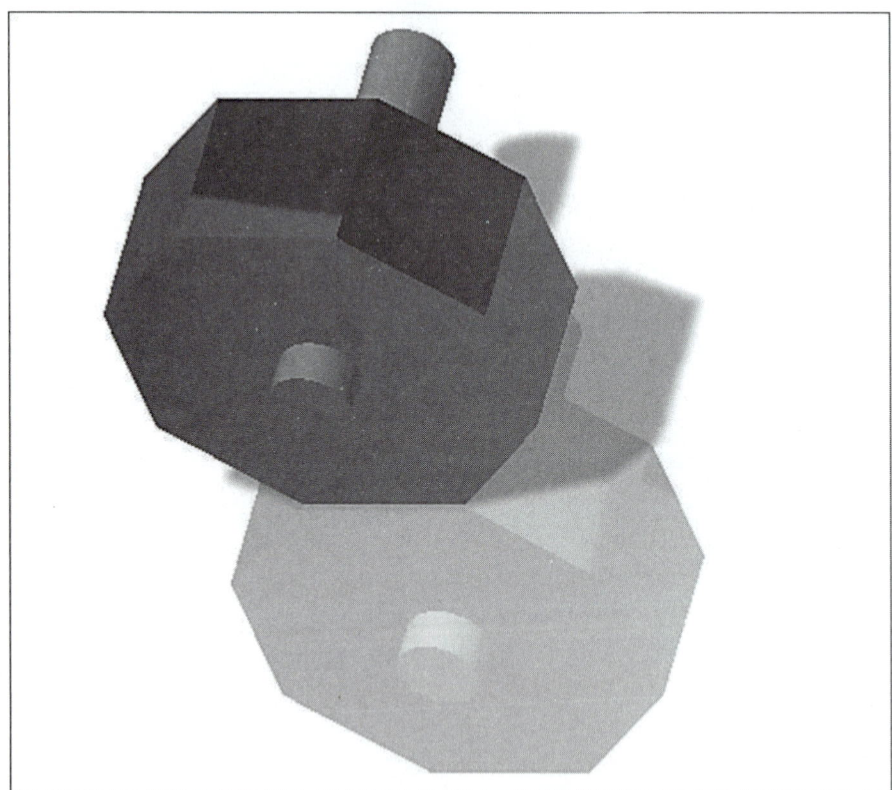

Figure 1.14(m) Assembled Pin *(the quality of your graphics card and graphics settings may prevent this display)*

You have completed your first Creo Parametric assembly. Next you will create a set of drawings for the Connector assembly and the Plate component.

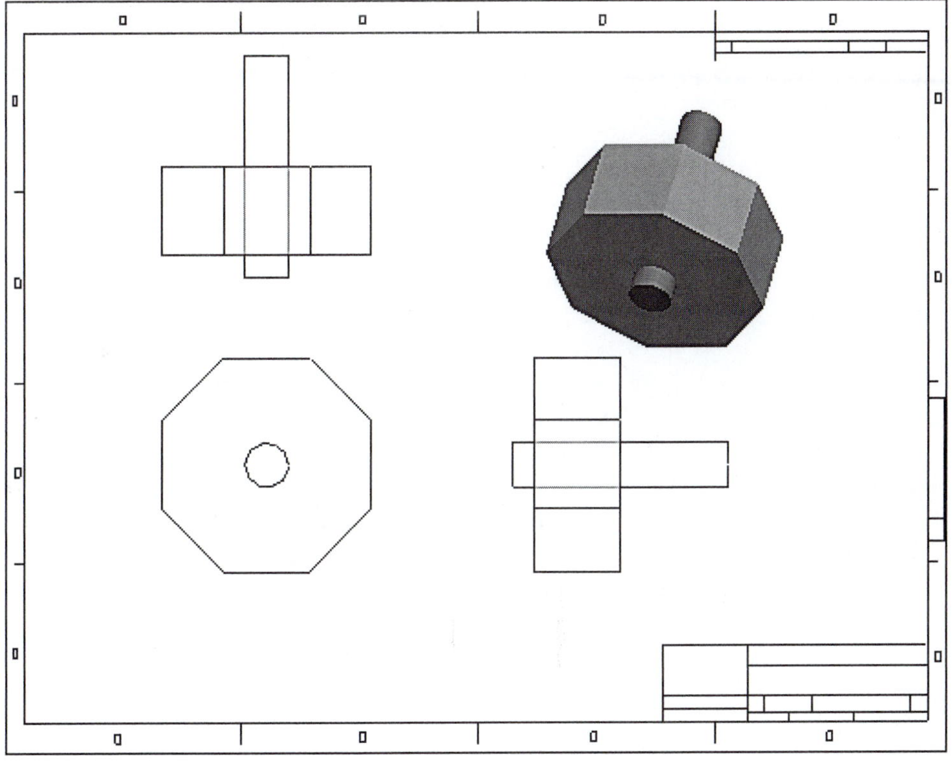

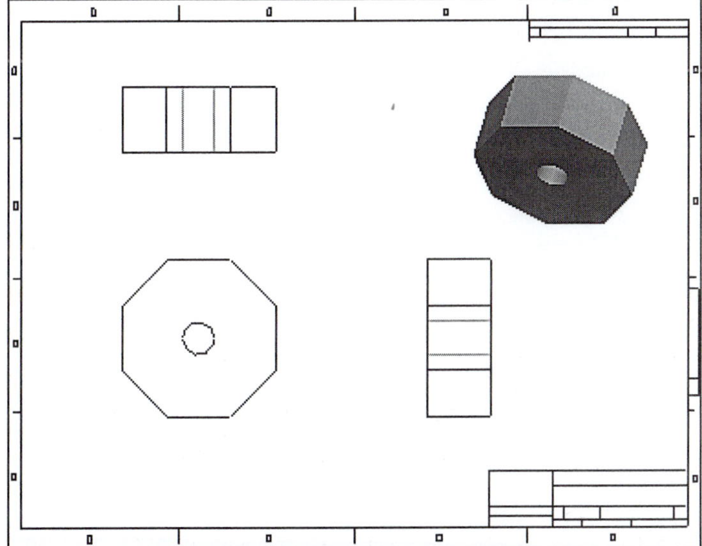

Figure 1.15 Assembly Drawing and Detail Drawing

Creating Drawings

Using the parts and the assembly just created, you will now create an assembly drawing of the Connector and a detail drawing of the Plate (Fig. 1.15).

The Drawing mode in PTC Creo Parametric 3.0 enables you to create and manipulate engineering drawings that use the 3D model (part or assembly) as a geometry source. You can pass dimensions, notes, and other elements of design between the 3D model and its views on a drawing.

Click: New > Type Drawing > Name-- *type* **connector** > Use default template [Fig. 1.16(a)] > **OK** [Fig. 1.16(b)] > Default Model **Browse** > **Preview** *on* > **connector.asm** [Fig. 1.16(c)]

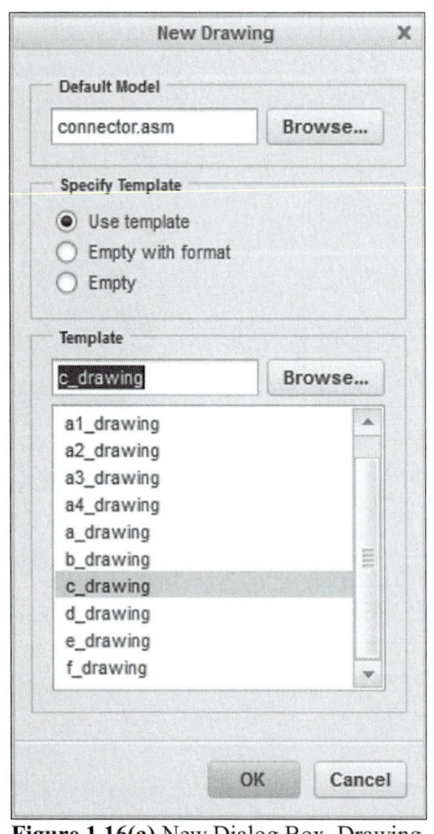

Figure 1.16(a) New Dialog Box, Drawing

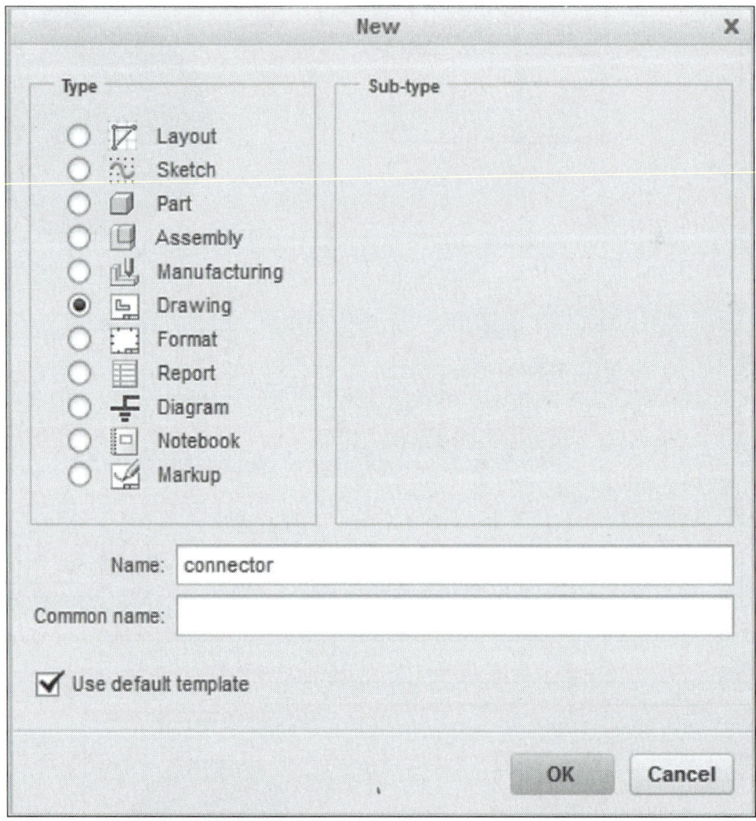

Figure 1.16(b) New Drawing Dialog Box

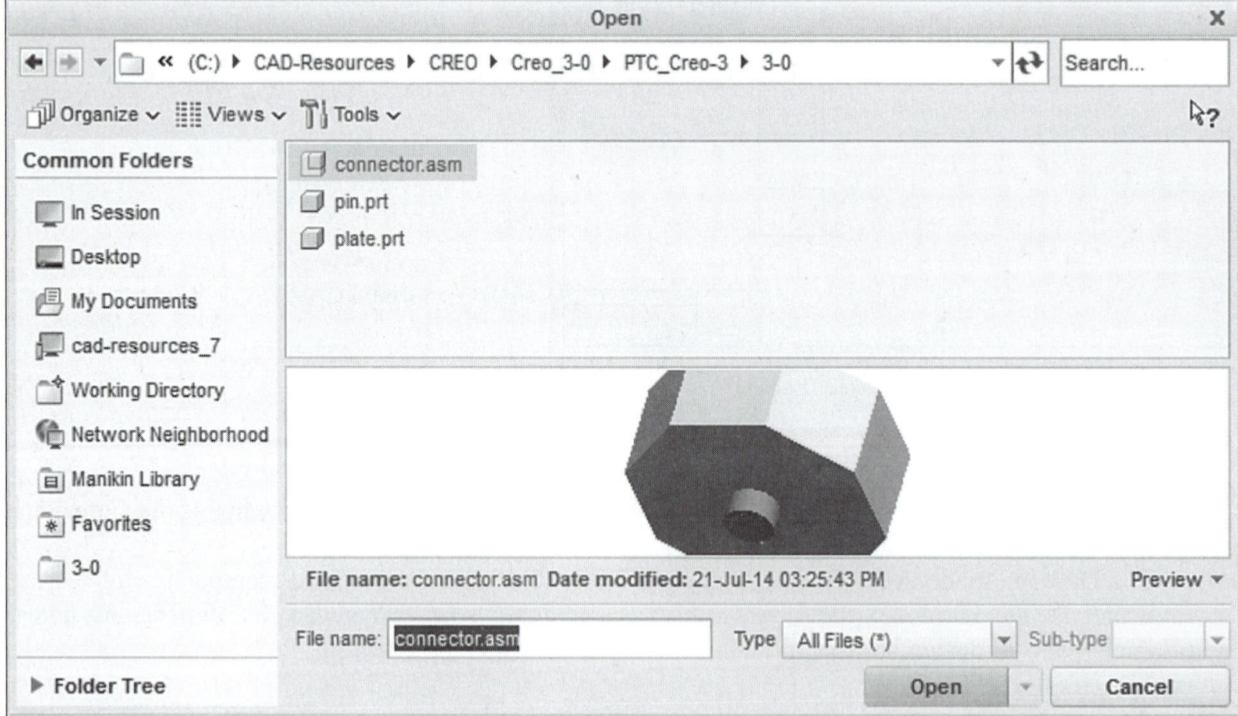

Figure 1.16(c) Connector Assembly

Click: **Open** > **OK** from the New Drawing dialog box [Fig. 1.16(d)] > **Layout** tab > in the Graphics Window, press and hold: **RMB** > **Sheet Setup** from the short cut menu [Fig. 1.16(e)]

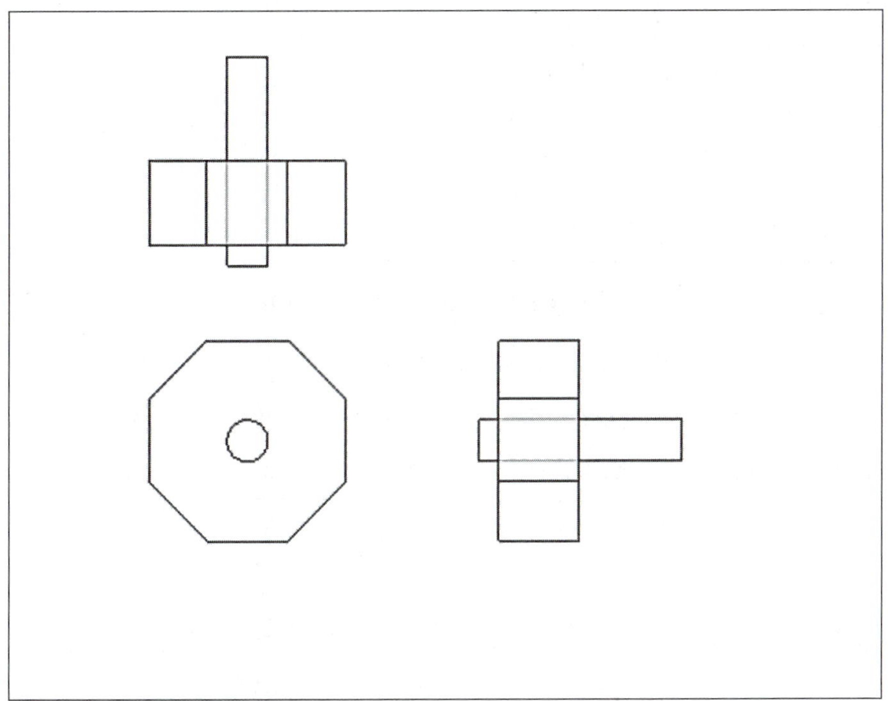

Figure 1.16(d) Drawing

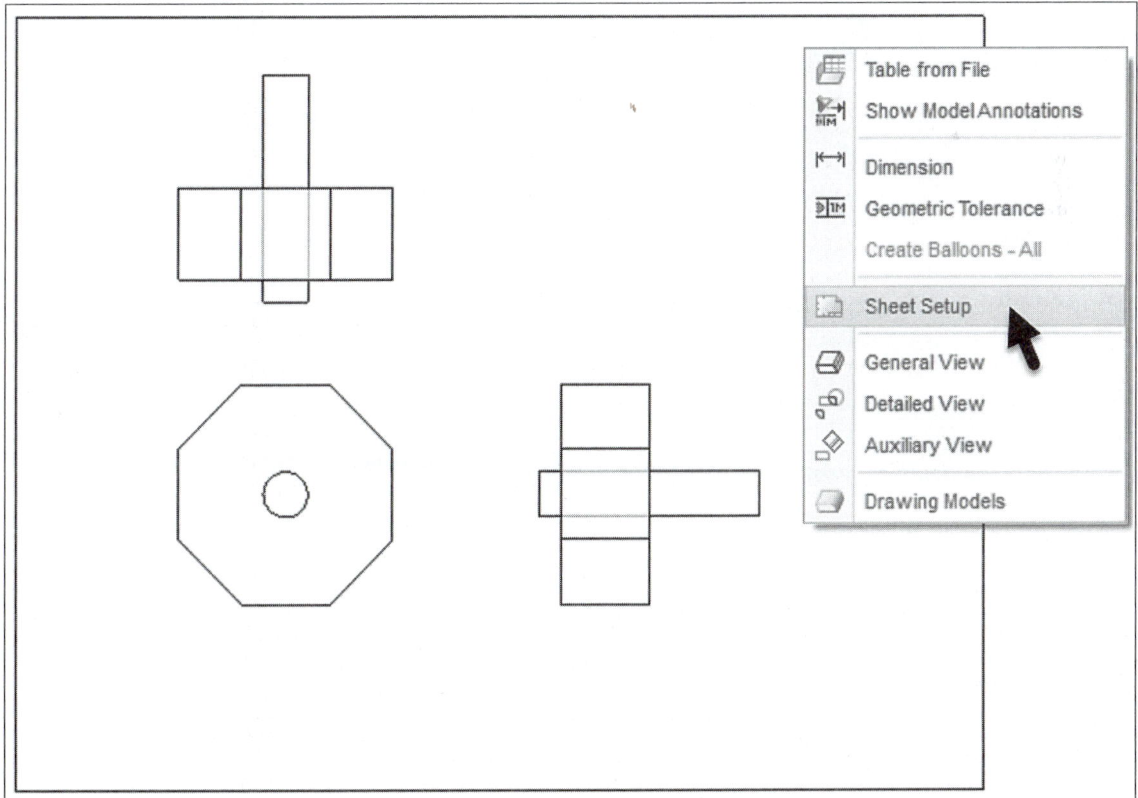

Figure 1.16(e) Drawing Sheet Setup

Sheet Setup dialog box displays [Fig. 1.16(f)] > click in the **C Size** field [Fig. 1.16(g)] > to see options, click: ▾ [Fig. 1.16(h)] > **Browse** [Fig. 1.16(i)] > **c.frm** [Fig. 1.16(j)] > **Open** [Fig. 1.16(k)] > **OK**

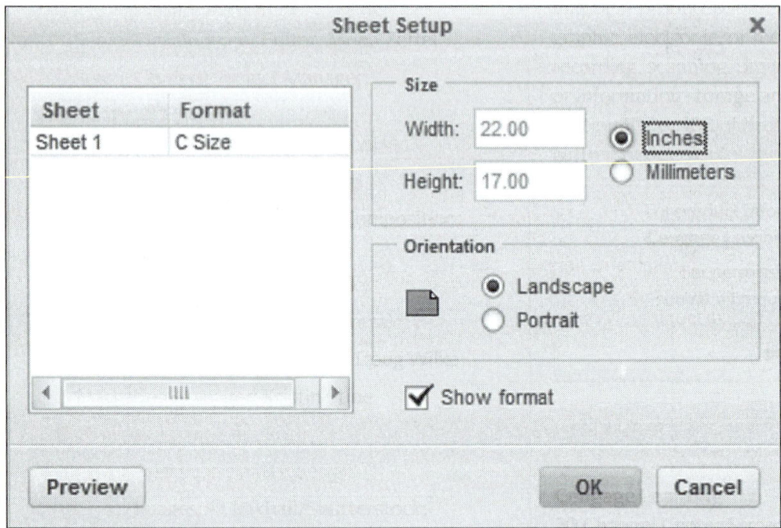

Figure 1.16(f) Sheet Setup Dialog Box

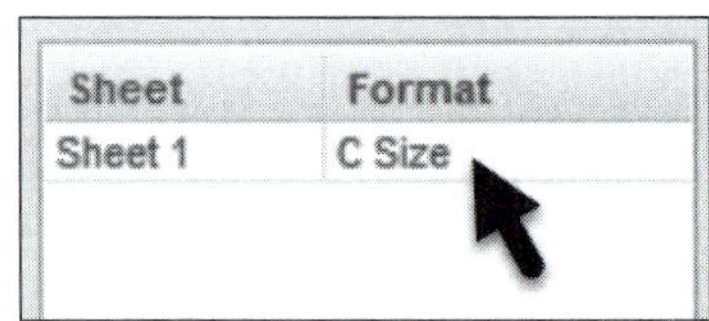

Figure 1.16(g) C Size Field

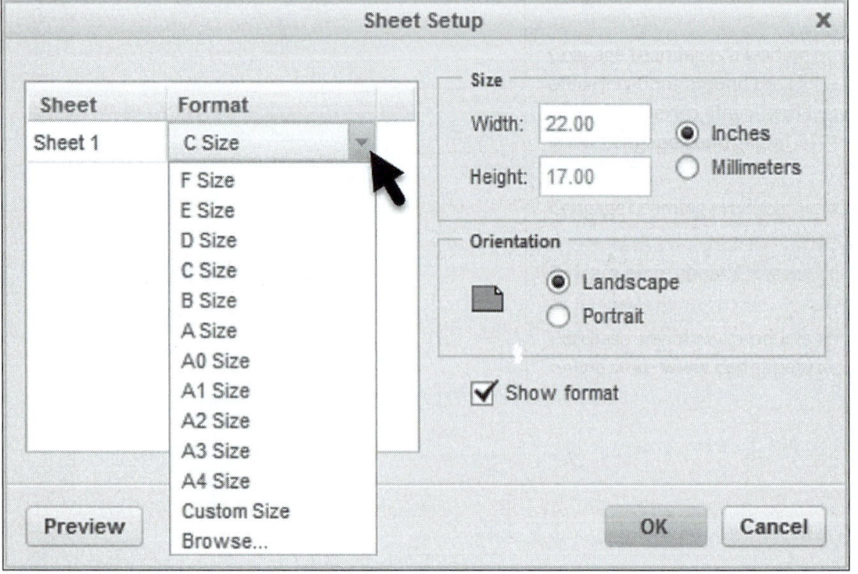

Figure 1.16(h) Format Options

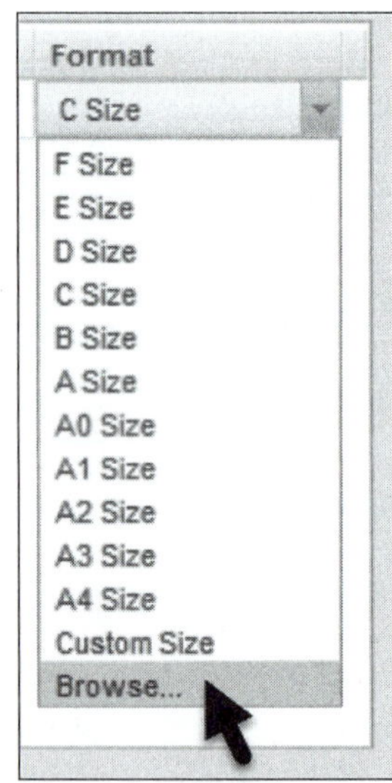

Figure 1.16(i) Browse

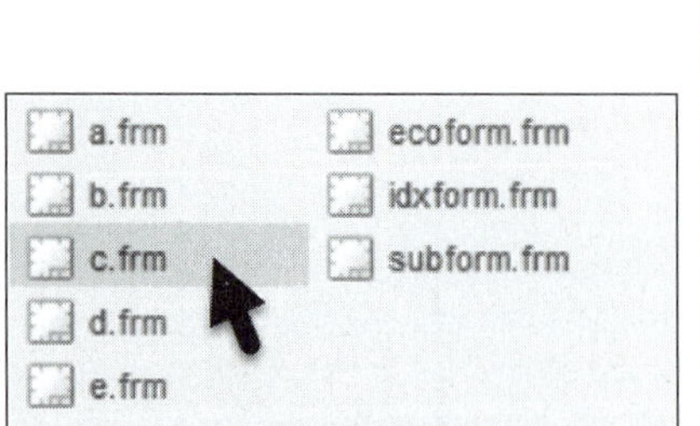

Figure 1.16(j) c.frm from the System Formats Library

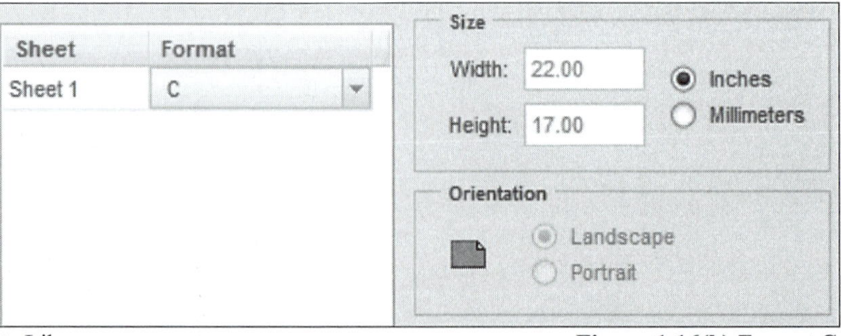

Figure 1.16(k) Format C

In the Graphics Window, press and hold: **RMB > General View** [Fig. 1.16(l)] > **No Combined State** *(selected by default)* [Fig. 1.16(m)] > **OK** > pick a position for the new view [Fig. 1.16(n)]

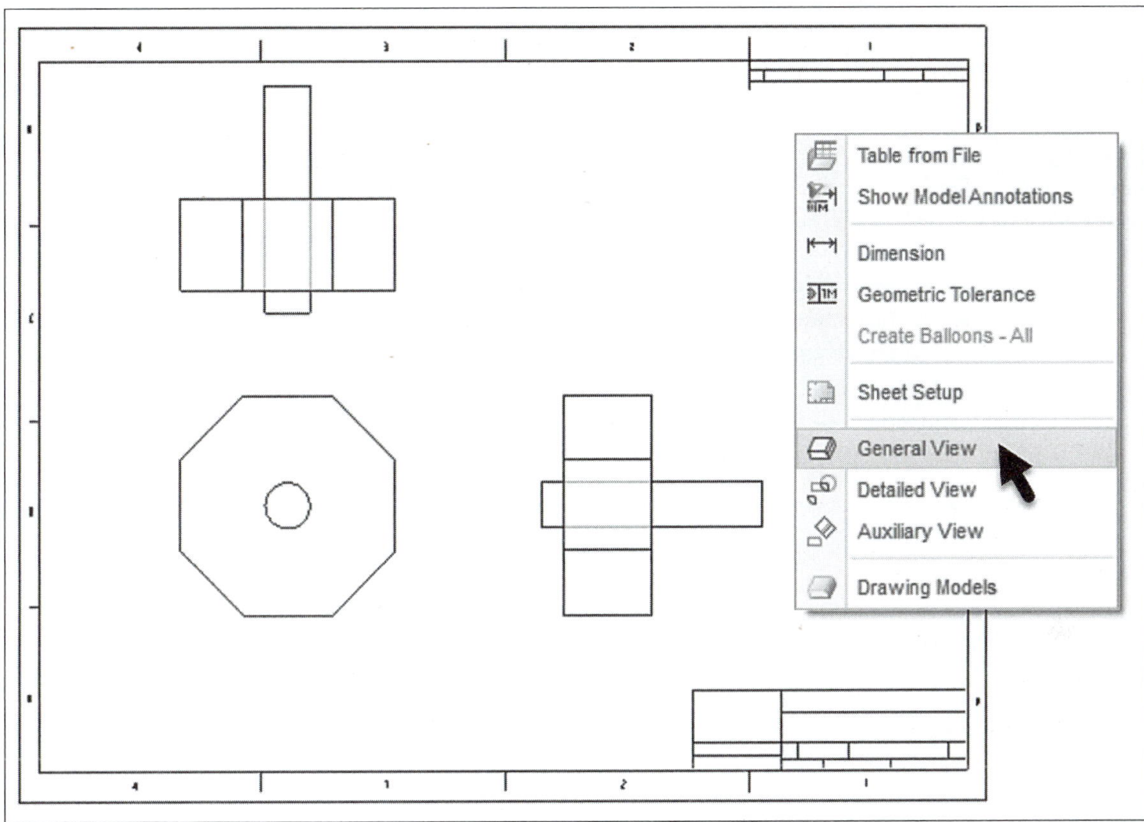

Figure 1.16(l) General View

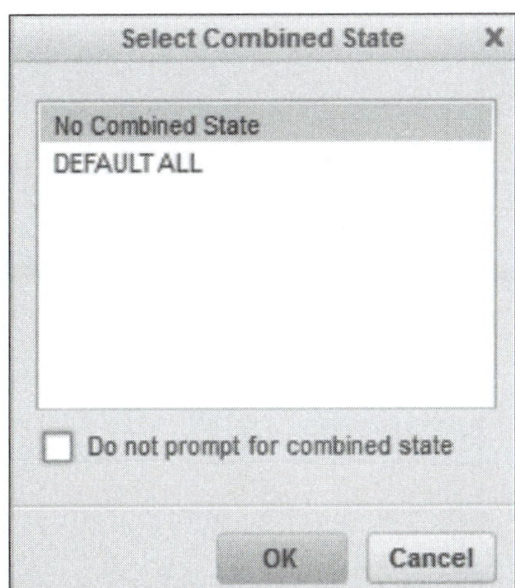

Figure 1.16(m) Select Combined State Dialog Box

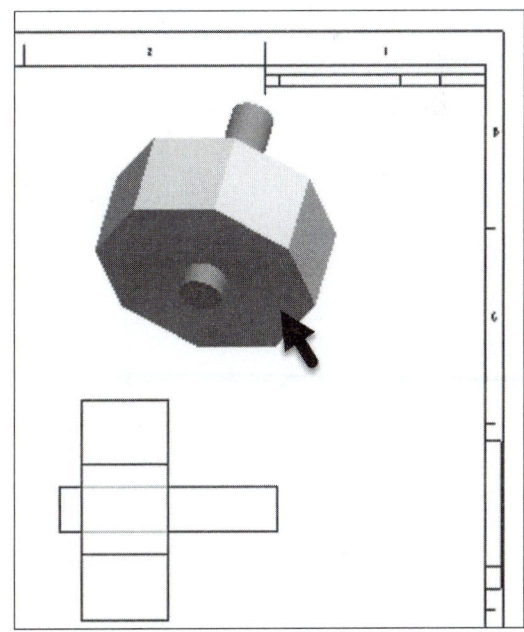

Figure 1.16(n) Select a Position for the New View

Click: **OK** [Fig. 1.16(o)] > **LMB** (in the Graphics Window to deselect) > **File** > **Save** > **OK** [Fig. 1.16(p)] > **File** > **Save As** > Type ▾ > **Zip File (*.zip)** *(scroll to the end of the list)* > **OK** > upload to your course interface or attach to an email and send to your instructor and/or yourself > **File** > **Close**

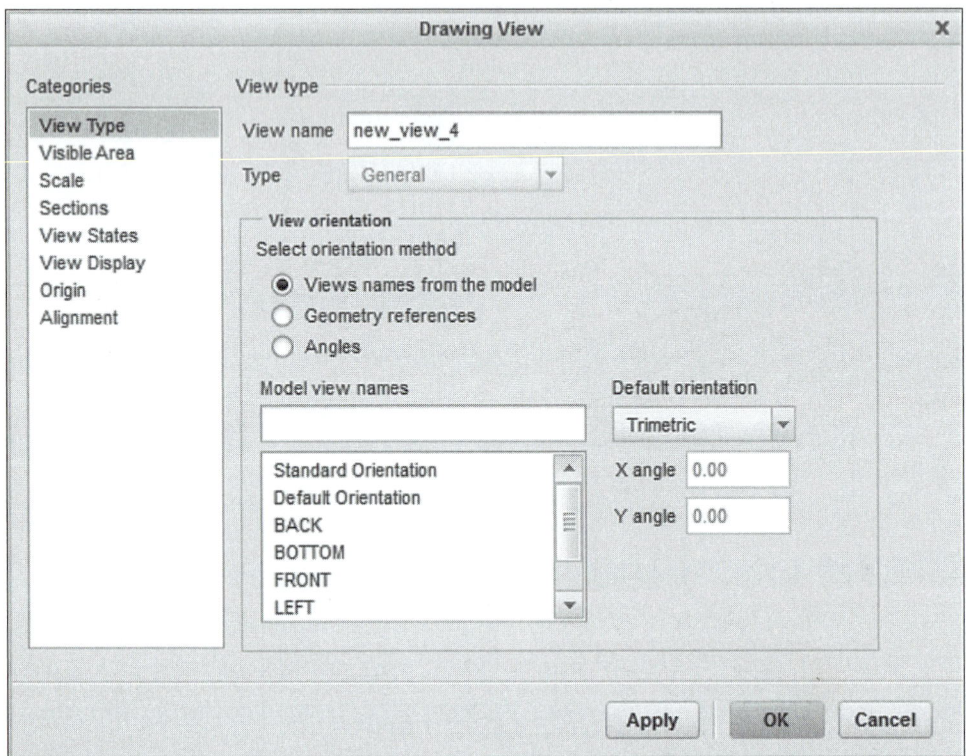

Figure 1.16(o) Drawing View Dialog Box

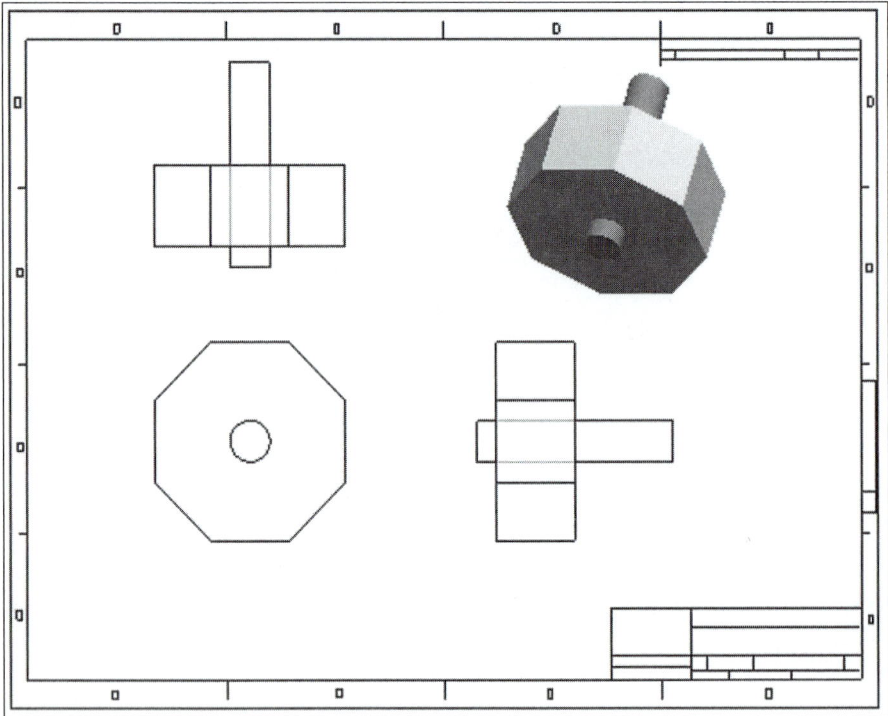

Figure 1.16(p) Completed Assembly Drawing

Click: New > Type Drawing > Name-- *type* **plate** > Use default template [Fig. 1.17(a)] > **OK** [Fig. 1.17(b)] > Default Model **Browse** > **Preview** *on* > **plate.prt** [Fig. 1.17(c)] > **Open**

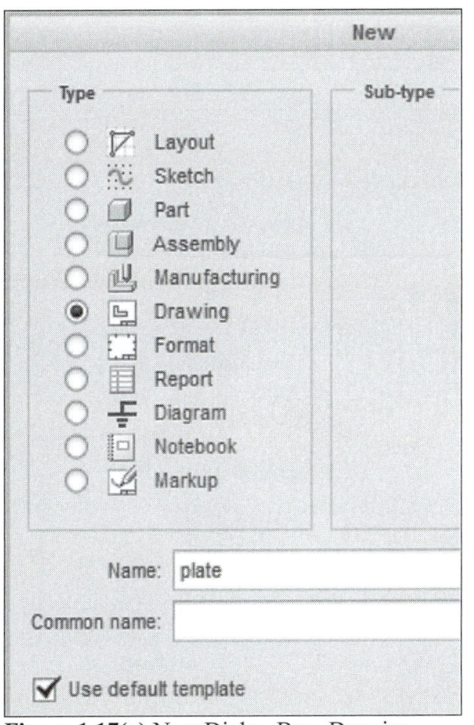

Figure 1.17(a) New Dialog Box, Drawing

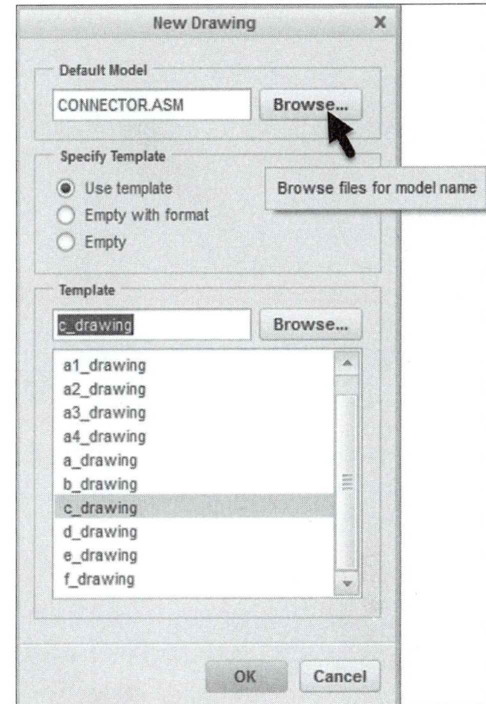

Figure 1.17(b) Browse

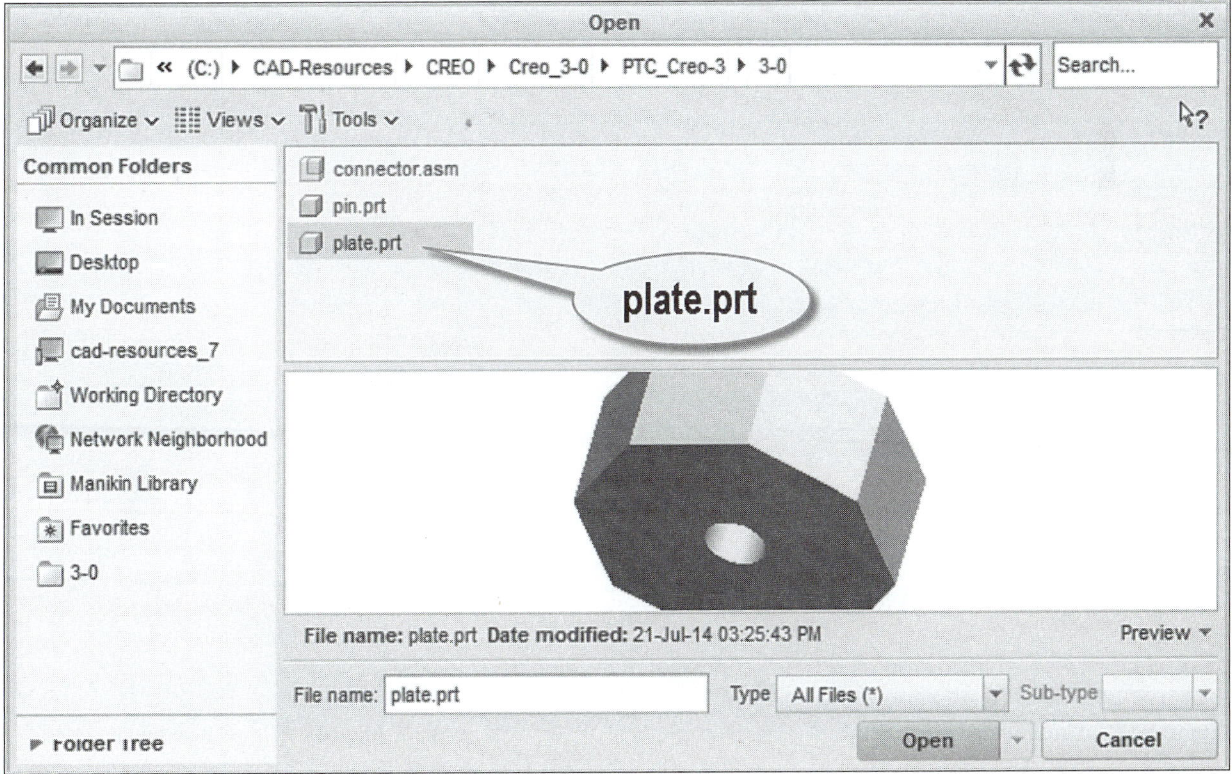

Figure 1.17(c) Plate Preview

Click: **OK** [Fig. 1.17(d)] drawing opens > in the Graphics Window, press and hold: **RMB > Sheet Setup** [Figs. 1.17(e-f)] > click in the **C Size** field [Fig. 1.17(g)] > ▾ to see options [Fig. 1.17(h)]

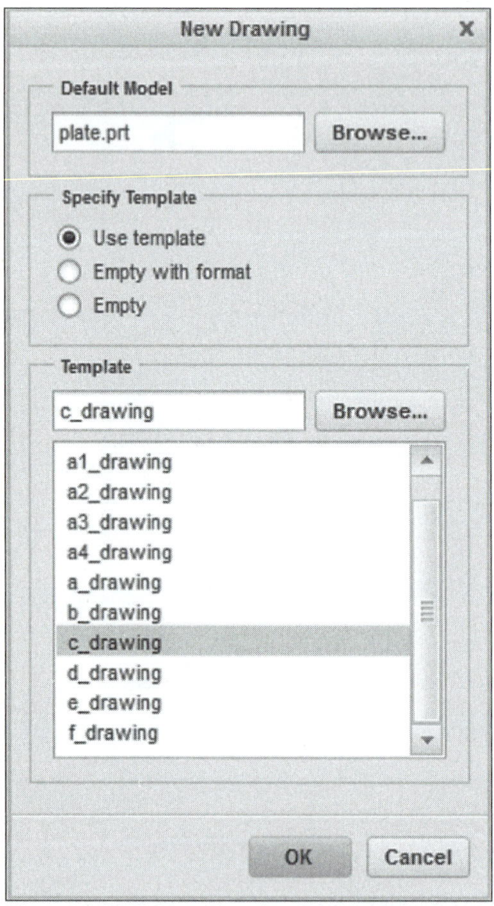

Figure 1.17(d) New Drawing Dialog Box

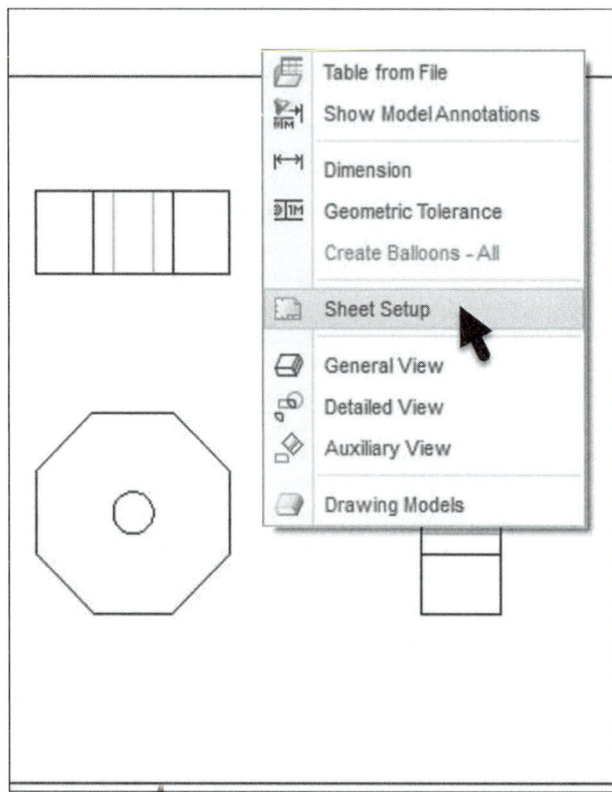

Figure 1.17(e) Plate Drawing, RMB > Sheet Setup

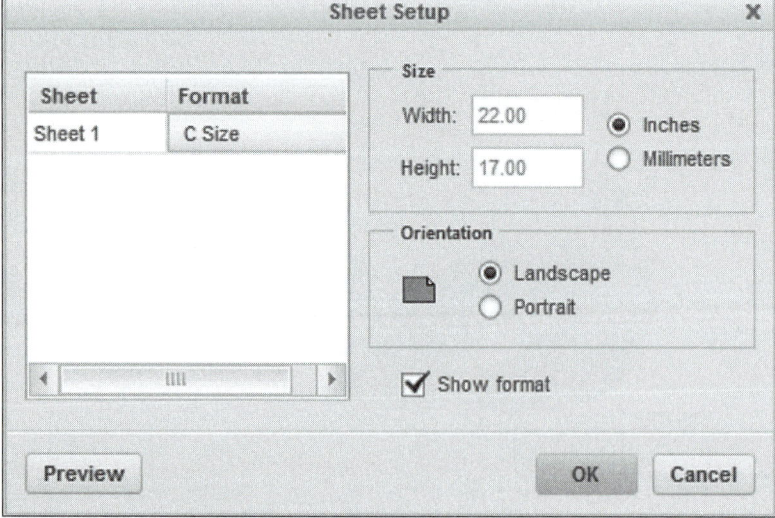

Figure 1.17(f) Sheet Setup Dialog Box

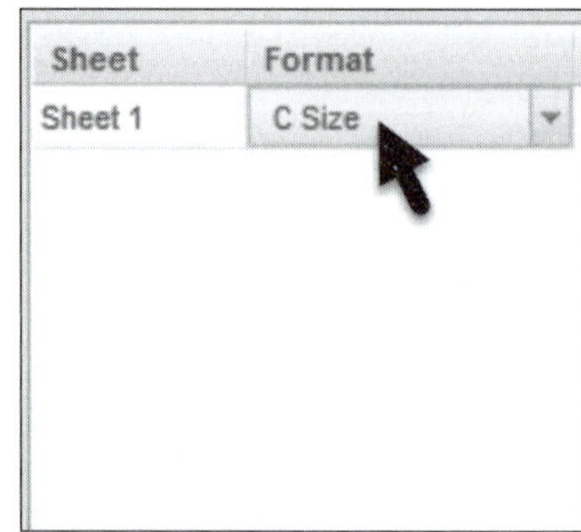

Figure 1.17(g) C Size Field

Click: **Browse** [Fig. 1.17(i)] > double-click on **c.frm** [Fig. 1.17(j)] > **OK** formatted drawing is displayed [Fig. 1.17(k)] > **Ctrl+S** > **Enter**

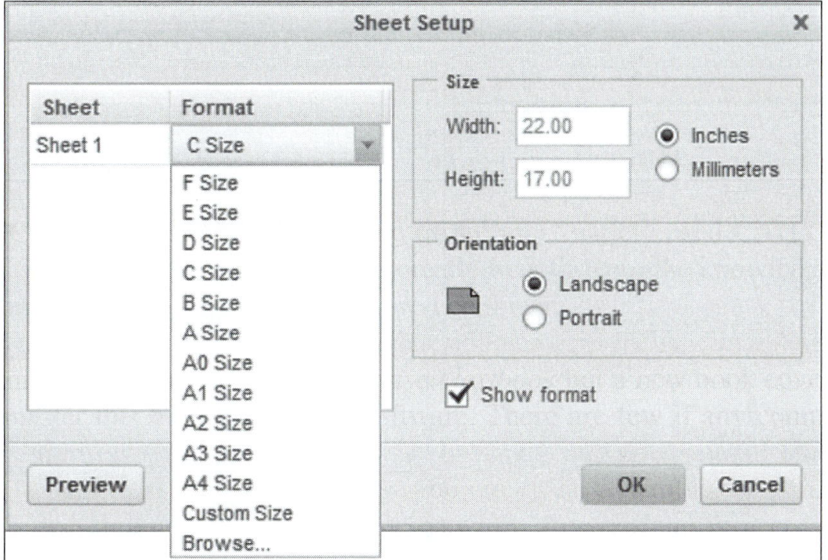

Figure 1.17(h) Sheet Setup Dialog Box, Format Options

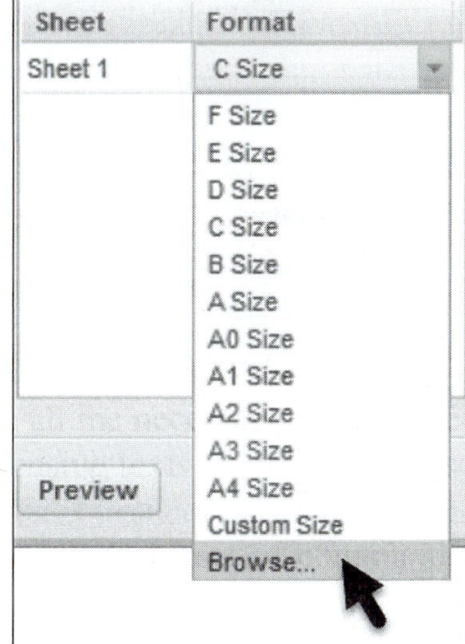

Figure 1.17(i) Browse

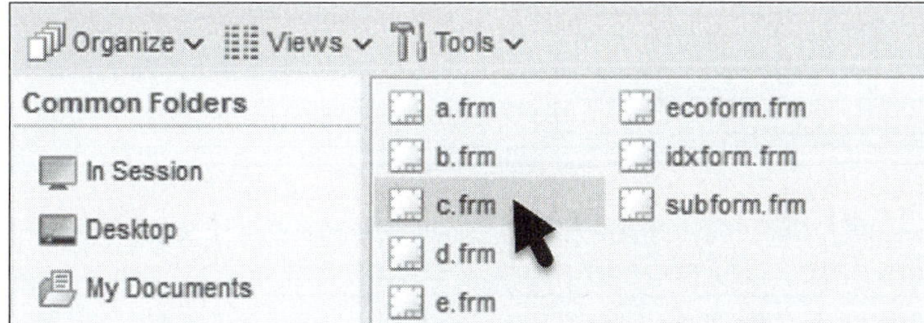

Figure 1.17(j) c.frm

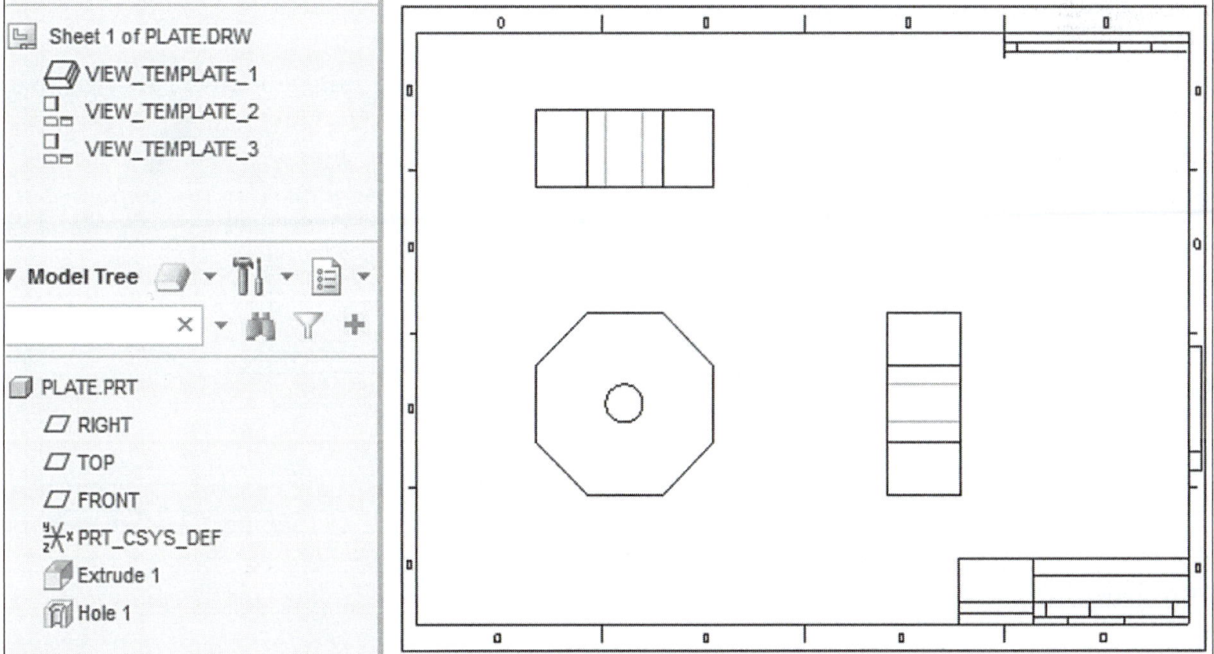

Figure 1.17(k) Formatted Drawing

In the Graphics Window, press and hold: **RMB** > **General View** [Fig. 1.17(l)] > **No Combined State** *(selected by default)* > **OK** > pick a position for the new view [Fig. 1.17(m)] > **OK** from the Drawing View dialog box [Fig. 1.17(n)] > **Refit**

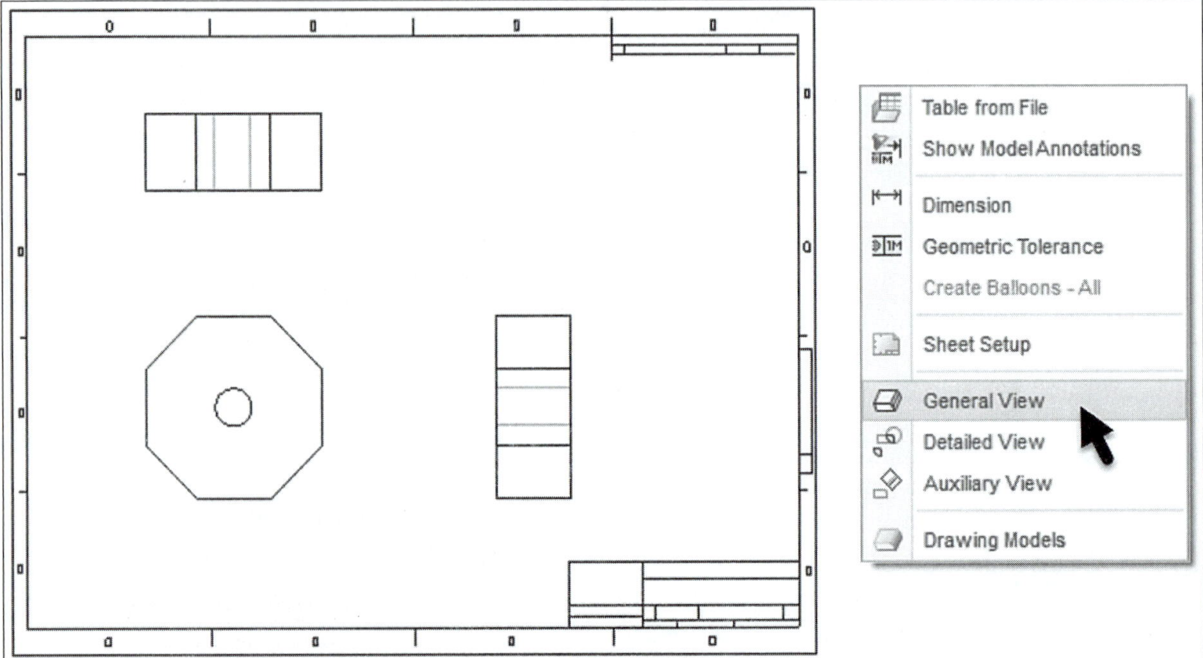

Figure 1.17(l) General View

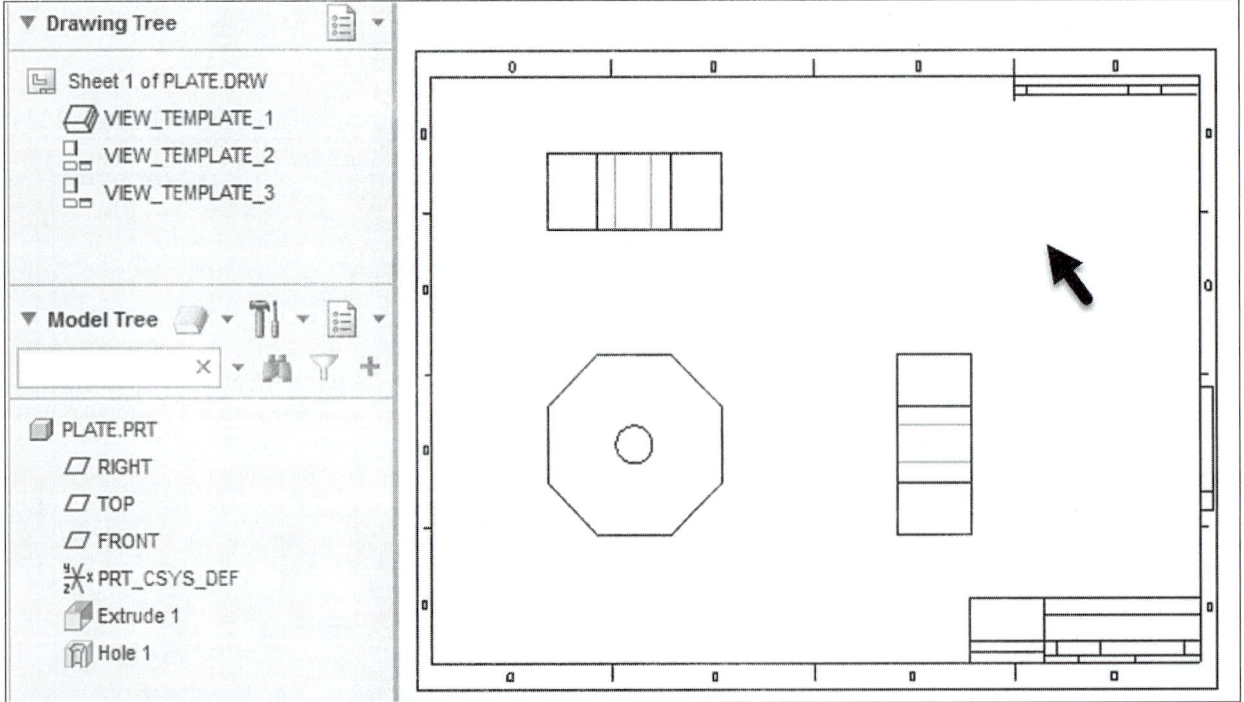

Figure 1.17(m) Select a Position for the New View

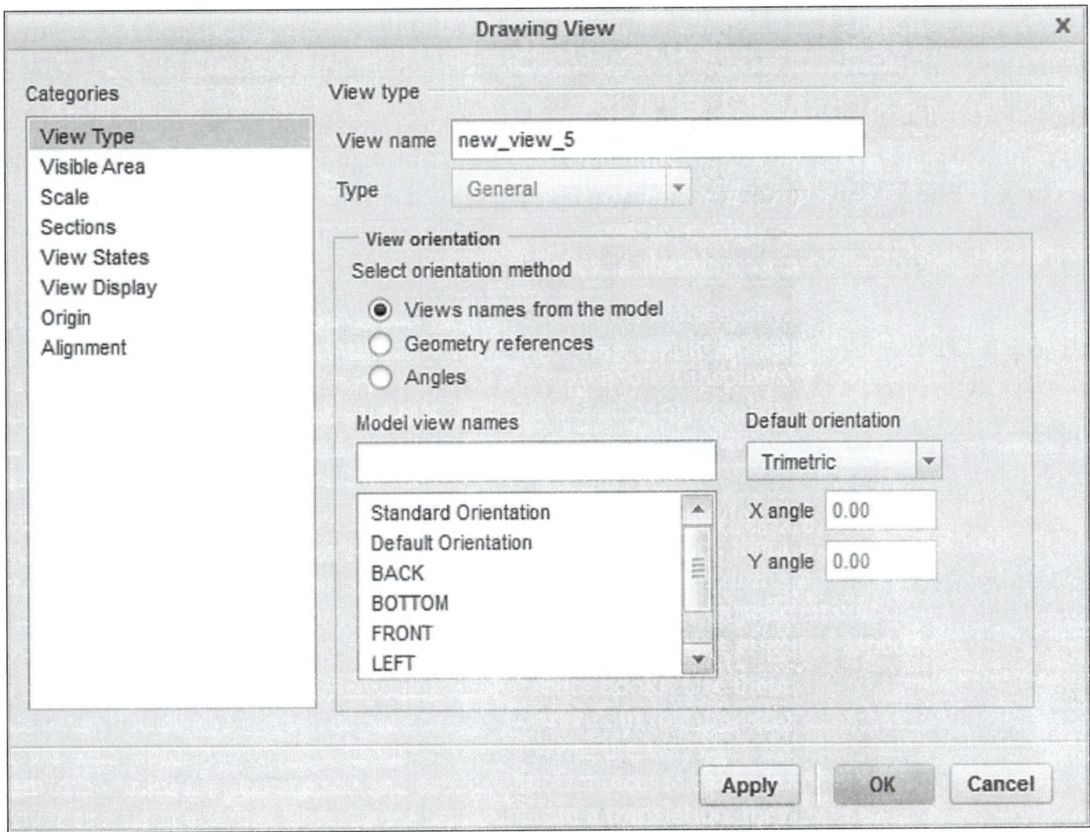

Figure 1.17(n) Drawing View Dialog Box

In the Graphics Window, press and hold: **RMB > Lock View Movement** *toggle off* [Fig. 1.17(o)]

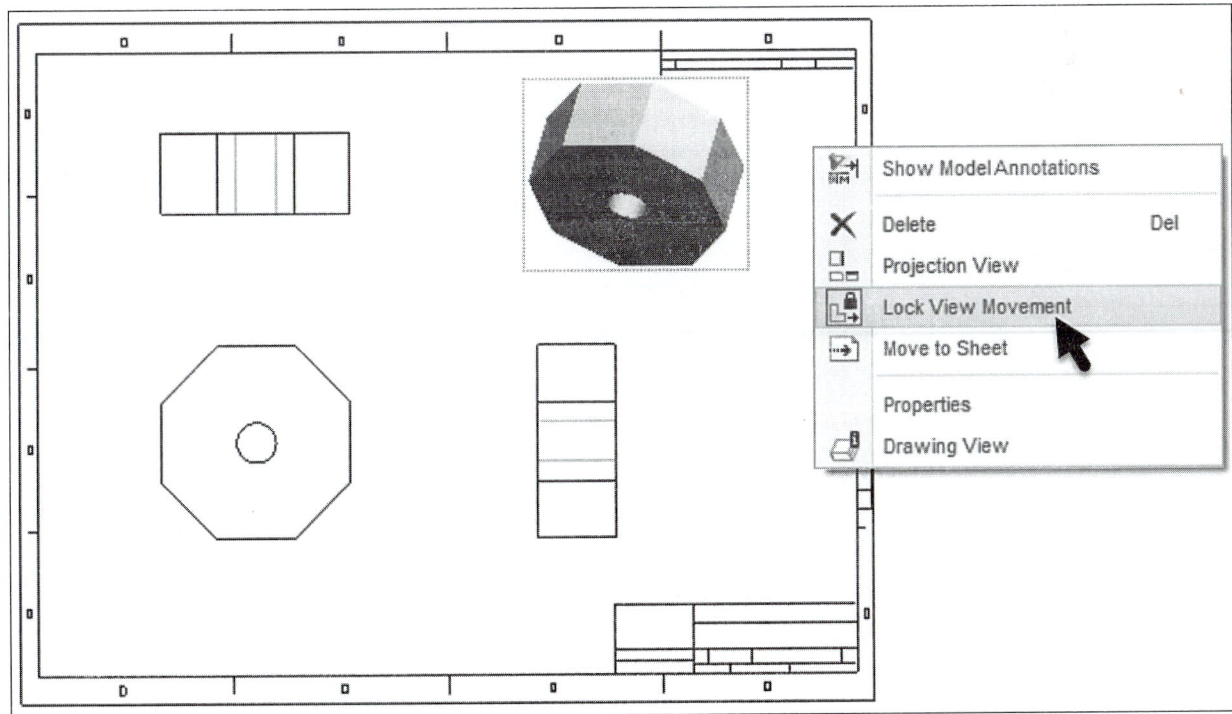

Figure 1.17(o) Lock View Movement off

Place the pointer on the *selected* view > press and hold down the **LMB** > move the pointer as desired [Fig. 1.17(p)] > release the **LMB** > **LMB** in the Graphics Window to deselect the view [Fig. 1.17(q)] > **Ctrl+S** > **Enter** > **File** > **Save As** > Type ▾ > **Zip File (*.zip)** > **OK** > upload to your course interface or attach to an email and send to your instructor or yourself > **File** > **Manage File** > **Delete Old Versions** > **Yes** > **File** > **Close** > **File** > **Manage Session** > **Erase Not Displayed** > **OK** > **File** > **Exit** > **Yes**

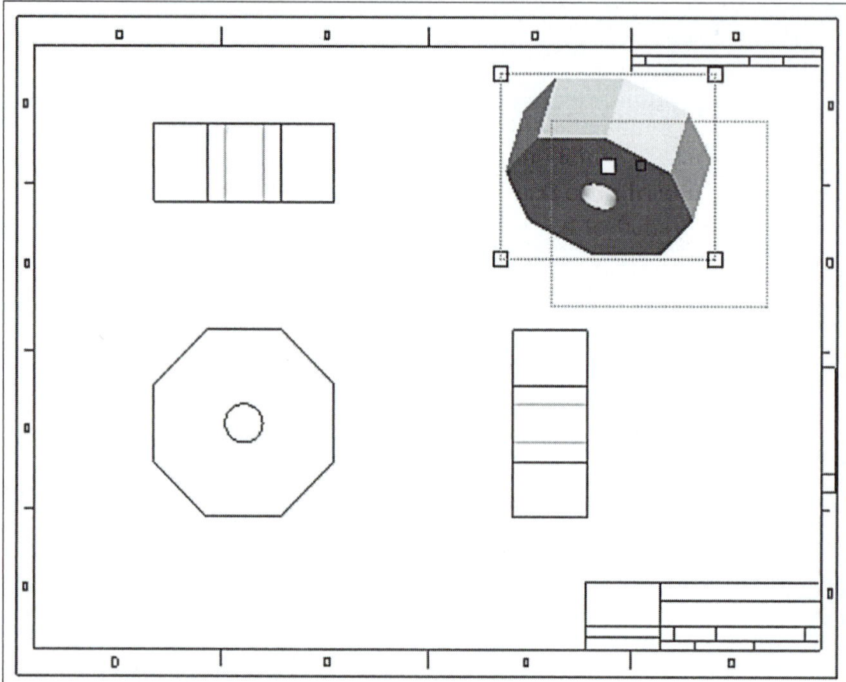

Figure 1.17(p) Move the View

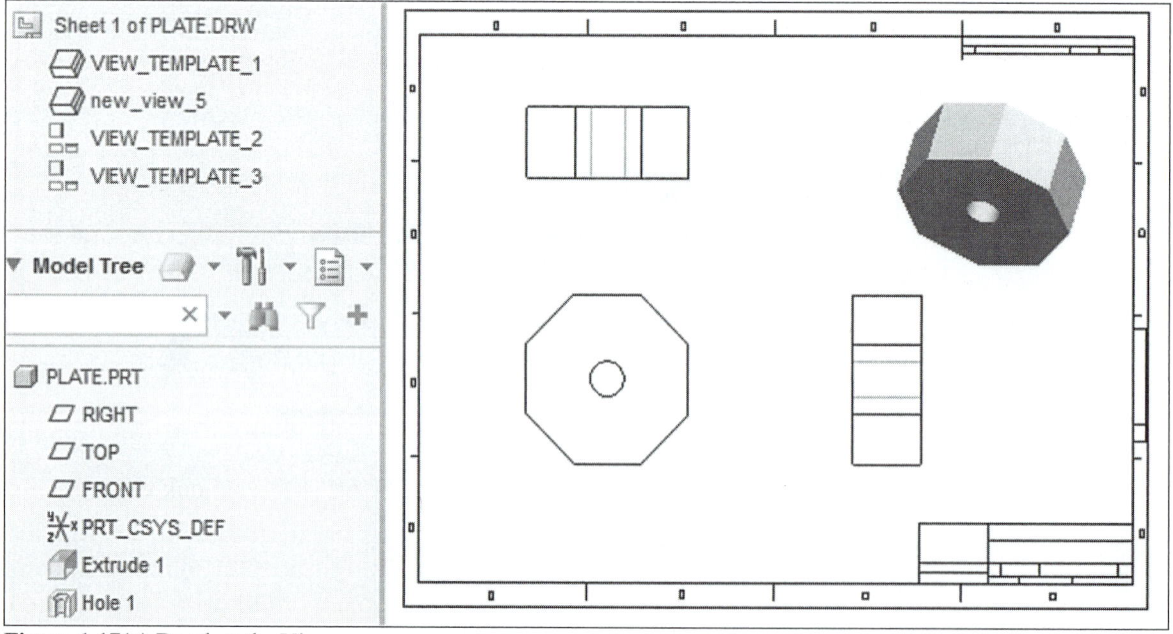

Figure 1.17(q) Deselect the View

You have completed the PTC Creo Parametric 3.0 overview. Spend some time exploring the website ***www.cad-resources.com*** for downloads, materials, and other CAD related information.

Lesson 2 PTC Creo Parametric 3.0

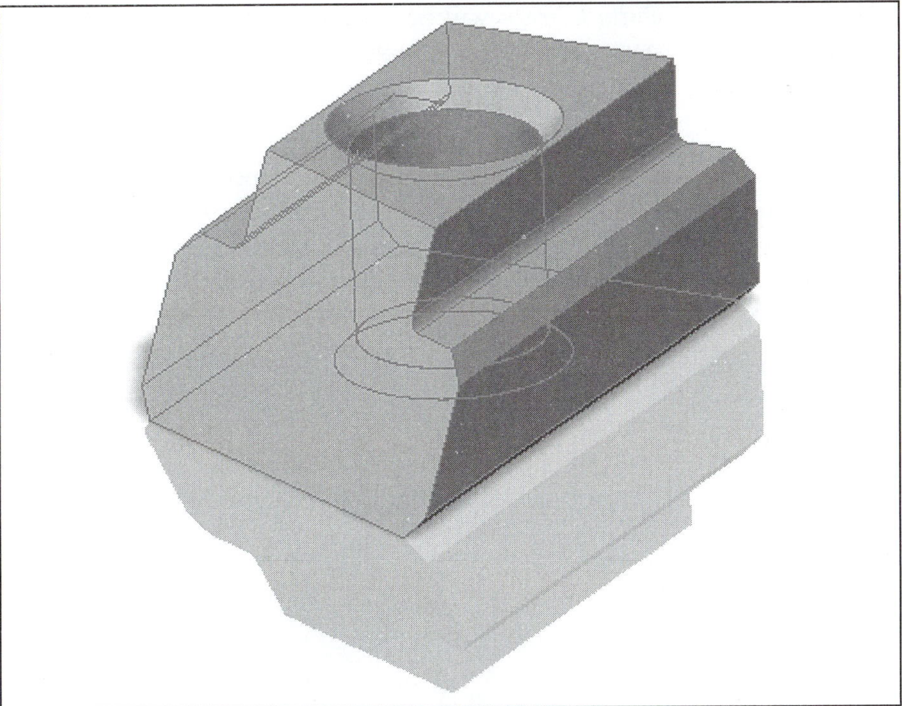

Figure 2.1 Part T-Nut

OBJECTIVES

- Understand and customize the **User Interface (UI)**
- **Download 3D CAD models** from the Creo Standard Parts Library
- Experience the **PTC Creo Parametric 3.0 Browser**
- Master the **File Functions**
- Become familiar with the **Help** center
- Use the **Command Locator**
- Be introduced to the **Display** and **View** capabilities
- Use **Mouse Buttons** to **Pan**, **Zoom**, and **Rotate** the object
- Change **Display Settings**
- Investigate an object with **Information Tools**
- Experience the **Model Tree** functionality

REFERENCES AND RESOURCES

For **Resources** go to **www.cad-resources.com** > click on the PTC Creo Parametric 3.0 book cover

- Lesson 2 Lecture
- Lesson 2 3D models embedded in a PDF
- Quick Reference Card
- Configuration Options

PTC Creo Parametric 3.0

This lesson will introduce you to the PTC Creo Parametric 3.0 working environment. An existing 3D CAD model (Fig. 2.1) from the Creo Standard Parts Library will be used to demonstrate the UI (user interface) and the general interaction required to master this CAD software.

PTC Creo Parametric 3.0 Interface

The PTC Creo Parametric 3.0 interface consists of a navigation window, an embedded Web browser, toolbars, information areas, and the graphics window (Fig. 2.2). Each Creo Parametric object opens in its own window. You can perform many operations from the Ribbon in multiple windows without canceling pending operations. Only one window is active at a time, but you can still perform some functions in the inactive windows. To activate a window simply press Ctrl+A.

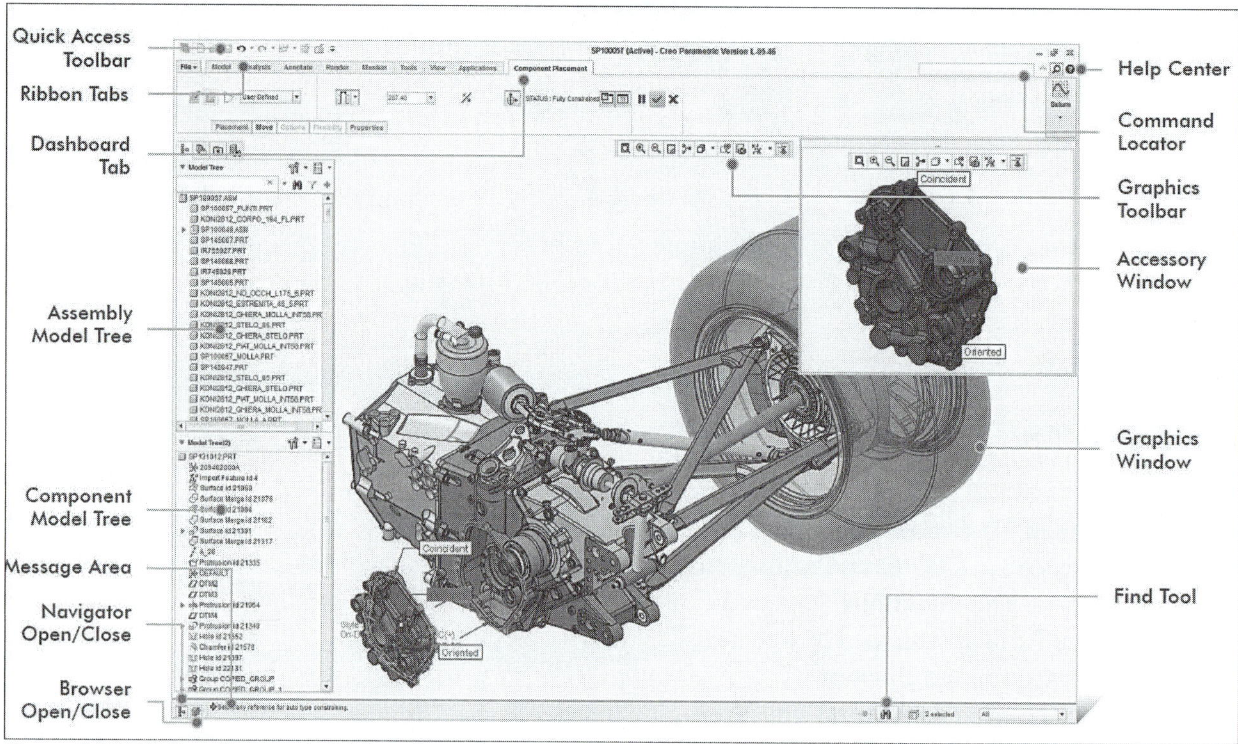

Figure 2.2 PTC Creo Parametric 3.0 Interface

The Creo Parametric window consists of the following elements:

- **The Browser** An embedded Web browser is located to the right of the navigator window. This browser provides you access to internal or external Web sites.

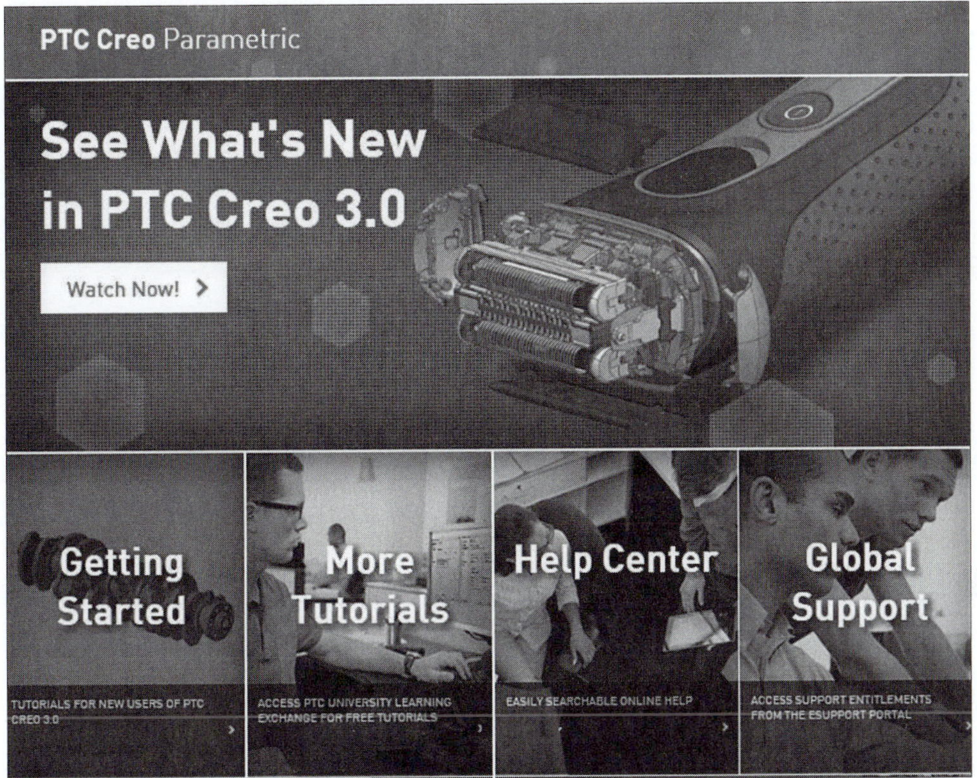

- **Graphics Window** The graphics window is the main working space (main window) for modeling.

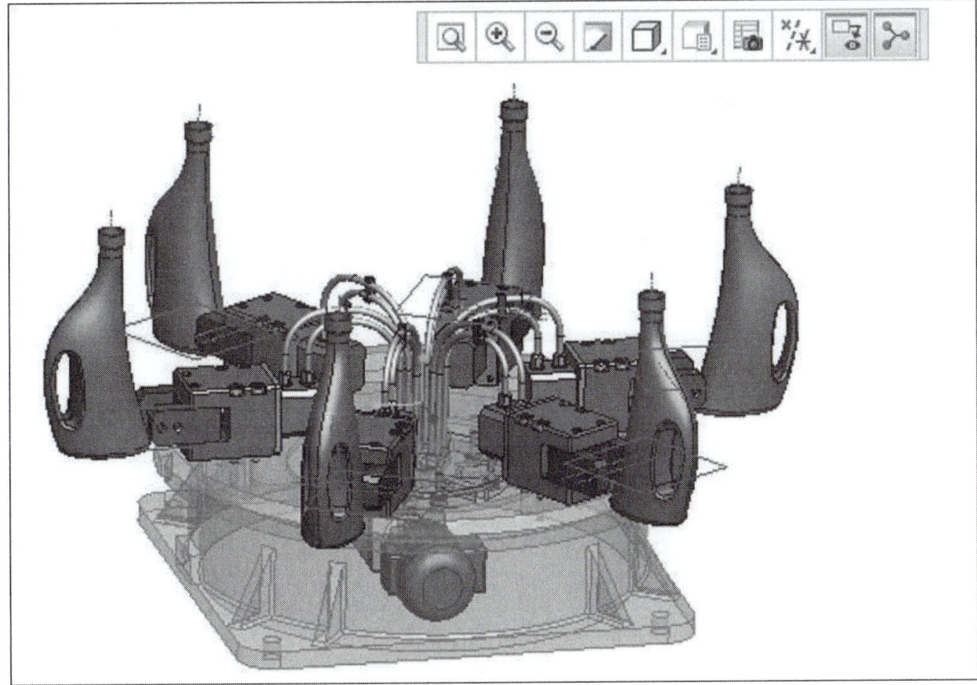

Navigation Area The navigation area is located on the left side of the Graphics Window. It includes tabs for the Model Tree and Layer Tree, Folder Browser, Favorites, History and Connections. on the status bar/message area (lower left corner of the window) controls the display of the navigator.

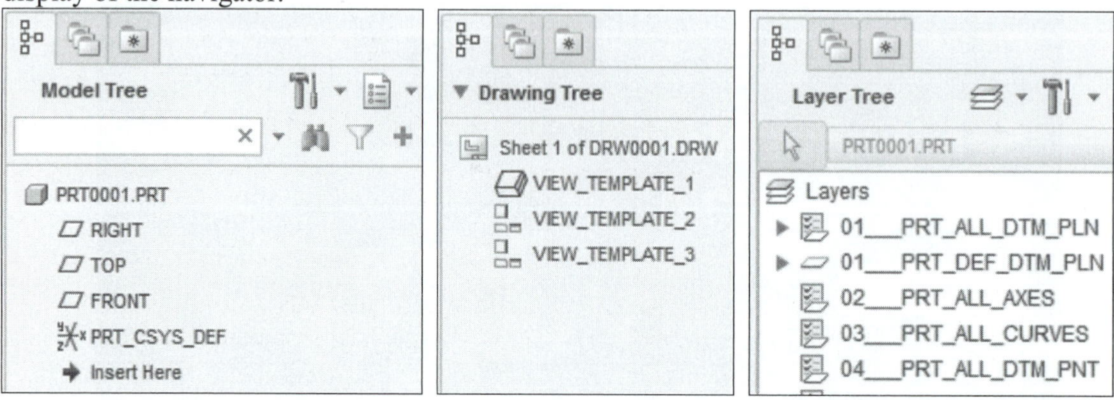

- **Quick Access Toolbar** The Quick Access toolbar is available regardless of which Ribbon tab is selected. By default it is located at the top of the window. It provides quick access to frequently used commands which are represented by icons or buttons, such as buttons for opening and saving files, undo, redo, regenerate, close window, switch windows, and so on. In addition, you can customize the Quick Access toolbar to include other frequently used buttons and cascading lists.

- **Graphics Toolbar** The Graphics toolbar contain icons to speed up access to commonly used View tab commands. By default, the Graphics toolbar consist of a row of buttons located directly under the Ribbon. The Graphics toolbar can be positioned on the top, left, bottom, and right of the graphics window, or in the status bar, or turned off. Graphics toolbar buttons can be added or removed to customize the layout.

- **File Menu** The File button in the upper-left corner of the window opens a menu that contains commands for managing files, models, preparing models for distribution, and for setting your Creo Parametric environment and configuration options.

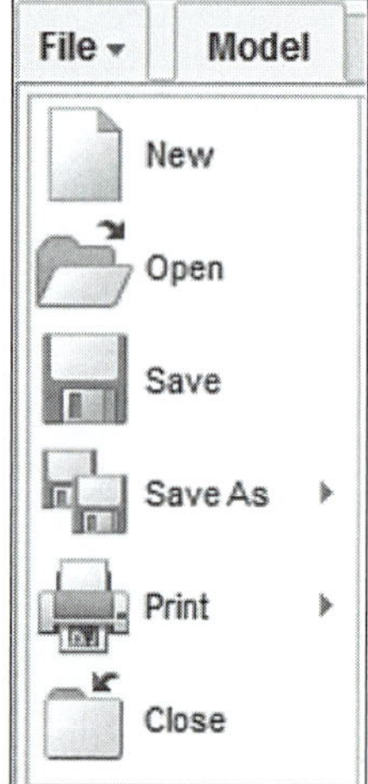

- **Ribbon** The Ribbon contains command buttons organized in a set of tabs. On each tab, the related buttons are grouped. You can minimize the Ribbon to make more space available on your screen. You can customize the Ribbon by adding, removing, or moving buttons. The following figure shows the different elements of a Ribbon.

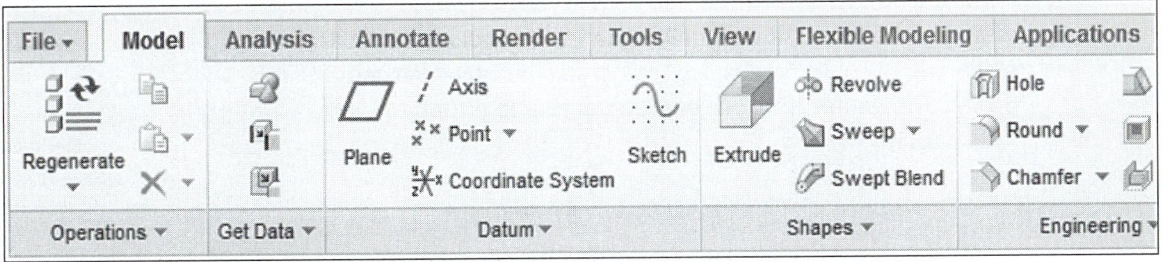

- **Tabs** Tabs are usually available when you are in a particular mode or application, or when you need them in a particular context. The Model, Analysis, Annotate, Render, Tools, View, Applications tabs are some commonly available tabs when you are in a mode. The Home tab is available when Creo Parametric has no model opened. Tabs related to a particular context open or close automatically when you activate or deactivate the context. Similarly, tabs related with a particular object open or close automatically when the related objects are selected or deselected, respectively. Tabs that contain tools of an application or controls of a tool have specific buttons to open and close them.

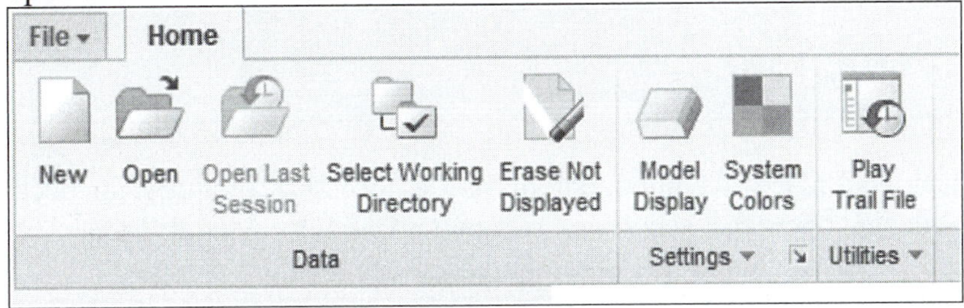

- **Screen Tips** The status bar messages also appear in small boxes near the menu option or dialog box item or toolbar button that the mouse pointer is passing over.

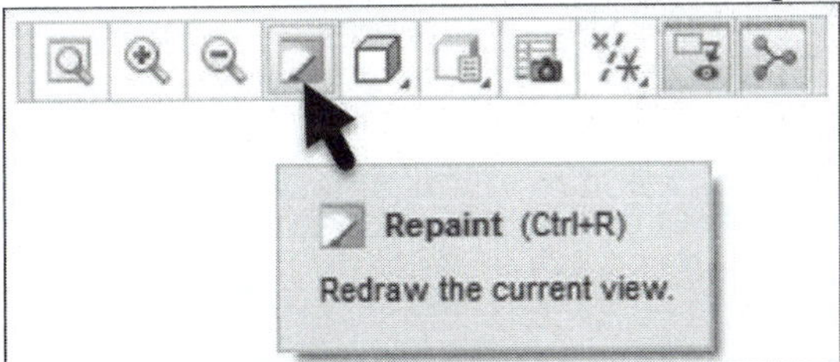

- **Message Area** Each window has a status bar/message area located at the bottom of the Creo Parametric window.

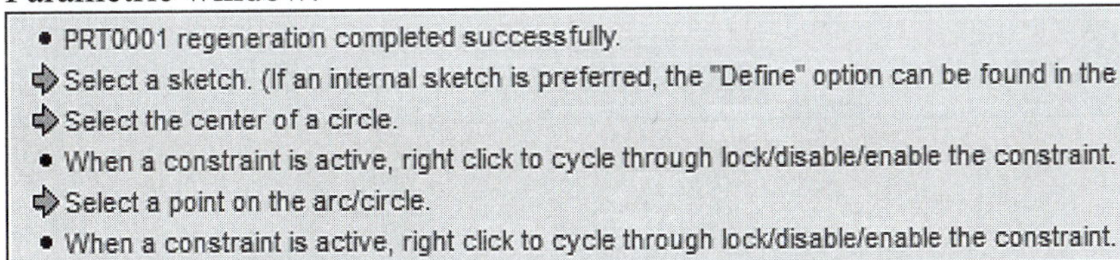

PRT0001 regeneration completed successfully.
Select a sketch. (If an internal sketch is preferred, the "Define" option can be found in the
Select the center of a circle.
When a constraint is active, right click to cycle through lock/disable/enable the constraint.
Select a point on the arc/circle.
When a constraint is active, right click to cycle through lock/disable/enable the constraint.

- **Command Locator** The Command Locator will quickly locate and initiate any command by opening the Ribbon and group associated with the command and listing optional locations.

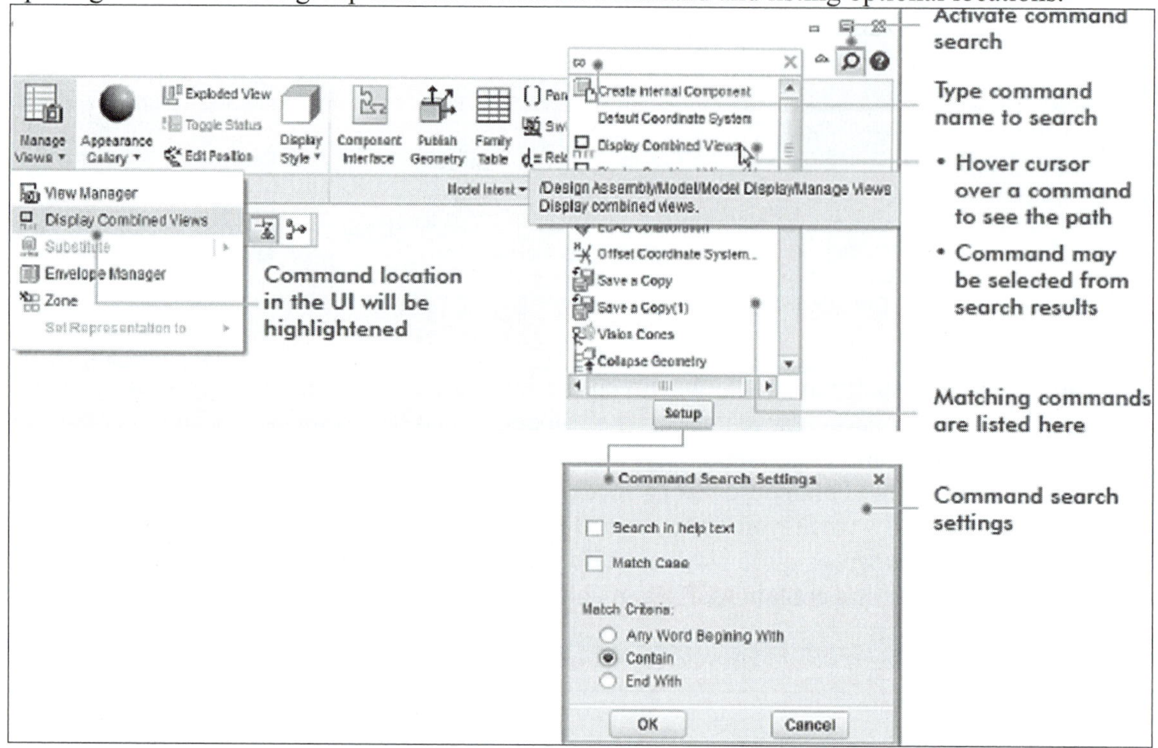

- **Accessory Window** When you place a component or use a tool that includes an accessory component or feature, you can open the accessory component and its Model Tree in a new window, even if it is already open in another window. An accessory window can be docked in the graphics window or opened separately.

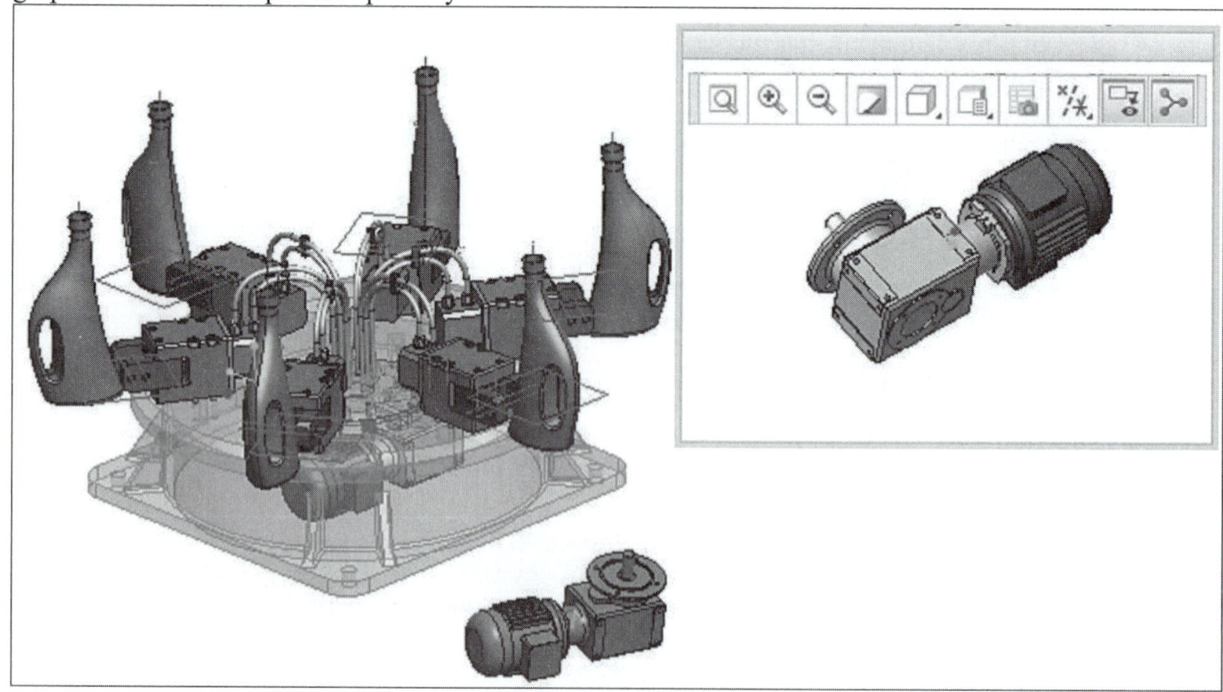

Lesson 2 STEPS

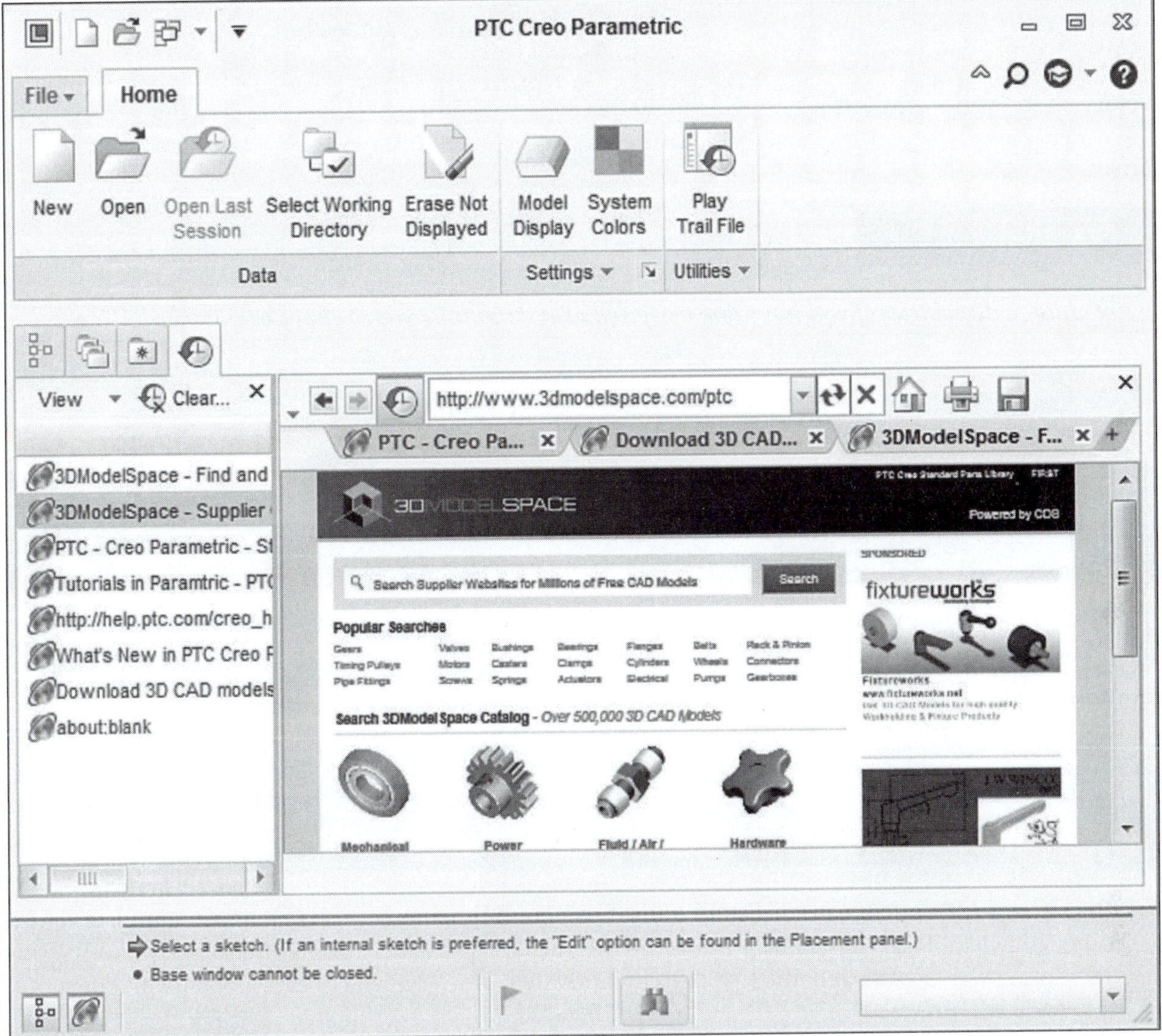

Figure 2.3 3DMODELSPACE Catalog (**Note: The PTC website may change names of certain items as they update their product library)**

Catalog Parts

In order to see and use the PTC Creo Parametric 3.0 UI, we must have an active object. You will download an existing 3D standard component using the Browser (Fig. 2.3). You will also experience using the Catalog of parts from PTC. You can access online catalogs from the **3DModelSpace** link. 3DModelSpace is available as a PTC Resource link in the Online Resources folder of Favorites tab in the Navigation area. The home page for the catalog lists catalogs of standard components.

You can browse catalogs, search for a component, view the details of the components, and download the selected component in Creo Parametric. You can also add a shortcut to a catalog. To access the online component catalogs, click the **Favorites** tab and expand the Personal Favorites folder. Online Resources is located within the Personal Favorites folder. Expand the Online Resources folder and select 3DModelSpace. The catalogs page opens in the browser and you can select a catalog to browse or you can right-click on a catalog to add a shortcut to it to be accessed from the Online Resources folder. You can also add the catalog shortcut to **Personal Favorites**.

Note: The PTC website may change names of certain items as they update their product library.

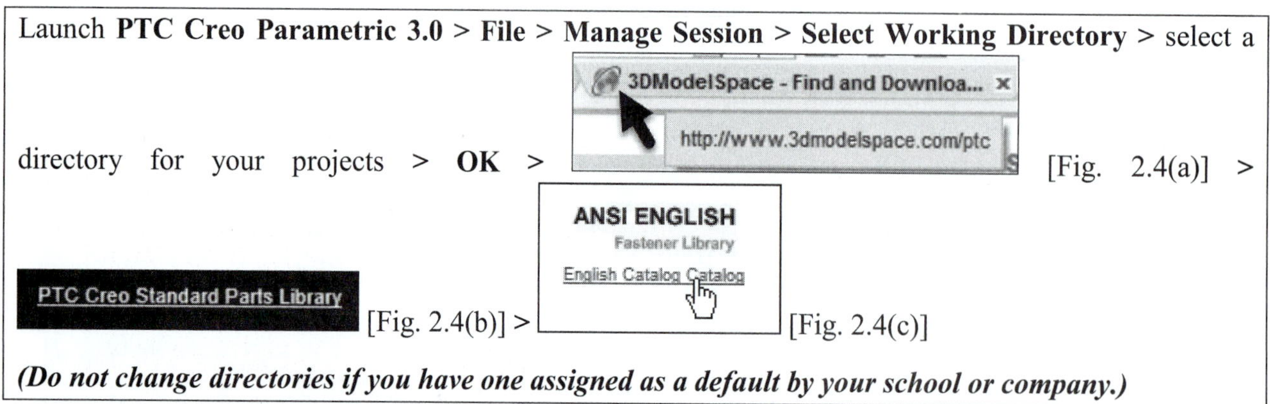

Launch **PTC Creo Parametric 3.0** > **File** > **Manage Session** > **Select Working Directory** > select a directory for your projects > **OK** > [Fig. 2.4(a)] > PTC Creo Standard Parts Library [Fig. 2.4(b)] > ANSI ENGLISH Fastener Library English Catalog Catalog [Fig. 2.4(c)]

(Do not change directories if you have one assigned as a default by your school or company.)

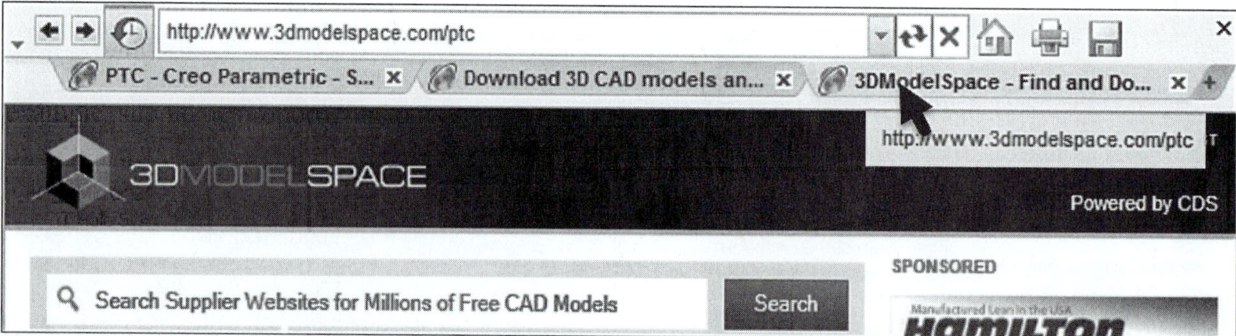

Figure 2.4(a) 3DModelSpace (*The PTC website may change names of certain items as they update their product library)*

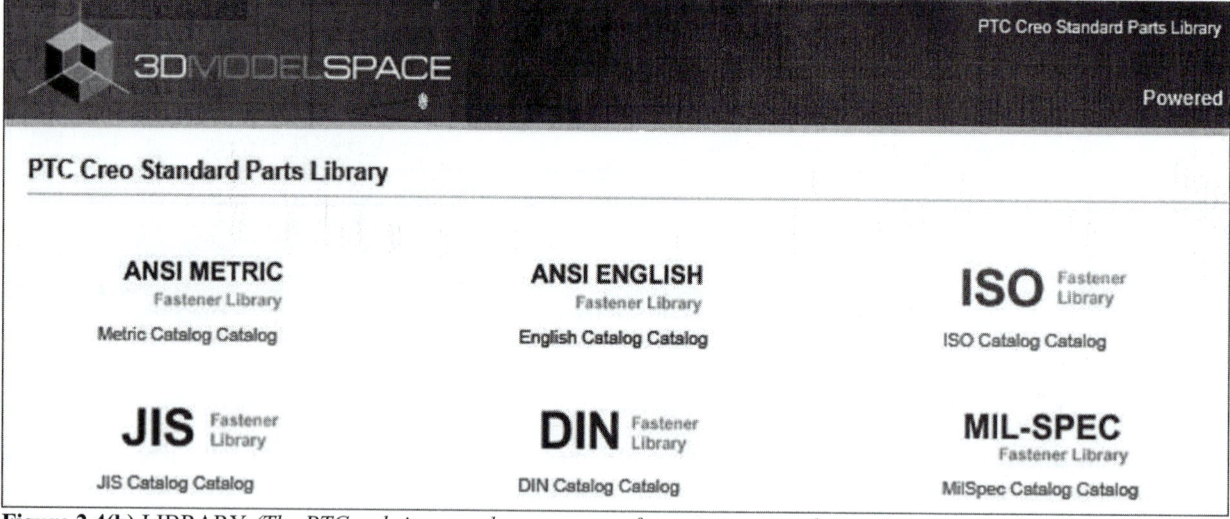

Figure 2.4(b) LIBRARY *(The PTC website may change names of certain items as they update their product library)*

PTC Creo Standard Parts Library

ANSI METRIC
Fastener Library
Metric Catalog Catalog

ANSI ENGLISH
Fastener Library
English Catalog Catalog

Figure 2.4(c) ANSI ENGLISH Fastener Library

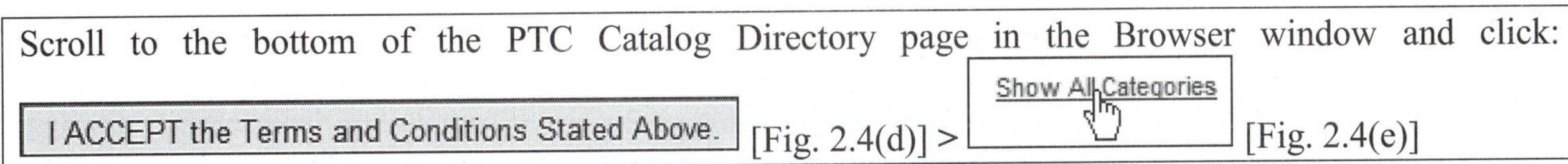

Scroll to the bottom of the PTC Catalog Directory page in the Browser window and click: I ACCEPT the Terms and Conditions Stated Above. [Fig. 2.4(d)] > Show All Categories [Fig. 2.4(e)]

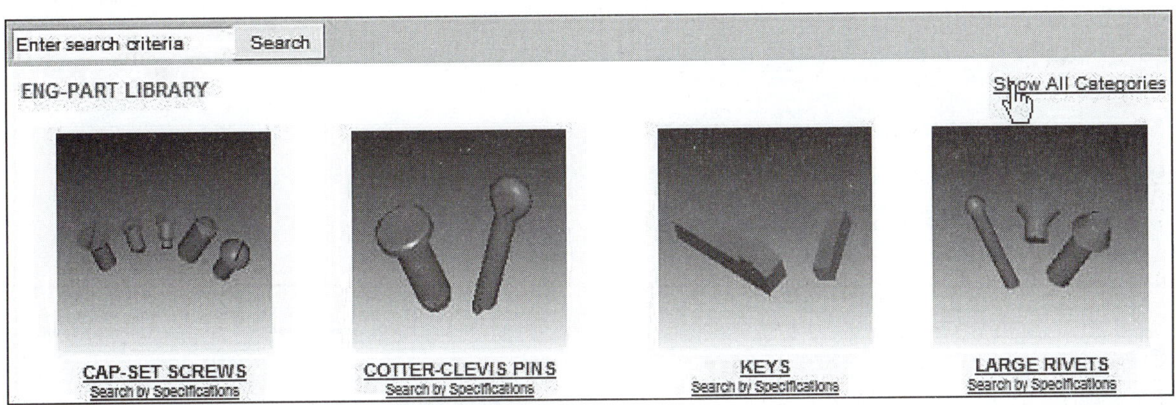

Figure 2.4(d) ENG-PART LIBRARY Category Images

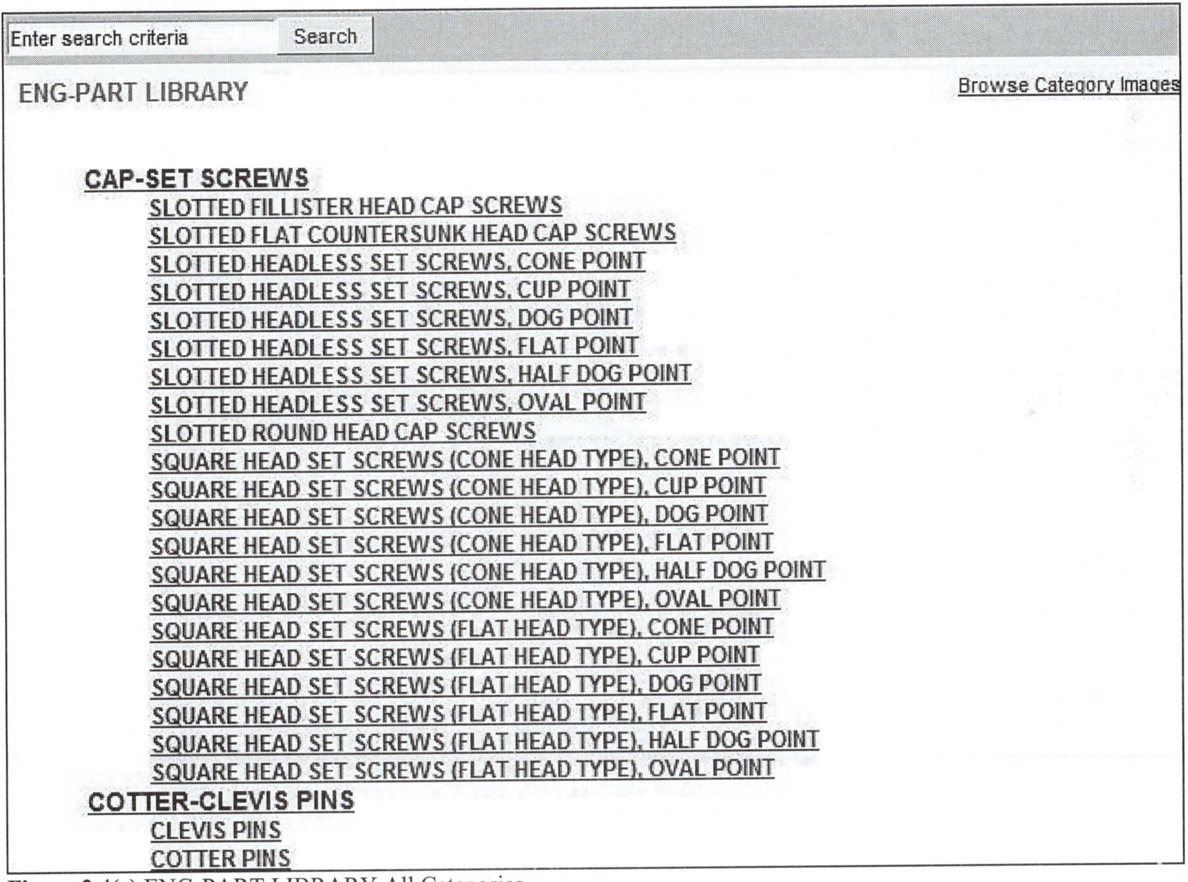

Figure 2.4(e) ENG-PART LIBRARY All Categories

TEE BOLTS-NUTS
STANDARD T-BOLTS
STANDARD T-NUTS
WOODRUFF KEYS

Scroll to the bottom of the list > STANDARD T-NUTS > **STANDARD T-NUTS** > **ASTN04** [Fig. 2.5(f)] > **Choose Format** > Select the appropriate version as needed

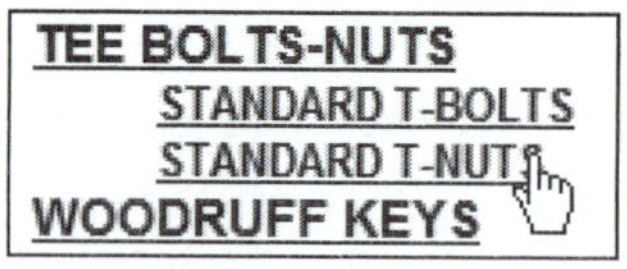

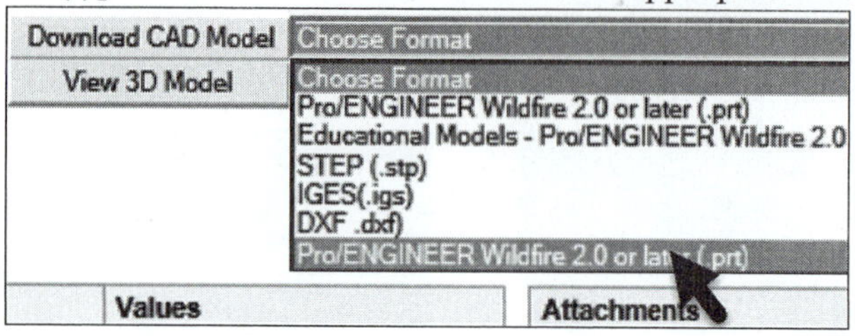

Download CAD Model | Choose Format
View 3D Model | Choose Format
Pro/ENGINEER Wildfire 2.0 or later (.prt)
Educational Models - Pro/ENGINEER Wildfire 2.0
STEP (.stp)
IGES(.igs)
DXF .dxf)
Pro/ENGINEER Wildfire 2.0 or later (.prt)
Values | Attachments

Pro/ENGINEER Wildfire 2.0 >

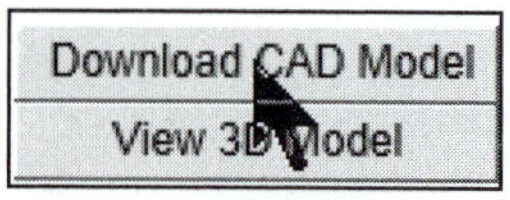

Download CAD Model
View 3D Model

> place mouse on the link

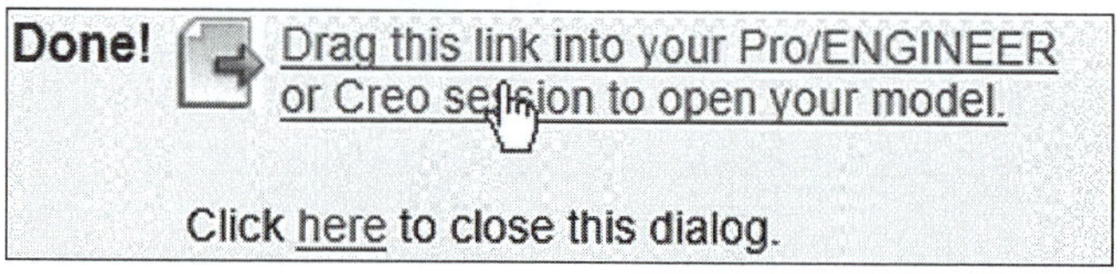

Done! Drag this link into your Pro/ENGINEER or Creo session to open your model.
Click here to close this dialog.

> press and hold down the **LMB** > drag and drop the link into the Graphics Window > **Open**

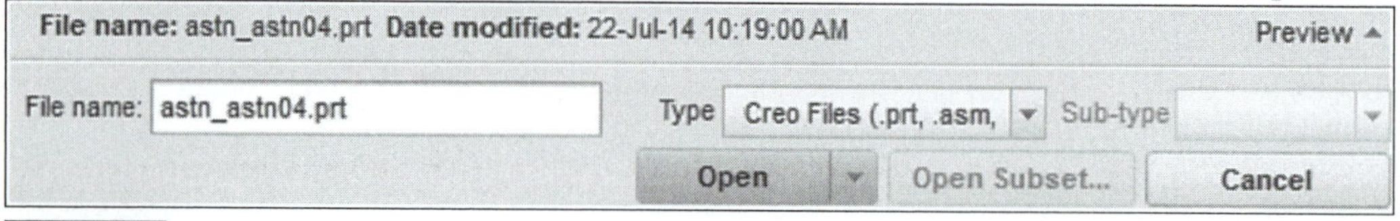

File name: astn_astn04.prt Date modified: 22-Jul-14 10:19:00 AM Preview
File name: astn_astn04.prt Type Creo Files (.prt, .asm, Sub-type
Open | Open Subset... | Cancel

[Fig. 2.5(g)]  > Toggle the display of the browser

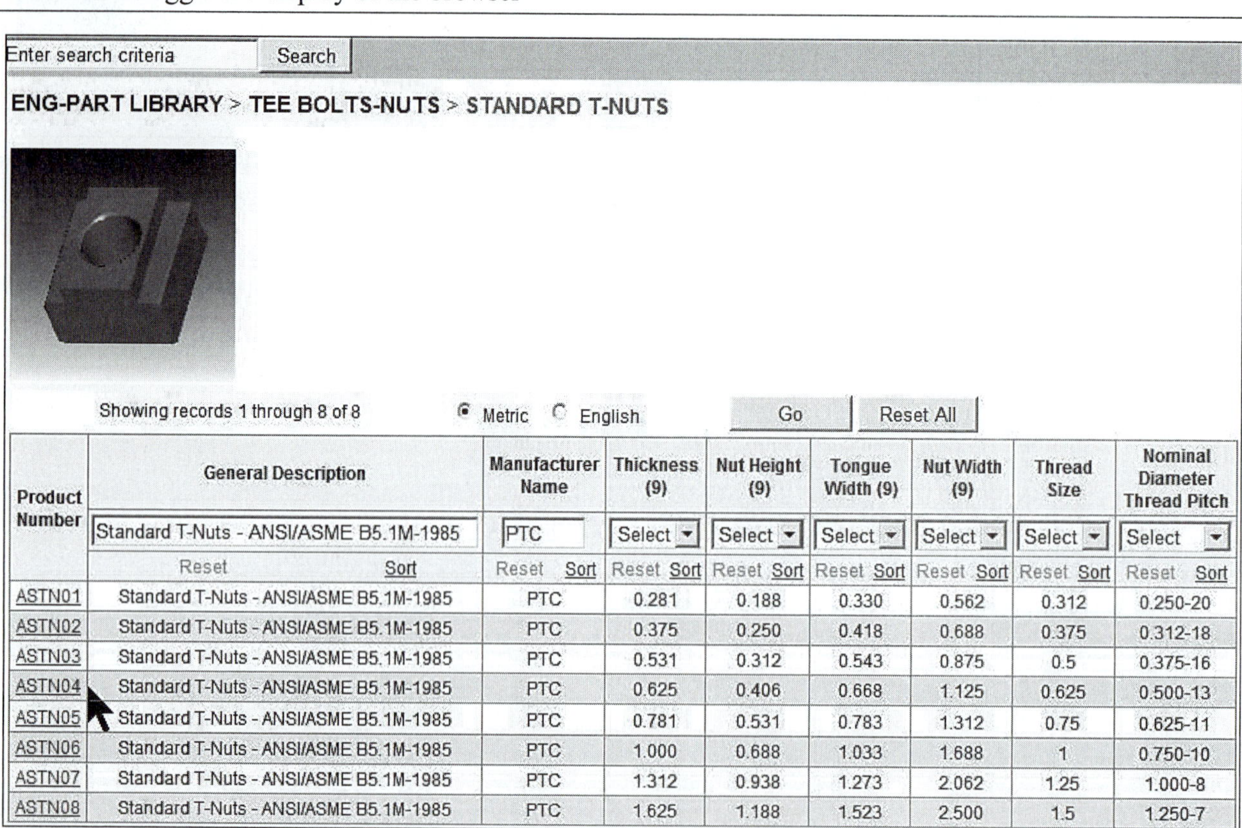

Enter search criteria | Search

ENG-PART LIBRARY > TEE BOLTS-NUTS > STANDARD T-NUTS

Showing records 1 through 8 of 8 — Metric / English — Go — Reset All

Product Number	General Description	Manufacturer Name	Thickness (9)	Nut Height (9)	Tongue Width (9)	Nut Width (9)	Thread Size	Nominal Diameter Thread Pitch
	Standard T-Nuts - ANSI/ASME B5.1M-1985	PTC	Select	Select	Select	Select	Select	Select
	Reset Sort	Reset Sort	Reset Sort	Reset Sort	Reset Sort	Reset Sort	Reset Sort	Reset Sort
ASTN01	Standard T-Nuts - ANSI/ASME B5.1M-1985	PTC	0.281	0.188	0.330	0.562	0.312	0.250-20
ASTN02	Standard T-Nuts - ANSI/ASME B5.1M-1985	PTC	0.375	0.250	0.418	0.688	0.375	0.312-18
ASTN03	Standard T-Nuts - ANSI/ASME B5.1M-1985	PTC	0.531	0.312	0.543	0.875	0.5	0.375-16
ASTN04	Standard T-Nuts - ANSI/ASME B5.1M-1985	PTC	0.625	0.406	0.668	1.125	0.625	0.500-13
ASTN05	Standard T-Nuts - ANSI/ASME B5.1M-1985	PTC	0.781	0.531	0.783	1.312	0.75	0.625-11
ASTN06	Standard T-Nuts - ANSI/ASME B5.1M-1985	PTC	1.000	0.688	1.033	1.688	1	0.750-10
ASTN07	Standard T-Nuts - ANSI/ASME B5.1M-1985	PTC	1.312	0.938	1.273	2.062	1.25	1.000-8
ASTN08	Standard T-Nuts - ANSI/ASME B5.1M-1985	PTC	1.625	1.188	1.523	2.500	1.5	1.250-7

Figure 2.5(f) ENG-PART LIBRARY > T-BOLTS-NUTS > STANDARD T-NUTS

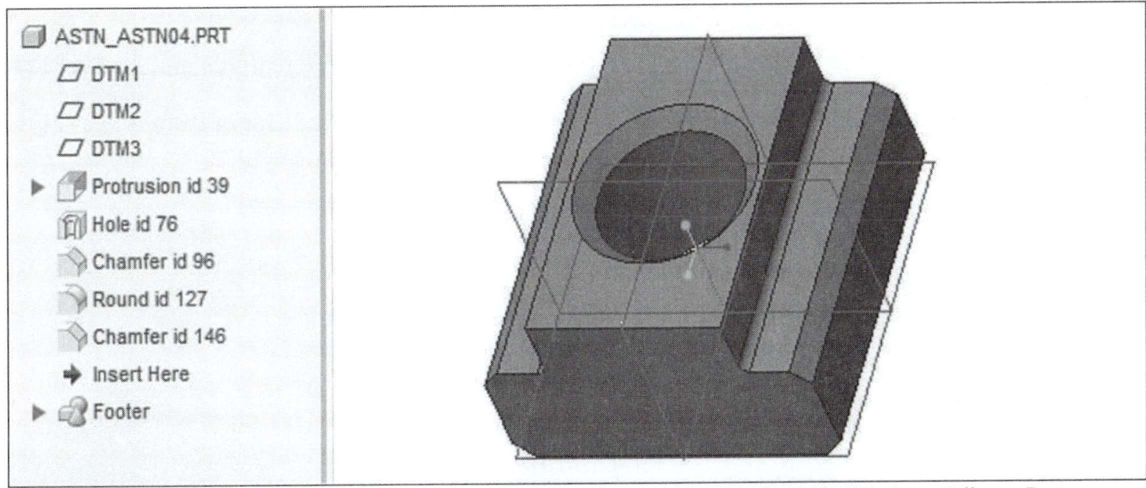

Figure 2.5(g) Drag and Drop the Link *(which is a Zip file of a part)* into the Graphics Window - Library Part

ALTERNATE MODEL- *If your system cannot access the Standard Parts Library; in the embedded web Browser > type: www.cad-resources.com > Enter*

http://www.cad-resources.com/
PTC - Creo Parametric -... Download 3D CAD models ... CAD Resources and Louis ...

> *click on the appropriate book cover icon > place your mouse cursor on the part icon* > *drag and drop the link into the Graphics Window >* ***OR*** *> click on the part icon > Save [Fig. 2.5(h)] > Save (if needed) > select the desired folder to save the zip file to > Save > Close (if needed) > Open a file manager window to the desired folder > drag and drop the zip file into your Graphics Window > Open (if needed) >* ***OR*** *> to the saved zip file, unzip or extract all > drag and drop the file into the Graphics Window*

File Download

Do you want to open or save this file?

Name: AA_astn-lesson-2.prt.zip
Type: WinZip File, 75.7 KB
From: **www.cad-resources.com**

Open Save Cancel

Always ask before opening this type of file

While files from the Internet can be useful, some files can potentially harm your computer. If you do not trust the source, do not open or save this file. What's the risk?

Figure 2.5(h) Zip File Download dialog box – *your dialog box may appear differently*

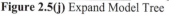

Click: close the Browser [Fig. 2.5(i)] > to expand the Model Tree features and Footer [Fig. 2.5(j)] > **RMB** on Interfaces **INTFC001** [Fig. 2.5(k)] > **Information** > **Feature Information** > close the Browser

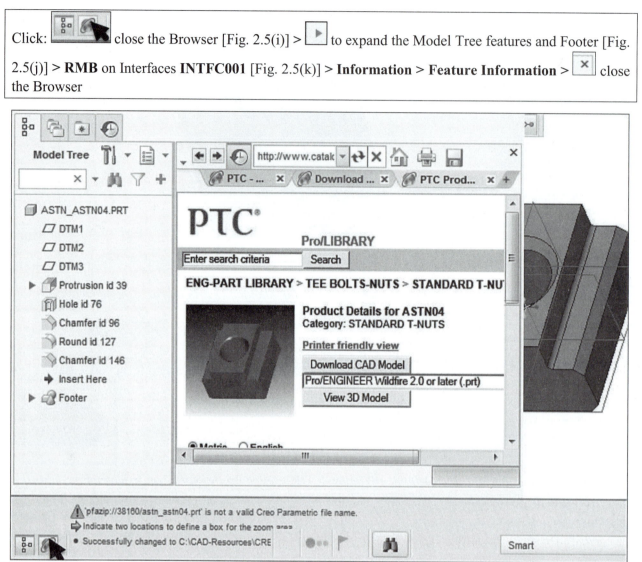

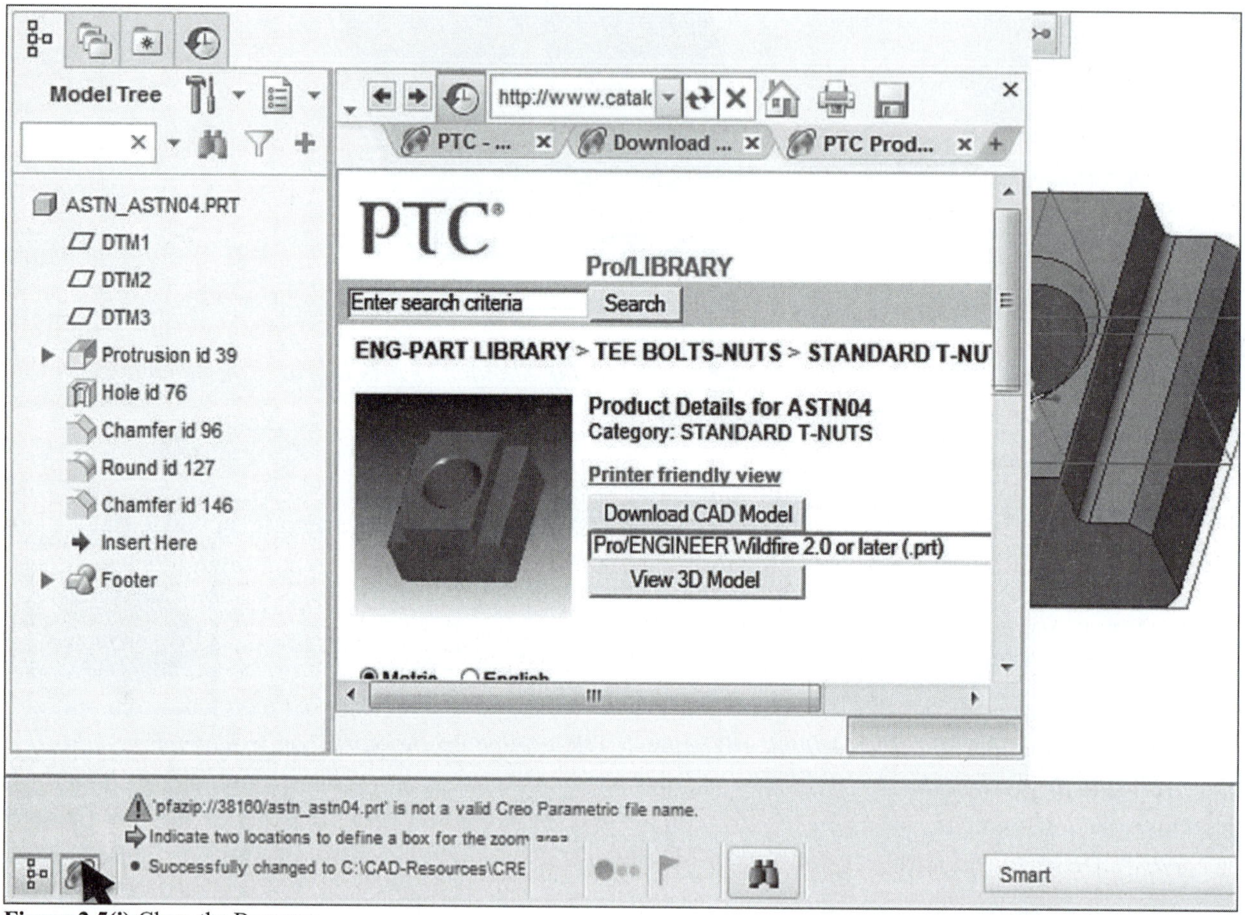

Figure 2.5(i) Close the Browser

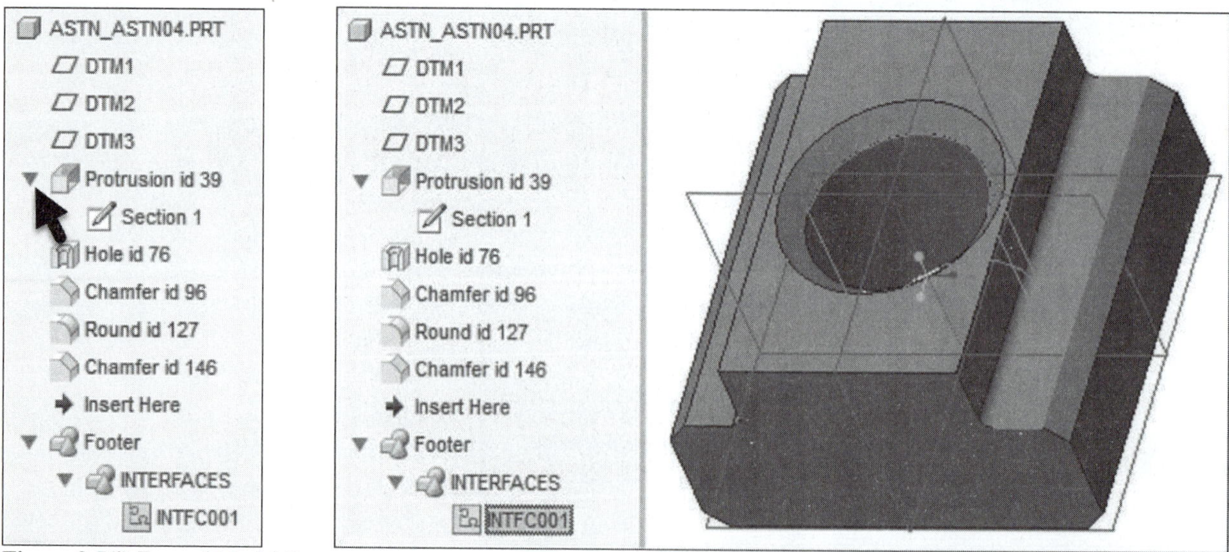

Figure 2.5(j) Expand Model Tree

Figure 2.5(k) INTERFACES

Click: **File** > **Save As** > New Name-- *type* **T-NUT** [Fig. 2.5(l)] > **OK** > **File** > **Close** > **Ctrl+O** open an existing model > **t-nut.prt** [Fig. 2.5(m)] > **Open** > ▶ to expand Model Tree items [Fig. 2.5(n)]

Model Name	ASTN_ASTN04.PRT
File Name	t-nut
Type	Part (*.prt)

Figure 2.5(l) Save As T-NUT

File Open

« (C:) ▸ CAD-Resources ▸ CREO ▸ Creo_3-0 ▸ PTC_Creo-3 ▸ 3-0

Search...

Organize ∨ Views ∨ Tools ∨

Common Folders: In Session, Desktop, My Documents, cad-resources_7, Working Directory, Network Neighborhood

Folder Tree

connector.asm
connector.drw
pin.prt
plate.drw
plate.prt
t-nut.prt

File name: t-nut.prt **Date modified:** 22-Jul-14 10:53:31 AM

Preview

File name: t-nut.prt Type Creo Files (.prt, .asm, Sub-type

Open Open Subset... Cancel

Figure 2.5(m) File Open Dialog Box, **t-nut.prt**

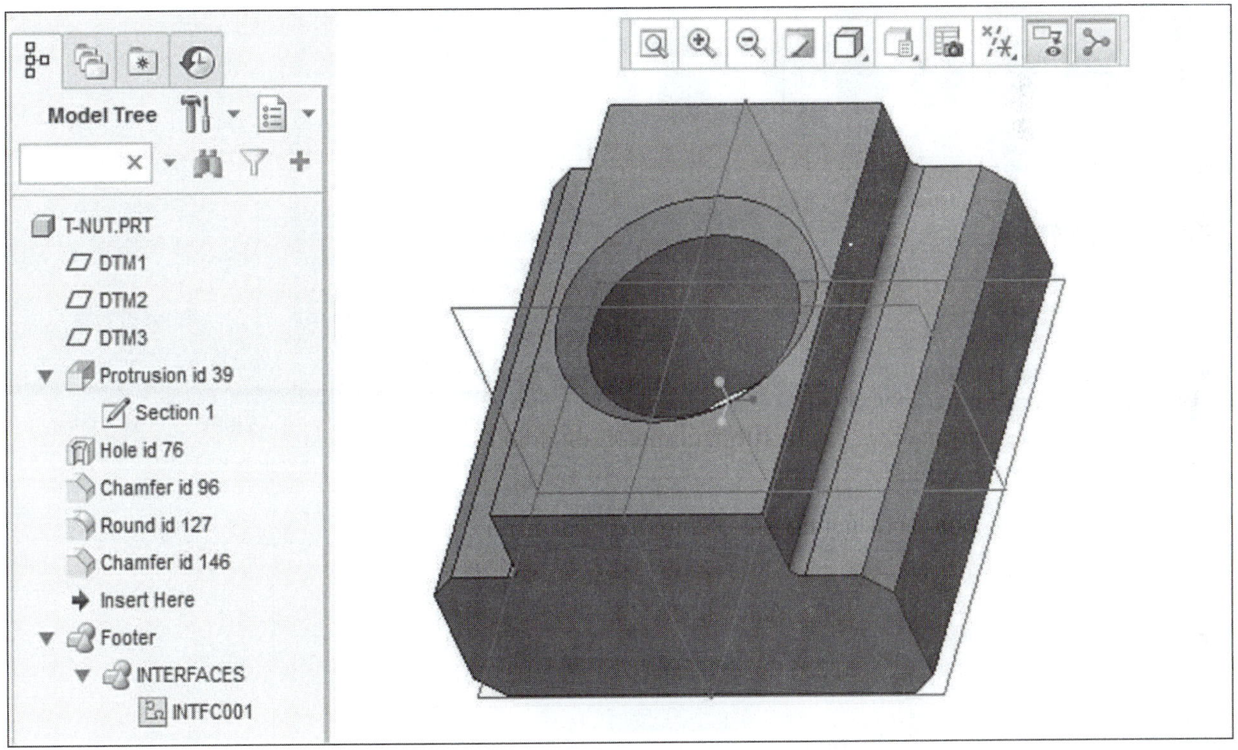

Figure 2.5(n) t-nut.prt

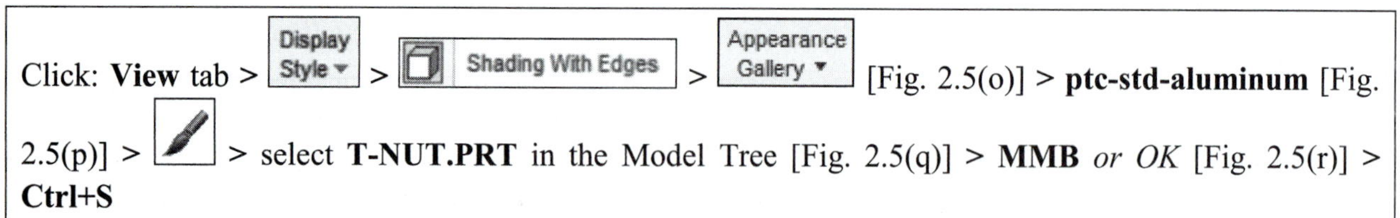

Click: **View** tab > Display Style > Shading With Edges > Appearance Gallery [Fig. 2.5(o)] > **ptc-std-aluminum** [Fig. 2.5(p)] > (edit icon) > select **T-NUT.PRT** in the Model Tree [Fig. 2.5(q)] > **MMB** *or OK* [Fig. 2.5(r)] > **Ctrl+S**

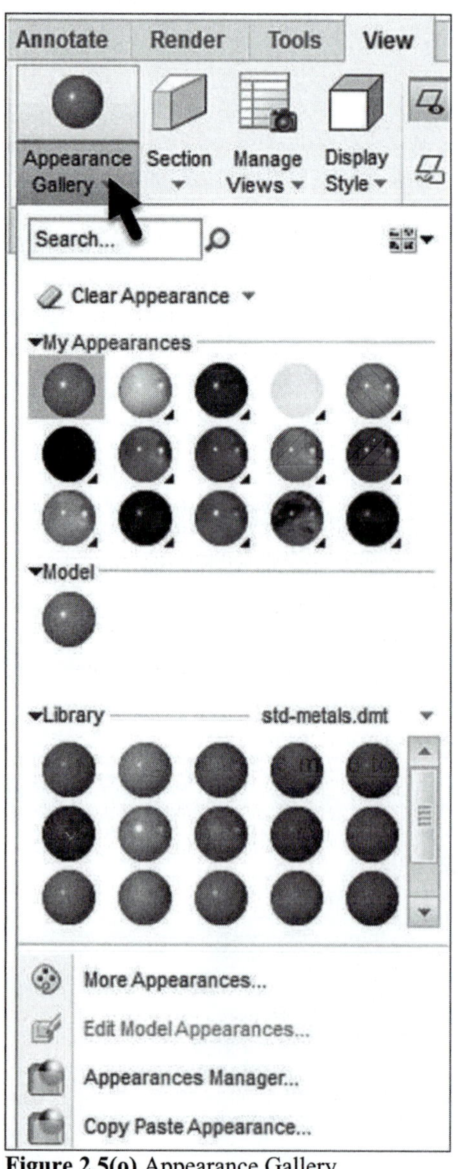

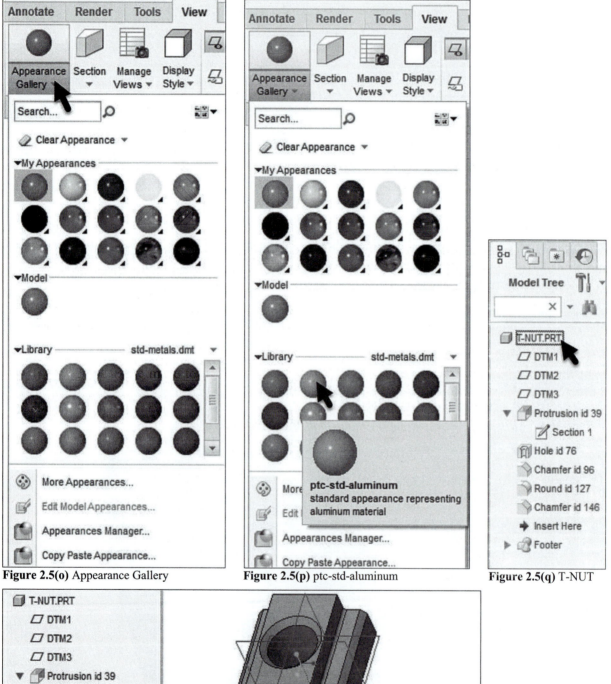

Figure 2.5(o) Appearance Gallery

Figure 2.5(p) ptc-std-aluminum

Figure 2.5(q) T-NUT

Figure 2.5(r) New Color

File Functions

The **File** tab provides options for opening, creating, saving, renaming, backing up files, and printing. Before using any File tool, make sure you have set your working directory to the folder where you wish to save objects for the project on which you are working. The Working Directory is a directory that you set up to contain PTC Creo Parametric 3.0 files. You can save a Creo Parametric file using **Save** [Fig. 2.6(a)] or **Save As** from the File tab. You can save files, make copies of files, or backups of files after creating them. The default folder to save or back up a file is determined as follows: The **My Documents** folder if you have not set a Working Directory. The **Working Directory** you have set for your current session. The folder you last accessed to open, save, save a copy, or back up your file. Each time you save a file, it creates a new version; ***nut.prt.1***, ***nut.prt.2***, etc. Save fifty times and you have fifty more files; ***nut.prt.50.*** To purge old versions: **File > Manage File > Delete Old Versions > Enter**. The last (highest number-last saved) version will be the remaining file and the one that opens the next time you access the file.

Save As [Fig. 2.6(b)] is the exportation of Creo Parametric files to different formats, and to save files as images. Since the name already exists in session, you cannot save or Rename a file using the same name as the original file name, even if you save the file in a different directory. Creo Parametric forces you to enter a unique file name by displaying the message: *An object with that name exists in session. Choose a different name.*

Creo Parametric file names are restricted to a maximum of 31 characters and have no spaces. A file can be a part, assembly, or drawing. Each is considered an "object" or "model".

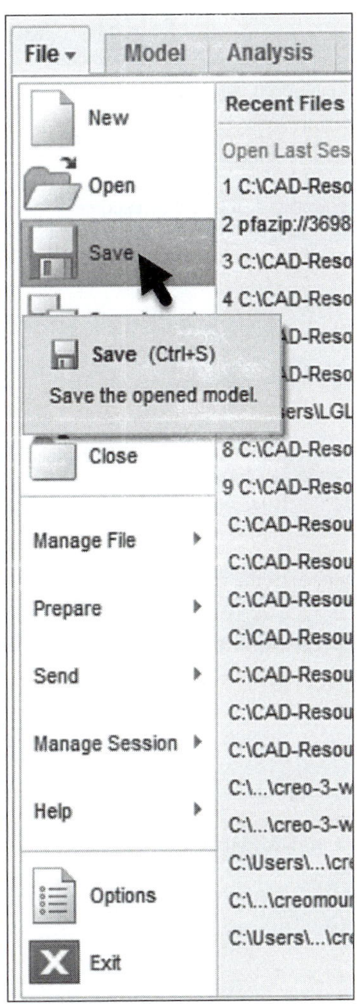

Figure 2.6(a) Save

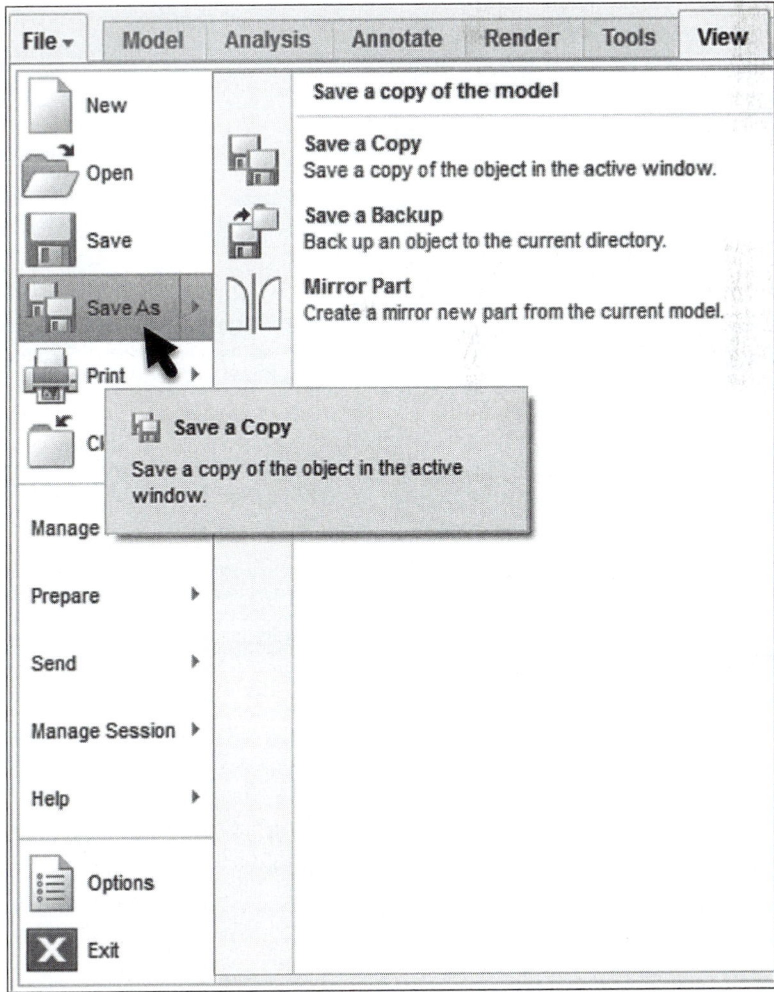

Figure 2.6(b) Save As

Help Center

Accessing the Help function [Fig. 2.7(a)] is one of the best ways to learn any CAD software. Use the Help Center [Fig. 2.7(b)] as often as possible to understand the tool or command you are using at the time and to expand your knowledge of the capabilities provided by PTC Creo Parametric. Use the **Help** menu to gain access to online information, release information, and customer service information.

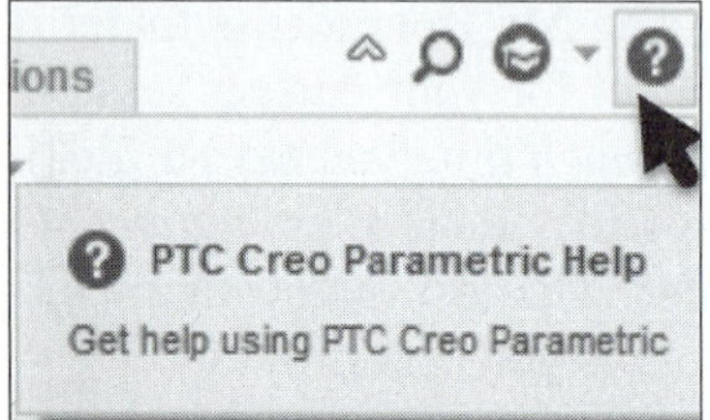

Figure 2.7(a) Activate the Help Center

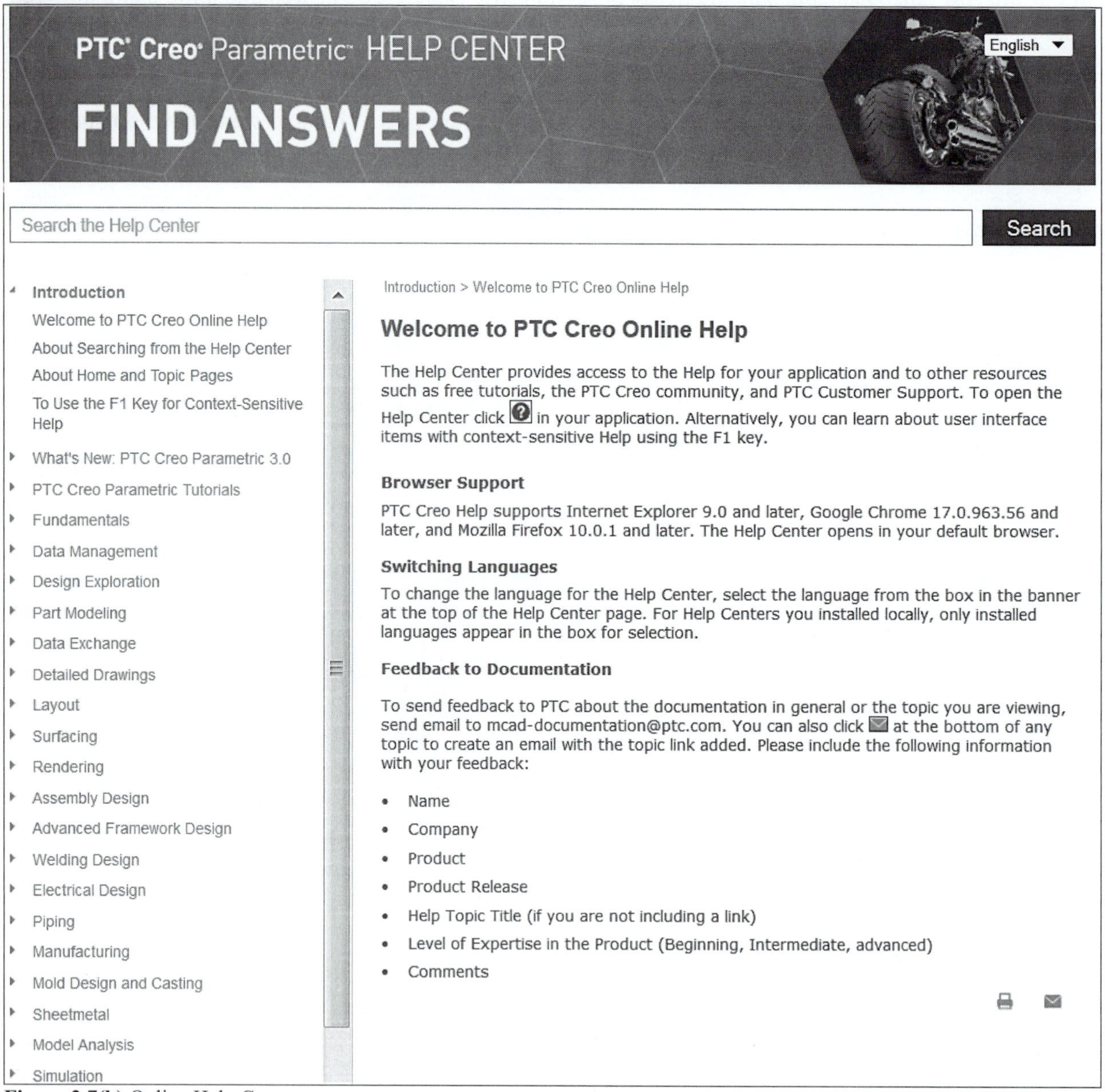

PTC® Creo® Parametric™ HELP CENTER

FIND ANSWERS

English

Search the Help Center | Search

- Introduction
  - Welcome to PTC Creo Online Help
  - About Searching from the Help Center
  - About Home and Topic Pages
  - To Use the F1 Key for Context-Sensitive Help
- What's New: PTC Creo Parametric 3.0
- PTC Creo Parametric Tutorials
- Fundamentals
- Data Management
- Design Exploration
- Part Modeling
- Data Exchange
- Detailed Drawings
- Layout
- Surfacing
- Rendering
- Assembly Design
- Advanced Framework Design
- Welding Design
- Electrical Design
- Piping
- Manufacturing
- Mold Design and Casting
- Sheetmetal
- Model Analysis
- Simulation

Introduction > Welcome to PTC Creo Online Help

Welcome to PTC Creo Online Help

The Help Center provides access to the Help for your application and to other resources such as free tutorials, the PTC Creo community, and PTC Customer Support. To open the Help Center click in your application. Alternatively, you can learn about user interface items with context-sensitive Help using the F1 key.

Browser Support

PTC Creo Help supports Internet Explorer 9.0 and later, Google Chrome 17.0.963.56 and later, and Mozilla Firefox 10.0.1 and later. The Help Center opens in your default browser.

Switching Languages

To change the language for the Help Center, select the language from the box in the banner at the top of the Help Center page. For Help Centers you installed locally, only installed languages appear in the box for selection.

Feedback to Documentation

To send feedback to PTC about the documentation in general or the topic you are viewing, send email to mcad-documentation@ptc.com. You can also click at the bottom of any topic to create an email with the topic link added. Please include the following information with your feedback:

- Name
- Company
- Product
- Product Release
- Help Topic Title (if you are not including a link)
- Level of Expertise in the Product (Beginning, Intermediate, advanced)
- Comments

Figure 2.7(b) Online Help Center

For **Context sensitive help**, move your pointer over the icon, dialog box, or menu command for which you want help and press the **F1** key on the keyboard. Click: **Model** tab > move the mouse pointer over  Boundary Blend > press **F1** [Fig. 2.8(a)] > **About the Boundary Blend Feature** [Fig. 2.8(b)] > repeat the process to investigate other commands

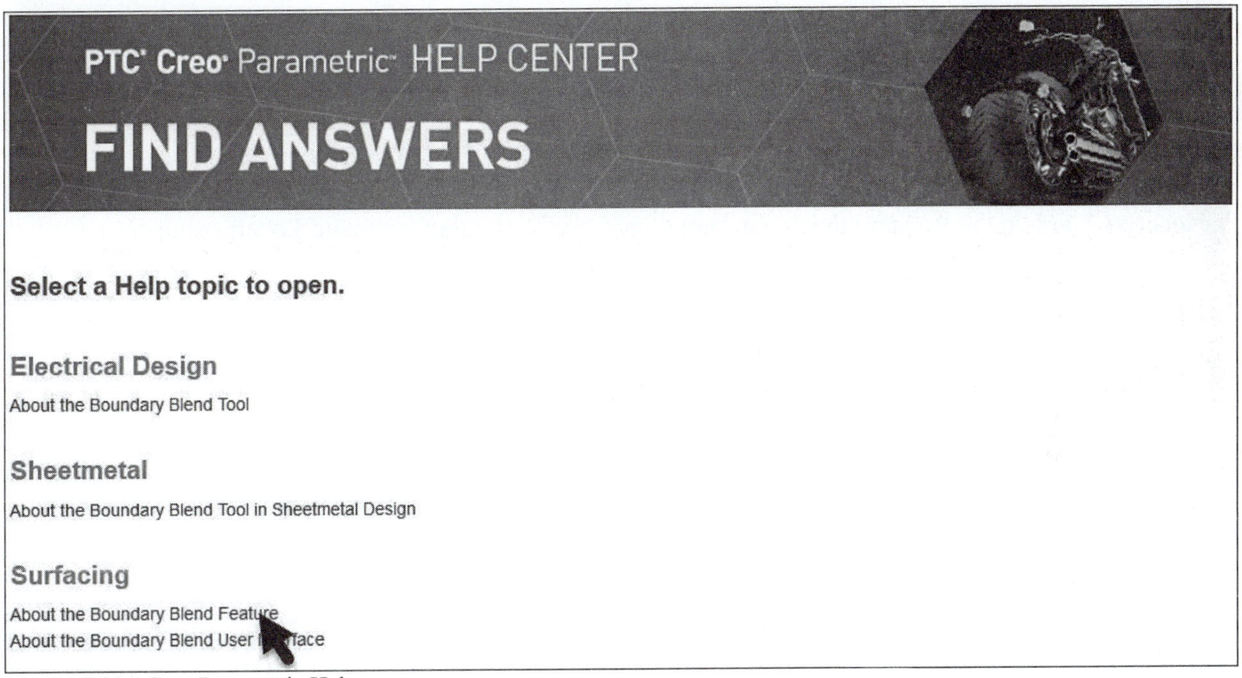

PTC Creo Parametric HELP CENTER

FIND ANSWERS

Select a Help topic to open.

Electrical Design

About the Boundary Blend Tool

Sheetmetal

About the Boundary Blend Tool in Sheetmetal Design

Surfacing

About the Boundary Blend Feature
About the Boundary Blend User Interface

Figure 2.8(a) Creo Parametric Help

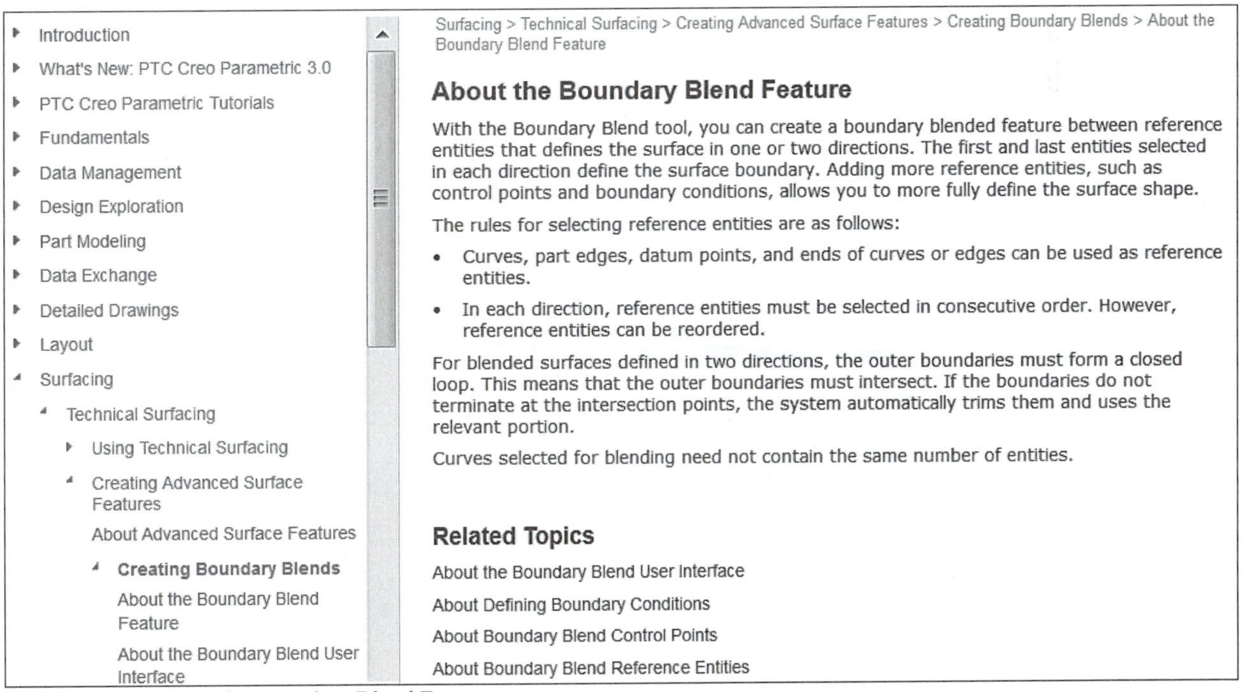

- Introduction
- What's New: PTC Creo Parametric 3.0
- PTC Creo Parametric Tutorials
- Fundamentals
- Data Management
- Design Exploration
- Part Modeling
- Data Exchange
- Detailed Drawings
- Layout
- Surfacing
  - Technical Surfacing
    - Using Technical Surfacing
    - Creating Advanced Surface Features
      - About Advanced Surface Features
      - **Creating Boundary Blends**
        - About the Boundary Blend Feature
        - About the Boundary Blend User Interface

Surfacing > Technical Surfacing > Creating Advanced Surface Features > Creating Boundary Blends > About the Boundary Blend Feature

About the Boundary Blend Feature

With the Boundary Blend tool, you can create a boundary blended feature between reference entities that defines the surface in one or two directions. The first and last entities selected in each direction define the surface boundary. Adding more reference entities, such as control points and boundary conditions, allows you to more fully define the surface shape.

The rules for selecting reference entities are as follows:

- Curves, part edges, datum points, and ends of curves or edges can be used as reference entities.
- In each direction, reference entities must be selected in consecutive order. However, reference entities can be reordered.

For blended surfaces defined in two directions, the outer boundaries must form a closed loop. This means that the outer boundaries must intersect. If the boundaries do not terminate at the intersection points, the system automatically trims them and uses the relevant portion.

Curves selected for blending need not contain the same number of entities.

Related Topics

About the Boundary Blend User Interface

About Defining Boundary Conditions

About Boundary Blend Control Points

About Boundary Blend Reference Entities

Figure 2.8(b) About the Boundary Blend Feature

Command Locator

The command search tool enables you to find known commands faster and preview the location of the command on the user interface. You can preview the location only if the command is located on the Ribbon, Quick Access toolbar, Graphics toolbar, or File menu. You can also run a command by clicking the command in the search list.

The tool displays commands not in the Ribbon under the **Commands not in the ribbon** category in the search list. The tool searches commands within the current mode.

Click:  **Command Search** in the upper top right-hand side of your **screen** [Fig. 2.9(a)] > Command Locator opens > *type* **View**  [Fig. 2.9(b)] > place pointer on **Perspective View** [Fig. 2.9(c)] > select  [Fig. 2.9(d)] > **Perspective View** in the Render tab Ribbon (toggle *off*) >  toggle Command Locator shut

Figure 2.9(a) Command Search

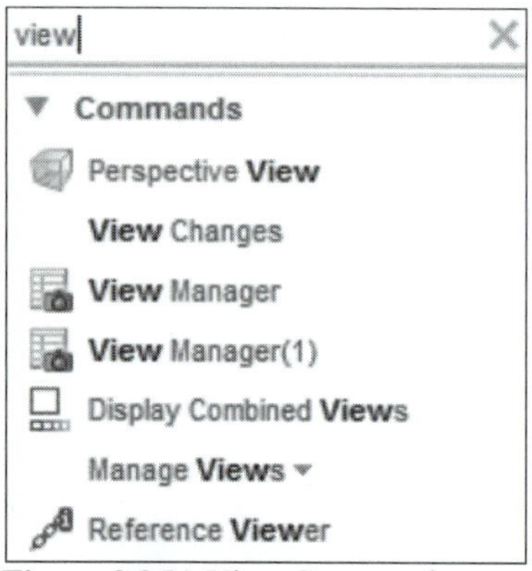

Figure 2.9(b) View Commands

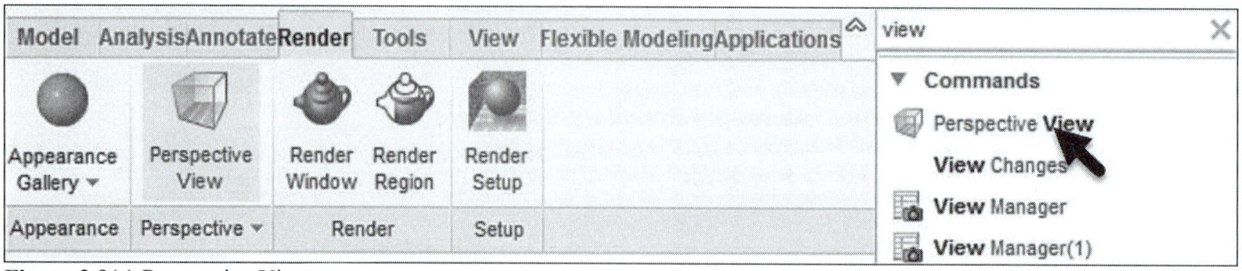

Figure 2.9(c) Perspective View

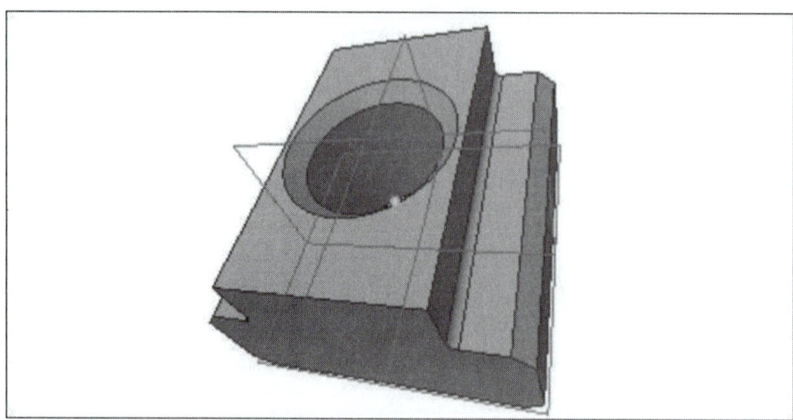

Figure 2.9(d) Perspective View

View and Display Functions

Using the **View** tab [Fig. 2.10(a)], you can adjust the model view, orient the view, hide and show entities, create and use advanced views, and set various model display options.

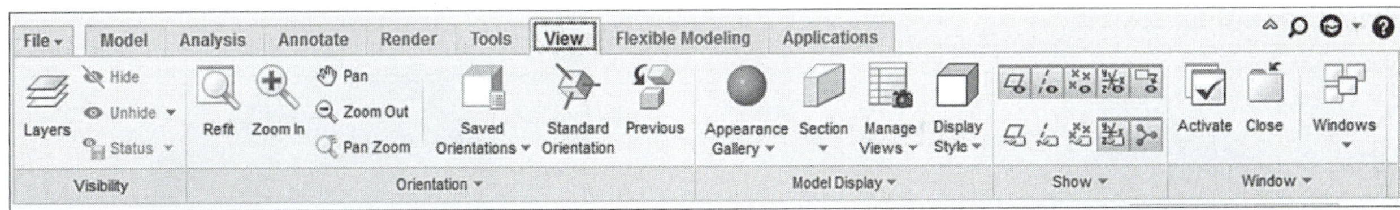

Figure 2.10(a) View tab

The following list includes some of the View operations you can perform:

- Orient the model view in the following ways, spin, pan, and zoom models and drawings, display the default orientation, revert to the previously displayed orientation, change the position or size of the model view, change the orientation (including changing the view angle in a drawing), and create new orientations
- Toggle display style; shading with edges, shading with reflections, shading, hidden line, no hidden line, and wireframe.
- Highlight items in the graphics window when you select them in the Model Tree
- Explode or un-explode an assembly view
- Repaint the graphics window
- Refit the model to the window after zooming in or out on the model
- Update drawings of model geometry
- Hide and unhide entities, and hide or show items during spin or animation

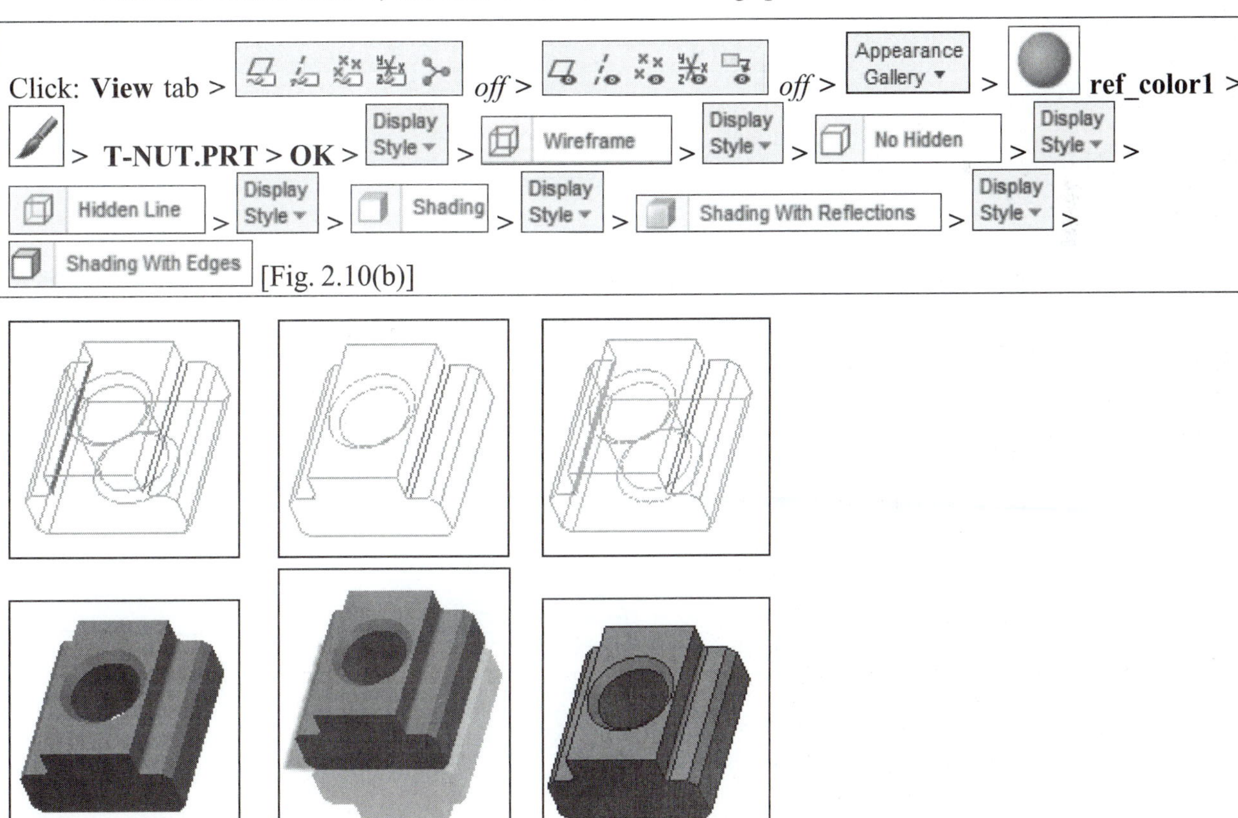

Click: **View** tab > *off* > *off* > Appearance Gallery > **ref_color1** > > **T-NUT.PRT** > **OK** > Display Style > Wireframe > Display Style > No Hidden > Display Style > Hidden Line > Display Style > Shading > Display Style > Shading With Reflections > Display Style > Shading With Edges [Fig. 2.10(b)]

Figure 2.10(b) Display Styles

Click: on > on > **Refit** > press and hold the **MMB** and rotate the model [Fig. 2.11(a)] > release the **MMB** > **Saved Orientations** > **Standard Orientation** [Fig. 2.11(b)]

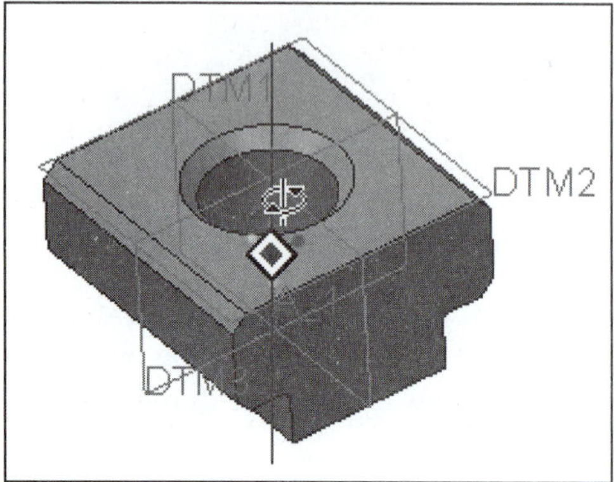

Figure 2.11(a) Rotated View

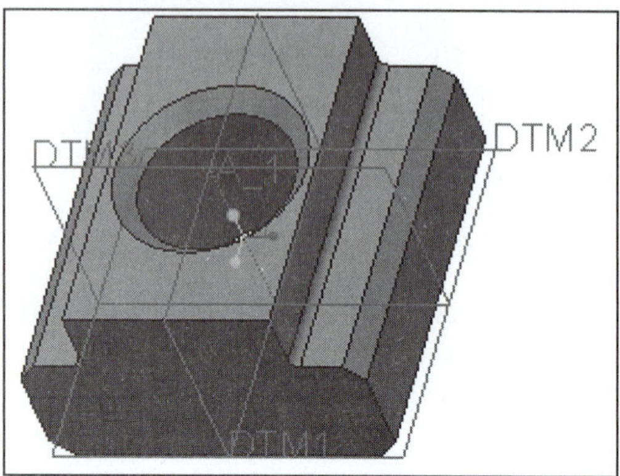

Figure 2.11(b) Standard Orientation

View Tools

Creo Parametric provides various view tools, including:

Zoom In Zoom in on the model, **Zoom Out** Zoom out from the model, **Refit** Adjust the zoom level to fully display the object on the screen, **Pan** Navigate from one location to another, **Reorient** Configure model orientation preferences, **View Normal**.

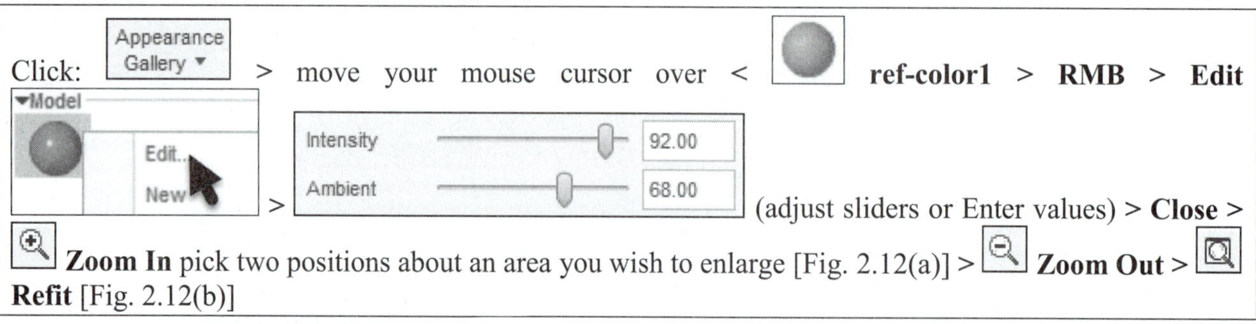

Click: > move your mouse cursor over < **ref-color1** > **RMB** > **Edit** > (adjust sliders or Enter values) > **Close** > **Zoom In** pick two positions about an area you wish to enlarge [Fig. 2.12(a)] > **Zoom Out** > **Refit** [Fig. 2.12(b)]

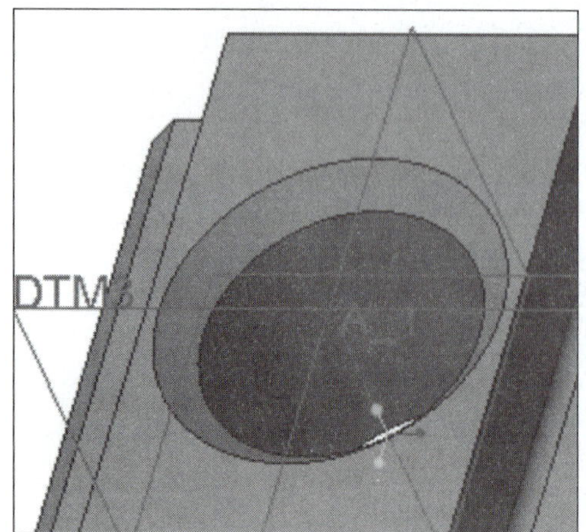
Figure 2.12(a) Zoom In

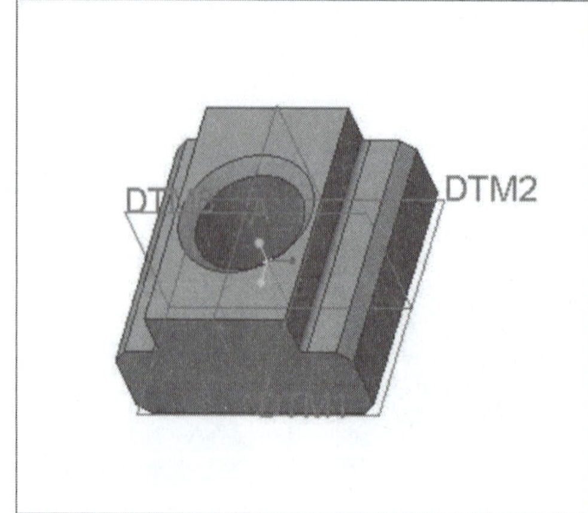

Figure 2.12(b) Refit

Using Mouse Buttons to Manipulate the Model

You can also dynamically reorient the model using the **MMB** by itself (**Spin**) or in conjunction with the **Shift** key (**Pan**) or **Ctrl** key (**Zoom, Turn**).

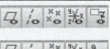

Click: *off* > hold down **Ctrl** key and **MMB** in the Graphics Window near the model and move the cursor up (zoom out) [Fig. 2.13(a)] > release the **Ctrl** key and **MMB** > hold down **Shift** key and **MMB** in the Graphics Window near the model and move the cursor about the screen (pan) [Fig. 2.13(b)] > release the **Shift** key and **MMB** > hold down **MMB** in the Graphics Window near the model and move the cursor around (spin) [Figs. 2.13(c-d)] > release the **MMB** > **Ctrl+D** > *on*

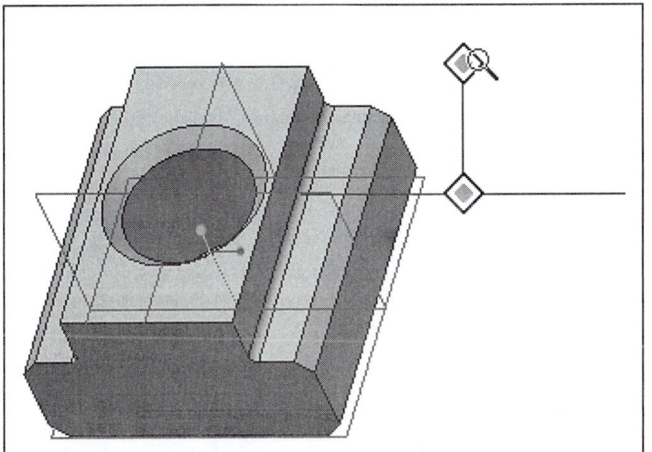

Figure 2.13(a) Zoom

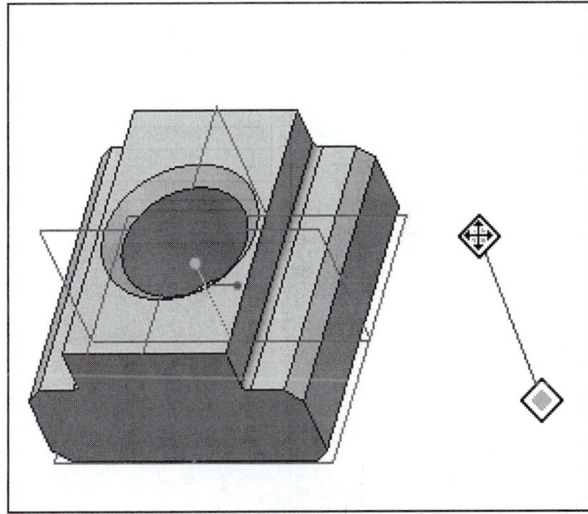

Figure 2.13(b) Pan

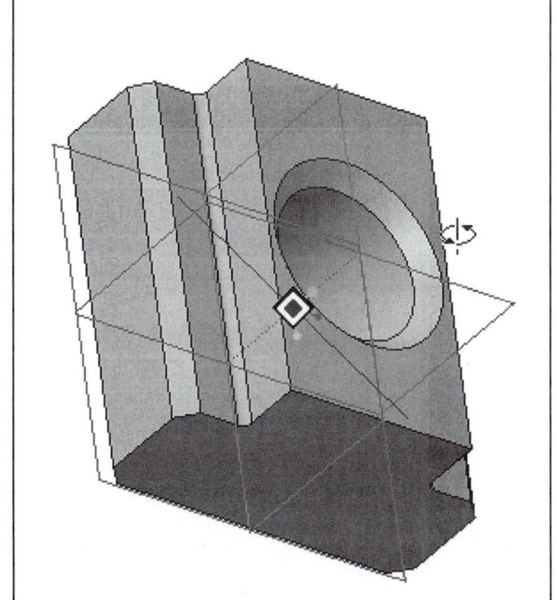

Figure 2.13(c) Spin

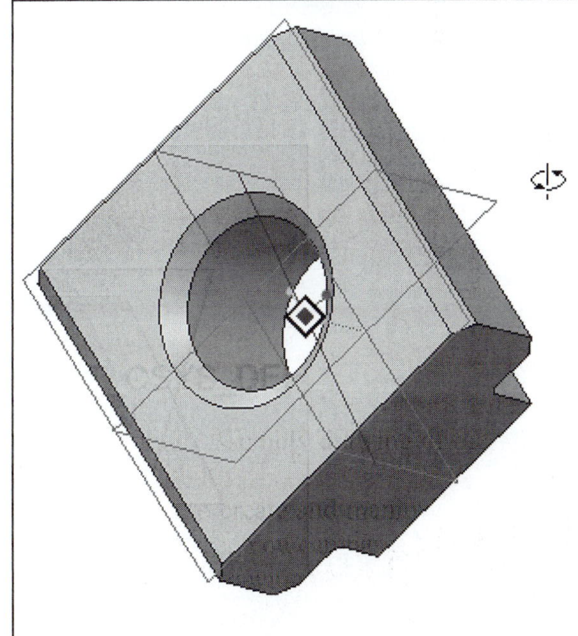

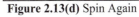

Figure 2.13(d) Spin Again

System Display Settings

You can make a number of changes to the default colors furnished by PTC Creo Parametric 3.0, customizing them for your own use:

- Define, save, and open color schemes
- Customize colors used in the user interface
- Redefine basic colors used in models
- Assign colors to be used by an entity
- Store a color scheme so you can reuse it
- Open a previously used color scheme

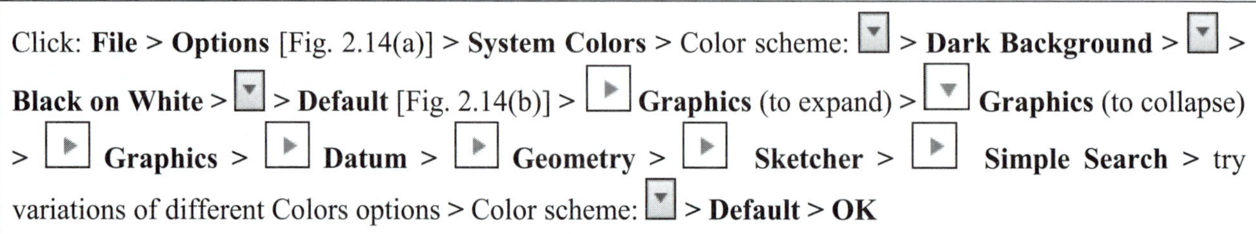

Click: **File** > **Options** [Fig. 2.14(a)] > **System Colors** > Color scheme: > **Dark Background** > > **Black on White** > > **Default** [Fig. 2.14(b)] > **Graphics** (to expand) > **Graphics** (to collapse) > **Graphics** > **Datum** > **Geometry** > **Sketcher** > **Simple Search** > try variations of different Colors options > Color scheme: > **Default** > **OK**

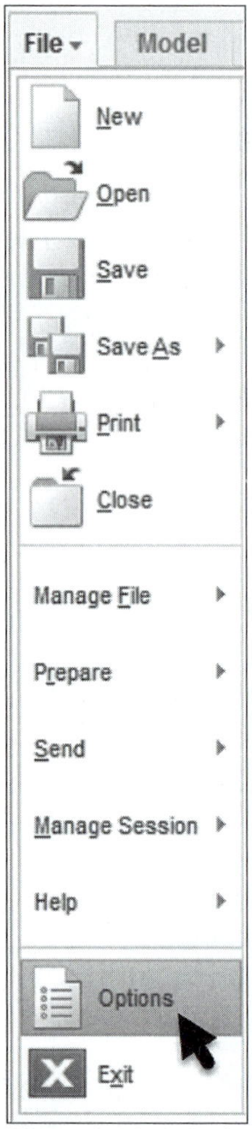

Figure 2.14(a) Options

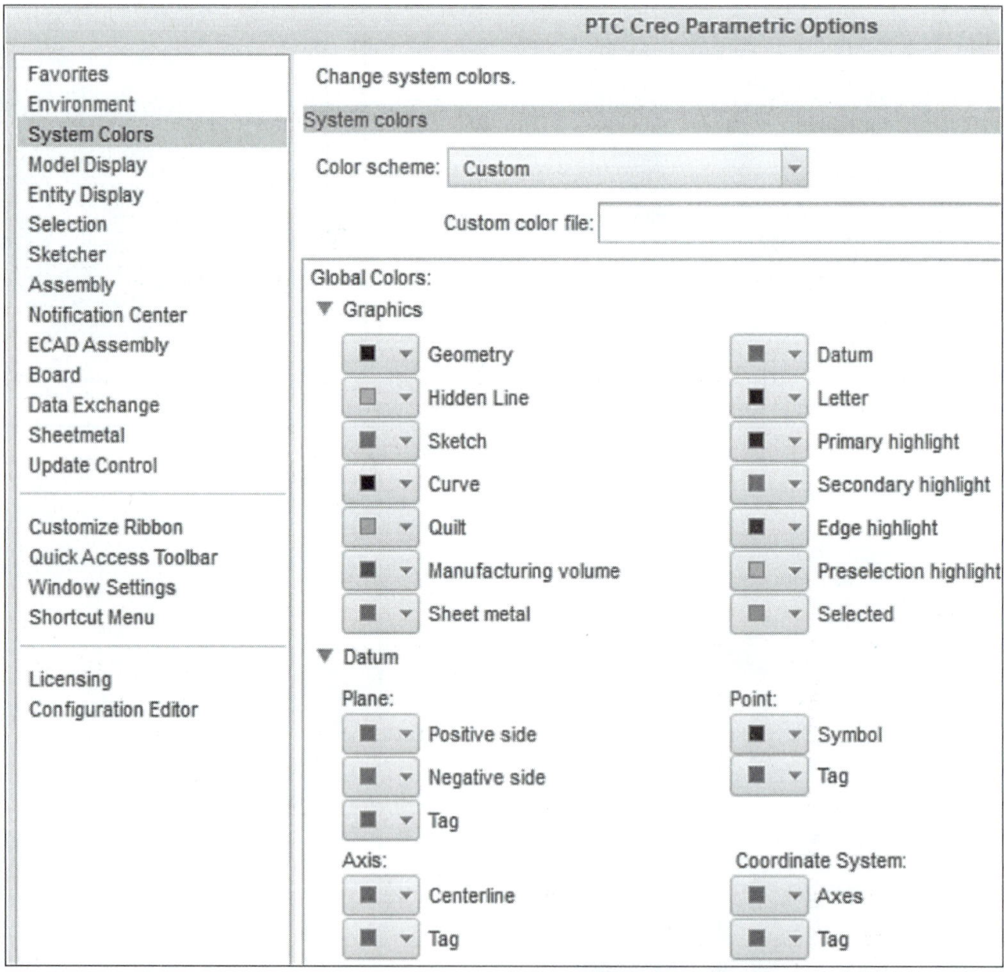

Figure 2.14(b) Expanded Global Options

Information Tools

At any time during the design process you can request model, feature, or other information. Picking on a feature in the graphics window or Model Tree and then **RMB > Information** will provide information about that feature in the Browser. This can also be accomplished by clicking on the feature name in the Model Tree and then **RMB > Information**. Both feature and model information can be obtained using this method. A variety of information can also be extracted using the **Tools** tab Ribbon.

Click: **Model** tab > in the Model Tree, pick once on the chamfer > **RMB > Information > Feature Information** [Figs. 2.15(a-b)] > × upper right corner of Browser or lower left of screen to close Browser > **File > Save > OK > File > Manage File > Delete Old Versions > Yes > Ctrl+D**

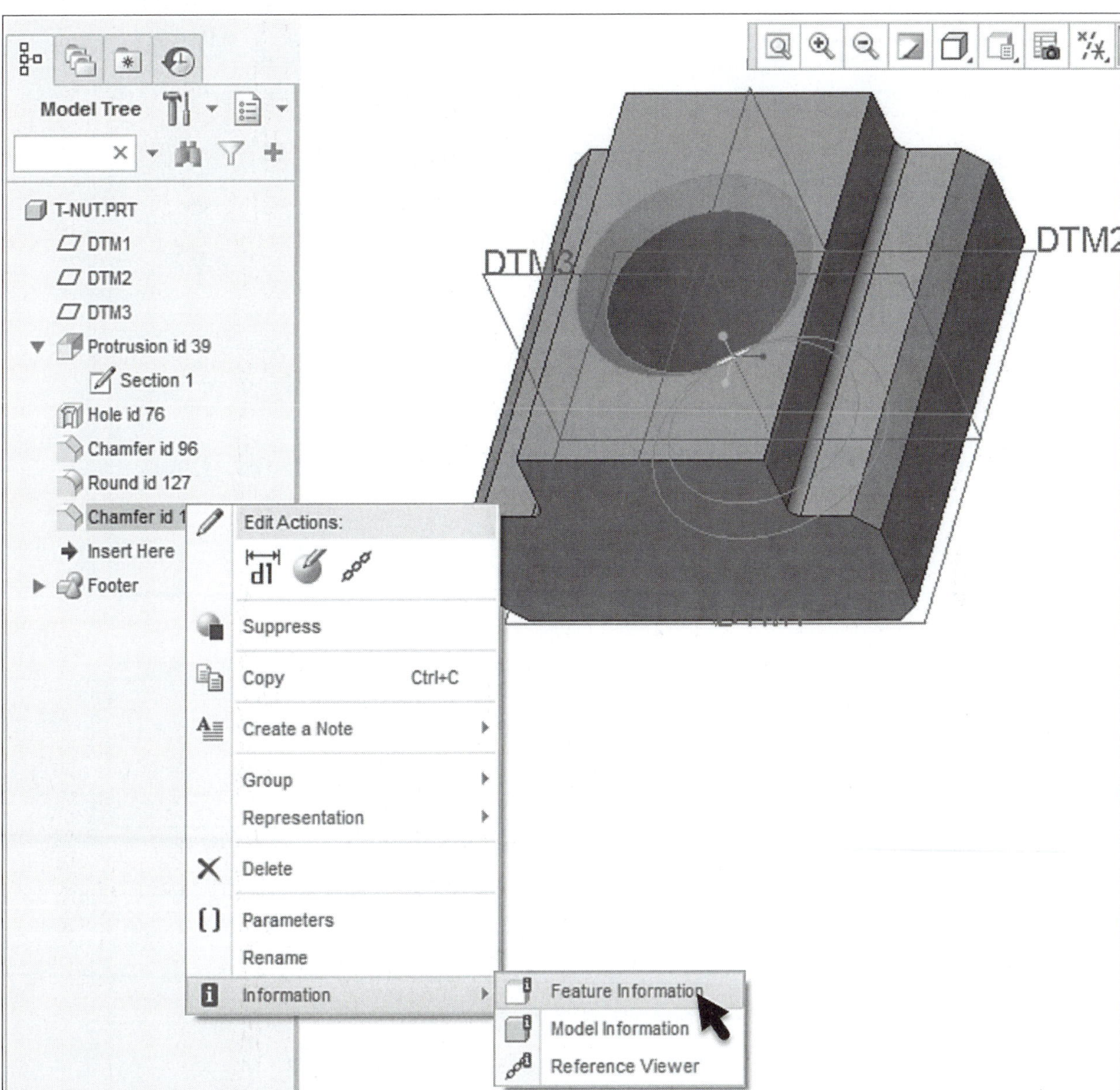

Figure 2.15(a) Information > Feature Information from the Model Tree

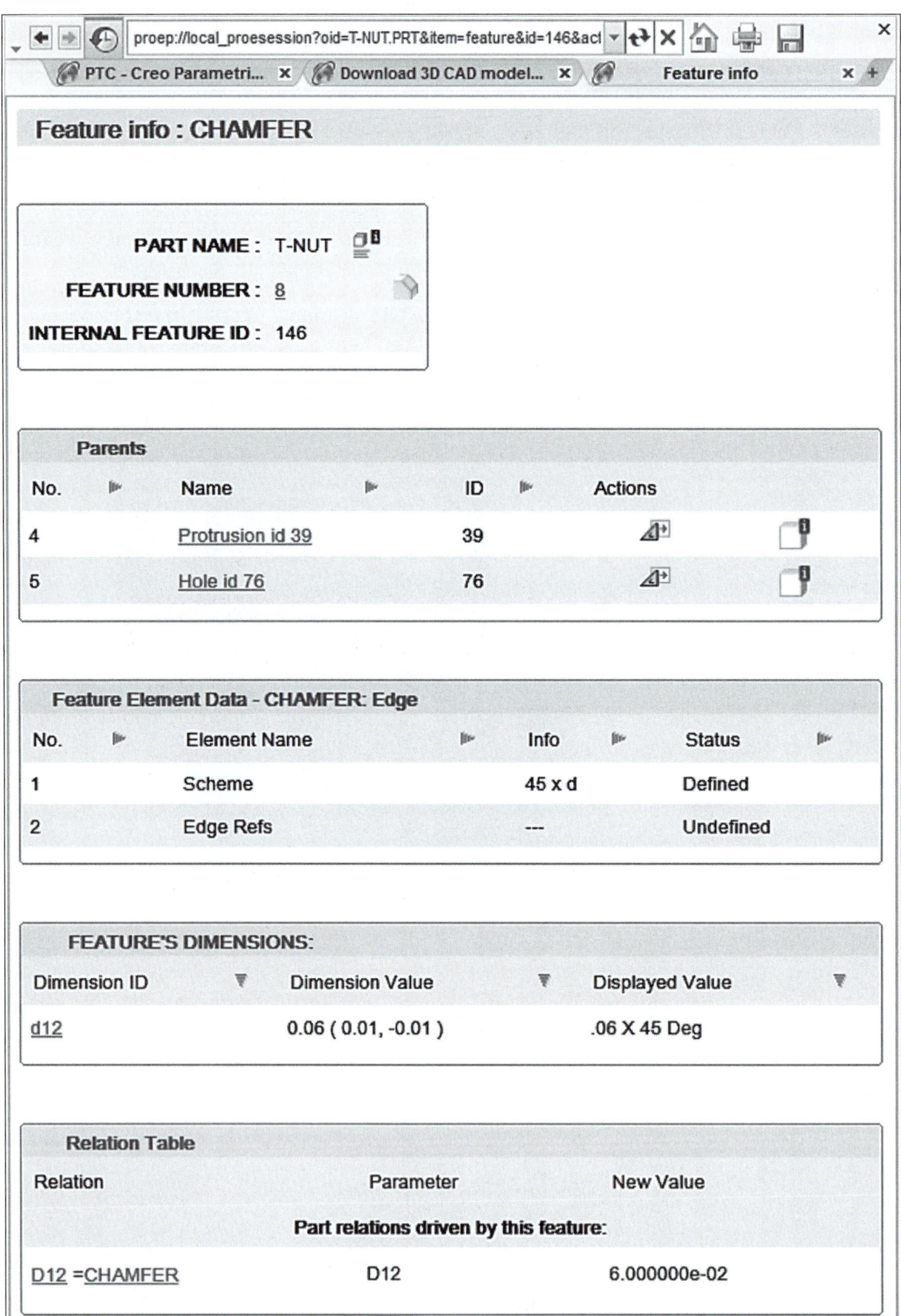

Figure 2.15(b) Feature Information: Chamfer, Displayed in the Browser

In the Model Tree, select: **T-NUT.PRT > RMB > Information > Model Information** [Figs. 2.15(c-d)] > ✕ close Browser

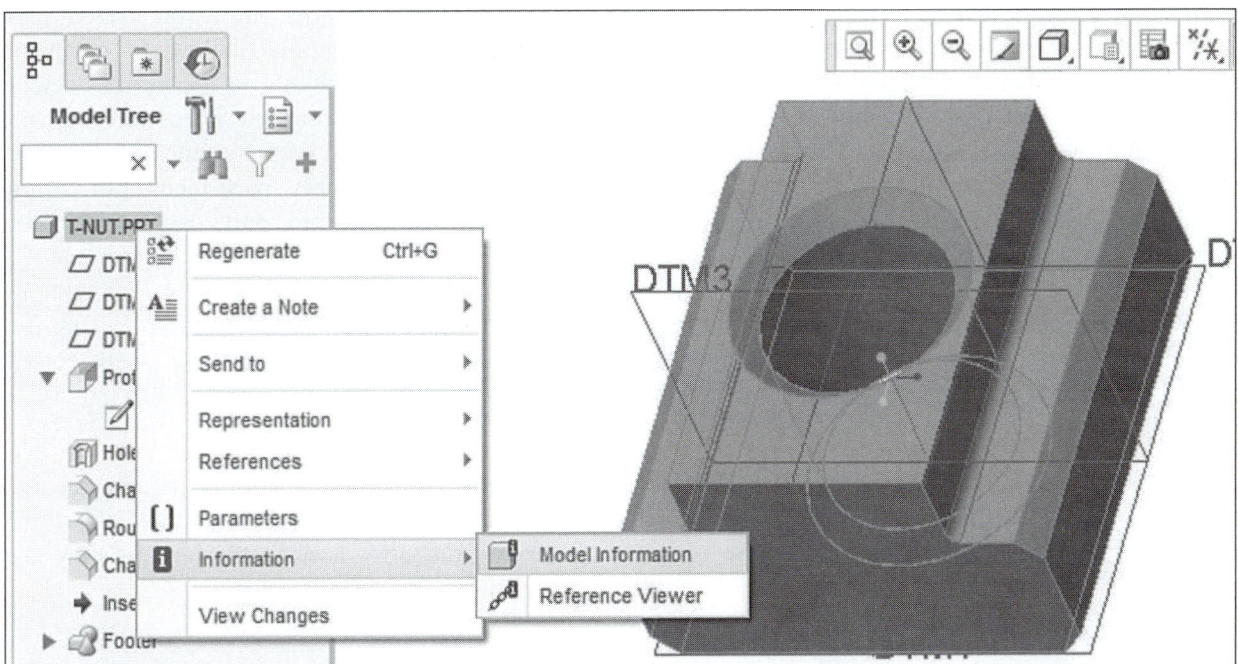

Figure 2.15(c) Model Information

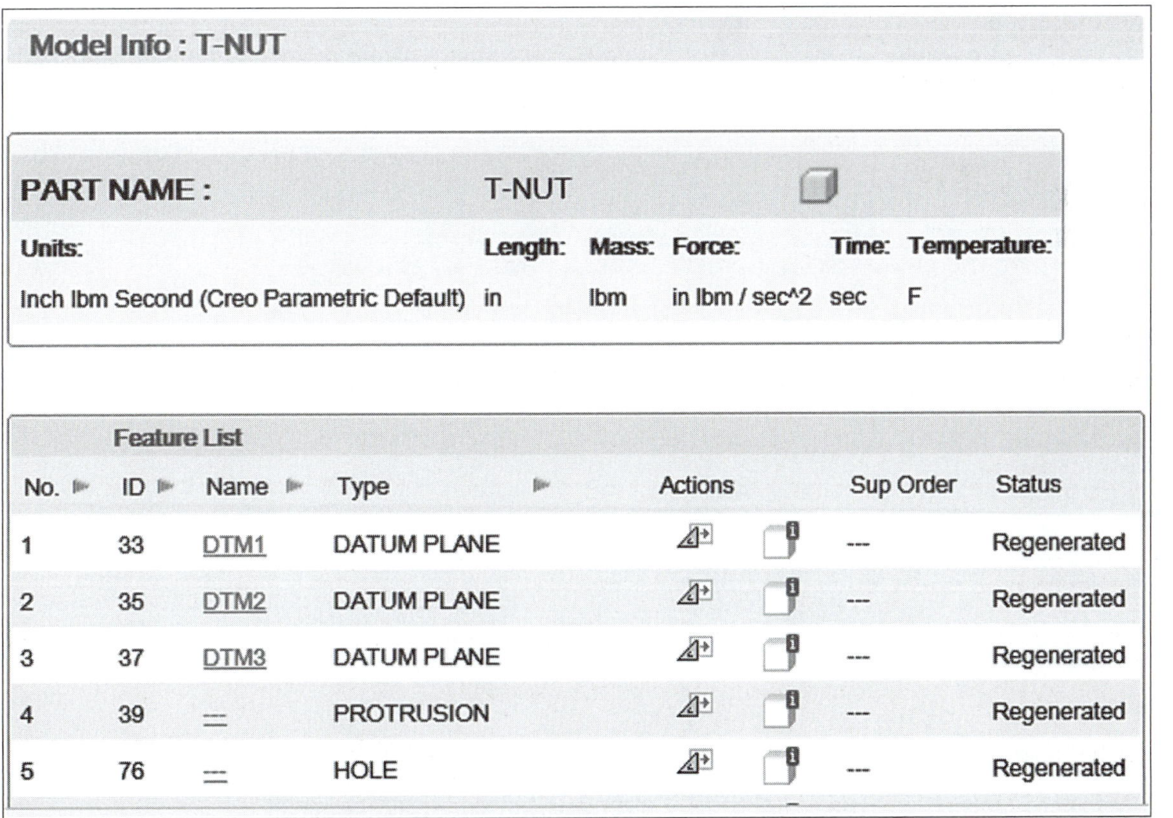

Model Info : T-NUT

PART NAME :	T-NUT				
Units:	Length:	Mass:	Force:	Time:	Temperature:
Inch lbm Second (Creo Parametric Default)	in	lbm	in lbm / sec^2	sec	F

Feature List

No.	ID	Name	Type	Actions	Sup Order	Status
1	33	DTM1	DATUM PLANE		---	Regenerated
2	35	DTM2	DATUM PLANE		---	Regenerated
3	37	DTM3	DATUM PLANE		---	Regenerated
4	39	---	PROTRUSION		---	Regenerated
5	76	---	HOLE		---	Regenerated

Figure 2.15(d) Information from the Model

The Model Tree

The **Model Tree** is a tabbed feature on the PTC Creo Parametric 3.0 Navigator that displays a list of every feature or part in the current part, assembly, or drawing.

The model structure is displayed in hierarchical (tree) format with the root object (the current part or assembly) at the top of its tree and the subordinate objects (parts or features) below. If you have multiple Creo Parametric windows open, the Model Tree contents reflect the file in the current active window. The Model Tree lists only the related feature and part level items in a current object and does not list the entities (such as edges, surfaces, curves, and so forth) that comprise the features. Each Model Tree item contains an icon that reflects its object type, for example, hidden, assembly, part, feature, or datum plane (also a feature). The icon can also show the display status for a feature, part, or assembly, for example, suppressed.

Selection in the Model Tree is object-action oriented; you select objects in the Model Tree without first specifying what you intend to do with them. You can select components, parts, or features using the Model Tree. You cannot select the individual geometry that makes up a feature (entities). To select an entity, you must select it in the graphics window.

With the **Settings** tab you can control what is displayed in the Model Tree. You can add informational columns to the Model Tree window, such as **Tree Columns** containing parameters and values, assigned layers, or feature name for each item. You can use the cells in the columns to perform context-sensitive edits and deletions. These options will be covered elsewhere in the text, as they are needed in the design process.

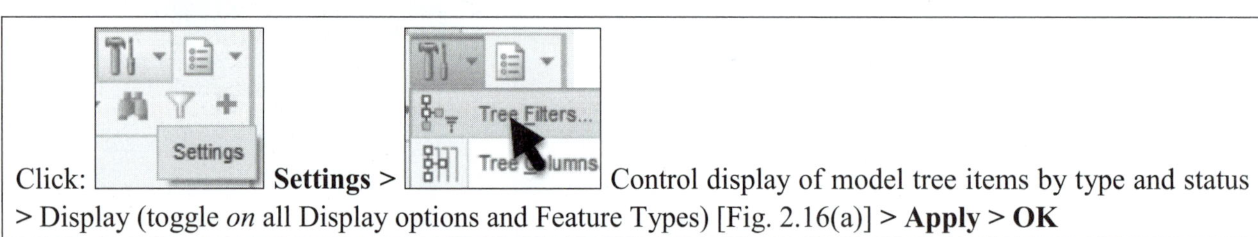

Click: **Settings** > Control display of model tree items by type and status > Display (toggle *on* all Display options and Feature Types) [Fig. 2.16(a)] > **Apply** > **OK**

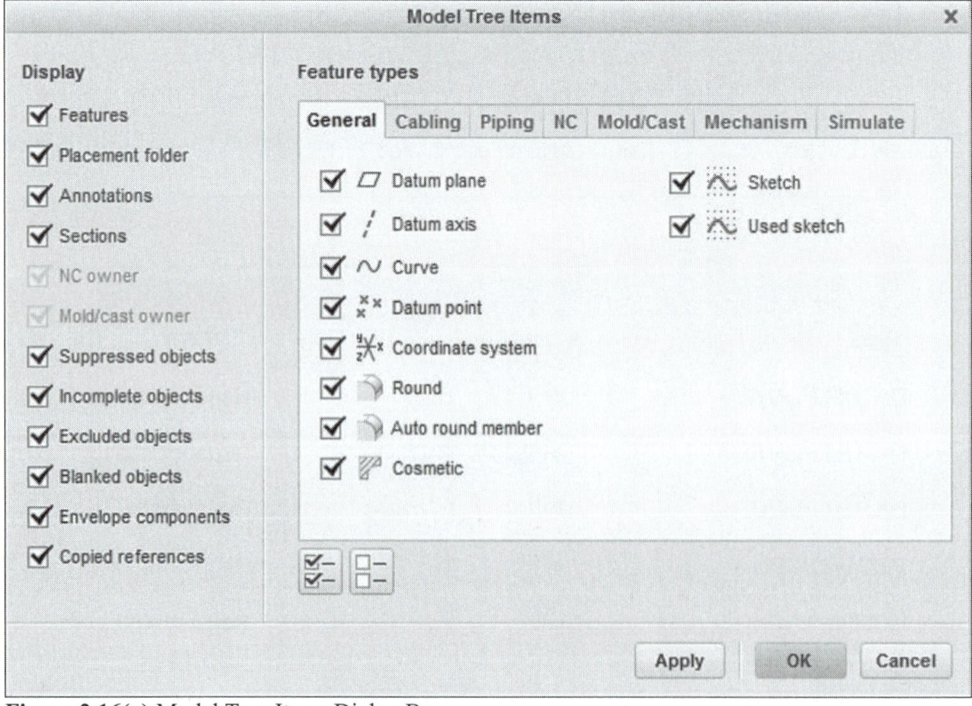

Figure 2.16(a) Model Tree Items Dialog Box

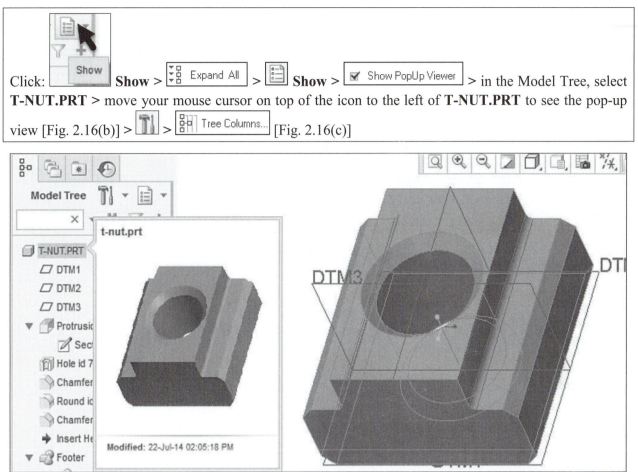

Click: Show Show > Expand All > Show > Show PopUp Viewer > in the Model Tree, select **T-NUT.PRT** > move your mouse cursor on top of the icon to the left of **T-NUT.PRT** to see the pop-up view [Fig. 2.16(b)] > > Tree Columns... [Fig. 2.16(c)]

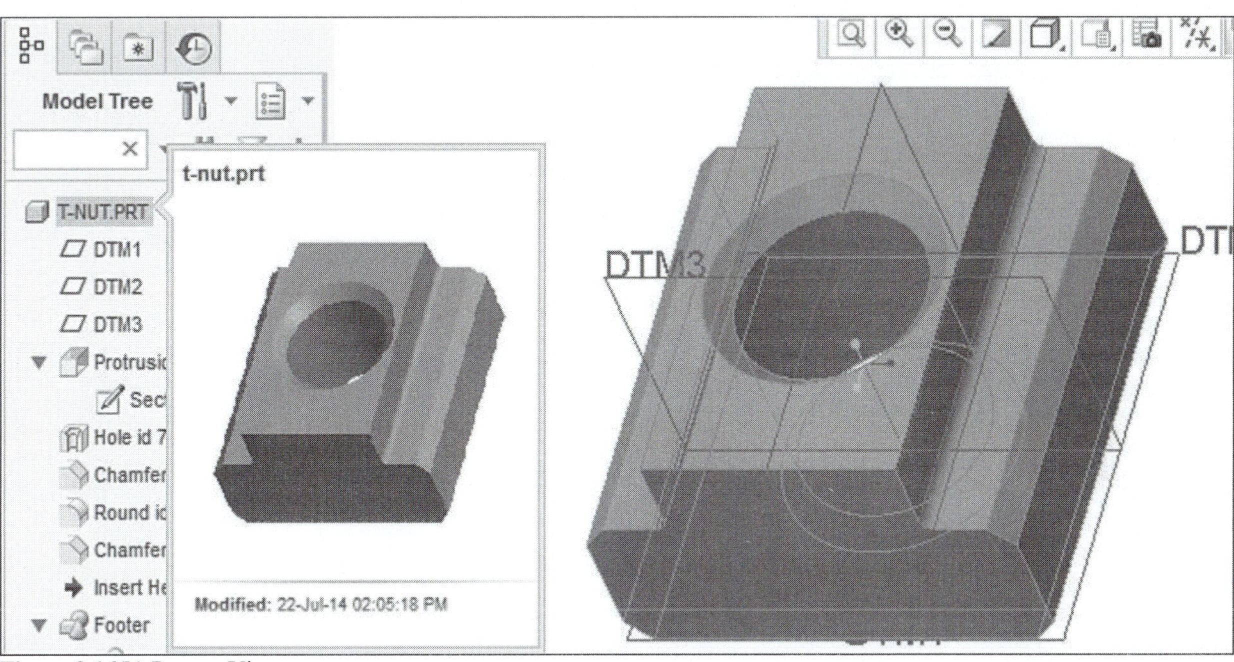

Figure 2.16(b) Pop-up Viewer

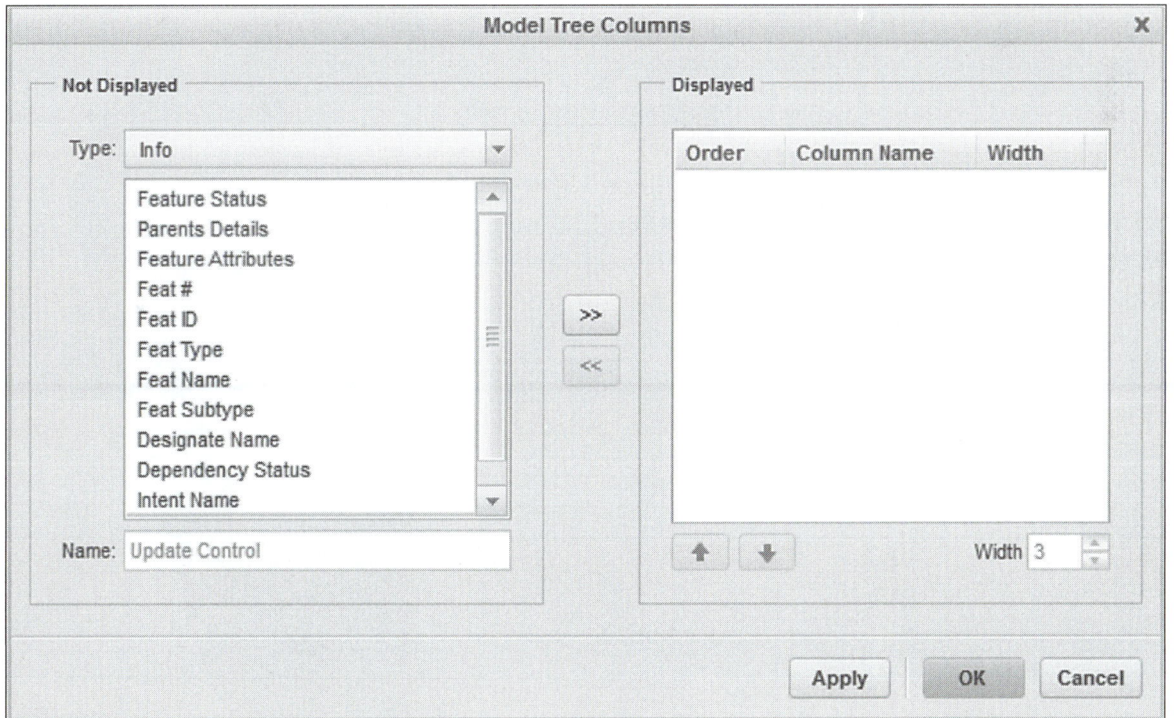

Figure 2.16(c) Model Tree Columns Dialog Box

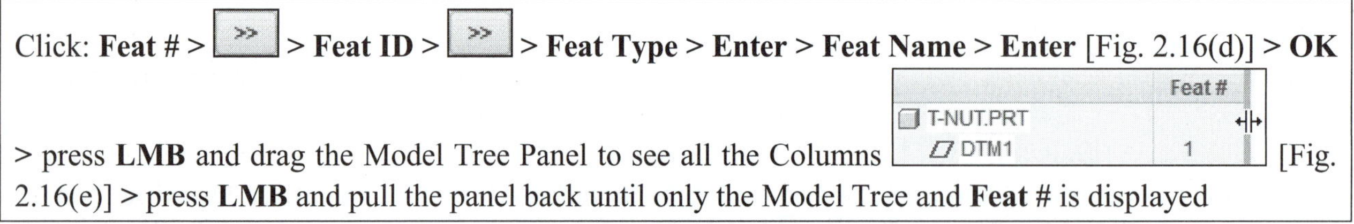

Click: **Feat #** > >> > **Feat ID** > >> > **Feat Type** > **Enter** > **Feat Name** > **Enter** [Fig. 2.16(d)] > **OK** > press **LMB** and drag the Model Tree Panel to see all the Columns [Fig. 2.16(e)] > press **LMB** and pull the panel back until only the Model Tree and **Feat #** is displayed

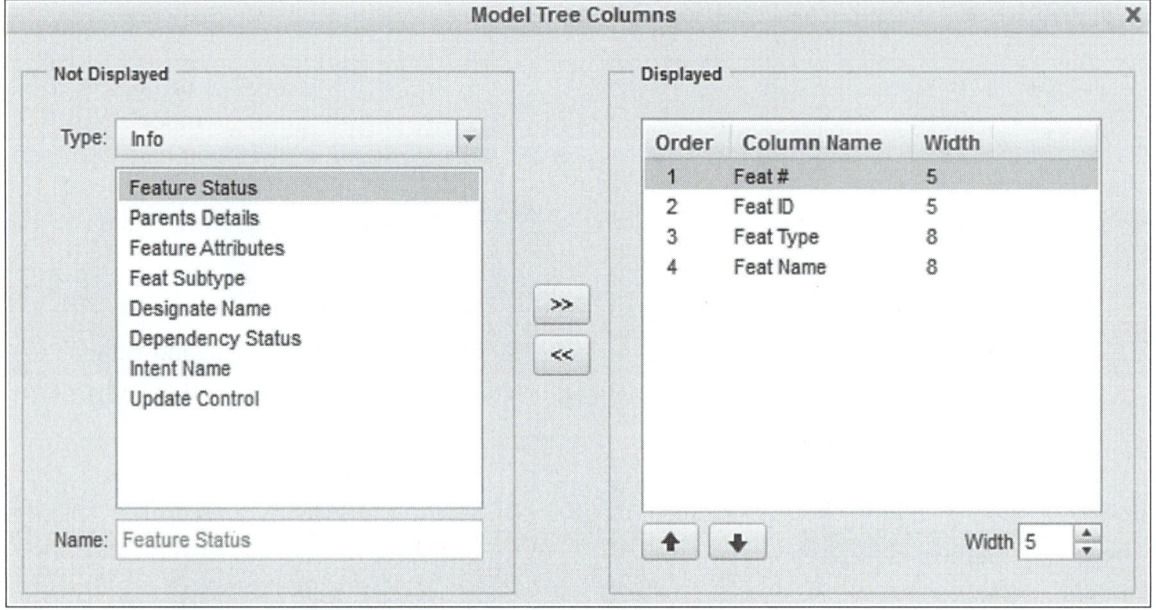

Figure 2.16(d) Adding Displayed Columns

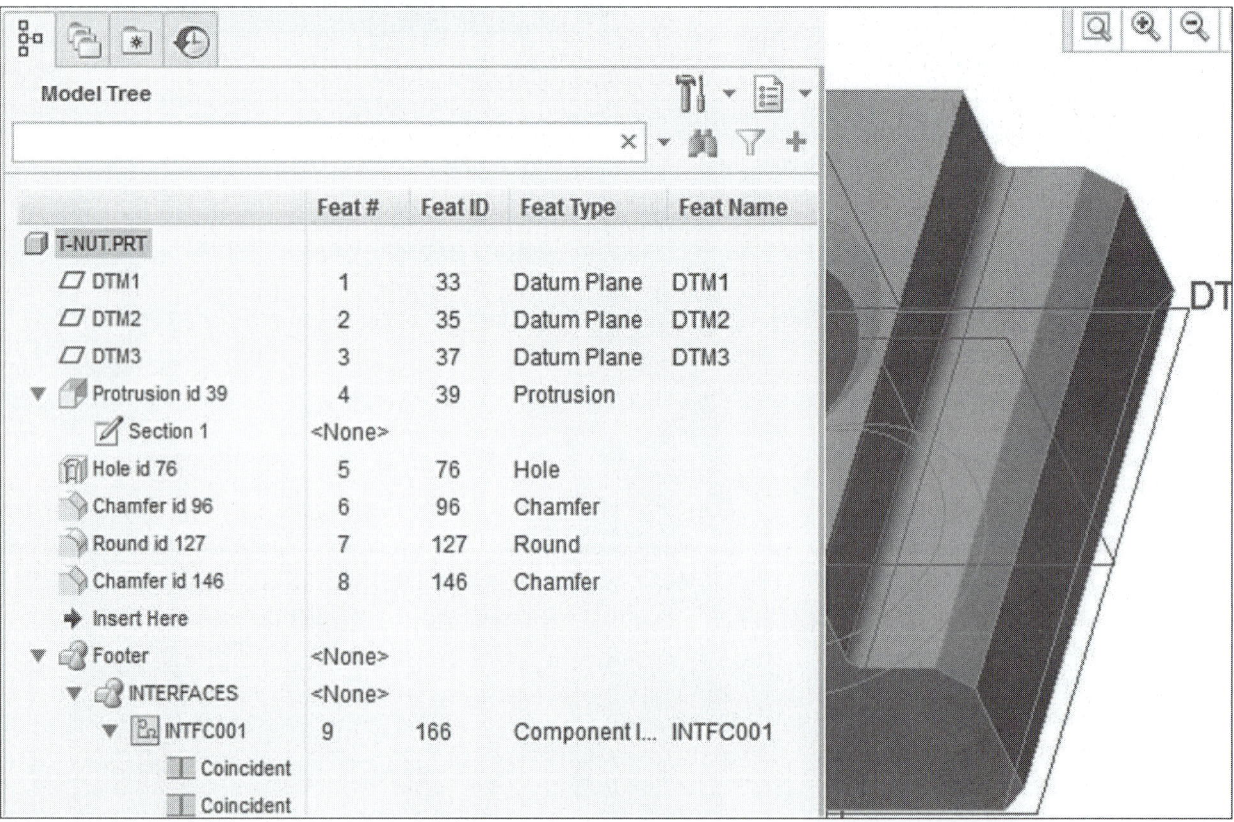

Figure 2.16(e) Model Tree and Model Tree Columns

Working on the Model

In Creo Parametric, you can select objects to work on from within the graphics window or in the Model Tree by using the mouse or the keyboard. The object types that are available for selection vary depending on whether you select an object from within the graphics window or in the Model Tree. You can select any type of object, including features, 3-D notes (model Annotations), parts, datum objects (planes, axes, curves, points, and coordinate systems), and geometry (edges and surfaces) from within the graphics window. Additionally, since the Model Tree displays only parts, components, and features, you can select only those object types from within the Model Tree.

Selection in both the graphics window and the Model Tree can be action-object or object-action oriented, depending on the process you choose within Creo Parametric to build your model. You can specify the action you want to perform on an object before you select the object, or you can select the object before you specify the action.

You can *dynamically* modify features from within the graphics window as you work. Features can be edited by selecting them from the graphics window directly, or from the Model Tree.

Click: **Refit** > pick on the hole in the Graphics Window [Fig. 2.17(a)] > press and hold **RMB** anywhere in the Graphics Window > **Delete** > **OK** in the Delete dialog box [Fig. 2.17(b)] > **Delete Rels** [Fig. 2.17(c)] > **Undo** from the Quick Access Toolbar [Fig. 2.17(d)]

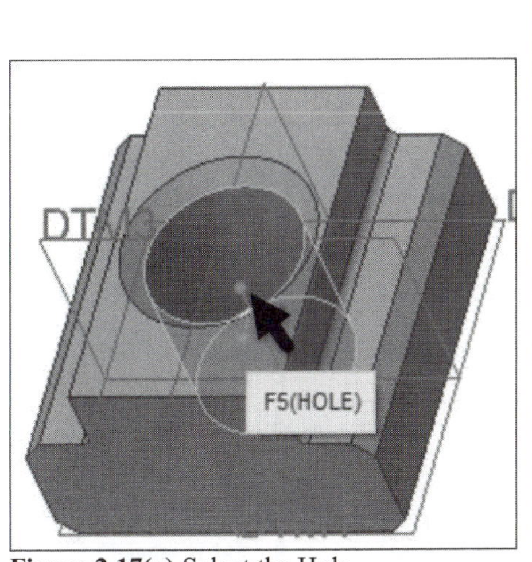

Figure 2.17(a) Select the Hole

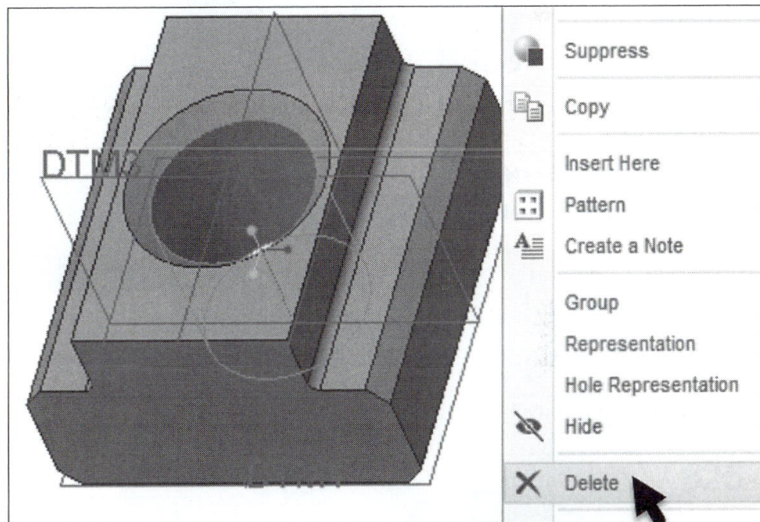

Figure 2.17(b) RMB > Delete

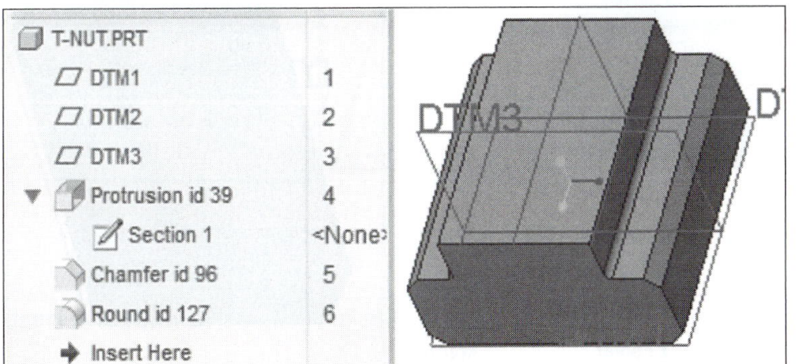

Figure 2.17(c) Hole is Deleted

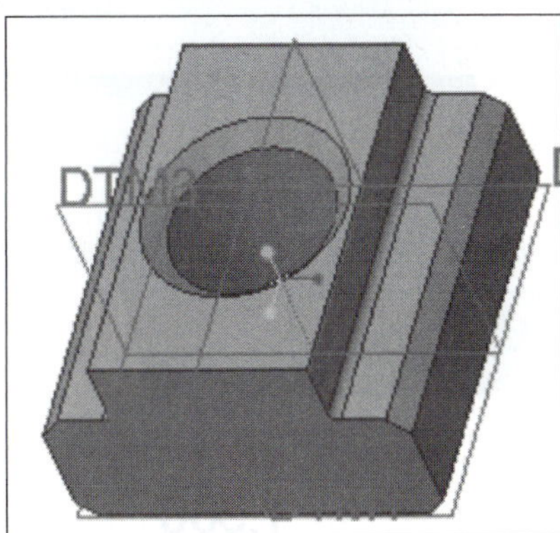

Figure 2.17(d) Undo, Hole Returns

Flexible Modeling

Flexible modeling allows you to experiment with your design without being committed to changes. You can make explicit modifications to selected geometry while ignoring pre-existing relationships. Use the buttons in the following groups to quickly select and manipulate geometry.

- **Recognize and Select**—selection only of the specified type of geometry.
- **Transform**—direct manipulation of the selected geometry.
- **Recognition**—recognizing patterns and symmetry so that when one member is modified, the modification is propagated to all pattern members or symmetrical geometry.
- **Edit Features**—editing of the selected geometry or surface.

The intention is that when you enter a tool, you can perform the required tasks using the right-click shortcut menus. The procedures for each tool use this approach since it is the fastest way to work and is sufficient for most tasks. Use the tabs to select rule–based surface sets, to view selections and references, and to change the default system selections. Detailed information about the individual tool tabs can be found in the About the User Interface topics for each tool.

Click: to close the Model Tree *(lower left-hand corner)* > **Flexible Modeling** tab [Fig. 2.20(a)] > **Ctrl+D** > Move **Move using Dragger** > select the front face [Fig. 2.20(b)] > place your mouse cursor on the *horizontal* circle, hold down the **LMB** and drag the face to rotate [Fig. 2.20(c)] > release the **LMB**

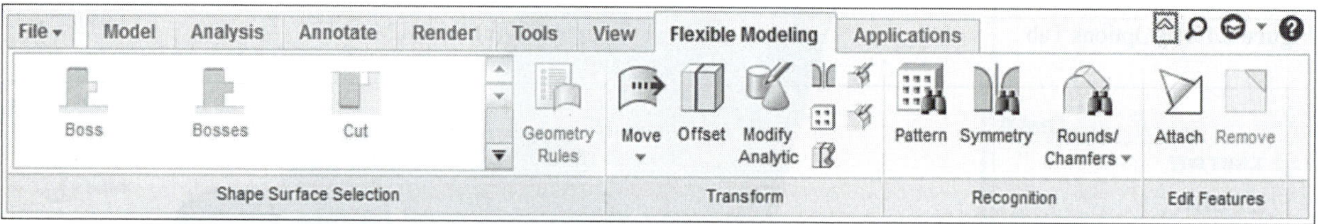

Figure 2.20(a) Flexible Modeling Tab

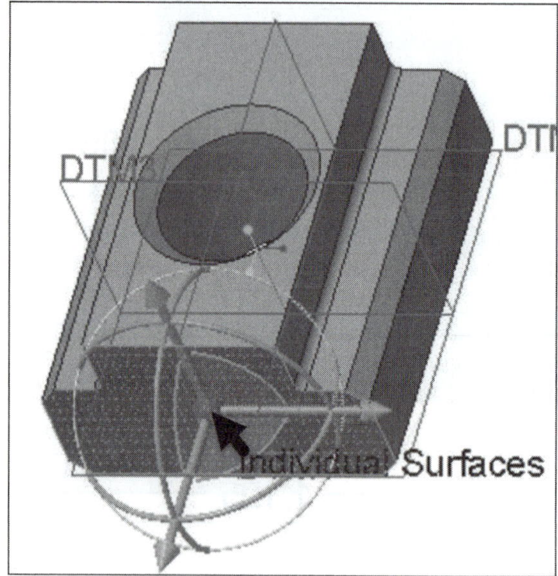

Figure 2.20(b) Select the Front Face

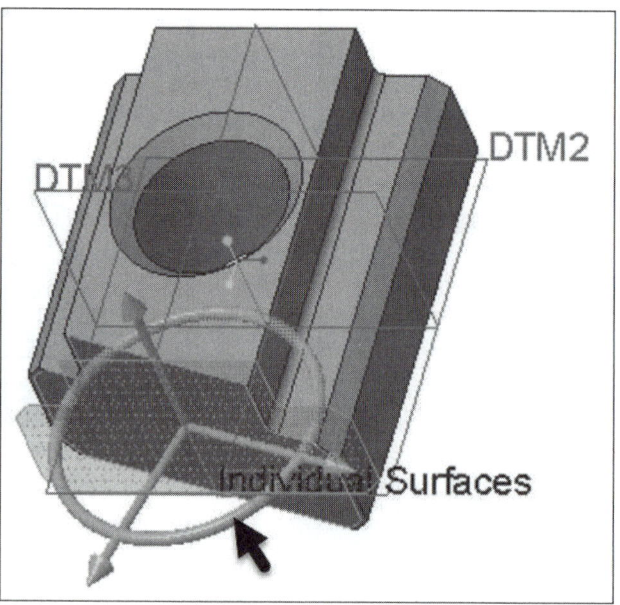

Figure 2.20(c) Rotate the Face with Green Circle

Drag the face by rotating the *vertical* circle [Fig. 2.20(d)] > pull the face with the *center* Ball [Fig. 2.20(e)] > spend some time and experiment with this tool > [Fig. 2.20(f)] > **Ctrl+S** > **File** > **Save As** > Type > **Zip File (*.zip)** > **OK** > upload your version of the part to your course interface or attach to an email and send to your instructor and/or yourself > **Close** > to open the Navigator/Model Tree *(lower left-hand corner)*

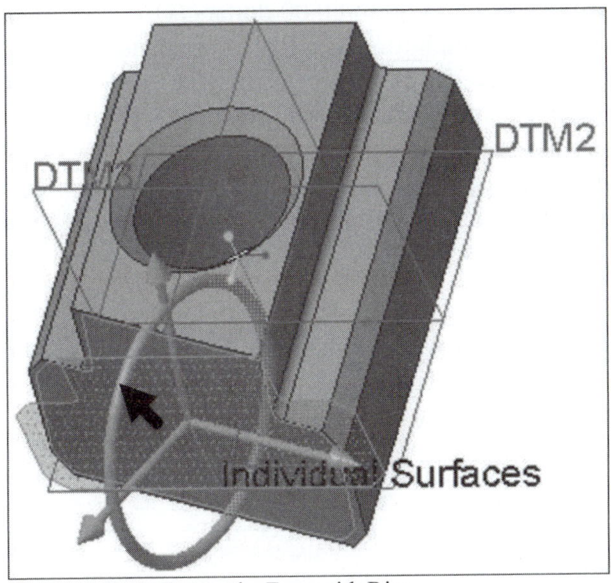

Figure 2.20(d) Rotate the Face with Ring

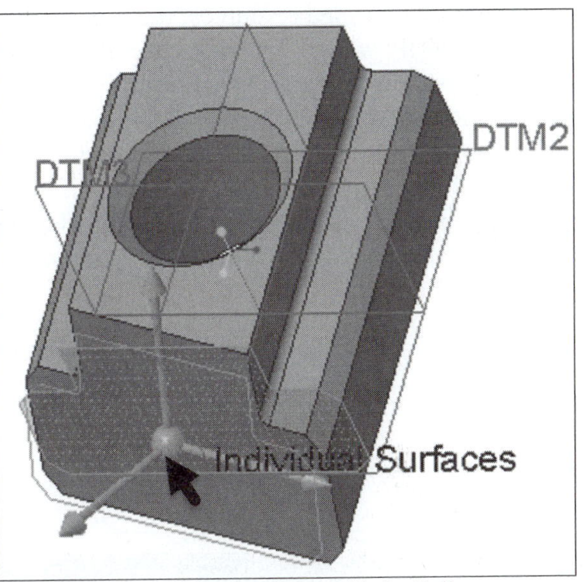

Figure 2.20(e) Pull the front Face with the Ball

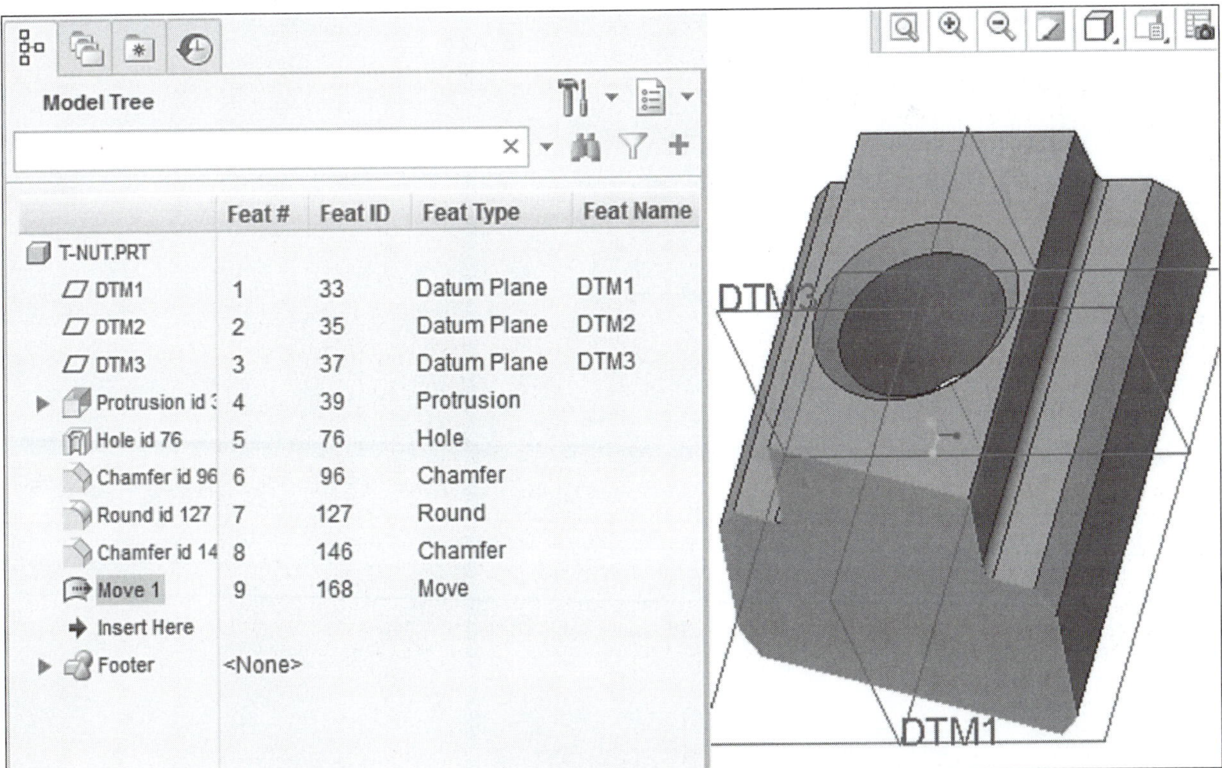

Model Tree
Feat # | Feat ID | Feat Type | Feat Name
T-NUT.PRT
DTM1 1 33 Datum Plane DTM1
DTM2 2 35 Datum Plane DTM2
DTM3 3 37 Datum Plane DTM3
Protrusion id 4 39 Protrusion
Hole id 76 5 76 Hole
Chamfer id 96 6 96 Chamfer
Round id 127 7 127 Round
Chamfer id 14 8 146 Chamfer
Move 1 9 168 Move
Insert Here
Footer <None>
DTM1

Figure 2.20(f) Flex Move Completed

Productivity Enhancements

Customizing your Creo Parametric screen interface and setting your mouse cursor to go to the default position are two productivity enhancements that can be employed to get your design created and documented quickly.

As far as input devices, the mouse is still a mouse, but- I encourage my students and trainees to buy a 5-Button mouse. Of course, many of you reading this already have a fancy Space Ball, etc. These input devices greatly improve your productivity as well as cut down on the amount of mouse picks, etc. But, for those with limited resources, the 5-button mouse for about 25.00 to 50.00 dollars (US) will be all that is needed to fly with Creo Parametric. Program your 4th and 5th button to initiate commands that you frequently use- Delete, Back, Undo, Recall Window, or whatever will reduce your need to move your mouse for the next input. I use the Back and Recall Window most since I want to quickly go back to previous Browser screens and also I flip programs when writing. Production oriented drafters-designers-engineers, will have different needs and thus different button assignments.

A simple way to eliminate much of your mouse movement by your hand and arm is to set the mouse cursor to always go to the default position after a pick. Every command pick will move the pointer to the default option on the next menu.

From your computer Desktop, click: **Start** > **Control Panel** > **Hardware and Sound** > **Mouse** > **Pointer Options** > Snap To

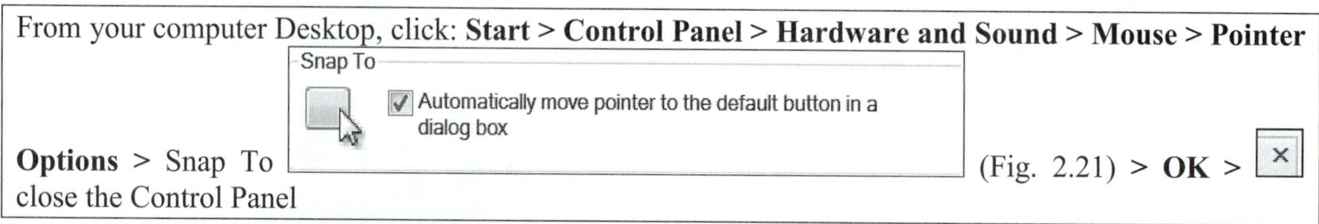

(Fig. 2.21) > **OK** > close the Control Panel

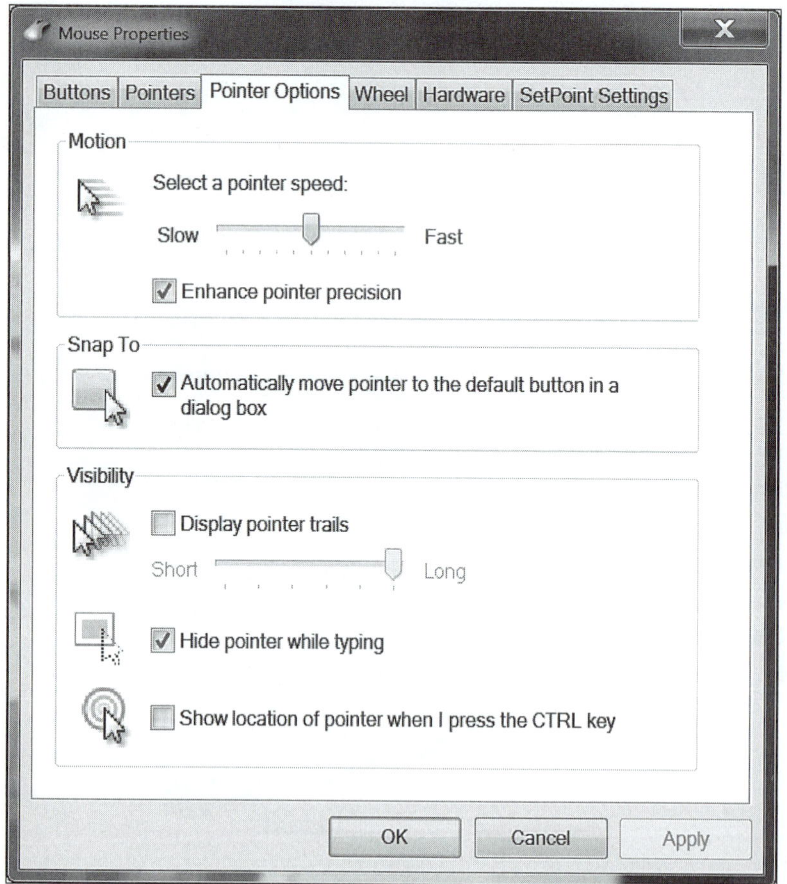

Figure 2.21 Mouse Properties *(the appearance of this dialog box may be different depending on your operating system)*

Customizing the User Interface (UI)

You can customize the Creo Parametric user interface, according to your needs or the needs of your group or company, to include the following:

- Create keyboard macros, called *mapkeys*, and add them to the menus and toolbars
- Add or remove existing toolbars
- Move or remove commands from the menus or toolbars
- Add options to the Menu Manager
- Blank (make unavailable) options in the Menu Manager
- Set default command choices for Menu Manager menus

Press: **Ctrl+O** > **t-nut.prt** > **Open** > **View** tab > [icons] *off* > move pointer anywhere in the Graphics Toolbar > **RMB** [Fig. 2.22(a)] > all options *on* [Fig. 2.22(b)] > **LMB** in Graphics Window > [toolbar] > [icons] > [icon] > [icons] > [toolbar]

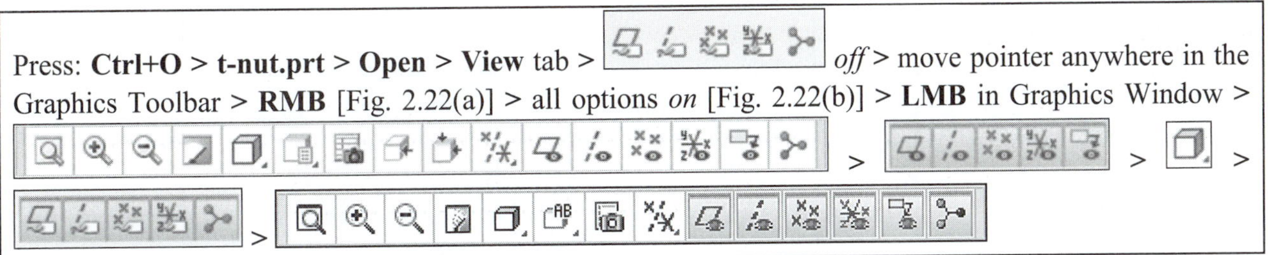

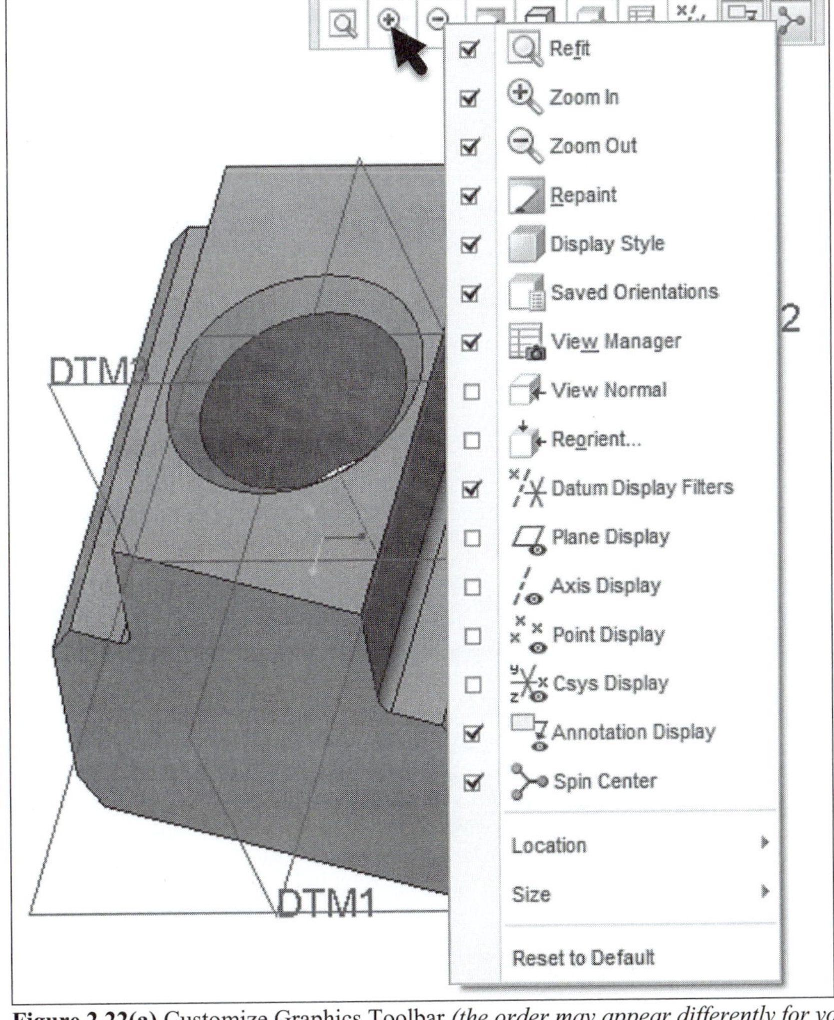

Figure 2.22(a) Customize Graphics Toolbar *(the order may appear differently for you)*

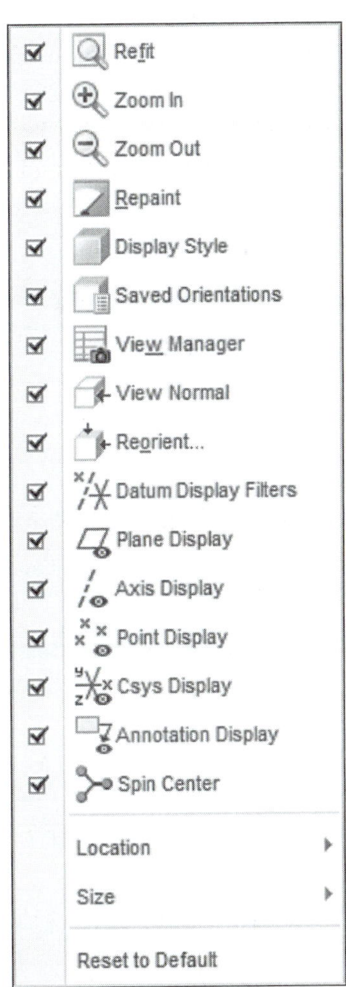

Figure 2.22(b) All Options On

To change the location of the Graphics Toolbar, click: **RMB** within the Graphics Toolbar > **Location** [Fig. 2.22(c)] > **Size** [Fig. 2.22(d)] > **Large** [Fig. 2.22(e)] > **RMB** within the Graphics Toolbar > **Reset to Default** [Fig. 2.22(f)]

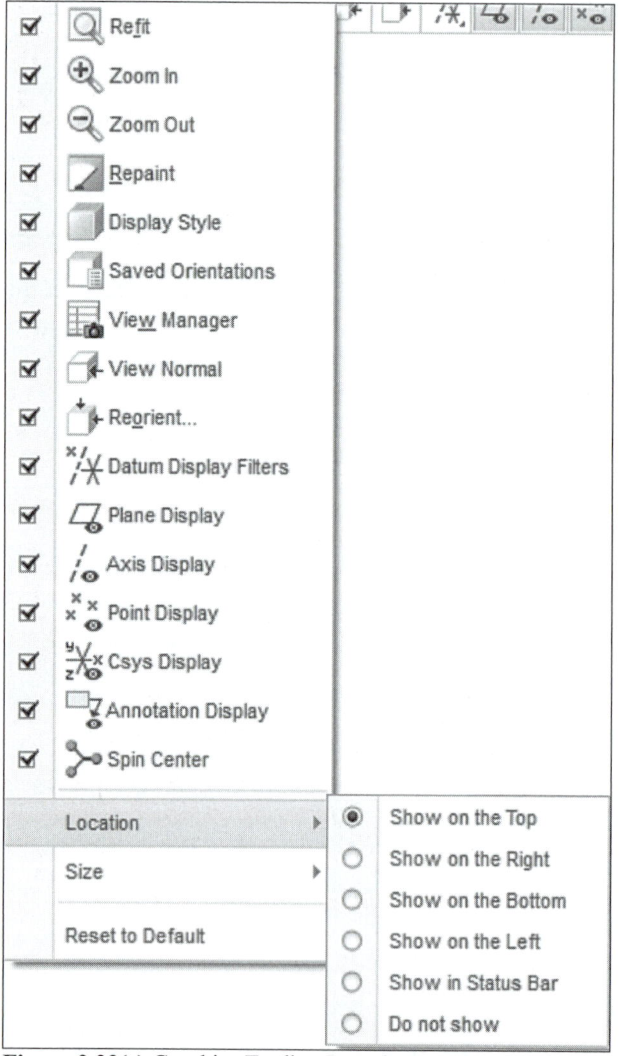

Figure 2.22(c) Graphics Toolbar Location

Figure 2.22(d) Graphics Toolbar Size

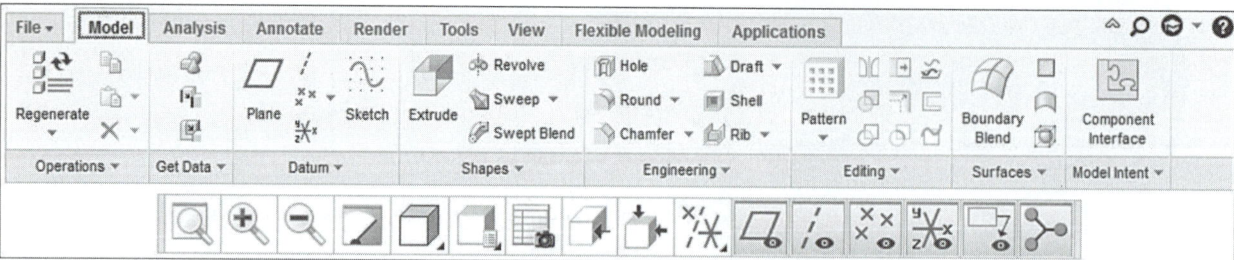

Figure 2.22(e) Graphics Toolbar Size Large *(the order may appear differently for you)*

Figure 2.22(f) Graphics Toolbar Reset to Default *(the order may appear differently for you)*

Place your pointer anywhere inside the **View** tab in the Ribbon [Fig. 2.23(a)] > **RMB** > check **Hide Command Labels** > 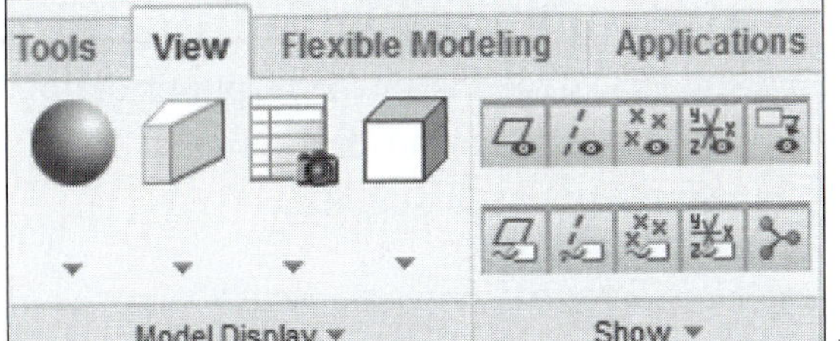

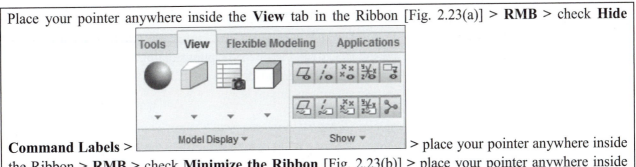

> place your pointer anywhere inside the Ribbon > **RMB** > check **Minimize the Ribbon** [Fig. 2.23(b)] > place your pointer anywhere inside the Ribbon tabs > **RMB** > uncheck **Minimize the Ribbon** > place your pointer anywhere inside the **View** tab > **RMB** > uncheck **Hide Command Labels** > place your pointer anywhere inside the Ribbon tabs > **RMB** > **Tabs** [Fig. 2.23(c)] leave all Tabs active > **LMB** anywhere in the Graphics Window

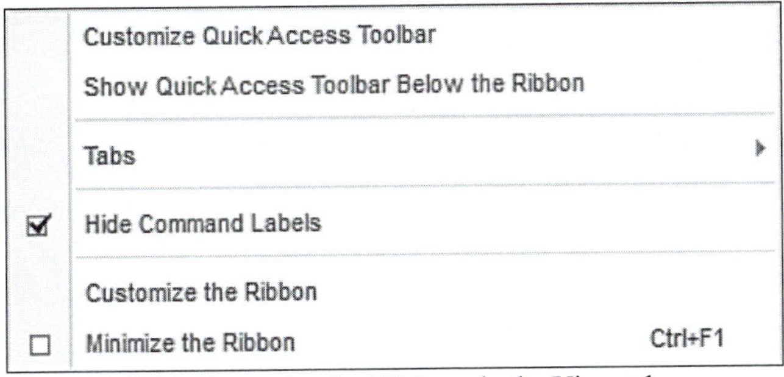

Figure 2.23(a) Customize the Ribbon via the View tab

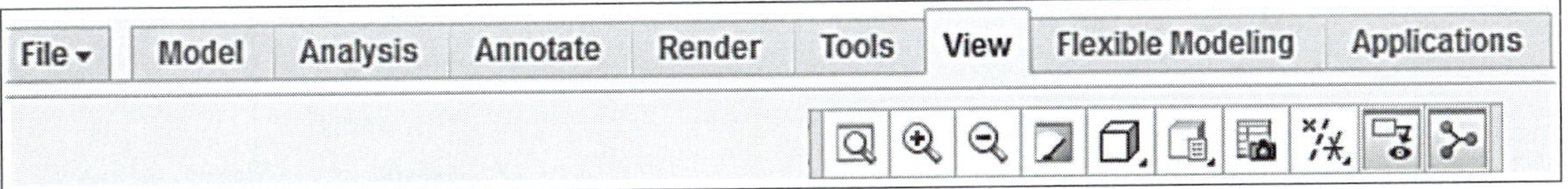

Figure 2.23(b) Minimized Ribbon

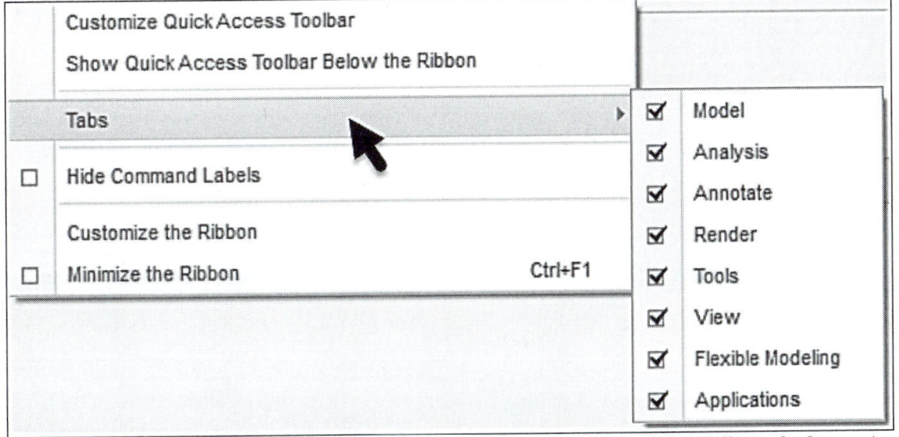

Figure 2.23(c) All Ribbon Tabs On *(the order and listing may appear differently for you)*

Click: **Model** tab > **Datum** Group *(Groups that have a ▾ indicate that there are overflow commands for that Group- more commands which are not typically placed on the Ribbon)* > **RMB** > **Groups** [Fig. 2.24(a)] > **Customize the Ribbon** > move the PTC Creo Parametric Options dialog box down if needed > **Datum** Group [Fig. 2.24(b)] > pick on **Curve** and drag and drop [Fig. 2.24(c)] into the Ribbon [Fig. 2.24(d)] > **OK** to close the PTC Creo Parametric Options dialog box

(After you are familiar with Creo Parametric's default UI, you can customize the interface based on your or your companies' productivity requirements Do not make changes now since the text and the PTC Help assume the default of commands and colors, etc.)

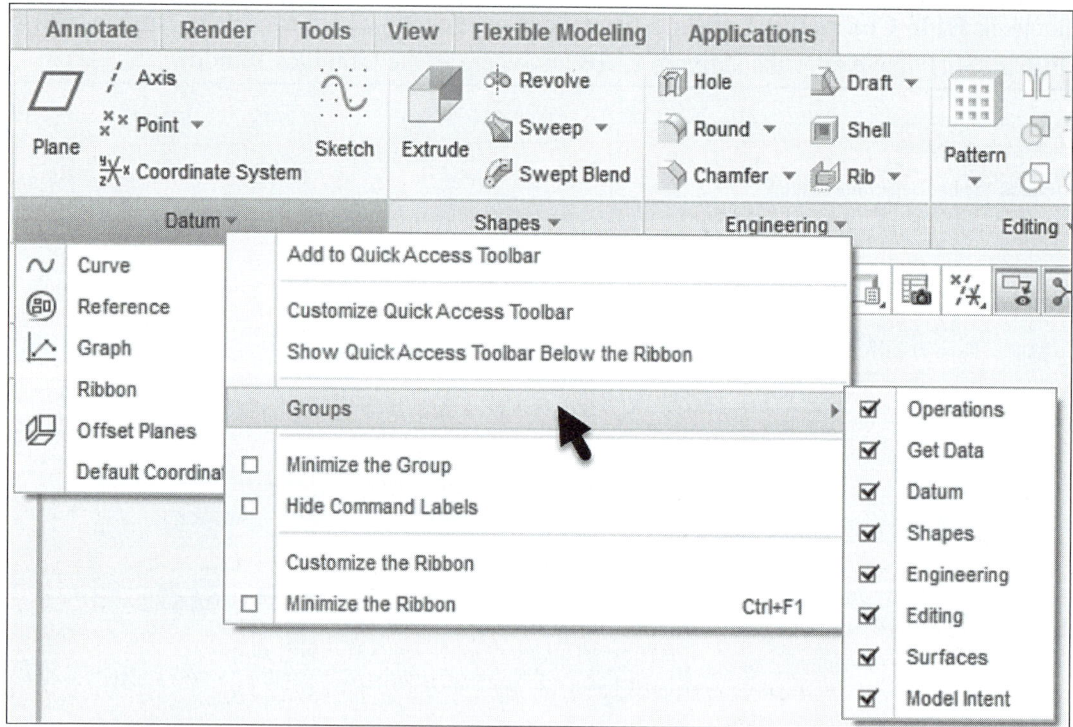

Figure 2.24(a) Groups

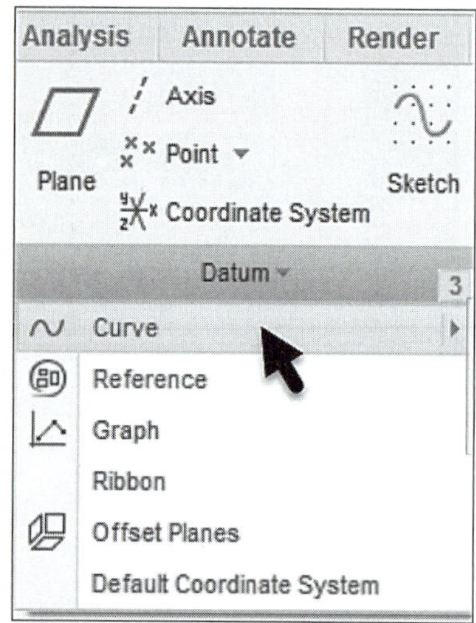

Figure 2.24(b) Click on Curve

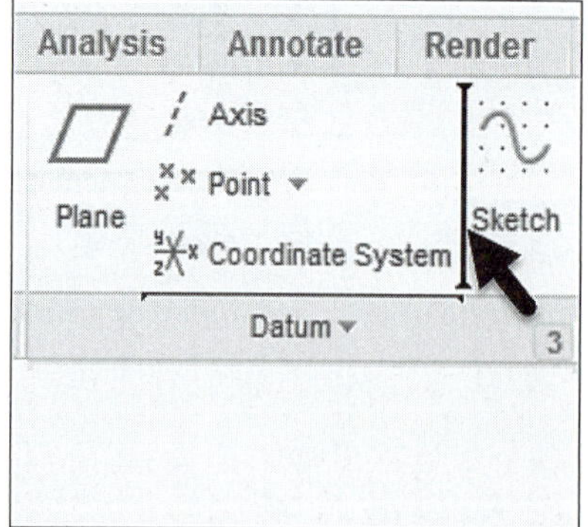

Figure 2.24(c) Drag and Drop Curve

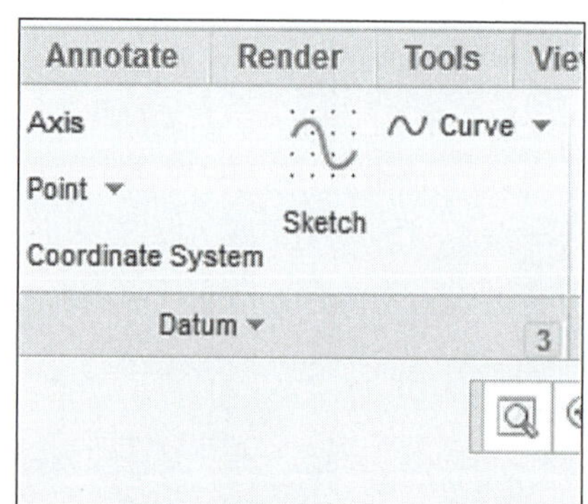

Figure 2.24(d) Curve

Click: **File > Options > Customize Ribbon > New Tab >** select **New Tab (Custom) > RMB > Rename** [Fig. 2.25(a)] > *type* **Production > OK >** select **New Group (Custom) > RMB > Rename >** *type* **Creo**

Production (Custom)
Creo (Custom)
> OK
File | Model | Production | Analysis | Annotate | Render | Tools

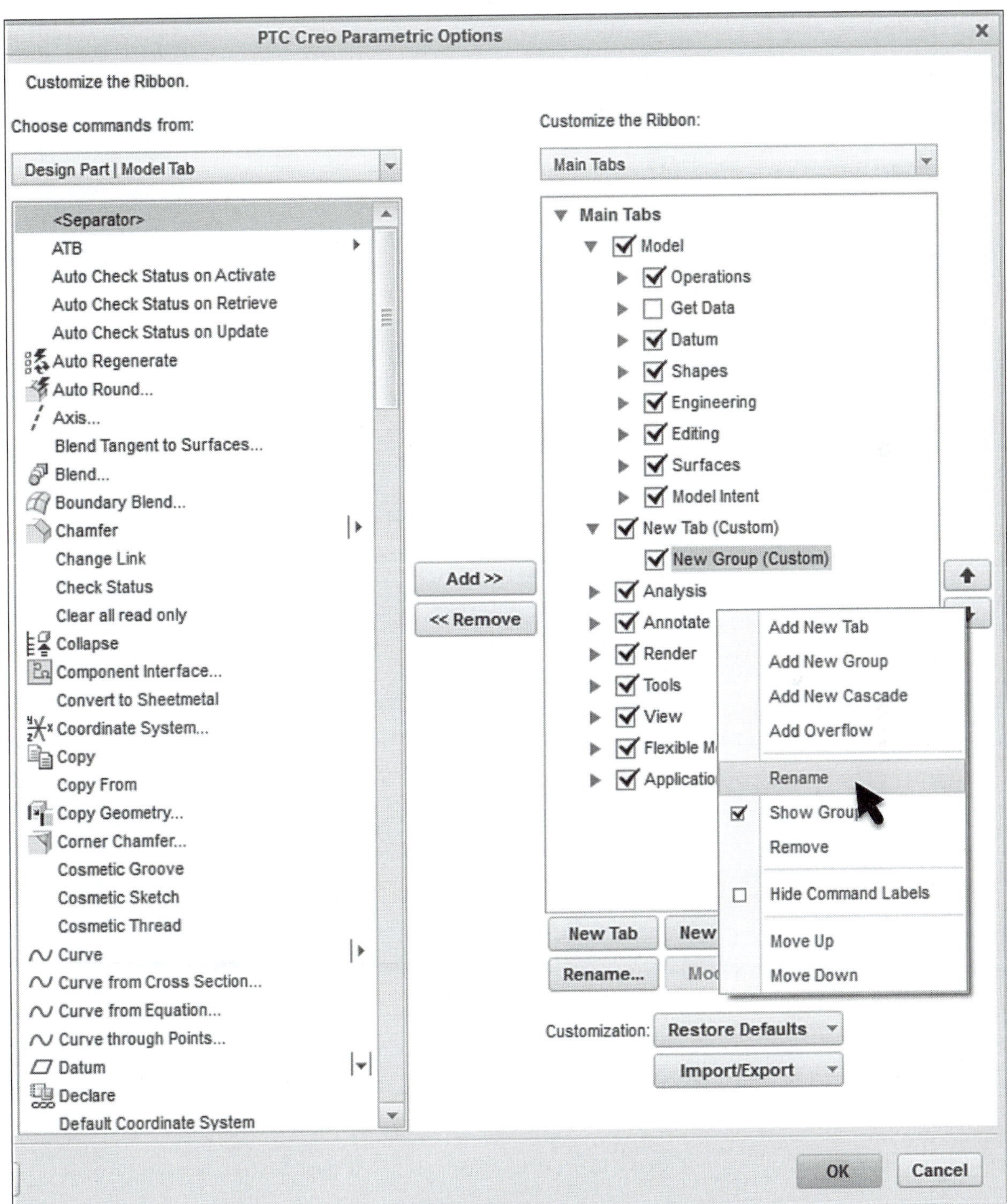

Figure 2.25(a) Customize Ribbon, New Tab > RMB > Rename

Click: **Edit Definition** > **Add** [Fig. 2.25(b)] > Choose commands from: **Design Part | Model Tab** > **All Commands** [Fig. 2.25(c)] > **Delete Old Versions** > **Add** [Fig. 2.25(d)] > **OK** > **Production** tab > **Delete Old Versions** [Fig. 2.25(e)] > **Yes** [Fig. 2.25(f)] > **Model** tab

(Customize your Production tab after you see what commands you would like to access faster.)

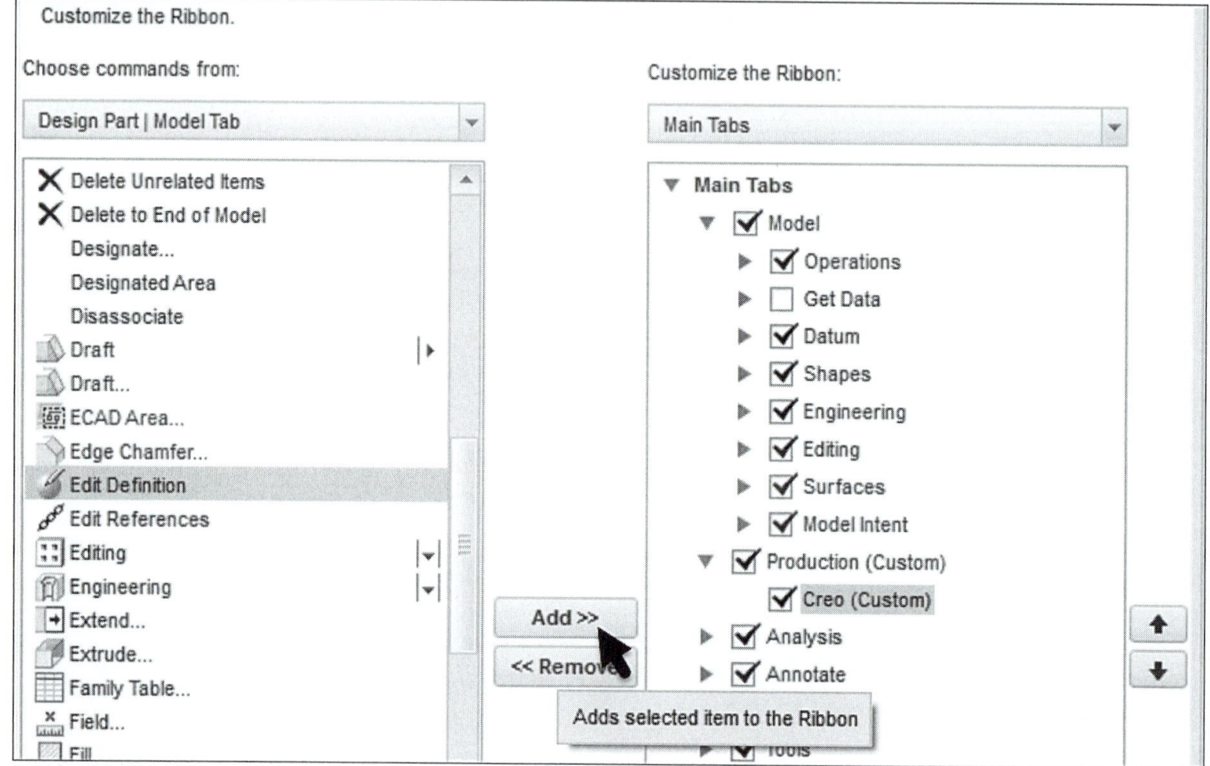

Figure 2.25(b) Edit Definition > Add

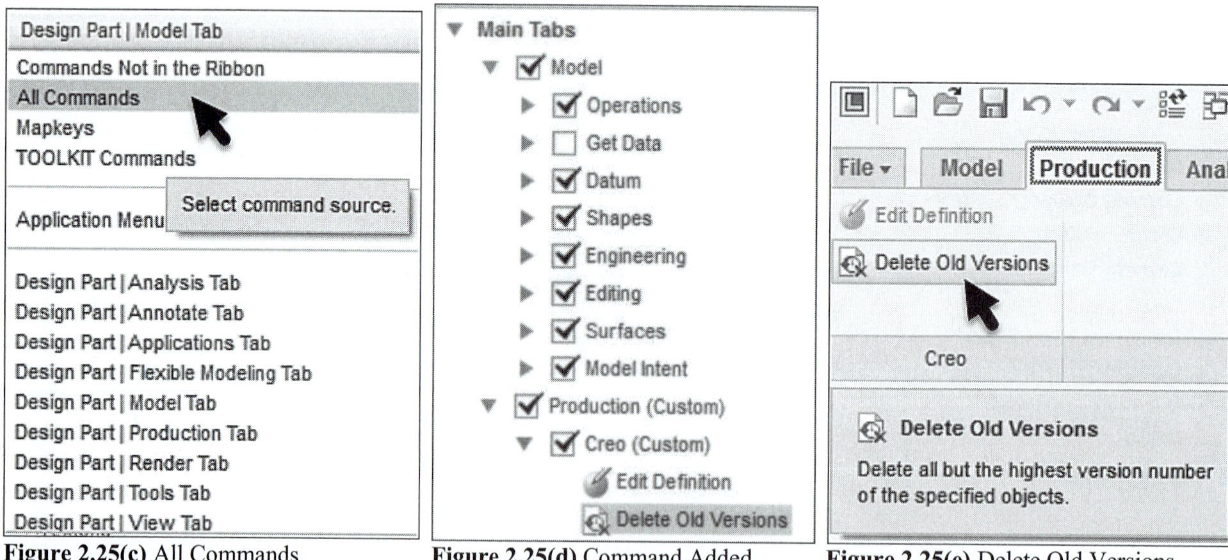

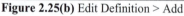

Figure 2.25(c) All Commands **Figure 2.25(d)** Command Added **Figure 2.25(e)** Delete Old Versions

Do you want to delete all old versions of C:\CAD-Resources\CREO\Creo_3-0\PTC_Creo-3\3-0\t-nut.prt.9?

Figure 2.25(f) Delete Old Versions Confirmation

Mapkeys

A **Mapkey** is a macro that maps frequently used command sequences to certain keyboard keys or sets of keys. Mapkeys are saved in the configuration file, and are identified with the option *mapkey*, followed by the identifier and then the macro. You can define a unique key or combination of keys which, when pressed, executes the mapkey macro. You can create a mapkey for virtually any task you perform frequently within PTC Creo Parametric 3.0. By adding mapkeys to your Ribbon or Toolbars, you can run mapkeys with a single mouse click or menu command and thus streamline your workflow.

Click: **File** > **Options** > **Environment** > **Mapkeys Settings** [Fig. 2.26(a)] > **New** from the Mapkeys dialog box [Fig. 2.26(b)] (Record Mapkey dialog box opens) > Key Sequence, *type* **$F3** > Name, *type* **Display** > Description, *type* **Change display** [Fig. 2.26(c)] > **Record**

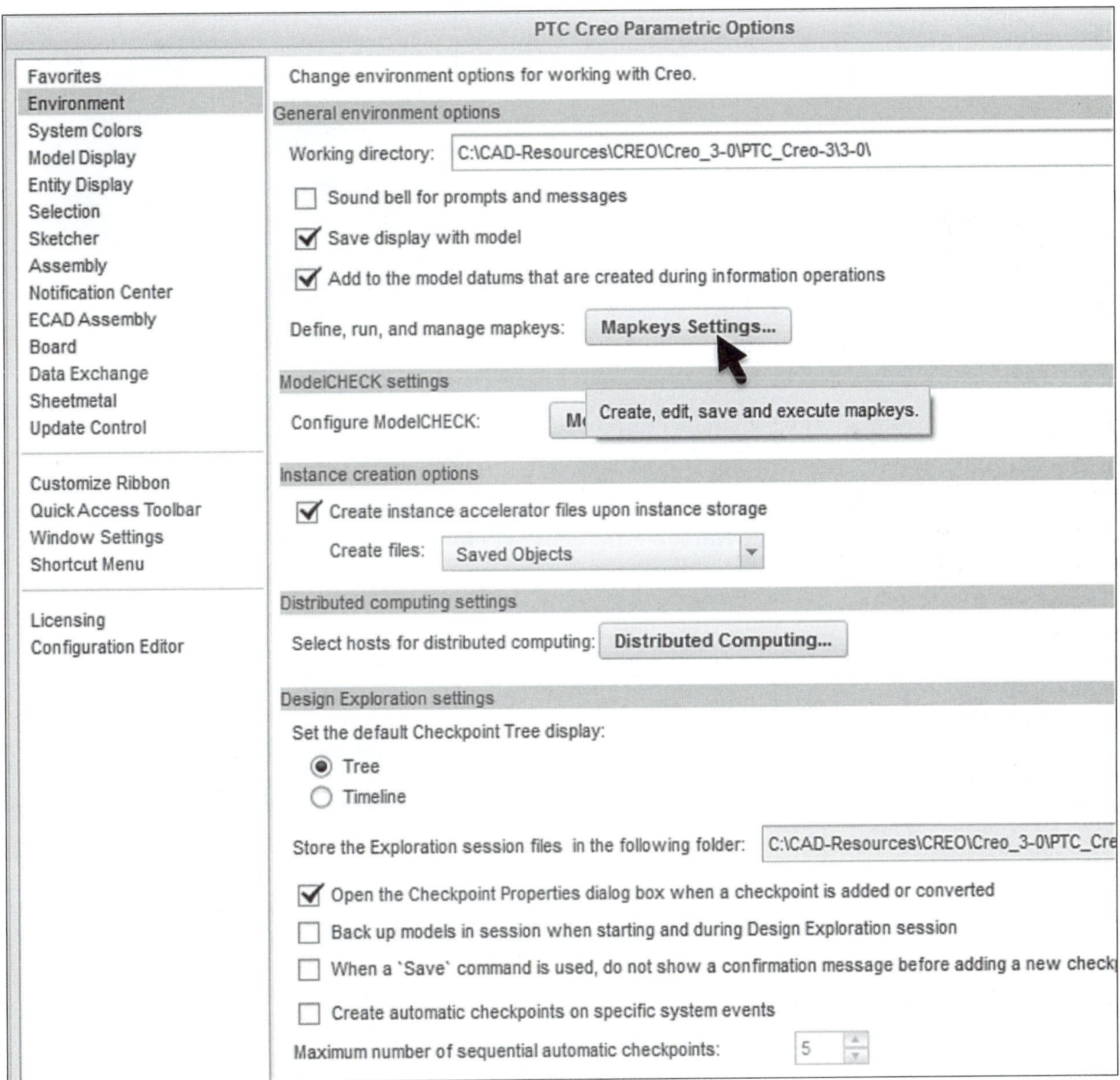

Figure 2.26(a) Creo Parametric Option Dialog Box > Environment > Mapkeys Settings

Click: **View** tab > toggle *off* > **File** > **Options** > **Model Display** > **Trimetric** > **Isometric** > **Fast hidden lines removal** > **OK** > **Apply** > **No** > **Stop** (from Record Mapkey dialog box) > **OK** (from Record Mapkey dialog box) > Save Mapkeys **Saved changed** (from Mapkeys dialog box) > File name: *type* **Creo_textbook.pro** > **OK** > **Close** (from Mapkeys dialog box) > press **F3** > press **F3** > **Ctrl+D** [Fig. 2.26(d)]

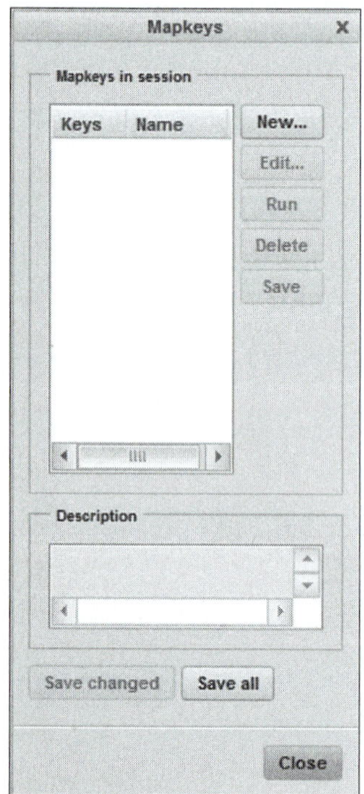

Figure 2.26(b) Mapkeys

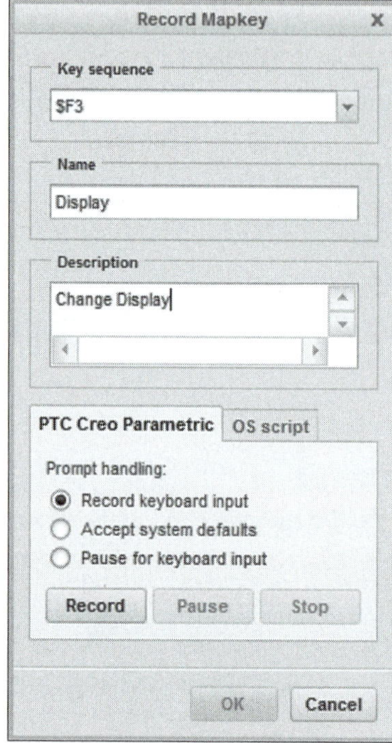

Figure 2.26(c) Record Mapkey

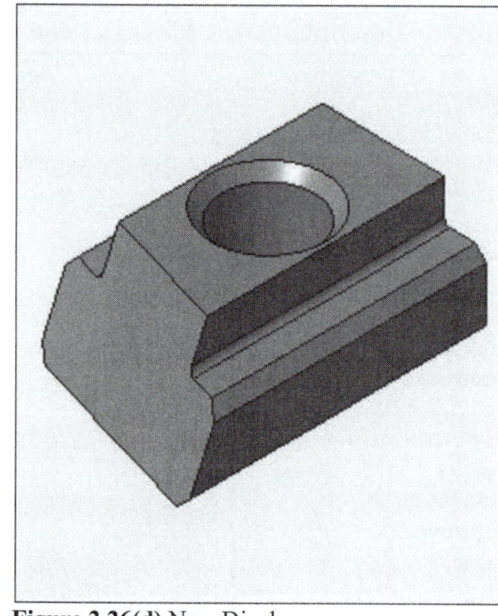

Figure 2.26(d) New Display

Click: **File** > **Options** > **Customize Ribbon** > **Design Part | Model Tab** >

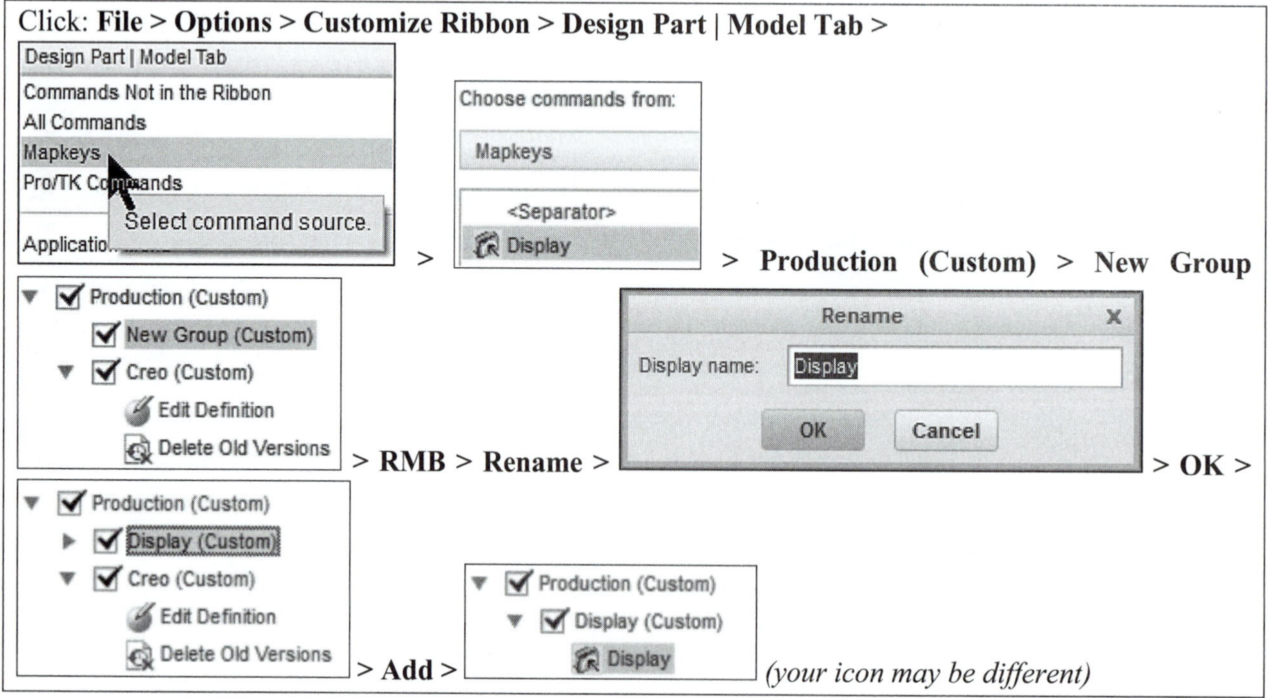

Click: **Modify > Choose from Existing** [Fig. 2.26(e)] > select the red diamond

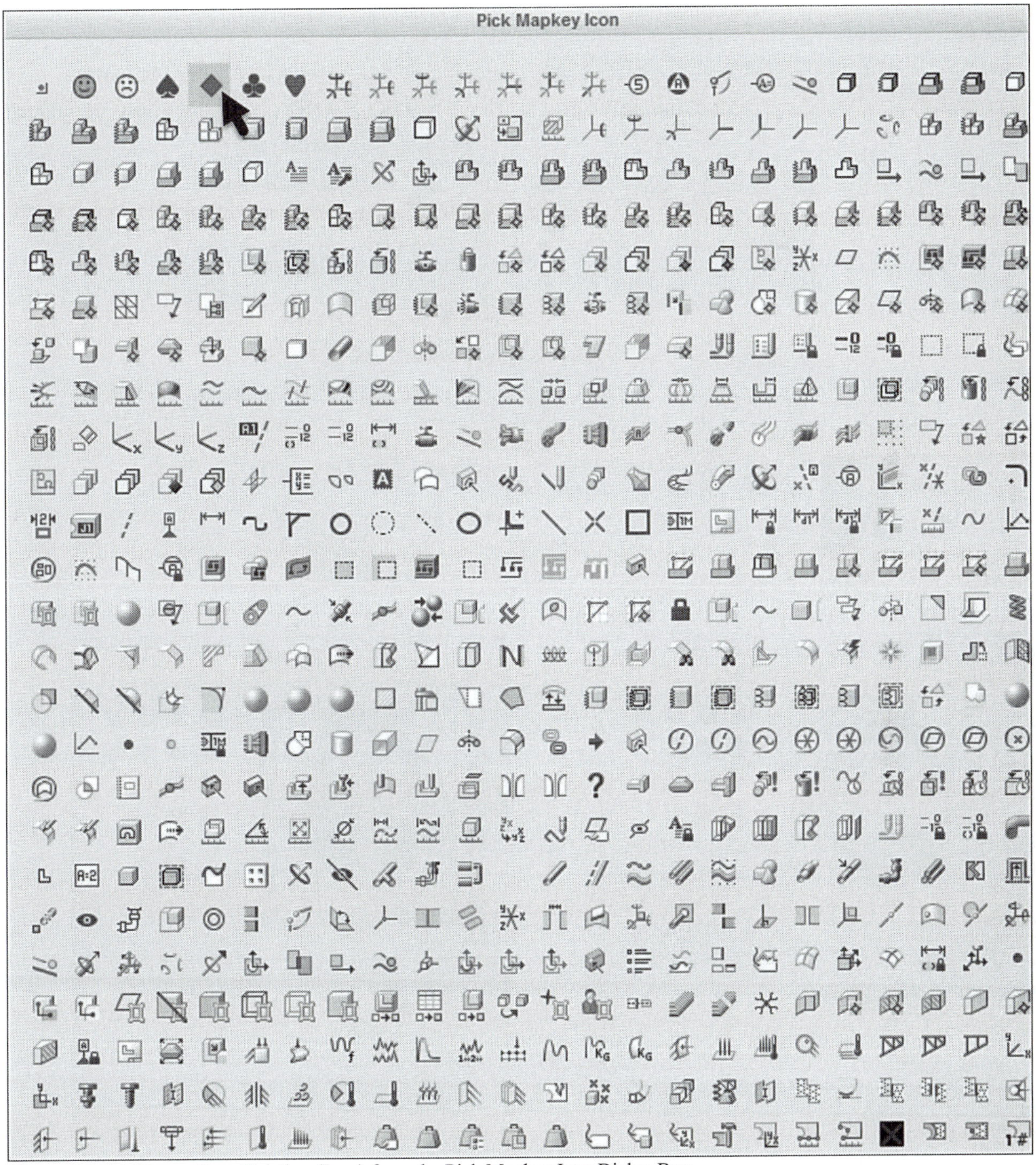

Figure 2.26(e) Choose from Existing (Icon) from the Pick Mapkey Icon Dialog Box

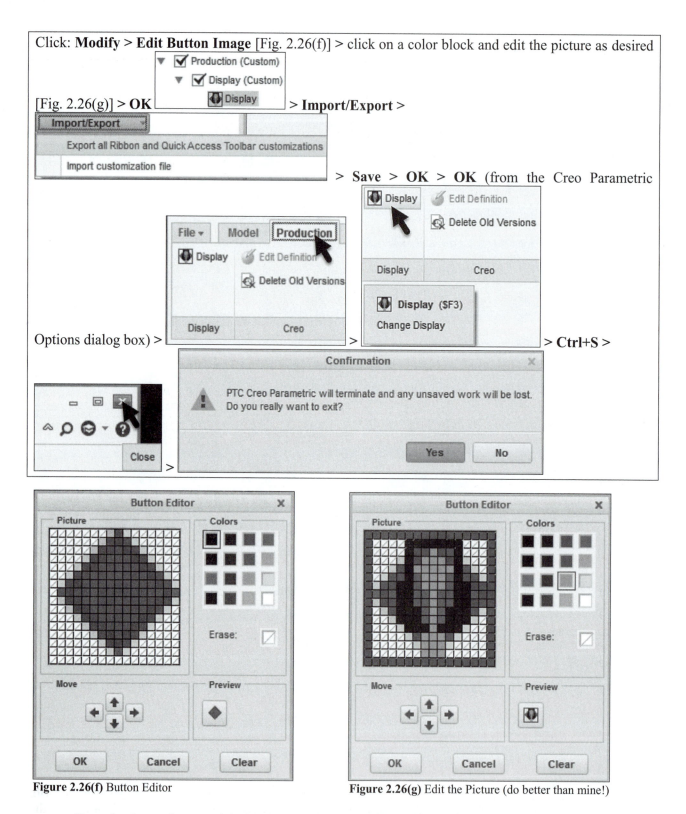

Click: **Modify** > **Edit Button Image** [Fig. 2.26(f)] > click on a color block and edit the picture as desired [Fig. 2.26(g)] > **OK** > **Import/Export** > **Save** > **OK** > **OK** (from the Creo Parametric Options dialog box) > > > **Ctrl+S** > >

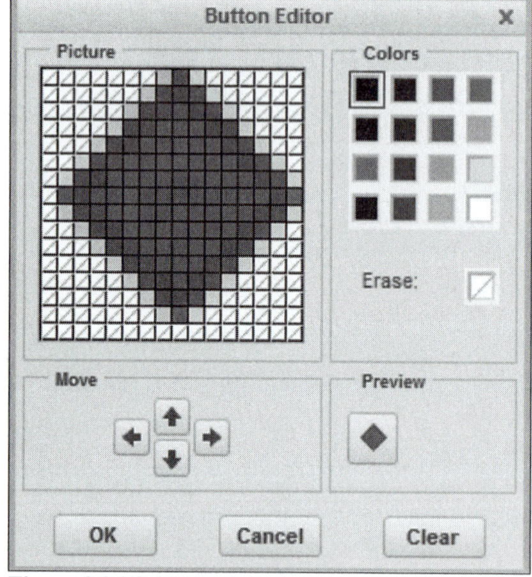

Figure 2.26(f) Button Editor

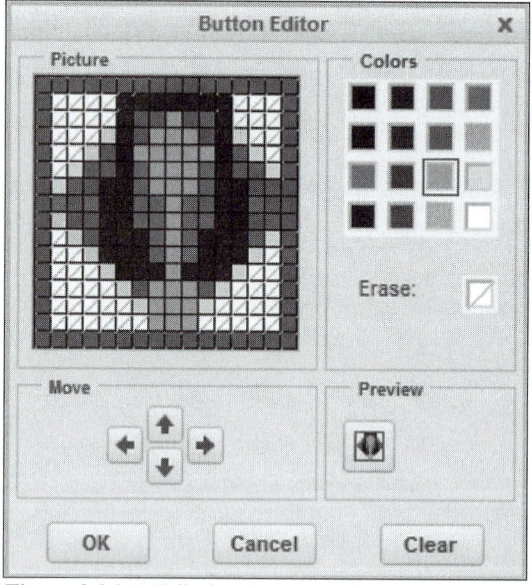

Figure 2.26(g) Edit the Picture (do better than mine!)

Download another model from the PTC Catalog or ***www.cad-resources.com*** and practice navigating the interface and using commands introduced in this Lesson.

Lesson 3 Direct Modeling

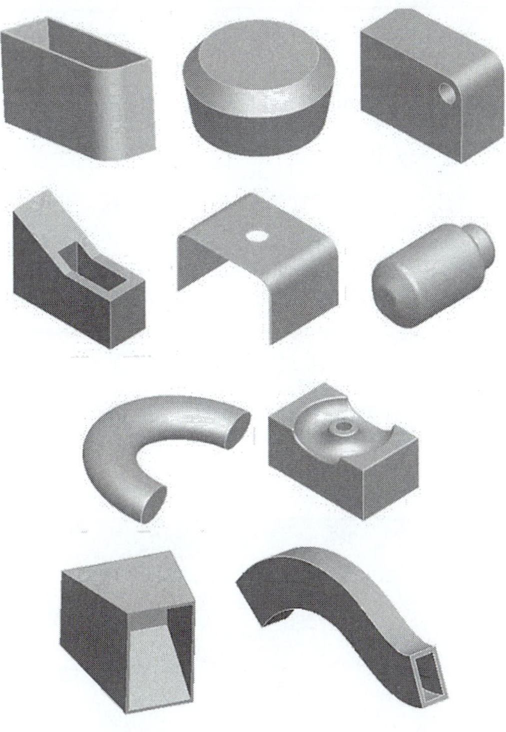

Figure 3.1 Quick Modeling Parts

OBJECTIVES

- Modeling simple parts quickly using **default selections**
- Sample **Feature Tools**
- Try out a variety of **Engineering Tools** including: **Hole**, **Shell**, **Round**, **Chamfer**, and **Draft**
- Sketch simple **sections**

REFERENCES AND RESOURCES

For **Resources** go to **www.cad-resources.com** > click on the PTC Creo Parametric 3.0 Book cover

- Lesson 3 Lecture
- Lesson 3 3D models embedded in a PDF
- Quick Reference Card
- Configuration Options

Modeling

The purpose of this lesson is to quickly introduce you to a variety of **Feature Tools** and **Engineering Tools**. You will model a variety of very simple parts (Fig. 3.1). Little or no explanation of the methodology or theory of the Tool or process will accompany the instructions. By using mainly default selections; you will create models that will display the power and capability of Creo Parametric.

In the next lesson, a detailed step-by-step systematic description will accompany all commands. Here, we hope to get you up and running on Creo Parametric without any belabored explanations.

Lesson 3 STEPS

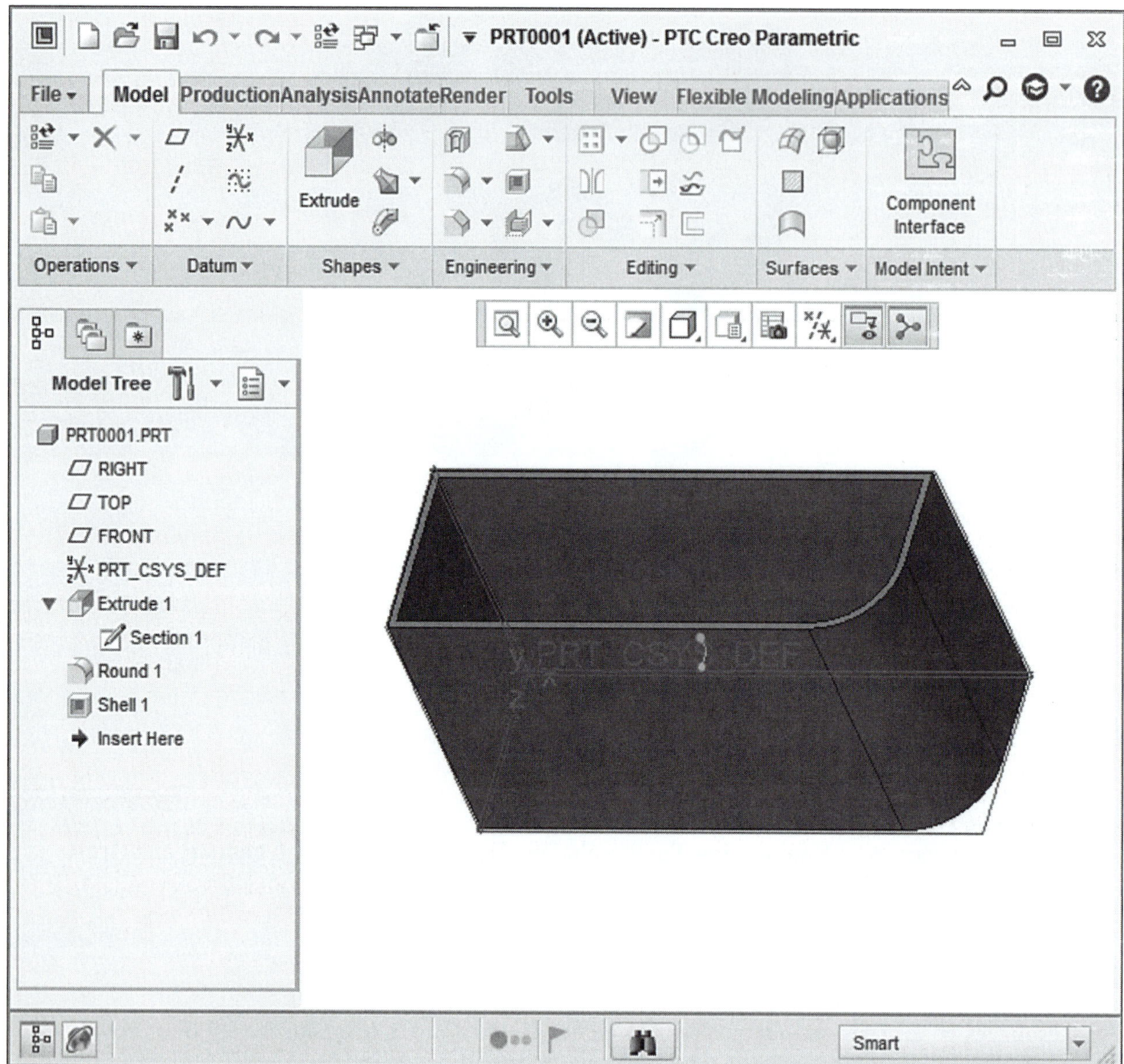

Figure 3.2(a) Part One – PRT0001.PRT

Part Model One (PRT0001.PRT) (Extrude)

This model [Fig. 3.2(a)] will introduce the **Extrude Tool** to create a simple box-shape, the **Round Tool** to add a round to one edge, and the **Shell Tool** to remove one surface and make the part walls a consistent thickness. In this direct modeling example, we will provide step-by-step illustrations. For subsequent parts, only important steps or illustrations displaying aspects of the command sequence that represent new material will be provided. The same applies to ***Tool Tips*** that appear as you pass the mouse pointer (cursor) over a button or icon. All parts have been created with out-of-the-box system settings for Creo Parametric including default templates, grid settings, and so forth.

(Do not change directories if you have one assigned as a default by your school or company.)

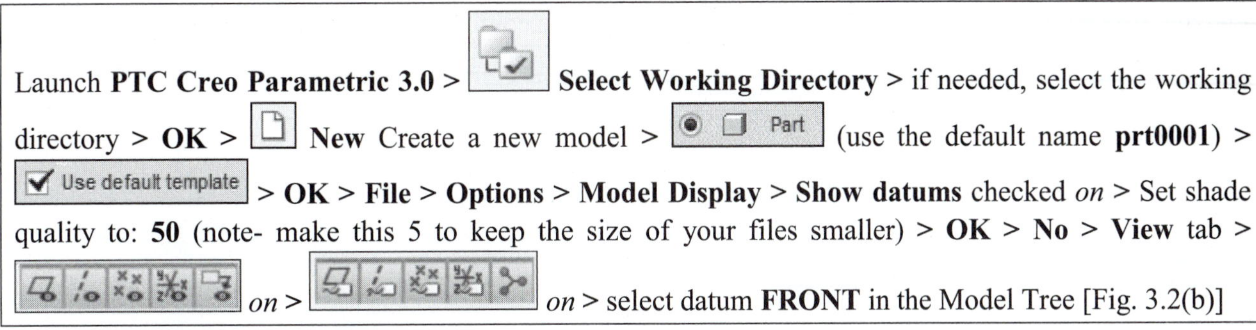

Launch **PTC Creo Parametric 3.0** > **Select Working Directory** > if needed, select the working directory > **OK** > **New** Create a new model > Part (use the default name **prt0001**) > Use default template > **OK** > **File** > **Options** > **Model Display** > **Show datums** checked *on* > Set shade quality to: **50** (note- make this 5 to keep the size of your files smaller) > **OK** > **No** > **View** tab > *on* > *on* > select datum **FRONT** in the Model Tree [Fig. 3.2(b)]

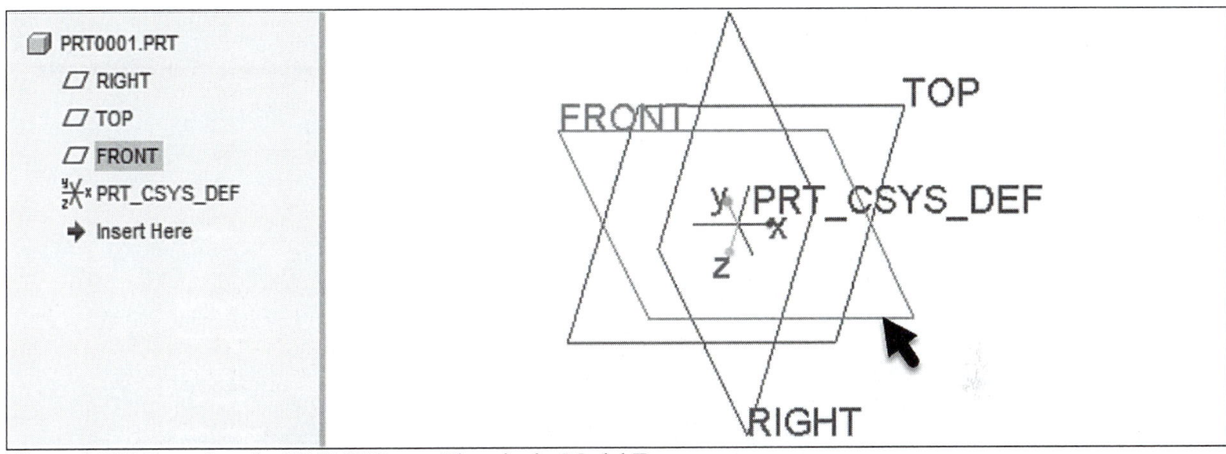

Figure 3.2(b) Pre-select the FRONT Datum Plane in the Model Tree

Click: **Model** tab > Extrude the pre-selected FRONT datum is the Sketch Plane, the sketch view remains in a 3D trimetric orientation > in the Graphics Window, press **RMB** > **Corner Rectangle** [Fig. 3.2(c)]

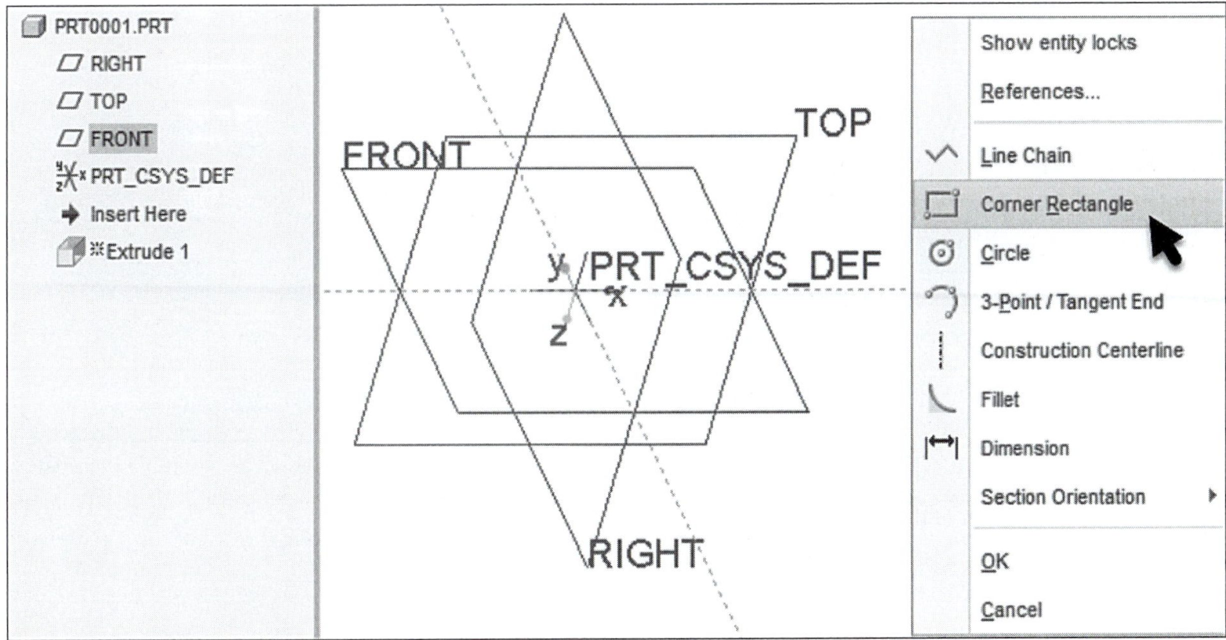

Figure 3.2(c) RMB > Corner Rectangle

Pick the first corner of the rectangle [Fig. 3.2(d)] > pick the second corner of the rectangle [Fig. 3.2(e)]

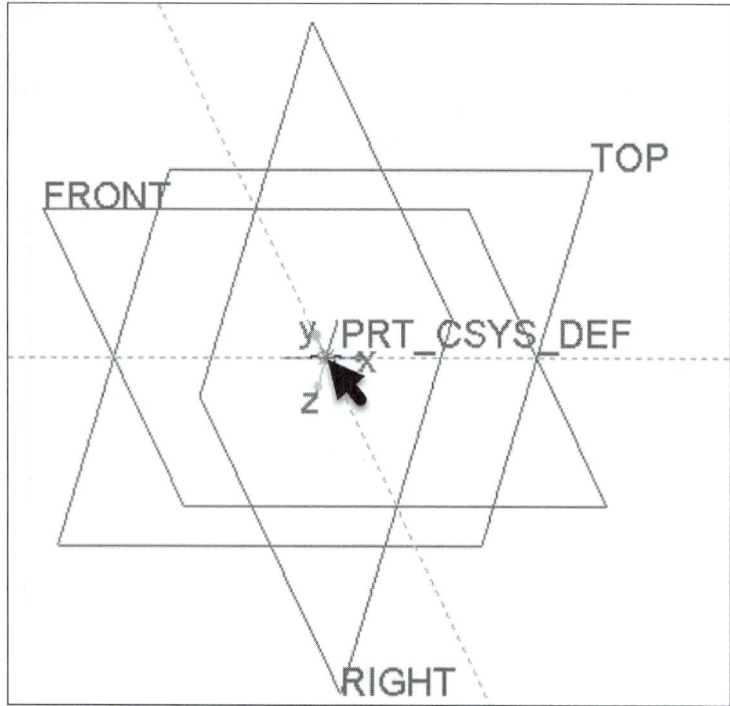

Figure 3.2(d) First Corner of the Rectangle at the origin

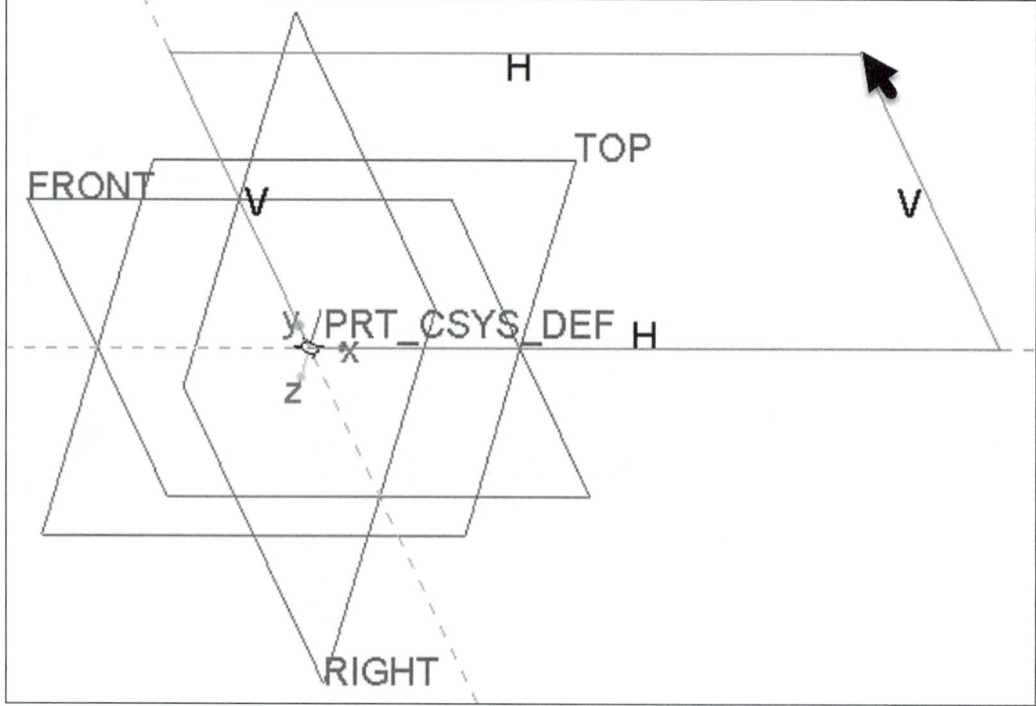

Figure 3.2(e) Second Corner of the Rectangle

Click: **MMB** to deselect the Corner Rectangle command > press **RMB > OK** [Fig. 3.2(f)]

(If your feature looks too thin, move a drag handle: place the pointer on the square drag handle > press and hold down the LMB > move the square drag handle forward to give the feature greater depth > release the LMB) The actual dimension does not need to be the same as shown. [Fig. 3.2(g)] **> Enter** to complete the Extrude **> Ctrl+S > OK**

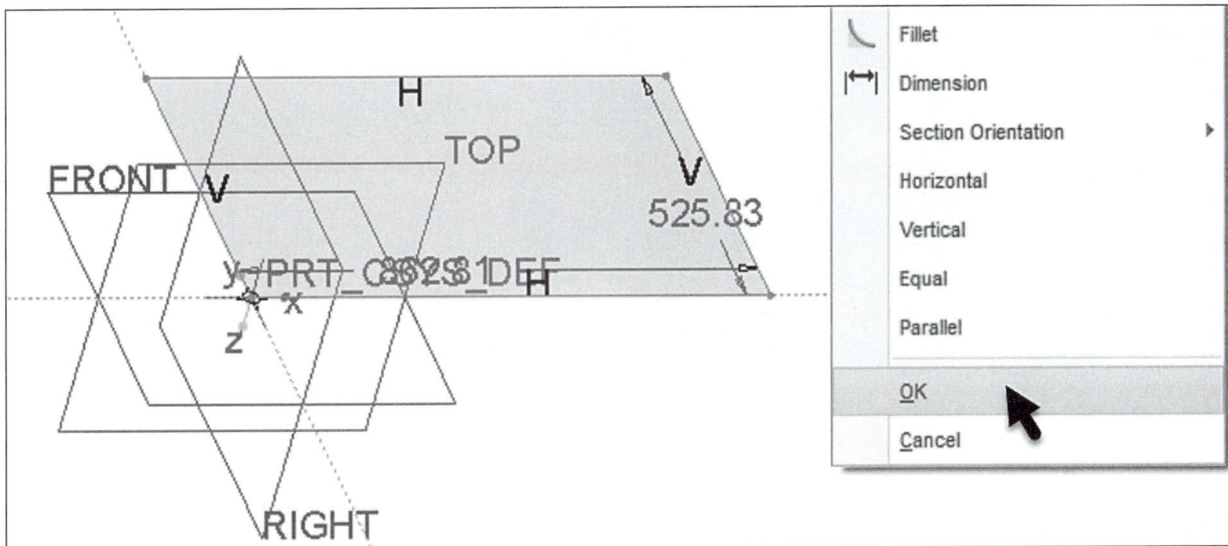

Figure 3.2(f) Shaded (Closed Loop) Rectangular Section

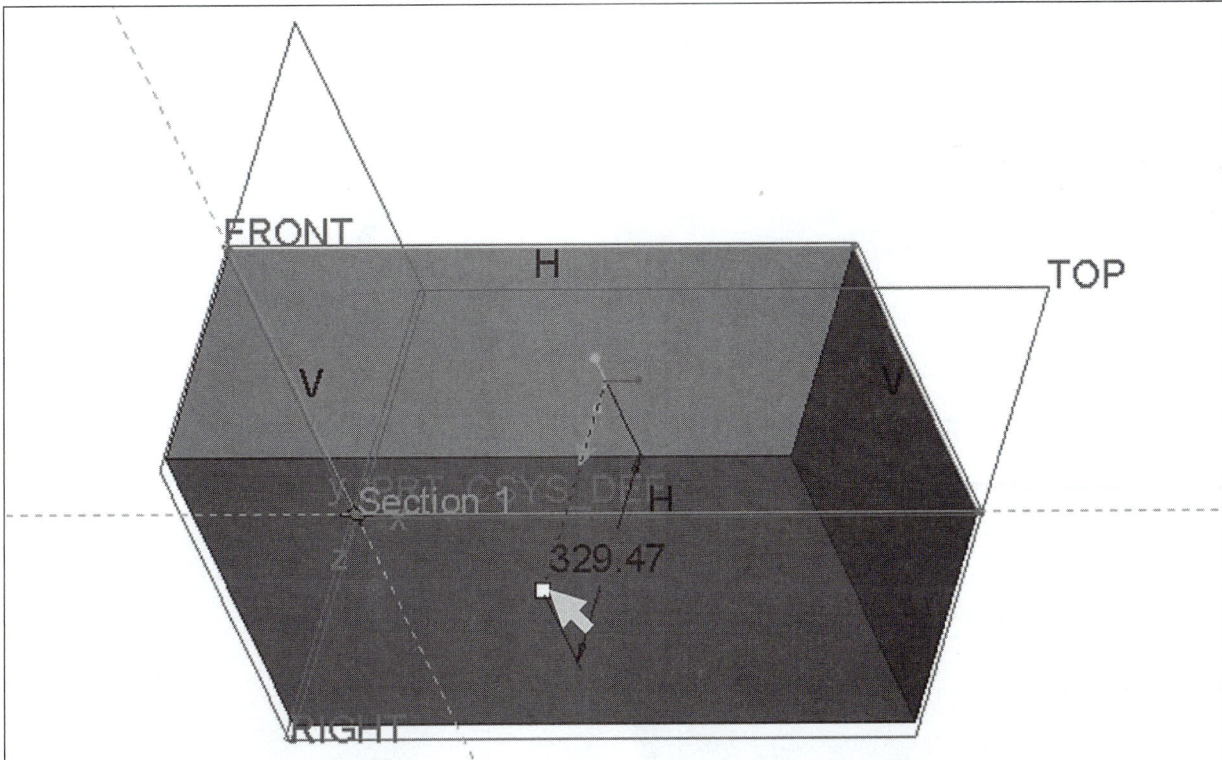

Figure 3.2(g) Depth Preview of Extrusion

With the extrusion selected, pick on the right front edge of the part [Fig. 3.2(h)] > press **RMB** > **Round** [Fig. 3.2(i)] > move a drag handle [Fig. 3.2(j)] > enlarge the round [Fig. 3.2(k)] > **MMB** to complete the Round

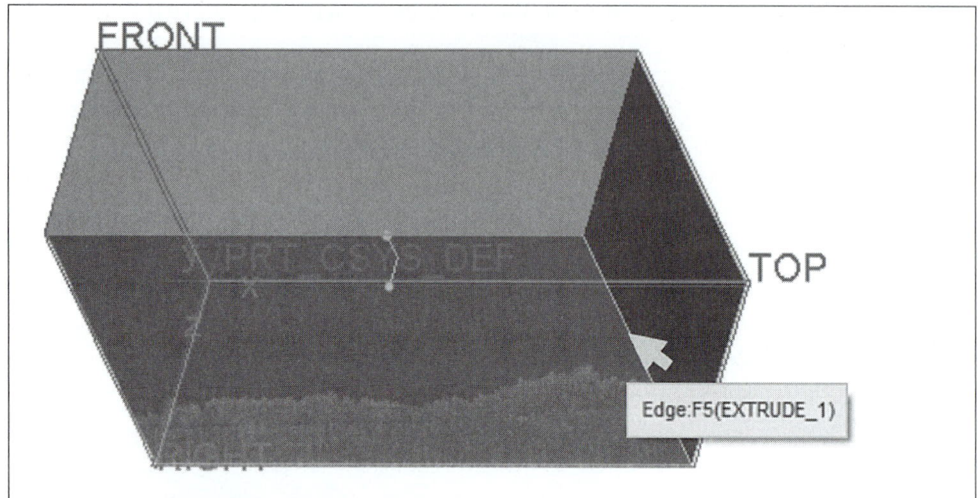

Figure 3.2(h) Pick on the Front Edge

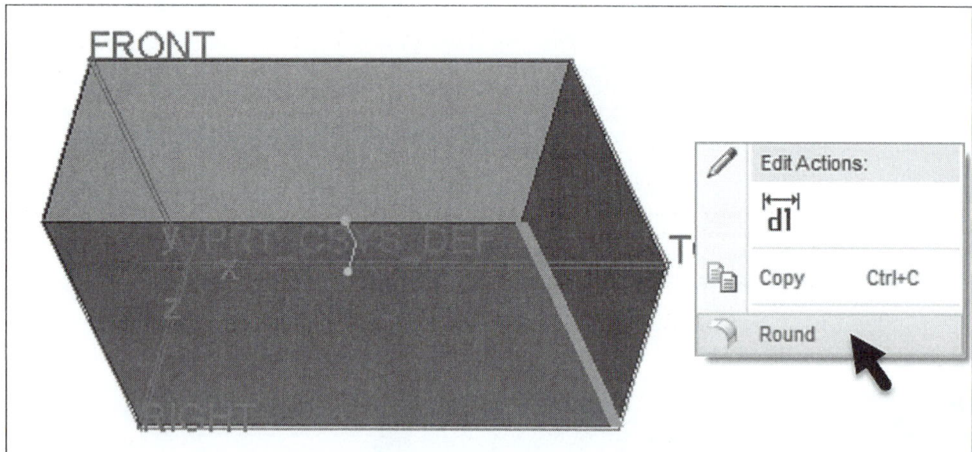

Figure 3.2(i) Round

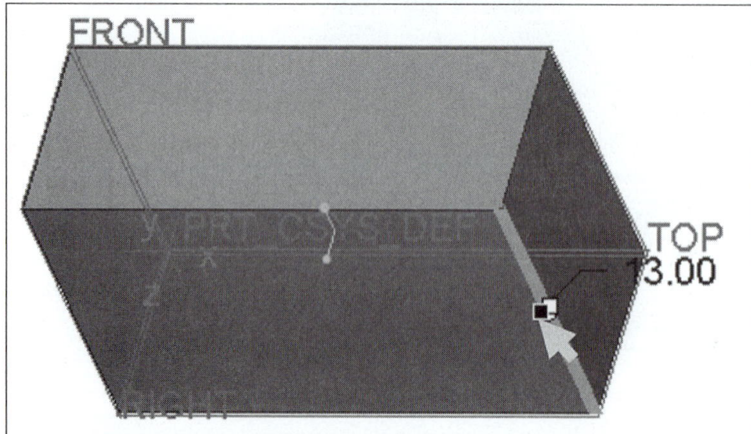

Figure 3.2(j) Move the Radius Drag Handle

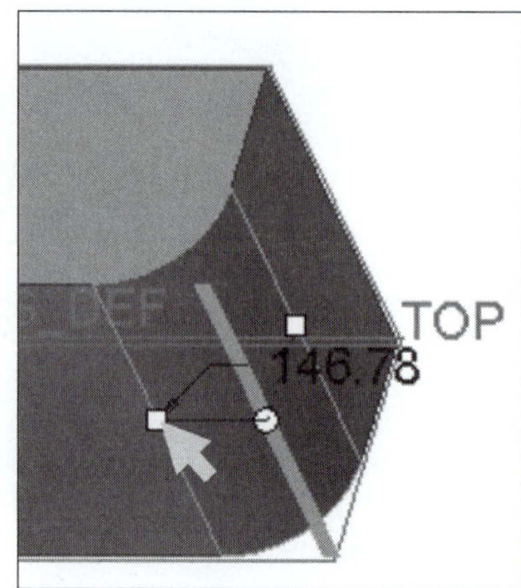

Figure 3.2(k) Enlarged Round

With the round feature *still selected*, pick on the top surface > slightly move the mouse pointer and pick again (the surface will highlight) [Fig. 3.2(l)] > Shell **Shell** [Fig. 3.2(m)] > **MMB** [Fig. 3.2(n)] > **Ctrl+S** > **File** > **Save As** > Type ▾ > **Zip File (*.zip)** > **OK** > **File** > **Close**

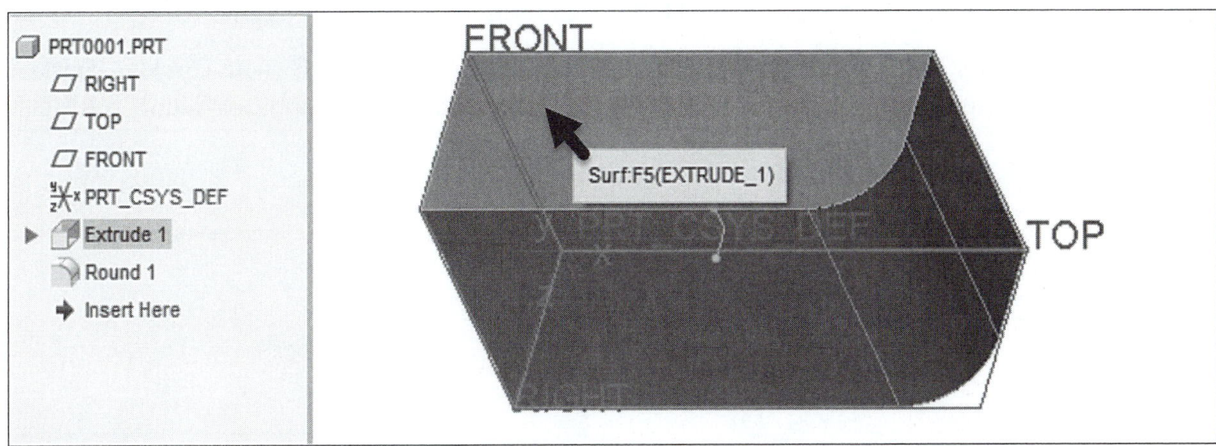

Figure 3.2(l) Selected Top Surface is Highlighted

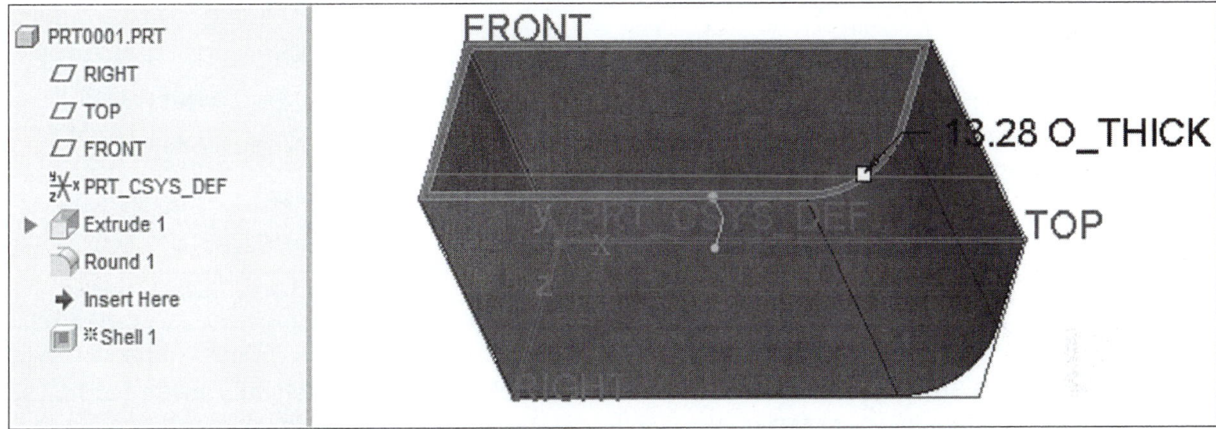

Figure 3.2(m) Shell Tool Applied

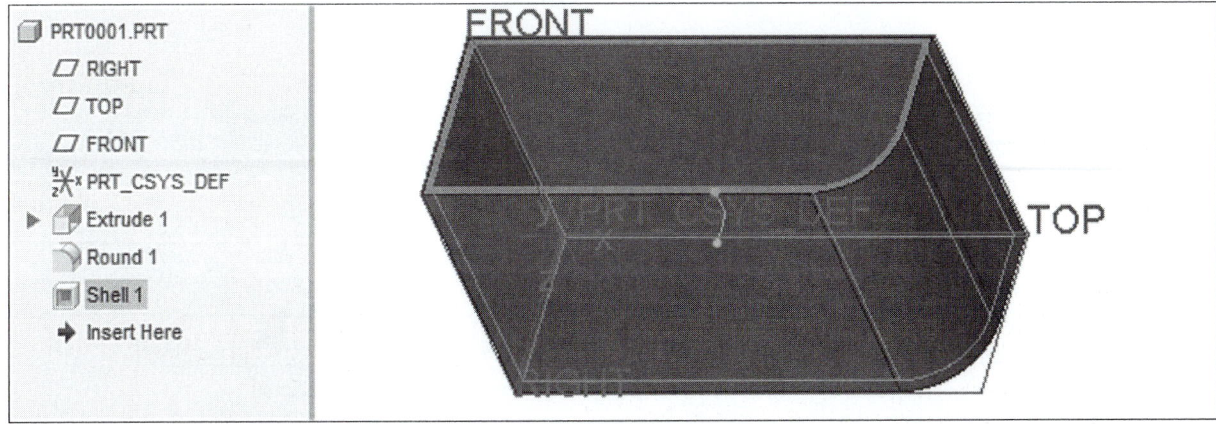

Figure 3.2(n) Completed Part

Part Model Two (PRT0002.PRT) (Draft)

Press: **Ctrl+N** > **Enter** (prt0002) > pick on the **TOP** datum from the Graphics Window > Extrude > in the Graphics Window, press **RMB** > Circle **Circle** [Fig. 3.3(a)] > press **RMB** (note Circle is checked *on*) [Fig. 3.3(b)] > release the **RMB** > slightly move the mouse pointer in the Graphics Window > **RMB**

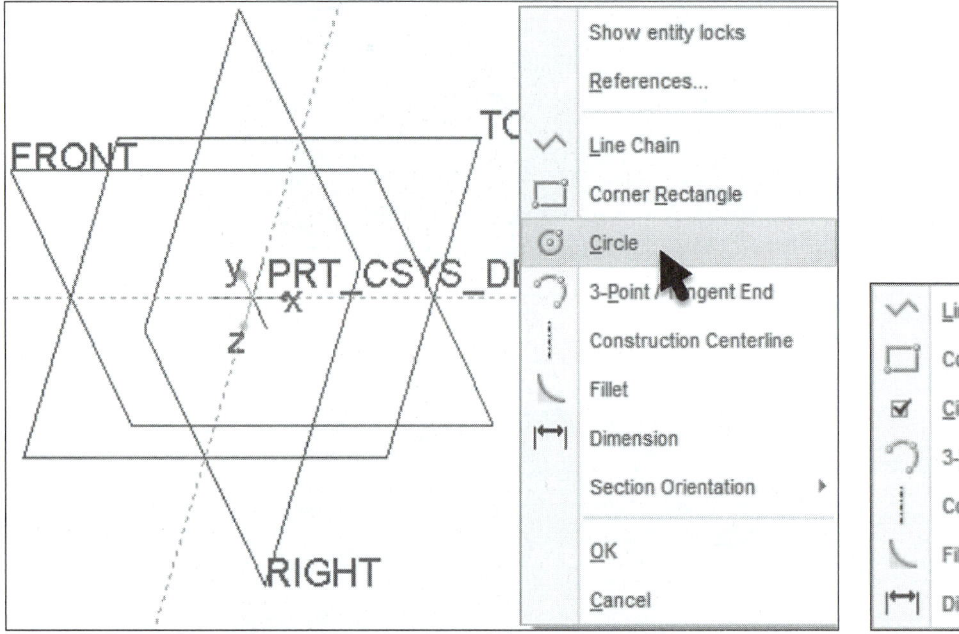

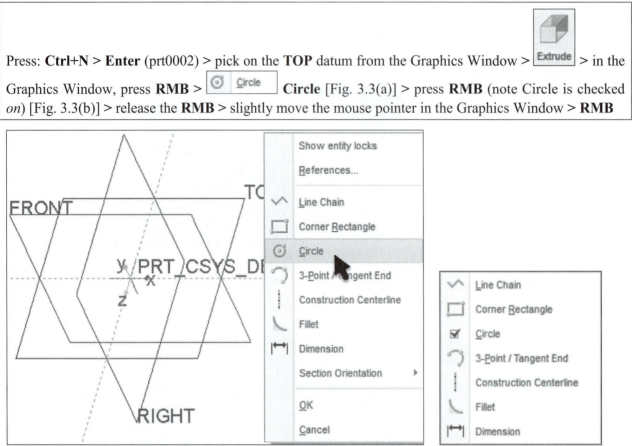

Figure 3.3(a) RMB > Circle

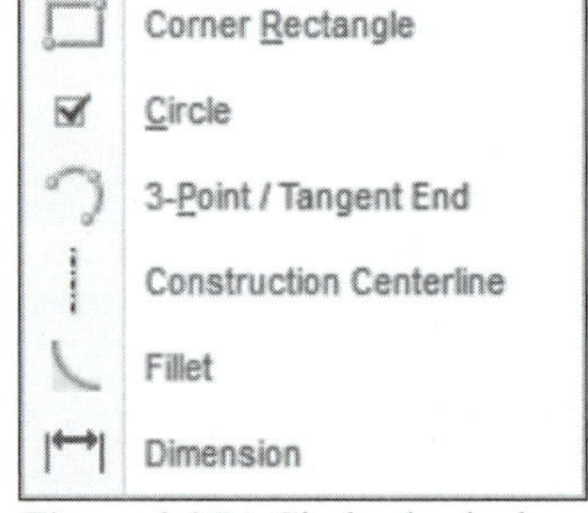

Figure 3.3(b) Circle checked on

Sketch a circle by picking at the origin [Fig. 3.3(c)] and then stretching the circle's diameter to a convenient size [Fig. 3.3(d)] > pick again to establish the circle's diameter > slightly move the mouse pointer in the Graphics Window > **MMB** to deselect the Circle command [Fig. 3.3(e)]

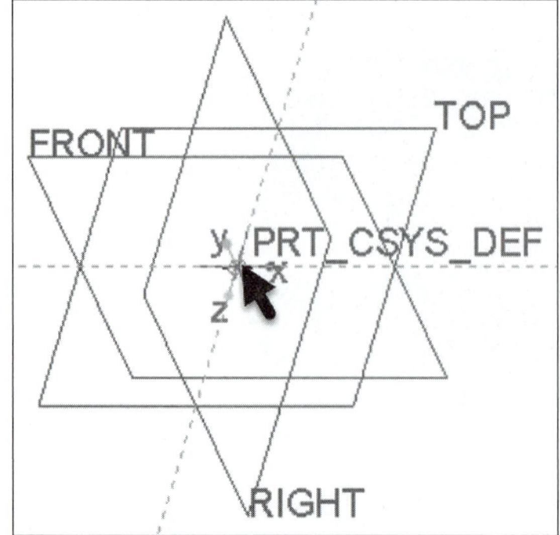

Figure 3.3(c) Pick the Center of the Circle

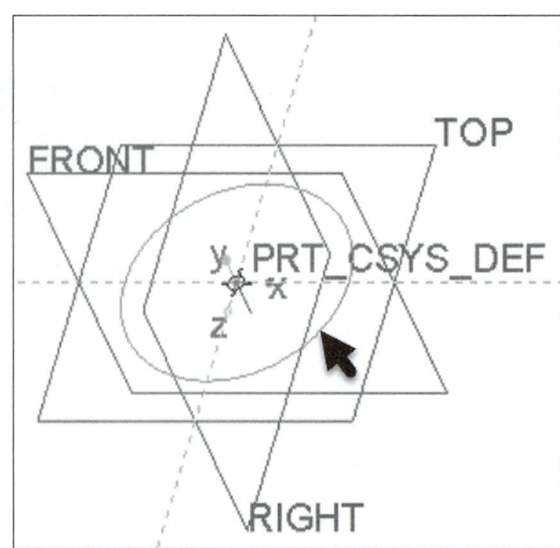

Figure 3.3(d) Pick to Establish the Circle's Diameter

Click: [✓] [Fig. 3.3(f)] > **Options** tab > **Add Taper** checked *on* > click in the taper value field > *type* **3** > **Enter** [Fig. 3.3(g)] > [Verify] **Verify** *on* > [toolbar] **Verify** *off* > **Options** tab > **Add Taper** checked *off* > [✓] > [Refit] > in the Graphics Window, **LMB** to deselect

(Note: There are angle limitations to adding a taper versus creating a draft.)

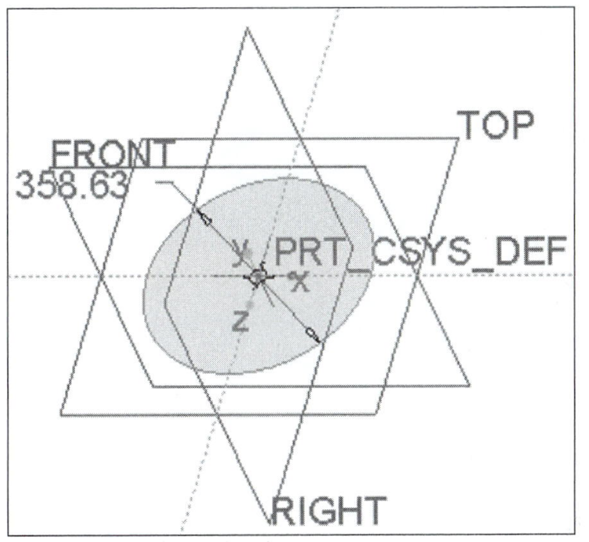

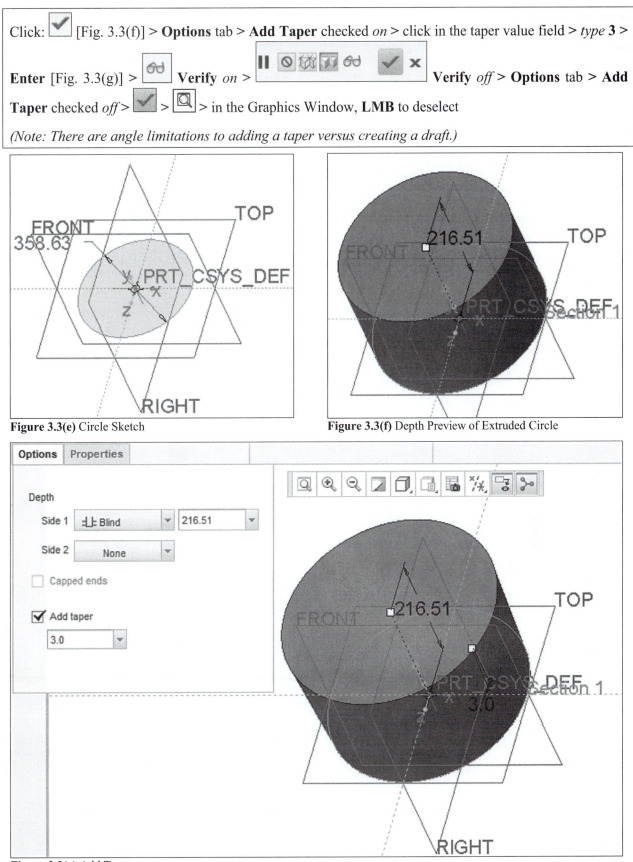

Figure 3.3(e) Circle Sketch

Figure 3.3(f) Depth Preview of Extruded Circle

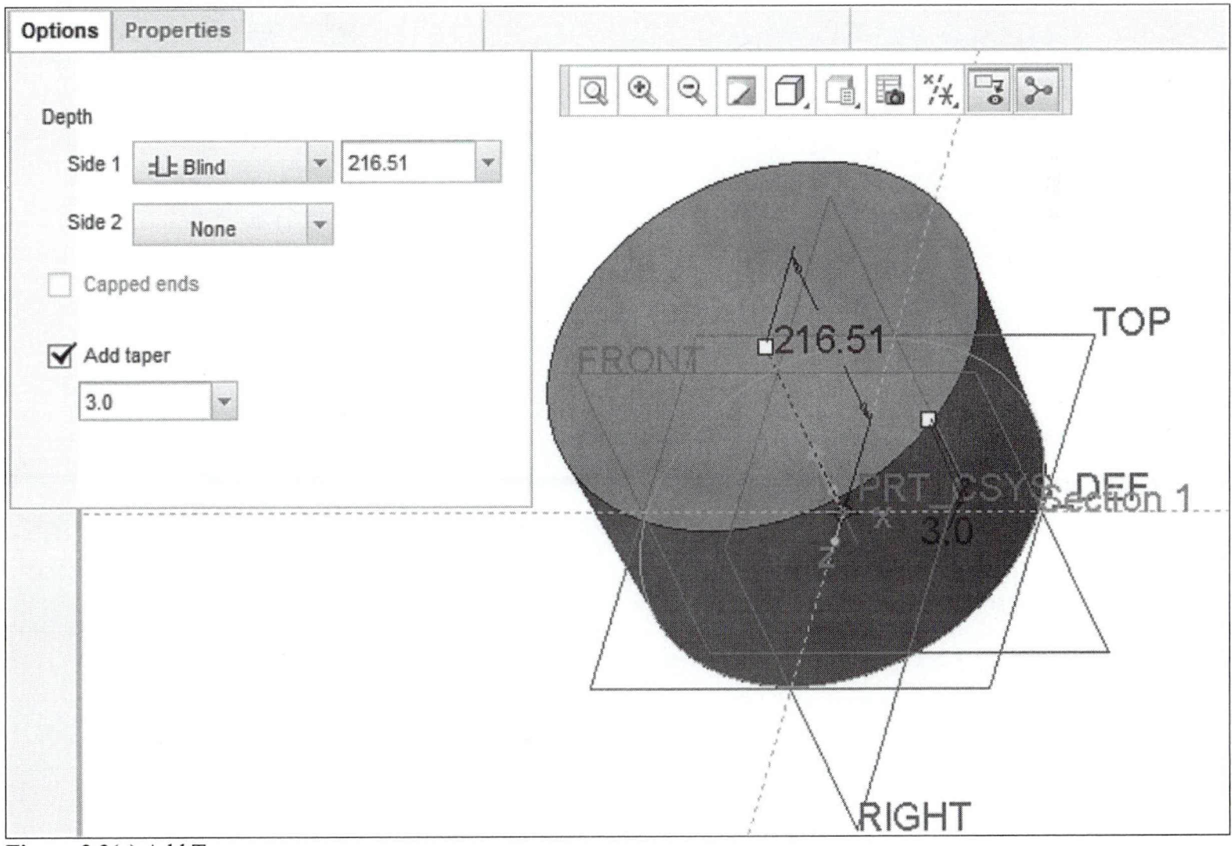

Figure 3.3(g) Add Taper

Click: Draft **Draft** > pick the front surface of the cylinder [Fig. 3.3(h)] > press **RMB** > **Draft Hinges** > pick the **TOP** datum as the Draft Hinge > double-click on the **1.0** degree dimension in the Graphics Window or click in the Angle value field > *type:* **10** [Fig. 3.3(i)] > **Enter** > > **Ctrl+S** > **OK**

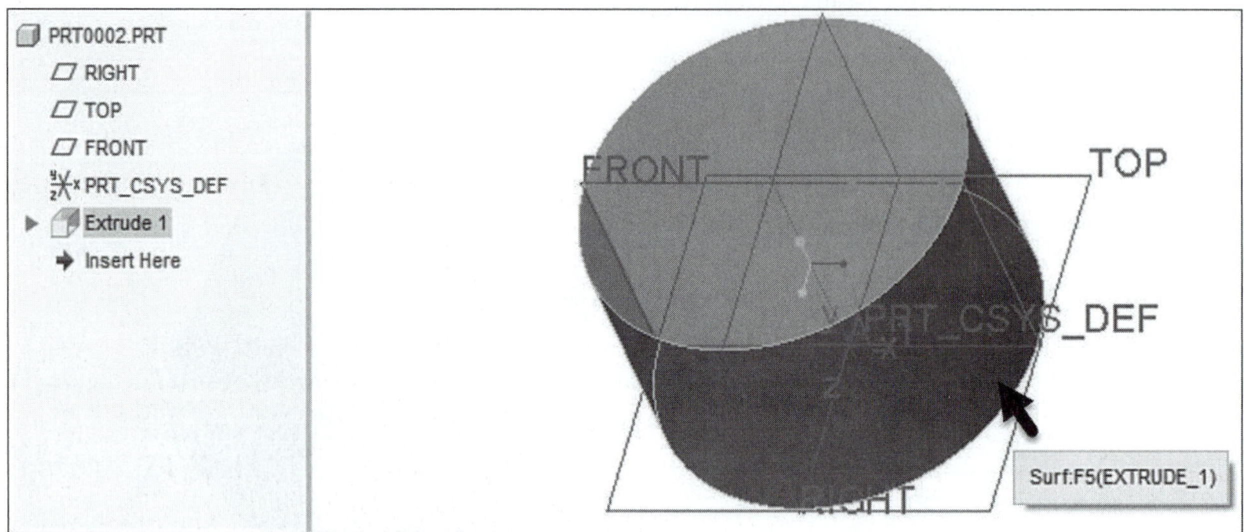

Figure 3.3(h) Vertical Surface of the Cylindrical Part Selected as the Draft Surface

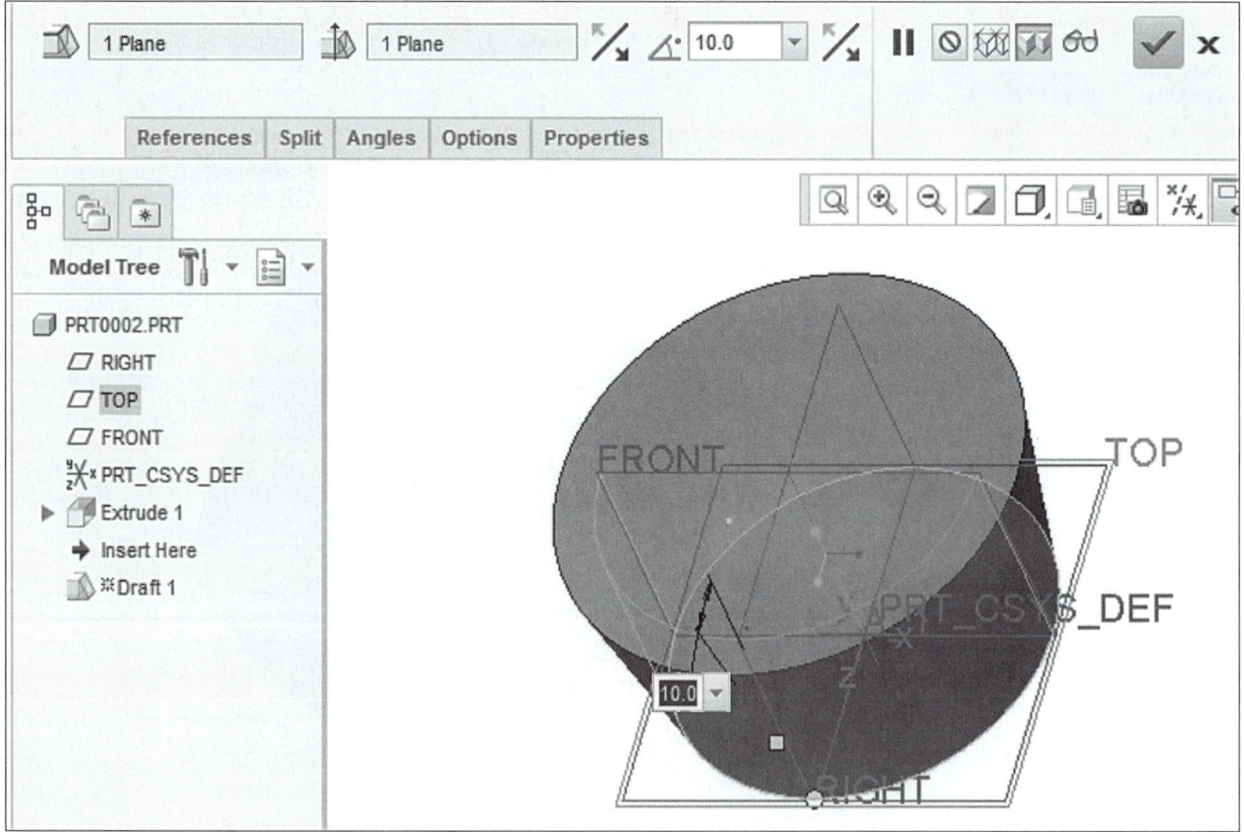

Figure 3.3(i) Draft Angle **10** Degrees

With the Draft feature *still selected*, pick on the top edge [Fig. 3.3(j)] > Chamfer **Chamfer** > click in the **D** value field > *type* **50** > **Enter** [Fig. 3.3(k)] > ✓ > **Ctrl+S**

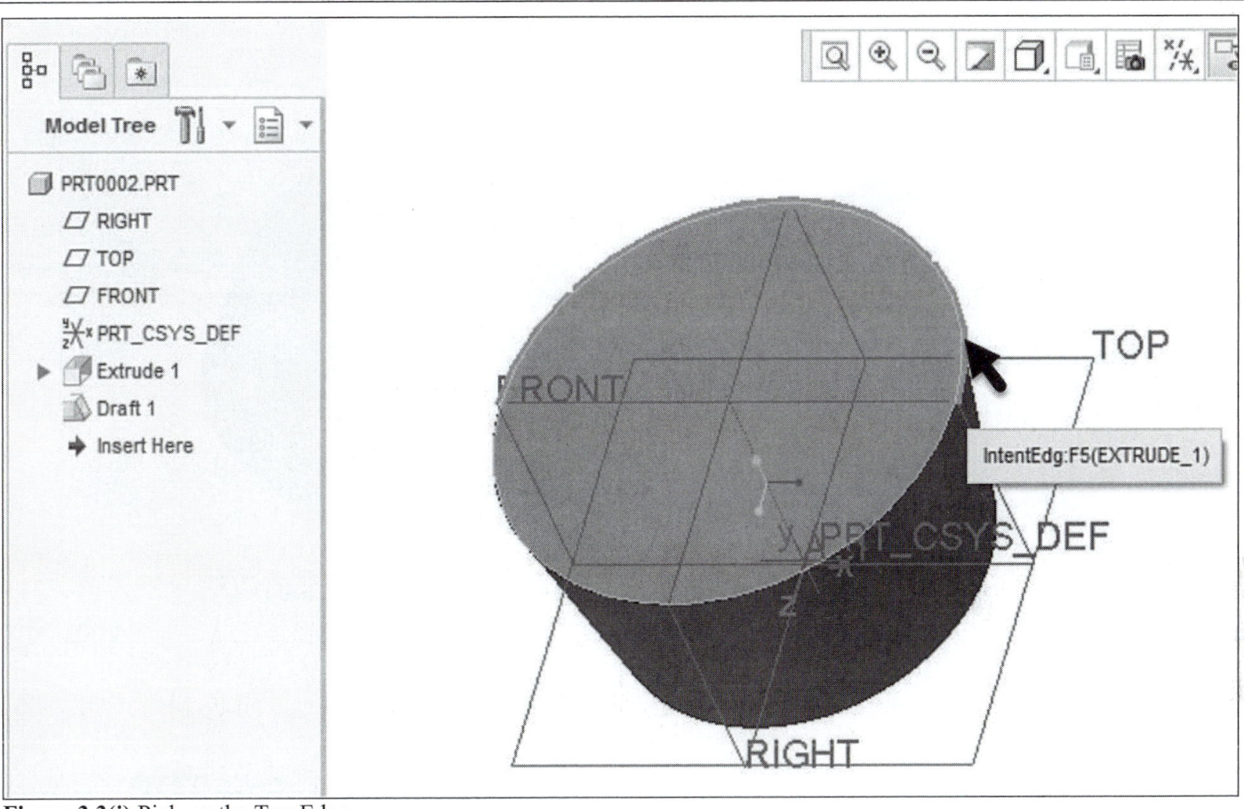

Figure 3.3(j) Pick on the Top Edge

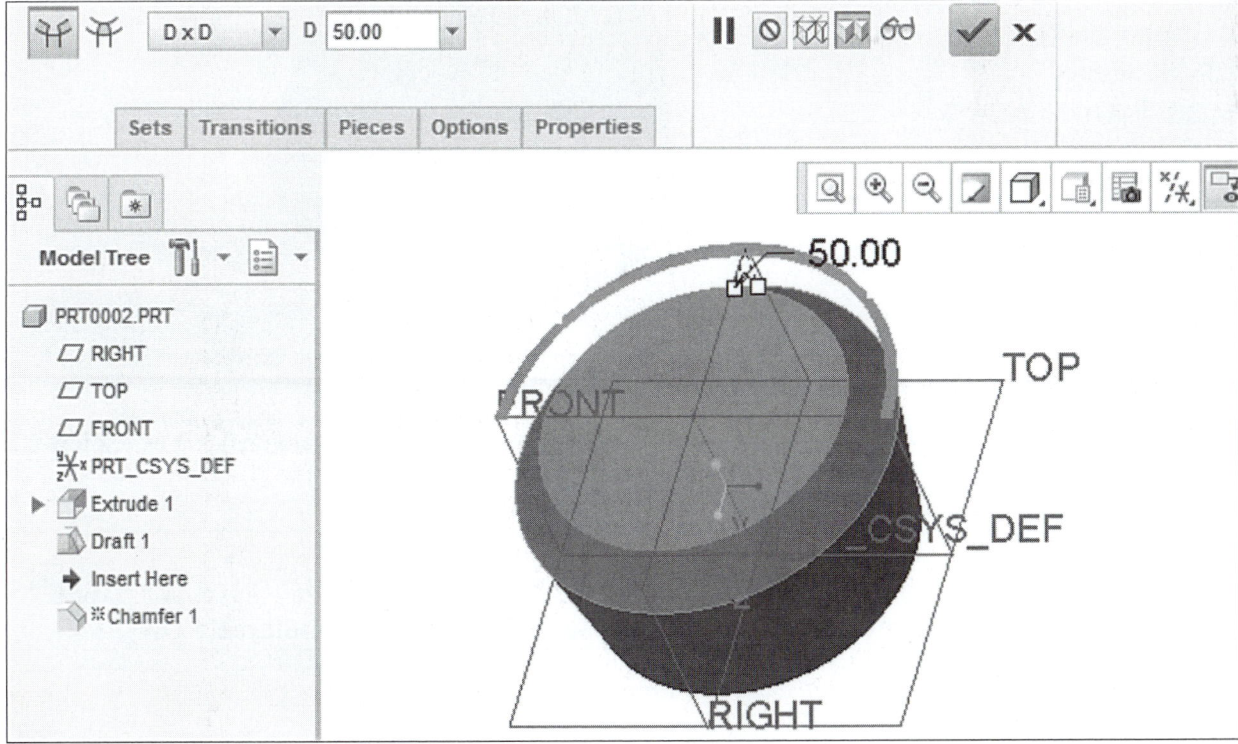

Figure 3.3(k) Chamfer Preview

Press and hold the **MMB** and spin the part to display the bottom surface [Fig. 3.3(l)] > release the **MMB** > pick on the bottom surface > slightly move the mouse pointer and pick the same surface again > Shell **Shell** [Fig. 3.3(m)] > ✓ [Fig. 3.3(n)] > **Ctrl+D** > in the Graphics Window, **LMB** to deselect > **Ctrl+S**

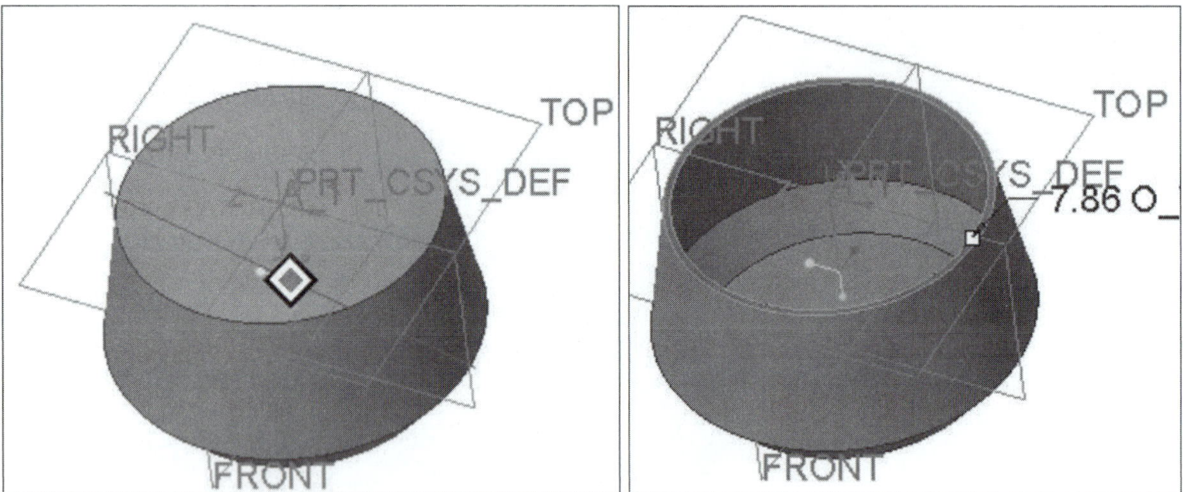

Figure 3.3(l) Spin the Model

Figure 3.3(m) Previewed Shell

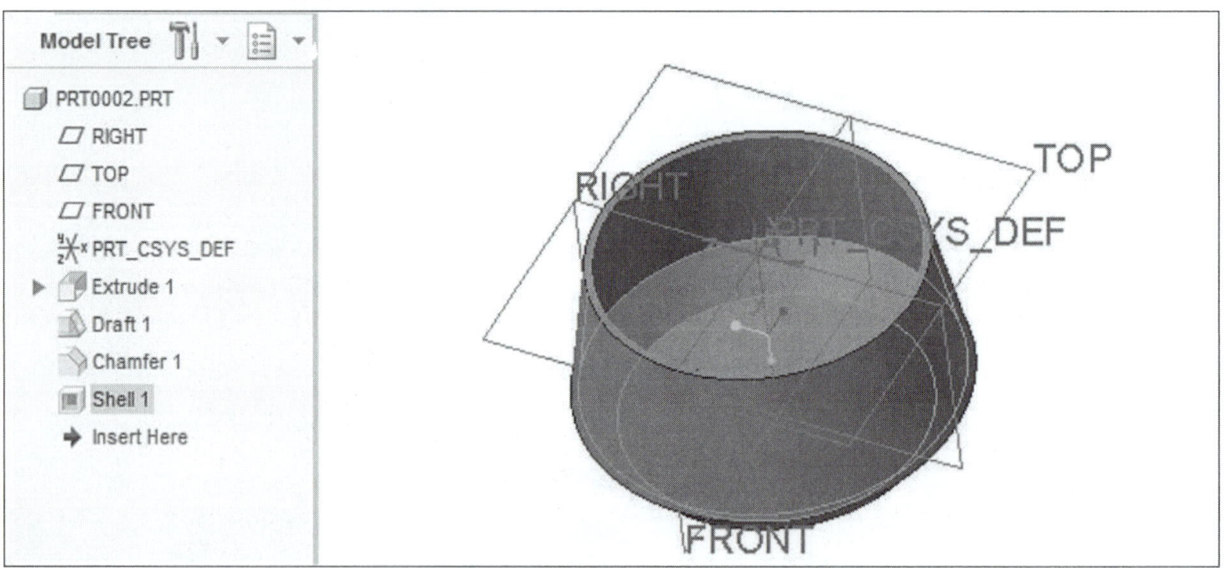

Figure 3.3(n) Completed Shell

Click: **File** > **Manage File** > **Delete Old Versions** (removes previously saved versions) > **Yes** > **View** tab > Display Style > Shading With Reflections > Display Style > Shading With Edges > **File** > **Save As** > Type > **Zip File (*.zip)** > **OK** > **File** > **Close** > **File** > **Manage Session** > **Erase Not Displayed** > **OK**

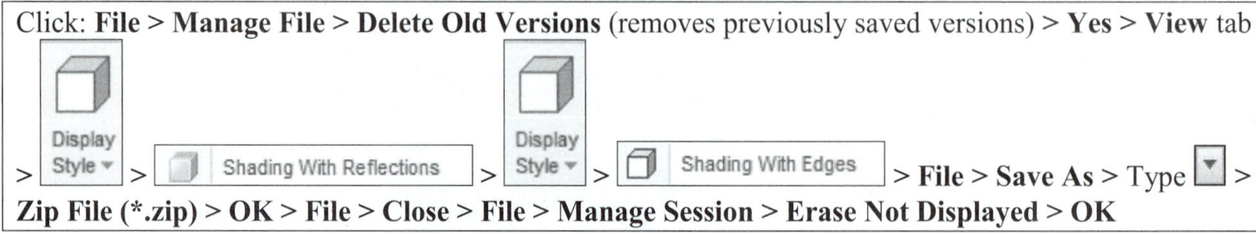

Part Model Three (PRT0003.PRT) (Hole)

Press: **Ctrl+N** (prt0003) > **OK** > preselect the **FRONT** datum > **RMB** in the Ribbon > **Customize the Ribbon** > move the Creo Parametric Options dialog down if needed > **RMB** on the Extrude button in the Shapes group of the Model tab Ribbon > **Small Button** > **OK** (in the Creo Parametric Options dialog box) > **Extrude** from the Ribbon > in the Graphics Window, press **RMB** > **Corner Rectangle** > sketch a rectangle by defining two corners > **MMB** > **LMB** to deselect [Fig. 3.4(a)] > press **RMB** > **Fillet** > pick the two lines that form the upper right-hand corner [Fig. 3.4(b)] > [Fig. 3.4(c)] > move the drag handle forward to increase the model's depth [Fig. 3.4(d)] > > **Ctrl+S** > **Enter**

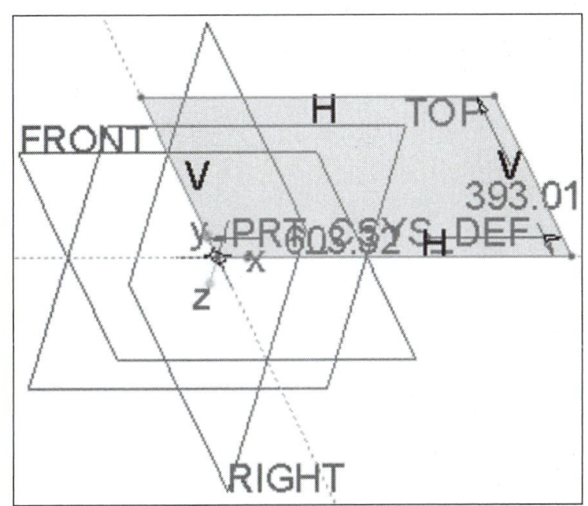

Figure 3.4(a) Sketch a Rectangle

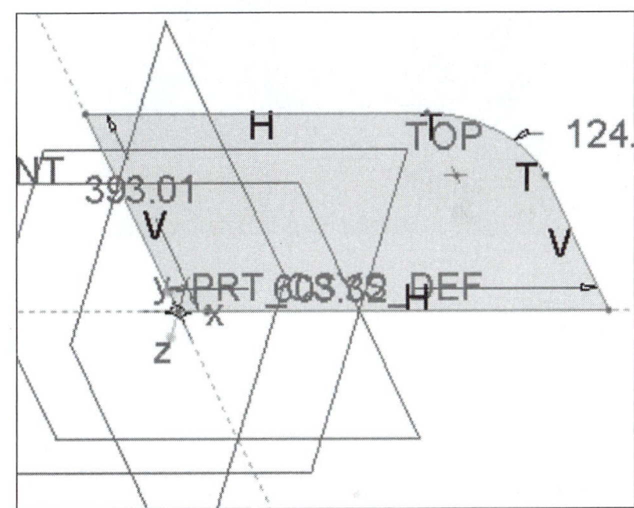

Figure 3.4(b) Create a Fillet

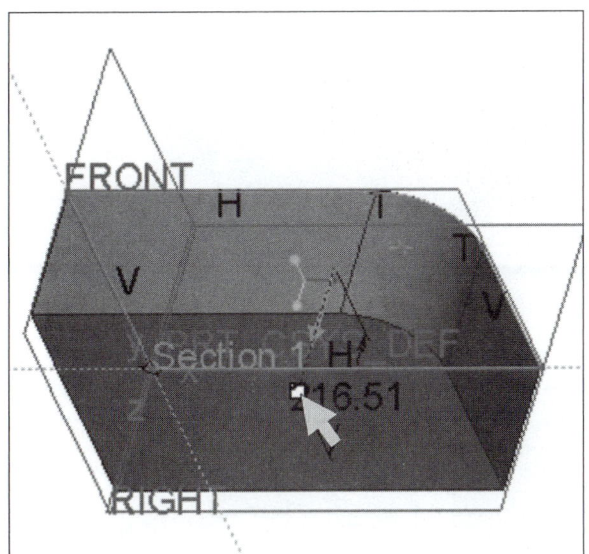

Figure 3.4(c) Depth Preview

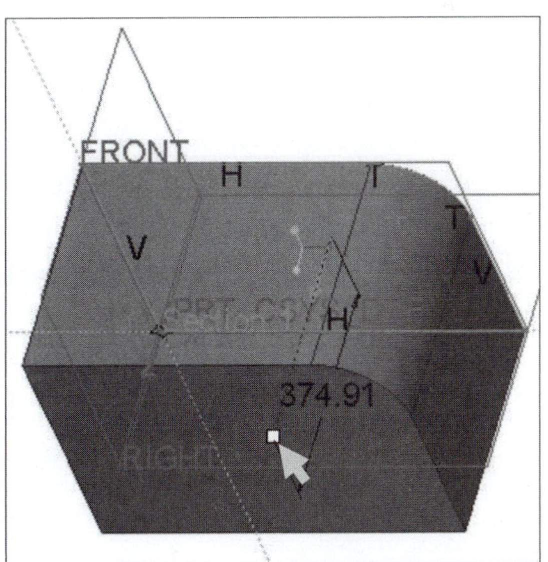

Figure 3.4(d) Move the Drag Handle to change the Depth

Click: **View** tab > Appearance Gallery > ptc-ceramic > pick on the part name (**PRT0003.PRT**) in the Model Tree > in the Graphics Window, click: **MMB** > **Model** tab > Axis **Axis** > pick on the cylindrical surface [Fig. 3.4(e)] > **OK** [Fig. 3.4(f)]

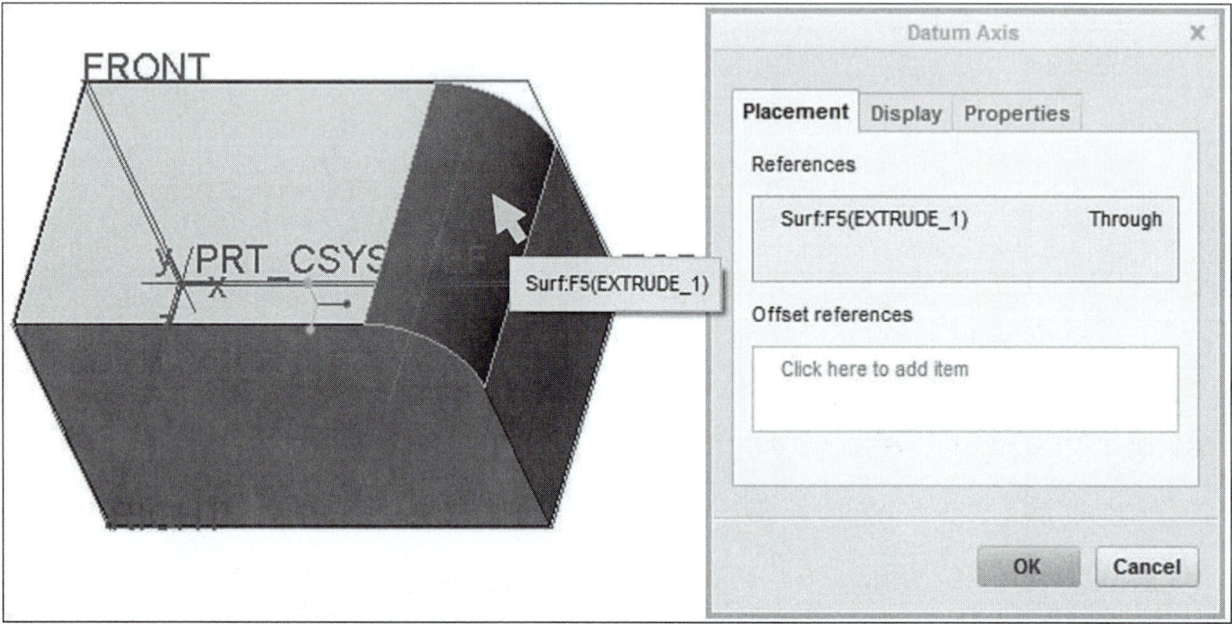

Figure 3.4(e) Pick on the Cylindrical Surface to Establish the References for a Datum Axis

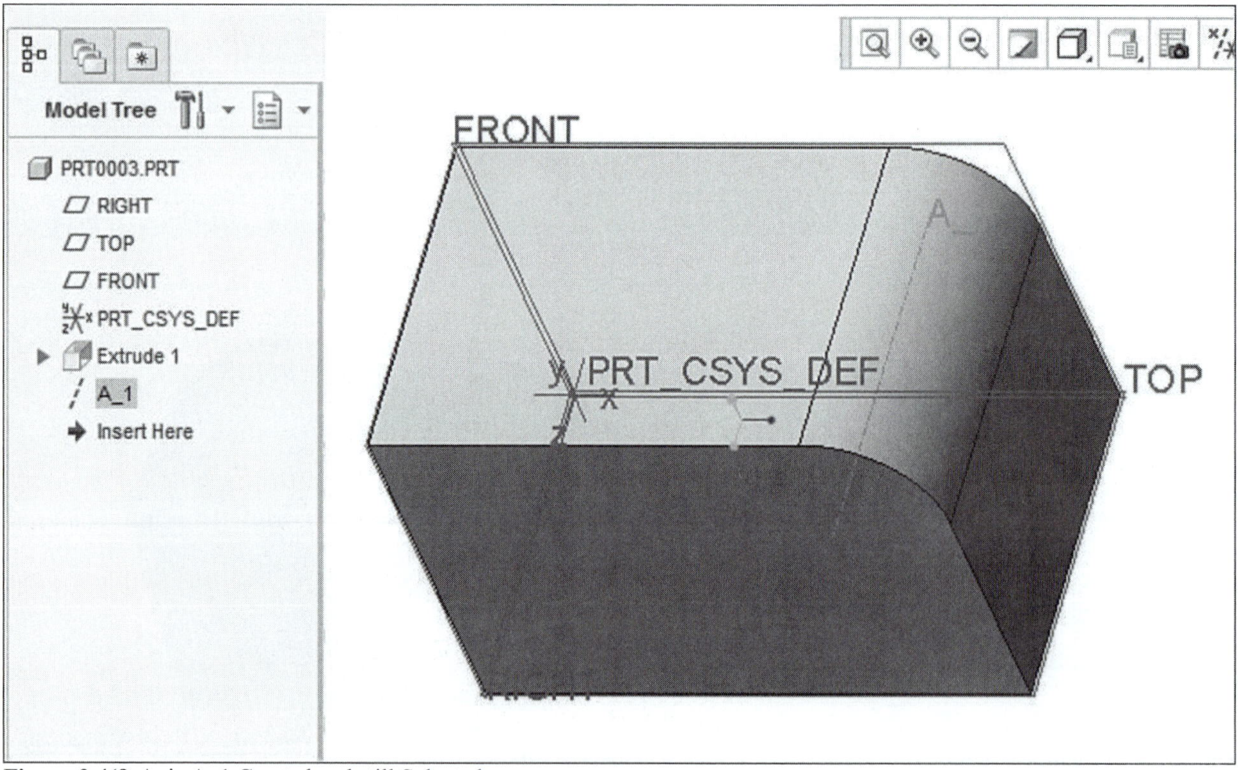

Figure 3.4(f) Axis A_1 Created and still Selected

With the datum axis *still highlighted/selected*, click: **Hole** [Fig. 3.4(g)] > expand depth options by opening slide-up panel > **Drill to intersect with all surfaces** > Placement tab > while holding down the **Ctrl** key, pick on the front ***surface*** *(not the FRONT Datum Plane)* [Fig. 3.4(h)] > release the **Ctrl** key

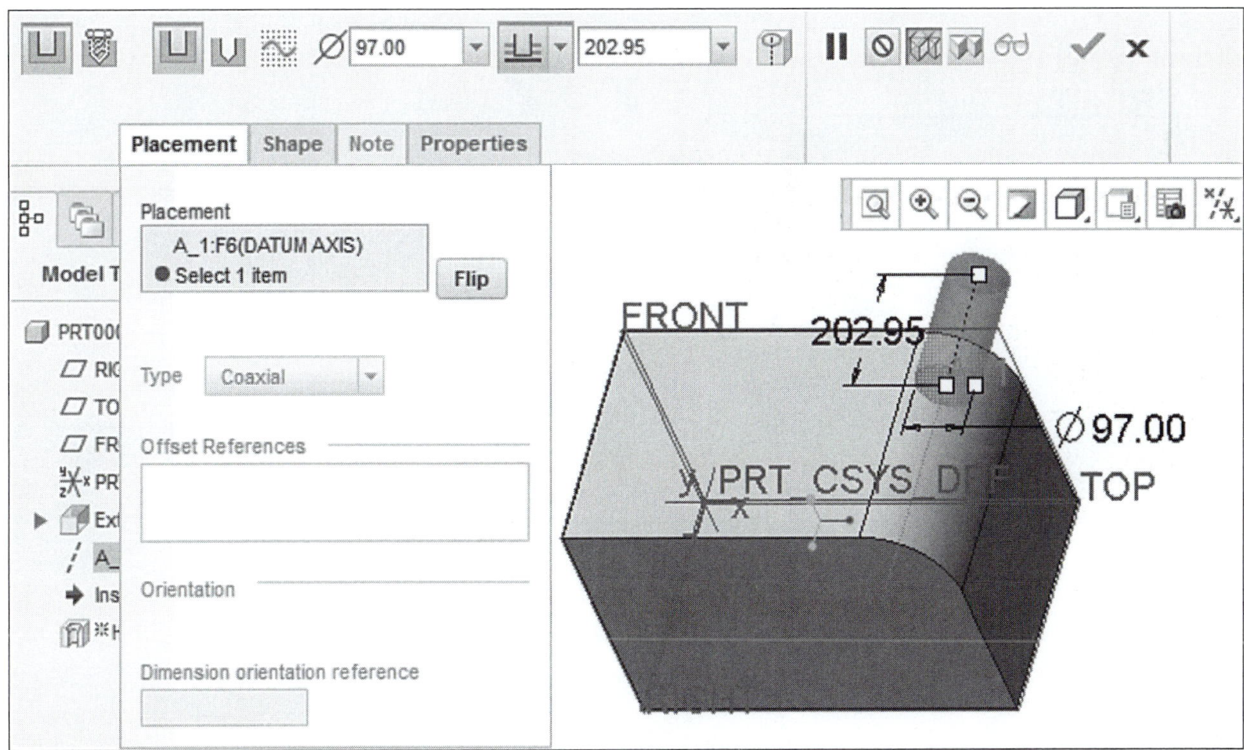

Figure 3.4(g) Hole Tool Display Preview

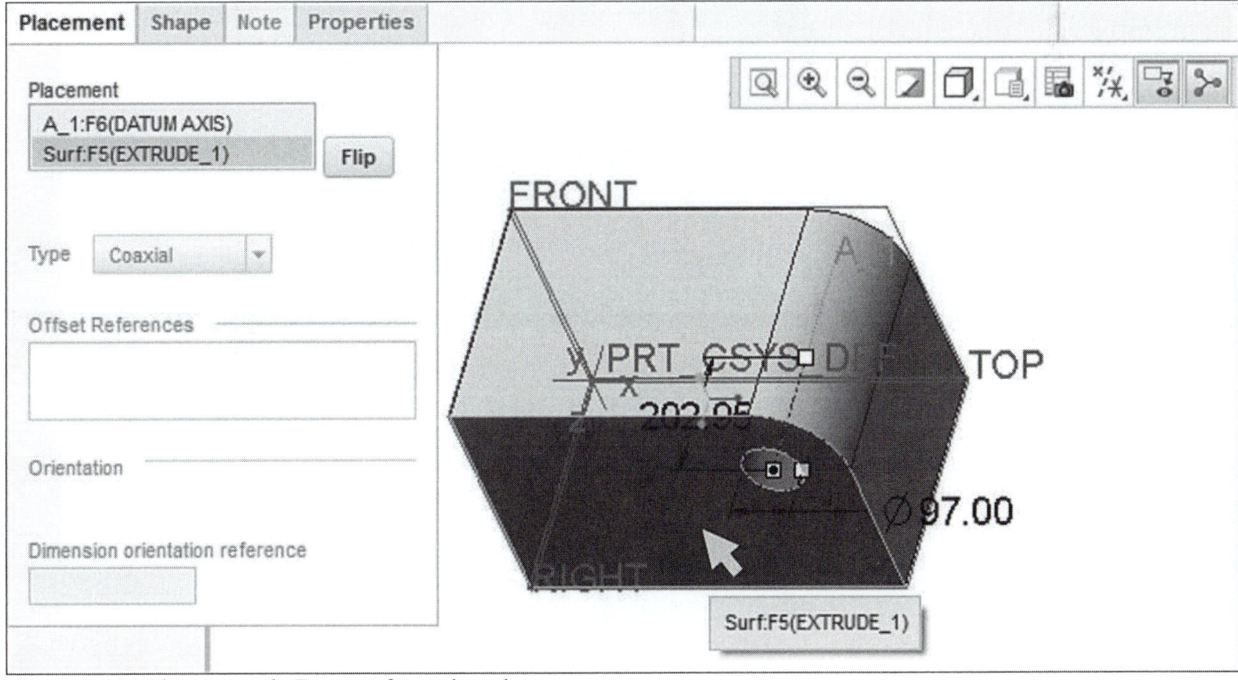

Figure 3.4(h) Placement tab, Front surface selected

Click: ✓ > with the hole *still selected*; in the Graphics Window, press: **RMB** > **Hole Representation** > **Convert to Lightweight** [Figs. 3.4(i-j)] > **Undo** > **View** tab > **Display Style** > **Shading with Reflections** [Fig. 3.4(k)] > **Display Style** > **Shading with Edges** > **Ctrl+S** > **File** > **Manage File** > **Delete Old Versions** > **Yes** > **File** > **Save As** > Type ▾ > **Zip File (*.zip)** > **OK** > **File** > **Close**

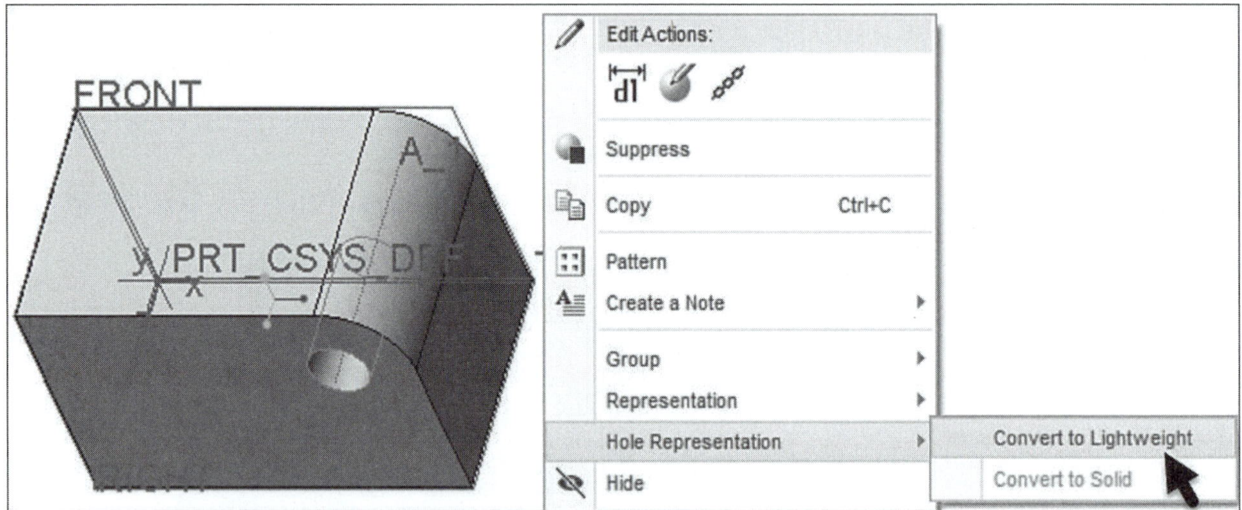

Figure 3.4(i) Convert the Hole Display to Lightweight *(your Hole number may be different)*

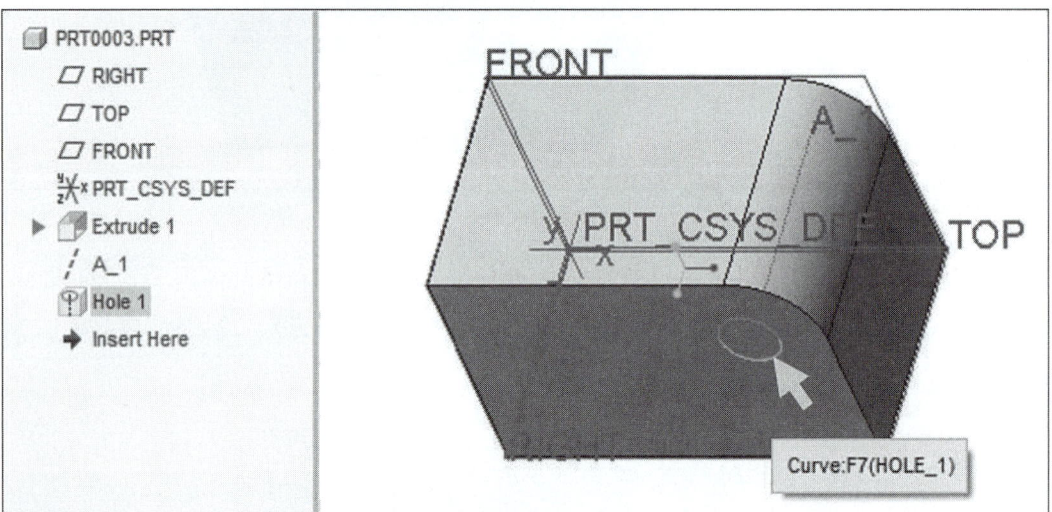

Figure 3.4(j) Lightweight Display of Hole. Note the change in the Model Tree Representation of the Hole Symbol

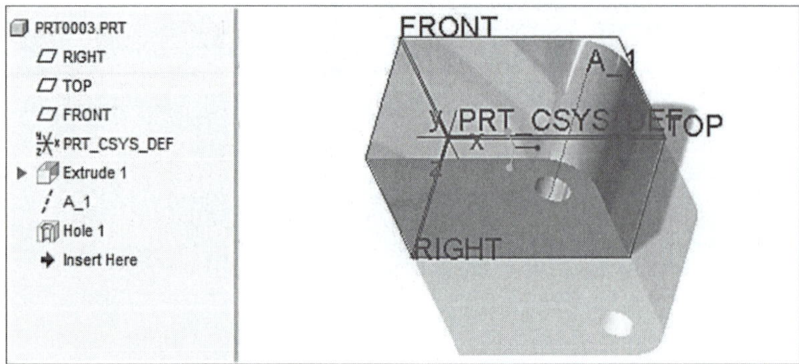

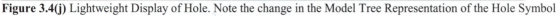

Figure 3.4(k) Shading with Reflections *(the quality of your graphics card and graphics settings may prevent this display)*

Part Model Four (PRT0004.PRT) (Cut)

Press: **Ctrl+N** (prt0004) > **Enter** > **View** tab > **Display Style** > **Shading with Edges** > **Model** tab > Extrude **Extrude** > Placement from the Dashboard > Define... > select the **FRONT** datum from the Model Tree > **Sketch** > **Sketch View** from the Graphics Toolbar > **Line Chain** > starting at the origin, sketch the outline using five lines forming a *closed section* [Fig. 3.5(a)] > **MMB** to end the current line > **MMB** to end the current tool and to display weak dimensions *[section will be shaded if done correctly]* > **Ctrl+D** > [Fig. 3.5(b)] > [Fig. 3.5(c)] > **Ctrl+S** > **OK** > in the Graphics Window, **LMB** to deselect

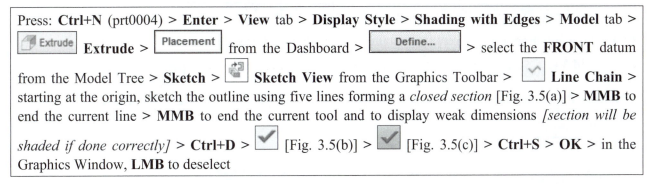

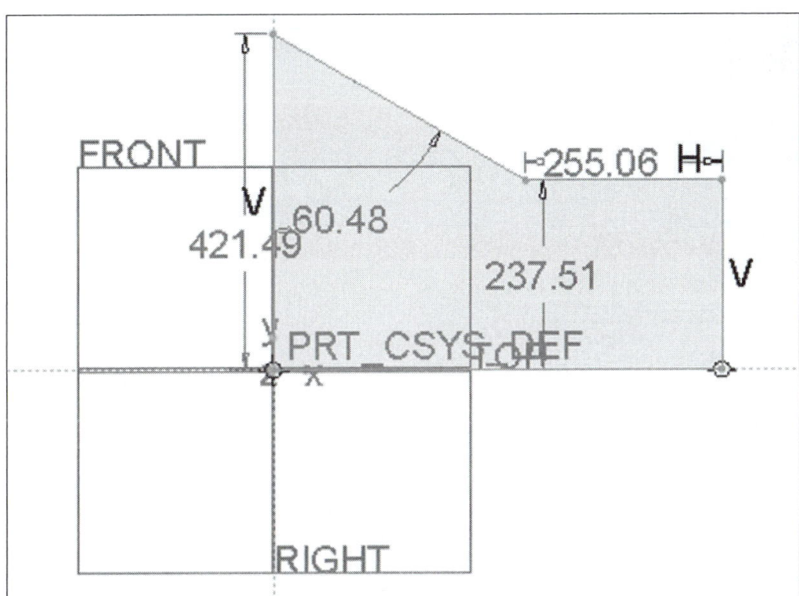

Figure 3.5(a) Sketch the Five Lines of the Enclosed Section *(your weak dimensions will be different)*

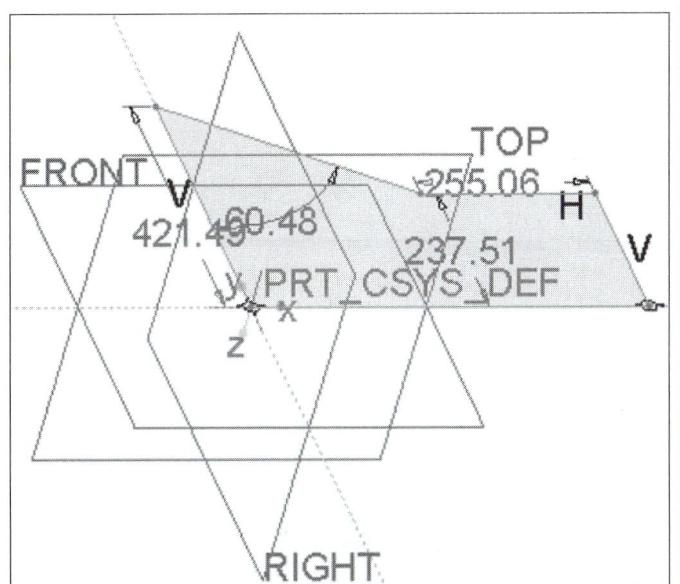

Figure 3.5(b) Sketch Preview in the Standard Orientation View

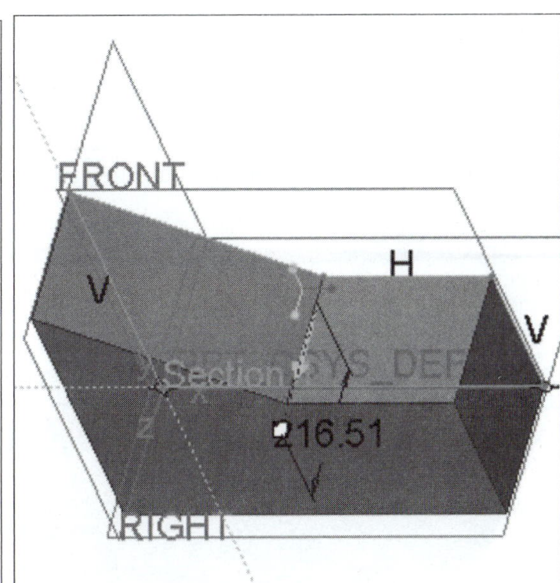

Figure 3.5(c) Depth Preview

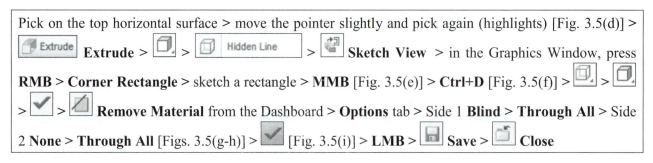

Pick on the top horizontal surface > move the pointer slightly and pick again (highlights) [Fig. 3.5(d)] > Extrude **Extrude** > > Hidden Line > **Sketch View** > in the Graphics Window, press **RMB** > **Corner Rectangle** > sketch a rectangle > **MMB** [Fig. 3.5(e)] > **Ctrl+D** [Fig. 3.5(f)] > > > > **Remove Material** from the Dashboard > **Options** tab > Side 1 **Blind** > **Through All** > Side 2 **None** > **Through All** [Figs. 3.5(g-h)] > [Fig. 3.5(i)] > **LMB** > **Save** > **Close**

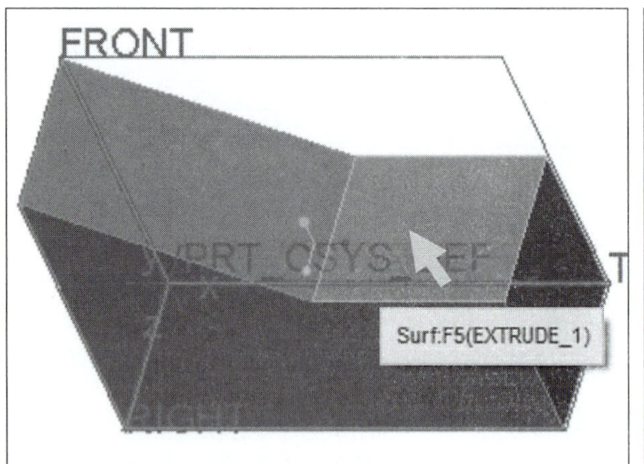

Figure 3.5(d) Select the Horizontal Surface

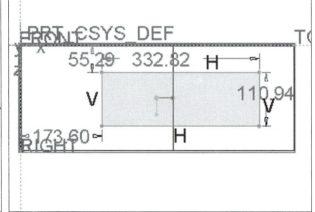

Figure 3.5(e) Sketch a Rectangle

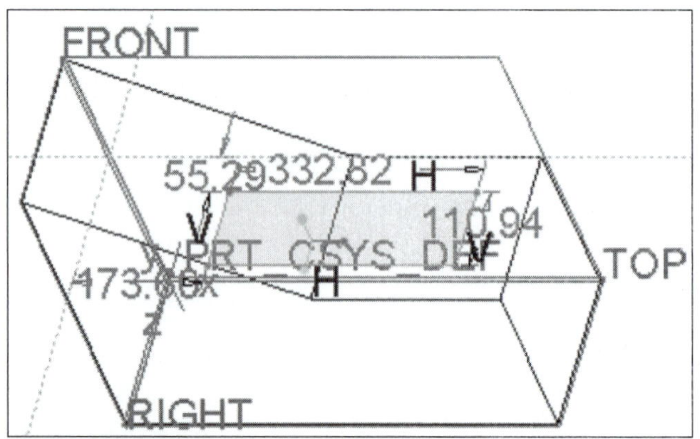

Figure 3.5(f) Sketch Displayed in Standard Orientation

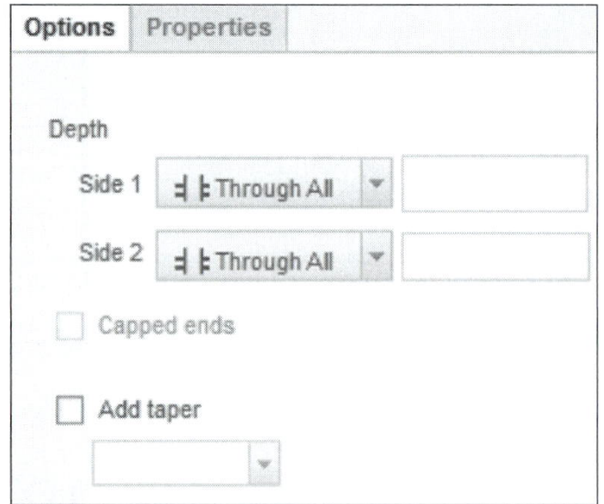

Figure 3.5(g) Through All

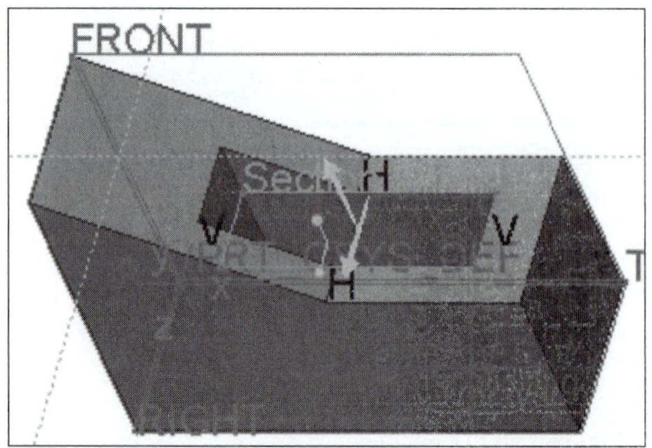

Figure 3.5(h) Preview of Cut

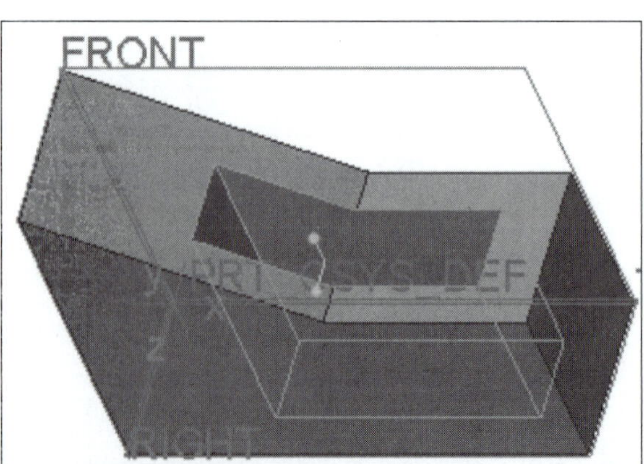

Figure 3.5(i) Completed Part

Part Model Five (PRT0005.PRT) (Mirror)

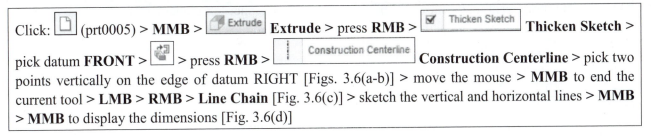

Click: (prt0005) > **MMB** > **Extrude** > press **RMB** > **Thicken Sketch** > pick datum **FRONT** > > press **RMB** > **Construction Centerline** > pick two points vertically on the edge of datum RIGHT [Figs. 3.6(a-b)] > move the mouse > **MMB** to end the current tool > **LMB** > **RMB** > **Line Chain** [Fig. 3.6(c)] > sketch the vertical and horizontal lines > **MMB** > **MMB** to display the dimensions [Fig. 3.6(d)]

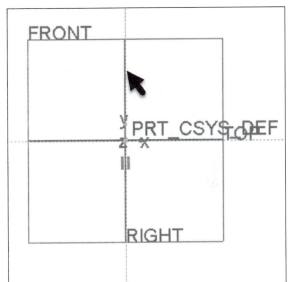

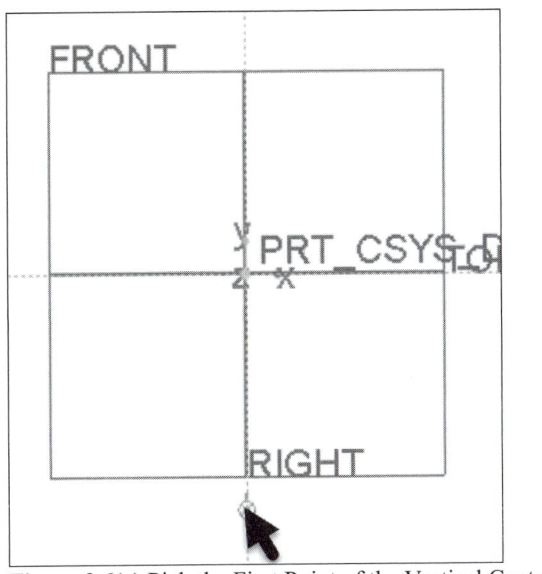

Figure 3.6(a) Pick the First Point of the Vertical Centerline

Figure 3.6(b) Pick the Second Point of the Vertical Centerline

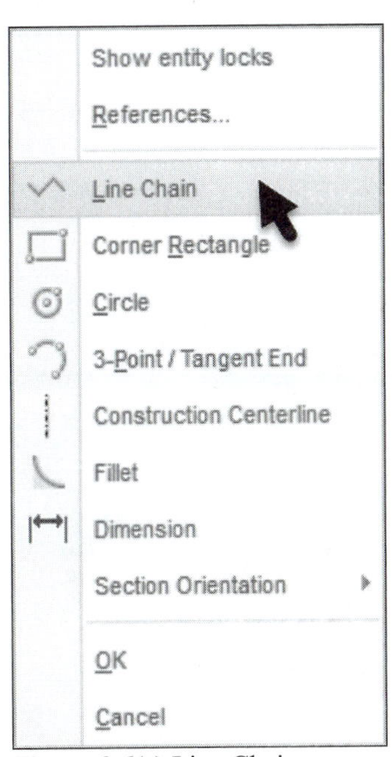

Figure 3.6(c) Line Chain

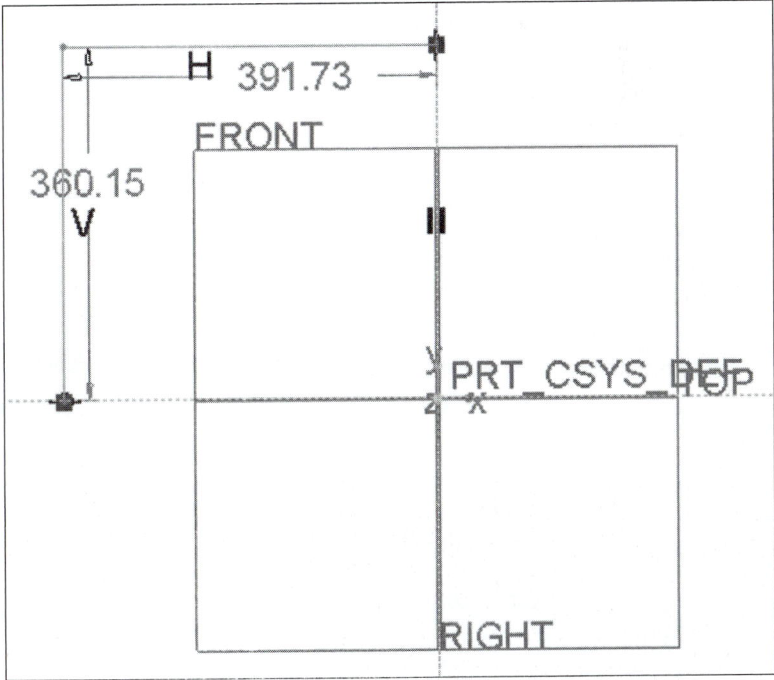

Figure 3.6(d) Sketch Two Lines

Click: **LMB** to deselect > press **RMB** > Fillet **Fillet** [Fig. 3.6(e)] > pick the lines near the corner > **MMB** to end the current tool > move the mouse pointer to the upper left corner of the Graphics Window > press **LMB** and hold while dragging a window until it incorporates the two lines and fillet [Fig. 3.6(f)] > release the **LMB** > **Mirror the selected entities** > pick the Centerline [Fig. 3.6(g)] > **Ctrl+D** > [Fig. 3.6(h)] > **Refit**

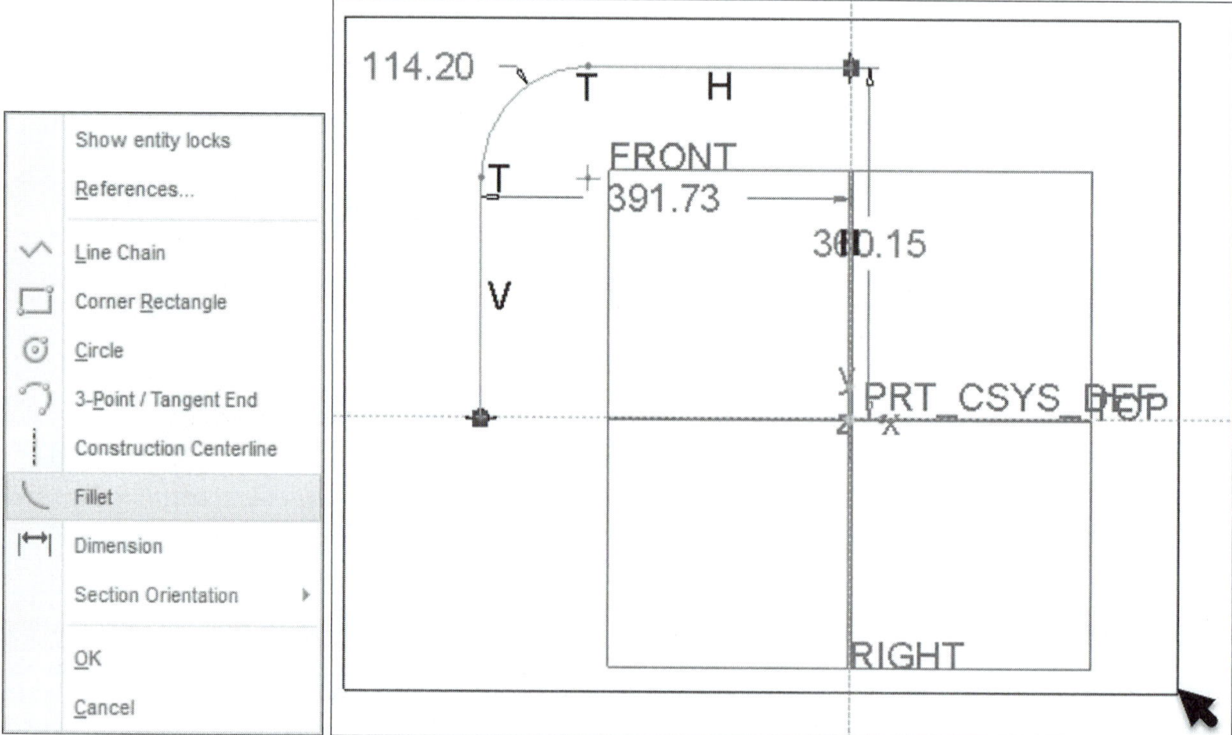

Figure 3.6(e) RMB > Fillet

Figure 3.6(f) Select Sketch Entities by Windowing

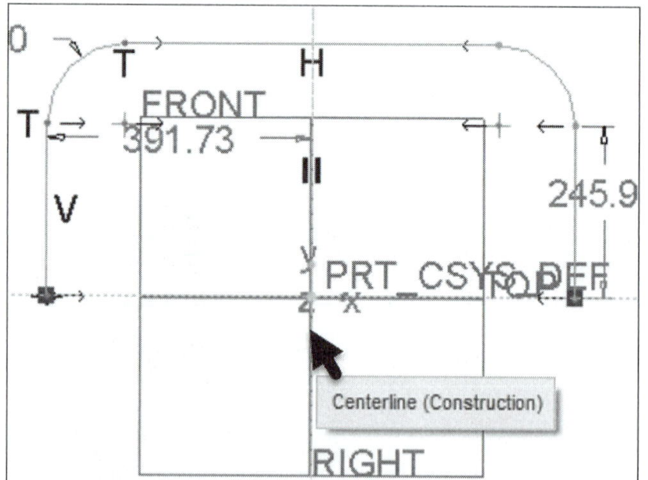

Figure 3.6(g) Mirrored Sketch

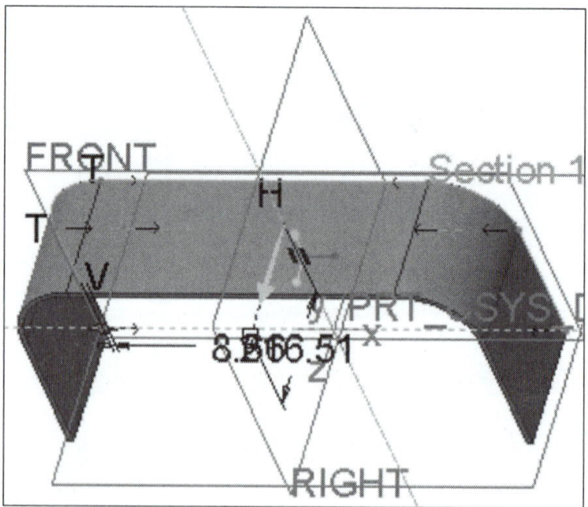

Figure 3.6(h) Depth Preview

Press and hold the **MMB** and rotate the model to see the drag **handle** > Place the mouse pointer on the drag handle > press **RMB** > **Symmetric** [Fig. 3.6(i)] > move a drag handle [Fig. 3.6(j)] > **Ctrl+D** > > **LMB** to deselect > press and hold the **Ctrl** key and select the **RIGHT** and **FRONT** datum planes in the Model Tree [Fig. 3.6(k)] > release the **Ctrl** key > **Axis** > Hole **Hole** > > > while holding down the **Ctrl** key, pick on the top face [Fig. 3.6(l)] > release the **Ctrl** key > **Enter** > Operations from left side of Ribbon below Regenerate > **Scale Model** > *type* **.01** > **Enter** > **Yes** > **Ctrl+D** > double-click on the model [Fig. 3.6(m)] > **Ctrl+S** > **OK** > **File** > **Close**

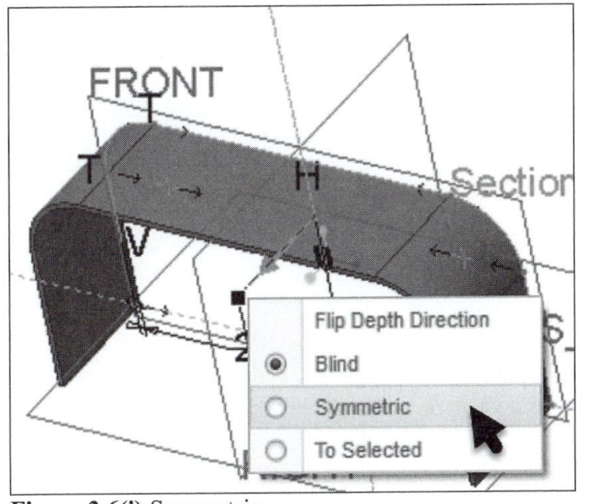

Figure 3.6(i) Symmetric

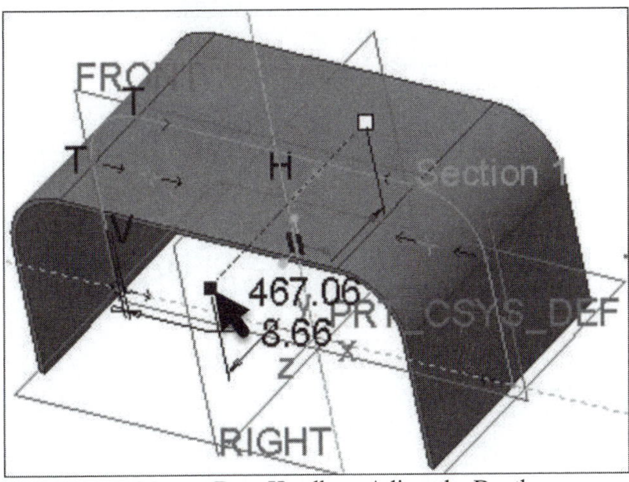

Figure 3.6(j) Move a Drag Handle to Adjust the Depth

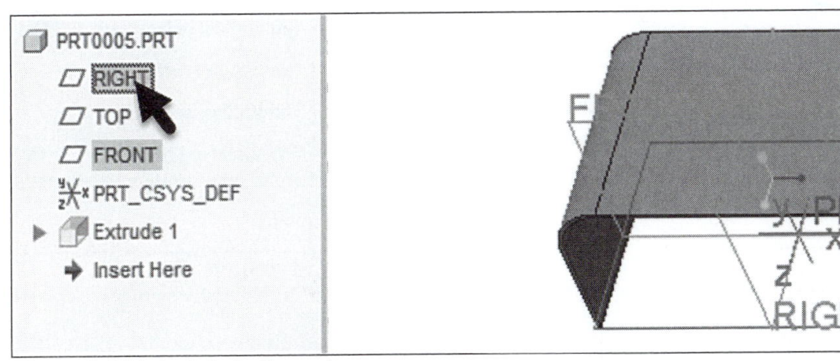

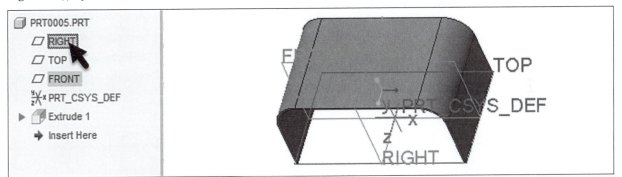

Figure 3.6(k) A_1 Axis Created

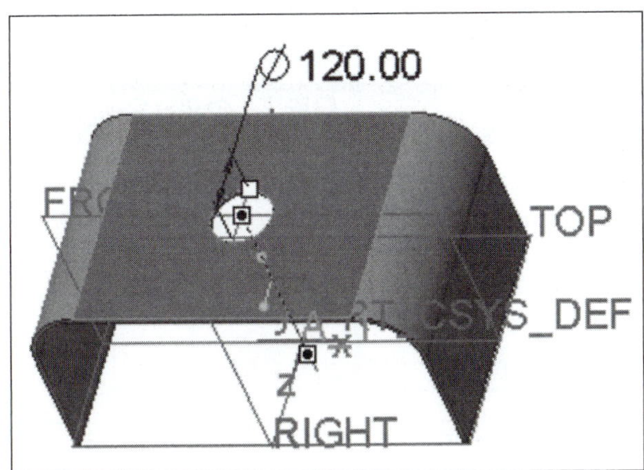

Figure 3.6(l) Hole Preview

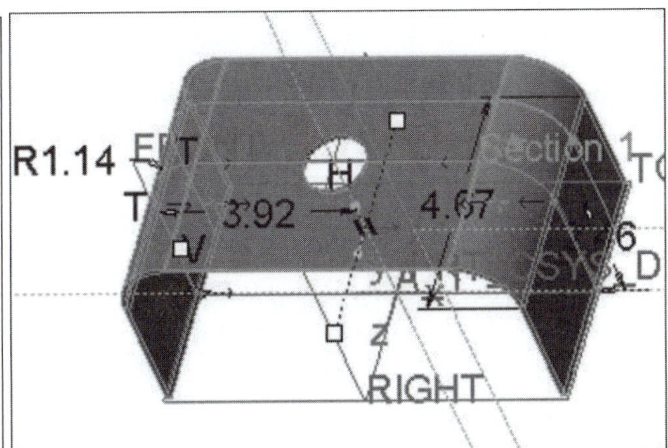

Figure 3.6(m) Scaled Model

Part Model Six (PRT0006.PRT) (Revolve)

Click: (prt0006) > **OK** > **View** tab > *on* > **Model** tab > pick datum **RIGHT** > Revolve > **Sketch View** > in the Graphics Window, press **RMB** > **Section Orientation** [Fig. 3.7(a)] > **Set horizontal reference** > pick on the Z-axis (edge of the TOP datum plane) [Figs. 3.7(b-c)]

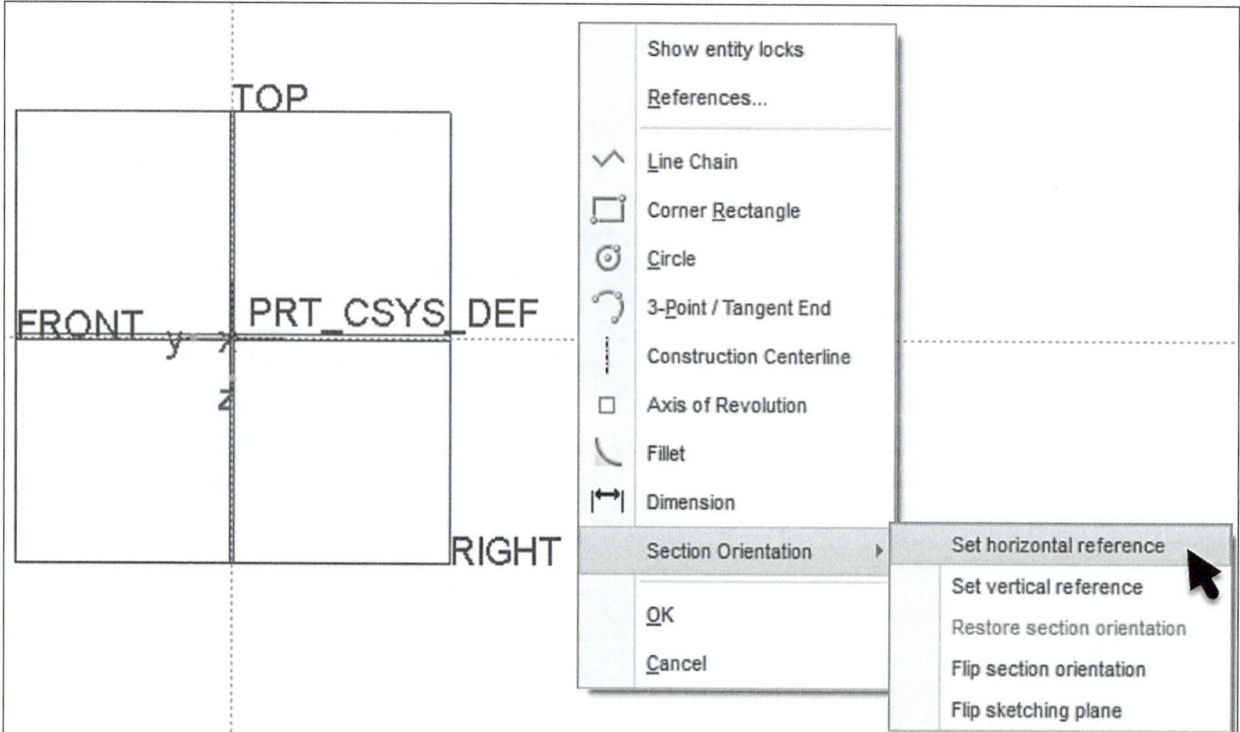

Figure 3.7(a) Section Orientation

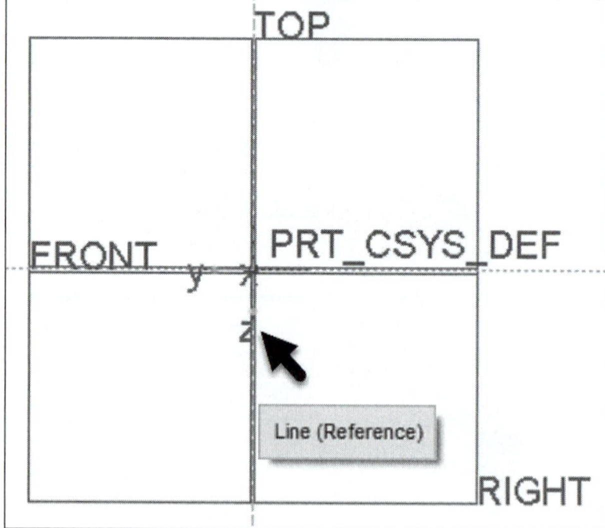

Figure 3.7(b) Select the Z-axis (Edge of Datum TOP)

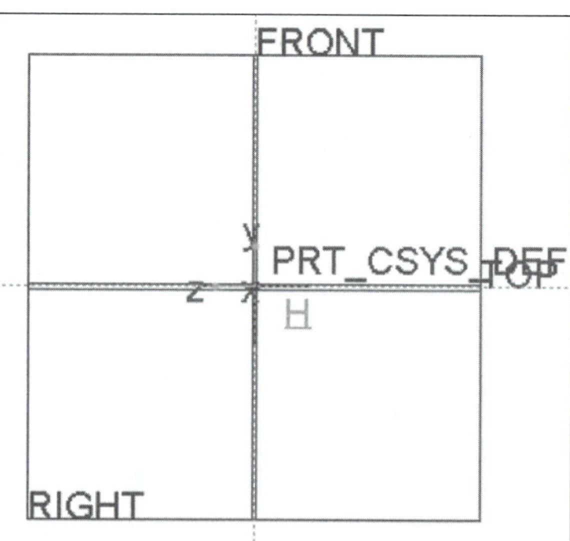

Figure 3.7(c) Horizontal Reference Changed

Press: **RMB** > **Axis of Revolution** > create horizontal axial centerline (on the edge of the TOP datum) [Fig. 3.7(d)] > press **RMB** > **Line Chain** > sketch the outline of the closed section *(Note that there are only sketched entities on one side of the Axis of Revolution)* > **MMB** to end the current line > **MMB** to end the current tool and to display the *weak* dimensions [Fig. 3.7(e)] > **Ctrl+D** > press **RMB** > **OK** [Fig. 3.7(f)]

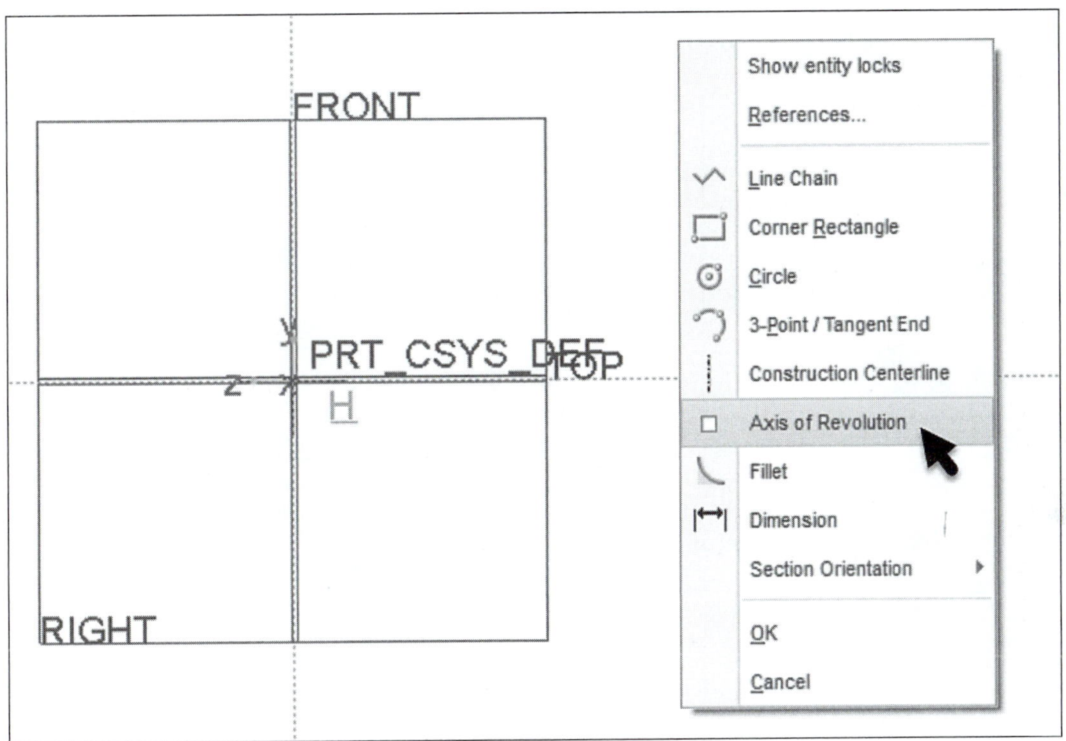

Figure 3.7(d) Designate the Axis of Revolution

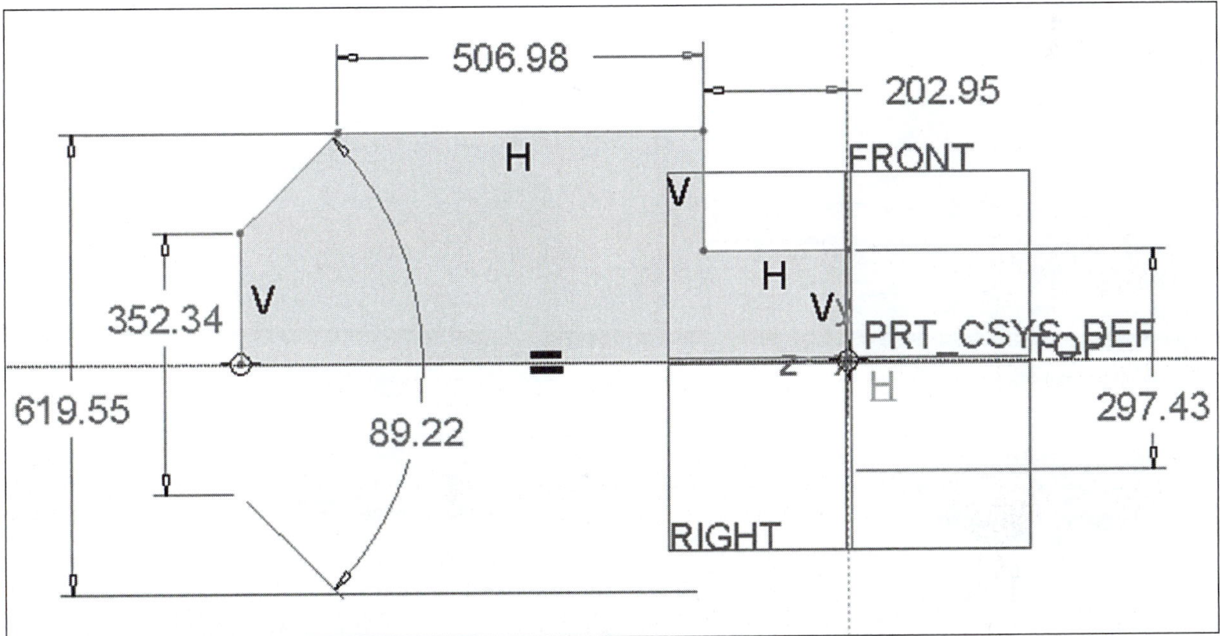

Figure 3.7(e) Completed Revolved Sketch- (your dimensions will be different) *(Note that there are only sketched entities on one side of the Axis of Revolution)*

Click: [Fig. 3.7(g)] > in the Graphics Window, **LMB** to deselect > slowly pick on the front edge of the part > slightly move the mouse > pick on the front edge again until it highlights > press and hold the **Ctrl** key > pick the other visible edges > release the **Ctrl** key > press **RMB** > **Round** [Fig. 3.7(h)] > Drag a handle to slightly adjust the radius value. The round surfaces will highlight if it is a valid size. (An invalid size will display only edges.) [Fig. 3.7(i)] > > in the Graphics Window, **LMB** to deselect

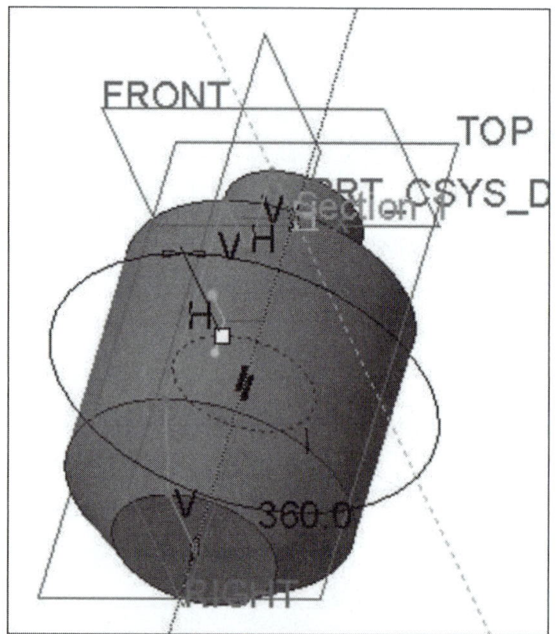

Figure 3.7(f) Revolve Preview

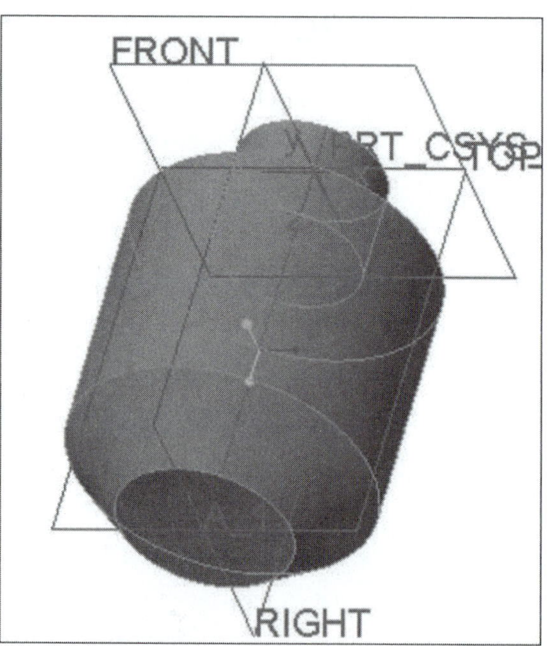

Figure 3.7(g) Completed Revolved Protrusion

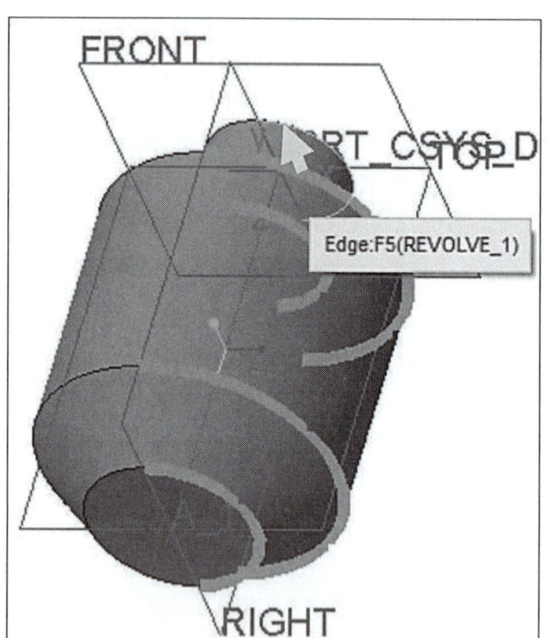

Figure 3.7(h) Select the Edges

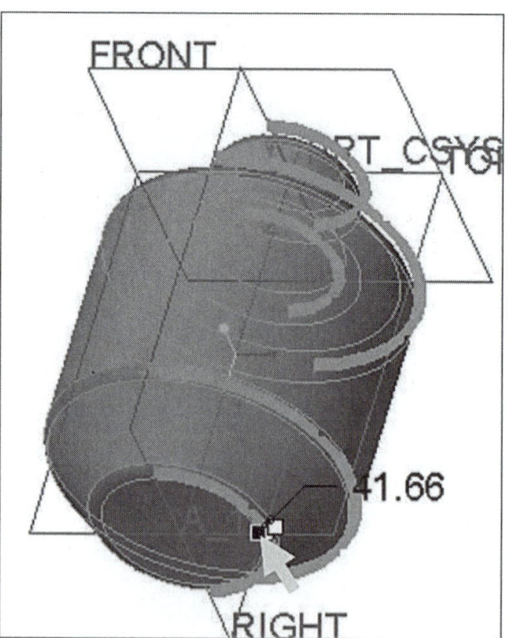

Figure 3.7(i) Move a Drag Handle to Adjust the Size

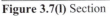

Click: View Manager from the Graphics Toolbar > **Sections** tab > **New** [Fig. 3.7(j)] > **Planar** > *type* **A** > **Enter** > select datum **RIGHT** [Fig. 3.7(k-l)] > > **Close** [Fig. 3.7(m)] > **File** > **Save** > **OK** > **File** > **Save As** > Type > **Zip File (*.zip)** > **OK** > **File** > **Close**

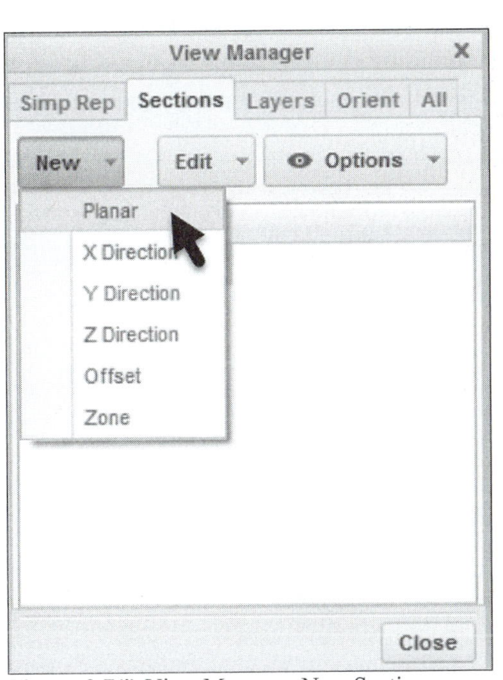

Figure 3.7(j) View Manager, New Section

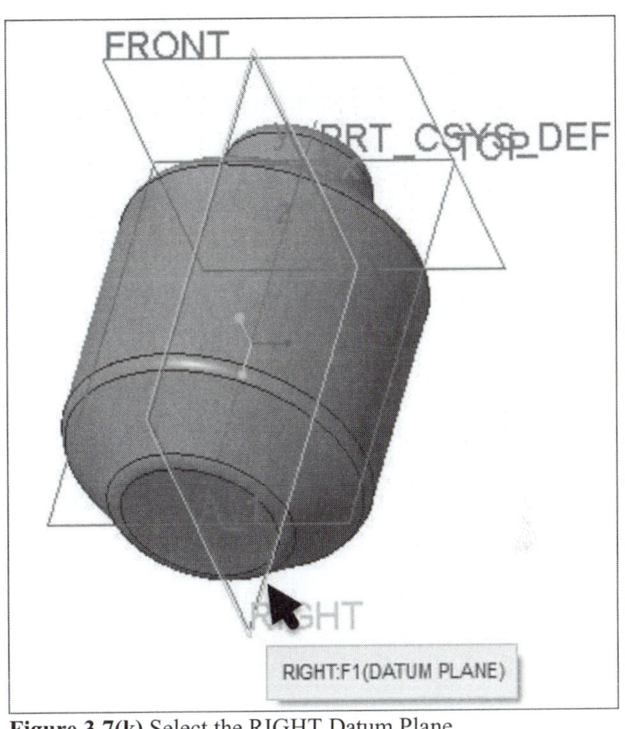

Figure 3.7(k) Select the RIGHT Datum Plane

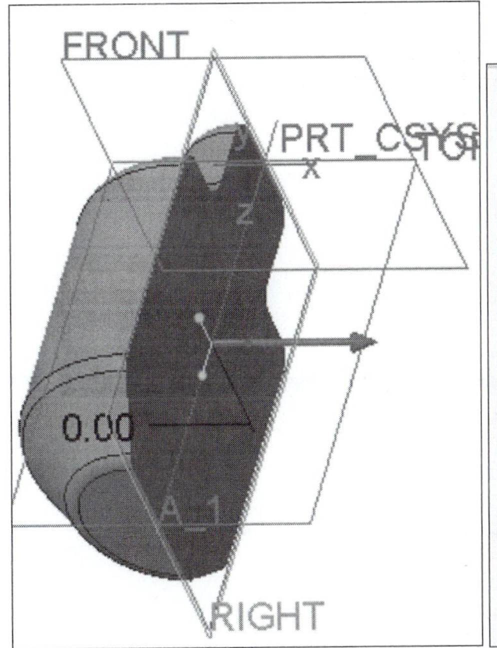

Figure 3.7(l) Section

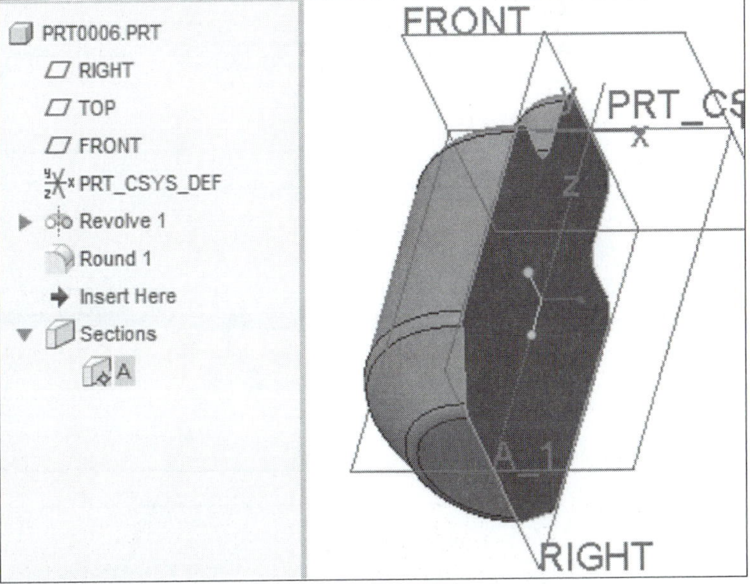

Figure 3.7(m) Section A

Part Model Seven (PRT0007.PRT) (Revolve Ellipse)

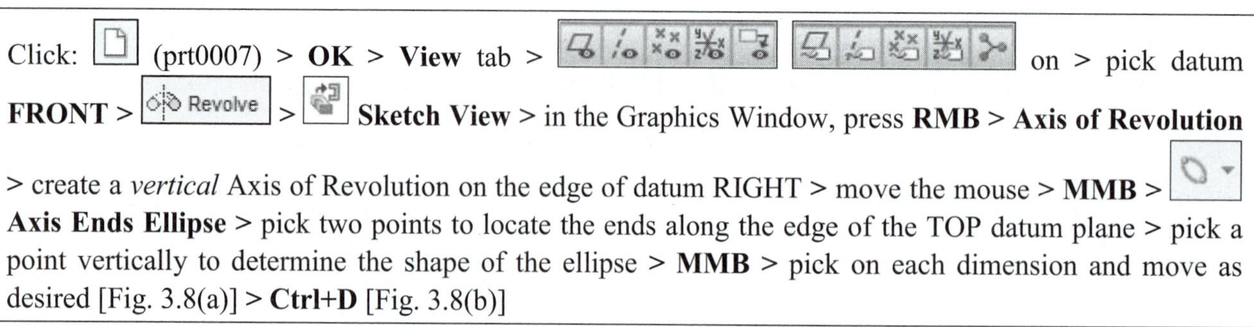

Click: (prt0007) > **OK** > **View** tab > on > pick datum **FRONT** > Revolve > **Sketch View** > in the Graphics Window, press **RMB** > **Axis of Revolution** > create a *vertical* Axis of Revolution on the edge of datum RIGHT > move the mouse > **MMB** > **Axis Ends Ellipse** > pick two points to locate the ends along the edge of the TOP datum plane > pick a point vertically to determine the shape of the ellipse > **MMB** > pick on each dimension and move as desired [Fig. 3.8(a)] > **Ctrl+D** [Fig. 3.8(b)]

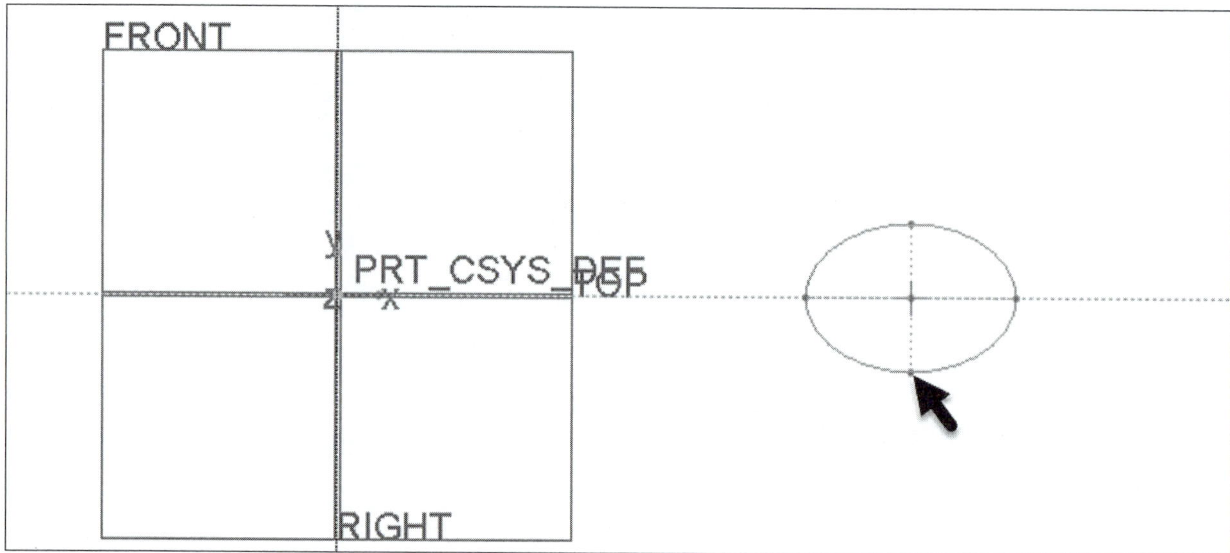

Figure 3.8(a) Sketch a Vertical Centerline and an Ellipse

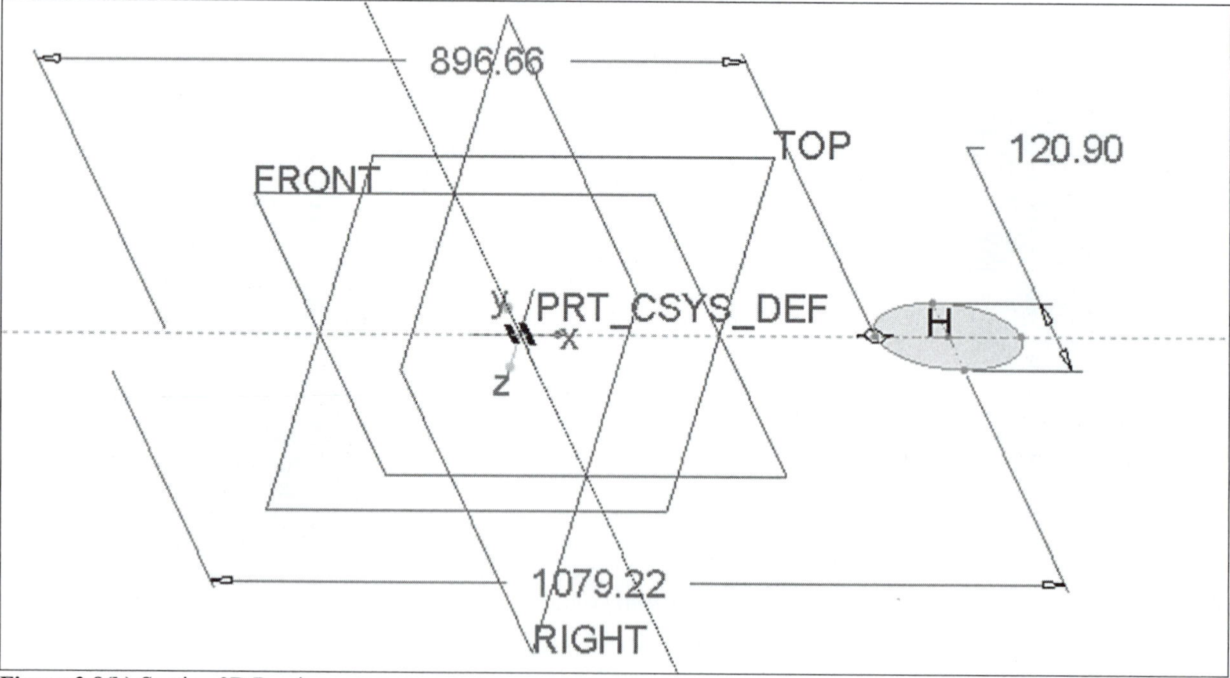

Figure 3.8(b) Section 3D Preview

Click: > **Ctrl+D** to resize > 360.00 > 180.00 > in the Graphics Window, press **RMB** > **Thicken Sketch** [Fig. 3.8(c)] > > **LMB** to deselect > **MMB** spin the model > close the Model Tree *(lower left-hand corner below the Navigator)* > **Tools** tab > **Model Information** > to adjust the Browser shade [Fig. 3.8(d)] > *(upper right corner of Browser)* or *(lower left of screen)* to close the Browser > open the Model Tree > **Ctrl+D** > **Ctrl+S** > **OK** > **File** > **Save As** > Type > **Zip File (*.zip)** > **OK** > **File** > **Close**

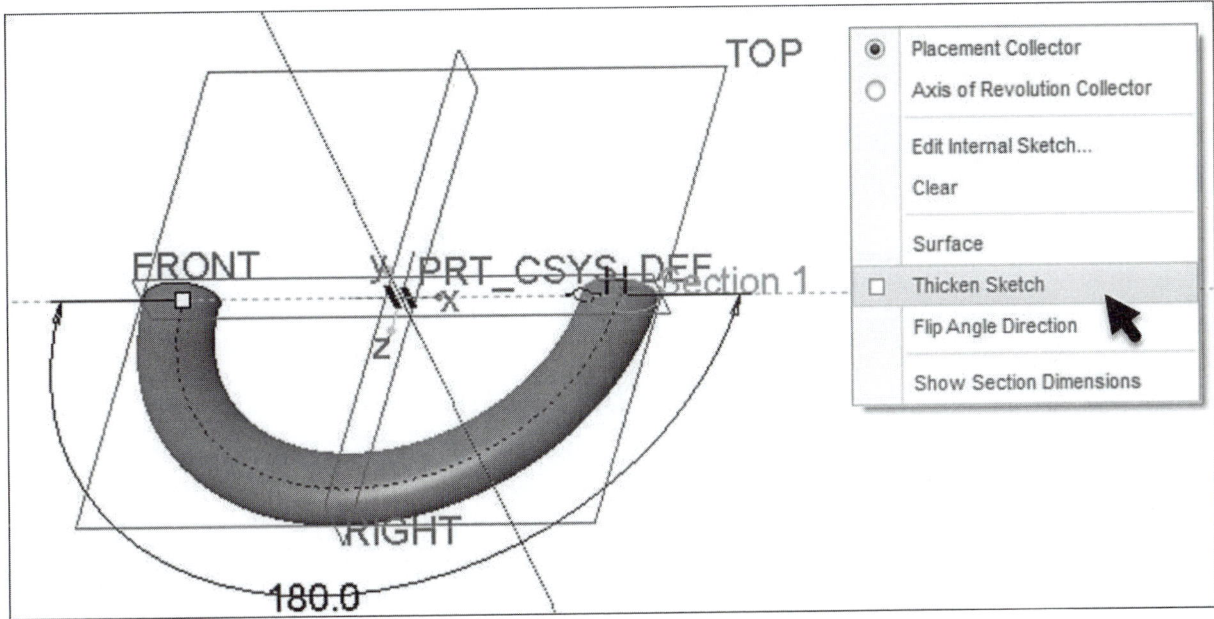

Figure 3.8(c) Revolve Preview

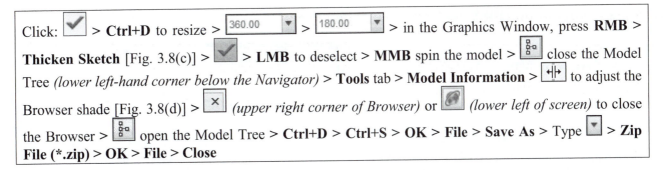

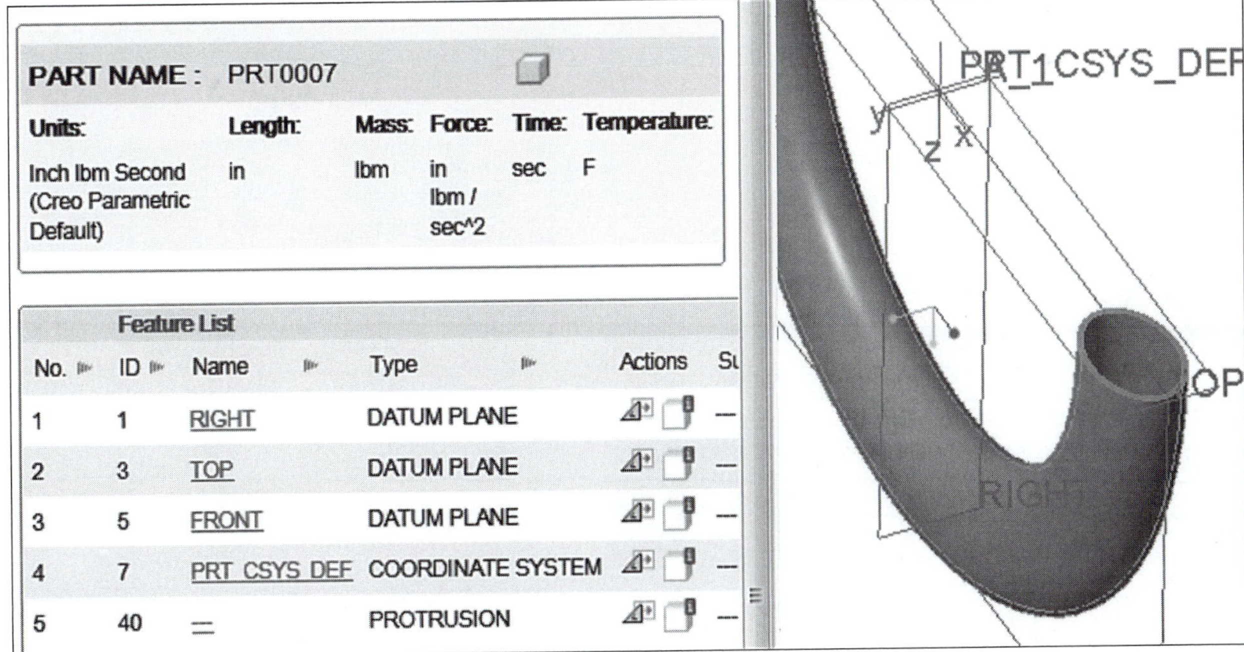

Figure 3.8(d) Completed Elliptical Torus

Part Model Eight (PRT0008.PRT) (Revolve Cut)

Press: **Ctrl+N** (prt0008) > **MMB** > pick datum **FRONT** > Extrude > > in the Graphics Window, press **RMB** > **Construction Centerline** > create a vertical centerline on the edge of datum RIGHT > move the mouse > **MMB** > **LMB** > press **RMB** > **Corner Rectangle** > sketch a symmetrical rectangle [Fig. 3.9(a)] > **MMB** > **Ctrl+D** > > > **Extrude on both sides** > pull one of the drag handles to increase the part's depth > **MMB** to complete the command > **LMB** to deselect

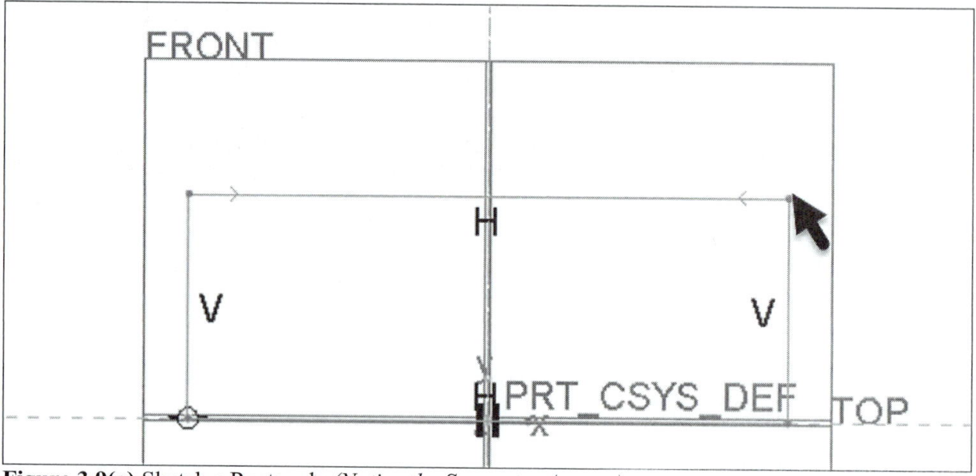

Figure 3.9(a) Sketch a Rectangle *(Notice the Symmetry Arrows)*

Click: Revolve > **Remove Material** from the Dashboard > Placement > Define... > Use Previous > **Sketch View** > in the Graphics Window, press **RMB** > **References** > **Select references for dimensioning and constraining** > pick the top edge/surface of the part [Fig. 3.9(b)] > **Close**

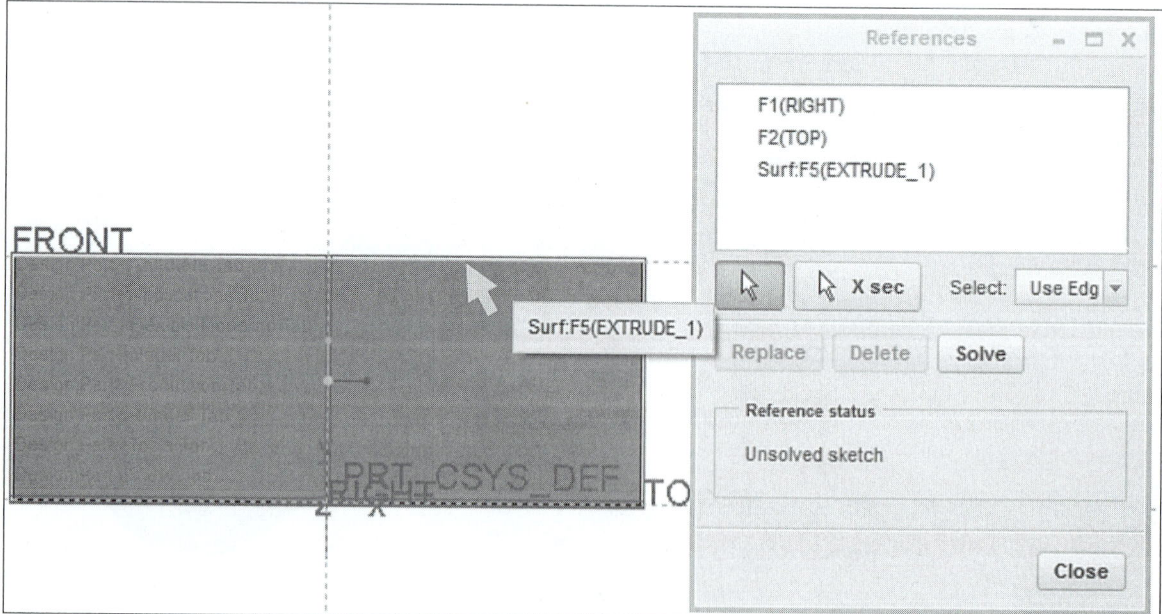

Figure 3.9(b) Select the Top Edge/Surface as a Reference

Press: **RMB** > **Axis of Revolution** > create a *vertical* Axis of Revolution on the edge of datum RIGHT > move the mouse > **MMB** > **LMB** > press **RMB** > **Circle** > pick on the top edge > pick a radius position [Fig. 3.9(c)] > move the mouse > **MMB** [Fig. 3.9(d)] > you can drag the circle diameter larger if necessary > **Ctrl+D** [Fig. 3.9(e)] > **MMB** [Fig. 3.9(f)]

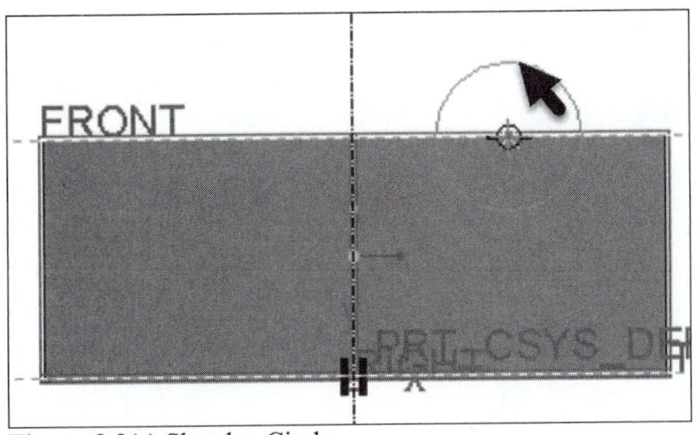

Figure 3.9(c) Sketch a Circle

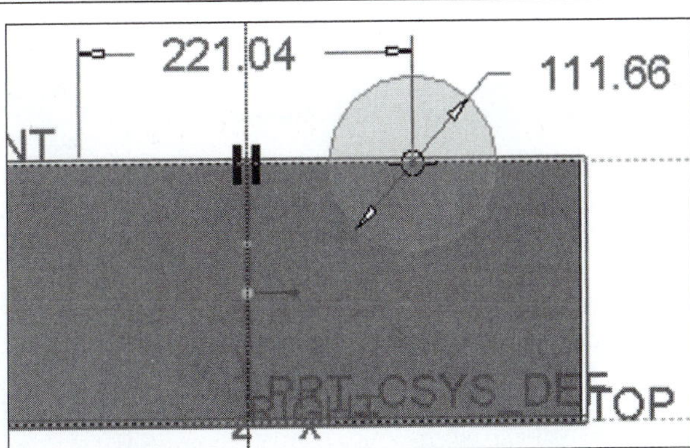

Figure 3.9(d) Completed Sketched Circle

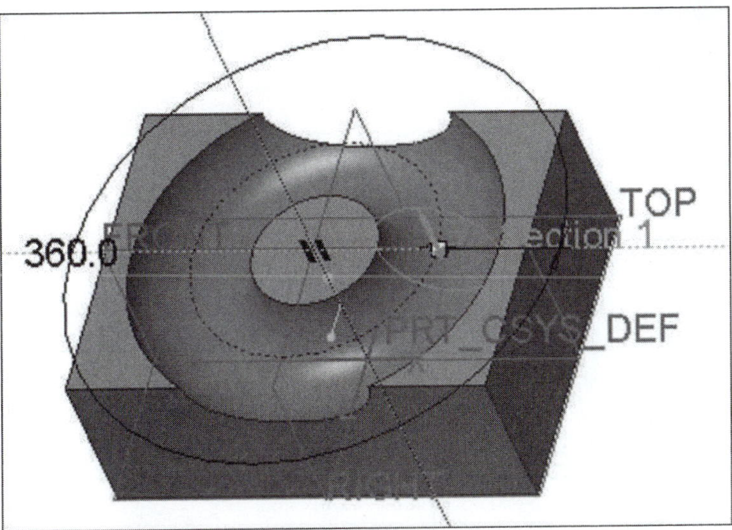

Figure 3.9(e) Revolved Cut Preview

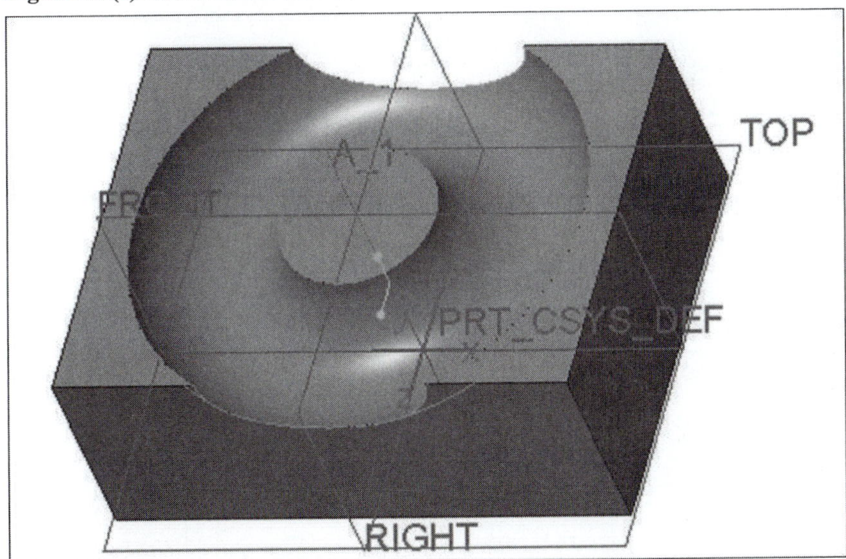

Figure 3.9(f) Completed Cut

Slowly pick on the edge of the part until it highlights > press and hold the **Ctrl** key > pick the other edges > release the **Ctrl** key > press **RMB** > **Round** [Fig. 3.9(g)] > Move a handle to adjust the radius value. Round surfaces will highlight for a valid size. [Fig. 3.9(h)] > **MMB** [Fig. 3.9(i)] > **LMB** to deselect > **Ctrl+S** > **OK**

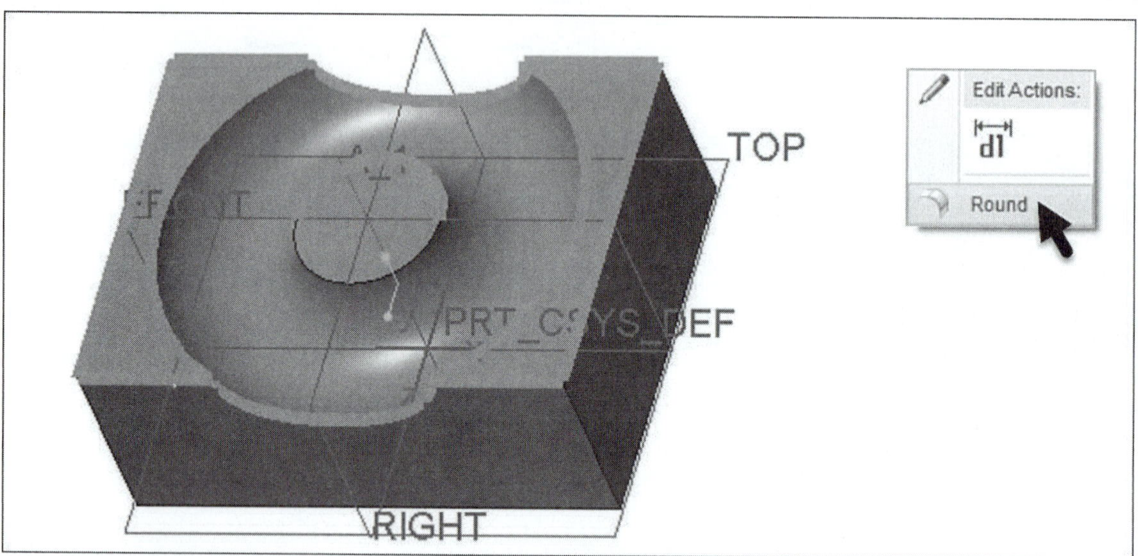

Figure 3.9(g) Select the Edges

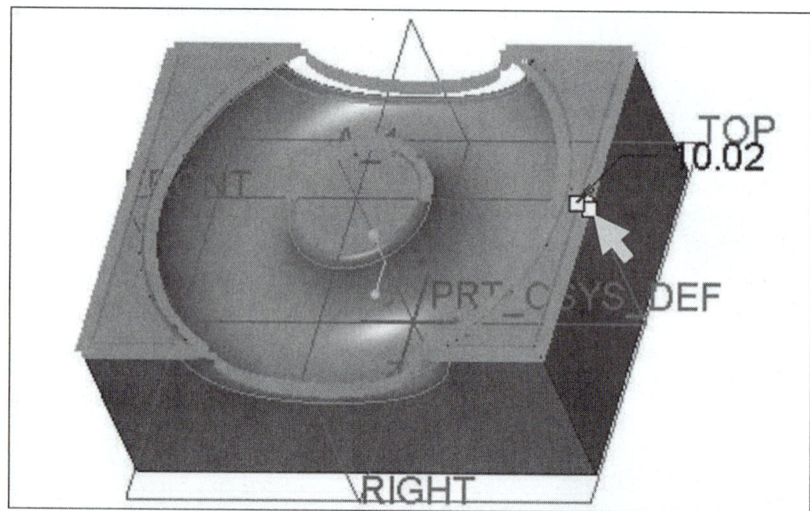

Figure 3.9(h) Round Preview

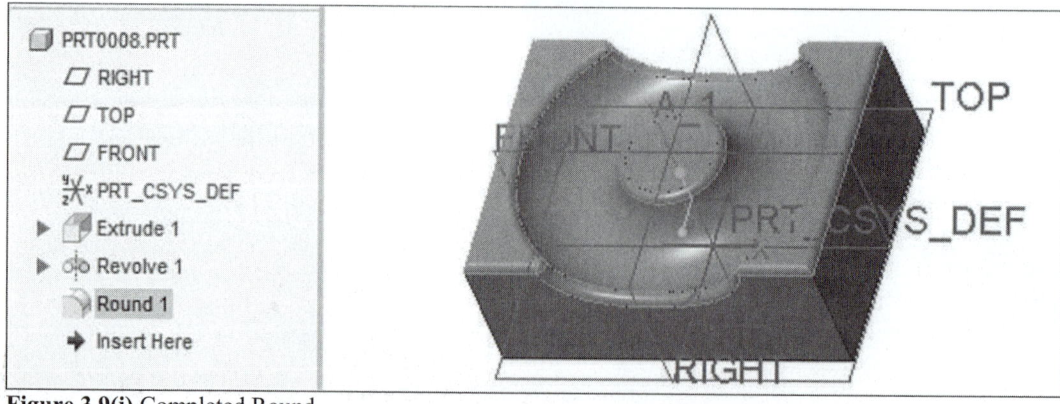

Figure 3.9(i) Completed Round

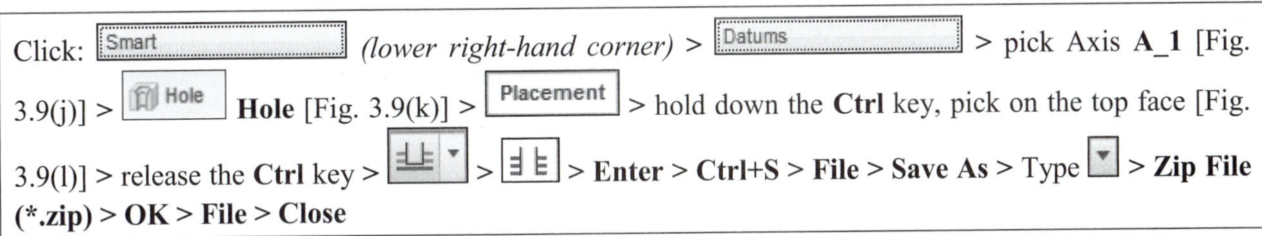

Click: Smart *(lower right-hand corner)* > Datums > pick Axis **A_1** [Fig. 3.9(j)] > Hole **Hole** [Fig. 3.9(k)] > Placement > hold down the **Ctrl** key, pick on the top face [Fig. 3.9(l)] > release the **Ctrl** key > > > **Enter** > **Ctrl+S** > **File** > **Save As** > Type > **Zip File (*.zip)** > **OK** > **File** > **Close**

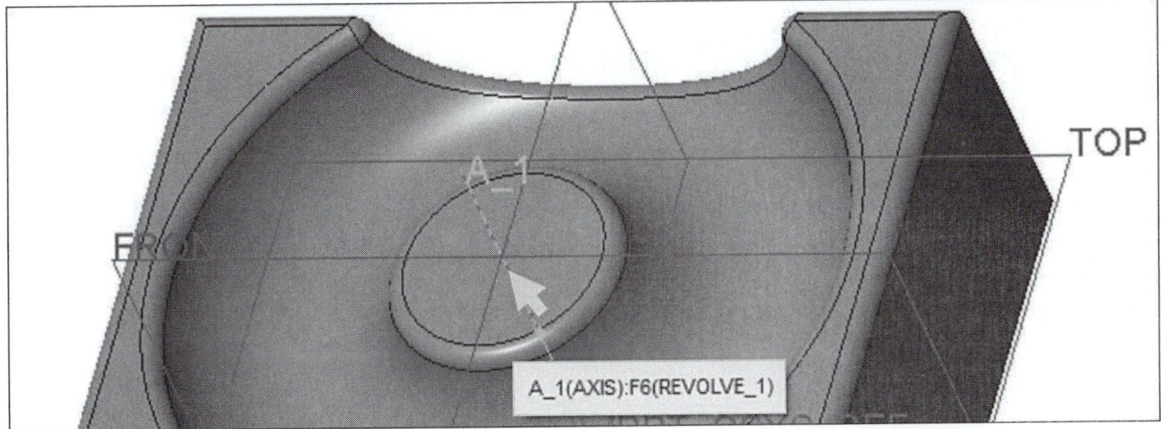

Figure 3.9(j) Select Axis **A_1** *(your axis label may be different)*

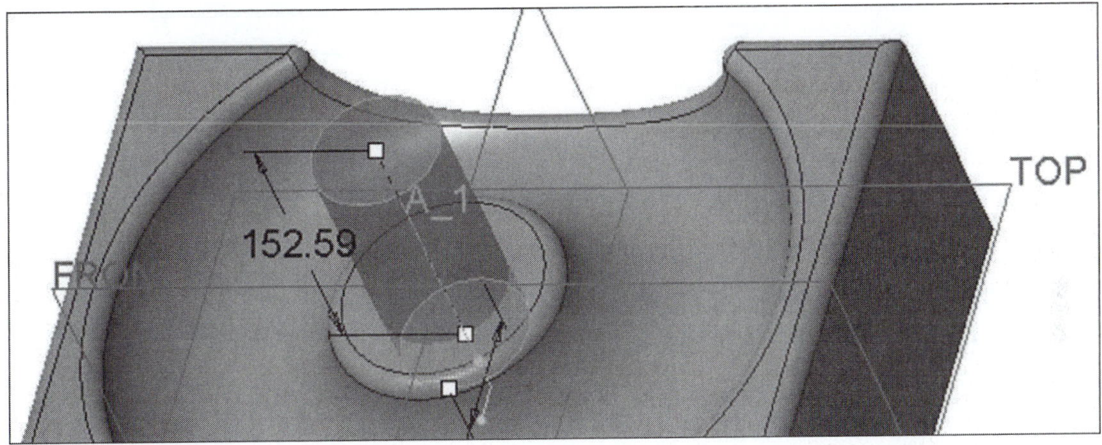

Figure 3.9(k) Preview

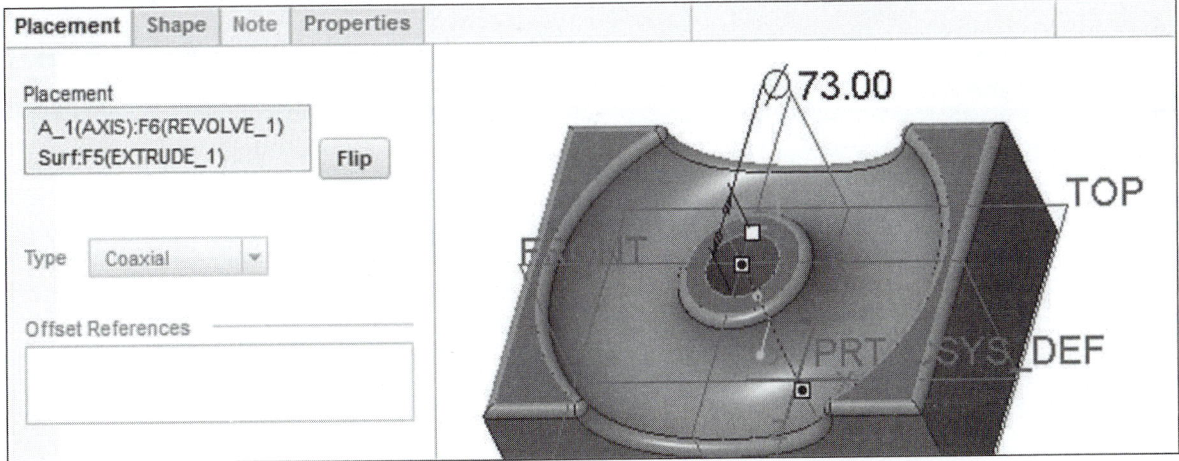

Figure 3.9(l) Select the Top Face of the Part

Part Model Nine (PRT0009.PRT) (Blend)

Click: (prt0009) > **Enter** > **Shapes** group of the Ribbon [Fig. 3.10(a)] > **Blend** > **Options** tab > Smooth > **Sections** tab > **Define** [Fig. 3.10(b)] > pick datum **FRONT** > **Sketch** > **Sketch View**

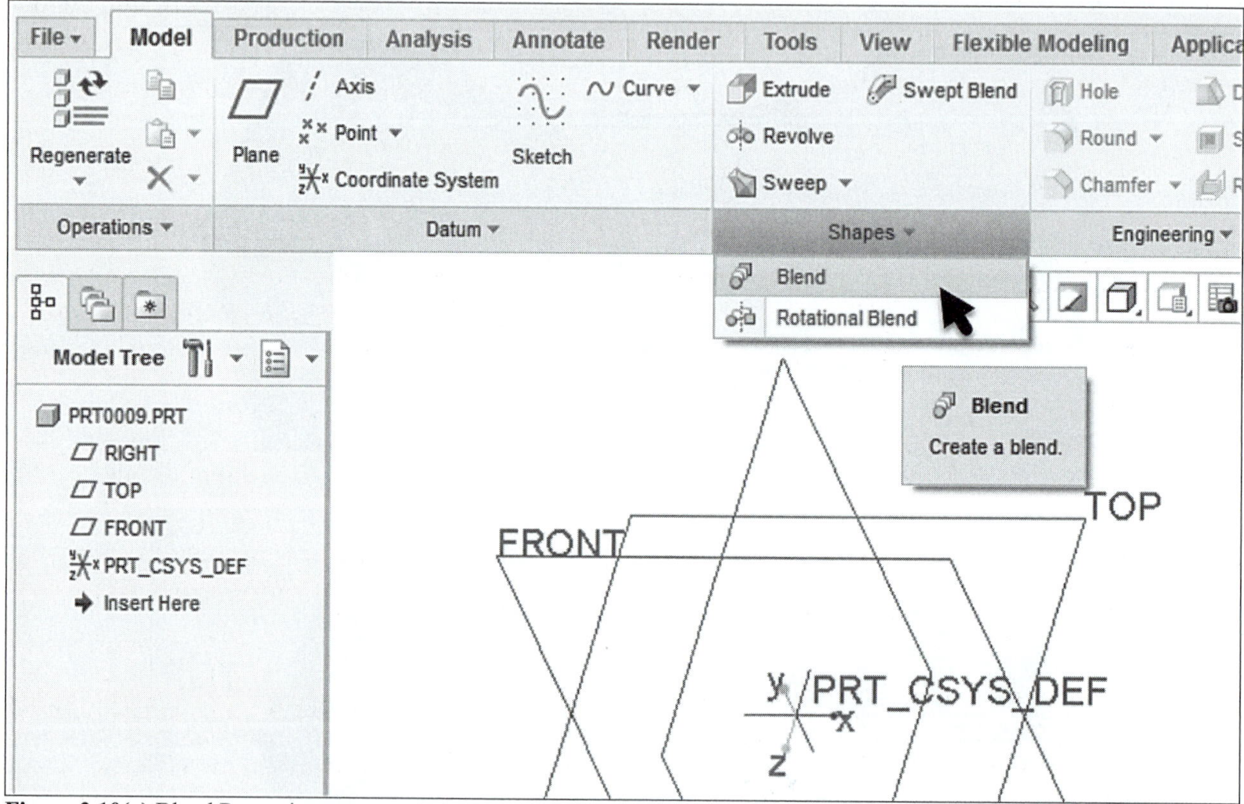

Figure 3.10(a) Blend Protrusion

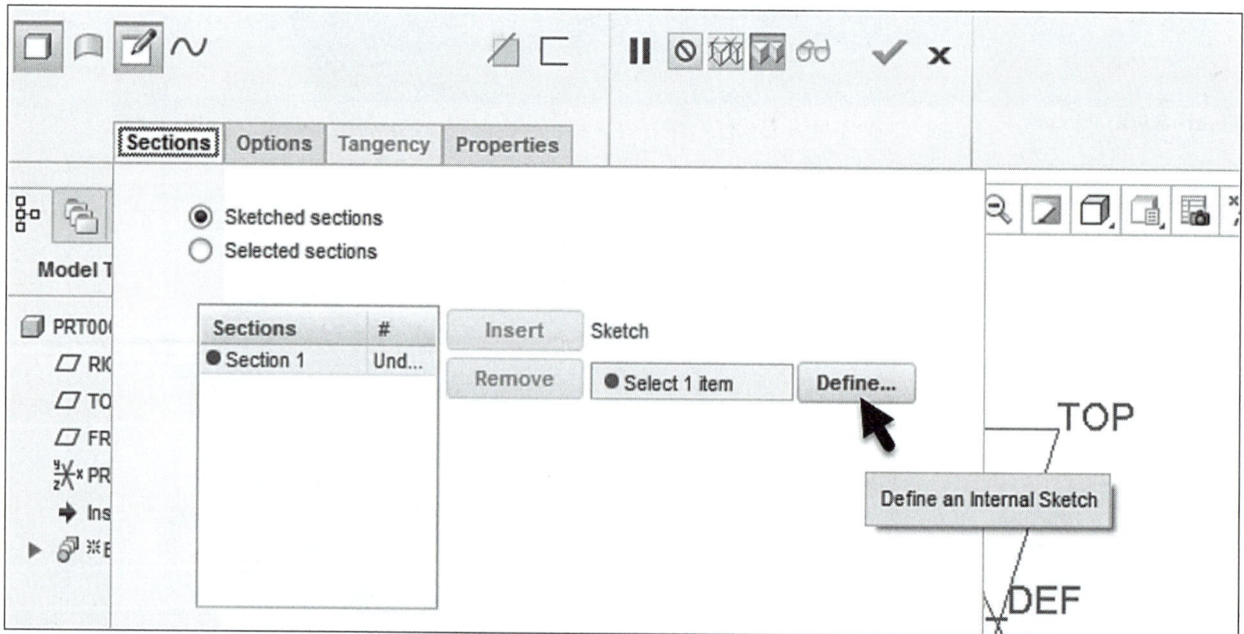

Figure 3.10(b) Blend Section Definition

Press **RMB** > **Construction Centerline** > create vertical and horizontal centerlines > move the mouse > **MMB** > **LMB** > press **RMB** > **Corner Rectangle** > start in the upper left-hand corner and pick two points (if you are careful, the rectangle will be symmetrical in both directions) [Fig. 3.10(c)] > **MMB** > **LMB** > > > **RMB** > **Corner Rectangle** > again, start in the upper left-hand corner and pick two points to create a symmetrical rectangle [Fig. 3.10(d)] > **MMB** > > **Ctrl+D** [Fig. 3.10(e)] >

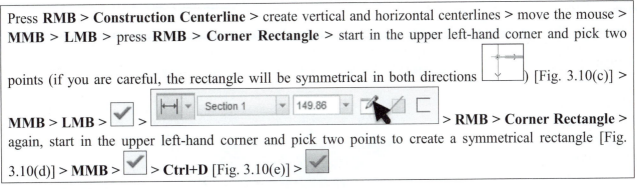

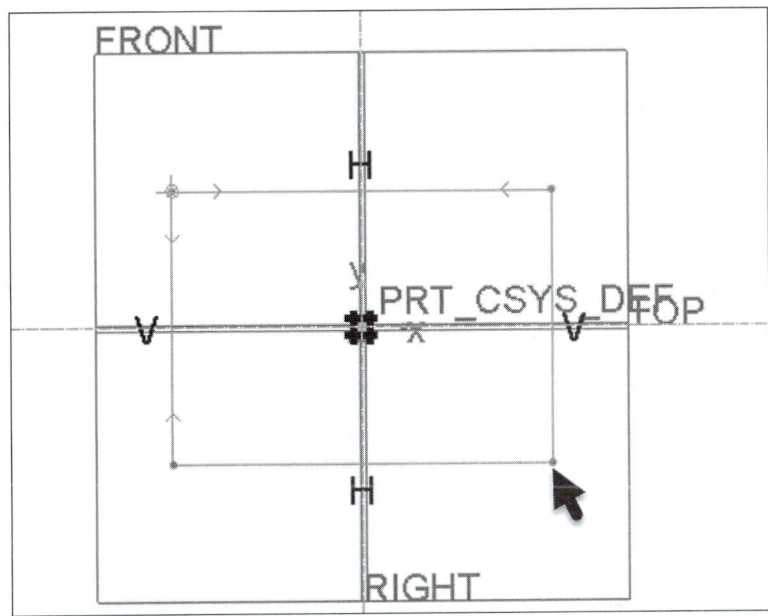

Figure 3.10(c) First Section Sketched Rectangle *(Symmetric in Both Directions)*

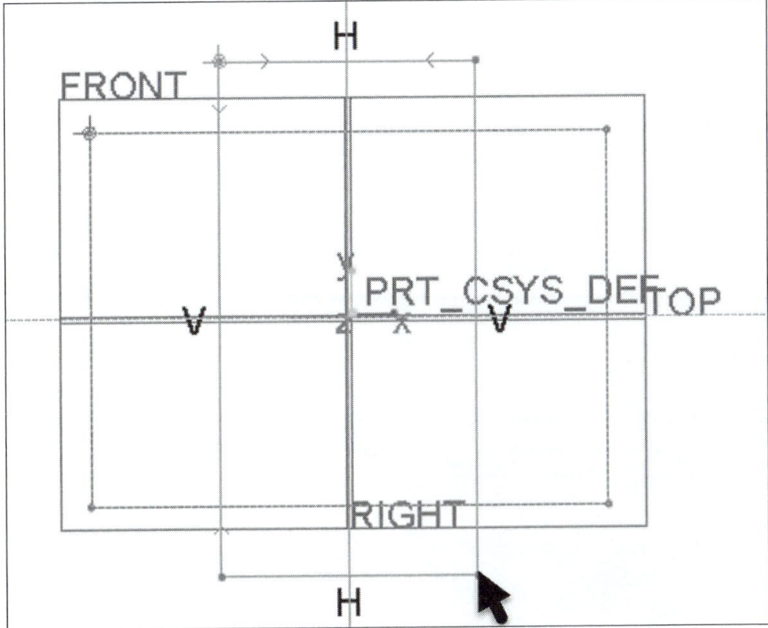

Figure 3.10(d) Second Section Sketched Rectangle *(Symmetric in Both Directions)*

With the blend protrusion *still selected*, press: **RMB** > [icon] **Edit** > place the mouse pointer on the drag handle [Fig. 3.10(f)] > press and hold down the **LMB** and move the drag handle [Fig. 3.10(g)] > release the **LMB** > move the mouse > **LMB** [Fig. 3.10(h)] > **Ctrl+D** > **Ctrl+S** > **Enter**

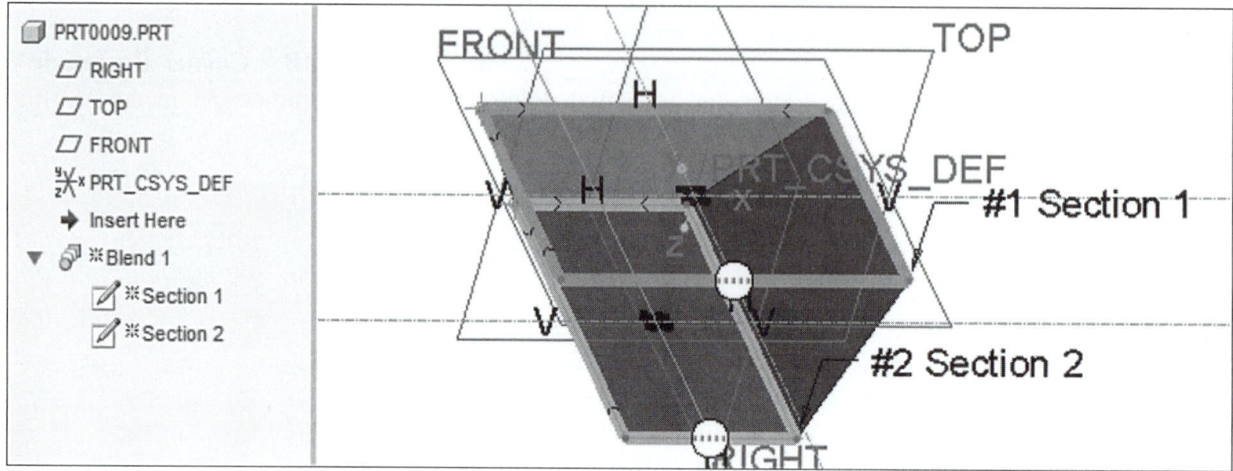

Figure 3.10(e) Edit

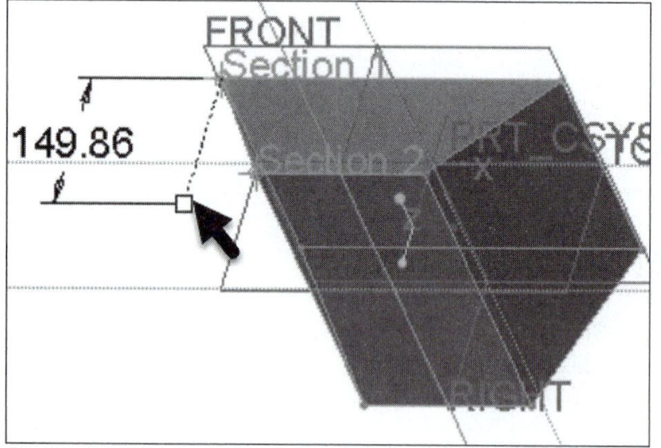

Figure 3.10(f) Depth Drag Handle

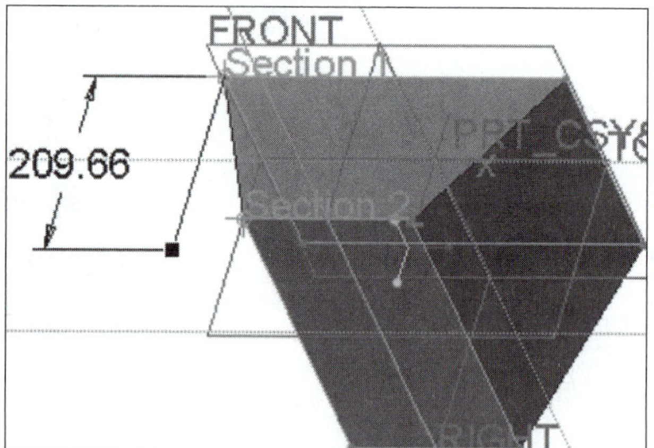

Figure 3.10(g) Move the Handle to Increase the Depth

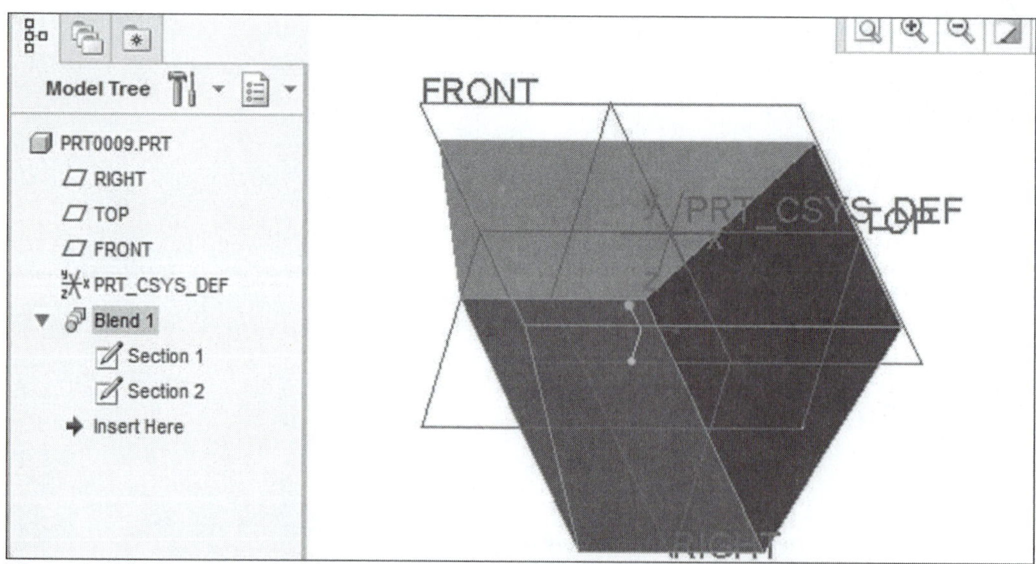

Figure 3.10(h) Regenerated Object

If you have the license, click: **Flexible Modeling** tab > **Move using Dragger** > pick on the front face [Fig. 3.10(i)] > press **LMB** on the *horizontal* circle, drag to rotate [Fig. 3.10(j)] > press **LMB** on the *vertical* circle, drag to rotate [Fig. 3.10(k)] > **Enter** > **Ctrl+S** > **File** > **Save As** > Type > **Zip File (*.zip)** > **OK** > **File** > **Close**

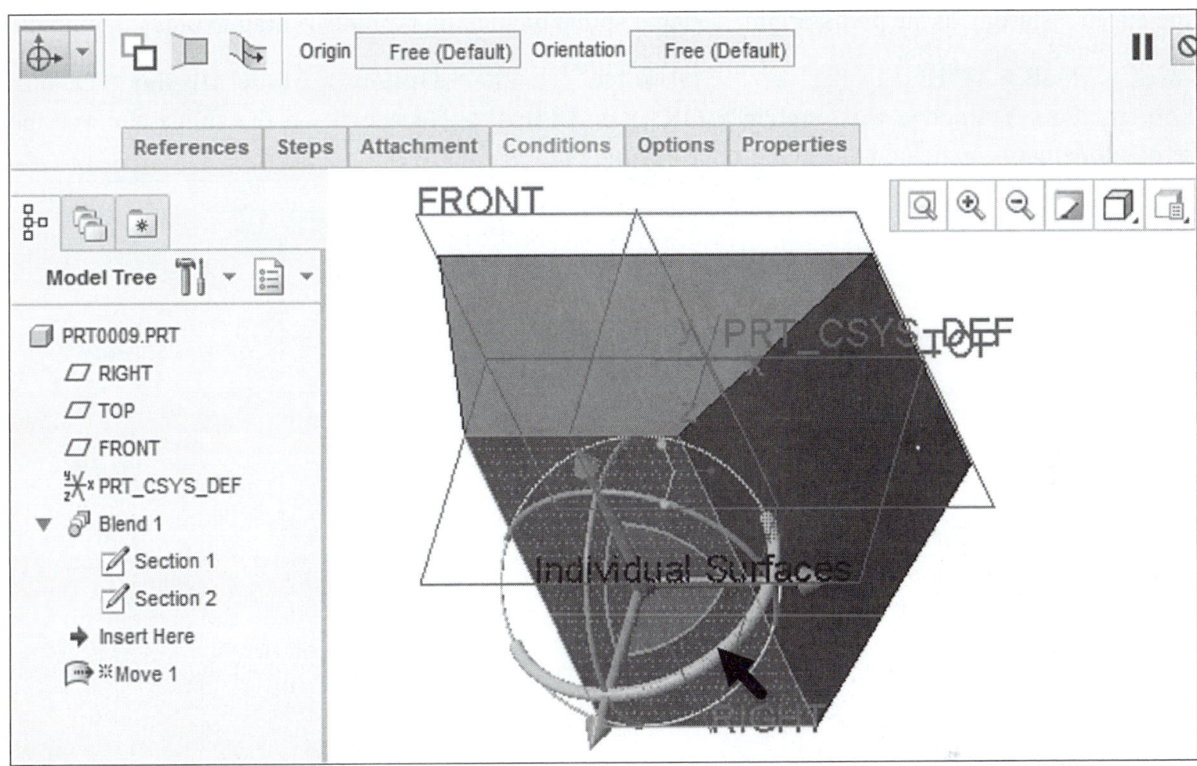

Figure 3.10(i) Set up Flexible Move

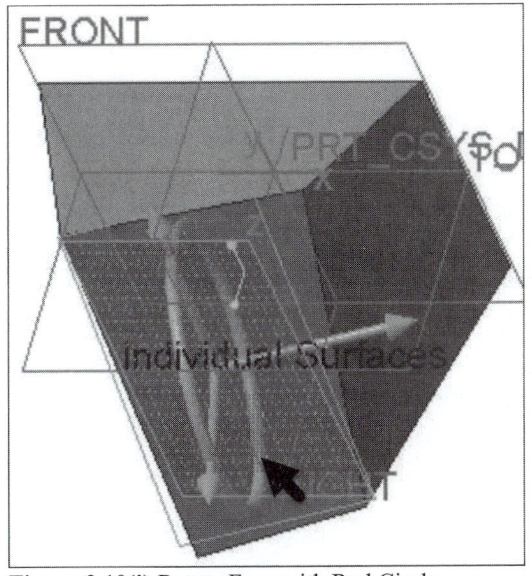

Figure 3.10(j) Rotate Face with Red Circle

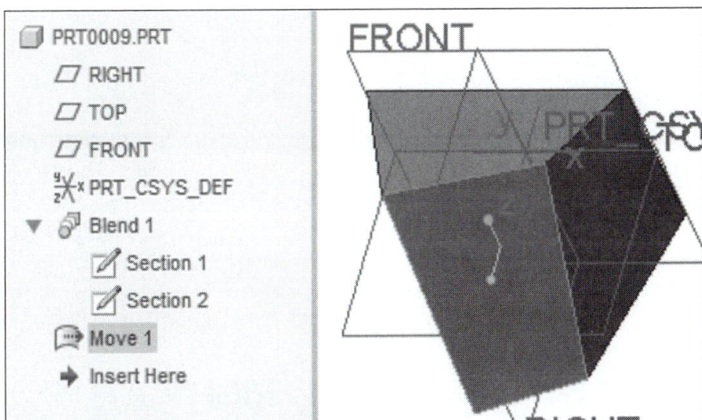

Figure 3.10(k) Move Completed

Part Model Ten (PRT0010.PRT) (Sweep)

For this part you will create the sketch before selecting a Shapes command. Press: **Ctrl+N** (prt0010) > **OK** > pick datum **FRONT** > Sketch from the Model tab Ribbon > **Sketch View** > **Spline** Create a spline curve > starting at the part's origin, sketch a spline by picking 6 points > drag to adjust the points as desired > **MMB** > **MMB** [Fig. 3.11(a)] > **Ctrl+D** > > **File** > **Options** > **Model Display** > Default model orientation: **Trimetric** > **Isometric** > **OK** [Fig. 3.11(b)] > **No** > pick on the spline > move the mouse pointer slightly > pick on the spline again *(spline thickens)* > **Ctrl+S** > **Enter**

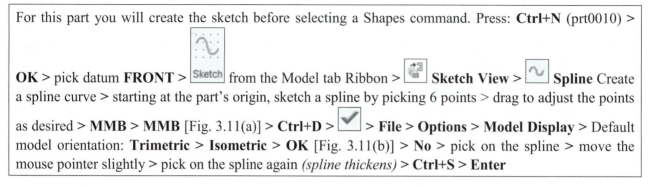

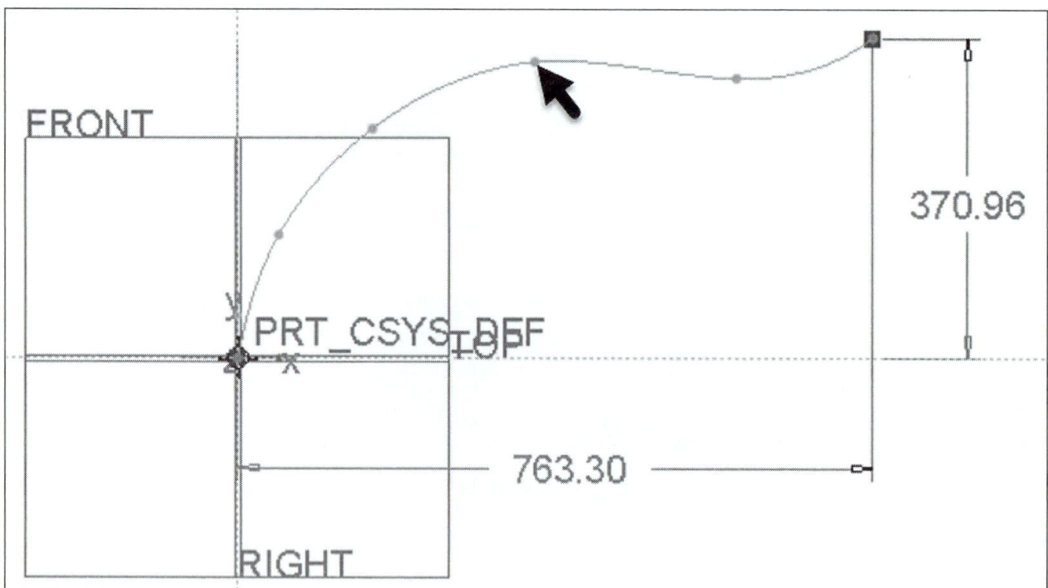

Figure 3.11(a) 6-point Spline

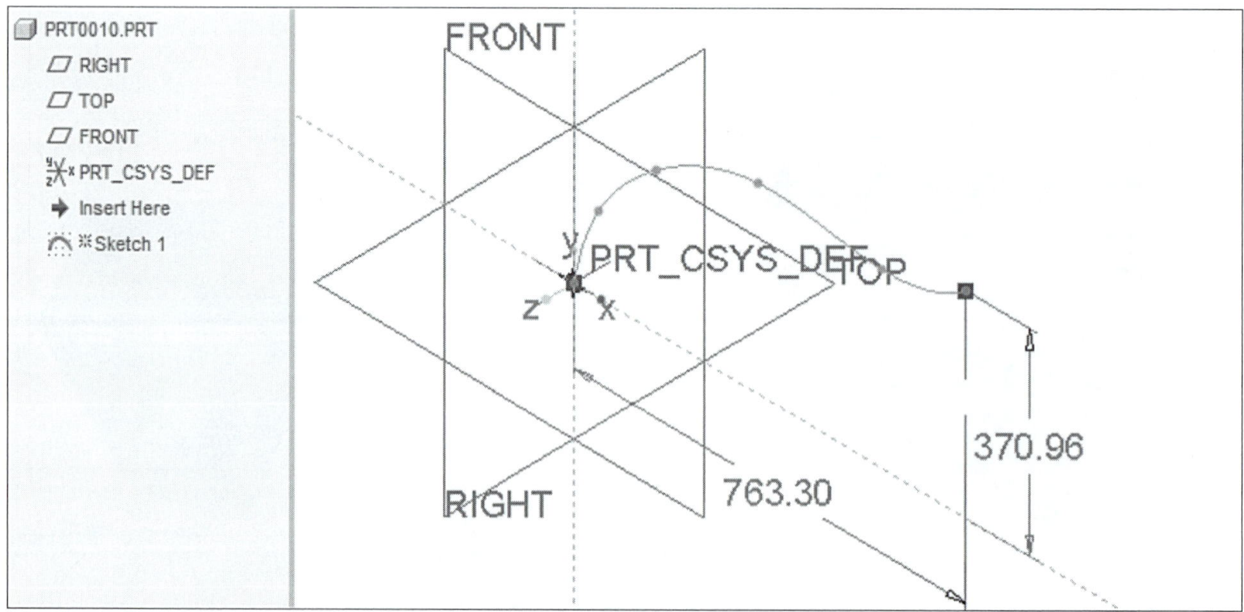

Figure 3.11(b) Select the Spline > Move the Mouse Pointer Slightly > Select the Spline Again

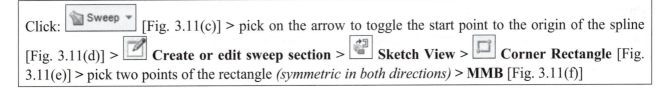

Click: Sweep [Fig. 3.11(c)] > pick on the arrow to toggle the start point to the origin of the spline [Fig. 3.11(d)] > **Create or edit sweep section** > **Sketch View** > **Corner Rectangle** [Fig. 3.11(e)] > pick two points of the rectangle *(symmetric in both directions)* > **MMB** [Fig. 3.11(f)]

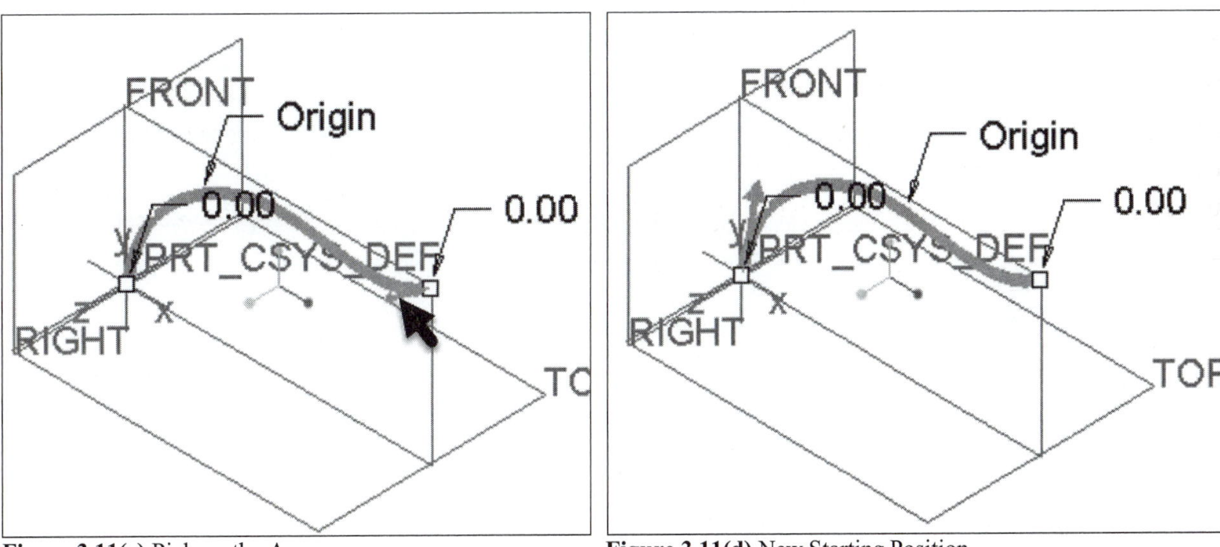

Figure 3.11(c) Pick on the Arrow

Figure 3.11(d) New Starting Position

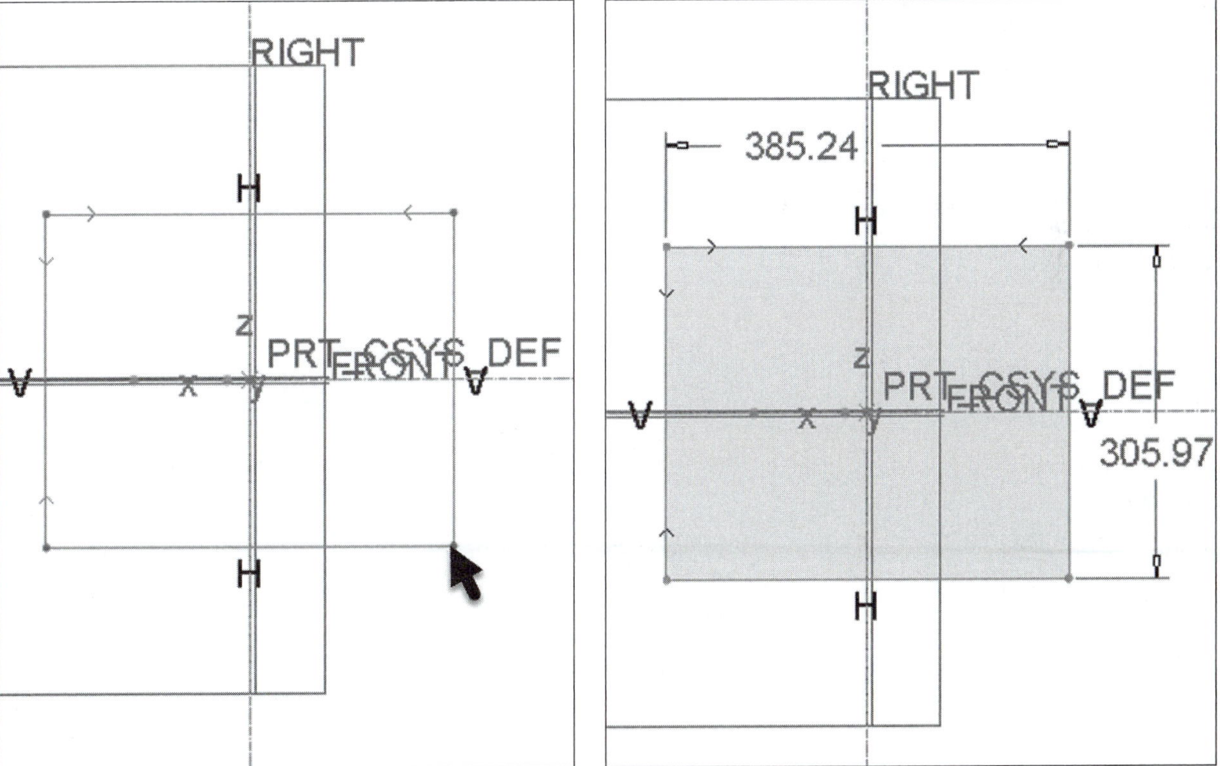

Figure 3.11(e) Sketch the Rectangle *(note symmetry)*

Figure 3.11(f) Dimensioned Sketch

Press: **Ctrl+D** > ✓ > ✓ > *if the sweep does not display- drag a sketch corner to a smaller size- it will display the feature when it is valid or if you get a Regeneration Failure* > **Cancel** from the Regeneration Failure dialog box [Fig. 3.11(g)] > **OK** from the Troubleshooter dialog box > ▶ [Fig. 3.11(h)]

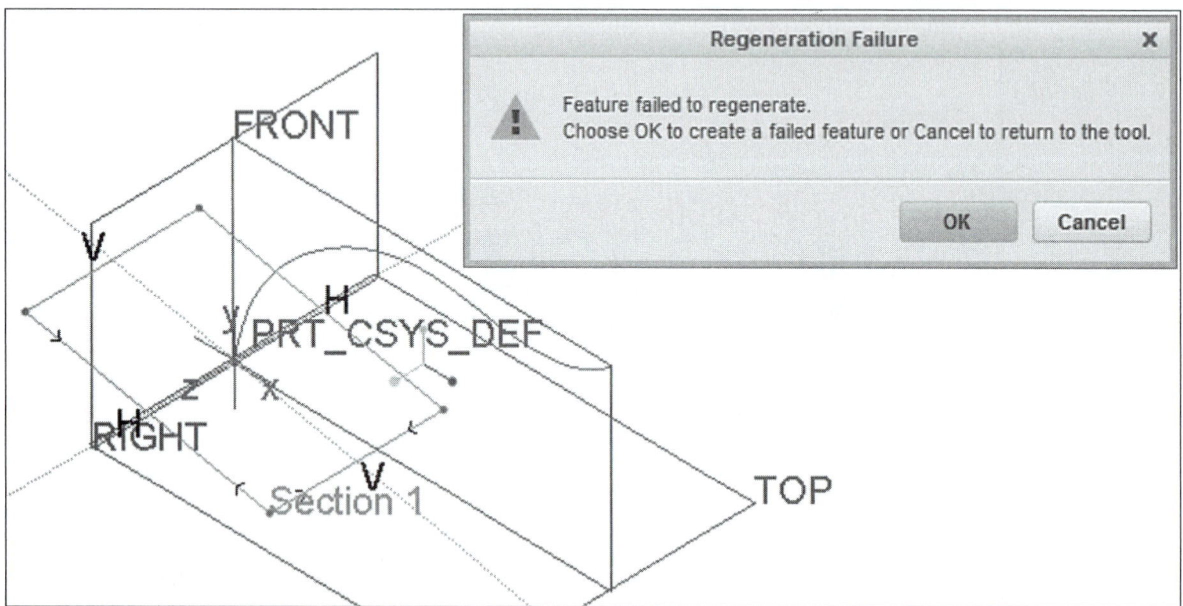

Figure 3.11(g) Regeneration Failure Dialog Box

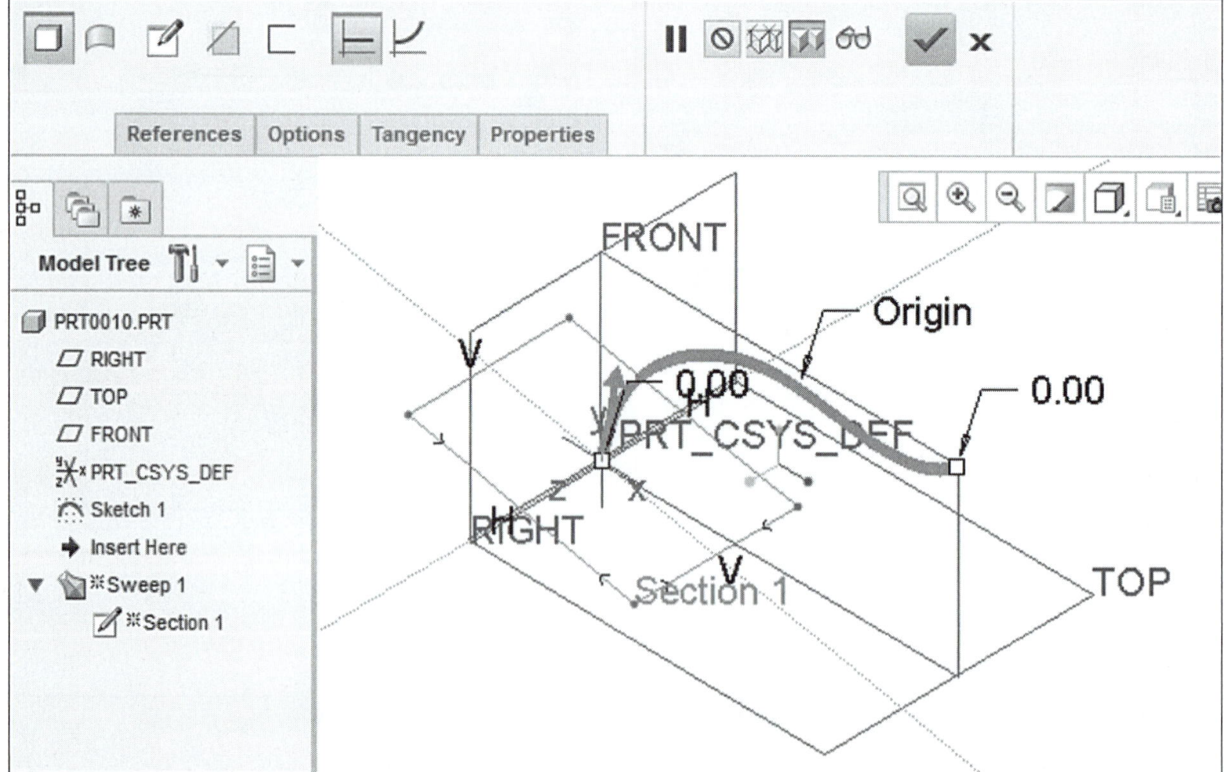

Figure 3.11(h) Editing the Feature

Press: **LMB** on one corner of the rectangular sketch and drag inward until a preview of the sweep displays [Fig. 3.11(i)] > release the **LMB** > ✓ > in the Graphics Window, **LMB** to deselect > select **Sketch 1** from the Model Tree *(your Sketch number may be different)* > **RMB** > **Edit the definition of the selected object** [Fig. 3.11(j)]

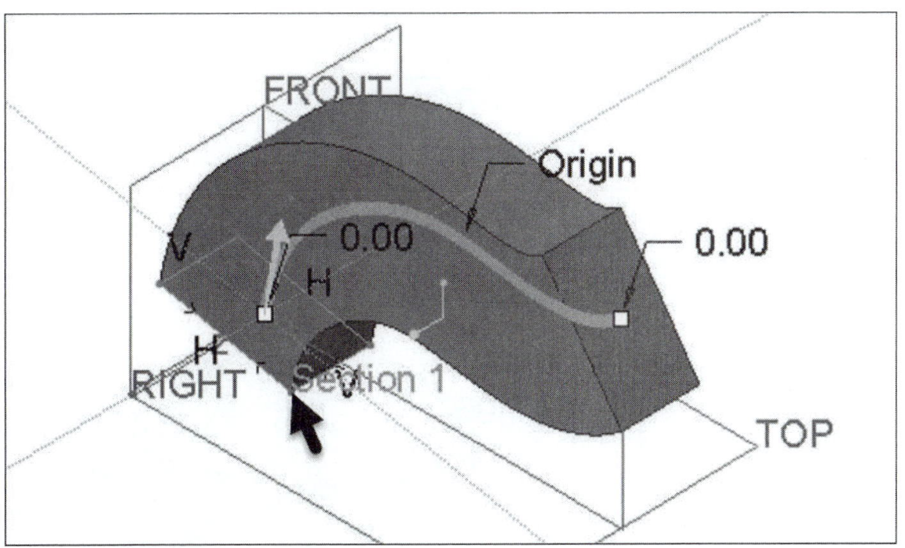

Figure 3.11(i) Feature Preview

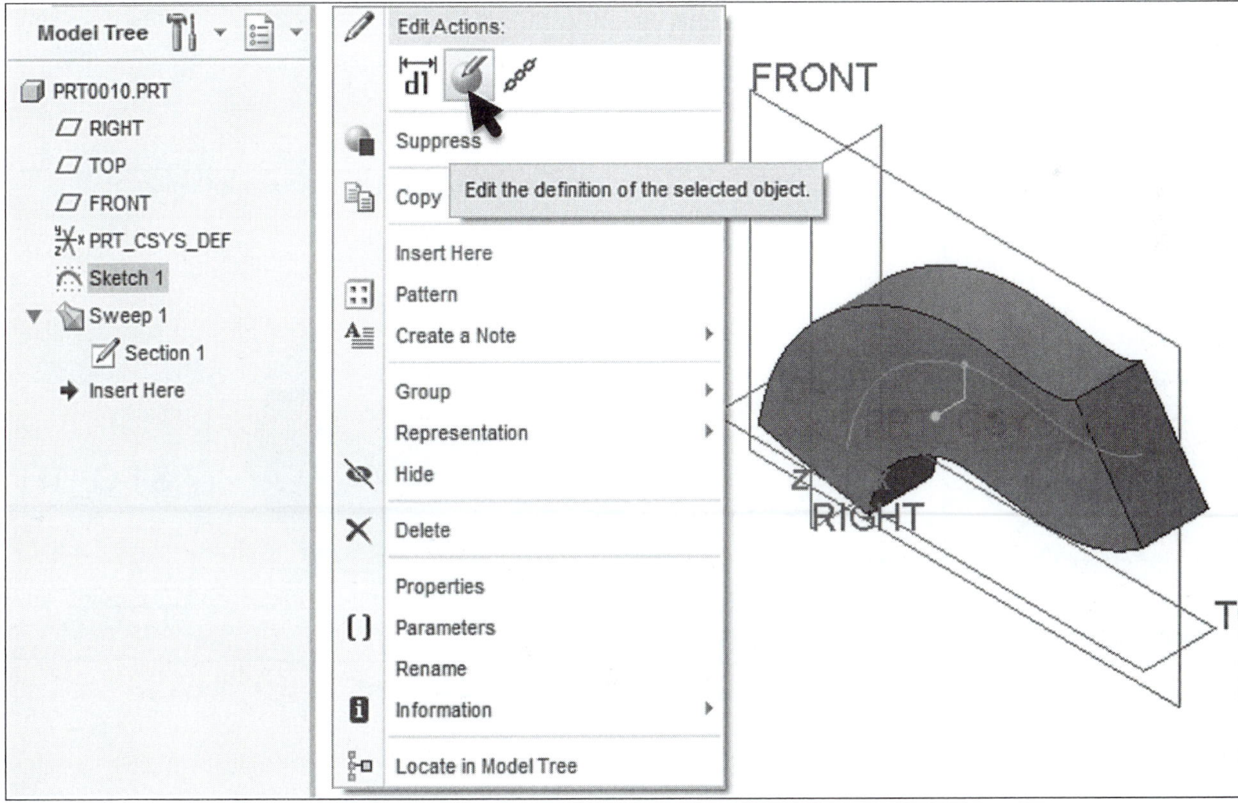

Figure 3.11(j) Edit Definition of the Sketch *(your Sketch number may be different)*

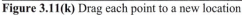

Press: **LMB** on each of the points of the spline and drag as desired [Fig. 3.11(k)] > press **RMB** > **OK** [Fig. 3.11(l)] > double-click on the sweep feature in the Graphics Window > press **LMB** on one of the rectangle's points and drag as desired > release the **LMB** > move the pointer slightly > **LMB** > **MMB** and rotate the part > Shell **Shell** > press and hold **Ctrl** > select both flat ends of the sweep > ✓ [Fig. 3.11(m)] > **Ctrl+D** > **Ctrl+S** > **File** > **Save As** > Type ▾ > **Zip File (*.zip)** > **OK** > **File** > **Close** > **File** > **Exit** > **Yes** > on your computer; open Windows Explorer, locate and put all ten part zip files into one (new) folder > zip that new folder > upload or email the new zip file as directed by your instructor.

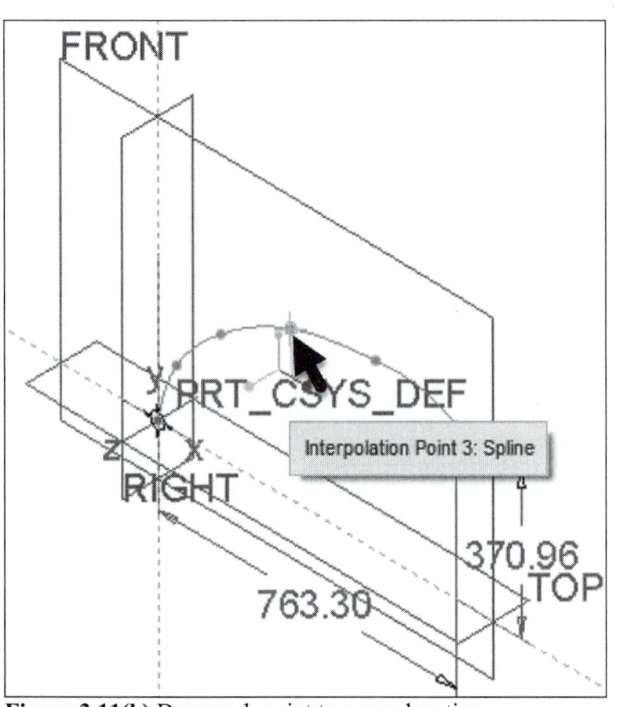

Figure 3.11(k) Drag each point to a new location

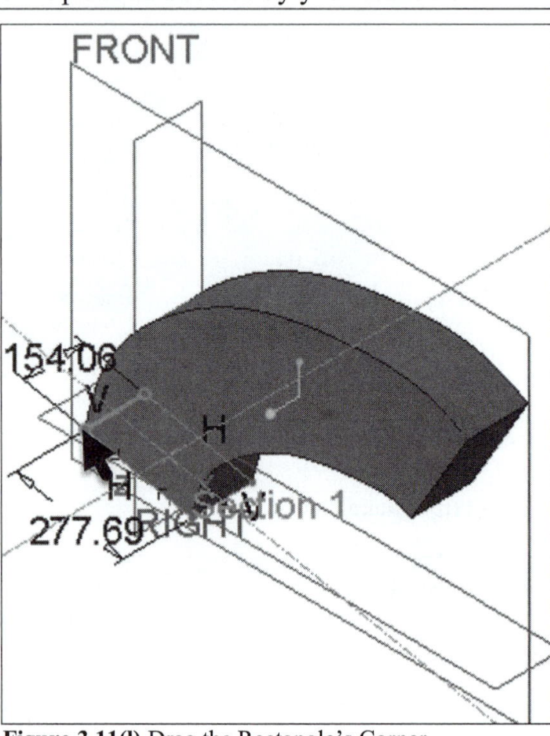

Figure 3.11(l) Drag the Rectangle's Corner

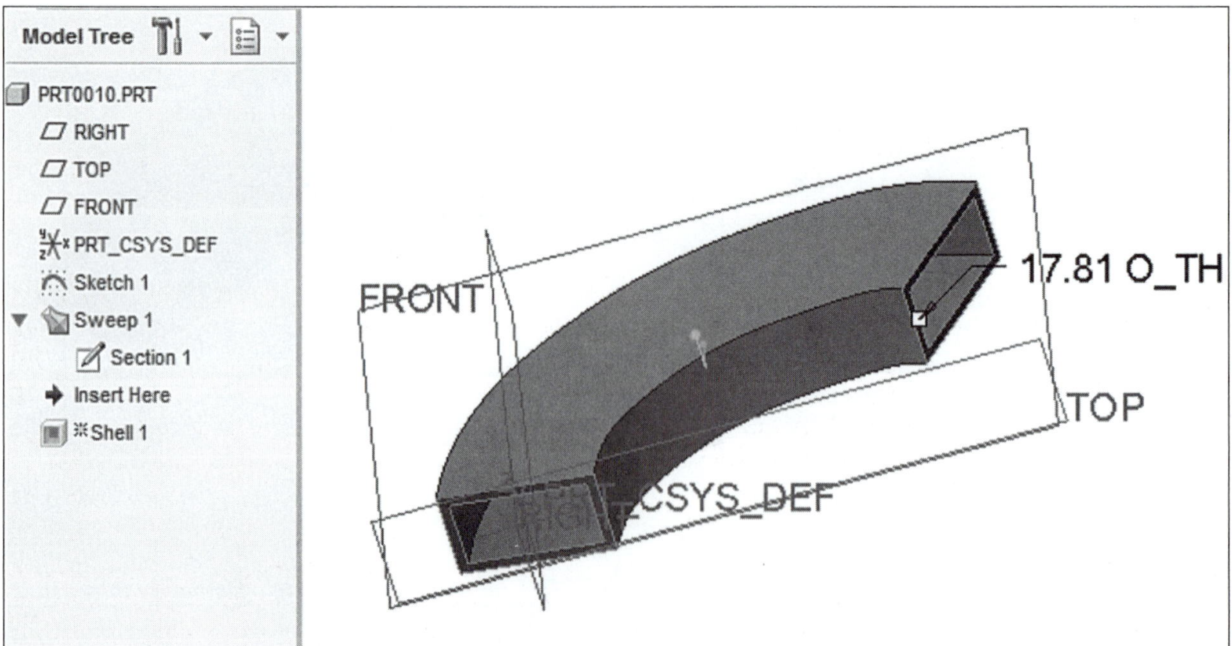

Figure 3.11(m) Dynamically Edit the Section

Lesson 4 Extrusions

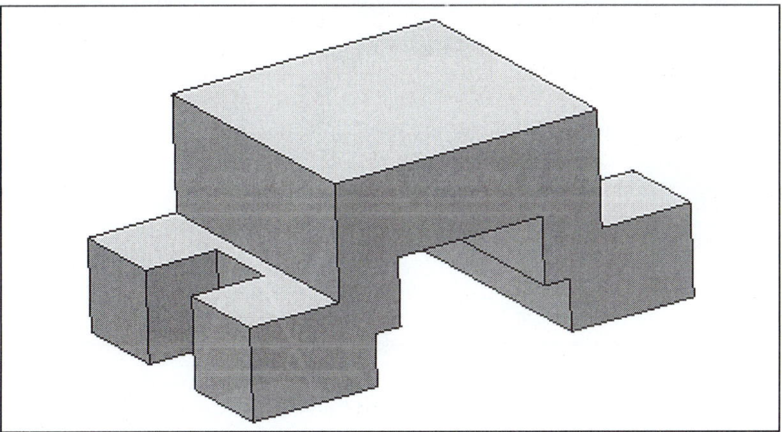

Figure 4.1 Clamp

OBJECTIVES

- Create a feature using an **Extruded** protrusion
- Understand **Setup** and **Environment** settings
- Define and set a **Material** type
- Create and use **Datum** features
- Sketch protrusion and cut feature geometry using the **Sketcher**
- Understand the feature **Ribbon**
- **Copy** a feature

REFERENCES AND RESOURCES

For **Resources** go to **www.cad-resources.com** > click on the PTC Creo Parametric 3.0 Book cover

- Lesson 4 Lecture
- Lesson 4 3D models embedded in a PDF
- Book Projects PDF
- Project Lectures
- Quick Reference Card
- Configuration Options

Extrusions

The design of a part using PTC Creo Parametric 3.0 starts with the creation of base features (normally datum planes), and a solid protrusion. Other protrusions and cuts are then added in sequence as required by the design. You can use various types of features as building blocks in the progressive creation of solid parts (Fig. 4.1). Certain features, by necessity, precede other more dependent features in the design process. Those dependent features rely on the previously defined features for dimensional and geometric references. The progressive design of features creates these dependent feature relationships known as *parent-child relationships*. The actual sequential history of the design is displayed in the Model Tree. The parent-child relationship is one of the most powerful aspects of Creo Parametric and parametric modeling in general. It is also very important as you modify a part. After a parent feature in a part is modified, all children are automatically modified to reflect the changes in the parent feature. It is therefore essential to reference feature dimensions so that Creo Parametric can correctly propagate design modifications throughout the model.

A **protrusion** is a part feature that adds or removes material. A protrusion is *always the first solid feature created.* This is usually the first feature created after a base feature of datum planes. Figure 4.2 shows four common protrusions.

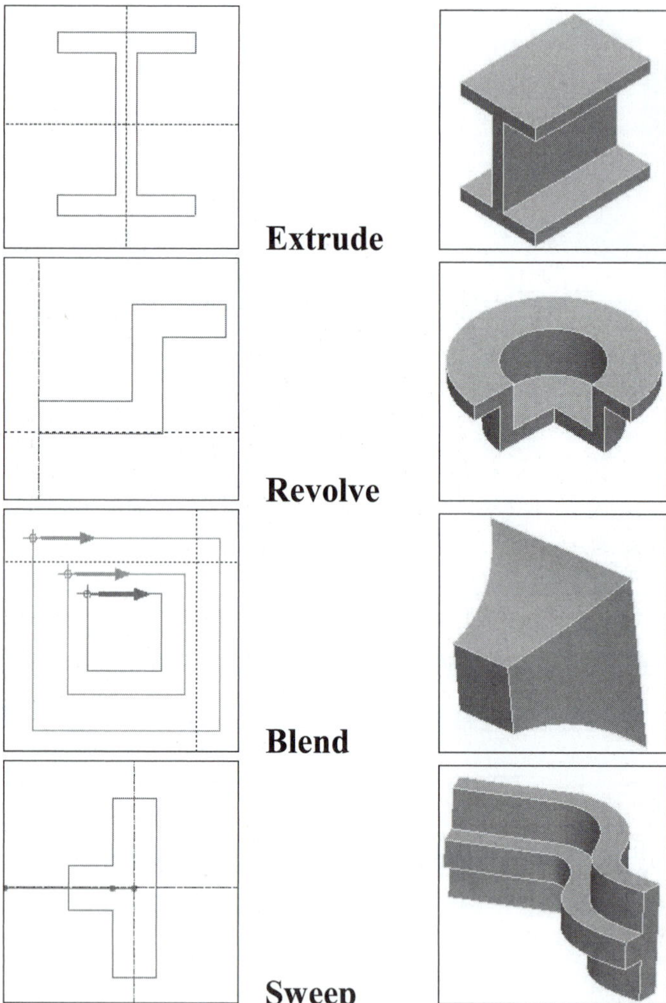

Figure 4.2 Basic Protrusions

The Design Process

It is tempting to directly start creating models. Nevertheless, in order to build value into a design, you need to create a product that can keep up with the constant design changes associated with the design-through-manufacturing process. Flexibility must be integral to the design. Flexibility is the key to a friendly and robust product design while maintaining design intent, and you can accomplish it through planning. To plan a design, you need to understand the overall function, form, and fit of the product. This understanding includes the following points:

- Overall size of the part
- Basic part characteristics
- The way in which the part can be assembled
- Approximate number of assembly components
- The manufacturing processes required to produce the part

Lesson 4 STEPS

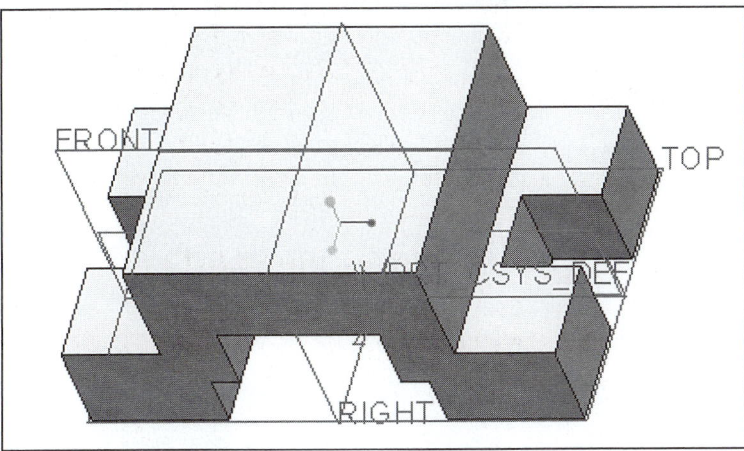

Figure 4.3 Clamp and Datum Planes

Clamp

The clamp in Figure 4.3 is composed of a protrusion and two cuts (one cut will be mirrored). A number of things need to be established before you actually start modeling. These include setting up the *environment*, selecting the *units*, and establishing the *material* for the part.

Before you begin any part, you must plan the design. The **design intent** will depend on a number of things that are out of your control and many that you can establish. Asking yourself a few questions will clear up the design intent you will follow: Is the part a component of an assembly? If so, what surfaces or features are used to connect one part to another? Will geometric tolerancing be used on the part and assembly? What units are being used in the design, SI or decimal inch? What is the part's material? What is the primary part feature? How should I model the part, and what features are best used for the primary protrusion (the first solid mass)? On what datum plane should I sketch to model the first protrusion? These and many other questions will be answered as you follow the systematic lesson part. However, you must answer many of the questions on your own when completing the *lesson project*, which does not come with systematic instructions.

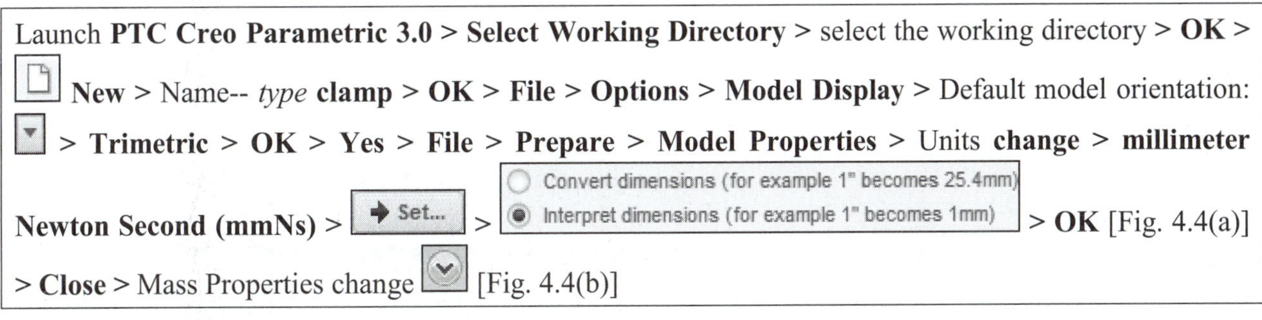

Launch **PTC Creo Parametric 3.0 > Select Working Directory** > select the working directory > **OK >** **New** > Name-- *type* **clamp > OK > File > Options > Model Display** > Default model orientation: > **Trimetric > OK > Yes > File > Prepare > Model Properties** > Units **change > millimeter Newton Second (mmNs)** > Set... > Convert dimensions (for example 1" becomes 25.4mm) / Interpret dimensions (for example 1" becomes 1mm) > **OK** [Fig. 4.4(a)] > **Close** > Mass Properties change [Fig. 4.4(b)]

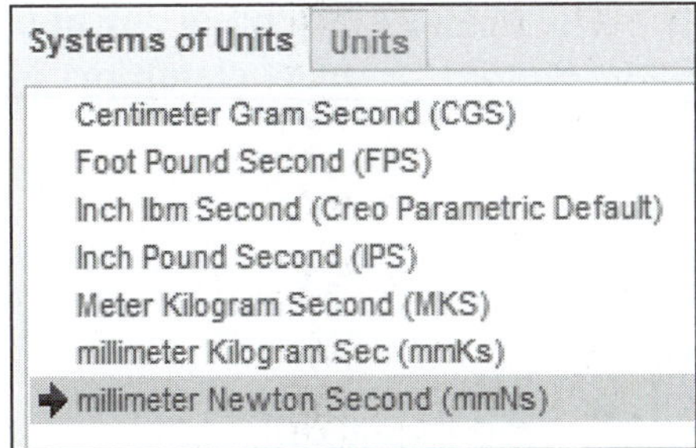

Figure 4.4(a) Units Manager Dialog Box

Model Properties			
Materials			
Material	Not assigned		
Units	millimeter Newton Second (mmNs)		
Accuracy	Relative 0.0012		
Mass Properties			
	Calculation source	Geometry	
	Origin for calculation	PRT_CSYS_DEF	
	Used density	0.0	
Relations, Parameters and Instances			
Relations	Not defined		
Parameters	2 defined		
Instance	Not defined		Active: Generic - CLAMP
Features and Geometry			
Tolerance	ANSI		
Names	4 defined		
Tools			
Flexible	Not defined		
Shrinkage	Not defined		
Simplified Representation	4 defined		Active: Master Rep
Pro/Program			
Interchange	Not defined		
Model Interfaces			
Reference Control	Default settings		
Detail Options			
Detail Options			

Figure 4.4(b) Model Properties Dialog Box

Click: Material **change** > **steel.mtl** > [Fig. 4.4(c)] > double-click on STEEL [Fig. 4.4(d)] > **Thermal** tab [Fig. 4.4(e)] > investigate other options and tabs > **Ok** > **OK** > **Close** > **Ctrl+S** > **Enter** *[you can end many commands by **Enter** or **OK** or **MMB** (Middle Mouse Button)]* > select **clamp.prt** in the Model Tree > **RMB** > **Information** > **Model Information**

PART NAME :	CLAMP				
MATERIAL FILENAME: STEEL					
Units:	**Length:**	**Mass:**	**Force:**	**Time:**	**Temperature:**
millimeter Newton Second (mmNs)	mm	tonne	N	sec	C

> close the Browser

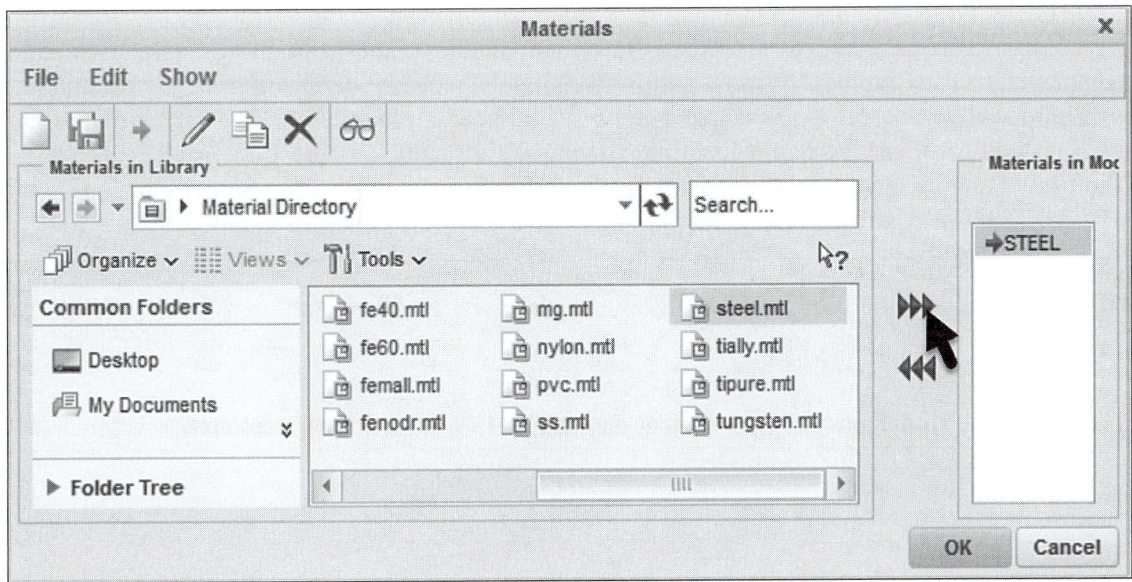

Figure 4.4(c) Material File

Figure 4.4(d) Material Definition, Structural Tab

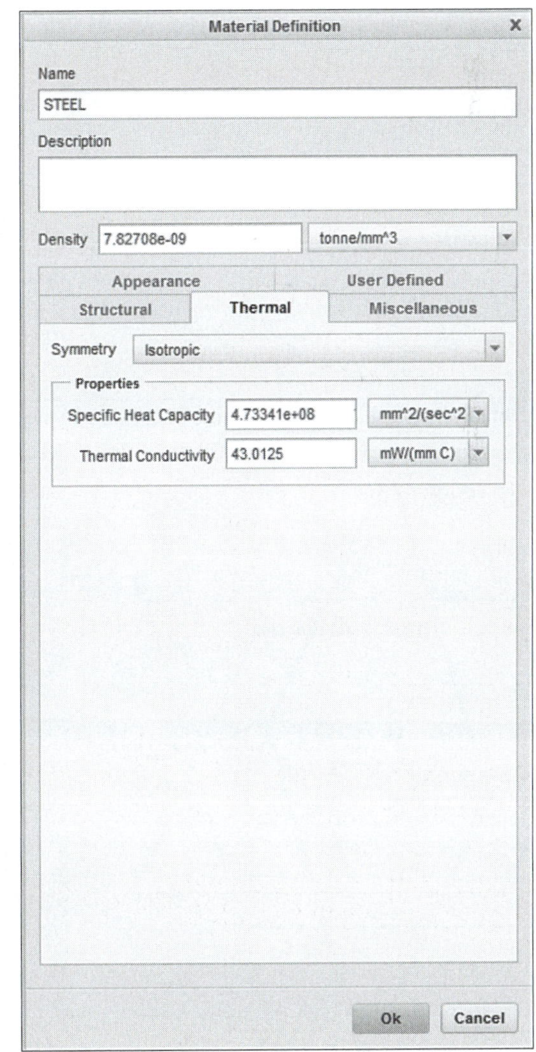

Figure 4.4(e) Material Definition, Thermal Tab

Since [Use default template] was selected, the default datum planes and the default coordinate system are displayed in the Graphics Window and in the Model Tree. *The **default datum planes** and the **default coordinate system** will be the first features on all parts and assemblies.* The datum planes are used to sketch on and to orient the part's features. Having datum planes as the first features of a part, instead of the first extrusion, gives the designer more flexibility during the design process. Picking on an item in the Model Tree will highlight/select that item on the model in the Graphics Window (Fig. 4.5).

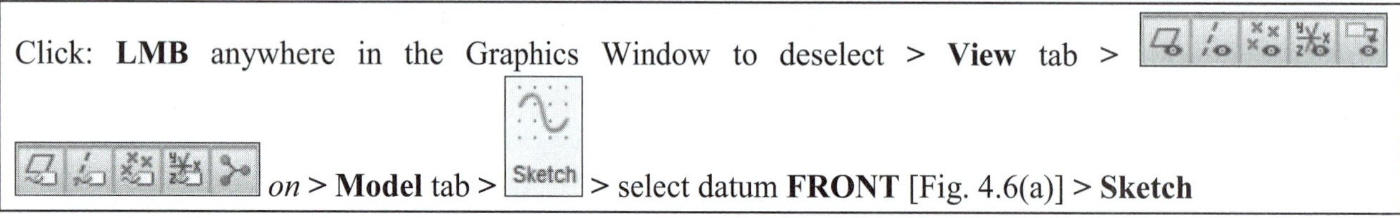

Click: **LMB** anywhere in the Graphics Window to deselect > **View** tab > *on* > **Model** tab > Sketch > select datum **FRONT** [Fig. 4.6(a)] > **Sketch**

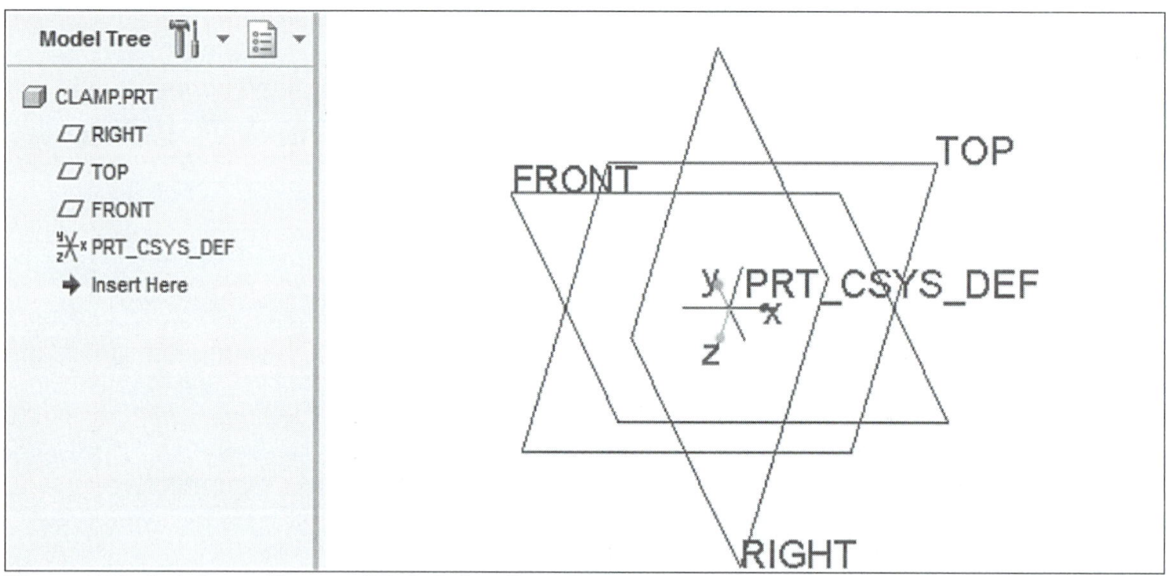

Figure 4.5 Default Datum Planes and Default Coordinate System

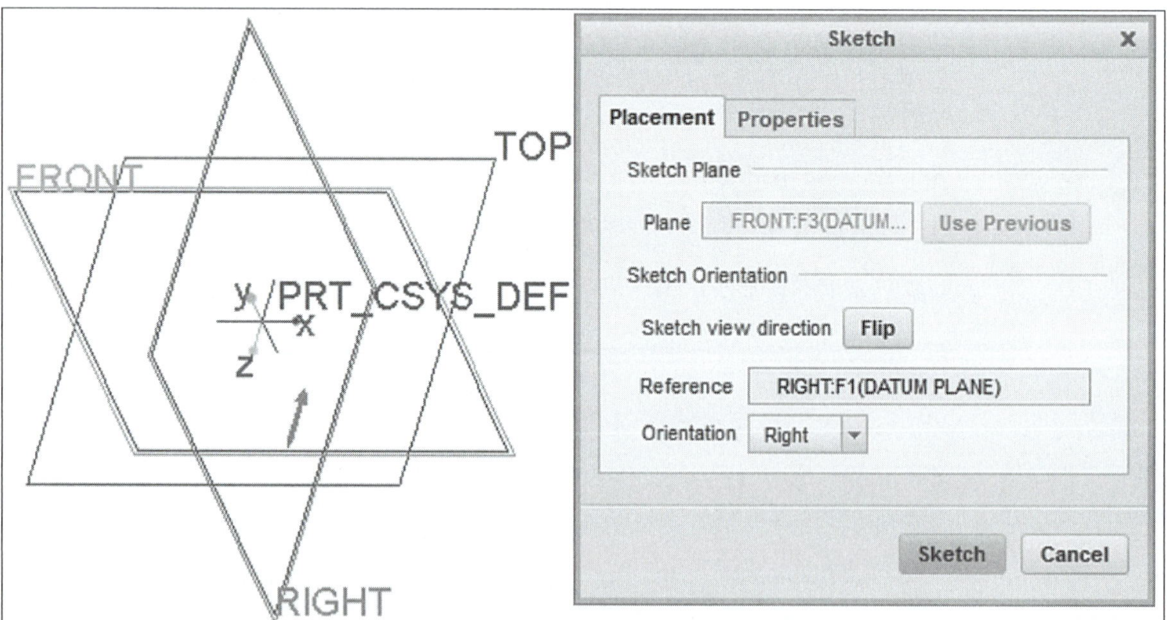

Figure 4.6(a) Sketch Dialog Box, Sketch Plane FRONT *(accept remaining default entries)*

Click: **Sketch View** > in the Graphics Window, press **RMB** > **References** [Fig. 4.6(b)] *(the RIGHT and TOP datum planes are the positional/dimensional references)* > **Close** > **File** > **Options** > **Sketcher** > ☑ Show the grid > ☑ Snap to grid > **OK** > **No** [Fig. 4.6(c)] > **Shift+MMB** (pan) and **Ctrl+MMB** (zoom) to reposition and resize the sketch as needed.

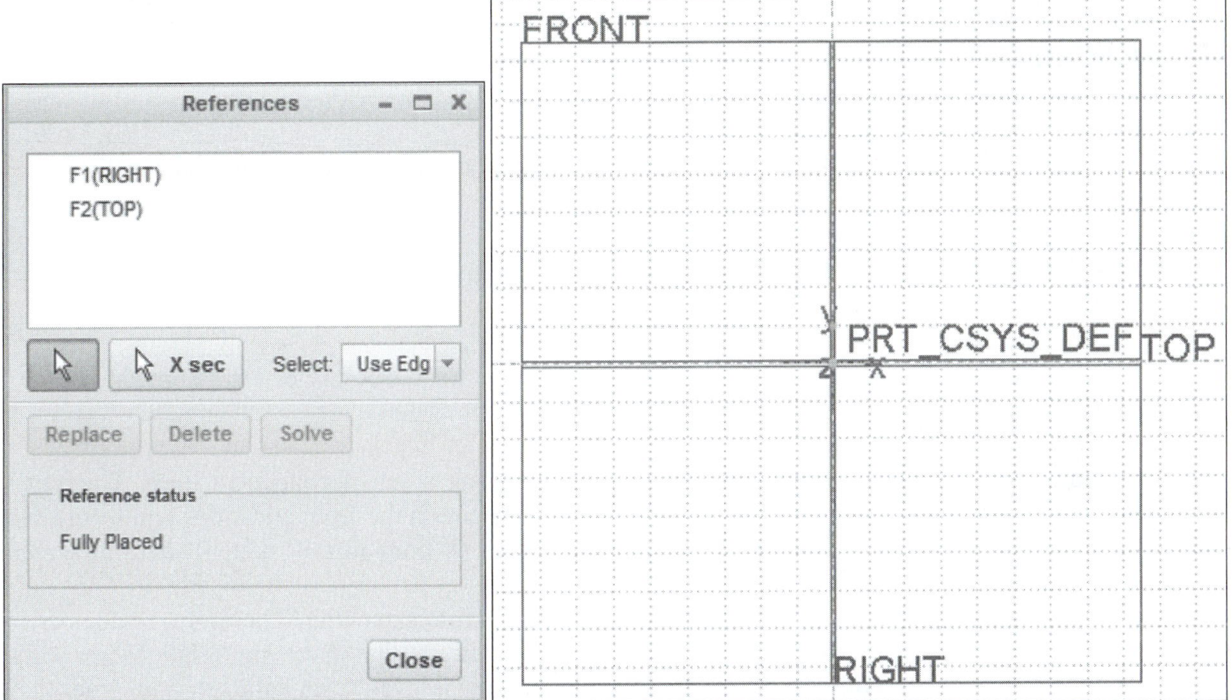

Figure 4.6(b) References Dialog Box

Figure 4.6(c) Grid On and Grid Snap are active

The sketch is now displayed and oriented in 2D [Fig. 4.6(c)]. The coordinate system is at the middle of the sketch, where datum RIGHT and datum TOP intersect. The X coordinate arrow points to the right and the Y coordinate arrow points up. The Z arrow is pointing toward you (out from the screen). The square box you see is the limited display of datum FRONT. This is similar to sketching on a piece of graph paper. Creo Parametric 3.0 is not coordinate-based software, so you need not enter geometry with X, Y, and Z coordinates.

Before starting the sketch, spend some time investigating the Sketch Ribbon [Fig. 4.6(d)].

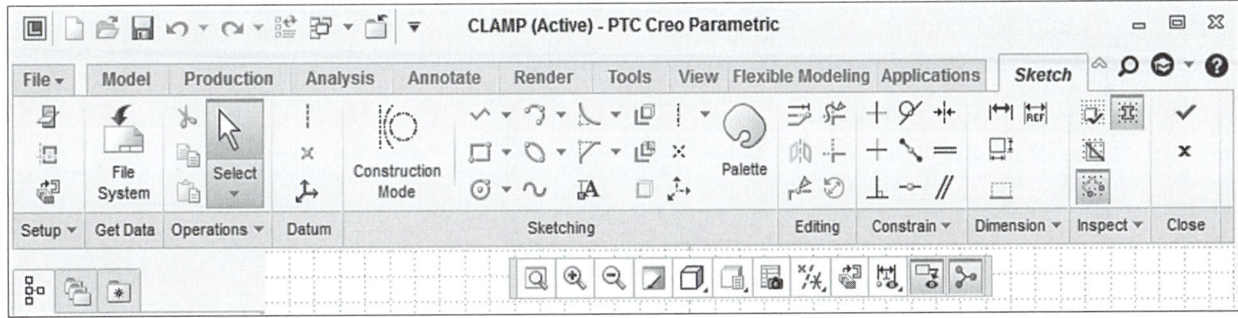

Figure 4.6(d) Sketch Ribbon *(shown here with "Hide Command Labels" checked, your Ribbon may appear differently)*

Because you checked Snap to grid, you can now sketch by simply picking grid points representing the part's geometry (outline). Because this is a sketch in the true sense of the word, you need only create geometry that *approximates* the shape of the feature; the sketch does not have to be accurate as far as size or dimensions are concerned. No two sketches will be the same between those using these steps, unless you count each grid space (which is unnecessary). Even with the grid snap off, Creo Parametric constrains the geometry according to rules, which include but are not limited to the following:

- **RULE:** Symmetry
  DESCRIPTION: Entities sketched symmetrically about a centerline are assigned equal values with respect to the centerline
- **RULE:** Horizontal and vertical lines
  DESCRIPTION: Lines that are approximately horizontal or vertical are considered exactly horizontal or vertical
- **RULE:** Parallel and perpendicular lines
  DESCRIPTION: Lines that are sketched approximately parallel or perpendicular are considered exactly parallel or perpendicular
- **RULE:** Tangency
  DESCRIPTION: Entities sketched approximately tangent to arcs or circles are assumed to be exactly tangent

The outline of the part's primary feature is sketched using a set of connected lines. The part's dimensions and general shape are provided in Figure 4.6(e). The cut on the front and sides will be created with separate sketched features. Sketch only one series of lines (8 lines in this sketch). ***Do not sketch lines on top of lines.***

It is important not to create any unintended constraints while sketching. Therefore, remember to exaggerate the sketch geometry and not to align geometric items that have no relationship. Creo Parametric is very smart: if you draw two lines at the same horizontal level, Creo Parametric assumes they are horizontally aligned. Two lines the same length will be constrained as so.

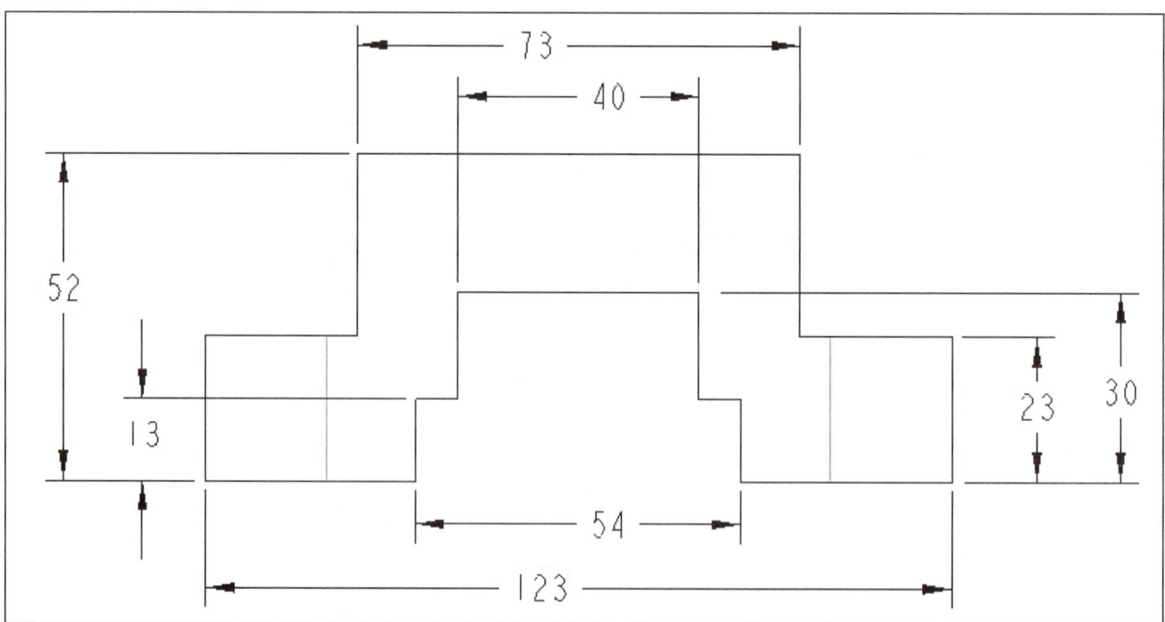

Figure 4.6(e) Front View of *Drawing* Showing Dimensions for the Clamp *(commands continue on the next page)*

With the cursor anywhere in the Graphics Window, but not on an object, press: **RMB** [Fig. 4.6(f)] > **Construction Centerline** > pick two vertical positions on the edge of datum RIGHT [Fig. 4.6(g)] > move the mouse > **MMB** > **LMB** > **Datum Display Filters** > **Select All** Toggle all *off* > **LMB** in the Graphics Window > press **RMB** > **Line Chain** > sketch eight lines *(watch the constraints as they appear- make each segment a different length and equal on one side of the centerline)* [Fig. 4.6(h)] > **MMB** to end the line sequence [Fig. 4.6(i)] > **MMB** > *If the sketch is not to your liking, click* ***Undo*** *and try again.*

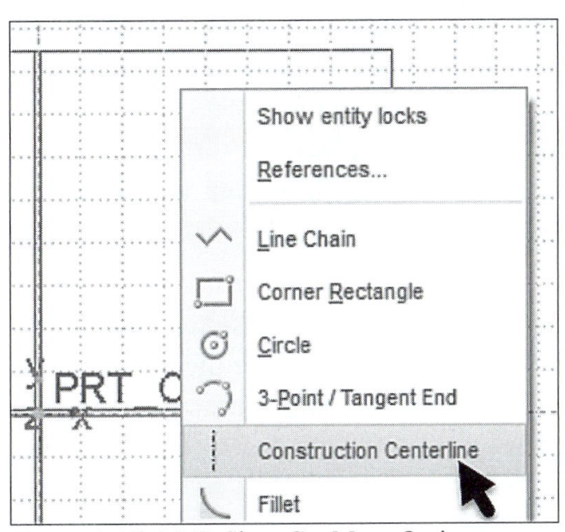

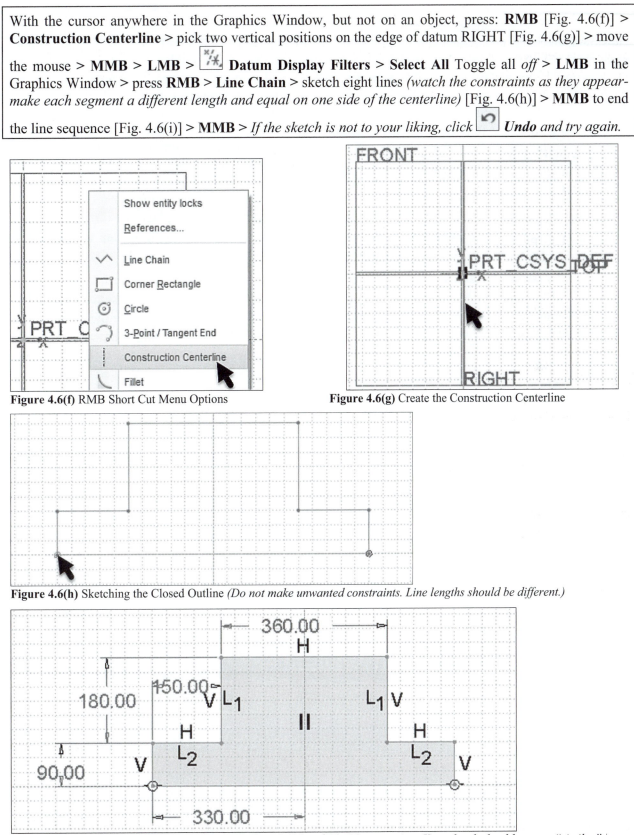

Figure 4.6(f) RMB Short Cut Menu Options

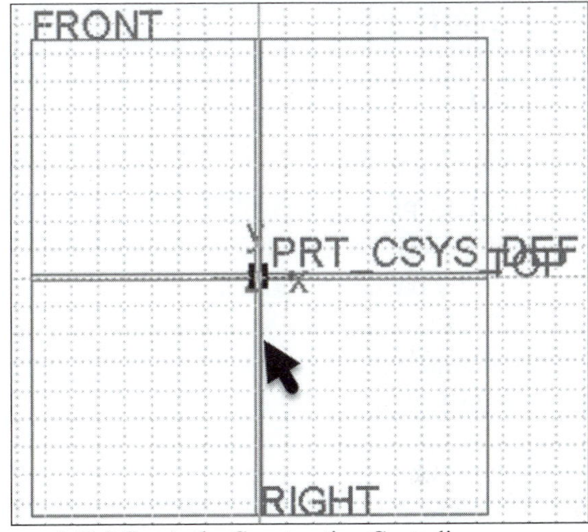

Figure 4.6(g) Create the Construction Centerline

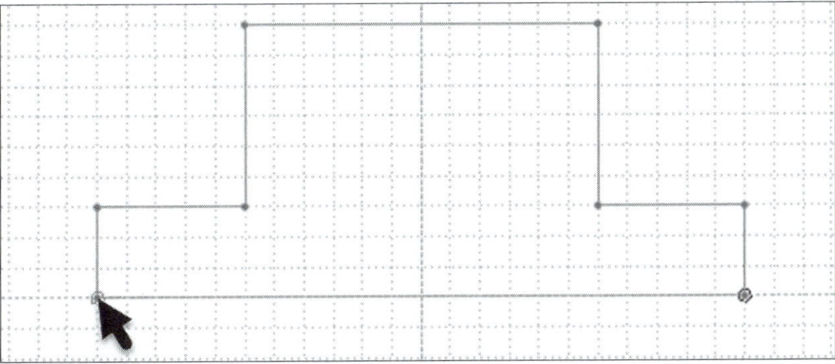
Figure 4.6(h) Sketching the Closed Outline *(Do not make unwanted constraints. Line lengths should be different.)*

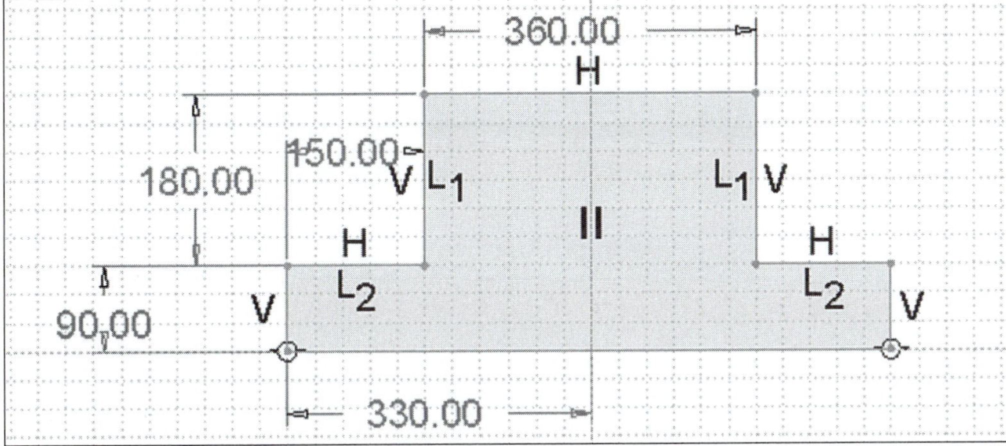

Figure 4.6(i) Default Dimensions Display *(Note the **L1=L1** and **L2=L2** constraints. Your sketch should appear "similar".)*

A sketcher constraint symbol L_1 V appears next to the entity that is controlled by that constraint. Sketcher constraints can be turned on or off (enabled or disabled) while sketching. Simply click the RMB as you sketch- before picking the position- and the constraint that is displaying will have a slash imposed over it. This will disable it for that entity. An **H** next to a line means horizontal; a **T** means tangent. Dimensions display, as they are needed according to the references selected and the constraints. Seldom are they the same as the required dimensioning scheme needed to manufacture the part. The dimensioning scheme is important, not the dimension value, which can be modified.

Place and create the dimensions as required. Do not be concerned with the perfect positioning of the dimensions, but in general, follow the spacing and positioning standards found in the **ASME Y14.5** standards. This saves you time when you create a drawing of the part. Dimensions placed at this stage of the design process are displayed on the drawing document by simply showing all the dimensions.

To dimension between two lines, simply pick the lines (LMB) and place the dimension value (MMB). To dimension a single line, pick on the line (LMB), and place the dimension with MMB.

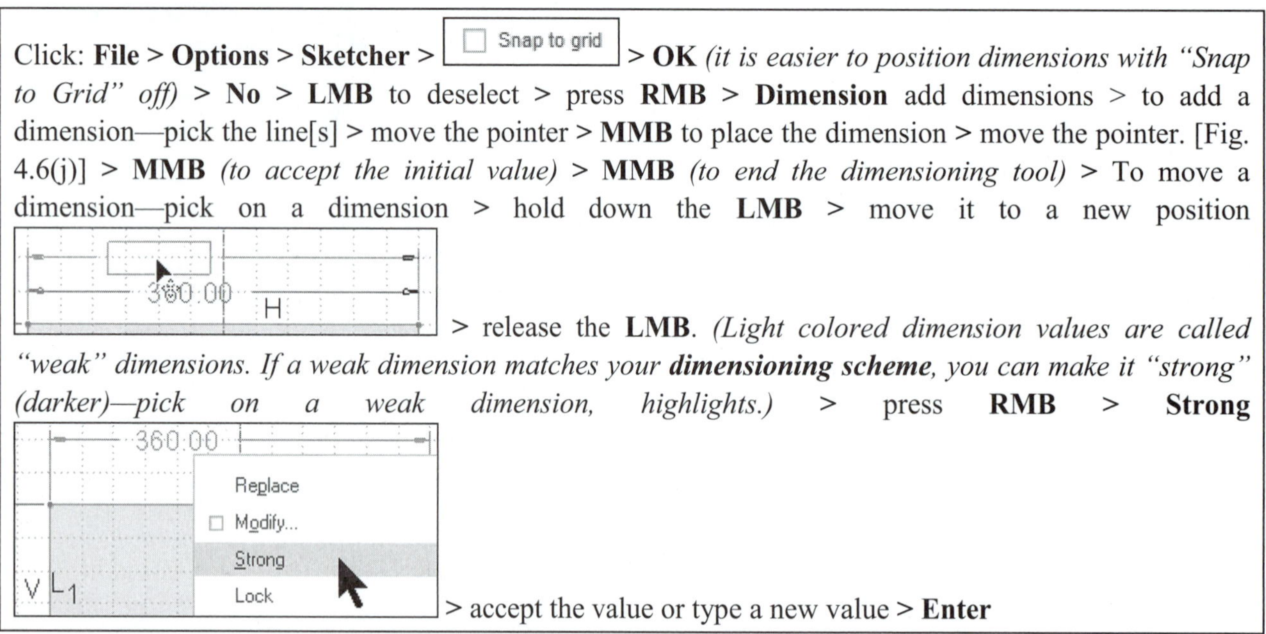

Click: **File > Options > Sketcher >** Snap to grid **> OK** *(it is easier to position dimensions with "Snap to Grid" off)* > **No** > **LMB** to deselect > press **RMB** > **Dimension** add dimensions > to add a dimension—pick the line[s] > move the pointer > **MMB** to place the dimension > move the pointer. [Fig. 4.6(j)] > **MMB** *(to accept the initial value)* > **MMB** *(to end the dimensioning tool)* > To move a dimension—pick on a dimension > hold down the **LMB** > move it to a new position > release the **LMB**. *(Light colored dimension values are called "weak" dimensions. If a weak dimension matches your **dimensioning scheme**, you can make it "strong" (darker)—pick on a weak dimension, highlights.)* > press **RMB** > **Strong** > accept the value or type a new value > **Enter**

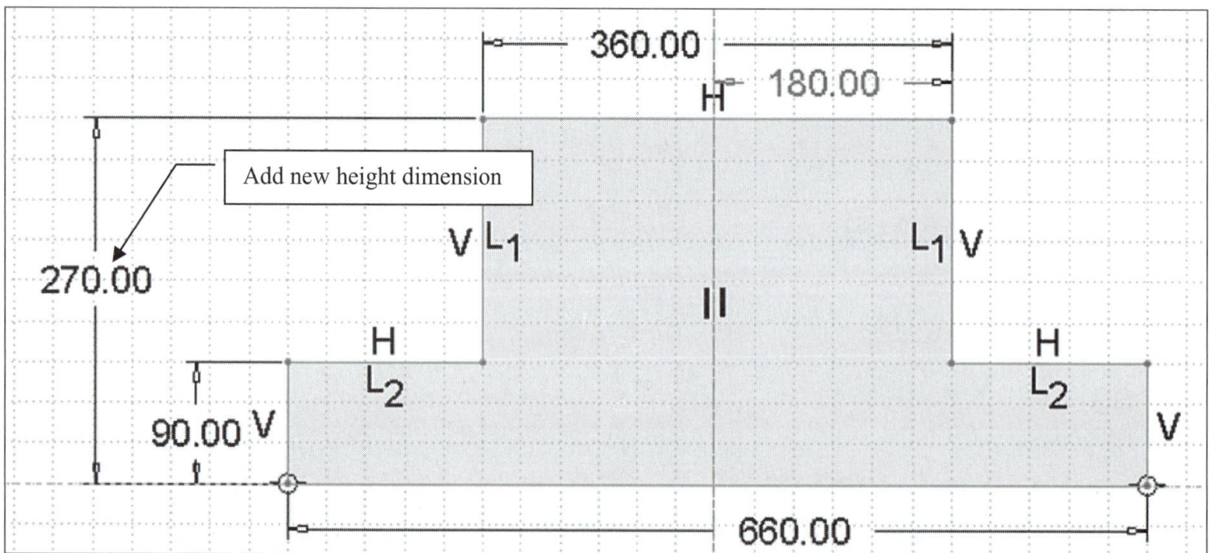

Figure 4.6(j) Re-dimensioned Sketch ***(you must have five similar dimensions- not the values)***

Next, control the sketch by adding symmetry constraints from the Sketch Ribbon; click: **Symmetric** [Fig. 4.6(k)] > pick the centerline and then pick two vertices (endpoints) to be symmetric [Fig. 4.6(l)] *(Note that because of symmetry, one of the weak dimensions is no longer needed and has been automatically deleted. Four dimensions remain.)* > **File** > **Options** > **Sketcher** > ☐ Show the grid > **OK** > **No** [Fig. 4.6(m)]

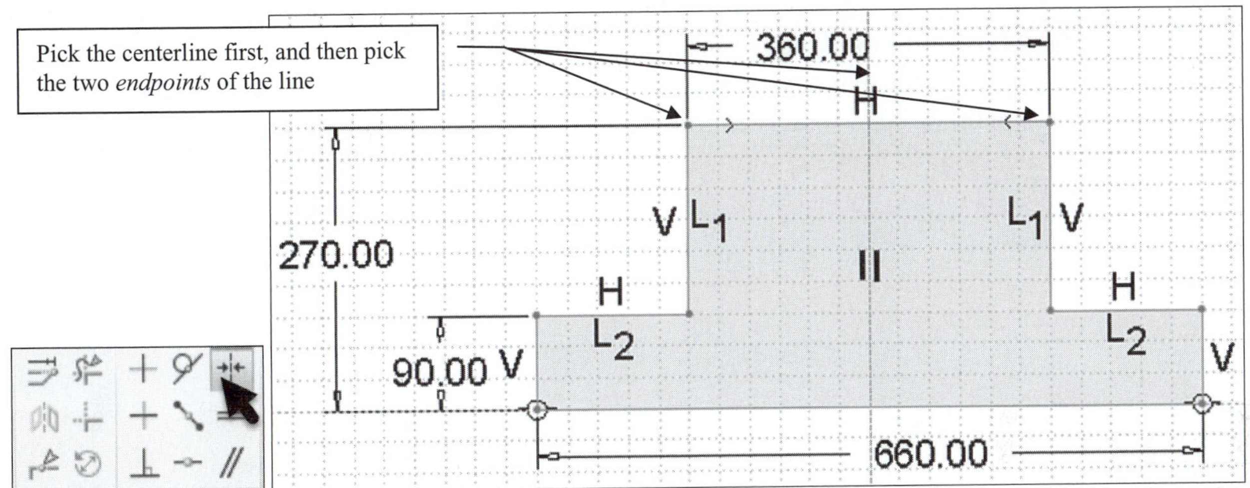

Figure 4.6(k) Constrain Palette

Figure 4.6(l) Adding Symmetry Constraint

Your original sketch *values* may be different from the example, but the final design values will be the same. ***DO NOT*** **CHANGE YOUR SKETCH DIMENSION VALUES TO THOSE IN FIGURE 4.6(m).**

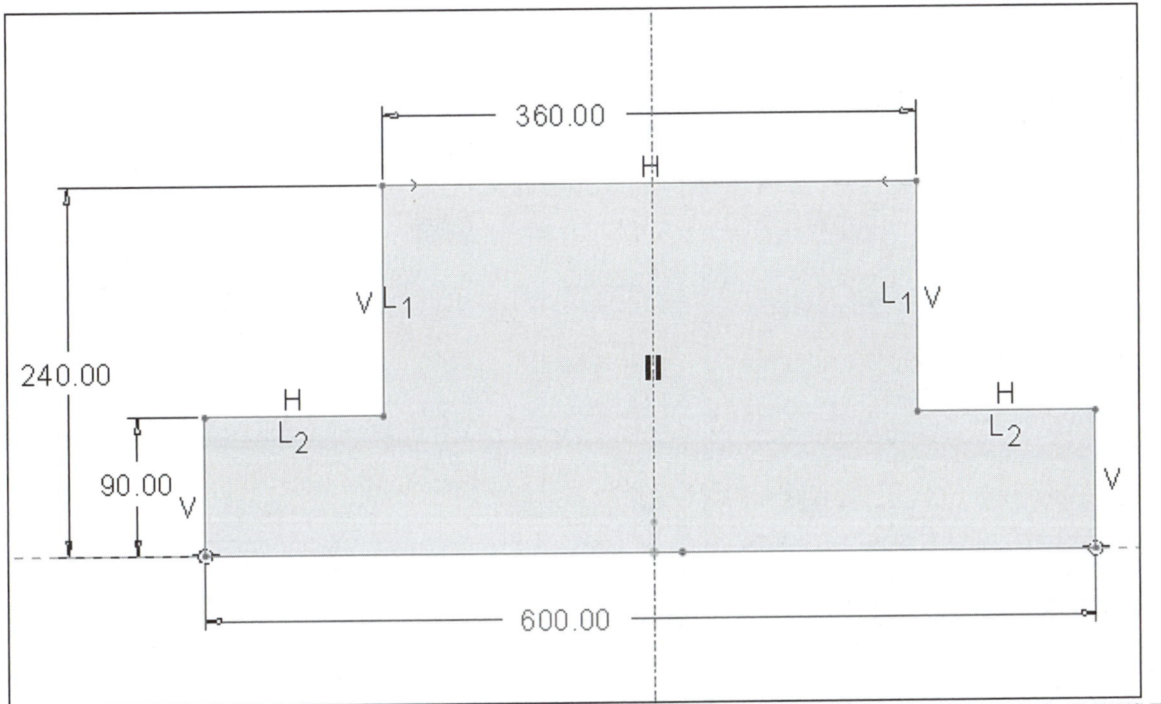

Figure 4.6(m) Sketch is Symmetrical *(your values may be different! – DO NOT CHANGE YOUR VALUES AT THIS TIME!)*

Again, do not change your *sketch* dimension values to these sketch values. Later, you will modify *your* sketch values to the required *design* dimensional values.

Click: Select > Window-in the sketch (place the cursor at one corner of the window; with the **LMB** depressed, drag the cursor to the opposite corner of the window and release the **LMB**) to capture all four dimensions, which will *highlight*. > press **RMB** > **Modify** [Fig. 4.6(n)] > Regenerate *(checked by default)* > Lock Scale > resize the Modify Dimensions dialog box *(so that you can see all four dimensions)* [Fig. 4.6(o)] > in the Modify Dimensions dialog box, locate the dimension for the overall width *(here it is 660, but your dimension value may be different)*, double-click in that length dimension value field and *type* in the design value at the prompt (**123**) > **Enter** > **LMB**

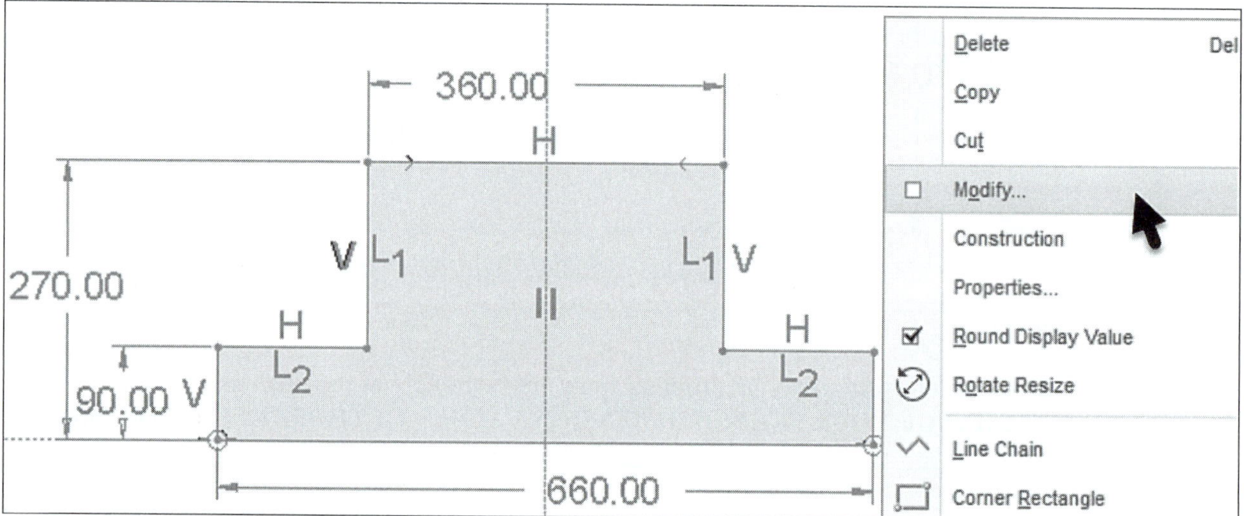

Figure 4.6(n) RMB > Modify Dimensions

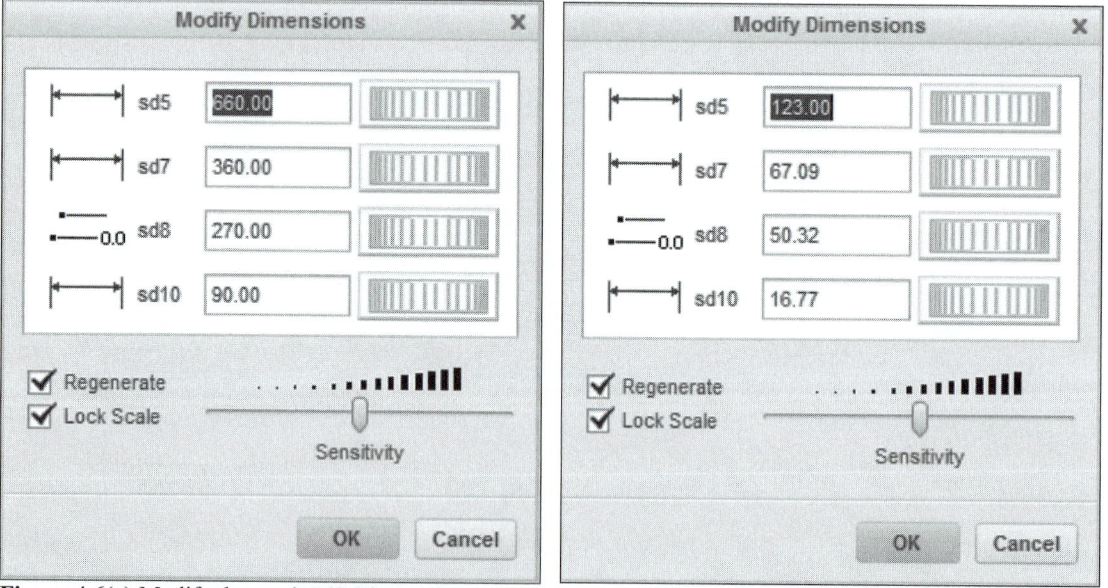

Figure 4.6(o) Modify the weak **660** Dimension to **123** *(your sketch weak dimension value may be different)*

Click: **OK** Regenerate the section and close the dialog [Fig. 4.6(p)] > double-click on another dimension on the sketch (in the Graphics Window) and modify the value > **Enter** > continue until all of the values are changed to the design sizes [Fig. 4.6(q)] *(your sketch **must** have these dimensions and dimension values)* > in the Graphics Window, **LMB** to deselect

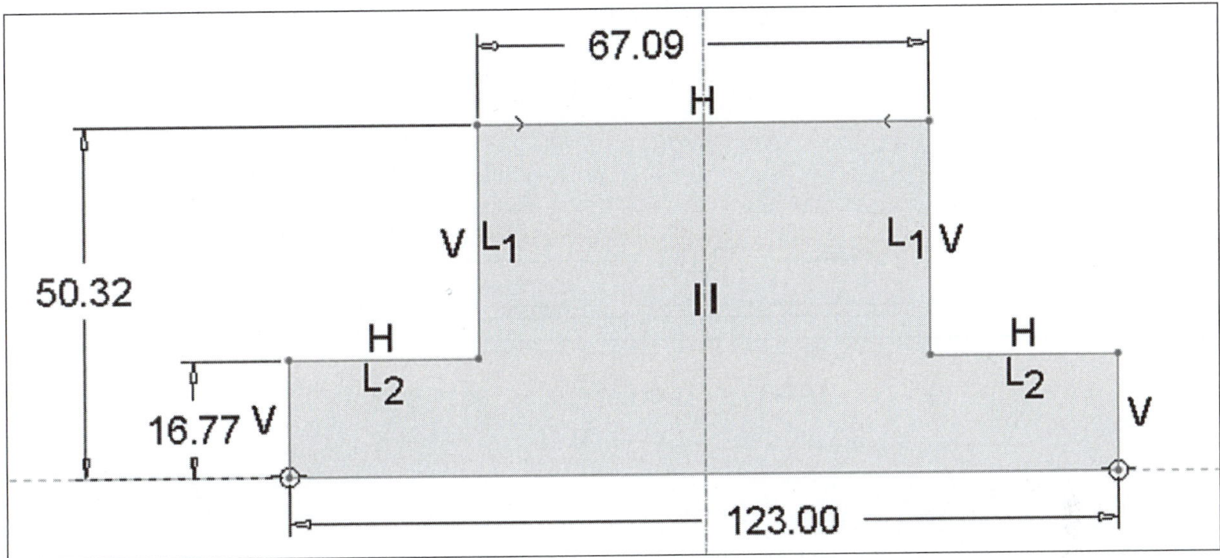

Figure 4.6(p) Modify each Dimension Individually

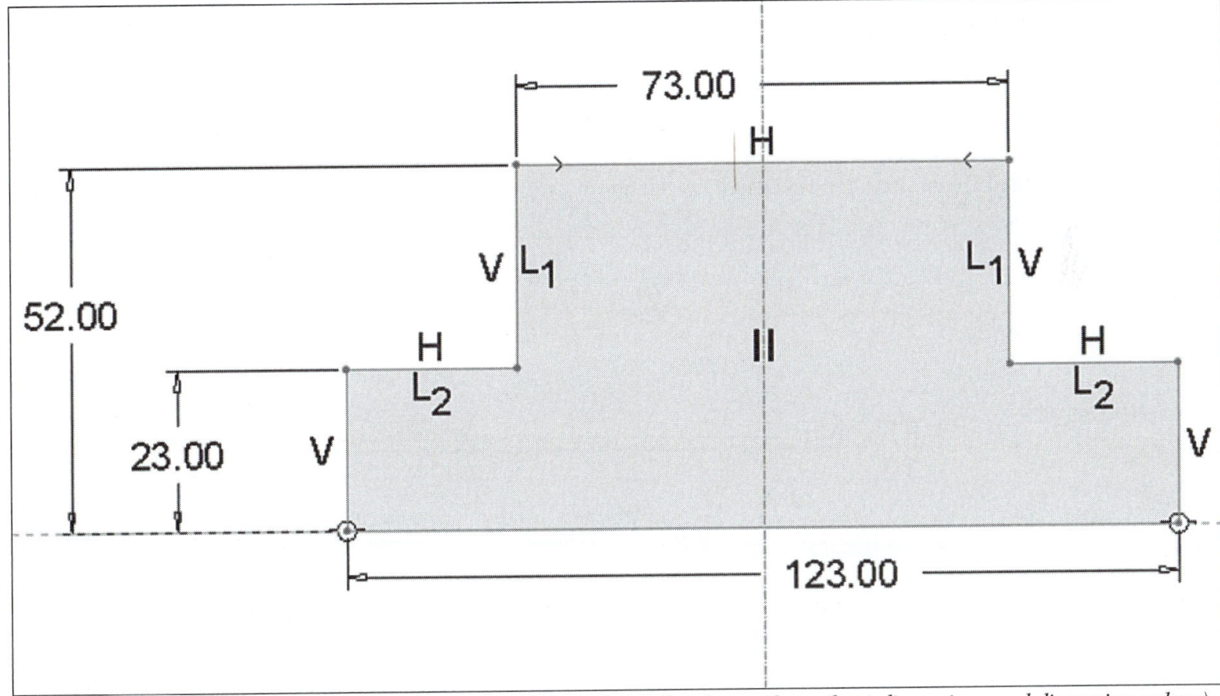

Figure 4.6(q) Modified Sketch showing the Design Values *(your sketch must have these dimensions and dimension values)*

Press: **Ctrl+D** [Fig. 4.6(r)] > **Datum Display Filters** from the Graphics Toolbar > **Select All** Toggle all datum display filters *on* > **LMB** in the Graphics Window > **View** tab on > **Sketch** tab > in the Graphics Window, press **RMB** > **OK** > **Save** from the Quick Access Toolbar > Sketch 1 remains *highlighted* in the Graphics Window and Model Tree, active and therefore *selected* [Fig. 4.6(s)].}

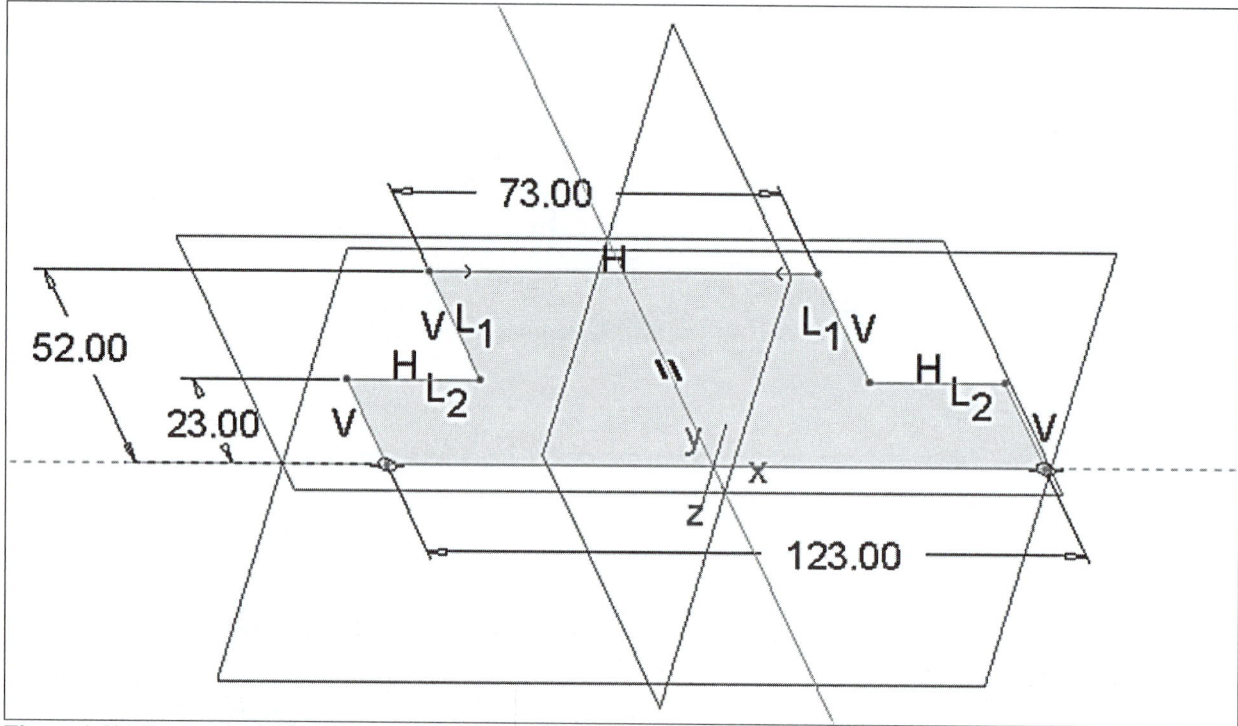

Figure 4.6(r) Regenerated Dimensions in the Standard Orientation View

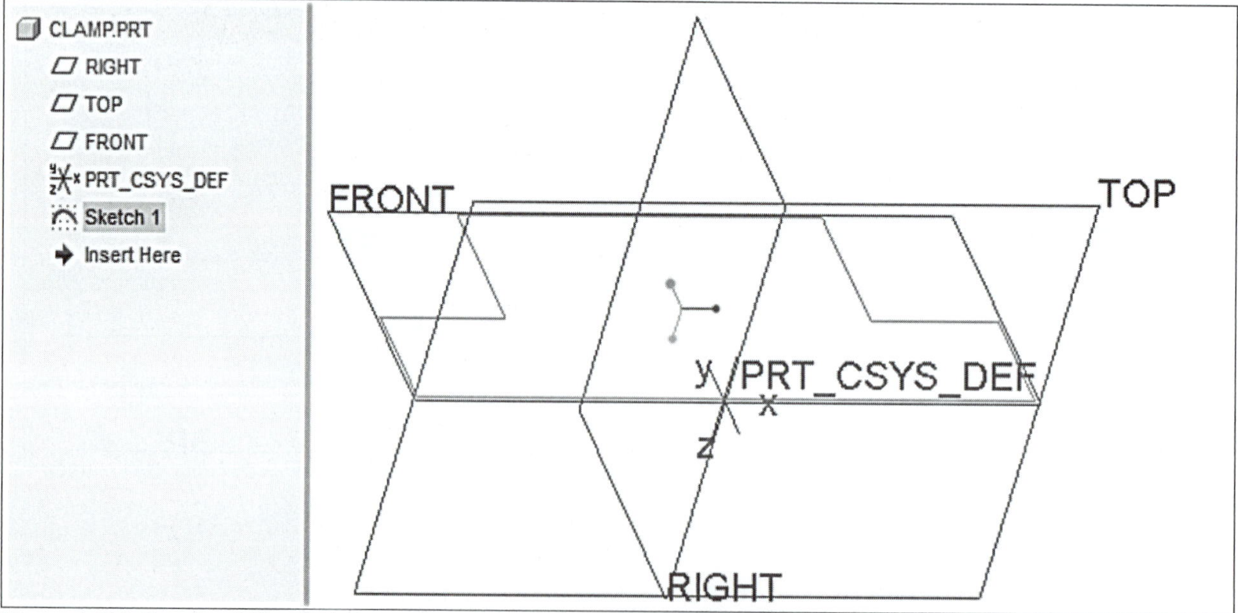

Figure 4.6(s) Completed Sketched Curve *(Datum Curve) [your Sketch number may be different]*

With the sketch *still selected*, click: Extrude **Extrude** [Fig. 4.7(a)] > double-click on the depth value on the model > *type* **70** [Fig. 4.7(b)] > **Enter** > place the pointer over the square drag handle *(it will change color)* > press **RMB** > **Symmetric** [Fig. 4.7(c)] > **Enter** [Fig. 4.7(d)] > **LMB** > **Ctrl+S**

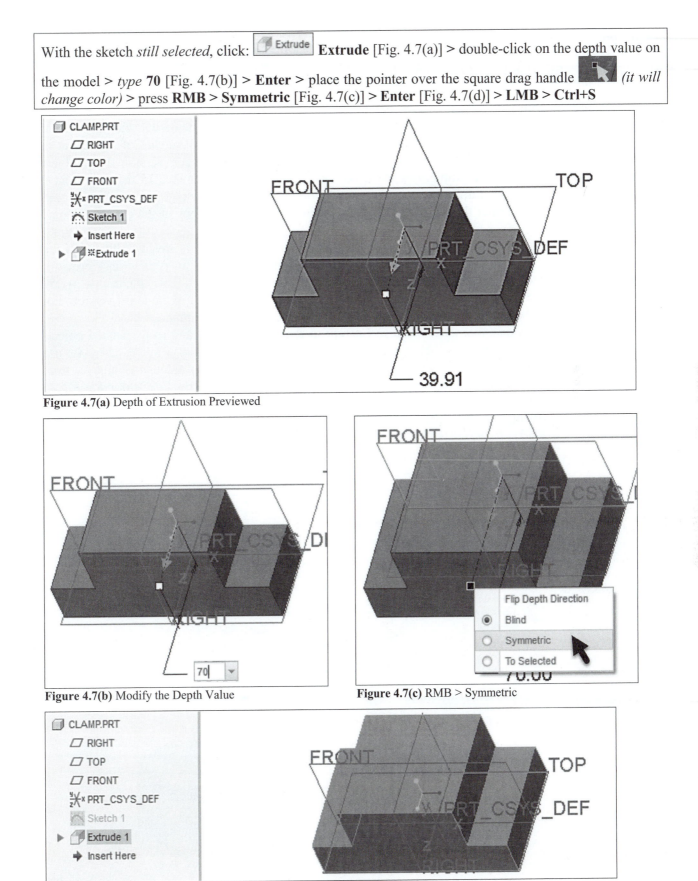

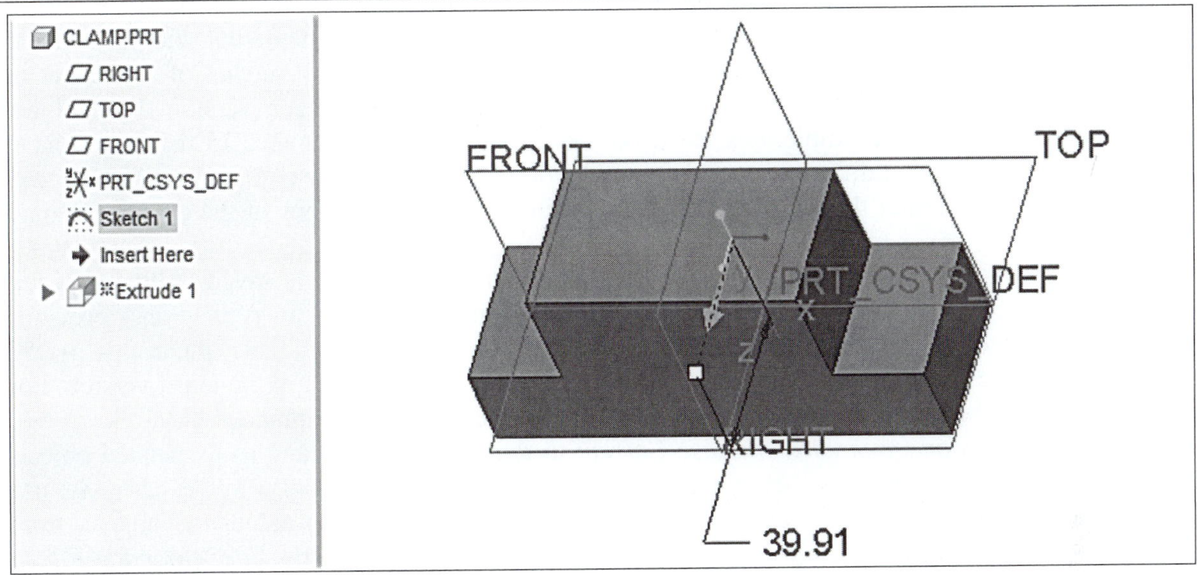

Figure 4.7(a) Depth of Extrusion Previewed

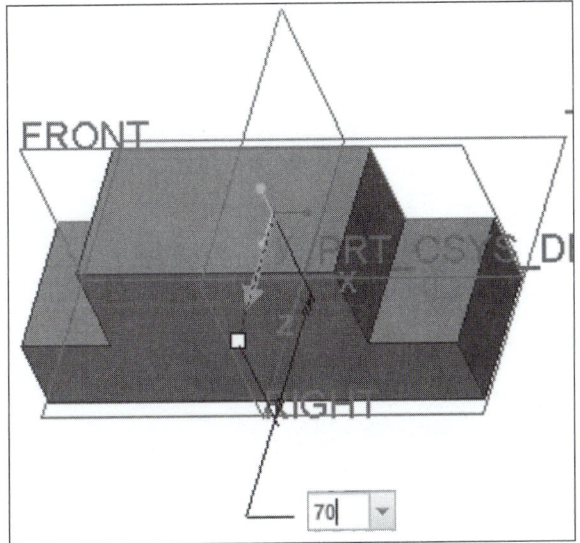

Figure 4.7(b) Modify the Depth Value

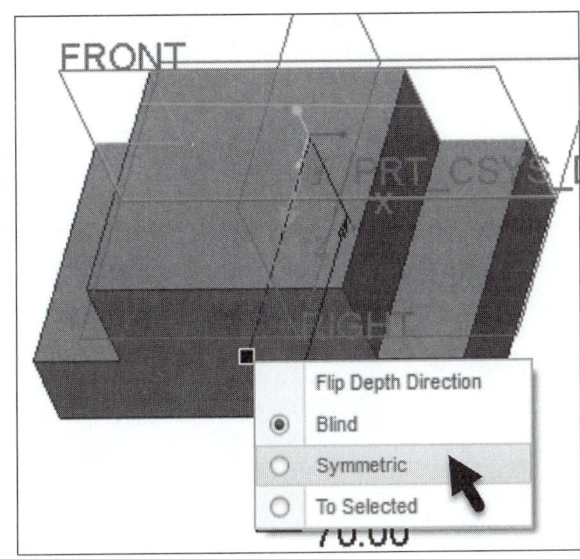

Figure 4.7(c) RMB > Symmetric

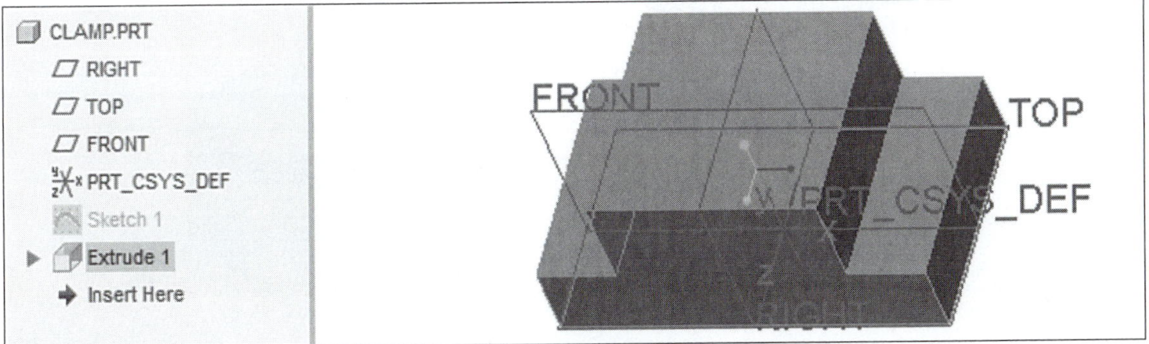

Figure 4.7(d) Completed Extrusion *(Sketch is hidden in the Model Tree)*

Click: **File > Options > Model Display > Trimetric > Isometric > OK > No > Ctrl+D > View** tab > **Display Style > Shading With Edges > Model** tab > [icon] from the Graphics Toolbar > **Ctrl+S**

Storing an object on the disk does not overwrite an existing object file. To preserve earlier versions, Creo Parametric saves the object to a new file with the same object name but with an updated version number. Every time you store an object using Save, you create a new version of the object in memory, and write the previous version to disk. Creo Parametric numbers each version of an object storage file consecutively (for example, box.sec.1, box.sec.2, box.sec.3). If you save 25 times, you have 25 versions of the object, all at different stages of completion. You can use *File > Manage File > Delete Old Versions* [Fig. 4.7(e)] after the *Save* command to eliminate previous versions of the object that may have been stored.

When opening an existing object file, you can open any version that is saved. Although Creo Parametric automatically retrieves the latest saved version of an object, you can retrieve any previous version by entering the full file name with extension and version number (for example, **partname.prt.5**). If you do not know the specific version number, you can enter a number relative to the latest version. For example, to retrieve a part from two versions ago, enter **partname.prt.3** *(or **partname.prt.-2**)*.

You use *File >Manage Session > Erase Current* to remove the object and its associated objects from memory. If you close a window before erasing it, the object is still in memory. In this case, you use *File >Manage Session > Erase Not Displayed* to remove the object and its associated objects from memory. This does not delete the object. It just removes it from active memory. *File >Manage File > Delete All Versions* removes the file from memory and from disk completely. You are prompted with a Delete All Confirm dialog box when choosing this command. Be careful not to delete needed files.

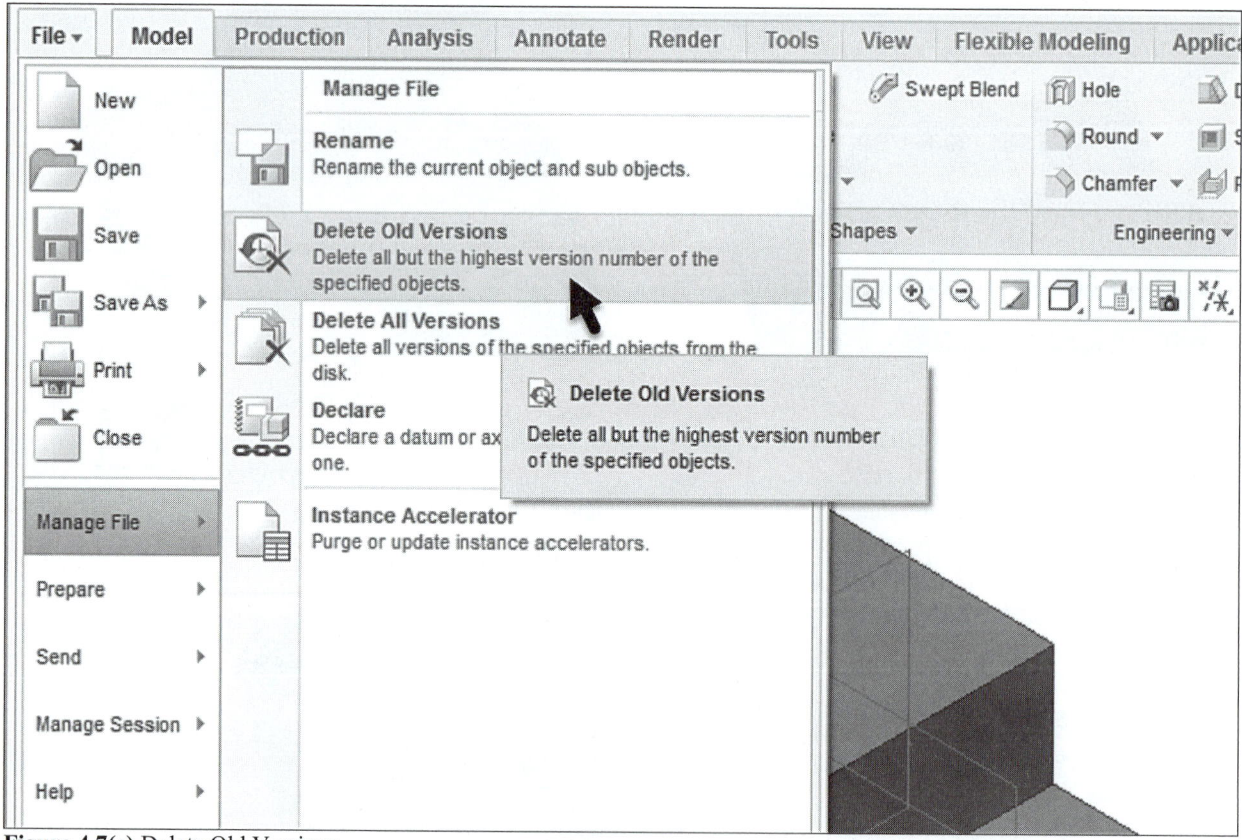

Figure 4.7(e) Delete Old Versions

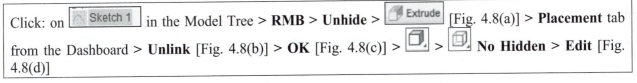

Click: on Sketch 1 in the Model Tree > **RMB** > **Unhide** > Extrude [Fig. 4.8(a)] > **Placement** tab from the Dashboard > **Unlink** [Fig. 4.8(b)] > **OK** [Fig. 4.8(c)] > > **No Hidden** > **Edit** [Fig. 4.8(d)]

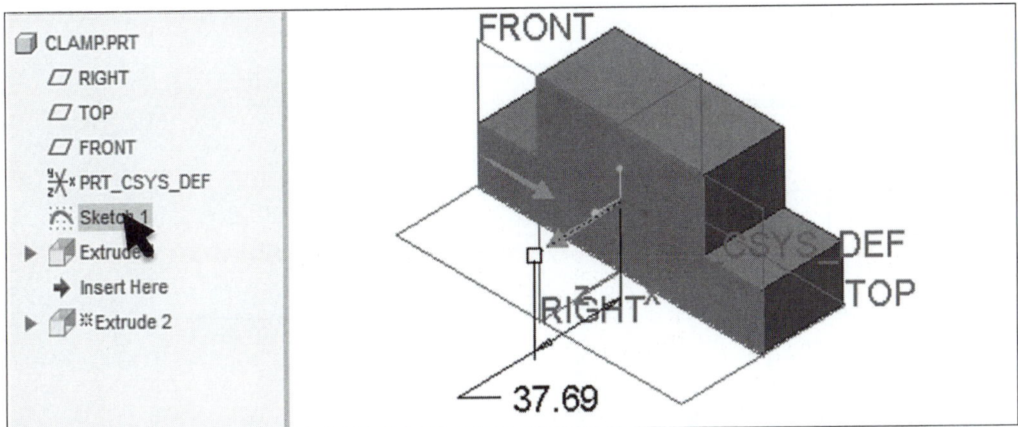

Figure 4.8(a) Click on the Sketch in the Model Tree > Extrude

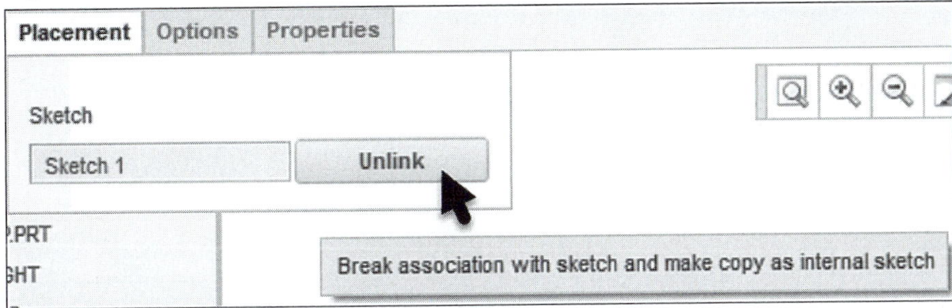

Figure 4.8(b) Unlink

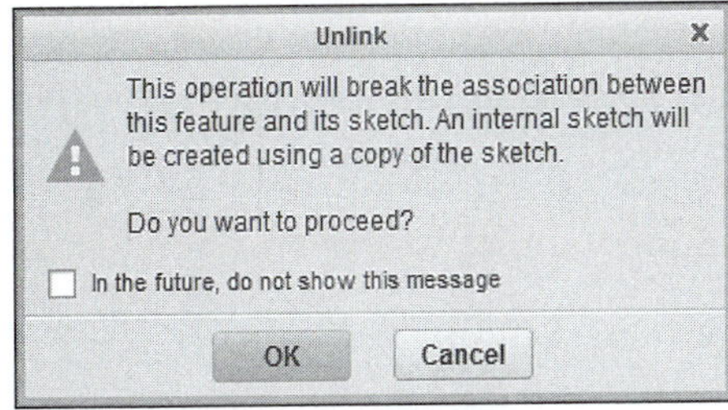

Figure 4.8(c) Unlink Dialog Box

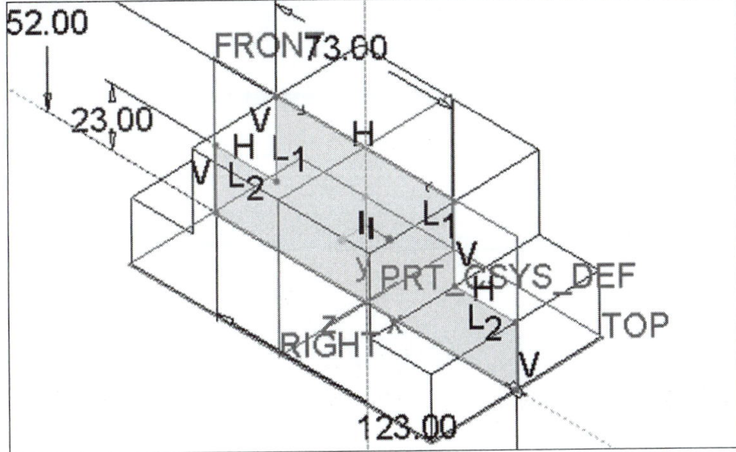

Figure 4.8(d) Internal Sketch Displayed

Click: **Sketch View** > drag the end points of the sketch to show the approximate size of the cut [Fig. 4.8(e)] > double-click on each value and modify to the design size > reposition the dimensions as needed [Fig. 4.8(f)] > **LMB** to deselect (only *four* dimensions remain > **Ctrl+D**

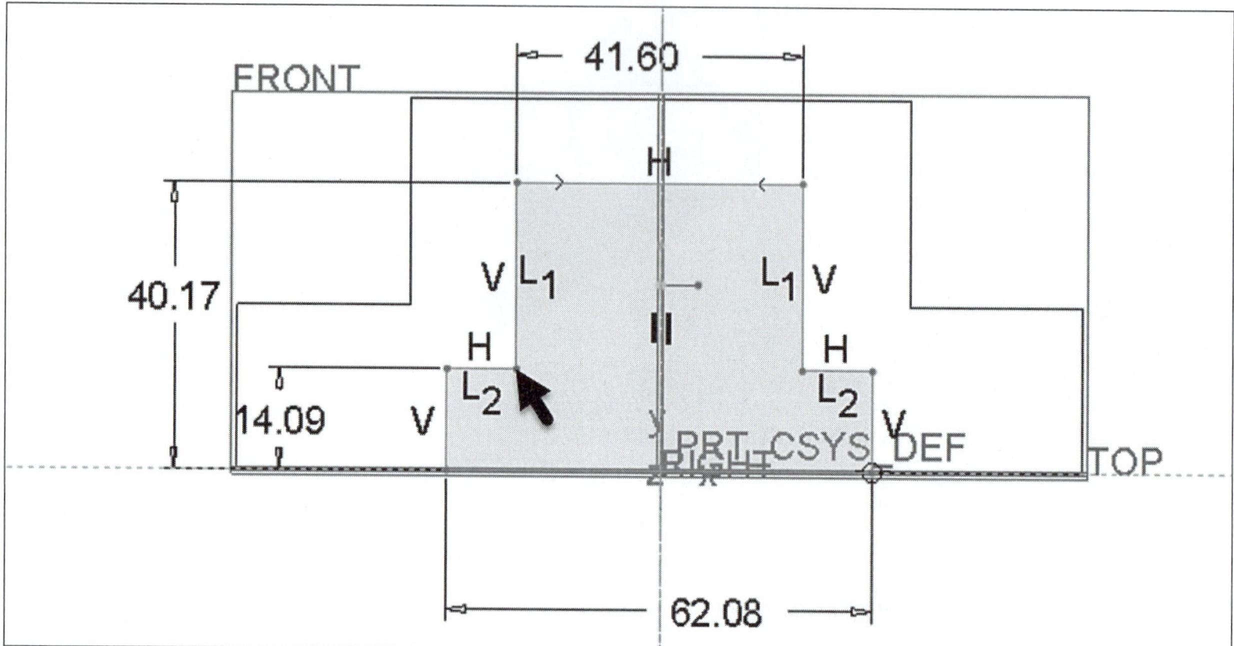

Figure 4.8(e) Drag Sketch Lines and End Points

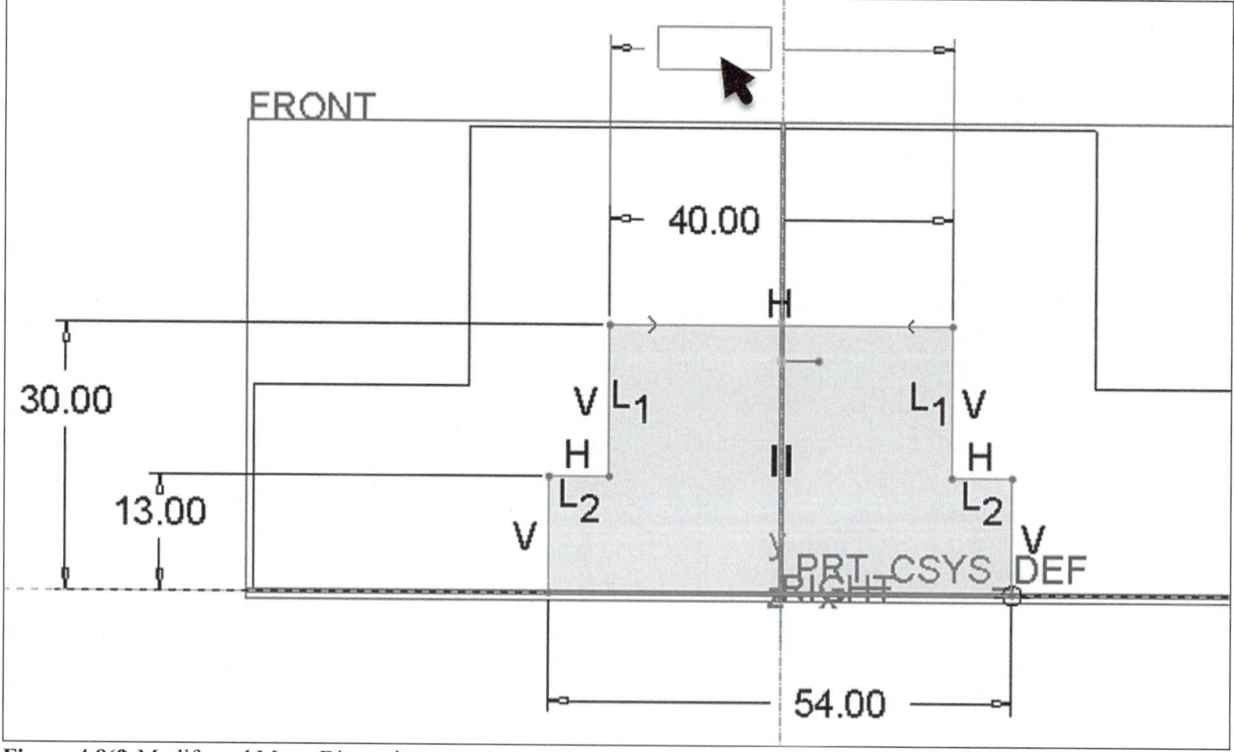

Figure 4.8(f) Modify and Move Dimensions.

Press: **RMB** > **OK** > [icon] > [icon] > press **RMB** > **Remove Material** should be checked > note the cut's direction arrows [Fig. 4.8(g)] > **Options** tab from the Dashboard > Side 1 **Blind** > **Through All** > Side 2 **None** > **Through All** [Fig. 4.8(h)] > **MMB** to complete the feature > **Ctrl+S** [Fig. 4.8(i)] > **LMB**

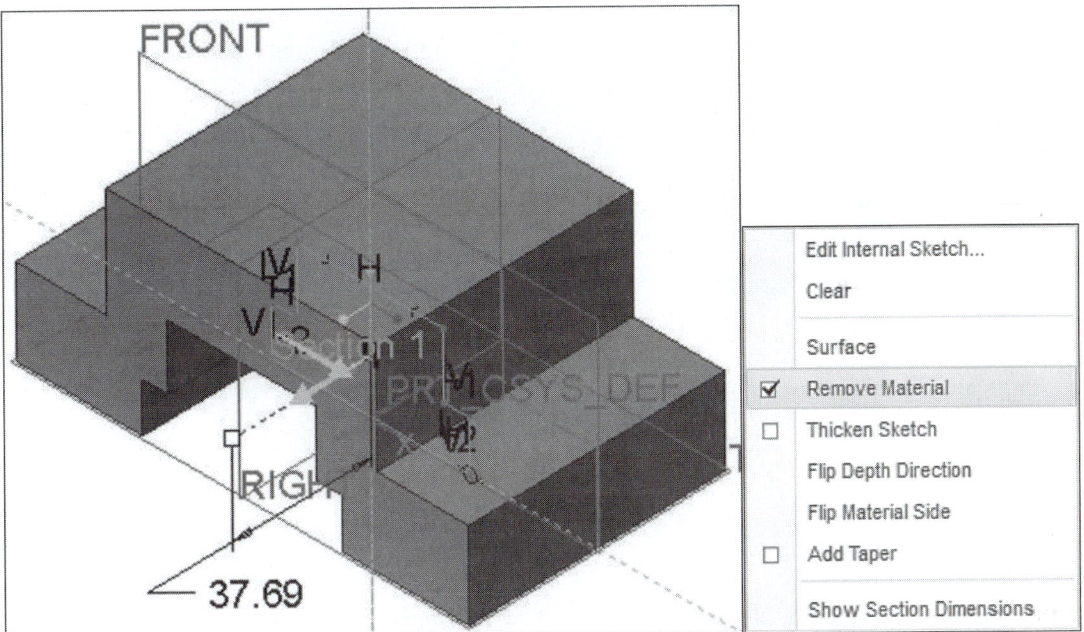

Figure 4.8(g) Cut Preview *(note the direction arrows – your arrows may be at different locations)*

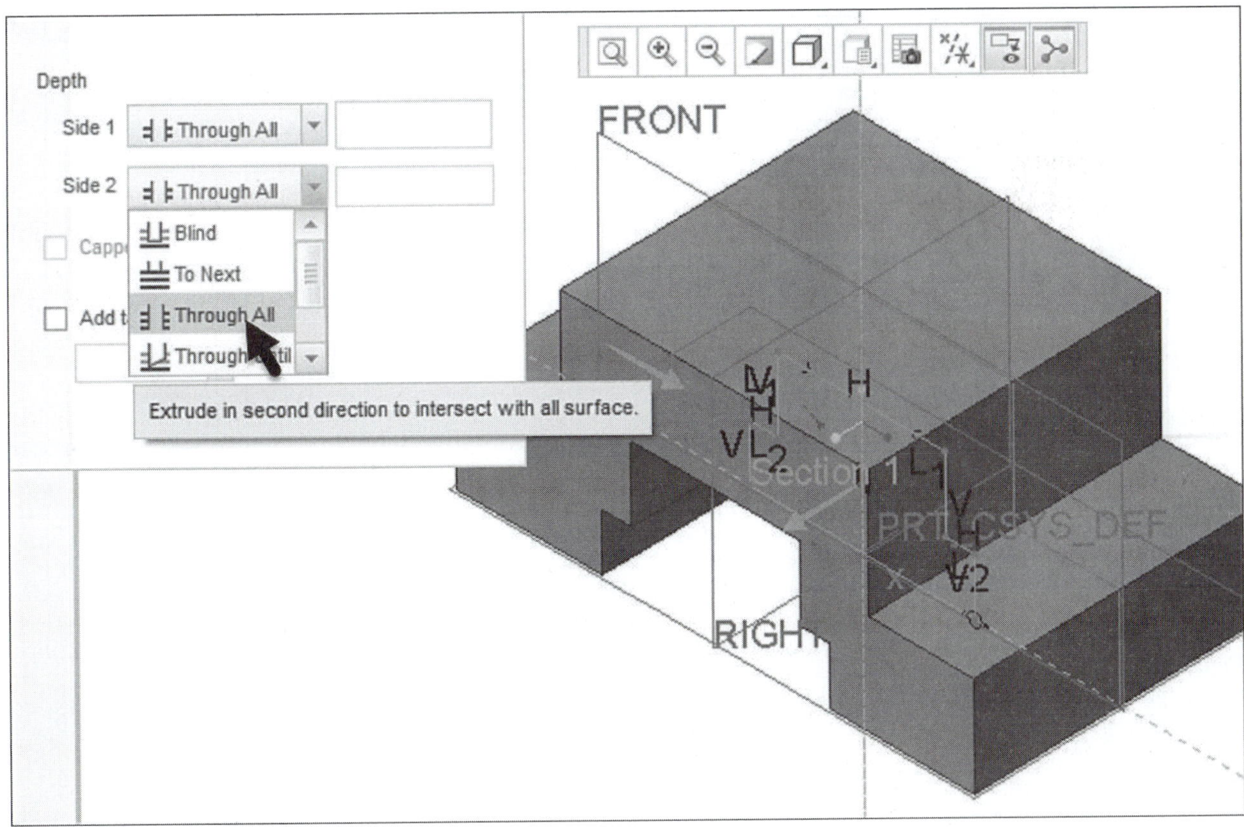

Figure 4.8(h) Options Depth

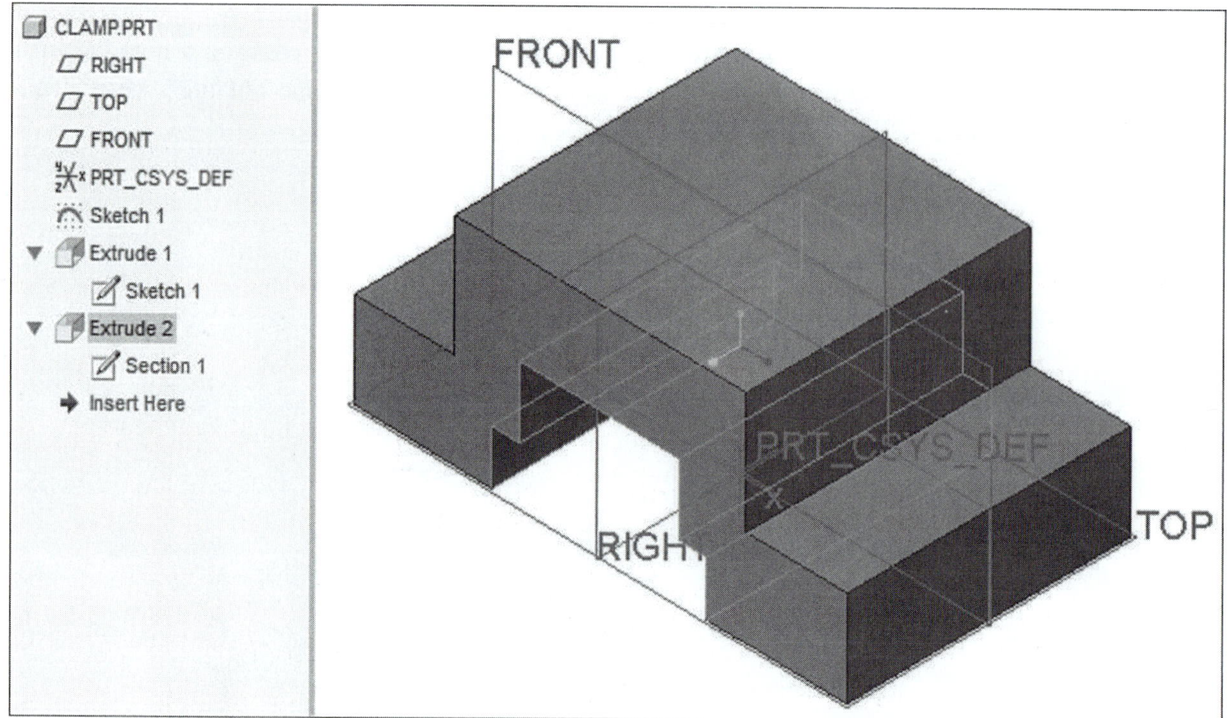

Figure 4.8(i) Completed Through All Cut

The next feature will be a **20 X 20** centered cut (Fig. 4.9). Because the cut feature is identical on both sides of the part, you can mirror and copy the cut after it has been created.

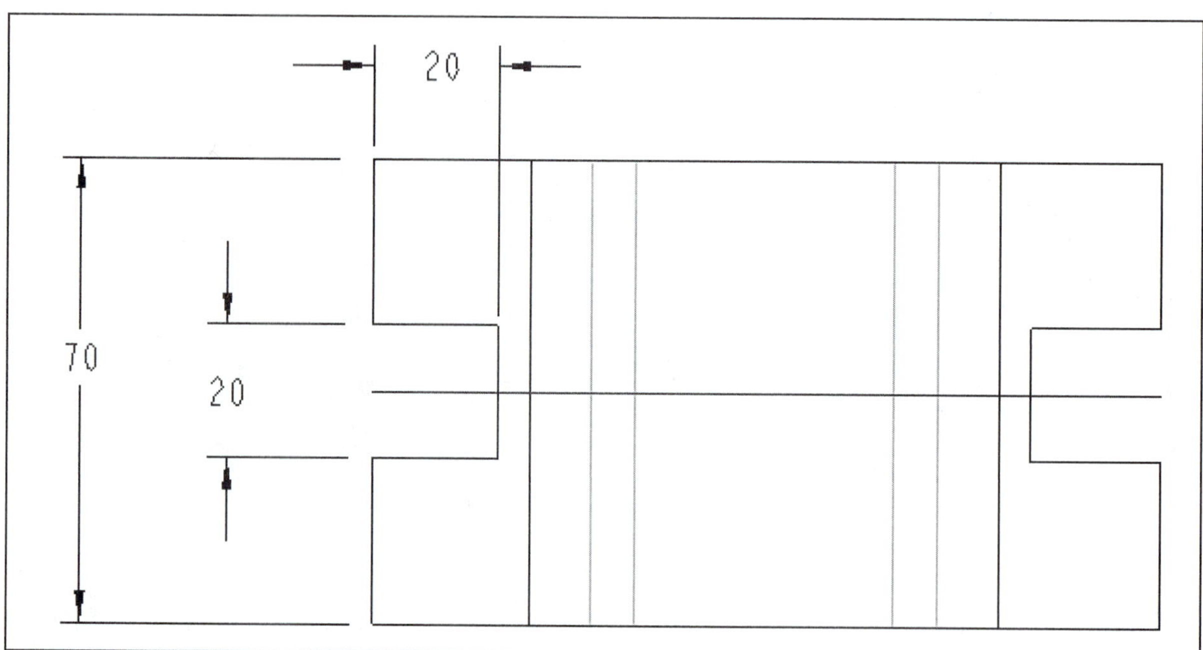

Figure 4.9 Top View of *Drawing* Showing Dimensions for the Cut

Click: Extrude **Extrude** > in the Graphics Window, press **RMB** > **Remove Material** > press **RMB** > **Define Internal Sketch** Sketch dialog box opens [Fig. 4.10(a)] > Sketch Plane--- Plane: select datum **TOP** from the model as the sketch plane > **Sketch** from the Sketch dialog box [Figs. 4.10(b-c)] > **Sketch View**

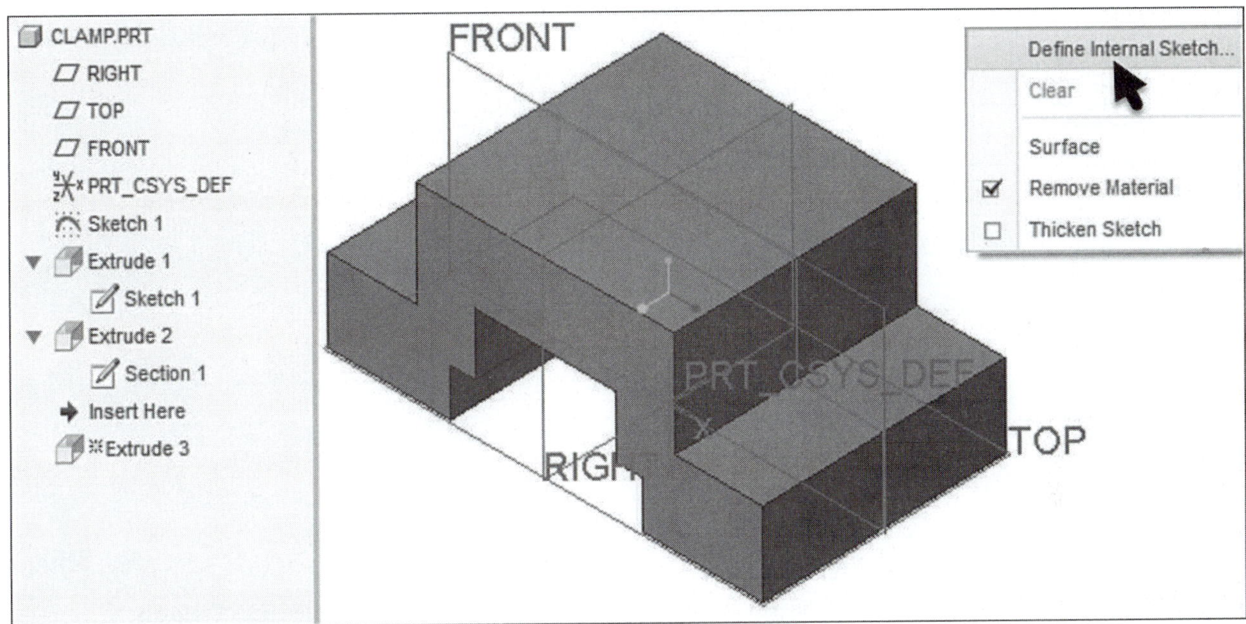

Figure 4.10(a) Define Internal Sketch

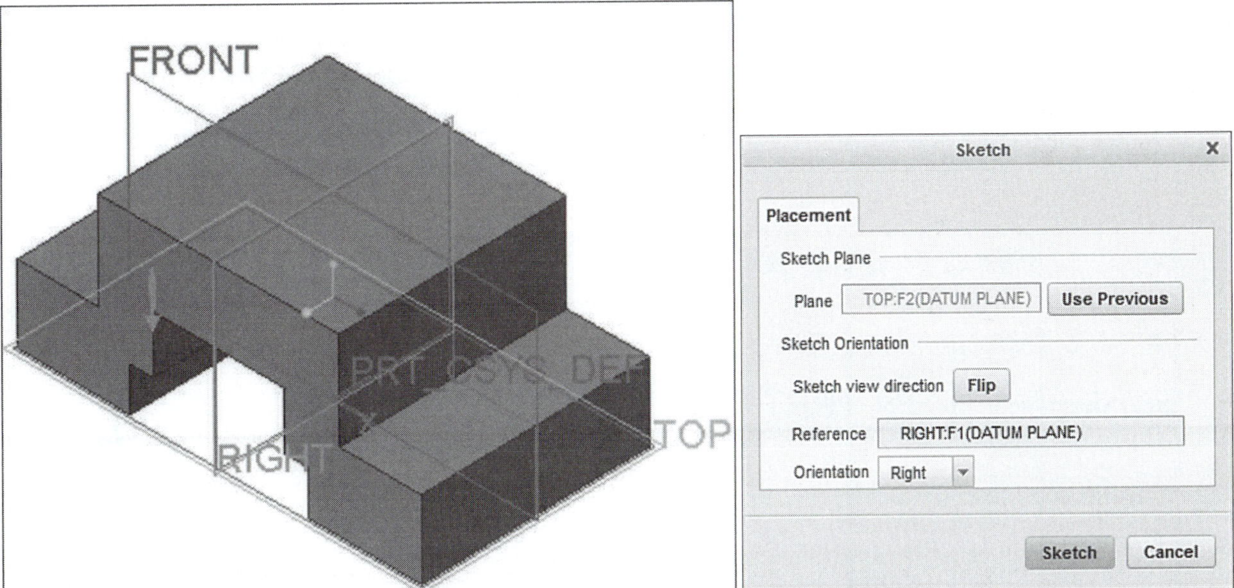

Figure 4.10(b) Top Datum Selected as Sketch Plane

Figure 4.10(c) Sketch Dialog Box

In the Graphics Window, press: **RMB** > **References** > pick the left edge/surface of the part [Fig. 4.10(d)] to add it to the References dialog box [Fig. 4.10(e)] > **Close** > check to see if the grid snap is *off*, click: **File** > **Options** > **Sketcher** > **Cancel** *(if both "Show the Grid" and "Snap to Grid" are unchecked) [If both "Show the Grid" and "Snap to Grid" are checked, then uncheck them > **OK** > No]*

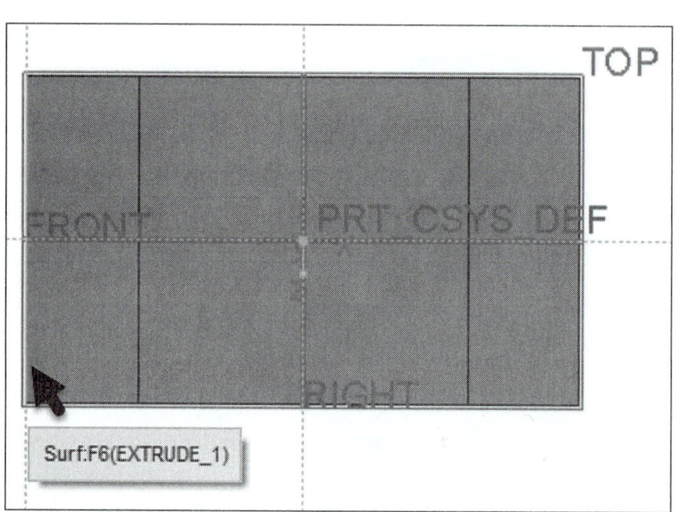

Figure 4.10(d) Add the left edge/surface of the part as a Reference

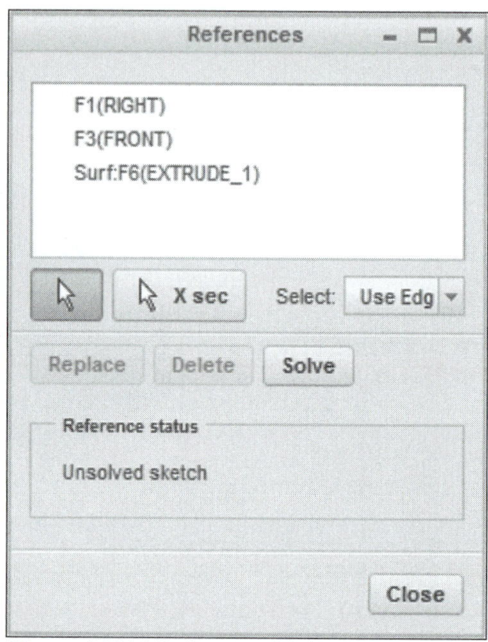

Figure 4.10(e) References Dialog Box

Click: **View** tab > **Display Style** > **Hidden Line** > **Sketch** tab > in the Graphics Window, press **RMB** > **Construction Centerline** [Fig. 4.10(f)] > create a *horizontal* centerline through the center of the part by picking two positions along the edge of the datum FRONT > move the mouse > **MMB** > **LMB**

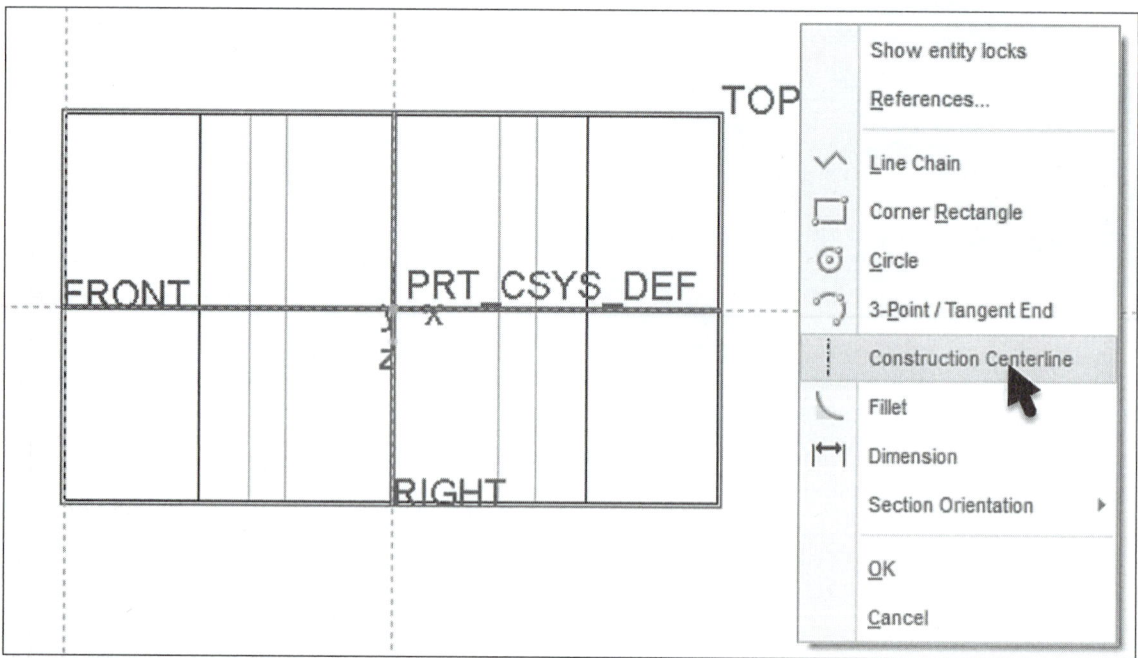

Figure 4.10(f) Horizontal Centerline

Press: **RMB > Corner Rectangle** [Fig. 4.10(g)] > Place the mouse on the left edge, click **LMB** to begin creating a rectangle/square. Move the pointer and make sure the legs of the rectangle are equal: **L = L**. [Fig. 4.10(h)] > **LMB** to complete the current rectangle > **MMB** to end the rectangle tool [Fig. 4.10(i)] > double-click on the dimension [Fig. 4.10(j)] > *type* **20** > **Enter** > move the mouse > **LMB** to deselect [Fig. 4.10(k)]

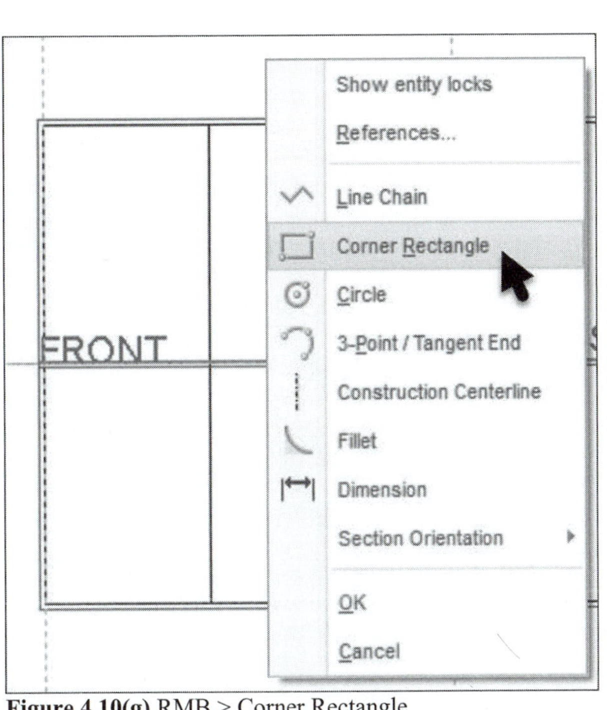

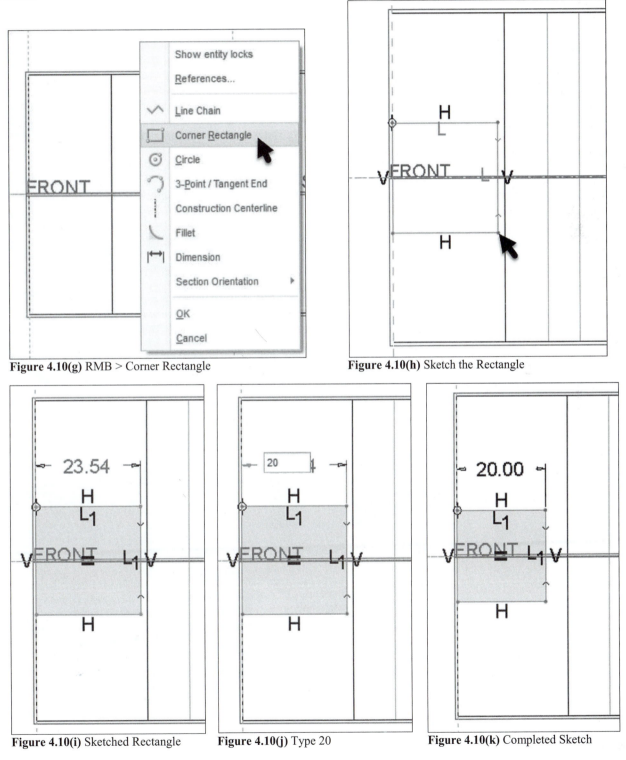

Figure 4.10(g) RMB > Corner Rectangle

Figure 4.10(h) Sketch the Rectangle

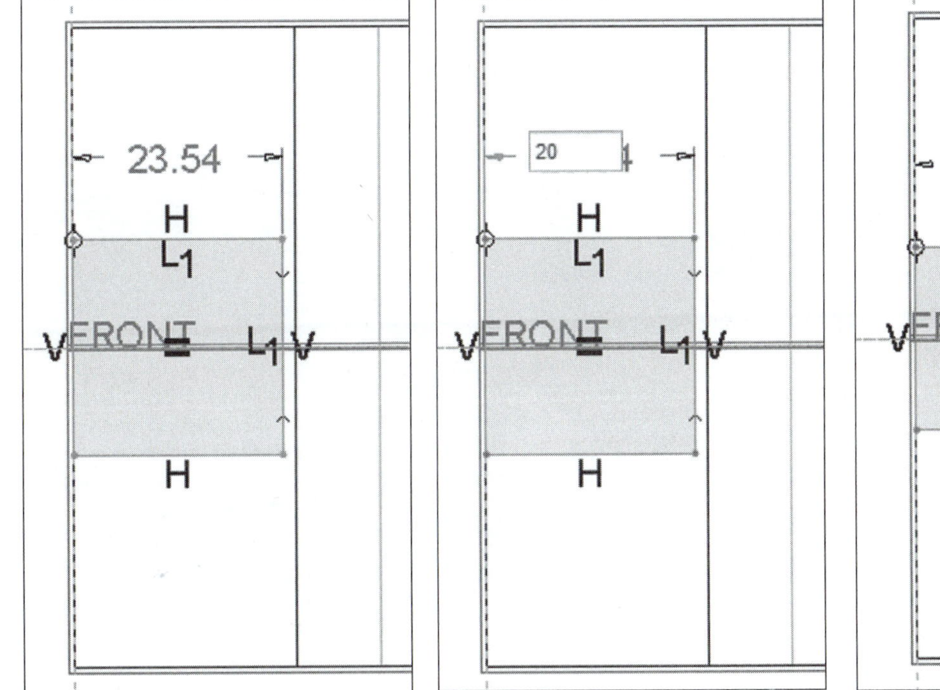

Figure 4.10(i) Sketched Rectangle

Figure 4.10(j) Type 20

Figure 4.10(k) Completed Sketch

Press: **Ctrl+D** > ✓ > ▢ from the Graphics Toolbar > **Shading With Edges** > ⇅ **Change depth direction of extrude to other side of sketch** *[or click on the arrow (vertical in this case) in the Graphics Window]* > **Options** tab > Side 1 **Blind** > **Through All** [Fig. 4.10(l)] > in the Graphics Window, press **MMB** to spin the model [Fig. 4.10(m)] > move the mouse > release the **MMB** > ✓ from the Dashboard > **File** > **Save** [Fig. 4.10(n)]

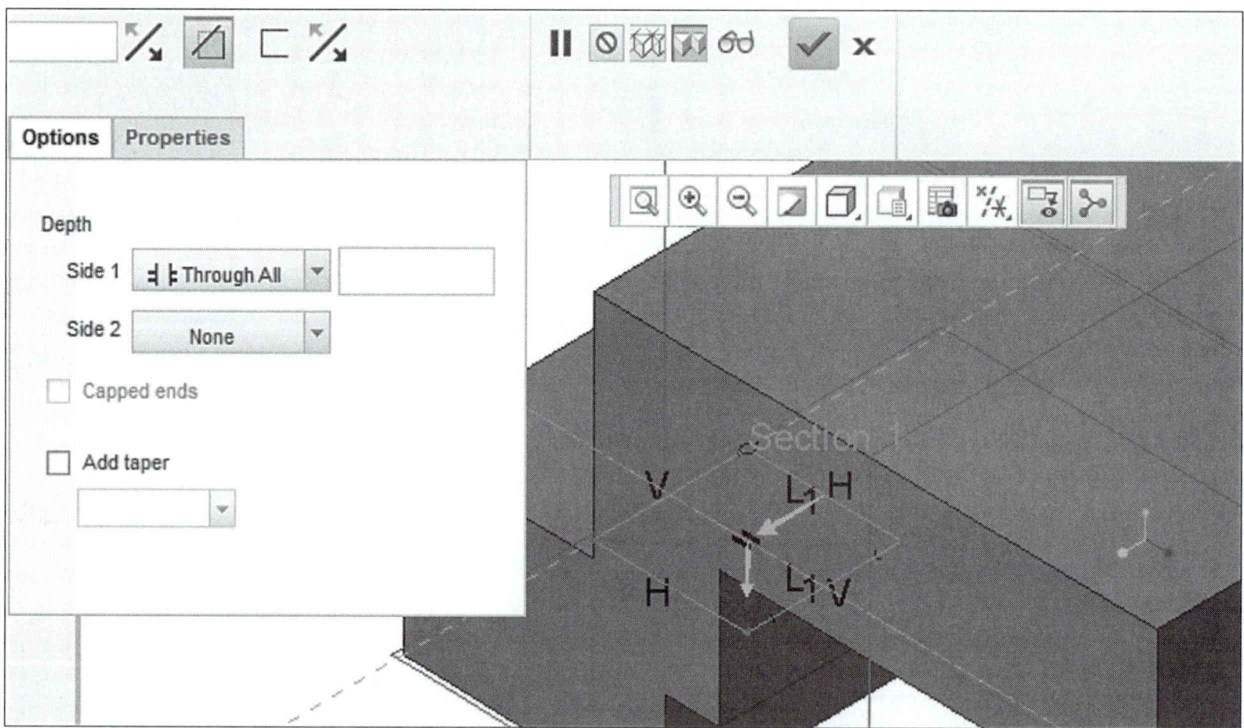

Figure 4.10(l) Through All

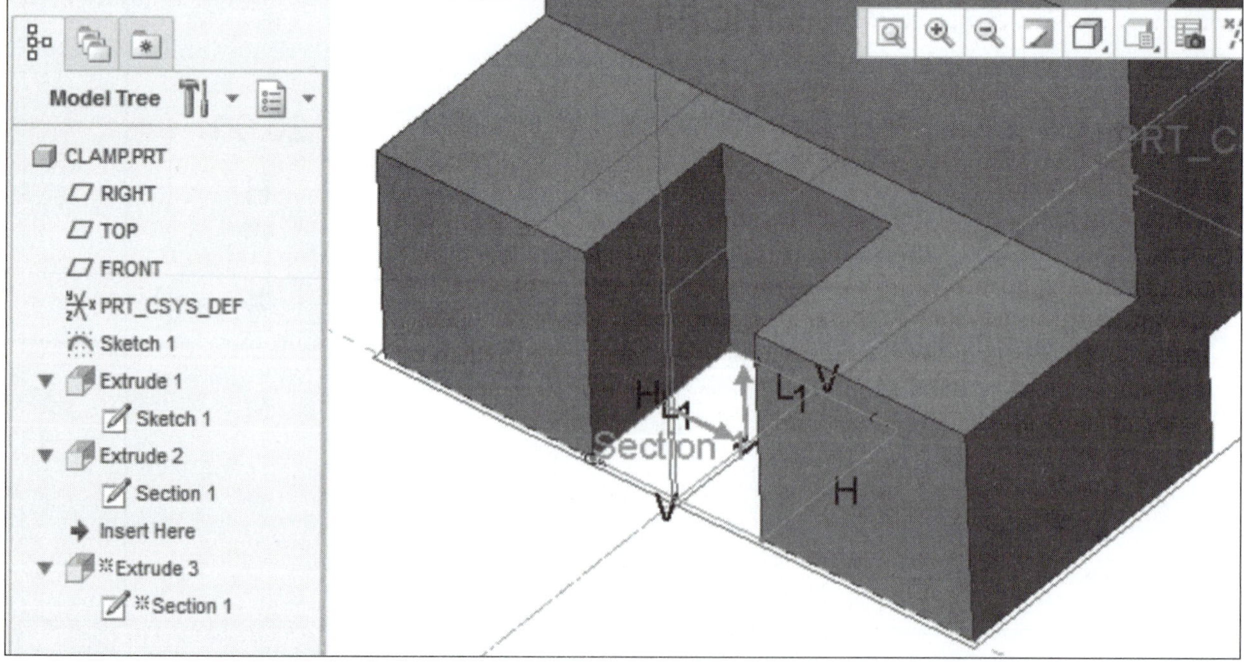

Figure 4.10(m) Cut Preview

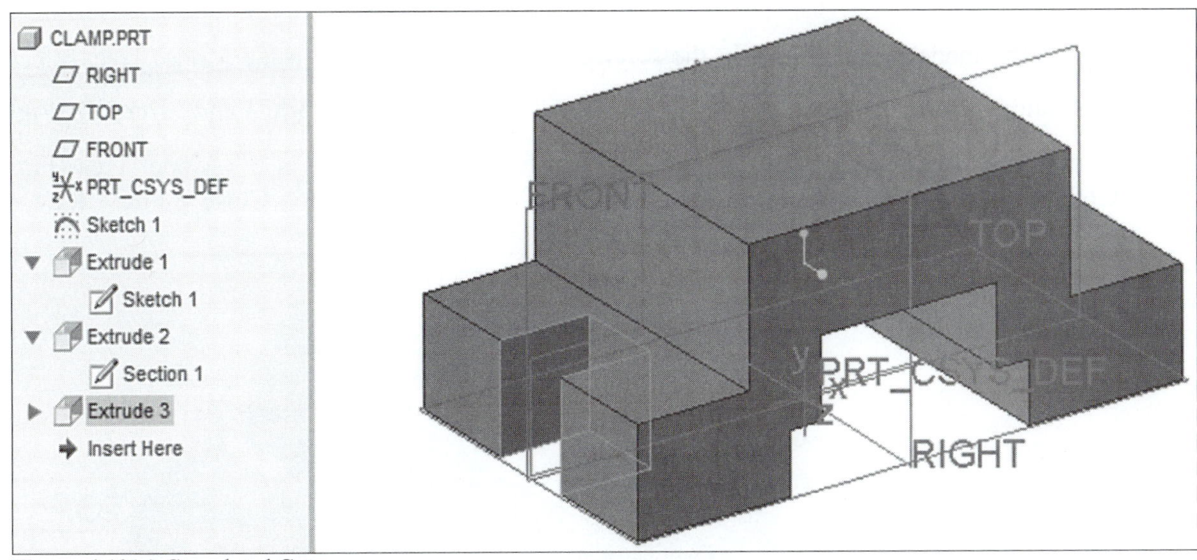

Figure 4.10(n) Completed Cut

With the cut still highlighted *[the extrude cut must be selected-highlighted for this tool to become active]*, from the Ribbon click: 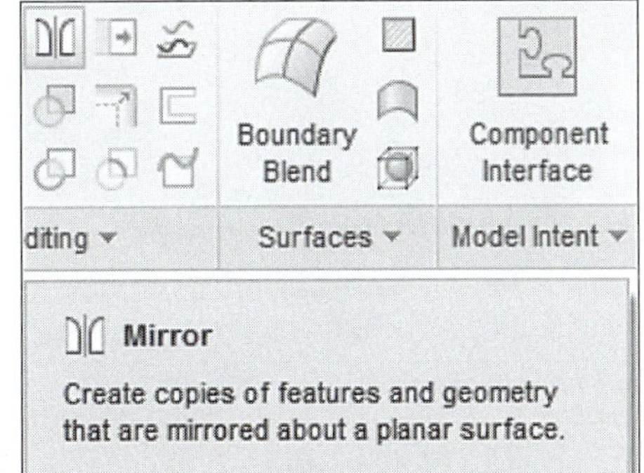

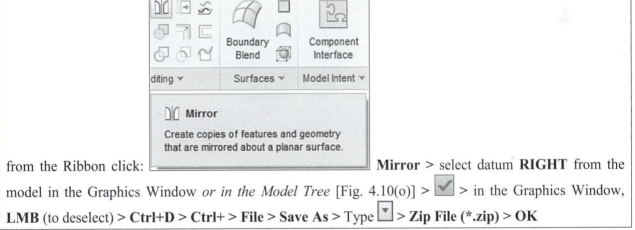

Mirror > select datum **RIGHT** from the model in the Graphics Window *or in the Model Tree* [Fig. 4.10(o)] > ✓ > in the Graphics Window, **LMB** (to deselect) > **Ctrl+D** > **Ctrl+** > **File** > **Save As** > Type ▾ > **Zip File (*.zip)** > **OK**

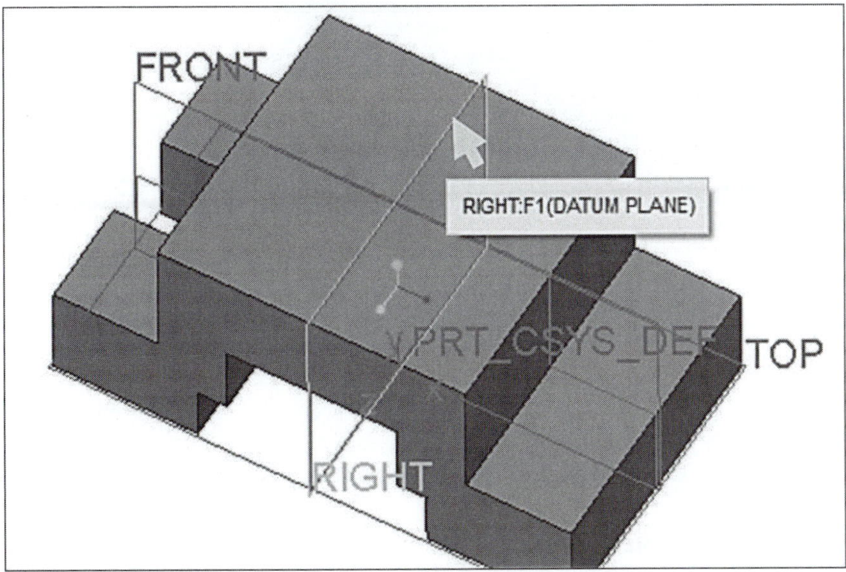

Figure 4.10(o) With the Extruded Cut Highlighted (Selected), Select the RIGHT Datum Plane

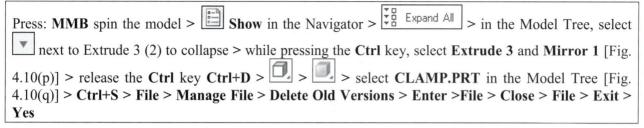

Press: **MMB** spin the model > **Show** in the Navigator > Expand All > in the Model Tree, select next to Extrude 3 (2) to collapse > while pressing the **Ctrl** key, select **Extrude 3** and **Mirror 1** [Fig. 4.10(p)] > release the **Ctrl** key **Ctrl+D** > > > select **CLAMP.PRT** in the Model Tree [Fig. 4.10(q)] > **Ctrl+S** > **File** > **Manage File** > **Delete Old Versions** > **Enter** >**File** > **Close** > **File** > **Exit** > **Yes**

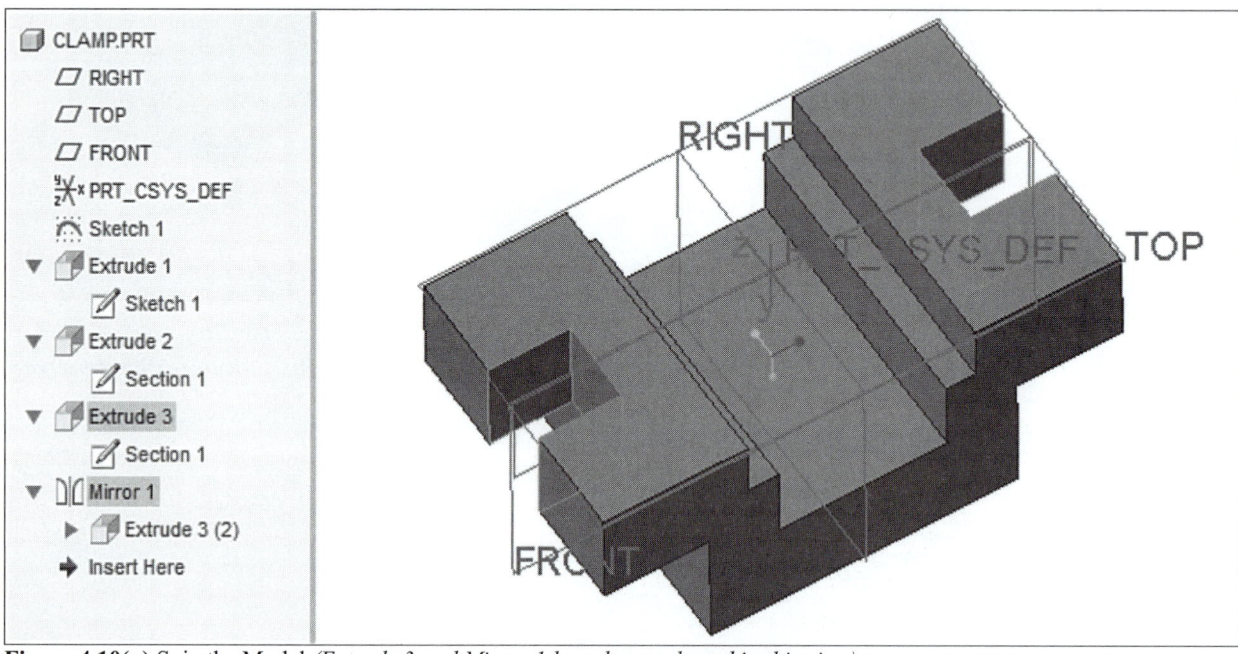

Figure 4.10(p) Spin the Model *(Extrude 3 and Mirror 1 have been selected in this view)*

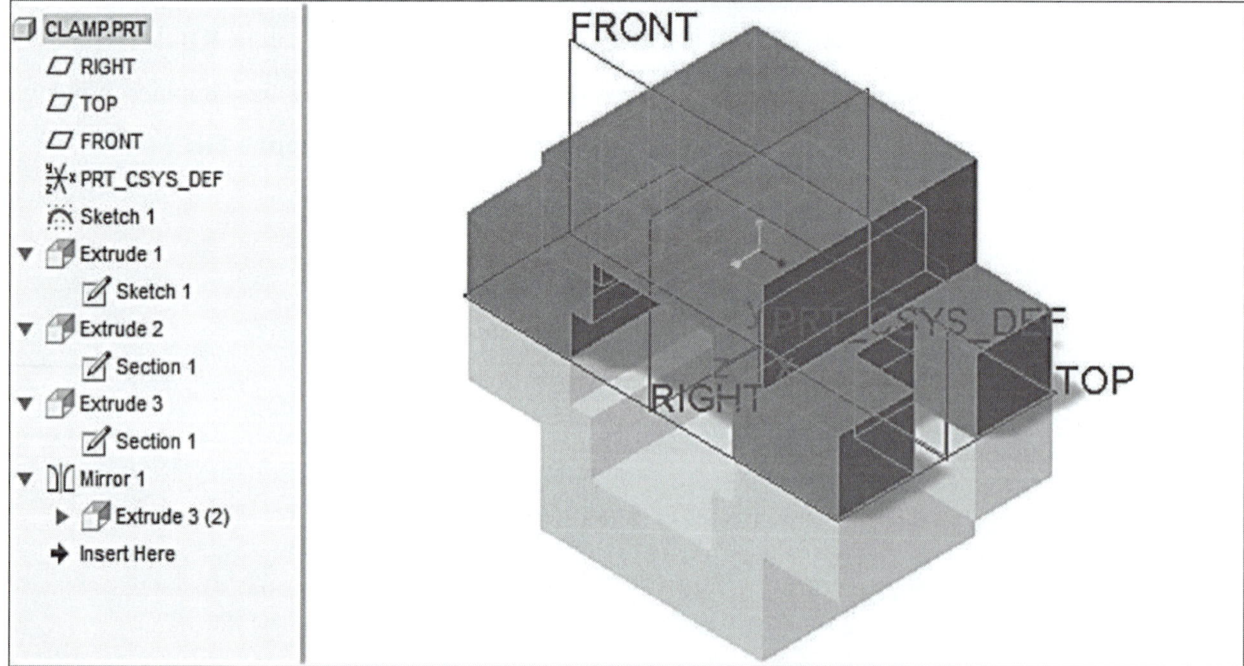

Figure 4.0(q) CLAMP.PRT Selected in the Model Tree

A complete set of extra projects are available at ***www.cad-resources.com > click on the image of your book cover > Lesson Projects.***

Lesson 5 Datums, Layers, and Sections

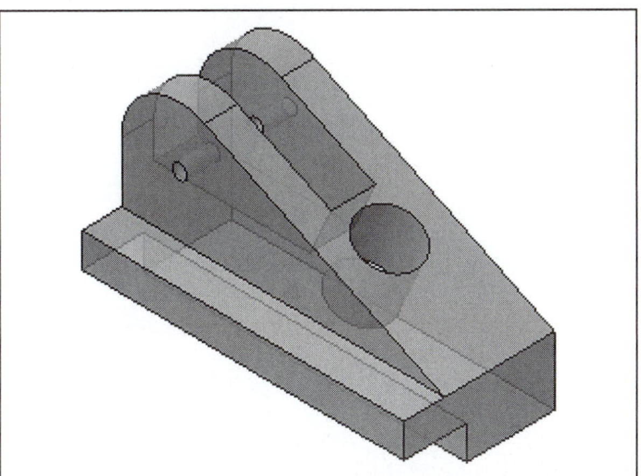

Figure 5.1 Anchor Model with Datum Features hidden on a layer

OBJECTIVES

- Create **Datums** to locate features
- Set datum planes for **geometric tolerancing**
- Learn how to change the **Color** and **Shading** of models
- Use **Layers** to organize features
- Use datum planes to establish **Model Sectioning**
- Add a simple **Relation** to control a feature
- Use **Info** command to extract **Relations** information

REFERENCES AND RESOURCES

For **Resources** go to **www.cad-resources.com** > click on the PTC Creo Parametric 3.0 Book cover

- Lesson 5 Lecture
- Lesson 5 3D models embedded in a PDF
- Book Projects PDF
- Project Lectures
- Quick Reference Card
- Configuration Options

DATUMS, LAYERS, and SECTIONS

Datums and **layers** are two of the most useful mechanisms for creating and organizing your design (Fig. 5.1). Features such as *datum planes* and *datum axes* are essential for the creation of all parts, assemblies, and drawings using PTC Creo Parametric 3.0.

Layers are an essential tool for grouping items and performing operations on them, such as selecting, hide/unhide, plotting, and suppressing. Any number of layers can be created. User-defined names are available, so layer names can be easily recognized.

Most companies have a layering scheme that serves as a *default standard* so that all projects follow the same naming conventions and objects/items are easily located by anyone with access. Layer information such as display status, is stored with each individual part, assembly, or drawing.

Lesson 5 STEPS

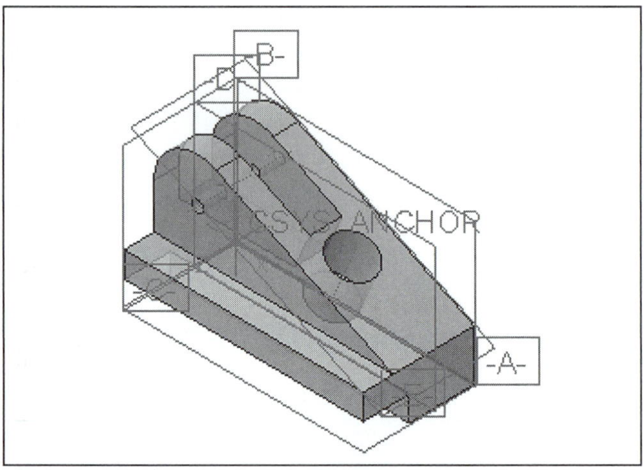

Figure 5.2 Anchor

Anchor

Though default datum planes have been sufficient in previous lessons, the Anchor (Fig. 5.2) incorporates the creation of user-defined datums and the assignment of datums to layers. The datum planes will be set as geometric tolerance features and put on a separate layer.

Launch **PTC Creo Parametric 3.0** > **Select Working Directory** > navigate to your directory if necessary > **OK** > **New** > Name-- *type* **anchor** > Use default template > **OK** > **File** > **Options** > **Model Display** > Default model orientation: **Trimetric** > **OK** > **No** *(if needed)* > **File** > **Prepare** > **Model Properties** > Units **change** > **Inch lbm Second** (Creo Parametric Default) > **Close** > Material **change** > double-click on **steel.mtl** > **OK** > **Close** > **View** tab > *on* > *on* > select the coordinate system on the model or in the Model Tree-- **PRT_CSYS_DEF** > press **RMB** [Fig. 5.3(a)] > **Rename** > *type* **csys_anchor** in the Model Tree > **Enter** *(automatically becomes uppercase)* [Fig. 5.3(b)] > **LMB**

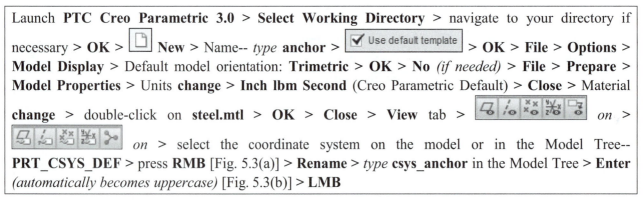

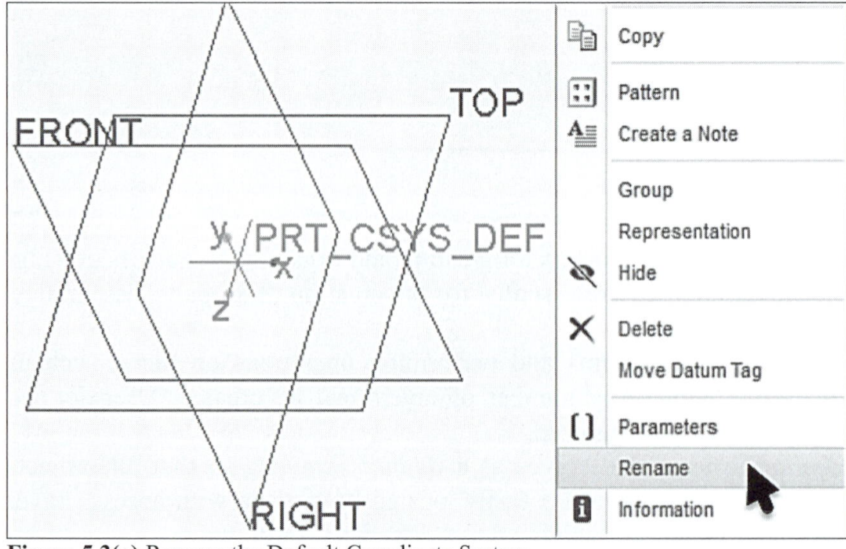

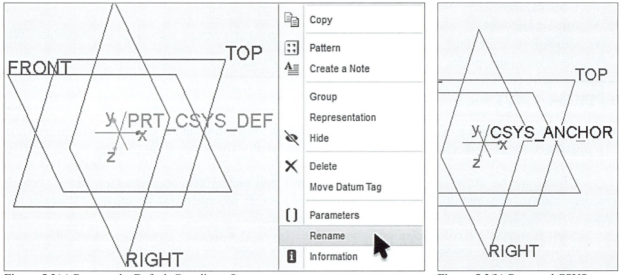

Figure 5.3(a) Rename the Default Coordinate System

Figure 5.3(b) Renamed CSYS

It is considered good practice to rename the default coordinate system to something similar to the components name. If an assembly has 25 components, it will have 25 default coordinate systems. Renaming the coordinate systems of each component will make them easier to visually identify.

The default datum planes and the default coordinate system are automatically placed on two layers each. For the datum planes; the *part default datum plane layer* and the *part all datum plane layer* [Fig 5.3(d)]. The coordinate system will also be layered in a similar fashion. The *part all datum coordinate system layer* for the coordinate system displays the new name created for the coordinate system.

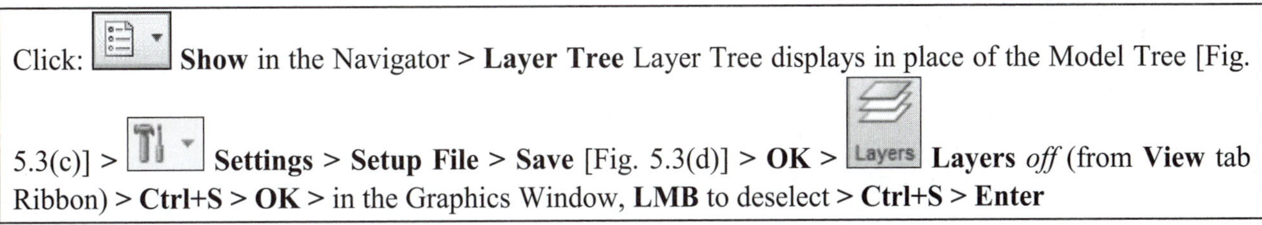

Click: **Show** in the Navigator > **Layer Tree** Layer Tree displays in place of the Model Tree [Fig. 5.3(c)] > **Settings** > **Setup File** > **Save** [Fig. 5.3(d)] > **OK** > **Layers** *off* (from **View** tab Ribbon) > **Ctrl+S** > **OK** > in the Graphics Window, **LMB** to deselect > **Ctrl+S** > **Enter**

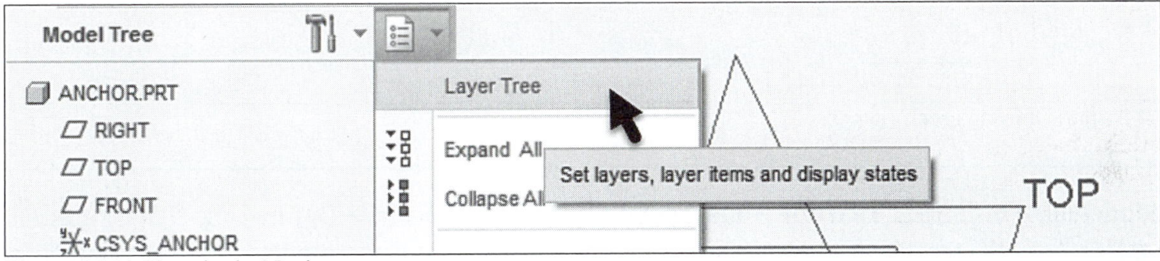

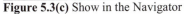

Figure 5.3(c) Show in the Navigator

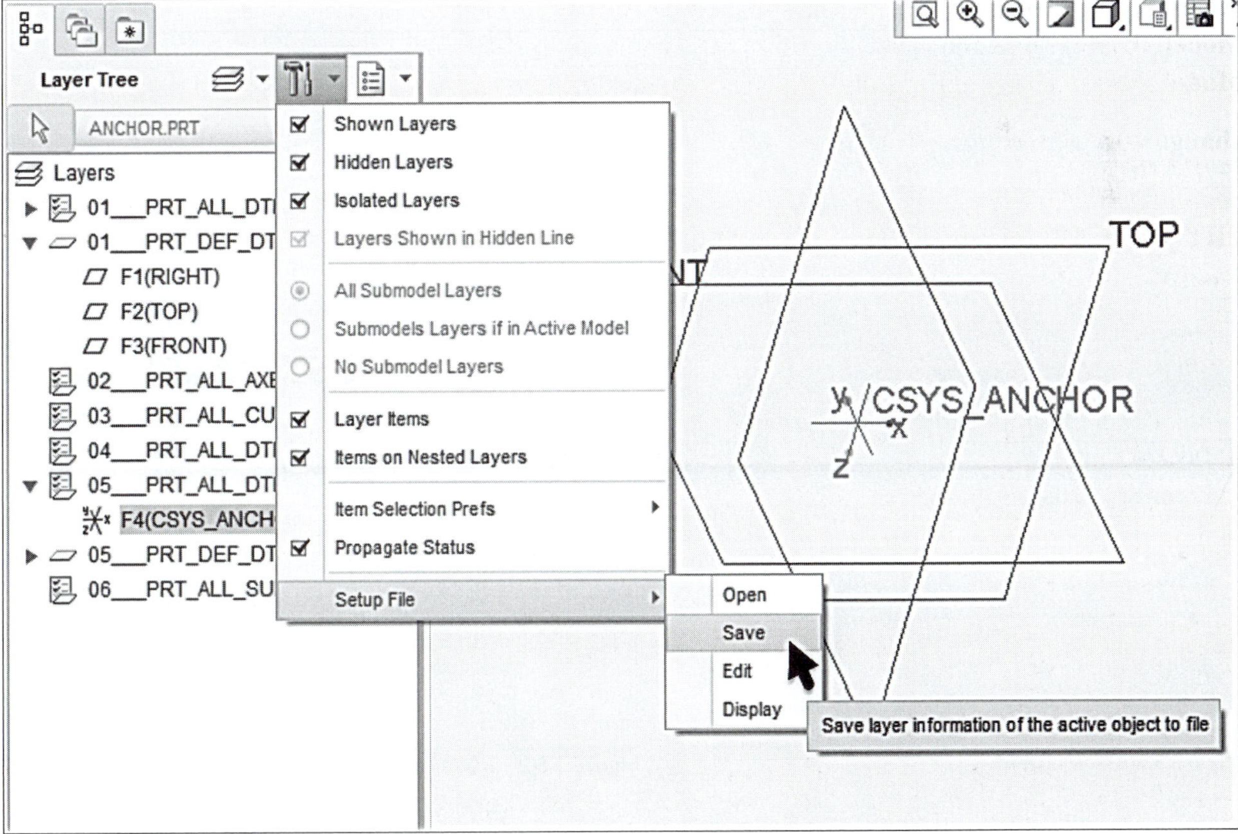

Figure 5.3(d) Layers Displayed in Layer Tree

The first feature will be sketched on the FRONT datum plane. For the protrusion, use only the **5.50**, **R1.00**, **1.125** (vertical), and **25°** dimensions (Fig. 5.4).

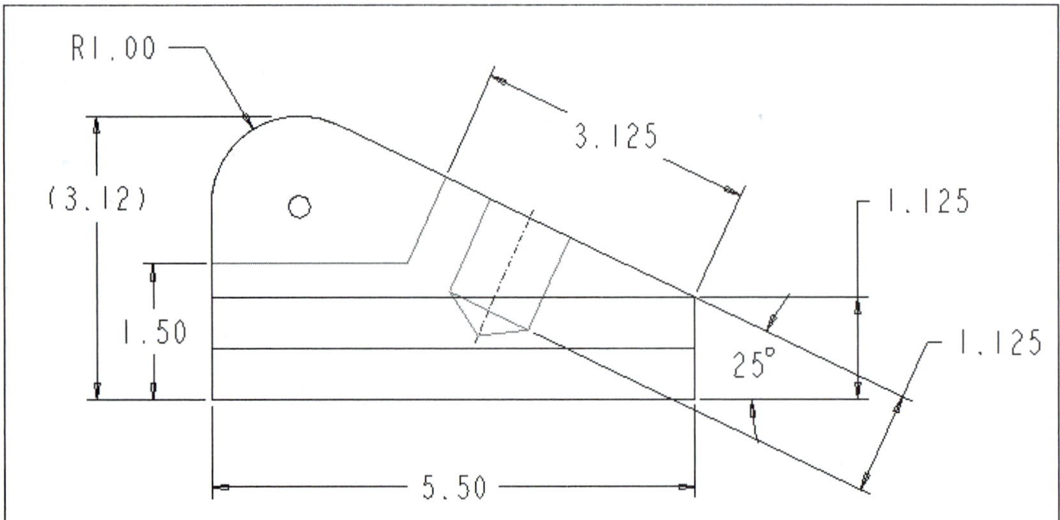

Figure 5.4 Anchor Drawing Front View

Click: **Model** tab > Extrude **Extrude** > **Placement** from the Dashboard > **Define** [Fig. 5.5(a)] > *in the Sketch dialog box:* Sketch Plane--- Plane: select datum **FRONT** from the model or Model Tree [Fig. 5.5(b)] > **Sketch** to close the Sketch dialog box > press **RMB** anywhere in the Graphics Window > **Line Chain**

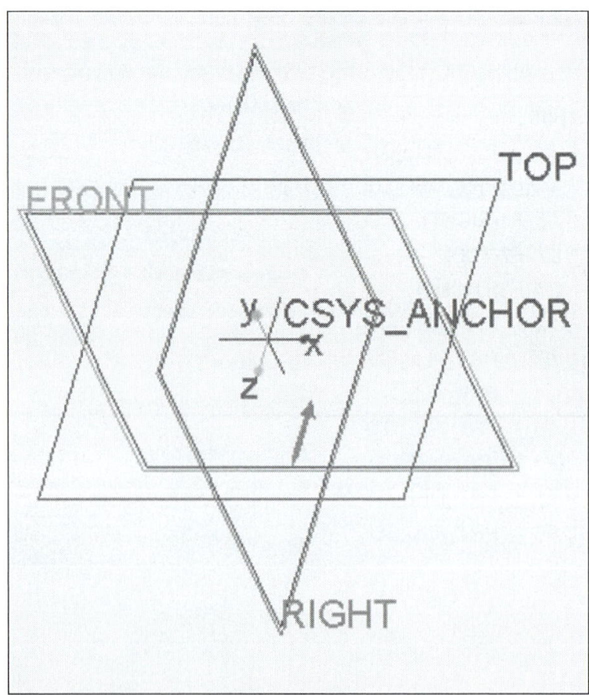

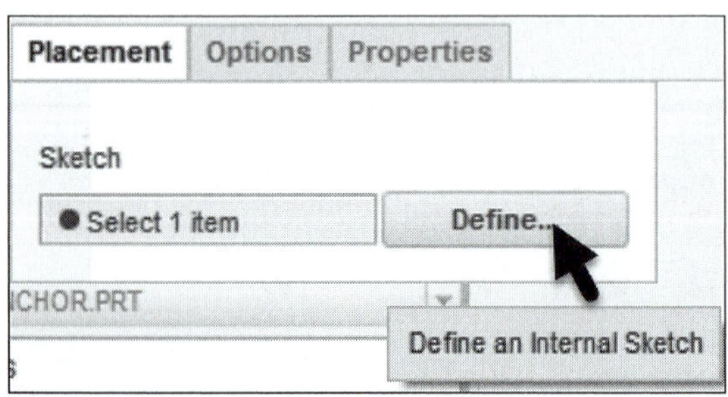

Figure 5.5(a) Dashboard

Figure 5.5(b) Select Datum FRONT

Starting at the coordinate system, sketch the four lines [Fig. 5.5(c)] > **MMB** > **MMB** > **LMB** > press **RMB** > **Fillet** > pick the left vertical line and the angled line to create a fillet [Fig. 5.5(d)] > **MMB** > **LMB** to deselect

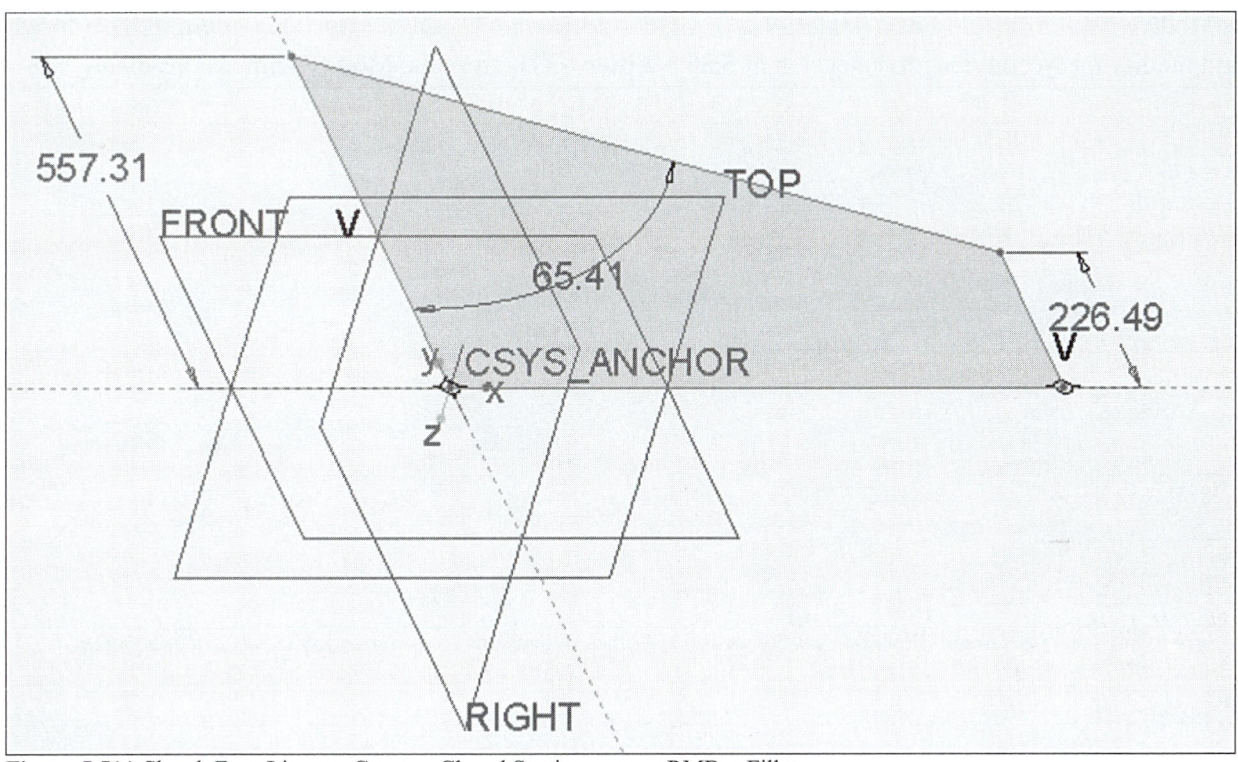

Figure 5.5(c) Sketch Four Lines to Create a Closed Section > press RMB > Fillet

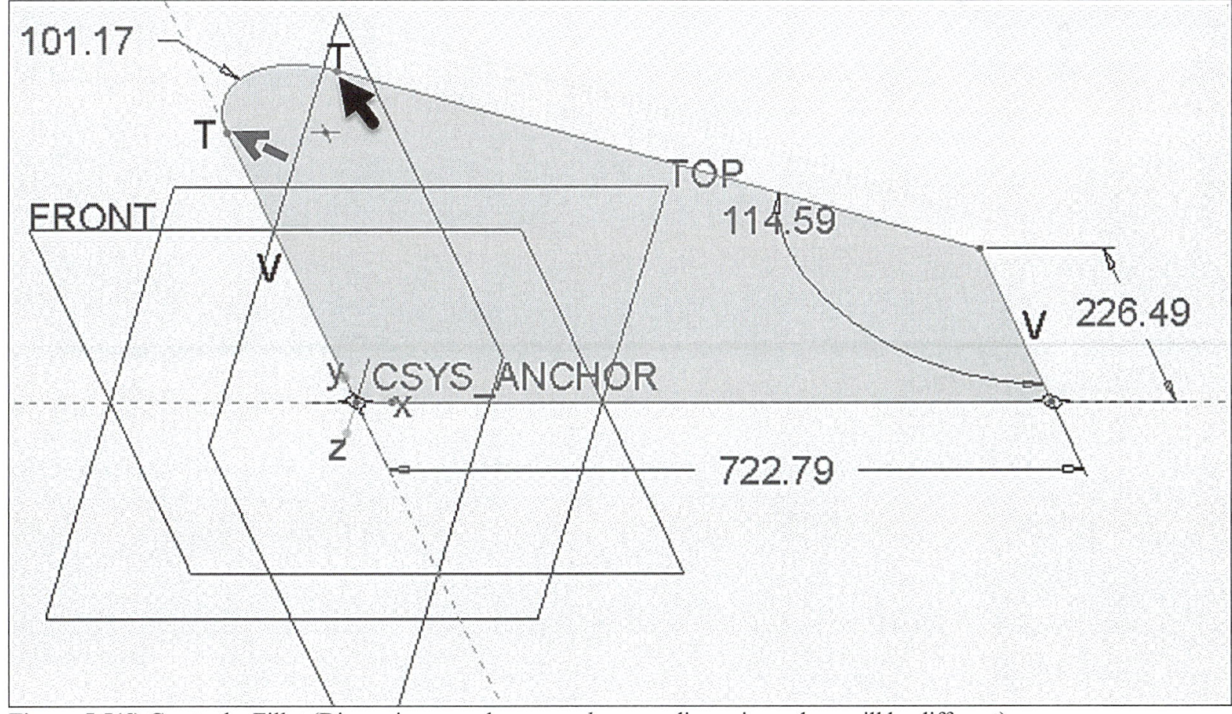

Figure 5.5(d) Create the Fillet (Dimensions are shown *weak* – your dimension values will be different)

Press: **RMB** > **Dimension** > pick on the horizontal line and then the angled line > move the pointer to the desired location for the dimension and click **MMB** to position the dimension value [Fig. 5.5(e)] > **Enter** *(or move the mouse > **MMB**)* to accept the initial value > **MMB** to end the current tool > reposition the dimensions > window-in dimensions *(or select all dimensions while pressing the **Ctrl** key)* > press **RMB** > **Modify** [Fig. 5.5(f)] > **Lock Scale** *on (if needed-- resize the Modify Dimensions dialog box)* > locate and modify the overall length dimension to **5.50** > **Enter** > **OK** from the Modify Dimensions dialog box

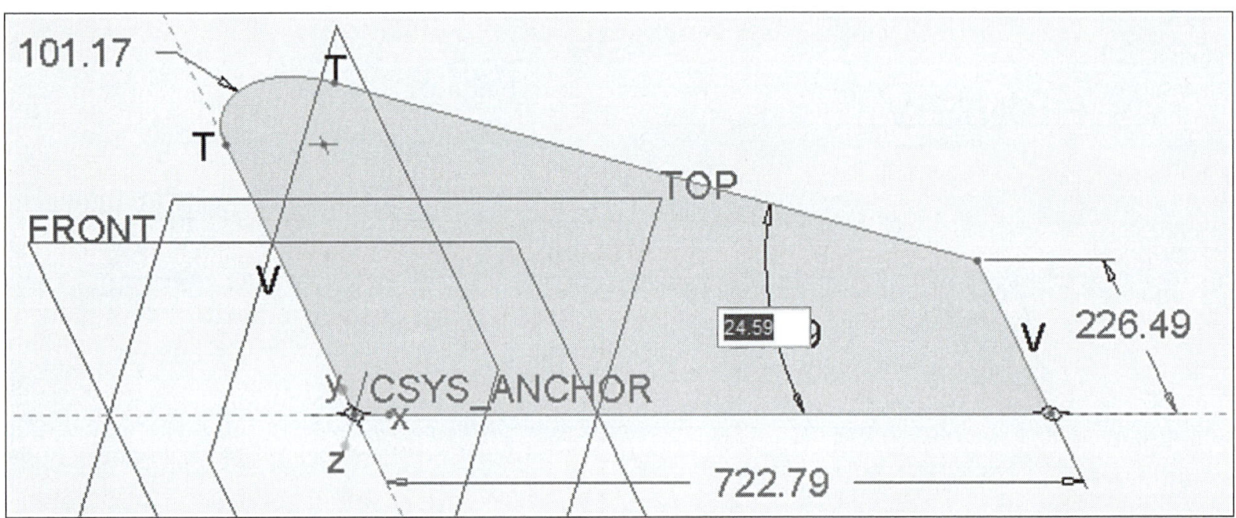

Figure 5.5(e) Create the Angle Dimension and Reposition the other Dimensions *(your dimension values will be different)*

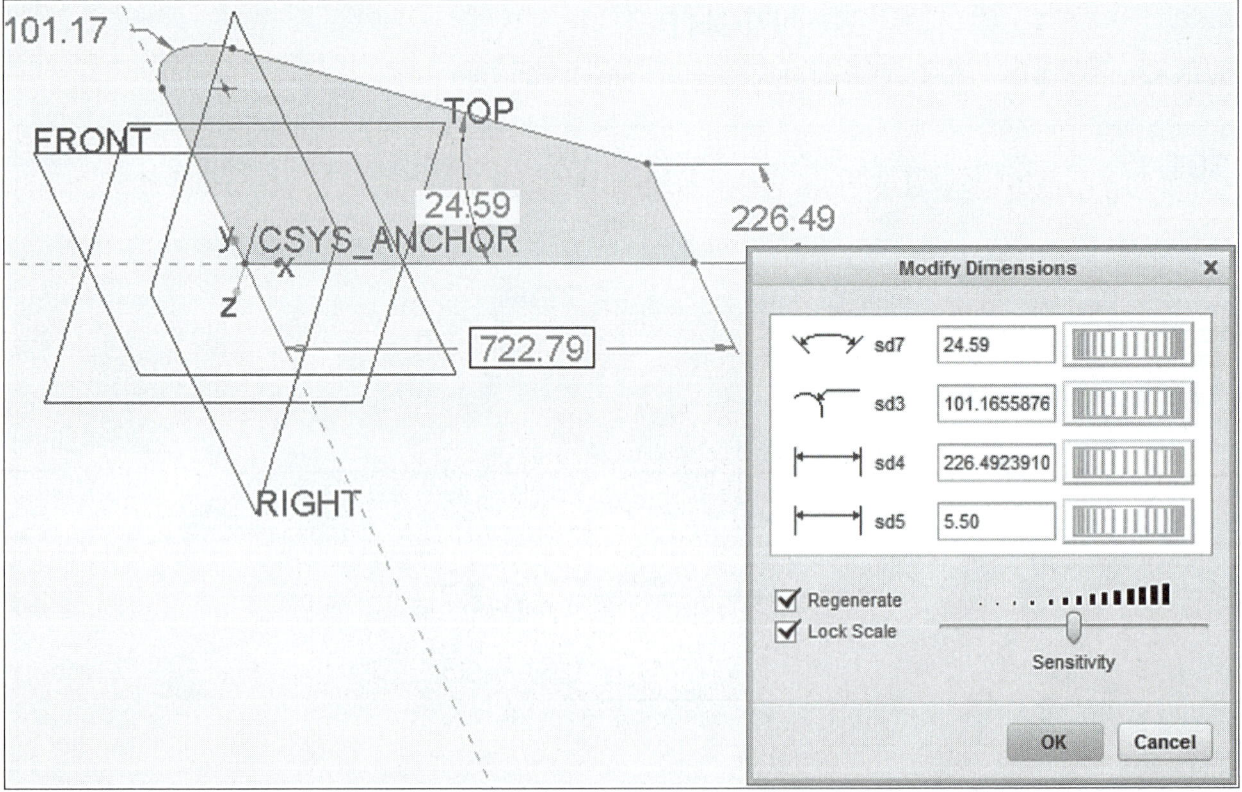

Figure 5.5(f) After Capturing the Dimensions in a Window, Modify the Overall Length Dimension Value

Double-click on each dimension value and modify to the design size [Fig. 5.5(g)] > reposition dimensions if needed > **LMB** to deselect > press **RMB** > **OK** > *type* **2.5625** in the depth value field [2.56250] > **Enter** [Fig. 5.5(h)] > [✓] > **Ctrl+D** > **Ctrl+S** > in the Graphics Window, **LMB** to deselect

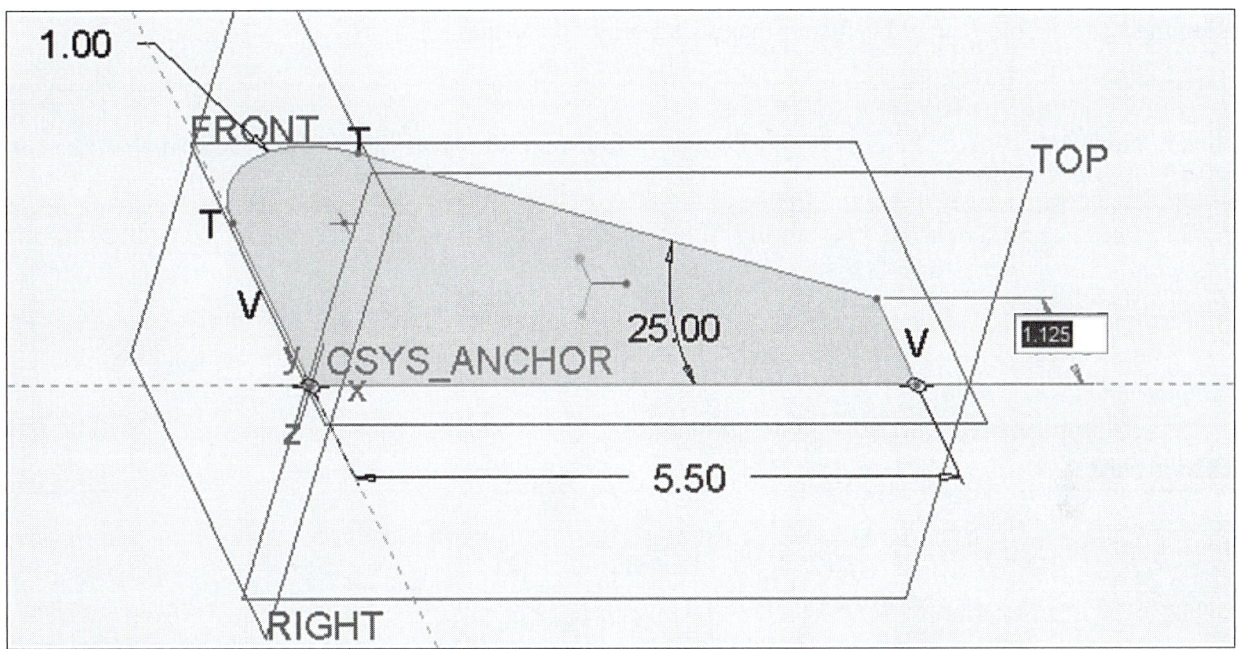

Figure 5.5(g) Modify the Dimensions to the Design Values

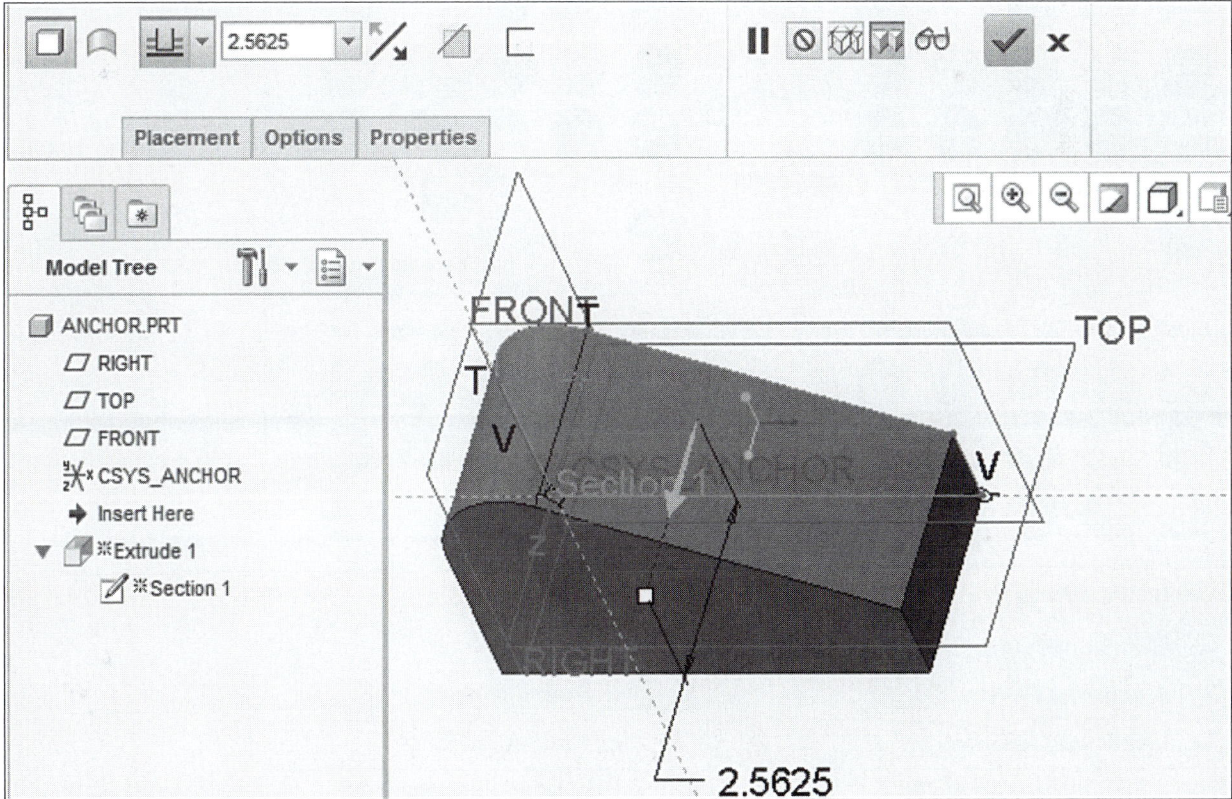

Figure 5.5(h) Modify the Depth to **2.5625** in the Dashboard

Colors

Colors are used to define material and light properties. Avoid the colors that are used as Creo Parametric defaults. Because feature and entity highlighting is currently defaulted to *green*, colors similar to *green* should be avoided. It is up to you to select colors that work with the type of project you are modeling. If the parts will be used in an assembly, each component should have a unique color scheme. In general, using light pastel colors would be better than using bright dark ones.

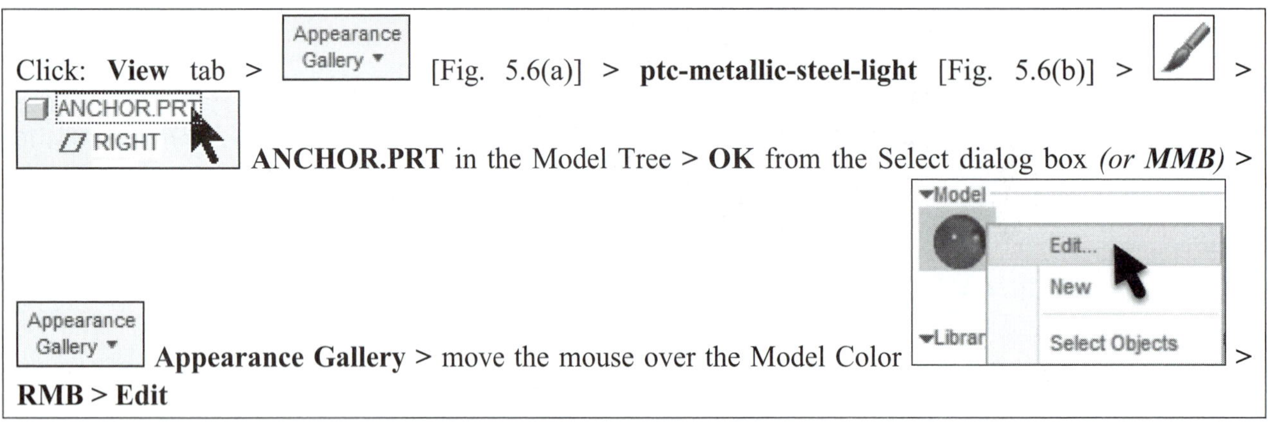

Click: **View** tab > Appearance Gallery [Fig. 5.6(a)] > **ptc-metallic-steel-light** [Fig. 5.6(b)] > > ANCHOR.PRT RIGHT **ANCHOR.PRT** in the Model Tree > **OK** from the Select dialog box *(or* ***MMB****)* > Appearance Gallery **Appearance Gallery** > move the mouse over the Model Color Model Edit... New Library Select Objects > **RMB** > **Edit**

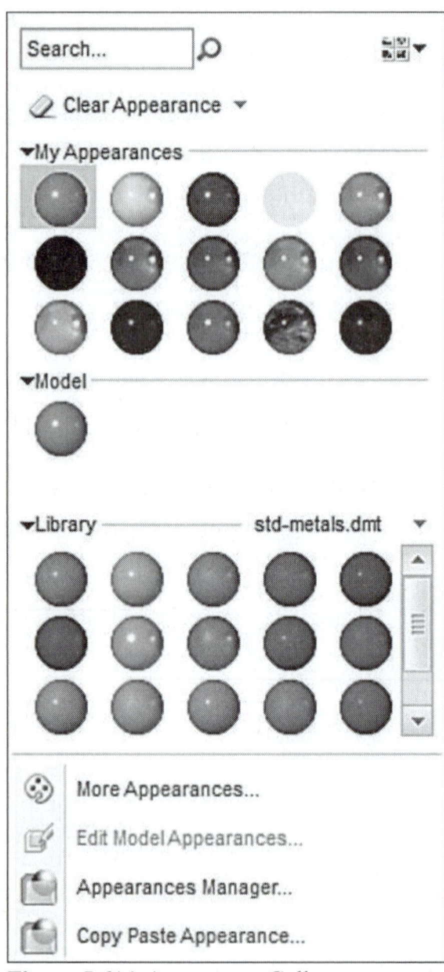

Figure 5.6(a) Appearance Gallery

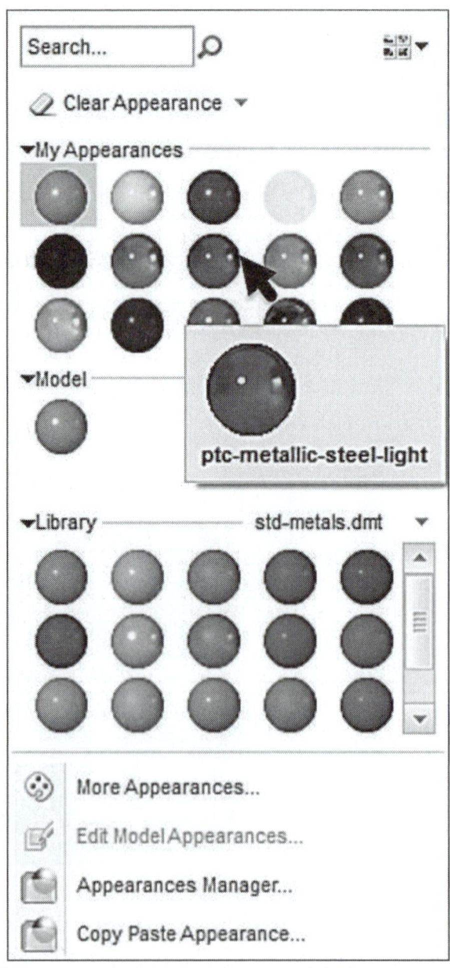

Figure 5.6(b) ptc-metallic- steel-light

Click on the color button in the Model Appearance Editor dialog, Properties, Color [Fig. 5.6(c)] *(color swatch)* to open the Color Editor dialog box [Fig. 5.6(d)] > move the **RGB/HSV Slider** bars to create a new color *(or type new values)* > you can also use the Color Wheel **Color Wheel** > or the Blending Palette **Blending Palette** > adjust the colors > **OK** from the Color Editor dialog box

Note that your color selection may be different depending on the PTC license and the graphics card.

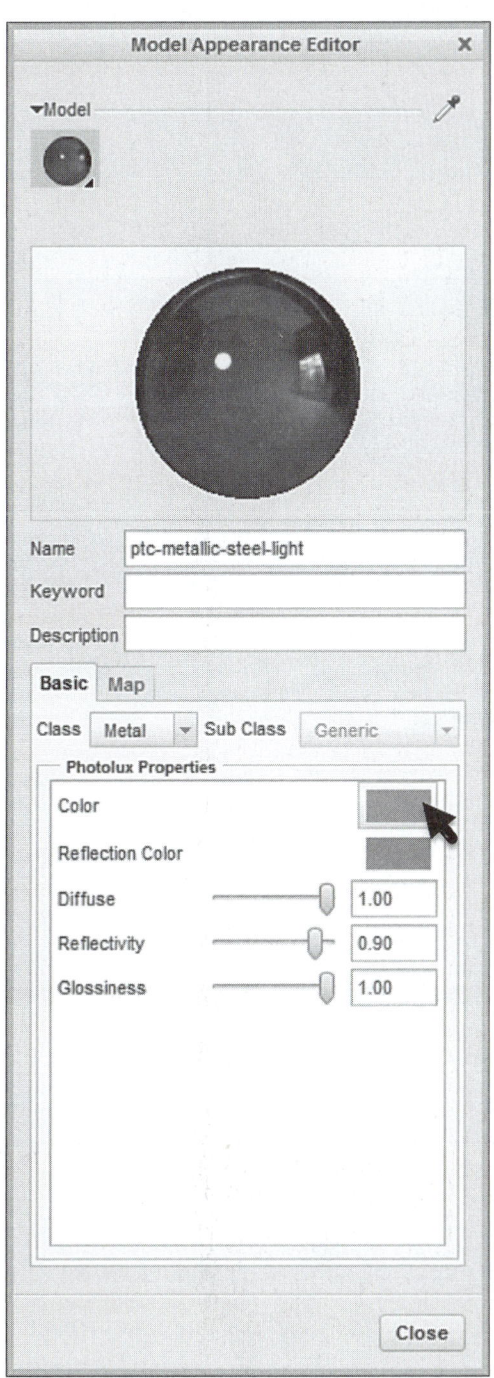

Figure 5.6(c) Model Appearance Editor

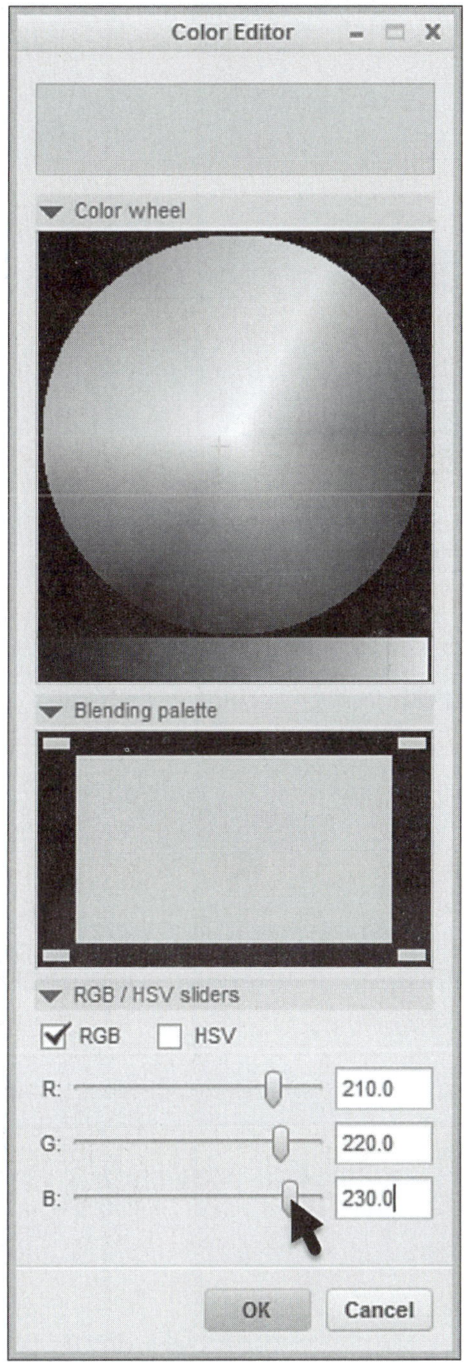

Figure 5.6(d) Color Editor

Click on the color button in the Model Appearance Editor dialog, Properties, Highlight Color [Fig. 5.6(e)] *(color swatch)* > adjust the **RGB/HSV Slider** bars to create a new *reflection* color *(or type new values)* > Color Wheel **Color Wheel** > Blending Palette **Blending Palette** > **OK** from the Color Editor > adjust the Properties slide bars from the Model Appearance Editor [Fig. 5.6(f)] > **Close**

Note that your color selection may be different depending on the PTC license and the graphics card.

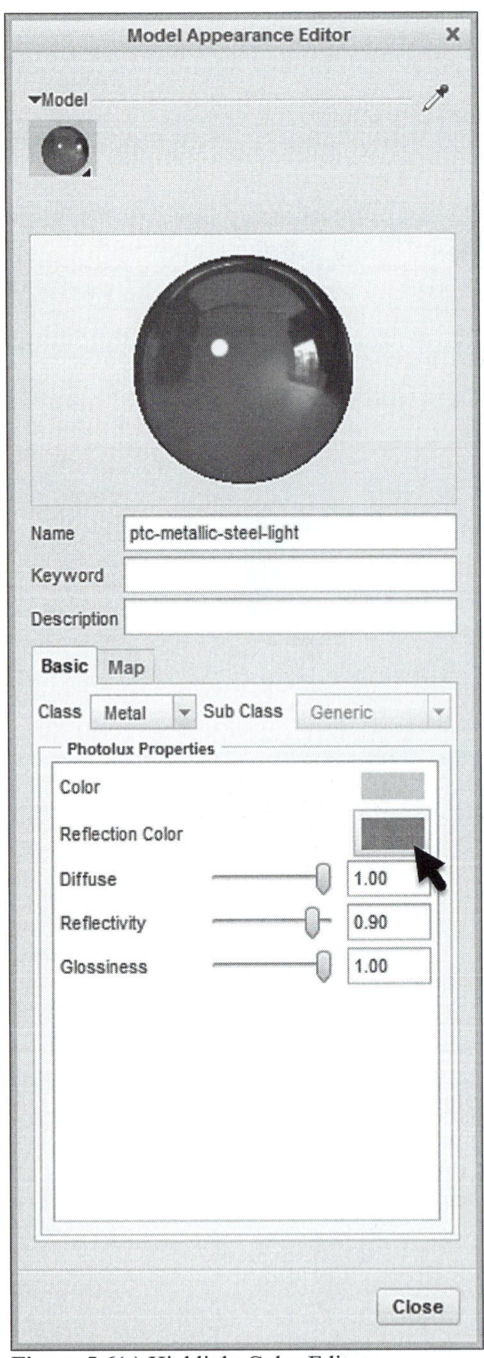

Figure 5.6(e) Highlight Color Editor

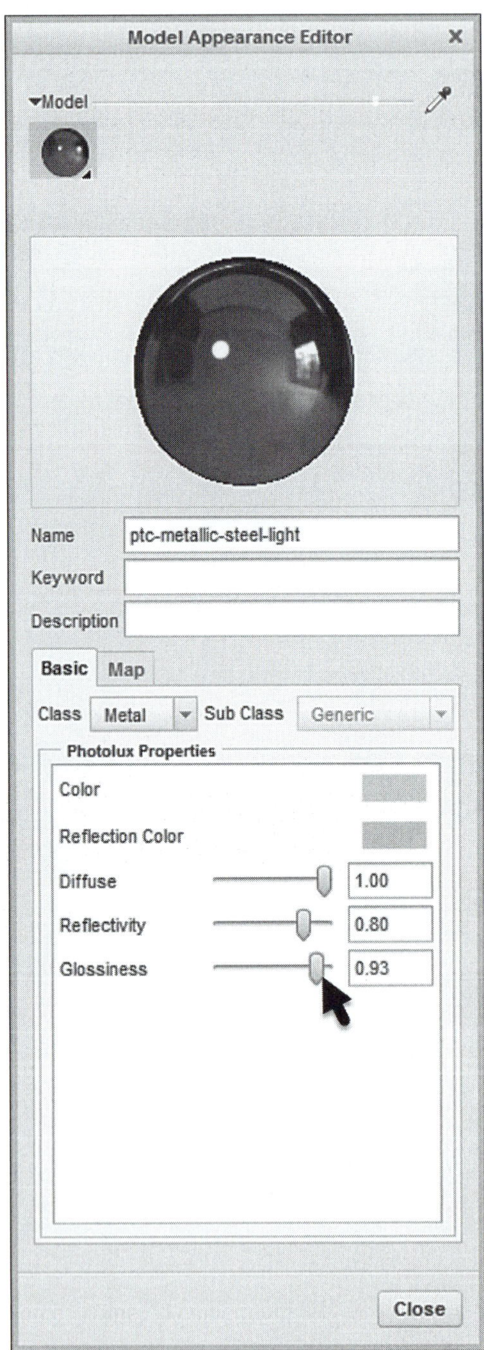

Figure 5.6(f) Adjust the Properties Slide Bars as Desired

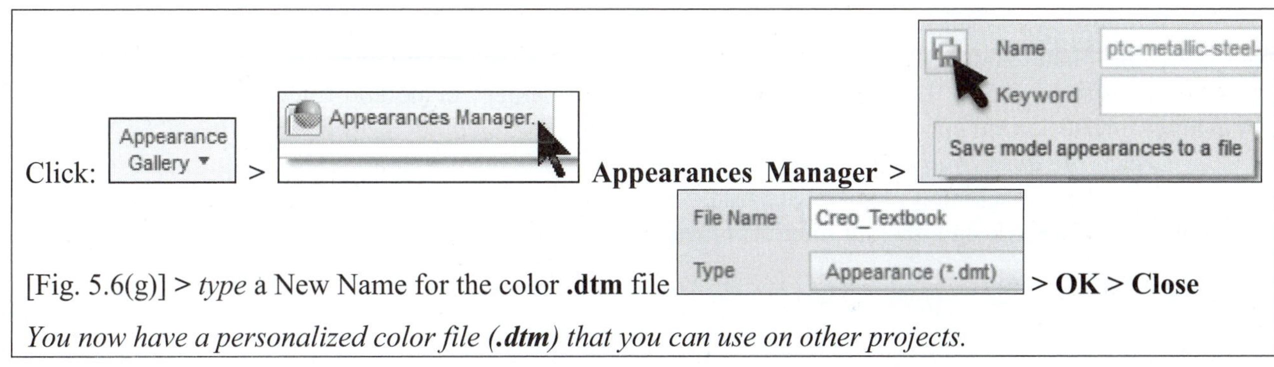

Click: Appearance Gallery ▾ > Appearances Manager **Appearances Manager >** Save model appearances to a file [Fig. 5.6(g)] > *type* a New Name for the color **.dtm** file > **OK > Close**

*You now have a personalized color file (**.dtm**) that you can use on other projects.*

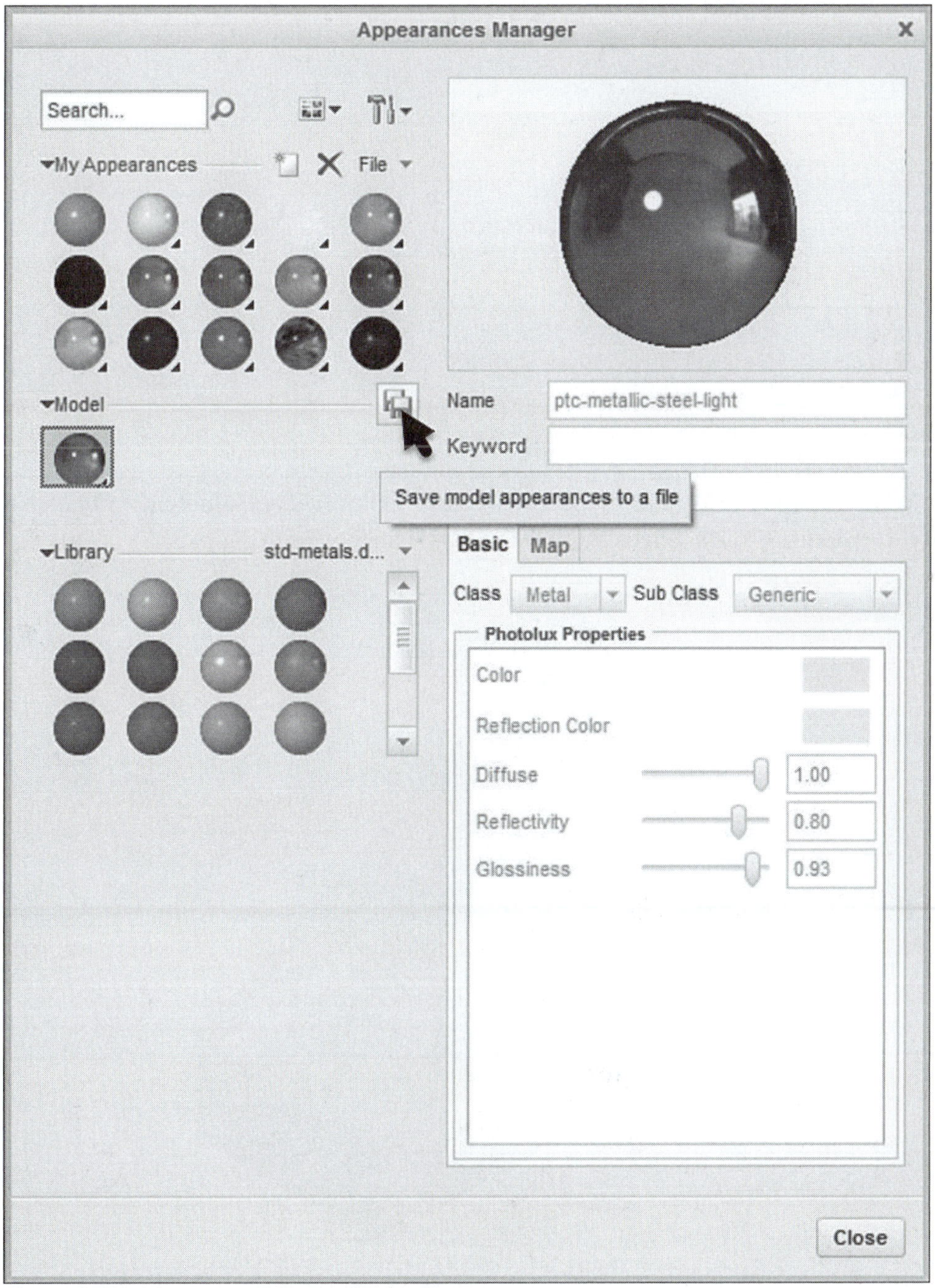

Figure 5.6(g) Appearances Manager Dialog Box

The next two features will remove material. For both of the cuts, use the RIGHT datum plane as the sketching plane, and TOP datum plane as the reference plane. Each cut requires just two lines, and two dimensions (Fig. 5.7). Use the first cut's sketching/placement plane and reference/orientation (**Use Previous**) for the second cut.

Create the cuts separately. It is correct engineering practice to separate the features into separate cuts as if you were machining the part. Separate features will allow ECO's (engineering change orders) to be implemented on a design without interfering with other non-related features.

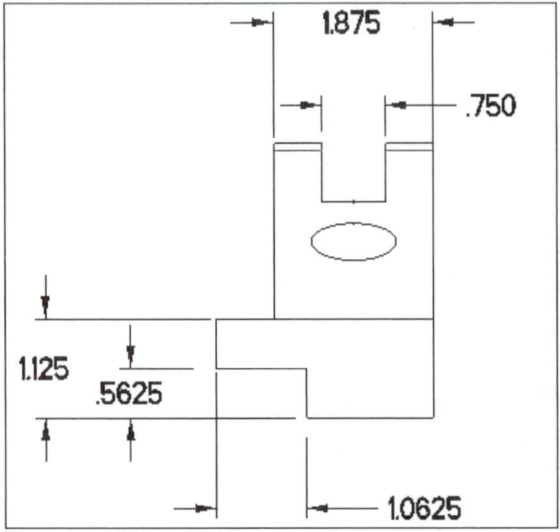

Figure 5.7 Side View showing Dimensions for the Cuts

Click: **File** > **Options** > **Model Display** > Default model orientation: **Trimetric** > **Isometric** > **OK** > **No** > **View** tab > **Display Style** > **Shading With Edges** > **Ctrl+D** > **Ctrl+S** [Fig. 5.8(a)]

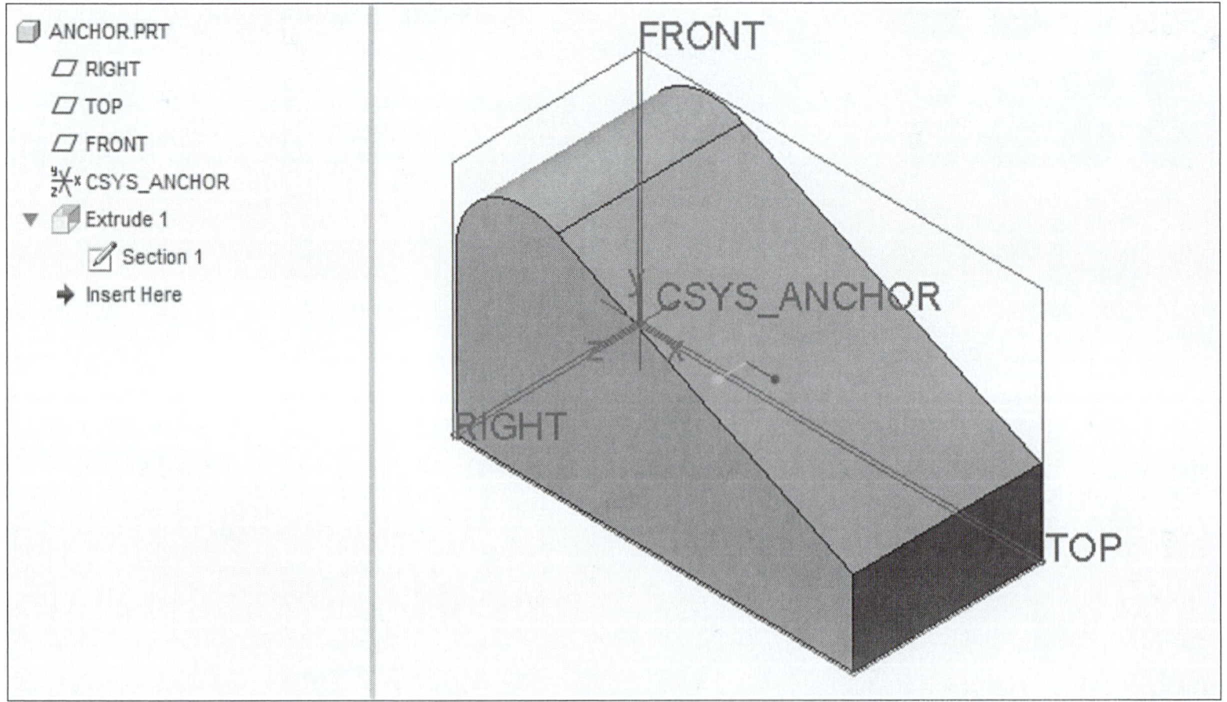

Figure 5.8(a) Standard Orientation Isometric, Shaded With Edges, Edited Color with Transparency on.

Click: **Model** tab > Extrude **Extrude** > in the Graphics Window, press **RMB** > **Remove Material** > press **RMB** > **Define Internal Sketch** > select datum **RIGHT** [Fig. 5.8(b)] for sketching [Fig. 5.8(c)] > **Sketch** *(or MMB)* to close the Sketch dialog box > **Sketch View** > in the Graphics Window, press **RMB** > **References** *(add a reference to align the cut with the top edge and the left edge of the part)* > pick on the *top edge/surface* of the part [Fig. 5.8(d)] > pick on the *left edge/surface* of the part [Fig. 5.8(e)] > **Solve** (notice the Reference status-- Fully Placed) [Fig. 5.8(f)] > **Close**

(If you select the wrong edge/surface, pick the incorrect references in the References dialog box. > Delete > Select the two edges/surfaces again.)

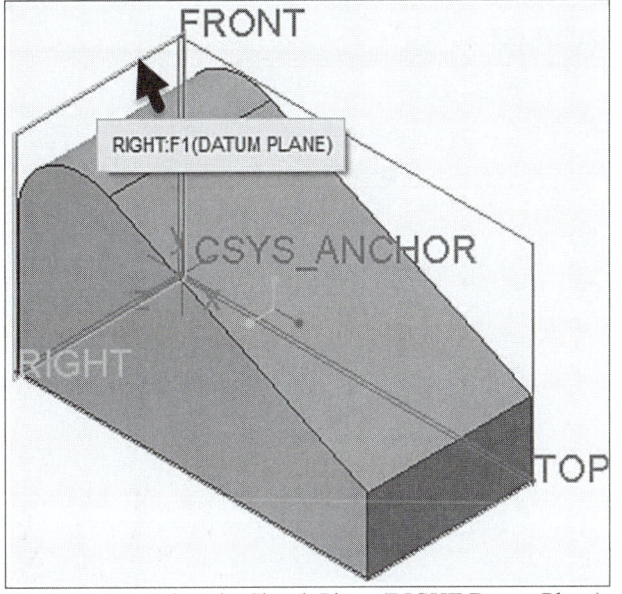

Figure 5.8(b) Select the Sketch Plane (RIGHT Datum Plane)

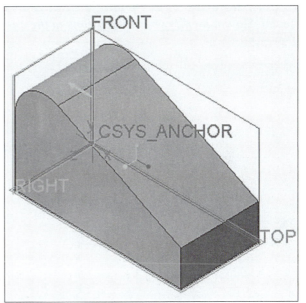

Figure 5.8(c) Sketch Viewing Direction

Figure 5.8(d) Add the top edge/surface

Figure 5.8(e) Add the left edge/surface

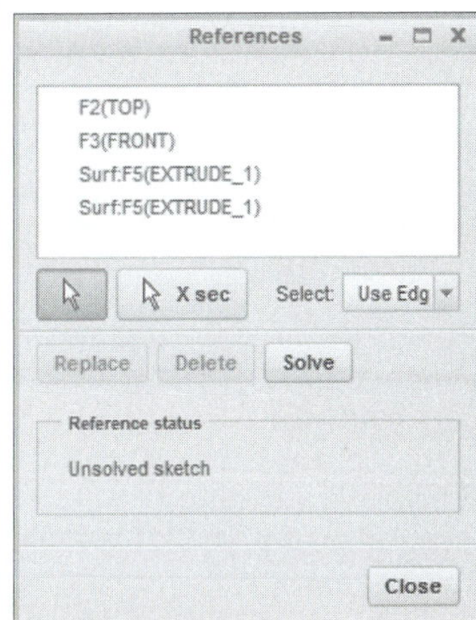

Figure 5.8(f) References Dialog Box

Press: [icon] > **Hidden Line** > **RMB** > **Line Chain** > sketch the two lines- start from the top edge with a *vertical* line and ending with the *horizontal* line locked to the left edge [Fig. 5.8(g)] > **MMB** > **MMB** [Fig. 5.8(h)] > **LMB** > press **RMB** > **Dimension** > ***add*** dimensions between the *horizontal* line and the bottom edge, and between the *vertical* line and the right edge [Fig. 5.8(i)], ***pick on the lines and edges not the endpoints*** > **MMB** (place the last dimension) > move the mouse > **MMB** (accept the default value) > **MMB** (end the current tool)

Note: "weak" (lighter colored) dimensions disappear as new dimensions are added. Created dimensions are "strong" dimensions (darker colored) and represent the design intent of the feature. These dimensions can be displayed on a drawing as driving-design dimensions to be used in manufacturing.

You must have the dimensioning *scheme* shown in Figure 5.8(i). Your dimension values will be different. ***Do not modify your dimensions to the ones shown here.***

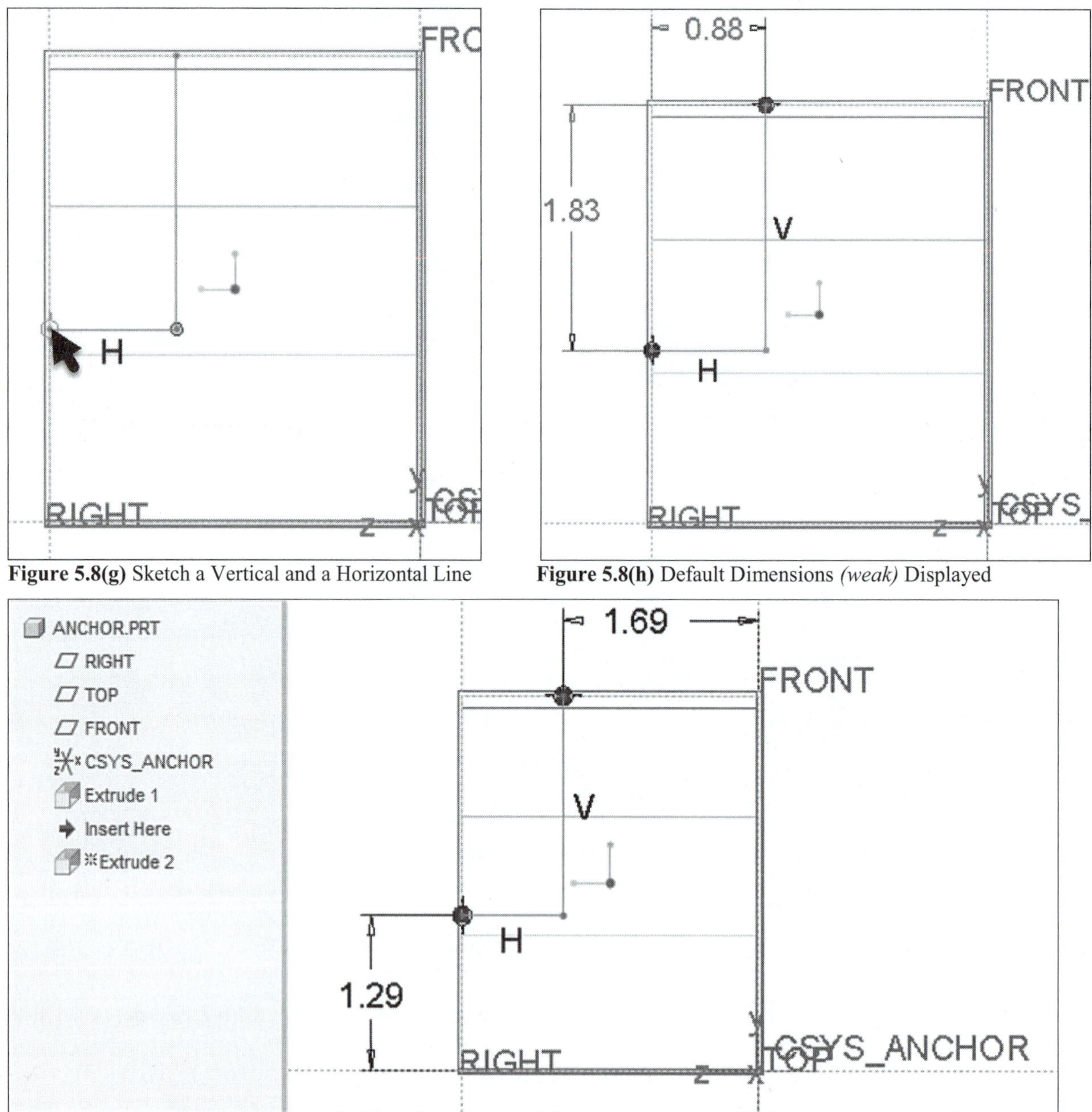

Figure 5.8(g) Sketch a Vertical and a Horizontal Line

Figure 5.8(h) Default Dimensions *(weak)* Displayed

Figure 5.8(i) New Dimensions *(strong)* ***(You* must have the dimensioning *scheme* shown here-** *your values will be different)*

The dimensional *design scheme* is correct for the cut, but the values are not the correct *design values*. Double-click on the *horizontal* dimension and change to the design value to **1.875**. [Fig. 5.8(j)] > **Enter** > repeat for the *vertical* dimension: **1.125** [Fig. 5.8(k)] > **LMB** in Graphics Window > **Ctrl+D** > [icon] > [icon] > press **RMB** > **OK** > if needed place the pointer on the depth direction arrow 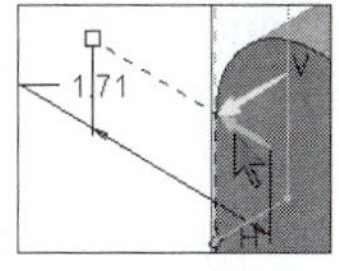 > **LMB** to select *(flips the depth direction arrow to the other side)* [Fig. 5.8(l)]

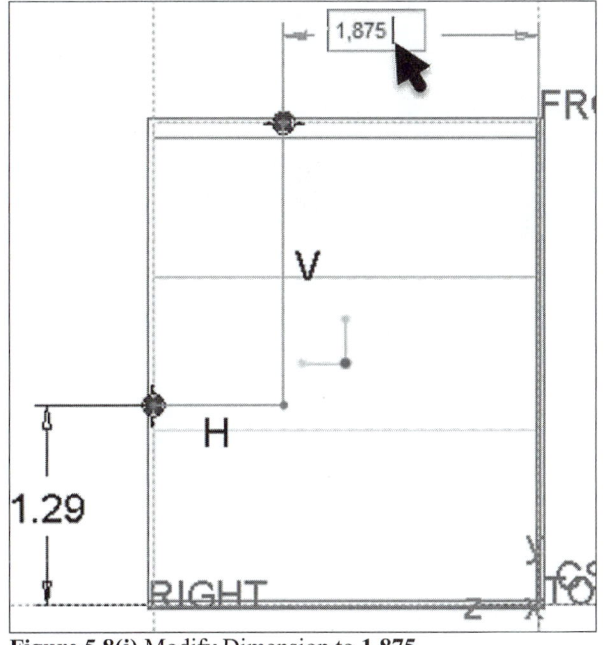

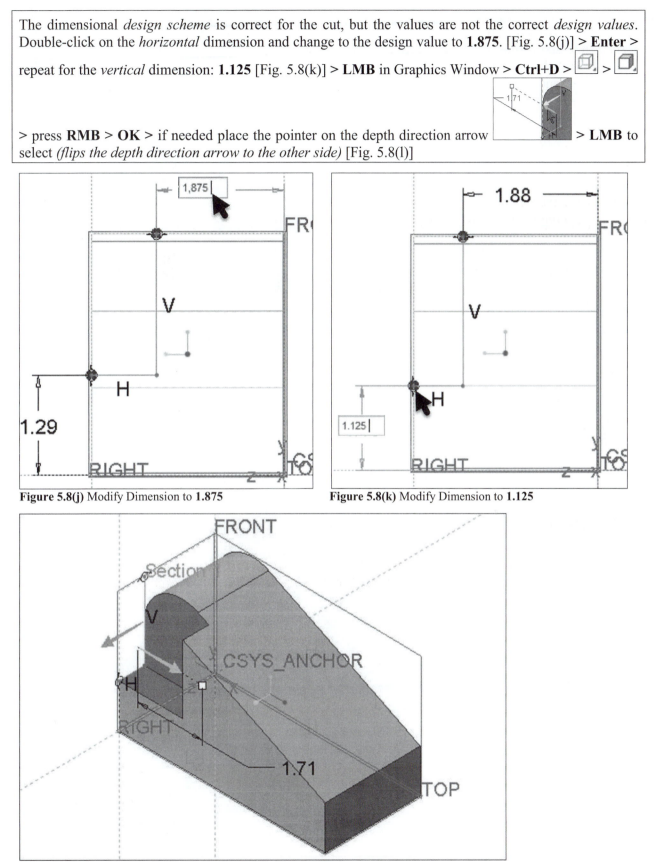

Figure 5.8(j) Modify Dimension to **1.875**

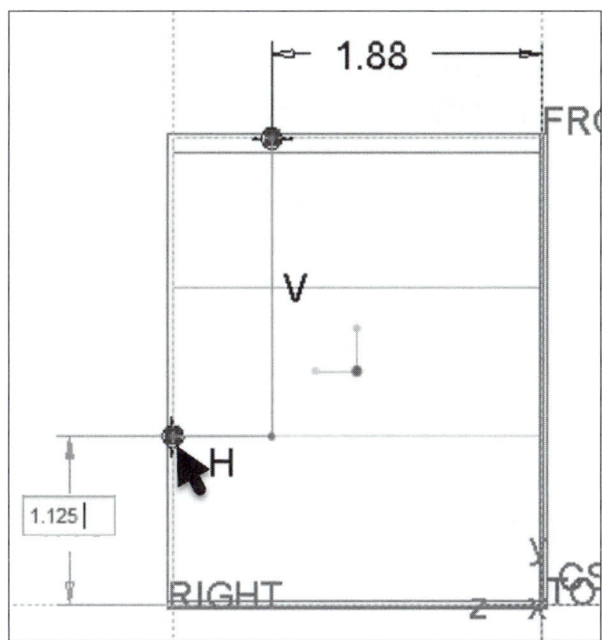

Figure 5.8(k) Modify Dimension to **1.125**

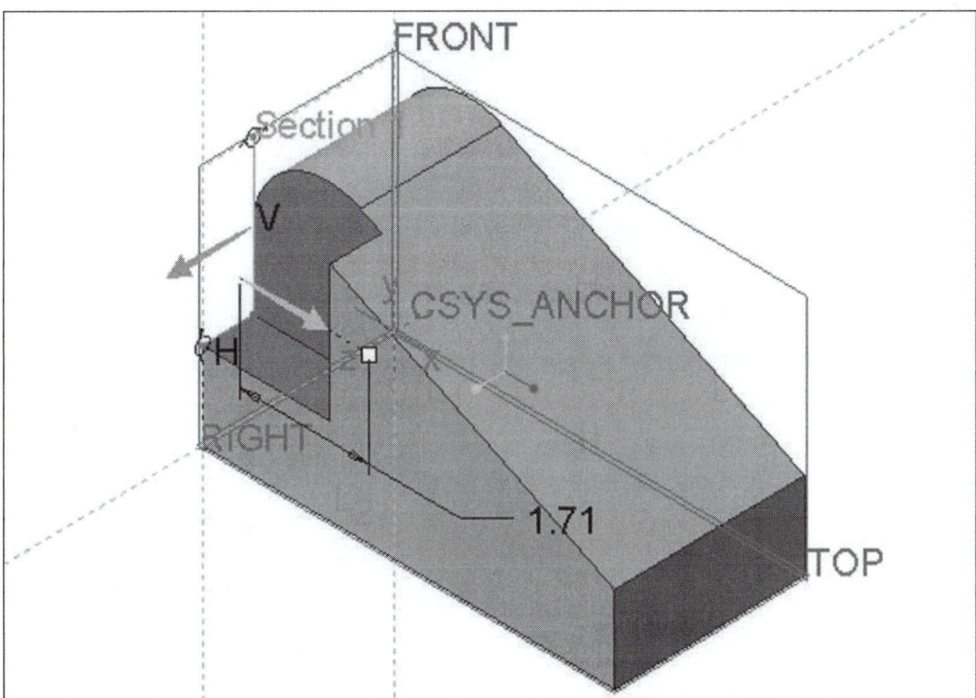

Figure 5.8(l) Extrusion Preview *(flip arrows if necessary)*

Place the pointer over the drag handle > press **RMB** [Fig. 5.8(m)] > **Through All** [Fig. 5.8(n)] > **Options** tab > **MMB** to complete the feature > in the Model Tree, click: ▸ (to expand) > ▸ > select Extrude 2 Section 1 [Fig. 5.8(o)] > in the Graphics Window, **LMB** to deselect > **Save**

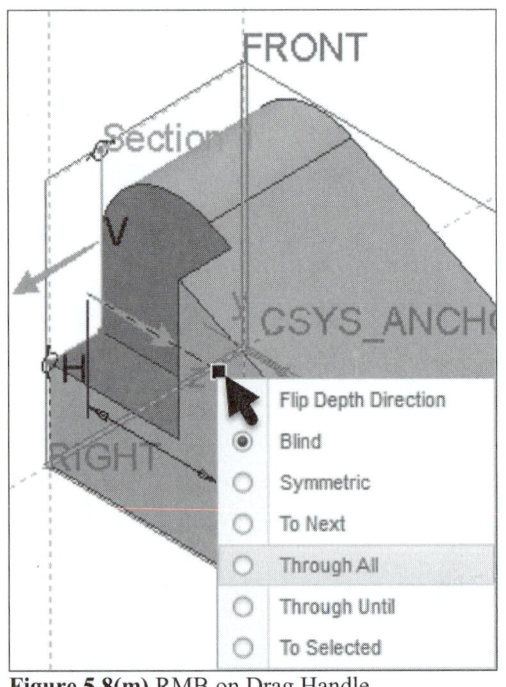

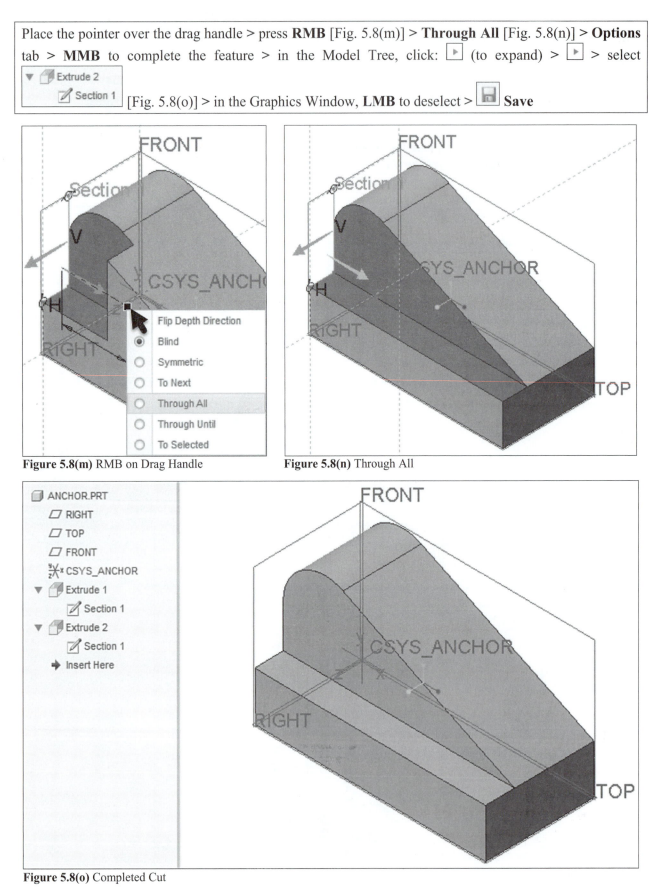

Figure 5.8(m) RMB on Drag Handle

Figure 5.8(n) Through All

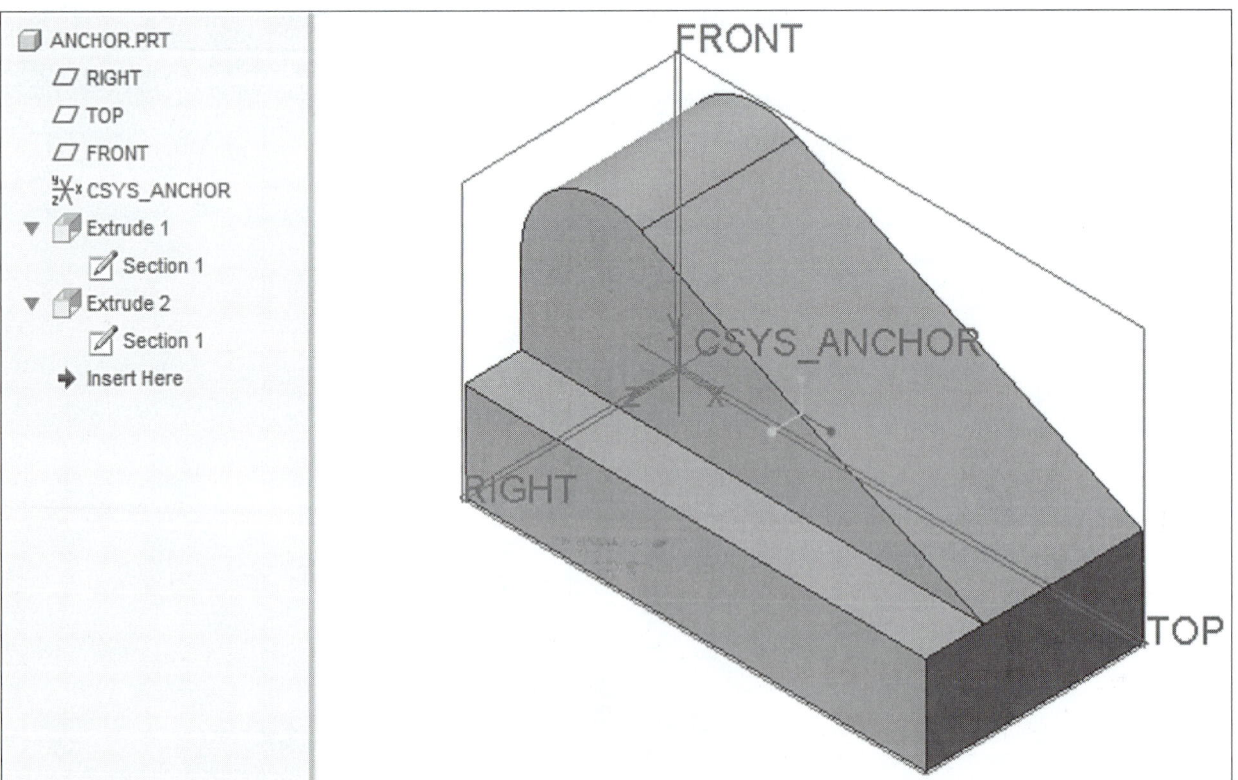

Figure 5.8(o) Completed Cut

Click: **File** > **Options** > **Sketcher** > Number of decimal places for dimension: 2 > **4** > **OK** > **Yes** > **OK** > > **Hidden Line** > > **Remove Material** from the Dashboard > in the Graphics Window, press **RMB** > **Define Internal Sketch** > **Use Previous** > **Sketch View** > **RMB** > **References** > add the left *vertical* edge/surface as a reference [Fig. 5.9(a)] > **Close** > **Line Chain** > sketch the vertical and then the horizontal lines > **MMB** > **MMB** [Fig. 5.9(b)] > **LMB** > reposition the default weak dimension scheme by moving the pointer on top of each dimension value, press and hold down the **LMB**, move the pointer to a new position, release the **LMB** to place the dimension [Fig. 5.9(c)]

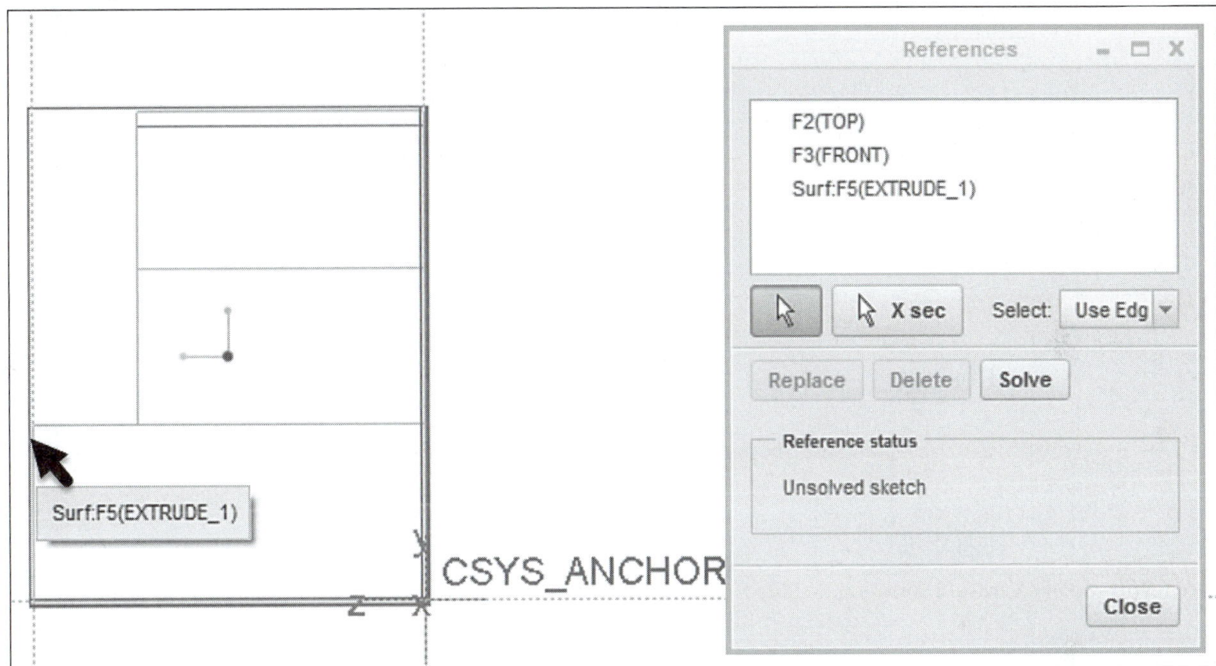

Figure 5.9(a) Add the Left Vertical Edge/Surface as a Reference

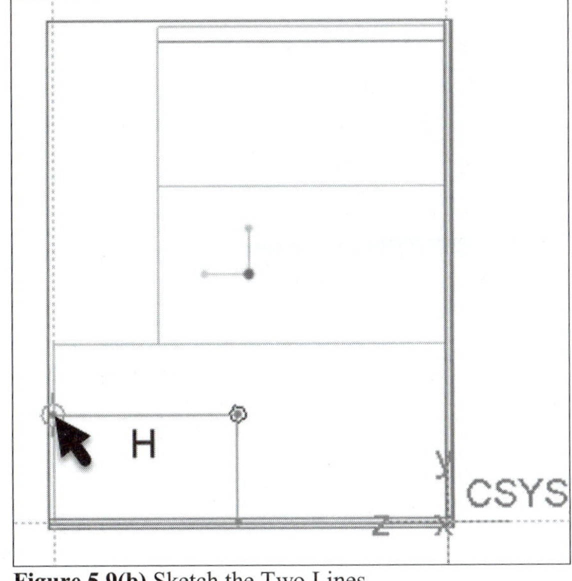

Figure 5.9(b) Sketch the Two Lines

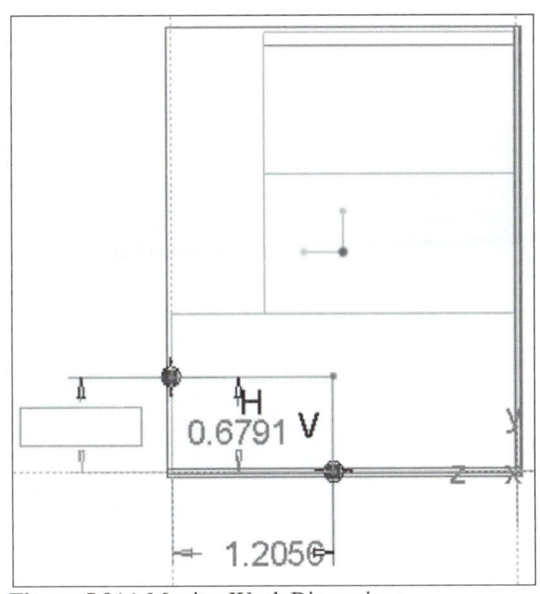

Figure 5.9(c) Moving Weak Dimensions

Double-click on the vertical dimension and *type* **.5625** [Fig. 5.9(d)] > **Enter** > double-click on the horizontal dimension and *type* **1.0625** [Fig. 5.9(e)] > **Enter** > **Ctrl+D** > > > from the Ribbon > > **Extrude to intersect with all surface** > flip the depth direction arrow if necessary > press **MMB** and spin the object to see the cut [Fig. 5.9(f)] > **Ctrl+D** > from the Dashboard [Fig. 5.9(g)] > **Ctrl+S** > **LMB** to deselect

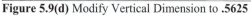

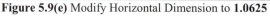

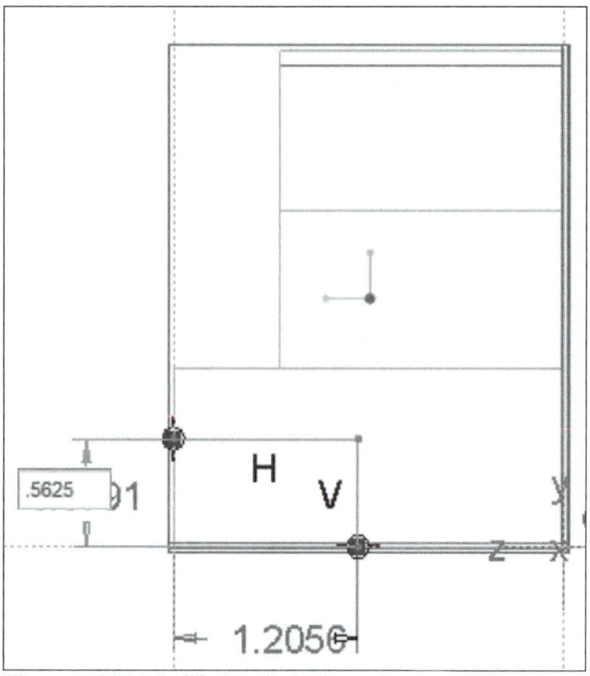

Figure 5.9(d) Modify Vertical Dimension to **.5625**

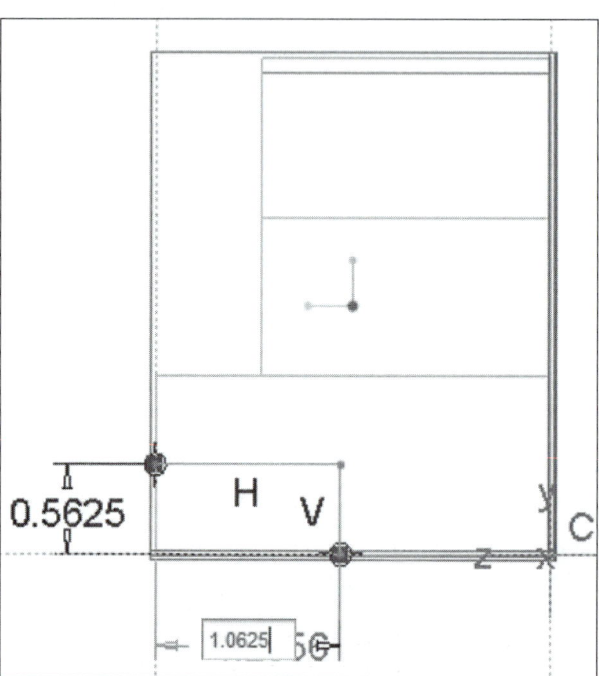

Figure 5.9(e) Modify Horizontal Dimension to **1.0625**

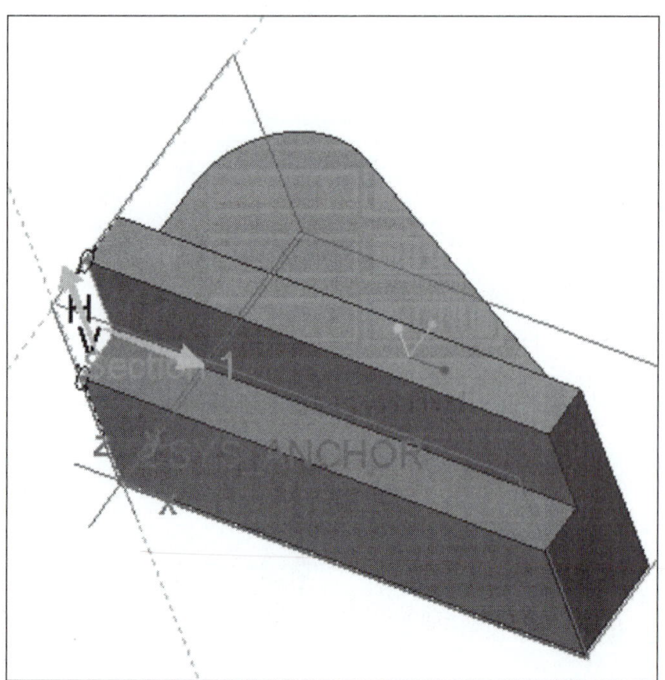

Figure 5.9(f) Preview of the Cut

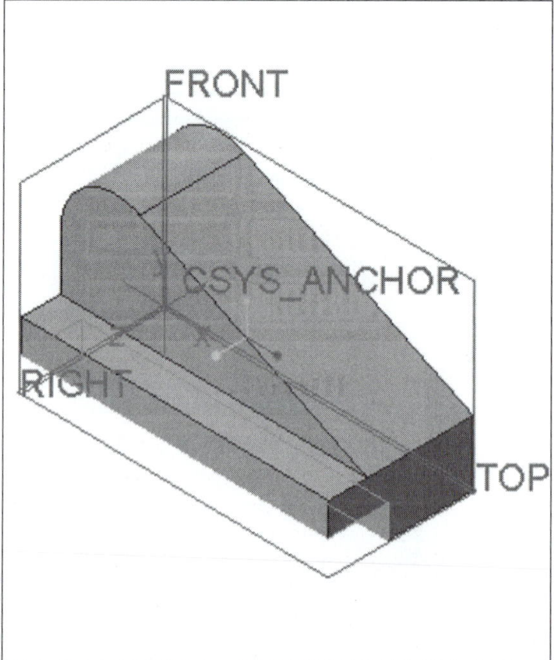

Figure 5.9(g) Completed Cut

For the next feature, you will need to create a new datum plane on which to sketch. New part datum planes are by default numbered sequentially, DTM1, DTM2, and so on. Also, a datum axis (A_1) will be created through the curved top of the part and used later for the axial location of a small hole. Holes and circular features automatically have axes when they are created. Features created with arcs, fillets and so on need to have an axis added if needed as a design reference.

Click: **Datum Display Filters** from the Graphics Toolbar > Csys Display *off* > Plane **Plane** Create a Datum Plane (from the Ribbon Datum Group) > References: pick datum **FRONT** as the plane to offset from > in the DATUM PLANE dialog box, Offset: Translation field, *type the distance* **1.875/2** *(1.875/2 = .9375, always have the CAD system do the math!)* [Fig. 5.10(a)] > **Enter** *(make sure that the value is .9375)* > **OK** > **LMB** to deselect > Axis > References: pick the curved surface [Fig. 5.10(b)] > **OK** > **LMB** to deselect *(you must deselect before the next feature is created)* > **Ctrl+S**

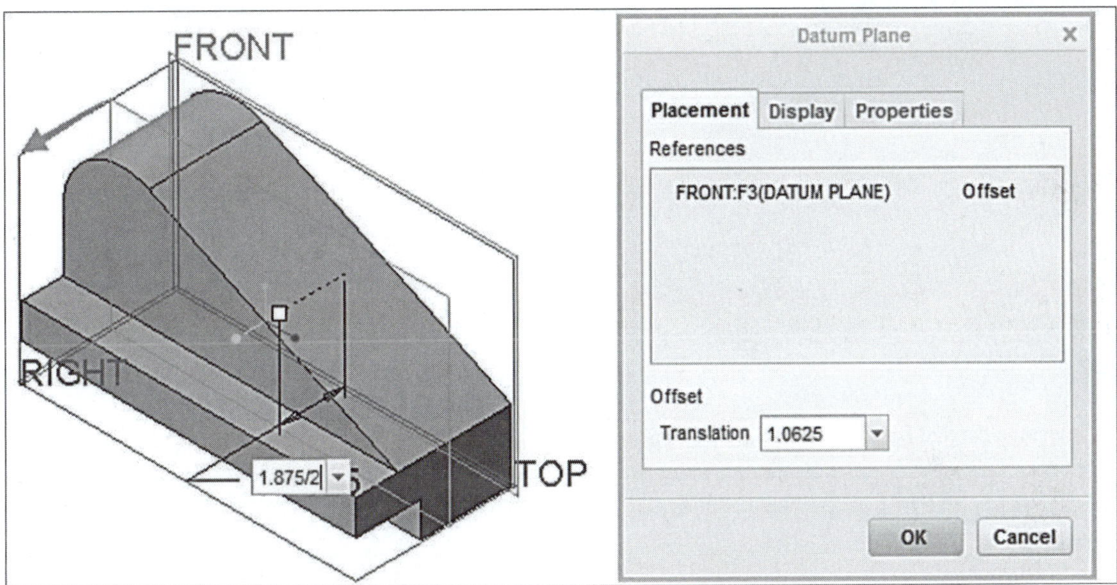

Figure 5.10(a) Offset Datum

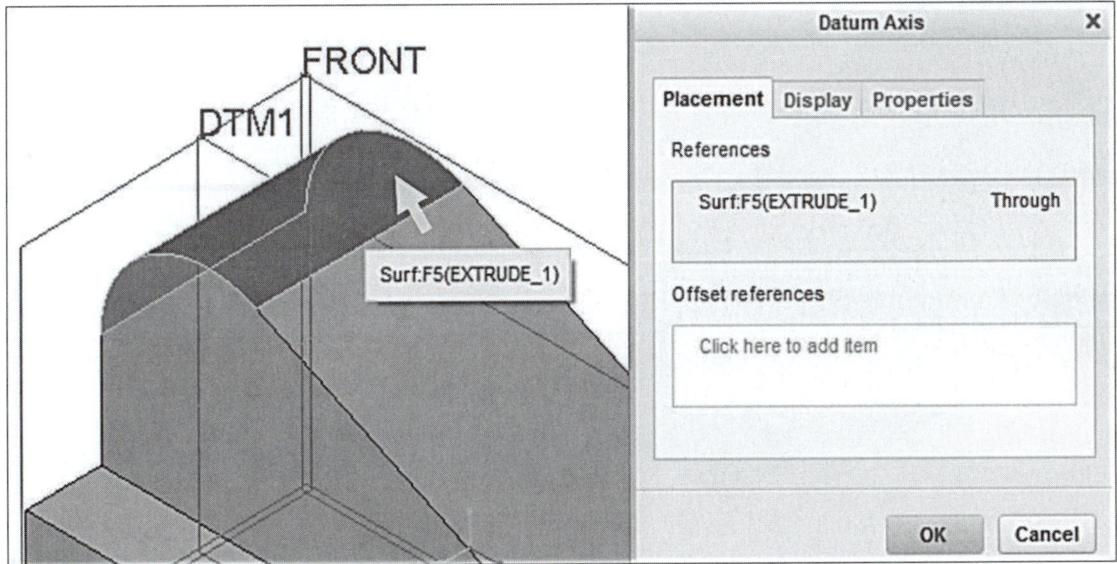

Figure 5.10(b) Datum Axis A_1 through the Center of the Curved Surface

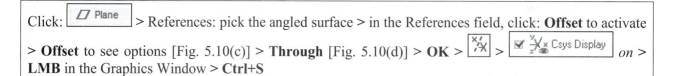

Click: [Plane] > References: pick the angled surface > in the References field, click: **Offset** to activate > **Offset** to see options [Fig. 5.10(c)] > **Through** [Fig. 5.10(d)] > **OK** > [icon] > [Csys Display] *on* > **LMB** in the Graphics Window > **Ctrl+S**

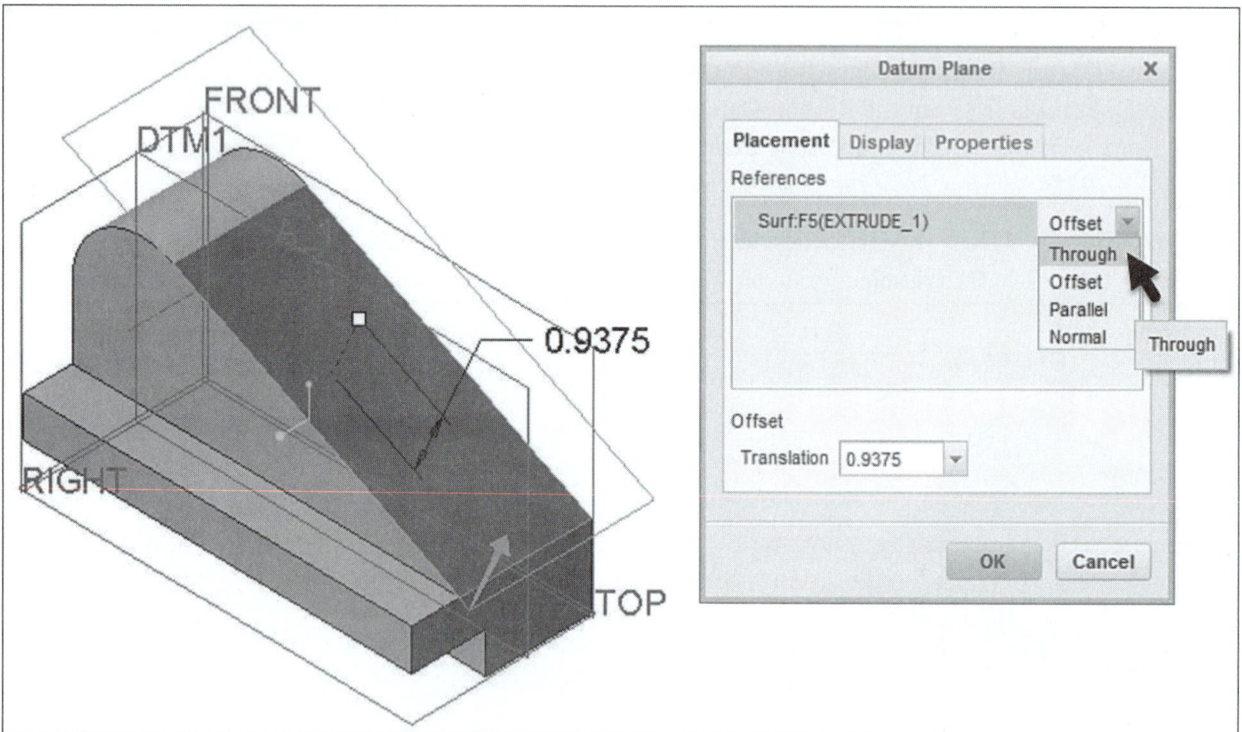

Figure 5.10(c) Select the Angled Surface, then select Through

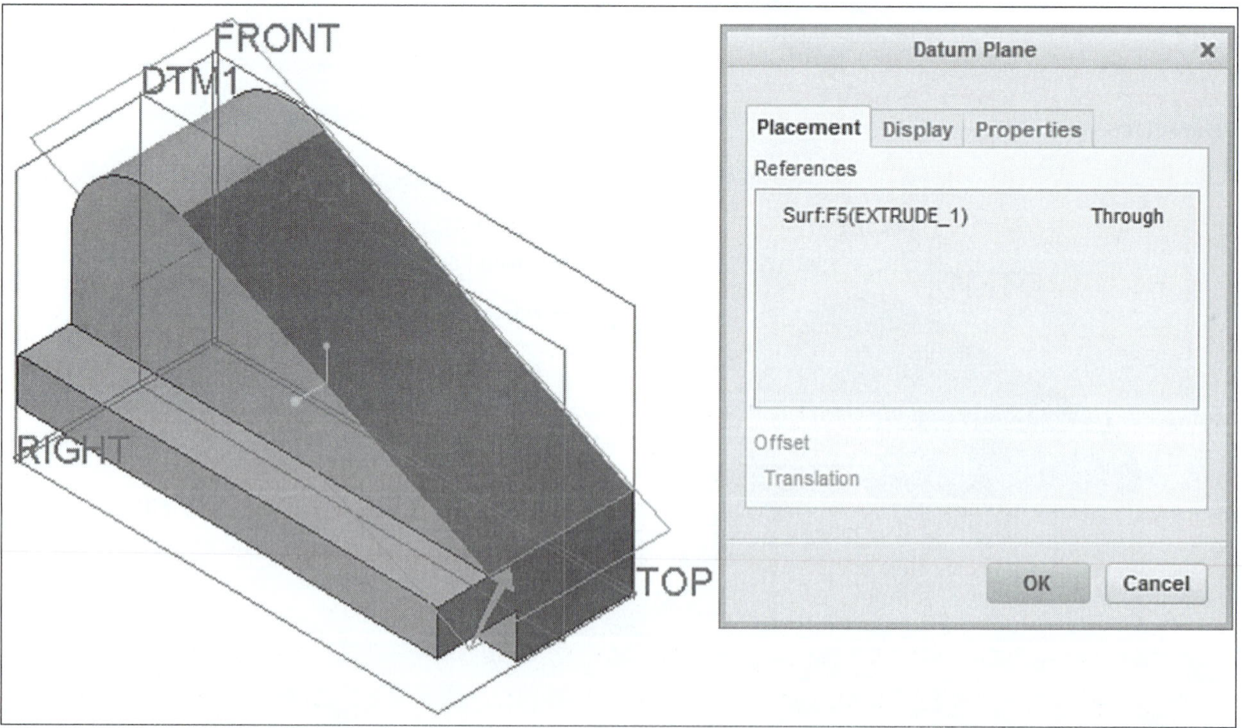

Figure 5.10(d) Through Datum Plane

Create a new layer and add the new datum planes and axis to it, click: **View** tab > Layers *(Layer Tree displays in place of the Model Tree in the Navigator)* > ▶ > ▶ [Fig. 5.11(a)] > Layer Tree > **New Layer** [Fig. 5.11(b)] > In the Layer Properties dialog box [Fig. 5.11(c)], Name: *type* **DATUM_FEATURES** *(do not press Enter)* as the name for new layer [Fig. 5.11(d)]

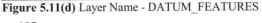

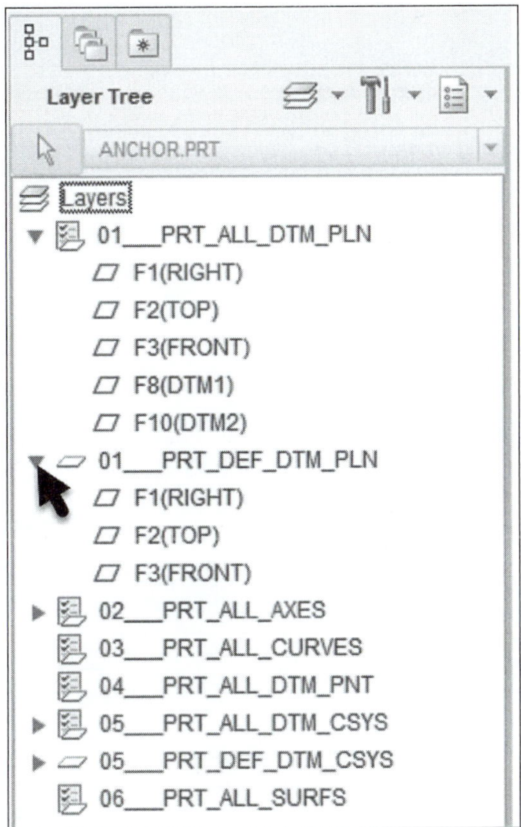

Figure 5.11(a) Layer Tree in the Navigator

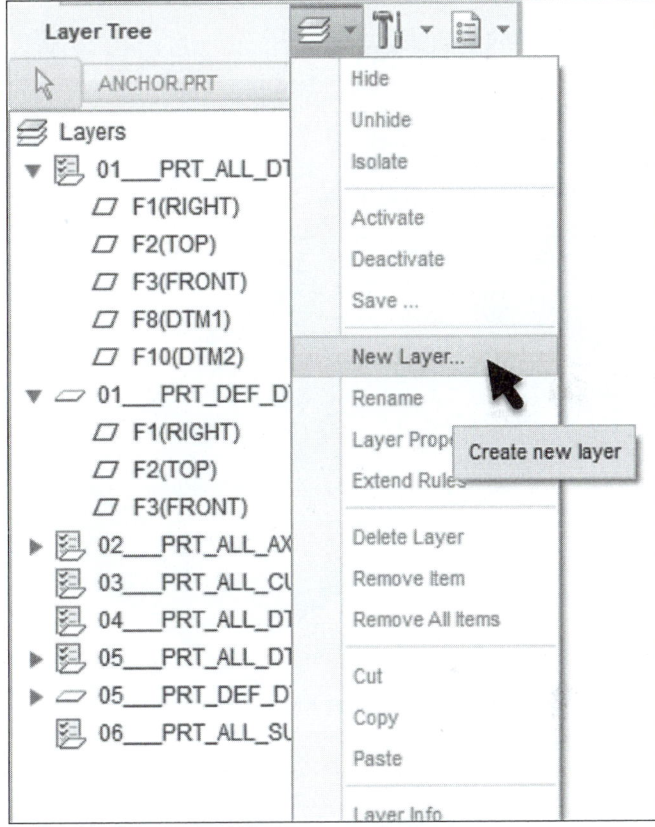

Figure 5.11(b) New Layer

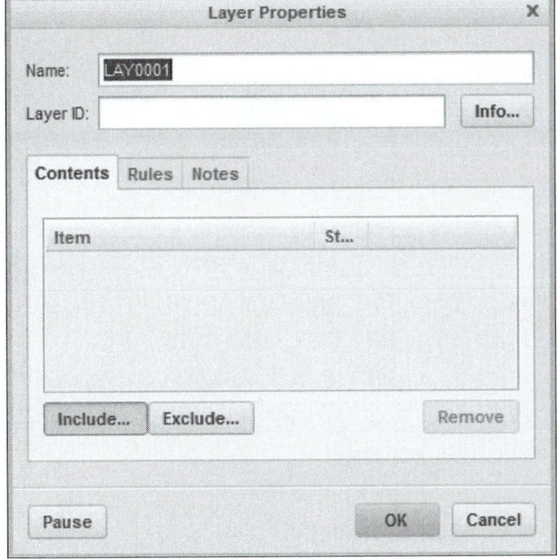

Figure 5.11(c) Layer Properties Dialog Box

Figure 5.11(d) Layer Name - DATUM_FEATURES

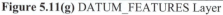

Select the two *new* datum planes from the model > with the cursor over the axis on the model, click **RMB** to toggle to the correct selection *(axis will highlight)* > select the axis with **LMB** [Fig. 5.11(e)] > **OK** [Fig. 5.11(f)] > ▸ expand layer to see items (DATUM_FEATURES: F8(DTM1), F9(A_1), F10(DTM2)) [Fig. 5.11(g)] > in the Navigator > **Model Tree** > **Ctrl+R** *(repaints/redraws the screen)* > **Ctrl+S**

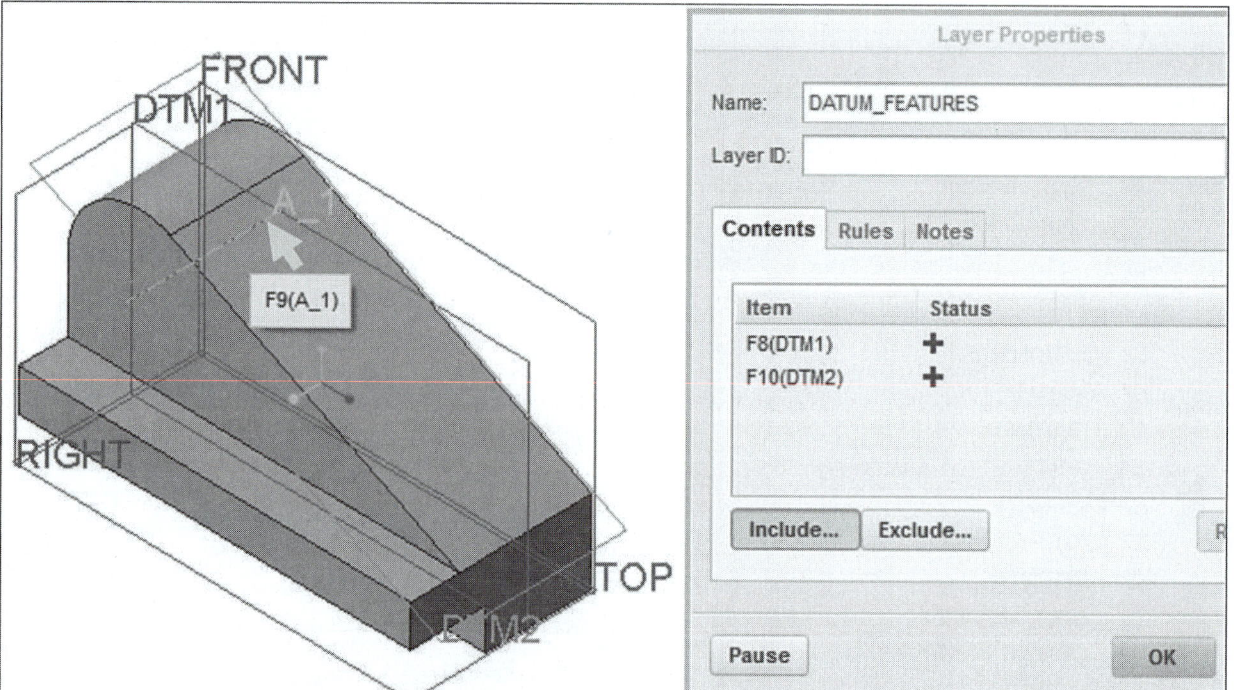

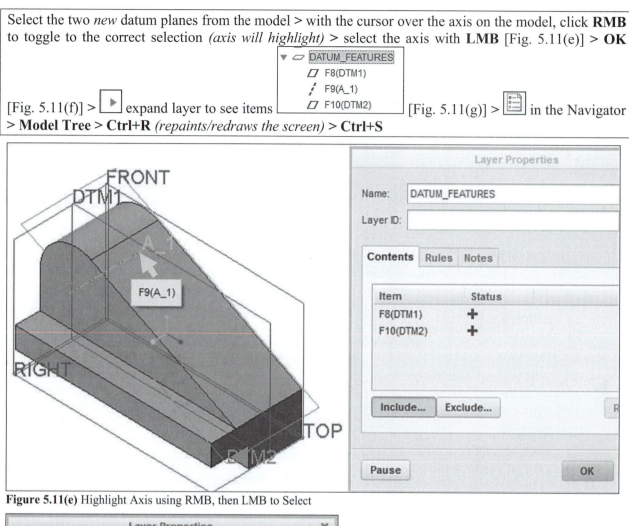

Figure 5.11(e) Highlight Axis using RMB, then LMB to Select

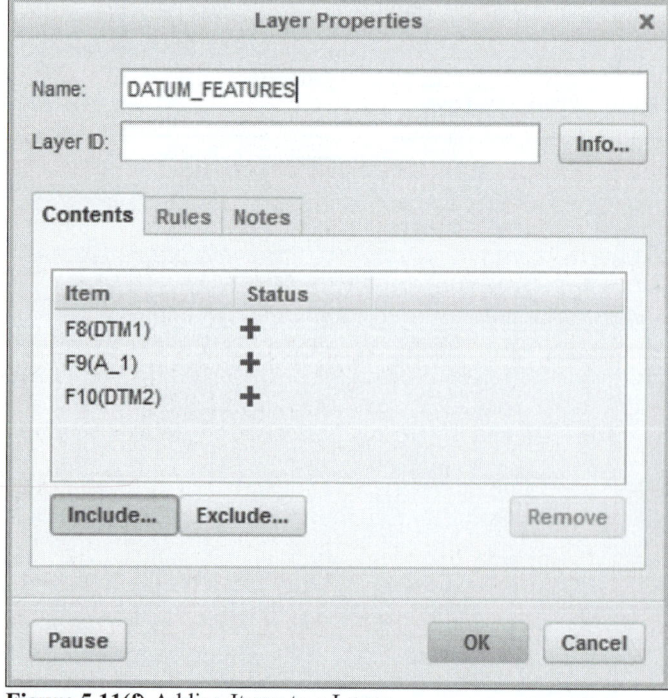

Figure 5.11(f) Adding Items to a Layer

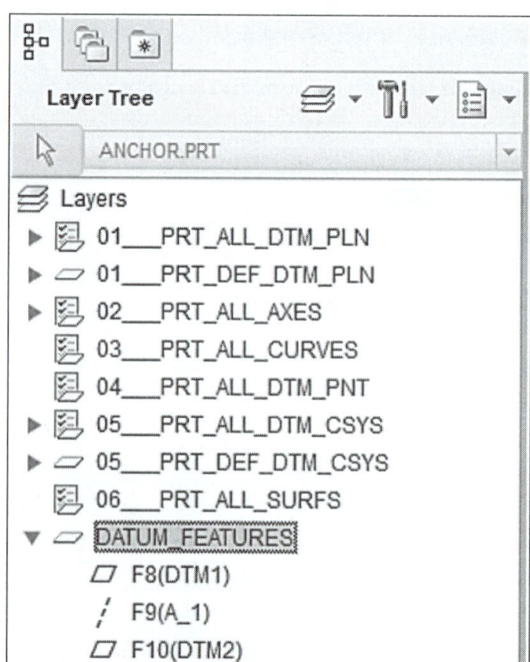

Figure 5.11(g) DATUM_FEATURES Layer

Geometric Tolerances

Before continuing with the modeling of the part, the datums used and created thus far will be "set" for geometric tolerancing. Geometric tolerances (GTOLs) provide a comprehensive method of specifying where on a part the critical surfaces are, how they relate to one another, and how the part must be inspected to determine if it is acceptable. They provide a method for controlling the location, form, profile, orientation, and run out of features. When you store a GTOL in a solid model, it contains parametric references to the geometry or feature it controls—its reference entity—and parametric references to referenced datums and axes. In Assembly mode, you can create a GTOL in a subassembly or a part. A GTOL that you create in Part or Assembly mode automatically belongs to the part or assembly that occupies the window; however, it can refer only to set datums belonging to that model itself, or to components within it. It cannot refer to datums outside of its model in some encompassing assembly, unlike assembly created features. You can add GTOLs in Part or Drawing mode, but they are reflected in all other modes. Creo Parametric treats them as annotations, and they are always associated with the model. Unlike dimensional tolerances, though, GTOLs do not affect part geometry. Before you can reference a datum plane or axis in a GTOL, you must set it as a reference. Creo Parametric encloses its name using the set datum symbol. After you have set a datum, you can use it in the usual way to create features and assemble parts. You enter the set datum command by picking on the datum plane or axis in the Model Tree or on the model itself > RMB > Properties. Creo Parametric encloses the datum name in a feature control frame. If needed, type a new name in the Name field of the Datum dialog box. Most datums will follow the alphabet, A, B, C, D, and so on. You can hide (not display) a set datum by placing it on a layer and then hiding the layer *(or selecting on the item in the Model Tree > RMB > Hide).*

Select datum **TOP** in the Model Tree > **RMB** > **Properties** [Figs. 5.12(a-b)]

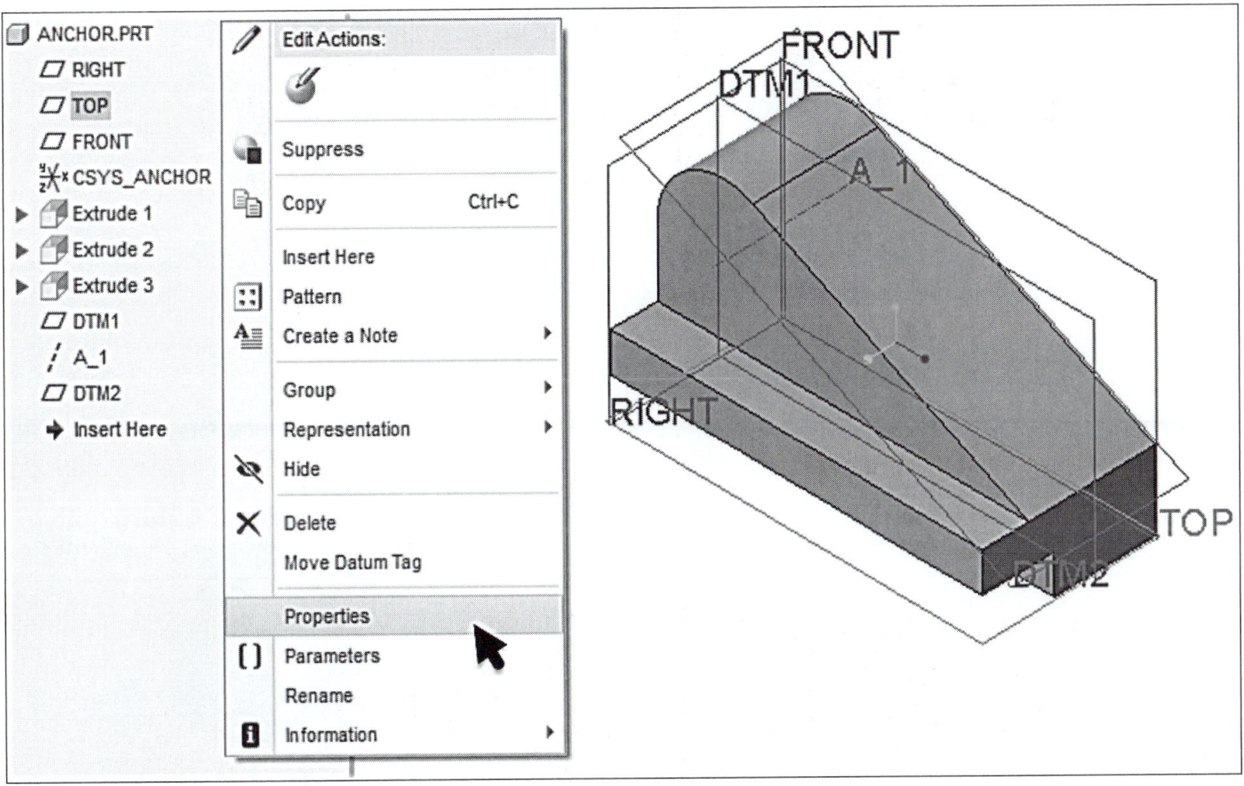

Figure 5.12(a) Datum Properties

Double-click in the Name field > *type* **A** > **Set** [Fig. 5.12(c)] > **OK** in the Datum dialog box > repeat the process and set datum FRONT as **B** and datum RIGHT as **C**. You can also set datums directly on the model. Pick **DTM1** on the model > press **RMB** > **Properties** > Name- *type* **D** > **Set** > **OK** > pick **DTM2** > press **RMB** > **Properties** > Name- *type* **E** > **Set** > **OK** [Fig. 5.12(d)] > **Ctrl+S** > **LMB**

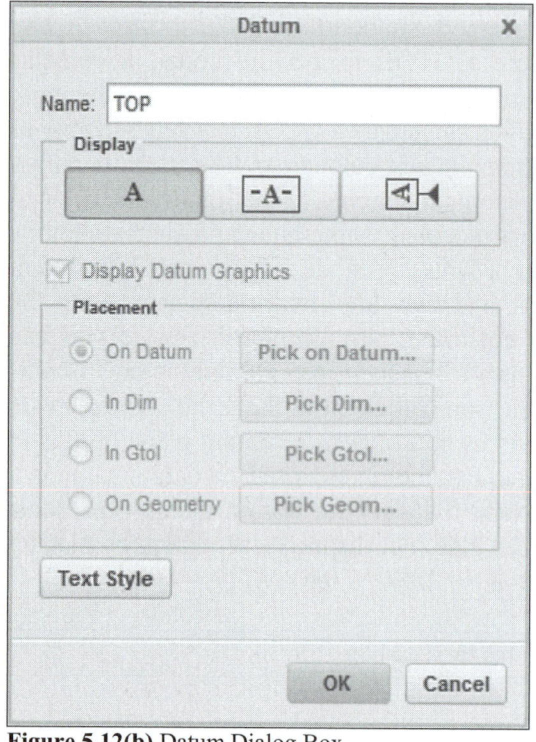

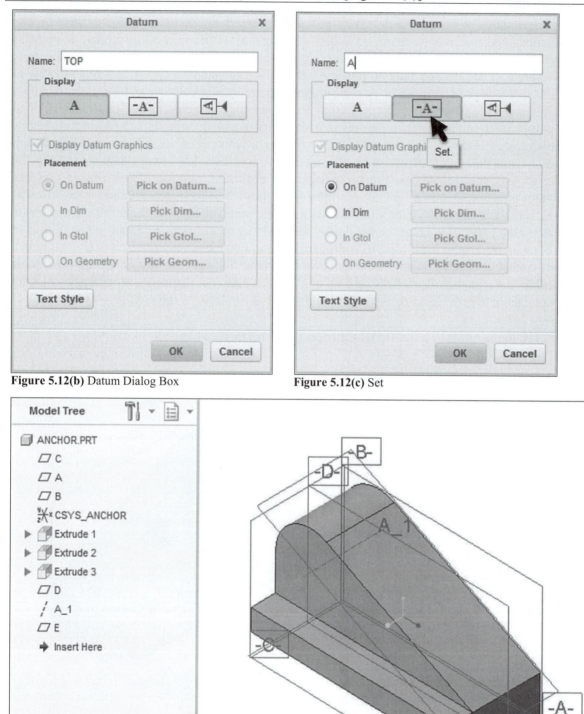

Figure 5.12(b) Datum Dialog Box

Figure 5.12(c) Set

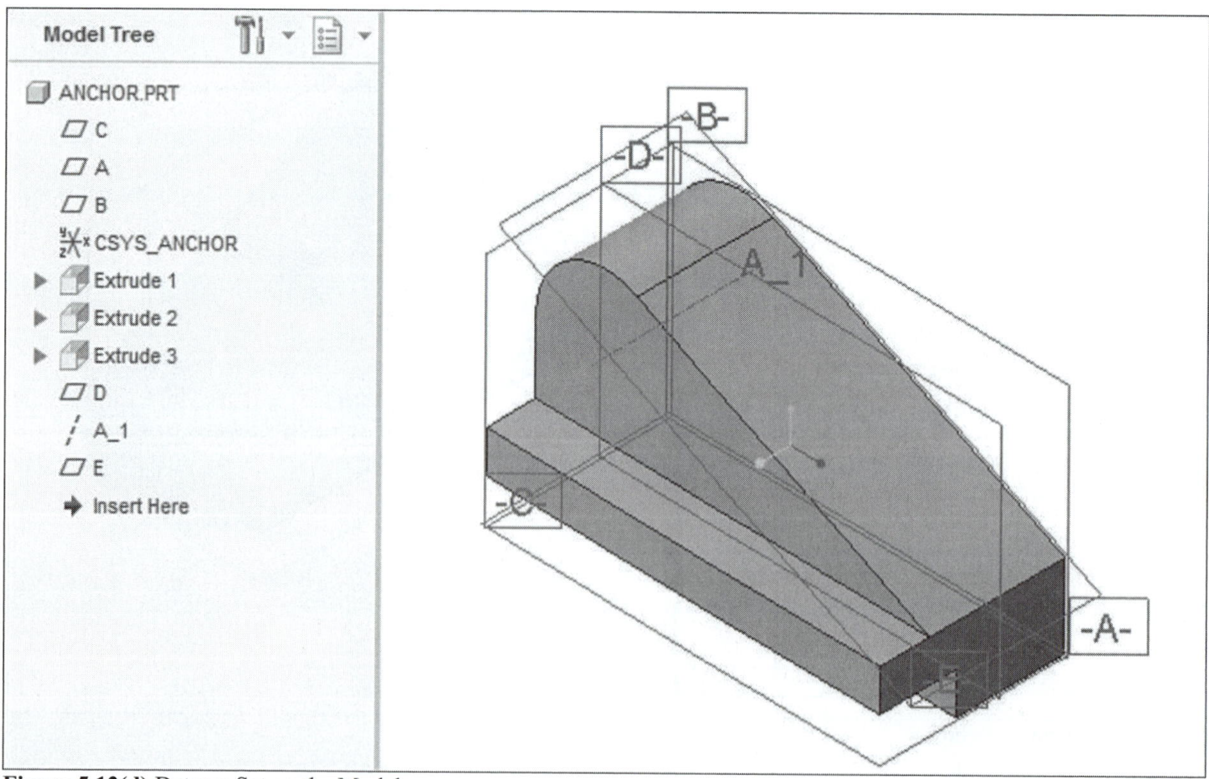

Figure 5.12(d) Datums Set on the Model

Create the slot on the top of the part by sketching on datum D and projecting the cut toward both sides. Use the Model Tree to select the appropriate datum planes.

Click: **Model** tab > Extrude > in the Graphics Window, press **RMB** > **Define Internal Sketch** > Plane: select datum **D** [Fig. 5.13(a)] > **Sketch** > add a references to align the cut with the angled edge of the part, click: **References** from the Ribbon Setup Group (*or RMB* > *References*) > pick on datum **E** > pick the small *vertical* surface on the right side of the part as the fourth reference [Fig. 5.13(b)] > **Close**

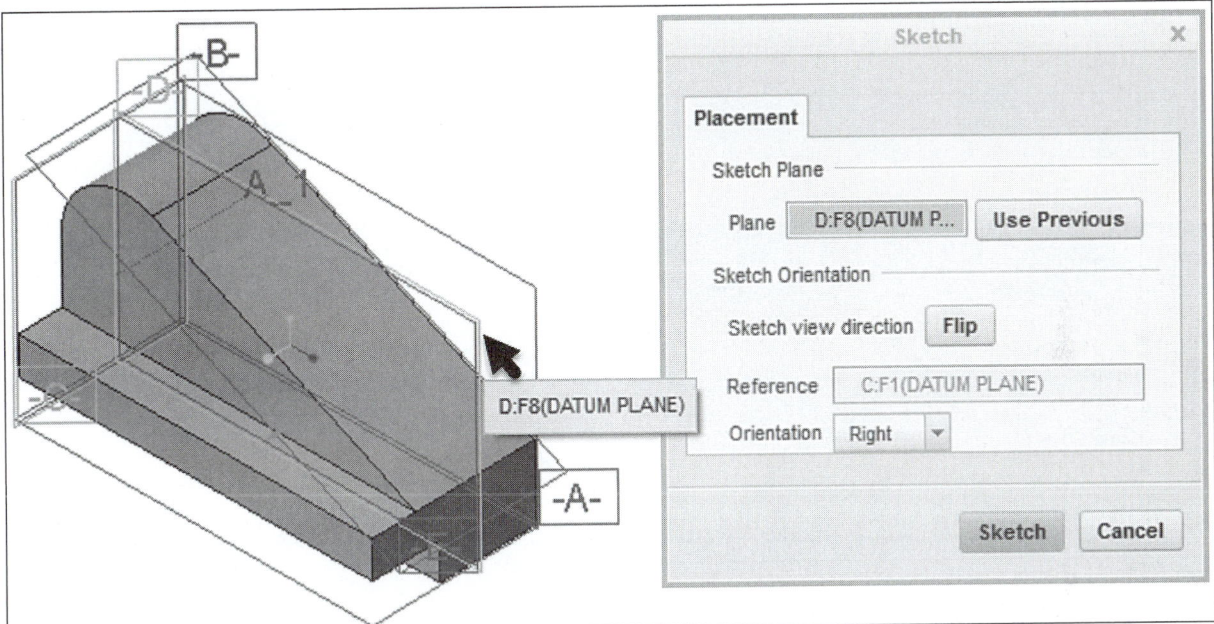

Figure 5.13(a) Sketch Plane and Sketch Orientation Reference

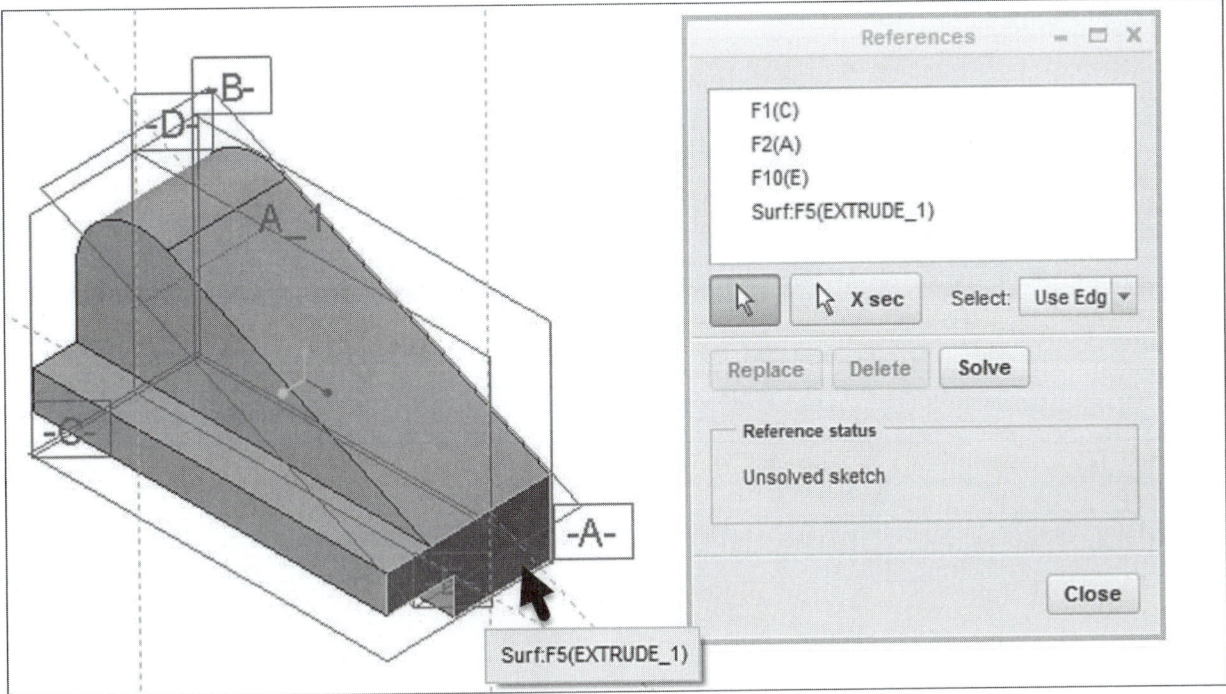

Figure 5.13(b) New References (Datum E and Small Right Vertical Surface)

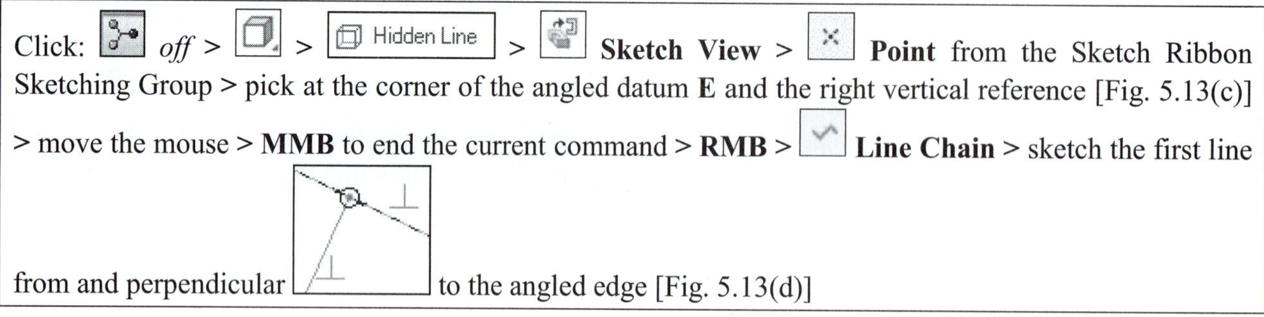

Click: off > > Hidden Line > **Sketch View** > **Point** from the Sketch Ribbon Sketching Group > pick at the corner of the angled datum **E** and the right vertical reference [Fig. 5.13(c)] > move the mouse > **MMB** to end the current command > **RMB** > **Line Chain** > sketch the first line from and perpendicular to the angled edge [Fig. 5.13(d)]

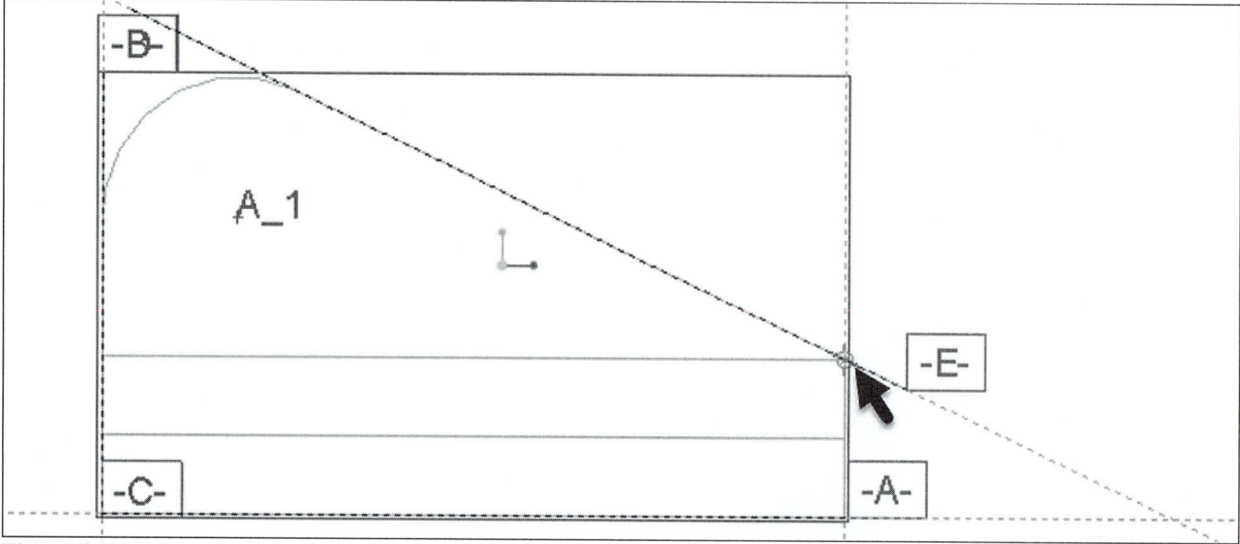

Figure 5.13(c) Construction Point

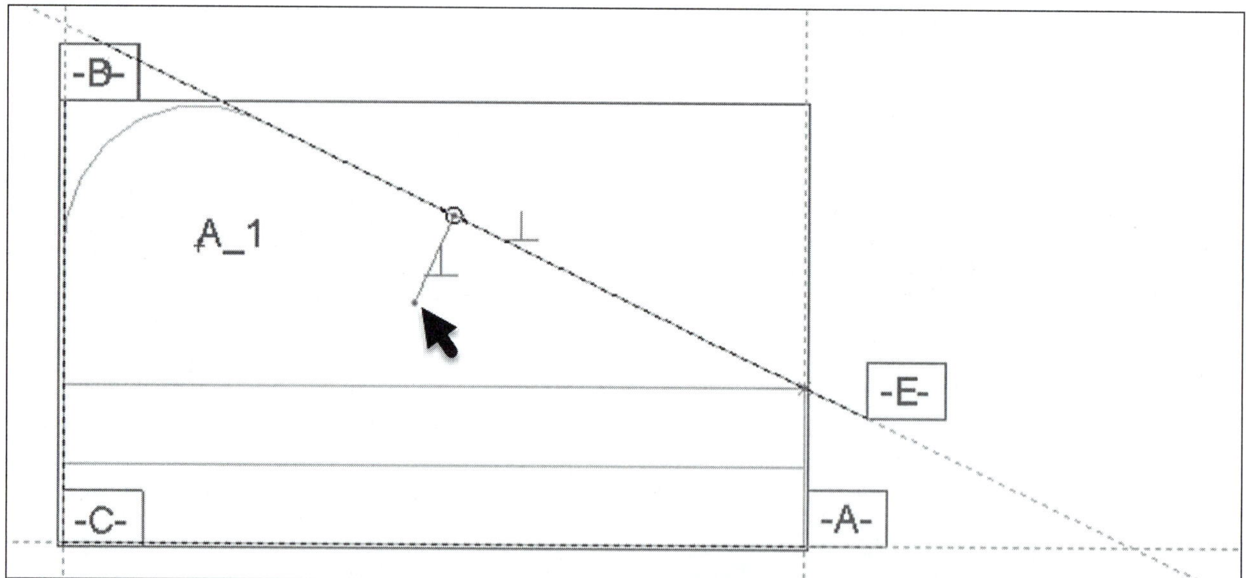

Figure 5.13(d) Create the first Line Perpendicular to Datum E

Sketch the next line horizontal (to the left vertical reference) [Fig. 5.13(e)] > **MMB** > **MMB** to end the line command [Fig. 5.13(f)] > **LMB** to deselect > move the dimension off of the object [Fig. 5.13(g)]

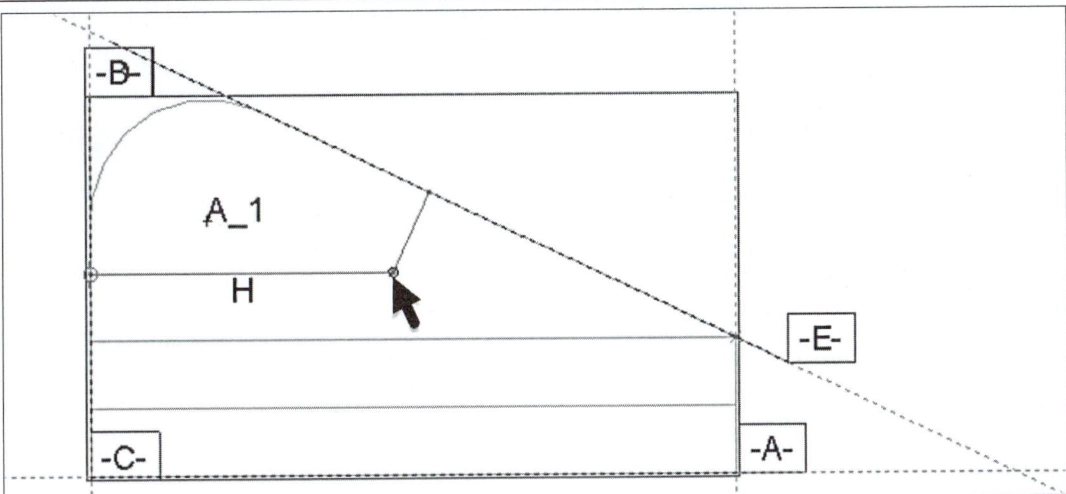

Figure 5.13(e) Create the Second Line Horizontal

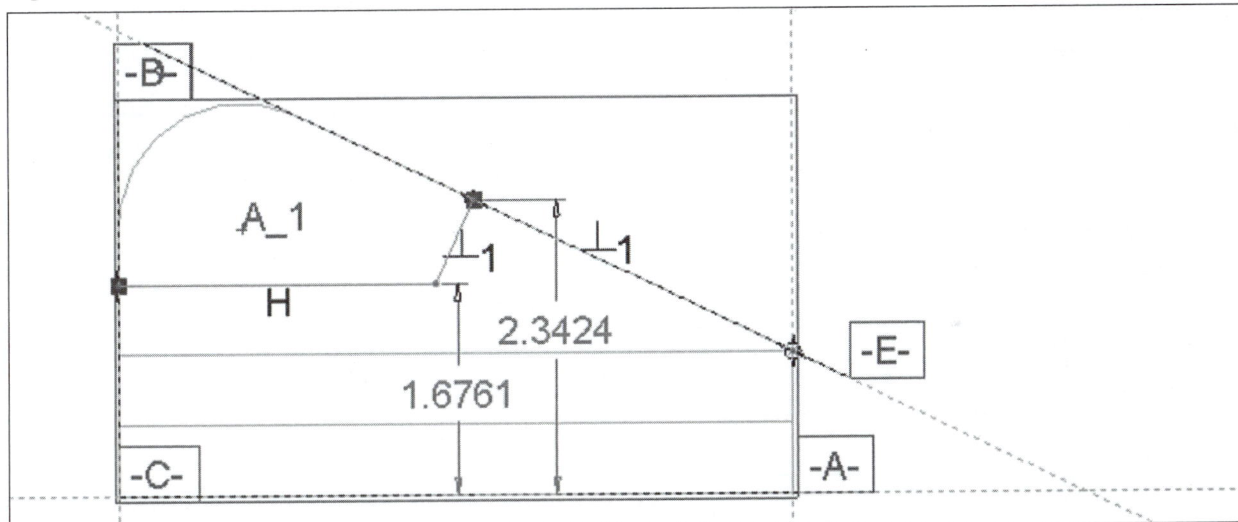

Figure 5.13(f) Completed Lines

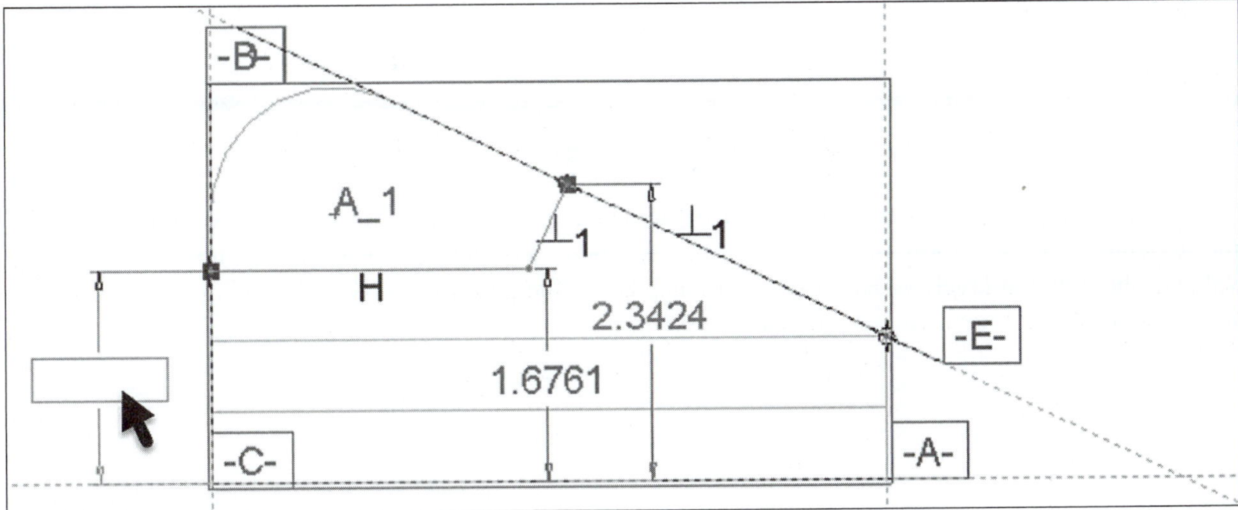

Figure 5.13(g) Move the Dimension

Double-click on the moved dimension > *type the design value of* **1.50** [Fig. 5.13(h)] > **Enter** > move the mouse off of the dimension > **LMB** to deselect > press **RMB** > **Dimension** > add a dimension by selecting angled line and then the construction point ***(do not pick endpoint to point)*** > **MMB** to place the dimension [Fig. 5.13(i)] > *type* **2.625** > **Enter** > **MMB** to end the dimensioning tool > **Ctrl+D** > ✔

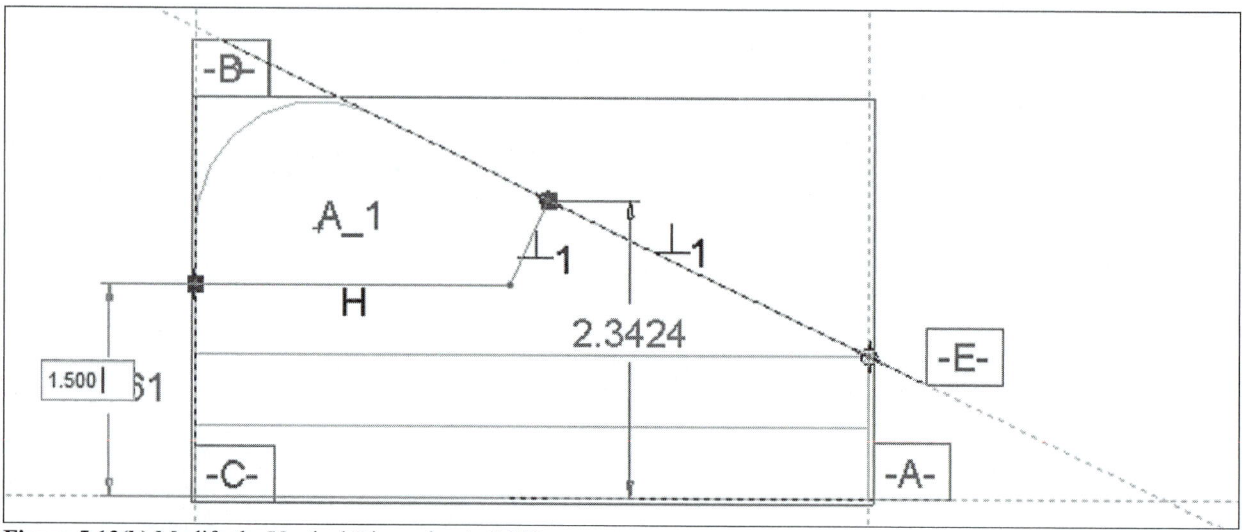

Figure 5.13(h) Modify the Vertical Dimension to **1.500**

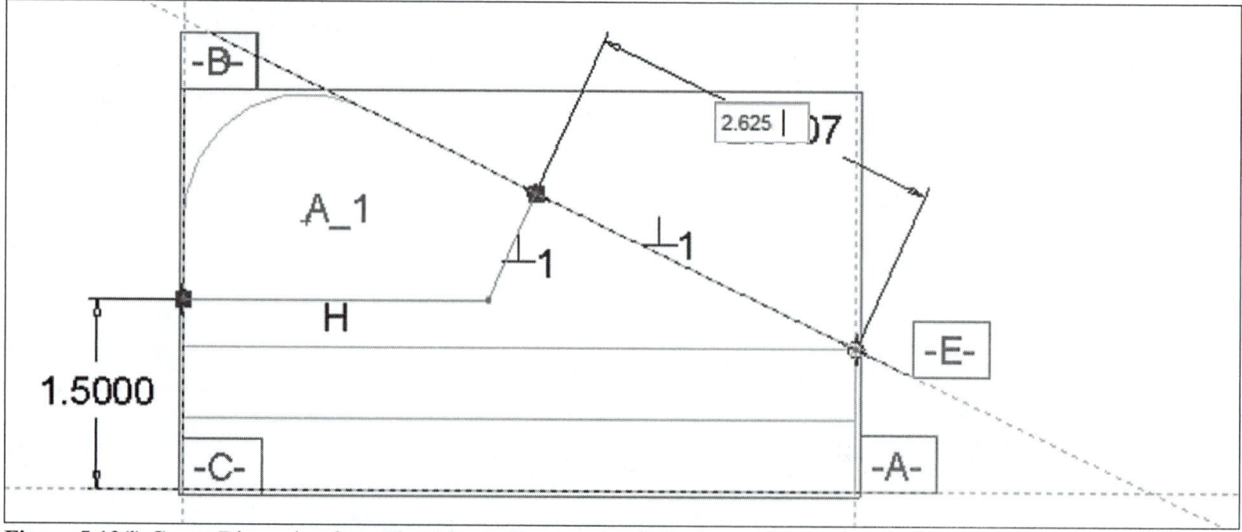

Figure 5.13(i) Create Dimension from First Line to Construction Point (not end point to point!)

The **2.625** value is different than that shown in Figure 5.4. When this part is used in a later lesson, the drawing detail will show the dimension modeled here. You will be instructed on how to modify the value at the drawing level instead of on the model. Since the part, assembly and drawing are associative, the component will regenerate with the new size.

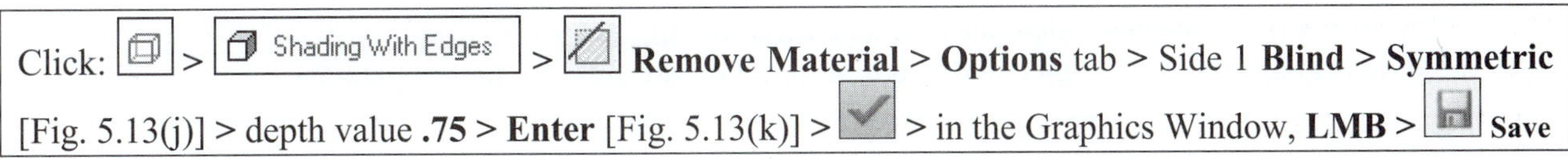

Click: > Shading With Edges > **Remove Material** > **Options** tab > Side 1 **Blind** > **Symmetric** [Fig. 5.13(j)] > depth value **.75** > **Enter** [Fig. 5.13(k)] > > in the Graphics Window, **LMB** > **Save**

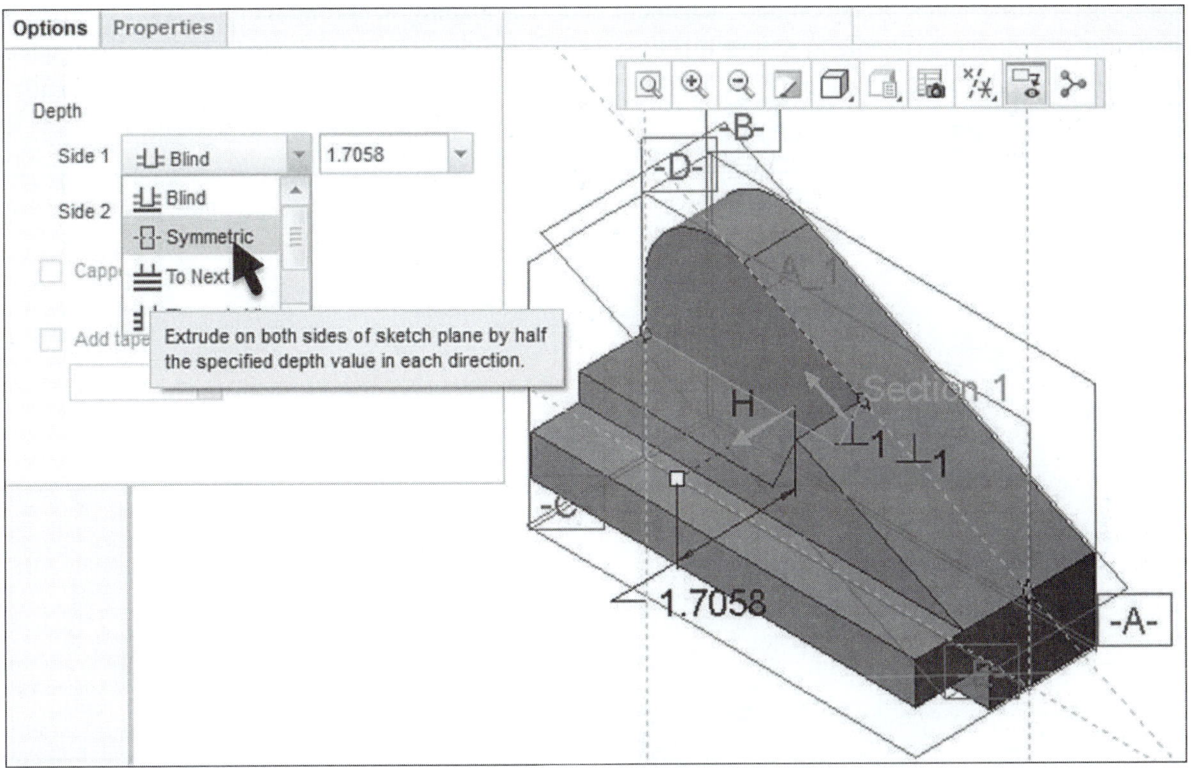

Figure 5.13(j) Symmetric Cut Preview

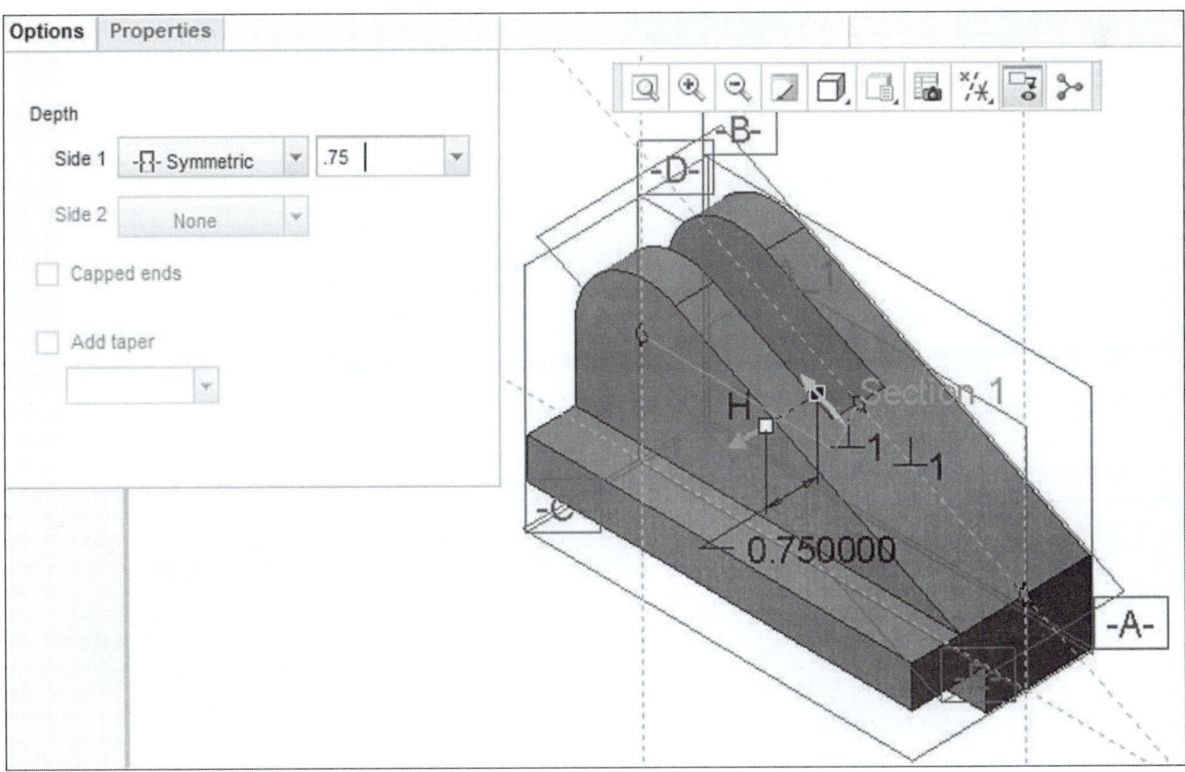

Figure 5.13(k) Symmetric Depth Value

The hole drilled in the angled surface appears to be aligned with Datum D. Upon inspection (Fig. 5.7), the hole is at a different distance and is not in line with the slot and datum plane.

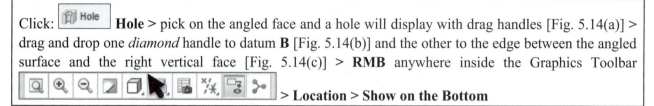

Click: Hole **Hole** > pick on the angled face and a hole will display with drag handles [Fig. 5.14(a)] > drag and drop one *diamond* handle to datum **B** [Fig. 5.14(b)] and the other to the edge between the angled surface and the right vertical face [Fig. 5.14(c)] > **RMB** anywhere inside the Graphics Toolbar > **Location** > **Show on the Bottom**

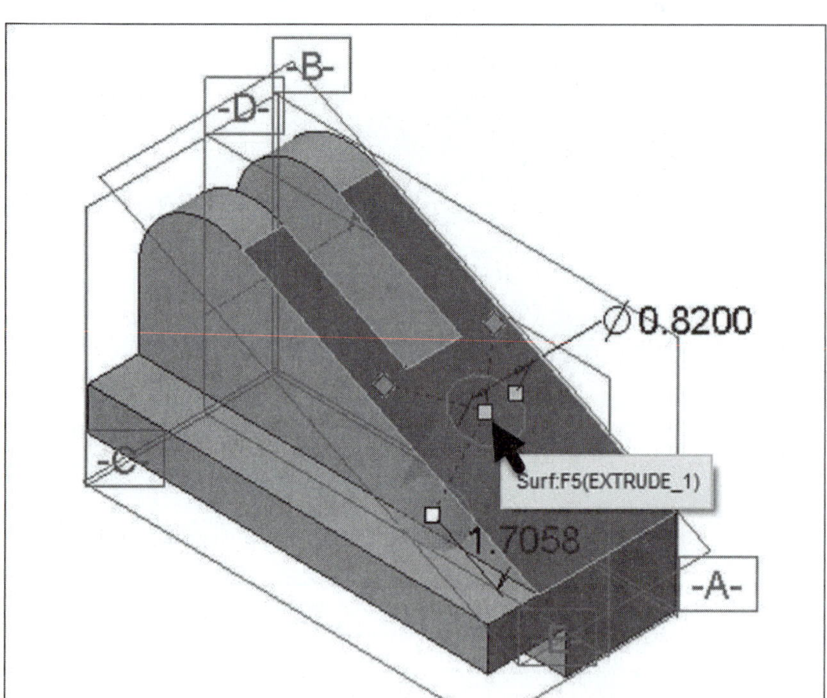

Figure 5.14(a) Hole Placement (two diamond placement drag handles and three square drag handles)

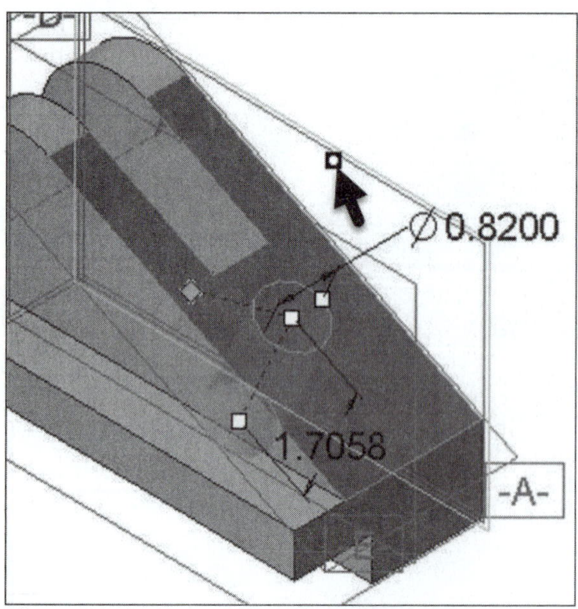

Figure 5.14(b) Drag and Drop Handle to Datum B

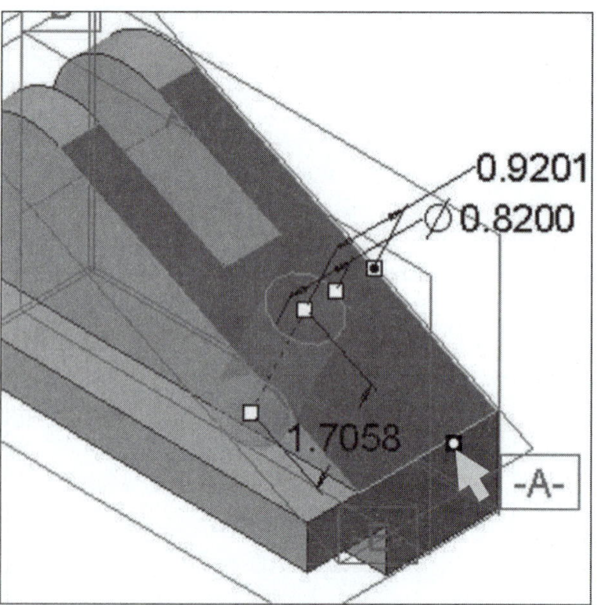

Figure 5.14(c) Drag and Drop other Handle to the Edge

Double-click on the dimension from the hole's center to datum **B** > *type* **.875** [Fig. 5.14(d)] > **Enter** > double-click on the dimension from the hole's center to the edge > *type* **2.0625** [Fig. 5.14(e)] > **Enter** > **Use standard hole profile as drill hole profile** in the Dashboard > Ø 1.0000 > 1.1250 > **Shape** tab [Fig. 5.14(f)] > ✓ > **Ctrl+D** > in the Graphics Window, **LMB** to deselect

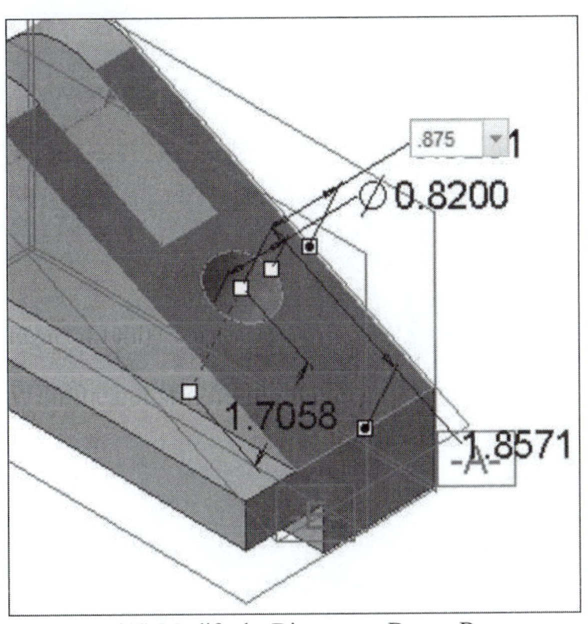

Figure 5.14(d) Modify the Distance to Datum B

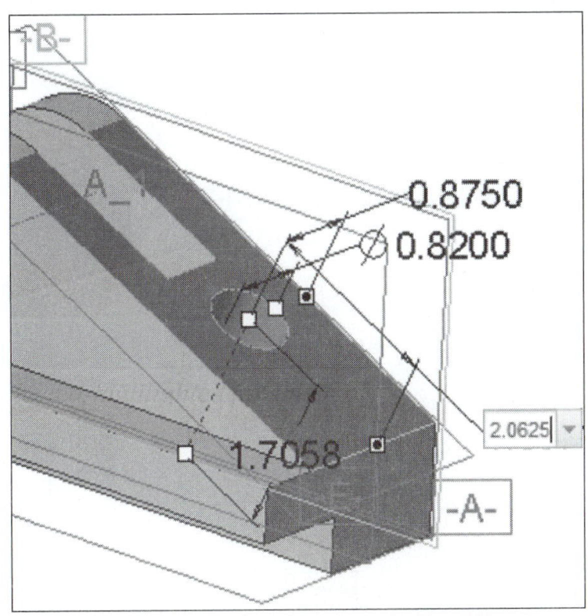

Figure 5.14(e) Modify the Distance to the Part Edge

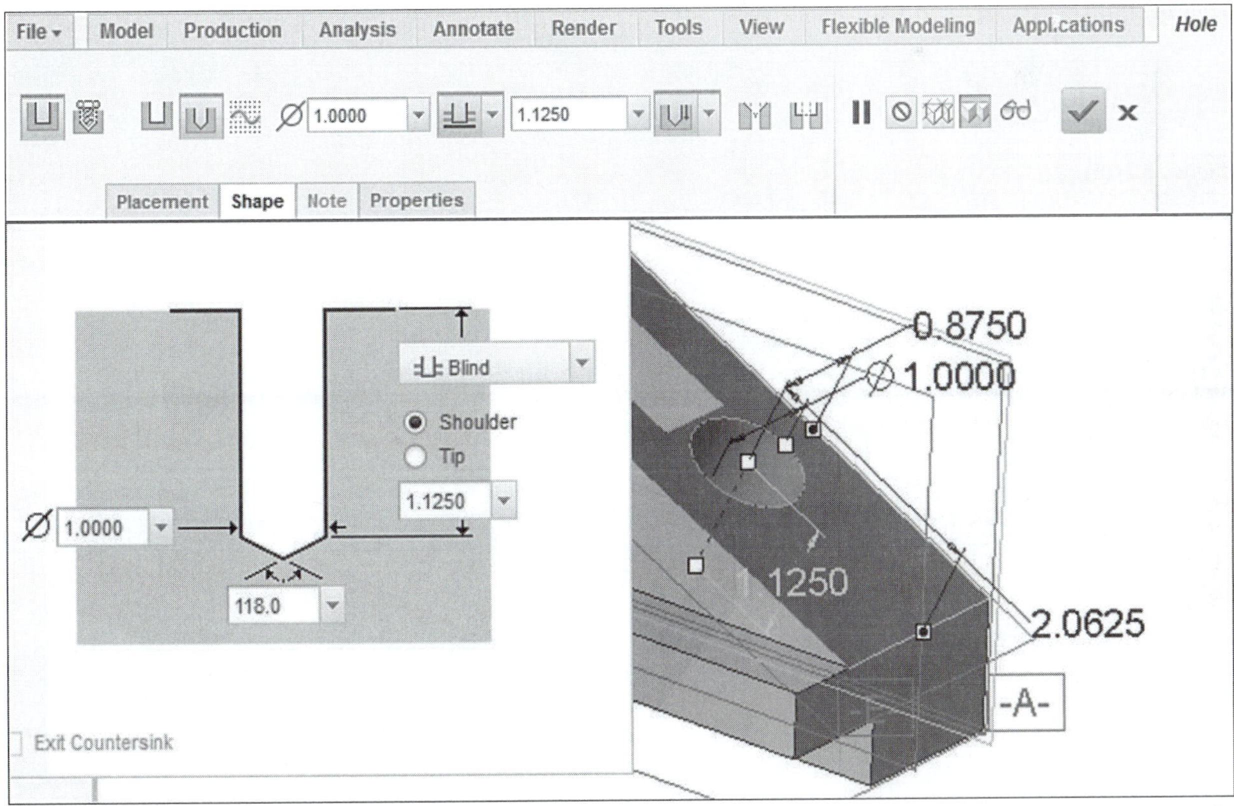

Figure 5.14(f) Shape Tab

Select axis **A_1** from the Model Tree > Hole **Hole** > hold down the **Ctrl** key and pick datum **D** from the model or the Model Tree > release the **Ctrl** key > **Placement** tab [Fig. 5.15(a)] > **Shape** tab > diameter ⌀ 0.25 **.250** > **Enter** > **Blind** > **Through All** > Side 2 **None** > **Through All** [Fig. 5.15(b)] > **MMB** to complete the command > **Ctrl+D** > **Ctrl+S** > in the Graphics Window, **LMB** to deselect

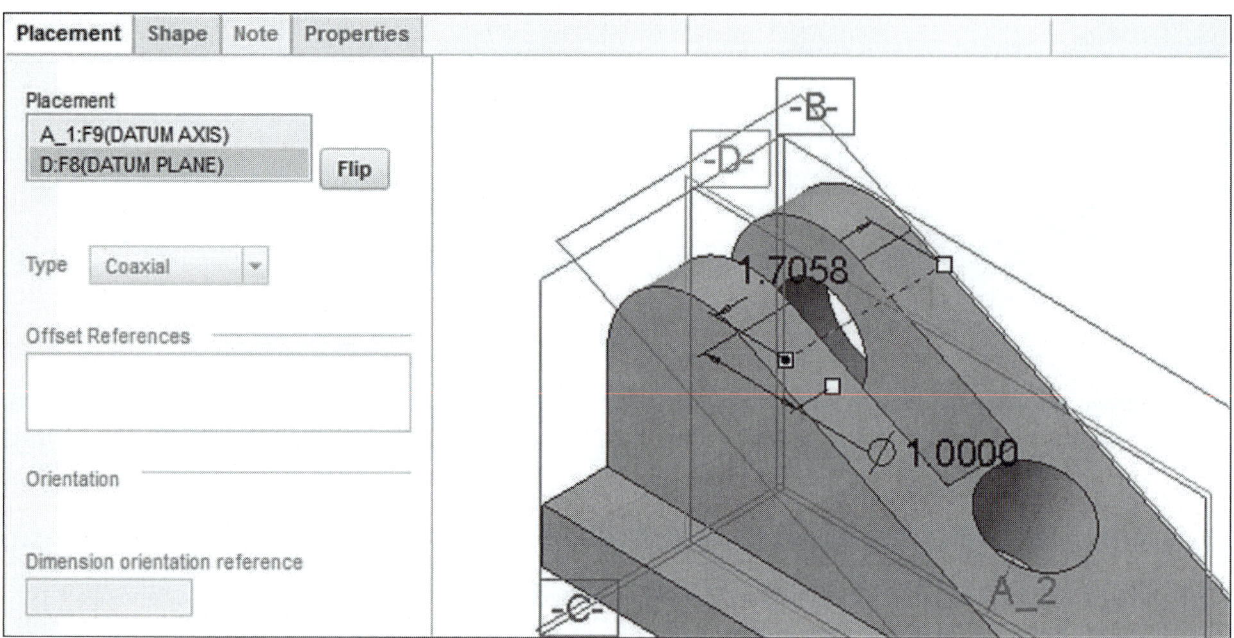

Figure 5.15(a) Placement Tab Placement References

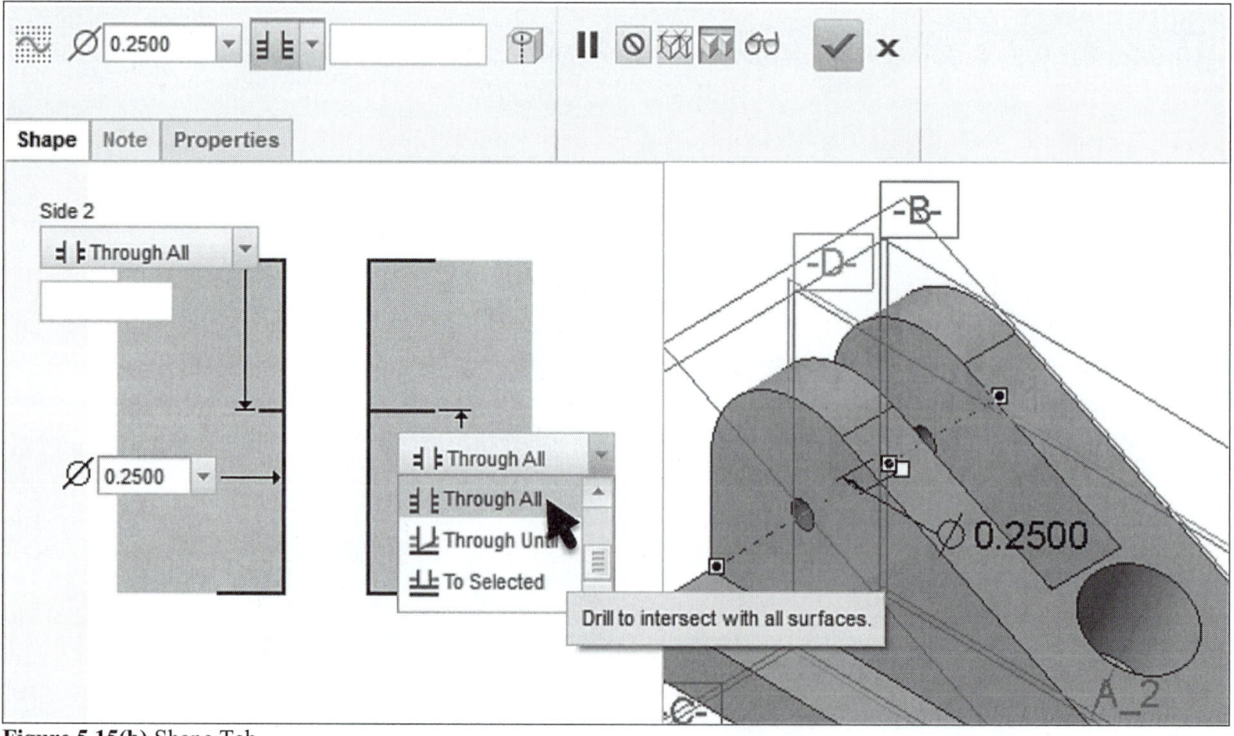

Figure 5.15(b) Shape Tab

Suppressing and Resuming Features using Layers

Next, you will create a new layer and add the two holes to it. The holes will then be selected in the Layer Tree and suppressed. Suppressed features are temporarily removed from the model along with their children (if any). In the example, you will notice that the holes and one axis will be suppressed. Since axis A_1 is the parent of the small hole, it will not be suppressed. Suppressing features is like removing them from regeneration temporarily. However, you can resume suppressed features as needed.

You can suppress features on a part to simplify the part model and decrease regeneration time. For example, while you work on one end of a shaft, it may be desirable to suppress features on the other end of the shaft. Similarly, while working on a complex assembly, you can suppress some of the features and components for which the detail is not essential to the current assembly process. Suppress features to do the following:

- Concentrate on the current working area by suppressing other areas
- Speed up a modification process because there is less to update
- Speed up the display process because there is less to display
- Temporarily remove features to try different design iterations

Unlike other features, the base feature cannot be suppressed. If you are not satisfied with the base feature, you can redefine the section of the feature, or start another part.

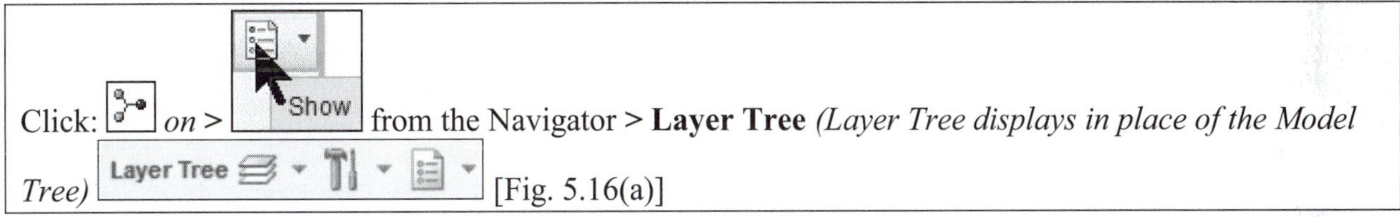

Click: *on >* from the Navigator > **Layer Tree** *(Layer Tree displays in place of the Model Tree)* [Fig. 5.16(a)]

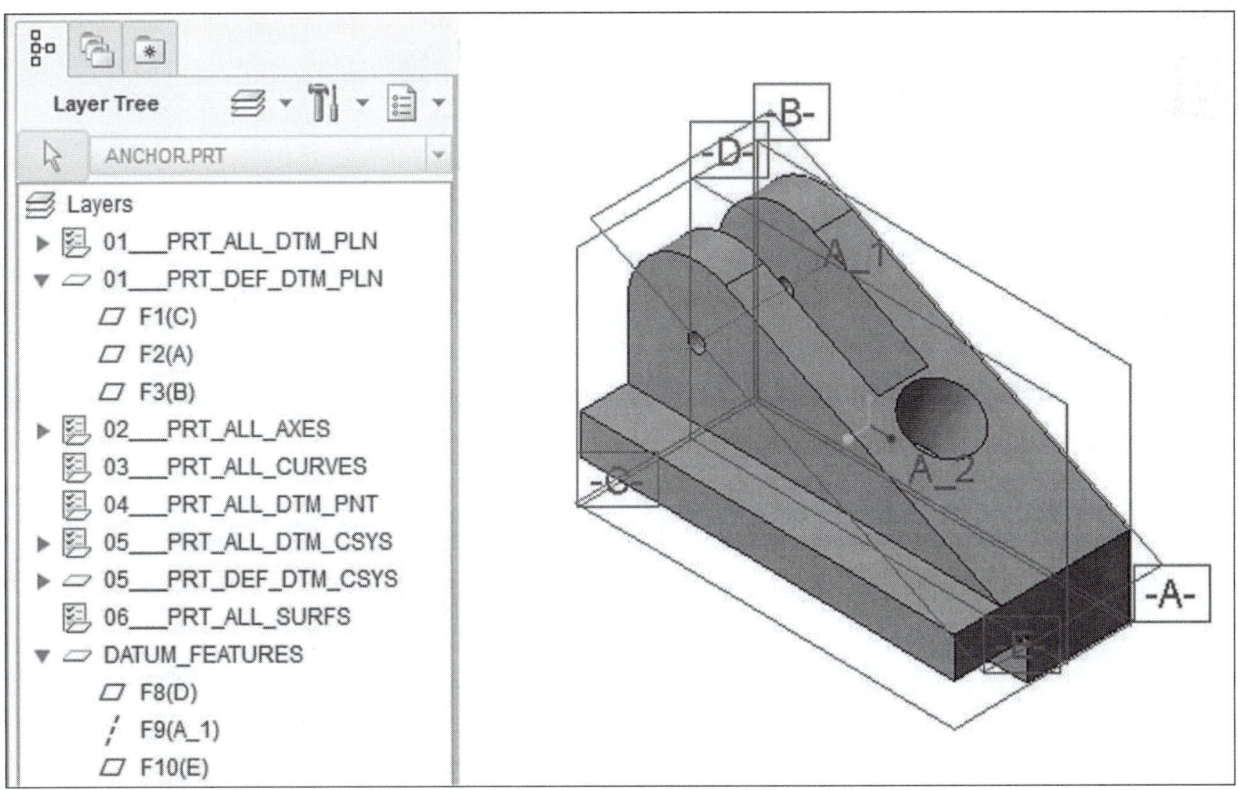

Figure 5.16(a) Layer Tree

Click: Layers in the Layer Tree > **RMB** > **New Layer** [Fig. 5.16(b)] > in the Layer Properties dialog box**,** Name: *type* **HOLES** as the name for the new layer > **Enter** > select the new **HOLES** layer in the Layer Tree > **RMB** > **Layer Properties** [Fig. 5.16(c)] *(you can avoid doing this by **not** pressing Enter after you type in a new layer name)* > Selection Filter (the lower right-hand side below the Graphics Window), click: **All** > **Feature** [Fig. 5.16(d)] > select the two holes from the model *(no need to hold down the Ctrl key)* [Fig. 5.16(e)] > **OK** > **LMB** in the Graphics Window > expand the HOLES layer > press and hold down the **Ctrl** key and pick on the two holes in the HOLES layer [Fig. 5.16(f)] > release the **Ctrl** key

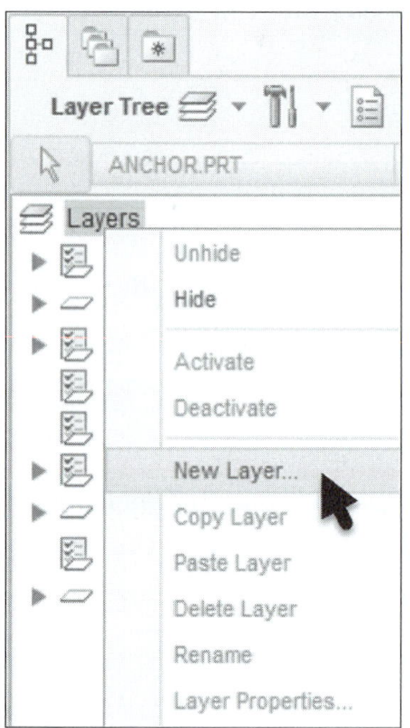

Figure 5.16(b) New Layer

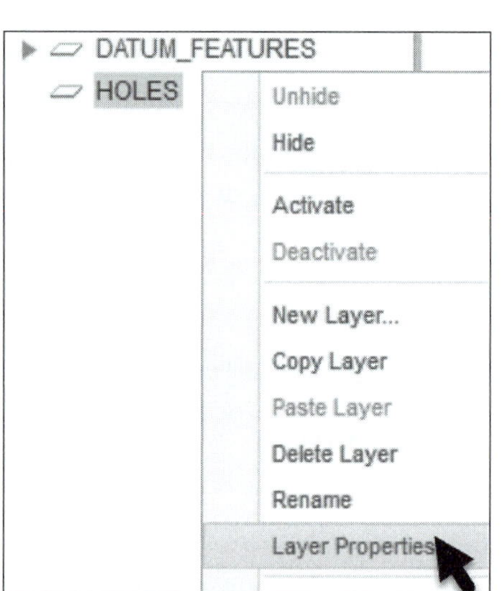

Figure 5.16(c) RMB > Layer Properties

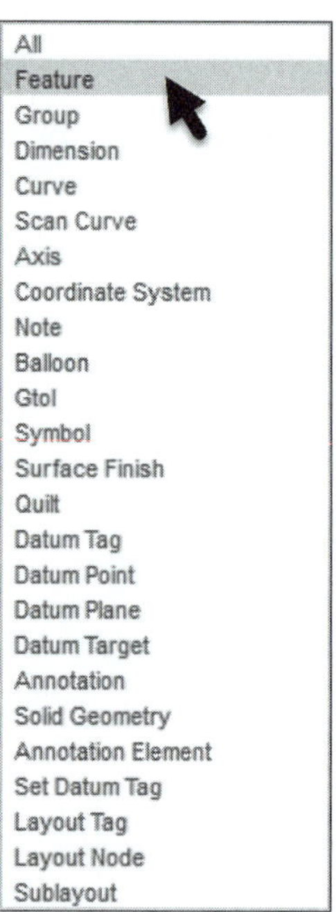

Figure 5.16(d) Filter

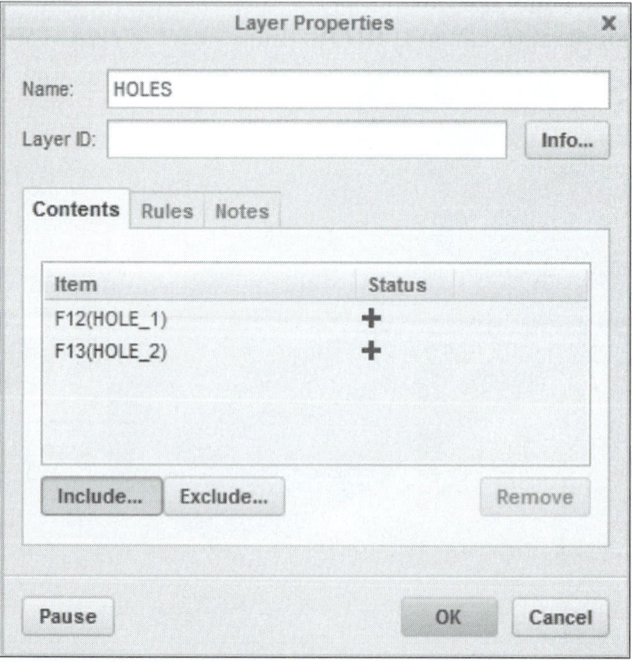

Figure 5.16(e) Holes Layer Items

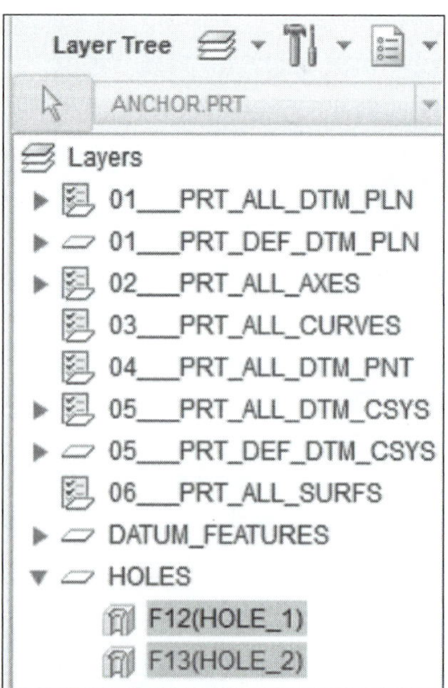

Figure 5.16(f) Holes Selected

Click: **Command Search** > *type* **suppress** > in the list, place the mouse on the icon (but do not click the **LMB**) Suppress [Fig. 5.16(g)] > to close > **Model** tab > **Operations** Group > next to Suppress > **Suppress** [Fig. 5.16(h)] > **OK** > **LMB** to deselect *F-1(HOLE_1) *F-1(HOLE_2) > > **Model Tree** > **Operations** Group > **Resume** > **Resume All** > in the Graphics Window, **LMB** to deselect [Fig. 5.16(i)]

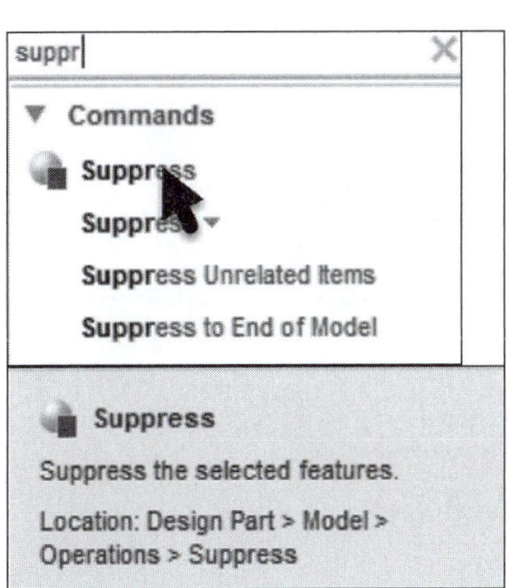

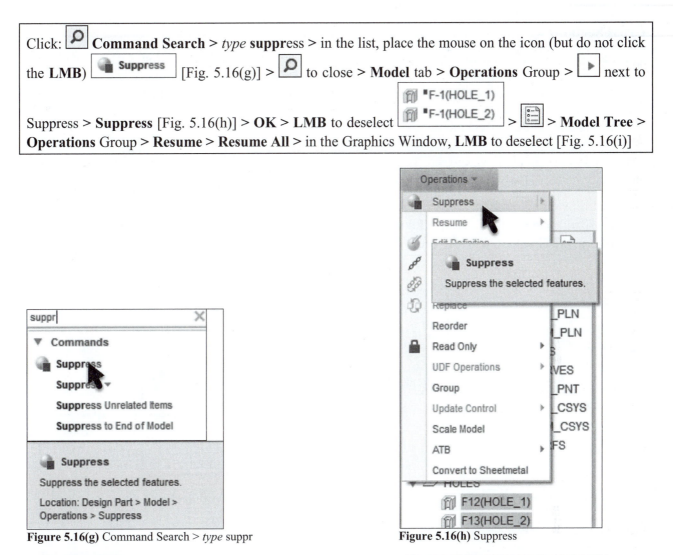

Figure 5.16(g) Command Search > *type* suppr

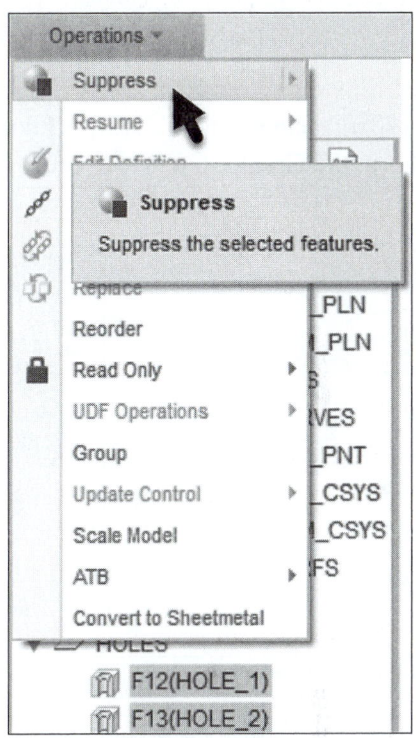

Figure 5.16(h) Suppress

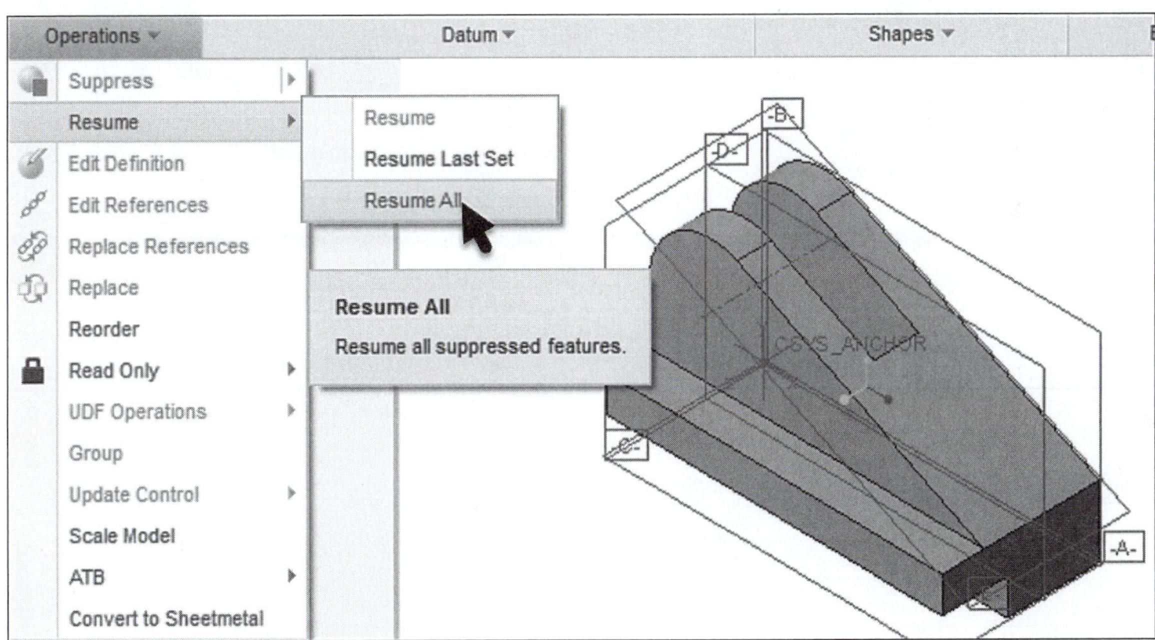

Figure 5.16(i) Resume All

Cross Sections

There are six types of cross sections: **Planar, X Direction**, **Y Direction**, **Z Direction**, **Offset**, and **Zone**. You will be creating a planar cross section. Creo Parametric can create standard planar cross sections of models (parts or assemblies), offset cross sections of models (parts or assemblies), planar cross sections of datum surfaces or quilts (Part mode only) and planar cross sections that automatically intersect all quilts and all geometry in the current model.

Click: **View** tab > **View Manager** View Manager dialog box displays [Fig. 5.17(a)] > **Sections** tab [Fig. 5.17(b)] > **New** [Fig. 5.17(c)] > **Planar** [Fig. 5.17(d)] > *type* **A** [Fig. 5.17(e)] > **Enter** > [Fig. 5.17(f)] > in the Graphics Window, press **RMB** > **Section Reference** [Fig. 5.17(g)]

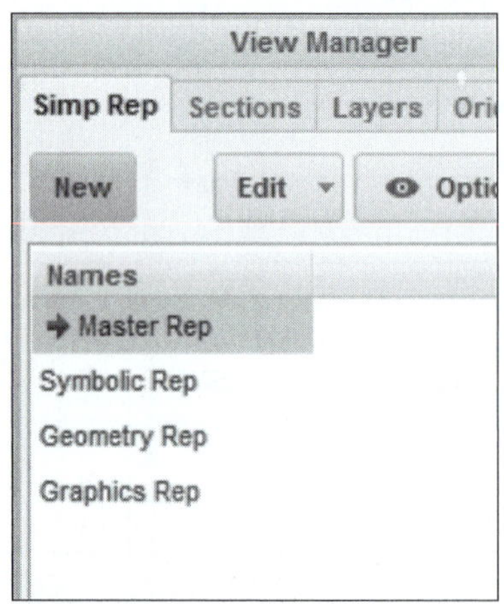

Figure 5.17(a) View Manager

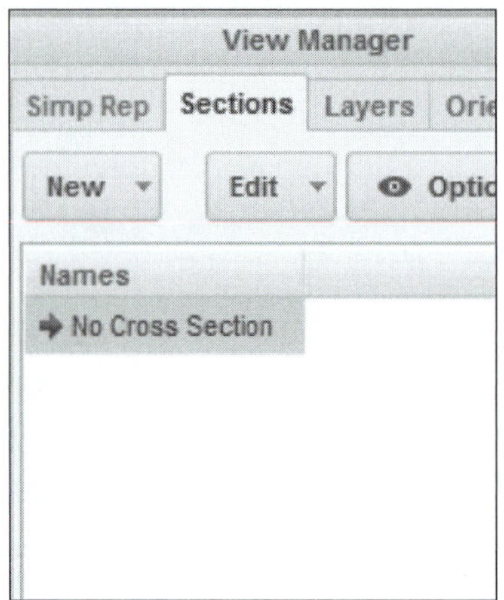

Figure 5.17(b) Sections Tab

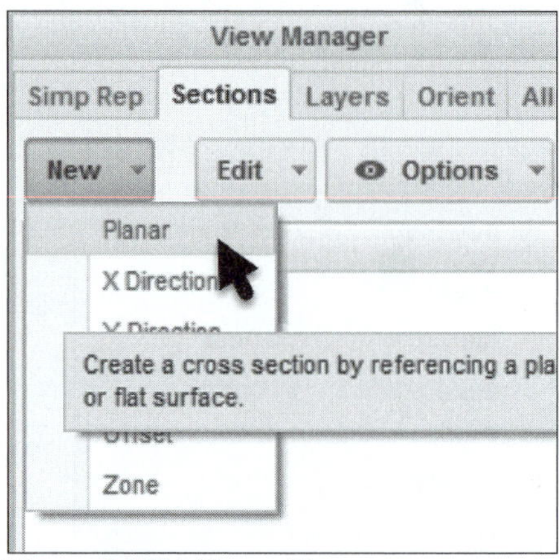

Figure 5.17(c) New > Planar

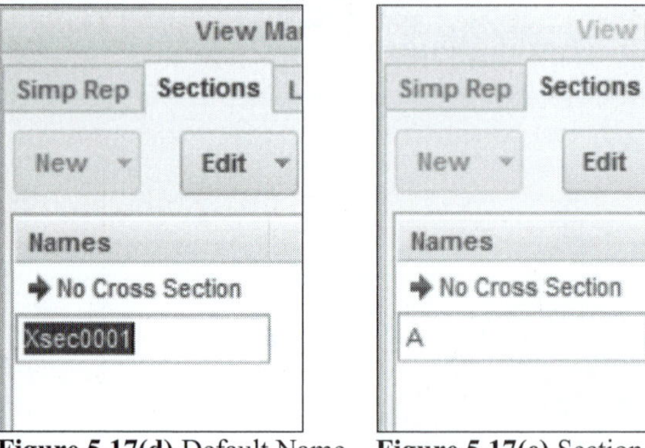

Figure 5.17(d) Default Name **Figure 5.17(e)** Section A

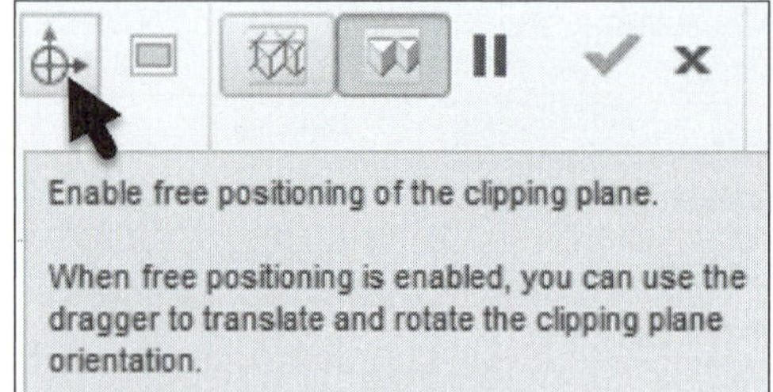

Figure 5.17(f) Activate Dragger

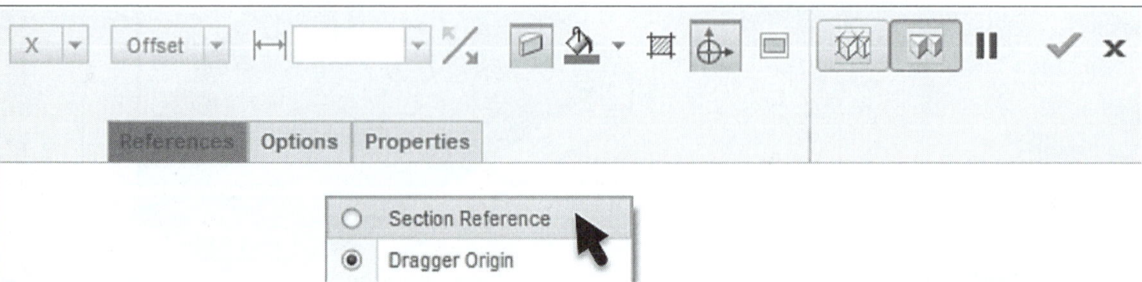

Figure 5.17(g) Section Reference

Select datum **D** [Fig. 5.17(h)] > [icon] > select a color [Fig. 5.17(i)] > [icon] > [icon] > [icon] > [icon] > [icon] > [icon] > **RMB** on [Names / No Cross Section / A] > **RMB** > **Edit Hatching** > **iso_alloy_steel** > **Apply** > **steel** > **Apply** > Angle field, *type* **150** > **Enter** > **Use a solid fill** > **Use hatch from the part** > **Apply** > [icon] **Close** > **RMB** on **No Cross Section** > **Activate** [Fig. 5.17(j)] > **Close** > in Model Tree, **RMB** on [icon] > uncheck **Show Section** > in the Graphics Window, **LMB** to deselect > **File** > **Save**

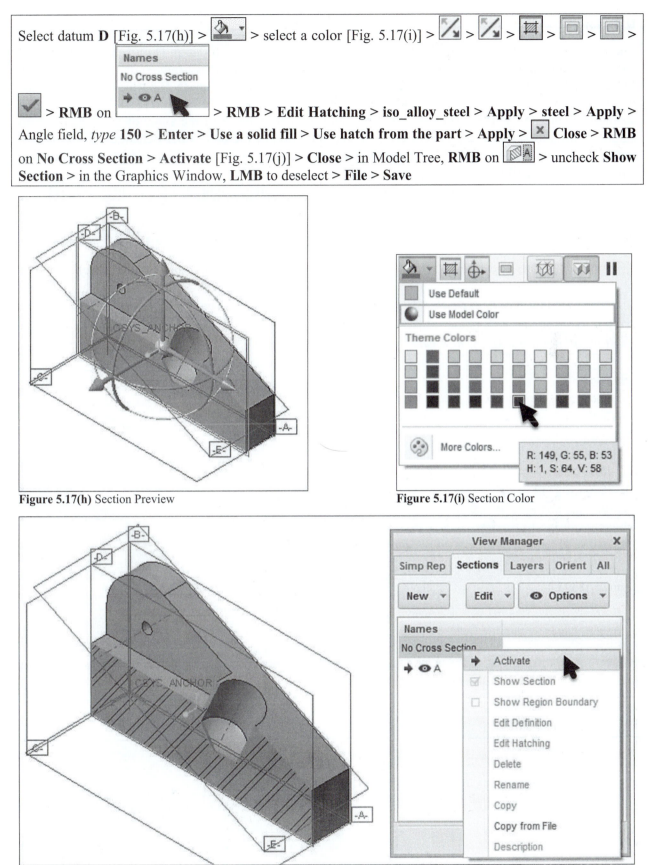

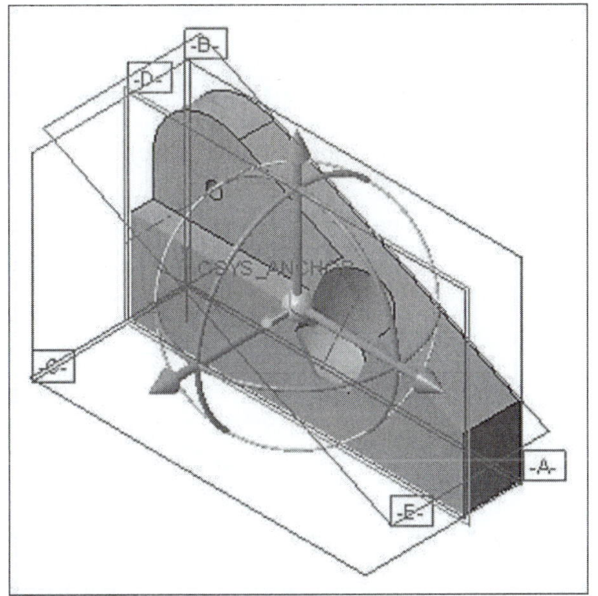

Figure 5.17(h) Section Preview

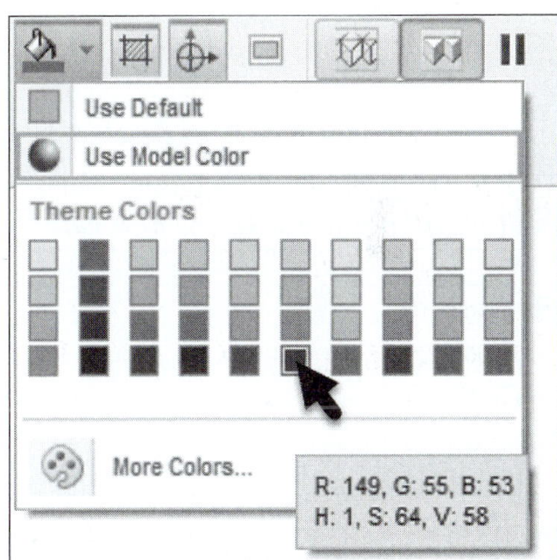

Figure 5.17(i) Section Color

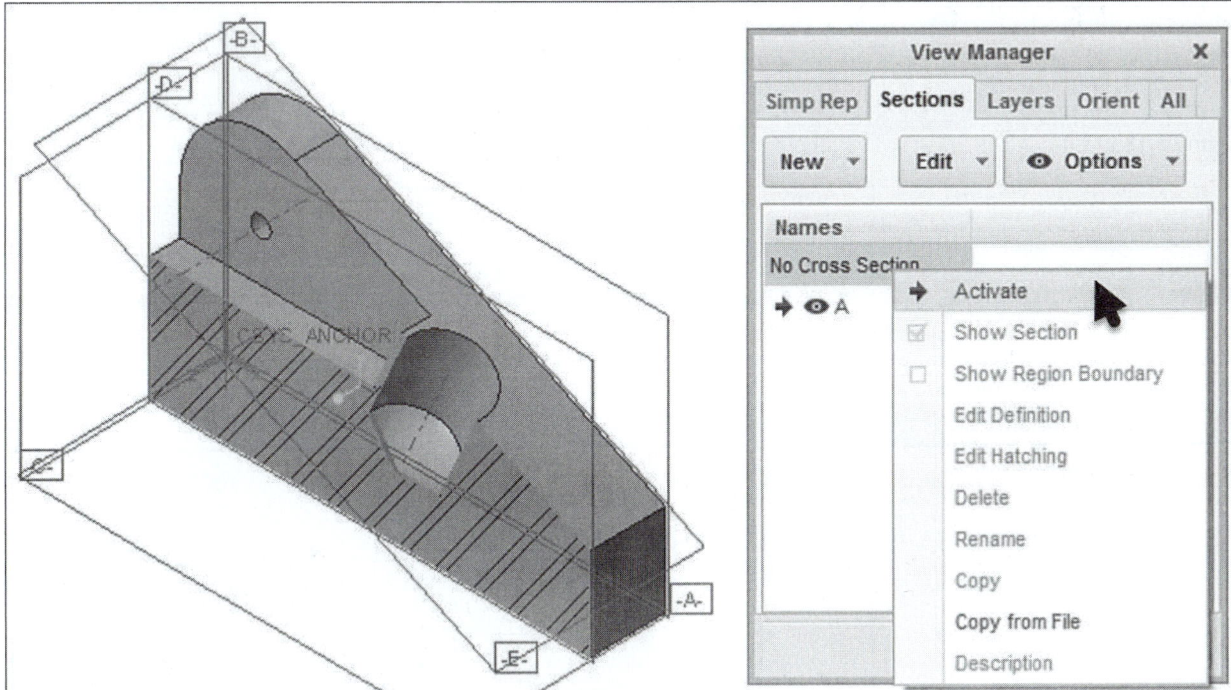

Figure 5.17(j) Activate No Cross Section

Relations

The section passes through datum D. The slot is the *child* of datum D. If datum D moves, so will the slot and the small hole (and X-Section A). In order to ensure that datum D stays centered on the upper portion of the part, you will need to create a relation to control the location of datum D. **Relations** will be covered in-depth in a later lesson. The first cut used a dimension from datum B for location (**1.875**). The relation should state that the distance from datum B to datum D will be one-half the value of the distance from datum B to the first cut. If the thickness of the upper portion of the part (**1.875**) changes, datum D will remain centered as will the slot and the X-Section. To start, you must first determine the feature dimension symbols (**d#**) required for the relation.

(READ! - Almost everyone will have different d# numbers than the ones displayed here)

Select the first cut (Extrude 2) from the Model Tree > **RMB** > **Edit** [Figs. 5.18(a-b)] > **Tools** tab > Switch Symbols > note symbol **d7** *(Note: your "d" values may be different)* [Fig. 5.18(c)] > in the Graphics Window, **LMB** to deselect > select datum **D** in the Model Tree > **RMB** > **Edit** > note symbol **d14** [Fig. 5.18(d)] *(Note: your "d" values may be different. If you have the wrong symbols, your relation will not work.* ***Use your d values not the ones in the text!****)*

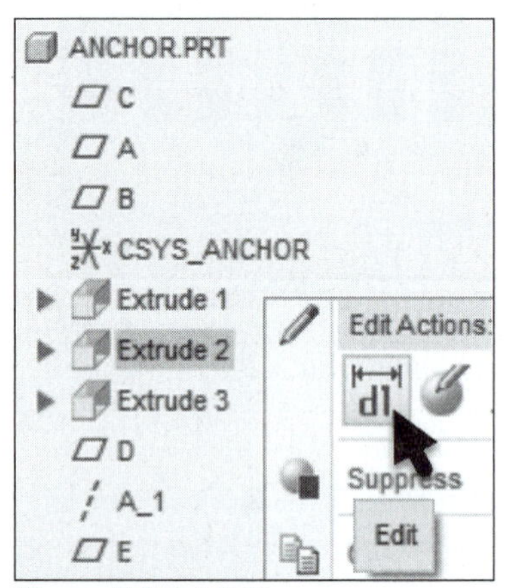

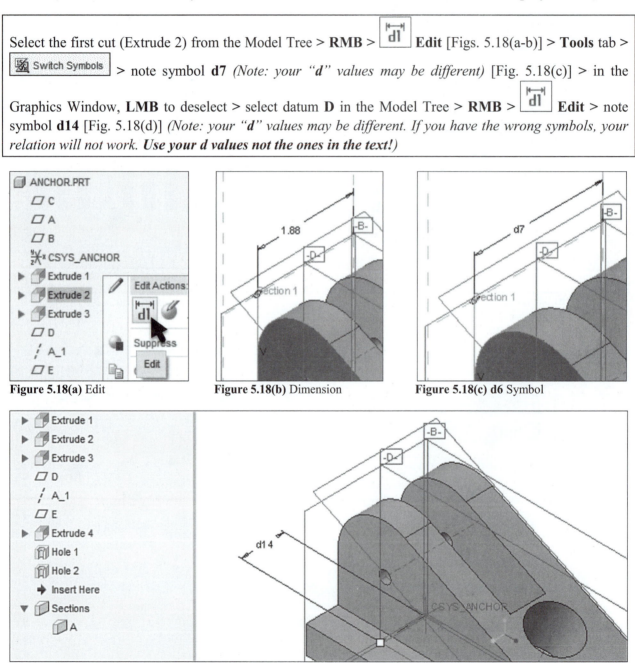

Figure 5.18(a) Edit

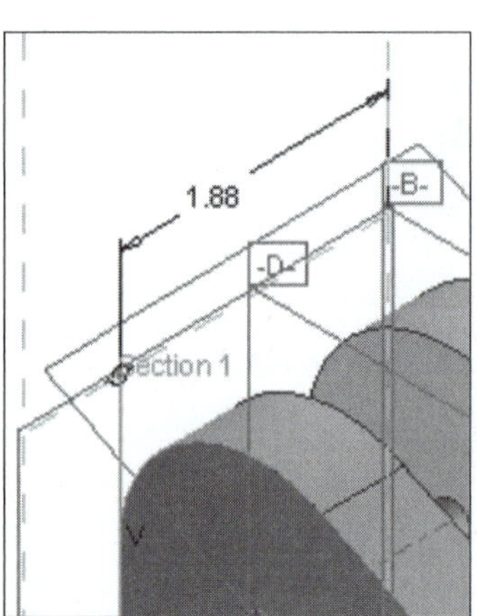

Figure 5.18(b) Dimension

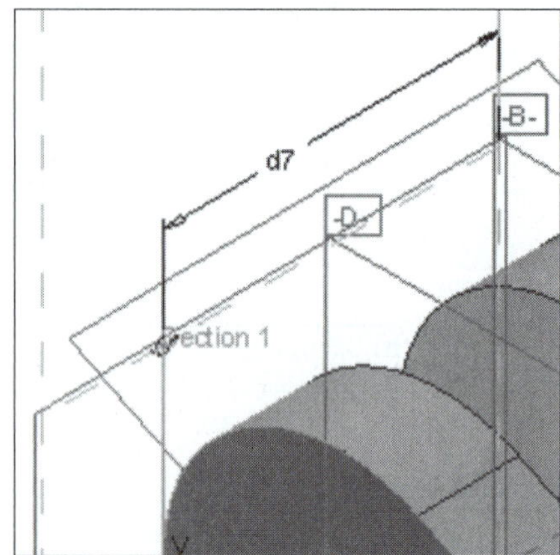

Figure 5.18(c) d6 Symbol

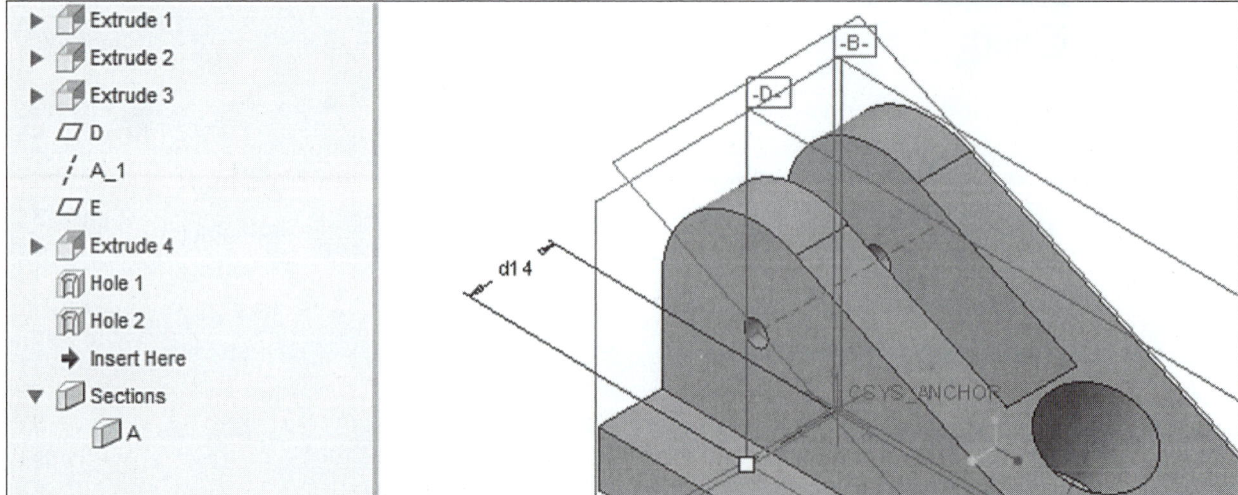

Figure 5.18(d) Using Edit to Display Datum D Dimension Symbol **d14** *(your "d" value may be different)*

Click: **d= Relations** from the Tools Ribbon > **Local Parameters** to see the parameters of the part > *type (or pick from the model)* **d14=d7/2** in the Relations field *(your "d" values may be different)* > **Execute/Verify Relations** [Fig. 5.18(e)] > **OK** from the Verify Relations dialog box > **OK** > **Switch Symbols** to display the dimension values > double click on the **.9375** dimension (in the information area at the bottom of the window: Dimension in ANCHOR is driven by relation d14=d7/2) > **Ctrl+S** > **LMB**

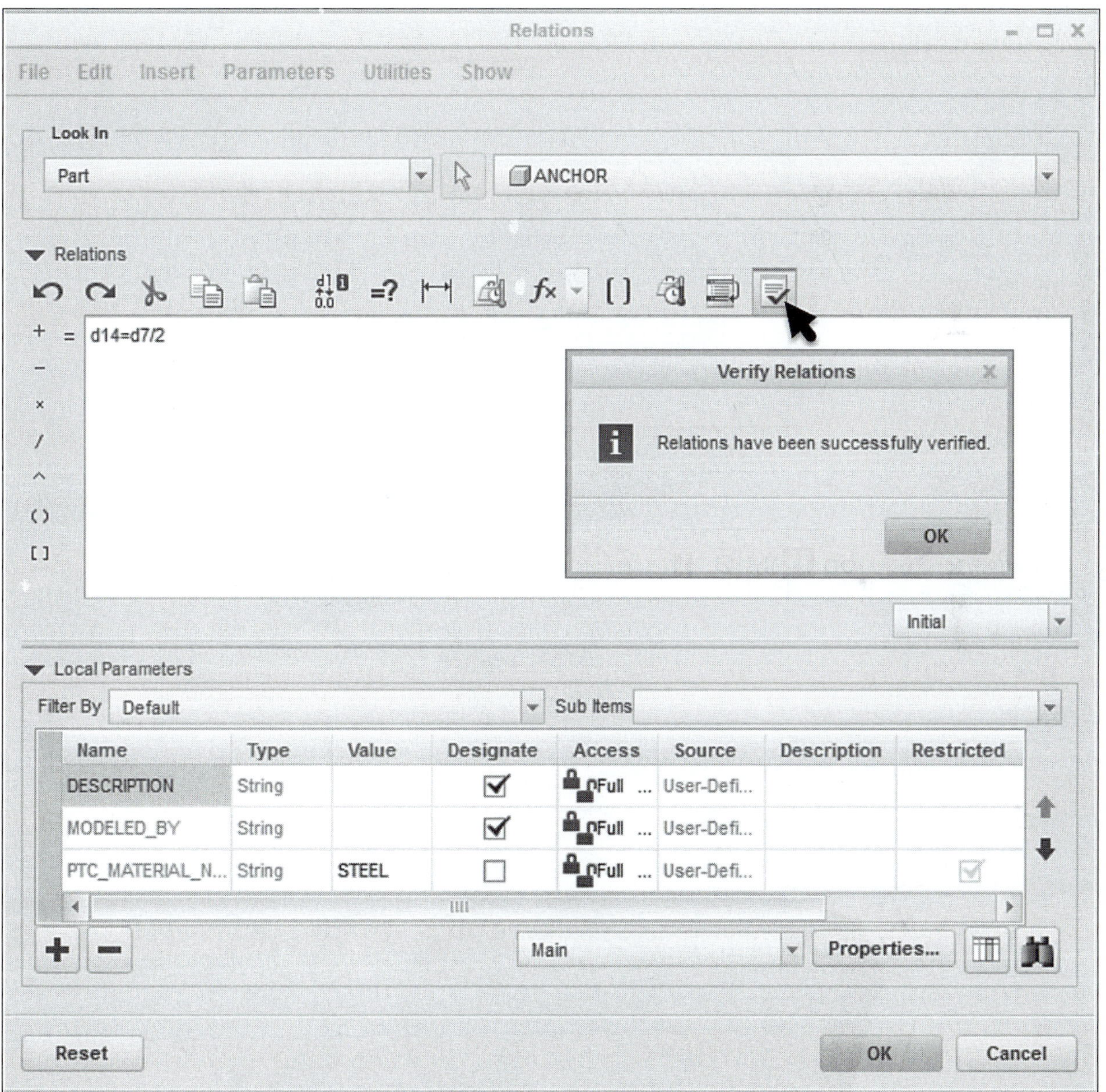

Figure 5.18(e) Relations **d11=d6/2** *(your **d** values may be different)*

The slot will now remain in the center of the upper protrusion regardless of changes in the protrusions width. In the real world, you will seldom encounter a situation where the project is designed and modeled without a "design change" or **ECO** (Engineering Change Order).

Let us assume that an ECO has been "issued" that states: *the location of the hole on the angled surface must be aligned with the center of the slot at all times.* A relation could be established to control the hole's position as was done with the slot; but instead of referencing the hole from Datum B with a dimension, we will *align* the holes offset reference from the D datum.

Select the large hole in the Model Tree or on the model > press **RMB** > **Edit the Definition** [Fig. 5.18(f)] > **Placement** tab > pick on datum **B (DATUM)** in the Offset References collector > **RMB** > **Remove** [Fig. 5.18(g)]

Offset References		
Edge:F6(EXT...	Offset	2.0625
● Select 1 item		

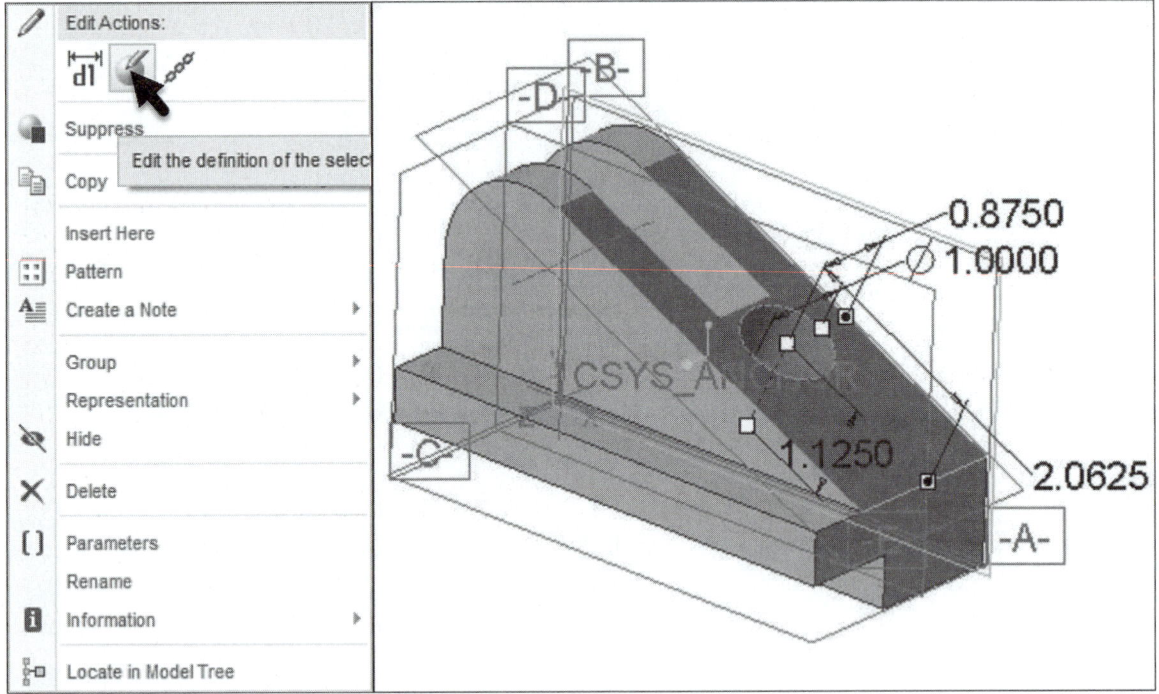

Figure 5.18(f) Edit the Definition

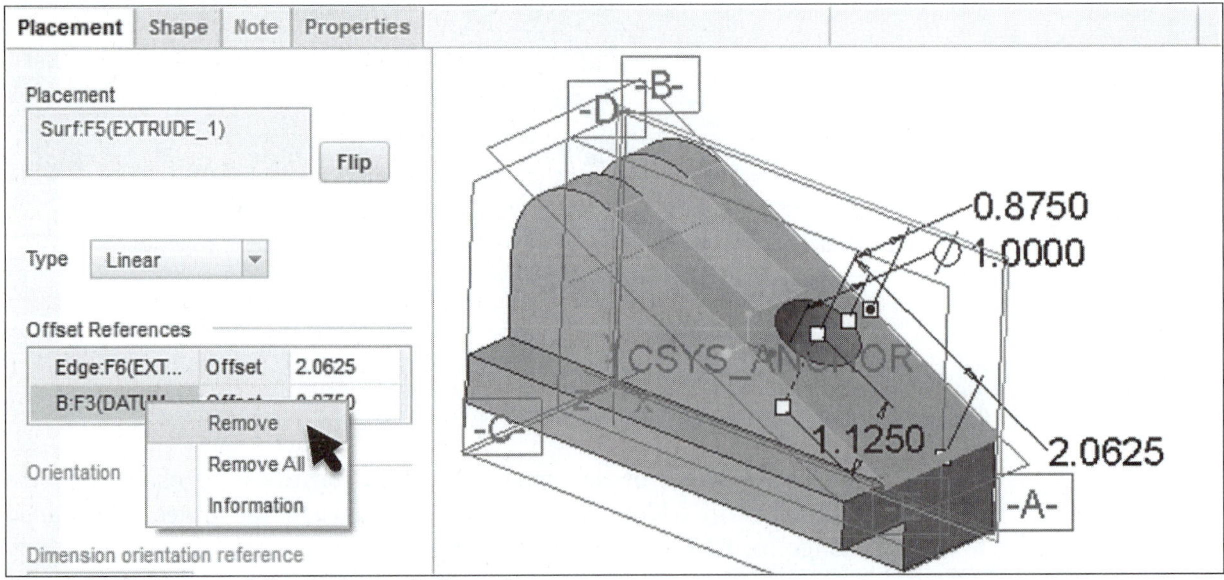

Figure 5.18(g) Remove Offset Reference B (Datum)

Press and hold down the **Ctrl** key and select **D (DATUM)** from the model or Model Tree [Fig. 5.18(h)] > release the **Ctrl** key > Offset References collector **Offset**, next to D (to activate) > **Offset** > **Align** [Fig. 5.18(i)] > **MMB** > **LMB** to deselect > **Ctrl+D** > **Ctrl+S**

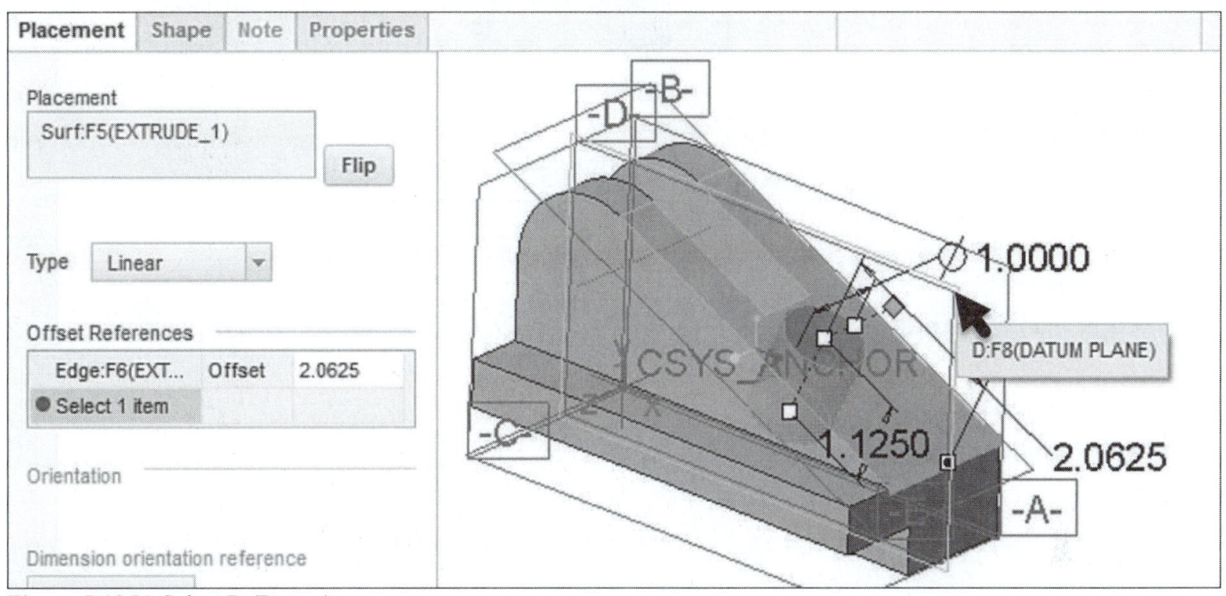

Figure 5.18(h) Select D (Datum)

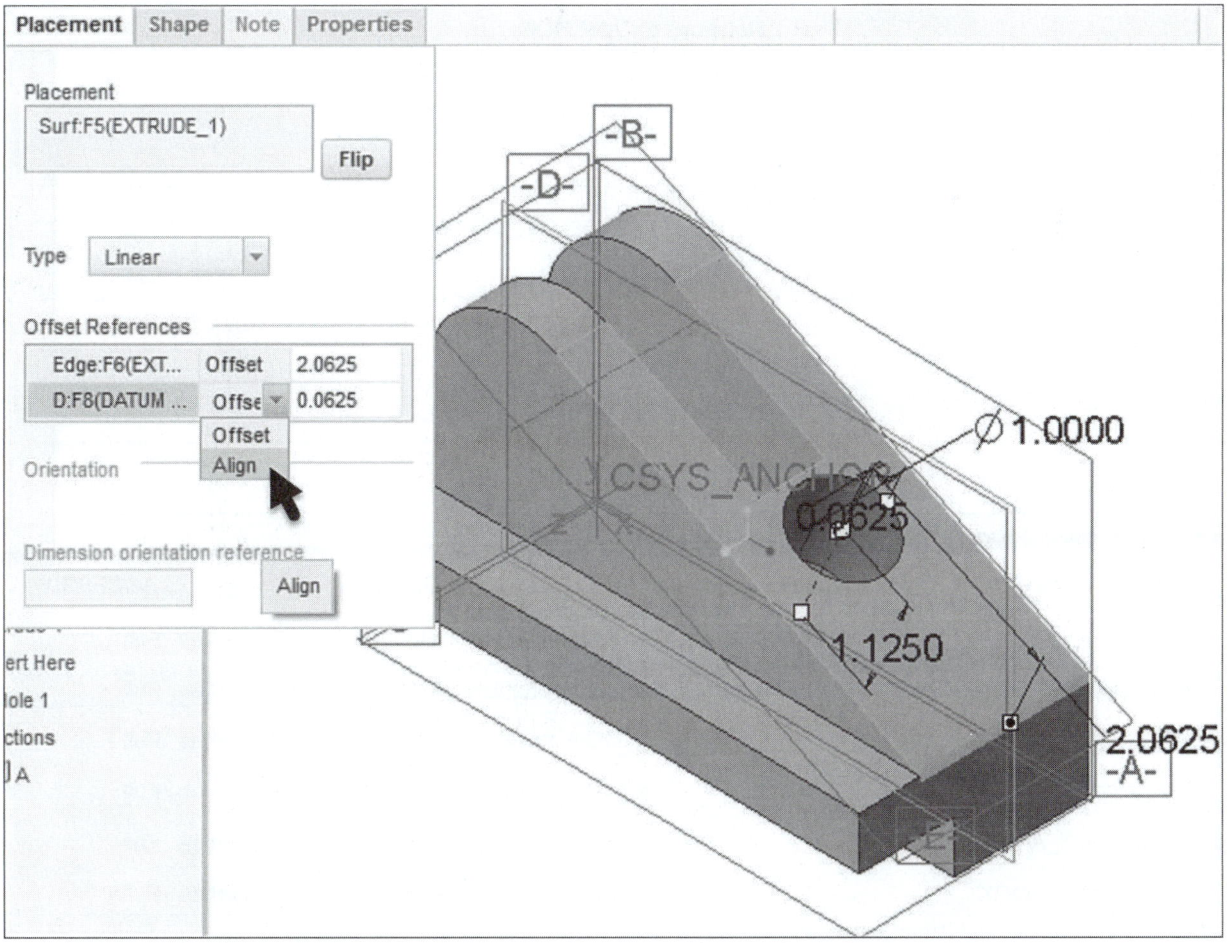

Figure 5.18(i) Align (Keep **3.0625** as the Offset distance from the edge for the first Offset Reference)

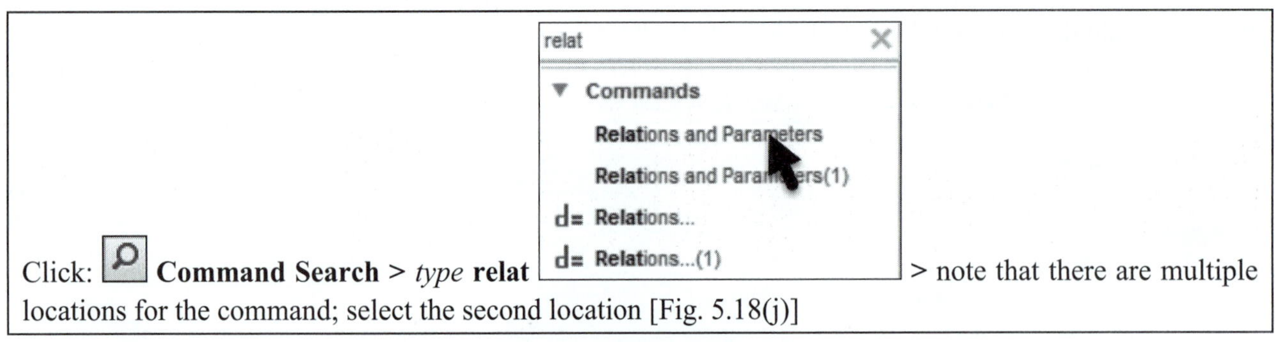

Click: **Command Search** > *type* **relat**

relat
Commands
Relations and Parameters
Relations and Parameters(1)
Relations...
Relations...(1)

> note that there are multiple locations for the command; select the second location [Fig. 5.18(j)]

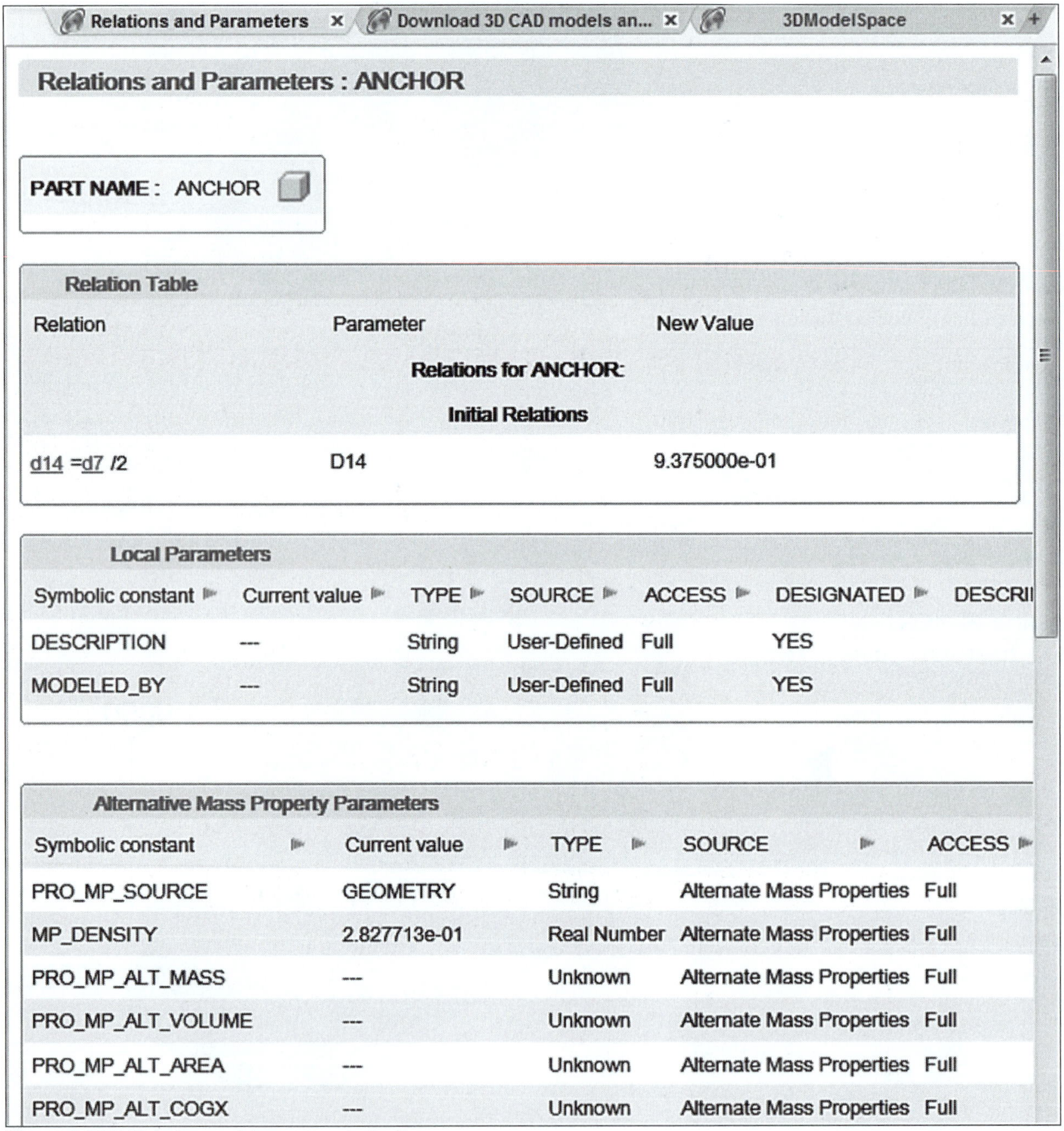

Relations and Parameters
Download 3D CAD models an...
3DModelSpace
Relations and Parameters : ANCHOR
PART NAME : ANCHOR
Relation Table
Relation
Parameter
New Value
Relations for ANCHOR:
Initial Relations
d14 =d7 /2
D14
9.375000e-01
Local Parameters
Symbolic constant
Current value
TYPE
SOURCE
ACCESS
DESIGNATED
DESCRI
DESCRIPTION --- String User-Defined Full YES
MODELED_BY --- String User-Defined Full YES
Alternative Mass Property Parameters
Symbolic constant
Current value
TYPE
SOURCE
ACCESS
PRO_MP_SOURCE GEOMETRY String Alternate Mass Properties Full
MP_DENSITY 2.827713e-01 Real Number Alternate Mass Properties Full
PRO_MP_ALT_MASS --- Unknown Alternate Mass Properties Full
PRO_MP_ALT_VOLUME --- Unknown Alternate Mass Properties Full
PRO_MP_ALT_AREA --- Unknown Alternate Mass Properties Full
PRO_MP_ALT_COGX --- Unknown Alternate Mass Properties Full

Figure 5.18(j) Relations and Parameters in the Browser

Select **d6**, and **d11** (your **d** values may be different) in the Browser [Fig. 5.18(k)] to display them in the Graphics Window > ✕ or [icon] to close the Browser > in the Graphics Window, **LMB > Ctrl+S**

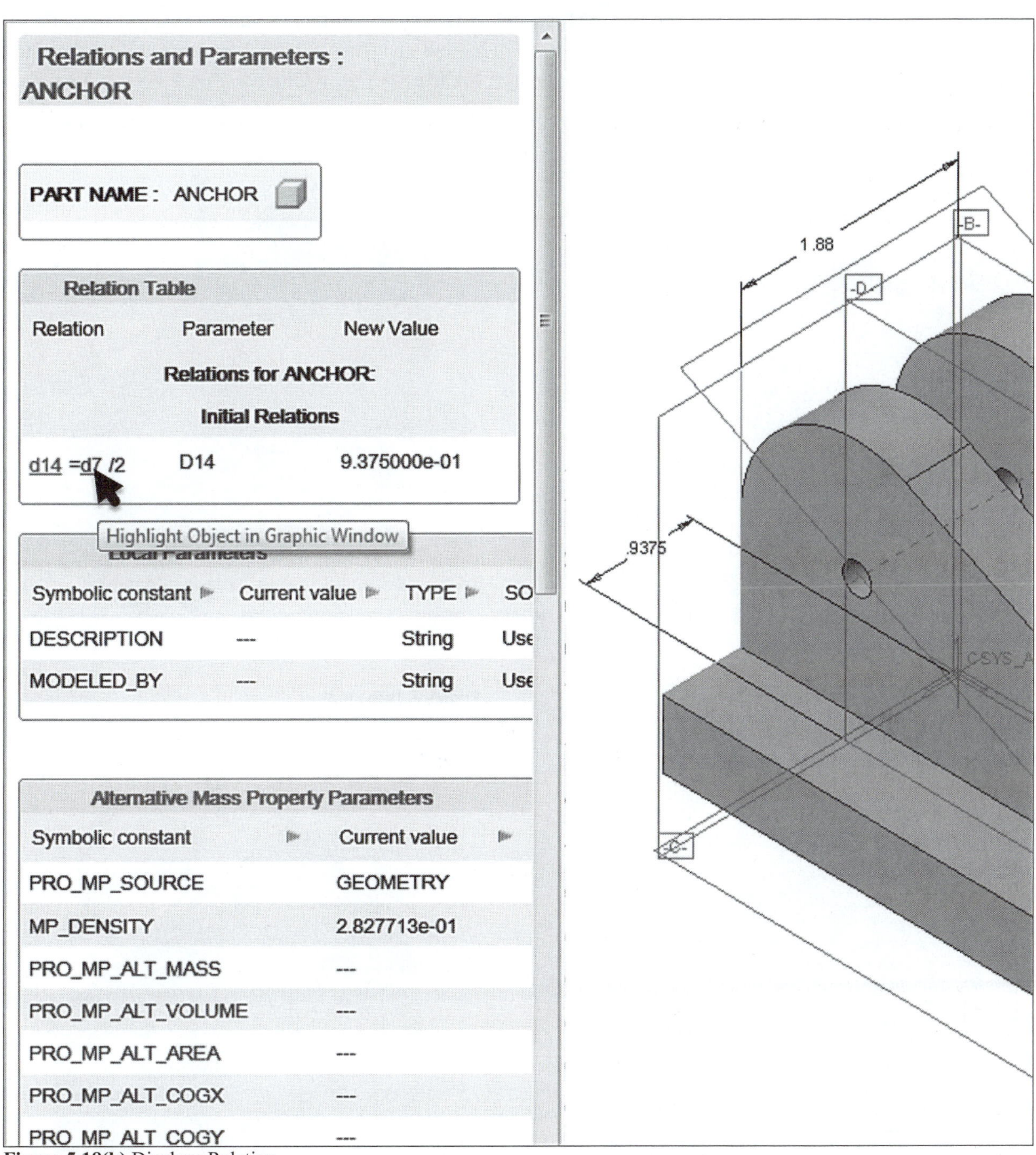

Figure 5.18(k) Displays Relation

Click: **File > Manage File > Delete Old Versions > Yes > File > Save As >** Type > **Zip File (*.zip) > OK >** upload to your course interface or attach to an email and send to your instructor and/or yourself >

Flexible Modeling tab *(if available)* > **Move using Dragger** > select the angled face [Fig. 5.19(a)] > pull the ball vertically [Fig. 5.19(b)] > rotate the *horizontal* arc [Fig. 5.19(c)] > rotate the *vertical* arc [Fig. 5.19(d)] > experiment with variations of the design > **Enter > File > Close > File > Exit > Yes**

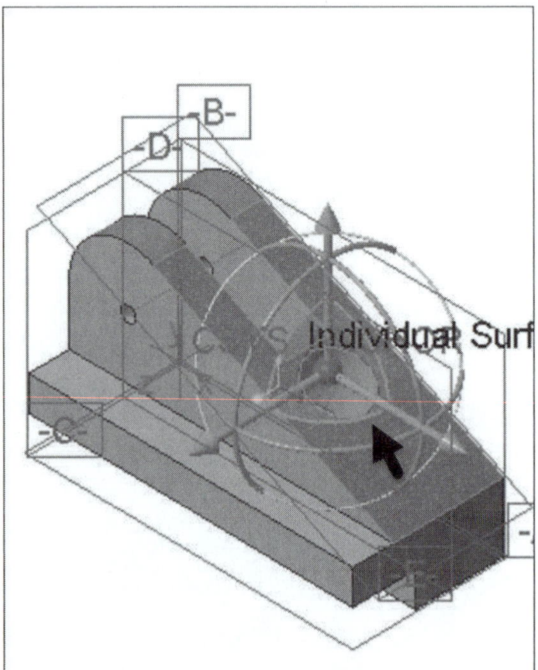

Figure 5.19(a) Flexible Modeling

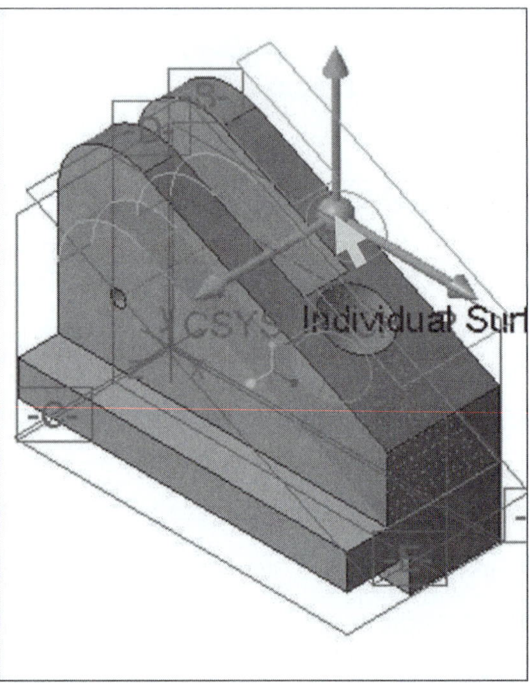

Figure 5.19(b) Pull

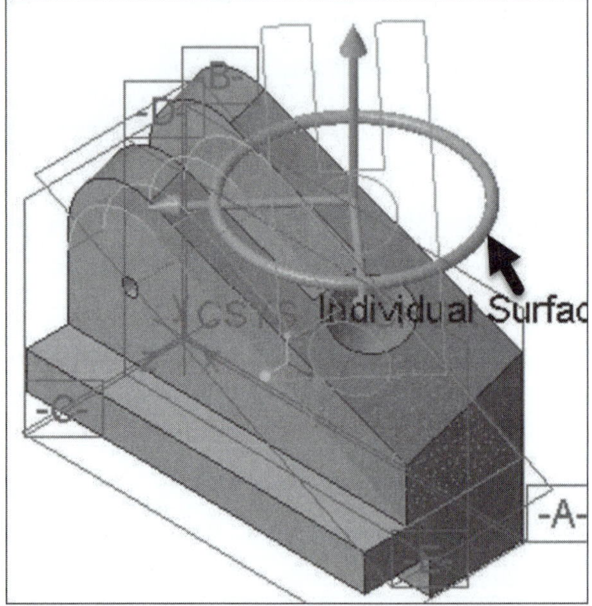

Figure 5.19(c) Rotate Horizontally

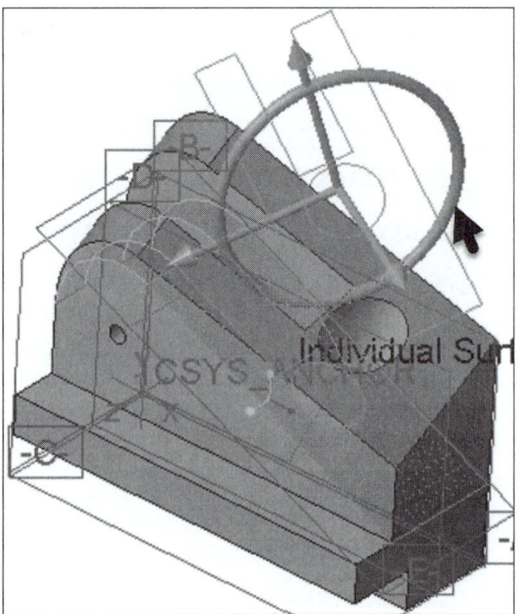

Figure 5.19(d) Rotate Vertically

The lesson is now complete. A complete set of Projects are available at ***www.cad-resources.com.***

Lesson 6 Revolved Features

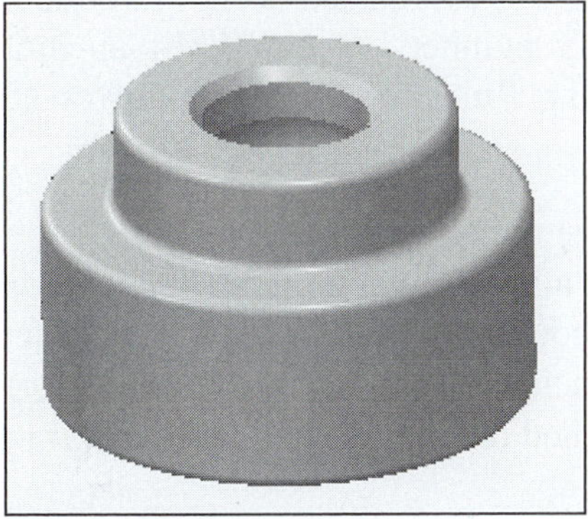
Figure 6.1(a) Clamp Foot

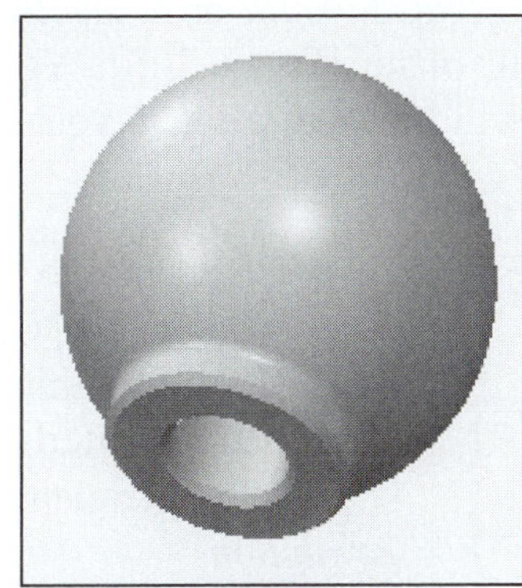
Figure 6.1(b) Clamp Ball

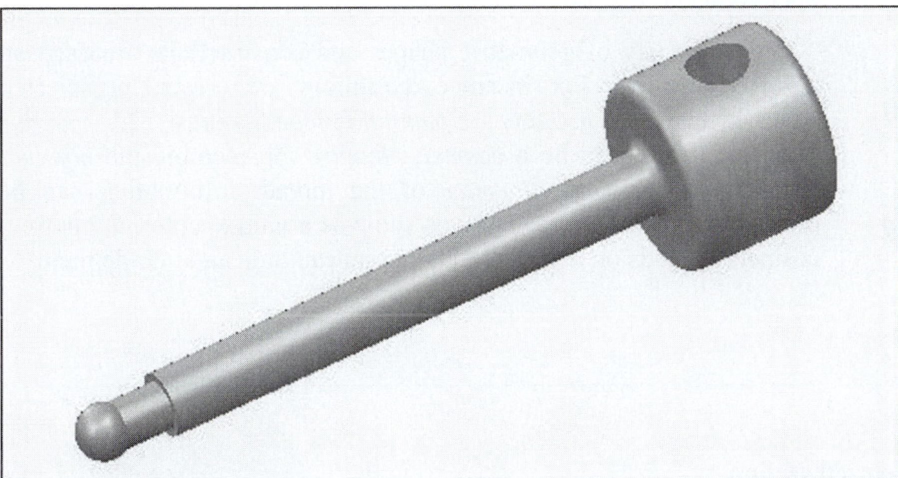
Figure 6.1(c) Clamp Swivel

OBJECTIVES

- Master the **Revolve Tool** [Figs. 6.1(a-c)]
- Create **Chamfers** along part edges
- Understand and use the **Navigation browser**
- Alter and set the **Items** and **Columns** displayed in the **Model Tree**
- Create standard **Tapped Holes**
- Create **Cosmetic Threads** and complete **tabular information** for threads
- Edit **Dimension Properties**
- Use the **Model Player** to extract information and dimensions
- Get a hard copy using the **Print** command

REFERENCES AND RESOURCES

For **Resources** go to **www.cad-resources.com** > click on the PTC Creo Parametric 3.0 Book cover

- Lesson 6 Lecture
- Lesson 6 3D models embedded in a PDF
- Book Projects PDF
- Project Lectures

REVOLVED FEATURES

The **Revolve Tool** creates a *revolved solid* or a *revolved cut* by revolving a sketched section around a centerline from the sketching plane [Fig. 6.2(a)]. You can have any number of centerlines in your sketch/section, but only one will be used to rotate your section geometry. Rules for sketching a revolved feature include:

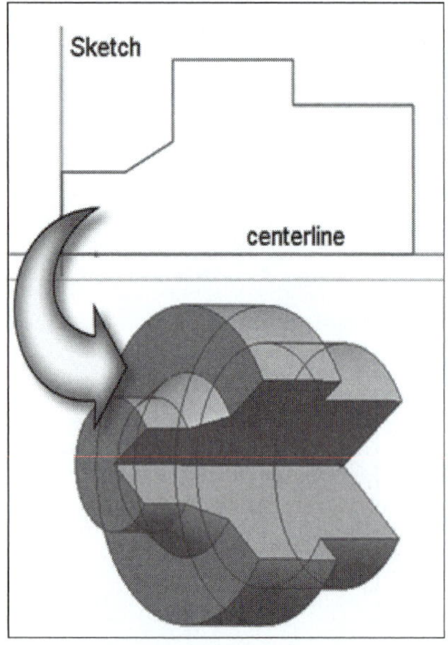

Figure 6.2(a) Revolved Section

- A section to be revolved must reference an axis
- By default, Creo Parametric allows only one *axis of revolution* per sketch (within a sketch, you may select from different construction centerlines to be designated as the one *axis of revolution*)
- The geometry must be sketched on only one side of the *axis of revolution*
- The section must be closed for a solid [Fig. 6.2(b)] but can be open for a cut or a protrusion with assigned thickness

A variety of geometric shapes and constructions are used on revolved features. For instance, **chamfers** are created at selected edges of the part. Chamfers are *pick-and-place* features.

Threads can be a *cosmetic feature* representing the *nominal diameter* or the *root diameter* of the thread. Information can be embedded in the feature. Threads show as a unique color. By putting cosmetic threads on a separate layer, you can hide and unhide them.

Revolved Solid Protrusion- closed section	
Revolved Protrusion with an assigned thickness- closed section	
Revolved Protrusion with an assigned thickness- open section	
Revolved Cut- open section	
Revolved Surface- open section	

Figure 6.2(b) Types of Revolved Features

Chamfers

Chamfers are created between abutting edges of two surfaces on the solid model. An edge chamfer removes a flat section of material from a selected edge to create a beveled surface between the two original surfaces common to that edge. Multiple edges can be selected.

There are four basic dimensioning schemes for edge chamfers (Fig. 6.3):

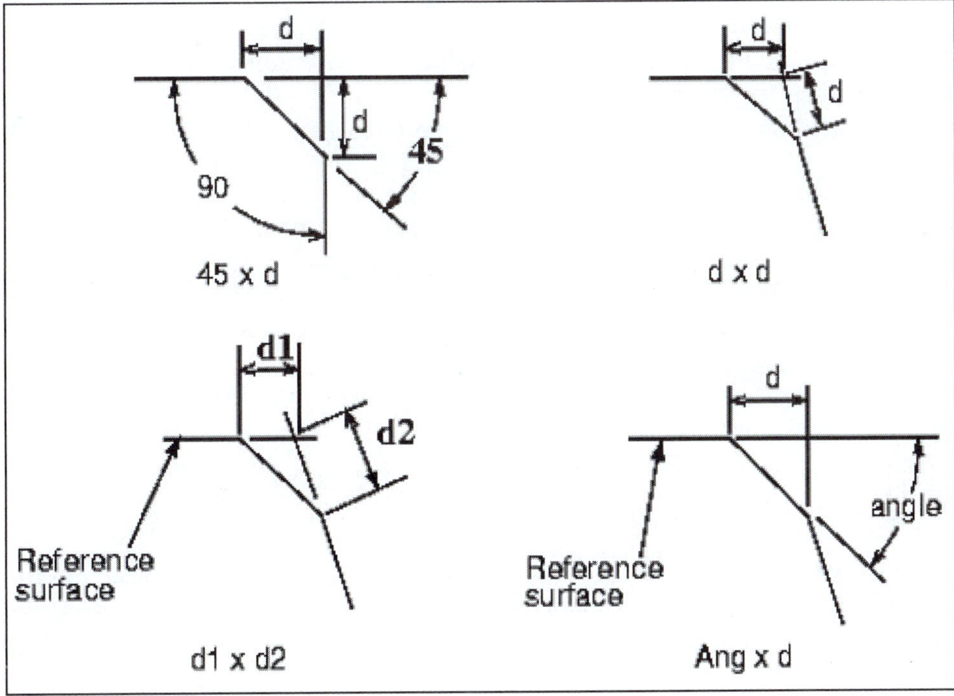

Figure 6.3 Chamfer Options

- **45 x d** Creates a chamfer that is at an angle of **45°** to both surfaces and a distance **d** from the edge along each surface. The distance is the only dimension to appear when edited. **45 x d** chamfers can be created only on an edge formed by the intersection of two *perpendicular* surfaces.
- **d x d** Creates a chamfer that is a distance **d** from the edge along each surface. The distance is the only dimension to appear when edited.
- **d1 x d2** Creates a chamfer at a distance **d1** from the selected edge along one surface and a distance **d2** from the selected edge along the other surface. Both distances appear along their respective surfaces.
- **Ang x d** Creates a chamfer at a distance **d** from the selected edge along one adjacent surface at an **Angle** to that surface.

Threads

Cosmetic threads are displayed with *magenta/purple* lines and circles. Cosmetic threads can be external or internal, blind or through. A cosmetic thread has a set of embedded parameters that can be defined at its creation or later, when the thread is added.

Standard Holes

Standard holes are a combination of sketched and extruded geometry. It is based on industry-standard fastener tables. You can calculate either the tapped or clearance diameter appropriate to the selected fastener. You can use Creo Parametric supplied standard lookup tables for these diameters or create your own. Besides threads, standard holes can be created with chamfers.

Navigation Window

Besides using the File command and corresponding options, the *Navigation window* can be used to directly access other functions.

As previously mentioned, the working directory is a directory that you set up to contain Creo Parametric files. You must have read/write access to this directory. You usually start Creo Parametric from your working directory. A new working directory setting is not saved when you exit the current Creo Parametric session. By default, if you retrieve a file from a non-working directory, rename the file and then save it, the renamed file is saved to the directory from which it was originally retrieved, if you have read/write access to that directory.

The navigation area is located on the left side (default) of the Creo Parametric main window. It includes tabs for the Model Tree and Layer Tree, **Folder browser**, **Favorites**, and **Drawing Tree** (in an active drawing).

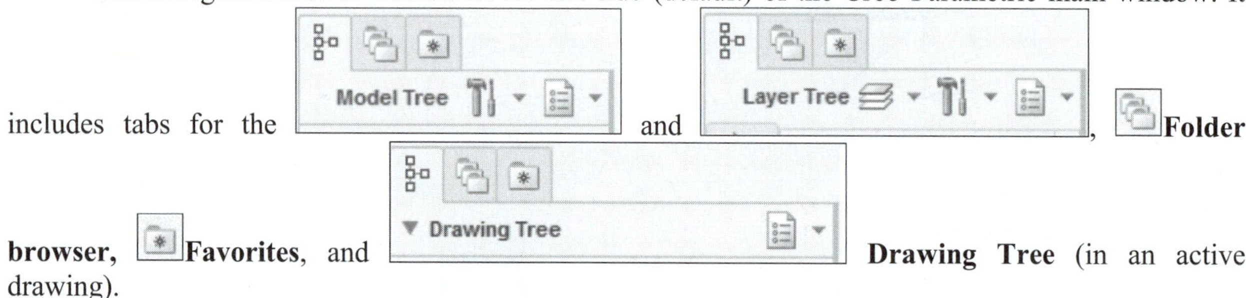

Folder Browser

The Folder browser (**Folder Browser**) (Fig. 6.4) is an expandable tree that lets you browse the file systems and other locations that are accessible from your computer. As you navigate the folders, the contents of the selected folder appear in the Creo Parametric 3.0 browser as a Contents page. The Folder browser contains top-level nodes for accessing file systems and other locations that are known to Creo Parametric:

Common Folders
- In Session
- Desktop
- My Documents
- cad-resources_7
- Working Directory
- Network Neighborhood
- Manikin Library
- Favorites
- 3-0

Figure 6.4 Folder Browser

In Session Creo Parametric 3.0 objects that have been retrieved into local memory.

Desktop Files and programs on the Desktop. Only Creo Parametric 3.0 items can be opened.

My Documents Files created and saved in Windows *My Documents* folder.

Working Directory Directory linked with the Select Working Directory command. All work will be accessed and saved in this location.

Network Neighborhood *(only for Windows)* The navigator shows computers on the networks to which you have access.

Manikin Library A Manikin model is a standard assembly that can be manipulated and positioned within the design scenario.

Favorites Saved folder locations for fast retrieval.

Manipulating Folders

To work with folders, you can use the **Browser** to accomplish most file management requirements. The Browser's **Views** drop-down menu includes a (simple) **List, Thumbnails,** and **Details**:

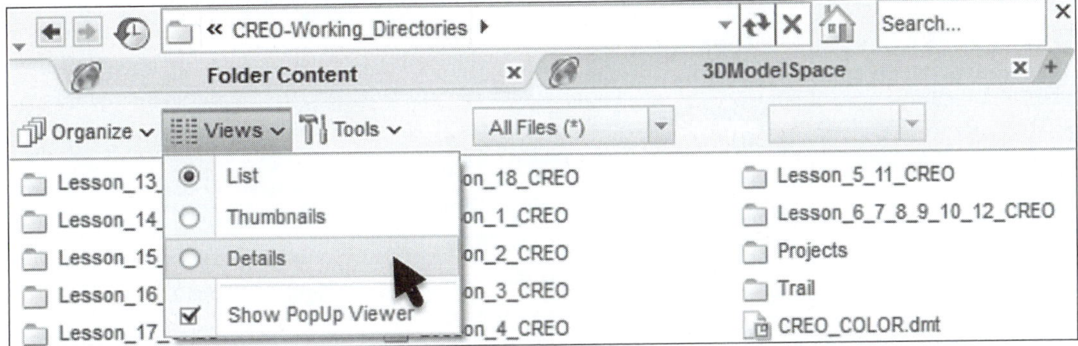

You can perform a variety of tasks from the Browser's **Organize** drop-down menu including **New Folder**, **Rename**, **Cut**, **Copy**, **Paste**, **Delete**, and **Add to common folders**:

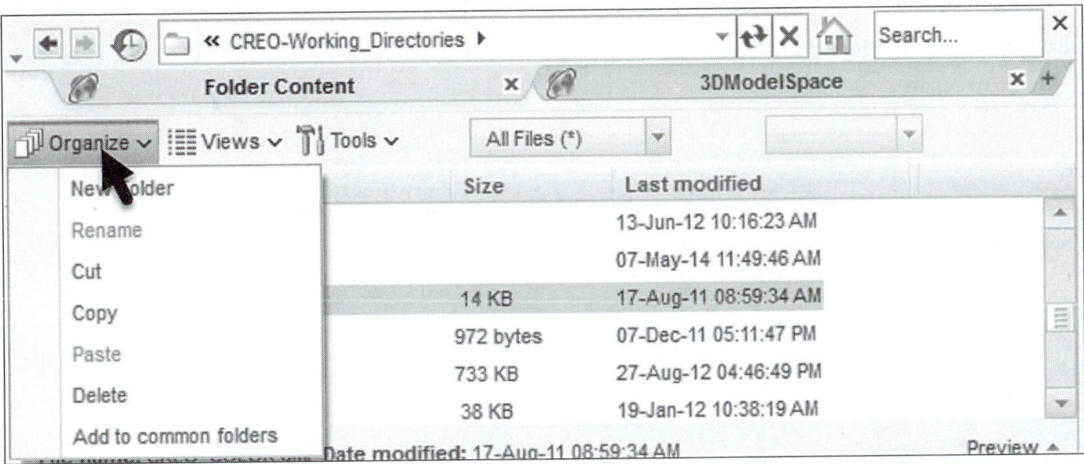

The Browser's **Tools** drop-down menu includes **Sort By**, go **Up One Level**, **Add to Favorites**, **Remove from Favorites**, **Organize Favorites**, show **All Versions**, **Show Instances,** and **Send to Mail Recipient (as Zipped Attachment)**:

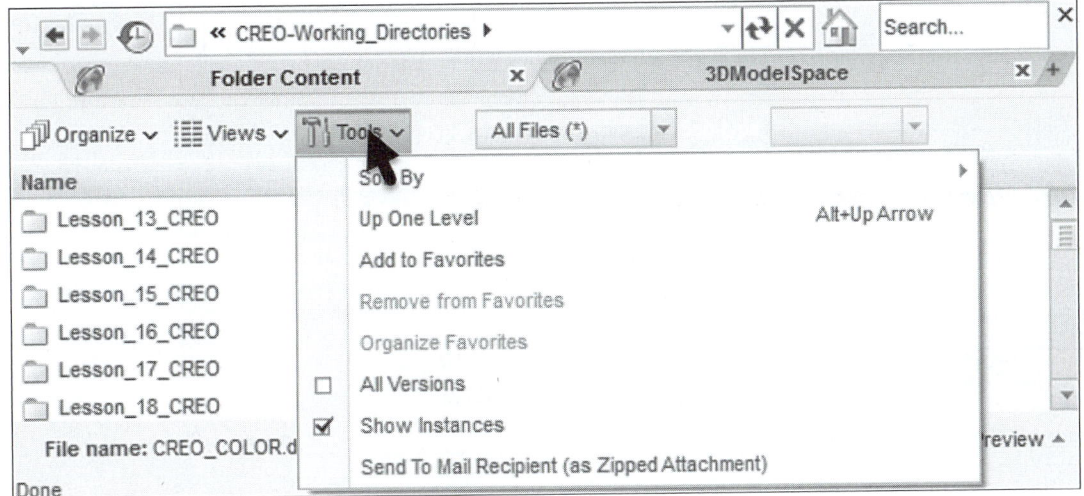

Do not change your working directory from your default system work folder unless instructed to do so.

From the Browser's **Type** field, you can limit the search to one of Creo Parametric's file types:

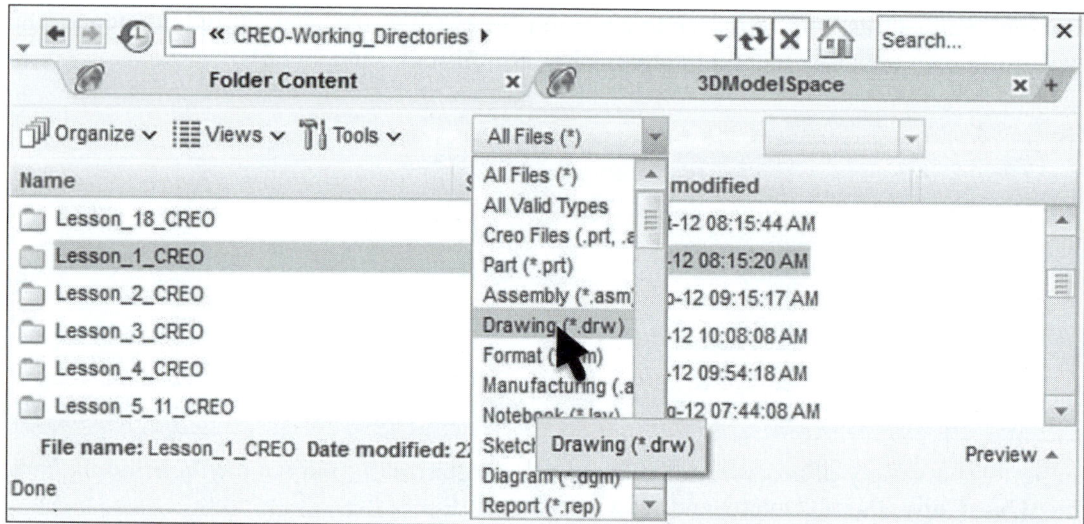

To set a working directory using the Browser, pick on the desired folder > ***RMB*** > ***Set Working Directory***:

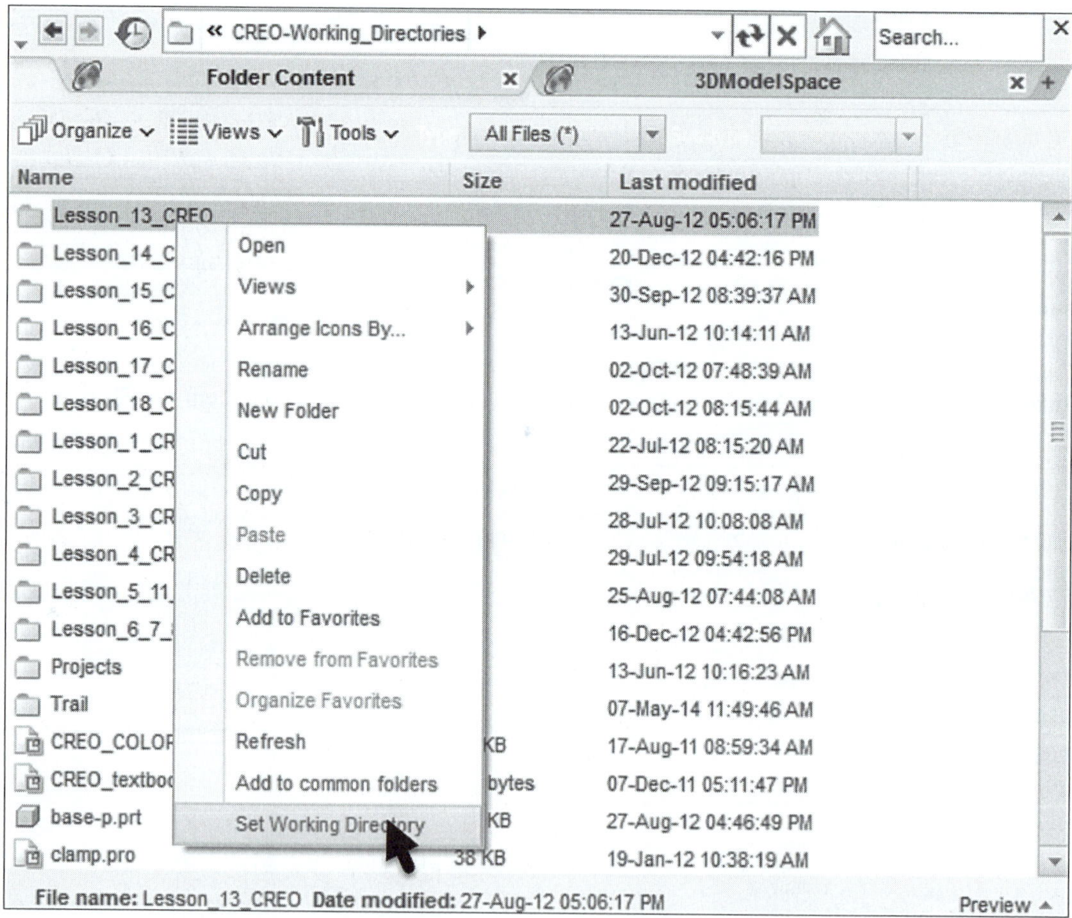

After you set (select) the working directory, pick on the object that you wish to preview:

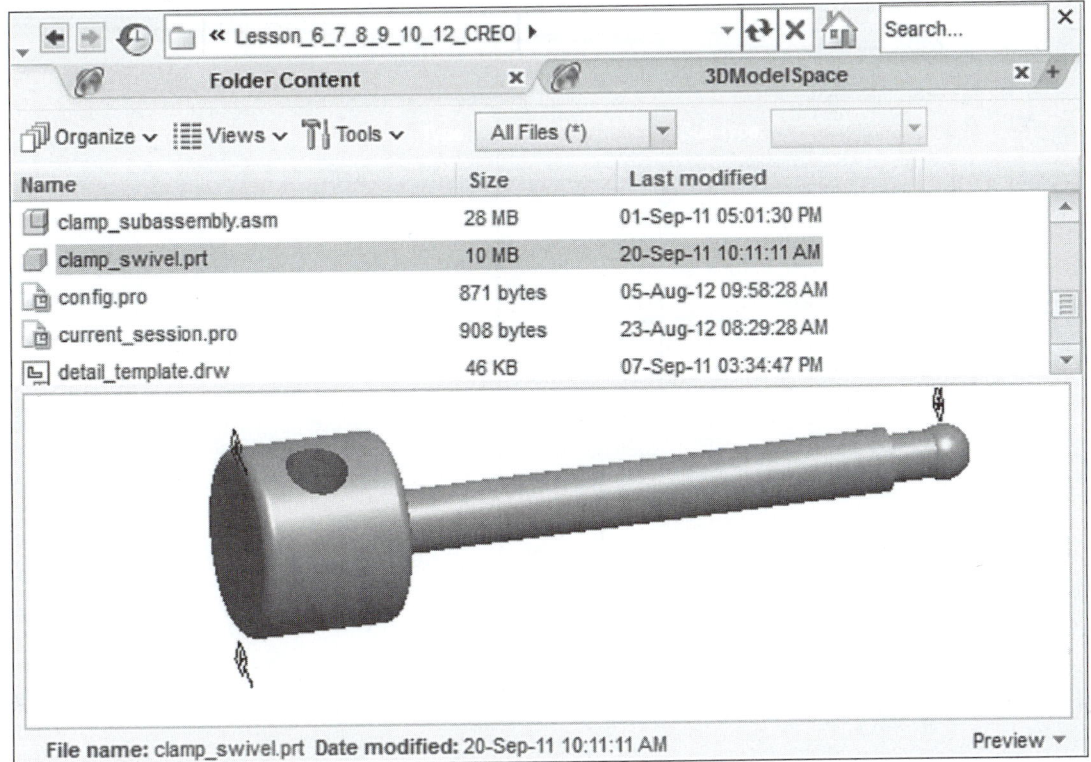

Within the preview window, dynamically reorient the model using the **MMB** by itself (**Spin**- press and hold the **MMB** as you move the mouse) or in conjunction with the **Shift** key (**Pan**) or **Ctrl** key (**Zoom, Turn**). A mouse with a thumbwheel (as the middle button) can be rolled to zoom in/out on the object. Double-clicking on the file name will open the part. After you complete this lesson you will have this same part in your working directory. From the Browser's toolbar you can also navigate to other directories or computer areas:

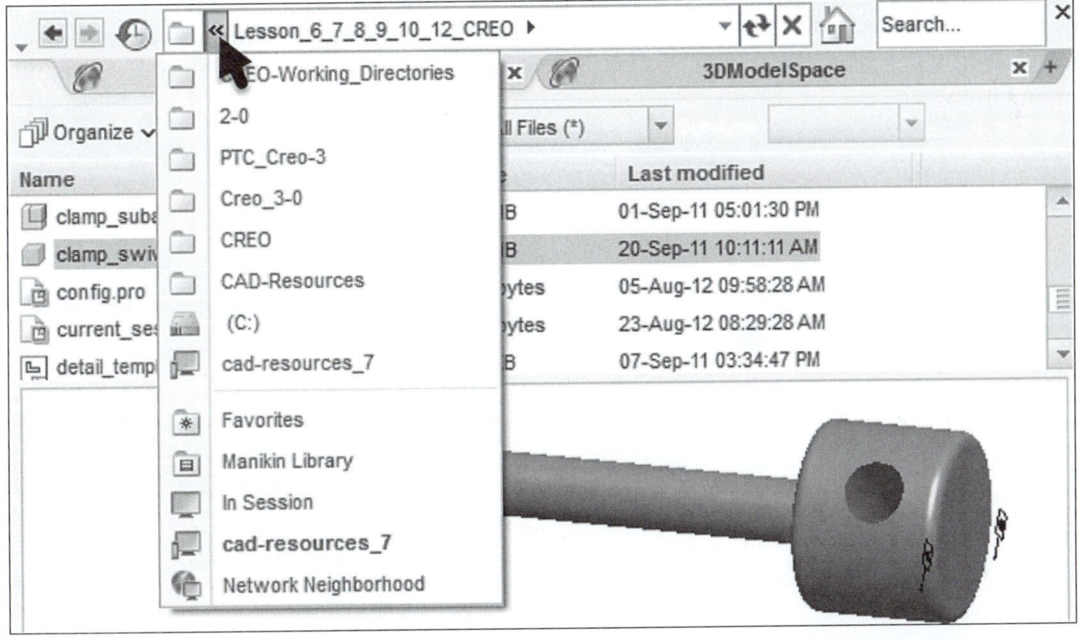

Do not change your working directory from your default work folder unless you are instructed.

Lesson 6 STEPS

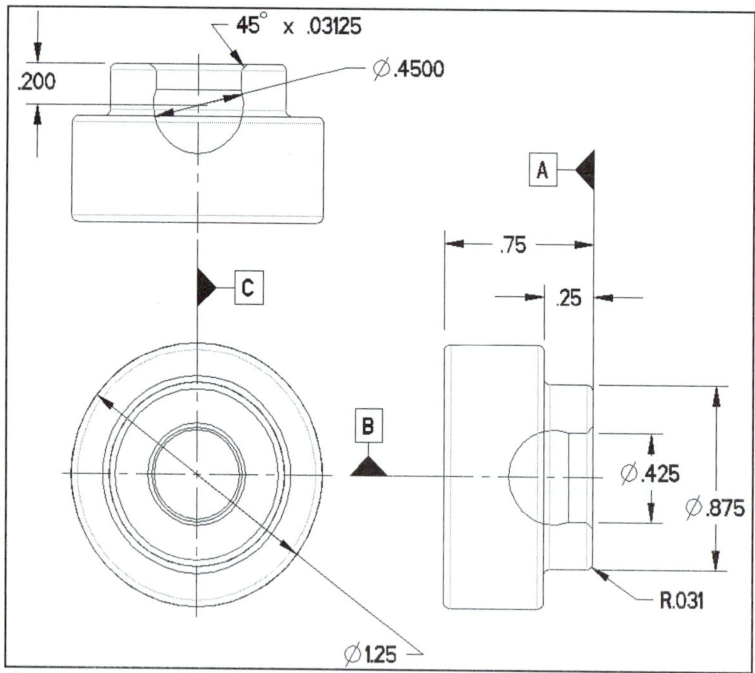

Figure 6.5 Clamp Foot Detail

Clamp Foot

The Clamp Foot (Fig. 6.5) is the first of three revolved parts to be created in this Lesson. The Clamp Foot, Clamp Ball, and Clamp Swivel are revolved parts needed for the Clamp assembly and drawings later in the text. This part requires a revolved extrusion (protrusion), a revolved cut, a chamfer and rounds.

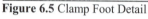

Launch **PTC Creo Parametric 3.0** > **Select Working Directory** from the Home Ribbon > navigate to your directory > **OK** > **Ctrl+N** > *type* **clamp_foot** > **OK** > **File** > **Prepare** > **Model Properties** > Units **change** > Inch lbm Second (Creo Parametric Default) > **Close** > **Close** > **File** > **Options** > **Configuration Editor** > **Import/Export** > **Import configuration file** > select **Creo_textbook.pro** [Fig. 6.6(a)] (your file name may be different) > **Open**

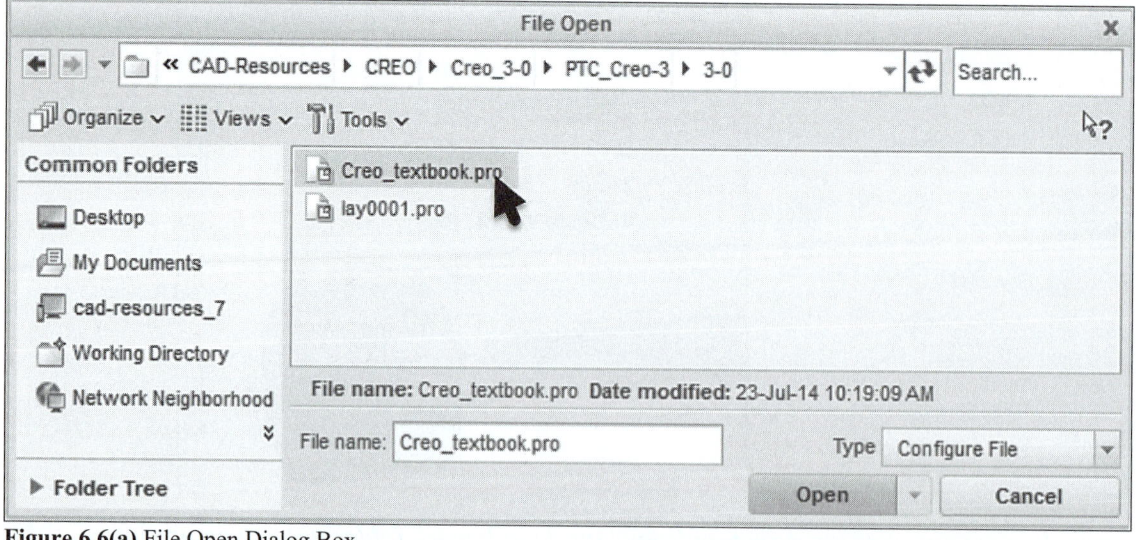

Figure 6.6(a) File Open Dialog Box

Click: **Show:** [Fig. 6.6(b)] > **All options** > Slide the vertical scroll bar *(right side of Creo Parametric Options dialog box)* down to the option *default_dec_places*. To the right of *default_dec_places*, click in the Value field > *type* **3** [Fig. 6.6(c)] > **Enter** default_dec_places 3 ✳ *(Status ✳ changed)*

PTC Creo Parametric Options
View and manage PTC Creo Parametric options.
Options
Sort: Alphabetical
Show: All Options
Name
Value
Status
Description
default_comp_skel_geom_scope
all *
Set the specified Shared Geometry reference control co
default_comp_skel_scope_setting
all *
Set the specified Accessible reference control conditions
default_dec_places
4
Sets the default number of decimal places (0-13) to whic
default_dep_copy_option
dep_dim *
Sets Full Dependent Copy as a default copy option.

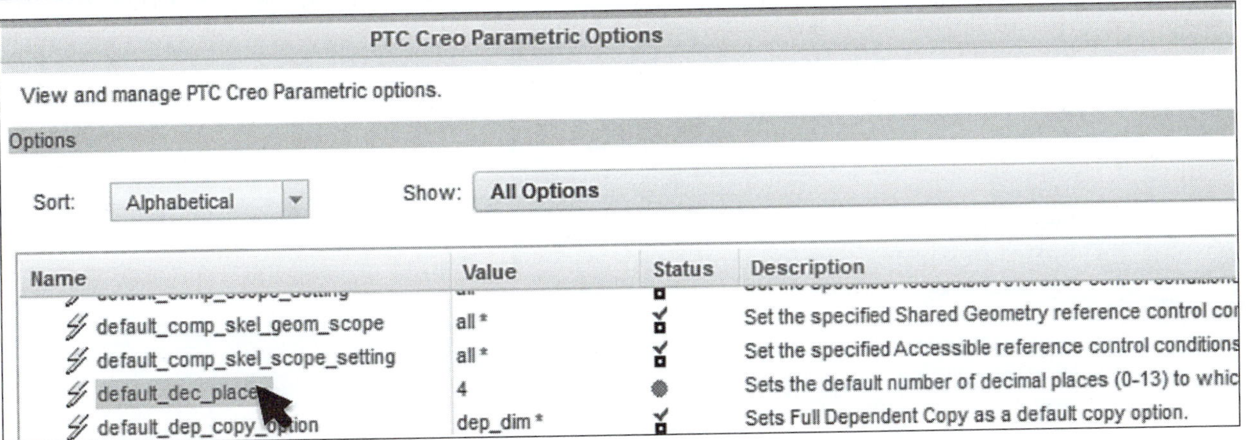

Figure 6.6(b) default_dec_places

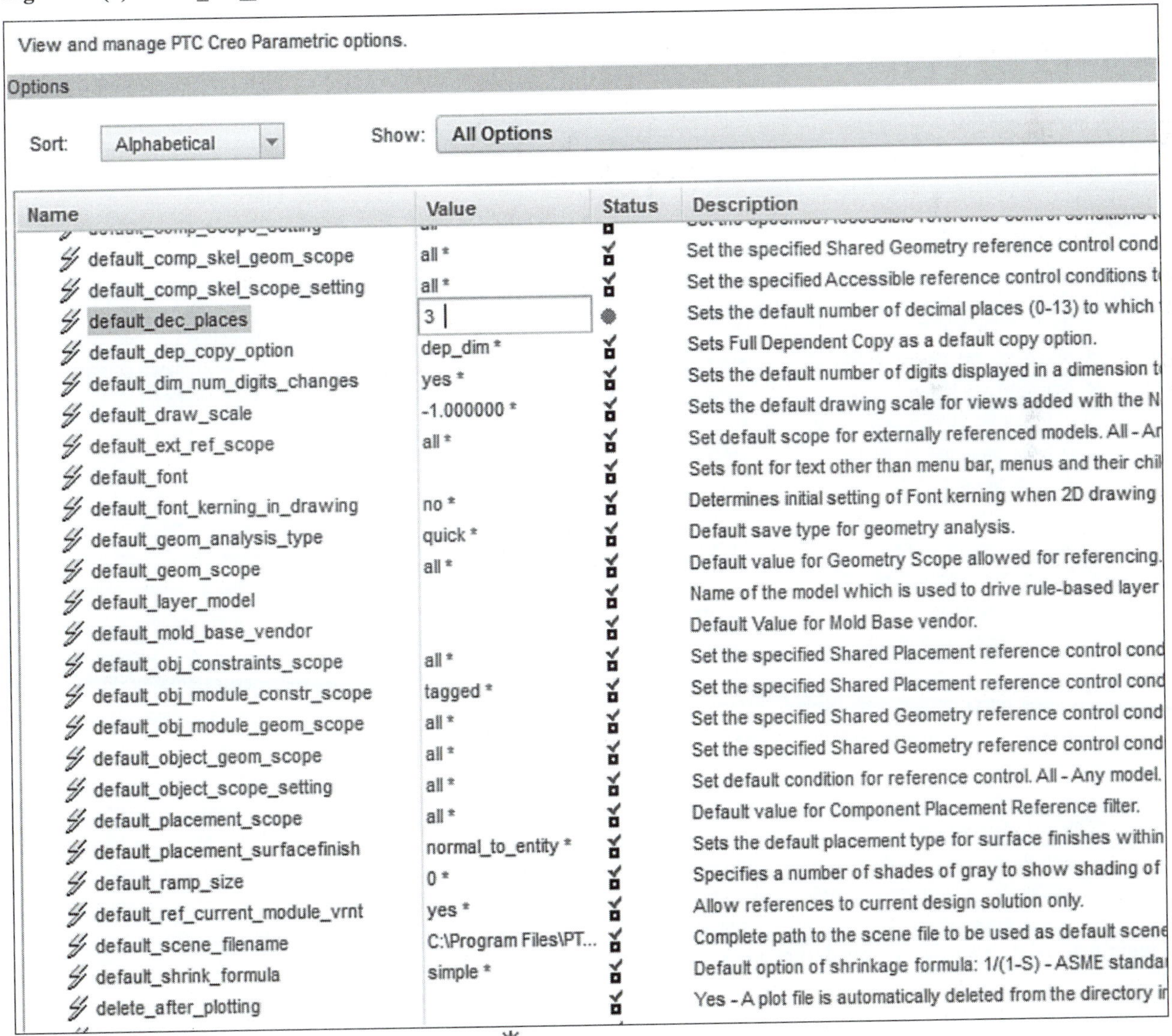

View and manage PTC Creo Parametric options.
Options
Sort: Alphabetical
Show: All Options
Name
Value
Status
Description
default_comp_skel_geom_scope
all *
Set the specified Shared Geometry reference control cond
default_comp_skel_scope_setting
all *
Set the specified Accessible reference control conditions t
default_dec_places
3
Sets the default number of decimal places (0-13) to which
default_dep_copy_option
dep_dim *
Sets Full Dependent Copy as a default copy option.
default_dim_num_digits_changes
yes *
Sets the default number of digits displayed in a dimension t
default_draw_scale
-1.000000 *
Sets the default drawing scale for views added with the N
default_ext_ref_scope
all *
Set default scope for externally referenced models. All - Ar
default_font
Sets font for text other than menu bar, menus and their chi
default_font_kerning_in_drawing
no *
Determines initial setting of Font kerning when 2D drawing
default_geom_analysis_type
quick *
Default save type for geometry analysis.
default_geom_scope
all *
Default value for Geometry Scope allowed for referencing.
default_layer_model
Name of the model which is used to drive rule-based layer
default_mold_base_vendor
Default Value for Mold Base vendor.
default_obj_constraints_scope
all *
Set the specified Shared Placement reference control cond
default_obj_module_constr_scope
tagged *
Set the specified Shared Placement reference control cond
default_obj_module_geom_scope
all *
Set the specified Shared Geometry reference control cond
default_object_geom_scope
all *
Set the specified Shared Geometry reference control cond
default_object_scope_setting
all *
Set default condition for reference control. All - Any model.
default_placement_scope
all *
Default value for Component Placement Reference filter.
default_placement_surfacefinish
normal_to_entity *
Sets the default placement type for surface finishes within
default_ramp_size
0 *
Specifies a number of shades of gray to show shading of
default_ref_current_module_vrnt
yes *
Allow references to current design solution only.
default_scene_filename
C:\Program Files\PT...
Complete path to the scene file to be used as default scene
default_shrink_formula
simple *
Default option of shrinkage formula: 1/(1-S) - ASME standa
delete_after_plotting
Yes - A plot file is automatically deleted from the directory i

Figure 6.6(c) default_dec_places Value: 3 Status: ✳

Click: **Add ...** > in the Add Option name field, *type*: **def_layer** *(you must type this with the keyboard,* ***but do not press Enter)*** [Fig. 6.6(d)] > Option value: ▼ > **layer_axis** > move the mouse pointer to the right side of the text: layer_axis, click in the Option value field > type *(space)* **datum_axes** > **OK** [Fig. 6.6(e)]

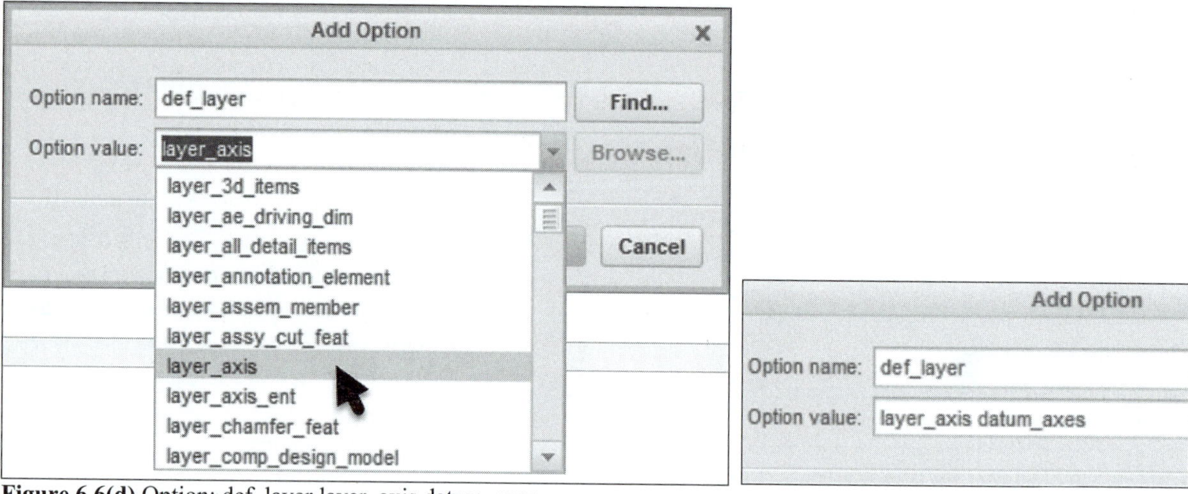

Figure 6.6(d) Option: def_layer layer_axis datum_axes

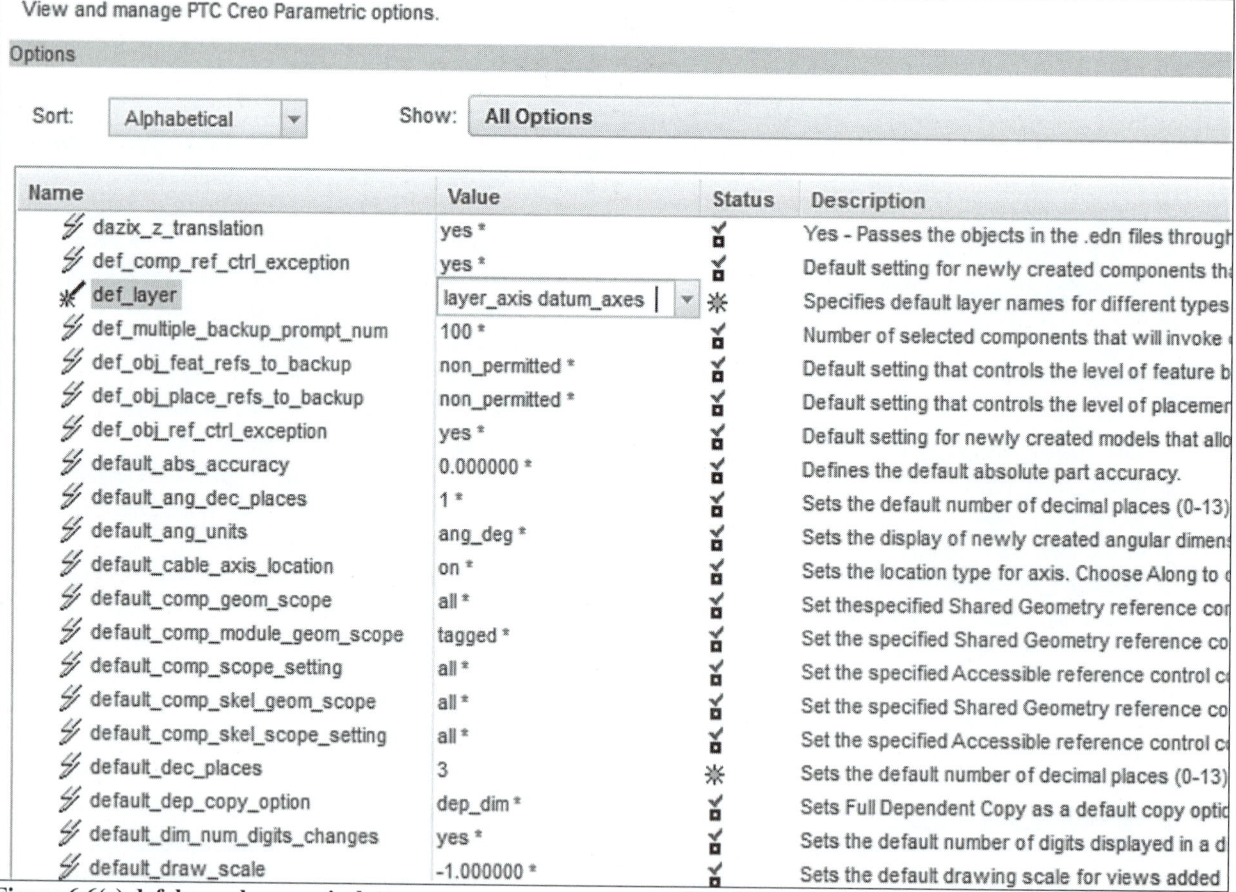

View and manage PTC Creo Parametric options.
Options
Sort: Alphabetical
Show: All Options
Name
Value
Status
Description
dazix_z_translation yes * Yes - Passes the objects in the .edn files through
def_comp_ref_ctrl_exception yes * Default setting for newly created components th
def_layer layer_axis datum_axes Specifies default layer names for different types
def_multiple_backup_prompt_num 100 * Number of selected components that will invoke
def_obj_feat_refs_to_backup non_permitted * Default setting that controls the level of feature b
def_obj_place_refs_to_backup non_permitted * Default setting that controls the level of placemen
def_obj_ref_ctrl_exception yes * Default setting for newly created models that allo
default_abs_accuracy 0.000000 * Defines the default absolute part accuracy.
default_ang_dec_places 1 * Sets the default number of decimal places (0-13)
default_ang_units ang_deg * Sets the display of newly created angular dimen
default_cable_axis_location on * Sets the location type for axis. Choose Along to
default_comp_geom_scope all * Set thespecified Shared Geometry reference co
default_comp_module_geom_scope tagged * Set the specified Shared Geometry reference co
default_comp_scope_setting all * Set the specified Accessible reference control c
default_comp_skel_geom_scope all * Set the specified Shared Geometry reference co
default_comp_skel_scope_setting all * Set the specified Accessible reference control c
default_dec_places 3 Sets the default number of decimal places (0-13)
default_dep_copy_option dep_dim * Sets Full Dependent Copy as a default copy optio
default_dim_num_digits_changes yes * Sets the default number of digits displayed in a d
default_draw_scale -1.000000 * Sets the default drawing scale for views added

Figure 6.6(e) def_layer, layer_axis datum_axes

Click: **Show** > Current Session [Fig. 6.6(f)] > **OK** [Fig. 6.6(g)] > **Yes** > select **Creo_textbook.pro** [Fig. 6.6(h)] (or your file name) > **OK** > **Ctrl+S** > **OK**

Name	Value	Status	Description
def_layer	layer_axis datum_axes	✳	Specifies default layer names for different types
default_dec_places	3	✳	Sets the default number of decimal places (0-13)
display	shadewithedges	●	Wireframe - All solid edges in white. Hiddenvis - H
drawing_setup_file	$PRO_DIRECTORY\text\pro...	●	Sets the default drawing setup file option values
edge_display_quality	very_high	●	Controls display quality of an edge for wireframe
format_setup_file	$PRO_DIRECTORY\text\pro...	●	Assigns a specified setup file to each drawing fo
mapkey	$F3 @MAPKEY_NAMEChan...	✳	Stores a keyboard macro. Use File > Options > En
orientation	trimetric *	●	Establishes the initial standard view orientation. U
pro_unit_length	unit_inch *	●	Sets the default units for new objects.

Figure 6.6(f) Showing current session options with def_layer selected *(your selected/highlighted name may be different)*

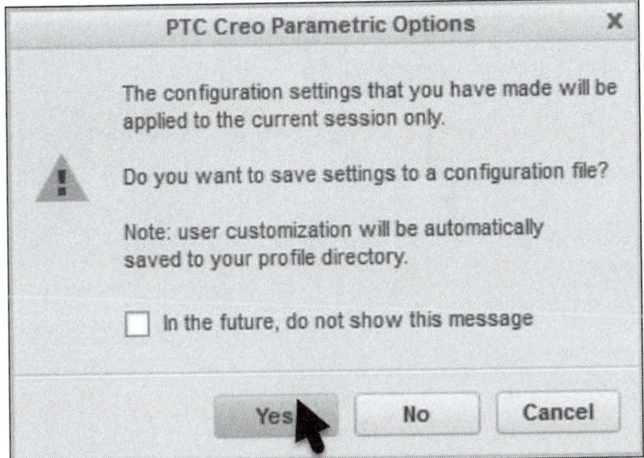

Figure 6.6(g) Creo Parametric Options Confirmation Dialog Box

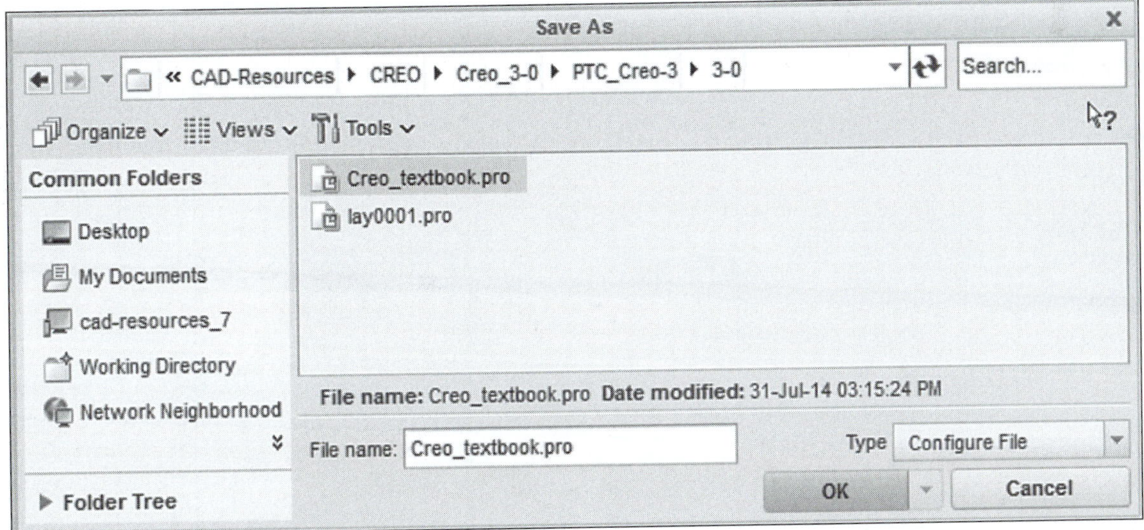

Figure 6.6(h) clamp.pro

Click: **View** tab > [icons] *on* > [icons] *on* > **Model** tab > Revolve > in the Graphics Window, press **RMB > Define Internal Sketch** > Sketch Plane--- Plane: select datum **RIGHT** [Fig. 6.7(a)] > Orientation **Left** > **Sketch** > [icon] **Sketch View** [Fig. 6.7(b)] *(To change the Graphics Toolbar location > RMB anywhere inside the Graphics Toolbar > Location > select an option. The Graphics Toolbar been relocated to the bottom of the Graphics Window here.)*

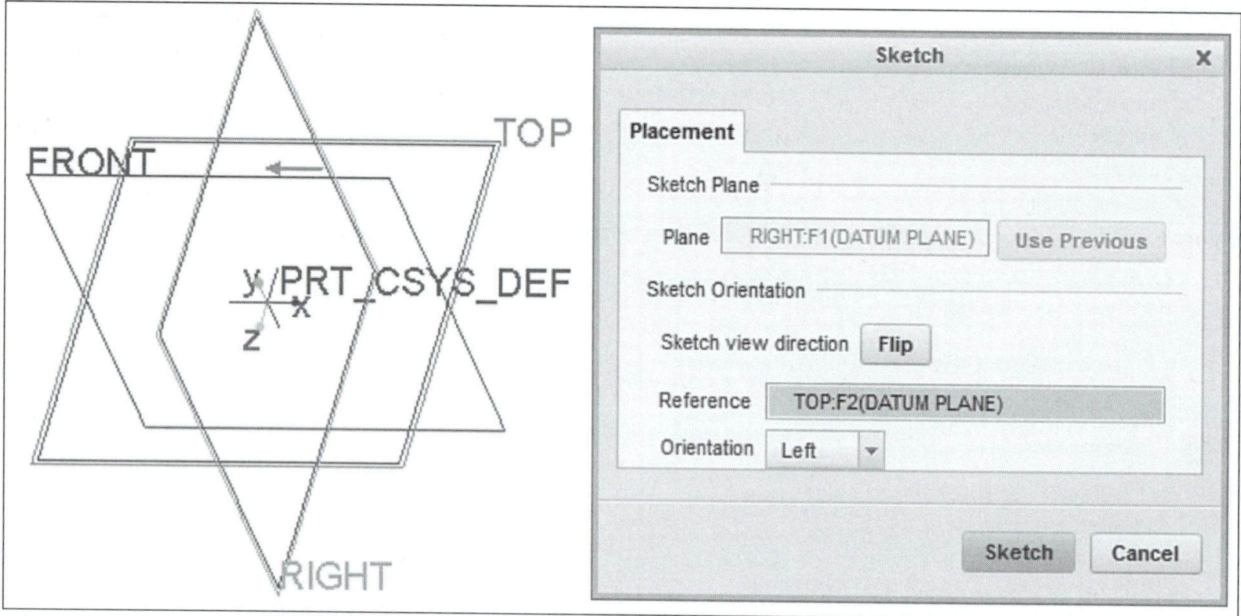

Figure 6.7(a) Sketch Plane and TOP:F2(DATUM PLANE) Reference

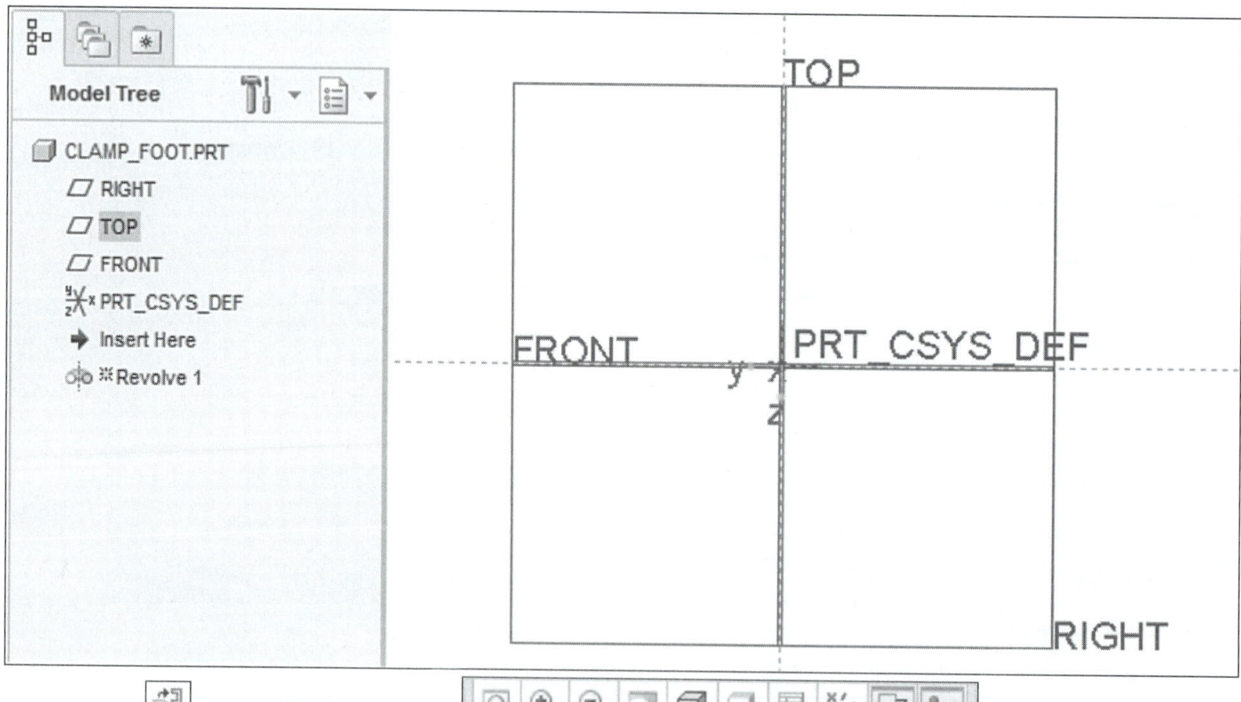

Figure 6.7(b) [icon] from the Graphics Toolbar

In the Graphics Window, press: **RMB > Axis of Revolution** sketch a *vertical* centerline through the default coordinate system [Fig. 6.7(c)] > move the mouse off of the centerline > **MMB** > press **RMB > Line Chain** [Fig. 6.7(d)] > sketch six lines > **MMB > MMB > LMB** [Fig. 6.7(e)] *(if incorrect, delete the lines and start again after clicking:* ***Undo****)*

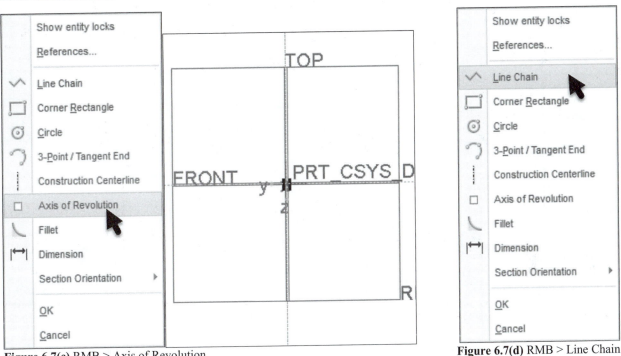

Figure 6.7(c) RMB > Axis of Revolution

Figure 6.7(d) RMB > Line Chain

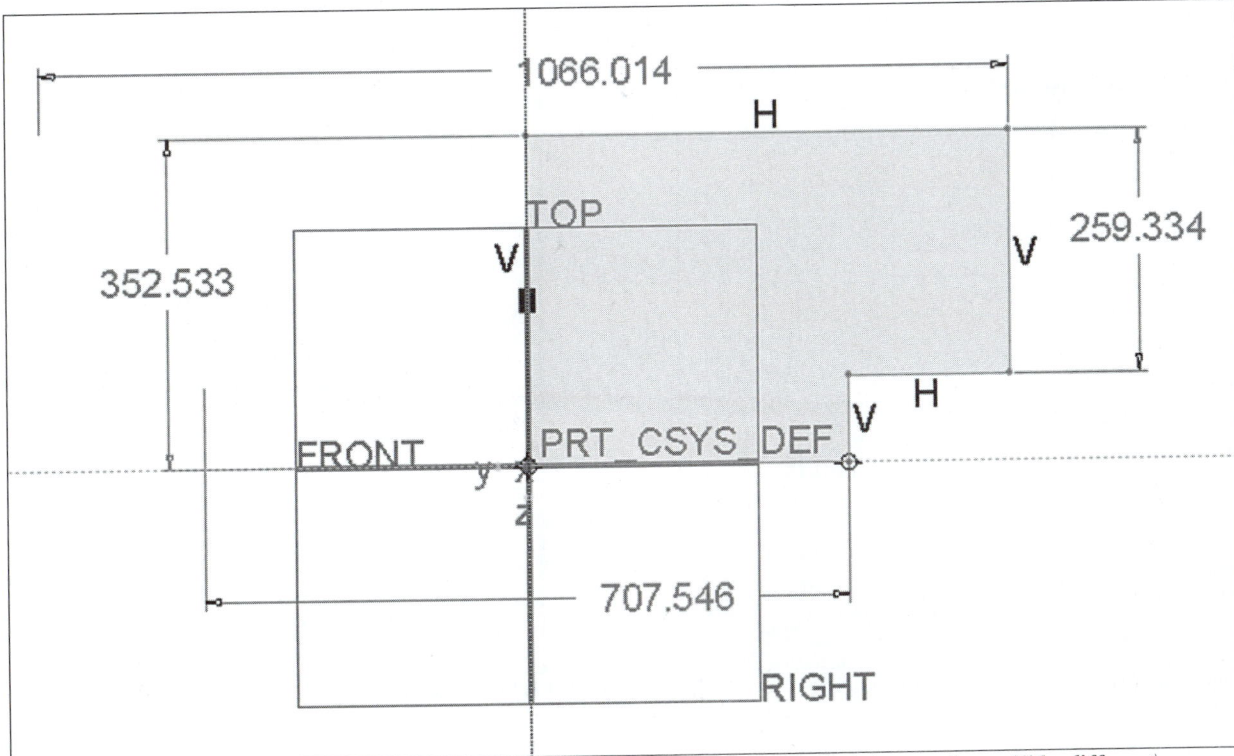

Figure 6.7(e) Sketch Six Lines to Form a Closed Section (on one side of the centerline) ***(your values will be different)***

Press **RMB** > **Dimension** > add the vertical dimension by picking both horizontal lines > **MMB** to place the dimension [Fig. 6.7(f)] > move the pointer > **MMB** > **MMB** > capture all of the sketch and dimensions with a window > press **RMB** > **Modify** > **Lock Scale** [Fig. 6.7(g)] > *type* **1.25** for the large diameter > **Enter** [Fig. 6.7(h)] > **OK**

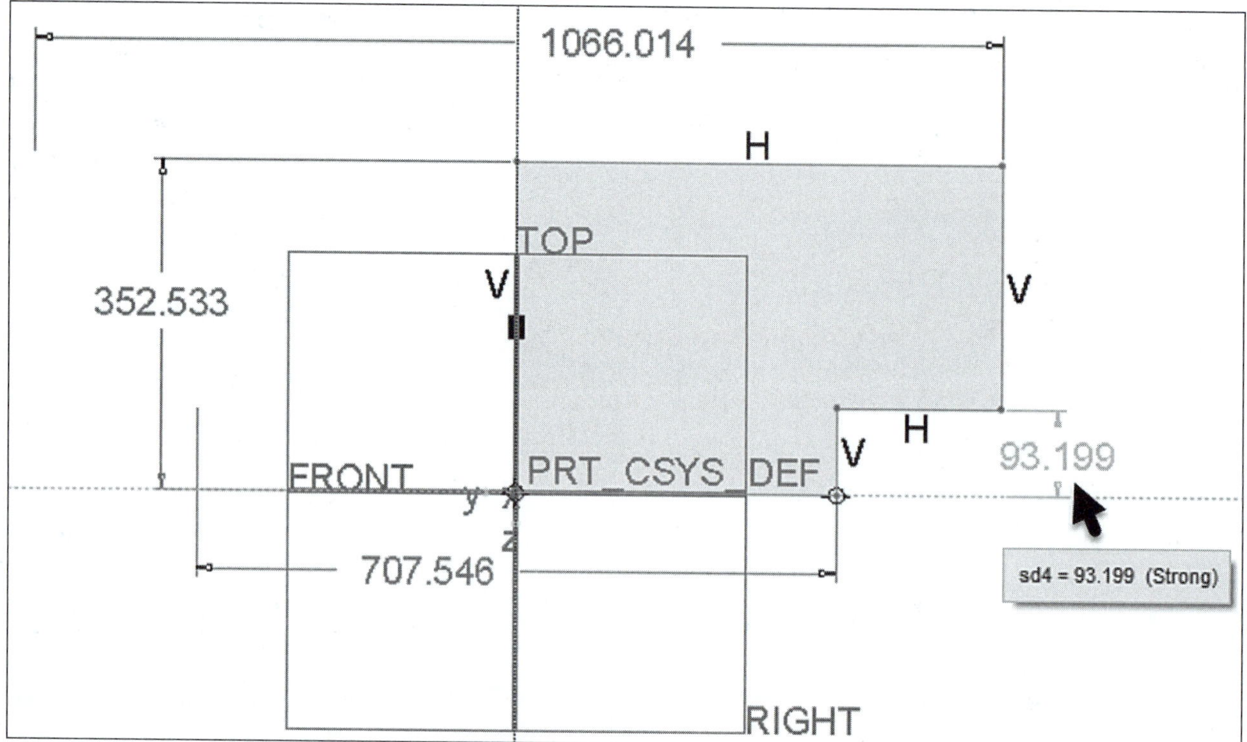

Figure 6.7(f) Sketcher Dimensioning *(your values will be different)*

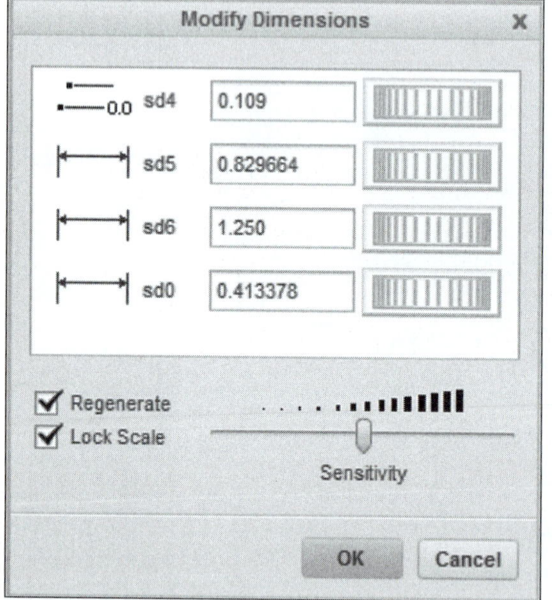

Figure 6.7(g) Modify Dimensions Dialog Box

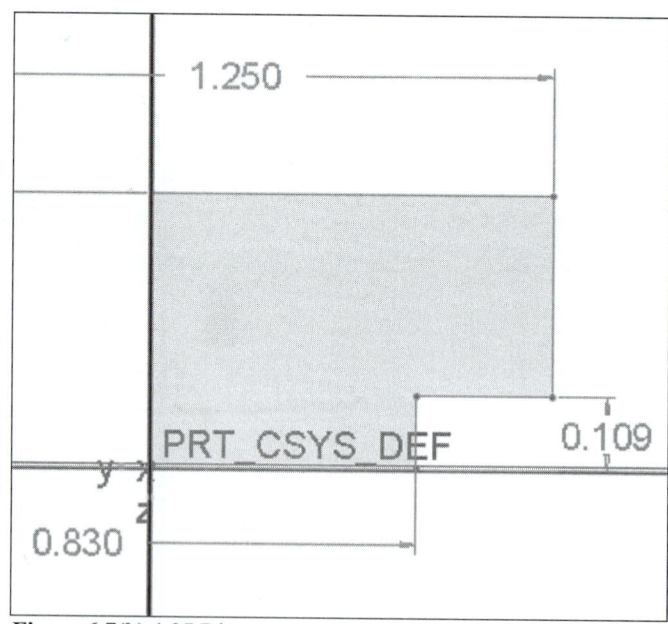

Figure 6.7(h) **1.25** Diameter

For each dimension, double-click on and change to the design value > **LMB** to deselect > reposition the dimensions as shown [Fig. 6.7(i)] > in the Graphics Window, press **RMB** > **OK** > in the Graphics Window, press **RMB** > **Show Section Dimensions** [Fig. 6.7(j)]

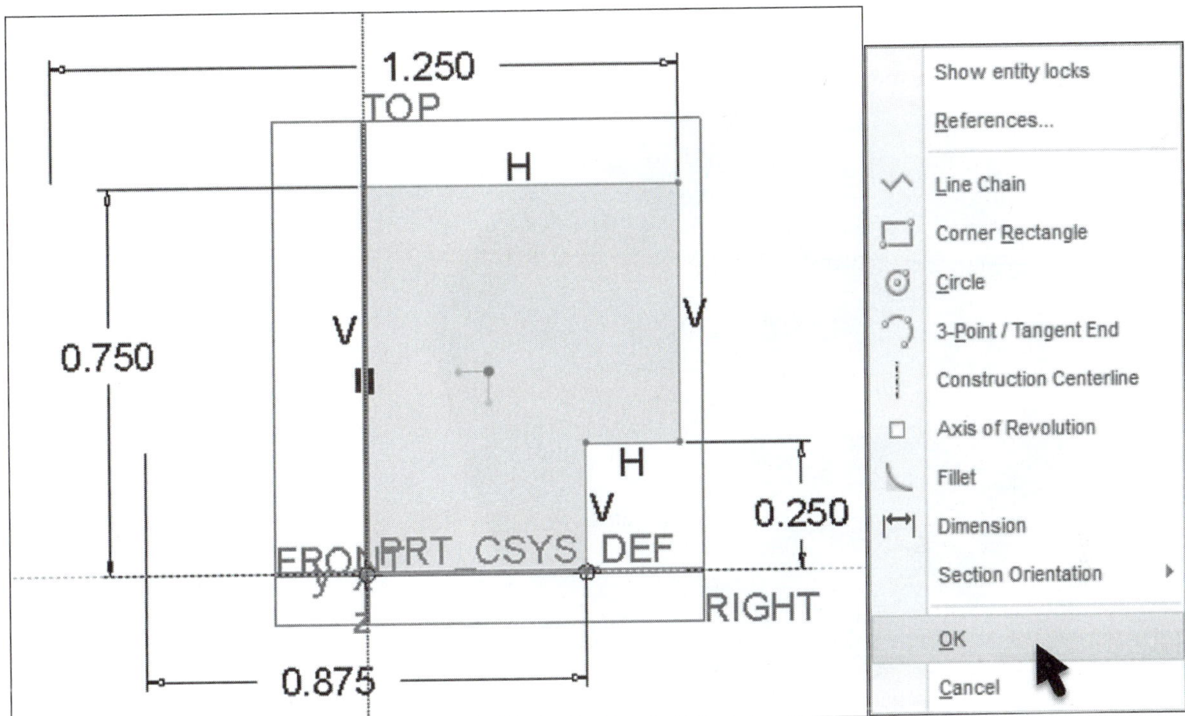

Figure 6.7(i) Design Value Dimensions

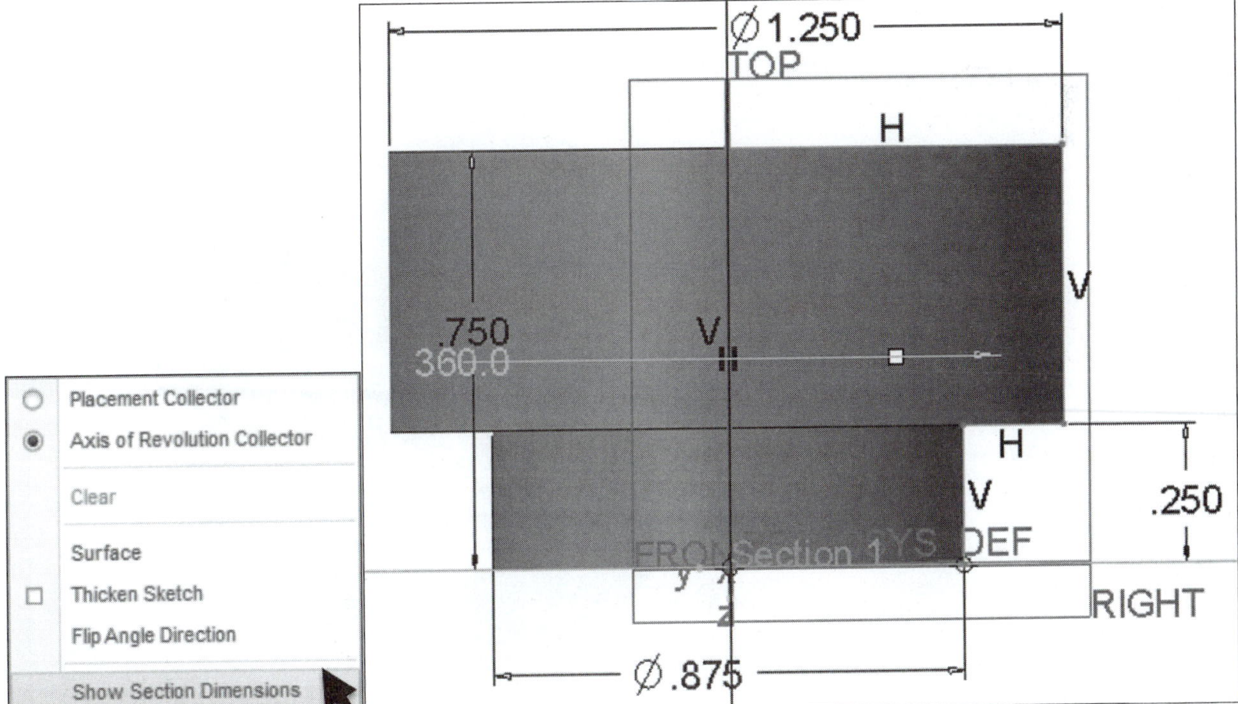

Figure 6.7(j) Show Section Dimensions

Press: **Ctrl+D** [Fig. 6.7(k)] > 🔍 **Zoom in** pick two corners > ✔ > 🔍 **Refit** > 💾 > **LMB** [Fig. 6.7(l)]

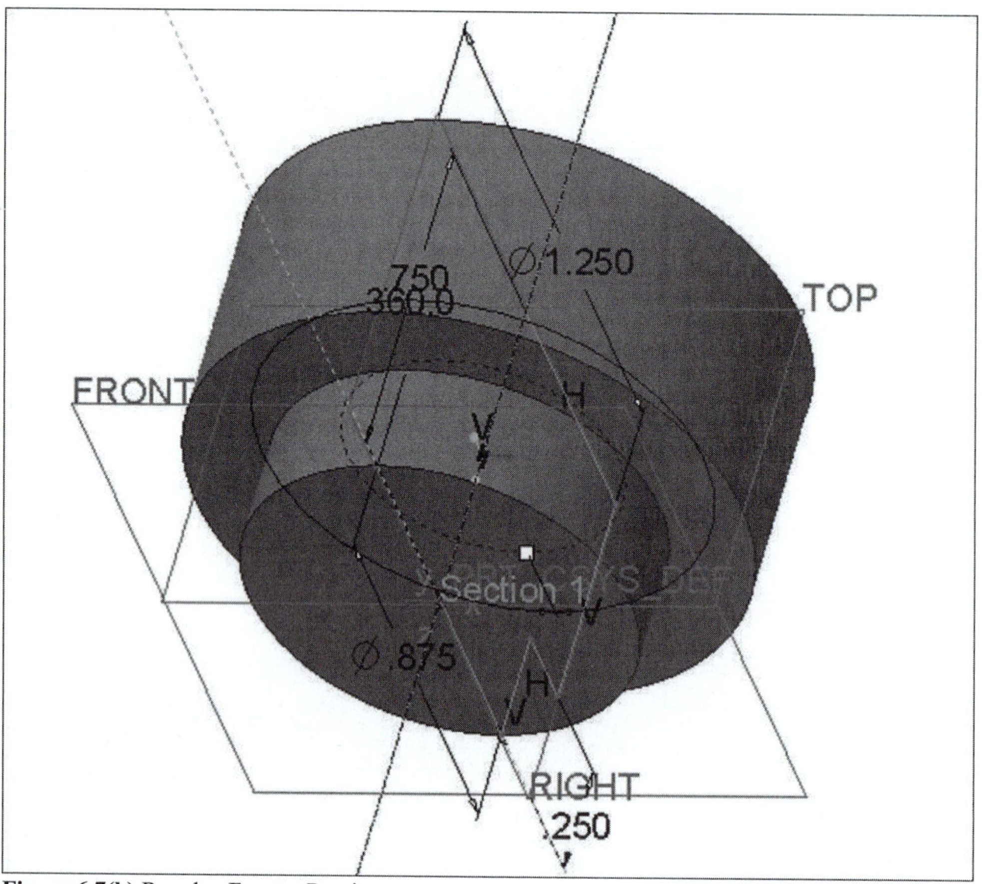

Figure 6.7(k) Revolve Feature Preview

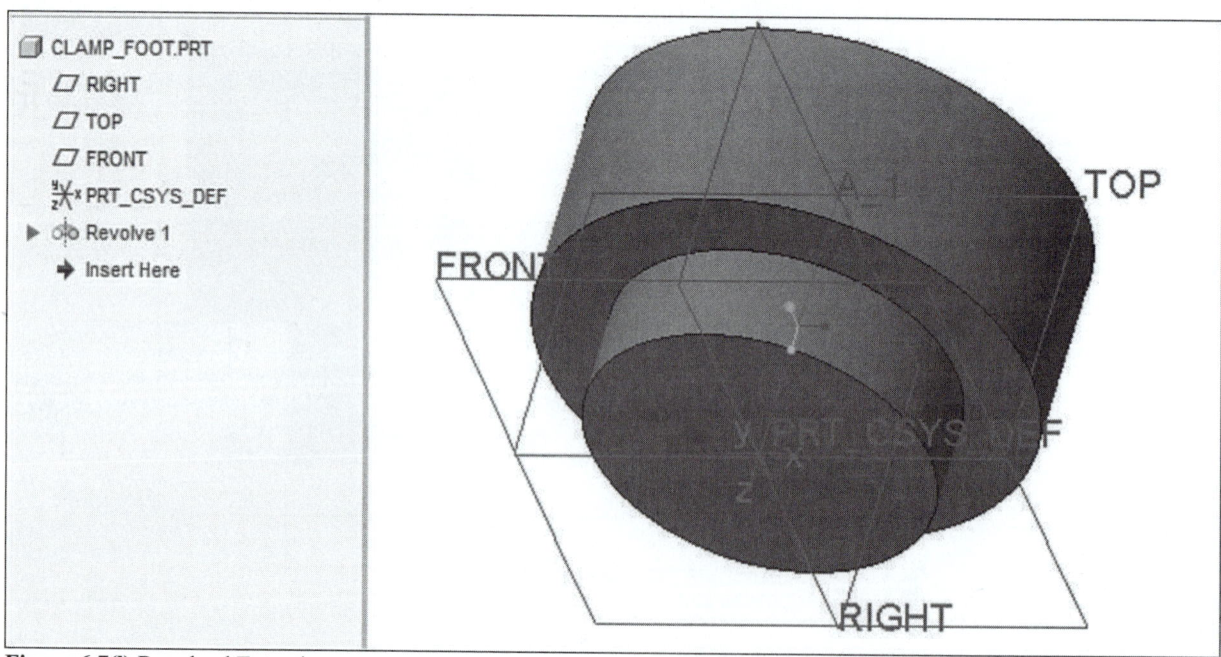

Figure 6.7(l) Revolved Extrusion

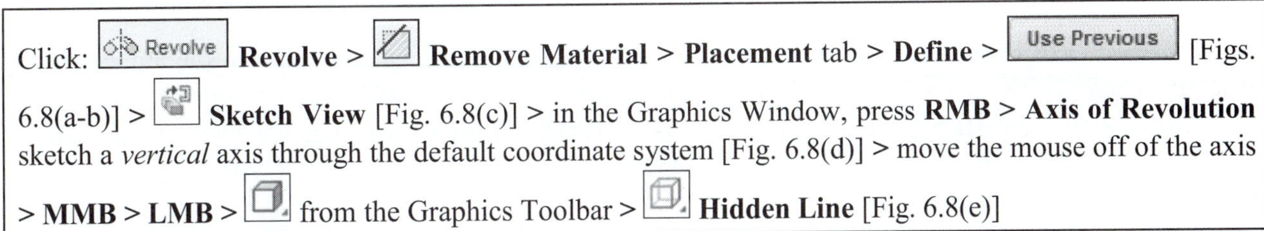

Click: Revolve **Revolve** > **Remove Material** > **Placement** tab > **Define** > Use Previous [Figs. 6.8(a-b)] > **Sketch View** [Fig. 6.8(c)] > in the Graphics Window, press **RMB** > **Axis of Revolution** sketch a *vertical* axis through the default coordinate system [Fig. 6.8(d)] > move the mouse off of the axis > **MMB** > **LMB** > from the Graphics Toolbar > **Hidden Line** [Fig. 6.8(e)]

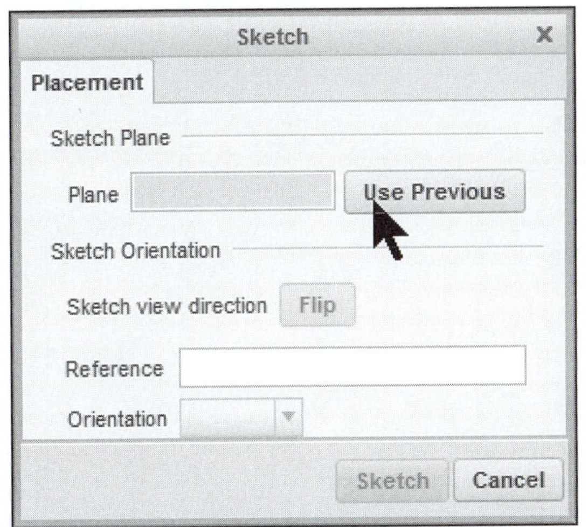

Figure 6.8(a) Use Previous

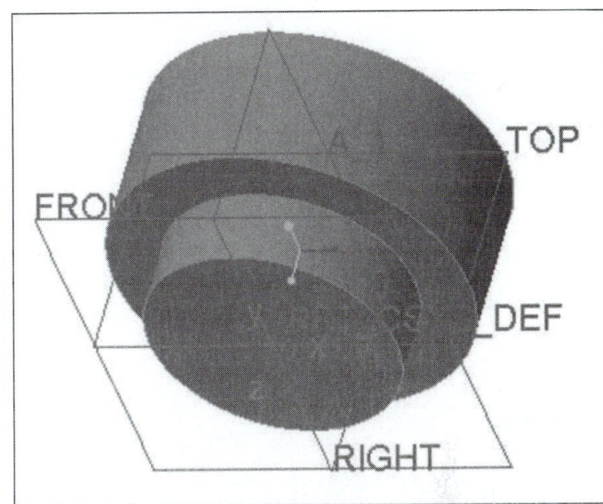

Figure 6.8(b) 3D Sketch View

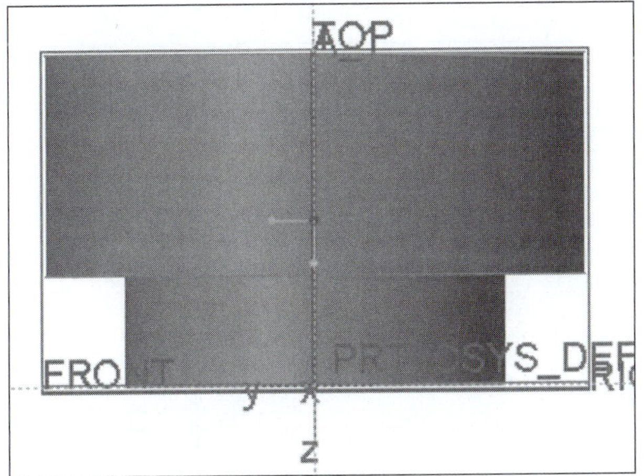

Figure 6.8(c) Sketch View

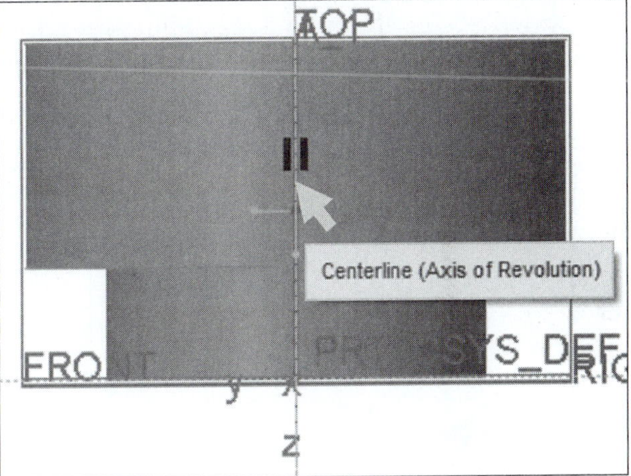

Figure 6.8(d) Sketch a Vertical Axis of Revolution

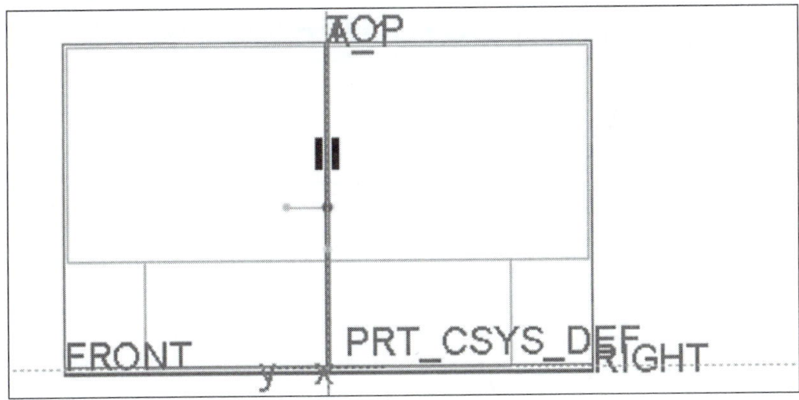

Figure 6.8(e) Hidden Line Display

From the Sketch Ribbon, click: ▾ next to Arc ▾ **Arc** > Center and Ends > sketch the arc *(LMB to place the center of the arc > LMB to place the start point of the arc > LMB to place the end point of the arc)* [Fig. 6.8(f)] > **MMB** > **LMB** [Fig. 6.8(g)] > press **RMB** > **Line Chain** sketch a *vertical* line ending at datum FRONT [Figs. 6.8(h-i)] > **MMB** > **MMB** to end the current tool > **LMB** in the Graphics Window

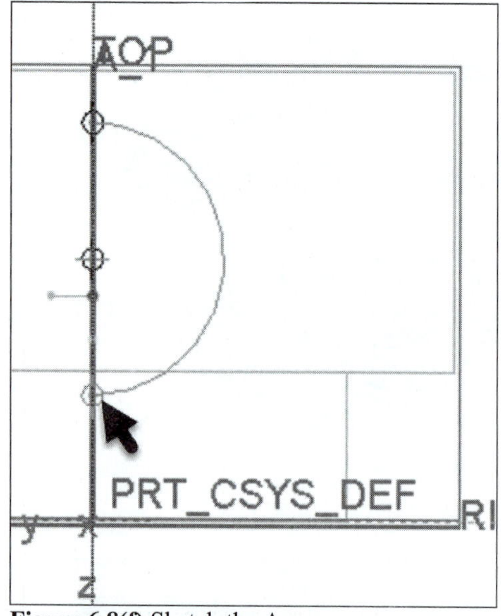

Figure 6.8(f) Sketch the Arc

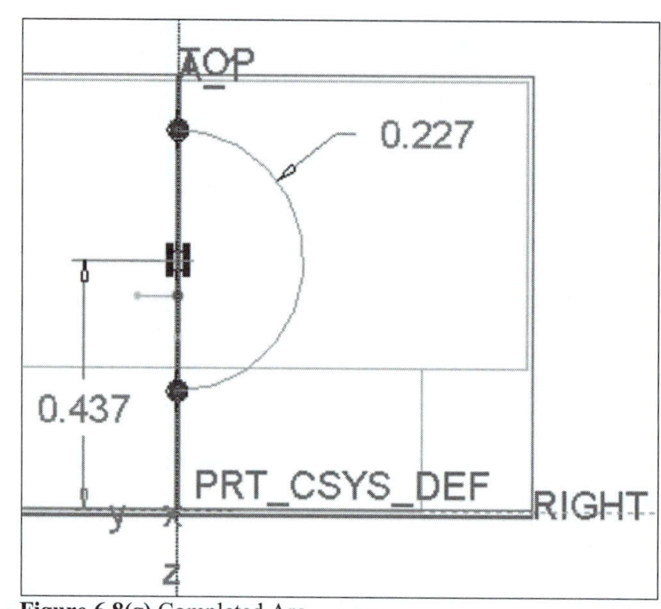

Figure 6.8(g) Completed Arc

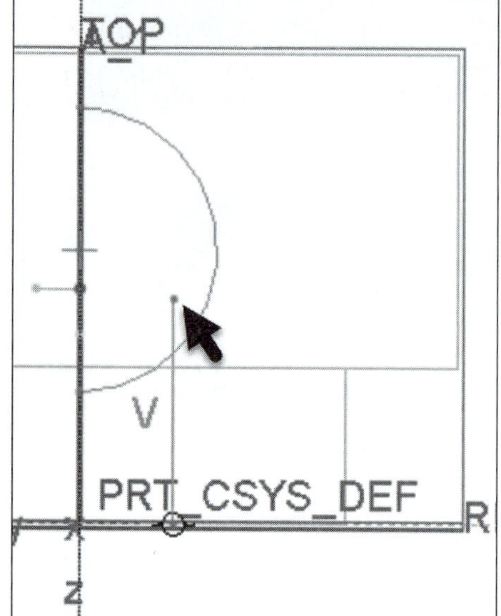

Figure 6.8(h) Sketch the Vertical Line

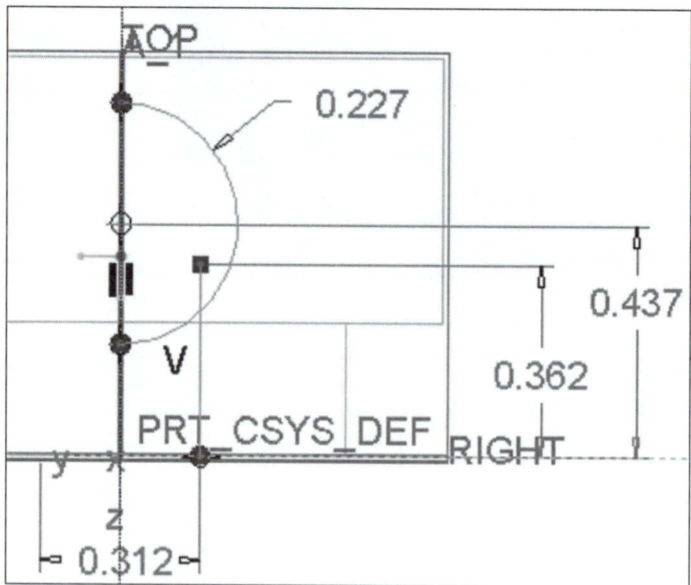

Figure 6.8(i) Completed Vertical Line

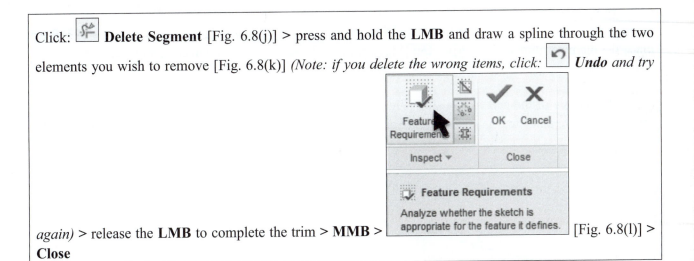

Click: **Delete Segment** [Fig. 6.8(j)] > press and hold the **LMB** and draw a spline through the two elements you wish to remove [Fig. 6.8(k)] *(Note: if you delete the wrong items, click:* ***Undo*** *and try again)* > release the **LMB** to complete the trim > **MMB** > [Fig. 6.8(l)] > **Close**

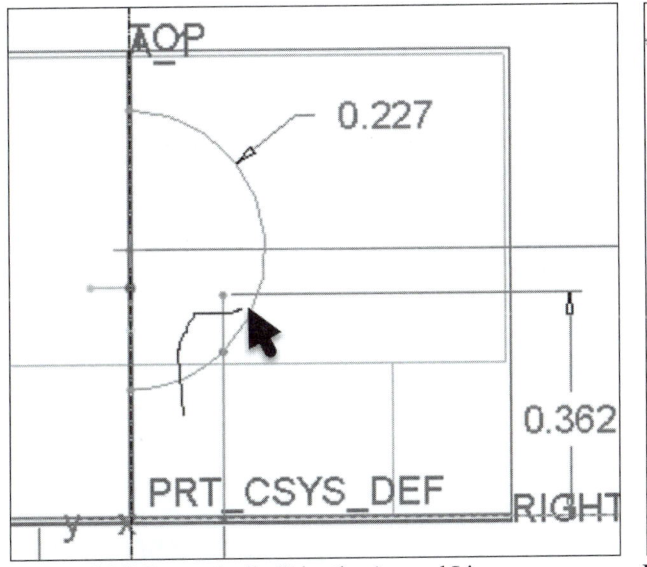

Figure 6.8(j) Dynamically Trim the Arc and Line

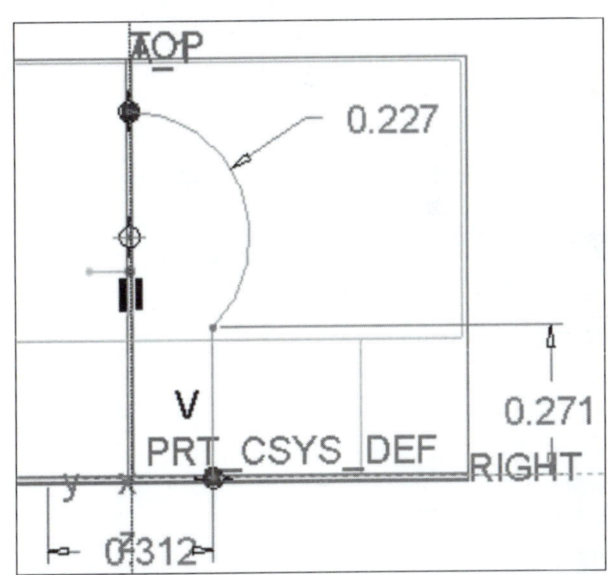

Figure 6.8(k) Trimmed Sketch

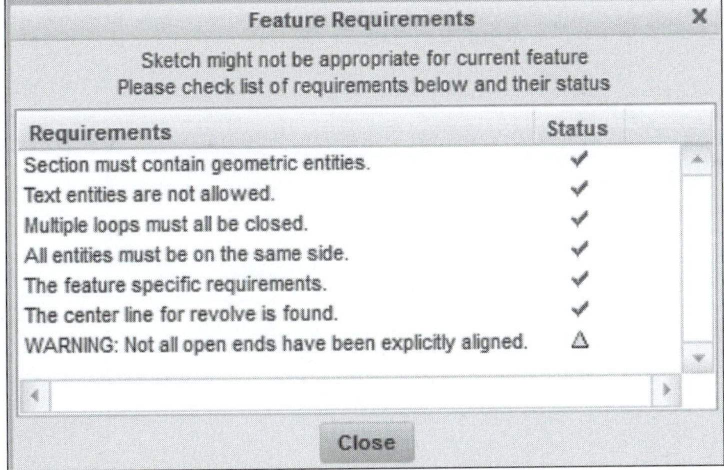

Figure 6.8(l) Feature Requirements

Click: [icon] > (Select All) > press **RMB** > **Dimension** > quickly *double-click* on the arc edge > **MMB** to place the diameter dimension > *type* **.450** [Fig. 6.8(m)] > **Enter** > add a dimension from the center of the arc to the bottom of the part > *type* **.200** [Fig. 6.8(n)] > **Enter** > Select to end the current tool > double-click on the hole diameter dimension > *type* **.425** > **Enter** > move the mouse off of the dimension > **LMB**

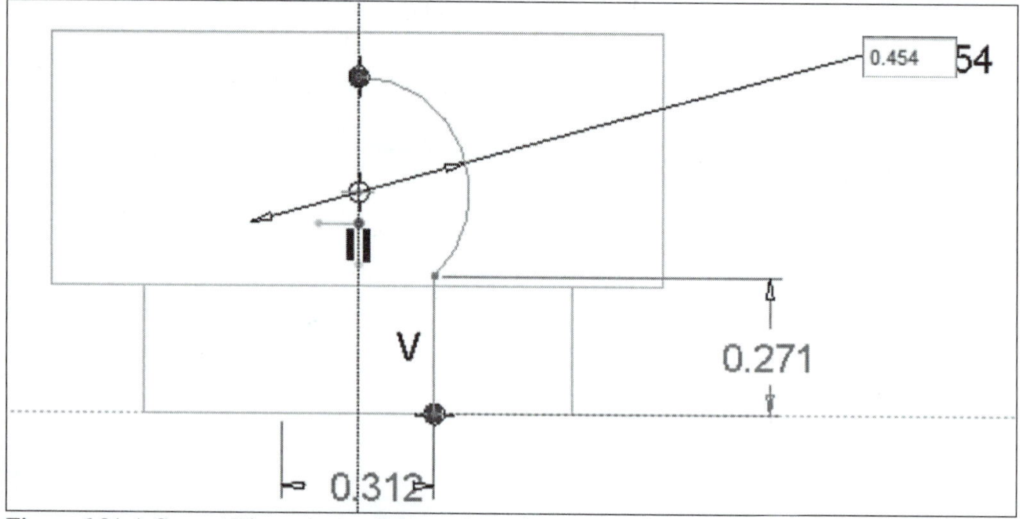

Figure 6.8(m) Correct Dimensioning Scheme *(your dimension values **will** be different)*

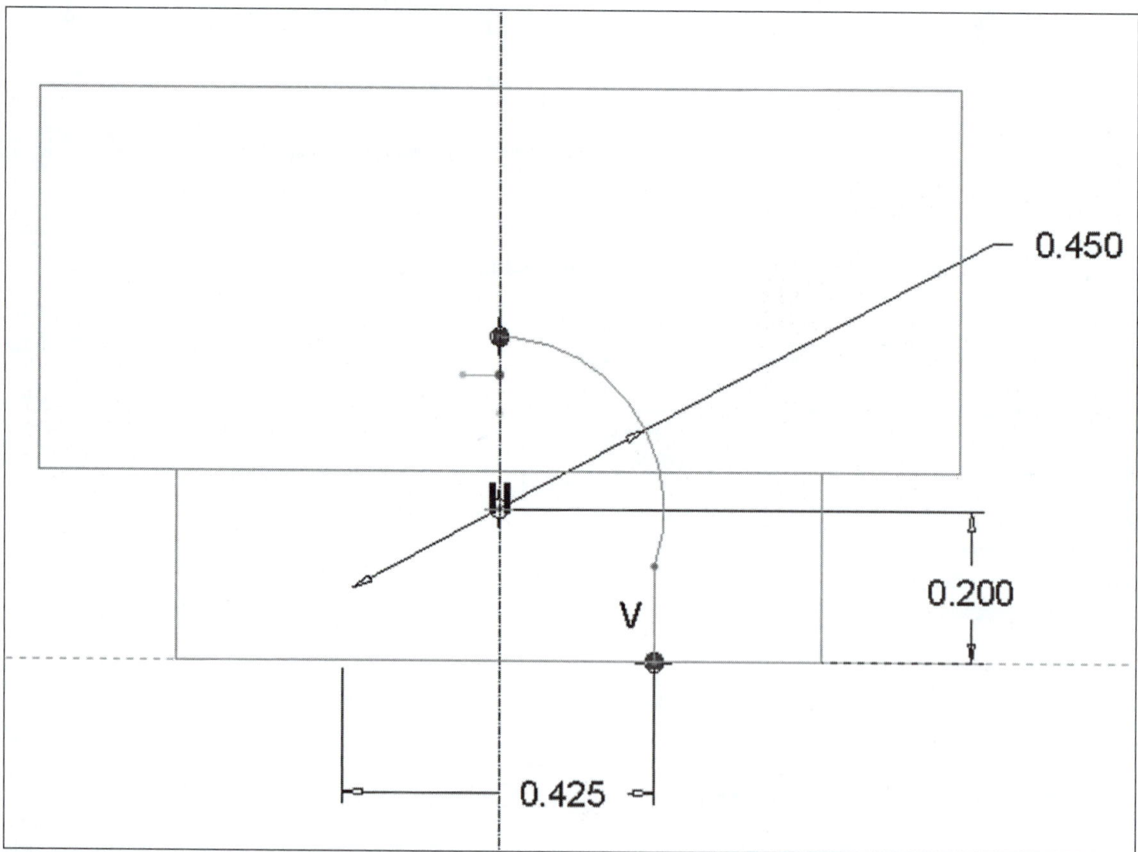

Figure 6.8(n) Modify Dimensions to **.450**, **.200**, and **.425** (these are the design values- see Figure 6.5)

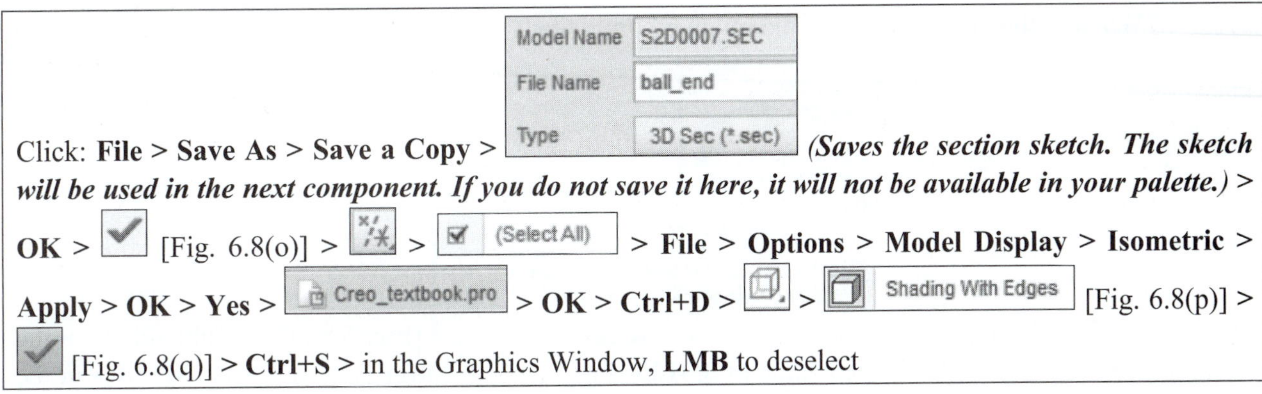

Click: **File > Save As > Save a Copy >** [Model Name: S2D0007.SEC; File Name: ball_end; Type: 3D Sec (*.sec)] ***(Saves the section sketch. The sketch will be used in the next component. If you do not save it here, it will not be available in your palette.) >***

OK > ✓ [Fig. 6.8(o)] **>** [icon] **>** (Select All) **> File > Options > Model Display > Isometric > Apply > OK > Yes >** Creo_textbook.pro **> OK > Ctrl+D >** [icon] **>** Shading With Edges [Fig. 6.8(p)] **>** ✓ [Fig. 6.8(q)] **> Ctrl+S >** in the Graphics Window, **LMB** to deselect

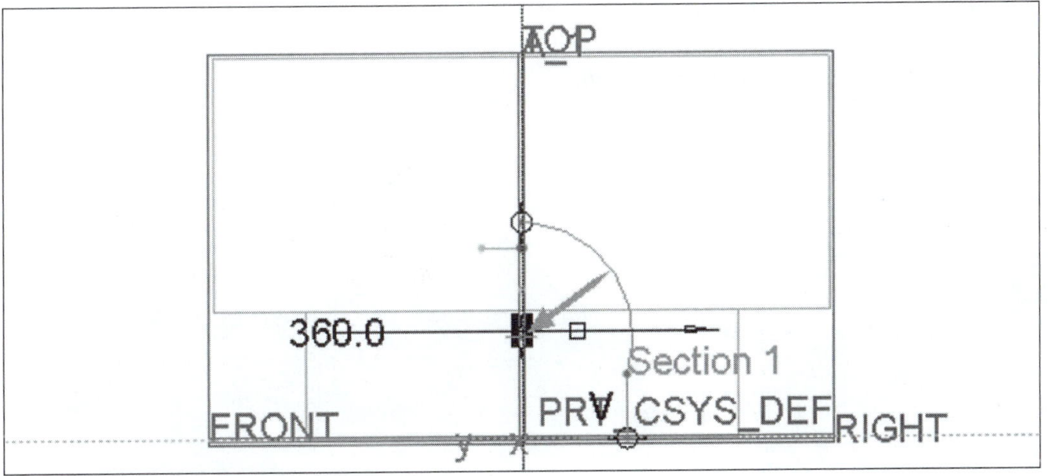

Figure 6.8(o) Previewed Revolved Cut

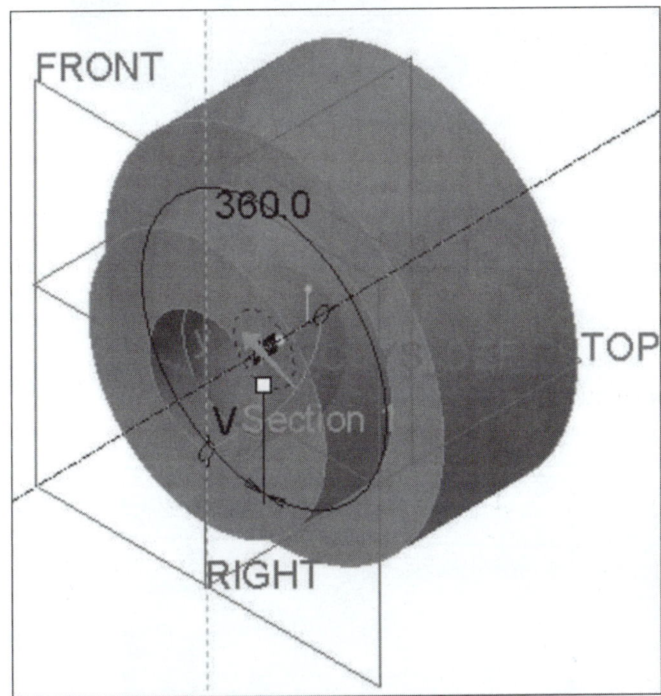

Figure 6.8(p) Revolved Cut Shading with Edges

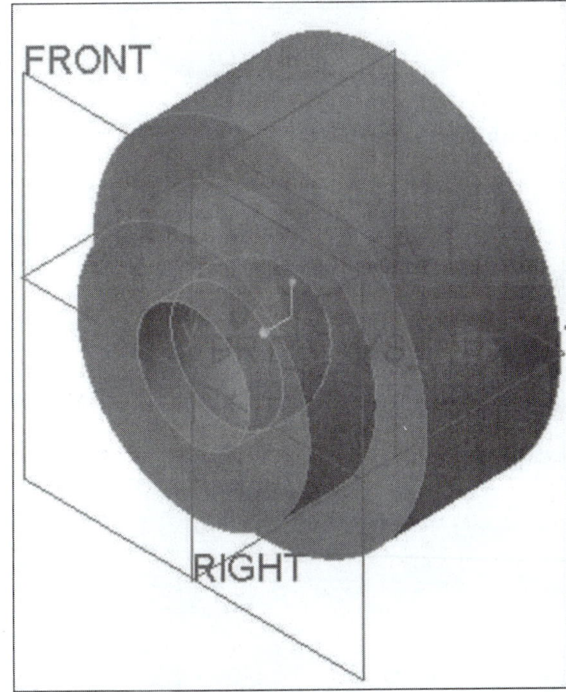

Figure 6.8(q) Completed Cut

Press: **MMB** spin the model > pick on the revolved protrusion > pick on the edge of the cut until selected *(highlights)* [Fig. 6.8(r)] > 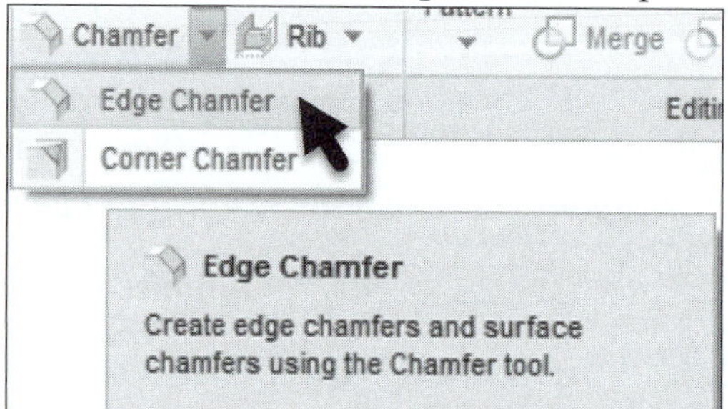

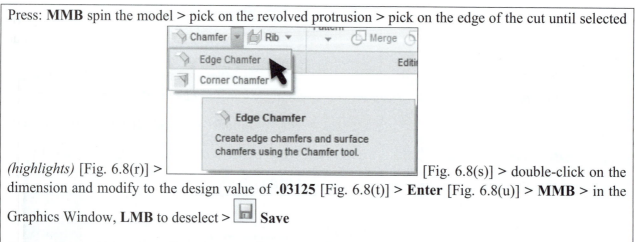

[Fig. 6.8(s)] > double-click on the dimension and modify to the design value of **.03125** [Fig. 6.8(t)] > **Enter** [Fig. 6.8(u)] > **MMB** > in the Graphics Window, **LMB** to deselect > **Save**

It is good practice to save after every new feature or edit.

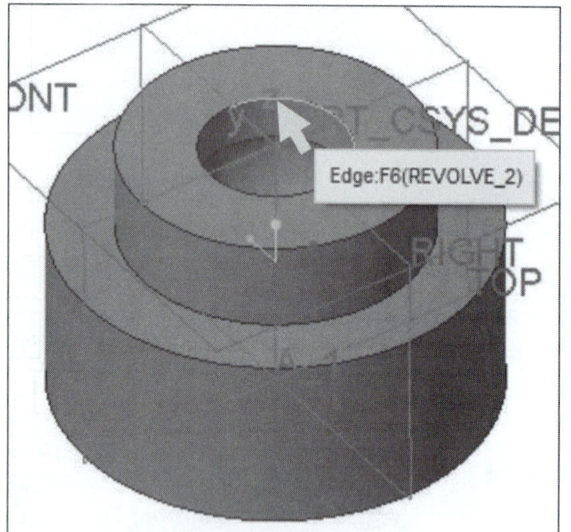

Figure 6.8(r) Pick on Edge

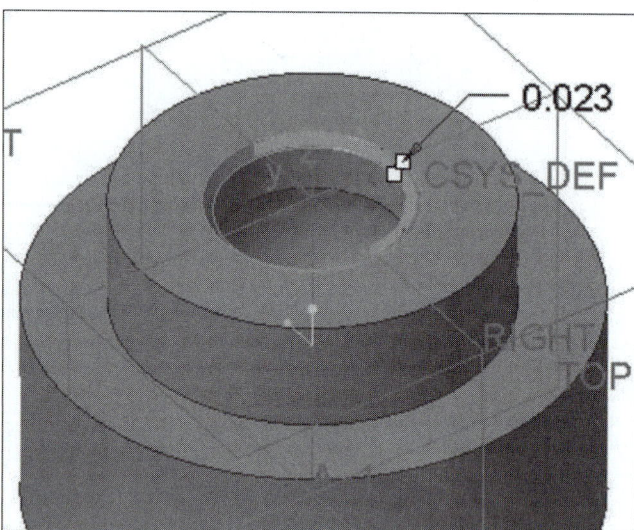

Figure 6.8(s) Chamfer Preview

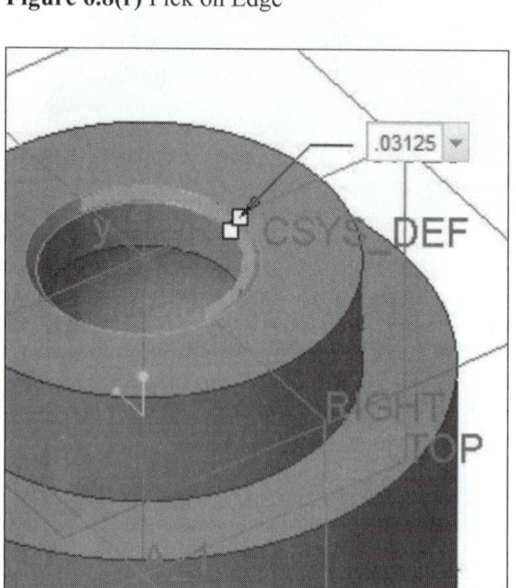

Figure 6.8(t) Chamfer Dimension

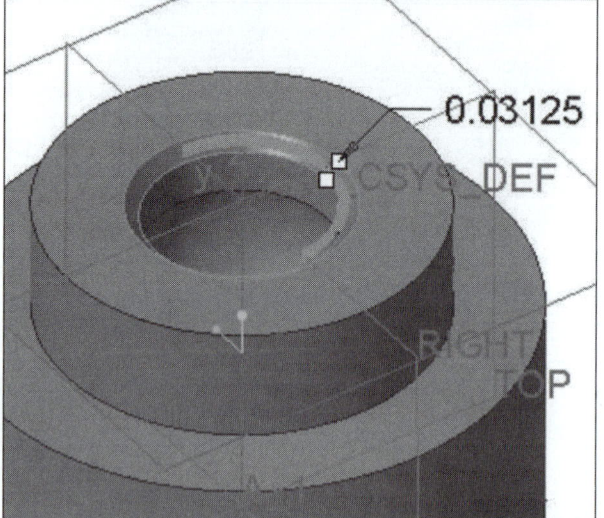

Figure 6.8(u) Chamfer

Select one edge, then press and hold the **Ctrl** key and pick on the remaining three edges > release the **Ctrl** key > press **RMB** > **Round** [Fig. 6.8(v)] > double-click on the dimension and modify to the design value of **.03125** [Fig. 6.8(w)] > **Enter** > **MMB** > **Ctrl+S** > **LMB** to deselect > **Ctrl+D**

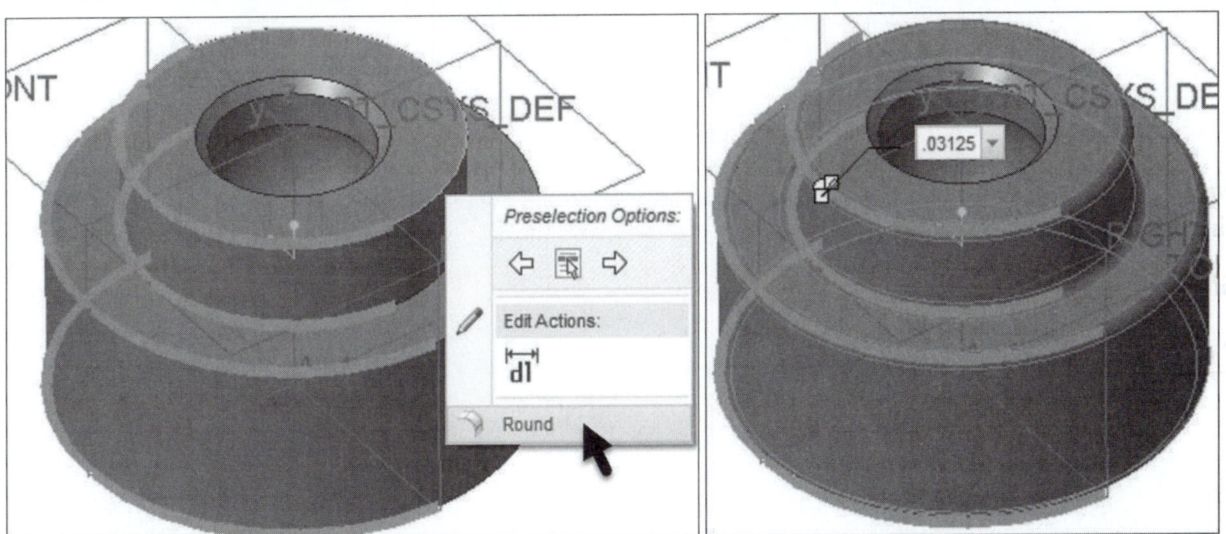

Figure 6.8(v) Round Edges

Figure 6.8(w) .03125 Round

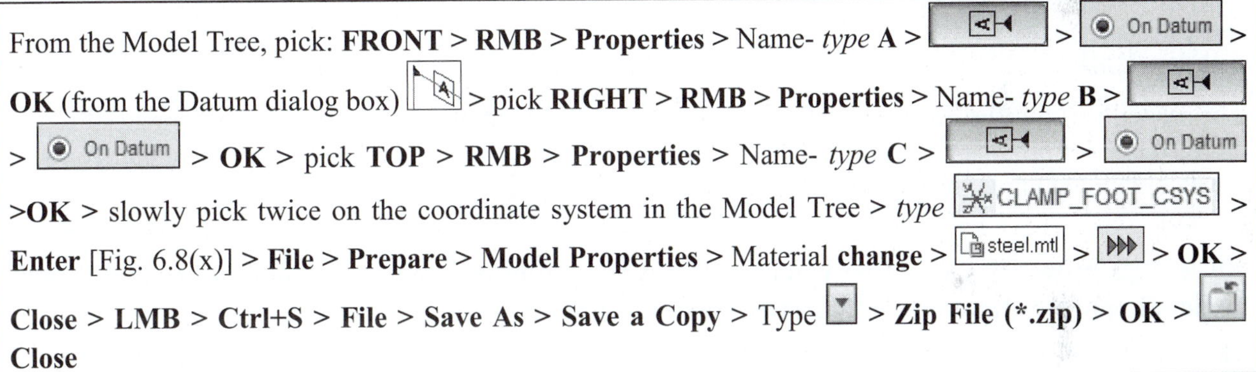

From the Model Tree, pick: **FRONT** > **RMB** > **Properties** > Name- *type* **A** > [icon] > On Datum > **OK** (from the Datum dialog box) [icon] > pick **RIGHT** > **RMB** > **Properties** > Name- *type* **B** > [icon] > On Datum > **OK** > pick **TOP** > **RMB** > **Properties** > Name- *type* **C** > [icon] > On Datum >**OK** > slowly pick twice on the coordinate system in the Model Tree > *type* CLAMP_FOOT_CSYS > **Enter** [Fig. 6.8(x)] > **File** > **Prepare** > **Model Properties** > Material **change** > steel.mtl > [icon] > **OK** > **Close** > **LMB** > **Ctrl+S** > **File** > **Save As** > **Save a Copy** > Type [icon] > **Zip File (*.zip)** > **OK** > [icon] **Close**

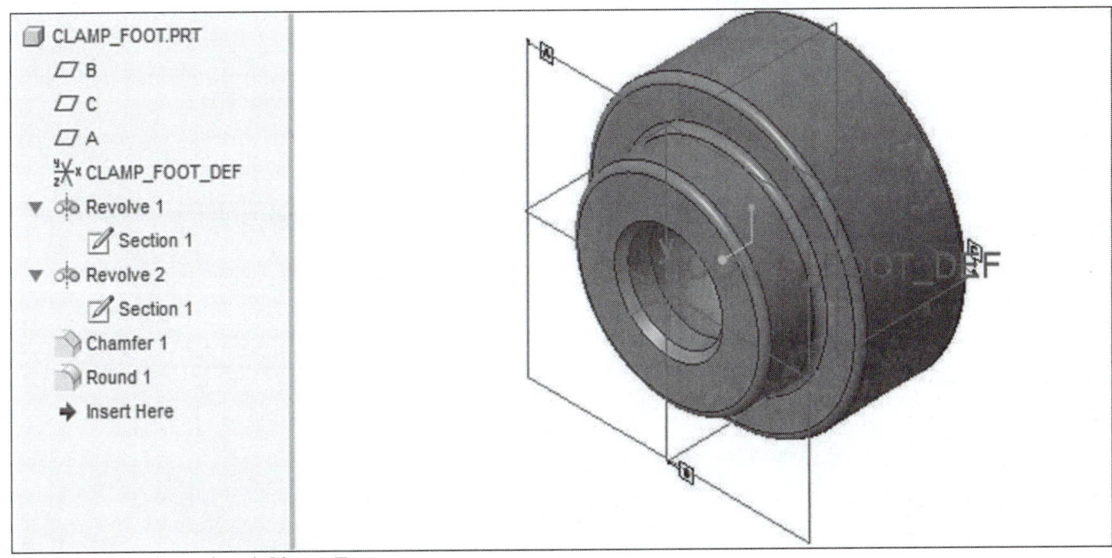

Figure 6.8(x) Completed Clamp Foot

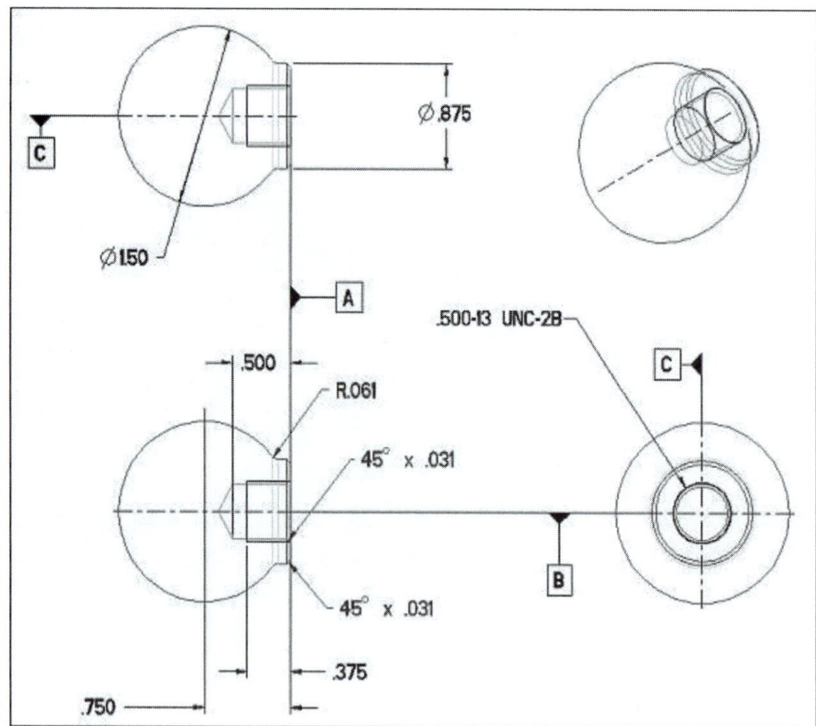

Figure 6.9 Clamp Ball Detail

Clamp Ball

The second part for this lesson is the Clamp Ball (Fig. 6.9). Much of the sketching is the same for this part as was done for the first revolved feature of the Clamp Foot. Instead of an internal revolved cut, you will create a standard hole. The Clamp Ball is made of *nylon*.

Click: New > Name-- *type* **clamp_ball** > **OK** > **File** > **Options** > **Configuration Editor** > **Import/Export** > **Import configuration file** > **Creo_textbook** [Fig. 6.10(a)] from your working directory (your saved file name may be different) > **Open** > **OK** > **No** > **File** > **Prepare** > **Model Properties** > Units **change** > **Inch lbm Second (Creo Parametric Default)** > **Close** > Material **change** > nylon.mtl > [Fig. 6.10(b)] > **OK** > **Close**

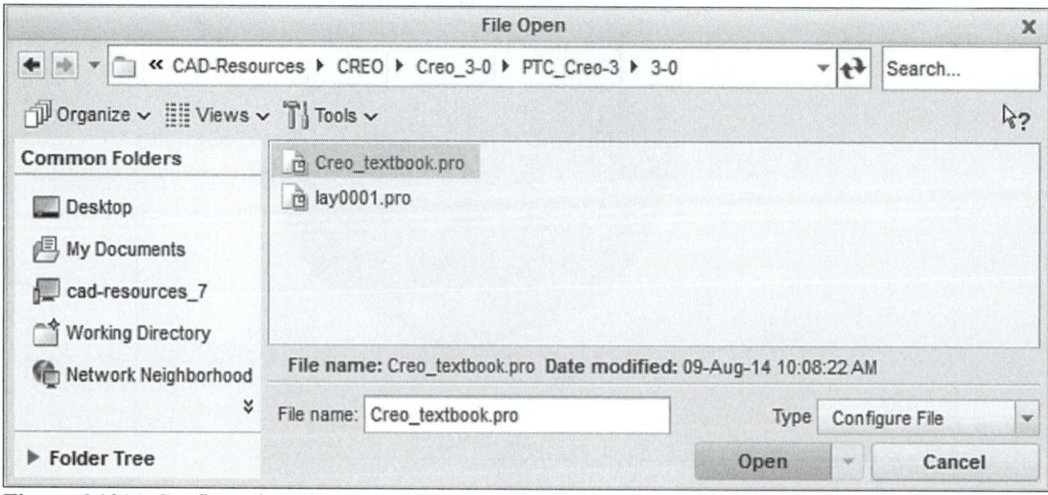

Figure 6.10(a) Configuration File

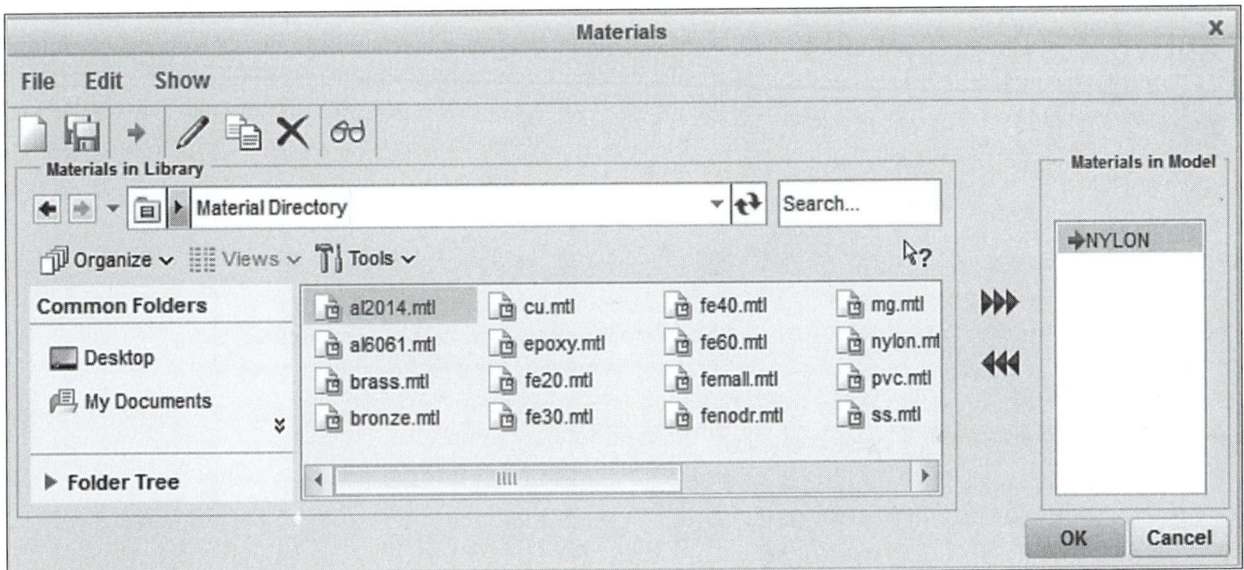

Figure 6.10(b) NYLON Material

Model Tree

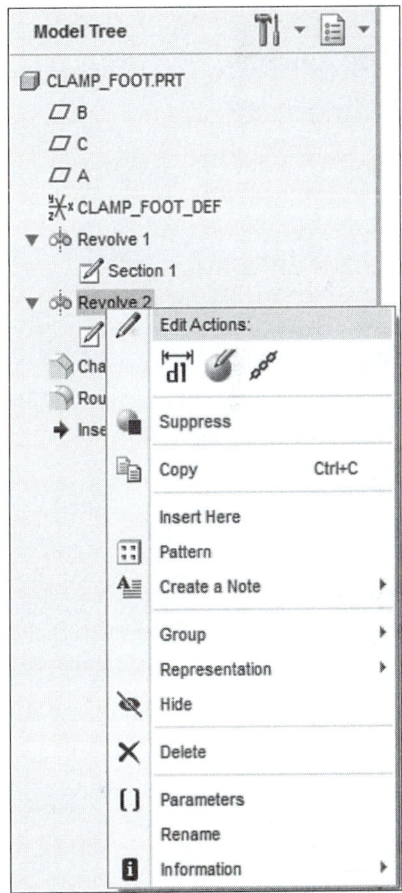

The Model Tree is a tabbed feature on the Creo Parametric navigator that contains a list of every feature (or component) in the current part, assembly, or drawing. The model structure is displayed in hierarchical (tree) format with the root object (the current feature, part or assembly) at the top of its tree and the subordinate objects (features, parts, or assemblies) below.

If you have multiple Creo Parametric windows open, the Model Tree contents reflect the file in the current "active" window. The Model Tree lists only the related feature- and part-level objects in a current file and does not list the entities (such as edges, surfaces, curves, and so forth) that comprise the features.

Each Model Tree item displays an icon that reflects its object type, for example, assembly, part, feature, or datum plane (also a feature). The icon can also show the display status for a feature, part, or assembly, for example, suppressed or hidden.

You can save the Model Tree as a .txt file. Selection in the Model Tree is object-action oriented; you select objects in the Model Tree without first specifying what you intend to do with them. Items can be added or removed from the Model Tree column display using Settings in the Navigator:

- Select objects, and perform object-specific operations on them using the shortcut menu.
- Filter the display by item type, for example, hiding or un-hiding datum features.
- Open a part within an assembly file by right-clicking the part in the Model Tree.
- Create or modify features and perform other operations such as deleting and redefining.
- Search the Model Tree for model properties or other feature information.
- Show the display status for an object, for example, suppressed or hidden.

Click: [Tools icon] > **Tree Filters** [Fig. 6.11(a)] > Display options *on* [Fig. 6.11(b)] > **Apply** > **OK**

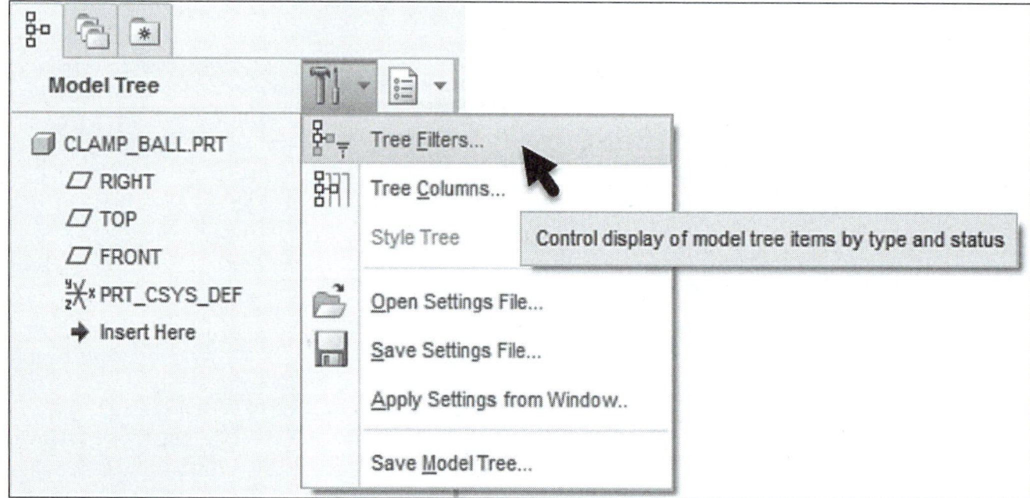

Figure 6.11(a) Tree Filters

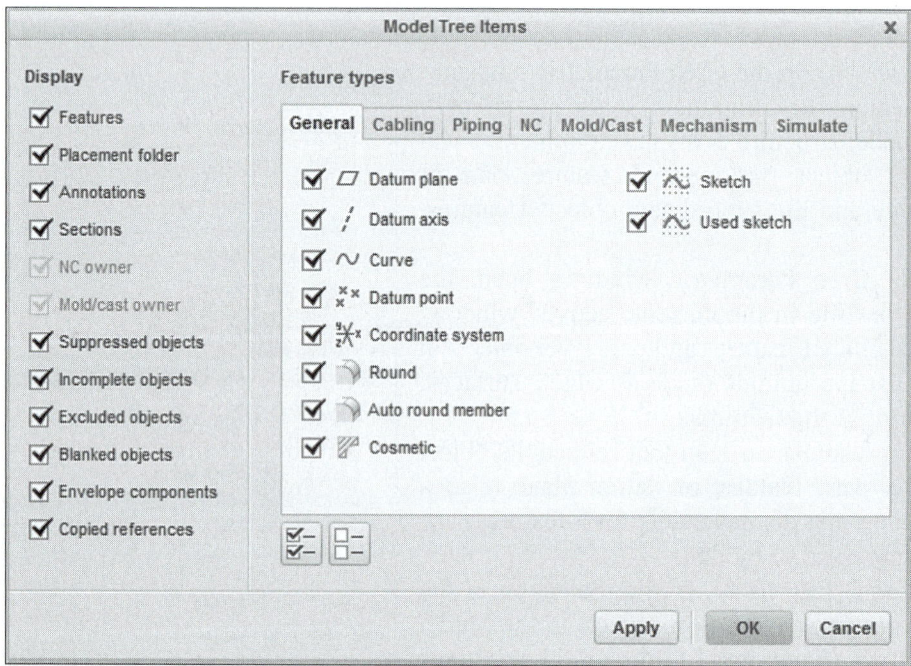

Figure 6.11(b) Model Tree Items Dialog Box

Click: [Tools icon] in the Navigator > **Tree Columns** [Fig. 6.12(a)] (the Model Tree Columns dialog opens)

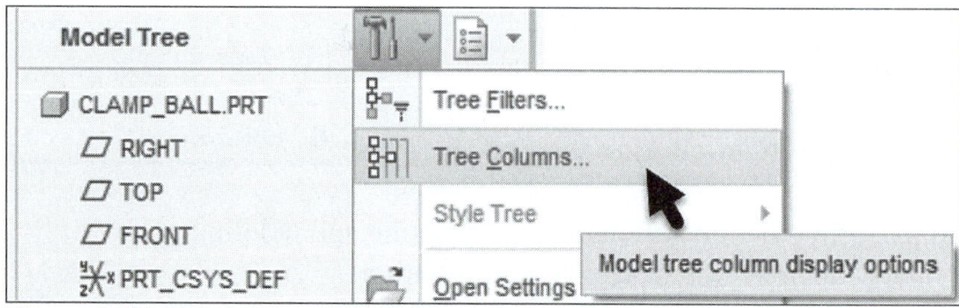

Figure 6.12(a) Model Tree Columns

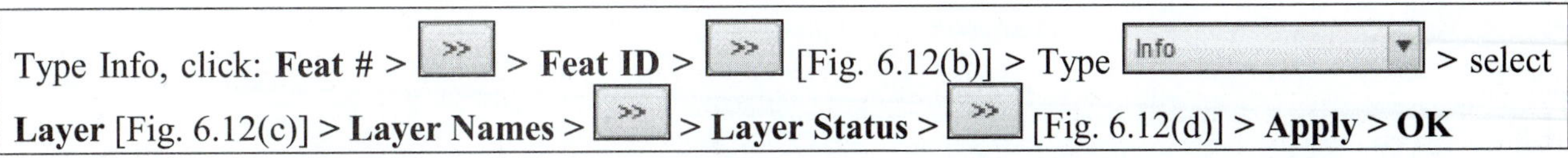
Type Info, click: **Feat #** > >> > **Feat ID** > >> [Fig. 6.12(b)] > Type Info > select **Layer** [Fig. 6.12(c)] > **Layer Names** > >> > **Layer Status** > >> [Fig. 6.12(d)] > **Apply** > **OK**

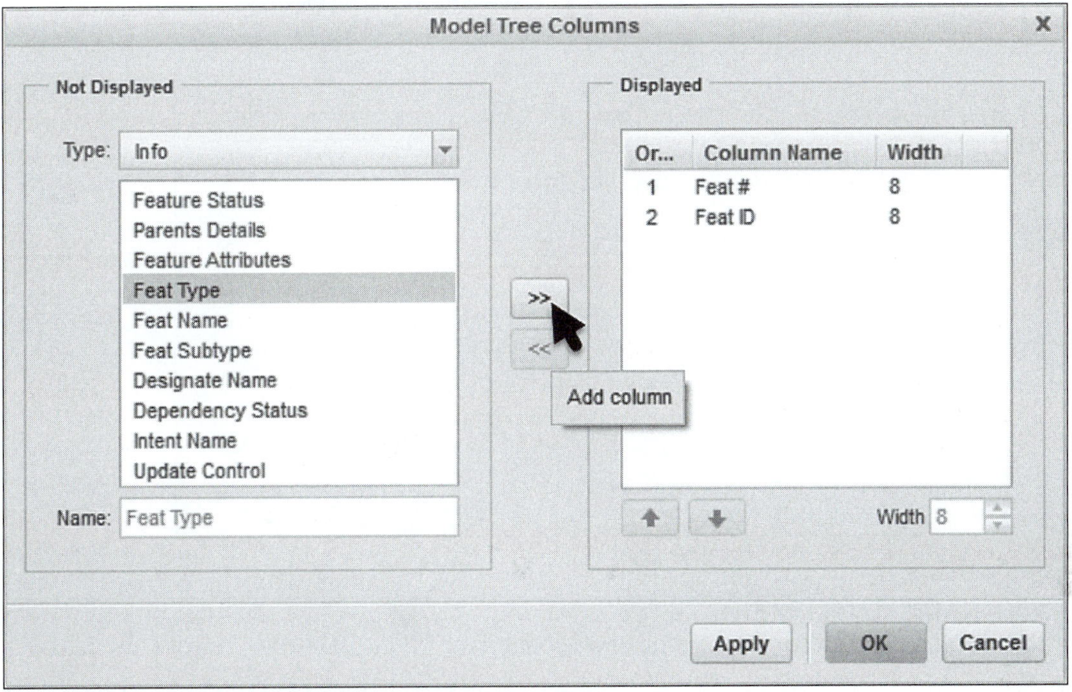

Figure 6.12(b) Model Tree Columns Dialog Box, Type: Info

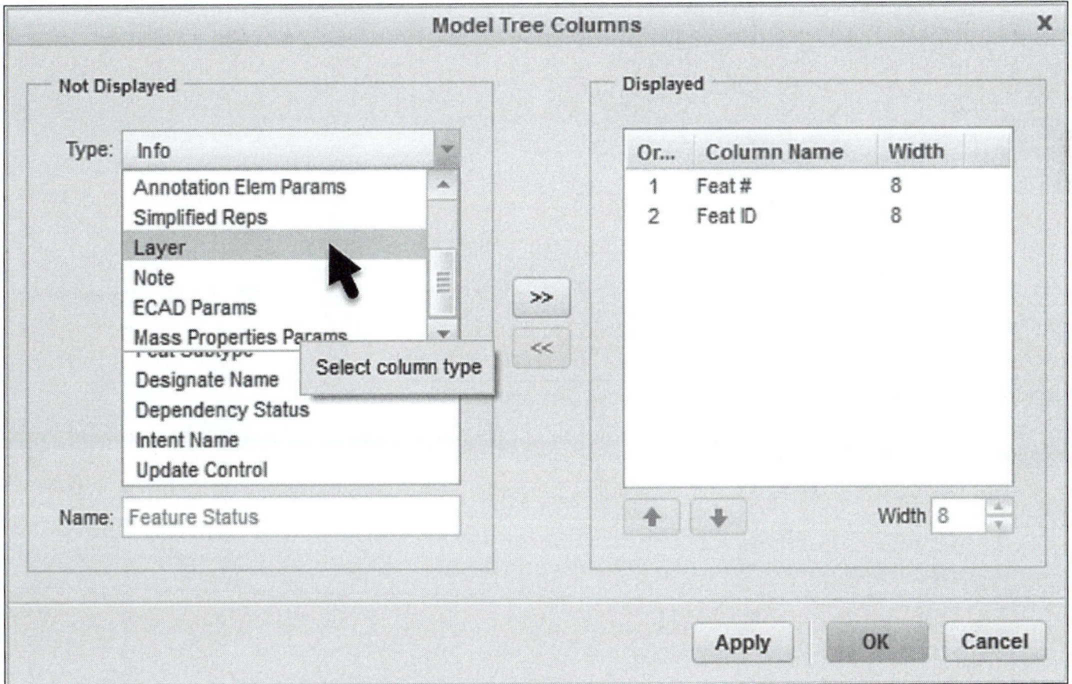

Figure 6.12(c) Model Tree Columns Dialog Box, Type: Layer

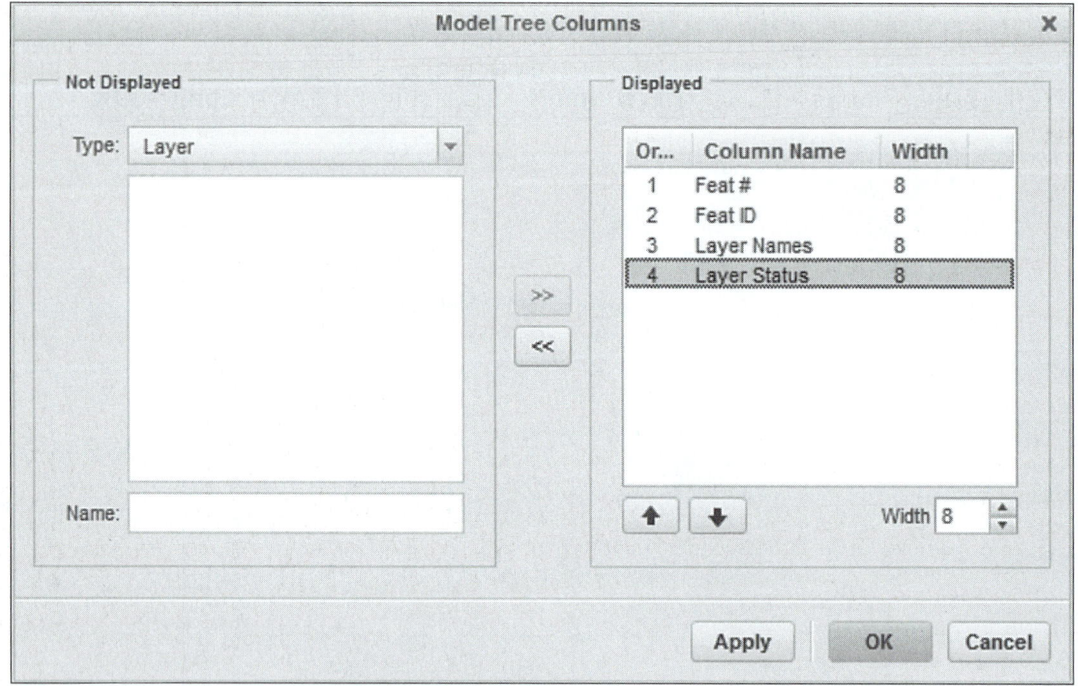

Figure 6.12(d) Model Tree Columns Dialog Box, Displayed List

Press: **LMB** on the sash and drag to expand the Model Tree [Fig. 6.12(e)] > release the **LMB** > adjust the width of each column by dragging the column divider Feat ID Layer Names > drag the sash to the left to decrease the Model Tree size [Fig. 6.12(f)]

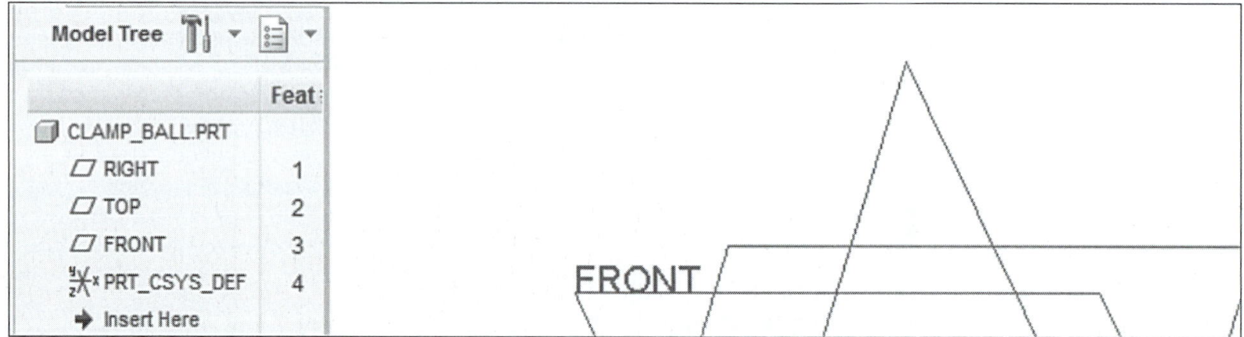

Figure 6.12(e) Expand the Model Tree and Adjust the Width of the Columns

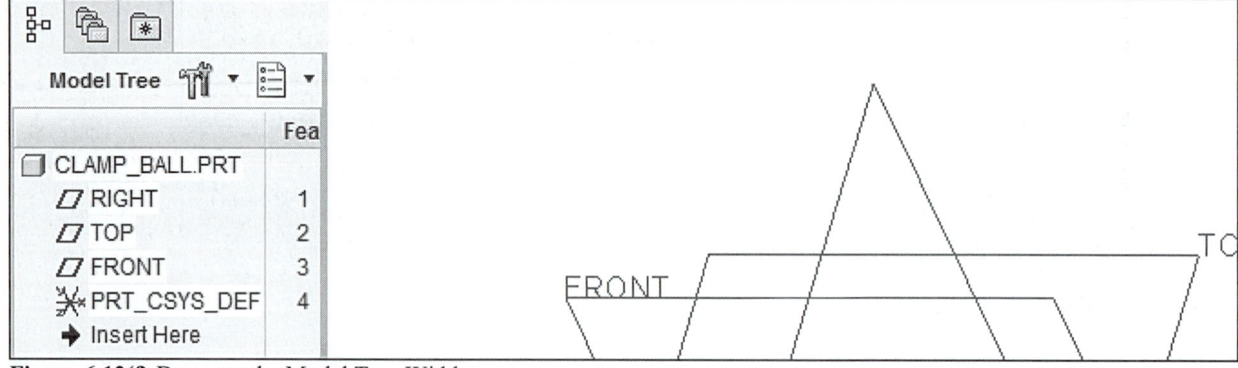

Figure 6.12(f) Decrease the Model Tree Width

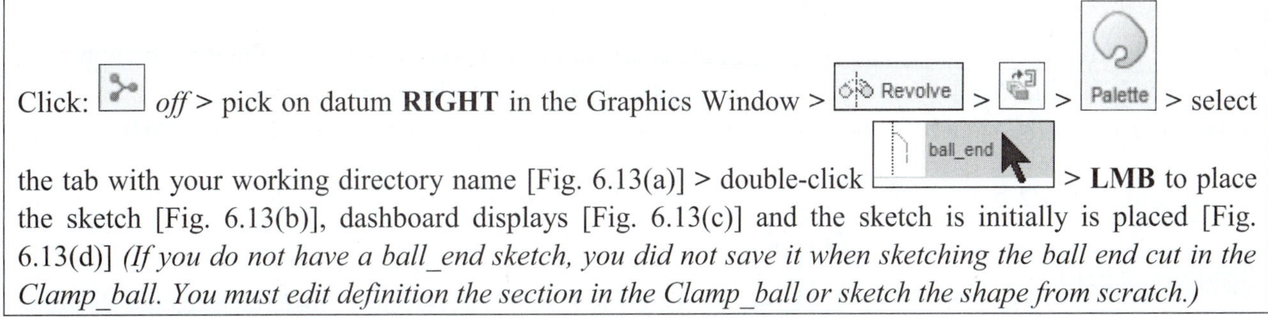

Click: [icon] *off* > pick on datum **RIGHT** in the Graphics Window > Revolve > [icon] > Palette > select the tab with your working directory name [Fig. 6.13(a)] > double-click ball_end > **LMB** to place the sketch [Fig. 6.13(b)], dashboard displays [Fig. 6.13(c)] and the sketch is initially is placed [Fig. 6.13(d)] *(If you do not have a ball_end sketch, you did not save it when sketching the ball end cut in the Clamp_ball. You must edit definition the section in the Clamp_ball or sketch the shape from scratch.)*

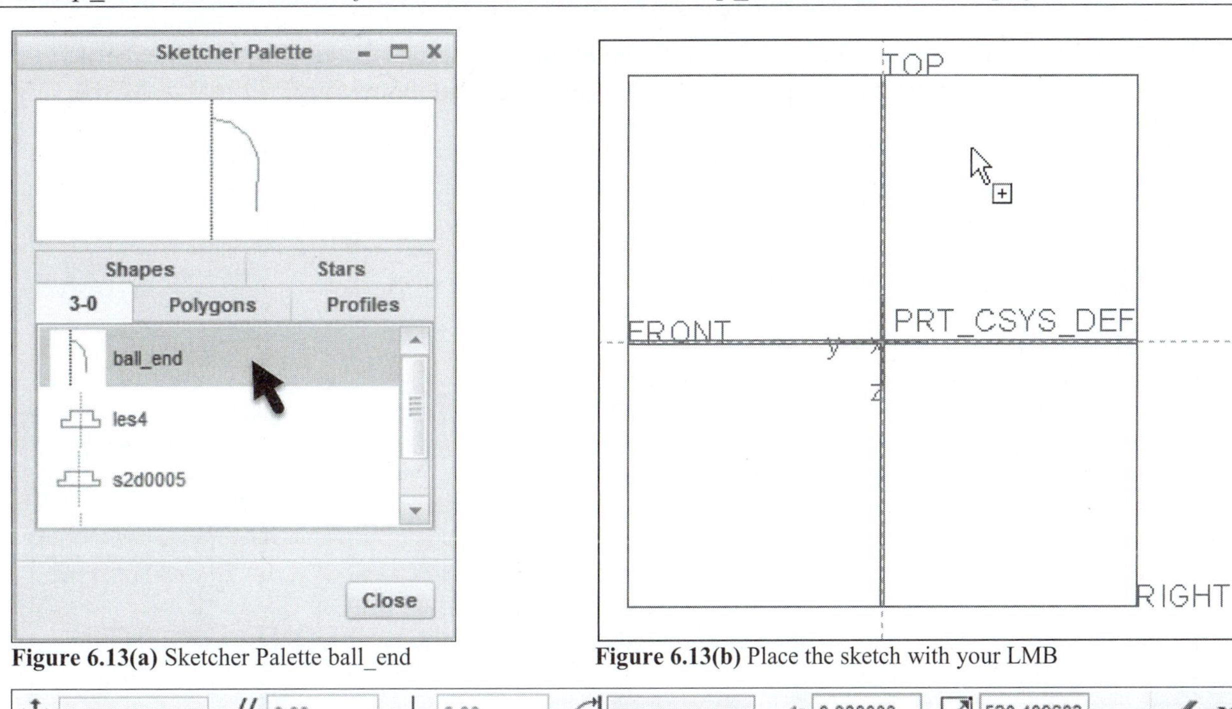

Figure 6.13(a) Sketcher Palette ball_end

Figure 6.13(b) Place the sketch with your LMB

Figure 6.13(c) Dashboard

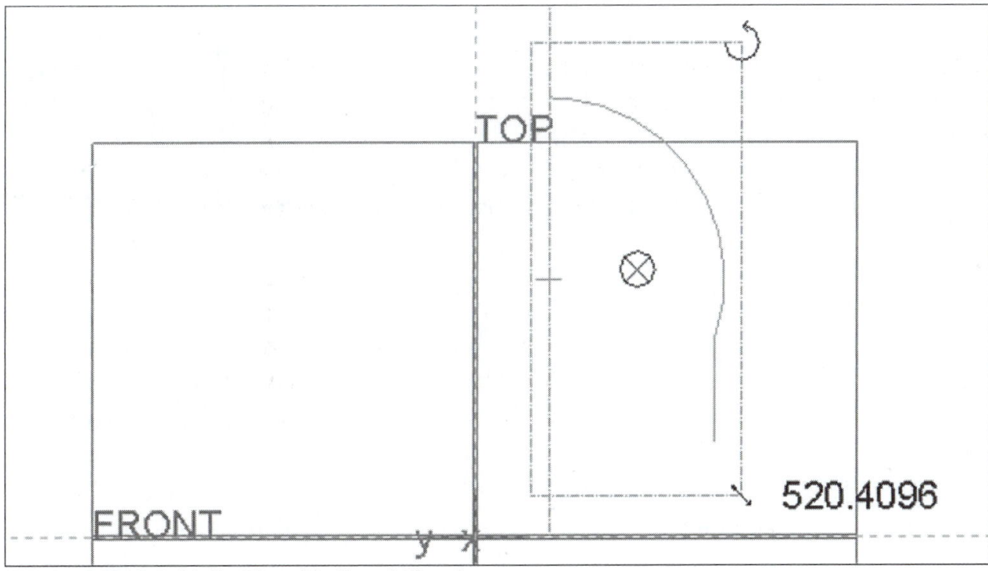

Figure 6.13(d) Initially Placed ball_end Sketch

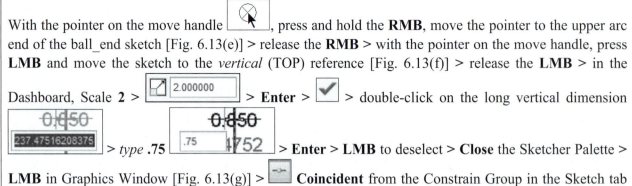

With the pointer on the move handle, press and hold the **RMB**, move the pointer to the upper arc end of the ball_end sketch [Fig. 6.13(e)] > release the **RMB** > with the pointer on the move handle, press **LMB** and move the sketch to the *vertical* (TOP) reference [Fig. 6.13(f)] > release the **LMB** > in the Dashboard, Scale **2** > > **Enter** > > double-click on the long vertical dimension > *type* **.75** > **Enter** > **LMB** to deselect > **Close** the Sketcher Palette > **LMB** in Graphics Window [Fig. 6.13(g)] > **Coincident** from the Constrain Group in the Sketch tab Ribbon > pick the **Line (Reference)** [Fig. 6.13(h)] > pick the **End Point** [Fig. 6.13(i)] > **MMB** [Fig. 6.13(j)]

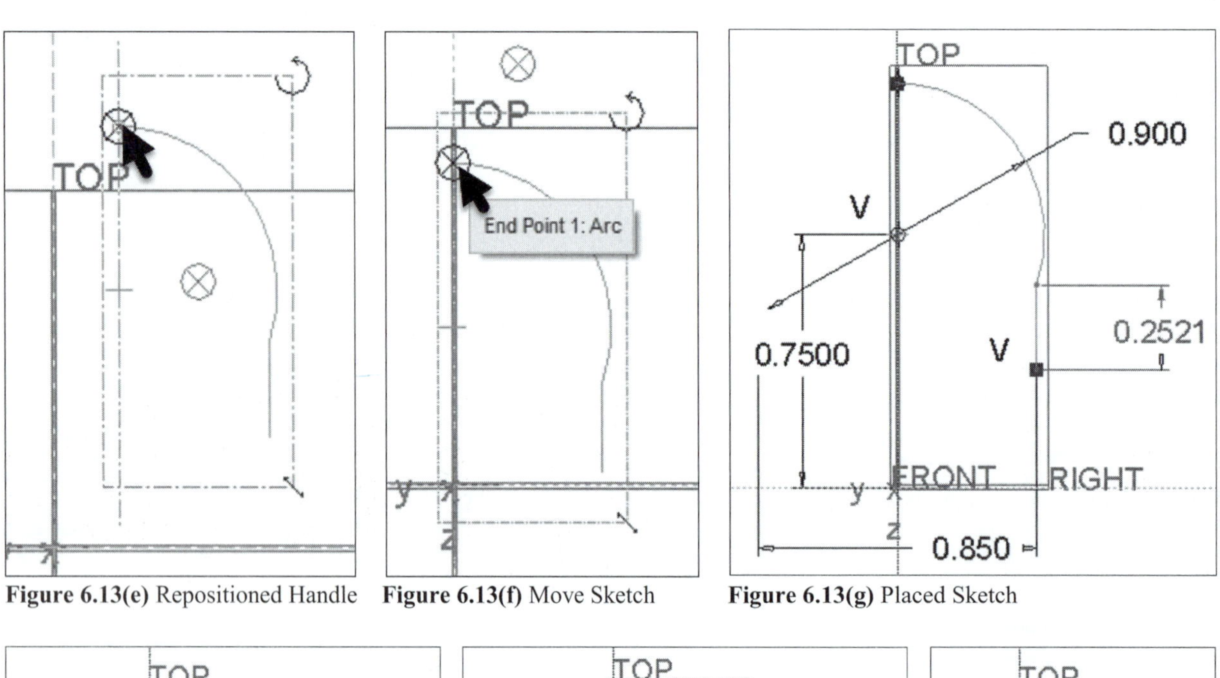

Figure 6.13(e) Repositioned Handle **Figure 6.13(f)** Move Sketch **Figure 6.13(g)** Placed Sketch

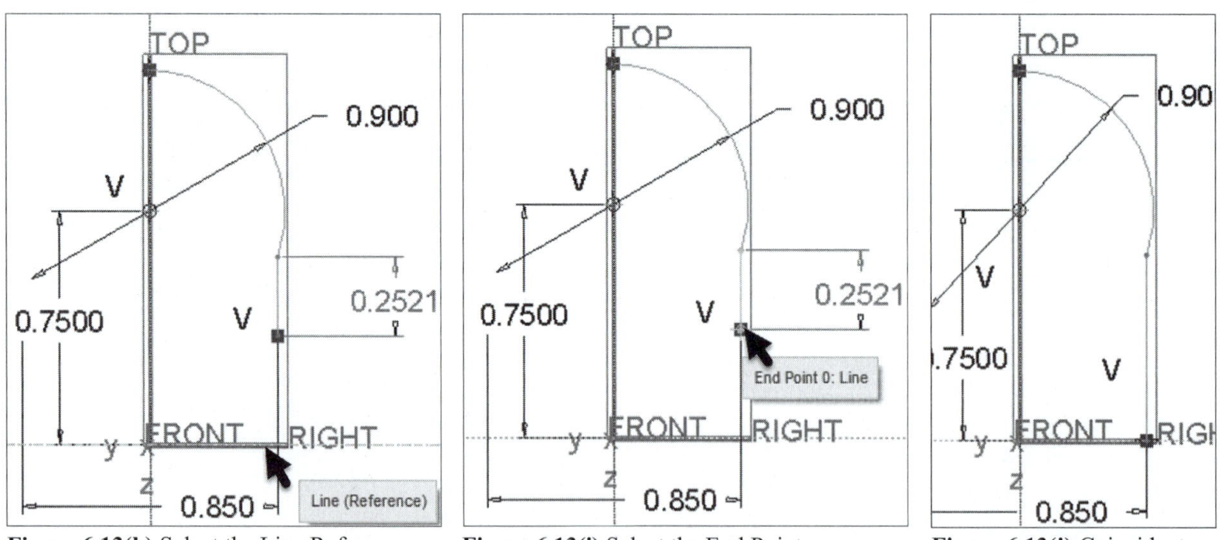

Figure 6.13(h) Select the Line Reference **Figure 6.13(i)** Select the End Point **Figure 6.13(j)** Coincident

Press: **RMB** > **Line Chain** sketch the *horizontal* and *vertical* lines to close the section (it will shade if done correctly) [Fig. 6.13(k)] > **MMB** > **MMB** > **LMB** > double-click on each dimension and modify to the design values (Fig. 6.9) of **1.500** diameter, **.875** diameter, and **.750** > **LMB** to deselect > reposition the dimensions [Fig. 6.13(l)] > > **Ctrl+D** [Fig. 6.13(m)] >

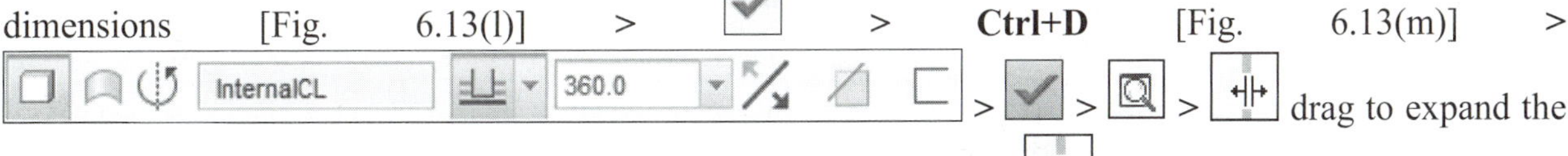

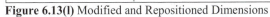

> > > drag to expand the Model Tree window (to the right) to see the columns > double-click to shut the Model Tree window > in the Graphics Window, **LMB** to deselect > **Ctrl+S** > **Enter** [Fig. 6.13(n)]

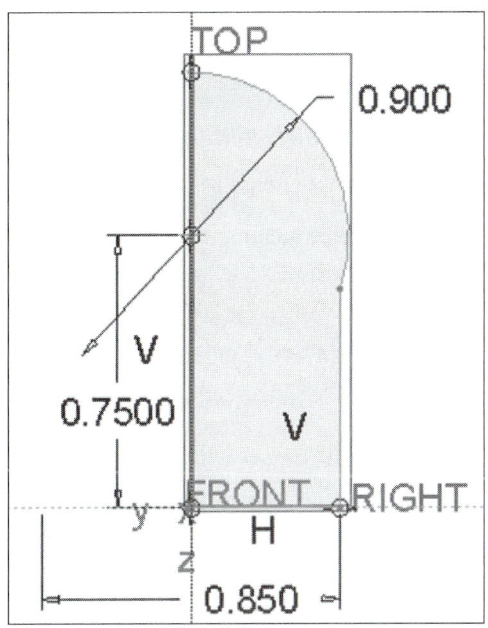

Figure 6.13(k) Closed Sketch

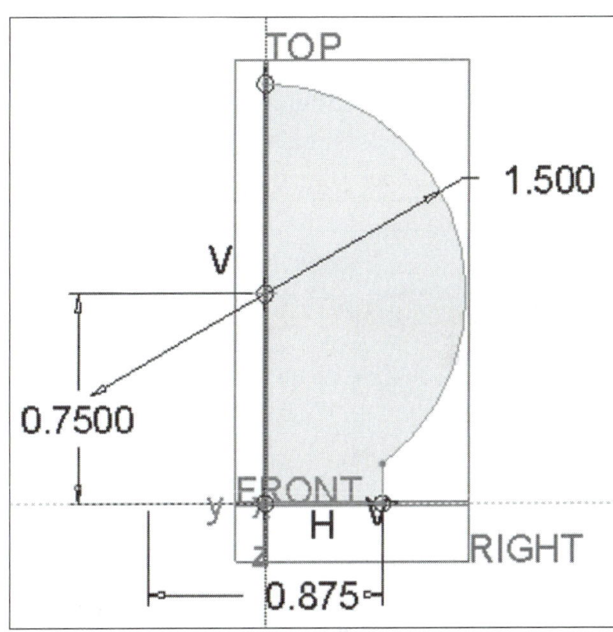

Figure 6.13(l) Modified and Repositioned Dimensions

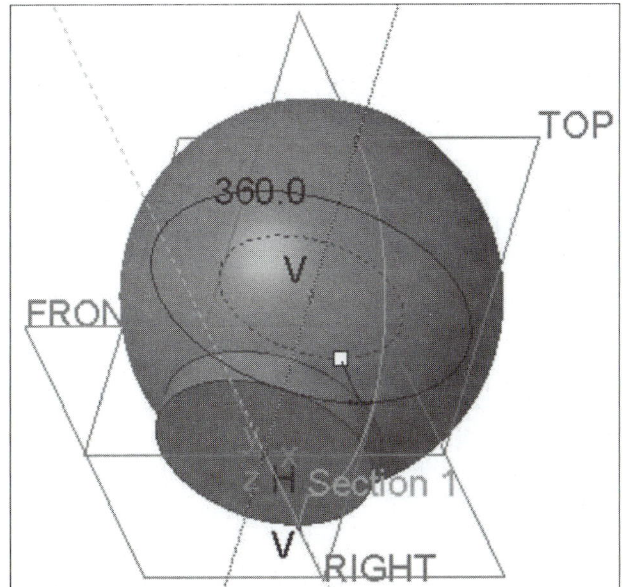

Figure 6.13(m) Previewed Revolved Feature

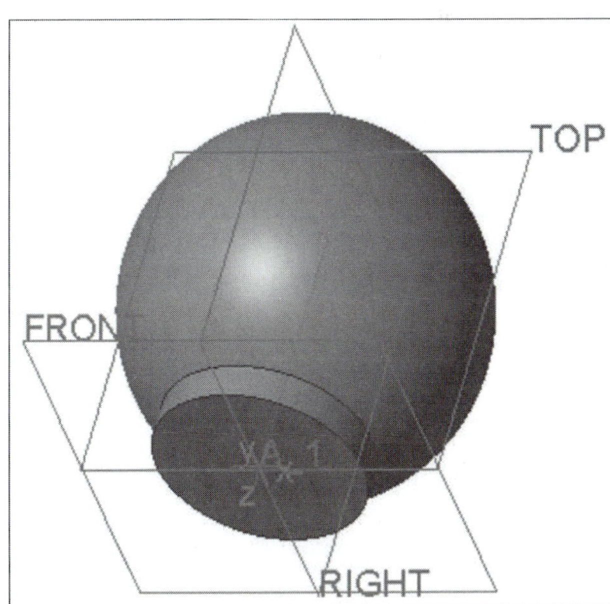

Figure 6.13(n) Completed Feature

Standard Holes

Hole charts are used to lookup diameters for a given fastener size. You can create custom hole charts and specify their directory location with the configuration file option *hole_parameter_file_path.* UNC, UNF and ISO hole charts are supplied with Creo Parametric. Create a standard **.500-13 UNC-2B** hole, **.375** thread depth, **.50** tap drill, and a standard chamfer.

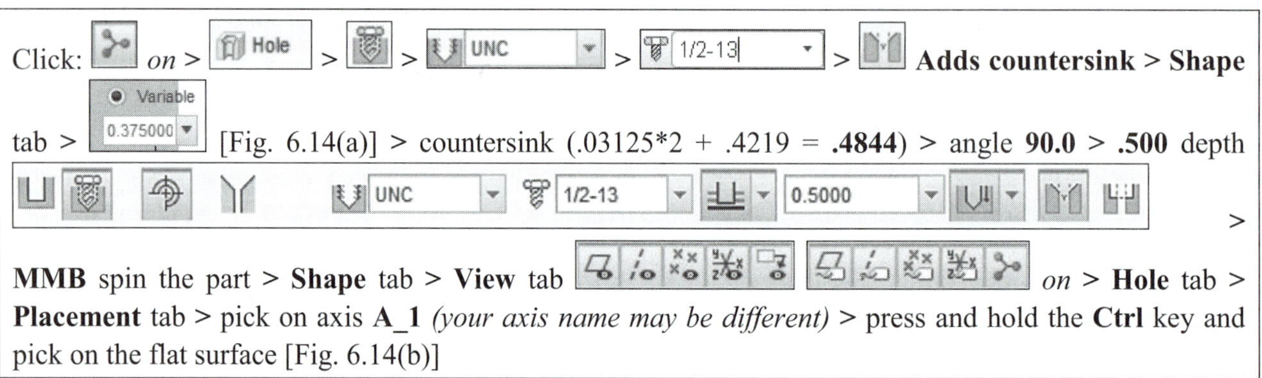

Click: *on* > Hole > > > UNC > 1/2-13 > **Adds countersink** > **Shape** tab > Variable 0.375000 [Fig. 6.14(a)] > countersink (.03125*2 + .4219 = **.4844**) > angle **90.0** > **.500** depth >

MMB spin the part > **Shape** tab > **View** tab *on* > **Hole** tab > **Placement** tab > pick on axis **A_1** *(your axis name may be different)* > press and hold the **Ctrl** key and pick on the flat surface [Fig. 6.14(b)]

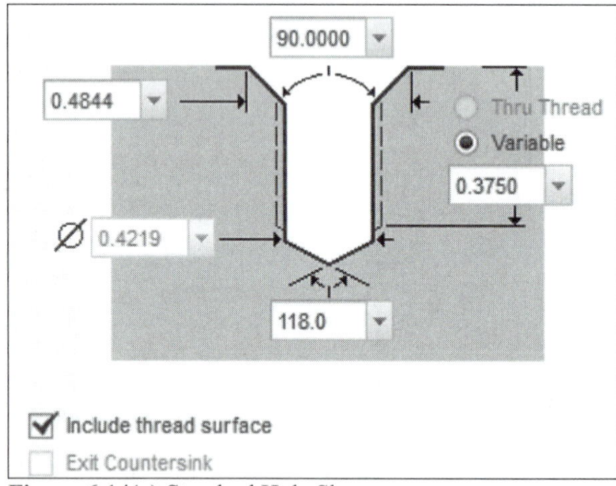

Figure 6.14(a) Standard Hole Shape

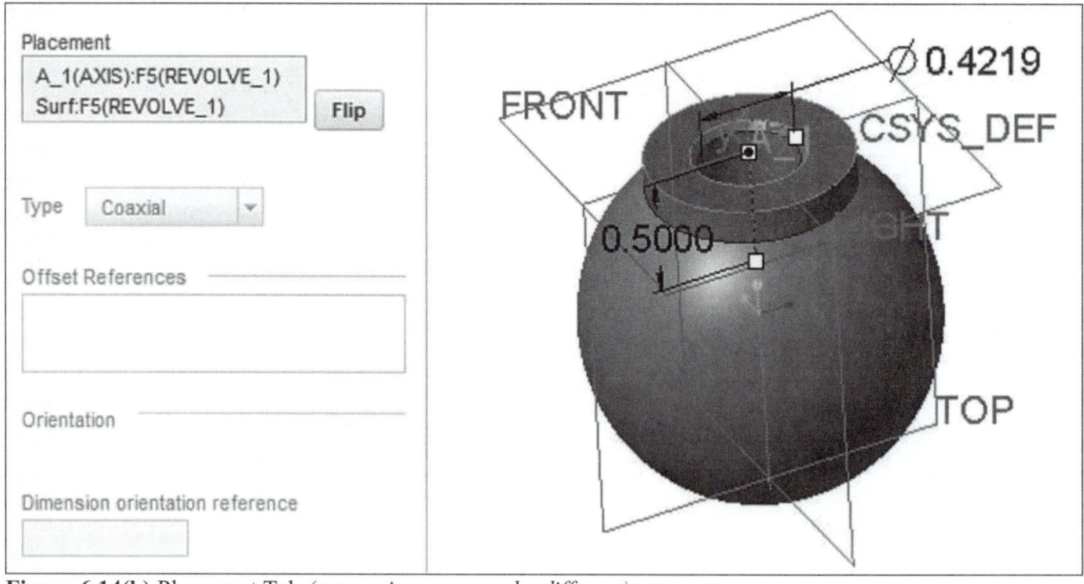

Figure 6.14(b) Placement Tab *(your axis name may be different)*

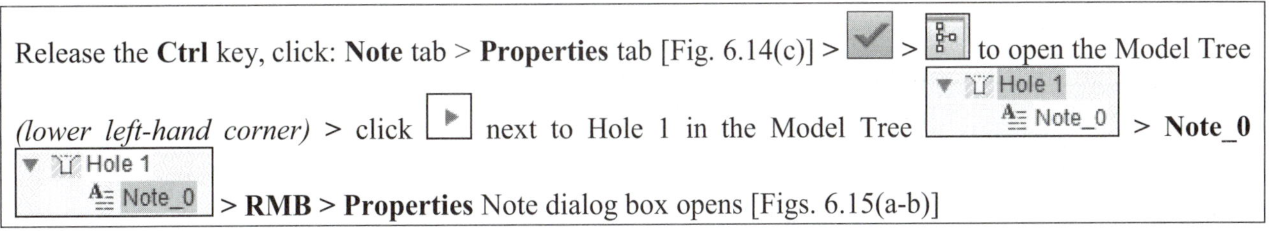

Release the **Ctrl** key, click: **Note** tab > **Properties** tab [Fig. 6.14(c)] > ✔ > to open the Model Tree *(lower left-hand corner)* > click ▶ next to Hole 1 in the Model Tree > **Note_0** > **RMB** > **Properties** Note dialog box opens [Figs. 6.15(a-b)]

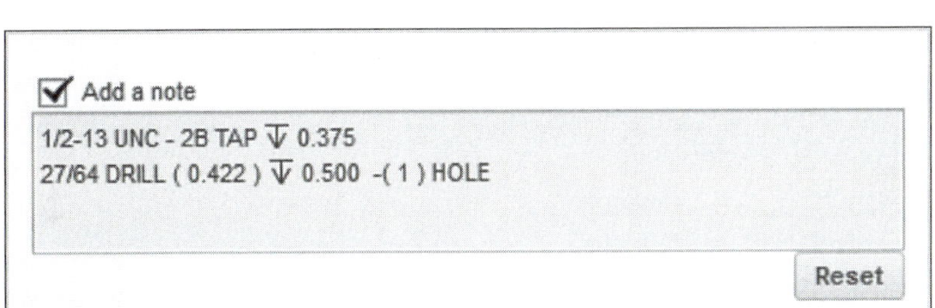

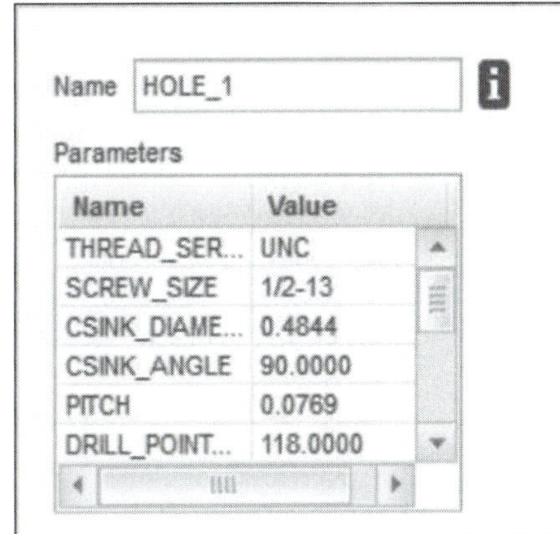

Figure 6.14(c) Note and Properties tabs

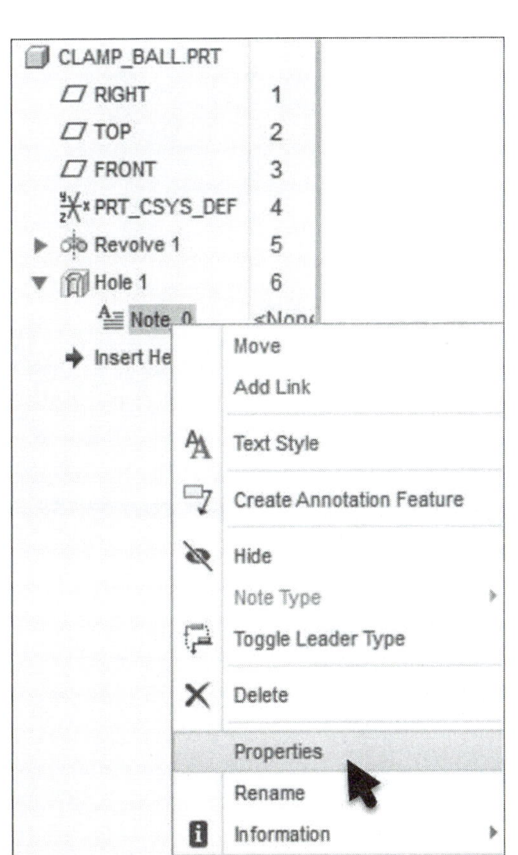

Figure 6.15(a) Note Properties

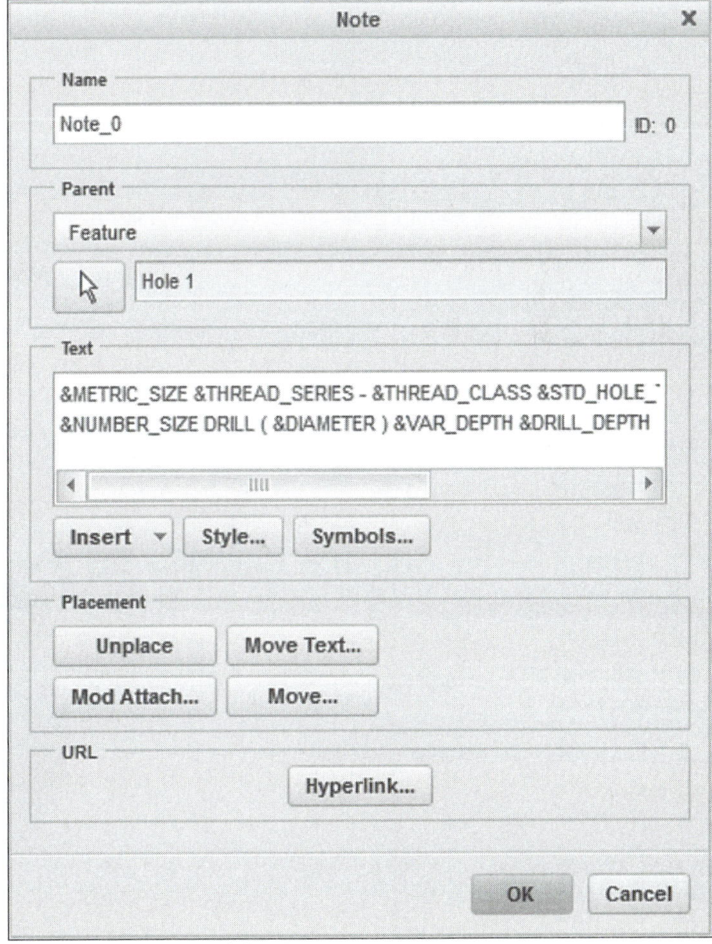

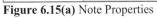

Figure 6.15(b) Note Dialog Box

In the Note Dialog Box, modify the Name and Text fields [Fig. 6.15(c)]. Move the mouse pointer within the Text field in front of &DIAMETER, click: **Symbols** > ⌀ add the diameter symbol [Fig. 6.15(d)] > **OK** from the Note dialog box > **MMB** spin the model [Fig. 6.15(e)] > **Ctrl+D** > **Ctrl+S**

Figure 6.15(c) Modified Note

Figure 6.15(d) Text Symbol Dialog Box

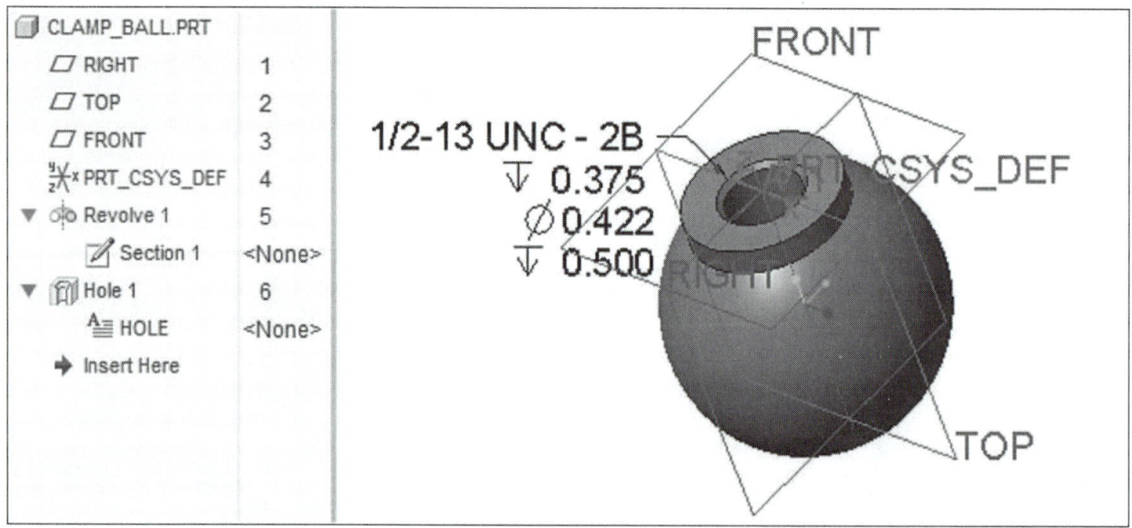

Figure 6.15(e) Annotation

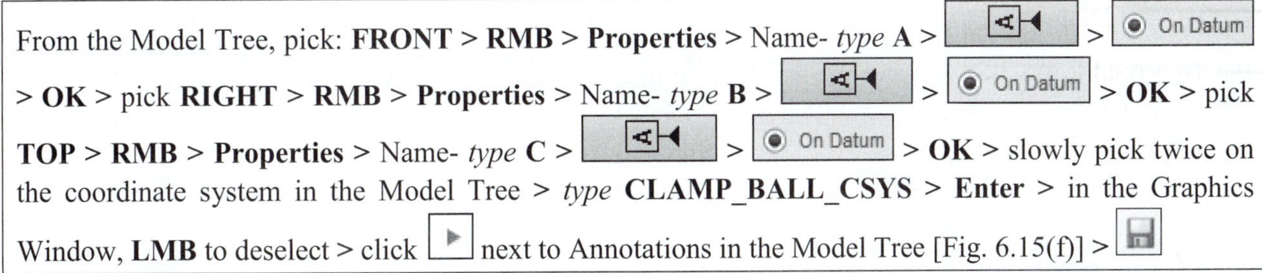

From the Model Tree, pick: **FRONT** > **RMB** > **Properties** > Name- *type* **A** > [icon] > On Datum > **OK** > pick **RIGHT** > **RMB** > **Properties** > Name- *type* **B** > [icon] > On Datum > **OK** > pick **TOP** > **RMB** > **Properties** > Name- *type* **C** > [icon] > On Datum > **OK** > slowly pick twice on the coordinate system in the Model Tree > *type* **CLAMP_BALL_CSYS** > **Enter** > in the Graphics Window, **LMB** to deselect > click [icon] next to Annotations in the Model Tree [Fig. 6.15(f)] > [icon]

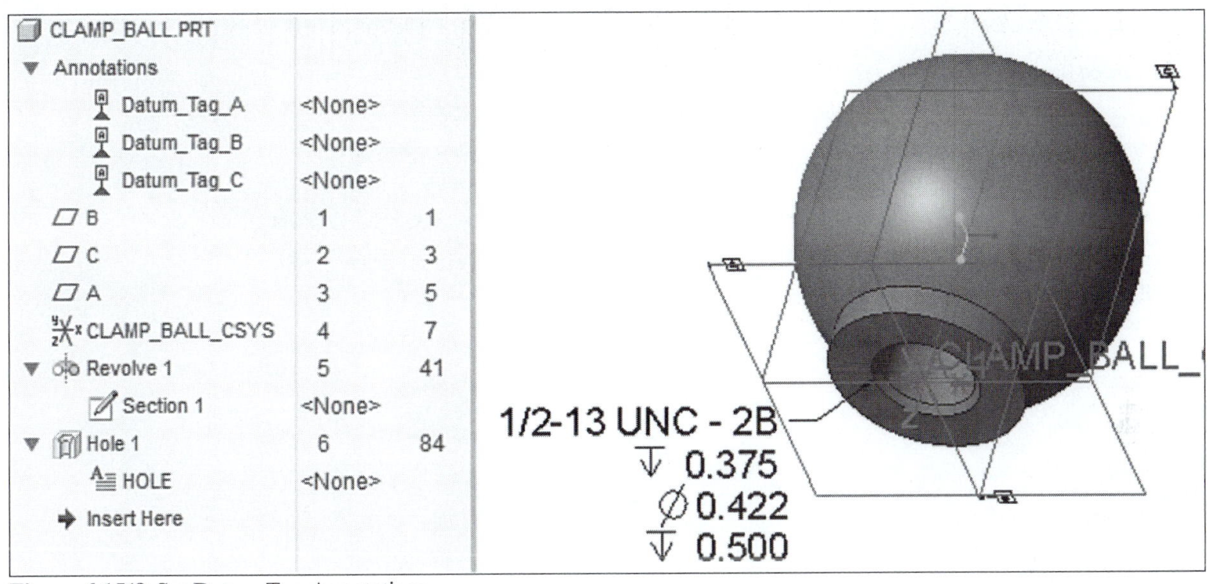

Figure 6.15(f) Set Datum Tag Annotations

Double-click on the model to display the dimensions for the protrusion > pick on the **∅.875** dimension > press **RMB** [Fig. 6.16(a)] > **Properties** Dimension Properties dialog box displays [Fig. 6.16(b)]

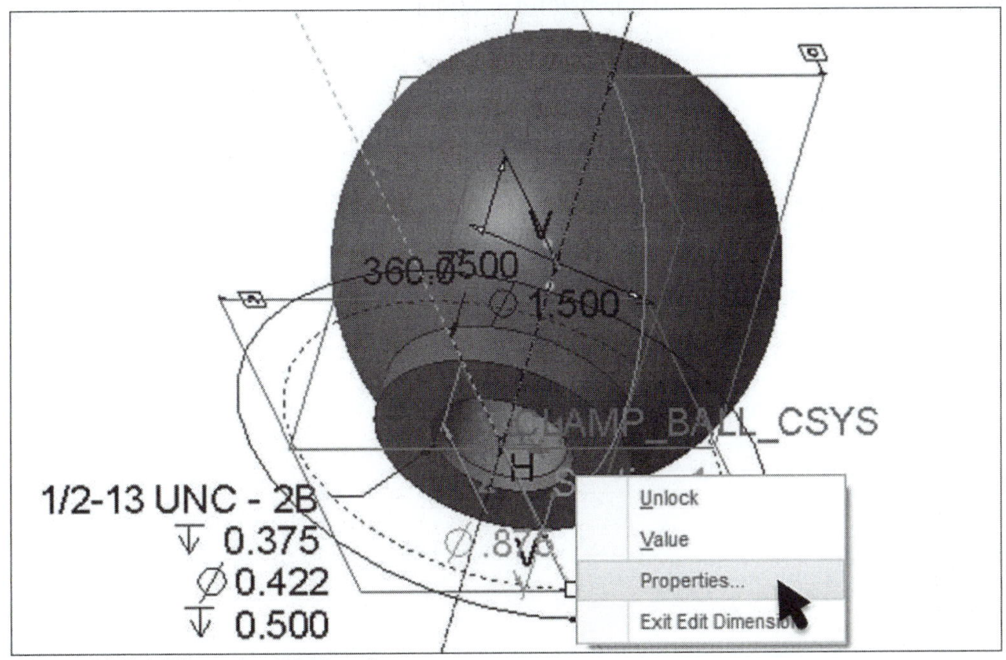

Figure 6.16(a) Dimension Displayed

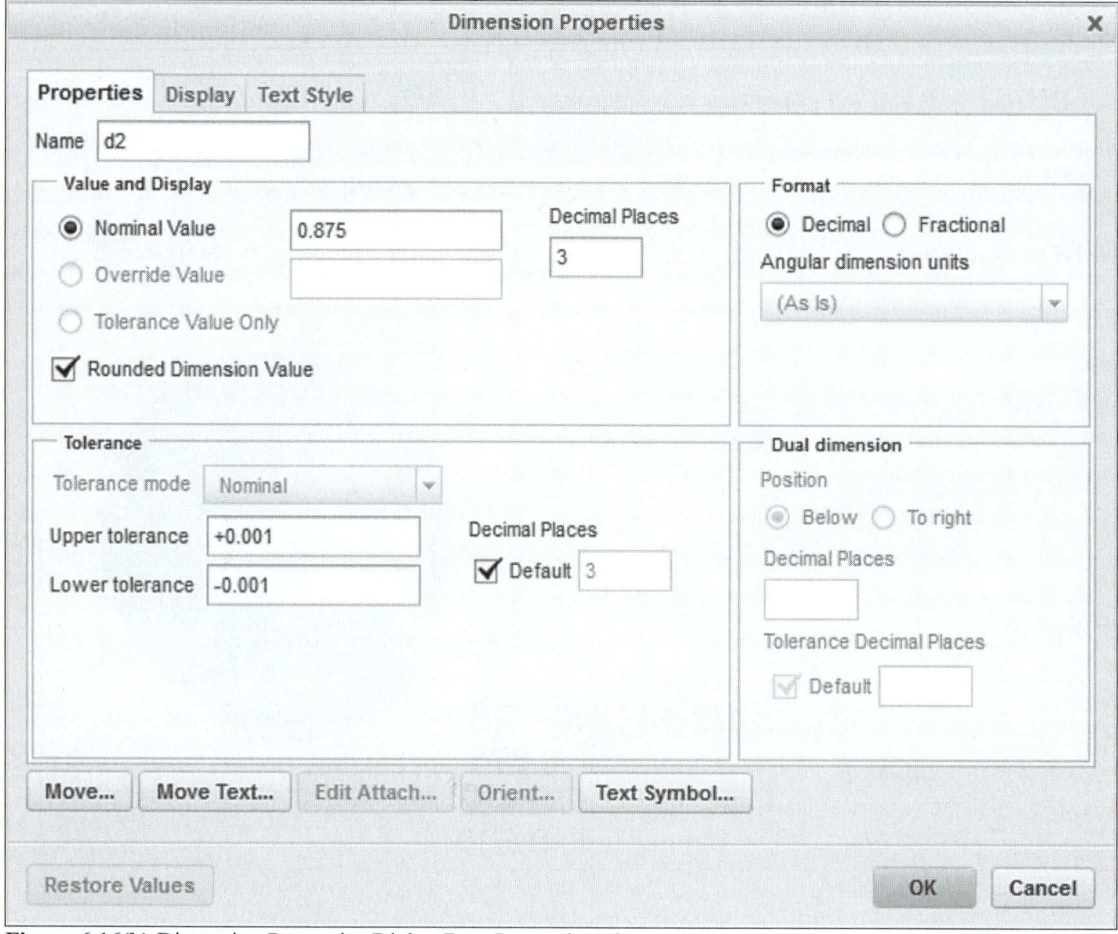

Figure 6.16(b) Dimension Properties Dialog Box, Properties tab

The **Display** tab [Fig. 6.16(c)] shows the parametric dimension symbol. The **Text Style** tab [Fig. 6.16(d)] provides options for Character, and Note/Dimension variations.

Figure 6.16(c) Display Tab

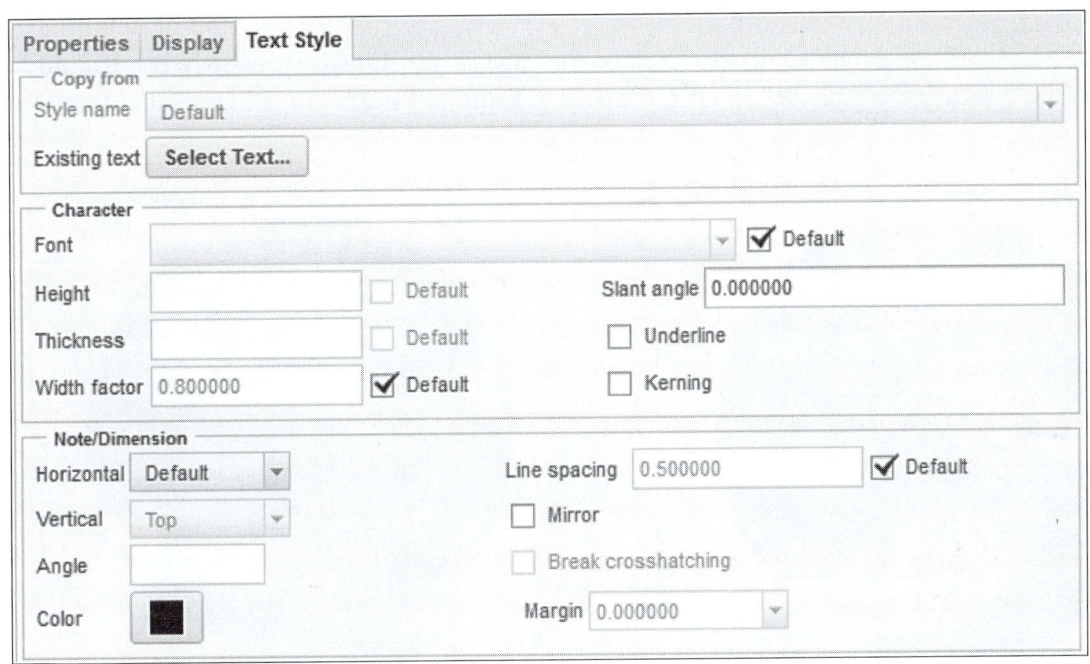

Figure 6.16(d) Dimension Properties, Text Style Tab

Click: **Move** > select a new position for the **Ø.875** dimension > **OK** > repeat to move the other dimensions as required *(pick on a dimension > press **RMB** > **Properties** > **Move** > select a new position > **OK**)* [Fig. 6.16(e)] > double-click **LMB** twice

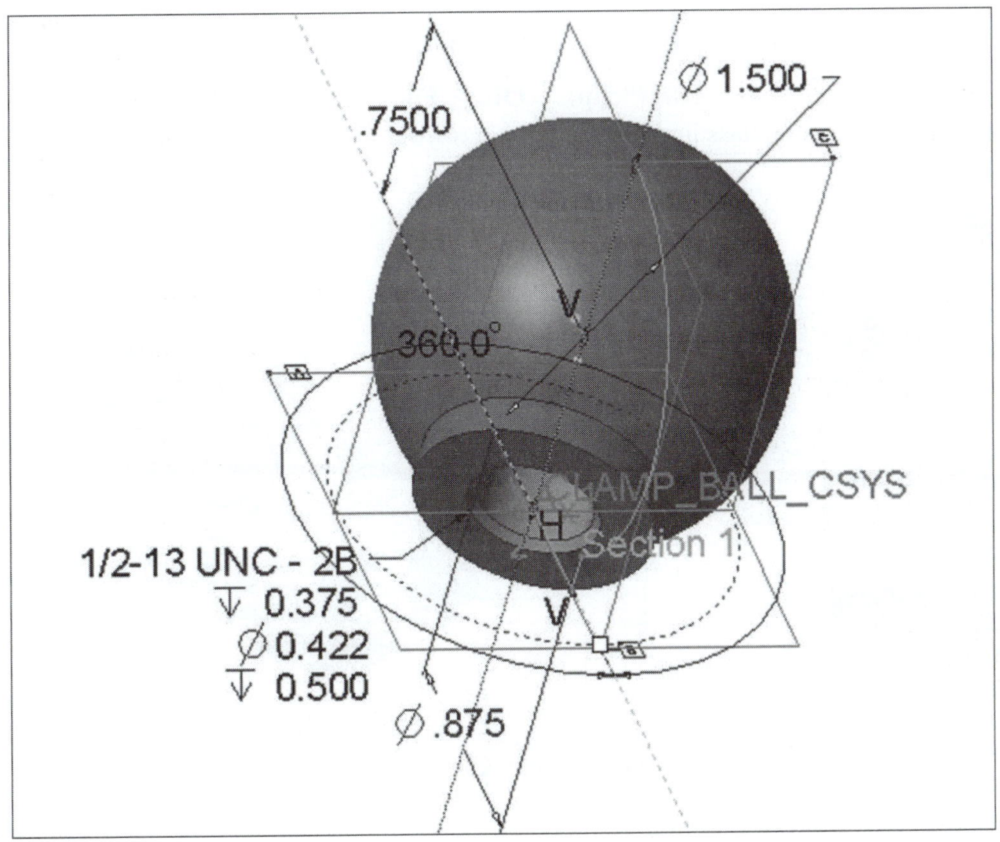

Figure 6.16(e) Moved Dimension

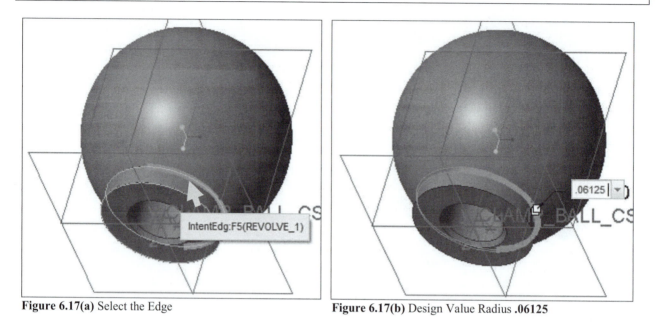

Click: **View** tab > [icon] off > pick on the model > pick on the edge [Fig. 6.17(a)] > press **RMB** > **Round** > double-click on the dimension > *type* **.06125** [Fig. 6.17(b)] > **Enter** > **Enter** > in the Graphics Window, **LMB** to deselect

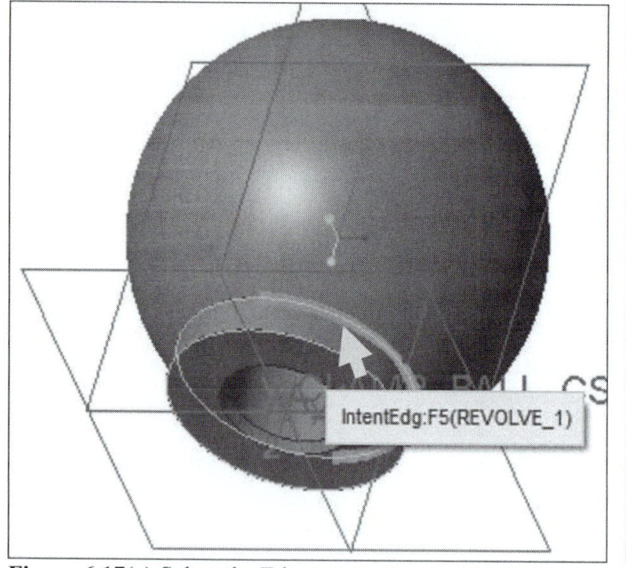

Figure 6.17(a) Select the Edge

Figure 6.17(b) Design Value Radius **.06125**

Click: **Model** tab > [Chamfer] **Chamfer** > pick on the edge [Fig. 6.18(a)] >

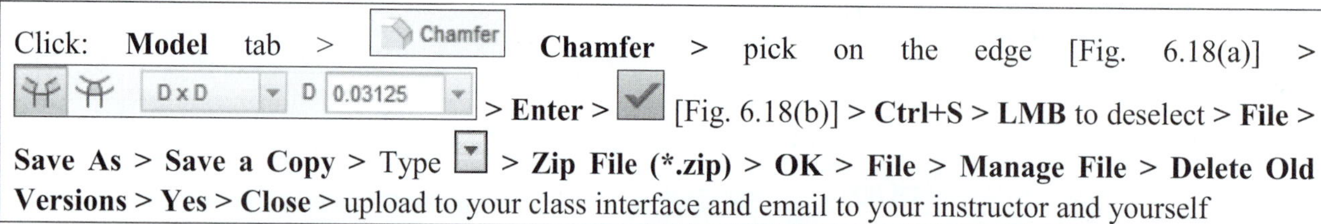

> **Enter** > [✓] [Fig. 6.18(b)] > **Ctrl+S** > **LMB** to deselect > **File** > **Save As** > **Save a Copy** > Type [▾] > **Zip File (*.zip)** > **OK** > **File** > **Manage File** > **Delete Old Versions** > **Yes** > **Close** > upload to your class interface and email to your instructor and yourself

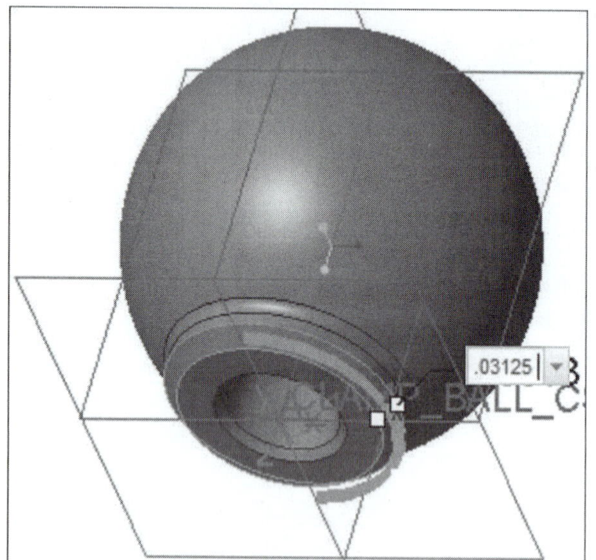

Figure 6.18(a) Chamfer **.03125**

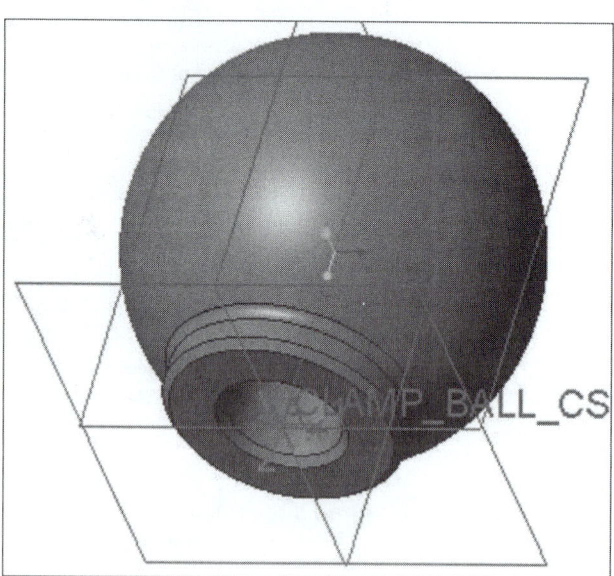

Figure 6.18(b) Completed Chamfer

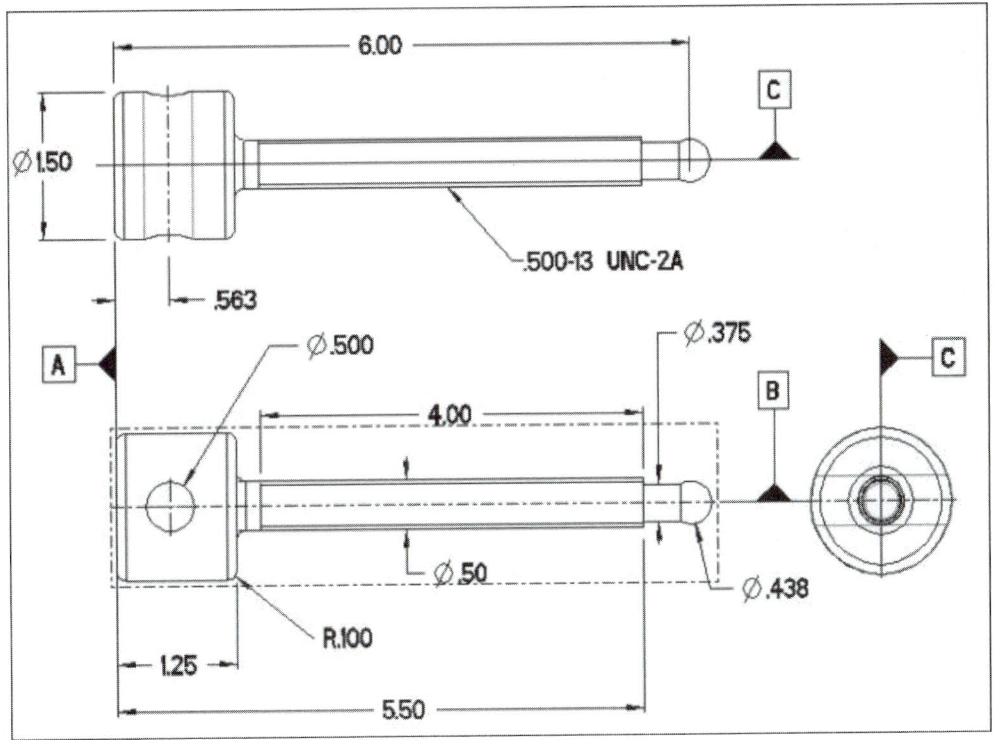

Figure 6.19 Clamp Swivel Detail

Clamp Swivel

The Clamp Swivel is the third part created by revolving one section about a centerline (Fig. 6.19). The Clamp Swivel is a component of the Clamp Assembly (Fig. 6.20).

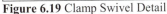

Click: **Create a new model** > Name-- *type* **clamp_swivel** > **OK**

(If you have just started a new session of PTC Creo Parametric 3.0, remember to load your configuration file > ***File > Options > Configuration Editor > Import/Export > Import configuration file*** *> click on* **Creo_textbook** *which was previously created and saved (your name could be different) >* ***Open > OK > No)***

> **File** > **Prepare** > **Model Properties** > Units **change** *(check the unit setting* Inch lbm Second*)* > **Close**

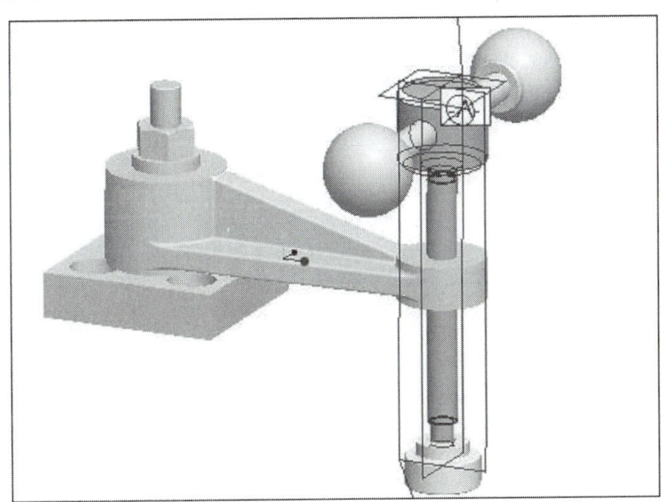
Figure 6.20 Clamp Swivel Highlighted in Clamp Assembly

Click: Material **change** > Create new material icon **Create new material** [Fig. 6.21(a)] > in Name field, *type* **STEEL_1066** > fill in the form [Fig. 6.21(b)] > **Save To Library…** > navigate to your working directory > File Name: **STEEl_1066 > OK**

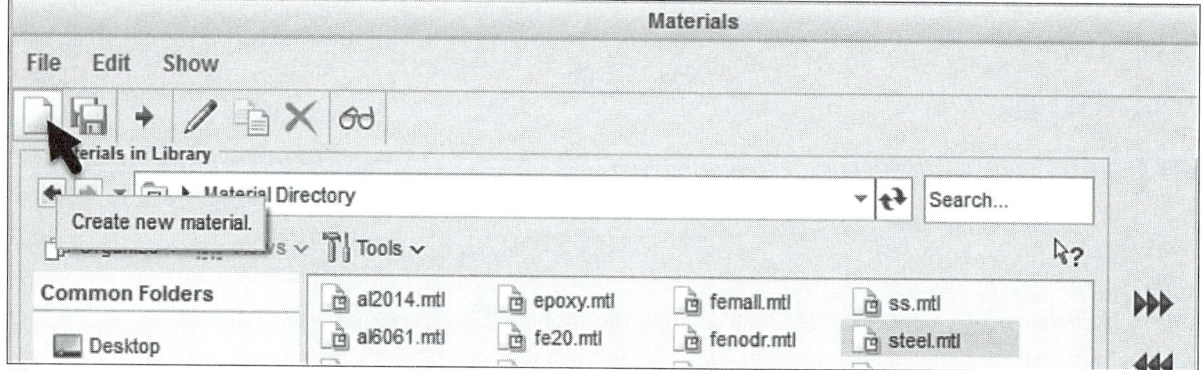

Figure 6.21(a) Materials Dialog Box

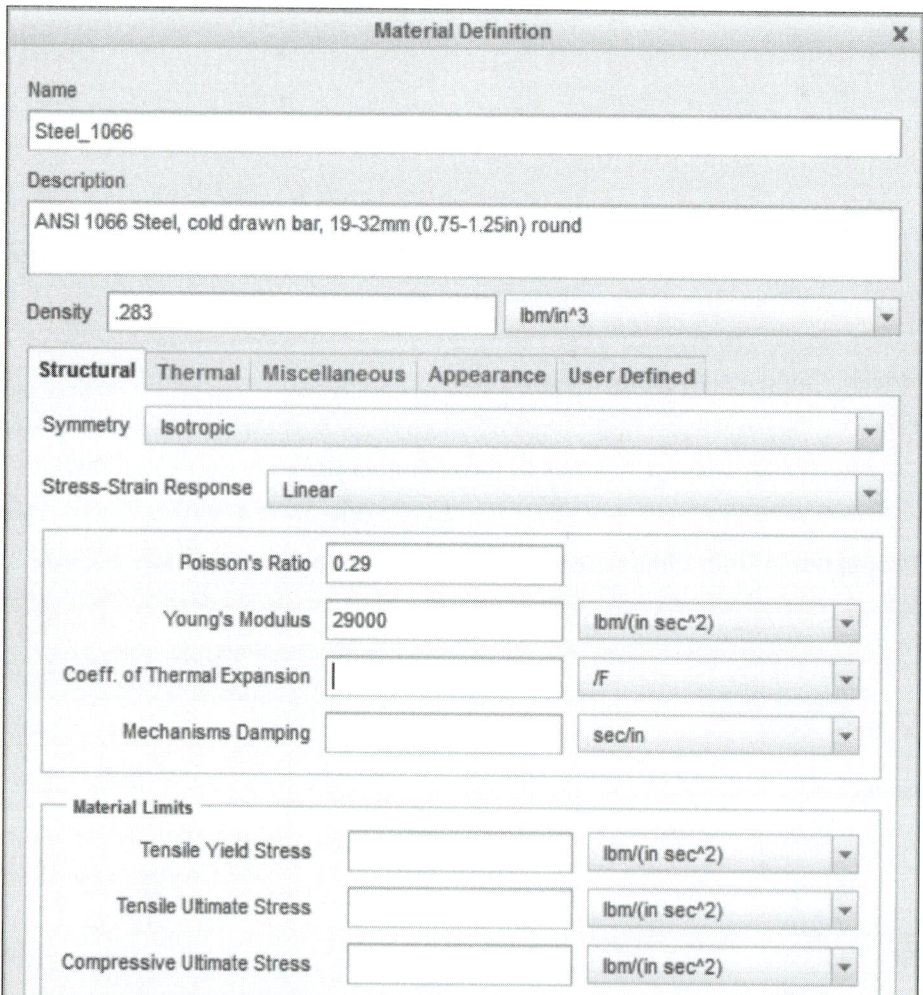

Figure 6.21(b) Material Definition Dialog Box

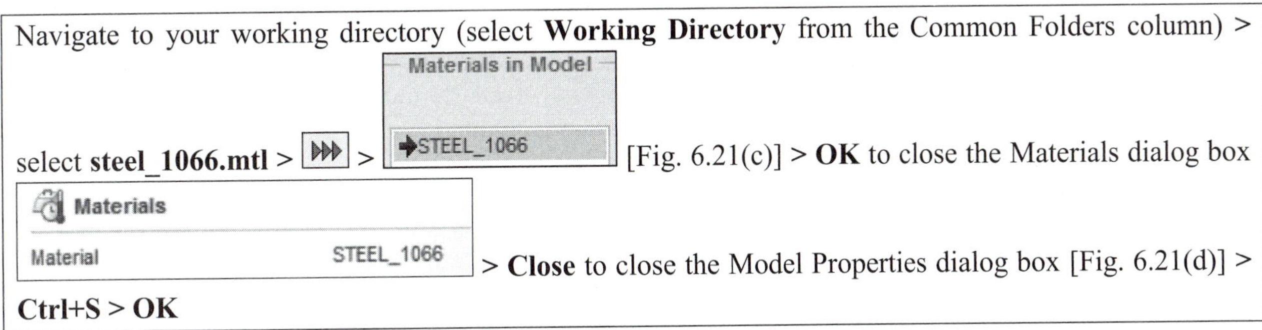

Navigate to your working directory (select **Working Directory** from the Common Folders column) > select **steel_1066.mtl** > ⏩ > Materials in Model ➔STEEL_1066 [Fig. 6.21(c)] > **OK** to close the Materials dialog box Materials Material STEEL_1066 > **Close** to close the Model Properties dialog box [Fig. 6.21(d)] > **Ctrl+S** > **OK**

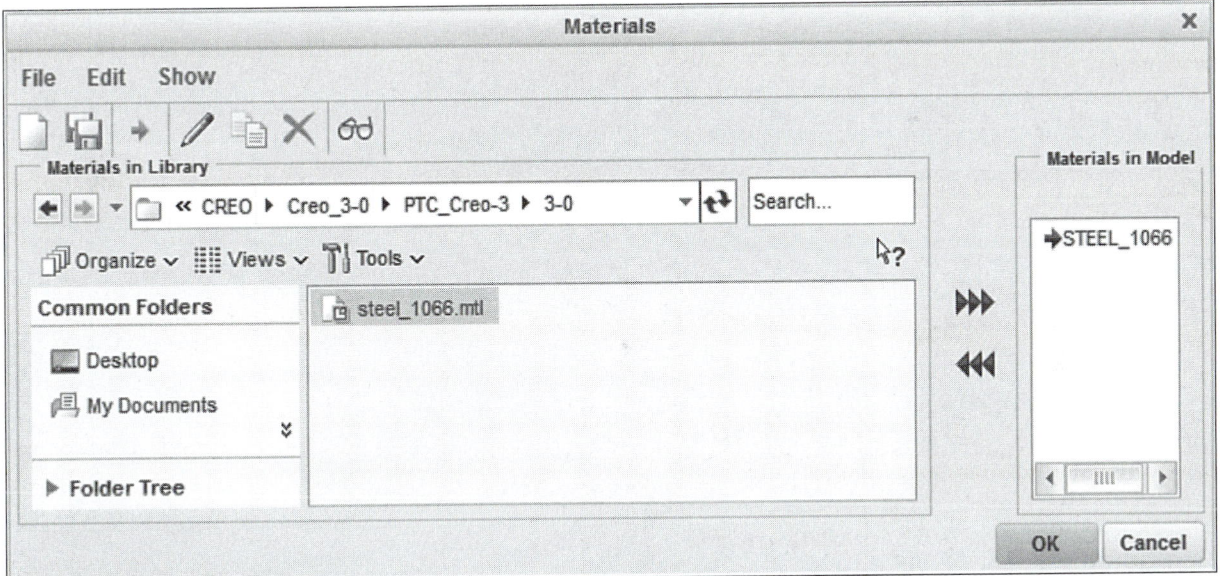

Figure 6.21(c) STEEL_1066

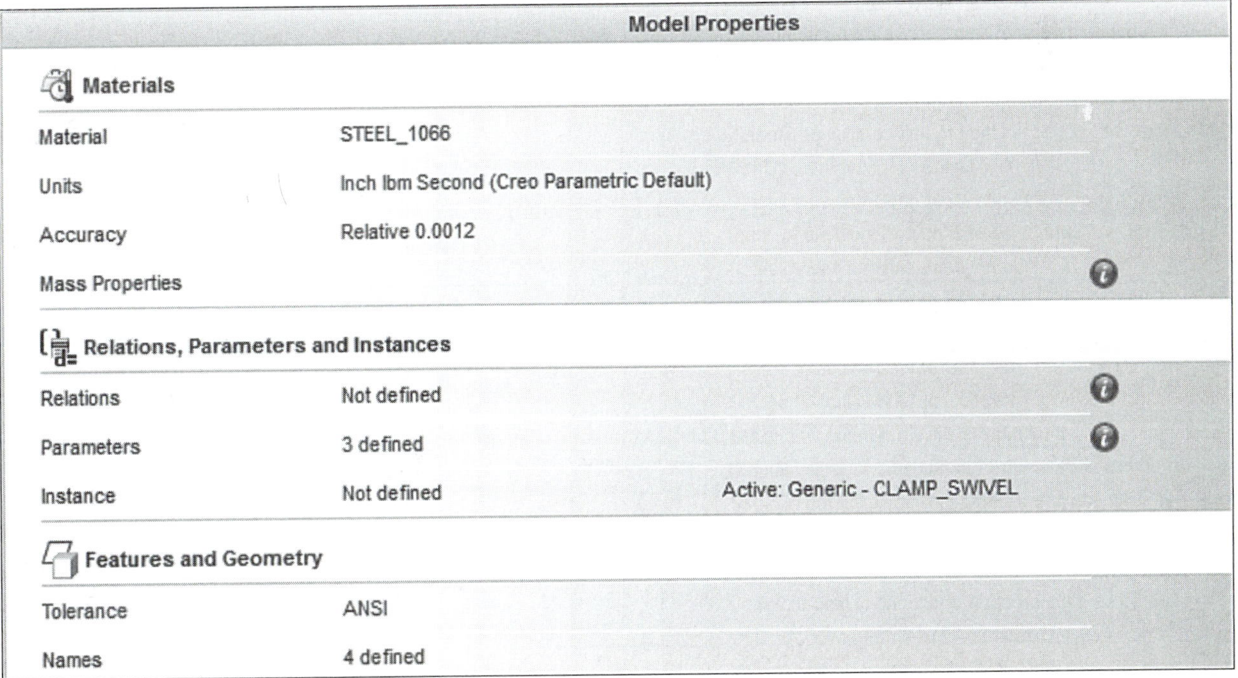

Figure 6.21(d) Model Properties

Rename the coordinate system to **CLAMP_SWIVEL_CSYS** > **Enter** > in the Graphics Window, **LMB** to deselect > **Tools** tab > **Model Information** [Fig. 6.21(e)] > [icon] to close the Browser [Fig. 6.21(f)] > [icon]

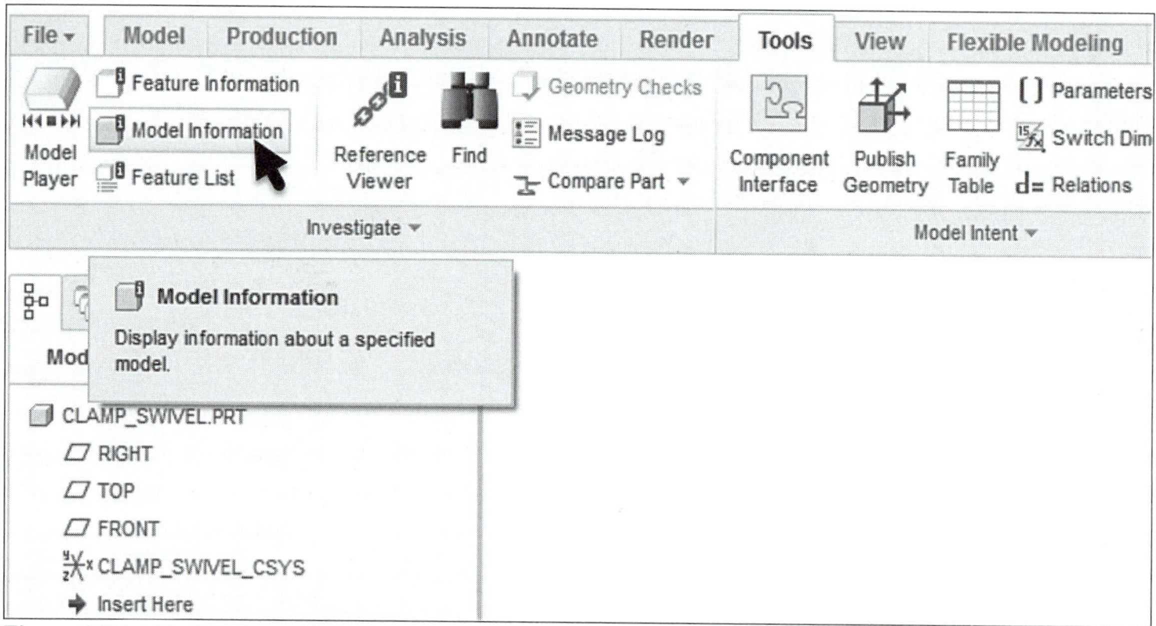

Figure 6.21(e) Model Information

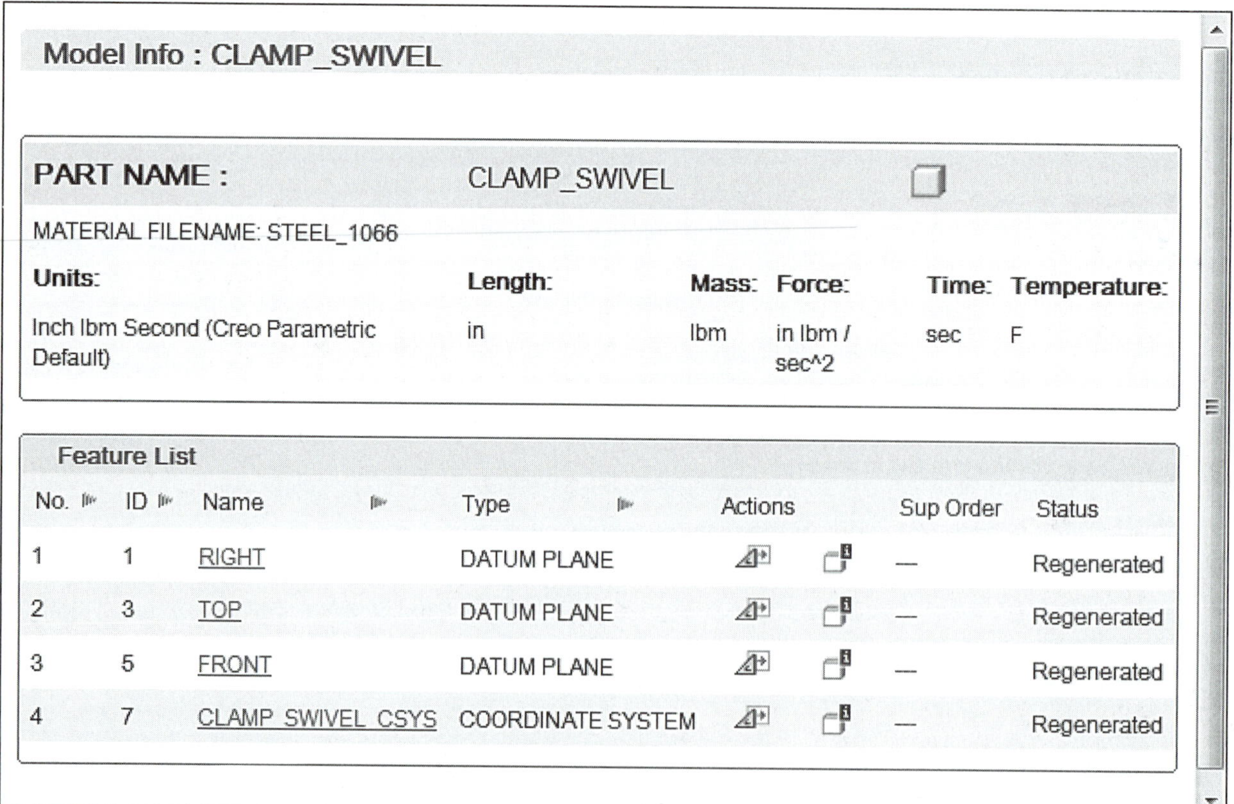

Figure 6.21(f) Model Info: MATERIAL FILENAME: STEEL_1066

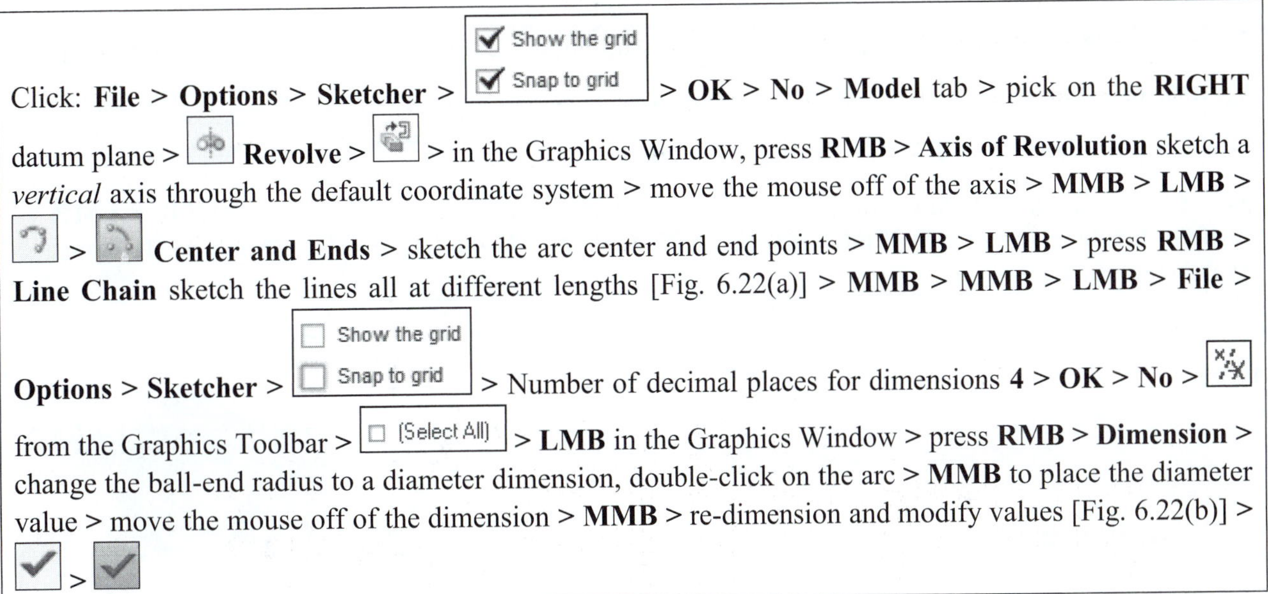

Click: **File** > **Options** > **Sketcher** > Show the grid, Snap to grid > **OK** > **No** > **Model** tab > pick on the **RIGHT** datum plane > **Revolve** > in the Graphics Window, press **RMB** > **Axis of Revolution** sketch a *vertical* axis through the default coordinate system > move the mouse off of the axis > **MMB** > **LMB** > > **Center and Ends** > sketch the arc center and end points > **MMB** > **LMB** > press **RMB** > **Line Chain** sketch the lines all at different lengths [Fig. 6.22(a)] > **MMB** > **MMB** > **LMB** > **File** > **Options** > **Sketcher** > Show the grid, Snap to grid > Number of decimal places for dimensions **4** > **OK** > **No** > from the Graphics Toolbar > (Select All) > **LMB** in the Graphics Window > press **RMB** > **Dimension** > change the ball-end radius to a diameter dimension, double-click on the arc > **MMB** to place the diameter value > move the mouse off of the dimension > **MMB** > re-dimension and modify values [Fig. 6.22(b)] > >

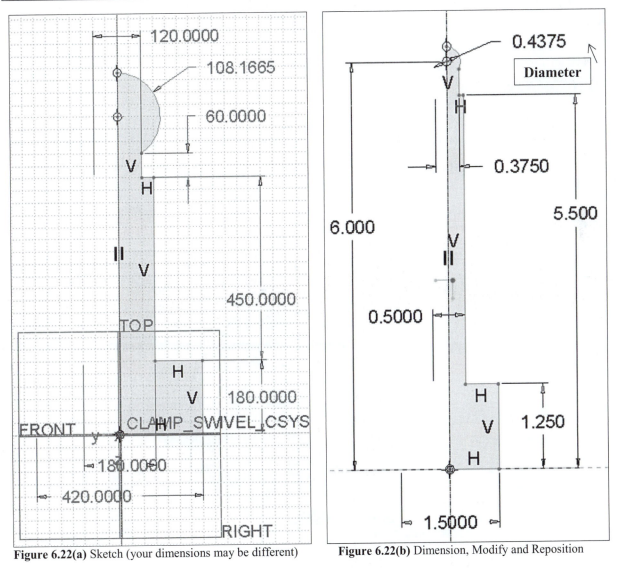

Figure 6.22(a) Sketch (your dimensions may be different)

Figure 6.22(b) Dimension, Modify and Reposition

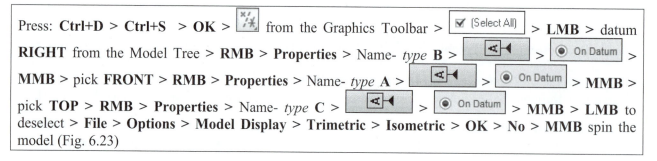

Press: **Ctrl+D** > **Ctrl+S** > **OK** > [icon] from the Graphics Toolbar > [☑ (Select All)] > **LMB** > datum **RIGHT** from the Model Tree > **RMB** > **Properties** > Name- *type* **B** > [icon] > [◉ On Datum] > **MMB** > pick **FRONT** > **RMB** > **Properties** > Name- *type* **A** > [icon] > [◉ On Datum] > **MMB** > pick **TOP** > **RMB** > **Properties** > Name- *type* **C** > [icon] > [◉ On Datum] > **MMB** > **LMB** to deselect > **File** > **Options** > **Model Display** > **Trimetric** > **Isometric** > **OK** > **No** > **MMB** spin the model (Fig. 6.23)

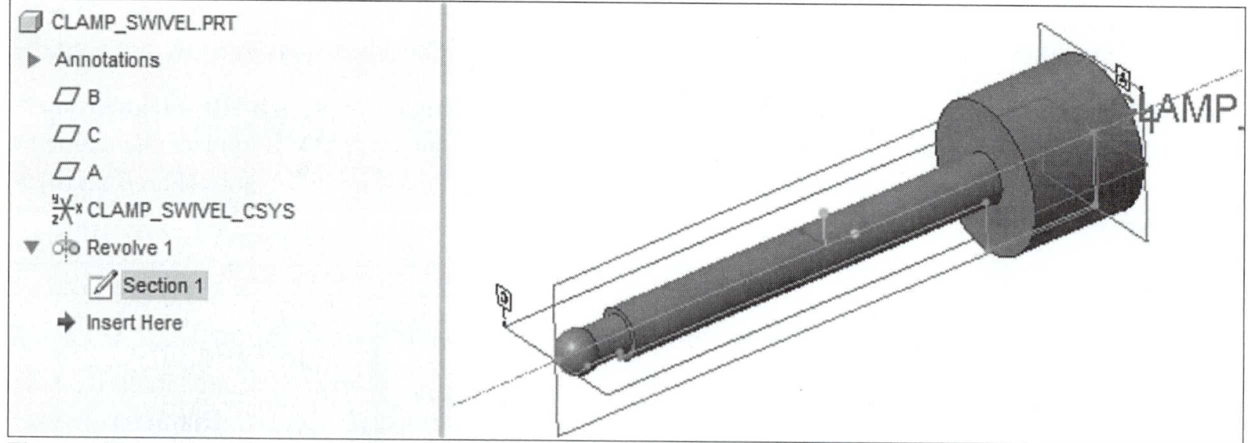

Figure 6.23 Completed Revolved Extrusion

Click: [Hole] **Hole** > pick on datum **C** > press **RMB** > **Offset References Collector** > **Placement** tab > [Flip] > [⌀ 0.5000] > **Enter** > [icon] > [icon] > pick datum **A** > Offset **.5625** > **Enter** > press and hold the **Ctrl** key and pick on datum **B** > release the **Ctrl** key > **Offset** > **Offset** >**Align** [Fig. 6.24(a)]

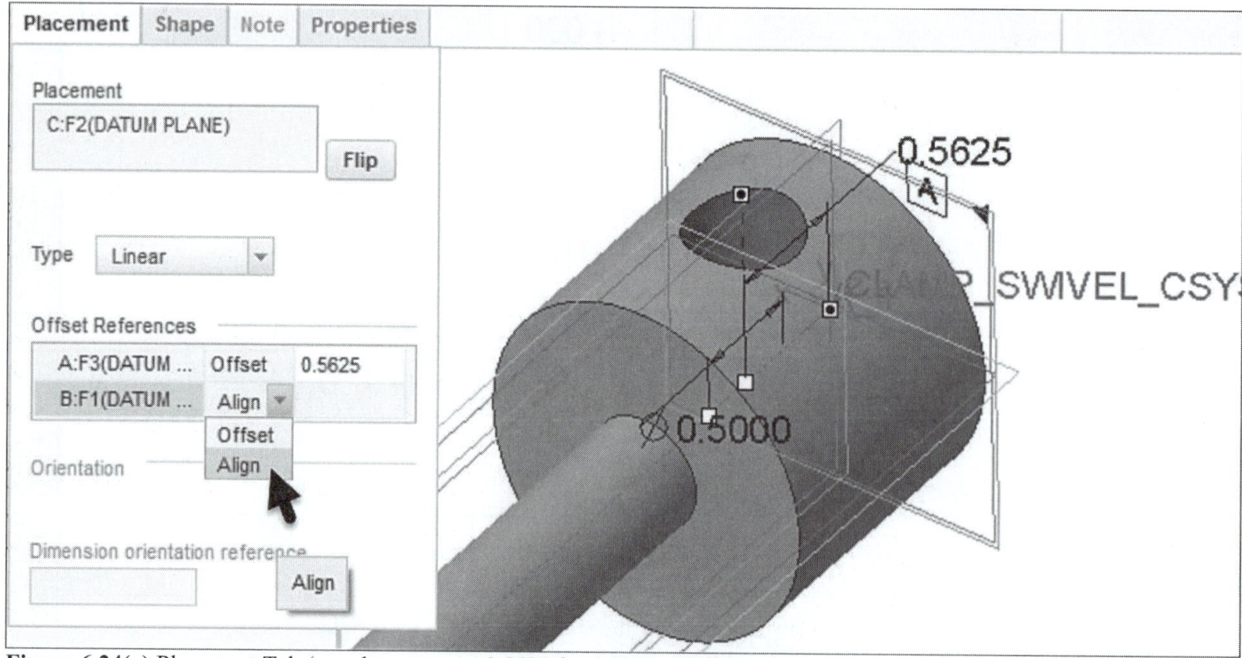

Figure 6.24(a) Placement Tab (may have to use -0.5625 for A Datum Offset value)

Click: **Shape** tab > Side 2: **None** > **Through All** [Fig. 6.24(b)] > > **LMB** in the Graphics Window > **Ctrl+S** > **LMB** to select the model > **LMB** to select one edge > press **RMB** > **Round** > hold down the **Ctrl** key and select the remaining two edges (Fig. 6.25) > release the **Ctrl** key > double-click on the dimension > *type* **.100** > **Enter** > **Enter** > **Ctrl+D** > **LMB** to deselect > **File** > **Save**

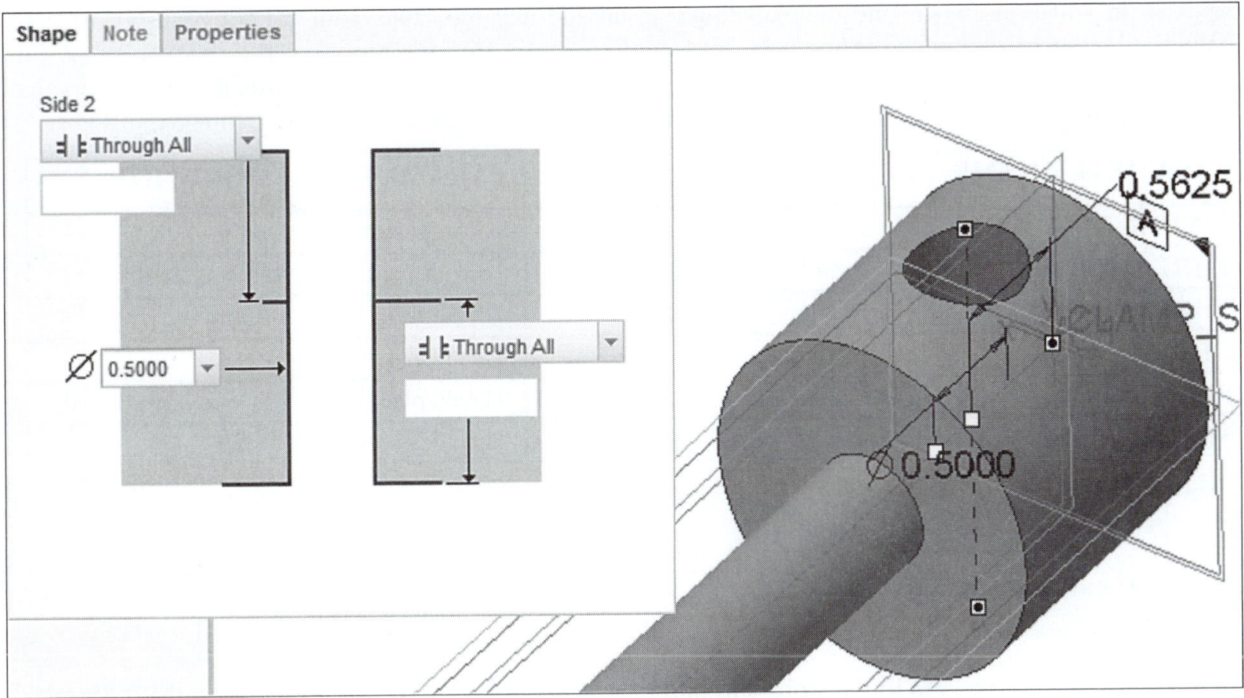

Figure 6.24(b) Shape Tab

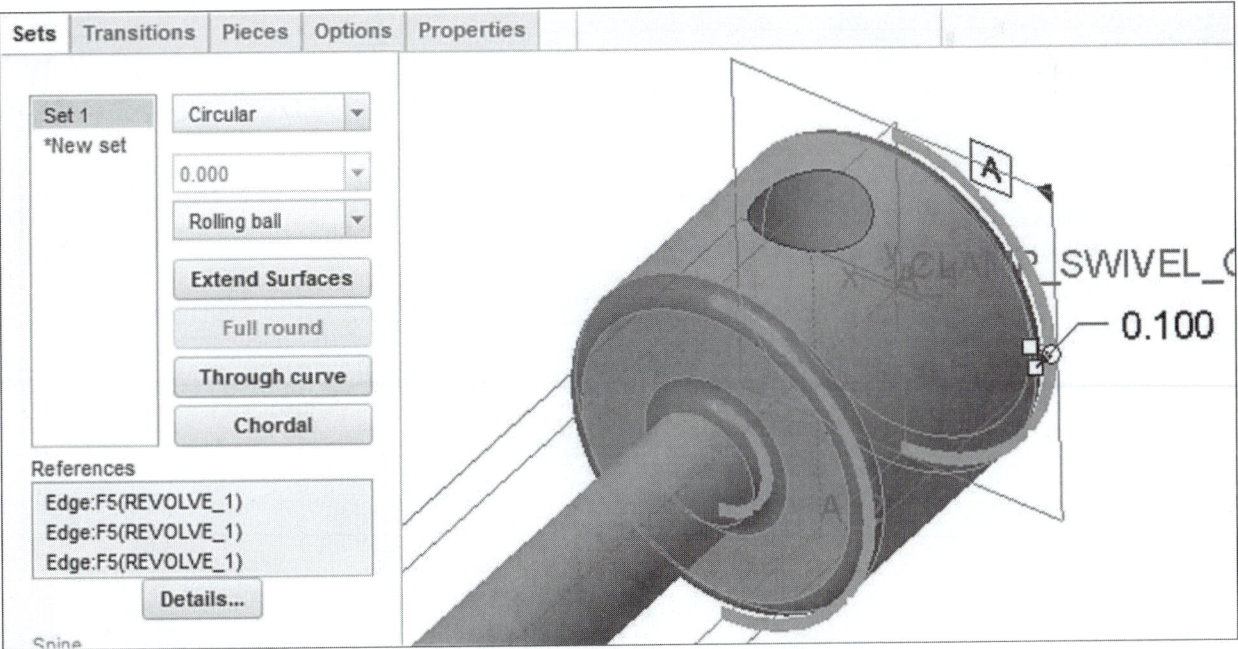

Figure 6.25 Rounds

Cosmetic Threads

A **cosmetic thread** is a feature that "represents" the diameter of a thread without having to show the actual threaded feature. Since a threaded feature is memory intensive, using cosmetic threads can save an enormous amount of visual memory on your computer. It is displayed in a unique default color. **Internal cosmetic threads** are created automatically when holes are created using the Hole Tool (Standard-Tapped). In situations where the internal thread is unique and the Hole Tool cannot be used, internal cosmetic threads can be added to the hole. **External cosmetic threads** represent the *root diameter*. For threaded shafts you must create the external cosmetic threads. The following table lists the parameters that can be defined for a cosmetic thread. In this table, "pitch" is the distance between two threads.

PARAMETER NAME	PARAMETER TYPE	PARAMETER DESCRIPTION
MAJOR_DIAMETER	Real Number	Thread major diameter
THREADS_PER_INCH	Real Number	Threads per inch (1/pitch)
THREAD FORM	String	Thread form
CLASS	String	Thread class
PLACEMENT	Character	Thread placement (A-external, B-internal)
METRIC	Yes No (True/False)	Thread is metric

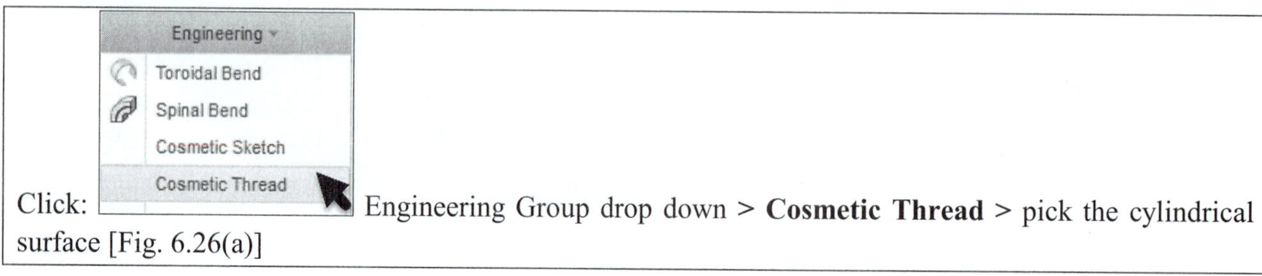

Click: Engineering Group drop down > **Cosmetic Thread** > pick the cylindrical surface [Fig. 6.26(a)]

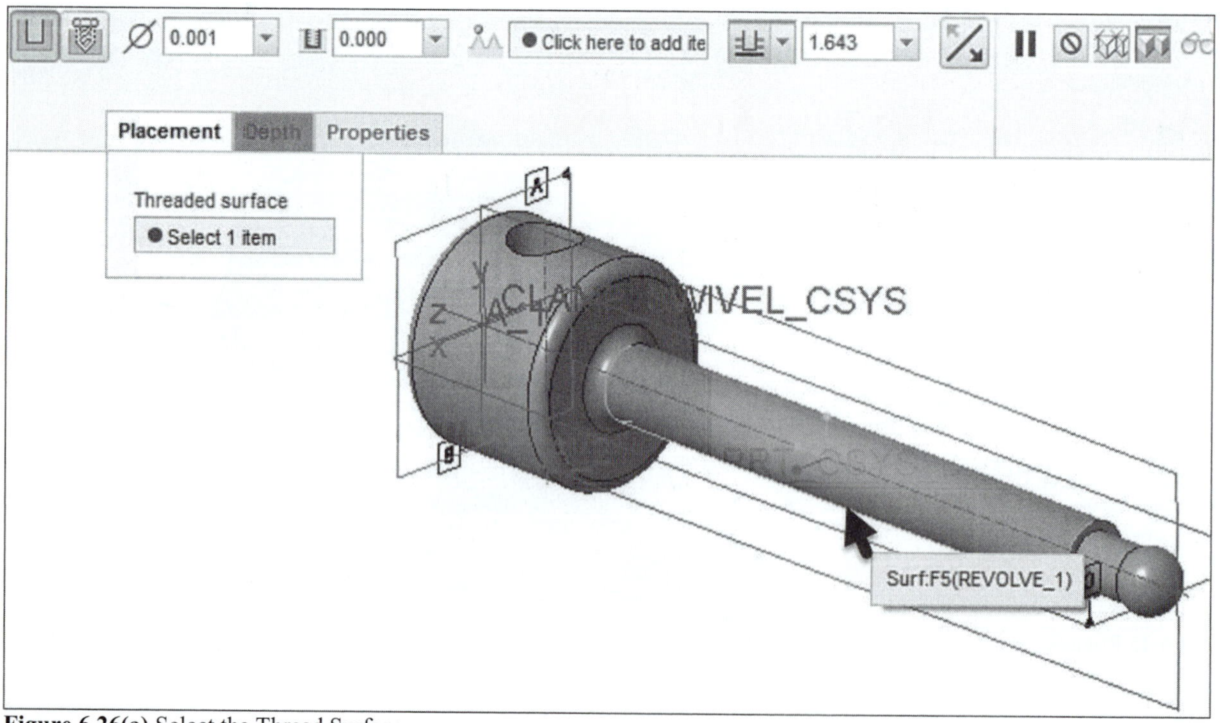

Figure 6.26(a) Select the Thread Surface

Pick the thread start surface--the flat edge lip surface [Fig. 6.26(b)] > in the Dashboard **4** 4.000 > **Enter** > ∅ 0.4485 thread root diameter: **.4485** [Fig. 6.26(c)] > **Enter** > **Properties** tab [Fig. 6.26(d)]

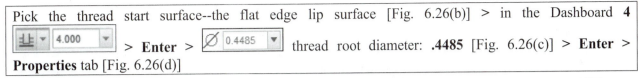

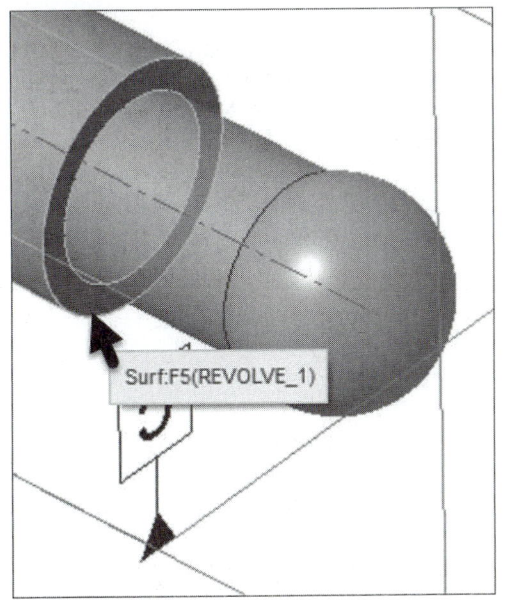

Figure 6.26(b) Select the Thread Start Surface

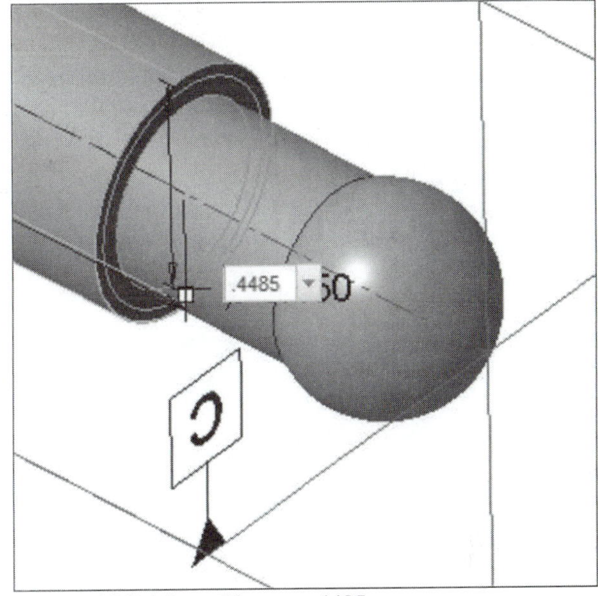

Figure 6.26(c) Root Diameter **.4485**

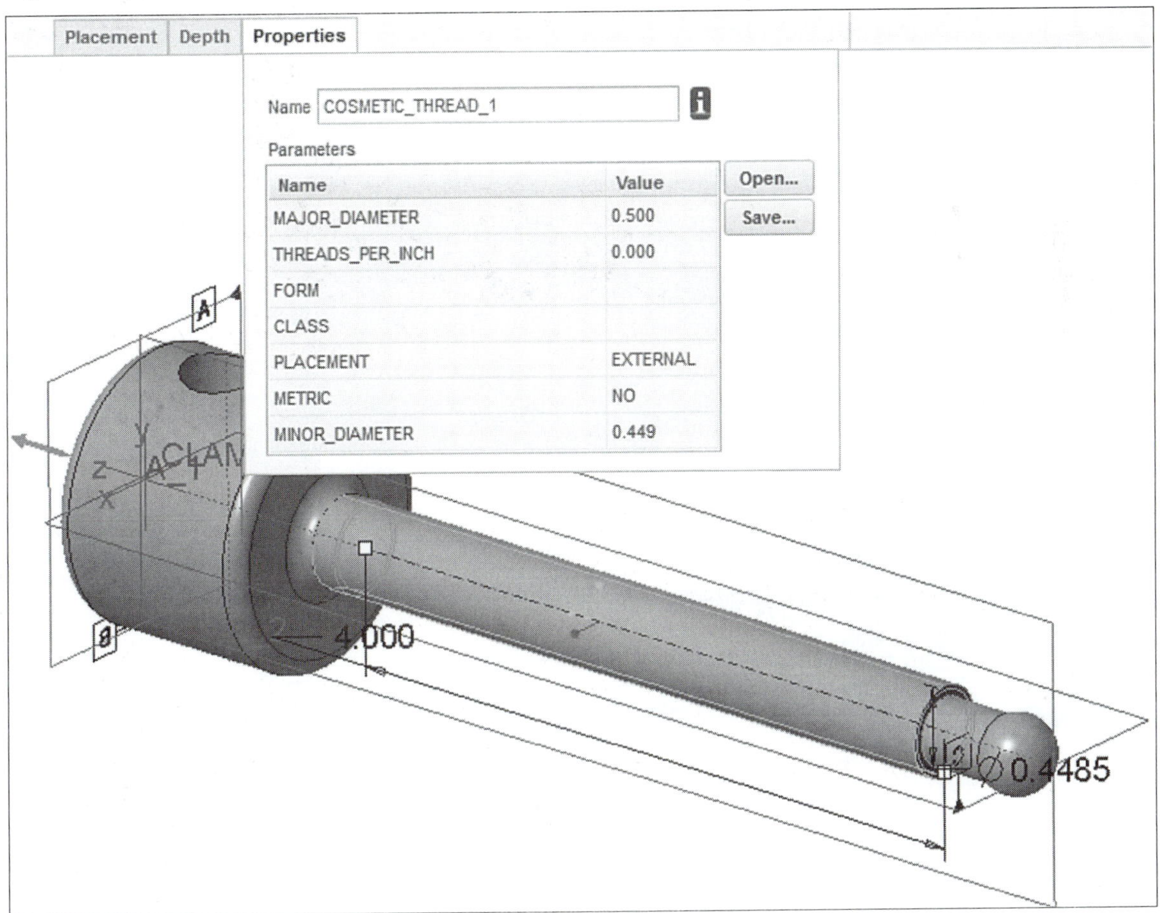

Name	Value
MAJOR_DIAMETER	0.500
THREADS_PER_INCH	0.000
FORM	
CLASS	
PLACEMENT	EXTERNAL
METRIC	NO
MINOR_DIAMETER	0.449

Figure 6.26(d) Properties Tab

After creating the cosmetic thread, edit the thread panel. The thread size of an external thread must be changed to the nominal size from the root diameter defaulted on the thread table. The task was to create an external cosmetic thread (**.500-13 UNC-2A**) using the **Ø.500** surface [Fig. 6.26(a)]. The thread started at the "neck" and extended **4.00** along the swivel's shaft [Fig. 6.26(b)]. The thread parameter [Fig. 6.26(e)] shows the major diameter as **0.500**.

Since you are cosmetically representing the *root diameter (.4485)* of the external thread on the model, the *thread diameter* is *smaller* than the *nominal (.500)* thread size.

> Click in each table field and change the values [Fig. 6.26(f)] > THREADS_PER_INCH **13** > FORM **UNC** > CLASS **2** > > in the Graphics Window, **LMB** to deselect > **Ctrl+S**

Name: COSMETIC_THREAD_1

Parameters

Name	Value
MAJOR_DIAMETER	0.500
THREADS_PER_INCH	0.000
FORM	
CLASS	
PLACEMENT	EXTERNAL
METRIC	NO
MINOR_DIAMETER	0.449

Open... Save...

Figure 6.26(e) Thread Parameters *(your display may differ depending on the number of decimal places set)*

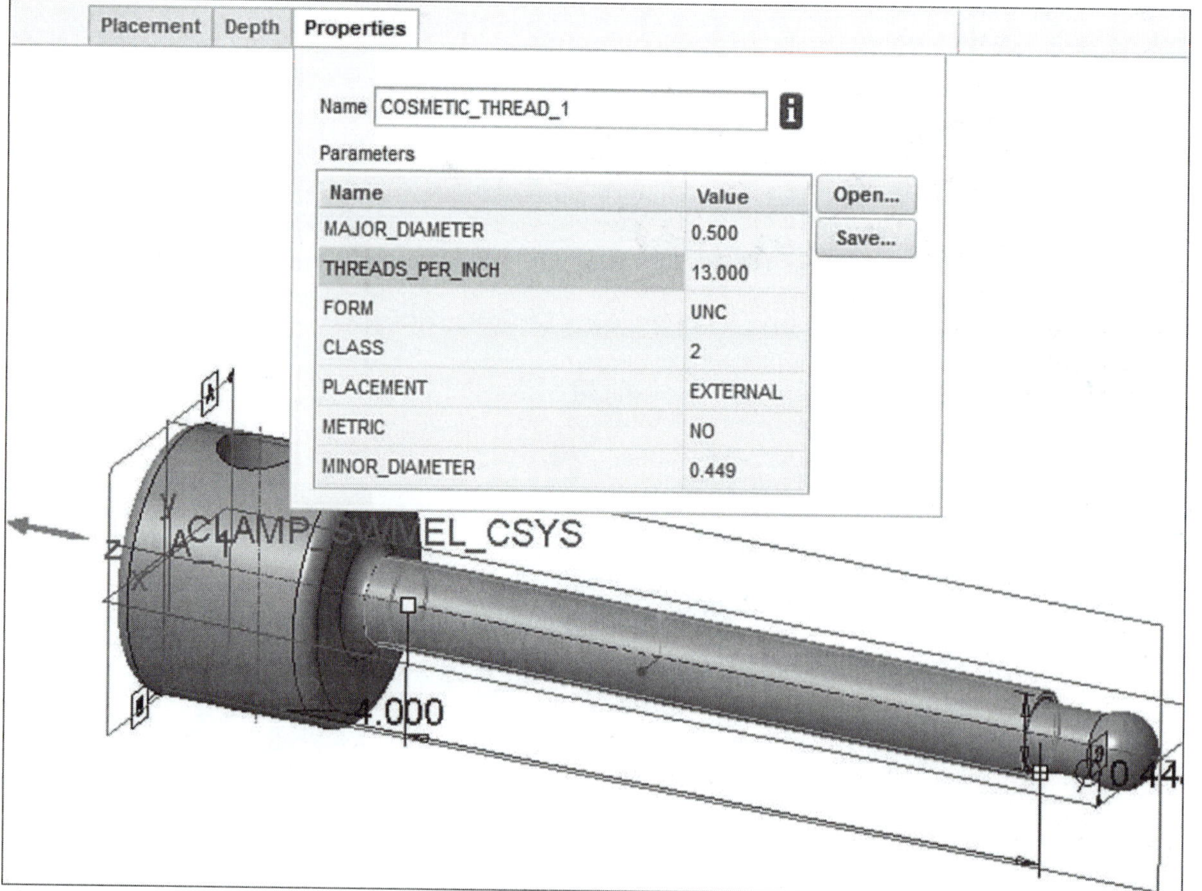

Figure 6.26(f) Edited Thread Parameters *(your display may differ depending on the number of decimal places set)*

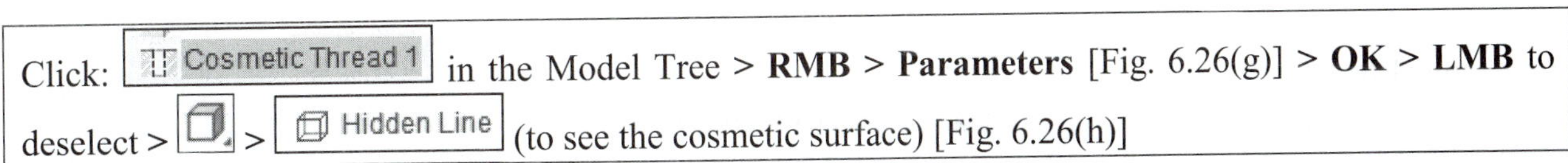

Click: Cosmetic Thread 1 in the Model Tree > **RMB** > **Parameters** [Fig. 6.26(g)] > **OK** > **LMB** to deselect > > Hidden Line (to see the cosmetic surface) [Fig. 6.26(h)]

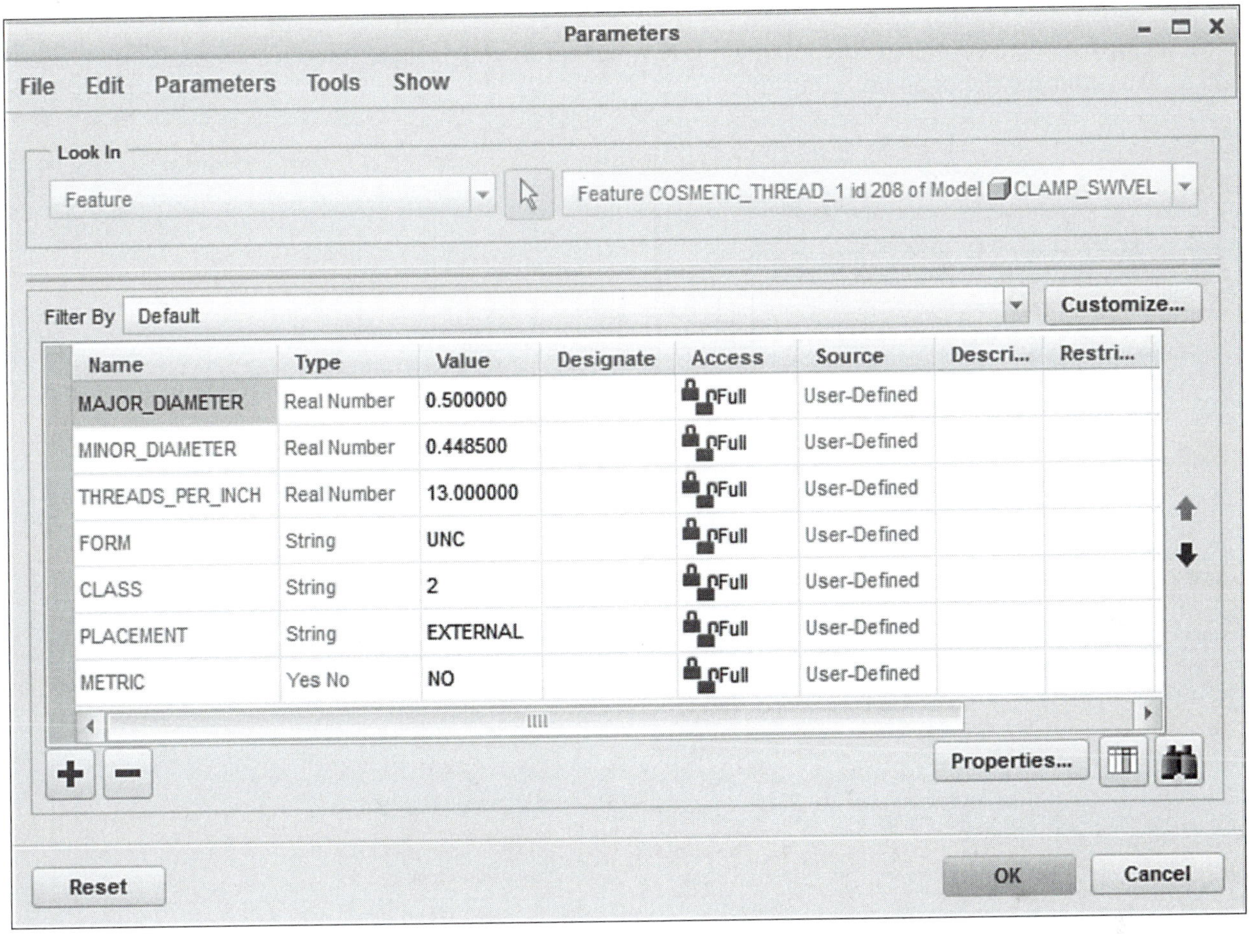

Parameters

File Edit Parameters Tools Show

Look In: Feature — Feature COSMETIC_THREAD_1 id 208 of Model CLAMP_SWIVEL

Filter By: Default — Customize...

Name	Type	Value	Designate	Access	Source	Descri...	Restri...
MAJOR_DIAMETER	Real Number	0.500000		Full	User-Defined		
MINOR_DIAMETER	Real Number	0.448500		Full	User-Defined		
THREADS_PER_INCH	Real Number	13.000000		Full	User-Defined		
FORM	String	UNC		Full	User-Defined		
CLASS	String	2		Full	User-Defined		
PLACEMENT	String	EXTERNAL		Full	User-Defined		
METRIC	Yes No	NO		Full	User-Defined		

Properties...

Reset OK Cancel

Figure 6.26(g) Thread Parameters

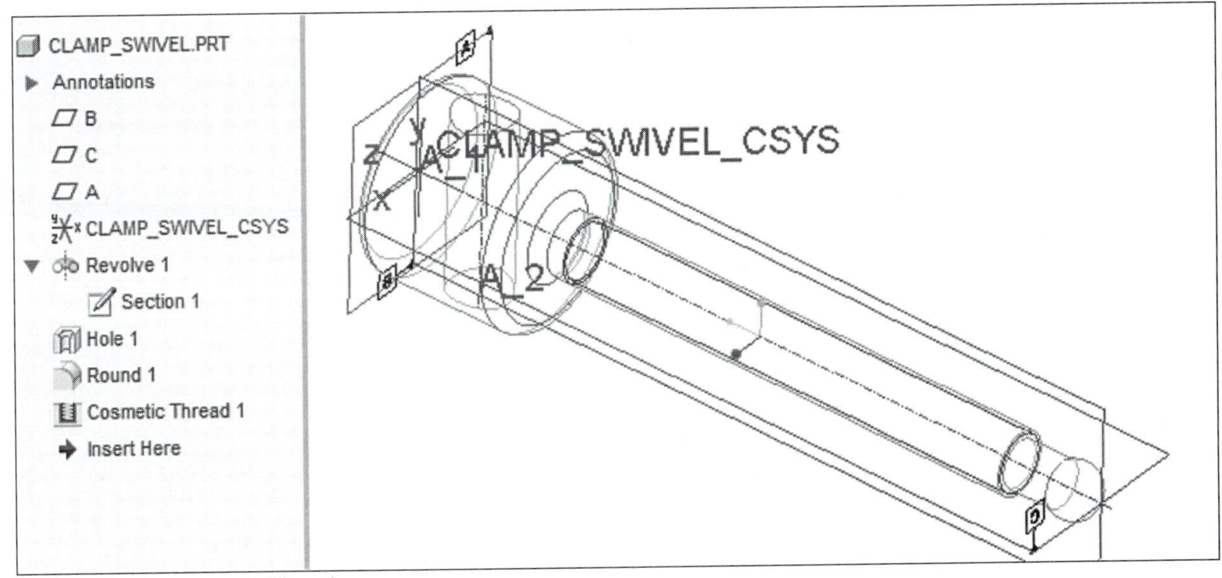

Figure 6.26(h) Cosmetic Thread

Select the cosmetic feature in the Model Tree > **RMB** > **Information** > **Feature Information** [Fig. 6.26(i)] > to close the Browser > **Ctrl+D** > in the Graphics Window, **LMB** to deselect > > **Shading With Edges** > > **File** > **Manage File** > **Delete Old Versions** > **Enter** > **File** > **Save As** > **Save a Copy** > Type > **Zip File (*.zip)** > **OK** > upload to your course interface or attach to an email and send to your instructor and/or yourself

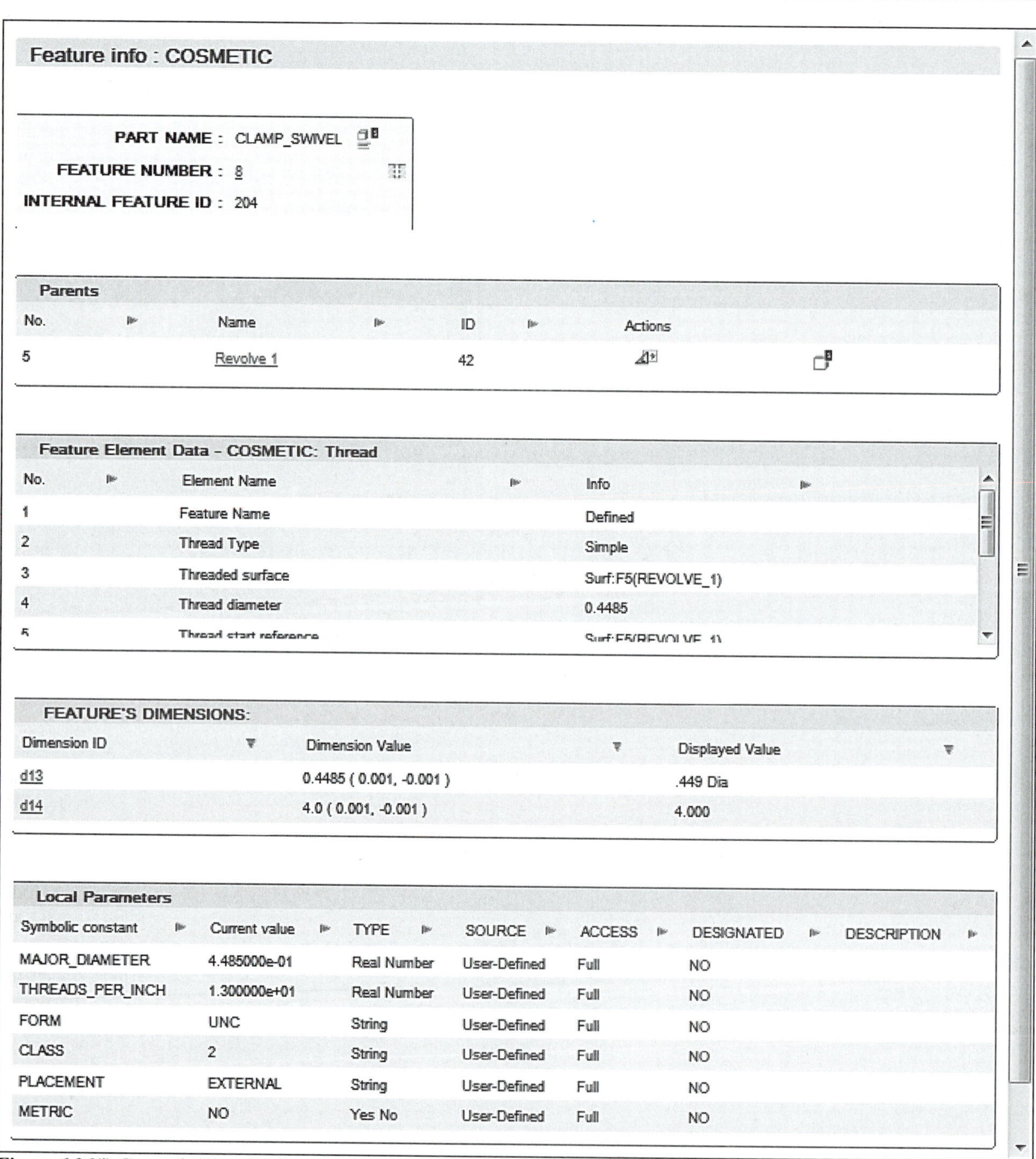

Feature info : COSMETIC

PART NAME : CLAMP_SWIVEL
FEATURE NUMBER : 8
INTERNAL FEATURE ID : 204

Parents

No.	Name	ID	Actions
5	Revolve 1	42	

Feature Element Data - COSMETIC: Thread

No.	Element Name	Info
1	Feature Name	Defined
2	Thread Type	Simple
3	Threaded surface	Surf:F5(REVOLVE_1)
4	Thread diameter	0.4485
5	Thread start reference	Surf:F5(REVOLVE_1)

FEATURE'S DIMENSIONS:

Dimension ID	Dimension Value	Displayed Value
d13	0.4485 (0.001, -0.001)	.449 Dia
d14	4.0 (0.001, -0.001)	4.000

Local Parameters

Symbolic constant	Current value	TYPE	SOURCE	ACCESS	DESIGNATED	DESCRIPTION
MAJOR_DIAMETER	4.485000e-01	Real Number	User-Defined	Full	NO	
THREADS_PER_INCH	1.300000e+01	Real Number	User-Defined	Full	NO	
FORM	UNC	String	User-Defined	Full	NO	
CLASS	2	String	User-Defined	Full	NO	
PLACEMENT	EXTERNAL	String	User-Defined	Full	NO	
METRIC	NO	Yes No	User-Defined	Full	NO	

Figure 6.26(i) Cosmetic Thread Information

Using the Model Player

The **Model Player** (Fig. 6.27) lets you observe how a part is built. You can:

- Move backward or forward through the feature-creation history of the model.
- Regenerate each feature in sequence, starting from the specified feature
- Display each feature as it is regenerated or rolled forward
- Update (regenerate all the features in) the entire display when you reach the desired feature
- Obtain information about the current feature (you can show dimensions, obtain regular feature information, investigate geometry errors, and enter Fix Model mode)

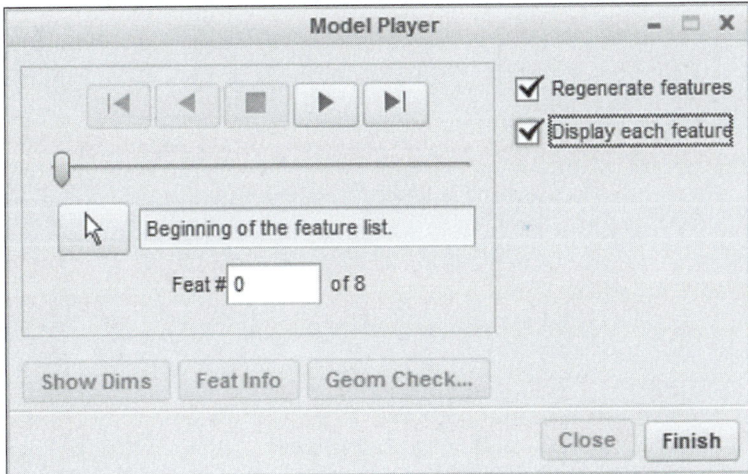

Figure 6.27 Model Player

You can select one of the following:

Regenerate features Regenerates each feature in sequence, starting from the specified feature
Display each feature Displays each feature in the graphics window as it is being regenerated
Compute CL (Available in Manufacturing mode only) When selected, the CL data is recalculated for each NC sequence during regeneration

Select one of the following commands:

Go to the beginning of the model moves immediately to the beginning of the model

Step backward through the model one feature at-a-time and regenerates the preceding feature

Stop play

Step forward through the model one feature at-a-time and regenerates the next feature

Go to the last feature in the model moves immediately to the end of the model (resume all)

Slider Bar Drag the slider handle to the feature at which you want model playback to begin. The features are highlighted in the graphics window as you move through their position with the slider handle. The feature number and type are displayed in the selection panel [such as #4 (COORDINATE SYSTEM)], and the feature number is displayed in the **Feat #** box.

Select feature from screen or model tree lets you select a starting feature from the graphics window or the Model Tree. Opens the **SELECT FEAT** and **SELECT** menus. After you select a starting feature, its number and ID are displayed in the selection panel, and the feature number is displayed in the **Feat #** box.

Feat # 5 of 5 Let's you specify a starting feature by typing the feature number in the box. After you enter the feature number, the model immediately rolls or regenerates to that feature.

To stop playback, click the **Stop play** button. Use the following commands for information:

- **Show Dims** Displays the dimensions of the current feature
- **Feat Info** Provides regular feature information about the current feature in an Information window
- **Geom Check** Investigates the geometry error for the current feature
- **Fix Model** Activates Resolve mode by forcing the current feature to abort regeneration
- **Close** Closes the Model Player and enters Insert mode at the current feature
- **Finish** Closes the Model Player and returns to the last feature in the model

Click: **File > Options > Model Display > Isometric > Trimetric > Apply > OK > No >** **Command Search >** *type* **Model Player >** [model player ▼ Commands Model Player...] > [Regenerate features, Display each feature] > **Go to the beginning of the model >** [Fig. 6.28(a)] > > > [Fig. 6.28(b)] > [Fig. 6.28(c)]

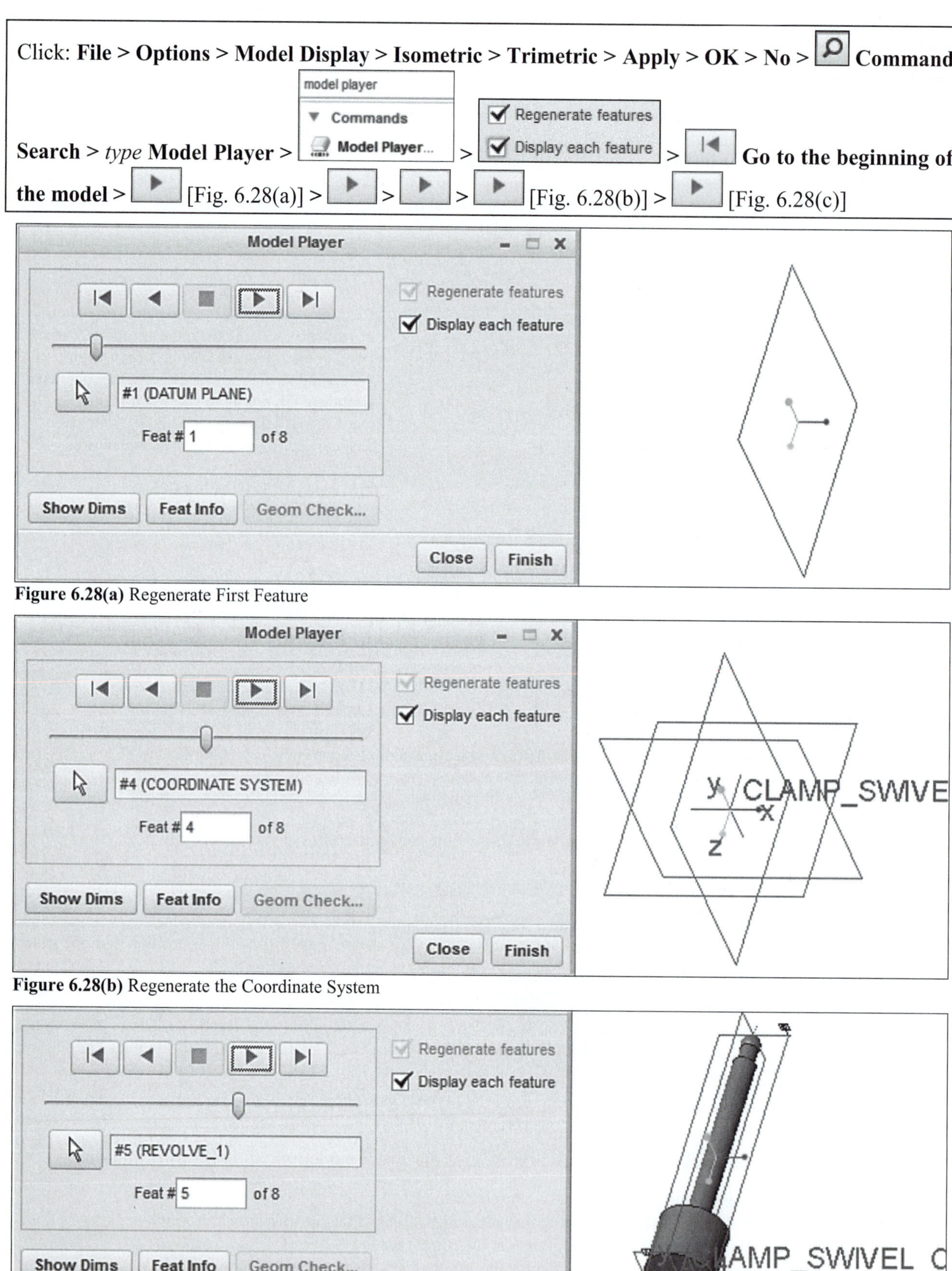

Figure 6.28(a) Regenerate First Feature

Figure 6.28(b) Regenerate the Coordinate System

Figure 6.28(c) Feature #5

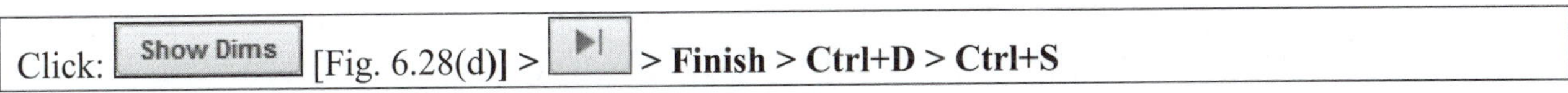

Click: Show Dims [Fig. 6.28(d)] > ▶| > Finish > Ctrl+D > Ctrl+S

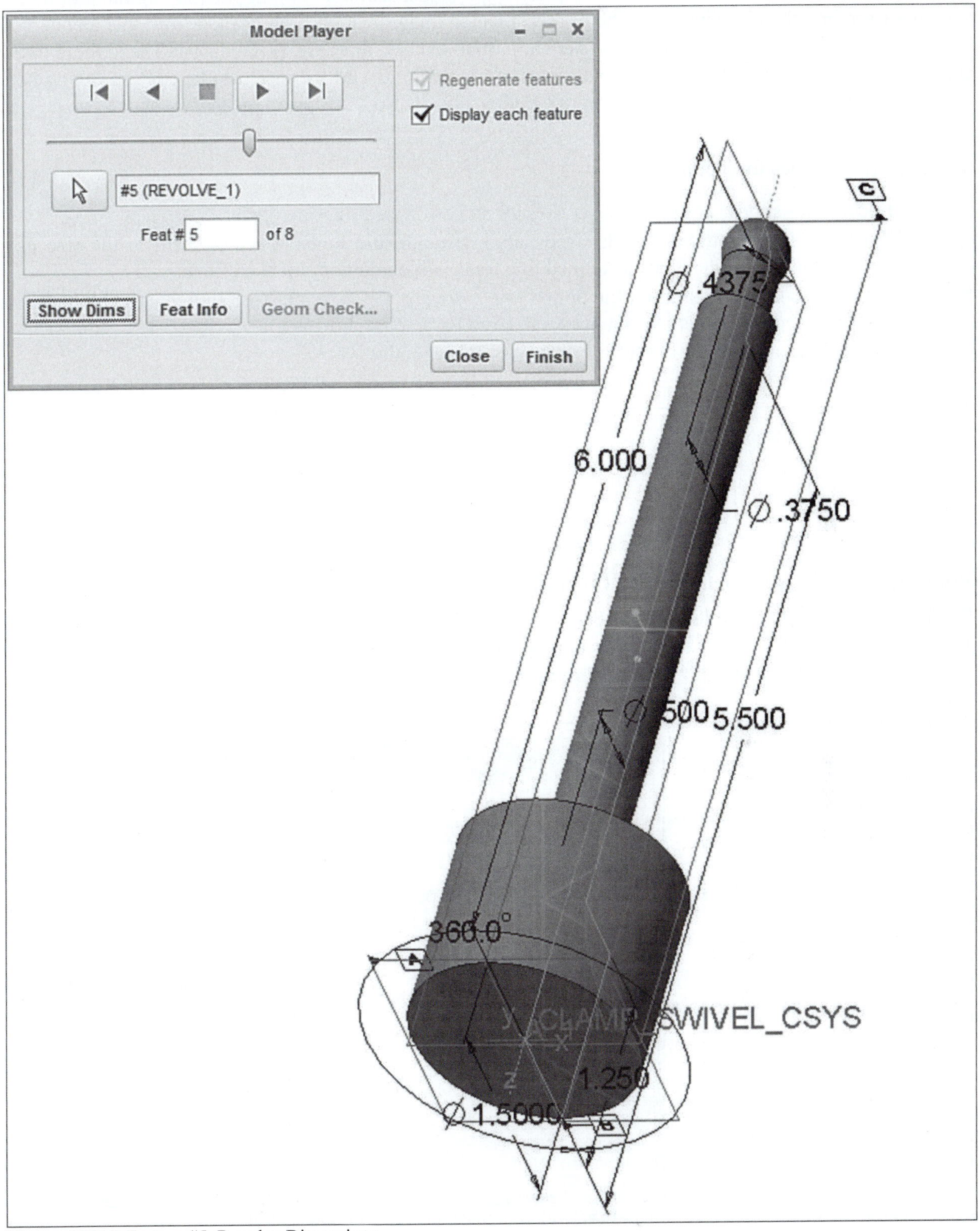

Figure 6.28(d) Feature #5, Revolve Dimensions

Printing and Plotting

From the File menu, you can print with the following options: scaling, clipping, displaying the plot on the screen, or sending the plot directly to the printer. Shaded images can also be printed from this menu. You can create plot files of the current object (sketch, part, assembly, drawing, or layout) and send them to the print queue of a plotter. You can configure your printer using the MS Printer Manager Print dialog box, available from the Creo Parametric 3.0 Print dialog box. If you are printing a non-shaded image, the Printer Configuration dialog box opens.

The following applies for plotting:

- Hidden lines appear as gray for a screen plot, but as dashed lines on paper.
- When Creo Parametric 3.0 plots Creo Parametric 3.0 line fonts, it scales them to the size of a sheet. It does not scale the user-defined line fonts, which do not plot as defined.
- You can use the configuration file option *use_software_linefonts* to make sure that the plotter plots a user-defined font exactly as it appears in Creo Parametric 3.0.
- You can plot a cross section from Part or Assembly mode.
- With a Pro/PLOT license, you can write plot files in a variety of formats.

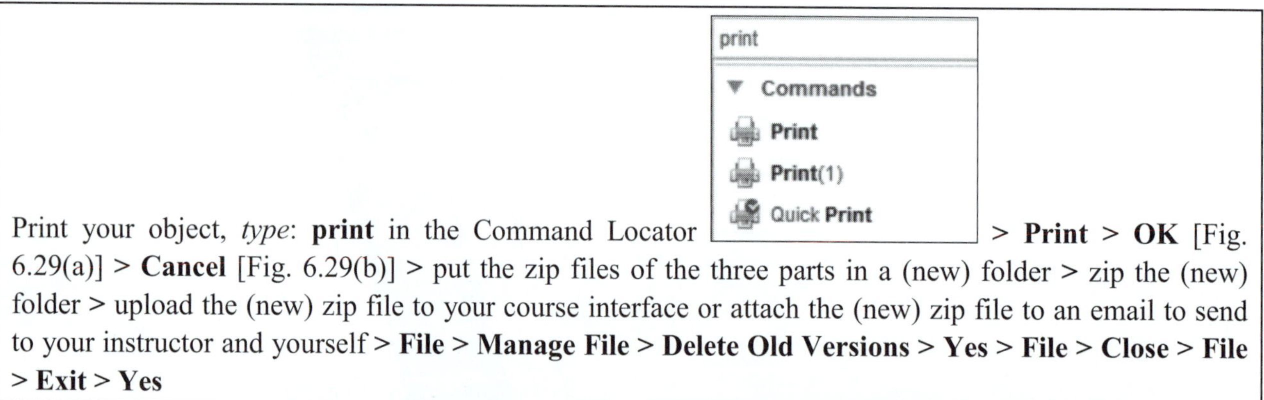

Print your object, *type*: **print** in the Command Locator > **Print** > **OK** [Fig. 6.29(a)] > **Cancel** [Fig. 6.29(b)] > put the zip files of the three parts in a (new) folder > zip the (new) folder > upload the (new) zip file to your course interface or attach the (new) zip file to an email to send to your instructor and yourself > **File** > **Manage File** > **Delete Old Versions** > **Yes** > **File** > **Close** > **File** > **Exit** > **Yes**

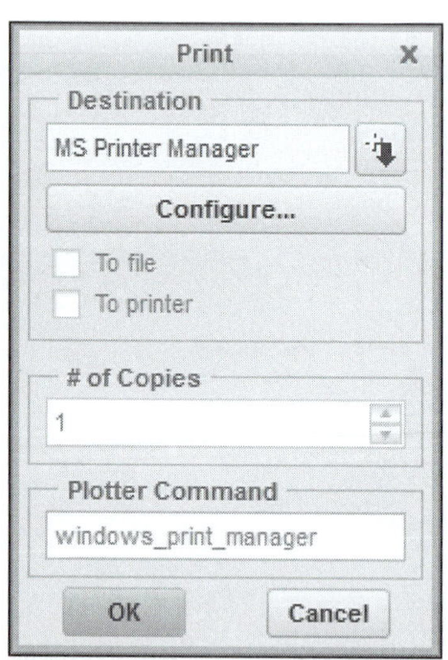

Figure 6.29(a) Print Dialog Box

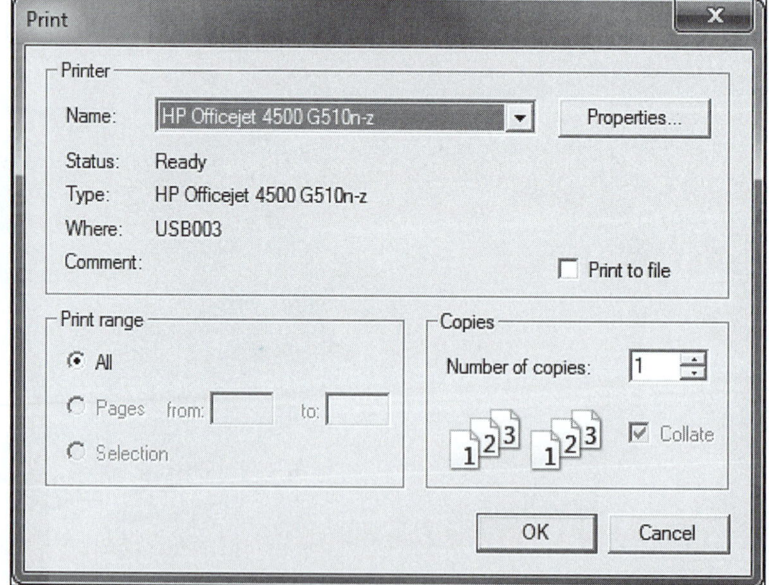

Figure 6.29(b) MS Printer Manager Print Dialog Box

A set of projects and lectures are available at ***www.cad-resources.com***.

Lesson 7 Feature Operations

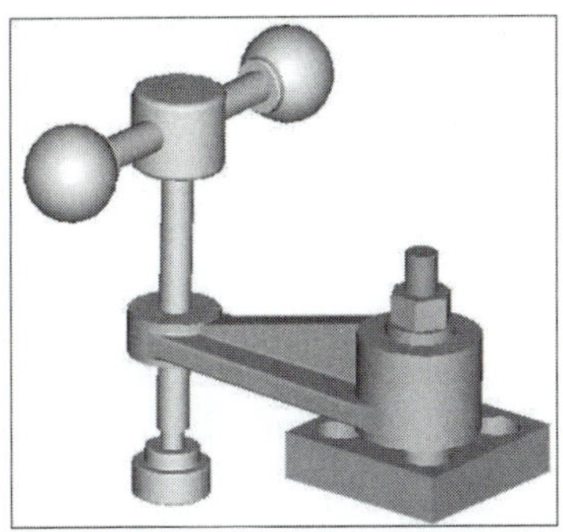

Figure 7.1 Swing Clamp Assembly

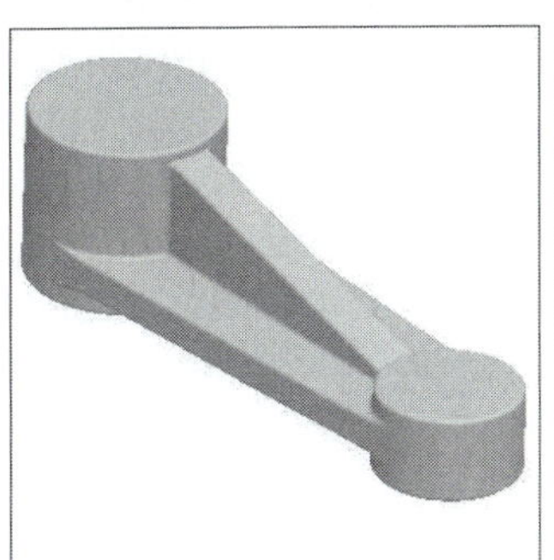

Figures 7.2(a-b) Clamp Arm (Casting- Workpiece) and Clamp Arm (Machined- Design Part)

OBJECTIVES

- Use **Copy**, **Paste**, and **Paste Special**
- Create **Ribs**
- Understand **Parameters** and **Relations**
- **Measure** geometry
- Solve **Failures**
- Create versions of the part using a **Family Table**

REFERENCES AND RESOURCES

For **Resources** go to **www.cad-resources.com** > click on the PTC Creo Parametric 3.0 Book cover

- Lesson 7 Lecture
- Lesson 7 3D models embedded in a PDF
- Book Projects PDF
- Project Lectures
- Quick Reference Card
- Configuration Options

FEATURE OPERATIONS

The Clamp Arm is used in the Swing Clamp Assembly (Fig. 7.1). This lesson will cover a wide range of PTC Creo Parametric 3.0 capabilities including: **Copy** and **Paste Special**, **Relations**, **Parameters**, **Failures**, **Family Tables**, and the **Rib Tool**. You will create two versions of the Clamp Arm; one with all cast surfaces [Fig. 7.2(a)] and the other with machined ends [Fig. 7.2(b)] (using a Family Table).

Ribs

A profile rib is a special type of protrusion designed to create a thin fin or web that is attached to a part. You always sketch a rib from a side view, and it grows about the sketching plane symmetrically or to either side. Because of the way ribs are attached to the parent geometry, they are always sketched as open sections. A trajectory rib is also available.

When sketching an open section, Creo Parametric may be uncertain about the side to which to add the rib. Creo Parametric adds all material in the direction of the arrow. If the incorrect choice is made, toggle the arrow direction by picking on the direction arrow on the screen.

A **profile ribs** must "see" material everywhere it attaches to the part; otherwise, it becomes an unattached feature. There are two types of profile ribs: straight [Fig. 7.3(a)] and rotational. The type is automatically set according to the attaching geometry (planar or curved).

Trajectory ribs [Fig. 7.3(b)] are most often used to strengthen plastic parts that include a base and a shell or other hollow area between pocket surfaces. The pocket surface and base must consist of solid geometry. Create a trajectory rib by sketching the rib path between pocket surfaces, or by selecting an existing sketch. The rib has a top and a bottom. The bottom is the end that intersects the part surface. The sketch plane that you select defines the top surface of the rib. The rib geometry's side surfaces extend to the next surface encountered. The rib sketch can contain open, closed, self-intersecting, or multiple loops.

A Trajectory Rib feature is one trajectory, which can include any number of segments in any shape. The feature can also include a round for each edge and a draft. You can define the draft, rounds, and rib width in the graphics window or on the Trajectory Rib tab. You can separate rounds as a standalone feature that can be redefined as another feature.

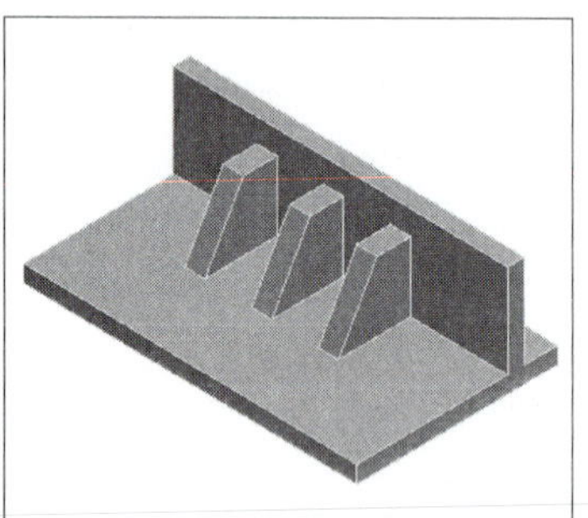

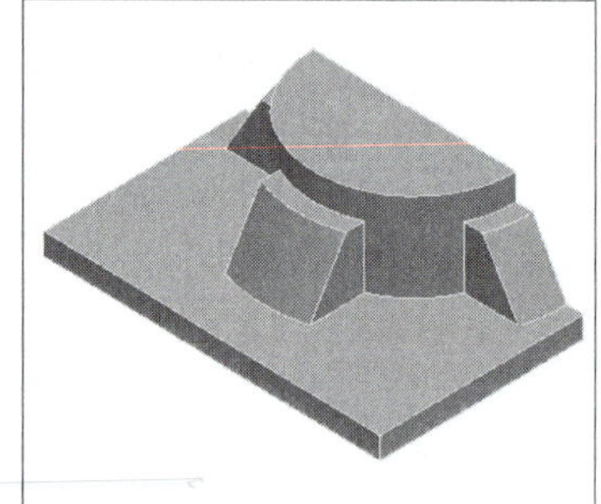

Figure 7.3(a) Straight and Rotational Ribs

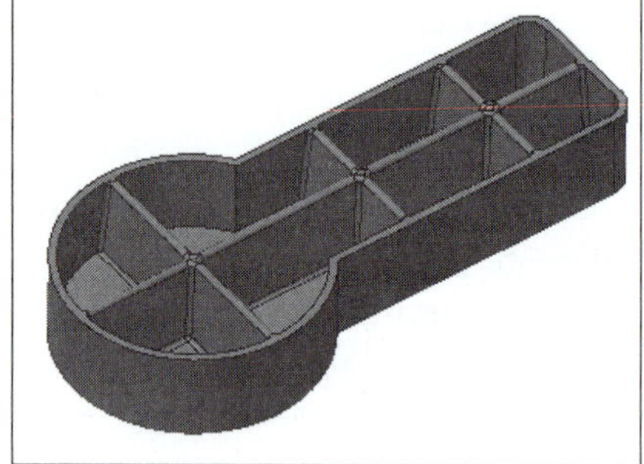

Figure 7.3(b) Trajectory Rib

Relations

Relations (also known as parametric relations) are user-defined equations written between symbolic dimensions and parameters. Relations capture design relationships within features or parts, or among assembly components, thereby allowing users to control the effects of modifications on models.

Relations can be used to control the effects of modifications on models, to define values for dimensions in parts and assemblies, and to act as constraints for design conditions (for example, specifying the location of a hole in relation to the edge of a part). They are used in the design process to describe conditional relationships between different features of a part or an assembly.

Relations can be used to provide a value for a dimension. However, they can also be used to notify you when a condition has been violated, such as when a dimension exceeds a certain value. There are two basic types of relations, equality and comparison.

An equality relation equates a parameter on the left side of the equation to an expression on the right side. This type of relation is used for assigning values to dimensions and parameters. The following are a few examples of equality relations:

d2 = 25.500 **d8 = d4/2** **d7 = d1+d6/2** **d6 = d2*(sqrt(d7/4.0+d4))**

A comparison relation compares an expression on the left side of the equation to an expression on the right side. This type of relation is commonly used as a constraint or as a conditional statement for logical branching. The following are examples of comparison relations:

d1 + d2 > (d3 + 5.5)	Used as a constraint
IF (d1 + 5.5) > = d7	Used in a conditional statement

Parameter Symbols

Four types of parameter symbols are used in relations:

- **Dimensions** These are dimension symbols, such as **d8**, **d12**.
- **Tolerances** These are parameters associated with ± symmetrical and plus-minus tolerance formats. These symbols appear when dimensions are switched from numeric to symbolic.
- **Number of Instances** These are integer parameters for the number of instances in a direction of a pattern.
- **User Parameter** These can be parameters defined by adding a parameter or a relation (e.g., **Volume = d3 * d4 * d5**).

Operators and Functions

The following operators and functions can be used in equations and conditional statements:

Arithmetic Operators

+	**Addition**
–	**Subtraction**
/	**Division**
*	**Multiplication**
^	**Exponentiation**
( )	**Parentheses for grouping** [for example, **(d0 = (d1–d2)*d3)**]

Assignment Operators

=	**Equal to**

The = (equals) sign is an assignment operator that equates the two sides of an equation or relation. When it is used, the equation can have only a single parameter on the left side.

Comparison Operators

Comparison operators are used whenever a TRUE/FALSE value can be returned. For example, the relation **d1** >= **3.5** returns TRUE whenever d1 is greater than or equal to **3.5**. It returns FALSE whenever **d1** is less than **3.5**. The following comparison operators are supported:

==	**Equal to**
>	**Greater than**
>=	**Greater than or equal to**
!=, <>,~=	**Not equal to**
<	**Less than**
<=	**Less than or equal to**
|	**Or**
&	**And**
~, !	**Not**

Mathematical Functions

The following operators can be used in relations, both in equations and in conditional statements. Relations may include the following mathematical functions:

cos ( )	**cosine**
tan ( )	**tangent**
sin ( )	**sine**
sqrt ( )	**square root**
asin ( )	**arc sine**
acos ( )	**arc cosine**
atan ( )	**arc tangent**
sinh ( )	**hyperbolic sine**
cosh ( )	**hyperbolic cosine**
tanh ( )	**hyperbolic tangent**

Failures

Sometimes model geometry cannot be constructed because features that have been modified or created conflict with or invalidate other features. This can happen when the following occurs:

- A protrusion is created that is unattached and has a one-sided edge.
- New features are created that are unattached and have one-sided edges.
- A feature is resumed that now conflicts with another feature (i.e. two chamfers on the same edge).
- The intersection of features is no longer valid because dimensional changes have moved the intersecting surfaces.
- A relation constraint has been violated.

After a feature fails, Creo Parametric will roll back the model to the last regenerated feature and display the children as red in the Model Tree. You can edit the definition of the failed feature at this juncture.

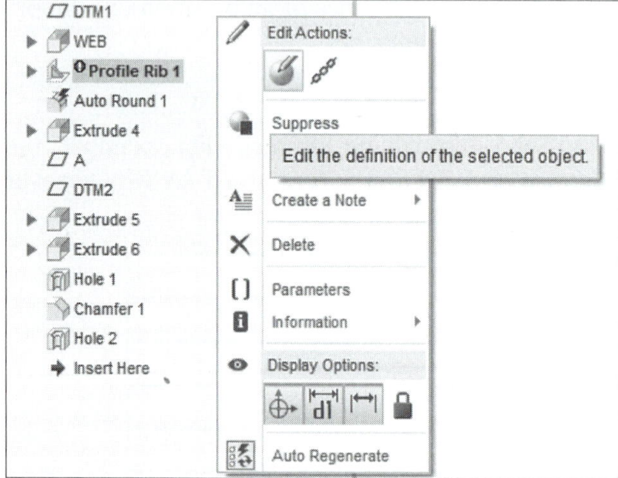

Family Tables

Family Tables are effective for two main reasons: they provide a beneficial tool, and they are easy to use. You need to understand the functionality of Family Tables, and you must understand when a Family Table is required and what circumstances should promote its use.

To determine whether a model is a candidate for a Family Table: establish whether the original and the variation would ever have to co-exist at the same time (both in the same assembly, both shown in the same drawing, both with an independent Bill of Materials) and whether they should be tied together (most of the same dimensions, features, and parameters). If so, the component is a candidate for the creation of a Family Table [Figs. 7.4(a-b)], otherwise, the model may be a candidate for copying to an independent model.

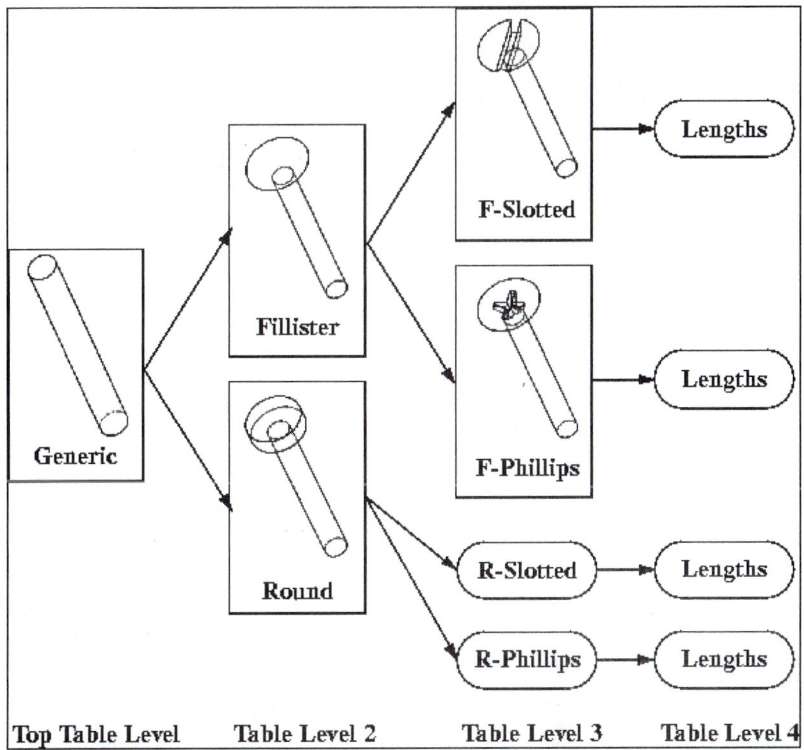

Figure 7.4(a) Screw Family

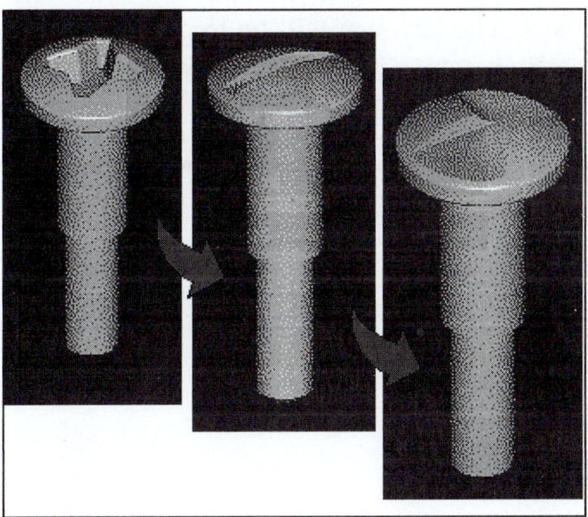
Figure 7.4(b) Screw Family

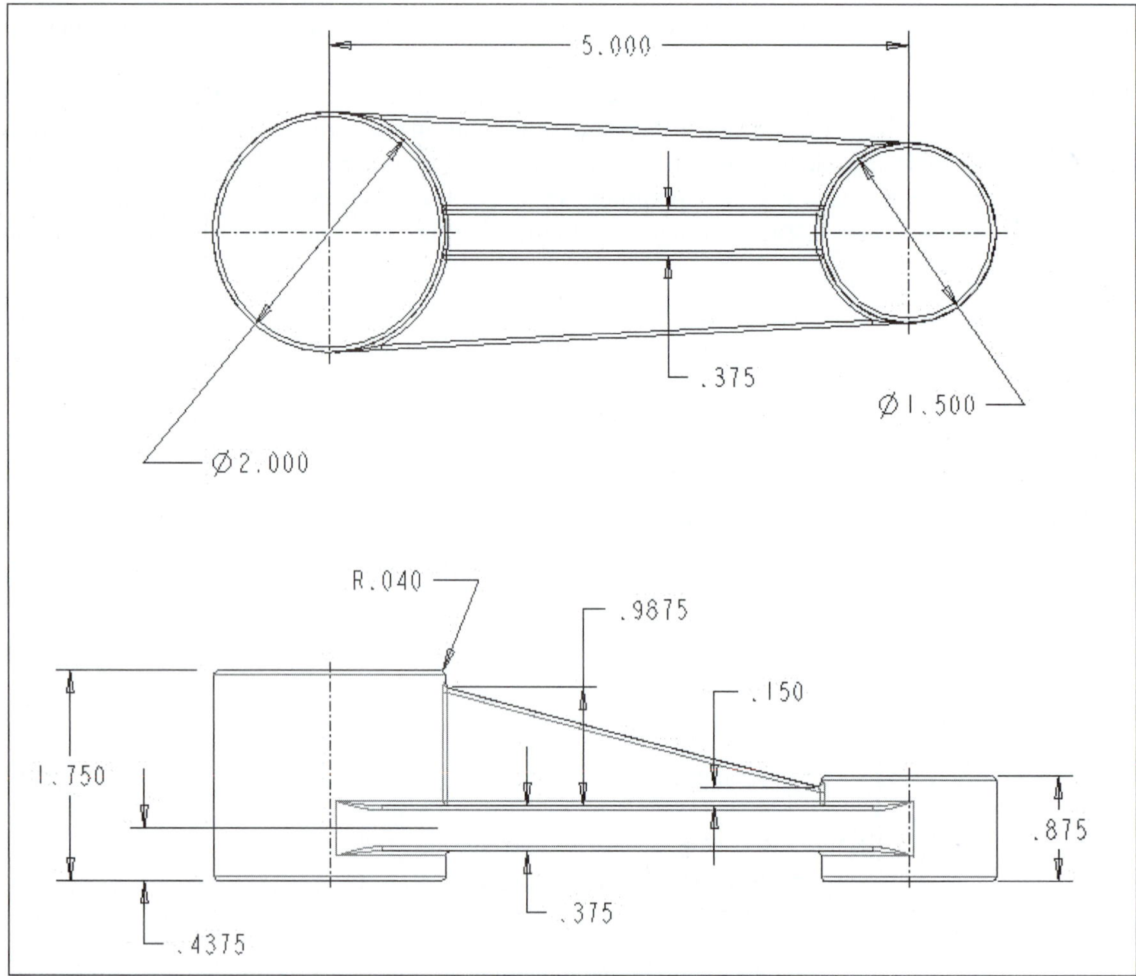

Figure 7.5 Clamp Arm Workpiece Drawing

Clamp Arm

The Clamp_Arm is modeled in two stages. Model the casting (Fig. 7.5) using the casting detail, and then use the machine detail to complete the last (machined) features. The last step will be to create a Family Table with an instance that suppresses the machined features. By having a *casting part* (***workpiece***) and a separate but almost identical *machined part* (***design part***) you can create an operation for machining and an NC sequence. During the manufacturing process, *you merge the workpiece into the design part* and create a ***manufacturing model***.

The difference between the two files is the difference between the volume of the workpiece/casting part and the volume of the design part/machined part. The removed volume can be seen as *material removal* when you are performing an **NC Check** operation on the manufacturing model. If the machining process gouges the part, the gouge will display the interference. The cutter location can also be displayed as an animated machining process.

Launch **PTC Creo Parametric 3.0** > **File** > **Manage Session** > **Select Working Directory** > select your working directory > **OK** > **File** > **New** > *type* **clamp_arm** > **OK** > **File** > **Options** > **Configuration Editor** > **Import/Export** > **Import configuration file** > **Creo_textbook.pro** *(open your previously created option file from Lesson 6)* > **Open** > **OK** > **No** > **File** > **Prepare** > **Model Properties** > Material **change** > **steel.mtl** > ▶▶▶ > **OK** > **Close** > change the coordinate system name to **clamp_arm_csys** > **Enter** > in the Graphics Window, **LMB** to deselect > [Save icon] > **OK** > **View** tab > [datum display icons] *on* > [datum display icons] *on* > [icon] *off* > **Model** tab > pick on datum **FRONT** to pre-select it > Extrude **Extrude** [Fig. 7.6(a)] > in the Graphics Window, press **RMB** > **Circle** > sketch a circle > move the mouse off of the circle > **MMB** to end the current tool > double-click on the diameter dimension > *type* **2.00** > **Enter** [Fig. 7.6(b)] > ✔ > click in the depth value field > *type* **1.75** > **Enter** [Fig. 7.6(c)] > ✔ > **Ctrl+S**

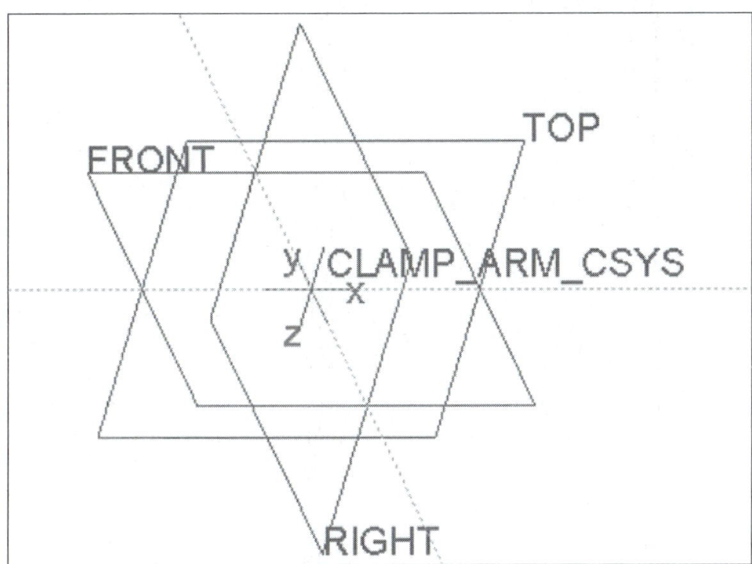

Figure 7.6(a) Sketch Plane (Display is set as Trimetric)

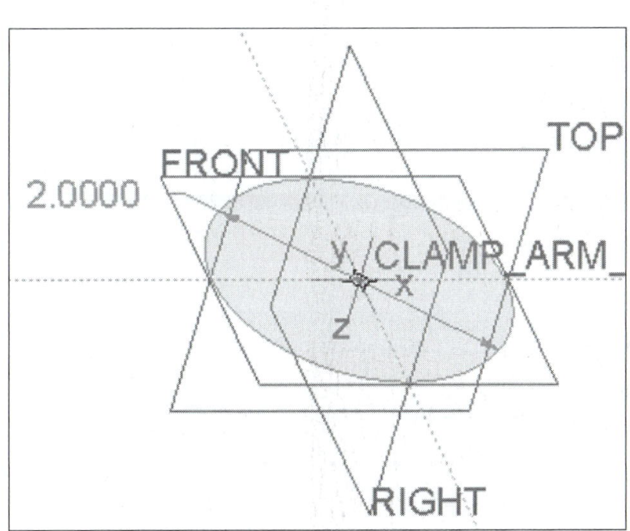

Figure 7.6(b) Completed Circle

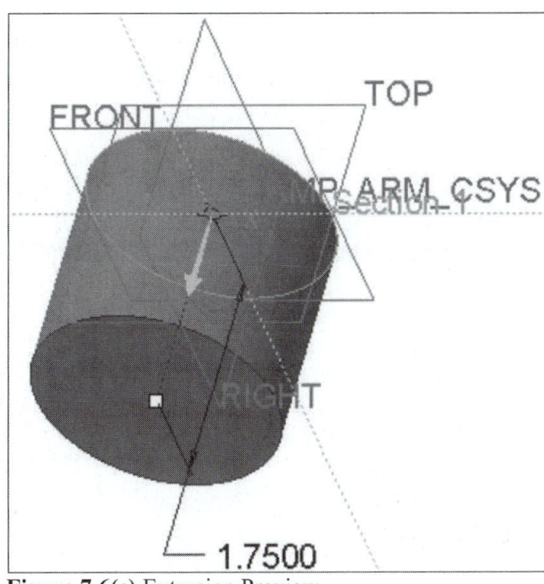

Figure 7.6(c) Extrusion Preview

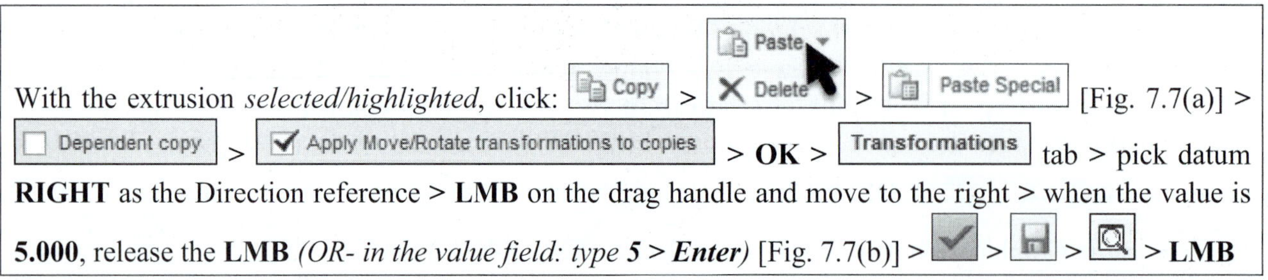

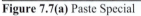

With the extrusion *selected/highlighted*, click: Copy > Paste / Delete > Paste Special [Fig. 7.7(a)] > Dependent copy > Apply Move/Rotate transformations to copies > **OK** > Transformations tab > pick datum **RIGHT** as the Direction reference > **LMB** on the drag handle and move to the right > when the value is **5.000**, release the **LMB** *(OR- in the value field: type* ***5*** *>* ***Enter)*** [Fig. 7.7(b)] > > > > **LMB**

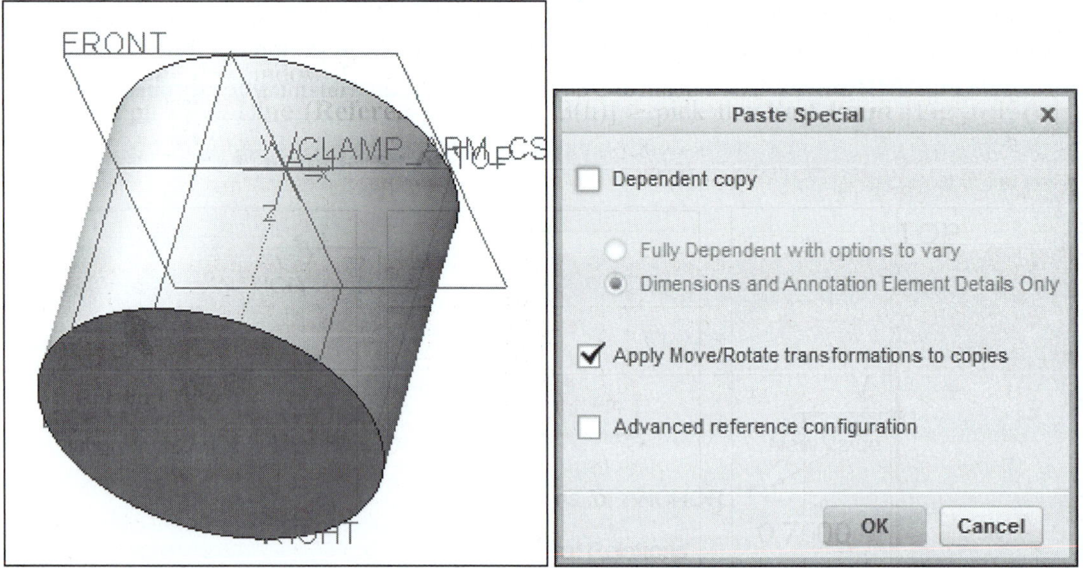

Figure 7.7(a) Paste Special

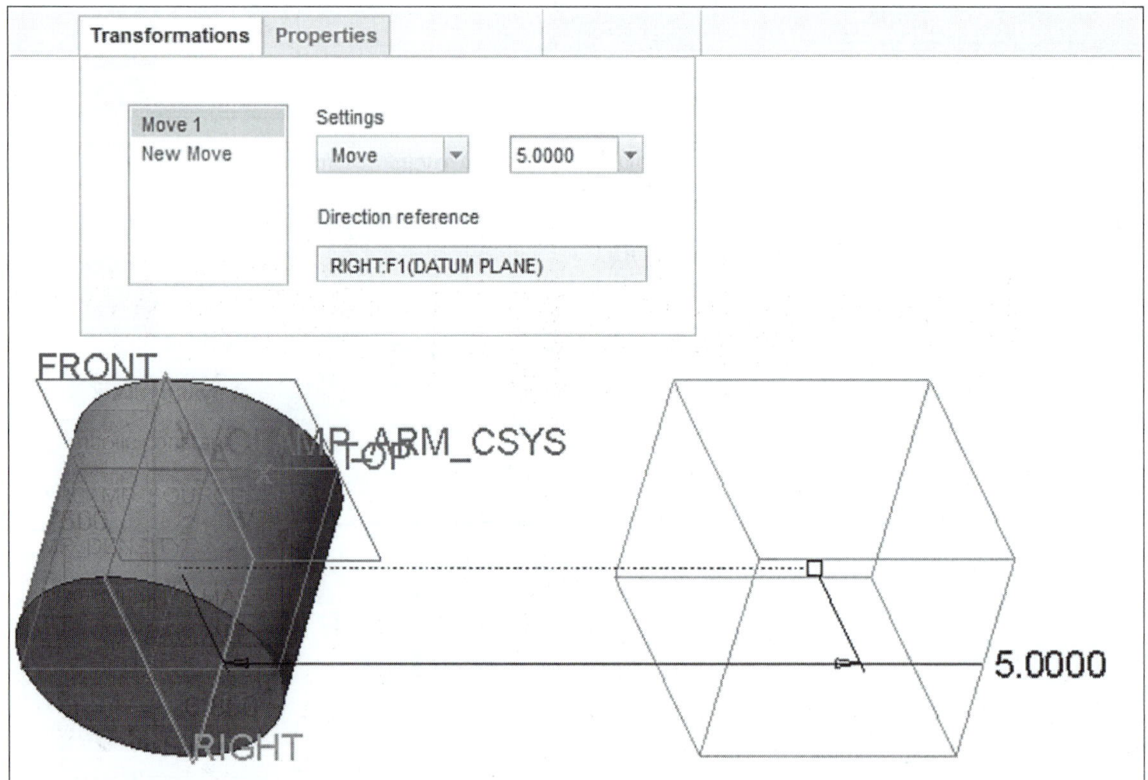

Figure 7.7(b) Transform the Copied Feature

Double-click on the copied feature > double-click on the height dimension > *type* **.875** > **Enter** > double-click on the diameter dimension > *type* **1.50** [Fig. 7.7(c)] > **Enter** > press **RMB** > **Exit Edit Dimension** > in the Graphics Window, **LMB** to deselect [Fig. 7.7(d)] > **Ctrl+S**

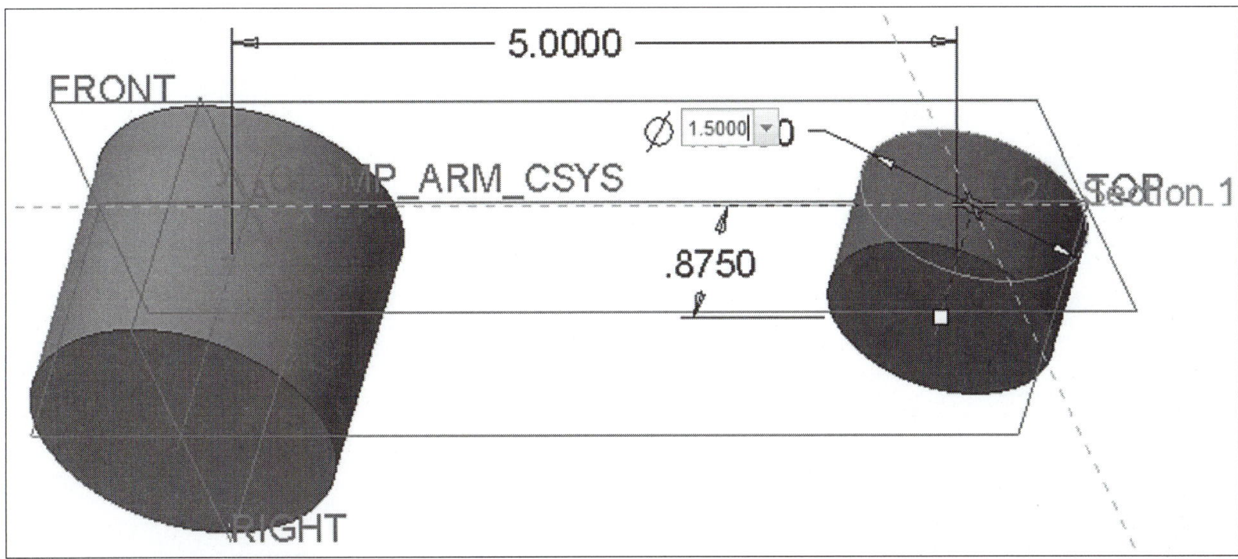

Figure 7.7(c) Modify the Dimensions *(double-click on each dimension and change to the design size)*

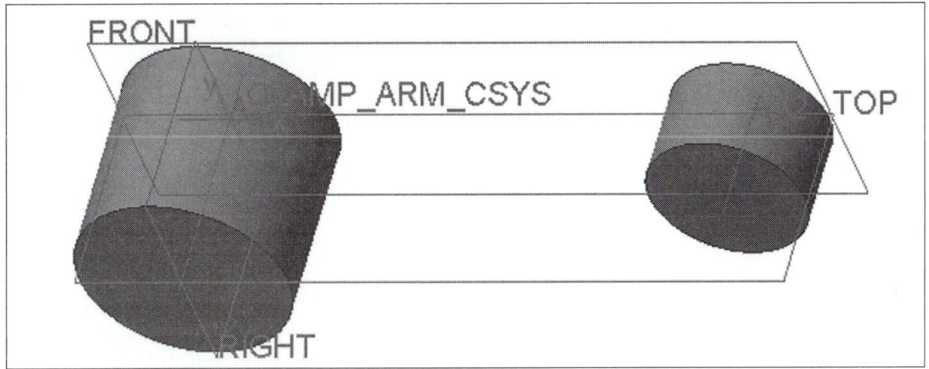

Figure 7.7(d) Regenerated Model

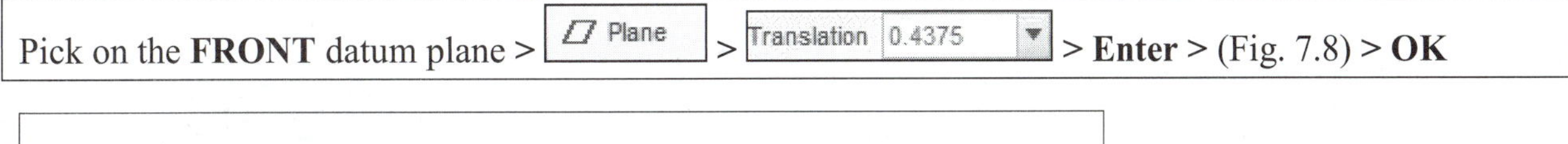

Pick on the **FRONT** datum plane > Plane > Translation 0.4375 > **Enter** > (Fig. 7.8) > **OK**

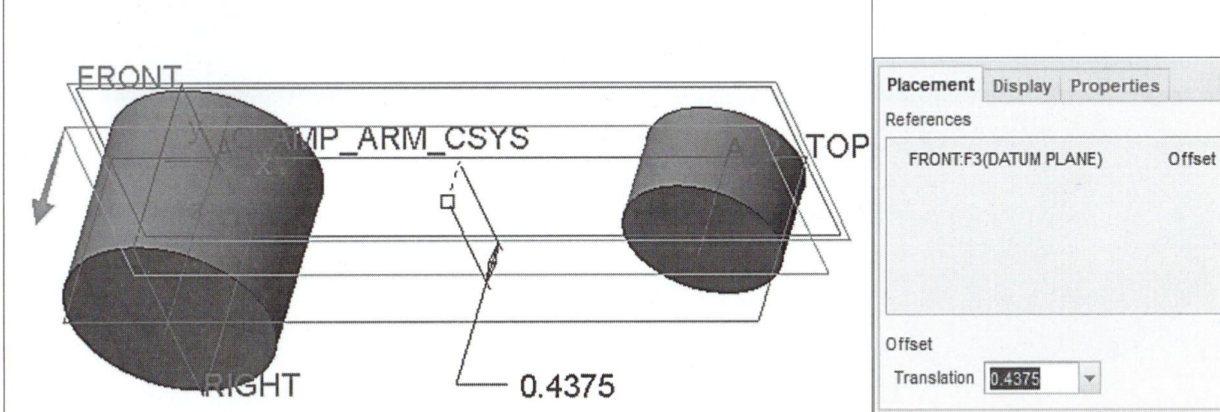

Figure 7.8 Offset Datum Plane

With datum **DTM1** *selected*, click: Extrude **Extrude** > **Sketch View** > **References** from the Ribbon Setup Group [Fig. 7.9 (a)] > select one arc from both circles [Fig. 7.9 (b] > **Solve** [Fig. 7.9 (c)] (Fully Placed) > **Ctrl+D** [Fig. 7.9(d)] > **Close** >

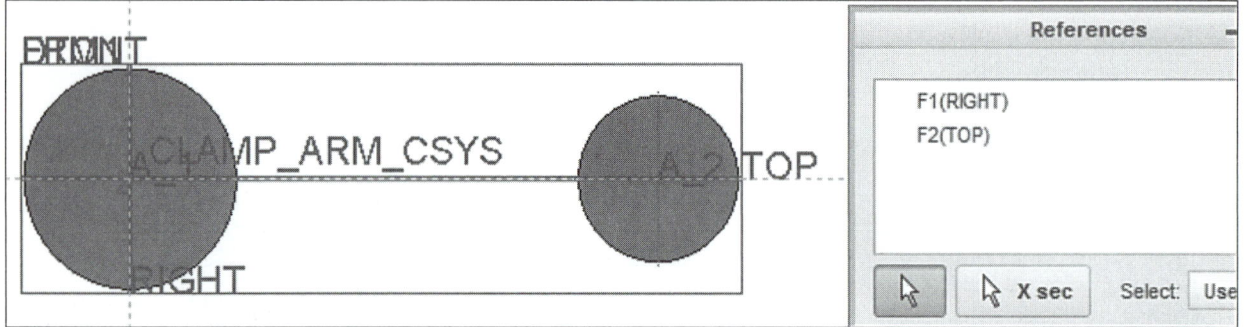

Figure 7.9(a) References

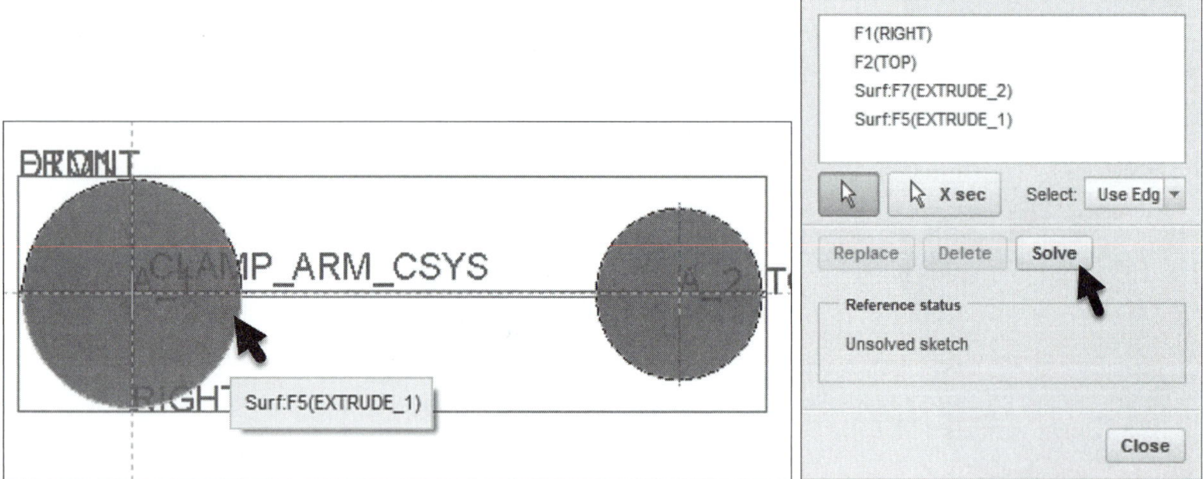

Figure 7.9(b) Add Reference

Figure 7.9(c) References Dialog Box

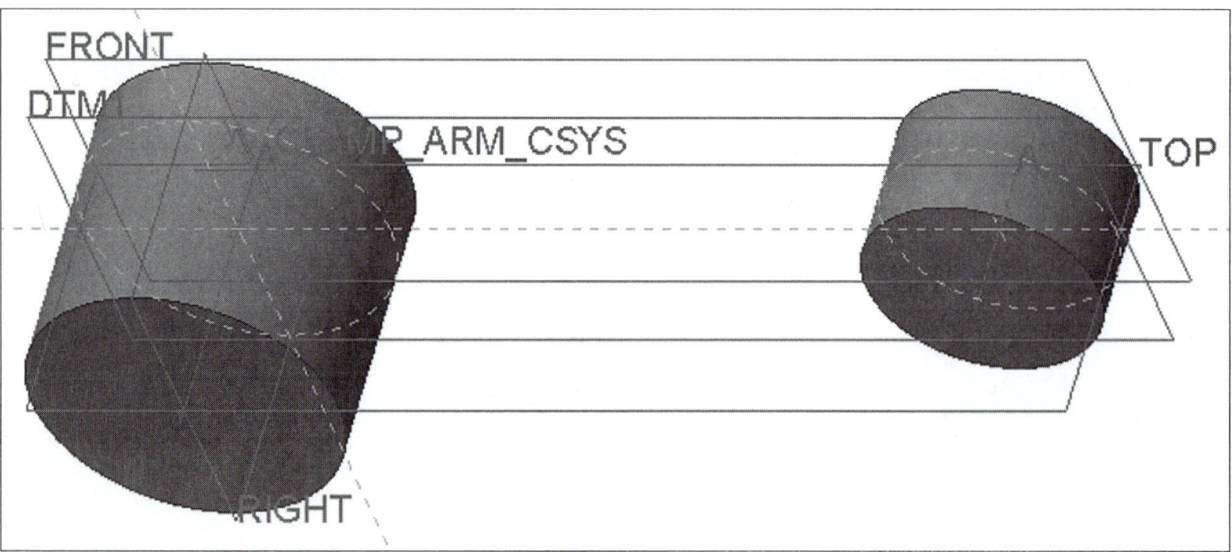

Figure 7.9(d) Four References

Press: **RMB** > **Construction Centerline** > create a horizontal centerline [Fig. 7.9(e)] > **MMB** > press **RMB** > **Circle** > pick on the center of a circular reference and then on its circular reference [Fig. 7.9(f)] > repeat on the opposite end [Fig. 7.9(g)] > **MMB** to end the Circle Tool > **Ctrl+D** [Fig. 7.9(h)]

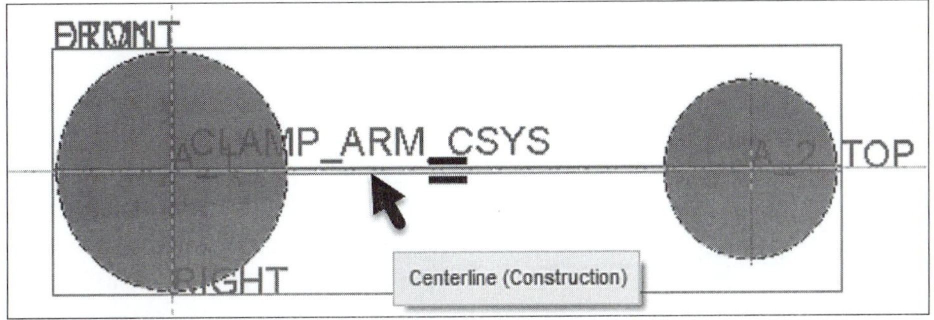

Figure 7.9(e) Add a Centerline

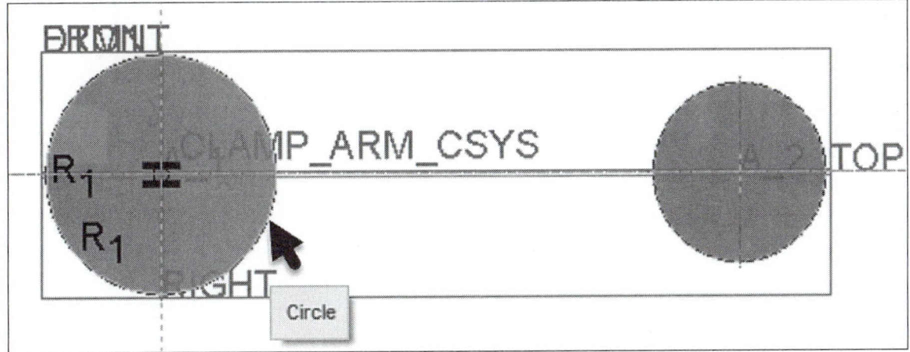

Figure 7.9(f) Create a Circle

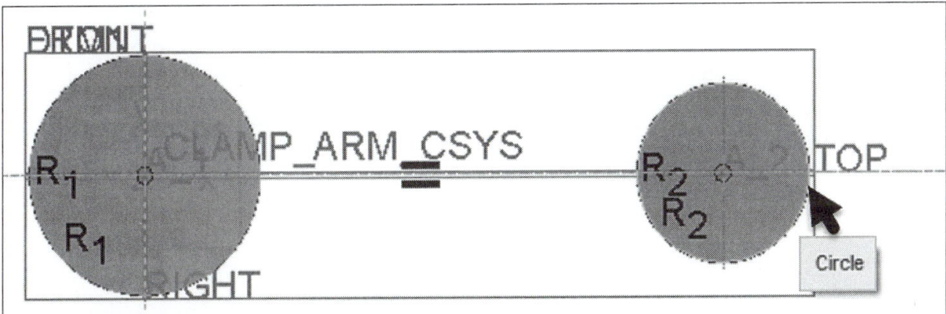

Figure 7.9(g) Create the Second Circle *(No weak dimensions will appear)*

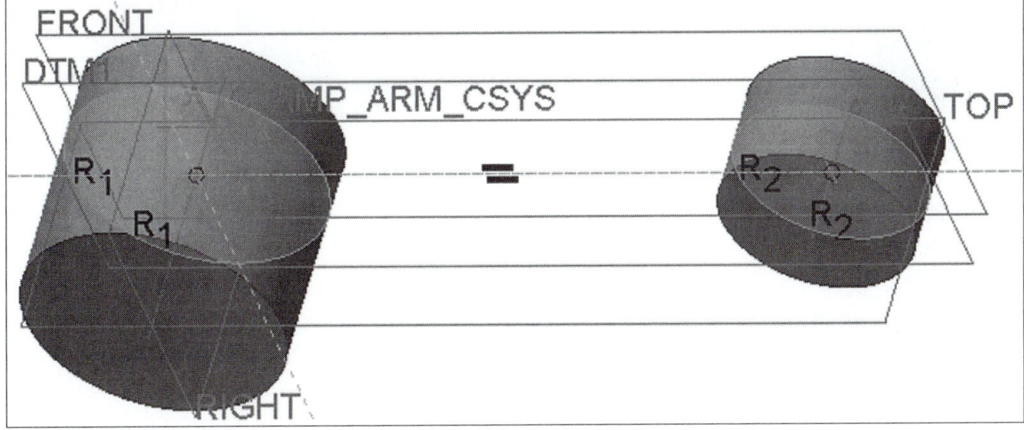

Figure 7.9(h) Completed Circles *(no dimensions should be displayed)*

Click: **Sketch View** > next to **Line Chain** > Line Tangent **Tangent Line** create a line tangent between the circular extrusions [Fig. 7.9(i)] > create the bottom tangent line [Fig. 7.9(j)] > **MMB** > **LMB** to deselect > **Ctrl+D** [Fig. 7.9(k)] > **Sketch** tab > **Sketch View**

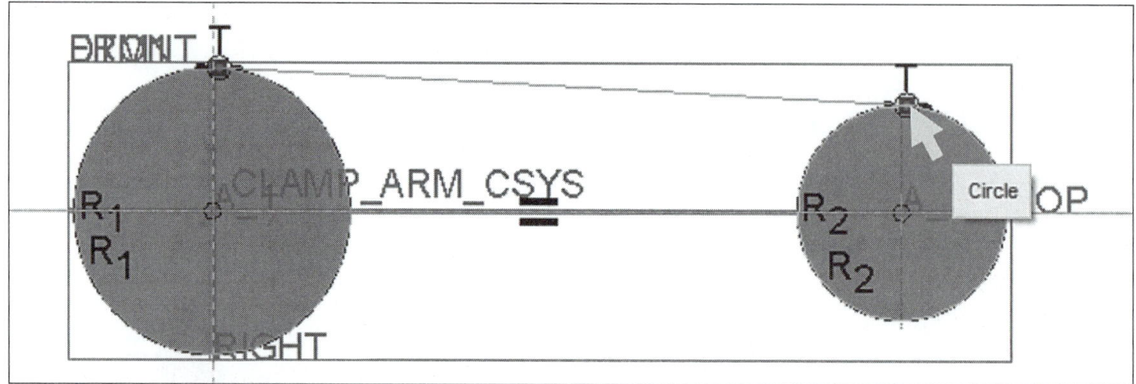

Figure 7.9(i) First Tangent Line

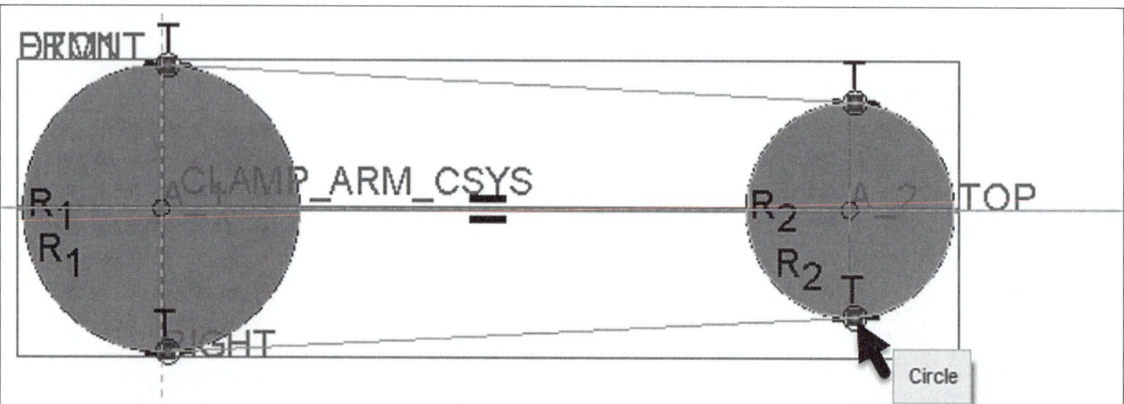

Figure 7.9(j) Second Tangent Line

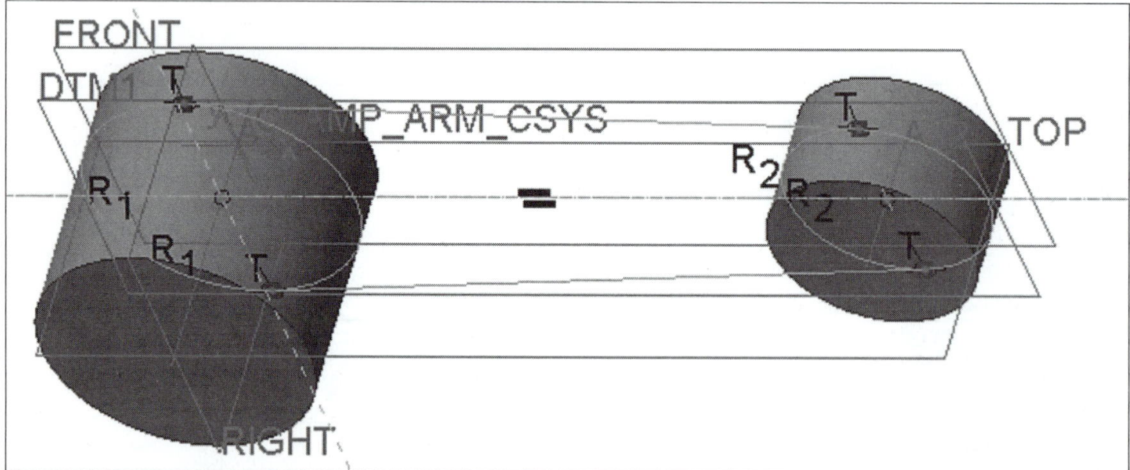

Figure 7.9(k) 3D Sketch *(T constraints should be displayed)*

Click: **View** tab > *off* > **Sketch** tab Delete Segment **Delete Segment** > press **LMB** and draw a spline through the unwanted entities on the right side [Fig. 7.9(l)] > release the **LMB** > draw a spline through the unwanted entities on the left side [Fig. 7.9(m)] > draw a spline through the tiny leftover piece on the lower-left side between the tangent position and the vertical reference [Fig. 7.9(n)] > repeat for the upper-left side [Fig. 7.9(o)]

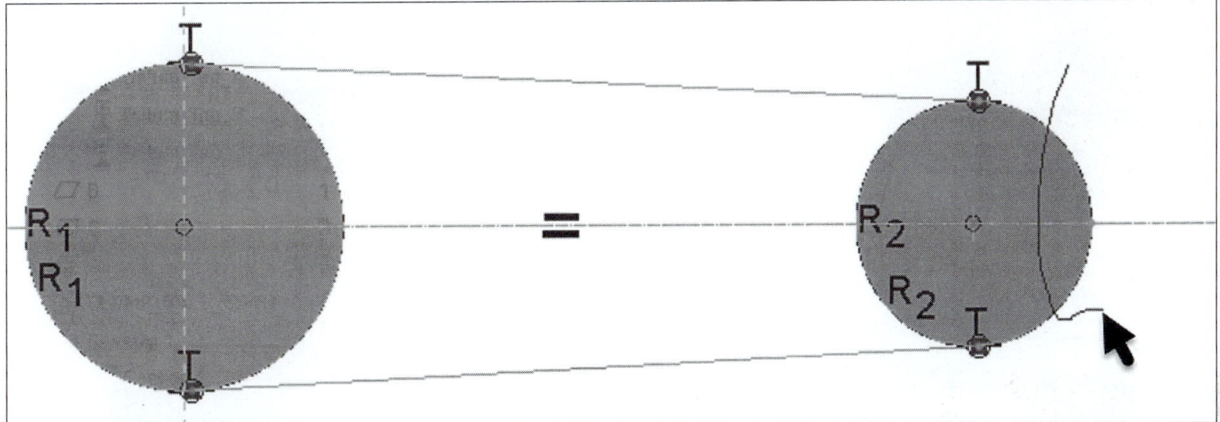

Figure 7.9(l) Trim Unwanted Entities on Right Side

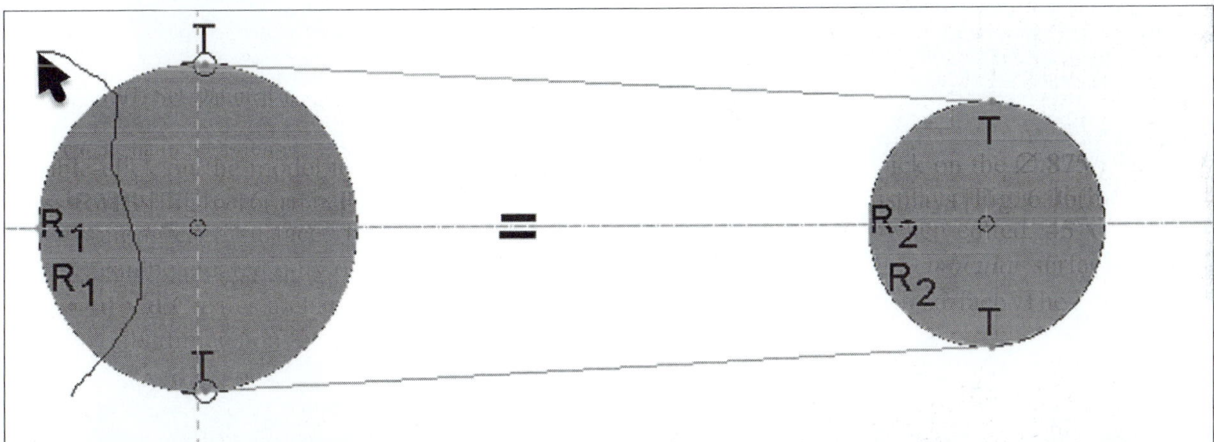

Figure 7.9(m) Trim Unwanted Entities on Left Side

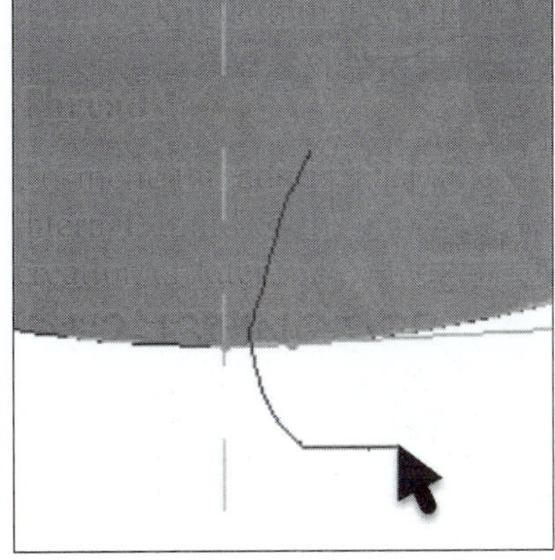

Figure 7.9(n) Trim Leftover on Bottom

Figure 7.9(o) Trim Leftover on Top

Press: **MMB** (section will shade if done correctly) [Fig. 7.9(p)] > **Ctrl+D** [Fig. 7.9(q)] > [Fig. 7.9(r)] > place the pointer over the drag handle > press **RMB** > **Symmetric** > modify depth value to **.375** 0.3750 > **Enter** [Fig. 7.9(s)]

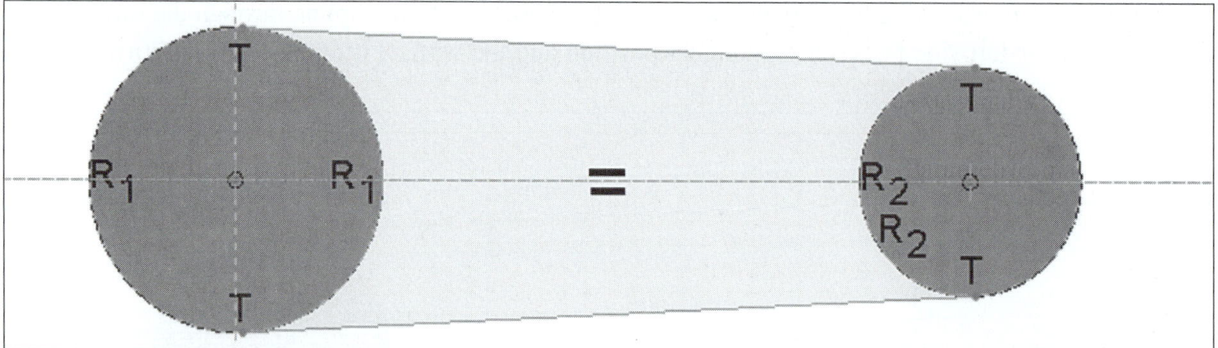

Figure 7.9(p) Completed Sketch [Note that there are no dimensions and four **T**'s (tangent constraints are displayed)]

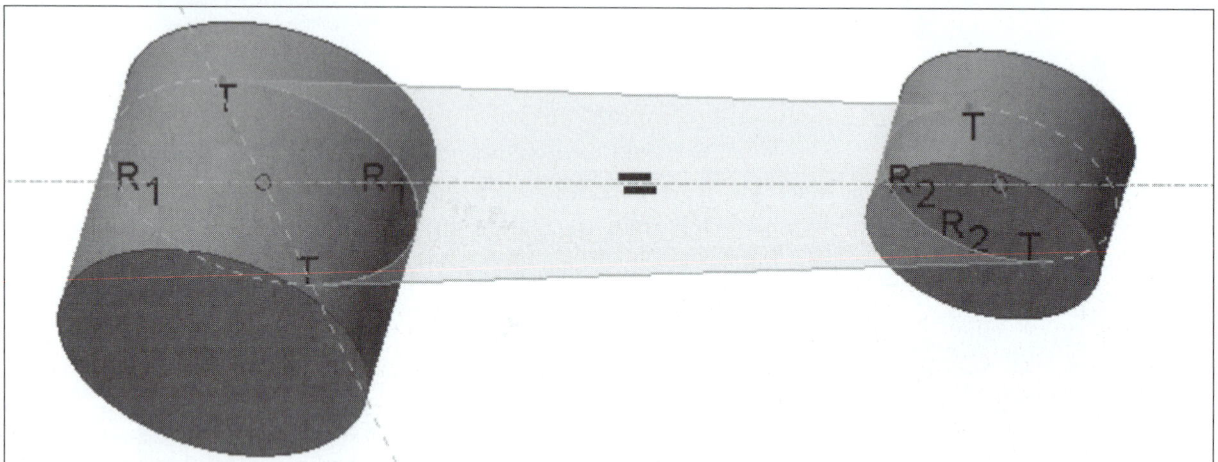

Figure 7.9(q) 3D Sketch

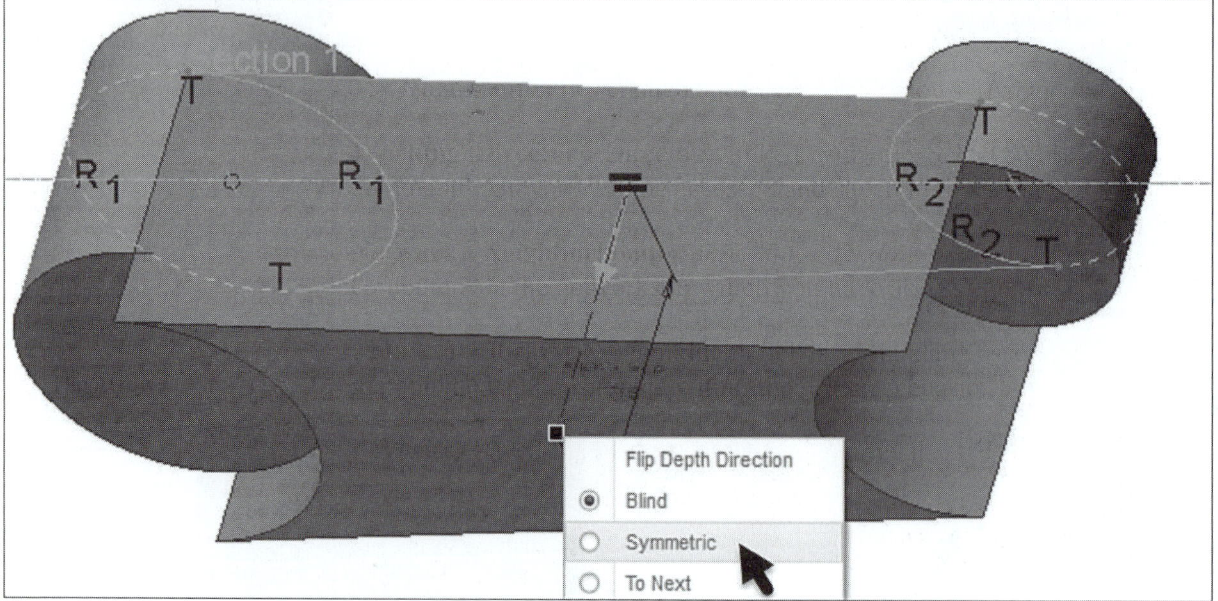

Figure 7.9(r) Preview

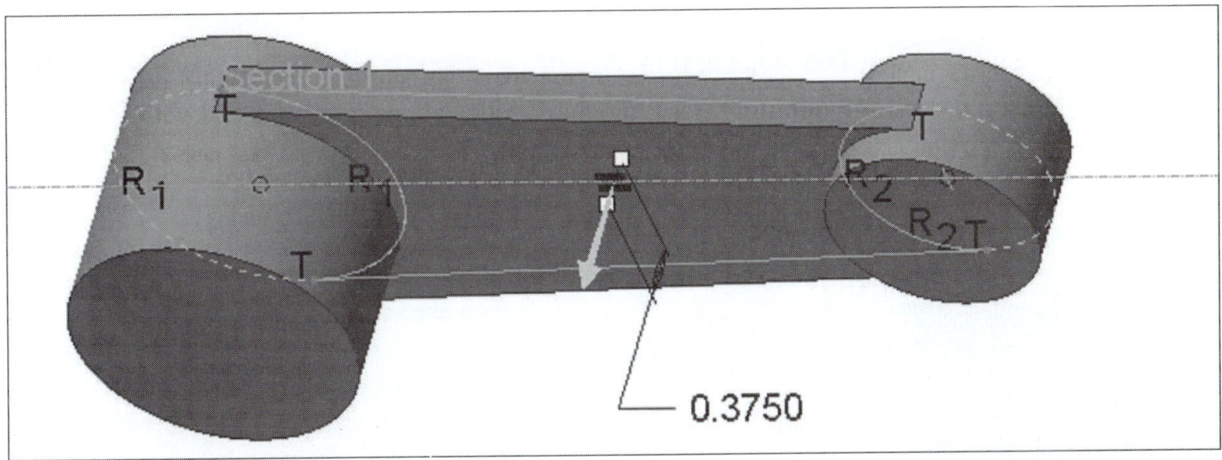

Figure 7.9(s) Preview Symmetric **.375** Web Feature

Click: [Fig. 7.9(t)] > **View** tab > *on* > *on* > **Ctrl+S** > in the Model Tree, click: to expand > select the sketch **Section 1** *(your name may be different)* for Extrude 3 [Fig. 7.9(u)] > in the Graphics Window, **LMB** to deselect

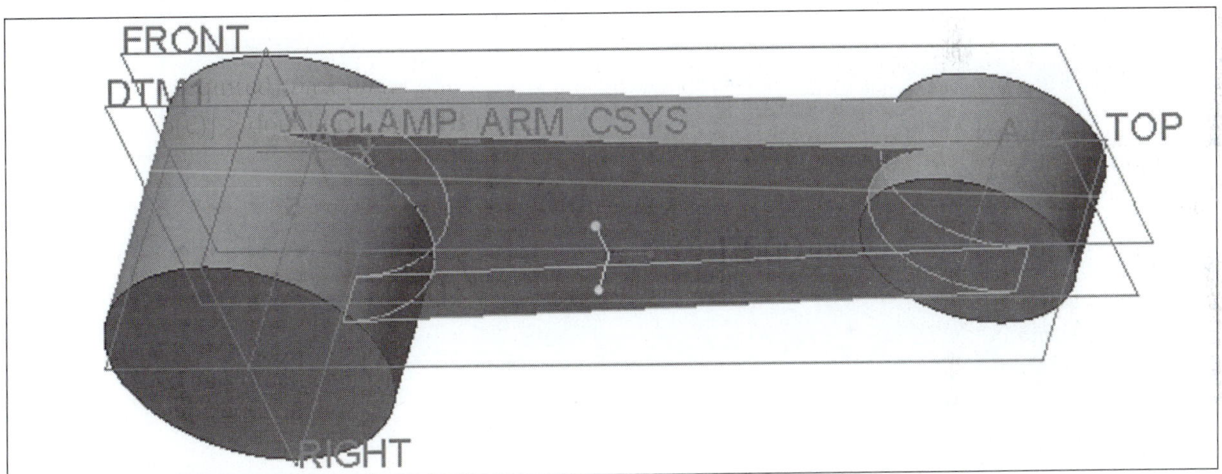

Figure 7.9(t) Completed Web Feature

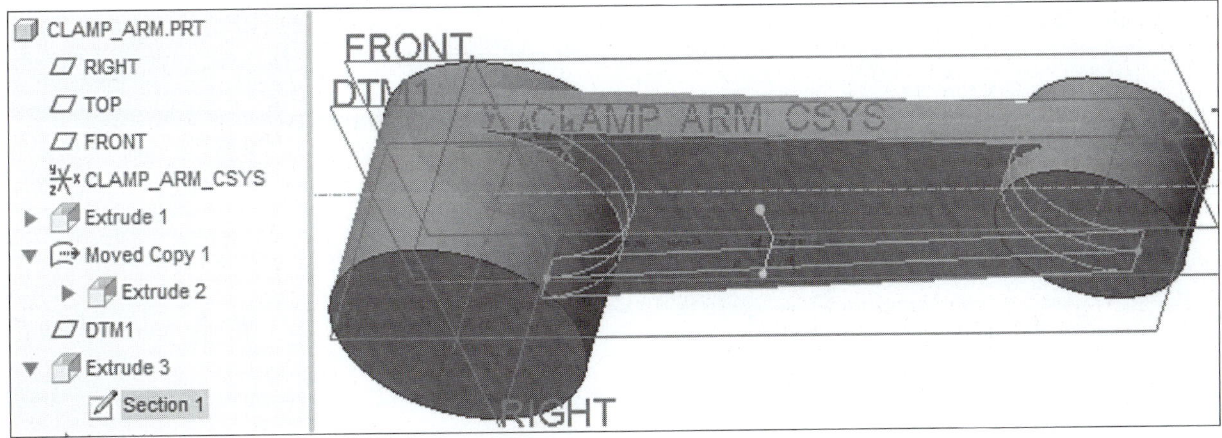

Figure 7.9(u) Model Tree

Click: **Model** tab > select datum **TOP** > [▼] next to [Rib] > [Profile Rib] **Profile Rib** > [icon] **Sketch View** > in the Graphics Window, press **RMB** > **Section Orientation** > **Set horizontal reference** [Fig. 7.10(a)] > pick on the **Line (Reference)** [Figs. 7.10(b-c)] > press **RMB** > **References** [Fig. 7.10(d)]

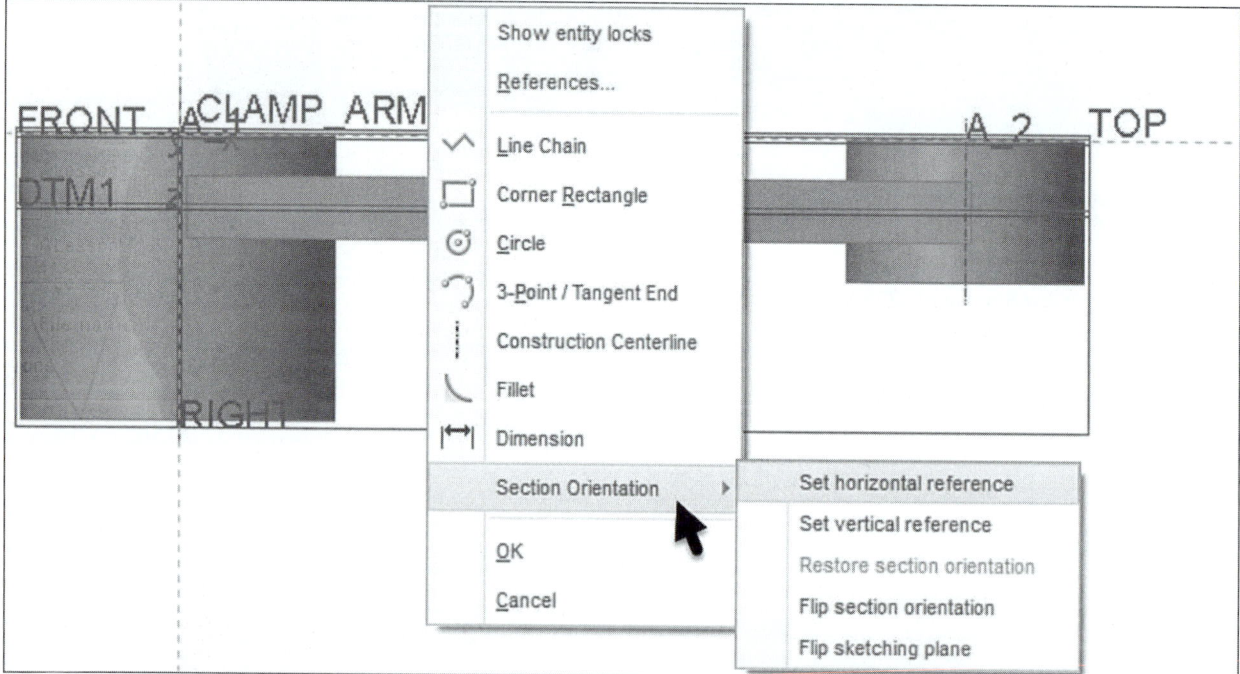

Figure 7.10(a) Profile Rib Tool and its Section Orientation

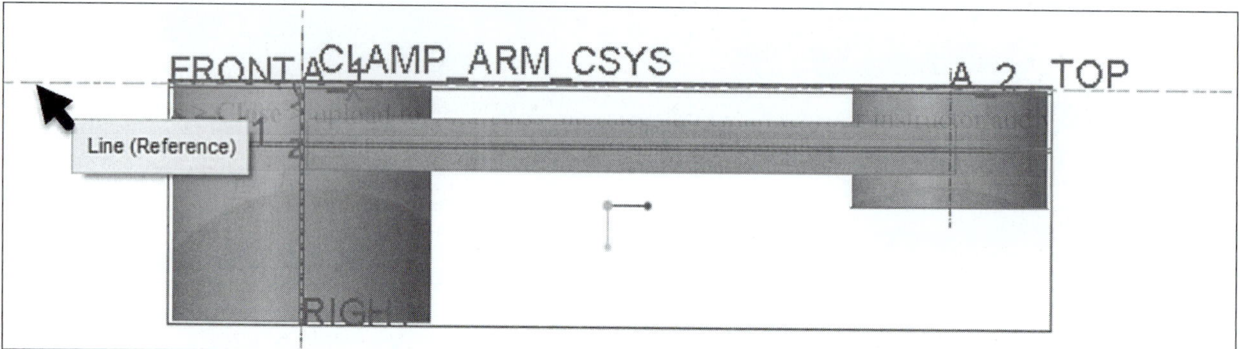

Figure 7.10(b) Flip the Viewing Direction

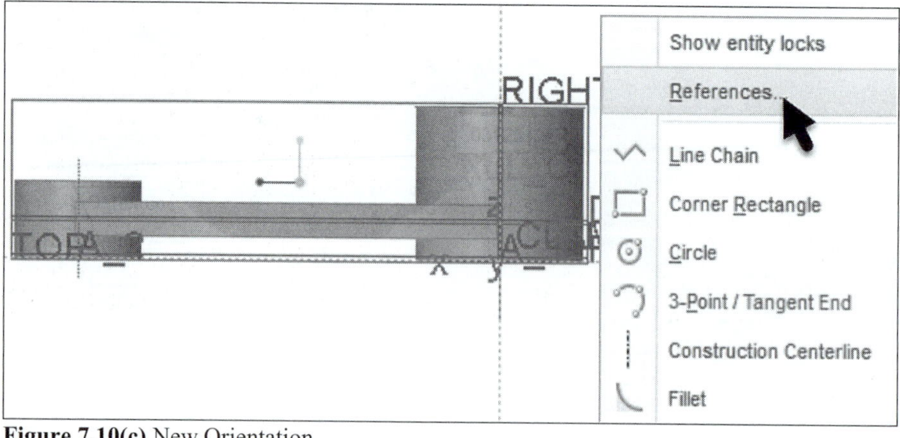

Figure 7.10(c) New Orientation

Figure 7.10(d) References

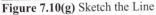

Click: [icon] > add three new references, pick on the inside vertical edge of both cylindrical extrusions and the top edge of the web extrusion [Figs. 7.10(e-f)] > **Solve** > **Close** > press **RMB** > **Line Chain** draw one angled line from one vertical reference to the other [Fig. 7.10(g)] (zoom in if necessary to pick the vertical surface/edge, not the endpoint) > move the pointer > **MMB** > **MMB** [Fig. 7.10(h)] > **LMB** to deselect

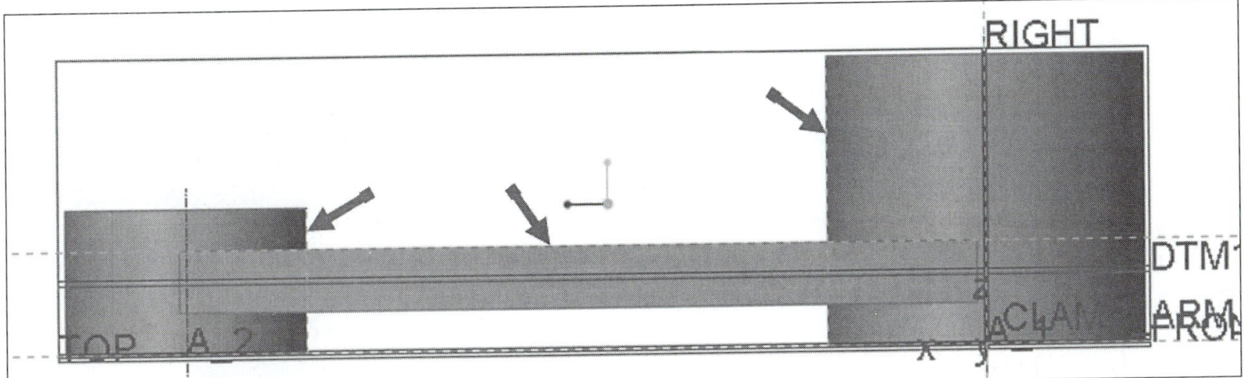

Figure 7.10(e) New References

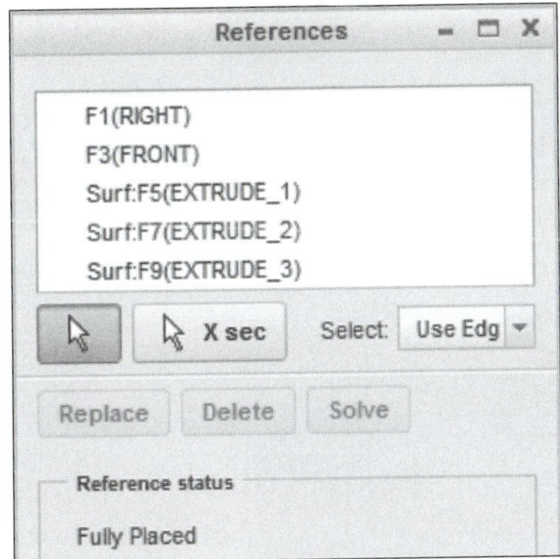

Figure 7.10(f) References Dialog Box

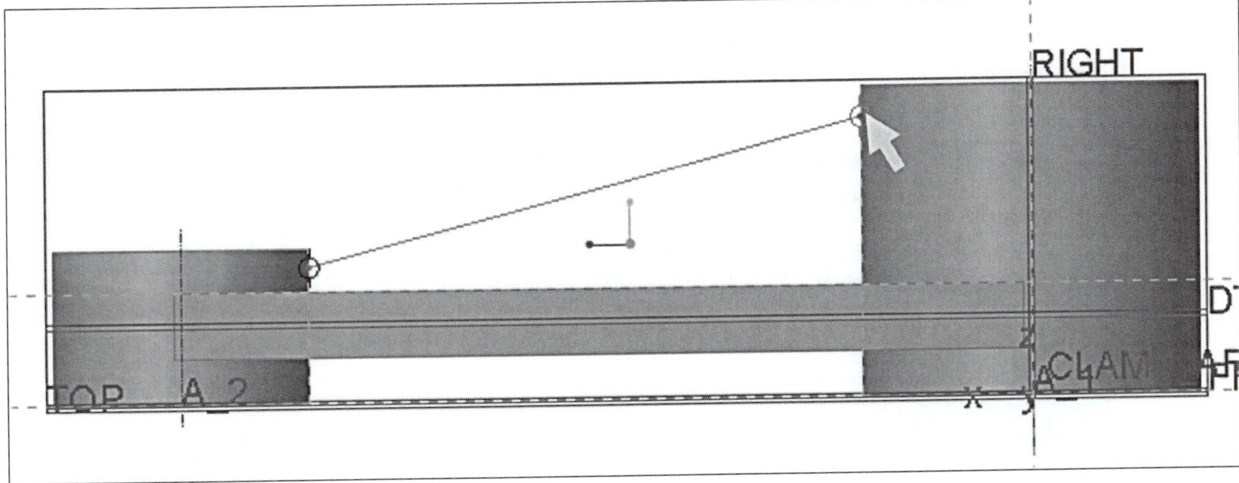

Figure 7.10(g) Sketch the Line

Press: **RMB** > **Dimension** add vertical dimensions from the edge reference to the end points of the angled line [Fig. 7.10(i)] > move the pointer > **MMB** > **MMB** > double-click on each dimension and modify the values (**.9875** and **.15**) [Fig. 7.10(j)] > **Ctrl+D** [Fig. 7.10(k)]

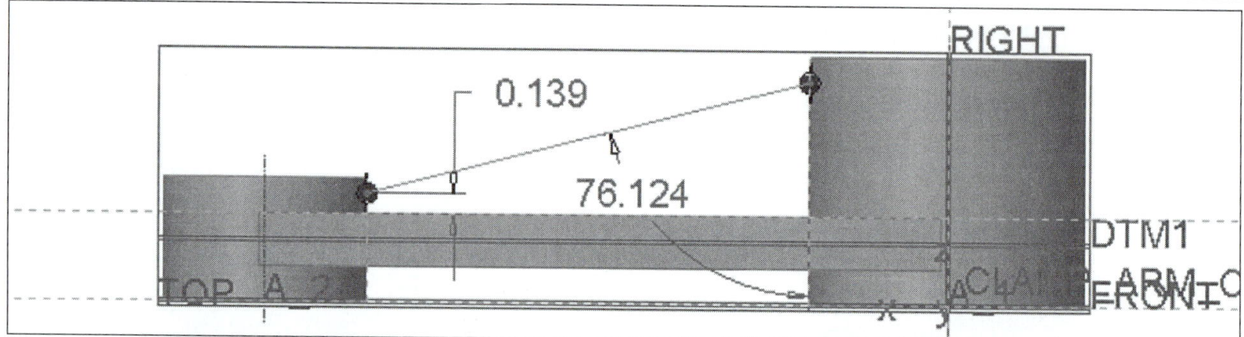

Figure 7.10(h) Dimensions as Sketched *(your values will be different)*

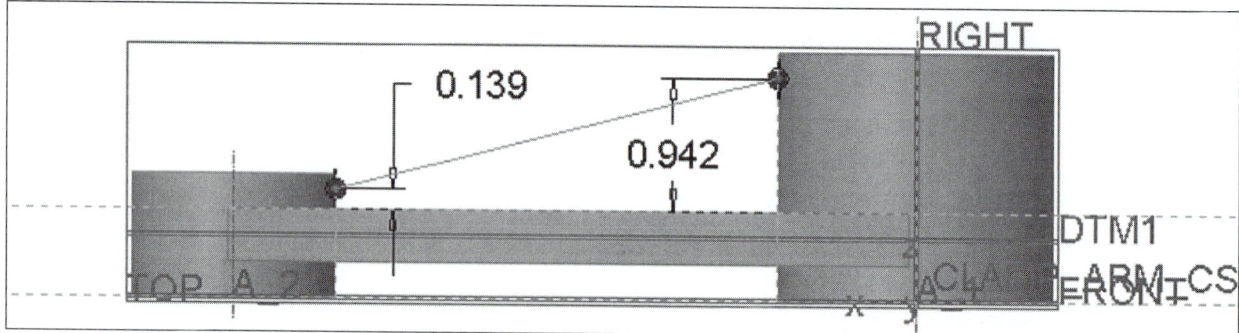

Figure 7.10(i) Create New dimensions *(your values will be different)*

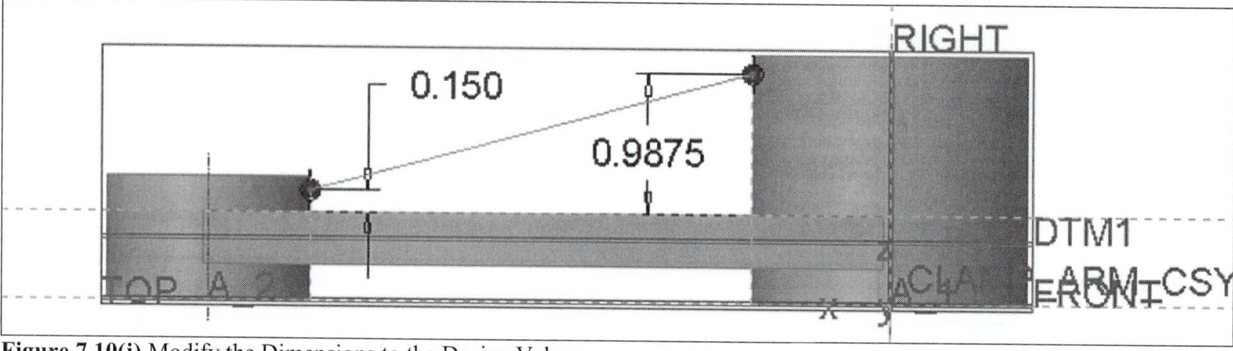

Figure 7.10(j) Modify the Dimensions to the Design Values

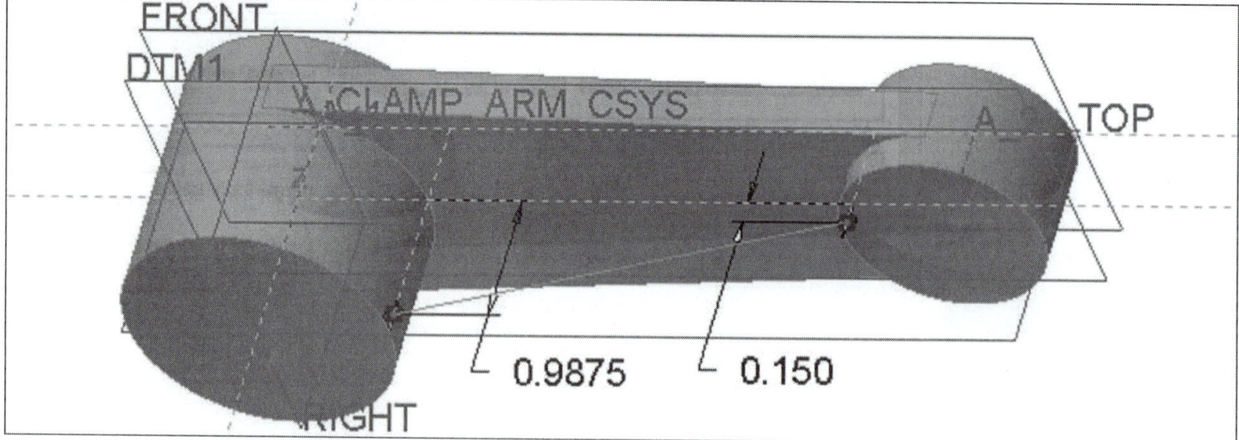

Figure 7.10(k) Rib Sketch in Standard Orientation

Click: ✓ > pick on the arrow to flip the direction of the rib creation towards the part [Fig. 7.10(l)] > press **LMB** on a drag handle > move the pointer until a rib thickness of **.375** is displayed [Fig. 7.10(m)] > release the **LMB** [0.375] *(or type in the value and press Enter)* > ✓ > **Ctrl+S** [Fig. 7.10(n)] > **LMB** in the Graphics Window to deselect

(Note: clicking [icon] *in the* ***Dashboard*** *of the* ***Profile Rib Tool*** *toggles the rib thickness from centered (both sides), to the right side, or to the left side of the sketch plane.)*

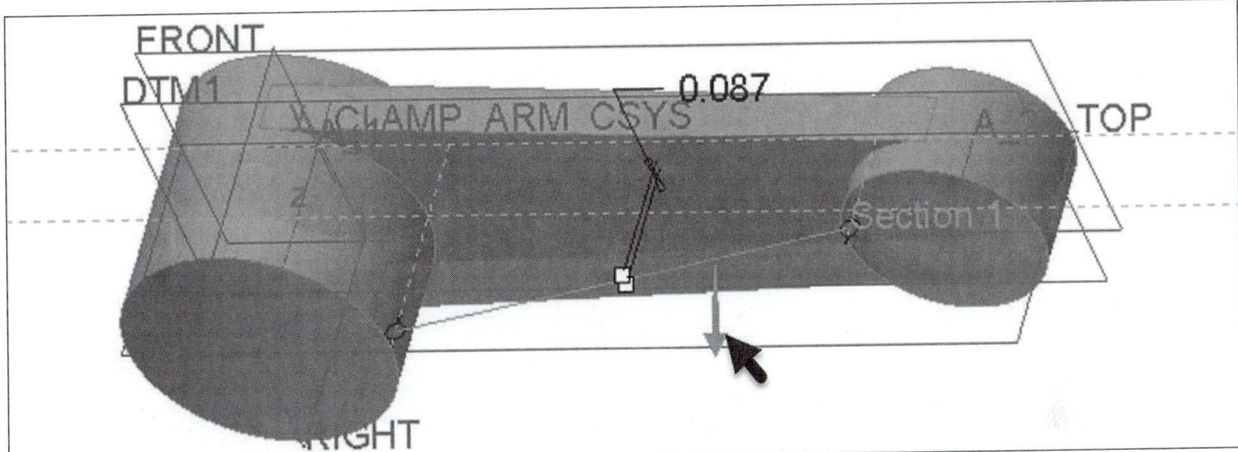

Figure 7.10(l) Flip Arrow (with the mouse pointer on the arrow, arrow color changes)

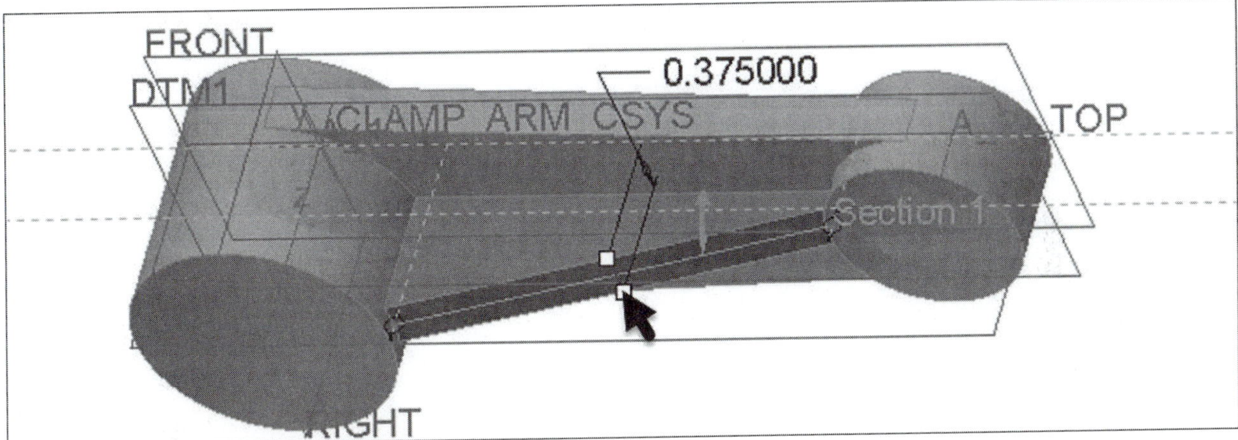

Figure 7.10(m) Previewed Rib

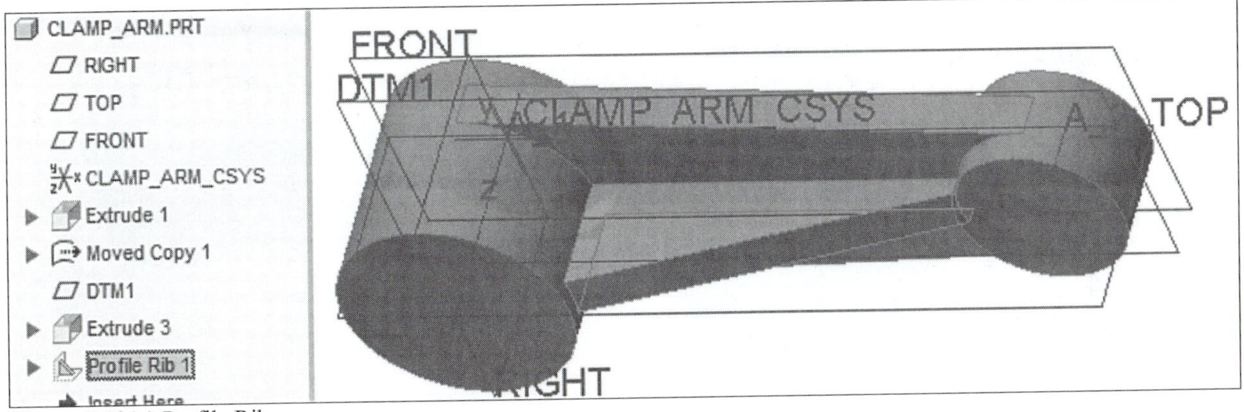

Figure 7.10(n) Profile Rib

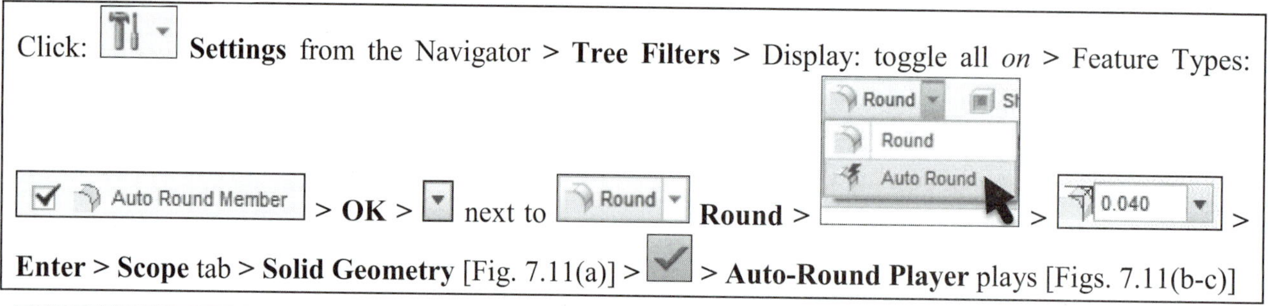

Click: **Settings** from the Navigator > **Tree Filters** > Display: toggle all *on* > Feature Types: Auto Round Member > **OK** > next to Round **Round** > Auto Round > 0.040 > **Enter** > **Scope** tab > **Solid Geometry** [Fig. 7.11(a)] > > **Auto-Round Player** plays [Figs. 7.11(b-c)]

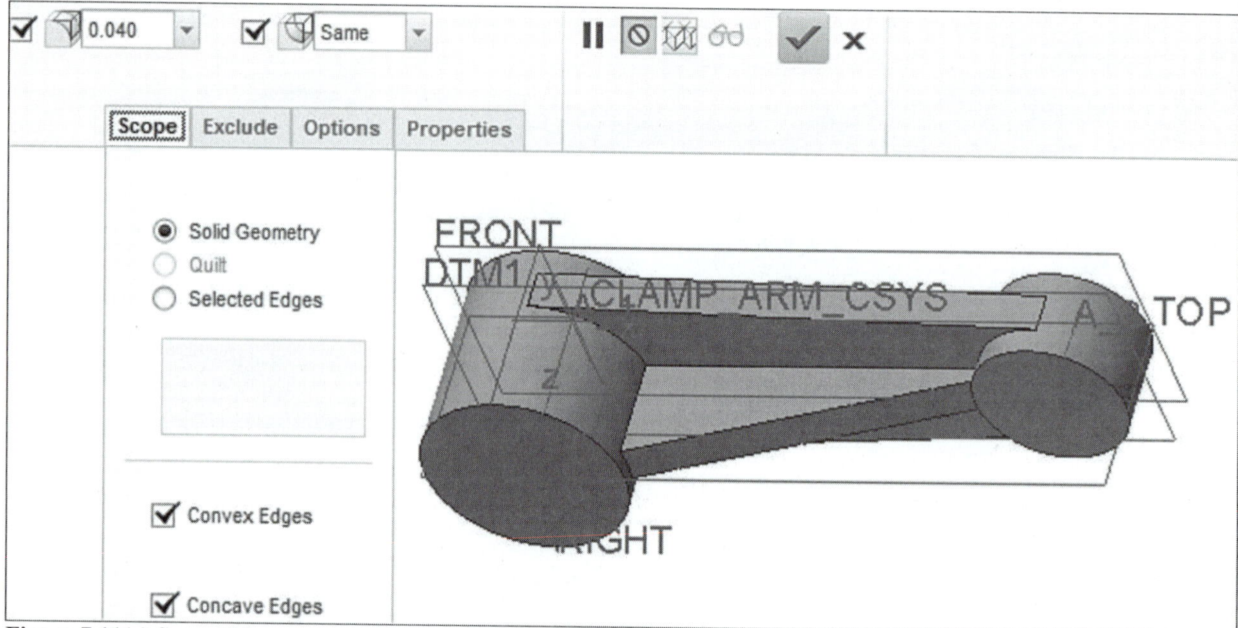

Figure 7.11(a) Scope Tab

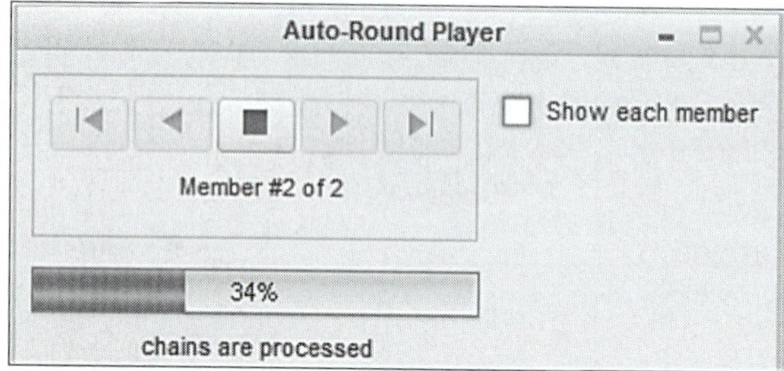

Figure 7.11(b) Auto-Round Player

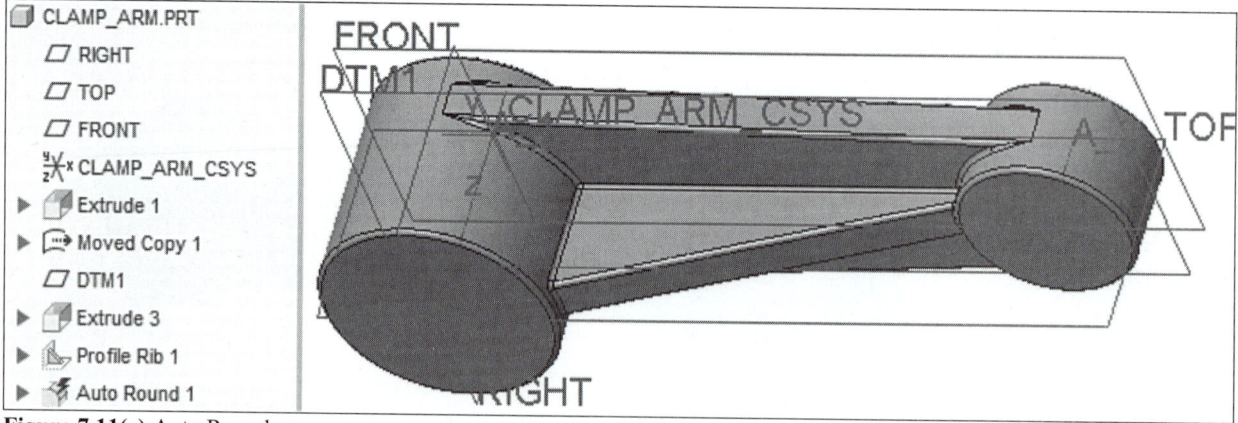

Figure 7.11(c) Auto Round

The "machined" features of the part can be created with cuts to "face" the ends of the cylindrical protrusions. A standard tapped hole, and a through hole with two countersinks are also added as "machined" features. You will be using the dimensions from the machine drawing (Fig. 7.12).

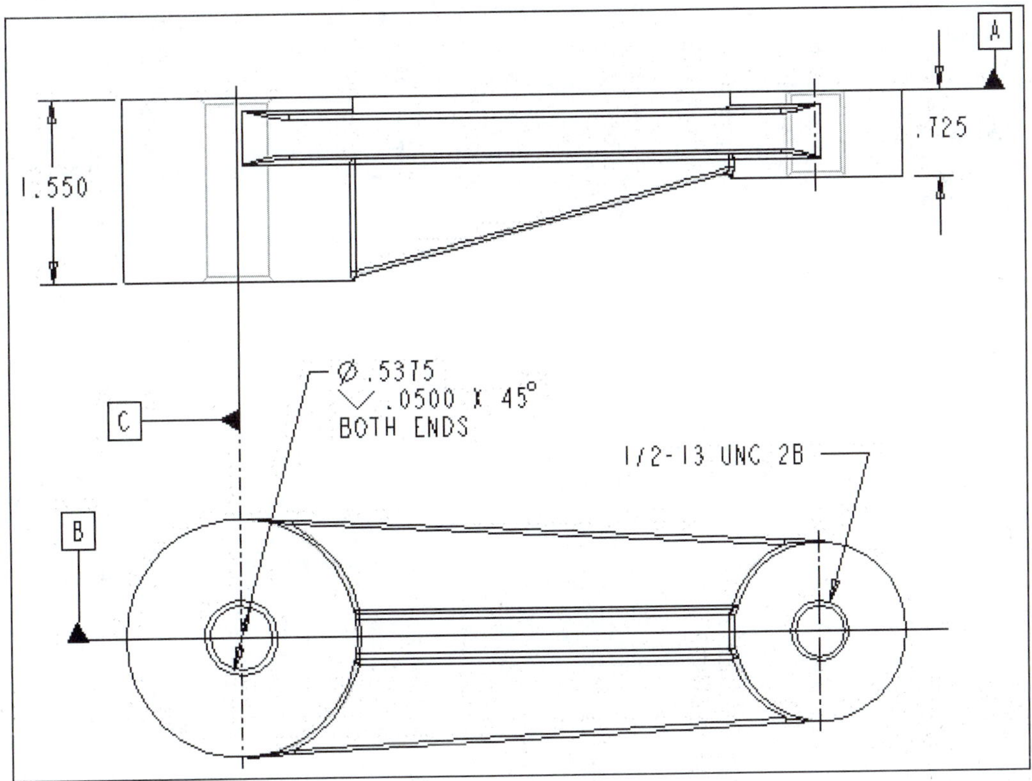

Figure 7.12 Clamp Arm Machining Drawing

Click: **LMB** to deselect > **Ctrl+S** > Extrude **Extrude** > in the Graphics Window, press **RMB** > **Remove Material** > press **RMB** > **Define Internal Sketch** > pick datum **TOP** > **Flip** [Fig. 7.13(a)] > **Sketch** > **Sketch View** > **View** tab > *off* > *off* > **Display Style** > **Shading**

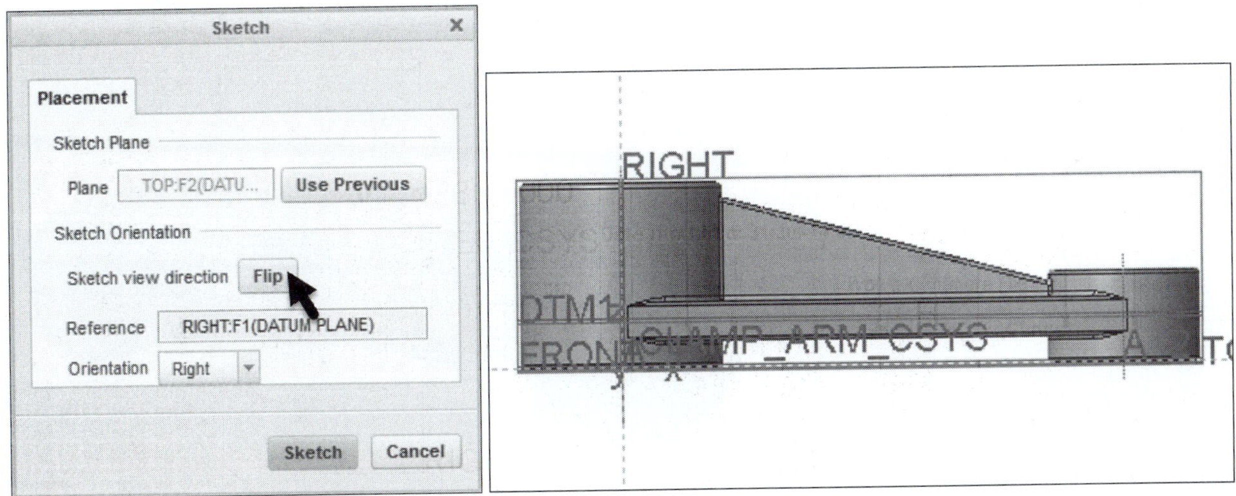

Figure 7.13(a) Flip Viewing Direction

Click: **Sketch** tab > in the Graphics Window, press **RMB** > **References** > pick the outside vertical edge of both cylindrical extrusions [Fig. 7.13(b)] > **Solve** > **Close** > press **RMB** > **Line Chain** > zoom in as needed > draw a horizontal line from one vertical reference to the other > move the pointer > **MMB** > **MMB** > **LMB** > double-click on the dimension > *type* **.100** [Fig. 7.13(c)] > **Enter** > move the pointer off of the dimension > **LMB** to deselect > press **RMB** > **OK** > **Ctrl+D** > press **RMB** > **Flip Material Side** > **Options** tab > Side 1: **Blind** > **Through All** > Side 2: **None** > **Through All** [Fig. 7.13(d)]

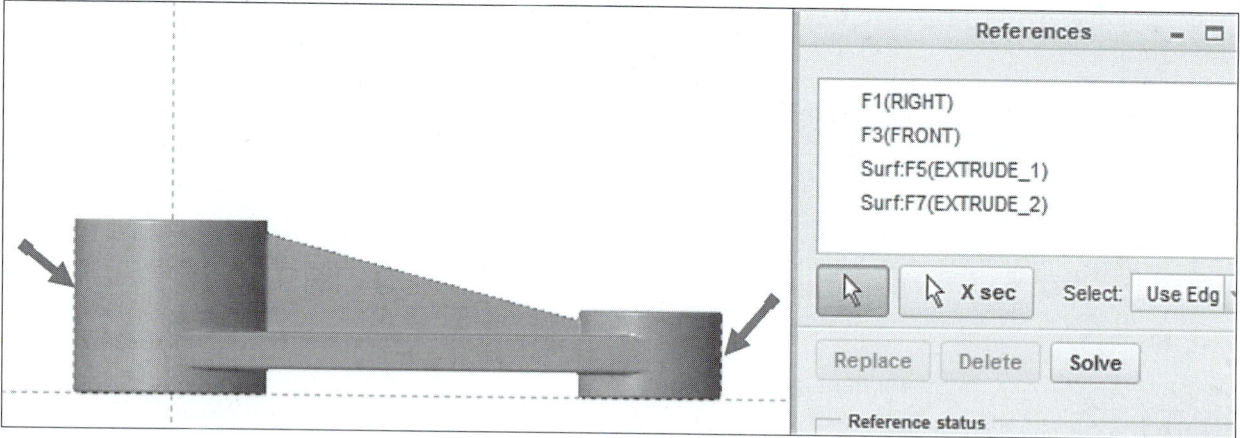

Figure 7.13(b) Add References

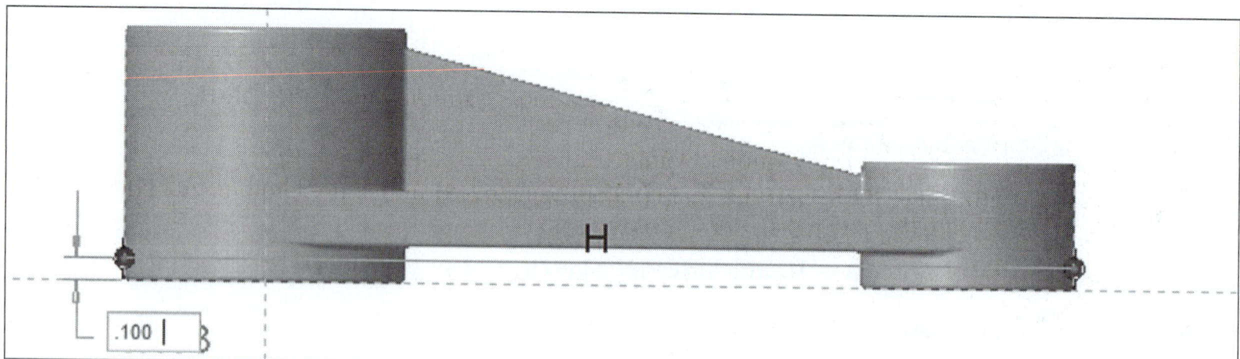

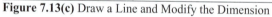

Figure 7.13(c) Draw a Line and Modify the Dimension

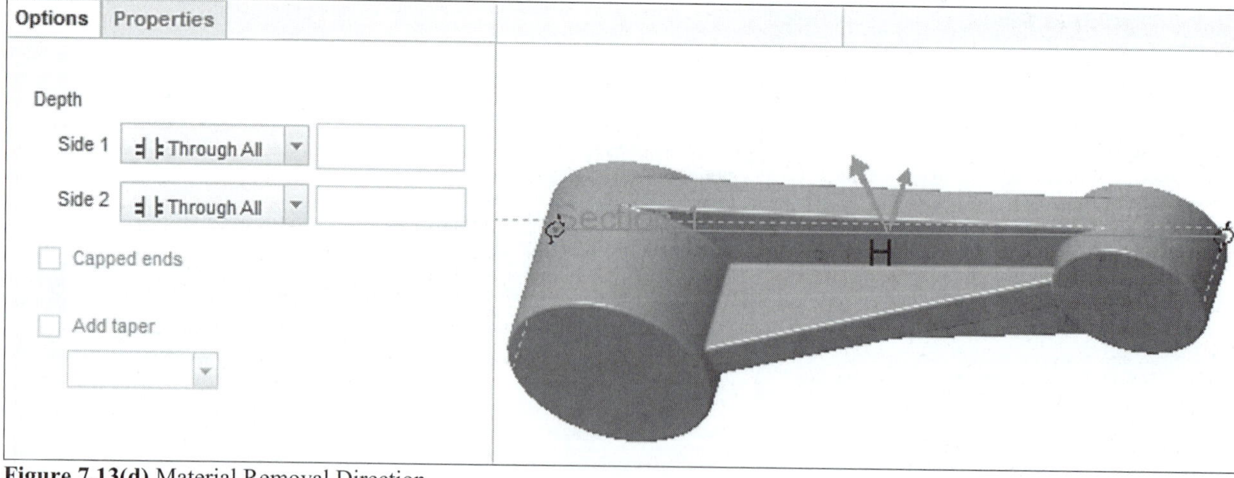

Figure 7.13(d) Material Removal Direction

Click: [icon] > **View** tab > [icon] *on* > [icon] *on* [Figs. 7.13(e-f)] > in the Graphics Window, **LMB** to deselect

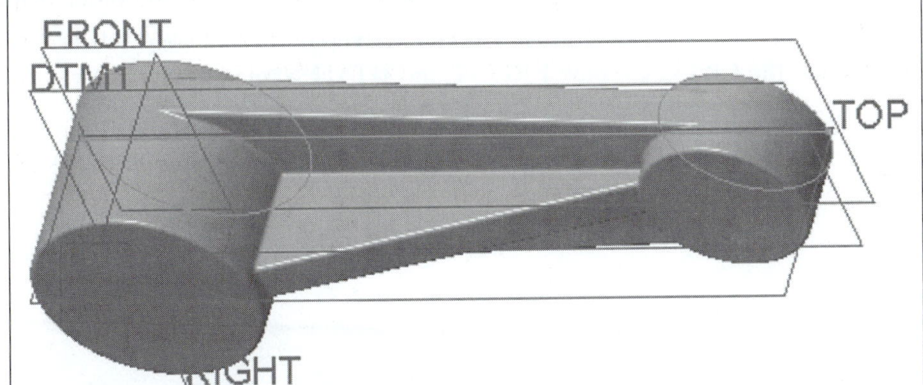

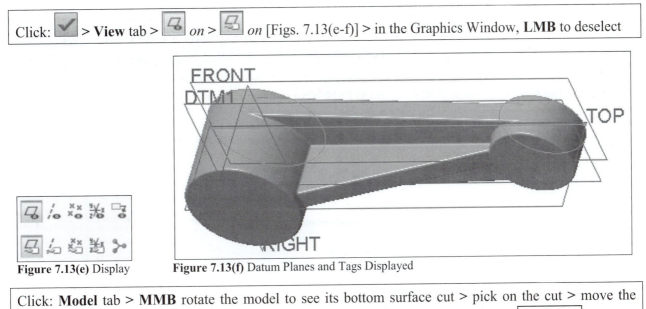

Figure 7.13(e) Display

Figure 7.13(f) Datum Planes and Tags Displayed

Click: **Model** tab > **MMB** rotate the model to see its bottom surface cut > pick on the cut > move the cursor a small distance and pick again (the cut *surface* will highlight) [Fig. 7.14(a)] > [Plane] > **Offset** > **Offset** > **Through** [Fig. 7.14(b)] > **OK** [Fig. 7.14(c)] > **LMB** to deselect > **Ctrl+D** > **Ctrl+S**

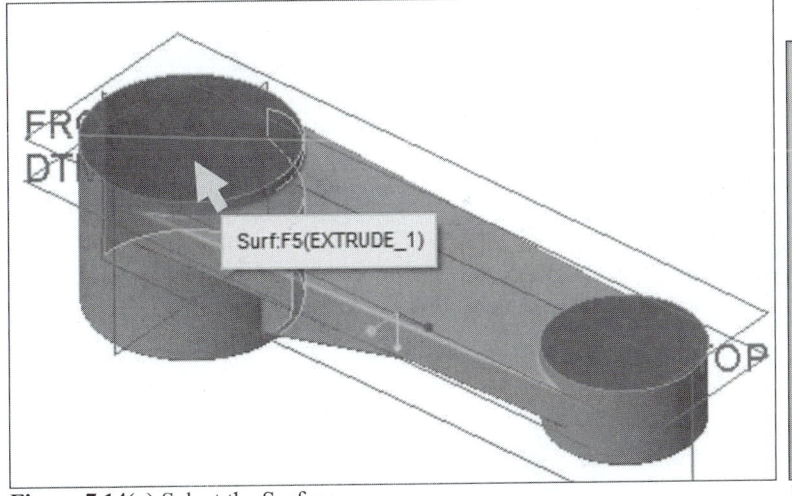

Figure 7.14(a) Select the Surface

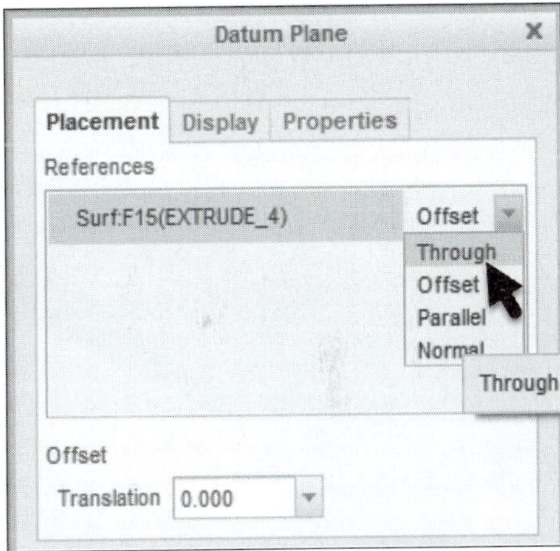

Figure 7.14(b) Datum Plane Through

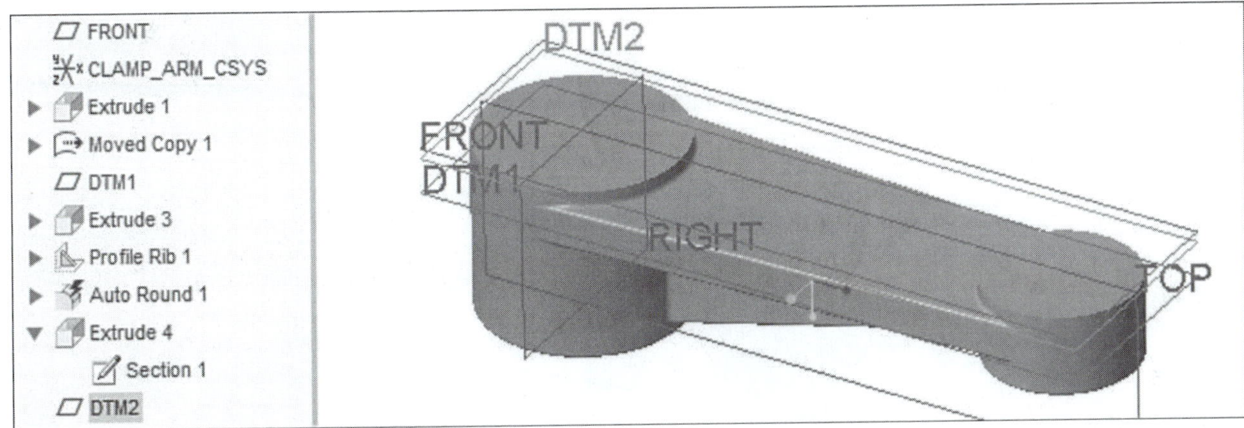

Figure 7.14(c) Completed Datum Plane

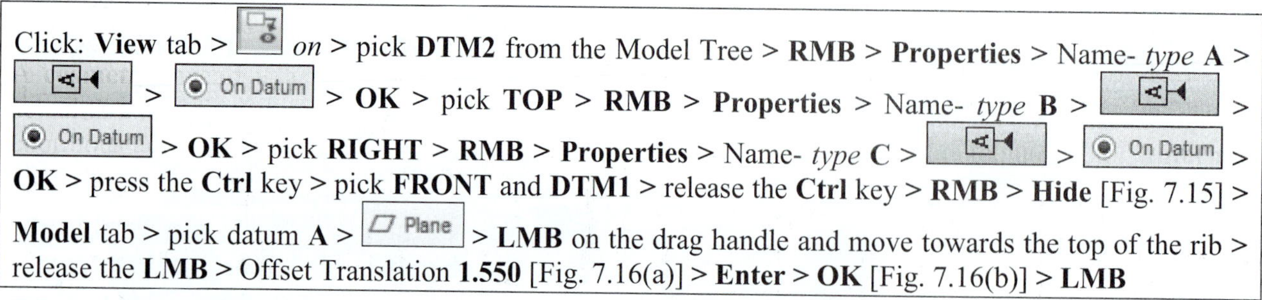

Click: **View** tab > [icon] *on* > pick **DTM2** from the Model Tree > **RMB** > **Properties** > Name- *type* **A** > [icon] > On Datum > **OK** > pick **TOP** > **RMB** > **Properties** > Name- *type* **B** > [icon] > On Datum > **OK** > pick **RIGHT** > **RMB** > **Properties** > Name- *type* **C** > [icon] > On Datum > **OK** > press the **Ctrl** key > pick **FRONT** and **DTM1** > release the **Ctrl** key > **RMB** > **Hide** [Fig. 7.15] > **Model** tab > pick datum **A** > Plane > **LMB** on the drag handle and move towards the top of the rib > release the **LMB** > Offset Translation **1.550** [Fig. 7.16(a)] > **Enter** > **OK** [Fig. 7.16(b)] > **LMB**

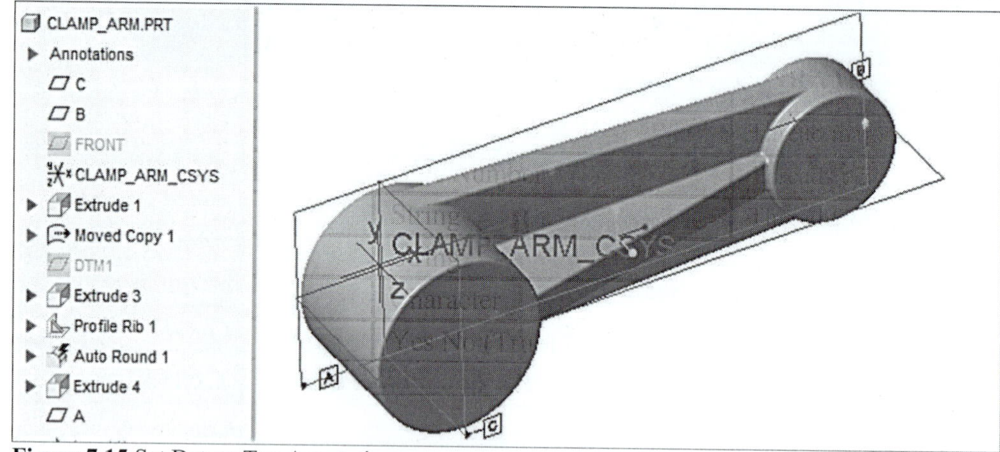

Figure 7.15 Set Datum Tag Annotations

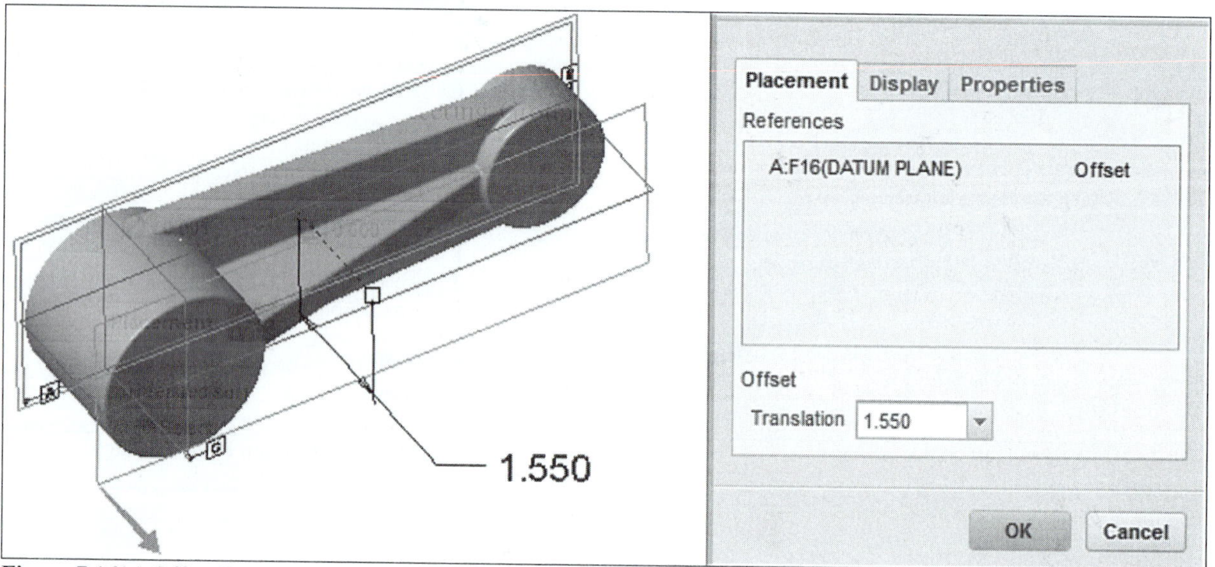

Figure 7.16(a) Offset Datum

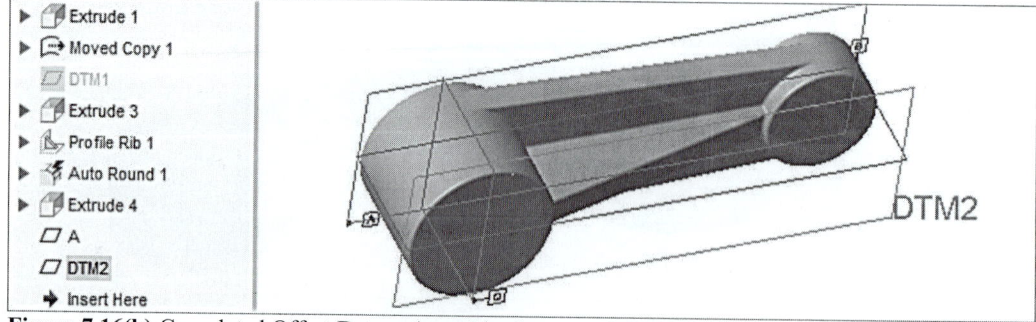

Figure 7.16(b) Completed Offset Datum *(your id values may be different)*

Click: Extrude **Extrude** > > **Placement** tab > **Define** > pick **DTM2** > Reference- pick datum **C** > Orientation **Right** [Fig. 7.17(a)] > **Sketch** to close the Sketch dialog box > **RMB** > **References** > pick on the circular surface of the large protrusion to add as a reference [Figs. 7.17(b)] > **Delete** the two extra references [Fig. 7.17(c)] > **Solve** > **Close** > > **No Hidden** [Fig. 7.17(d)]

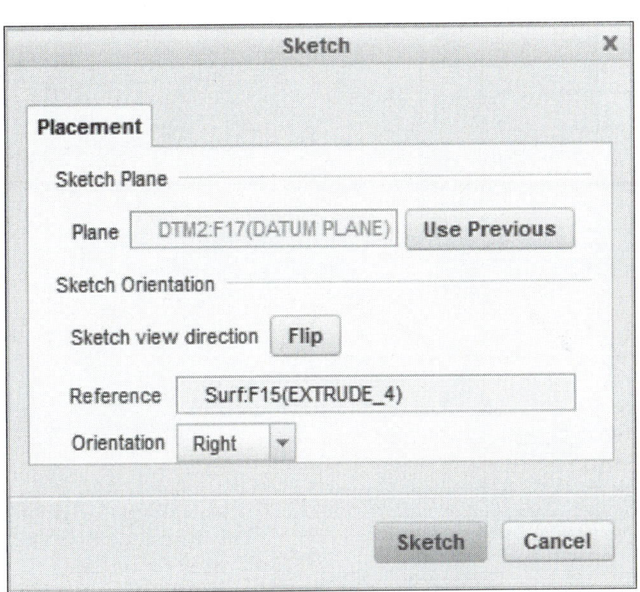

Figure 7.17(a) Sketch Dialog Box

Figure 7.17(b) References

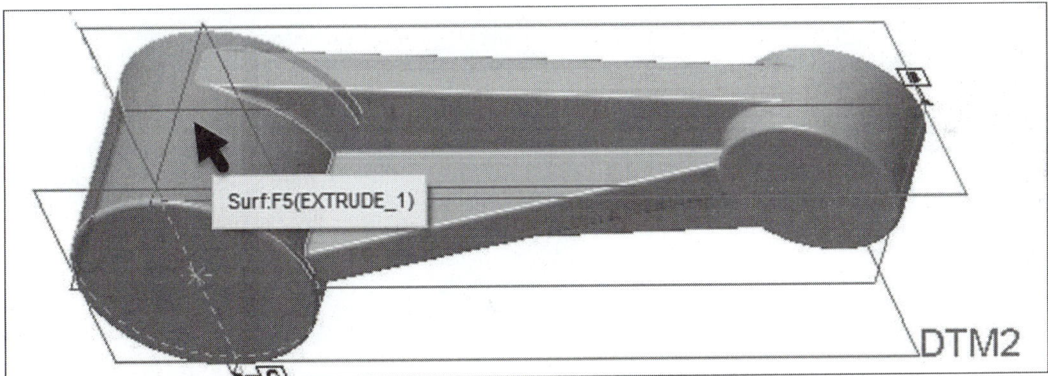

Figure 7.17(c) Select the Cylindrical Surface

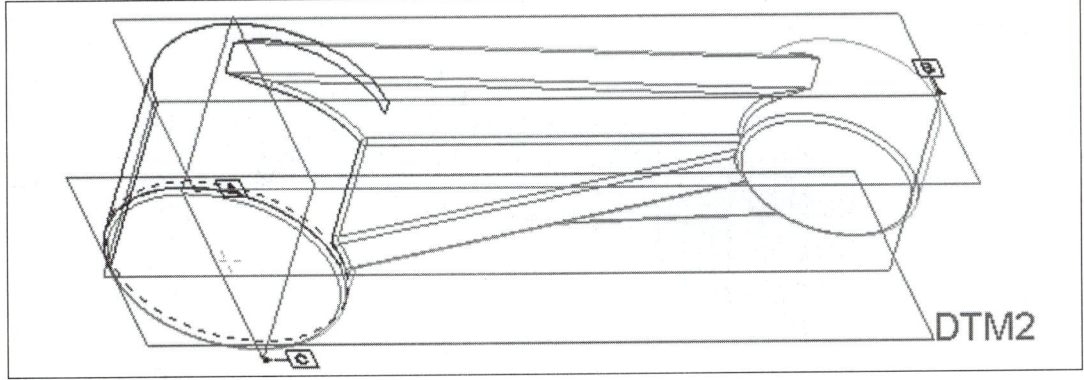

Figure 7.17(d) Circular Reference

Press: **RMB > Circle >** sketch a circle by picking the center [Fig. 7.17(e)] and one edge (on surface reference) [Fig. 7.17(f)] *(the circle is locked into the reference, no dimensions will be displayed; if you get dimensions, undo and redo the circle)* > **MMB** [Fig. 7.17(g)] > > > > if needed click on the arrow to flip the direction of the cut [Fig. 7.17(h)] > > **Shading** [Fig. 7.17(i)] > > **LMB** to deselect > **Shading With Edges** [Fig. 7.17(j)] > **Ctrl+D > Ctrl+S**

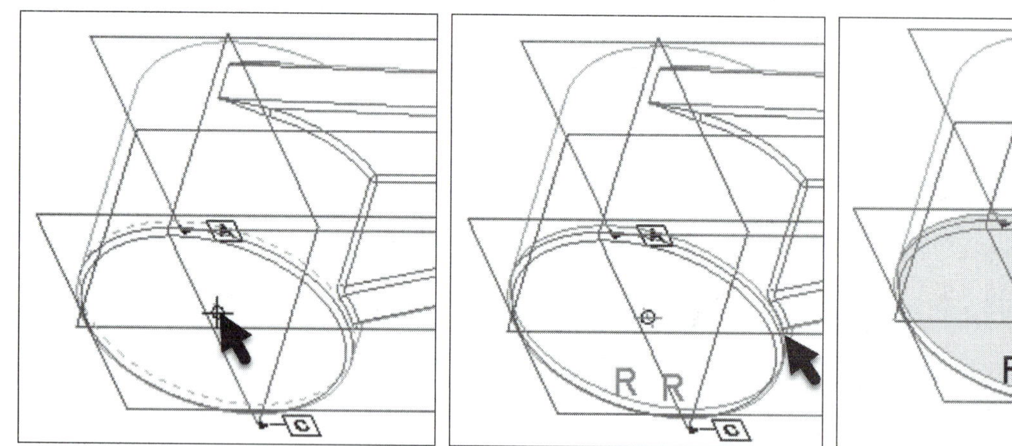

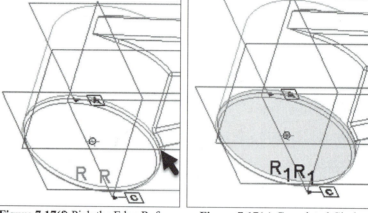

Figure 7.17(e) Pick the Center **Figure 7.17(f)** Pick the Edge Reference **Figure 7.17(g)** Completed Circle

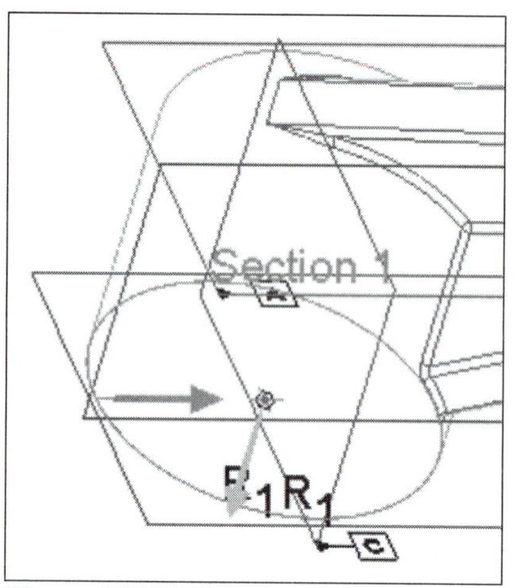

Figure 7.17(h) Cut

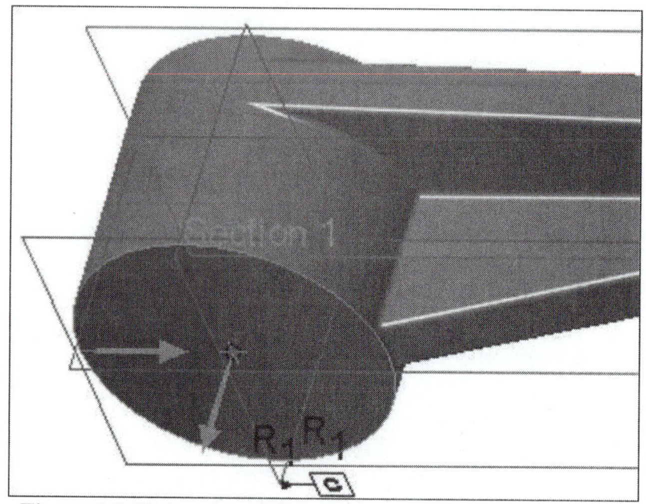

Figure 7.17(i) Previewed Cut

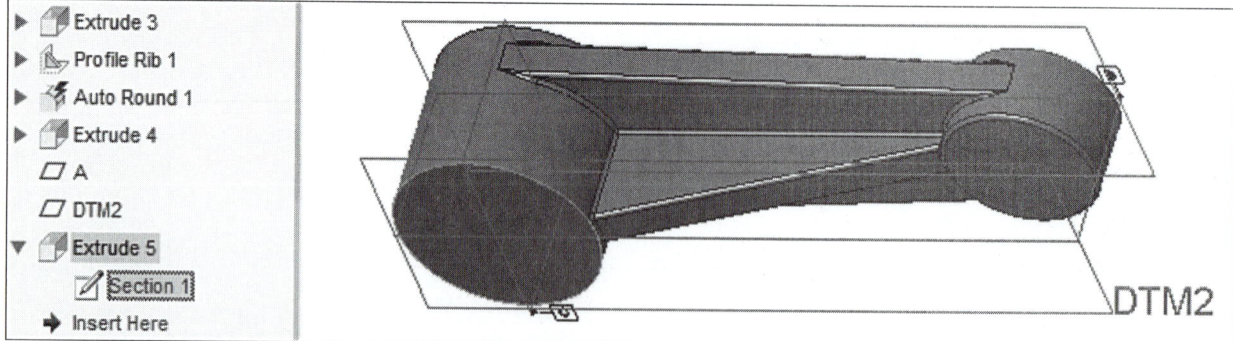

Figure 7.17(j) Completed Cut

Click: **DTM2** in the Model Tree > **RMB** > **Hide** > Extrude > > > **Remove Material** > **Placement** tab > **Define** > Datum > **Plane** > pick datum **A** > **LMB** on the drag handle and move > double-click on the dimension > *type* **.725** > **Enter** [Fig. 7.17(k)] > **OK** > click on **Datum B** Reference **B:F2(DATUM PLANE)** > Orientation **Top** [Fig. 7.17(l)] > **Sketch** > press **RMB** > **References** > hold down the **Ctrl** key and select **F1** and **F2** > **Delete** > > pick on the circular surface of the small protrusion [Figs. 7.17(m-n)] > **Solve** > **Close**

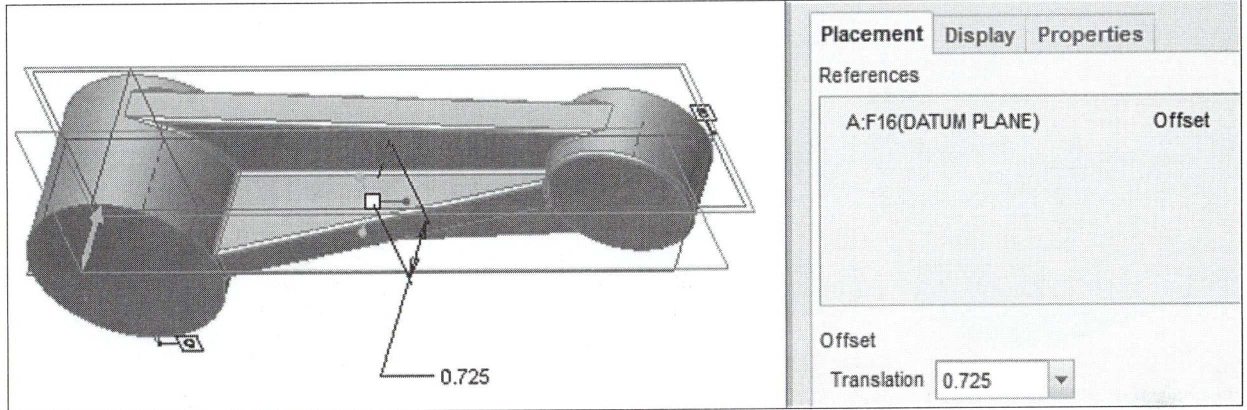

Figure 7.17(k) Creating an Internal Datum Plane Offset from Datum A for Sketching

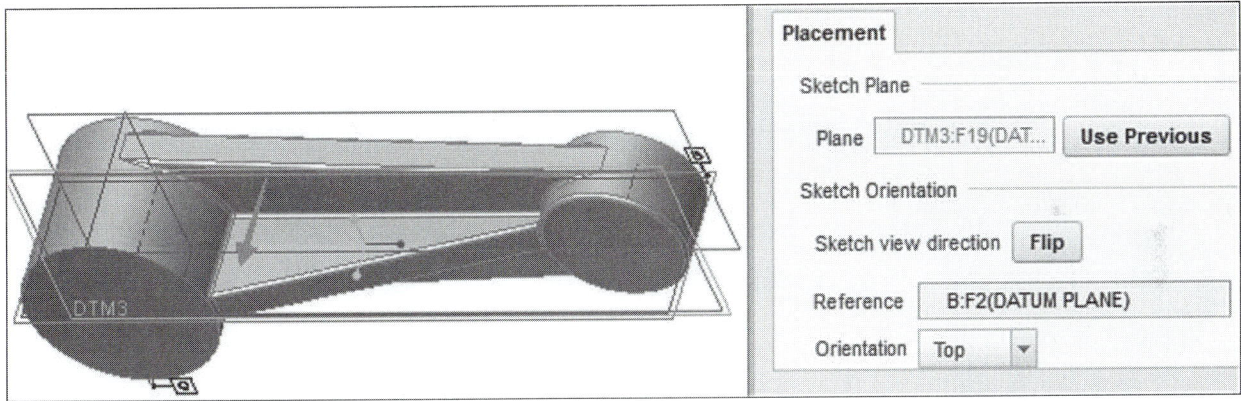

Figure 7.17(l) Sketch Plane and Orientation

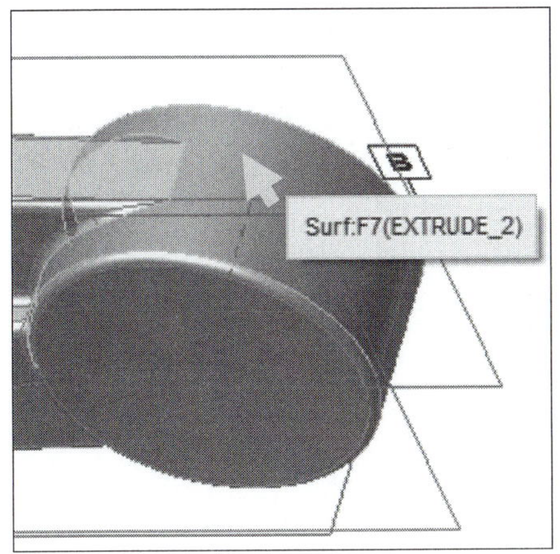

Figure 7.17(m) Select the Surface

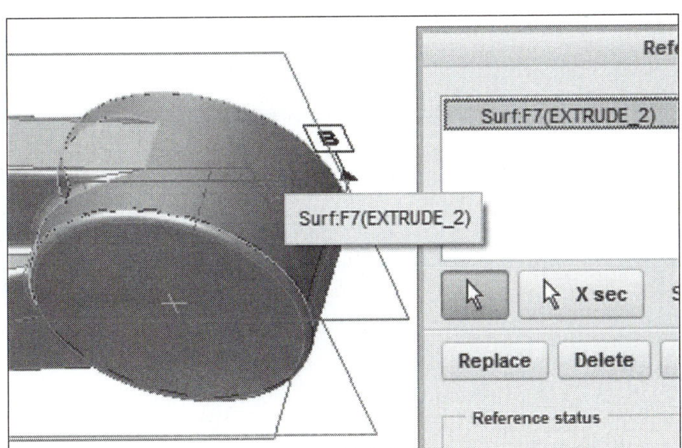

Figure 7.17(n) Reference

Click: Center and Point > sketch a circle by picking the center [Fig. 7.17(o)] and the surface reference [Fig. 7.17(p)] *(the circle is locked into references, no dimensions will be displayed; if you get dimensions, undo and redo the circle)* > **MMB** [Fig. 7.17(q)] > [Fig. 7.17(r)] > > in the Graphics Window, **LMB** to deselect > > Select **Extrude 6** and **DTM3** in the Model Tree [Fig. 7.17(s)] > **LMB** in Graphics Area > **Ctrl+D** > **Ctrl+S**

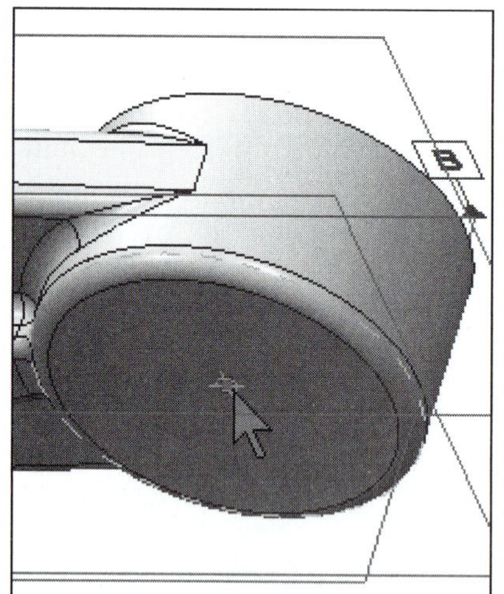

Figure 7.17(o) Pick Center

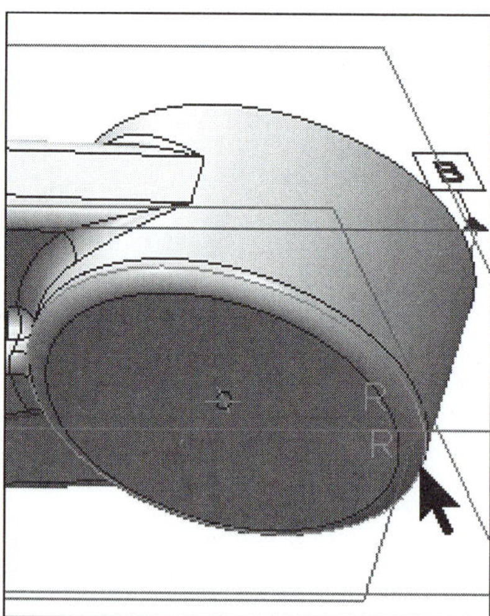

Figure 7.17(p) Pick Reference

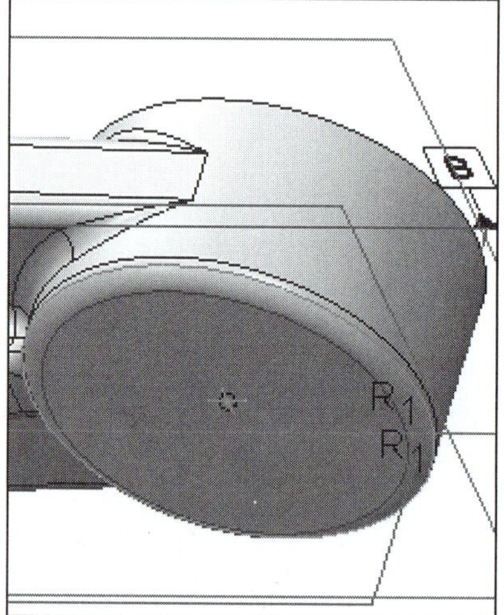

Figure 7.17(q) Shaded Section

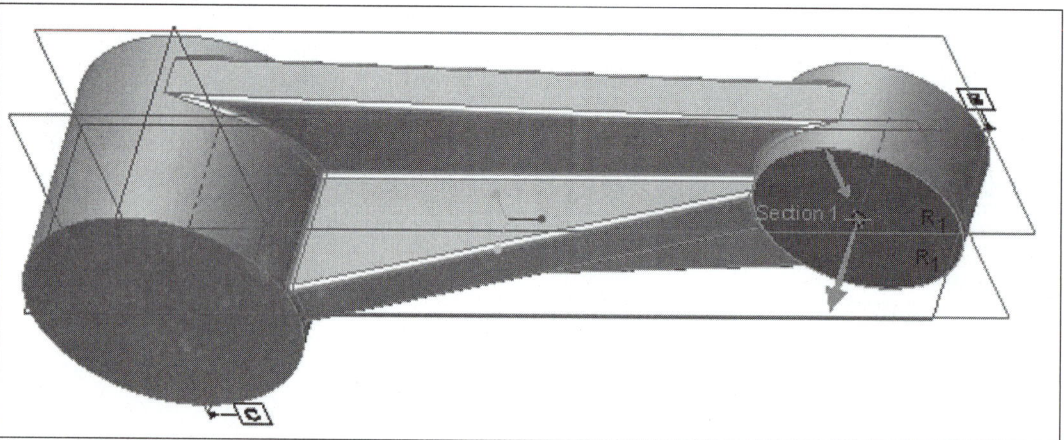

Figure 7.17(r) Previewed Through All Cut

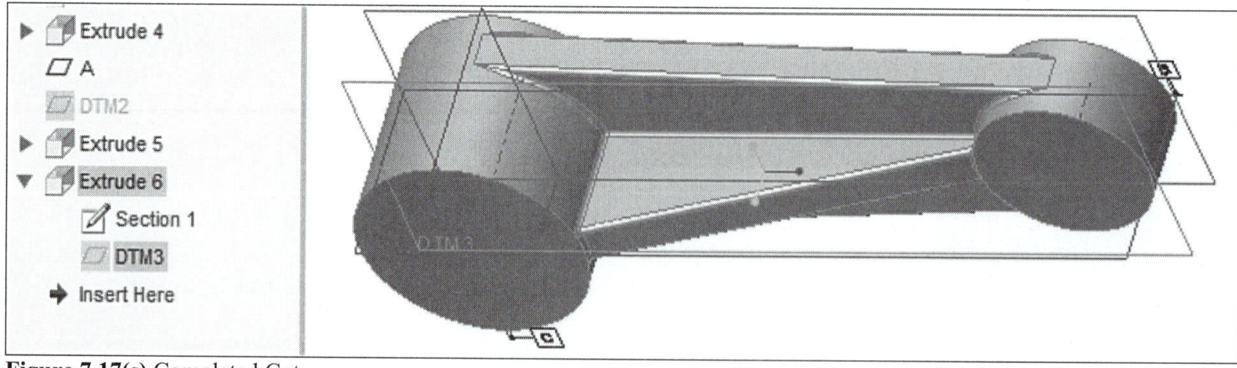

Figure 7.17(s) Completed Cut

Measuring Geometry

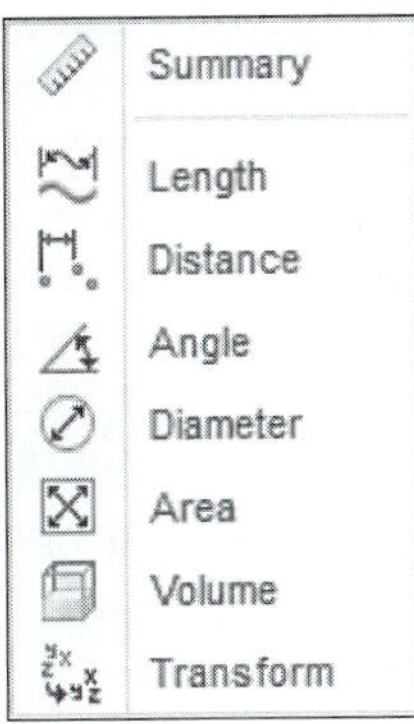

Using analysis measure, you can measure model geometry with one of the following commands:

- **Length** Displays the length of the curve or edge
- **Distance** Displays the distance between two entities
- **Angle** Displays the angle between two entities
- **Diameter** Displays the diameter of the surface
- **Radius** Displays the radius of the surface
- **Area** Displays the area of the selected surface, quilt, facets, or an entire model
- **Volume** Displays the volume of the object
- **Transform** Displays the transformation matrix between two coordinate systems.

Click: **Analysis** tab > **Measure** > **Distance** > pick the top surface of the large circular protrusion [Fig. 7.18(a)] > press and hold **Ctrl** > place the cursor over the bottom and *click* **RMB** until it highlights [Fig. 7.18(b)] > **LMB** to select (**1.55000**) [Fig. 7.18(c)] > **Clear all Selections** > repeat the process to measure the shorter end (**.725**) >

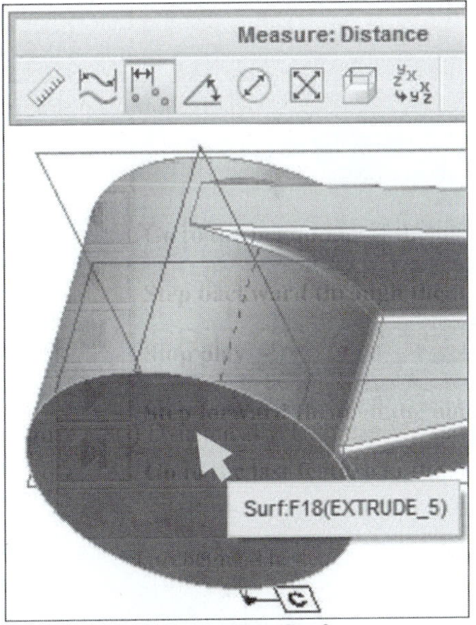

Figure 7.18(a) Select Top Surface

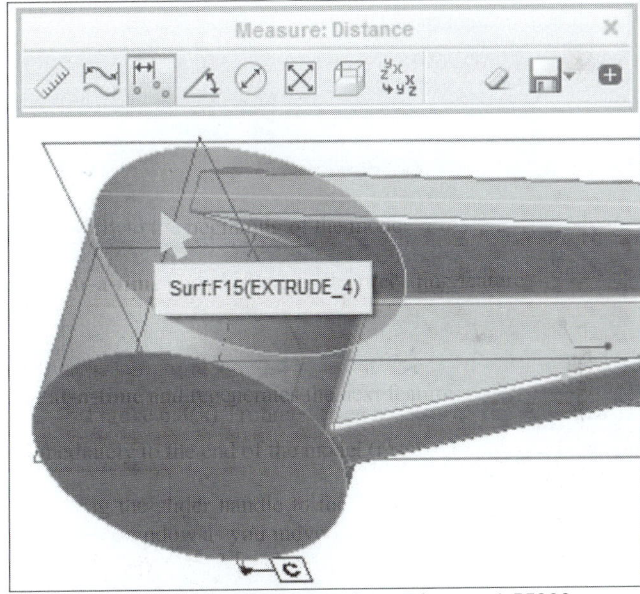

Figure 7.18(b) Select Bottom Surface- Distance **1.55000**

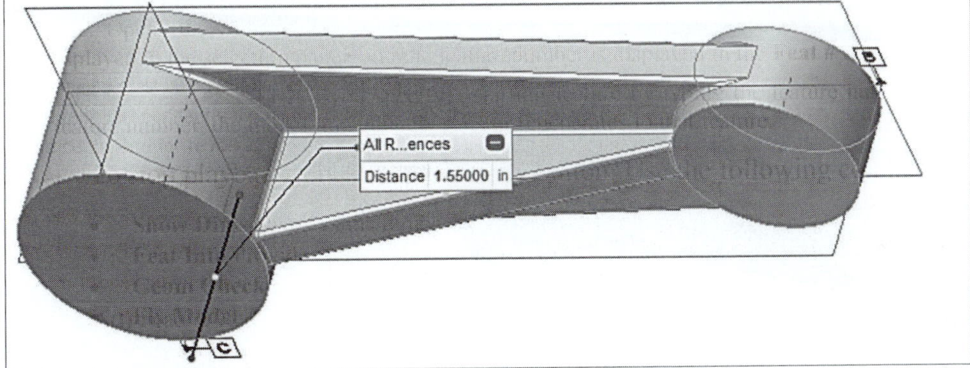

Figure 7.18(c) Distance **1.55000**

Editing the Model

During the design of a component there are modifications made to the design. The ability to make changes without causing failures is important. "Flexing" the model; changing and editing dimension values to see if the model integrity withstands these modifications, establishes your designs robustness.

Click: **Auto Round 1** feature in the Model Tree > **RMB** > dl **Edit** > double-click on the dimension value > *type* **.08** > **Enter** > move the pointer off of the dimension > **LMB** to complete the edit > move the pointer > **LMB** to regenerate and deselect > **OK** > double-click on the **Rib** > double-click on the **.9875** dimension value > *type* **.90** > **Enter** > press **RMB** > **Exit Edit Dimension** > **LMB** > double-click on the **Rib** > drag a handle until the value is **1.000** [Fig. 7.19(a)] > release the **LMB**[Figs. 7.19(b-c)] > double-click **LMB** > (Undo: Edit Value / Undo: Edit Value / Undo: Edit Value) to the design values > **Measure** > **Diameter** > select the round >

Surf:...ND_1)		
Diameter	0.0800000	in
Radius	0.0400000	in

> **MMB**

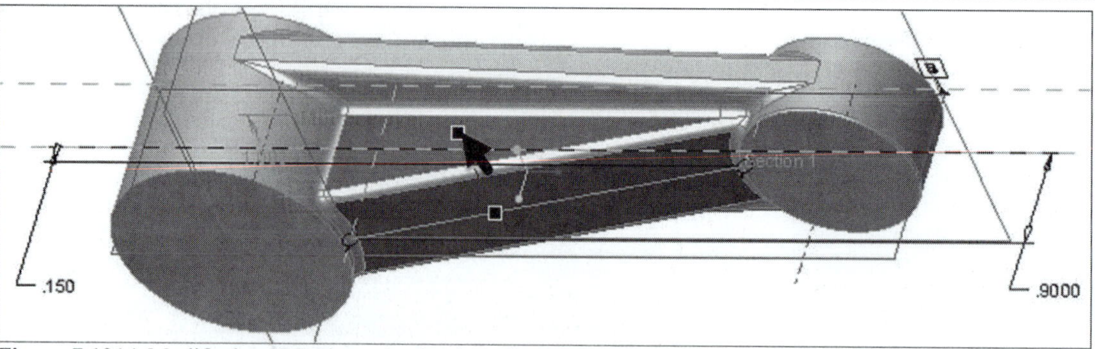

Figure 7.19(a) Modify the **.375** Dimension by dragging a handle to **1.00**

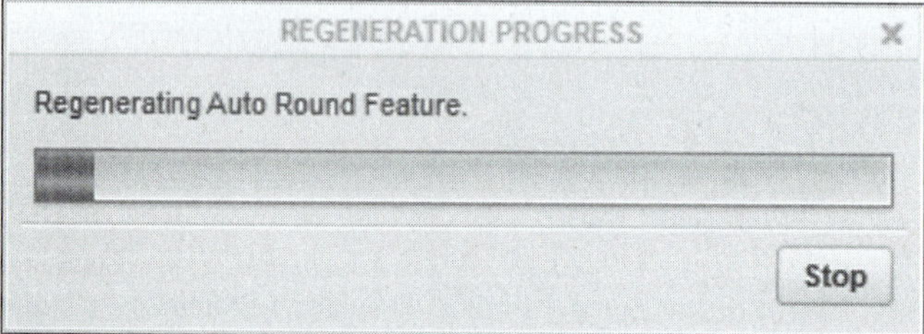

Figure 7.19(b) Regenerate

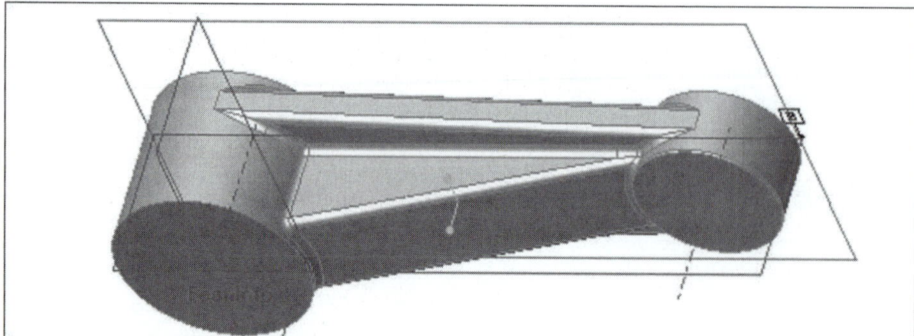

Figure 7.19(c) Flexed Part

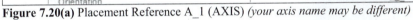

Click: **Ctrl+S** > **View** tab > [icon] *on* > [icon] *on* > **Model** tab > [Hole] **Hole** > [icon] > [icon] > [Ø 0.5375000] > **Placement** tab > pick the Axis [Fig. 7.20(a)] > hold down the **Ctrl** key > pick the top surface [Fig. 7.20(b)] > release the **Ctrl** key > **Shape** tab [Fig. 7.20(c)] > **Enter** > **LMB** to deselect

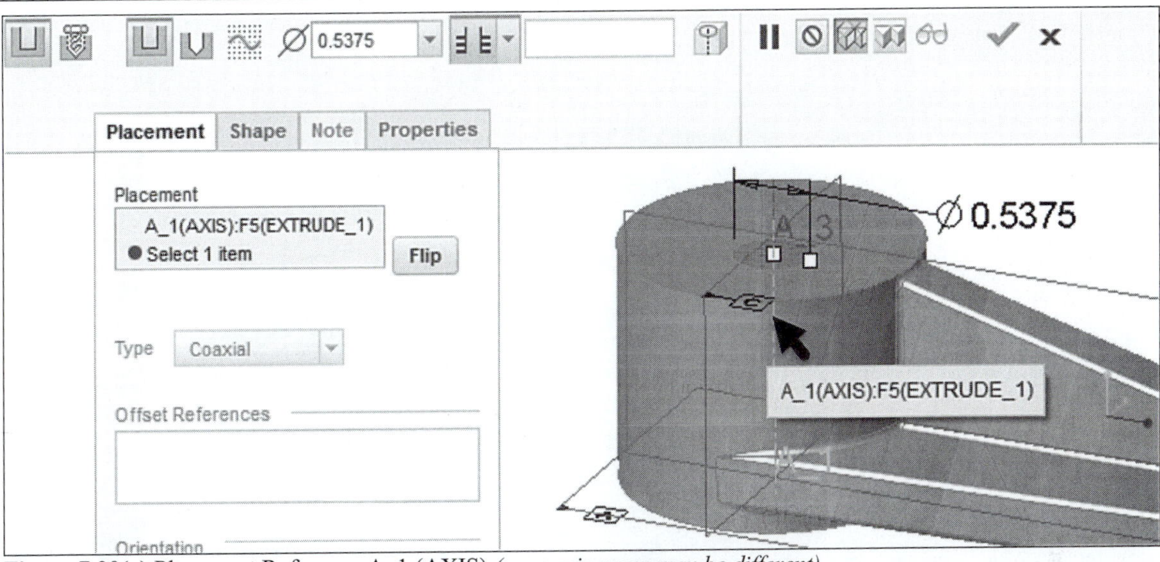

Figure 7.20(a) Placement Reference A_1 (AXIS) *(your axis name may be different)*

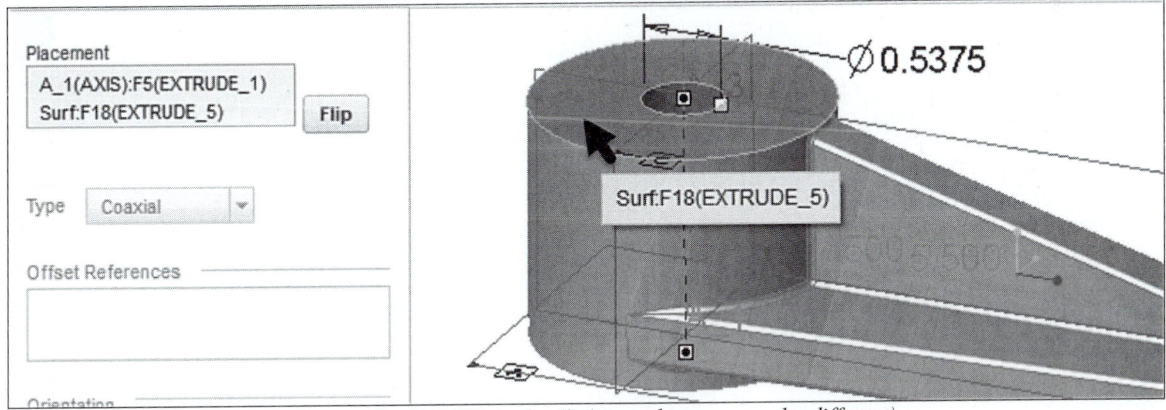

Figure 7.20(b) Placement Reference Surf:F18(Extrude_5) *(your id names may be different)*

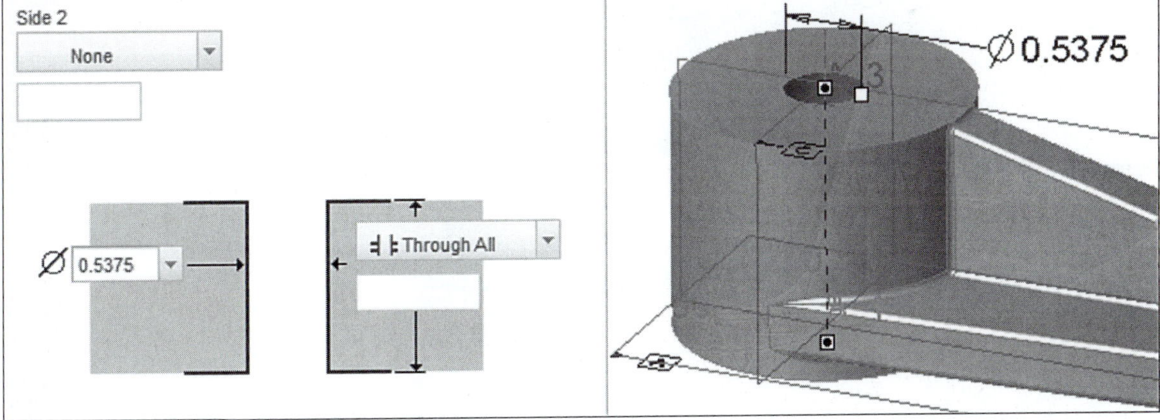

Figure 7.20(c) Shape Tab

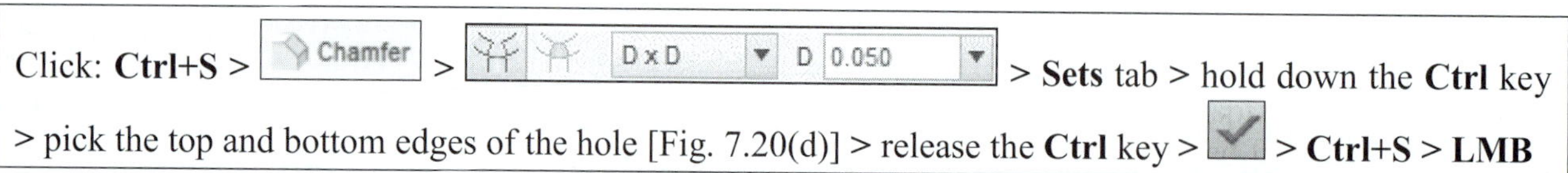
Click: **Ctrl+S** > Chamfer > D x D D 0.050 > **Sets** tab > hold down the **Ctrl** key > pick the top and bottom edges of the hole [Fig. 7.20(d)] > release the **Ctrl** key > > **Ctrl+S** > **LMB**

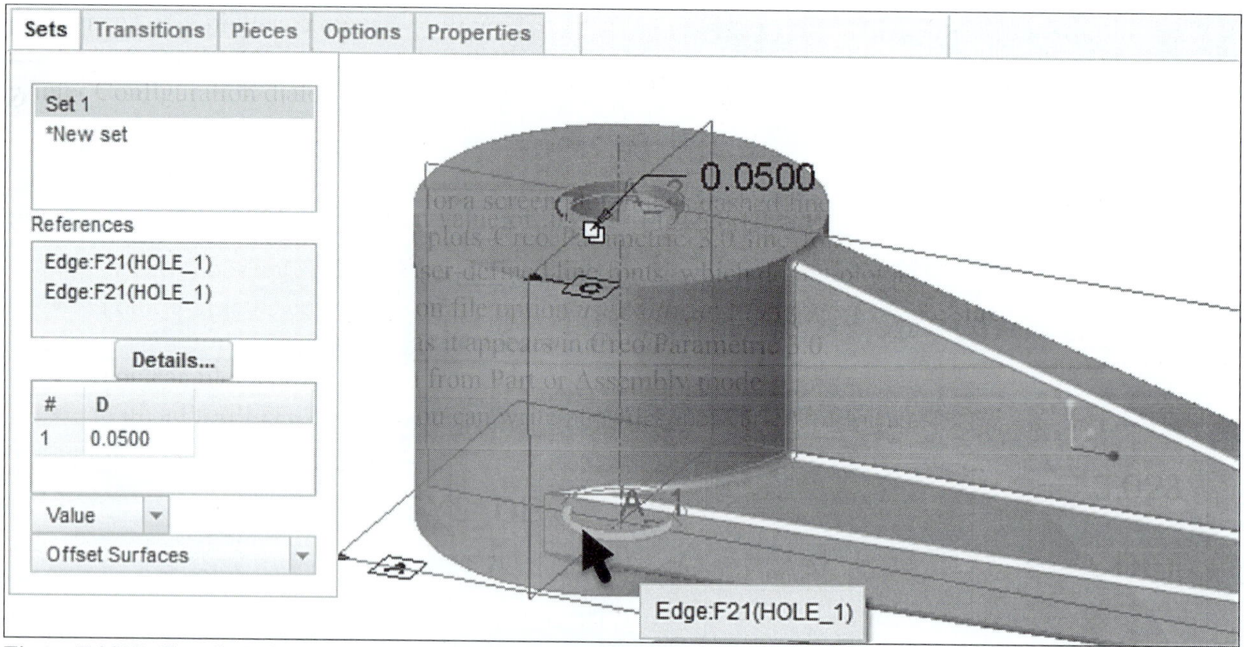

Figure 7.20(d) Chamfer Edges

Click: Hole **Hole** > pick axis **A_2** [Fig. 7.21(a)]

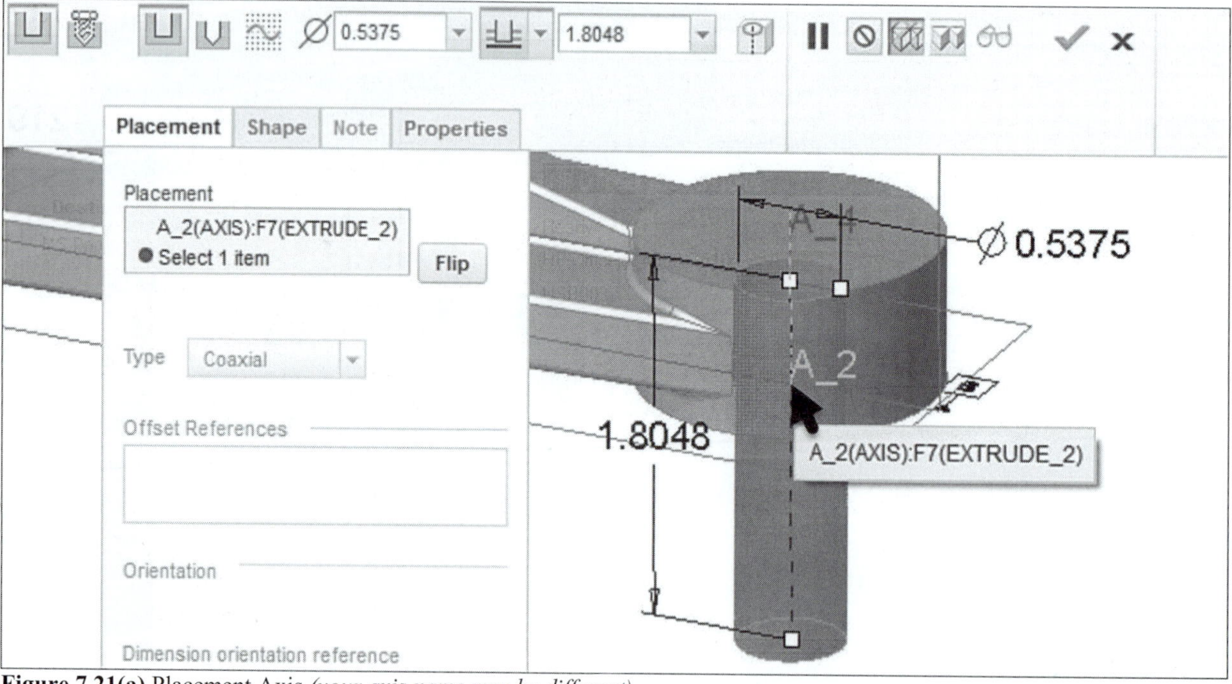

Figure 7.21(a) Placement Axis *(your axis name may be different)*

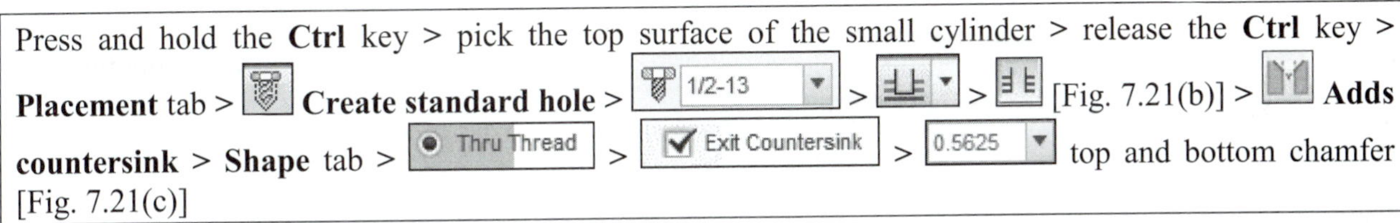

Press and hold the **Ctrl** key > pick the top surface of the small cylinder > release the **Ctrl** key > **Placement** tab > **Create standard hole** > 1/2-13 > > [Fig. 7.21(b)] > **Adds countersink** > **Shape** tab > Thru Thread > Exit Countersink > 0.5625 top and bottom chamfer [Fig. 7.21(c)]

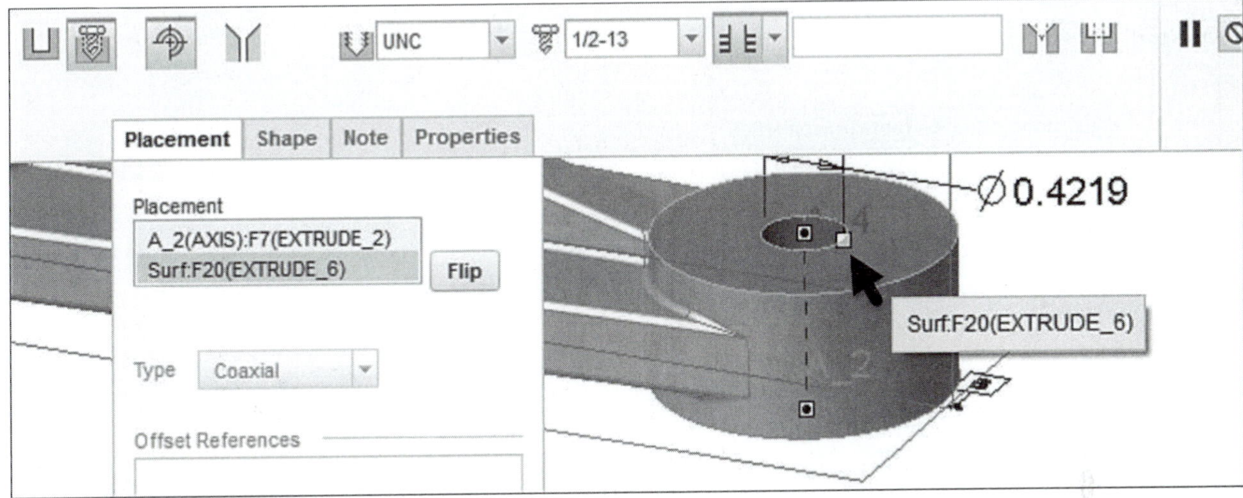

Figure 7.21(b) Placement Surface

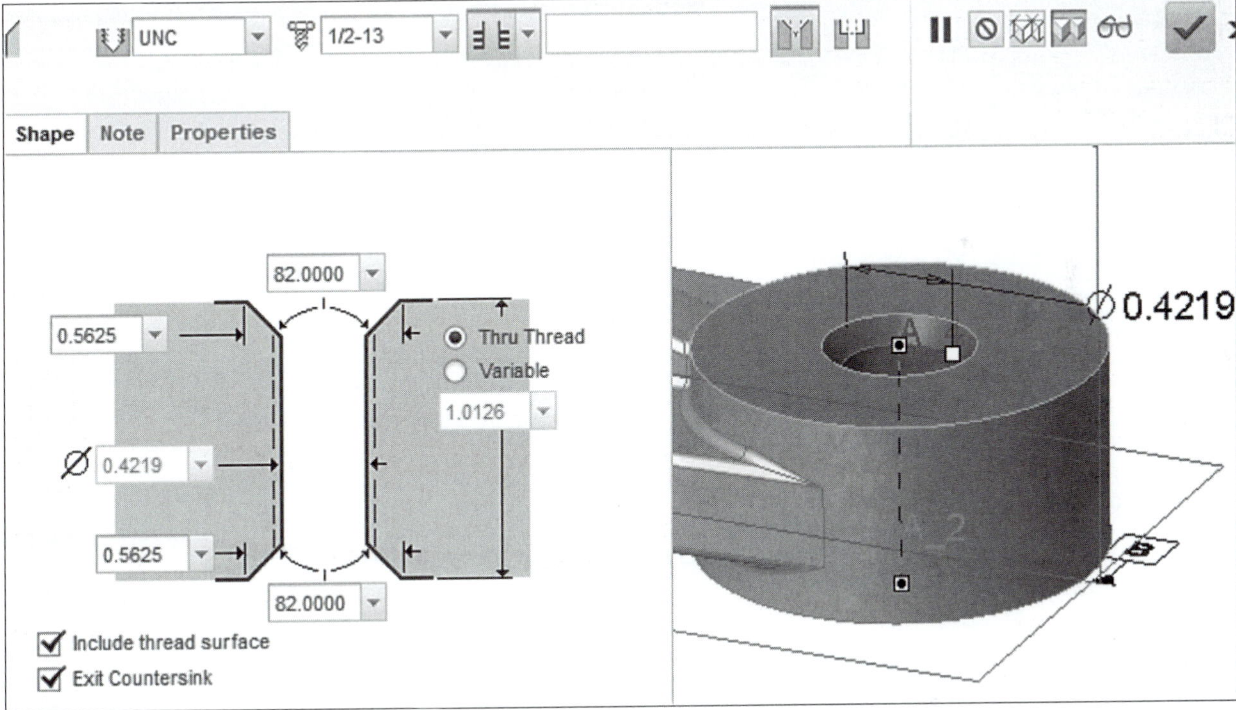

Figure 7.21(c) Shape Tab

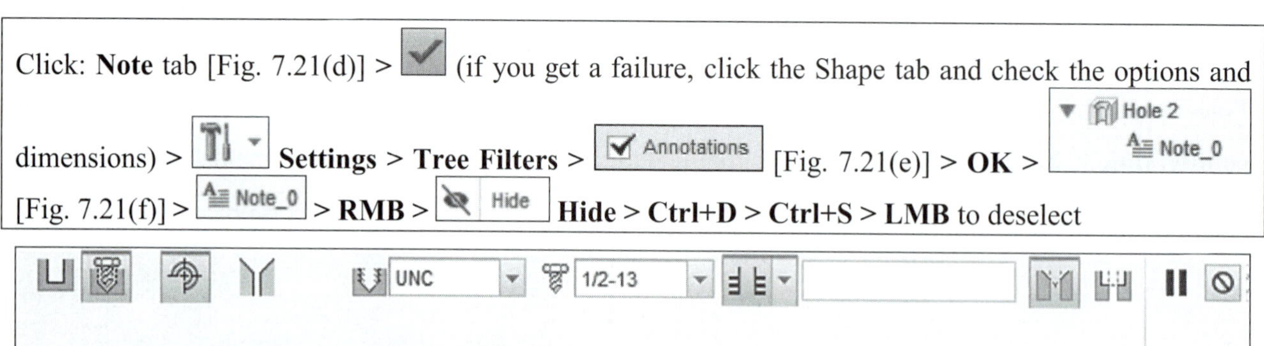

Click: **Note** tab [Fig. 7.21(d)] > ✓ (if you get a failure, click the Shape tab and check the options and dimensions) > **Settings** > **Tree Filters** > ✓ Annotations [Fig. 7.21(e)] > **OK** > Hole 2 / Note_0 [Fig. 7.21(f)] > Note_0 > **RMB** > Hide **Hide** > **Ctrl+D** > **Ctrl+S** > **LMB** to deselect

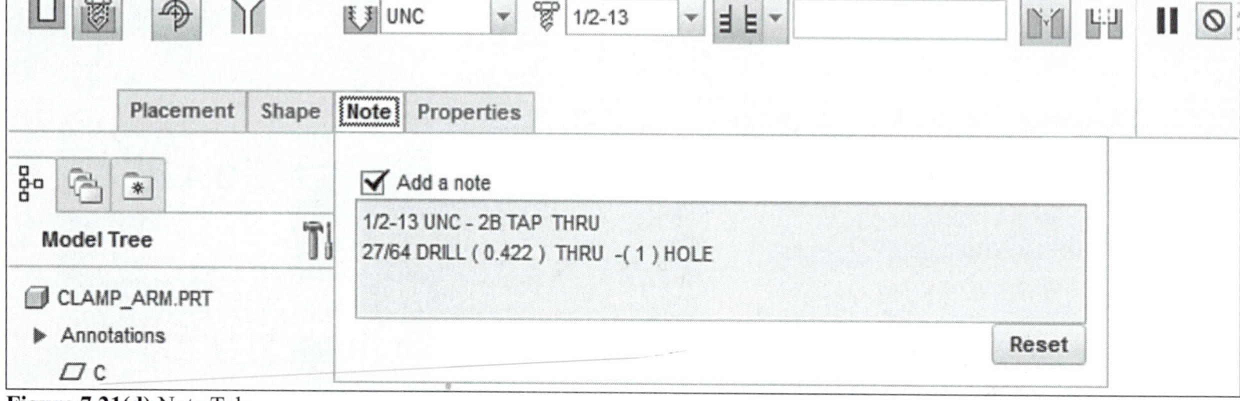

Figure 7.21(d) Note Tab

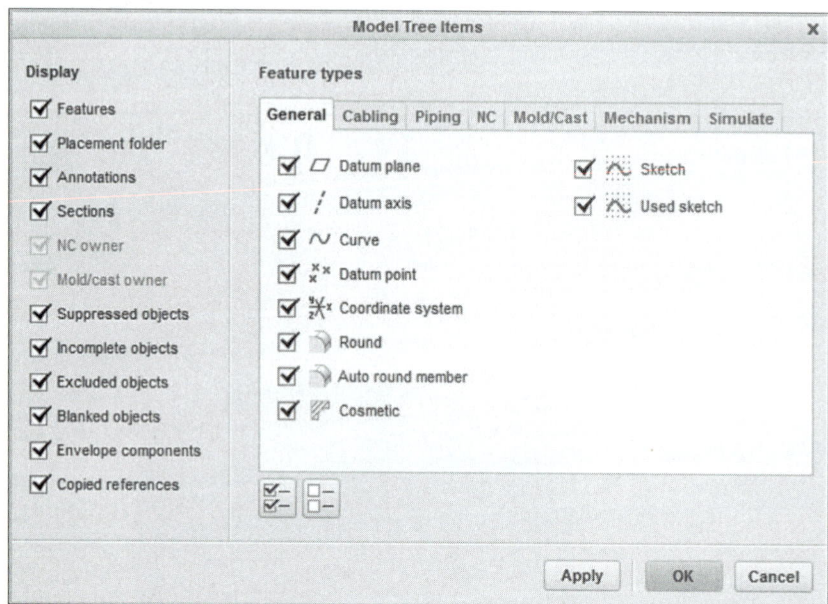

Figure 7.21(e) Model Tree Items

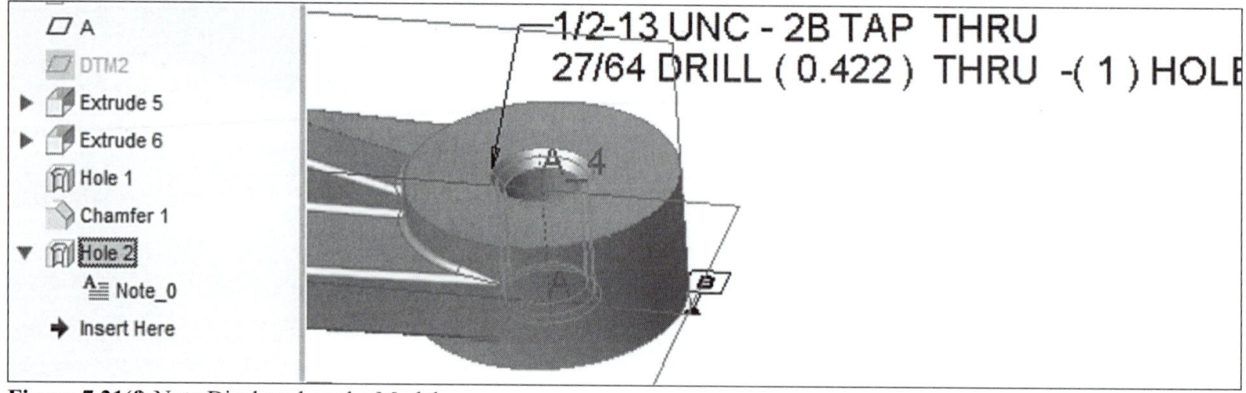

Figure 7.21(f) Note Displayed on the Model

Slowly click twice on the **Extrude 3** feature in the Model Tree > *type* **WEB** as the new name > **Enter** > in the Graphics Window, **LMB** to deselect > Write a relation to keep the thickness of the rib the same as that for the web. With the **Ctrl** key pressed, pick on the WEB and the rib in the Model Tree > release the **Ctrl** key > **RMB** > dl **Edit** > place the mouse pointer over each dimension [Fig. 7.22(a)] and note the **d** values [Fig. 7.22(b)] > **Tools** tab > d= Relations > *type (or pick from the model)* **d10=d8** [Fig. 7.22(c)] *(your "d" values may be different)* > **Execute/Verify** > **OK** > **OK** > **Ctrl+D** > **Ctrl+S** > **LMB**

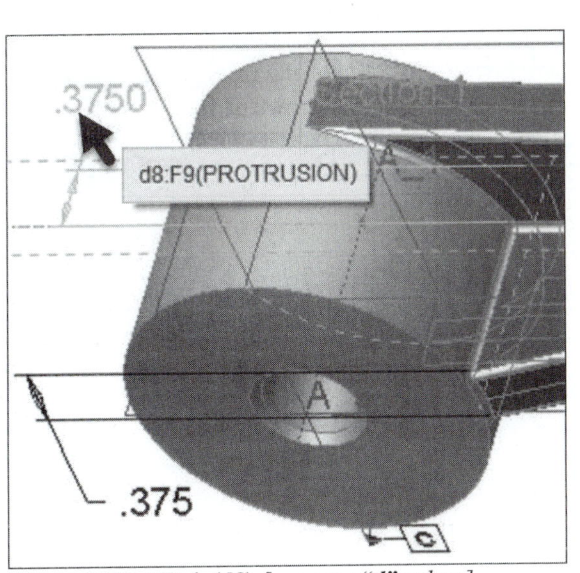

Figure 7.22(a) Web (**d8**) *[use your "d" values]*

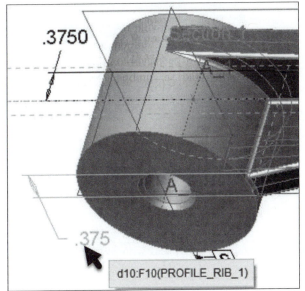

Figure 7.22(b) Rib (**d10**) *[use your "d" values]*

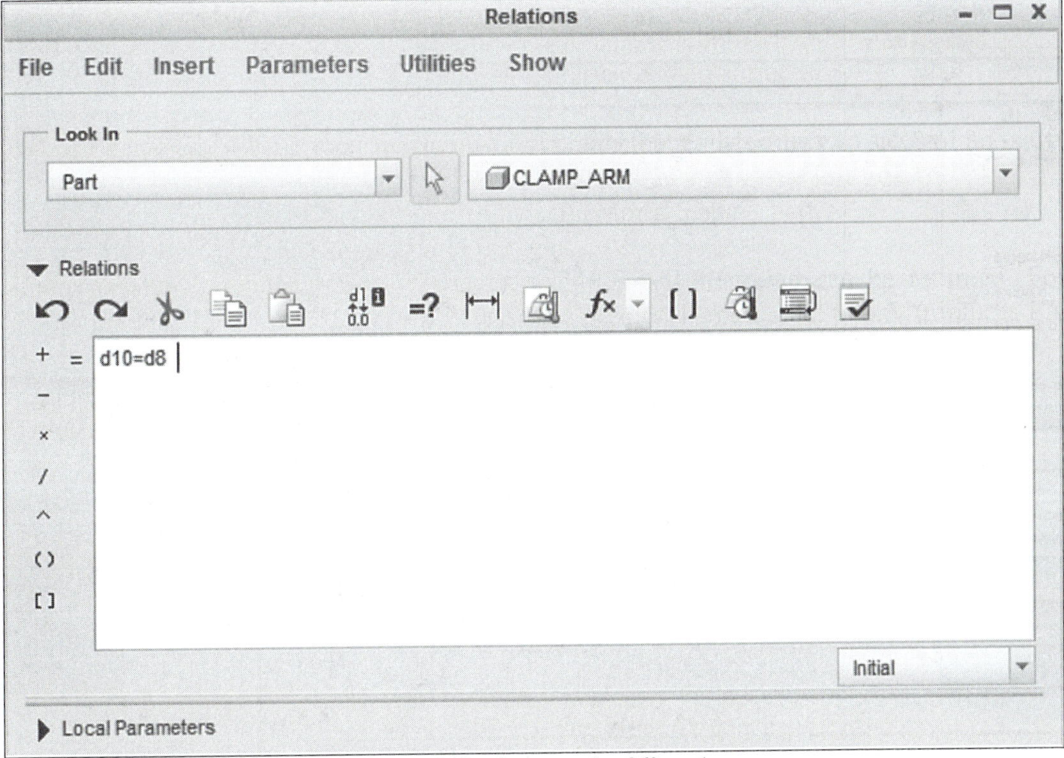

Figure 7.22(c) Relations Dialog Box *(your "d" values may be different)*

Pick on **WEB** in the Model Tree > **RMB** > **Edit** > double-click on the **.375** dimension > *type* **.60** [Fig. 7.22(d)] > **Enter** [Fig. 7.22(e)] > move the pointer into the Graphics Window > **LMB** [Fig. 7.22(f)] > slightly move the pointer > **LMB** > **Cancel** [Fig. 7.22(g)] > repeat and modify the value to **.20** > **Enter** > move the pointer > **LMB** > move the pointer > **LMB** [Fig. 7.22(h)] > **Undo**

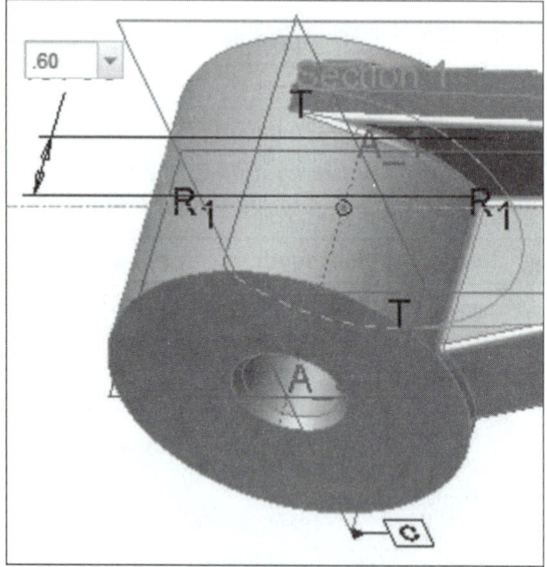

Figure 7.22(d) Edit the Web Dimension to **.60**

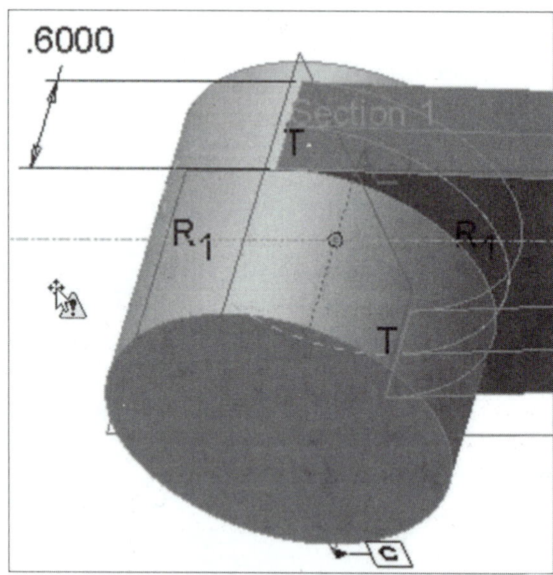

Figure 7.22(e) Failed Feature

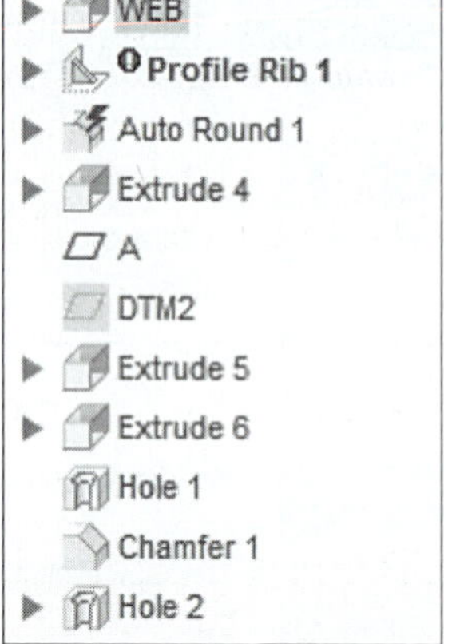

Figure 7.22(f) Failed Features in Red

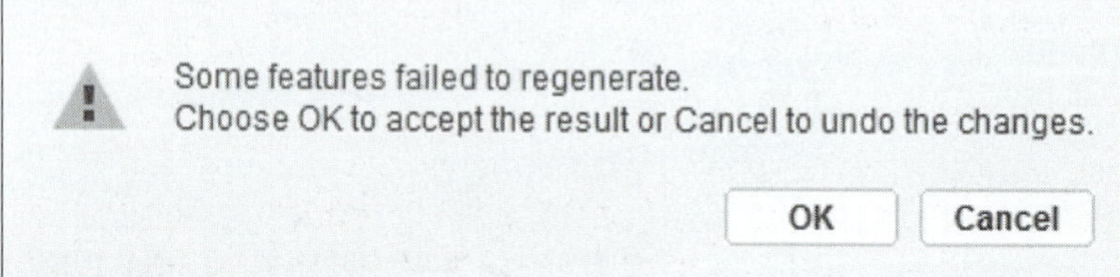

Figure 7.22(g) Failure Diagnostics Dialog Box

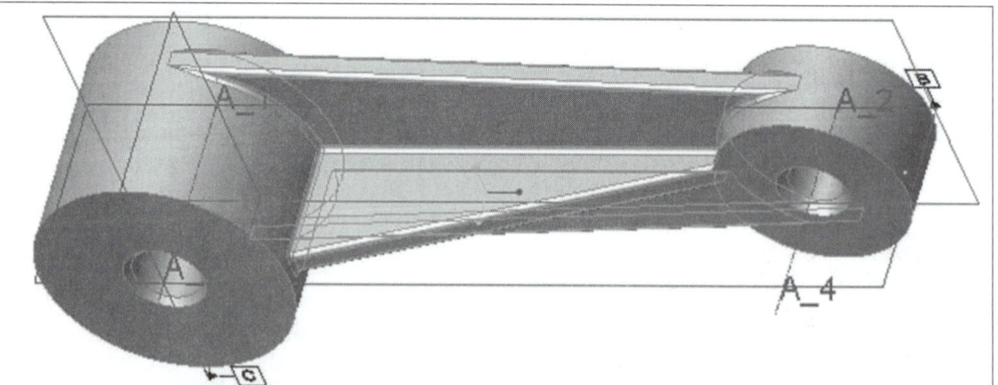

Figure 7.22(h) .20 Dimension driving the WEB and the Rib Size

Family Tables

Family Tables are used any time a part or assembly has several unique iterations developed from the original model. The iterations must be considered as separate models, not just iterations of the original model. In this lesson, you will add instances of a family table to create a machined part and a casting of the Clamp Arm (a version without machined cuts or holes).

You will be creating a Family Table from the generic model. The (base) model is the **Generic**. Each variation is referred to as an **Instance**. When you create a Family Table, Creo Parametric 3.0 allows you to *select dimensions,* which can vary between instances. You can also *select features* to add to the Family Table. Features can vary by being suppressed or resumed in an instance. When you are finished selecting items (e.g., dimensions, features, and parameters), the Family Table is automatically generated.

When adding features to the table; enter an **N** to suppress the feature, or a **Y** to resume the feature. Each instance must have a unique name.

Family tables are spreadsheets, consisting of columns and rows. *Rows* contain instances and their corresponding values; *columns* are used for items. The column headings include the *instance name* and the names of all of the *dimensions, parameters, features, members,* and *groups* that were selected to be in the table. The Family Table dialog box is used to create and modify family tables.

Family tables include:

- The base object (generic object or *generic*) on which all members of the family are based.
- Dimensions, parameters, feature numbers, user-defined feature names, and assembly member names that are selected to be table-driven (*items*).
    - o **Dimensions** are listed by name (for example, **d125**) with the associated symbol name (if any) on the line below it (for example, depth).
    - o **Parameters** are listed by name (dim symbol).
    - o **Features** are listed by feature number with the associated feature type (for example, [cut]) or feature name on the line below it. The generic model is the first row in the table. Only modifying the actual part, suppressing, or resuming features can change the table entries belonging to the generic; *you cannot change the generic model by editing its row entries in the family table.*
- Names of all family members (*instances*) created in the table and the corresponding values for each of the table-driven items

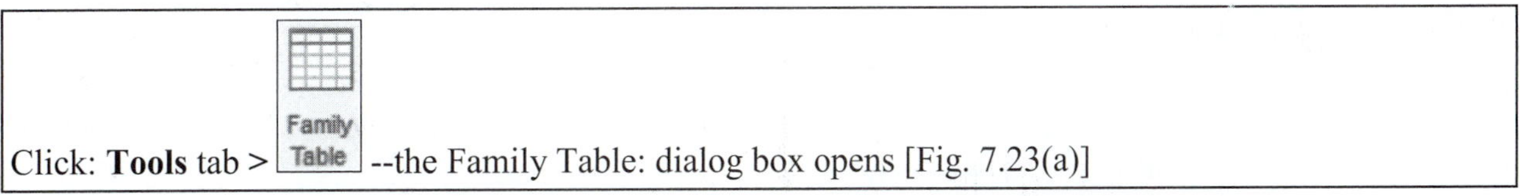

Click: **Tools** tab > Family Table --the Family Table: dialog box opens [Fig. 7.23(a)]

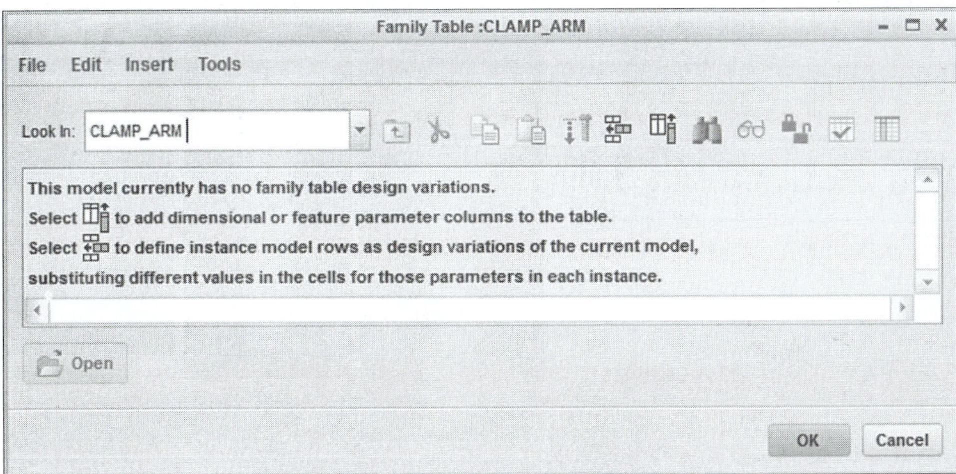

Figure 7.23(a) Family Table Dialog Box

Click: **Add/delete the table columns** > Feature from the Add Item options > select the cuts and holes from the model or the Model Tree [Fig. 7.23(b)] *(the order in which the items are listed will determine the default order in which they will appear in the table – default column order)* > **OK** [Fig. 7.23(c)] > **Insert a new instance at the selected row** > pick on the name of the new instance **CLAMP_ARM_INST** > *type* **CLAMP_ARM_DESIGN** [Fig. 7.23(d)] > **Enter** *(adds a new instance)*

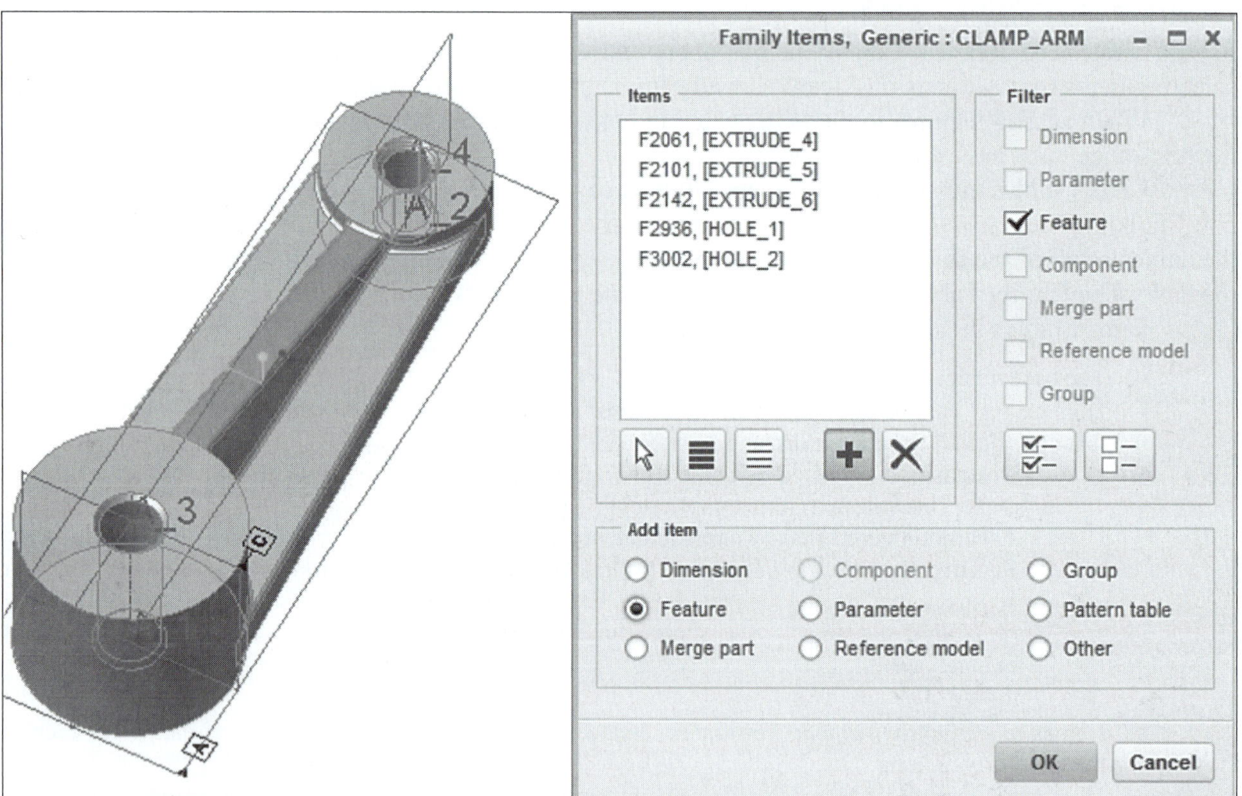

Figure 7.23(b) Family Items Dialog Box, Adding Features

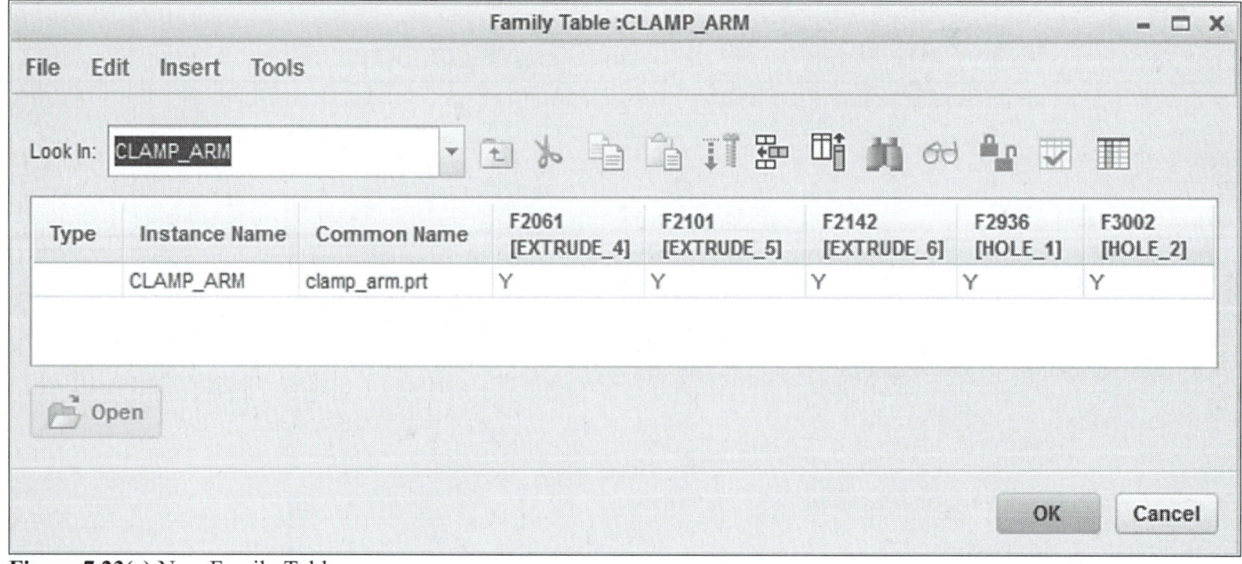

Figure 7.23(c) New Family Table

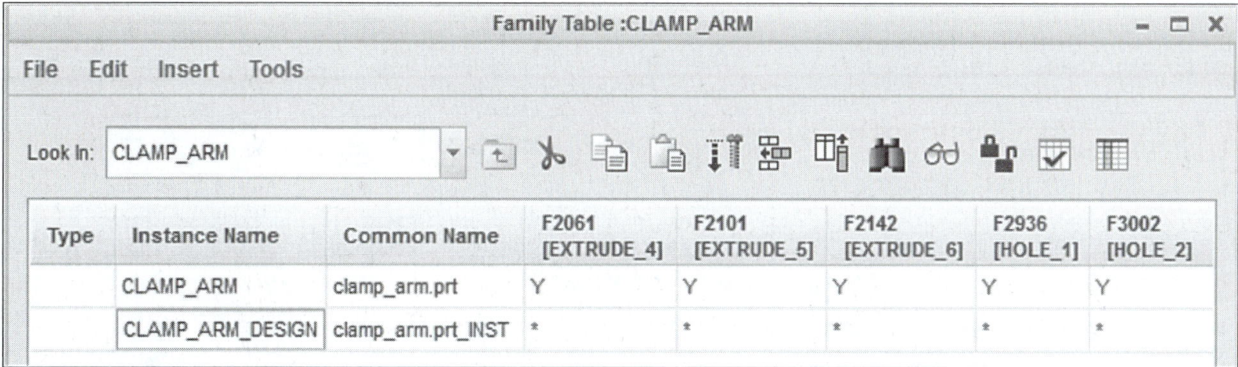

Figure 7.23(d) Add an Instance

The second instance **CLAMP_ARM_INST** should be highlighted, *type* **CLAMP_ARM_WORKPIECE** [Fig. 7.23(e)] > pick in the cell of the first feature and change to **N** (not used) [Fig. 7.23(f)] > change all cells for the CLAMP_ARM_WORKPIECE to **N** [Fig. 7.23(g)]

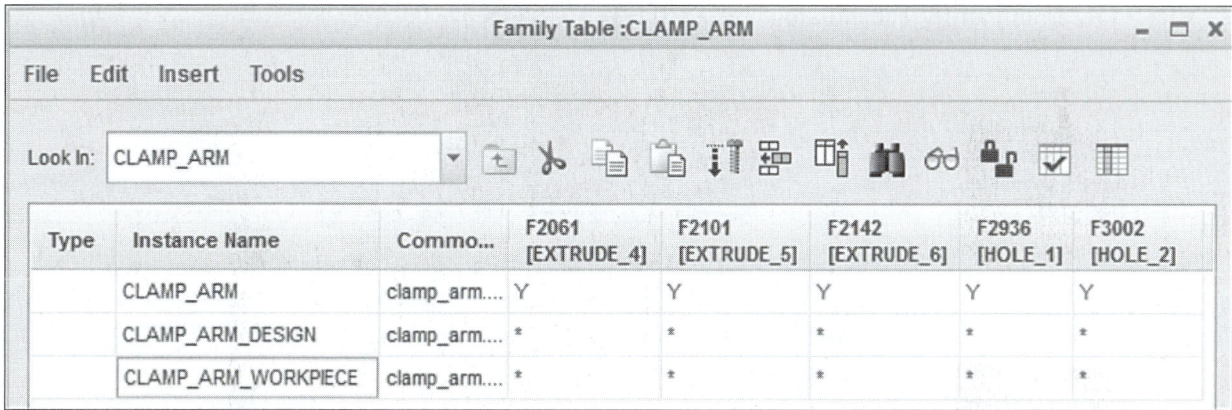

Figure 7.23(e) Add a Second Instance

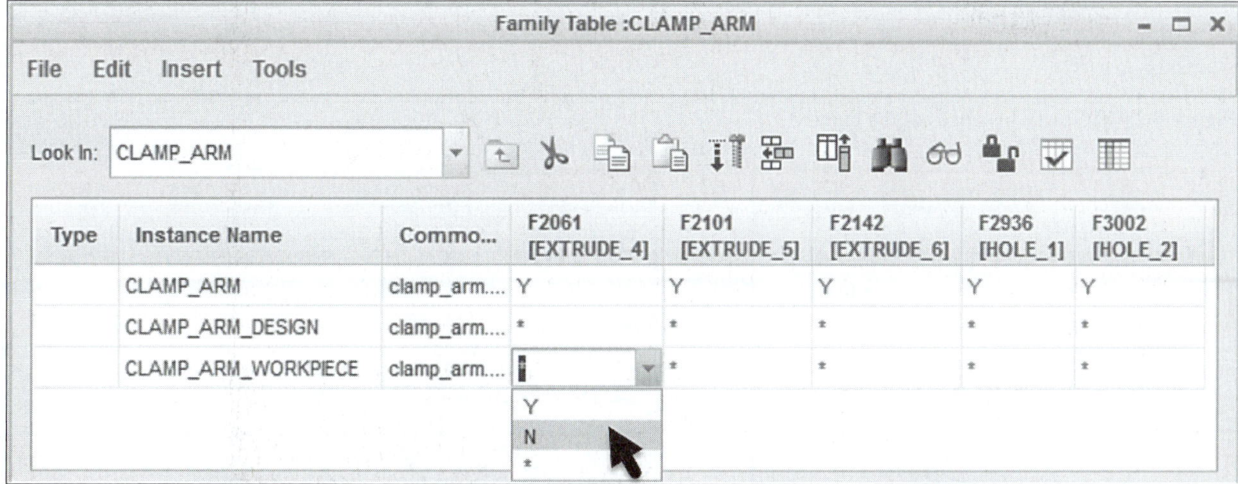

Figure 7.23(f) Change Feature to N (not used) *(pick on the drop down arrow for the list of choices)*

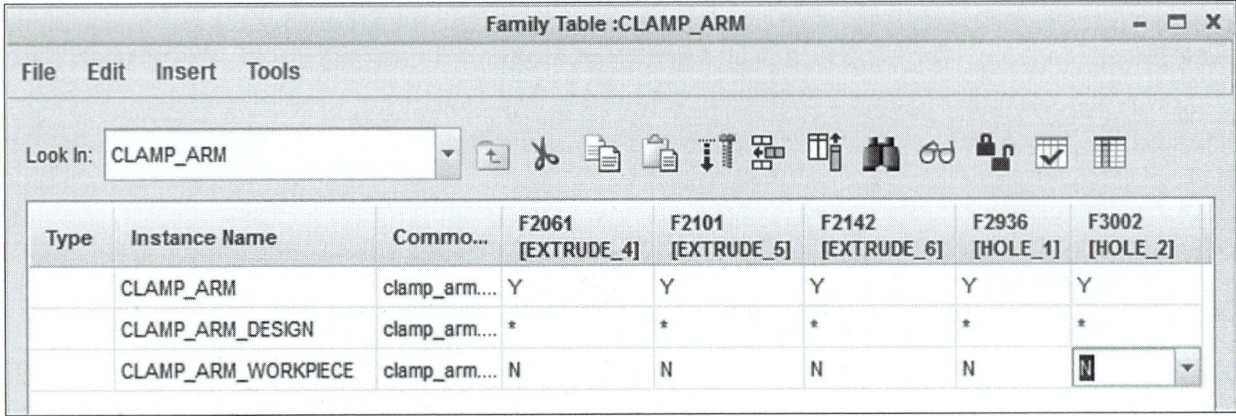

Figure 7.23(g) N for All Machined Features

Note: If your school or company uses the parameter "Common Name", enter the appropriate data.

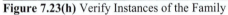

Click: **Verify instances of the family** [Fig. 7.23(h)] > **VERIFY** [Fig. 7.23(i)] > **CLOSE** [Fig. 7.23(j)] > **OK** from Family Table dialog box > **Ctrl+D** > **Ctrl+S**

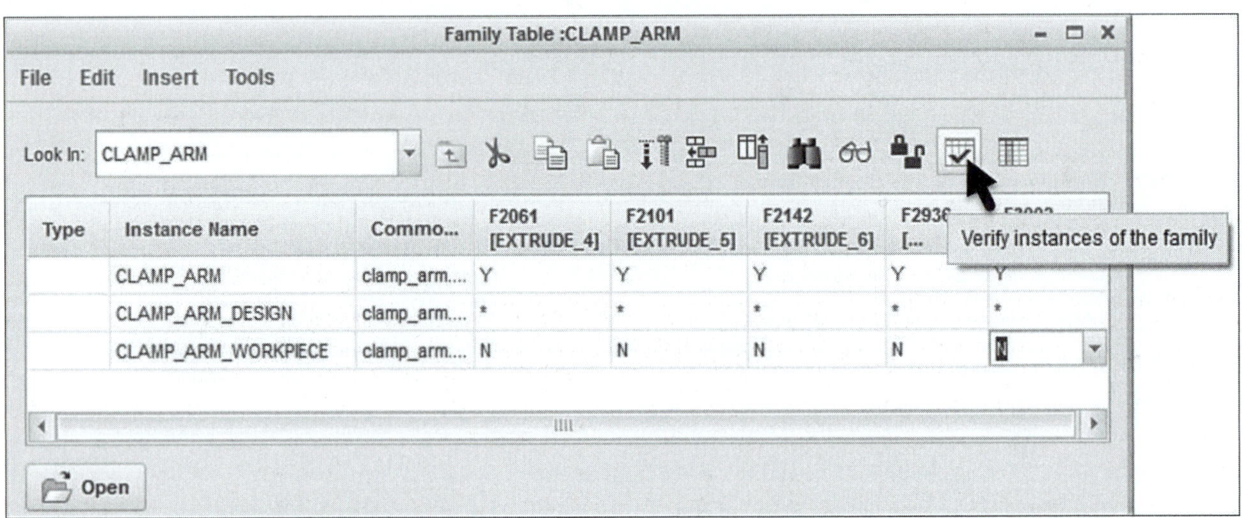

Figure 7.23(h) Verify Instances of the Family

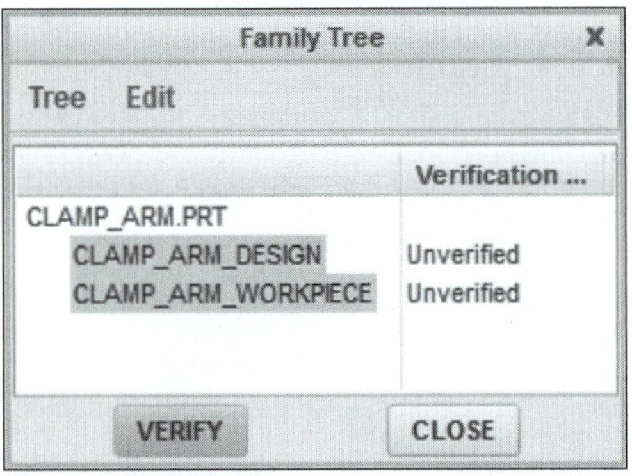

Figure 7.23(i) Verify

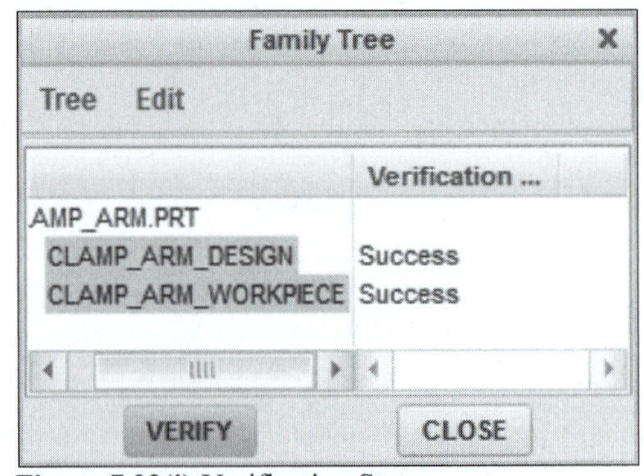

Figure 7.23(j) Verification Status

A Family Table controls whether a feature is present or not for a given design instance, not whether a feature is displayed. The Generic is the base model [Fig. 7.23(k)] and is typically not a member of an assembly.

Click: **Tools** tab > **Family Table** > **CLAMP_ARM_DESIGN** > Open [Fig. 7.23(l)] > **Close** > **Family Table** > **CLAMP_ARM_WORKPIECE** > Open [Fig. 7.23(m)] > **Close** (the Generic will be the only object on your screen) > **Ctrl+S**

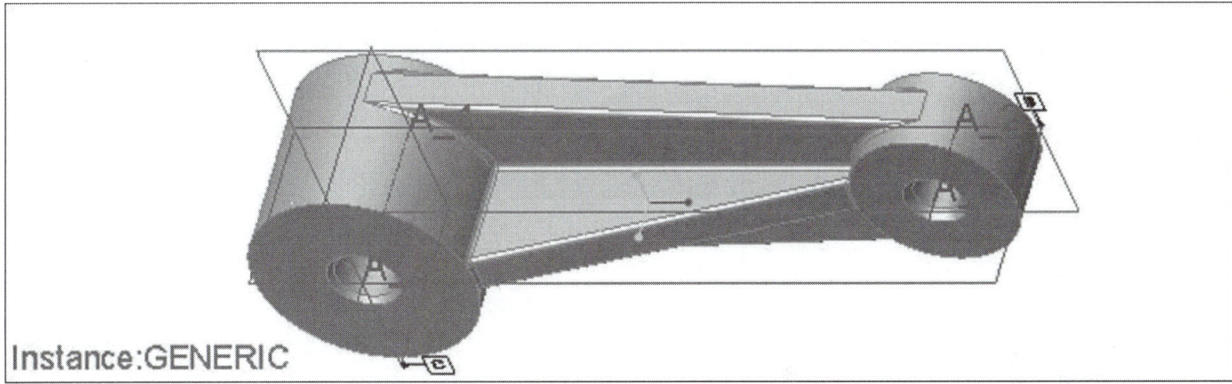

Figure 7.23(k) Instance: GENERIC

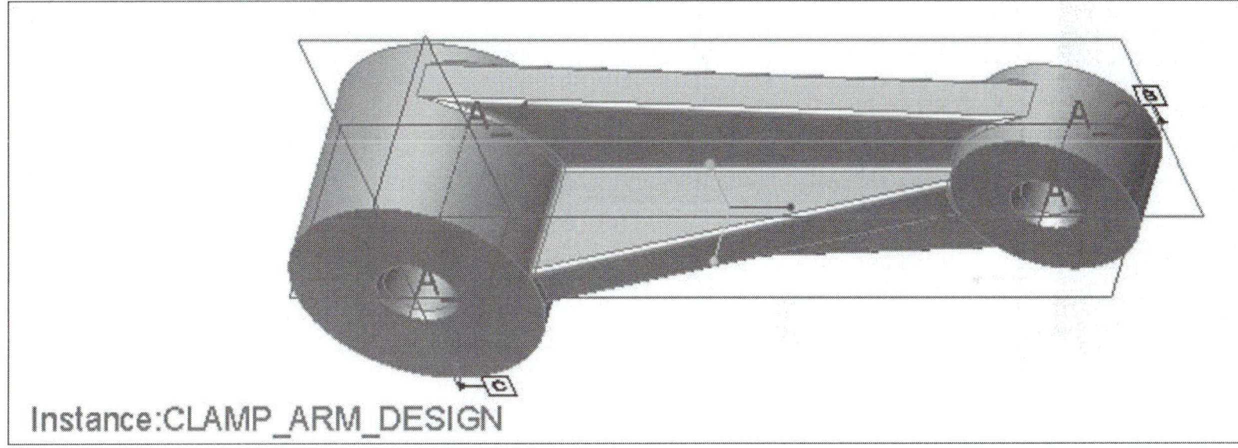

Figure 7.23(l) Instance: CLAMP_ARM_DESIGN

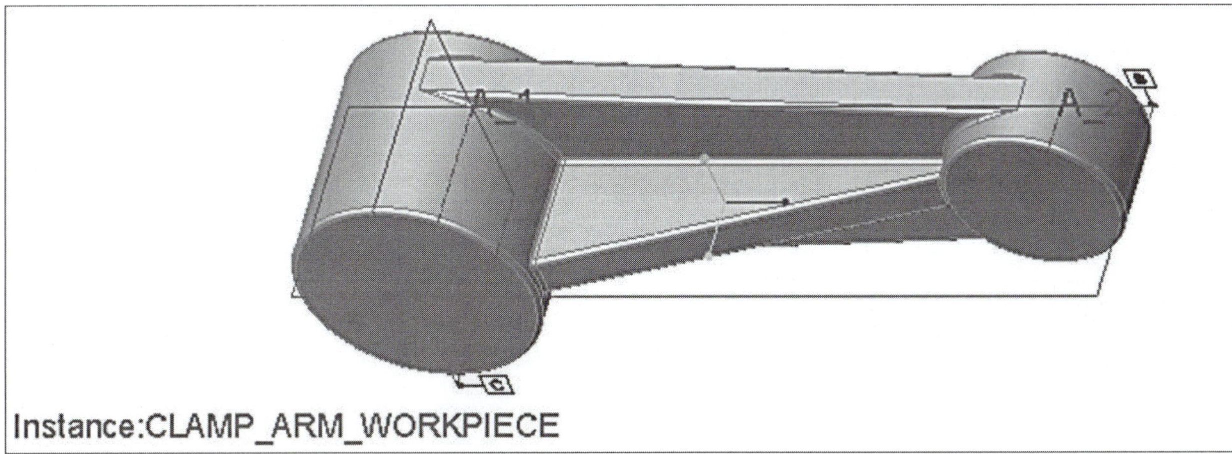

Figure 7.23(m) Instance: CLAMP_ARM_WORKPIECE

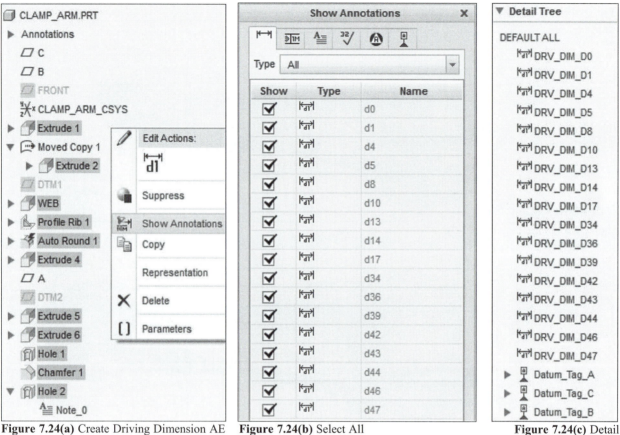

Click: > > **Annotate** tab > select the features in the Model Tree [Fig. 7.24(a)] > **RMB > Show Annotations >** [Fig. 7.24(b)] > **Apply > Cancel** [Fig. 7.24(c)] > pick on a dimension and move to a new position > select dimensions for the threaded hole > **RMB > Delete > LMB >** pick on > press **RMB > Flip** ( ) [Fig. 7.24(d)]

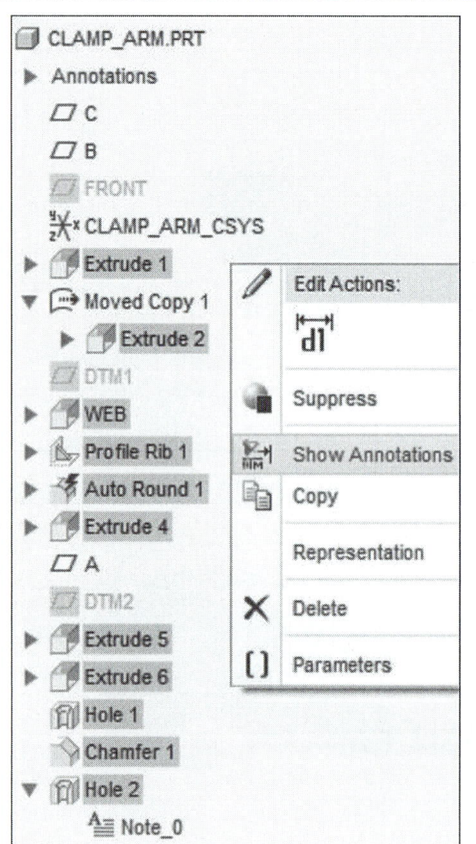

Figure 7.24(a) Create Driving Dimension AE Tree

Show Annotations

Type All

Show	Type	Name
✓		d0
✓		d1
✓		d4
✓		d5
✓		d8
✓		d10
✓		d13
✓		d14
✓		d17
✓		d34
✓		d36
✓		d39
✓		d42
✓		d43
✓		d44
✓		d46
✓		d47

Figure 7.24(b) Select All

Detail Tree

DEFAULT ALL
DRV_DIM_D0
DRV_DIM_D1
DRV_DIM_D4
DRV_DIM_D5
DRV_DIM_D8
DRV_DIM_D10
DRV_DIM_D13
DRV_DIM_D14
DRV_DIM_D17
DRV_DIM_D34
DRV_DIM_D36
DRV_DIM_D39
DRV_DIM_D42
DRV_DIM_D43
DRV_DIM_D44
DRV_DIM_D46
DRV_DIM_D47
Datum_Tag_A
Datum_Tag_C
Datum_Tag_B

Figure 7.24(c) Detail

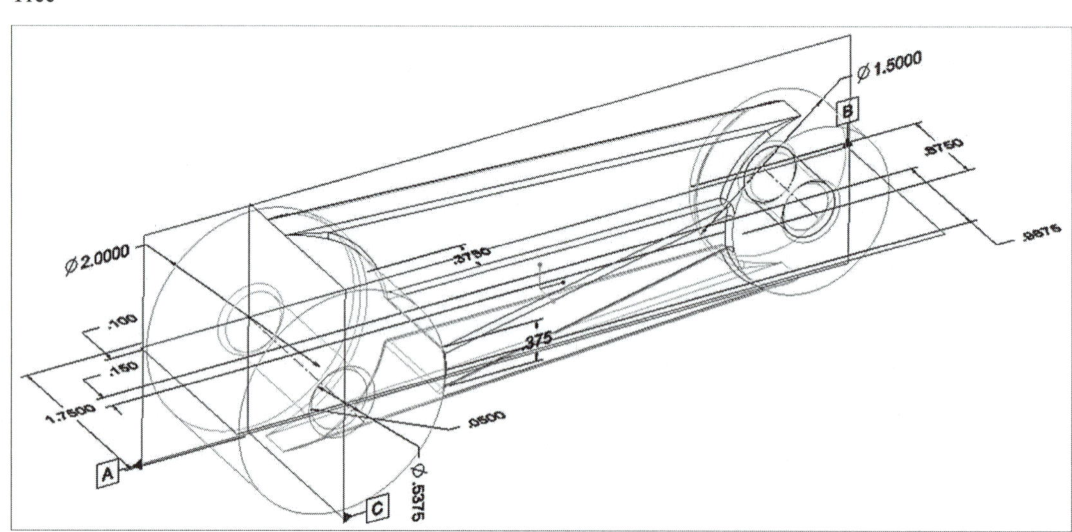

Figure 7.24(d) Repositioned Dimensions

Click: FRONT [Fig. 7.24(e)] > **Ctrl+D** > **Ctrl+S** > **Ok** (if needed, from Conflicts dialog box) > **Tools** tab > [ ] Parameters **Parameters** [Fig. 7.25] > **OK** > **File** > **Manage File** > **Delete Old Versions** > **Yes** > **File** > **Save As** > **Save a Copy** > Type ▾ > **Zip File (*.zip)** > **OK** > upload the zip file to your class interface system and email to yourself and your instructor > **File** > **Close**

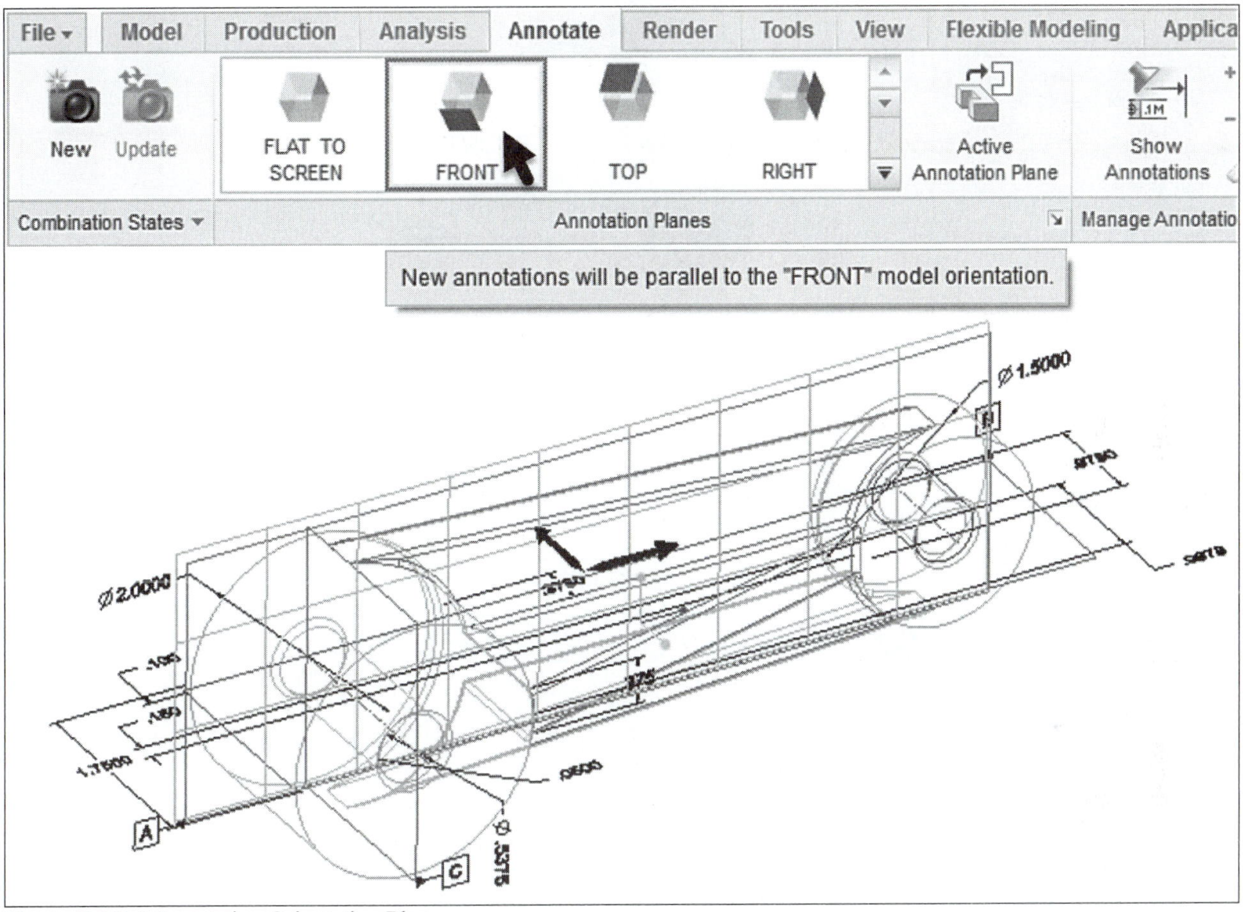

Figure 7.24(e) Annotation Orientation Plane

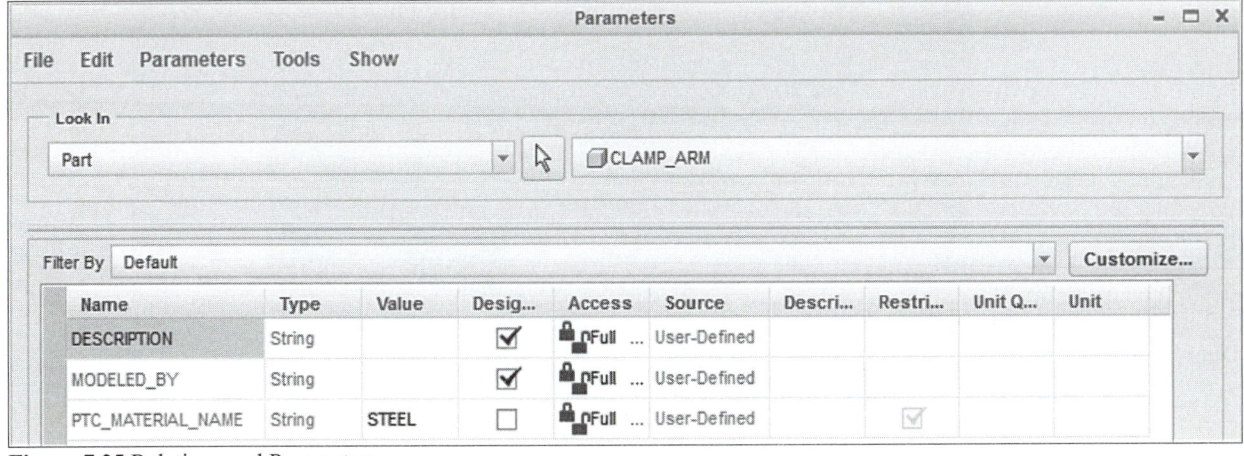

Figure 7.25 Relations and Parameters

You now have two separate models, a casting (workpiece) and a machined part (design part). During the manufacturing process, the workpiece is merged (assembled) into the design part thereby creating a manufacturing model [Fig. 7.26(a)]. The difference between the two objects is the difference between the volume of the casting and the volume of the machined part. The manufacturing model is used to machine the part [Fig. 7.26(b)]. If you have the manufacturing module you could now virtually "machine" the part.

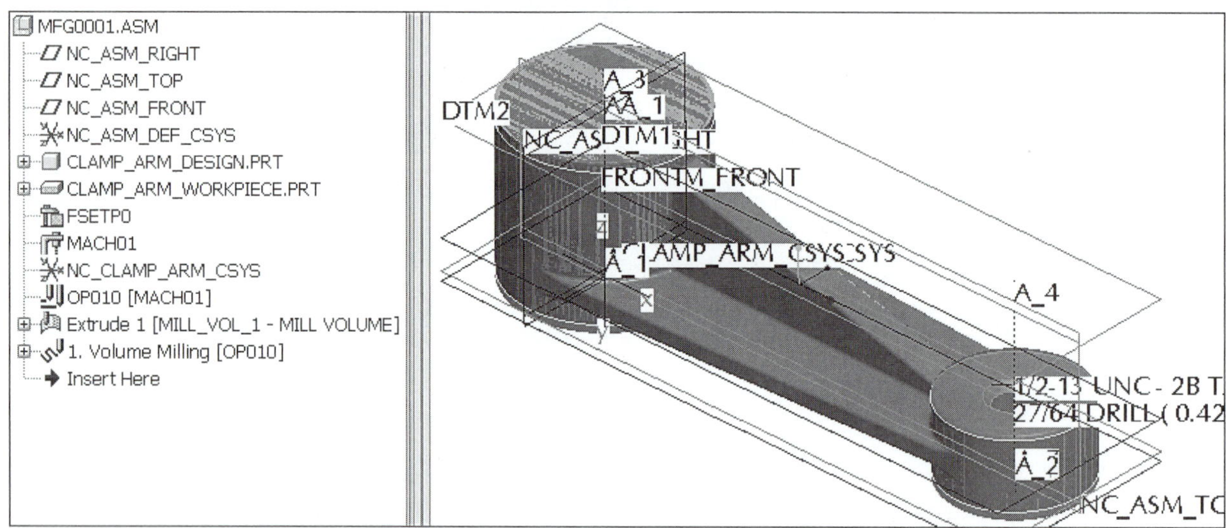

Figure 7.26(a) Manufacturing Model

Figure 7.26(b) Facing

A complete set of extra projects are available at ***www.cad-resources.com***

Lesson 8 Assemblies

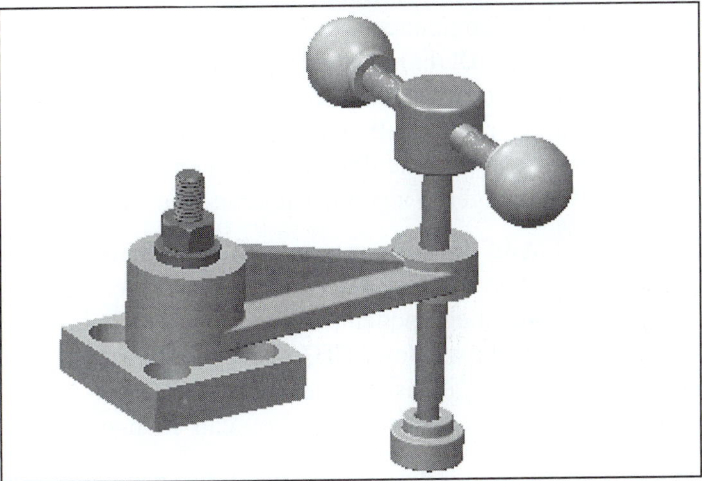

Figure 8.1(a) Swing Clamp Assembly

OBJECTIVES

- **Assemble** components to create an assembly
- Create a **subassembly**
- Understand and use a variety of **Assembly Constraints**
- **Modify** a component constraint
- Check for **clearance** and **interference**

REFERENCES AND RESOURCES

For **Resources** go to **www.cad-resources.com** > click on the PTC Creo Parametric 3.0 Book cover

- Lesson 8 Lecture
- Lesson 8 3D models embedded in a PDF
- Book Projects PDF
- Project Lectures

ASSEMBLY CONSTRAINTS

Assembly mode allows you to place together components and subassemblies to create an assembly [Fig. 8.1(a)]. Assemblies [Fig. 8.1(b)] can be modified, reoriented, documented, or analyzed. An assembly can be assembled into another assembly, thereby becoming a subassembly.

Figure 8.1(b) Swing Clamp (CARRLANE at **www.carrlane.com**)

Placing Components

The 3D Dragger is a graphical tool for precision handling of assembly components. The *axes* are called **draggers**. Pull or rotate the draggers to make changes to the components position and orientation. The draggers enable movement in one or more *degrees of freedom* (DOFs). They support linear, planar, free, translation, and angular movements. You can pull a dragger along a 2D or 3D trajectory. When you move a geometric entity using a dragger, the relative distances from the start point of the entity appear in the graphics window as you pull the dragger. You can hold down the SHIFT key while pulling a dragger to snap the dragger to a geometric reference.

To assemble components, use: **Add a component to the assembly**. After selecting a component from the Open dialog box, the dashboard (Fig. 8.2) opens and the component appears in the assembly window. Alternatively, you can select a component from a browser window and drag it into the graphics window. If there is an assembly in the window, PTC Creo Parametric 3.0 will begin to assemble the component into the current assembly. Using icons in the dashboard, you can specify the screen window in which the component is displayed while you position it. You can change window options at any time using: **Show component in a separate window while specifying constraints** or **Show component in the assembly window while specifying constraints**.

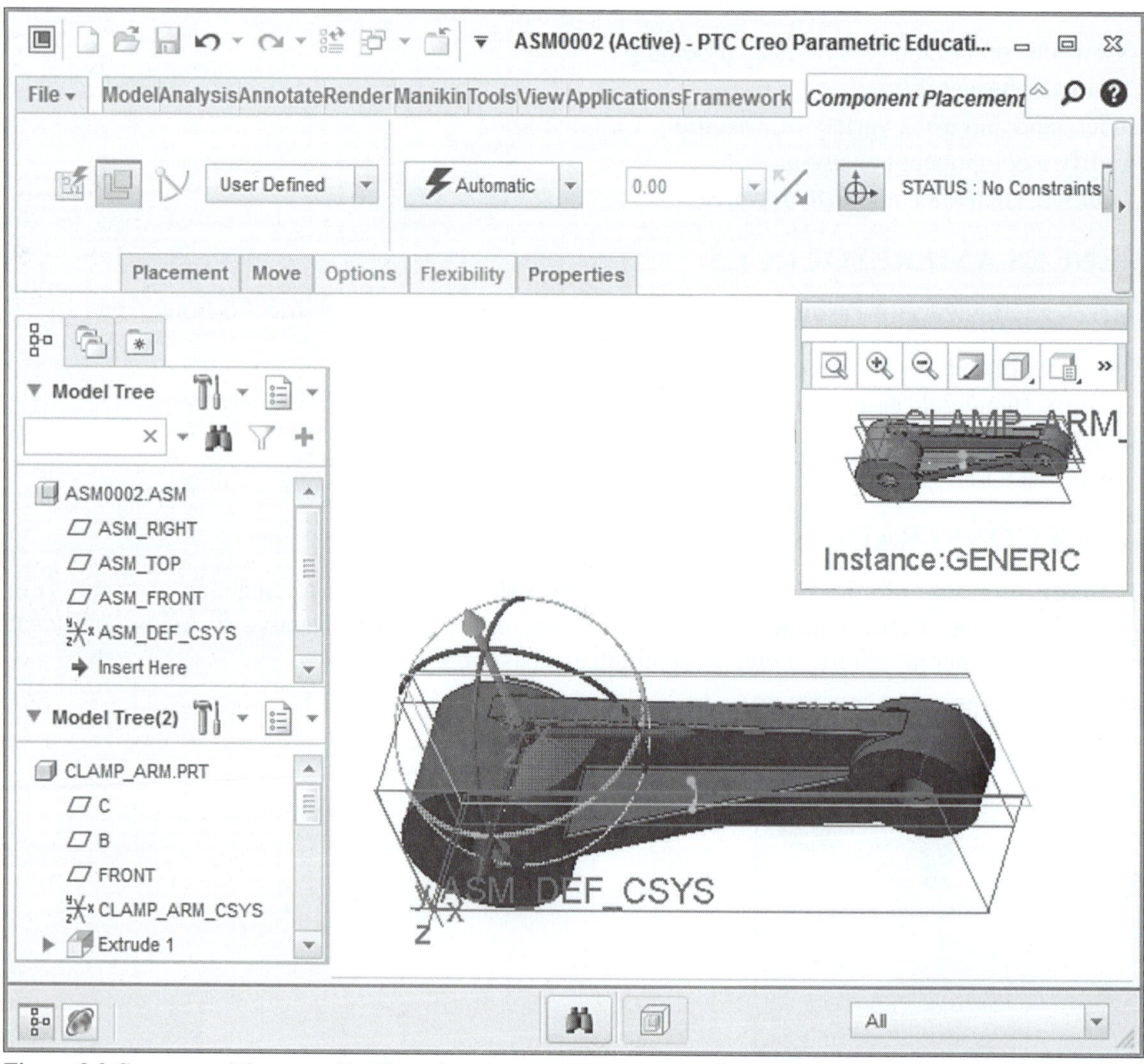

Figure 8.2 Component Placement Dashboard

Lesson 8 STEPS- Part One: Bottom-Up Design

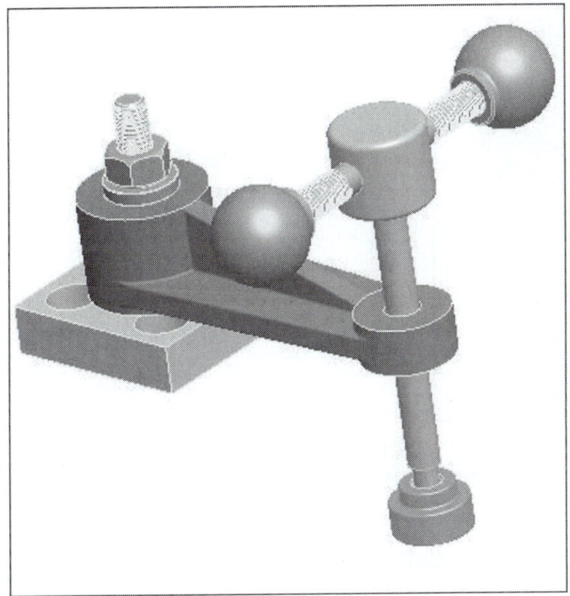

Figure 8.3(a) Swing Clamp Main Assembly

Figure 8.3(b) Swing Clamp Sub-Assembly

Swing Clamp Assembly

The parts required in this lesson are from this text. *If you have not modeled these parts previously, please do so before you start the following systematic instructions.* The other components required for the assembly are standard *off-the-shelf* hardware items that you can get by accessing a company catalog. The **Flange Nut**, the **3.50 Double-ended Stud**, and the **5.00 Double-ended Stud** are standard items. The **Clamp Plate** component is the first component of the main assembly and will be modeled later when completing the main assembly [Fig. 8.3(a)] using the *top-down design* approach.

Because you will be creating the sub-assembly [Fig. 8.3(b)] using the *bottom-up design* approach, all the components must be available before any assembling starts. *Bottom-up design* means that existing parts are assembled, one by one, until the assembly is complete. The assembly starts with a set of default datum planes and a coordinate system. The parts are constrained to the datum features of the assembly. The sequence of assembly will determine the parent-child relationships between components.

Top-down design is the design of an assembly where one or more component parts are created in Assembly mode as the design unfolds. Some existing parts are available, such as standard components and a few modeled parts. The remaining design evolves during the assembly process. The main assembly will involve creating one part using the *top-down design* approach.

Regardless of the design method, the assembly default datum planes and coordinate system should be on their own separate *assembly layer*. Each part should also be placed on separate assembly layers; the part's datum features should already be on *part layers*.

Before starting the assembly, you will be modeling each part or retrieving *standard parts* and saving them under unique names into *your* working directory. *Unless instructed to do so,* ***do not use the library parts directly in the assembly***. Start this process by retrieving the standard parts [Figs. 8.3(c-e)].

The three standard parts are also available at *http://www.cad-resources.com* > *pick on the appropriate book cover > download the commercial or academic versions > drag and drop the (each) file into an empty Graphics Window*

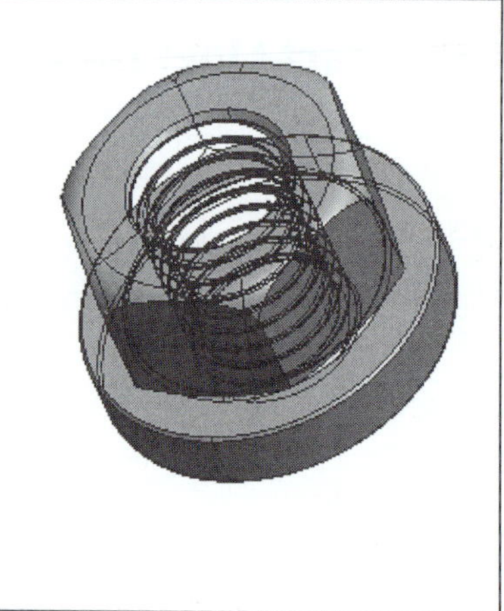

Figures 8.3(c-e) Standard Parts from CarrLane

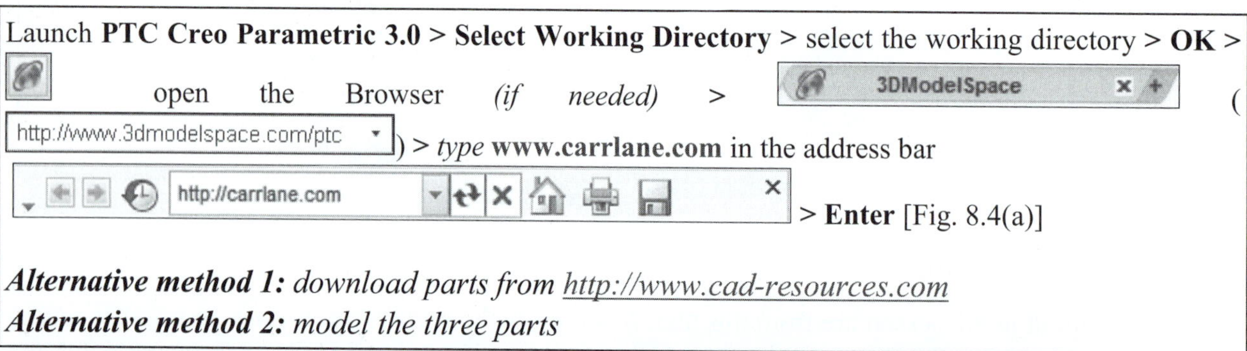

Launch **PTC Creo Parametric 3.0 > Select Working Directory >** select the working directory **> OK >** open the Browser *(if needed)* > 3DModelSpace (http://www.3dmodelspace.com/ptc) > *type* **www.carrlane.com** in the address bar http://carrlane.com > **Enter** [Fig. 8.4(a)]

Alternative method 1: *download parts from http://www.cad-resources.com*
Alternative method 2: *model the three parts*

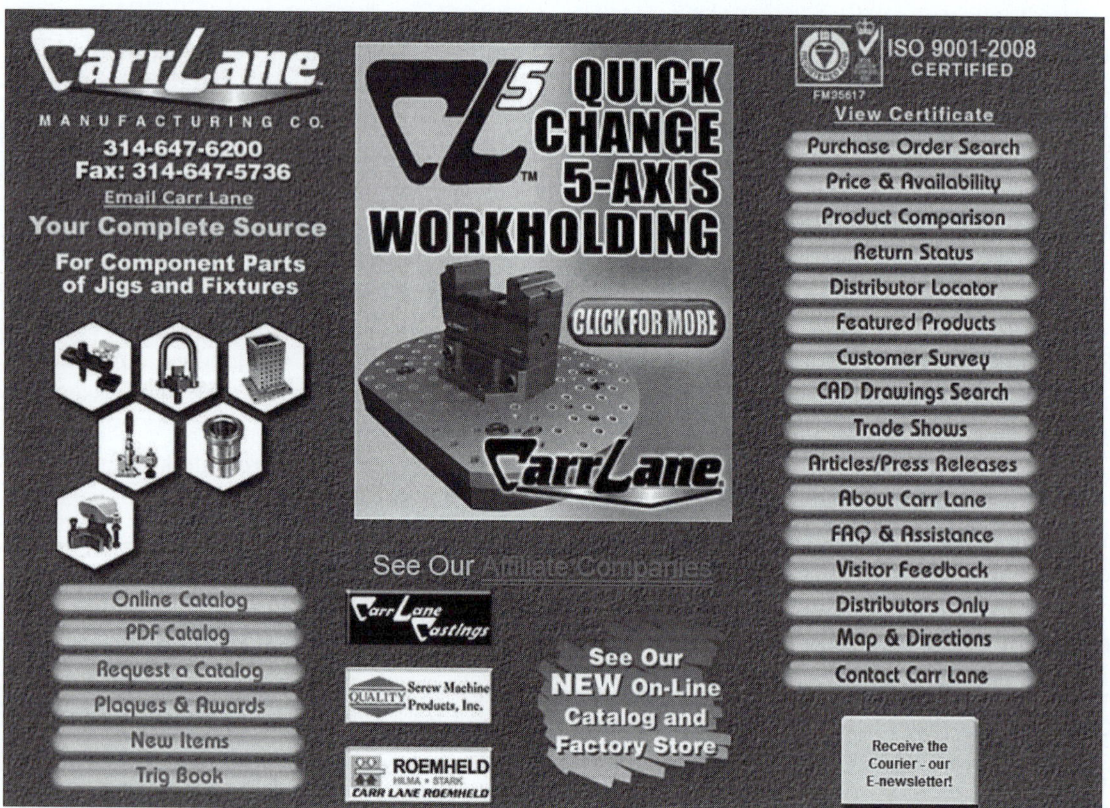

Figure 8.4(a) Carr Lane (The Website may have since been updated)

Click: **Online Catalog** Online Catalog > **Clamps and Accessories** Clamps and Accessories >

STUDS, BOLTS, AND T NUTS
STUDS, BOLTS, AND T NUTS
> **Clamping Studs**
Clamping Studs
[Fig. 8.4(b)]

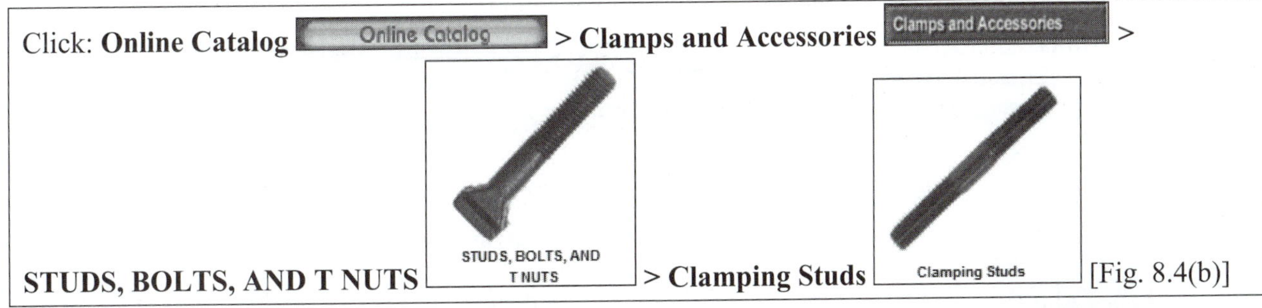

CarrLane
USA Catalog
General Information
Alignment Pins
Hoist Rings
Handles/Knobs/Screw Clamps
Spring-Loaded Devices
Supports, Rests, and Feet
Locators
Clamps and Accessories
Plug Gages
Threaded Inserts
Jig and Fixture Bases
Chuck Jaws
Modular Fixturing
Toggle Clamps
Drill-Jig Bushings
Power Workholding
SEARCH
New Items
Find a Carr Lane Distributor
View My Shopping Cart
CAD Drawings Search
Part Number Conversion
Trig Book
Carr Lane Home Page
Features
Dimensions
CLAMPS AND ACCESSORIES » STUDS, BOLTS, AND T NUTS » Clamping Studs
CLAMPING STUDS
FEATURES: High-strength general-purpose Clamping Stud, for use with a T Nut, or for installation directly into a fixture plate. Available in steel or stainless steel. Made in USA.
SIZES (STEEL): Available in the following thread sizes, each in a choice of many lengths — see Dimensions page for part numbers:
1/4-20 M6
5/16-18 M8
3/8-16 M10
1/2-13 M12
5/8-11 M16
3/4-10 M20
7/8-9 —
1"-8 —
SIZES (STAINLESS): Available in the following thread sizes, each in a choice of many lengths — see Dimensions page for part numbers:
3/8-16
1/2-13
5/8-11
MATERIAL (STEEL): 1144 steel, 115,000 psi min tensile strength, 100,000 psi min yield strength, black oxide finish. DFARS and RoHS compliant.
MATERIAL (STAINLESS): 300-series stainless, passivated. DFARS and RoHS compliant.
For smaller stainless steel sizes, see Fixture Studs.
Steel
& Stainless Steel
CarrLane
MANUFACTURING CO.
en_447
Figure 8.4(b) Carr Lane CLAMPING STUDS

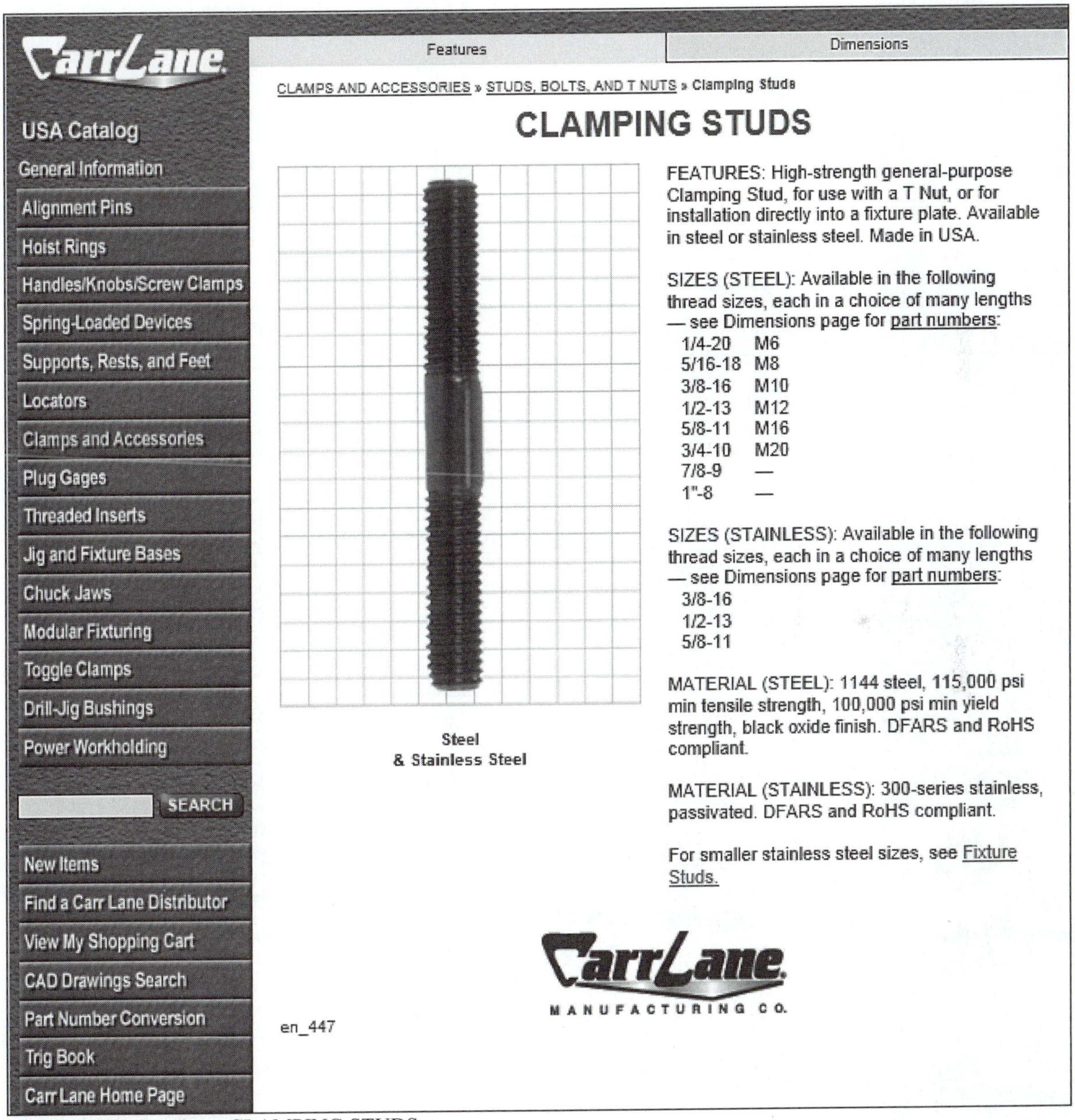

Click: **Dimensions** [Dimensions] [Fig. 8.4(c)] > select [CL-1/2-13X3.50-STUD]

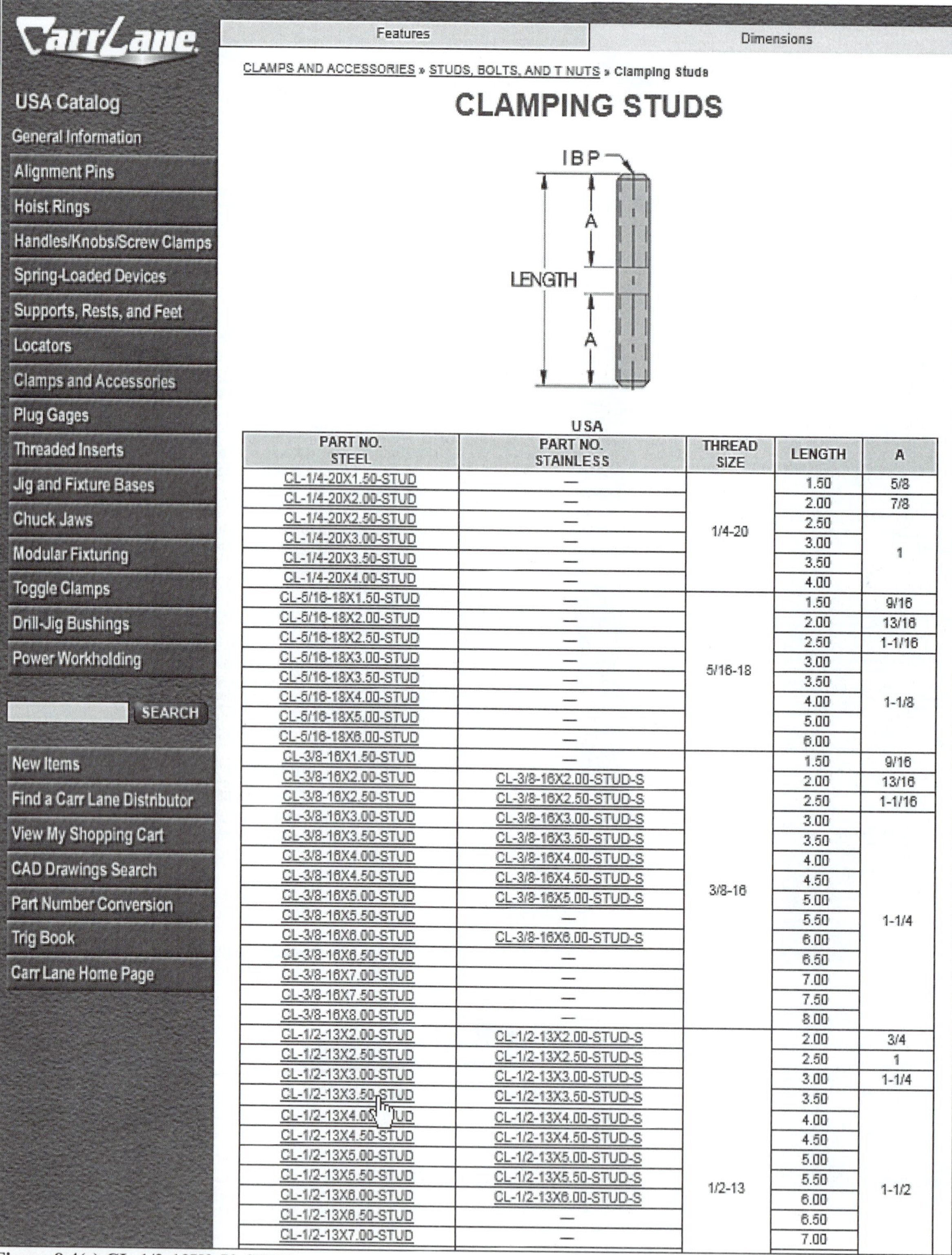

PART NO. STEEL	PART NO. STAINLESS	THREAD SIZE	LENGTH	A
CL-1/4-20X1.50-STUD	—	1/4-20	1.50	5/8
CL-1/4-20X2.00-STUD	—		2.00	7/8
CL-1/4-20X2.50-STUD	—		2.50	1
CL-1/4-20X3.00-STUD	—		3.00	
CL-1/4-20X3.50-STUD	—		3.50	
CL-1/4-20X4.00-STUD	—		4.00	
CL-5/16-18X1.50-STUD	—	5/16-18	1.50	9/16
CL-5/16-18X2.00-STUD	—		2.00	13/16
CL-5/16-18X2.50-STUD	—		2.50	1-1/16
CL-5/16-18X3.00-STUD	—		3.00	1-1/8
CL-5/16-18X3.50-STUD	—		3.50	
CL-5/16-18X4.00-STUD	—		4.00	
CL-5/16-18X5.00-STUD	—		5.00	
CL-5/16-18X6.00-STUD	—		6.00	
CL-3/8-16X1.50-STUD	—	3/8-16	1.50	9/16
CL-3/8-16X2.00-STUD	CL-3/8-16X2.00-STUD-S		2.00	13/16
CL-3/8-16X2.50-STUD	CL-3/8-16X2.50-STUD-S		2.50	1-1/16
CL-3/8-16X3.00-STUD	CL-3/8-16X3.00-STUD-S		3.00	1-1/4
CL-3/8-16X3.50-STUD	CL-3/8-16X3.50-STUD-S		3.50	
CL-3/8-16X4.00-STUD	CL-3/8-16X4.00-STUD-S		4.00	
CL-3/8-16X4.50-STUD	CL-3/8-16X4.50-STUD-S		4.50	
CL-3/8-16X5.00-STUD	CL-3/8-16X5.00-STUD-S		5.00	
CL-3/8-16X5.50-STUD	—		5.50	
CL-3/8-16X6.00-STUD	CL-3/8-16X6.00-STUD-S		6.00	
CL-3/8-16X6.50-STUD	—		6.50	
CL-3/8-16X7.00-STUD	—		7.00	
CL-3/8-16X7.50-STUD	—		7.50	
CL-3/8-16X8.00-STUD	—		8.00	
CL-1/2-13X2.00-STUD	CL-1/2-13X2.00-STUD-S	1/2-13	2.00	3/4
CL-1/2-13X2.50-STUD	CL-1/2-13X2.50-STUD-S		2.50	1
CL-1/2-13X3.00-STUD	CL-1/2-13X3.00-STUD-S		3.00	1-1/4
CL-1/2-13X3.50-STUD	CL-1/2-13X3.50-STUD-S		3.50	1-1/2
CL-1/2-13X4.00-STUD	CL-1/2-13X4.00-STUD-S		4.00	
CL-1/2-13X4.50-STUD	CL-1/2-13X4.50-STUD-S		4.50	
CL-1/2-13X5.00-STUD	CL-1/2-13X5.00-STUD-S		5.00	
CL-1/2-13X5.50-STUD	CL-1/2-13X5.50-STUD-S		5.50	
CL-1/2-13X6.00-STUD	CL-1/2-13X6.00-STUD-S		6.00	
CL-1/2-13X6.50-STUD	—		6.50	
CL-1/2-13X7.00-STUD	—		7.00	

Figure 8.4(c) CL-1/2-13X3.50-STUD

Click: **Interactive 3D Viewer** [Fig. 8.4(d)] (download/activate the viewer (plugin) if necessary) > **CAD Downloads** > 3D Download Formats: > **Pro/E** *(Creo)* **Part/Assembly (*.prt)** > **Go** [Fig. 8.4(e)]

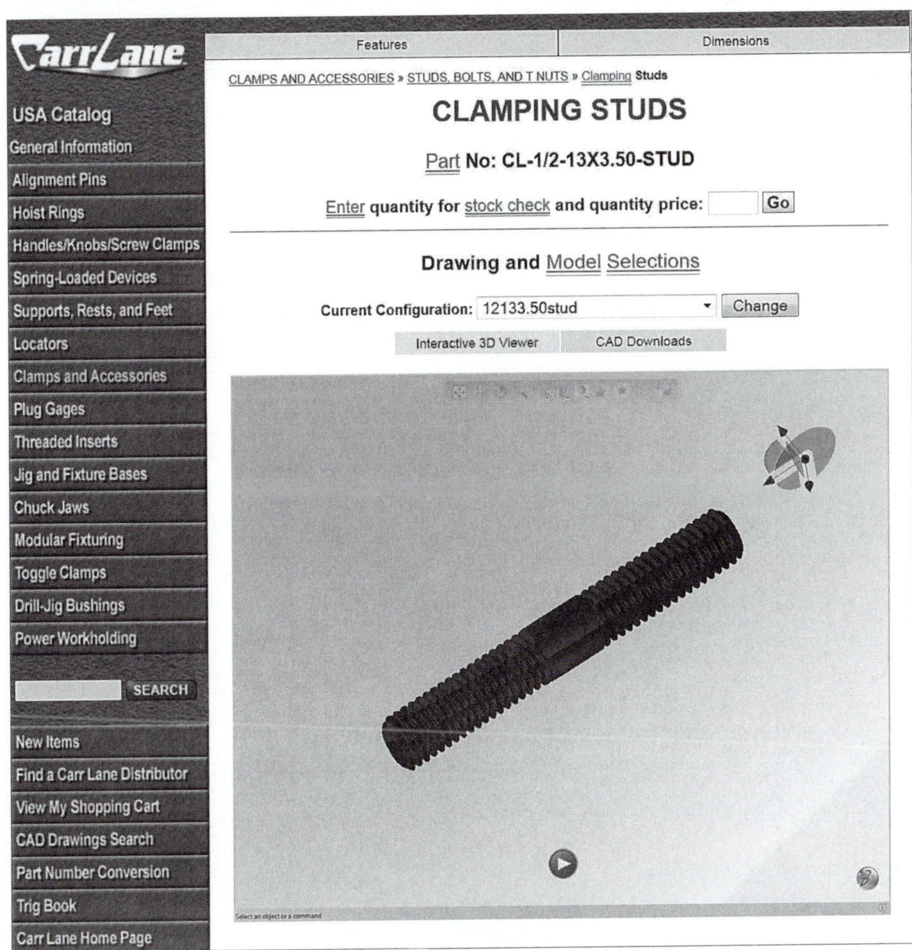

Figure 8.4(d) 3D Viewer (download the viewer if desired or allowed)

CLAMPING STUDS

Part No: CL-1/2-13X3.50-STUD

Enter quantity for stock check and quantity price: Go

Drawing and Model Selections

Current Configuration: 12133.50stud Change

Interactive 3D Viewer | CAD Downloads

The file you requested is ready to download.

Part Number: 12133.50STUD
Format: Pro/E Part/Assembly
Type: 3D

Download Drawing

If you experience trouble downloading, right click on the "Download Drawing" button and select "Save Target As..." (or "Save Link As..." in Mozilla Firefox). Then open in appropriate program.

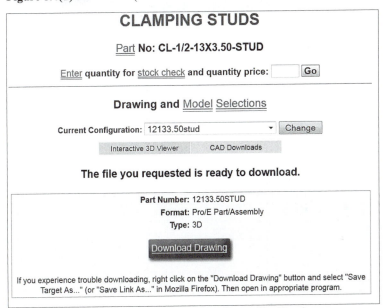

Figure 8.4(e) Pro/E *(Creo Parametric)* Part/Assembly (*.prt)

Click: **Download Drawing** > Save ▾ > **Save as** (Save As dialog box opens) *[if this does not work, right click on* **Download Drawing** *and select "Save Target As..."]* > navigate to your Working Directory > in File name field; *type* (a new name) **Carrlane-12-13350_STUD.prt > Enter** > double-click on the Browser sash to close > **File > Open > Carrlane-12-13350_STUD.prt > Open > View** tab > **Appearance Gallery >** ▾ next to Clear Appearance > **Clear All Appearances** [Fig. 8.4(f)] > **Yes** > change the coordinate system name to **350_STUD > Enter** [Fig. 8.4(g)] > in the Graphics Window, **LMB** to deselect > **Ctrl+D > Ctrl+S >** **Close** *(you may also model the stud [Fig. 8.4(h)])*

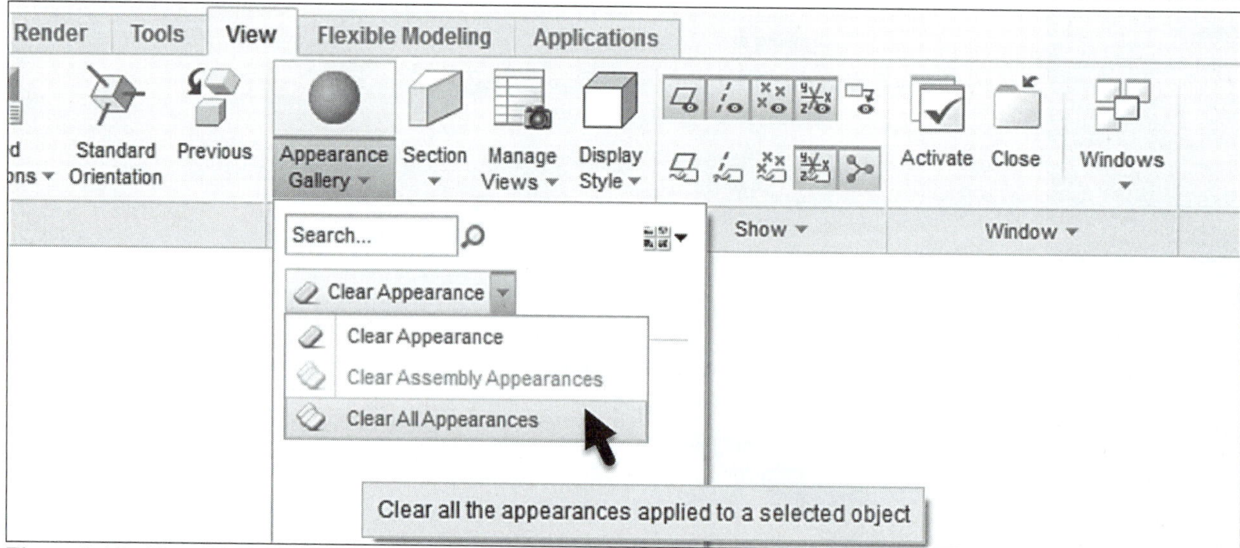

Figure 8.4(f) Clear All Appearances

Figure 8.4(g) New Coordinate System Name: **350_STUD**

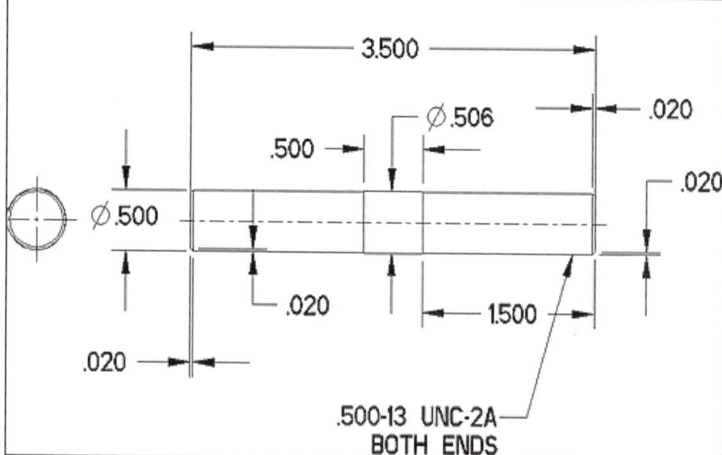

Figure 8.4(h) Model your own stud (without threads) if the catalog is not available

Click: [icon] open the Browser > [icon] > [icon] until you get to the CLAMPING STUDS Dimensions page > select CL-1/2-13X5.00-STUD > **CAD Downloads** > 3D Download Formats: **Pro/E** *(Creo Parametric 3.0)* **Part/Assembly (*.prt)** > **Go** > **Download Drawing** > Save [icon] > **Save as** > navigate to your Working Directory > File name **Carrlane-12-13500_STUD.prt** > **Save** > [icon] close Browser > [icon] > **Views** tab > **Thumbnails** [Fig. 8.4(i)] > **Carrlane-12-13500_STUD.prt** > **Open** > **View** tab > **Appearance Gallery** > [icon] next to Clear Appearance > **Clear All Appearances** > **Yes** > rename the coordinate system to **500_STUD** [Fig. 8.4(j)] > **Ctrl+D** > **Ctrl+S** > **File** > **Close** *(model the stud if needed [Fig. 8.4(k)])*

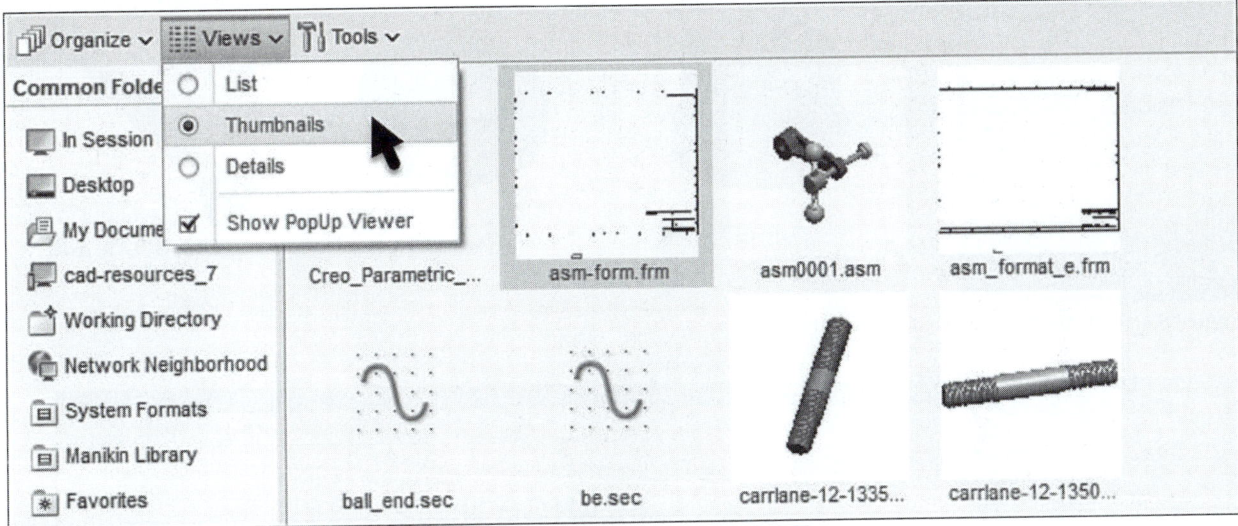

Figure 8.4(i) File > Open > Views Tab > Thumbnails > Carrlane-12-13500_STUD.prt

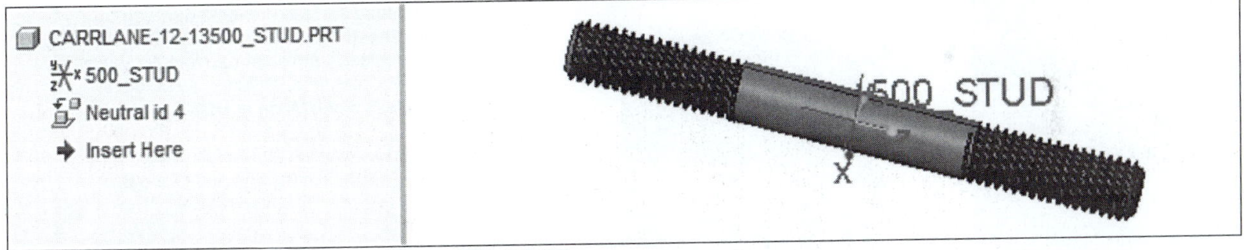

Figure 8.4(j) New Coordinate System Name: **500_STUD**

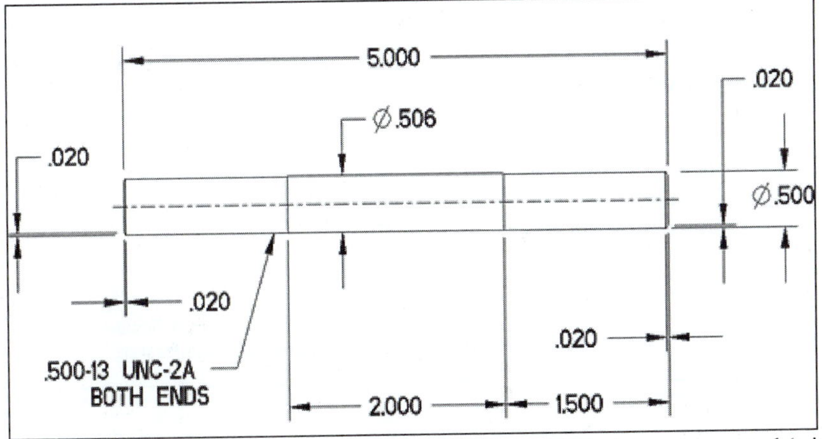

Figure 8.4(k) Double-ended Stud **Ø.500** by **5.00** Length. Model your own stud (without threads) if the catalog is not available.

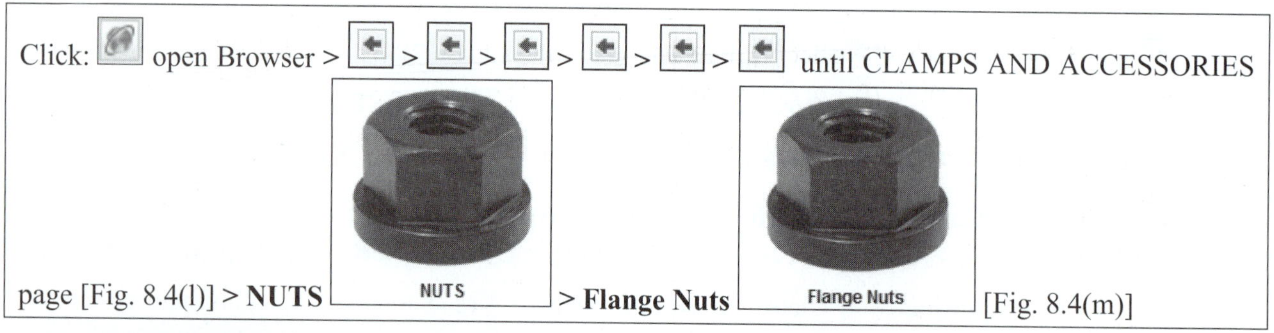

Click: open Browser > > > > > > until CLAMPS AND ACCESSORIES page [Fig. 8.4(l)] > **NUTS** > **Flange Nuts** [Fig. 8.4(m)]

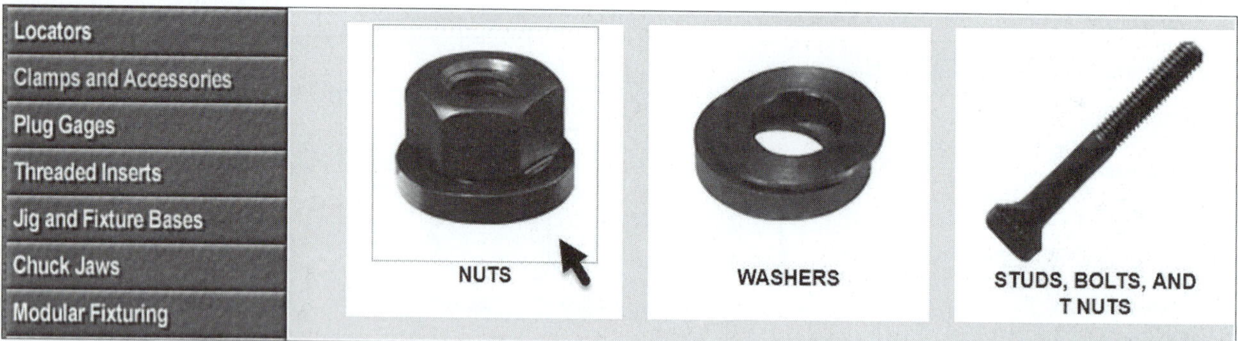

Figure 8.4(l) CLAMPS AND ACCESSORIES

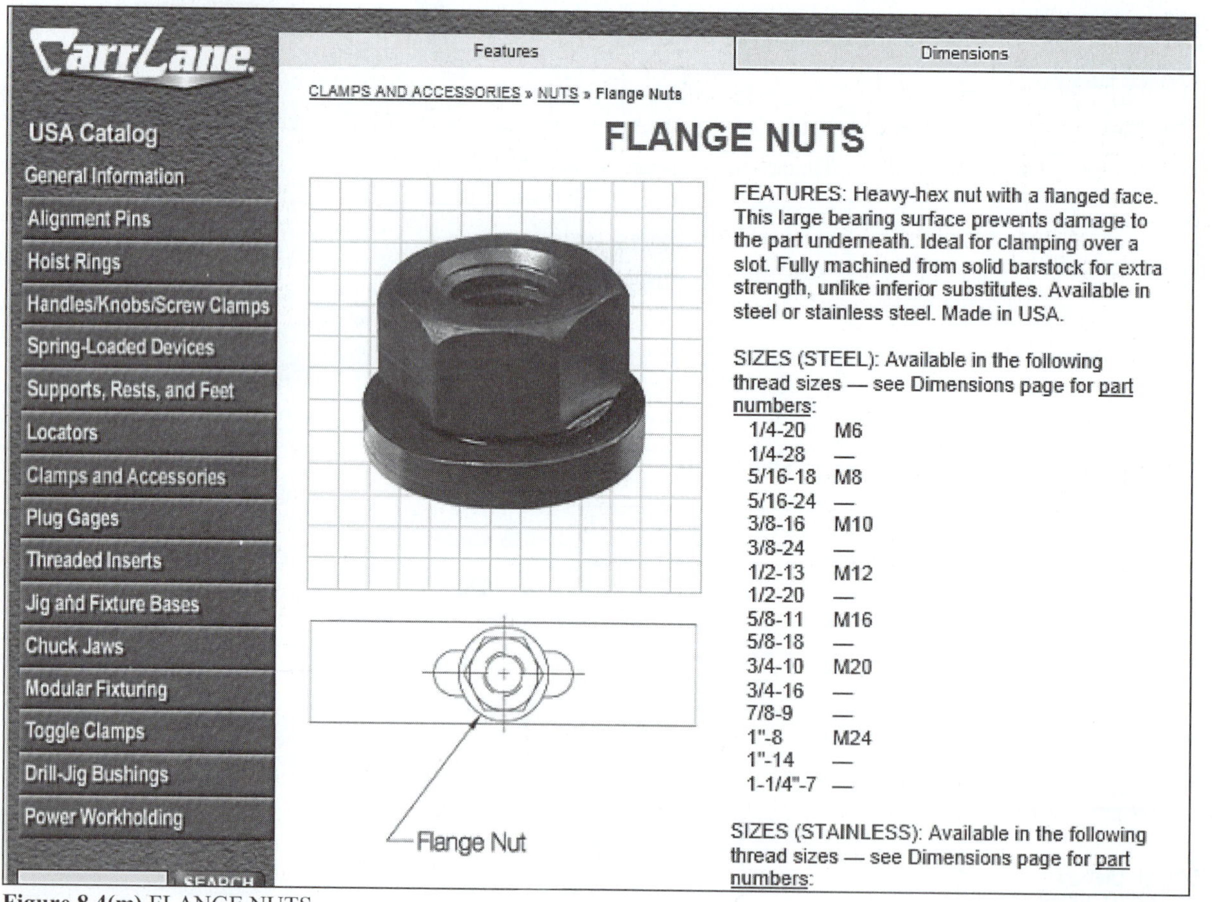

Figure 8.4(m) FLANGE NUTS

Click: **Dimensions** [Fig. 8.4(n)] > select CL-3-FN > **Interactive 3D Viewer** > **CAD Downloads** > 3D Download Formats: **Pro/E** *(Creo Parametric 3.0)* **Part/Assembly** > **Go** > **Download Drawing** > Save > **Save as** > navigate to your Working Directory > File name, *type* **carrlane_500_fn.prt** > **Save** > double-click on the Browser sash to close > **Open** > **carrlane_500_fn.prt** > **Open** > **View** tab > **Appearance Gallery** > next to Clear Appearance > **Clear All Appearances** > **Yes** > rename the coordinate system to **CARRLANE_500_FN** [Fig. 8.4(o)] > **Enter** > in the Graphics Window, **LMB** to deselect > **Ctrl+D** > **Ctrl+S** > **File** > **Close** *(you may also model the Flange Nut [Fig. 8.4(p)])*

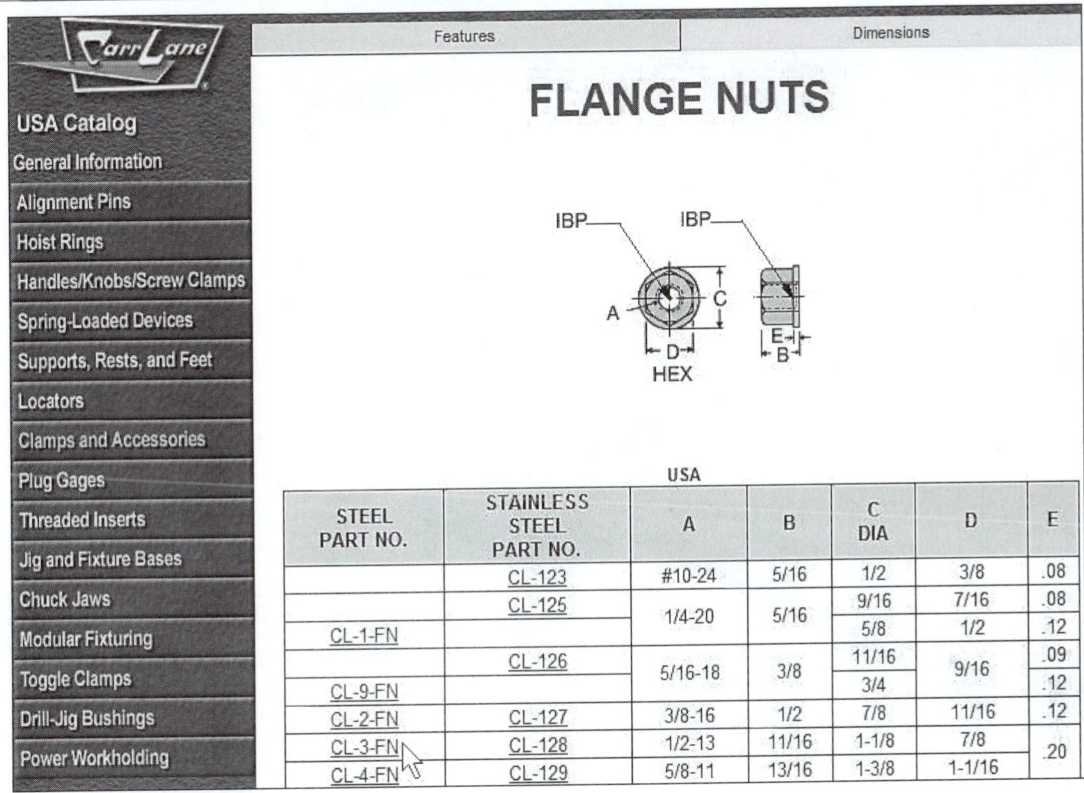

STEEL PART NO.	STAINLESS STEEL PART NO.	A	B	C DIA	D	E
	CL-123	#10-24	5/16	1/2	3/8	.08
	CL-125	1/4-20	5/16	9/16	7/16	.08
CL-1-FN				5/8	1/2	.12
	CL-126	5/16-18	3/8	11/16	9/16	.09
CL-9-FN				3/4		.12
CL-2-FN	CL-127	3/8-16	1/2	7/8	11/16	.12
CL-3-FN	CL-128	1/2-13	11/16	1-1/8	7/8	.20
CL-4-FN	CL-129	5/8-11	13/16	1-3/8	1-1/16	

Figure 8.4(n) CL-3-FN

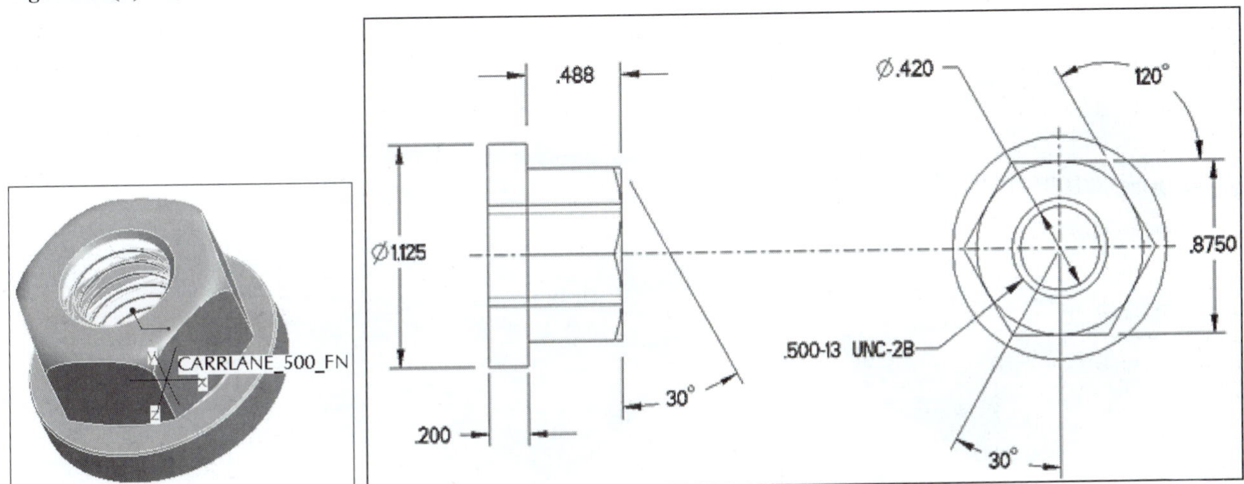

Figure 8.4(o) Flange Coordinate System Name

Figure 8.4(p) Flange Nut Dimensions for Modeling.

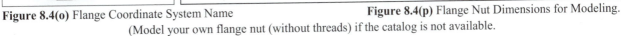

(Model your own flange nut (without threads) if the catalog is not available.

Open the models for the Clamp_Arm_Design [Fig. 8.5(a)], the Clamp_Swivel [Fig. 8.5(b)], the Clamp_Ball [Fig. 8.5(c)] and the Clamp_Foot [Fig. 8.5(d)]. > Review the components and standard parts for unique **colors**, **layering**, **coordinate system naming**, and **set datum planes**. > Close all windows. Components remain "in session".

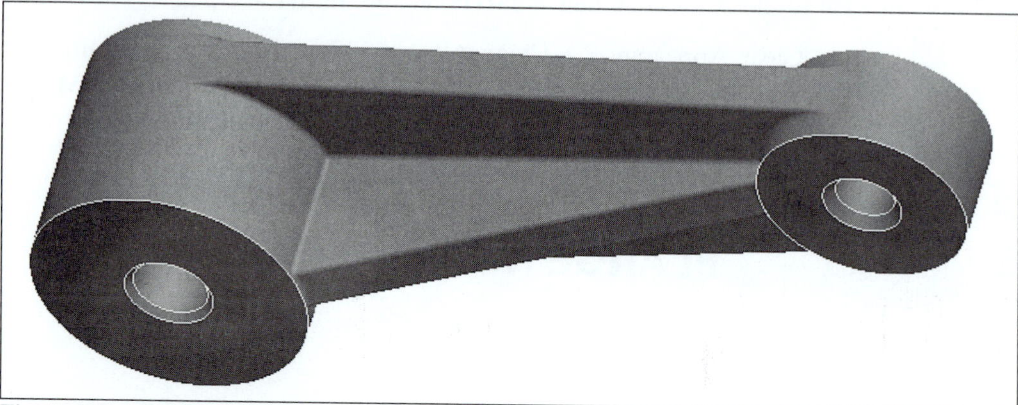

Figure 8.5(a) Clamp_Arm_Design

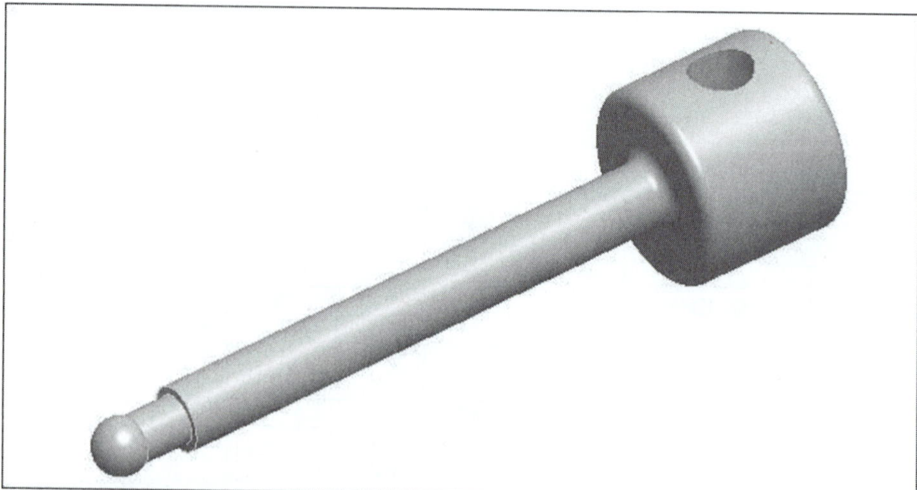

Figure 8.5(b) Clamp_Swivel

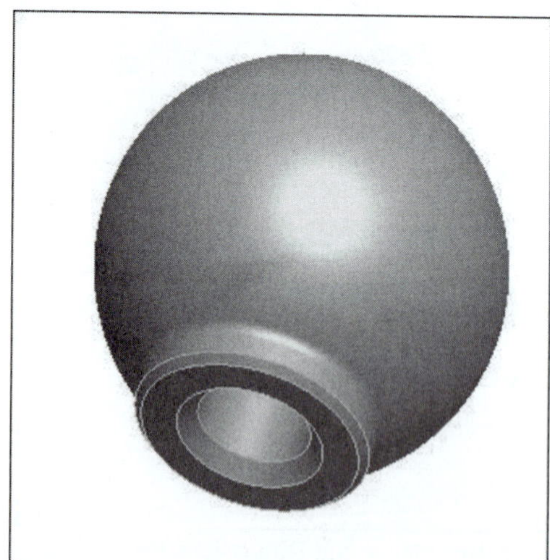

Figure 8.5(c) Clamp_Ball

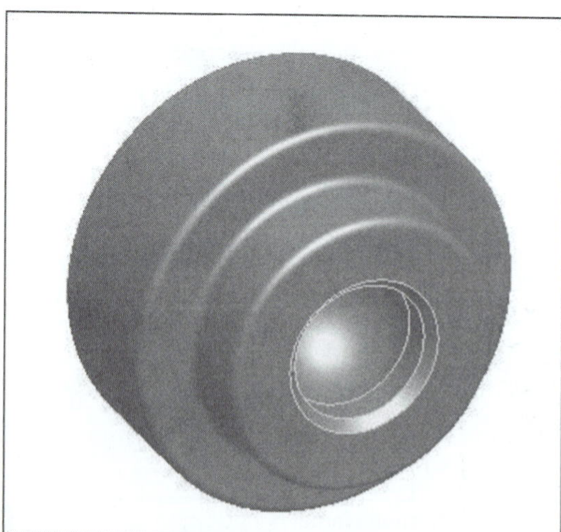

Figure 8.5(d) Clamp_Foot

You now have eight components (two identical Clamp_Ball components are used) required for the assembly. The Clamp_Plate (Fig. 8.6) will be created using *top-down design* procedures when you start the main assembly. *All parts must be in the same working directory used for the assembly.*

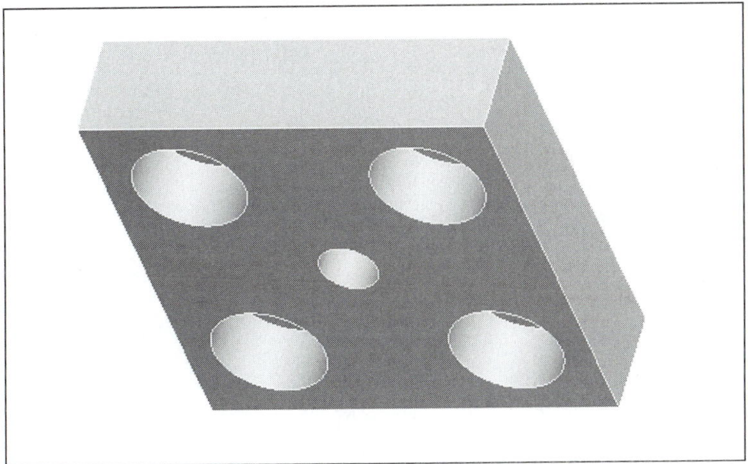

Figure 8.6 Clamp_Plate

A subassembly will be modeled first. The main assembly is created second. The subassembly will be added to the main assembly to complete the project. Note: the assembly and the components can have different units. Therefore, you must check and correctly set the assembly units before creating or assembling components or sub-assemblies.

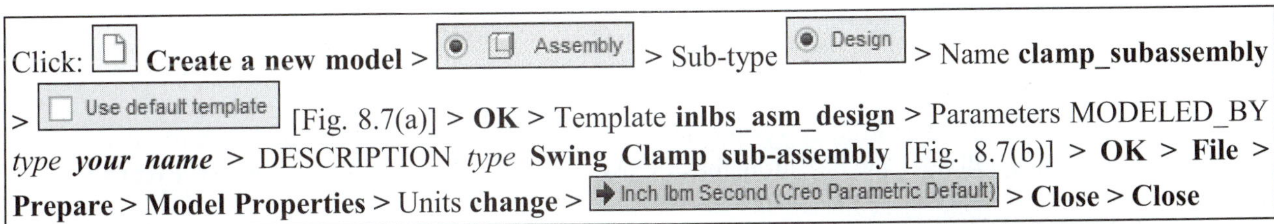

Click: **Create a new model** > Assembly > Sub-type Design > Name **clamp_subassembly** > Use default template [Fig. 8.7(a)] > **OK** > Template **inlbs_asm_design** > Parameters MODELED_BY *type* ***your name*** > DESCRIPTION *type* **Swing Clamp sub-assembly** [Fig. 8.7(b)] > **OK** > **File** > **Prepare** > **Model Properties** > Units **change** > Inch lbm Second (Creo Parametric Default) > **Close** > **Close**

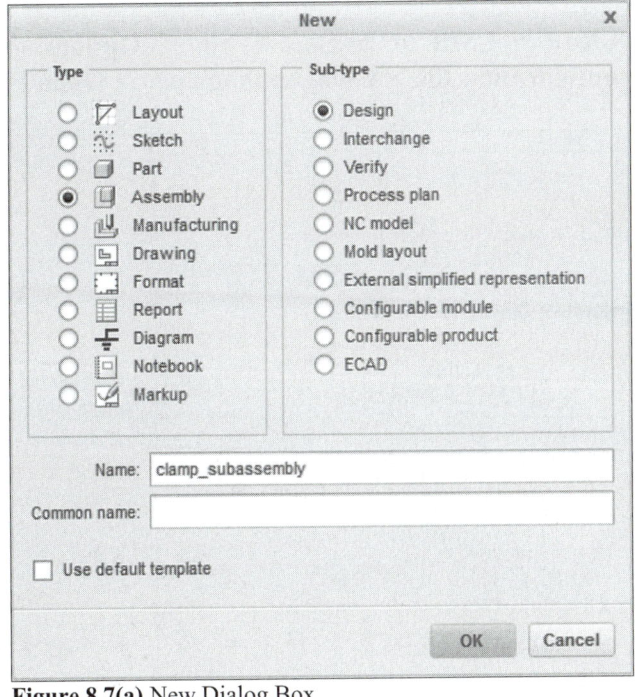

Figure 8.7(a) New Dialog Box

Figure 8.7(b) New File Options Dialog Box

Click: **Settings** from the Navigator > Tree Filters... > check all Display options *on* [Fig. 8.7(c)] > **Apply** > **OK** > **Ctrl+S** > **OK** > **View** tab > *on* > *on*

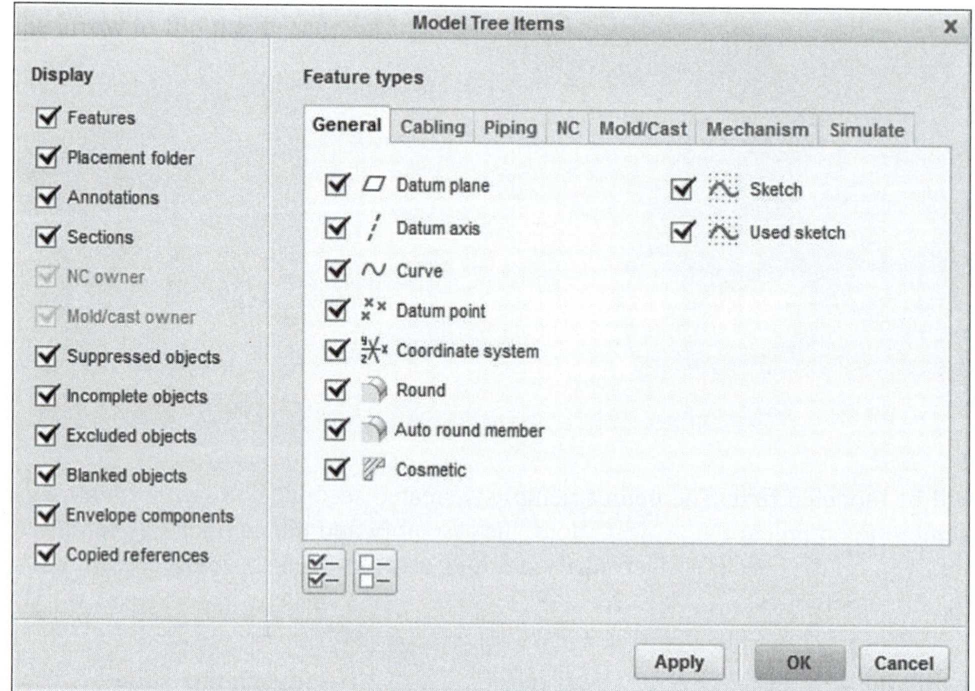

Figure 8.7(c) Model Tree Items Dialog Box

Datum planes and the coordinate system are created per the template provided by Creo Parametric 3.0. The datum planes will have the default names, **ASM_RIGHT**, **ASM_TOP**, and **ASM_FRONT**.

Change the coordinate system name: slowly double-click on ASM_DEF_CSYS in the Model Tree > *type the new name* SUB_ASM_CSYS > **Enter** [Fig. 8.7(d)] > **LMB** to deselect > **File** > **Options** > **Configuration Editor** > **Import/Export** > **Import configuration file** > **Creo_textbook.pro** > **Open** > **OK** > **No**

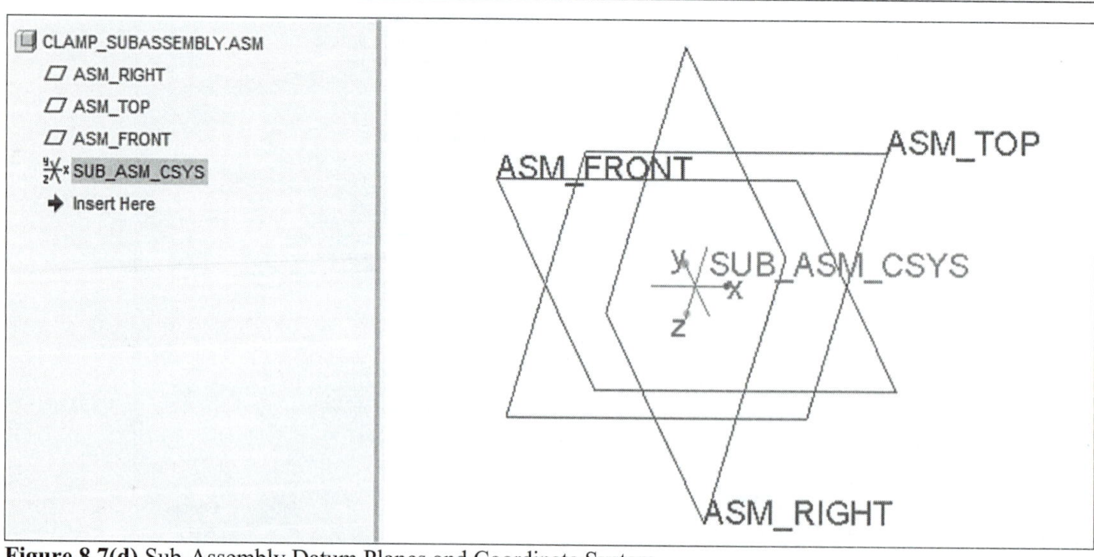

Figure 8.7(d) Sub-Assembly Datum Planes and Coordinate System

Regardless of the design methodology, the assembly datum planes and coordinate system should be on their own separate *assembly layer*. Each part should also be placed on separate assembly layers; the part's datum features should already be on *part layers*. Look over the default template for assembly layering.

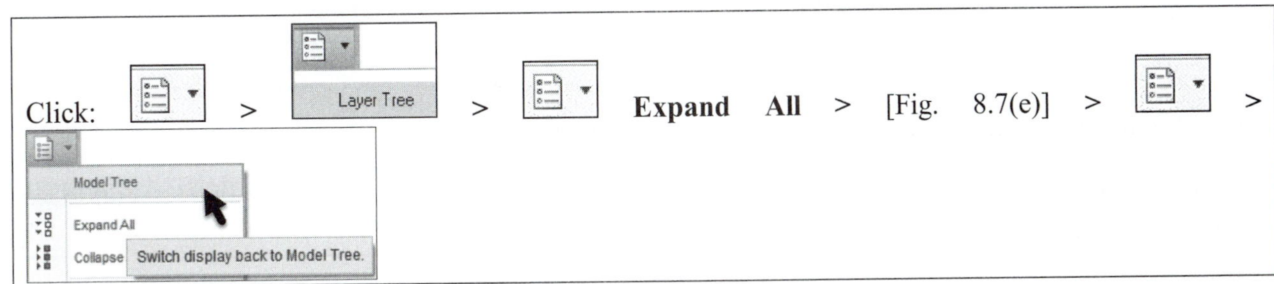

Layer Tree
CLAMP_SUBASSEMBLY.ASM (TOP MODEL, ACTIVE)
Layers
01_ALL_DTM_PLN
CLAMP_SUBASSEMBLY layer 01__ASM_ALL_DTM_PLN.
F1(ASM_RIGHT)
F2(ASM_TOP)
F3(ASM_FRONT)
CLAMP_SUBASSEMBLY layer 01___PRT_ALL_DTM_PLN.
CLAMP_SUBASSEMBLY layer 01__ASM_DEF_DTM_PLN.
F1(ASM_RIGHT)
F2(ASM_TOP)
F3(ASM_FRONT)
CLAMP_SUBASSEMBLY layer 01___PRT_DEF_DTM_PLN.
01__ASM_ALL_DTM_PLN
F1(ASM_RIGHT)
F2(ASM_TOP)
F3(ASM_FRONT)
01__ASM_DEF_DTM_PLN
F1(ASM_RIGHT)
F2(ASM_TOP)
F3(ASM_FRONT)
01___PRT_ALL_DTM_PLN
01___PRT_DEF_DTM_PLN
02_ALL_AXES
CLAMP_SUBASSEMBLY layer 02__ASM_ALL_AXES.
CLAMP_SUBASSEMBLY layer 02___PRT_ALL_AXES.
02__ASM_ALL_AXES
02___PRT_ALL_AXES
03_ALL_CURVES
CLAMP_SUBASSEMBLY layer 03__ASM_ALL_CURVES.
CLAMP_SUBASSEMBLY layer 03___PRT_ALL_CURVES.
03__ASM_ALL_CURVES
03___PRT_ALL_CURVES
04_ALL_DTM_PNT
CLAMP_SUBASSEMBLY layer 04__ASM_ALL_DTM_PNT.
CLAMP_SUBASSEMBLY layer 04___PRT_ALL_DTM_PNT.
04__ASM_ALL_DTM_PNT
04___PRT_ALL_DTM_PNT
05_ALL_DTM_CSYS
CLAMP_SUBASSEMBLY layer 05__ASM_ALL_DTM_CSYS.
F4(SUB_ASM_CSYS)
CLAMP_SUBASSEMBLY layer 05__ASM_DEF_DTM_CSYS.
F4(SUB_ASM_CSYS)

ASM_TOP
ASM_FRONT
y SUB_ASM_CSYS
x
z
ASM_RIGHT

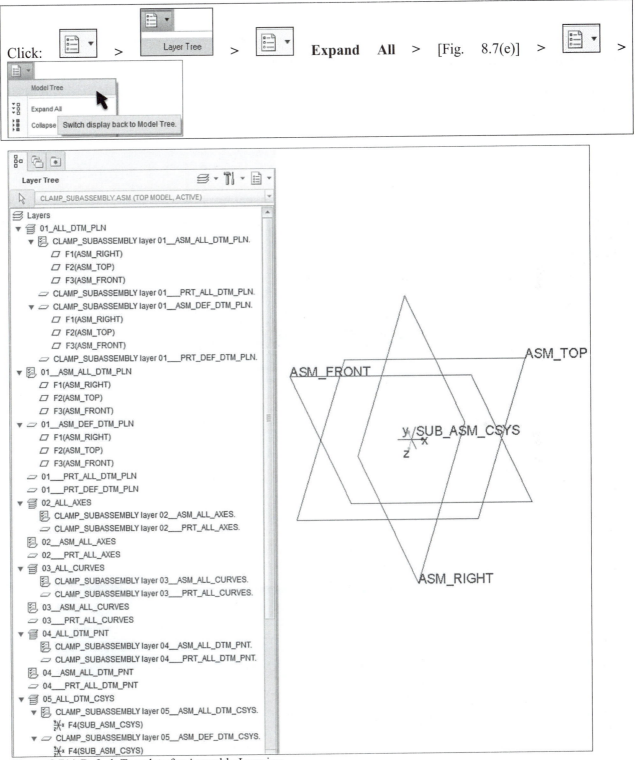

Figure 8.7(e) Default Template for Assembly Layering

The first component to be assembled to the subassembly is the Clamp_Arm. The simplest and quickest method of adding a component to an assembly is to match the coordinate systems. The first component assembled is usually where this *constraint* is used, because after the first component is established, few if any of the remaining components are assembled to the assembly coordinate system (with the exception of *top-down design*) or, for that matter, other parts' coordinate systems. Make sure all your models are in the same working directory before you start the assembly process.

Click: **Model** tab > Assemble > **clamp_arm.prt** > **Views** tab > **List** > **Preview** *on* [Fig. 8.8(a)] > **Open** > **CLAMP_ARM_DESIGN** [Fig. 8.8(b)] > **Open** > **Hide 3D Dragger** > **View** tab > *off* [Fig. 8.8(c)] > **Component Placement** tab

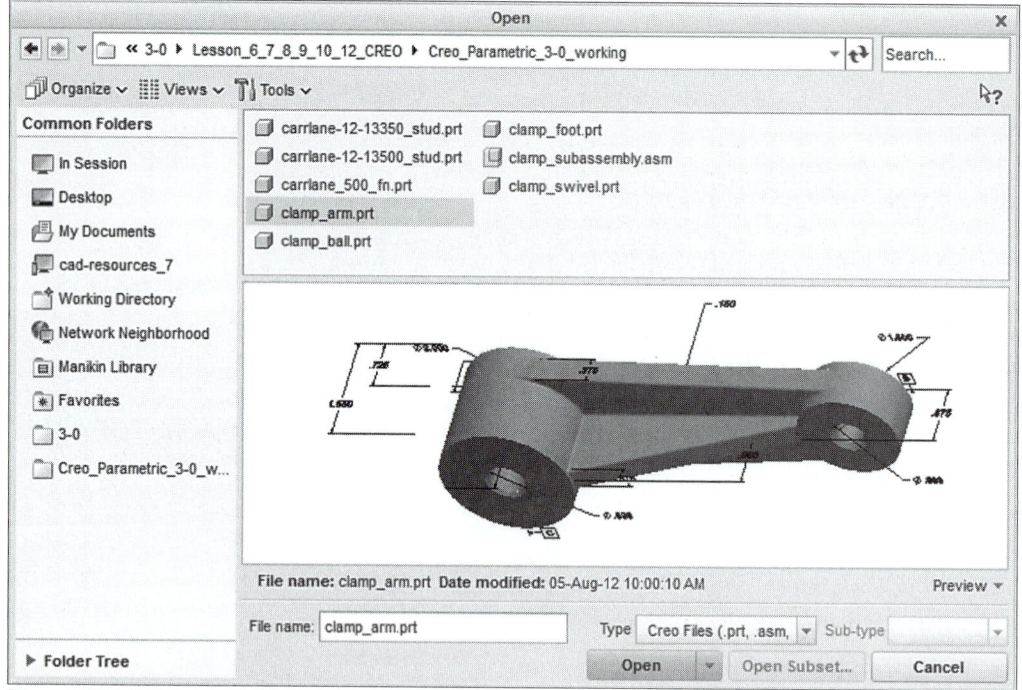

Figure 8.8(a) Previewed Clamp_Arm Component

Figure 8.8(b) Clamp_Arm_Design

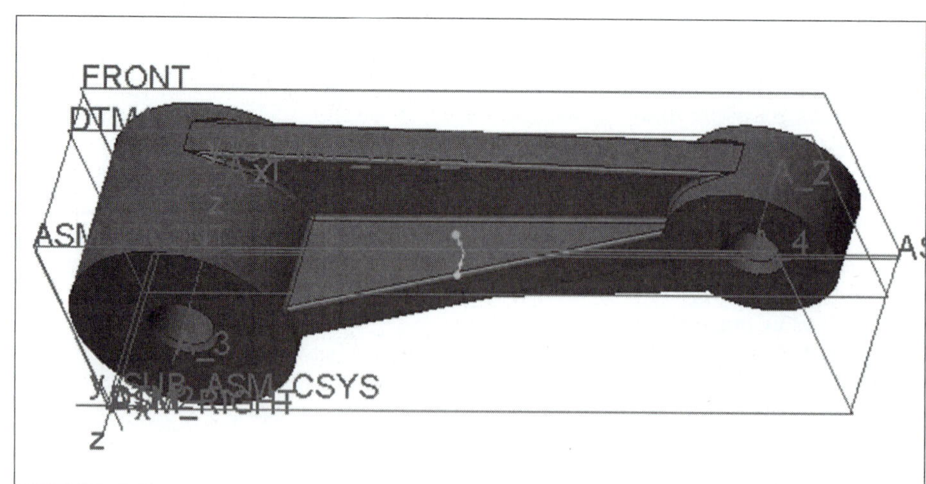

Figure 8.8(c) Clamp_Arm_Design in the Graphics Window

When an assembly is complicated, or it is difficult to select constraining geometry, a separate window aids in the selection process. For simple assemblies (as is the Clamp Assembly and Clamp Subassembly) working in the assembly window is more convenient. Also, since this is the first component being added to the assembly, a separate window is not needed.

In the Graphics Window, press: **RMB** > **Default Constraint** [Fig. 8.8(d)] (this option puts the component at the default position on the assembly model, which is the same as using the Coincident constraint) > ✓ > 💾 > **File** > **Options** > **Model Display** > Default Model orientation: **Trimetric** > **Isometric** > **OK** > **No** > in the Model Tree, pick: ▶ (▼ CLAMP_ARM_DESIGN<CLAMP_ARM>.PRT) > ▶ (▼ Placement) > ▶ (▼ Set1) *(your Set # may be different)* (Default) [Fig. 8.8(e)]

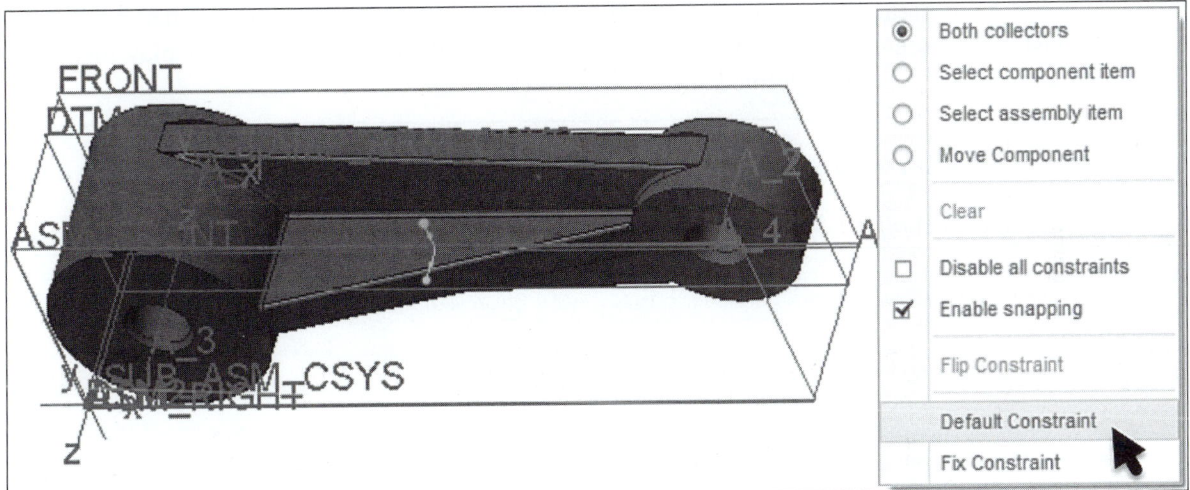

Figure 8.8(d) Default Constraint

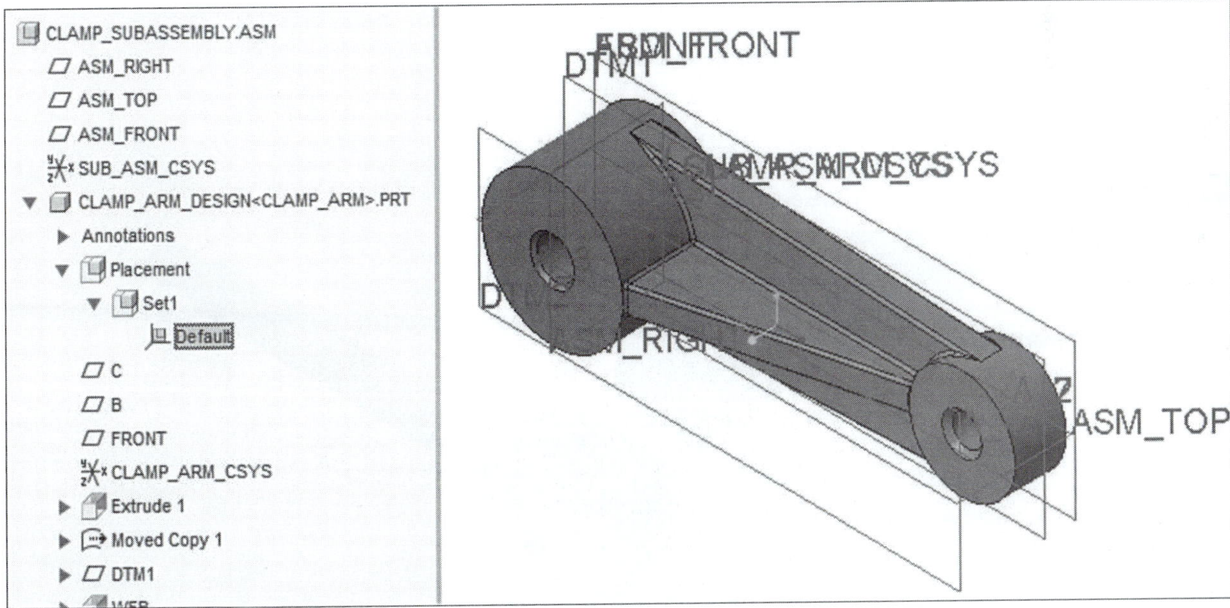

Figure 8.8(e) Expanded Placement Folder

The Clamp_Arm_Design has been assembled to the default location. This means to align the default Creo Parametric-created coordinate system of the component to the default Creo Parametric-created coordinate system of the assembly. Creo Parametric places the component at the assembly origin. By using the constraint, Coincident and selecting the assembly and then the component's coordinate systems would have accomplished the same thing, but with more picks.

Component placement is based on placement definition sets. These sets determine how and where the component relates to the assembly. The sets are either user-defined or predefined. A user-defined constraint set has zero or more constraints (a packaged component may have no constraints). Predefined constraint sets have a predefined number of constraints. Placement of a component in an assembly is determined by the constraints in all sets defined. A single set of constraints can define placement of a component. If constraints from one set conflict with constraints from another set, the placement status becomes invalid. The constraints must be redefined or removed until placement status becomes valid. Constraints can be added or deleted at will in a user-defined constraint set, there are no predefined constraints. Each type of predefined constraint set (also called a connection) has a predefined number of constraints. Constraint sets are displayed in the Placement folder of the Model Tree. Display hierarchy follows the order in which they were defined.

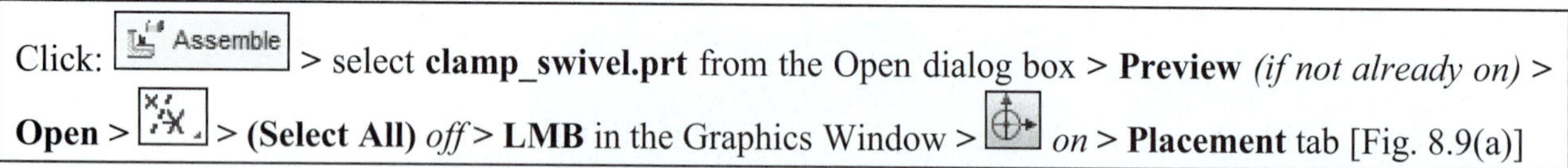

Click: Assemble > select **clamp_swivel.prt** from the Open dialog box > **Preview** *(if not already on)* > **Open** > > **(Select All)** *off* > **LMB** in the Graphics Window > *on* > **Placement** tab [Fig. 8.9(a)]

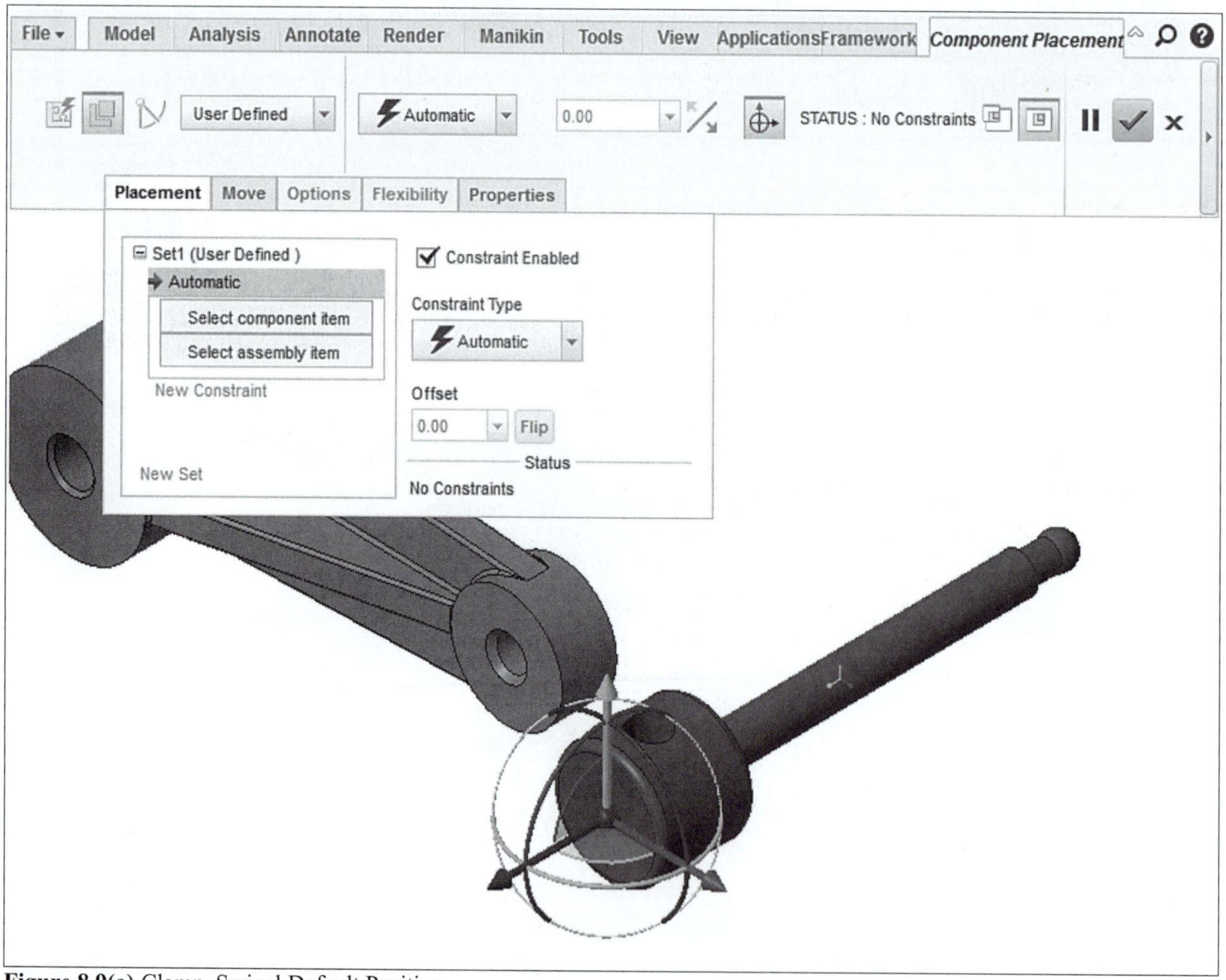

Figure 8.9(a) Clamp_Swivel Default Position

Click: **Placement** tab (closed) > pick on the cylindrical surface of the Clamp_Swivel [Fig. 8.9(b)] > pick on the hole surface of the Clamp_Arm_Design [Fig. 8.9(c)] (constraint becomes Coincident) > **View** tab > **Plane Display** *on* > **Plane Tag Display** *off* > **Component Placement** tab

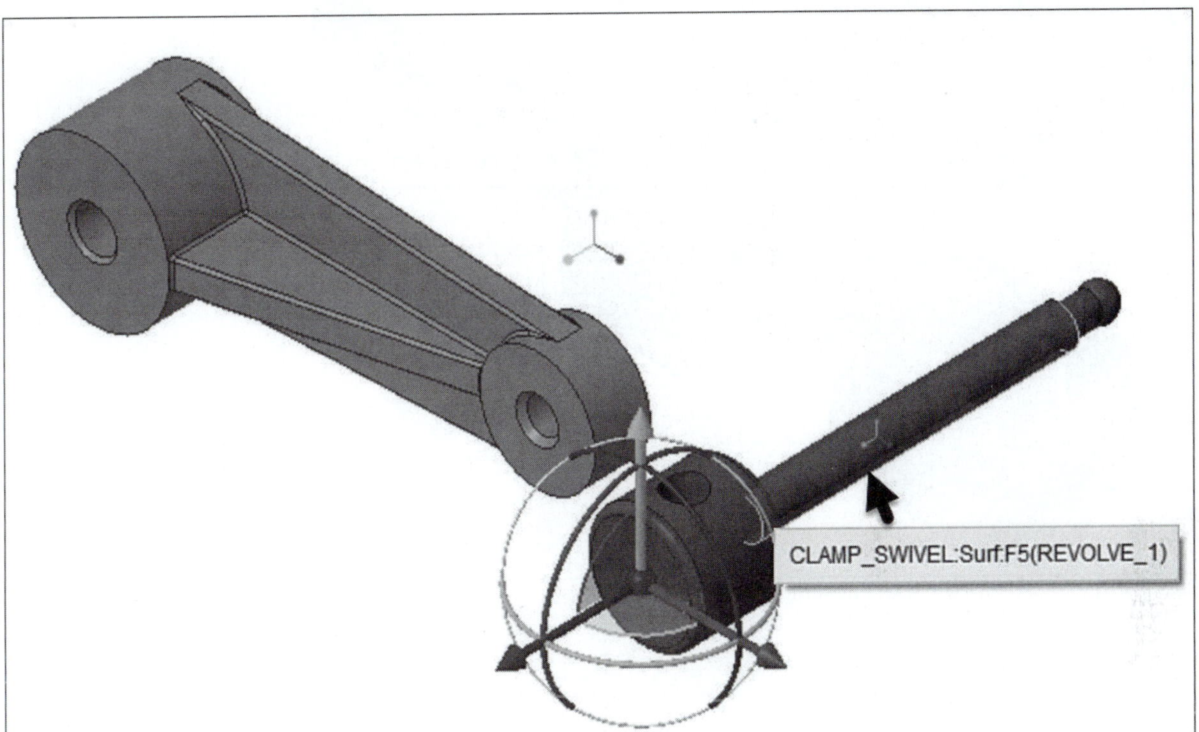

Figure 8.9(b) Select on the Clamp_Swivel Surface

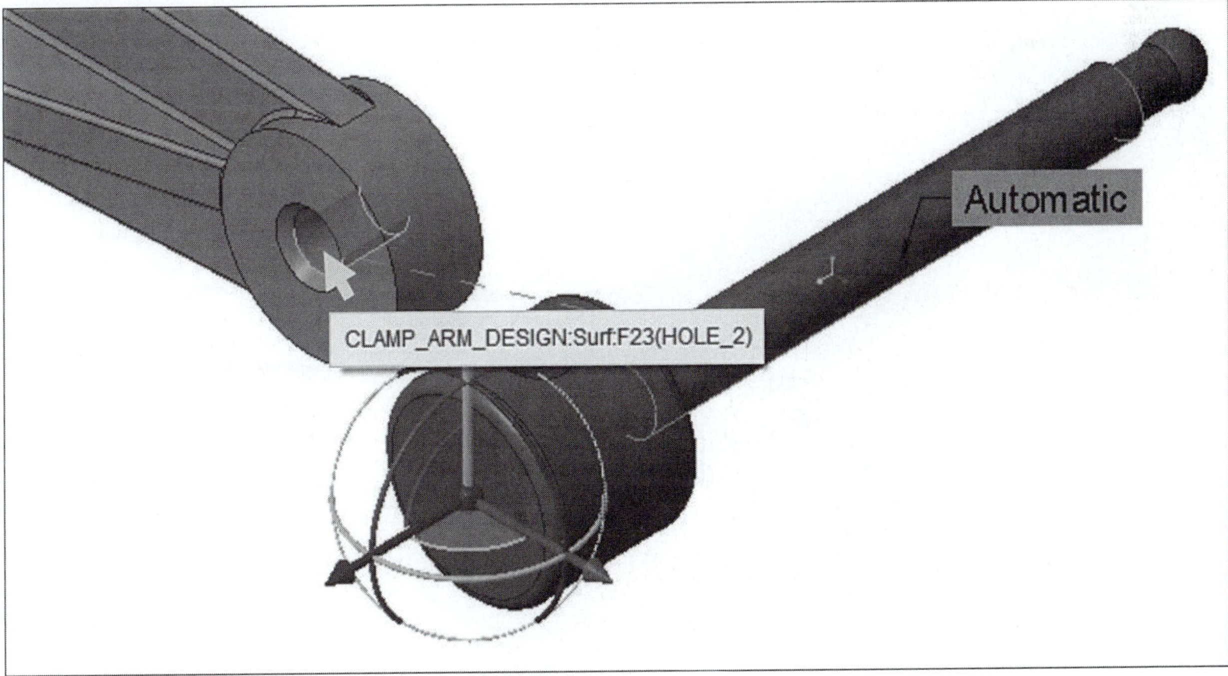

Figure 8.9(c) Select on the Clamp_Arm_Design Hole Surface

Place the pointer on the Dragger Axis, press and hold: **LMB** on the Dragger Axis [Fig. 8.9(d)] > pull forward > release the **LMB** > press **RMB** > **New Constraint** [Fig. 8.9(e)]

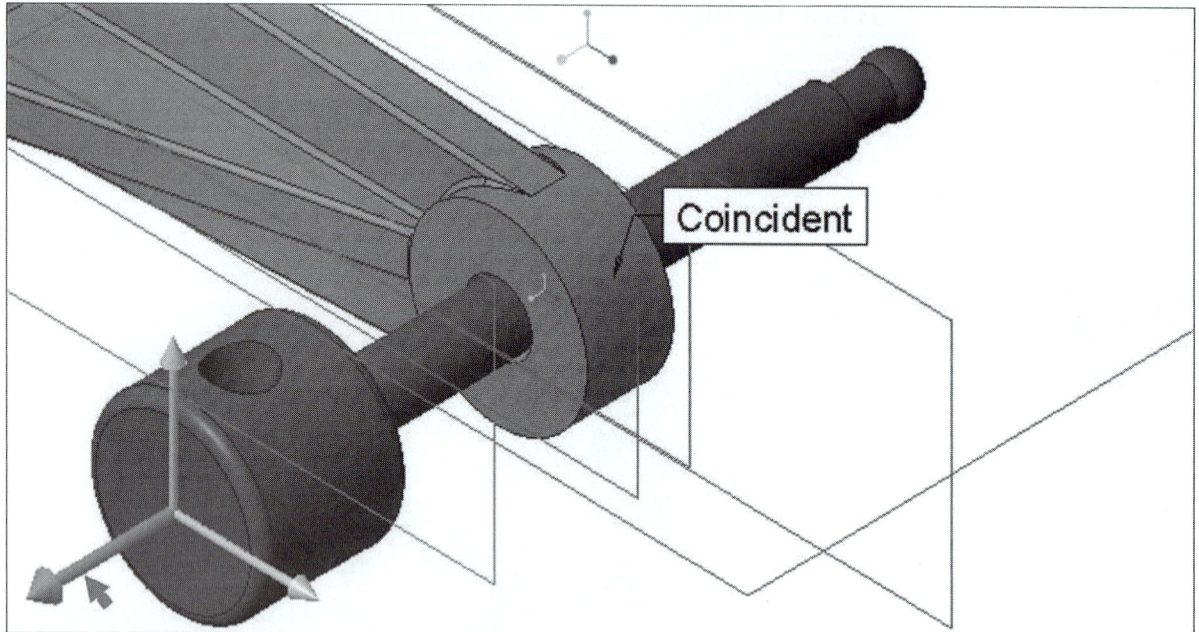

Figure 8.9(d) Pull the Dragger Axis Forward

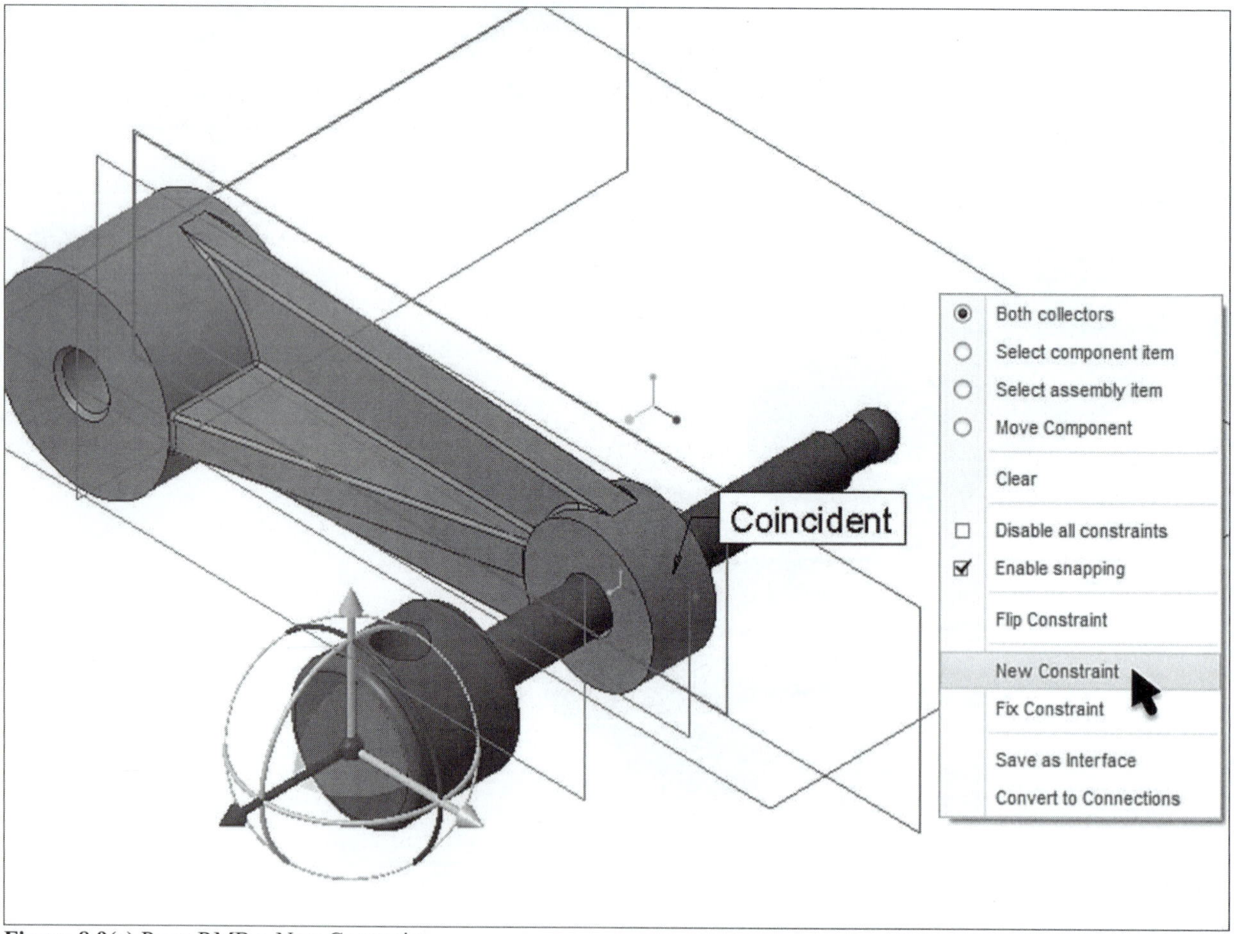

Figure 8.9(e) Press RMB > New Constraint

Click: Hide 3D Dragger > Placement tab > pick the surface of the Clamp_Arm_Design [Fig. 8.9(f)] > click the **RMB** to toggle to the desired surface (or rotate the model to see the bottom surface of the Swivel) > pick the surface of the Clamp_Swivel [Fig. 8.9(g)]

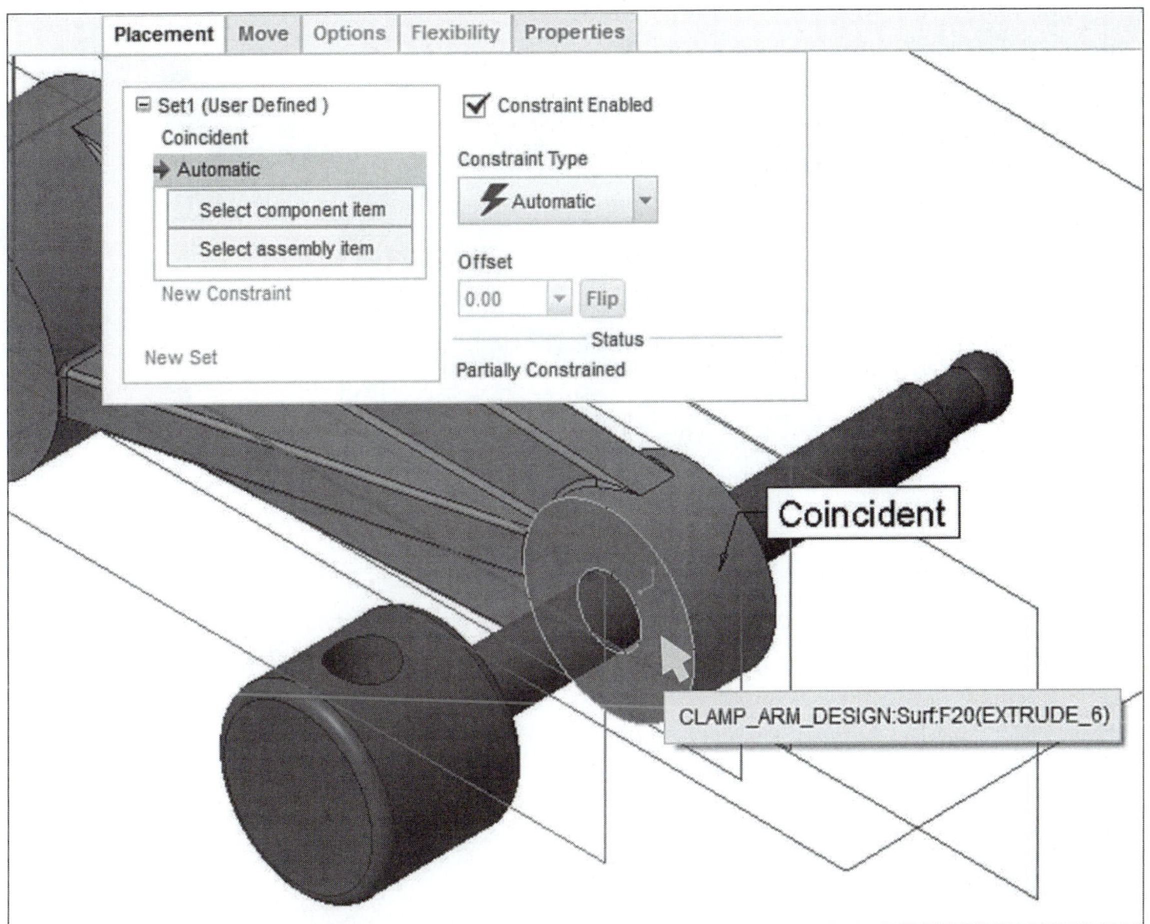

Figure 8.9(f) Select on the Top Surface of the Small Circular Extrusion of the Clamp_Arm_Design

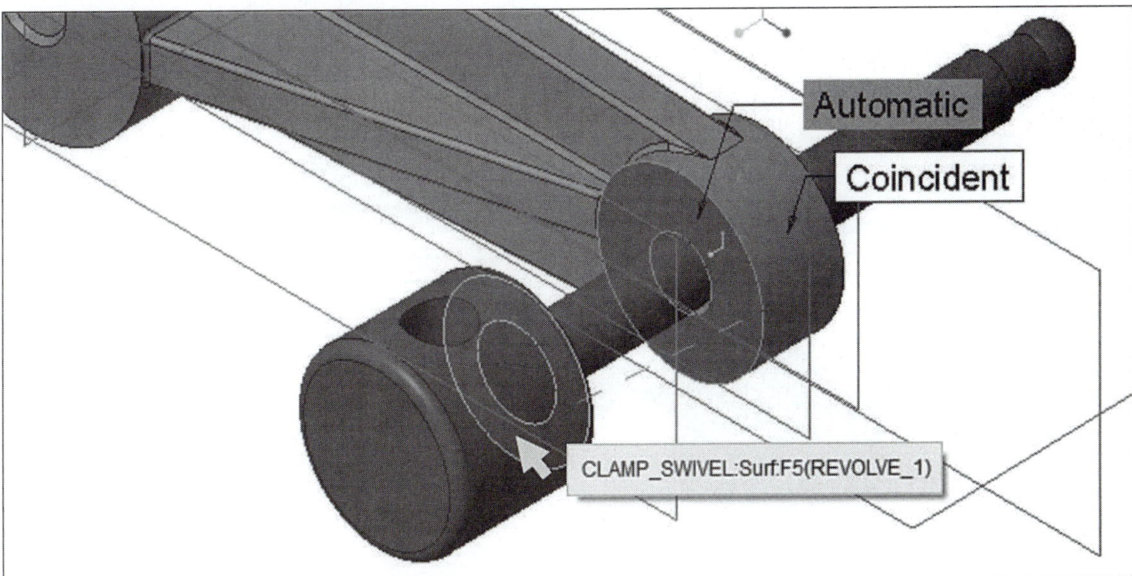

Figure 8.9(g) RMB > Select on the Underside Surface of the Clamp_Swivel

Click: Coincident > Distance > **LMB** on the handle and drag the Clamp_Swivel until it is **1.50** offset from the Clamp_Arm_Design surface [Fig. 8.9(h)] > **Placement** tab (closed) [Fig. 8.9(i)]

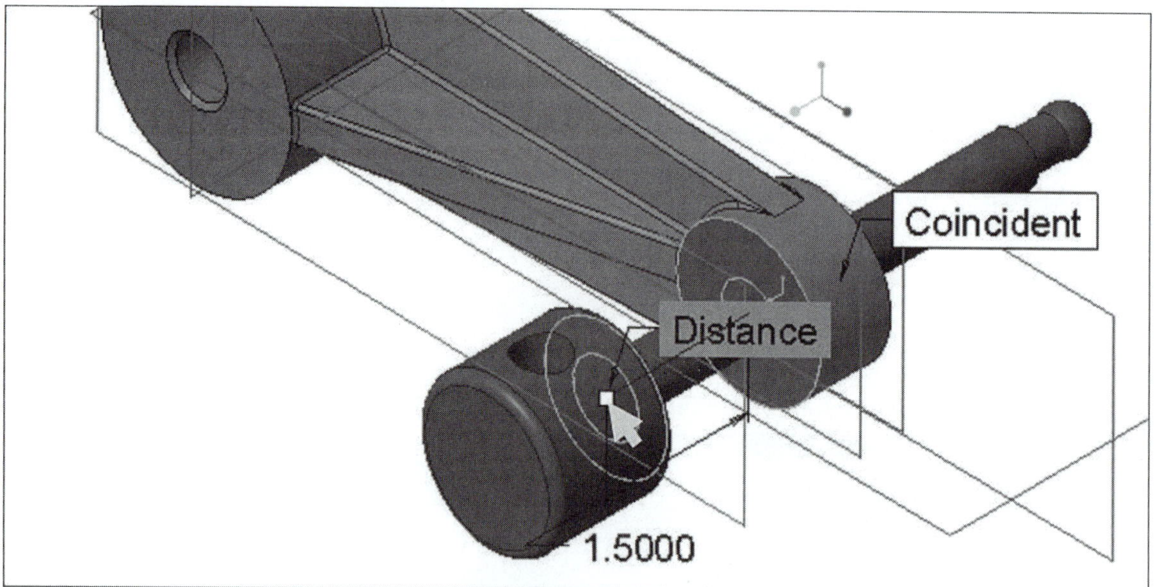

Figure 8.9(h) Drag the handle until **1.50** for the (Mate) Offset Distance

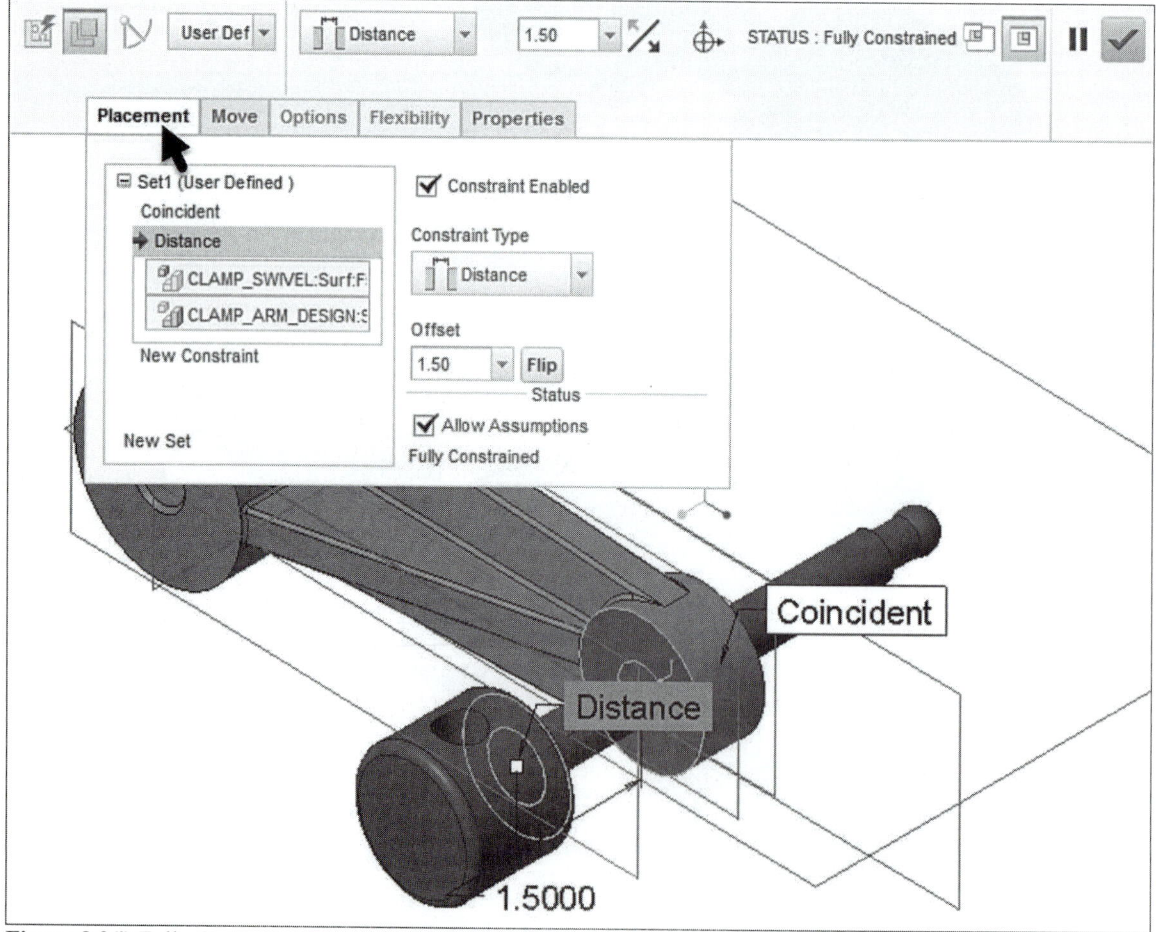

Figure 8.9(i) Fully Constrained Component

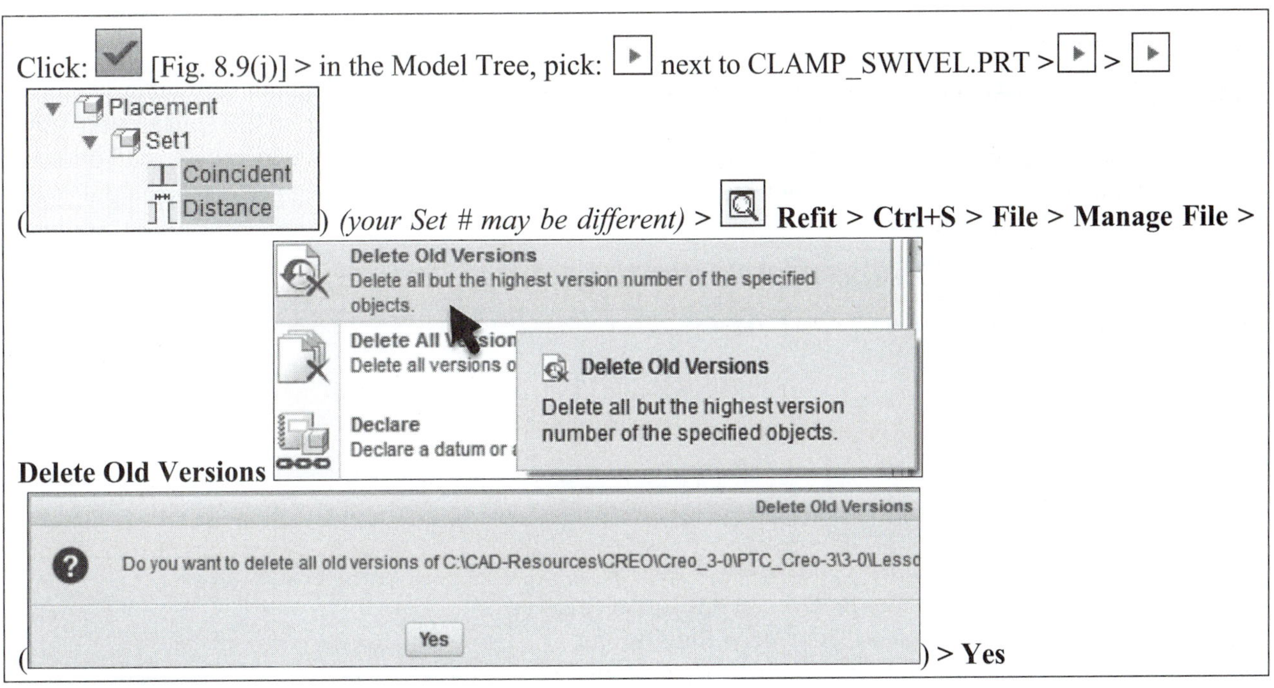

Click: ✓ [Fig. 8.9(j)] > in the Model Tree, pick: ▶ next to CLAMP_SWIVEL.PRT > ▶ > ▶ (Placement > Set1 > Coincident, Distance) *(your Set # may be different)* > **Refit > Ctrl+S > File > Manage File > Delete Old Versions** (Delete Old Versions: Delete all but the highest version number of the specified objects.) > **Yes**

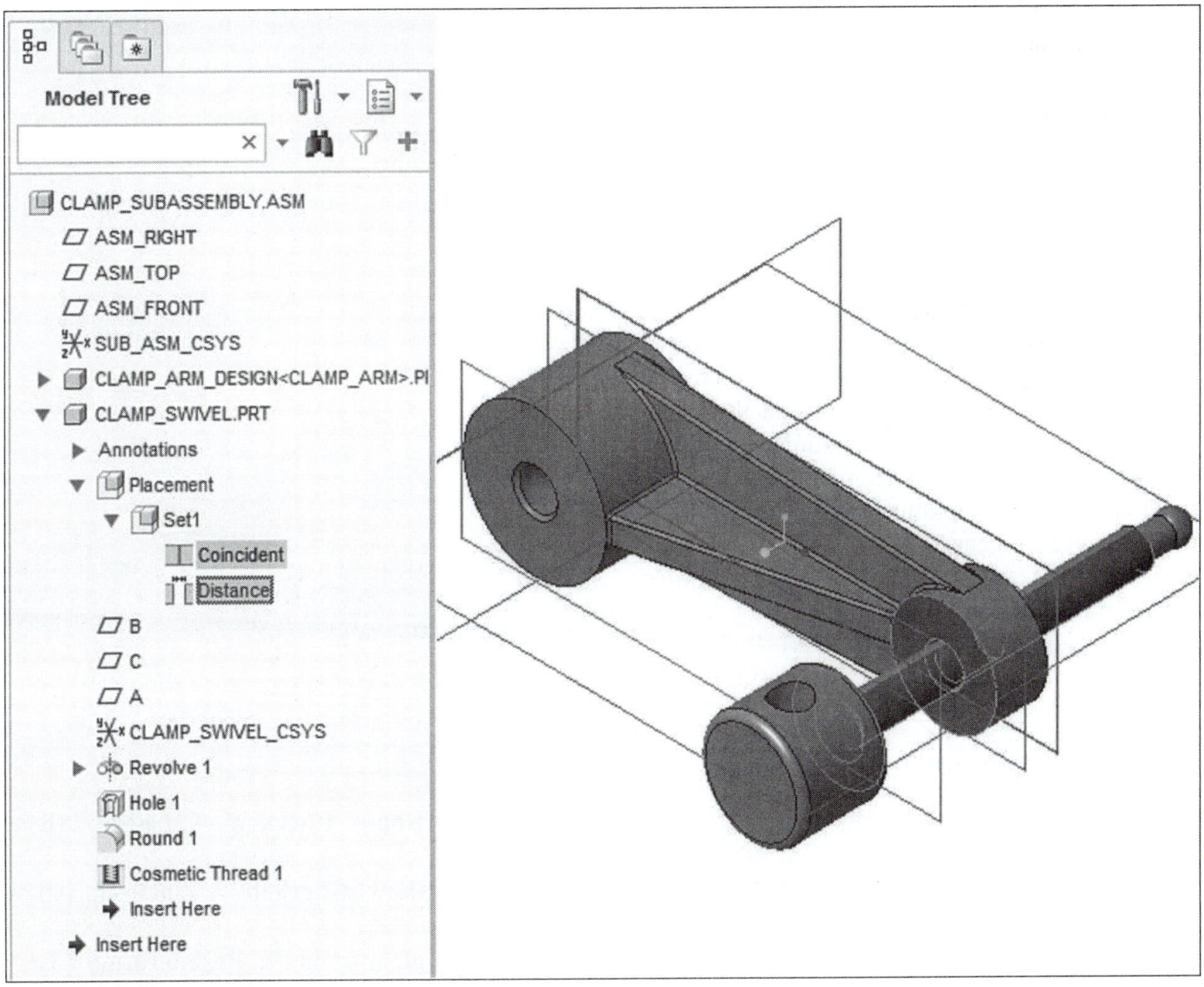

Figure 8.9(j) Placement Status

Regenerating Models

You can use **Regenerate** to find bad geometry, broken parent-child relationships, or any other problem with a part feature or assembly component. In general, it is a good idea to regenerate the model every time you make a change, so that you can see the effects of each change in the graphics window as you build the model. By regenerating often, it helps you stay on course with your original design intent by helping you to resolve failures as they happen.

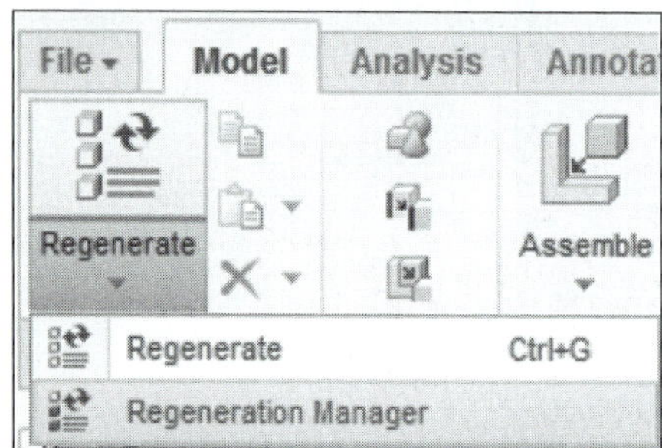

When Creo Parametric regenerates a model, it recreates the model feature by feature, in the order in which each feature was created, and according to the hierarchy of the parent-child relationship between features.

In an assembly, component features are regenerated in the order in which they were created, and then in the order in which each component was added to the assembly. Creo Parametric regenerates a model automatically in many cases, including when you open, save, or close a part or assembly or one of its instances, or when you open an instance from within a Family Table. You can also use the Regenerate command to manually regenerate the model.

Regenerate, lets you recalculate the model geometry, incorporating any changes made since the last time the model was saved. If no changes have been made, Creo Parametric 3.0 informs you that the model has not changed since the last regeneration. The **Regeneration Manager** opens the Regeneration Manager dialog box if there are features or components that have been changed that require regeneration. A column next to the Regeneration List indicates each entries Status (Regenerated or Unregenerated).

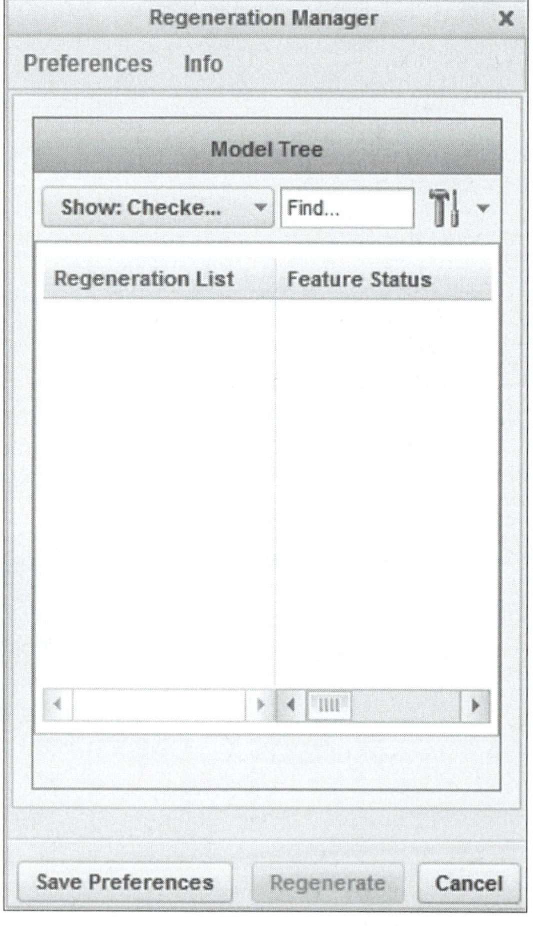

In the Regeneration Manager dialog box you can:

- Select all features/components for regeneration, select **Show Checked** > **Check All** > **Regenerate**.
- Omit all features/components from regeneration, select **Show Checked** > **Uncheck All** > **Regenerate**.
- Determine the reason an object requires regeneration, select an entry in the **Regeneration List** > **RMB** > **Feature Info**.

Double-click on the Clamp_Swivel > double-click on the **1.50** dimension > *type* **1.75** [Fig. 8.10(a)] > **Enter** > **Regenerate** *(upper left on Model Ribbon)* > Regeneration Manager Regeneration Manager dialog box opens [Fig. 8.10(b)] > **Show Checked** > **Check All** > **Regenerate** [Fig. 8.10(c)] > **Undo** (dimension is now **1.50**) > double-click on the Clamp_Swivel > in the Graphics Window, **LMB** to deselect > **Ctrl+S**

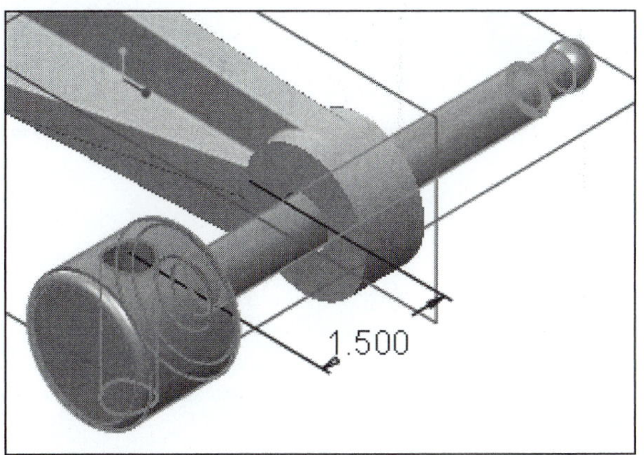

Figure 8.10(a) Double-click on the Clamp_Swivel

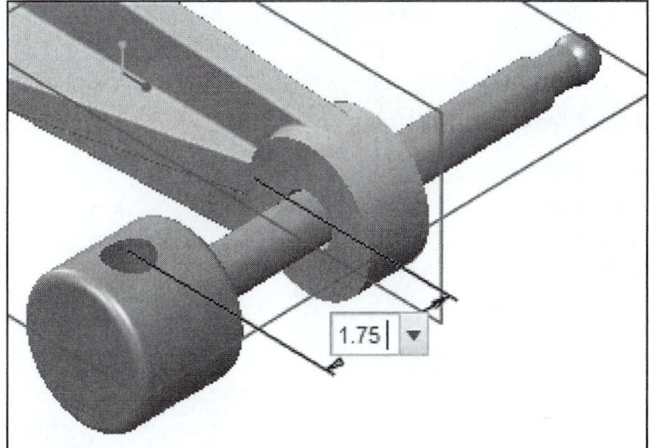

Figure 8.10(b) Modify the Offset Value to **1.75**

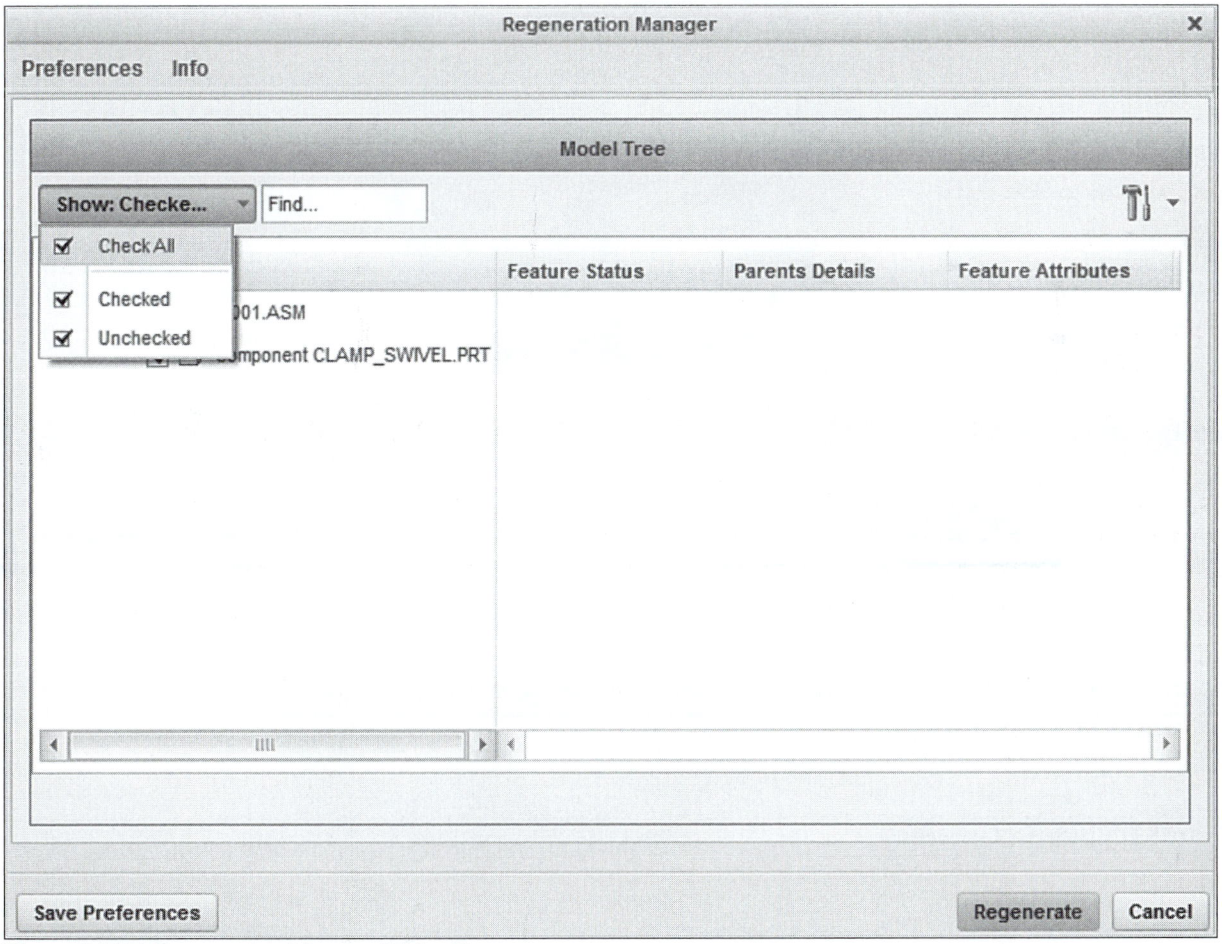

Figure 8.10(c) Regeneration Manager Dialog Box

The next component to be assembled is the Clamp_Foot. Click: Assemble > select the **clamp_foot.prt** from the Open dialog box > **Preview** *on* > **MMB** rotate the model in the Preview Window > **RMB** to see options (Fig 8.11) > **No Hidden** > **RMB** > **Shaded** > **Open**

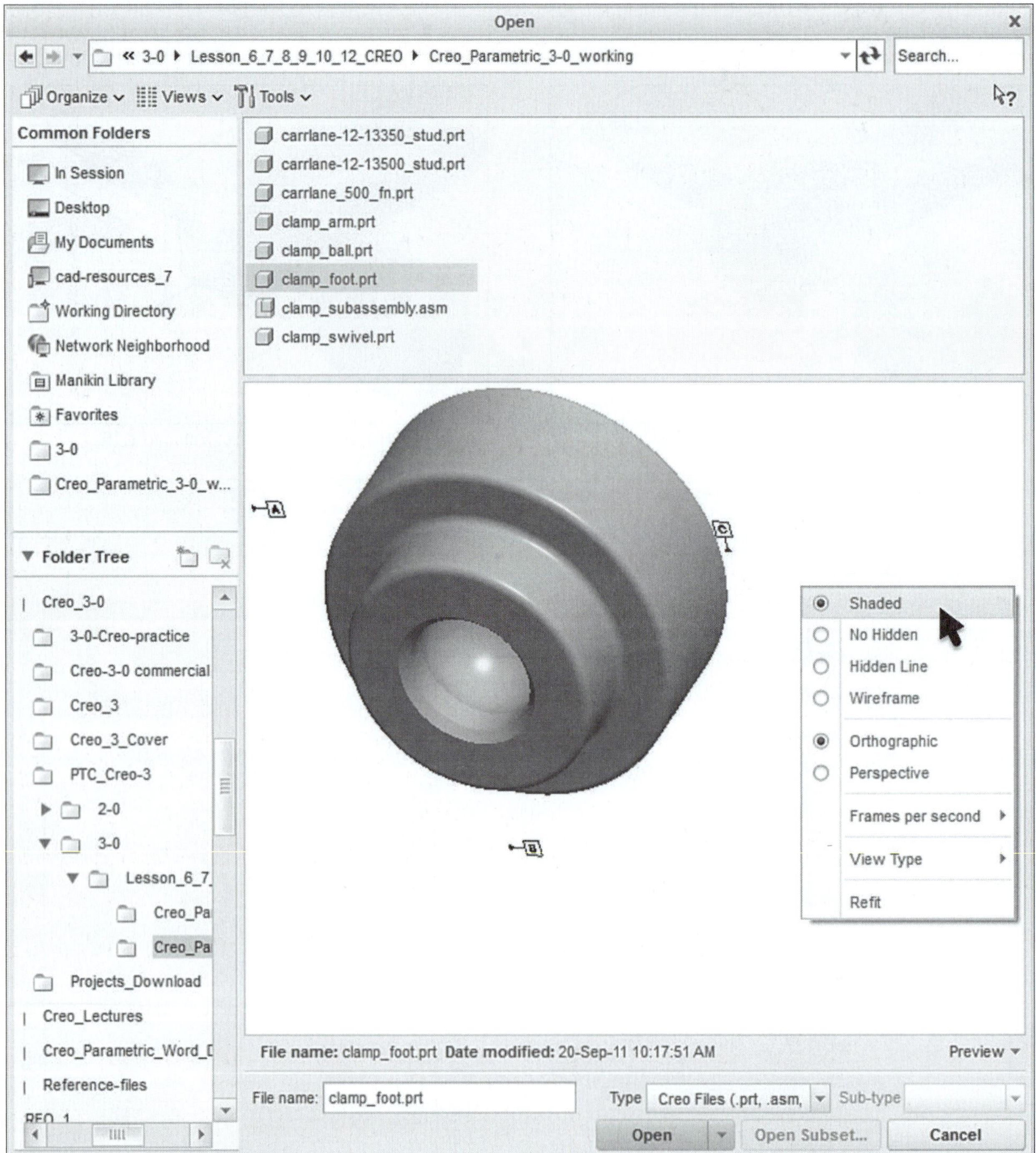

Figure 8.11 Clamp_Foot Preview

Click: **Show component in a separate window while specifying constraints** > **Show 3D Dragger** [Fig. 8.12(a)] > use the colored axes draggers to reposition the component model

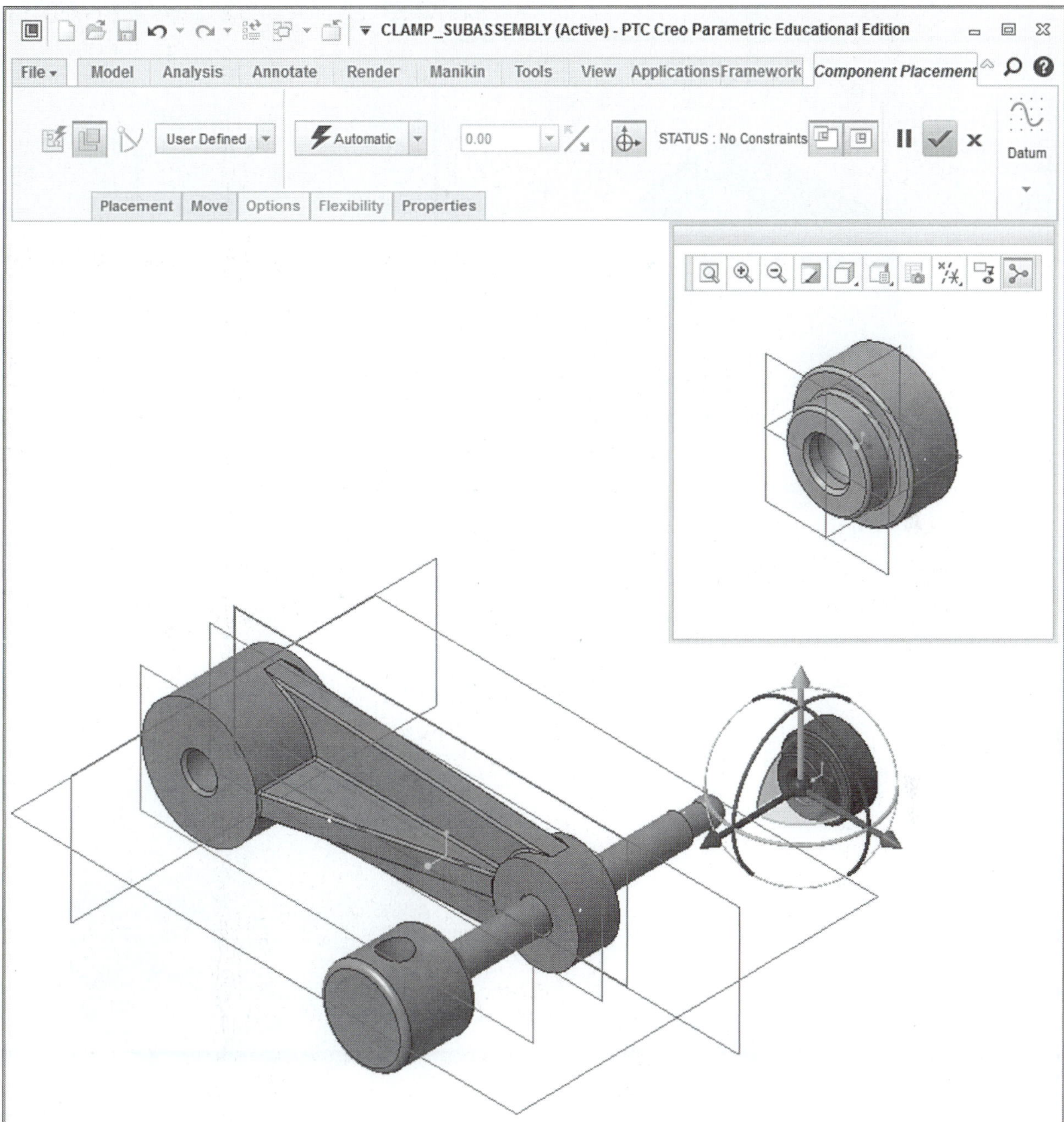

Figure 8.12(a) Component in a Separate Window while Specifying Constraints

Click: Hide 3D Dragger > resize your windows as desired > from the Accessory (Component) Window, pick the ***internal*** spherical surface of the Clamp_Foot [Fig. 8.12(b)] > pick the ***external*** spherical surface of the Clamp_Swivel [Fig. 8.12(c)] > Coincident from the Dashboard > Centered

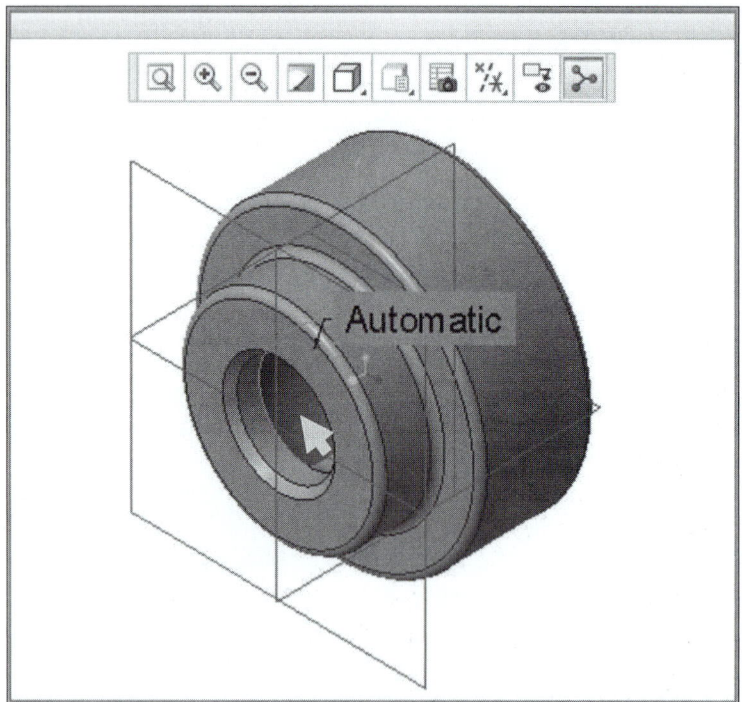

Figure 8.12(b) Select the Internal Spherical Surface of the Clamp_Foot

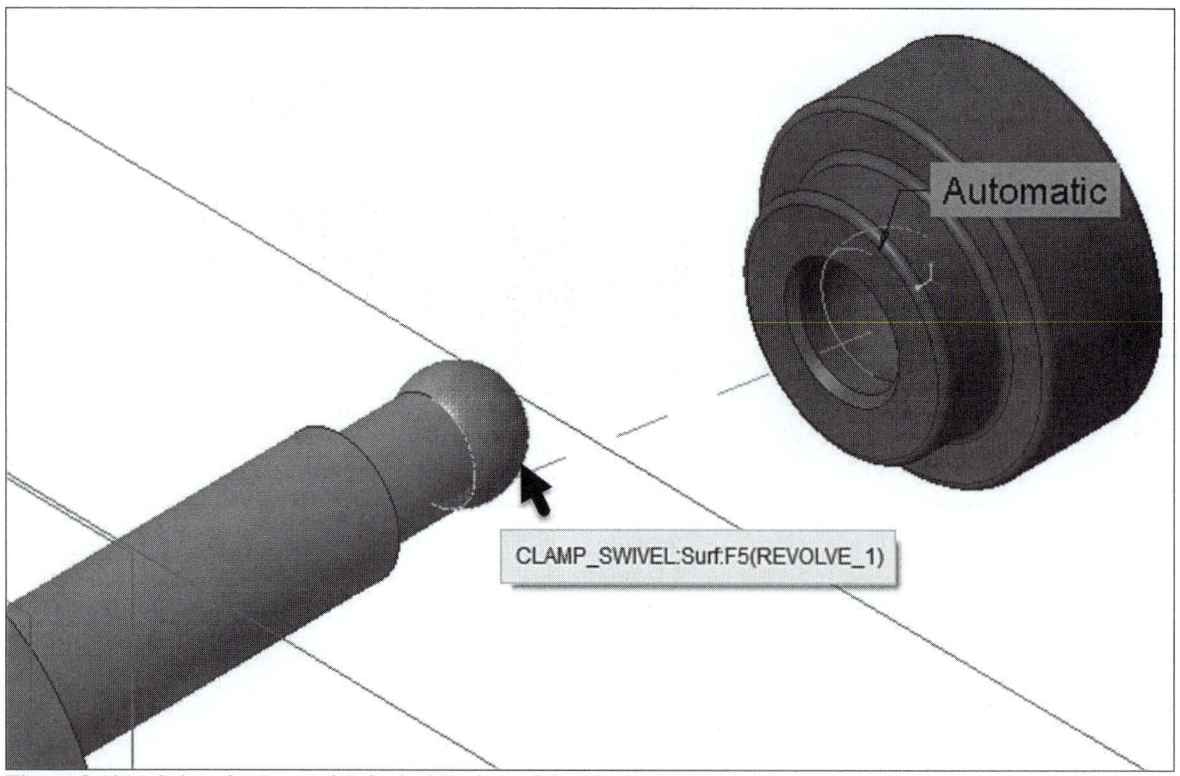

Figure 8.12(c) Select the External Spherical Surface of the Clamp_Swivel

From the Accessory (Component) Window, pick the surface of the Clamp_Foot [Fig. 8.12(d)] > from the assembly window, click: **RMB** to toggle to the desired surface (or rotate the model) > pick the surface of the Clamp_Arm_Design [Fig. 8.12(e)]

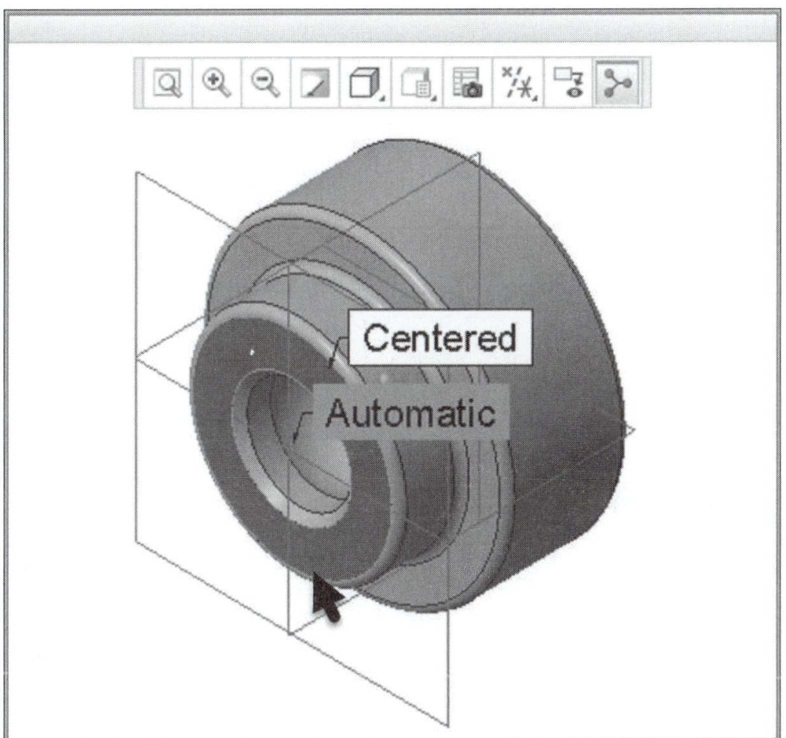

Figure 8.12(d) Select the External Cylindrical Surface of the Clamp_Foot

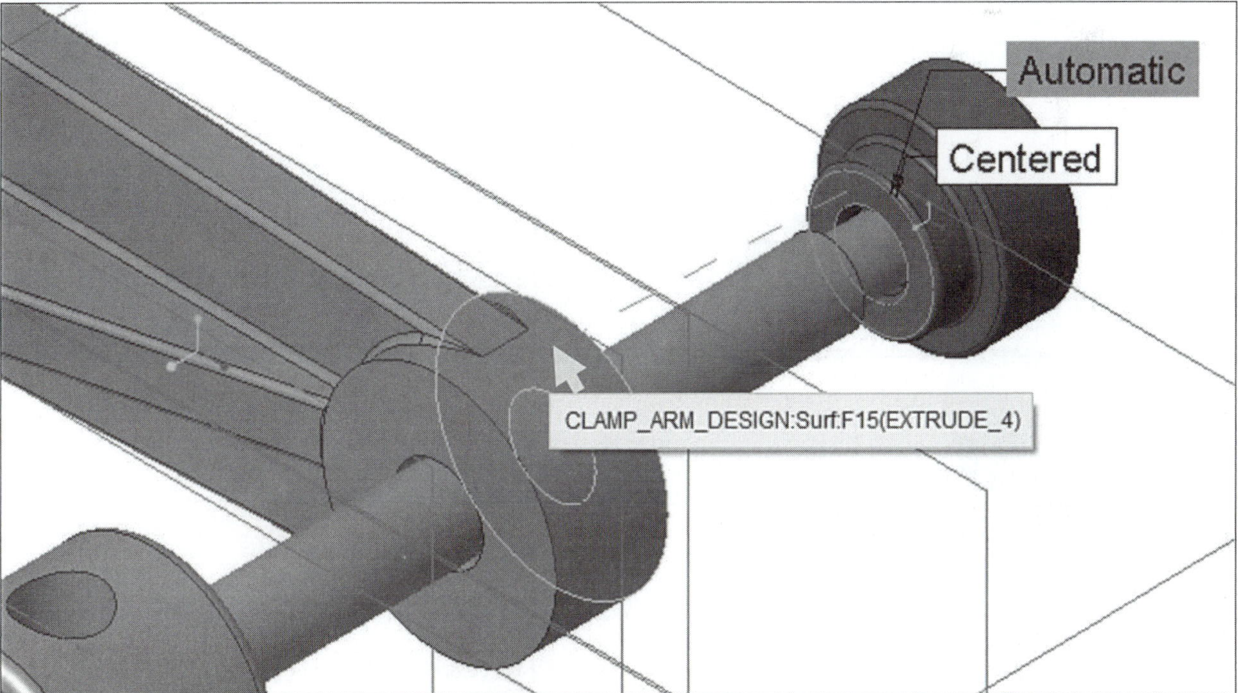

Figure 8.12(e) Select the External Cylindrical Surface of the Clamp_Arm_Design

Click: [icon] *off* > **Placement tab** > [Coincident] > [Parallel] > from the Graphics Window, press **RMB** [Fig. 8.12(f)] > **Flip Constraint** [Fig. 8.12(g)] > **Flip** from the Placement tab [Fig. 8.12(h)] > [✔ Allow Assumptions / Fully Constrained]

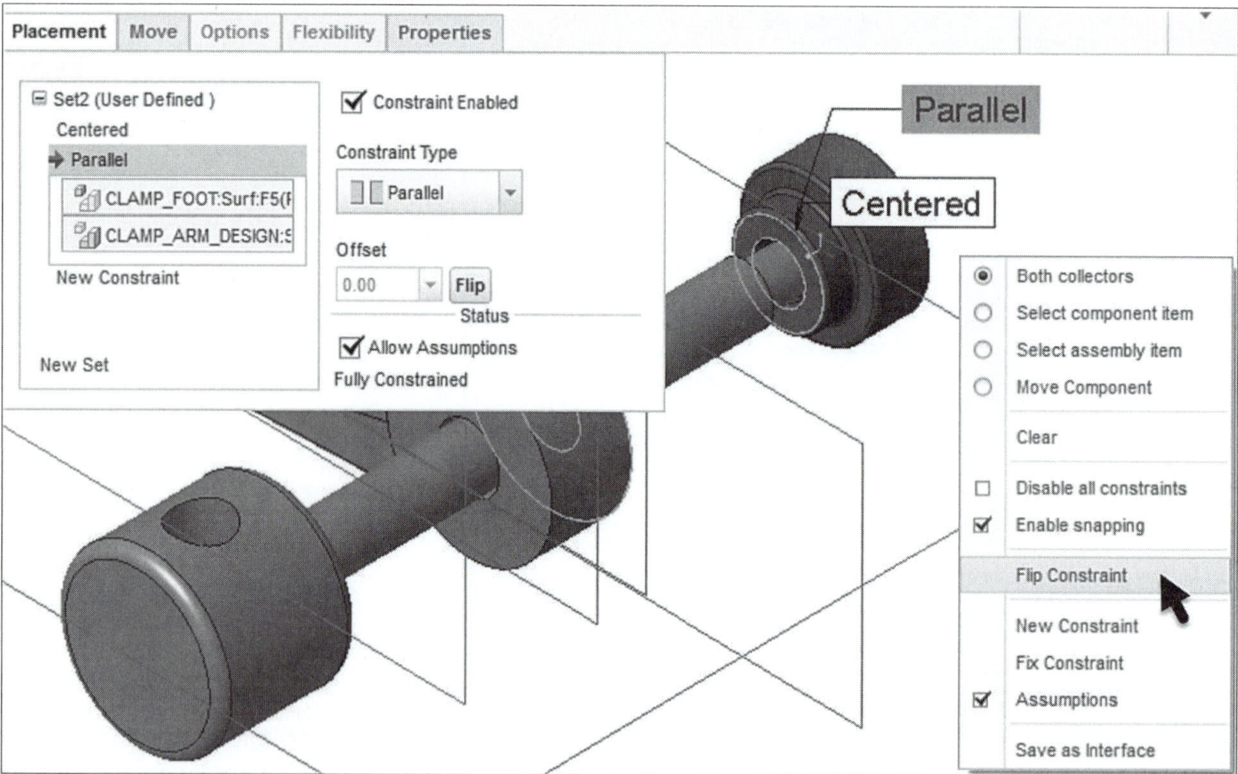

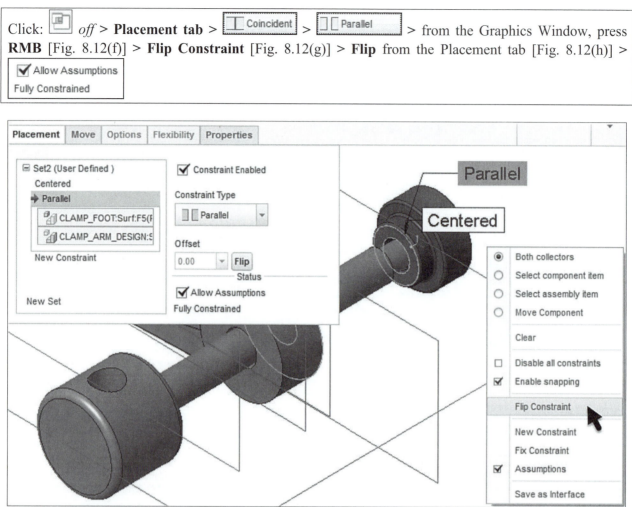

Figure 8.12(f) Flip Constraint

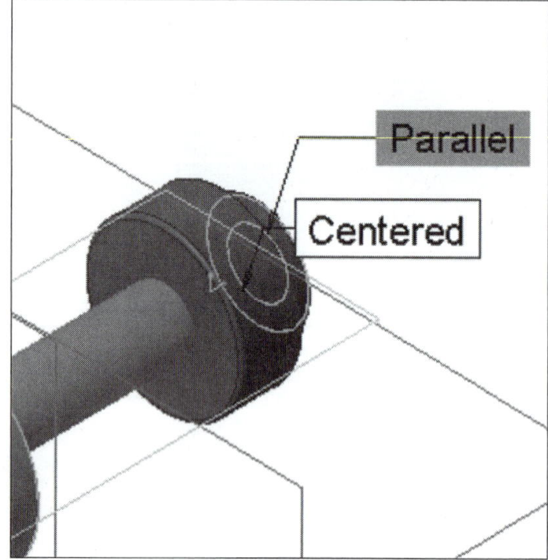

Figure 8.12(g) Flip Constraint

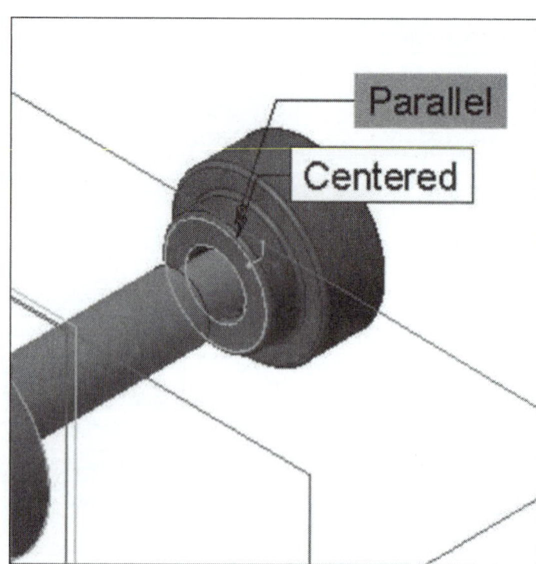

Figure 8.12(h) Original Orientation

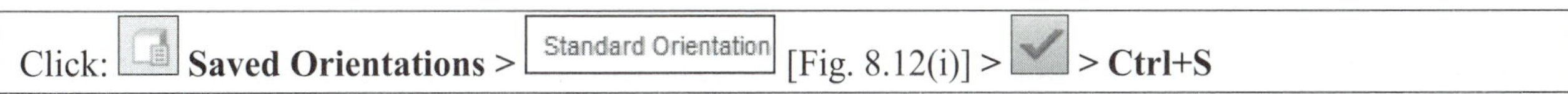

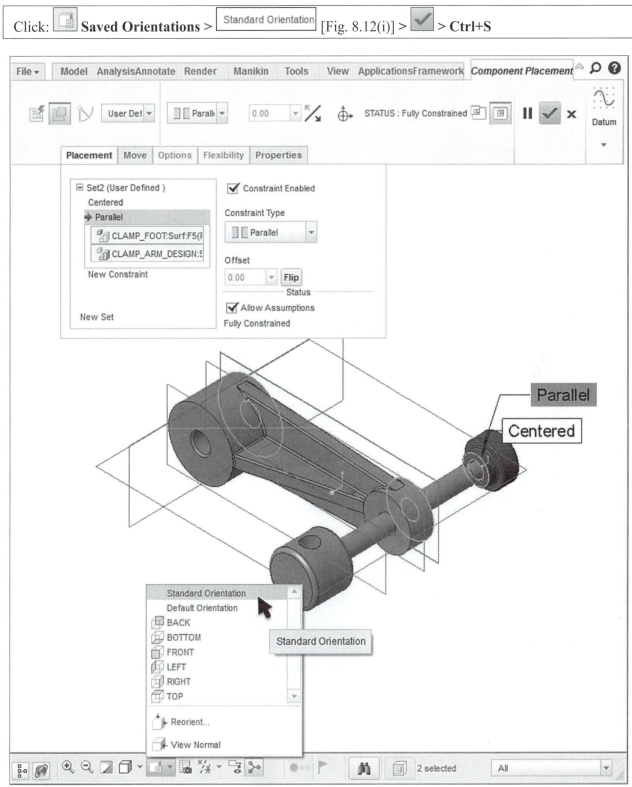

Click: **Saved Orientations** > Standard Orientation [Fig. 8.12(i)] > **Ctrl+S**

Figure 8.12(i) Fully Constrained Component

Assemble the longer **5.00** stud. Click: Assemble > select **Carrlane-12-13500_STUD.prt > Open >** **Show component in a separate window** *off* > **Show 3D Dragger** > rotate the part using the colored arc draggers [Fig. 8.13(a)] (you can rotate and move the component as desired [Fig. 8.13(b)])

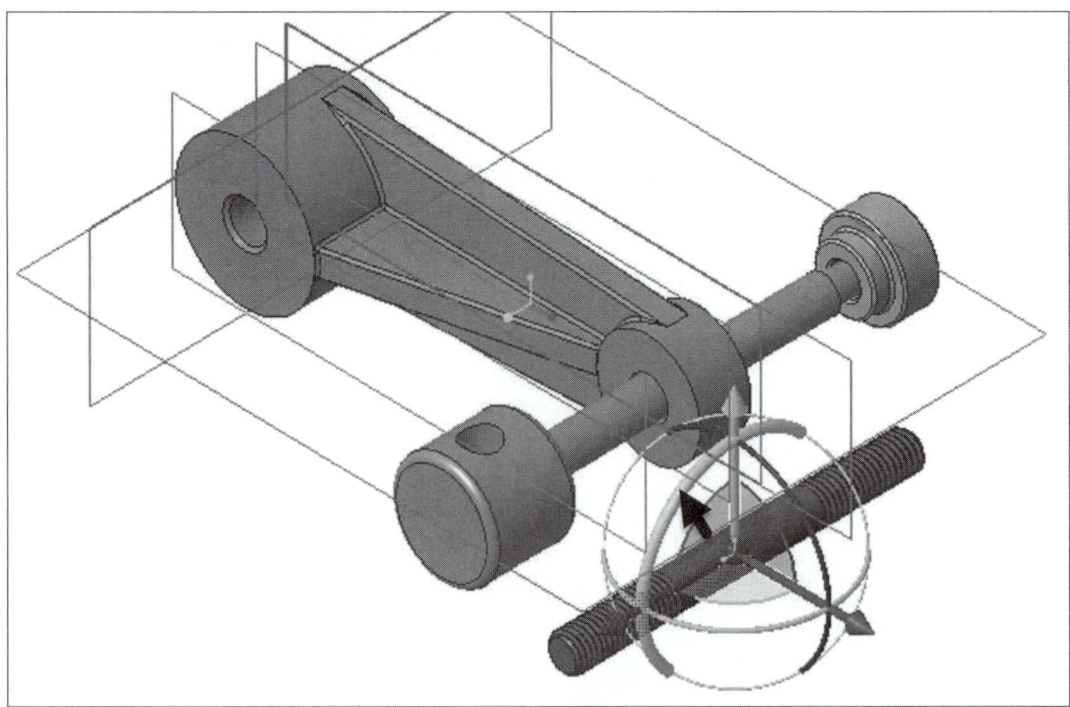

Figure 8.13(a) Press and hold down the LMB on the Arc

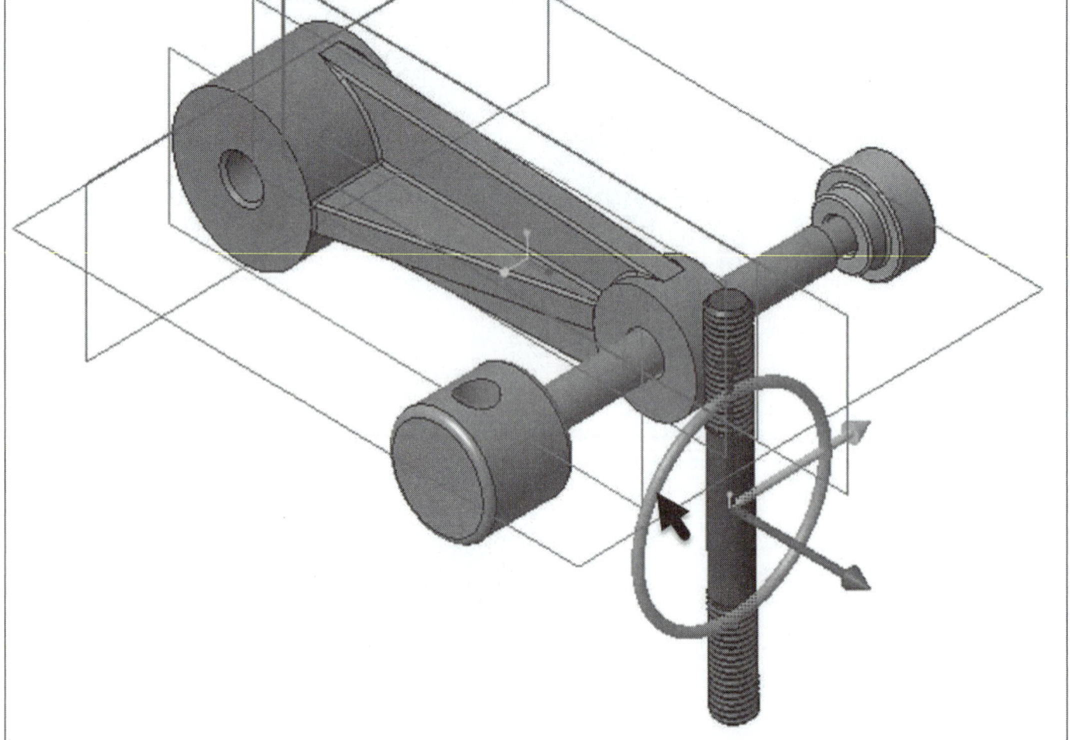

Figure 8.13(b) Move the Mouse Pointer to Rotate the Part > at the Desired Orientation, Release the LMB

Click: Hide 3D Dragger > pick the external cylindrical surface of the Carrlane-12-13500_STUD [Fig. 8.13(c)] > pick the internal cylindrical surface of the hole of the Clamp_Swivel [Fig. 8.13(d)]

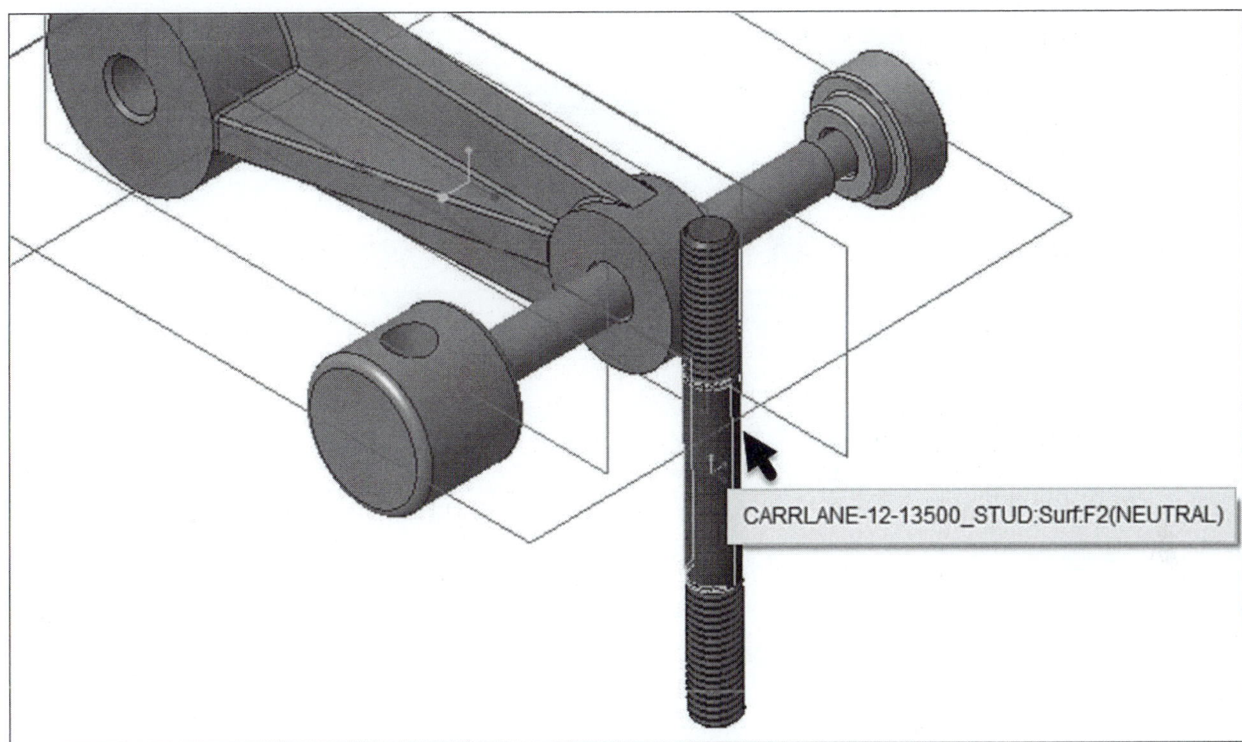

Figure 8.13(c) Select the Stud's Cylindrical Surface

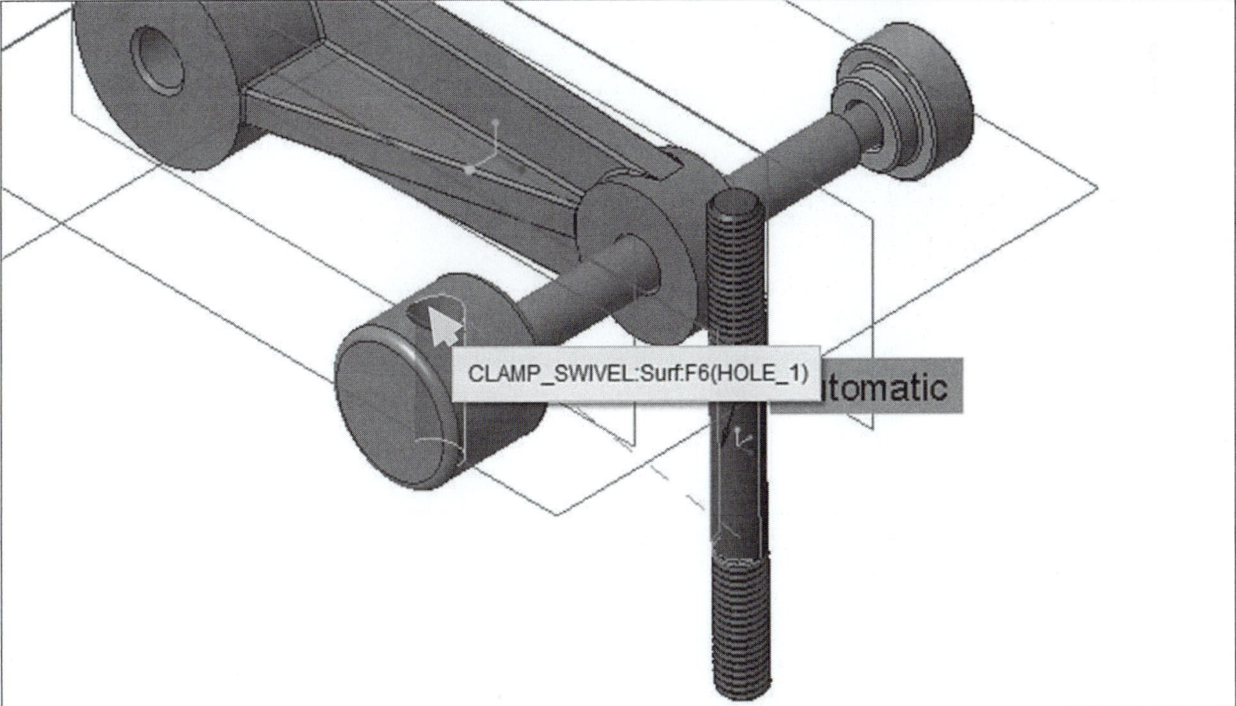

Figure 8.13(d) Select the Swivel's Hole's Cylindrical Surface

Pick the end surface of the Carrlane-12-13500_STUD [Fig. 8.14(a)] > pick datum **C** of the Clamp_Swivel [Fig. 8.14(b)] *(The hidden datum plane will highlight as you pass your cursor over it. If you get the wrong selection: RMB > Clear and try again.)*

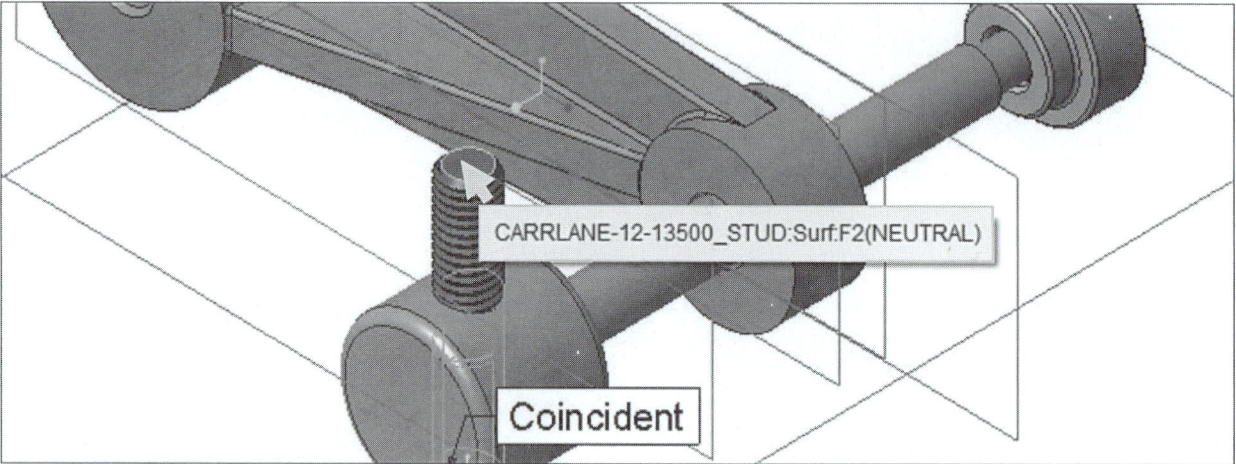

Figure 8.14(a) Select the End Surface

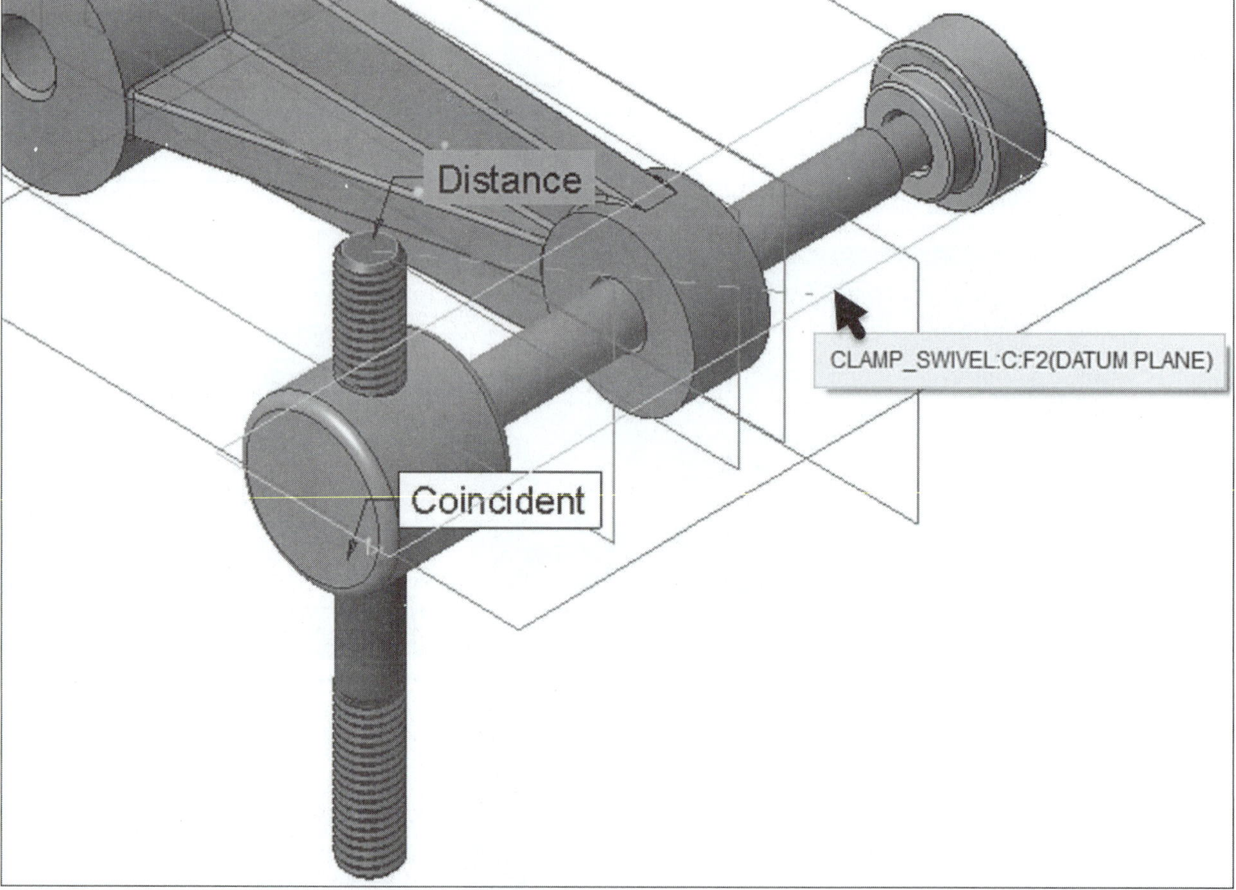

Figure 8.14(b) Select the Datum

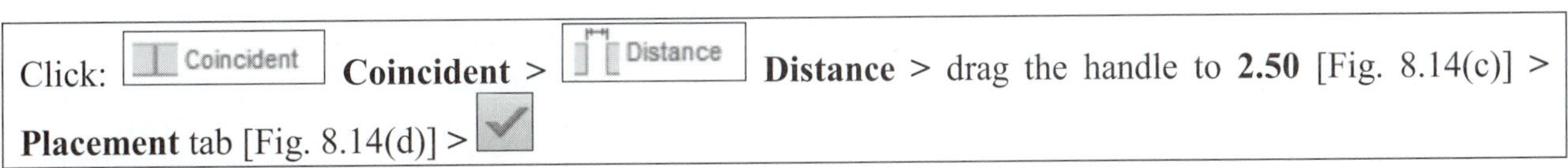

Click: Coincident **Coincident** > Distance **Distance** > drag the handle to **2.50** [Fig. 8.14(c)] > **Placement** tab [Fig. 8.14(d)] >

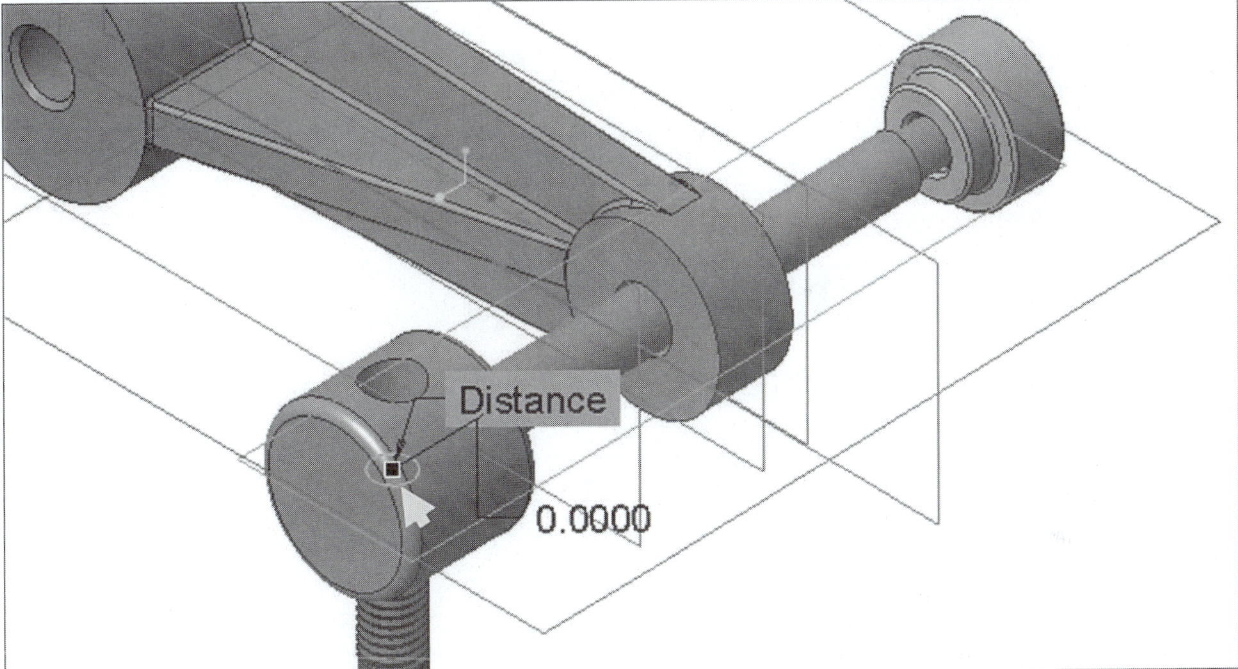

Figure 8.14(c) Constraint Type changed to Distance *(may already be Distance, depending on the initial location of the stud)*

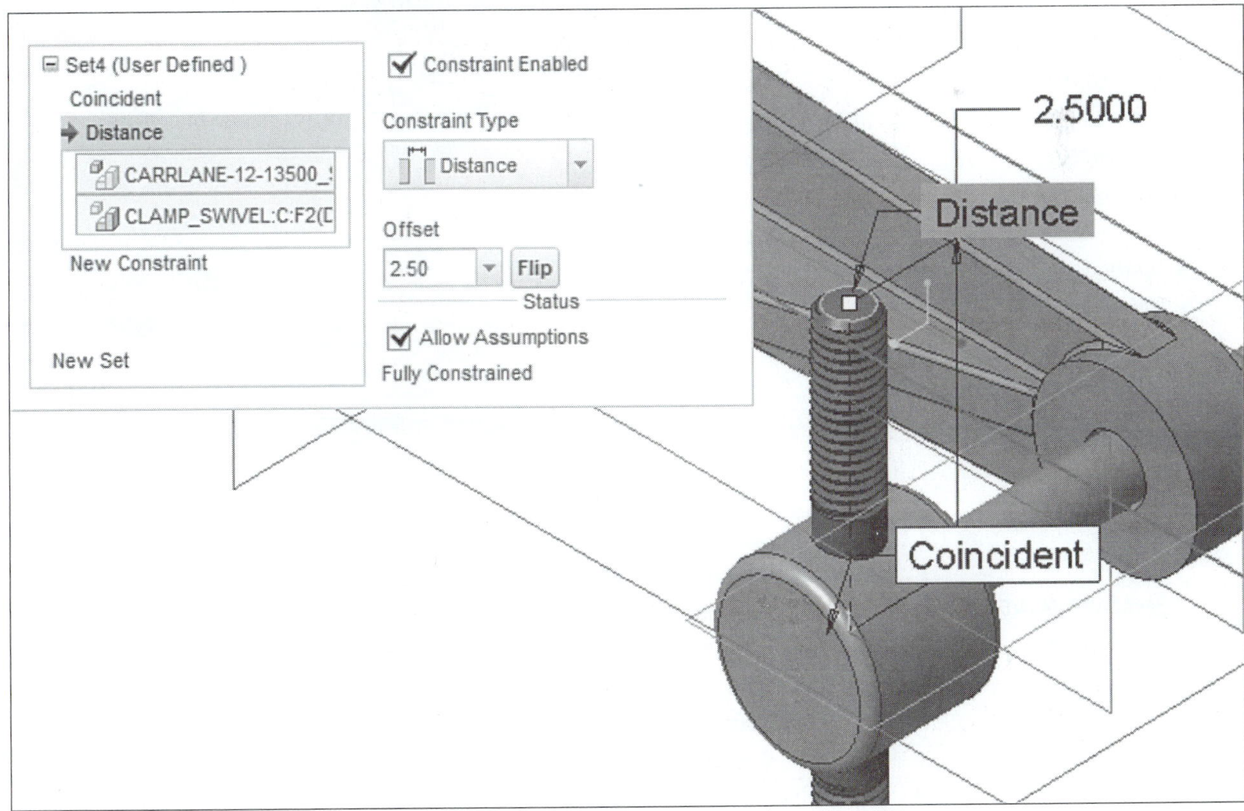

Figure 8.14(d) Press LMB on the Handle and Drag to an Offset of **2.50** > Release the LMB

Click: **Ctrl+S** > **Settings** > Tree Columns... > use >> to add the columns names to be Displayed [Fig. 8.15(a)] > **Apply** > **OK** > resize the Model Tree and Model Tree column widths [Fig. 8.15(b)]

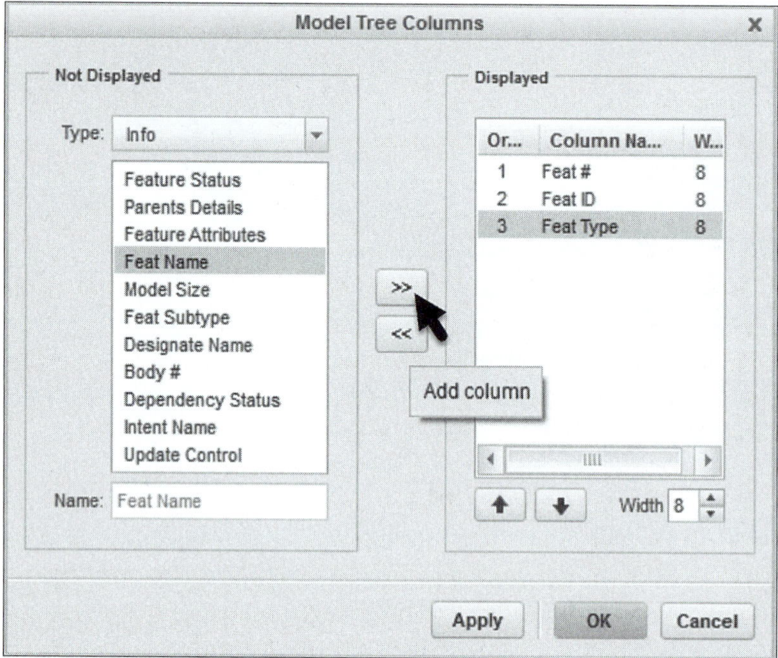

Figure 8.15(a) Model Tree Columns Dialog Box

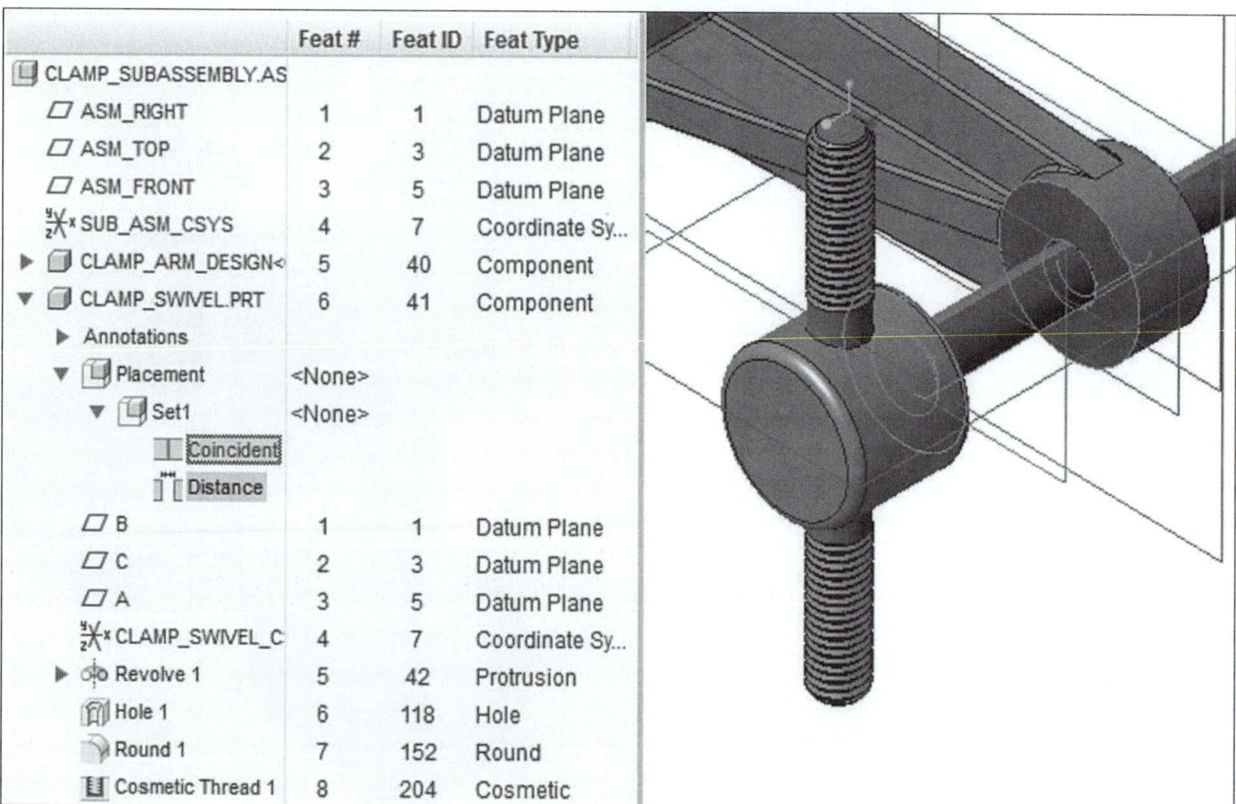

Feat # Feat ID Feat Type
CLAMP_SUBASSEMBLY.AS
ASM_RIGHT 1 1 Datum Plane
ASM_TOP 2 3 Datum Plane
ASM_FRONT 3 5 Datum Plane
SUB_ASM_CSYS 4 7 Coordinate Sy...
CLAMP_ARM_DESIGN< 5 40 Component
CLAMP_SWIVEL.PRT 6 41 Component
Annotations
Placement <None>
Set1 <None>
Coincident
Distance
B 1 1 Datum Plane
C 2 3 Datum Plane
A 3 5 Datum Plane
CLAMP_SWIVEL_C 4 7 Coordinate Sy...
Revolve 1 5 42 Protrusion
Hole 1 6 118 Hole
Round 1 7 152 Round
Cosmetic Thread 1 8 204 Cosmetic

Figure 8.15(b) Model Tree with Adjusted Columns

The Clamp_Ball handles are the last components of the Clamp_Subassembly. Reduce the Model Tree size to show only the Feat # column. Click: Assemble > select **clamp_ball.prt** > **Preview** *on* > **MMB** spin and zoom the Preview Window [Fig. 8.16(a)] > **Open** > **MMB** to spin the model as shown [Fig. 8.16(b)]

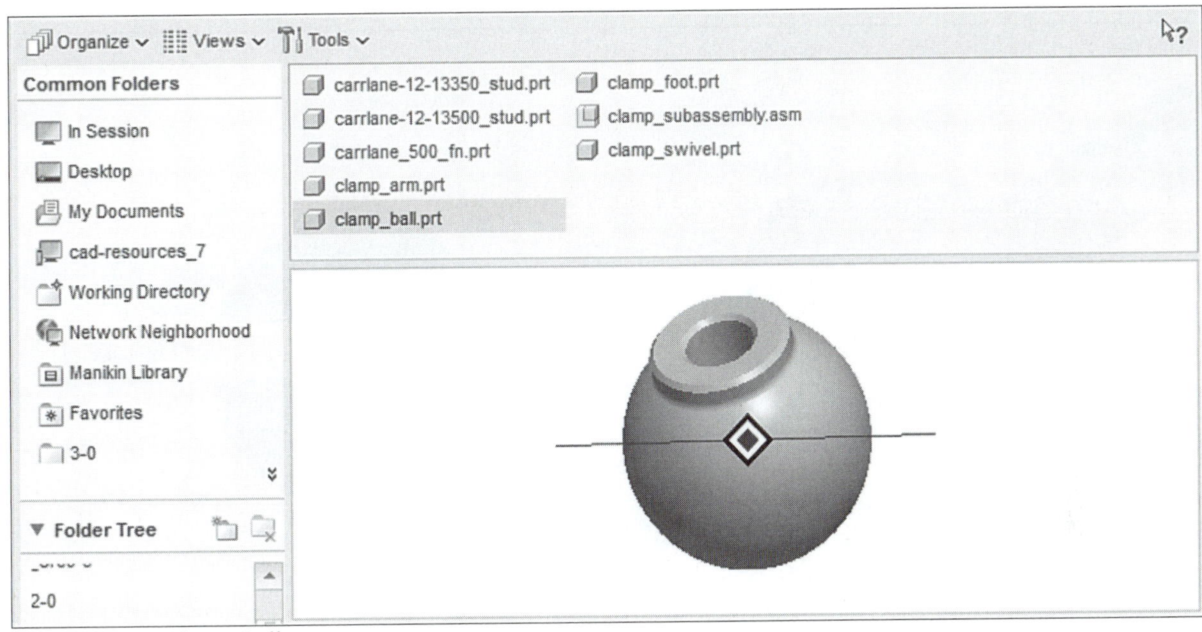

Figure 8.16(a) Clamp_Ball

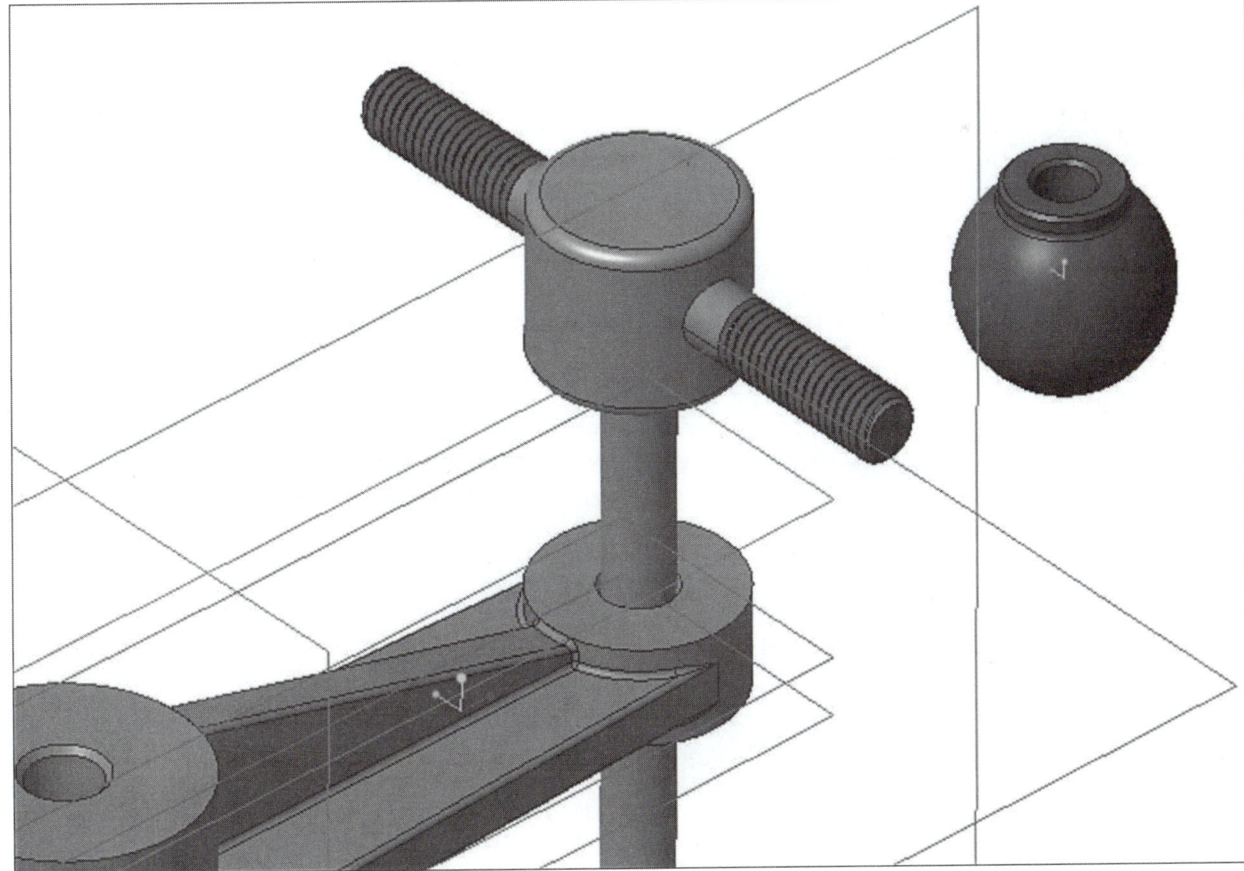

Figure 8.16(b) Spin the Subassembly

To move the new component independently in the assembly window, press and hold **Ctrl+Alt > MMB** on the Clamp_Ball and rotate > **Ctrl+Alt > RMB** pan as needed > release the **RMB** and **Ctrl+Alt** keys > **Hide 3D Dragger** > pick on the internal cylindrical surface of the hole of the Clamp_Ball [Fig. 8.16(c)] > pick on the external cylindrical surface of the Carrlane-12-13500_STUD [Fig. 8.16(d)]

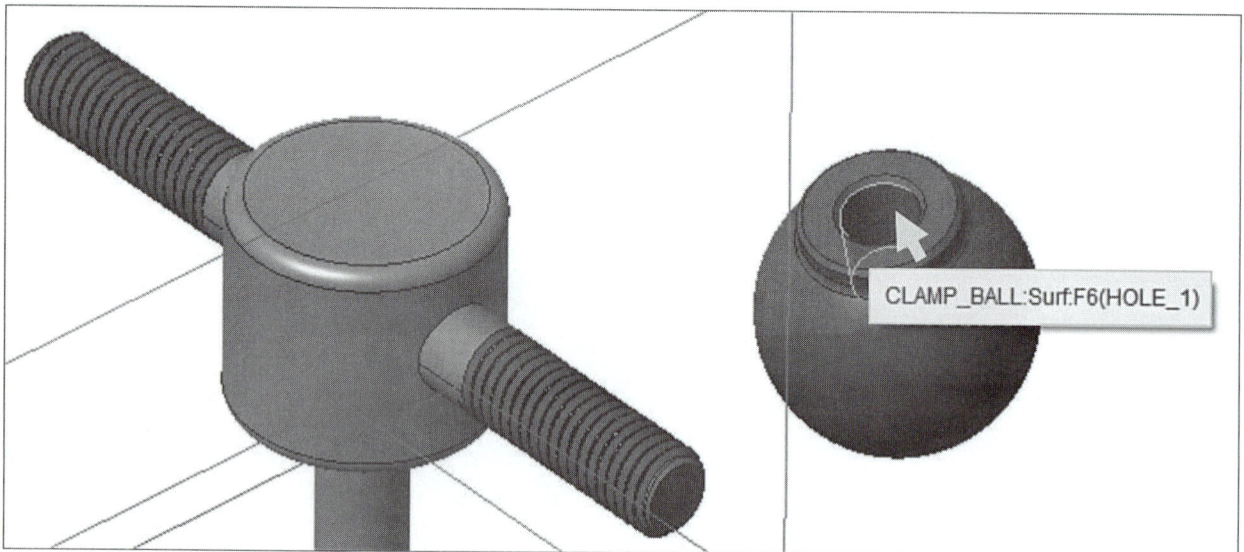

Figure 8.16(c) Select the Hole's Surface

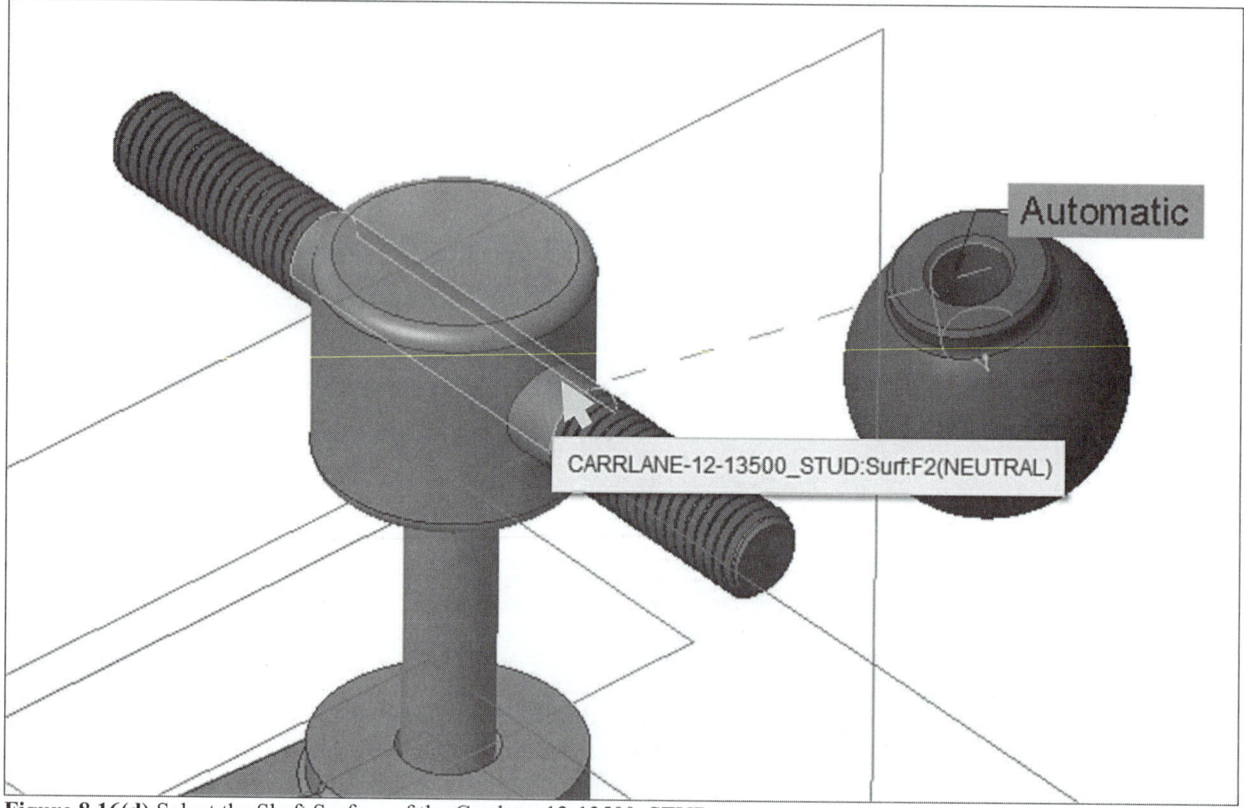

Figure 8.16(d) Select the Shaft Surface of the Carrlane-12-13500_STUD

Click: **Placement** tab in the Dashboard > **New Constraint** > **Shift** key > **RMB** pan as needed > pick the end surface of the Carrlane-12-13500_STUD [Fig. 8.16(e)] > pick the flat end of the Clamp_Ball [Fig. 8.16(f)] (**RMB** to toggle to the desired surface if needed)

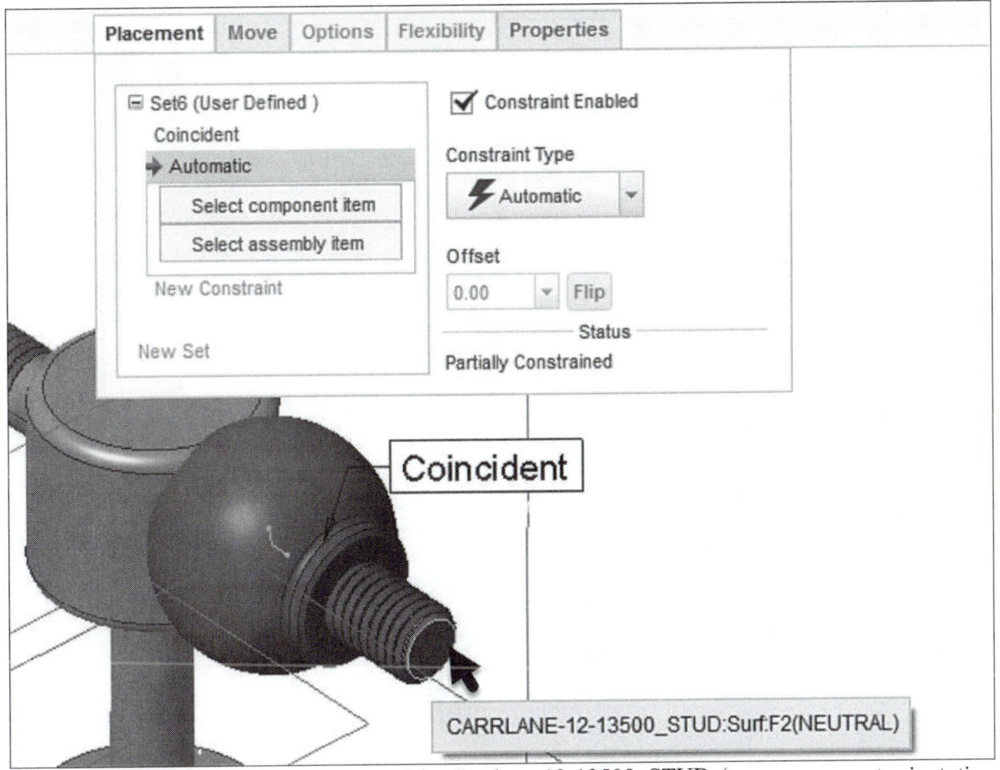

Figure 8.16(e) Select the End Surface of the Carrlane-12-13500_STUD *(your component orientation may appear differently)*

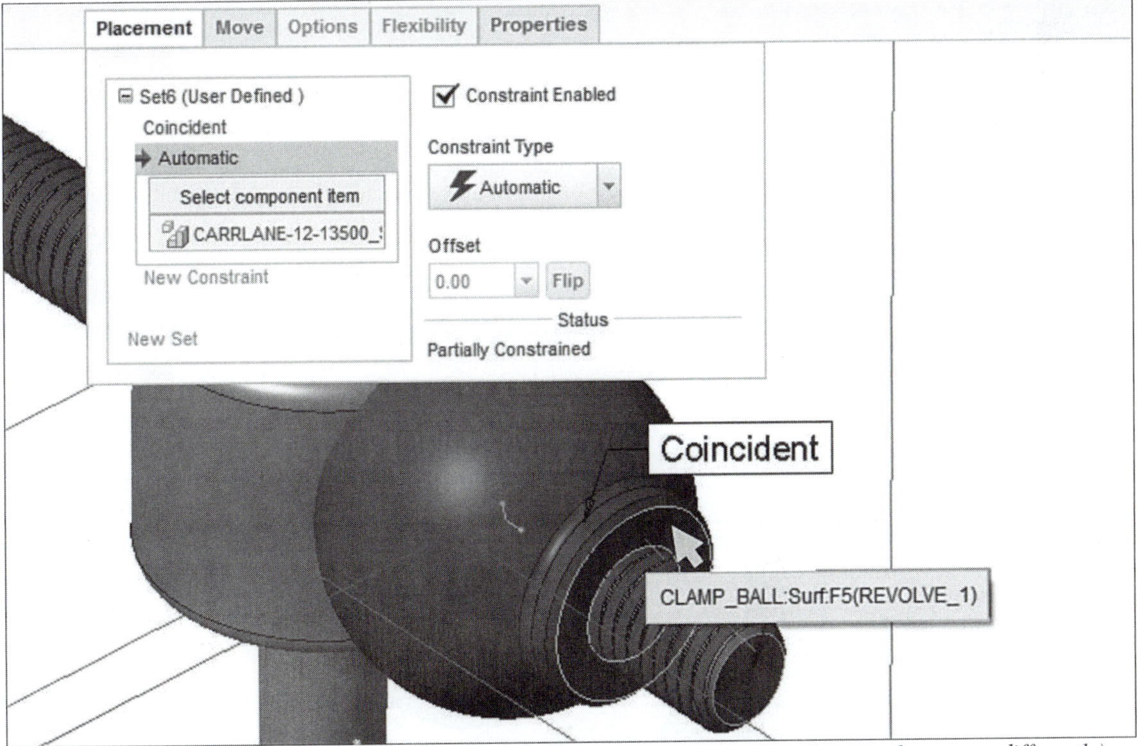

Figure 8.16(f) Select the End Surface of the Clamp_Ball *(your component orientation may be appear differently)*

Click: Offset **Flip** (changes the constraint type from Align to Mate) *(Flip as needed)* [Fig. 8.16(g)] > Coincident > Distance (may already be Distance, based on the initial location of the component) > drag the handle toward the Clamp_Swivel until **.50** [Fig. 8.16(h)] > ✓ > **Ctrl+S**

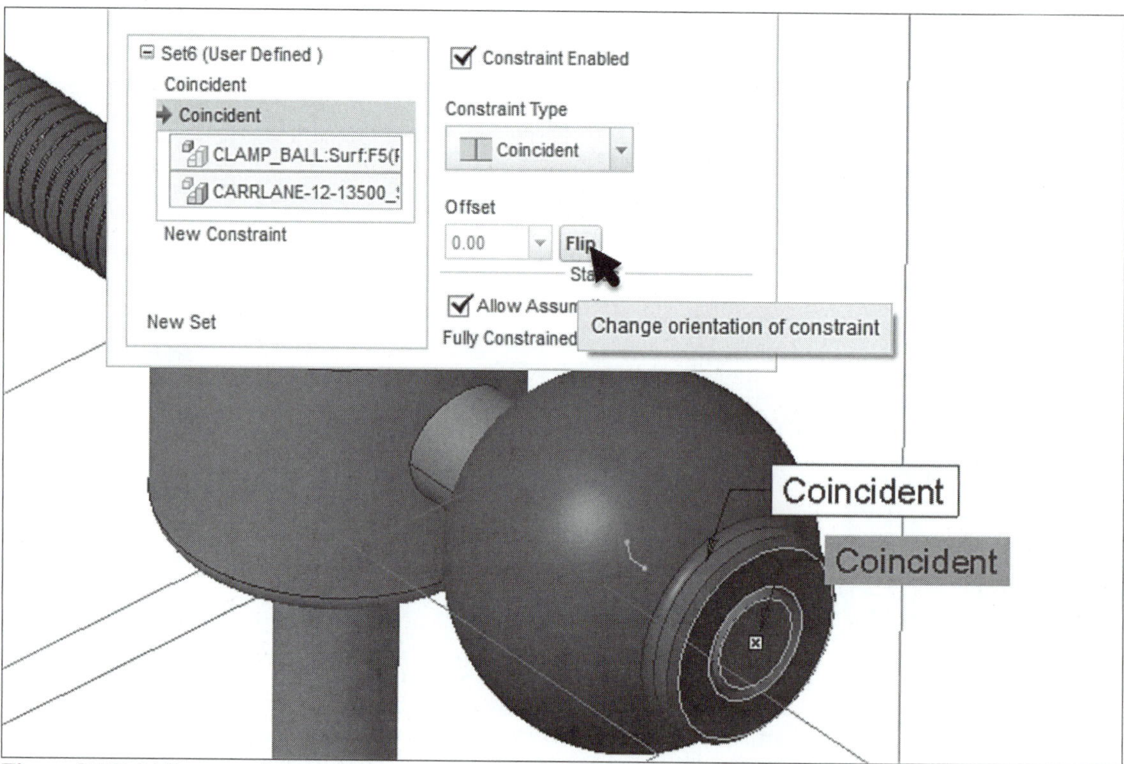

Figure 8.16(g) Flip as needed

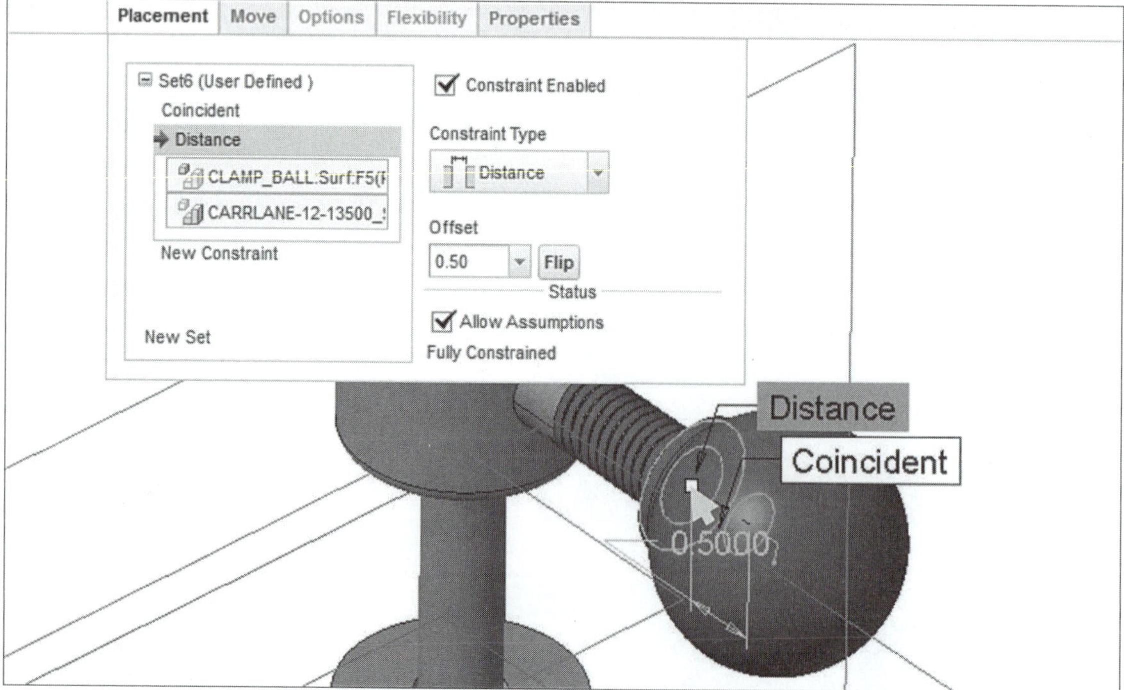

Figure 8.16(h) Move the Drag Handle so that the Clamp_Ball is Offset **.500**

MMB to spin the model > pick on the **Clamp_Ball** component to select it *(highlights)* [Fig. 8.16(i)] > > **Hidden Line** > **Ctrl+C** (copy) > **Ctrl+V** (paste) > pick the opposite end cylindrical surface of the Carrlane-12-13500_STUD as the new reference [Fig. 8.16(j)]

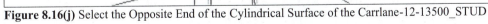

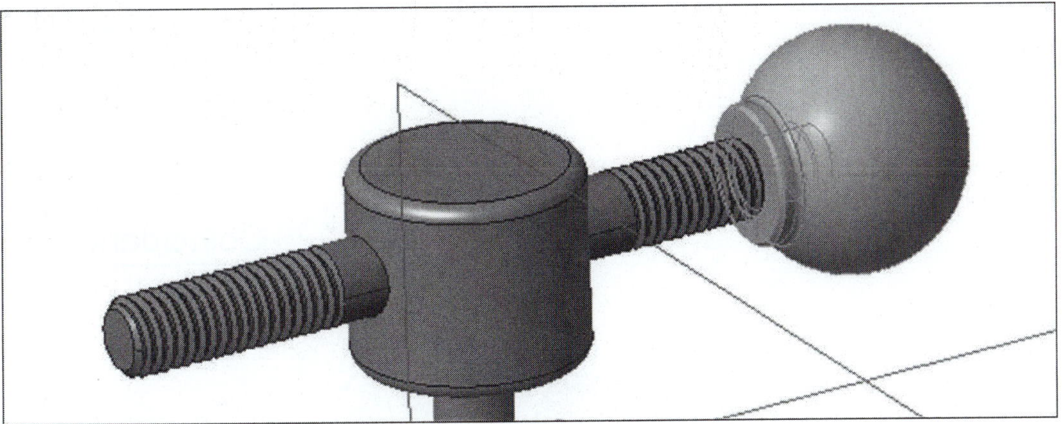

Figure 8.16(i) First Clamp_Ball Selected *(highlighted)*

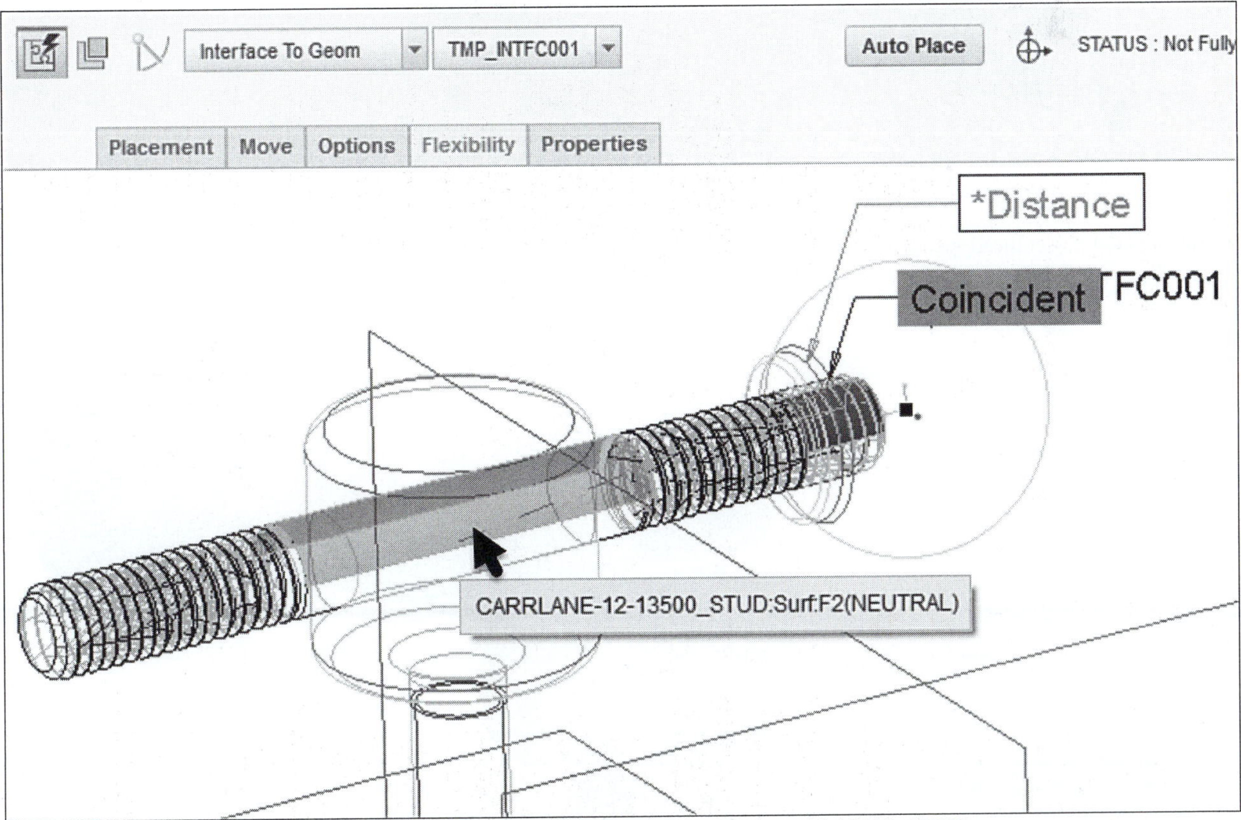

Figure 8.16(j) Select the Opposite End of the Cylindrical Surface of the Carrlane-12-13500_STUD

Pick the opposite end (flat) surface of the Carrlane-12-13500_STUD [Fig. 8.16(k)] > [icon] > [icon] **Shading** > press **RMB** > **Flip Constraint** > in the Graphics Window; double-click on the **.500** dimension > *type* **.4375** (to modify the distance from the end of the shaft so that it does not bottom-out) [Fig. 8.16(l)] > **Enter** > [icon] > **Ctrl+S** > in the Graphics Window, **LMB** to deselect

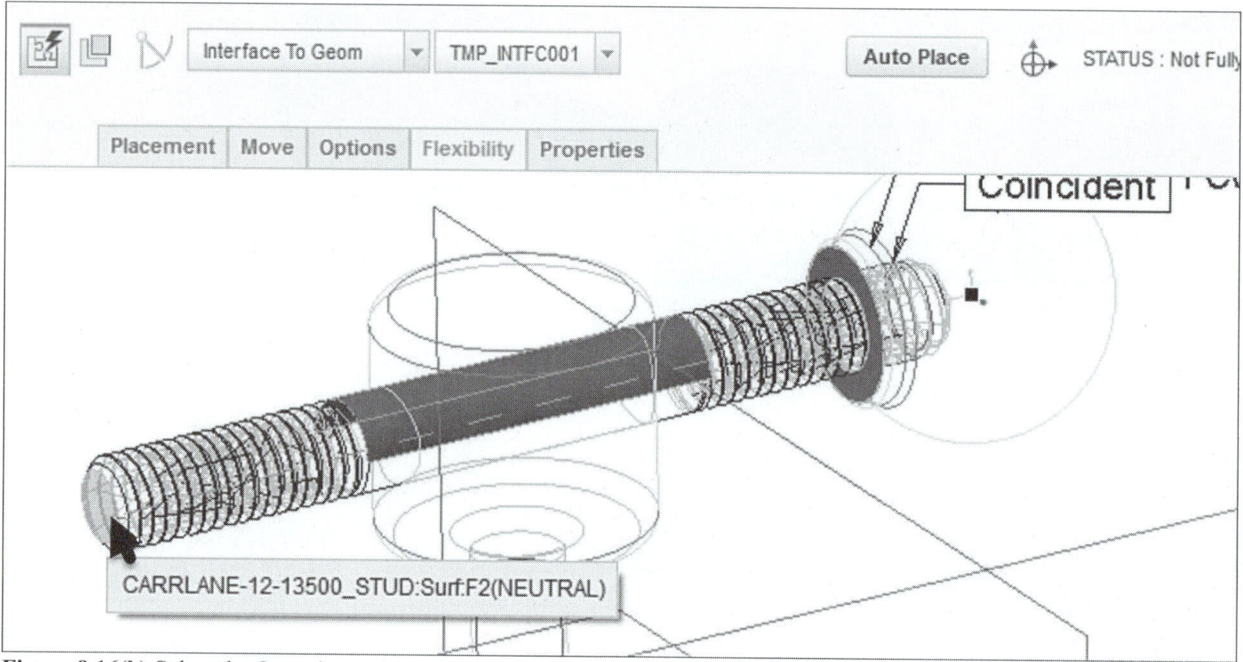

Figure 8.16(k) Select the Opposite End (Flat) Surface of the Carrlane-12-13500_STUD

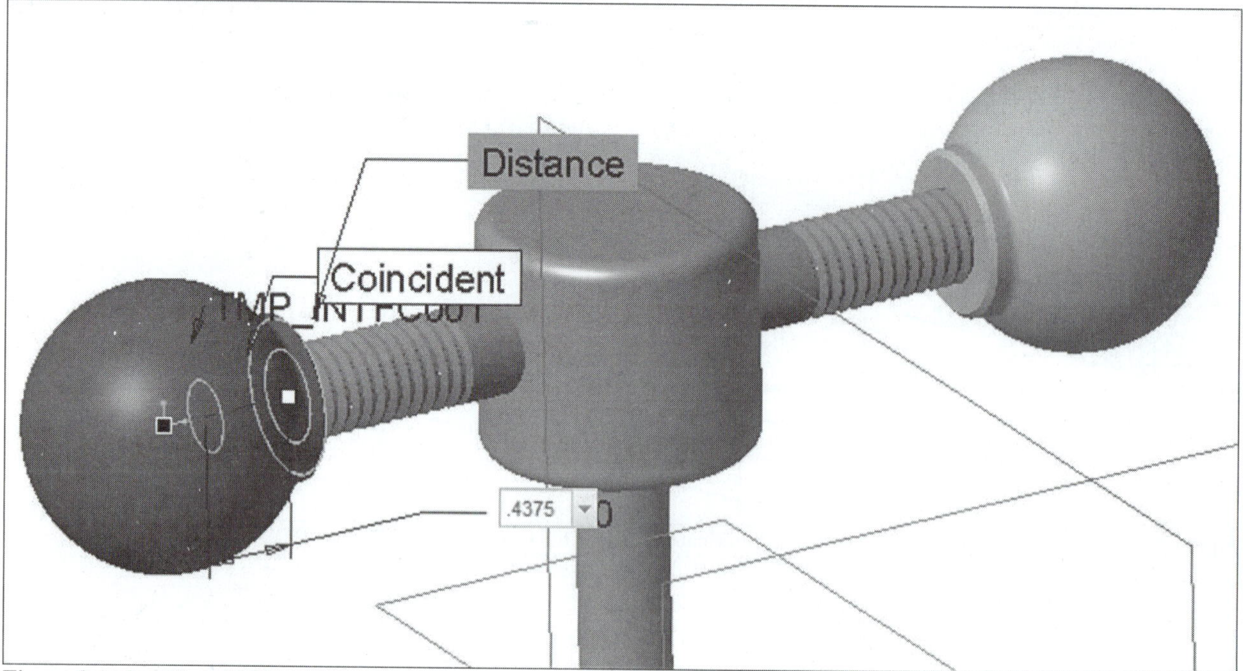

Figure 8.16(l) Offset **.4375**

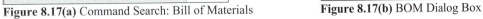

Double-click on the first Clamp_Ball and modify the offset distance from **.500** to **.4375** > **Enter** > **Regenerate** > **Ctrl+D** > **Command Search** > *type* **bill** > **Bill of Materials** [Fig. 8.17(a)] > Top Level [Fig. 8.17(b)] > **OK** [Fig. 8.17(c)] > close the Browser

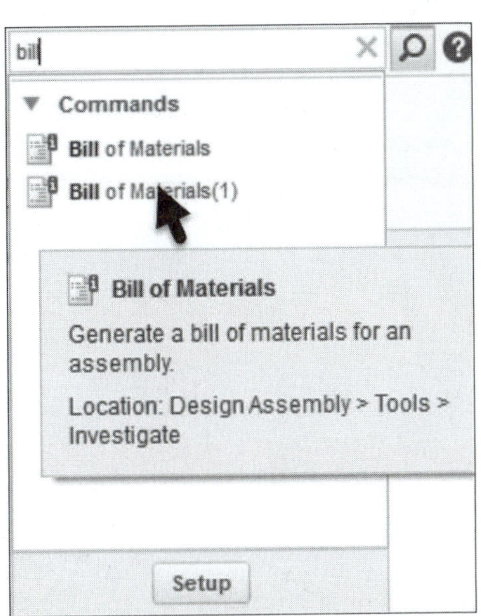

Figure 8.17(a) Command Search: Bill of Materials

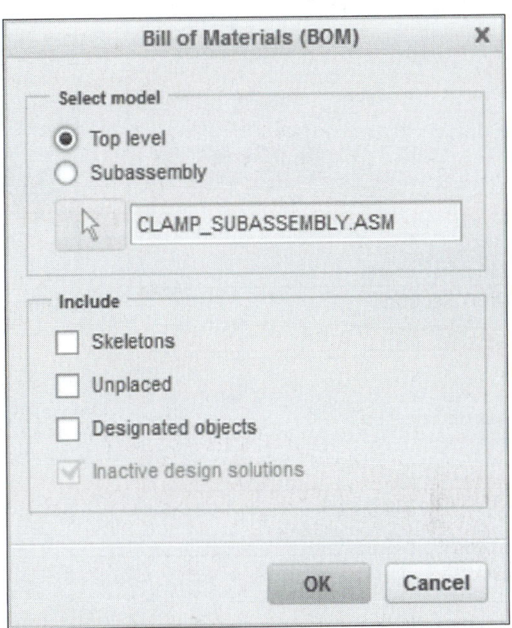

Figure 8.17(b) BOM Dialog Box

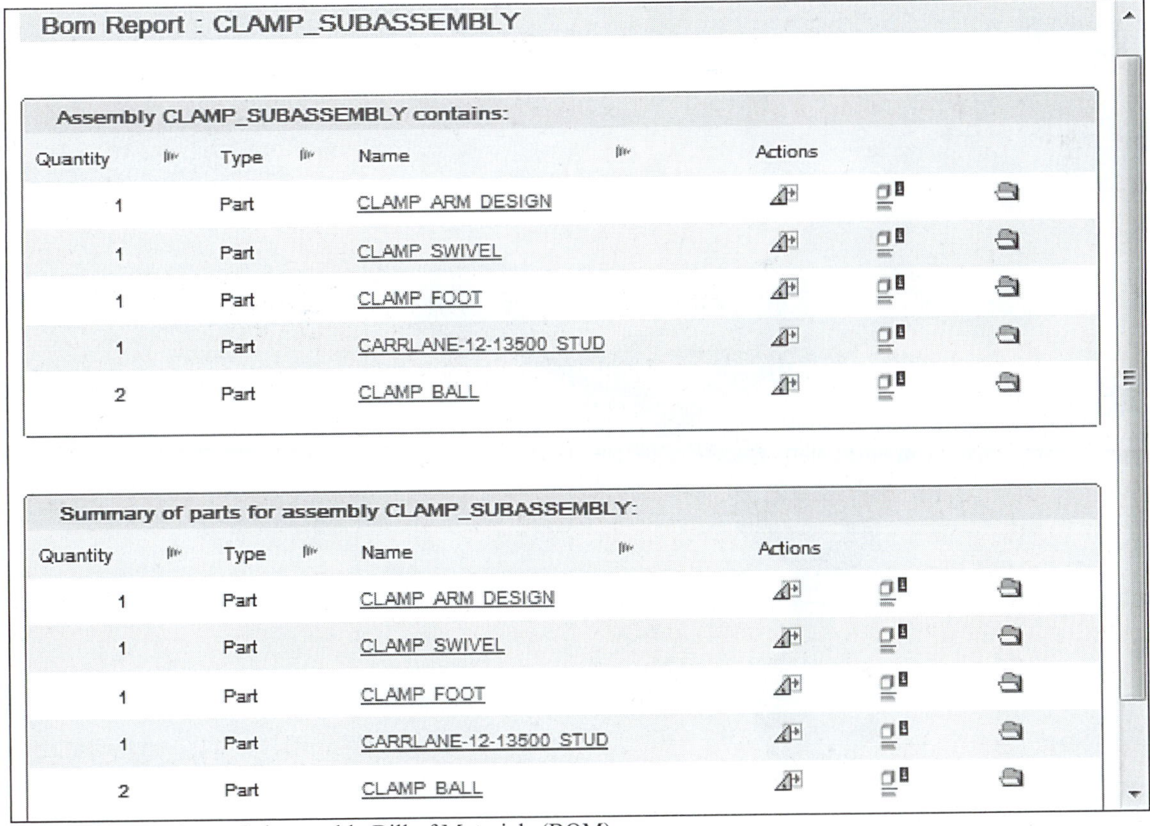
Bom Report : CLAMP_SUBASSEMBLY

Assembly CLAMP_SUBASSEMBLY contains:

Quantity	Type	Name	Actions
1	Part	CLAMP_ARM_DESIGN	
1	Part	CLAMP_SWIVEL	
1	Part	CLAMP_FOOT	
1	Part	CARRLANE-12-13500_STUD	
2	Part	CLAMP_BALL	

Summary of parts for assembly CLAMP_SUBASSEMBLY:

Quantity	Type	Name	Actions
1	Part	CLAMP_ARM_DESIGN	
1	Part	CLAMP_SWIVEL	
1	Part	CLAMP_FOOT	
1	Part	CARRLANE-12-13500_STUD	
2	Part	CLAMP_BALL	

Figure 8.17(c) Clamp_Subassembly Bill of Materials (BOM)

Ctrl+D > **Ctrl+S** > **View** tab > [icons] *on* > [icon] **View Manager** from the Status Bar > **Sections** tab [Fig. 8.18(a)] > **New** > **Planar** [Fig. 8.18(b)] > *type* **A** [Fig. 8.18(c)] > **Enter** > pick datum **ASM_TOP** [Fig. 8.18(d)] > [✓] [Fig. 8.18(e)]

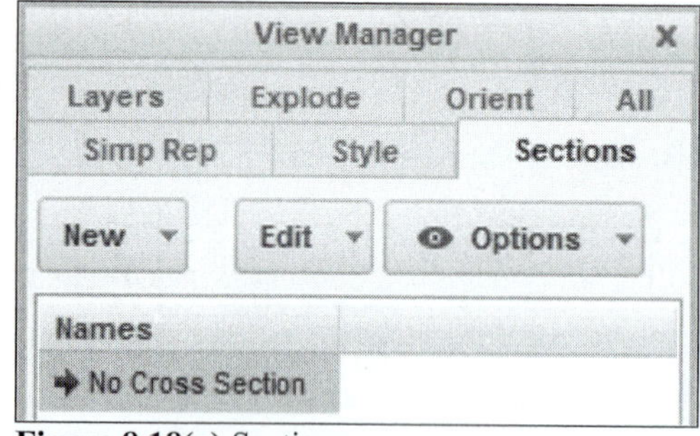

Figure 8.18(a) Sections

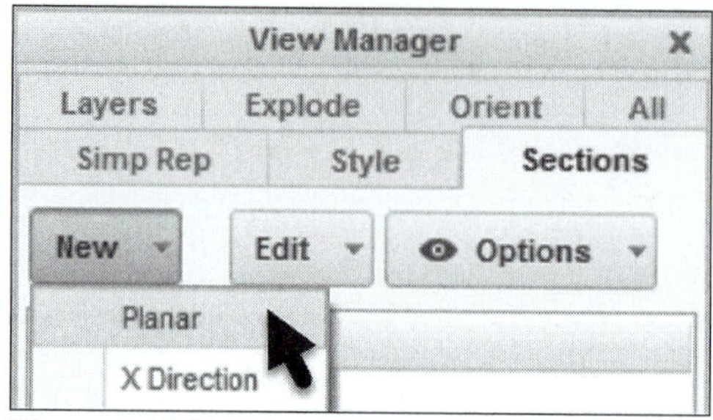

Figure 8.18(b) New > Planar

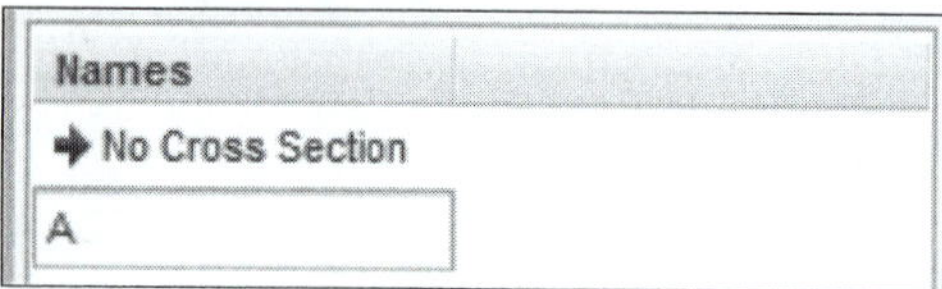

Figure 8.18(c) A is the New Cross Section Name

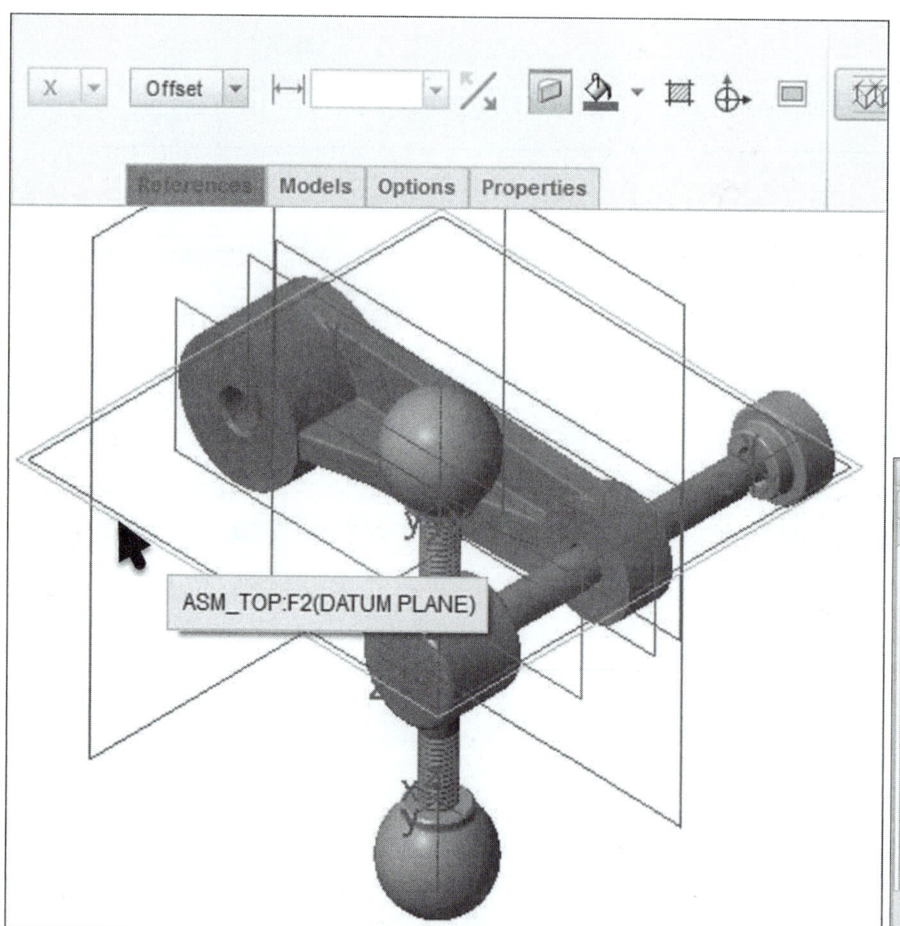

Figure 8.18(d) Select Assembly Datum Top

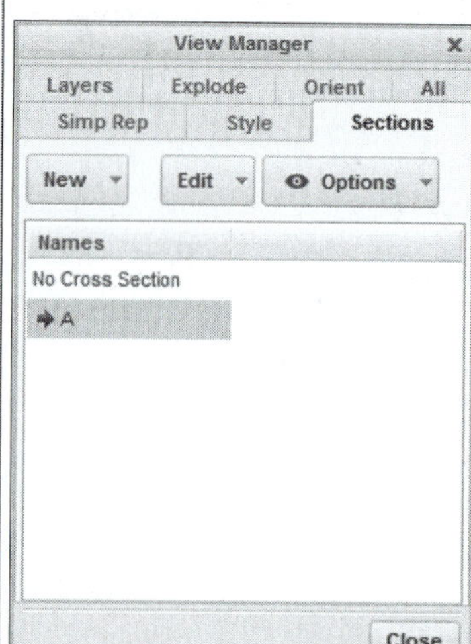

Figure 8.18(e) Section A

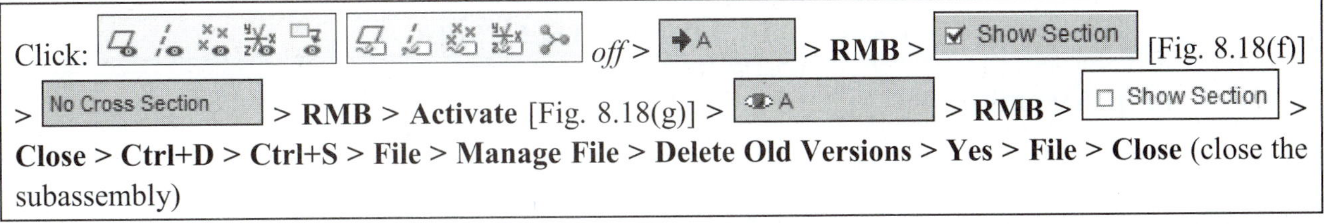

Click: off > ➜A > **RMB** > ☑ Show Section [Fig. 8.18(f)] > No Cross Section > **RMB** > **Activate** [Fig. 8.18(g)] > A > **RMB** > ☐ Show Section > **Close** > **Ctrl+D** > **Ctrl+S** > **File** > **Manage File** > **Delete Old Versions** > **Yes** > **File** > **Close** (close the subassembly)

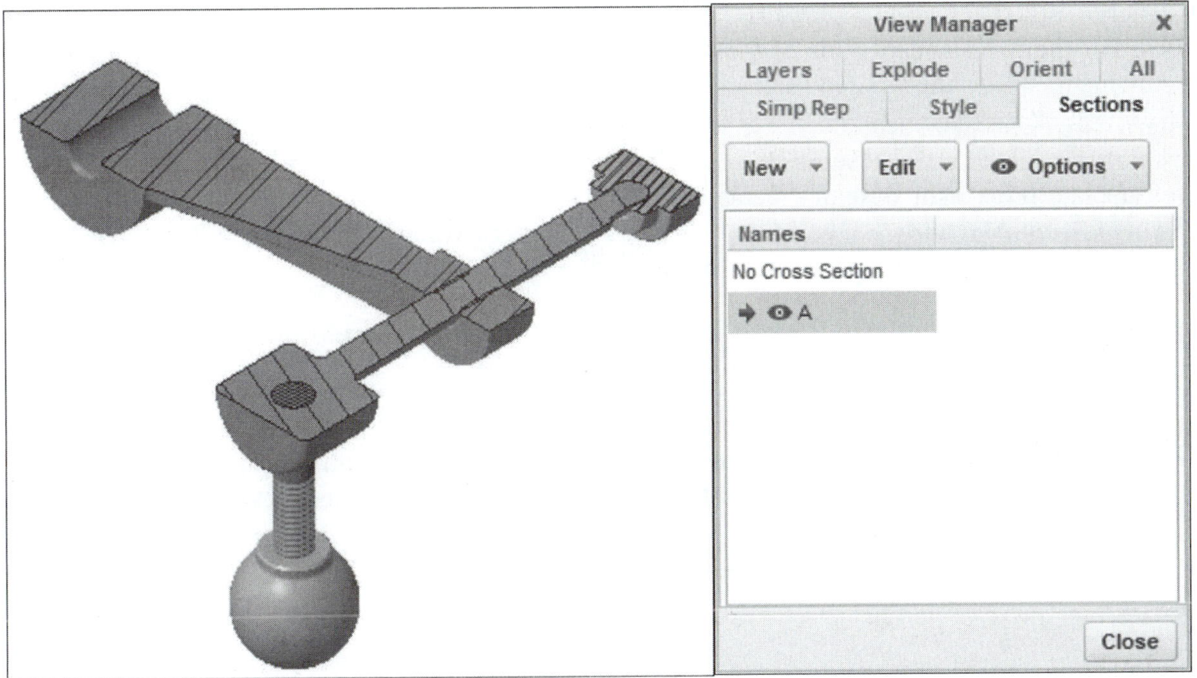

Figure 8.18(f) Show Section A Cross Hatching

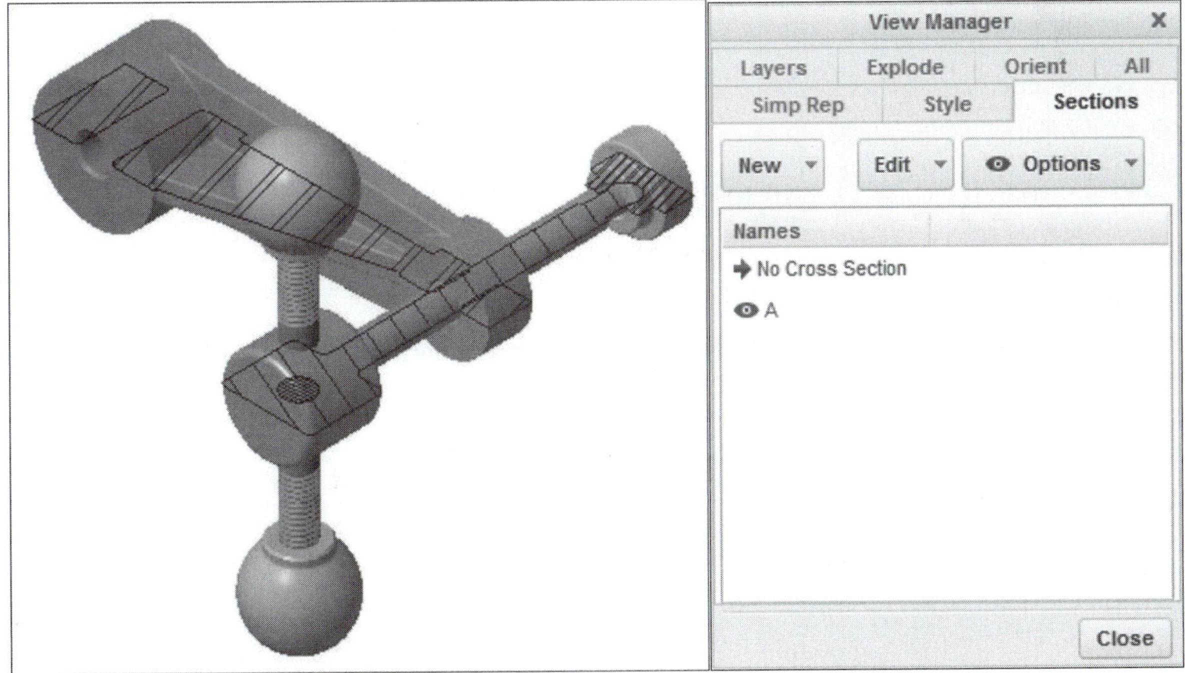

Figure 8.18(g) No Cross Section Set Active

Part Two: Top-Down Design

Swing Clamp Assembly

The first features for the main assembly will be the default datum planes and coordinate system. The first part assembled on the main assembly will be the Clamp_Plate, which will be created using *top-down design*; where the assembly is active, and the component is created within the assembly mode. The subassembly is still *"in session-in memory"* even though it does not show on the screen after its window is closed. For the two standard parts of the main assembly, you will use specific constraints instead of using Automatic, which allows Creo Parametric 3.0 to default to an assumed constraint.

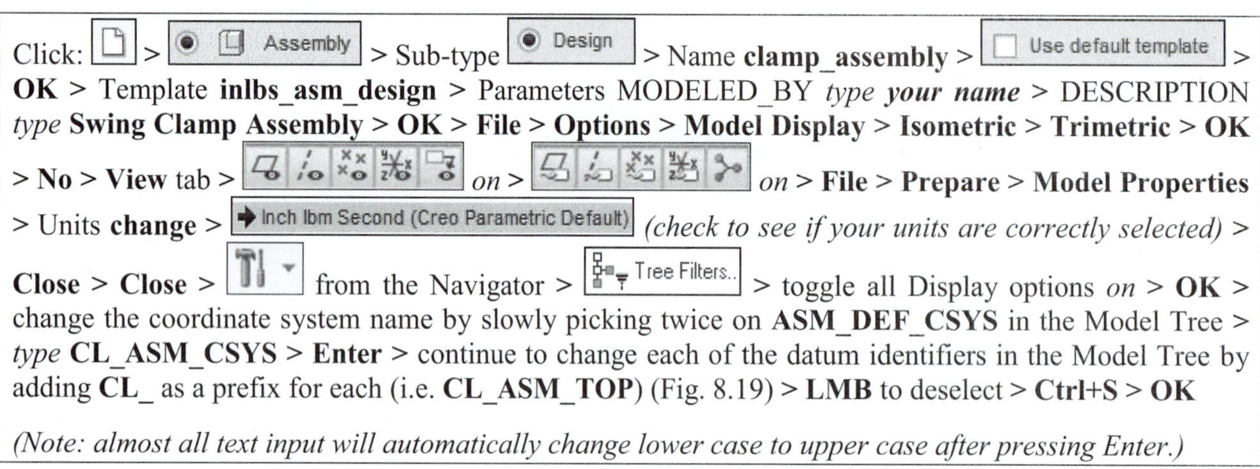

Click: [icon] > Assembly > Sub-type Design > Name **clamp_assembly** > Use default template > **OK** > Template **inlbs_asm_design** > Parameters MODELED_BY *type* ***your name*** > DESCRIPTION *type* **Swing Clamp Assembly** > **OK** > **File** > **Options** > **Model Display** > **Isometric** > **Trimetric** > **OK** > **No** > **View** tab > [icons] *on* > [icons] *on* > **File** > **Prepare** > **Model Properties** > Units **change** > Inch lbm Second (Creo Parametric Default) *(check to see if your units are correctly selected)* > **Close** > **Close** > [icon] from the Navigator > Tree Filters.. > toggle all Display options *on* > **OK** > change the coordinate system name by slowly picking twice on **ASM_DEF_CSYS** in the Model Tree > *type* **CL_ASM_CSYS** > **Enter** > continue to change each of the datum identifiers in the Model Tree by adding **CL_** as a prefix for each (i.e. **CL_ASM_TOP**) (Fig. 8.19) > **LMB** to deselect > **Ctrl+S** > **OK**

(Note: almost all text input will automatically change lower case to upper case after pressing Enter.)

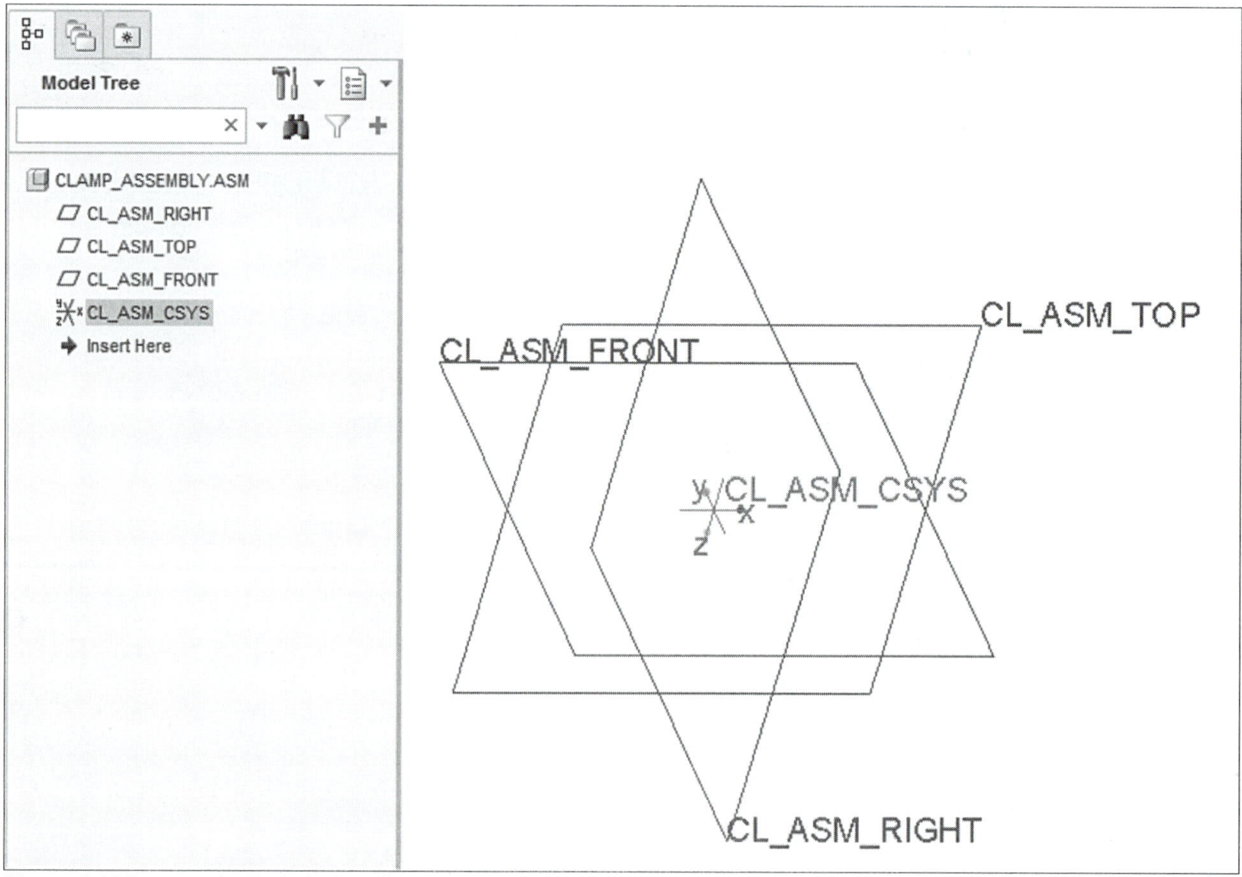

Figure 8.19 Rename the Default Coordinate System and Datum Planes

Creating Components in the Assembly Mode

The Clamp_Subassembly is now complete. The subassembly will be added to the assembly after the Clamp_Plate is created and assembled. Using [Create] and the Component Create dialog box (Fig. 8.20) you can create different types of components: parts, subassemblies, skeleton models, and bulk items. You cannot reroute components created in the Assembly mode.

The following methods allow component creation in the context of an assembly without requiring external dependencies on the assembly geometry:

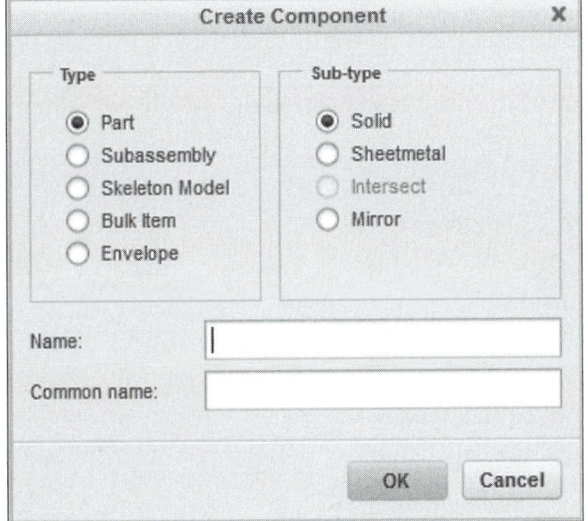

Figure 8.20 Component Create Dialog Box

- Create a component by copying another component or existing start part or start assembly
- Create a component with default datums
- Create an empty component
- Create the first feature of a new part; this initial feature is dependent on the assembly
- Create a part from an intersection of existing components
- Create a mirror copy of an existing part or subassembly
- Create Solid or Sheetmetal components
- Mirror Components

Main Assembly, Top-Down Design

The Clamp_Plate component is the first component of the main assembly. The Clamp_Plate is a new part. You will be modeling the plate *"inside"* the assembly using *top-down design*. The drawing in Figure 8.21 provides the dimensions necessary to model the Clamp_Plate.

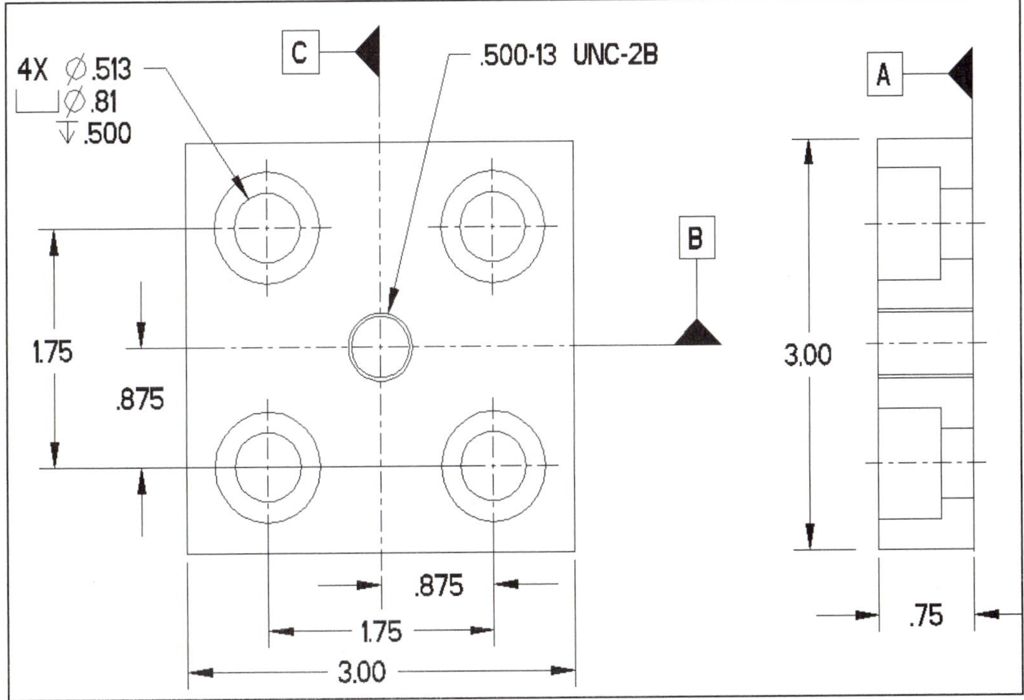

Figure 8.21 Clamp_Plate Detail Drawing

Click: **Model** tab > Create **Create a component in assembly mode** Component Create dialog box displays > Type Part > Sub-type Solid > Name **clamp_plate** [Fig. 8.22(a)] > **OK** Creation Options dialog box displays [Fig. 8.22(b)] > Locate Default Datums > Align Csys To Csys > **OK** > pick **CL_ASM_CSYS** from the Graphics Window, CLAMP_PLATE.PRT displays in the Model Tree, *the small symbol/icon* *indicates that this component is now active* > expand and highlight the items in the Model Tree for the CLAMP_PLATE.PRT [Fig. 8.23(a)] > **LMB** in the Graphics Window to deselect

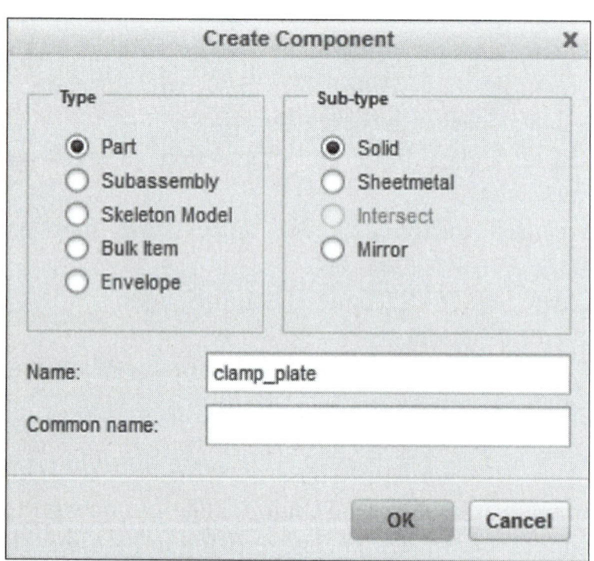

Figure 8.22(a) Component Create Dialog Box

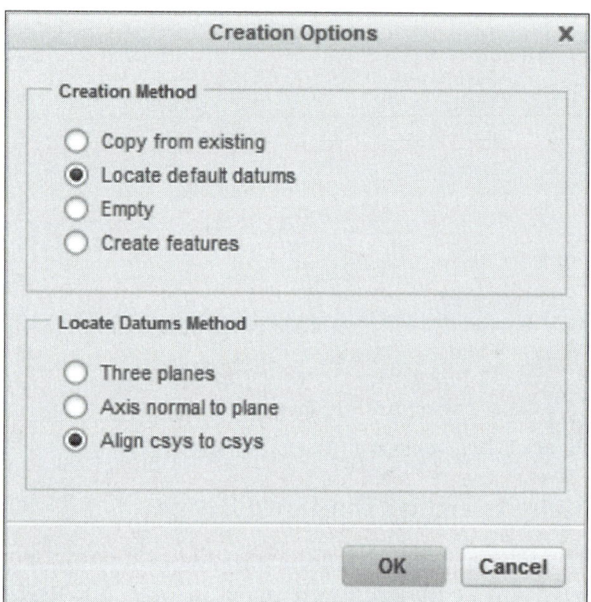

Figure 8.22(b) Creation Options Dialog Box

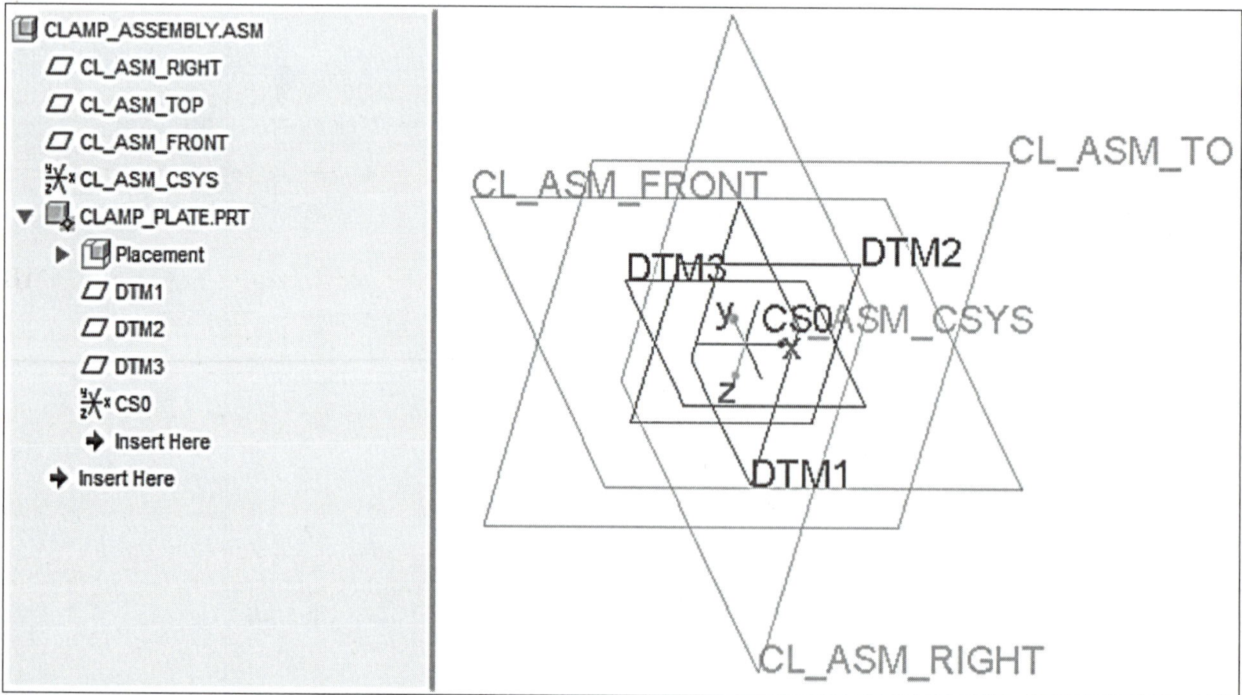

Figure 8.23(a) New Component Default Datums and Coordinate System Created and Assembled

Assembly Tools are now *unavailable* in the Model Ribbon. The current component is the Clamp_Plate. *You are now effectively in Part mode*, except that you can see the assembly features and components.

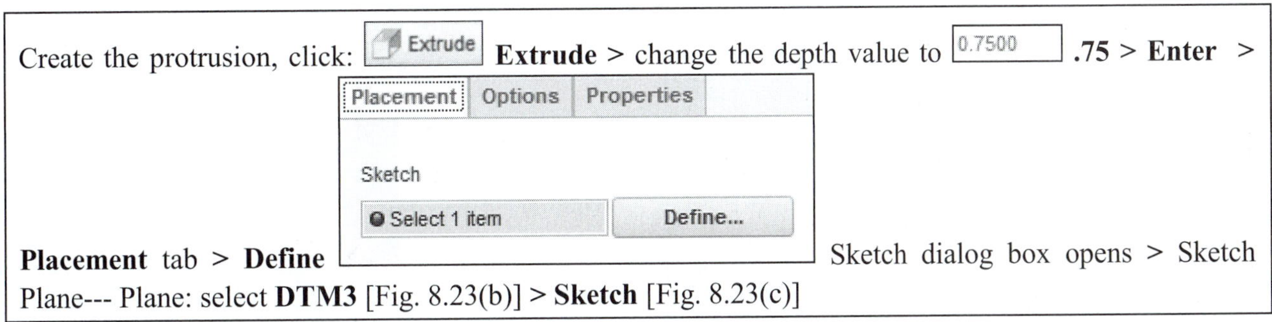

Create the protrusion, click: **Extrude** > change the depth value to **.75** > **Enter** > **Placement** tab > **Define** Sketch dialog box opens > Sketch Plane--- Plane: select **DTM3** [Fig. 8.23(b)] > **Sketch** [Fig. 8.23(c)]

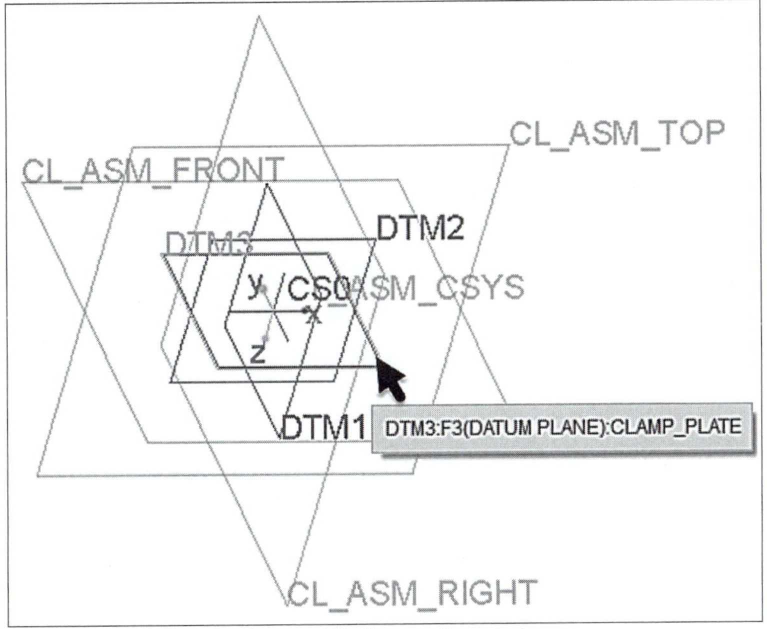

Figure 8.23(b) Sketch Plane and Reference Orientation

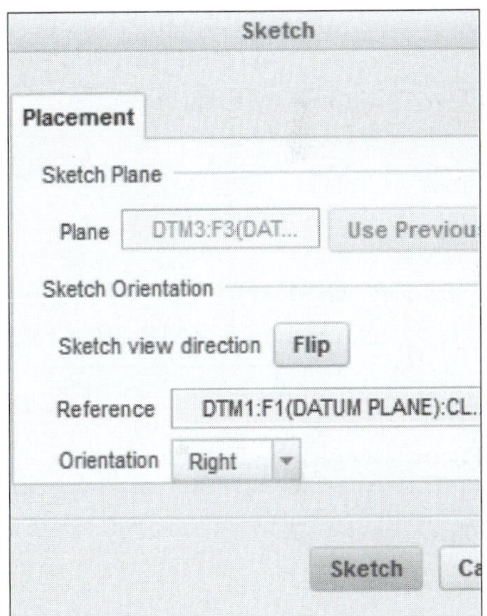

Figure 8.23(c) Sketch Dialog Box

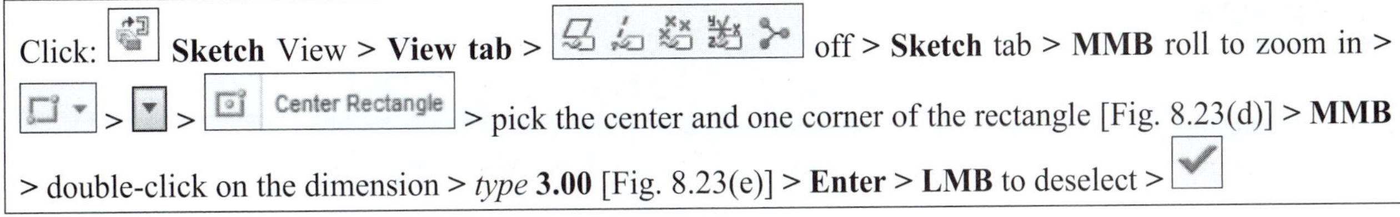

Click: **Sketch** View > **View tab** > off > **Sketch** tab > **MMB** roll to zoom in > > > > pick the center and one corner of the rectangle [Fig. 8.23(d)] > **MMB** > double-click on the dimension > *type* **3.00** [Fig. 8.23(e)] > **Enter** > **LMB** to deselect >

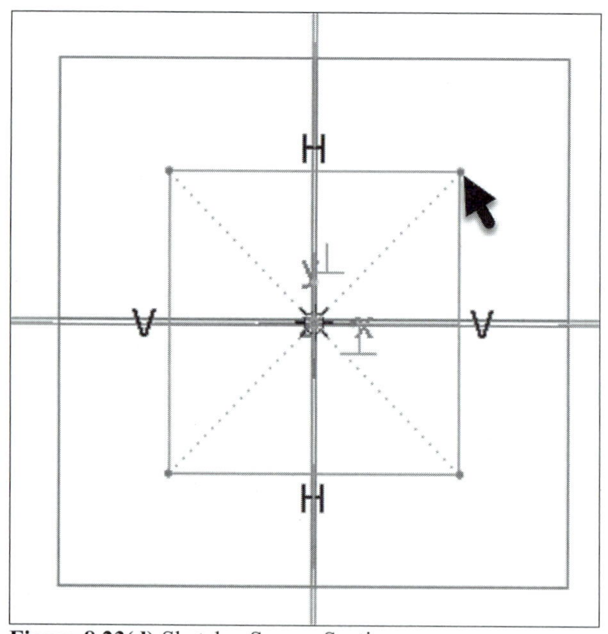

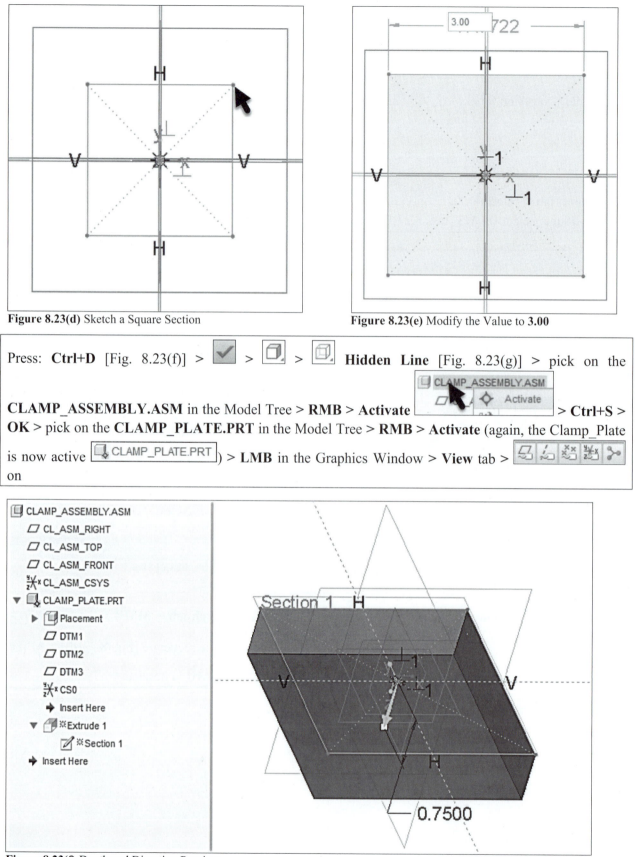

Figure 8.23(d) Sketch a Square Section

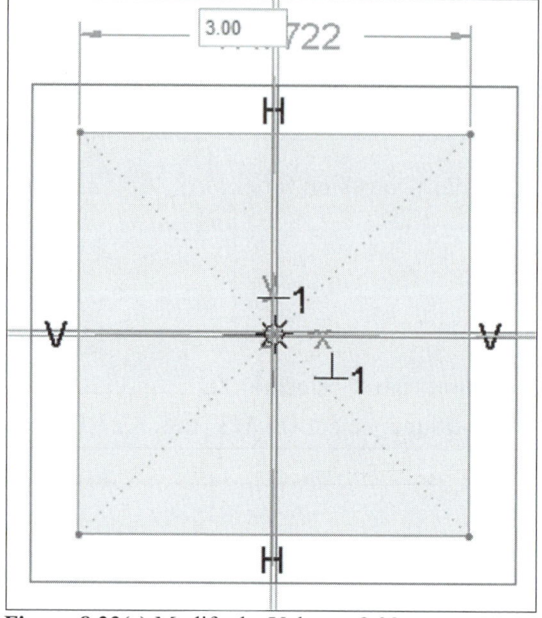

Figure 8.23(e) Modify the Value to **3.00**

Press: **Ctrl+D** [Fig. 8.23(f)] > ✓ > ▢ > ▢ **Hidden Line** [Fig. 8.23(g)] > pick on the **CLAMP_ASSEMBLY.ASM** in the Model Tree > **RMB** > **Activate** [CLAMP_ASSEMBLY.ASM / Activate] > **Ctrl+S** > **OK** > pick on the **CLAMP_PLATE.PRT** in the Model Tree > **RMB** > **Activate** (again, the Clamp_Plate is now active [CLAMP_PLATE.PRT]) > **LMB** in the Graphics Window > **View** tab > [icons] on

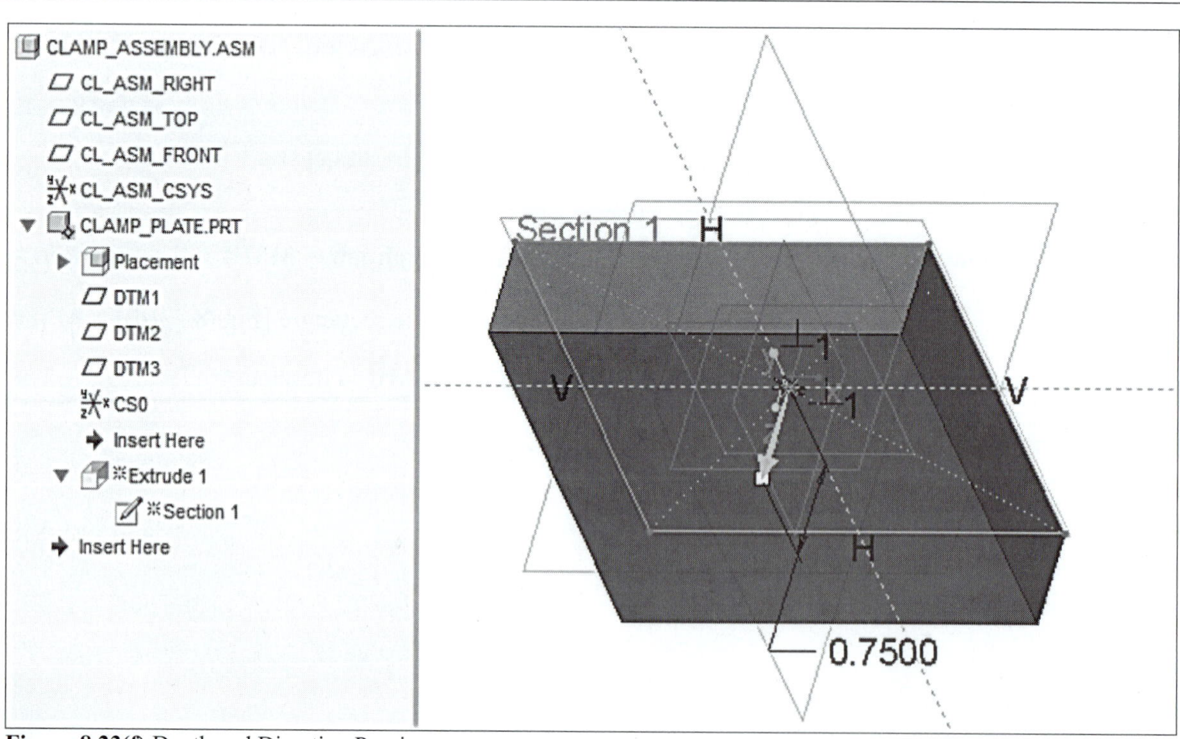

Figure 8.23(f) Depth and Direction Preview

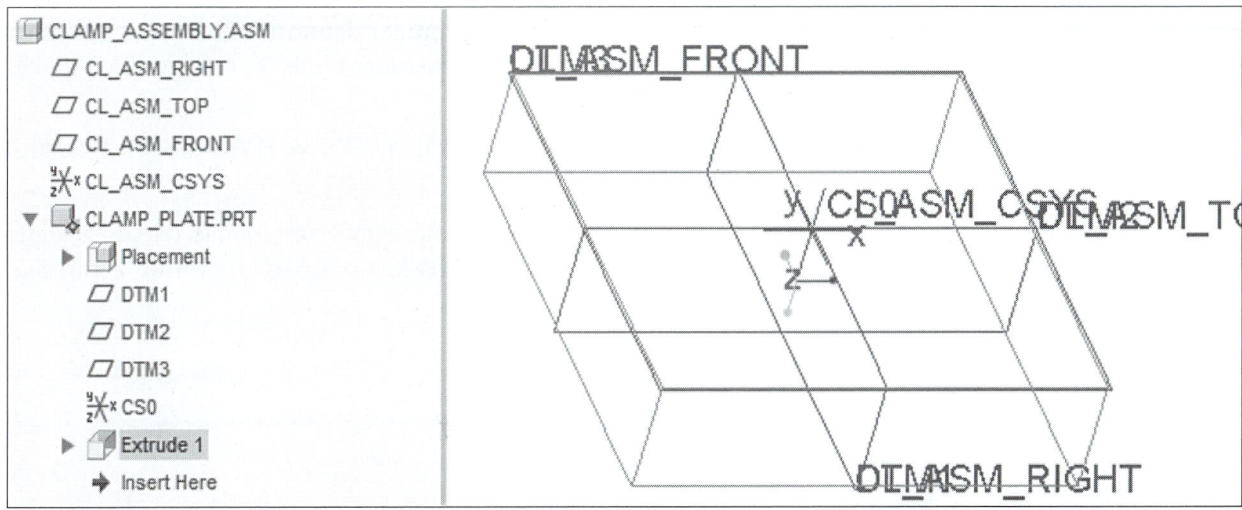

Figure 8.23(g) Completed Protrusion

Throughout the lesson, follow the exactly steps as given. Drag the handles to the appropriate references and let Creo Parametric select the references automatically ***even if you think the selections are incorrect***. Later, we will address these inconsistencies, but if you correct them on your own, steps described later will not work (or be needed).

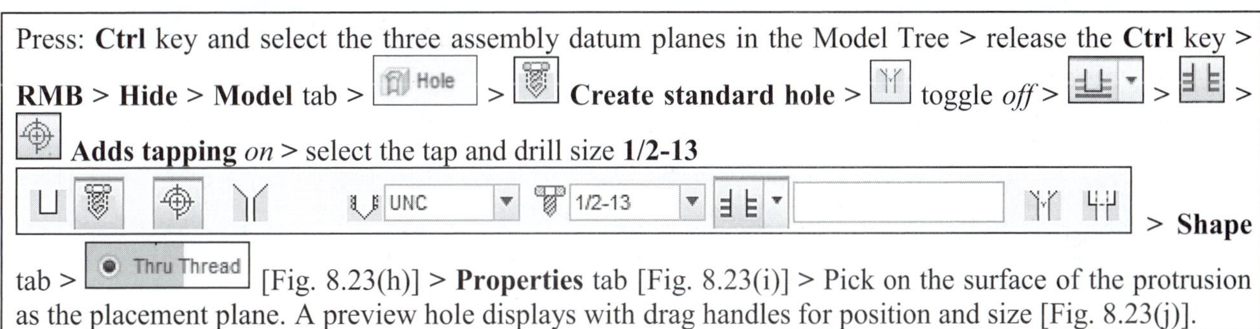

Press: **Ctrl** key and select the three assembly datum planes in the Model Tree > release the **Ctrl** key > **RMB** > **Hide** > **Model** tab > Hole > **Create standard hole** > toggle *off* > > > **Adds tapping** *on* > select the tap and drill size **1/2-13**

> **Shape** tab > Thru Thread [Fig. 8.23(h)] > **Properties** tab [Fig. 8.23(i)] > Pick on the surface of the protrusion as the placement plane. A preview hole displays with drag handles for position and size [Fig. 8.23(j)].

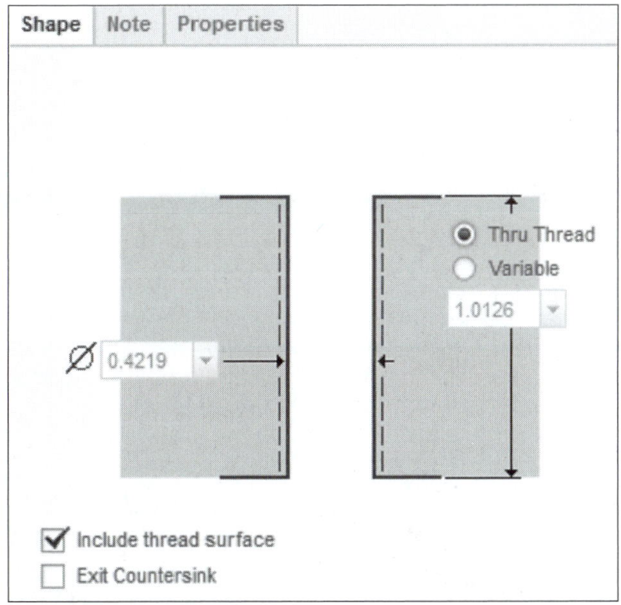

Figure 8.23(h) Hole Shape

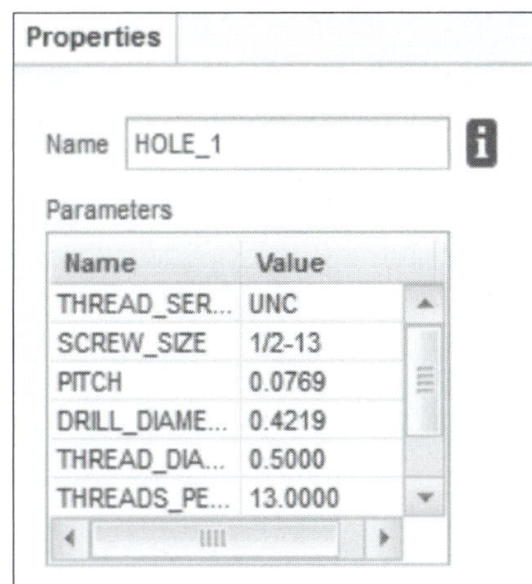

Figure 8.23(i) Hole Properties

To establish the Offset References, move one drag handle to the vertical datum (**DTM1**) and the other drag handle to the horizontal datum (**DTM2**) respectively > **Placement** tab > change *both* linear dimensions to **Align** DTM1:F1(DAT... Align DTM2:F2(DAT... Align [Fig. 8.23(k)] > ✓ > **LMB** in the Graphics Window > **Ctrl+D** > Hole 1 Note_0 [Fig. 8.23(l)] > 💾 > **LMB** to deselect > press the **Ctrl** key > select the three assembly datum planes in the Model Tree > release the **Ctrl** key > **RMB** > **Unhide** > **LMB** to deselect

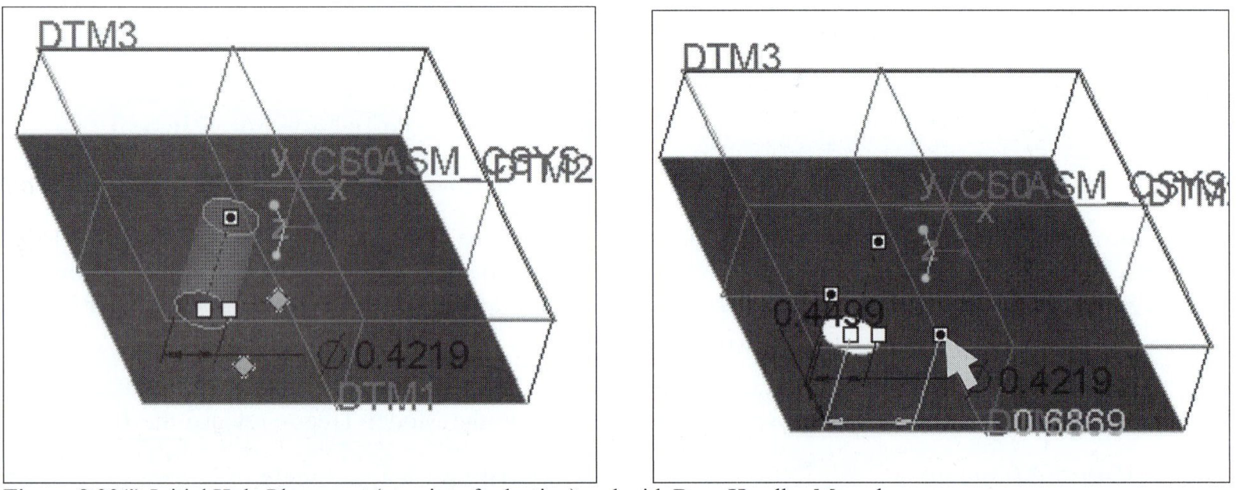

Figure 8.23(j) Initial Hole Placement (at point of selection) and with Drag Handles Moved

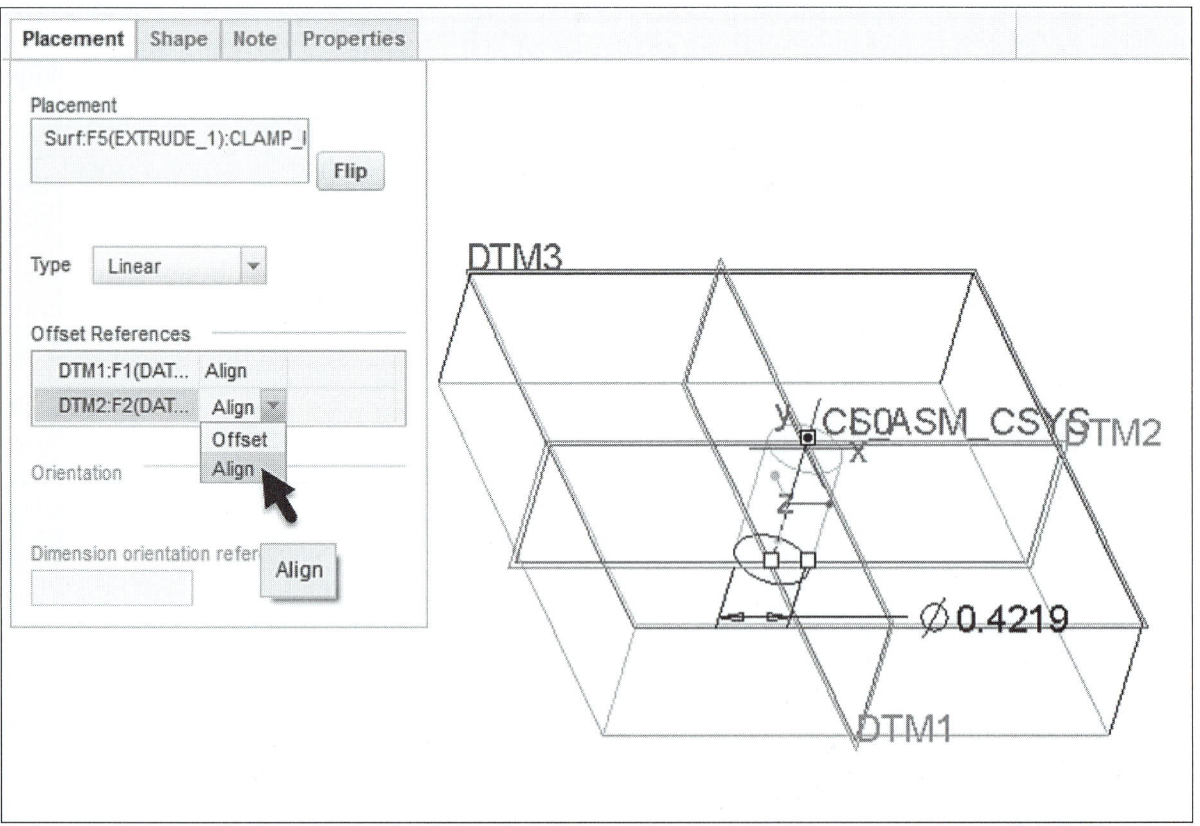

Figure 8.23(k) Hole Offset References Set to Align

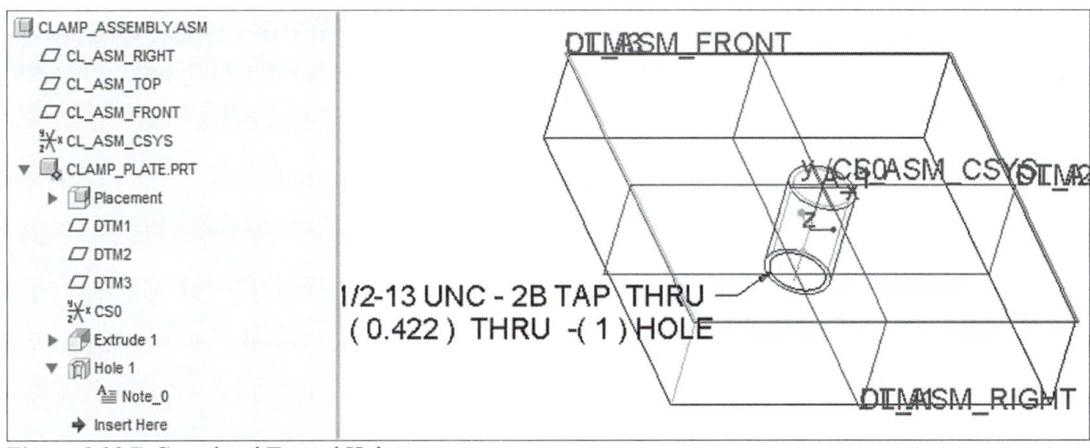

Figure 8.23(l) Completed Tapped Hole

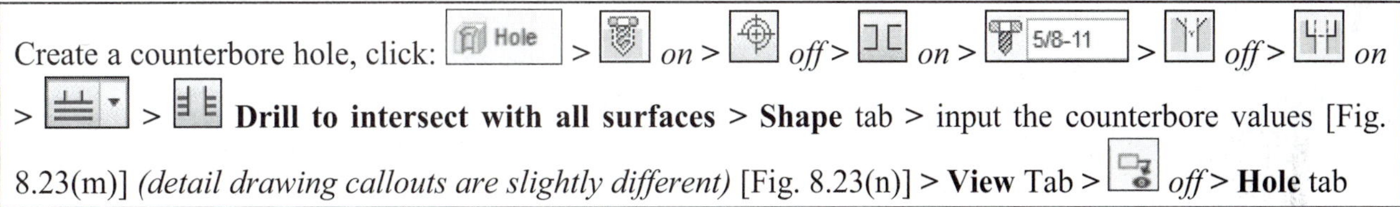

Create a counterbore hole, click: Hole > *on* > *off* > *on* > 5/8-11 > *off* > *on* > > **Drill to intersect with all surfaces** > **Shape** tab > input the counterbore values [Fig. 8.23(m)] *(detail drawing callouts are slightly different)* [Fig. 8.23(n)] > **View** Tab > *off* > **Hole** tab

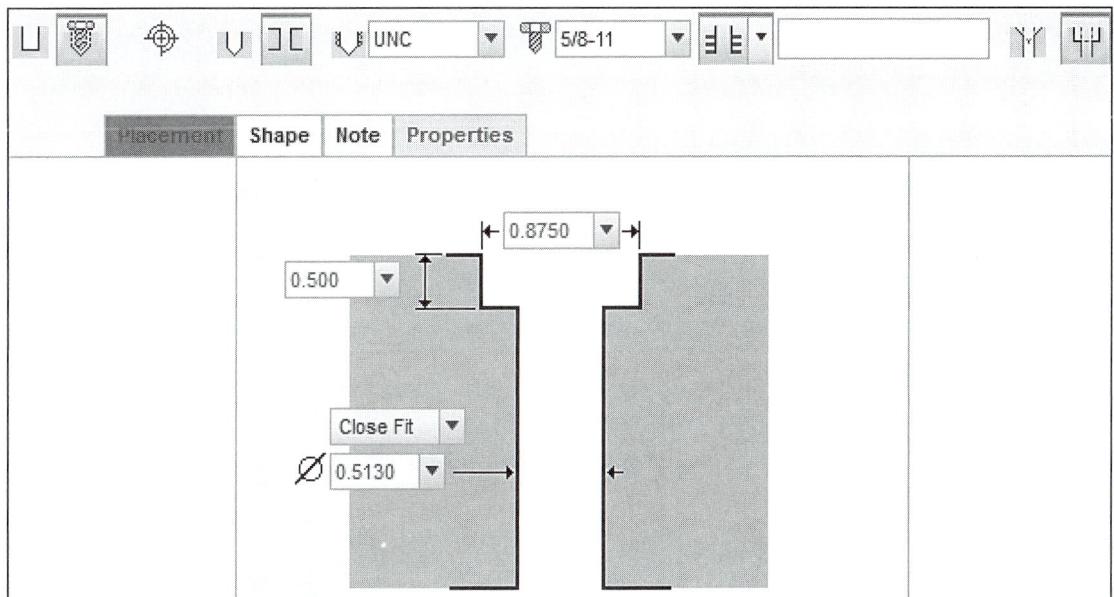

Figure 8.23(m) Counterbore Specifications

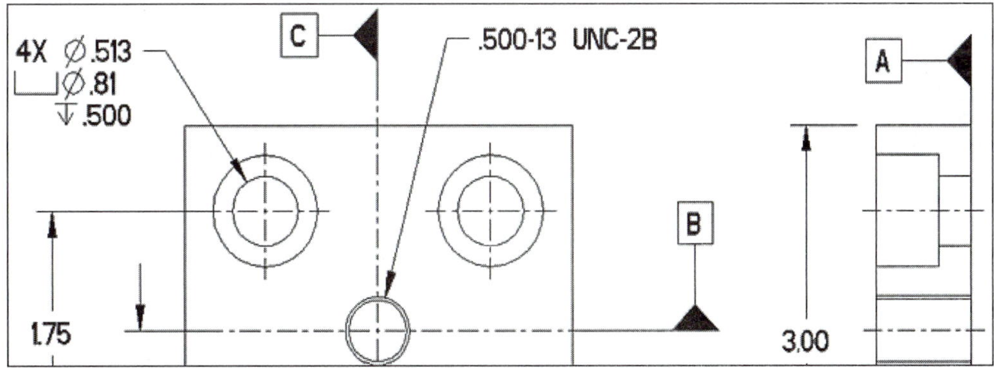

Figure 8.23(n) Hole Callout on the Detail Drawing

Pick on the front surface to place the counterbore hole [Fig. 8.23(o)] > press **RMB** > **Offset References Collector** > **Placement** tab > pick on **CL_ASM_RIGHT** from the model or Model Tree > hold down the **Ctrl** key and pick on **CL_ASM_TOP** from the model or the Model Tree [Fig 8.23(o)] > release the **Ctrl** key > modify both linear dimensions to **.875** [Fig. 8.23(p)] > ✓ > **LMB** in the Graphics Window

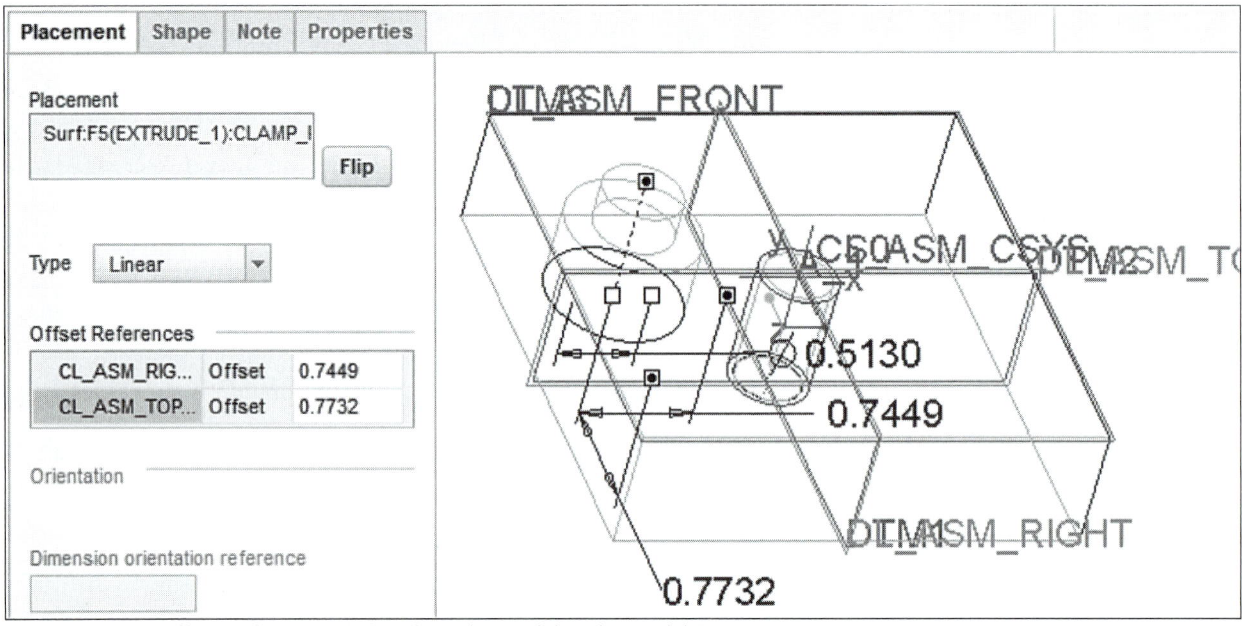

Figure 8.23(o) Offset References- Select from the Model or the Model Tree

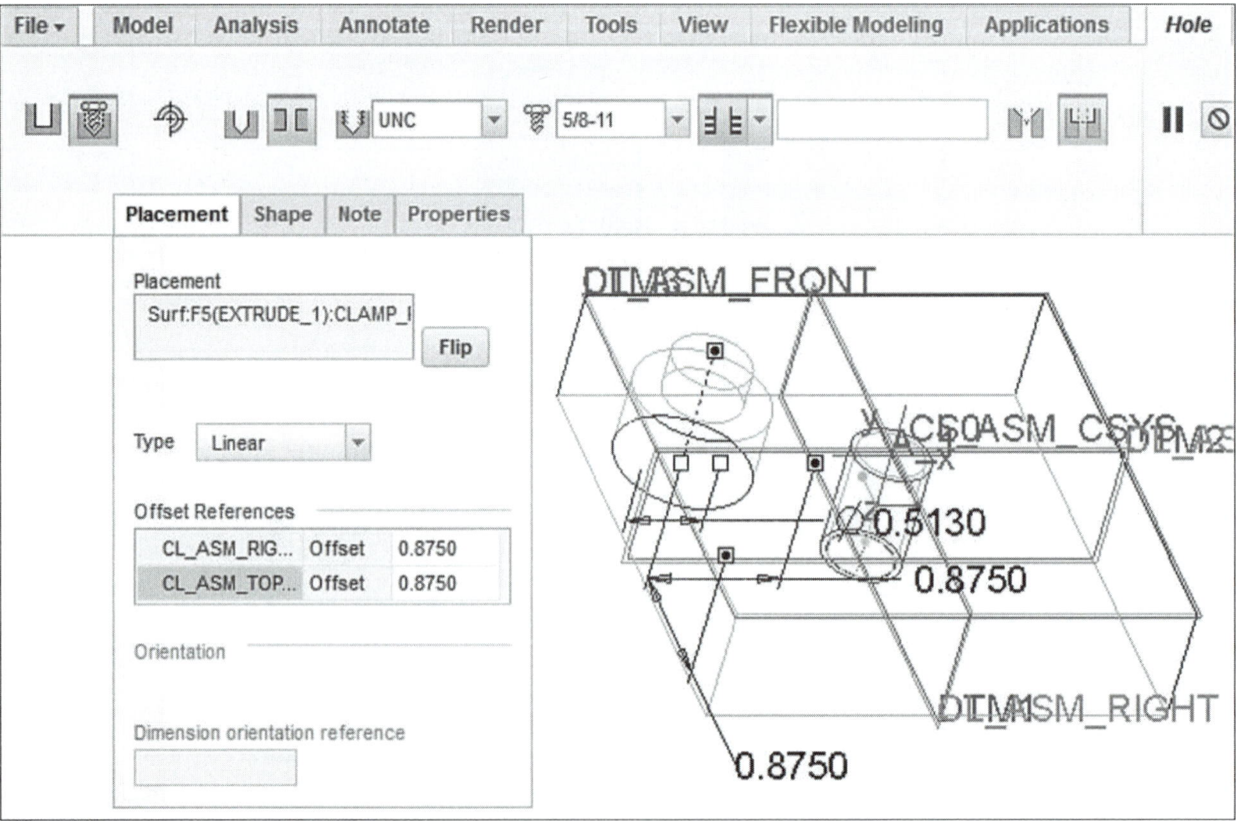

Figure 8.23(p) .875 Offset Dimensions *(you may need to use negative .875)*

The holes "seem" correct, but it is important to check for external references. Pick on **Hole 1** from the Model Tree > **RMB** > **Information** > **Reference Viewer** > **Reference Filters** tab > ◉ References > ☑ Components in path [Fig. 8.23(q)] [DTM1 and DTM2 are Internal (Local) References] > **Close** > pick on **Hole 2** from the Model Tree > **RMB** > **Information** > **Reference Viewer** > **Reference Filters** tab > expand [Fig. 8.23(r)] (CL_ASM_RIGHT and CL_ASM_TOP are External References) > **Close**

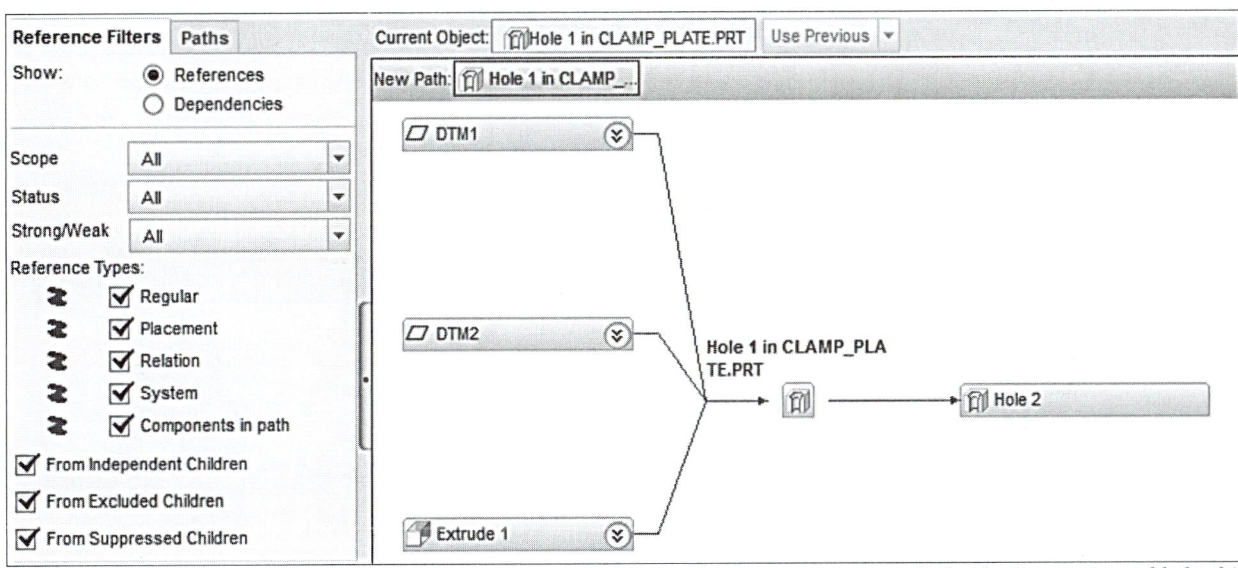

Figure 8.23(q) DTM1 and DTM2 as the HOLE 1 Internal References *(place pointer on top of a leader line to view added info)*

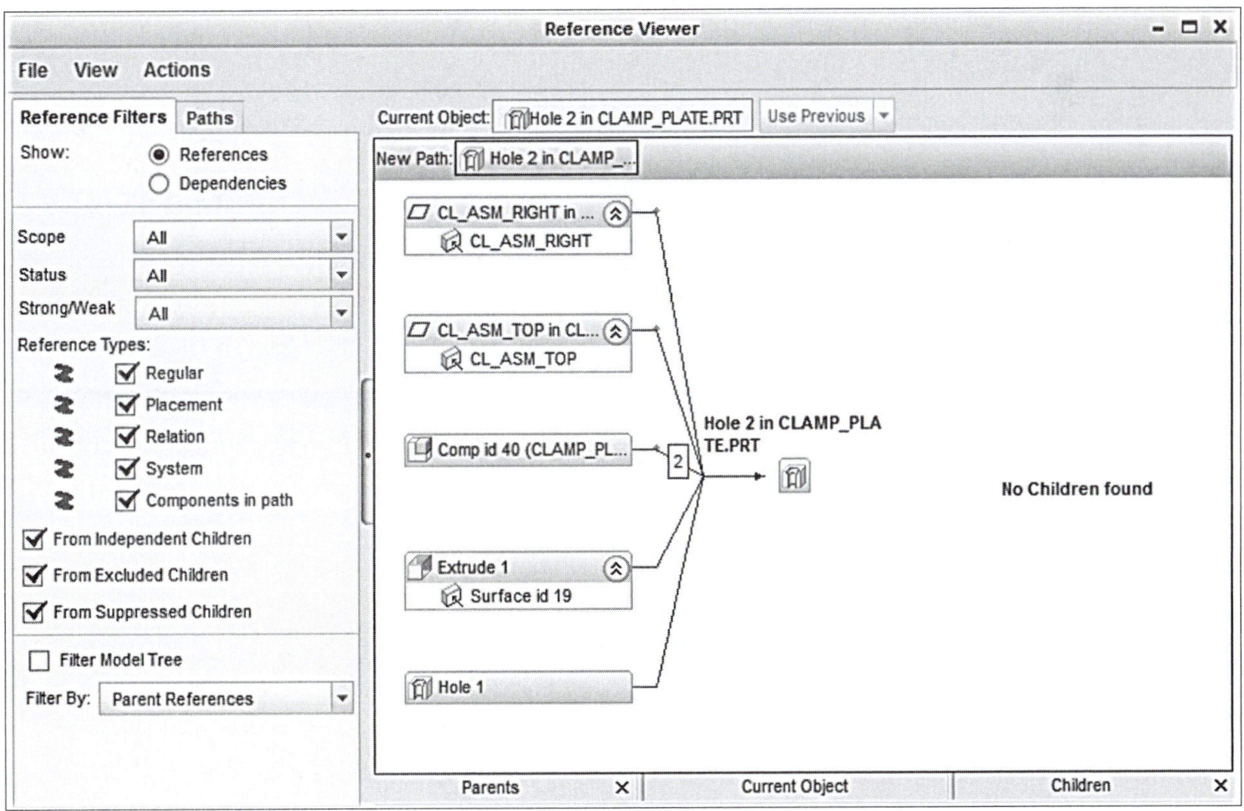

Figure 8.23(r) Expand the References to see the CL_ASM_RIGHT and CL_ASM_TOP External References

If you look back at Figure 8.23(o) you will see that the assembly datums were used to place the Hole 2 instead of DTM1 and DTM2. The following commands will show you how to edit the references. If your references are correct- do not change them. If both of your holes have external references, then use the following steps to reference the Hole 2 from DTM1 and DTM2.

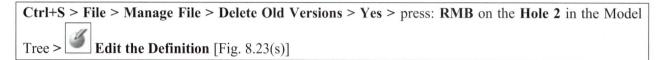

Ctrl+S > **File** > **Manage File** > **Delete Old Versions** > **Yes** > press: **RMB** on the **Hole 2** in the Model Tree > **Edit the Definition** [Fig. 8.23(s)]

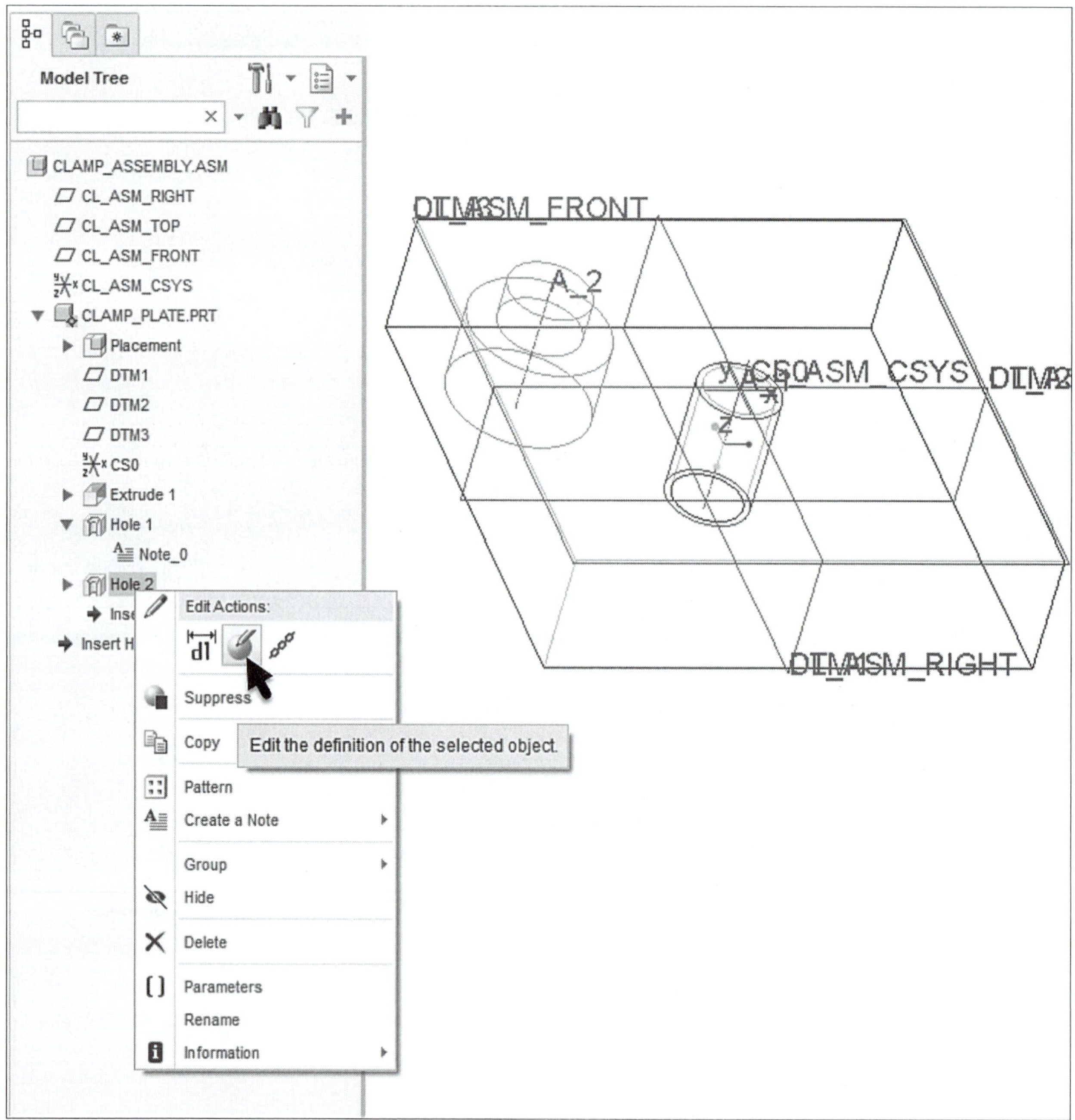

Figure 8.23(s) Edit Definition

Click: **Placement** tab > click in the Offset References field > **RMB** > **Remove All** [Fig. 8.23(t)] > place the cursor over datum CL_ASM_TOP to highlight > click **RMB** *[do not hold down the RMB, unless you want to "Pick From List" (accessing the query select list) or choose one of the other options]* and DTM2 highlights [Fig. 8.23(u)] > pick on **DTM2**

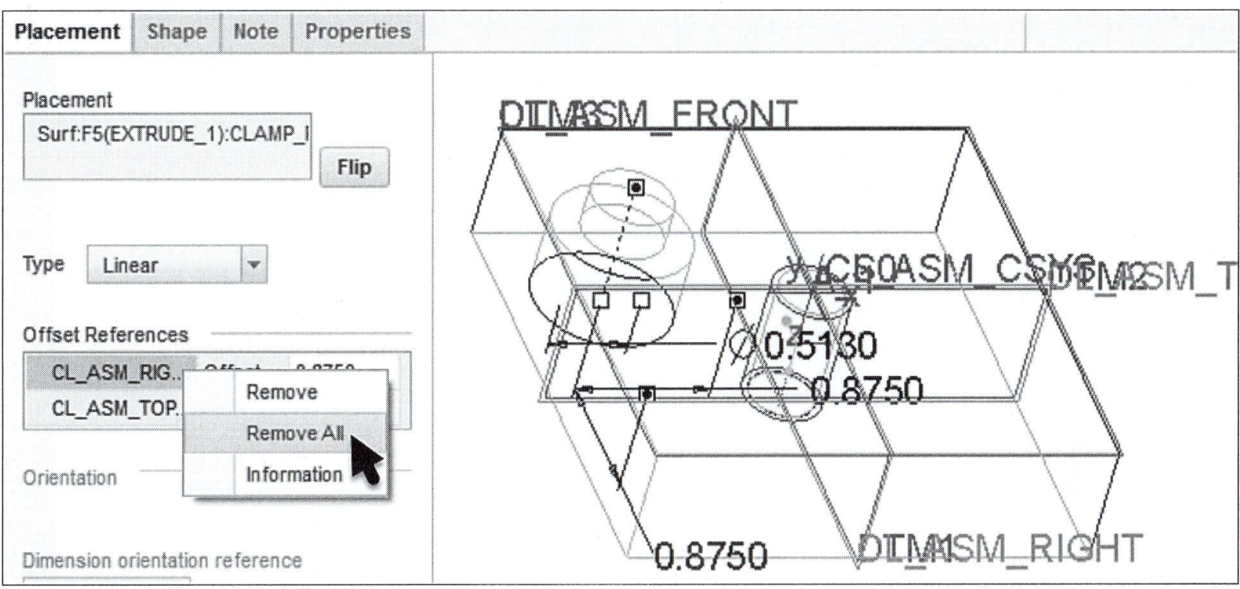

Figure 8.23(t) Remove All Offset References

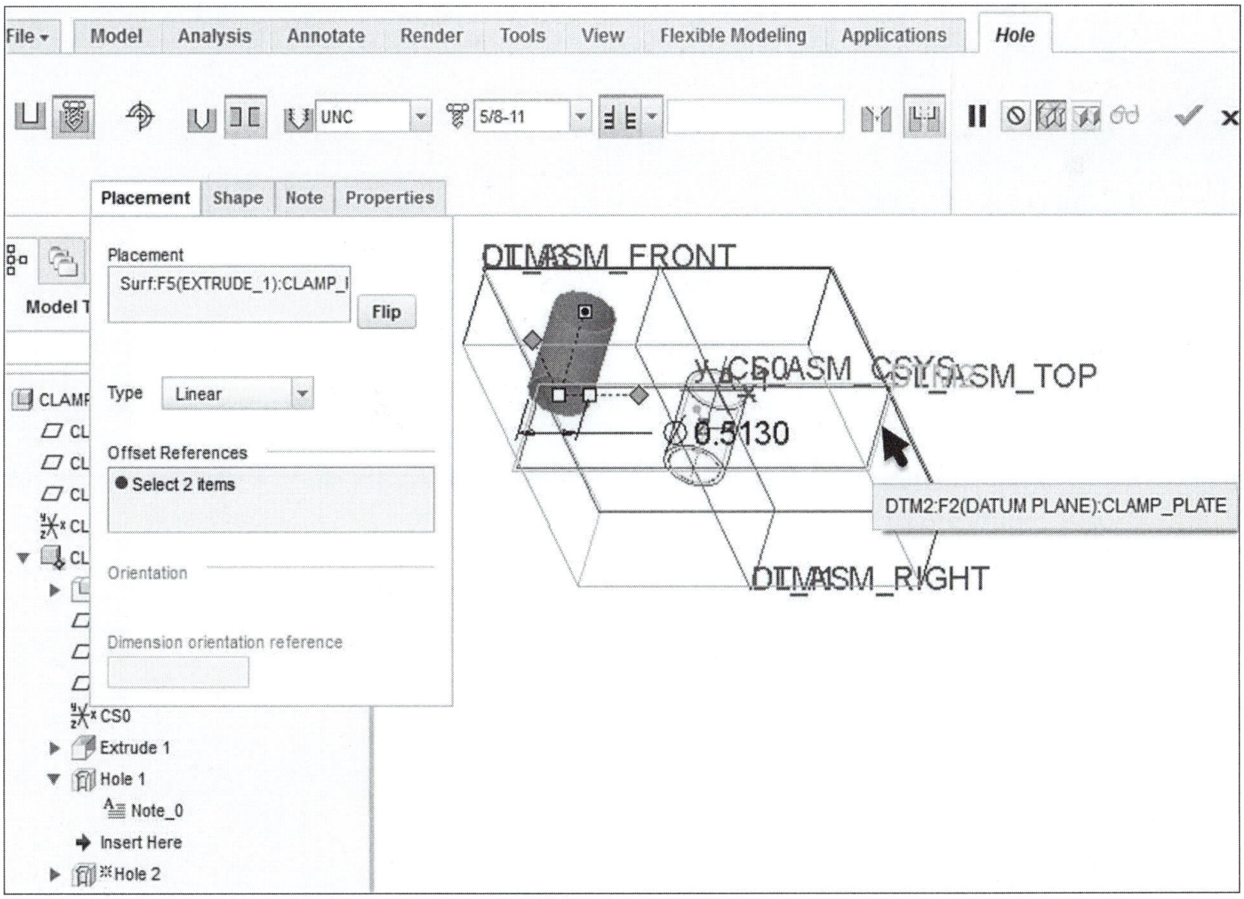

Figure 8.23(u) Place Pointer on CL_ASM_TOP > RMB > select DTM2

Press and hold the **Ctrl** key and place the cursor over CL_ASM_RIGHT (highlights) > click **RMB** (DTM1 highlights) [Fig. 8.23(v)] > pick on **DTM1** [Fig. 8.23(w)] *(both datums could also have been selected from the Model Tree)* > release the **Ctrl** key > **.875** is still the Offset distance for both References > ✓ to complete the command > **Ctrl+D** > **Ctrl+S** > in the Graphics Window, **LMB** to deselect

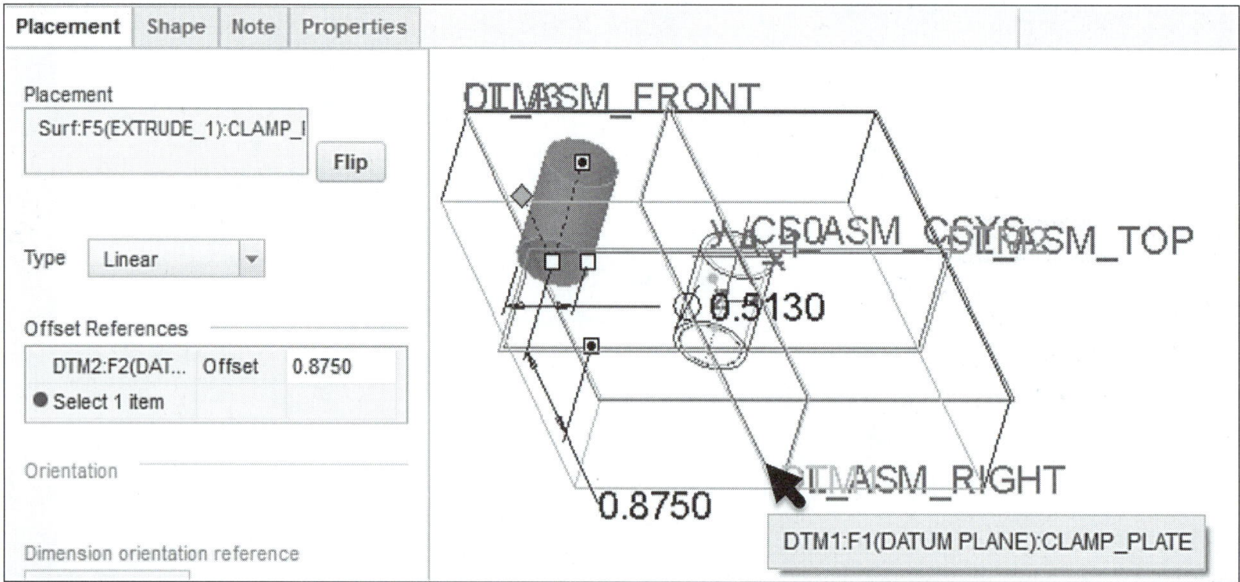

Figure 8.23(v) Place Pointer on CL_ASM_RIGHT > RMB > select DTM1

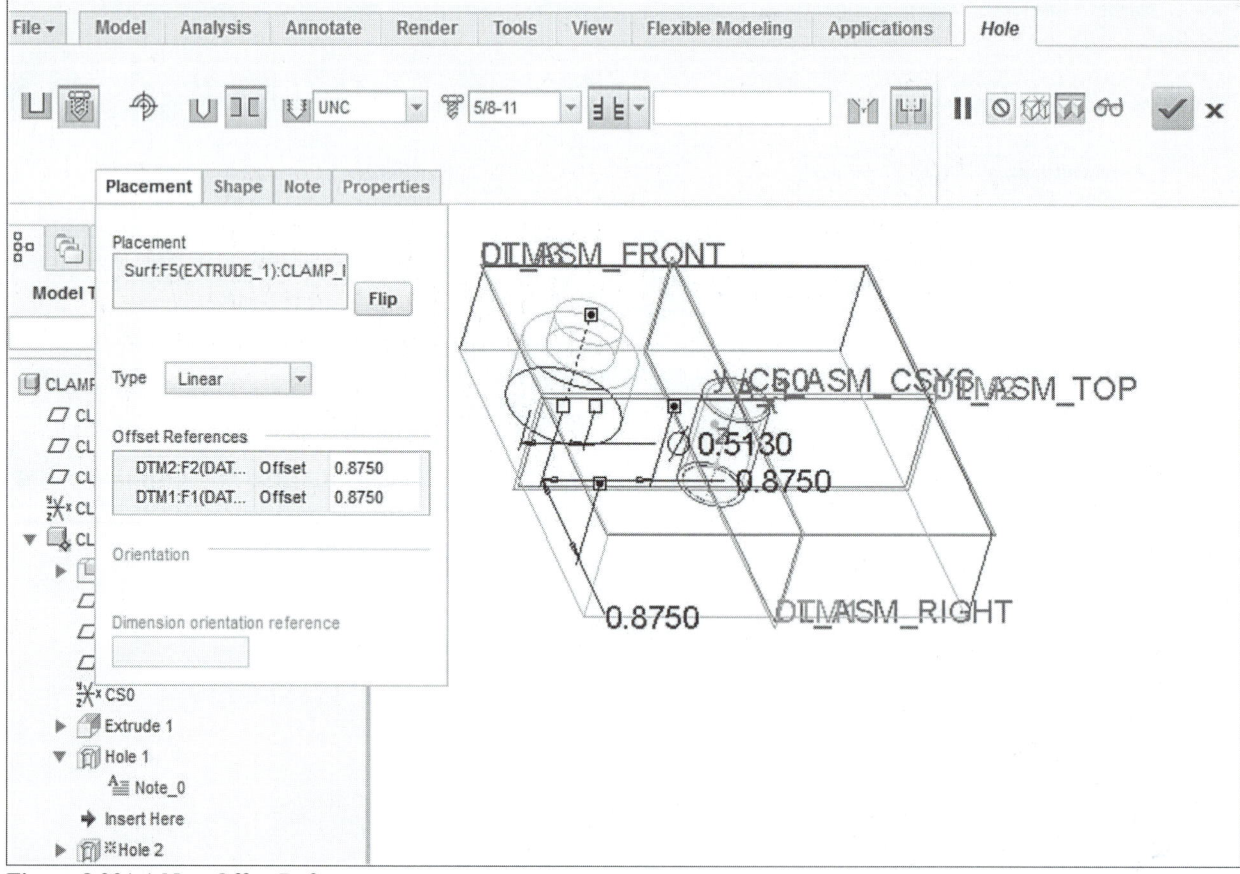

Figure 8.23(w) New Offset References

Pick on **Hole 2** from the Model Tree or model > **Tools** tab > **Reference Viewer** > **Reference Filters** tab > **References** *on* > [expand icon] expand the references [Fig. 8.23(x)] > **Close** > **LMB** to deselect > **Model** tab [*If necessary*, repeat the previous process to select the correct references (DTM2 and DTM1) for the tapped hole in the center of the part [Fig. 8.23(y)]. The hole will be aligned with both references.]

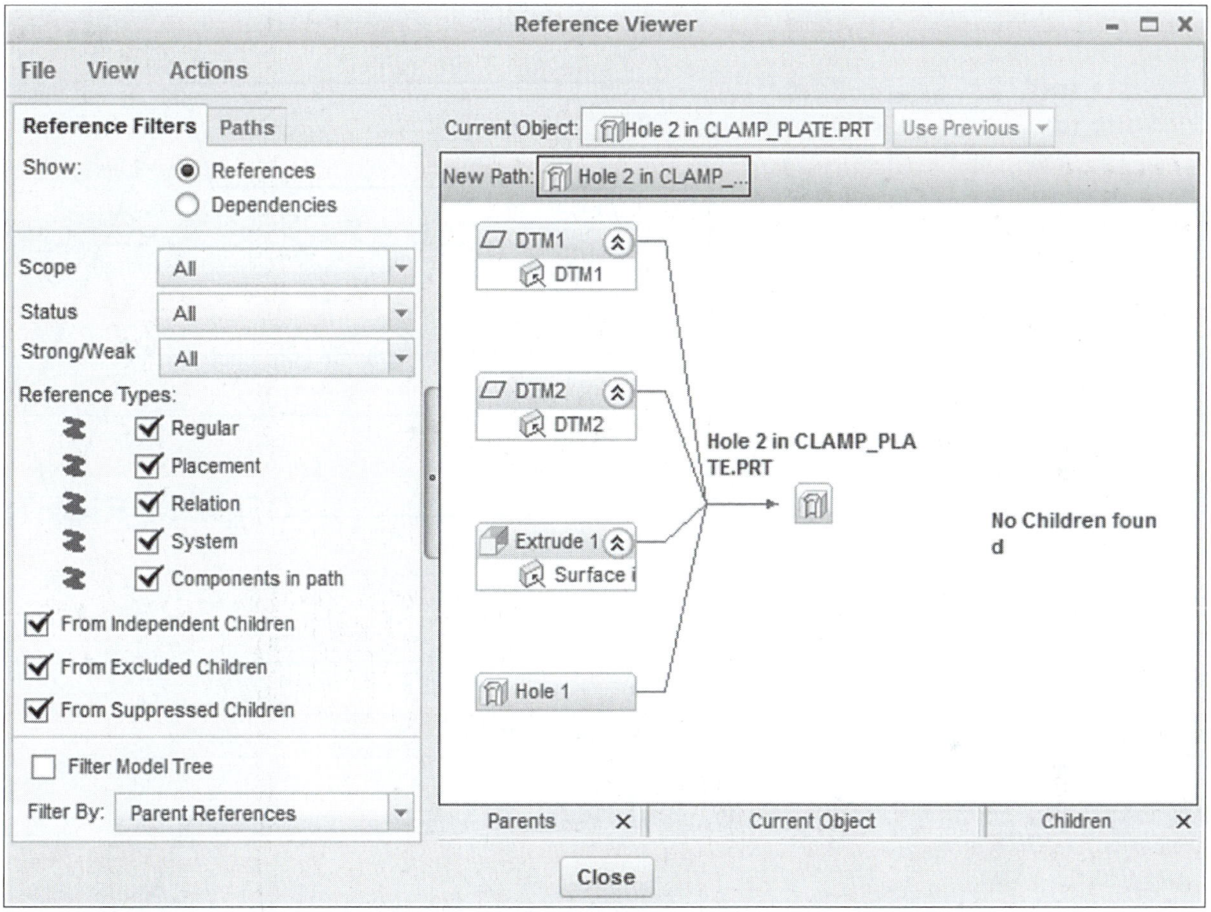

Figure 8.23(x) Reference Viewer (Holes are now Offset from the Datum Planes of the Clamp_Plate)

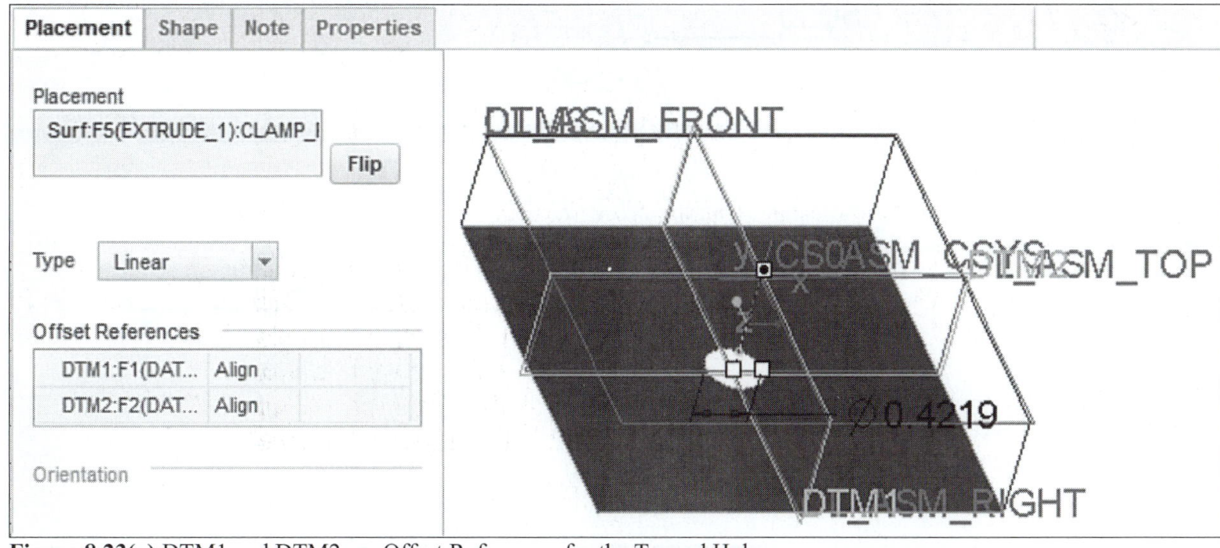

Figure 8.23(y) DTM1 and DTM2 are Offset References for the Tapped Hole

Ctrl+D > *(if the CLAMP_PLATE.PRT is not active, click on* ***CLAMP_PLATE.PRT*** *in the Model Tree >* ***RMB*** *>* ***Activate)*** > double-click on the counterbore hole > double-click on the counterbore diameter dimension > *type* **.8125** > **Enter** > move the mouse > **LMB** > move the mouse > **LMB** > **Ctrl+G** to Regenerate > pick on the counterbore hole in the Model Tree > **RMB** > **Pattern** [Fig. 8.24(a)] > **Dimensions** tab > pick on the *horizontal* **.875** dimension > *type* **–1.75** [Fig. 8.24(b)] > **Enter**

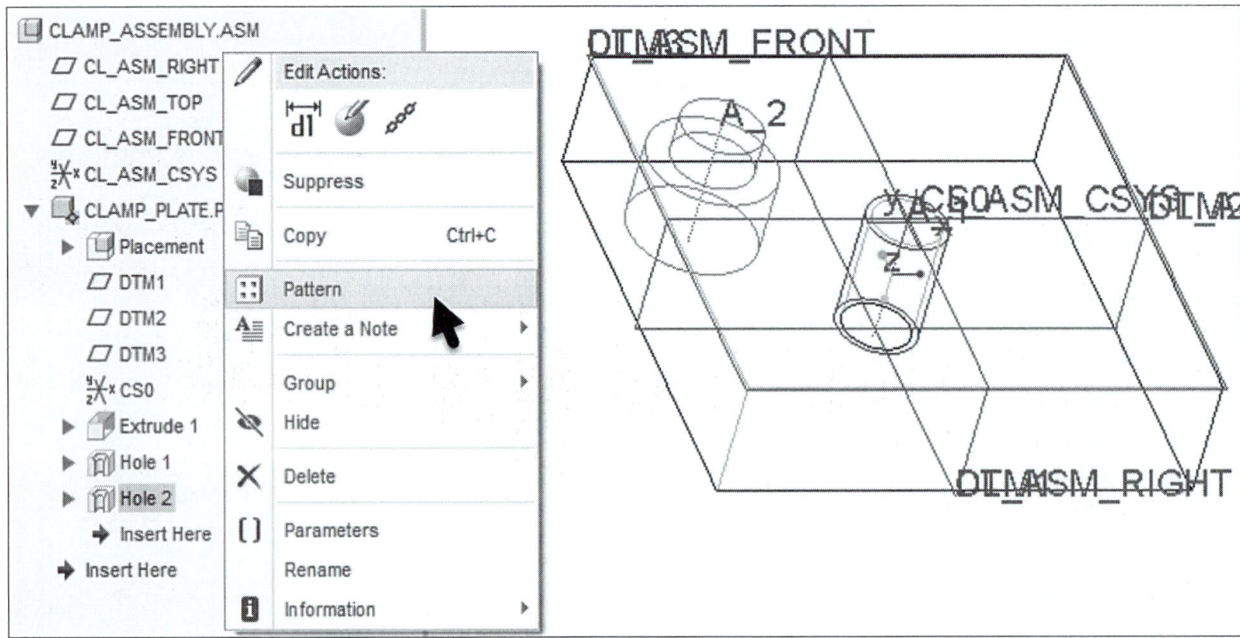

Figure 8.24(a) Pattern the Counterbore Hole

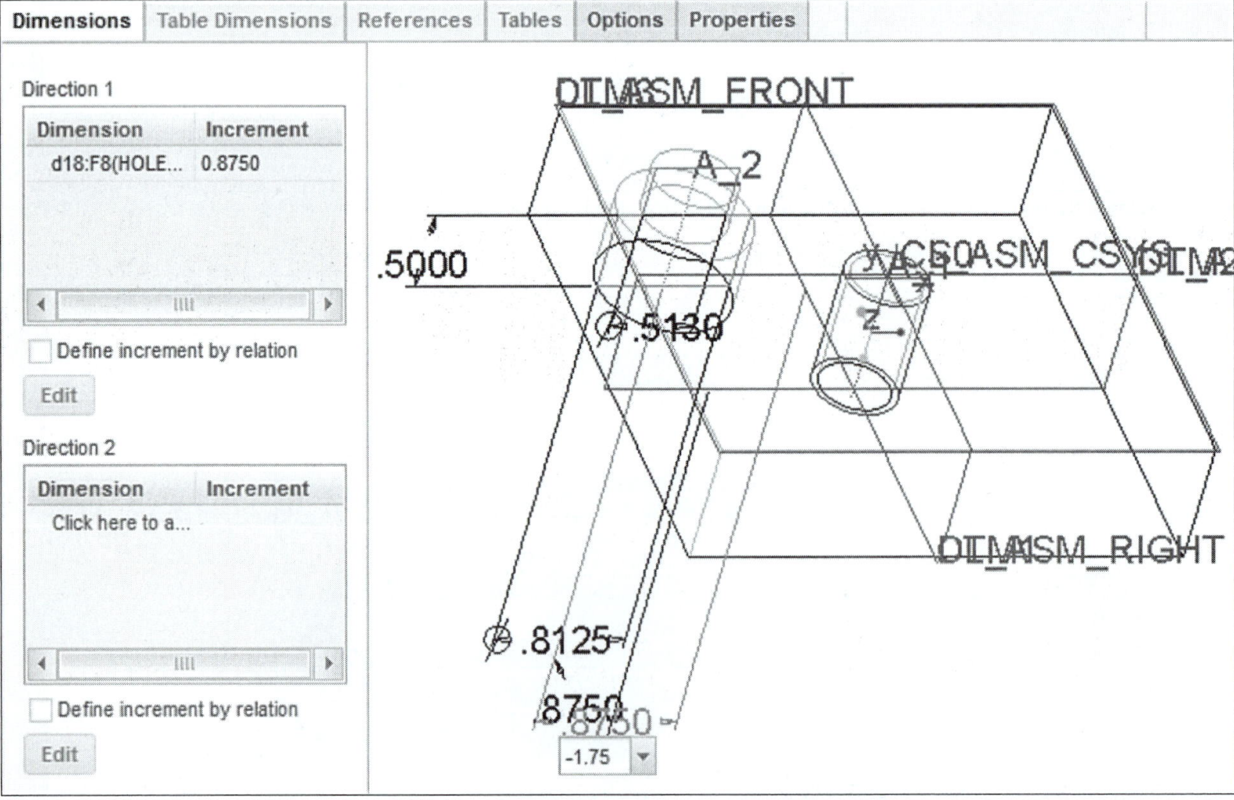

Figure 8.24(b) Direction 1 Dimension **–1.75**, Two Items

Press: **RMB > Direction 2 Dimensions** *(or click in Direction 2 collector box, pick* ***Select Items****)* > pick on the vertical **.875** dimension [Fig. 8.24(c)] > A "combo box" opens in the Graphics Window, with the dimension increment initially equal to the dimension value. Highlight the value (if needed) and *type* **–1.75** [Fig. 8.24(d)] > **Enter** > > **Regenerate** > **Save > LMB** to deselect

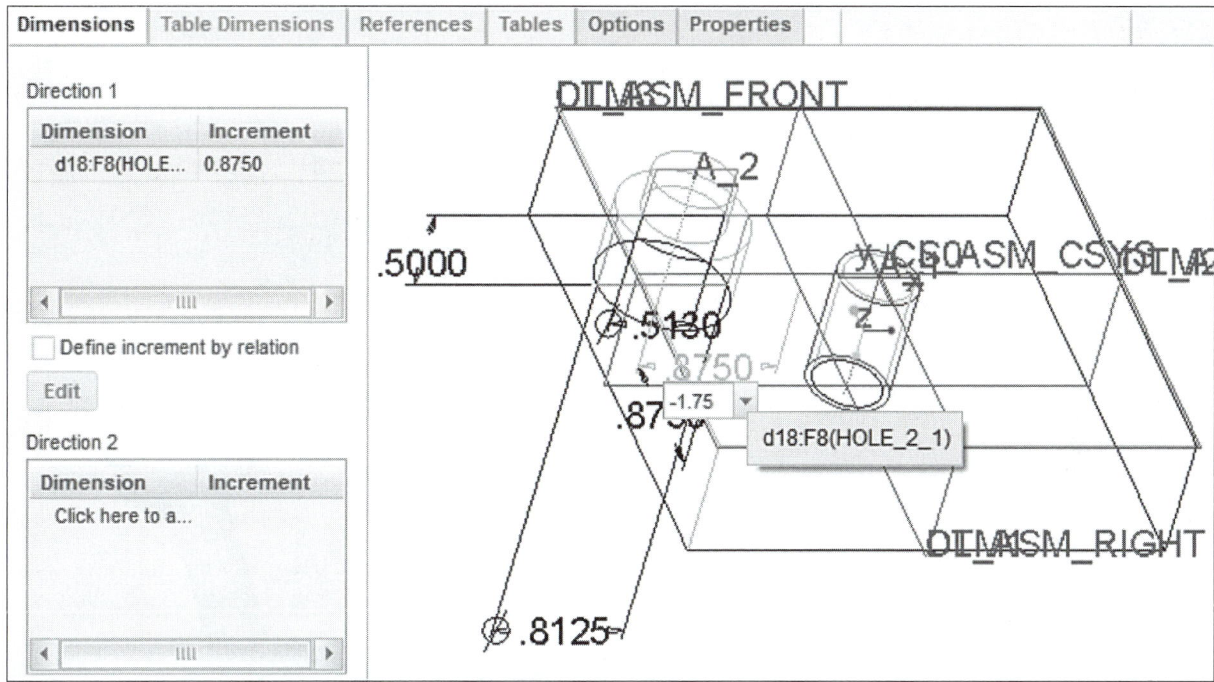

Figure 8.24(c) Direction 2 Dimension **–1.75**, Two Items

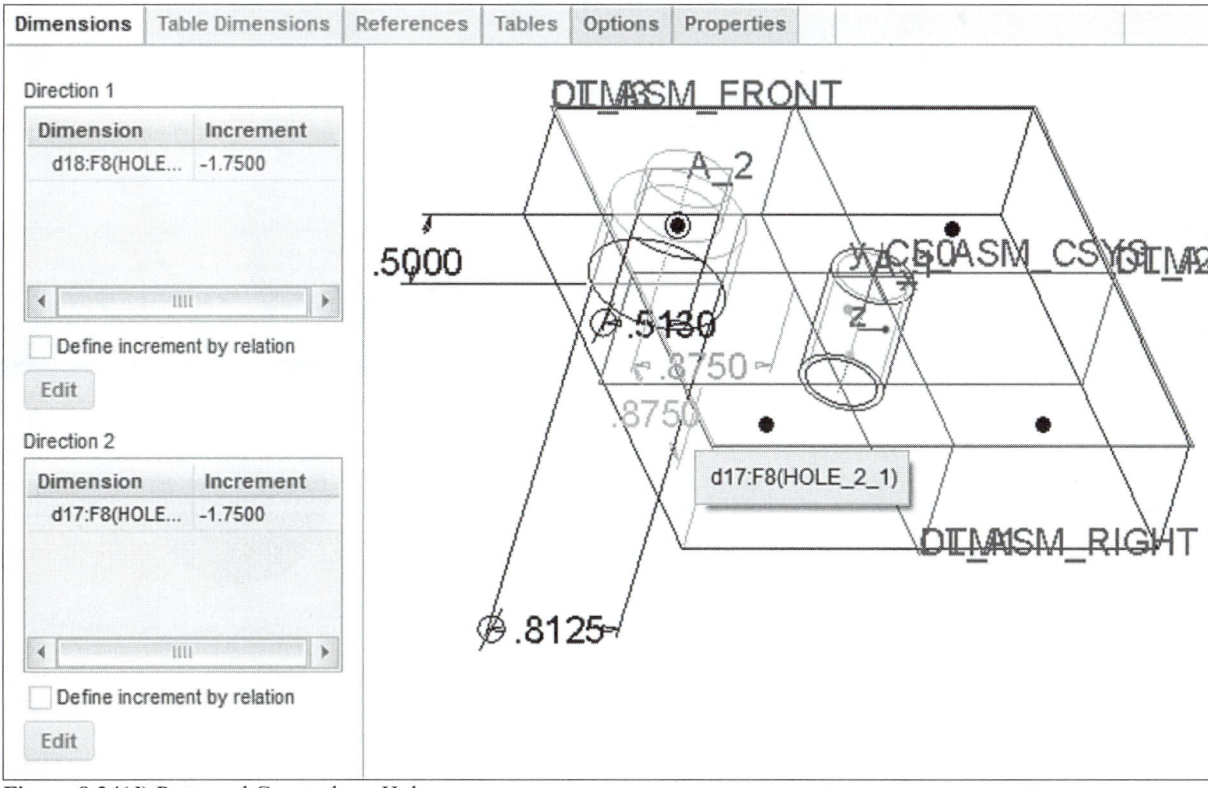

Figure 8.24(d) Patterned Counterbore Holes

Click: **CLAMP_ASSEMBLY.ASM** from the Model Tree > **RMB** > **Activate** > **View** tab > > > **Model** tab > Assemble > select **clamp_subassembly.asm** > **Preview** *on* [Fig. 8.25(a)] > **Open** > **File** > **Options** > **Model Display** > **Trimetric** > **Isometric** > **OK** > **No** > **Ctrl+D**

*(Follow the steps **exactly** as provided)*

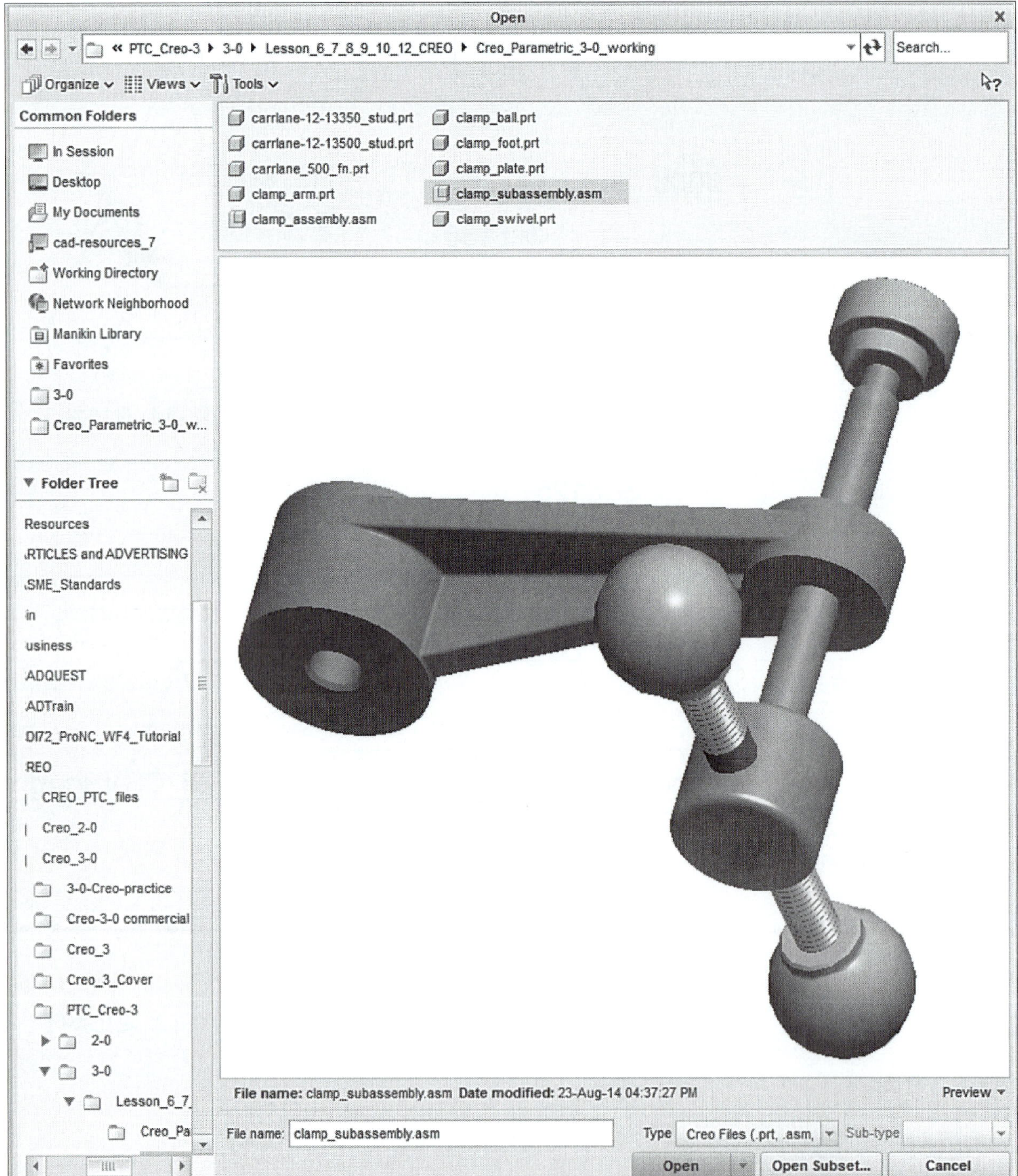

Figure 8.25(a) Preview Clamp_Subassembly

Click: Hide 3D Dragger > Automatic > Coincident [Fig. 8.25(b)] > **MMB** spin the model > pick the bottom surface of the Clamp_Arm_Design (of the Clamp_Subassembly) [Fig. 8.25(c)]

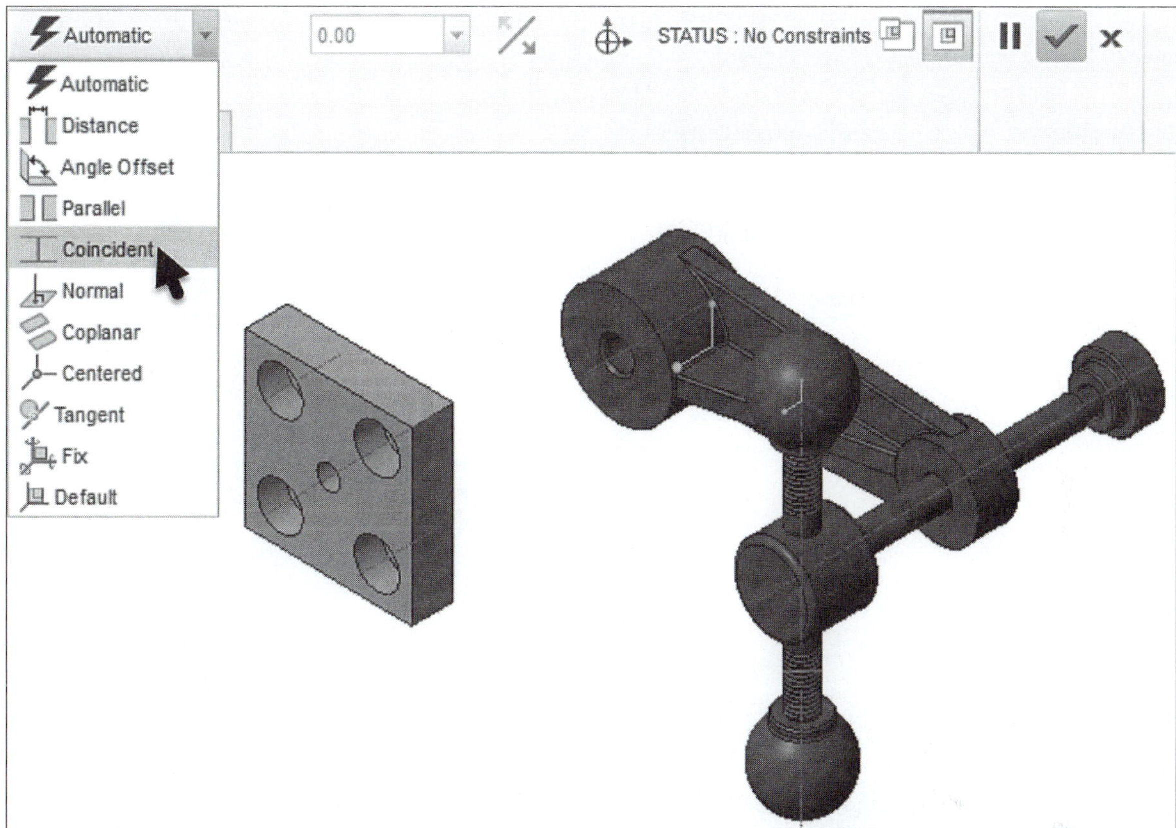

Figure 8.25(b) Assembling the Clamp_Subassembly

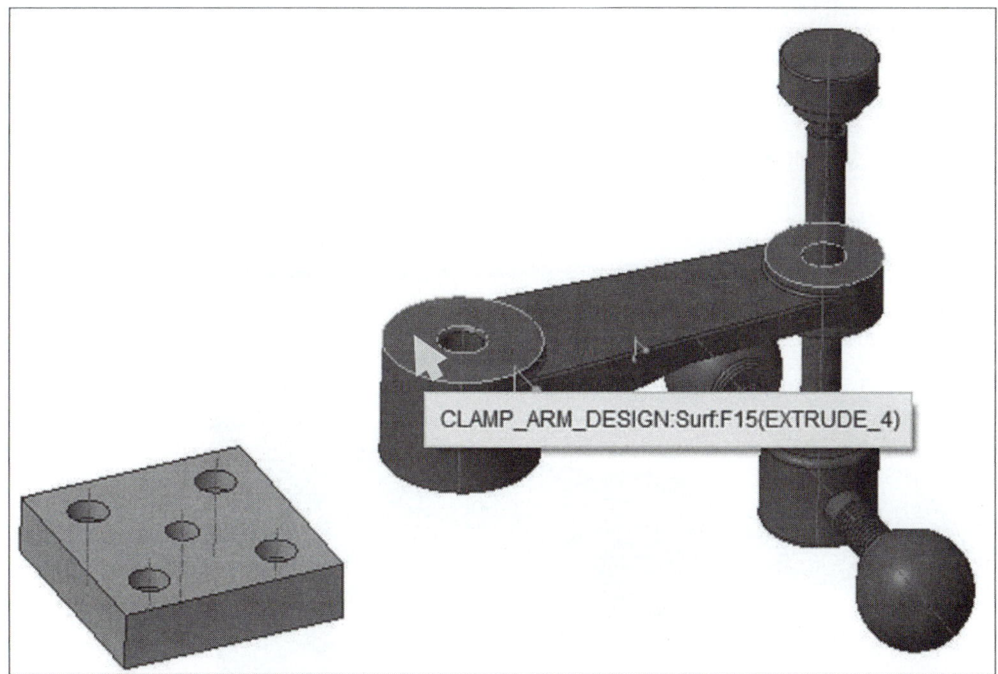

Figure 8.25(c) Bottom Surface of the Clamp_Arm_Design is selected as the Component Reference

Click: **Placement** tab > pick the non-counterbore surface of the Clamp_Plate [Fig. 8.25(d)] > **Flip** [Fig. 8.25(e)]

Make sure you follow the selections exactly- even if they "seem" wrong.

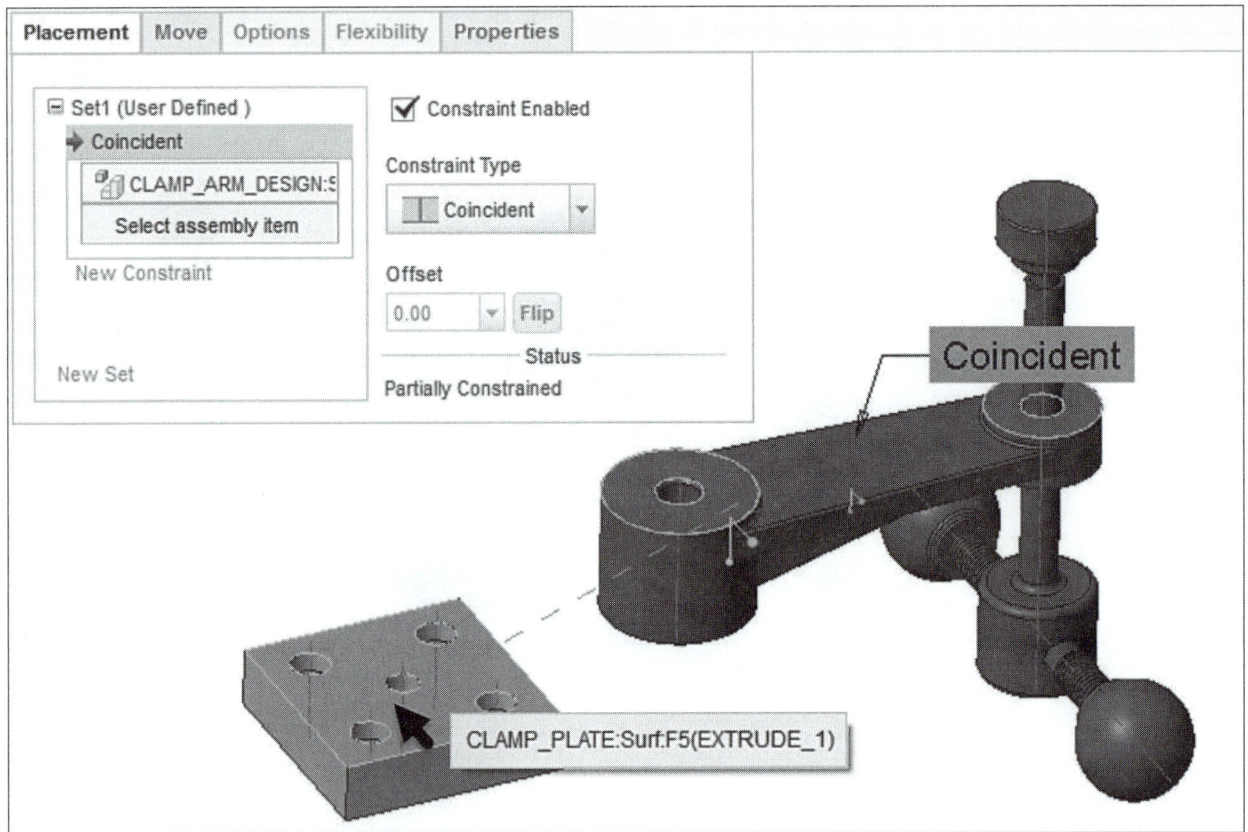

Figure 8.25(d) Non-counterbore Surface of the Clamp_Plate is highlighted as the Assembly Reference

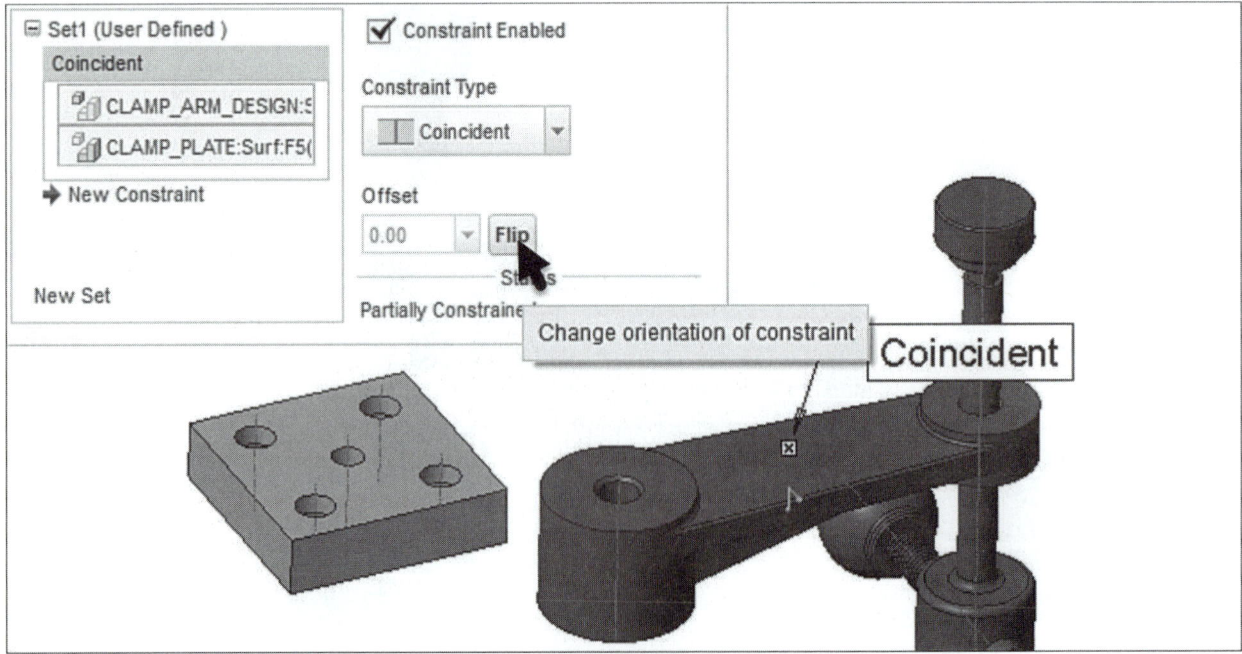

Figure 8.25(e) Coincident

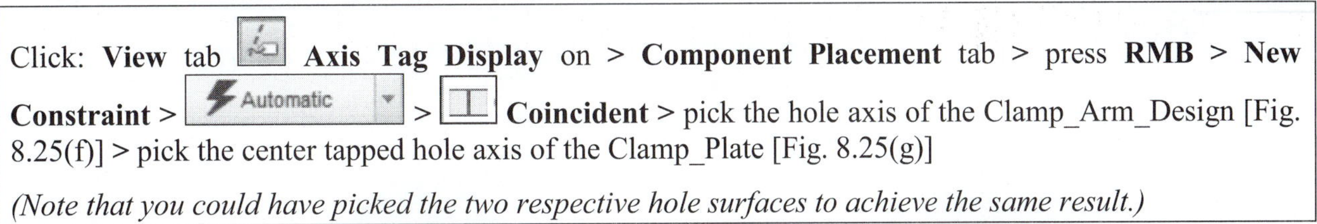

Click: **View** tab **Axis Tag Display** on > **Component Placement** tab > press **RMB** > **New Constraint** > Automatic > **Coincident** > pick the hole axis of the Clamp_Arm_Design [Fig. 8.25(f)] > pick the center tapped hole axis of the Clamp_Plate [Fig. 8.25(g)]

(Note that you could have picked the two respective hole surfaces to achieve the same result.)

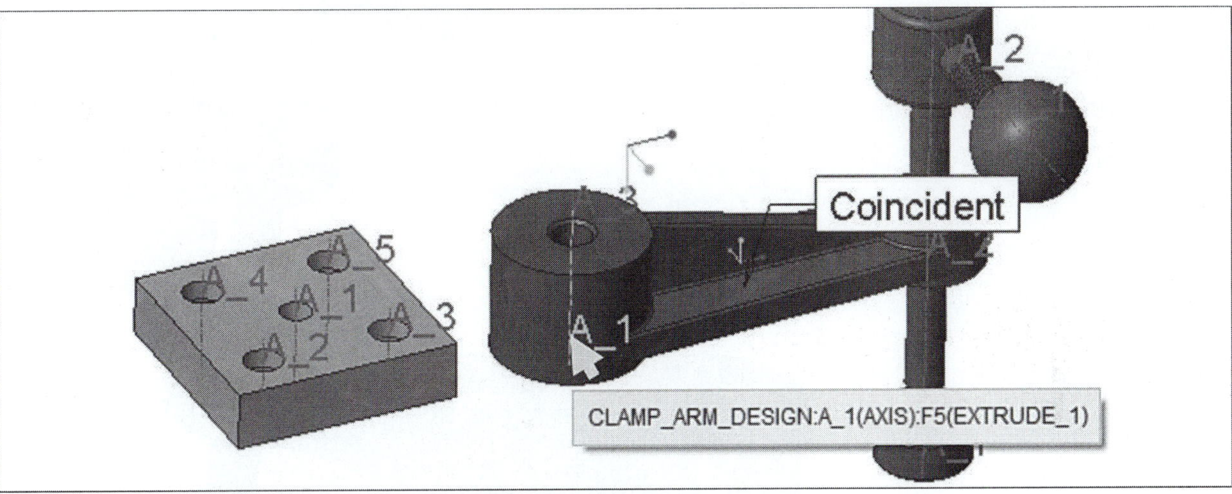

Figure 8.25(f) Select the Axis on the Clamp_Arm_Design

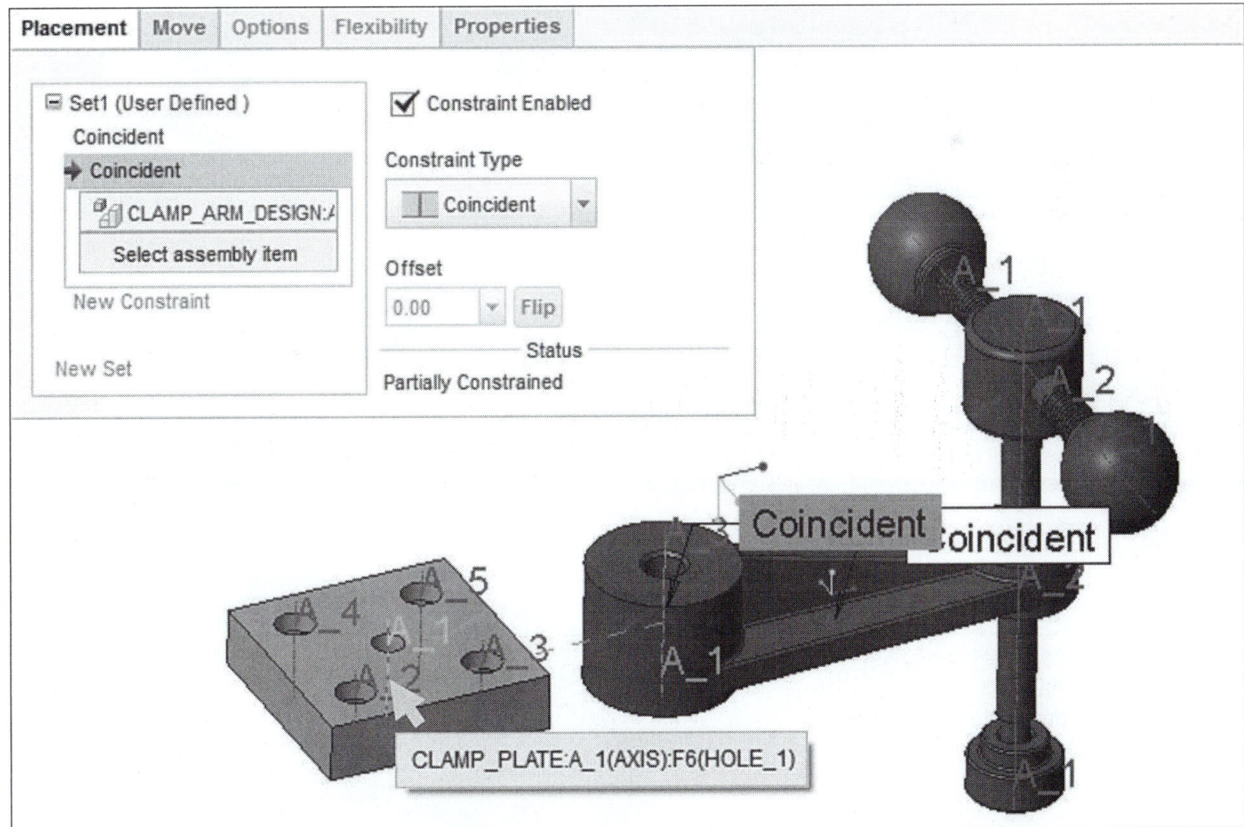

Figure 8.25(g) Select the Tapped Hole Axis on the Clamp_Plate

Click: [Fig. 8.25(h)] > select **CLAMP_PLATE.PRT** [Fig. 8.25(i)] > **Ctrl+D** > **Ctrl+S**

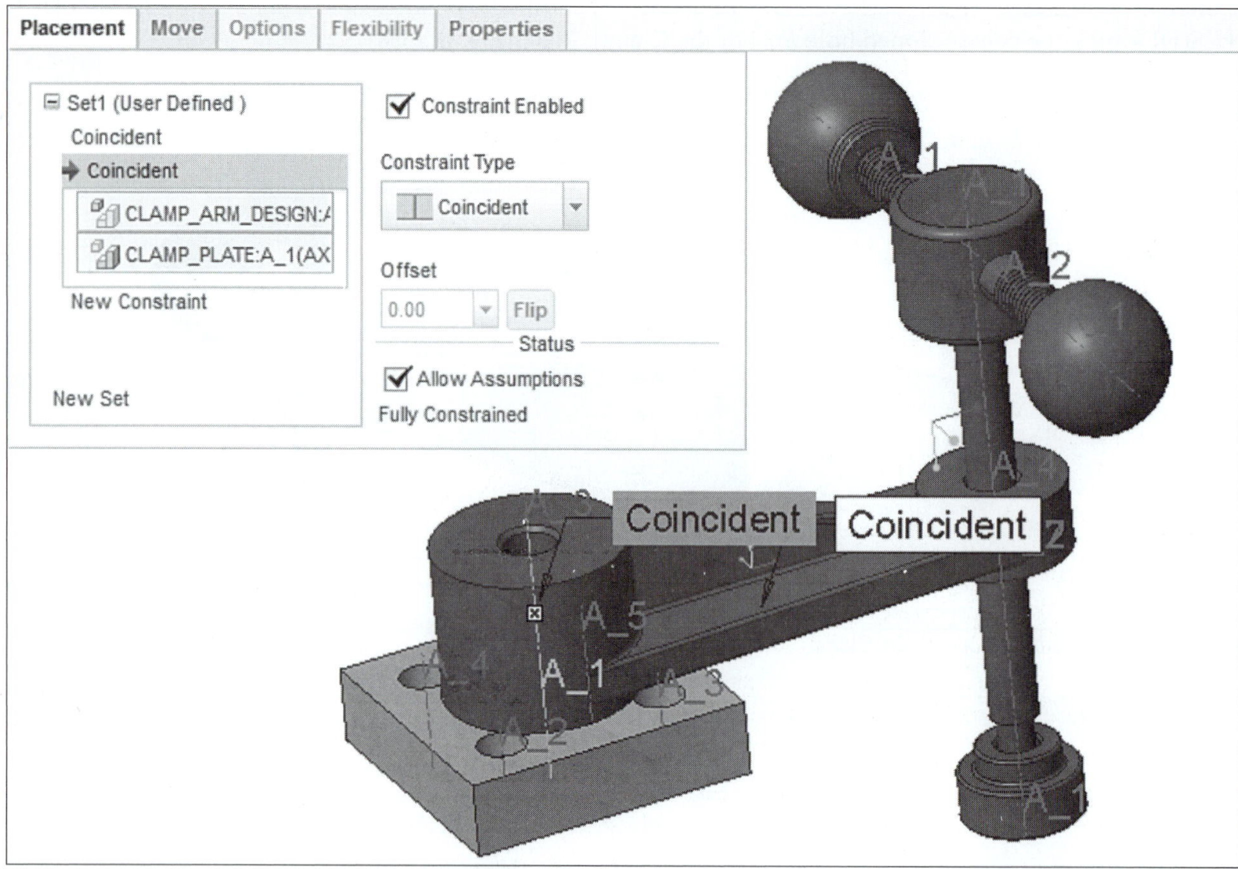

Figure 8.25(h) Fully Constrained Clamp_Subassembly

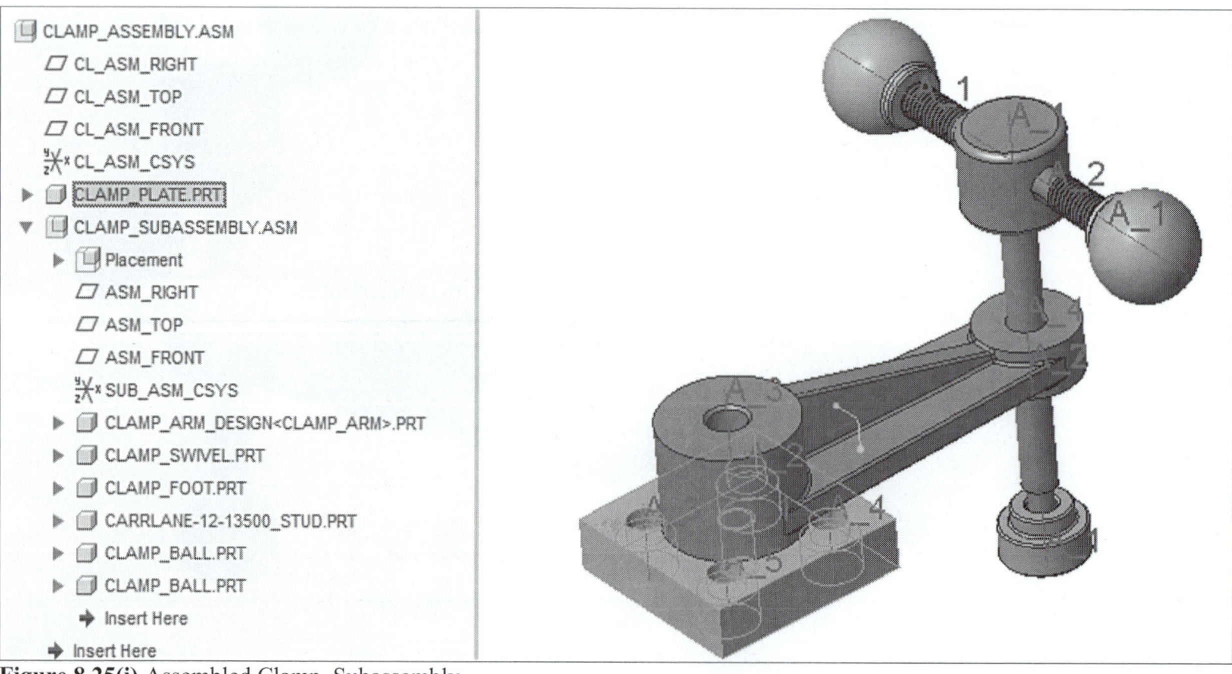

Figure 8.25(i) Assembled Clamp_Subassembly

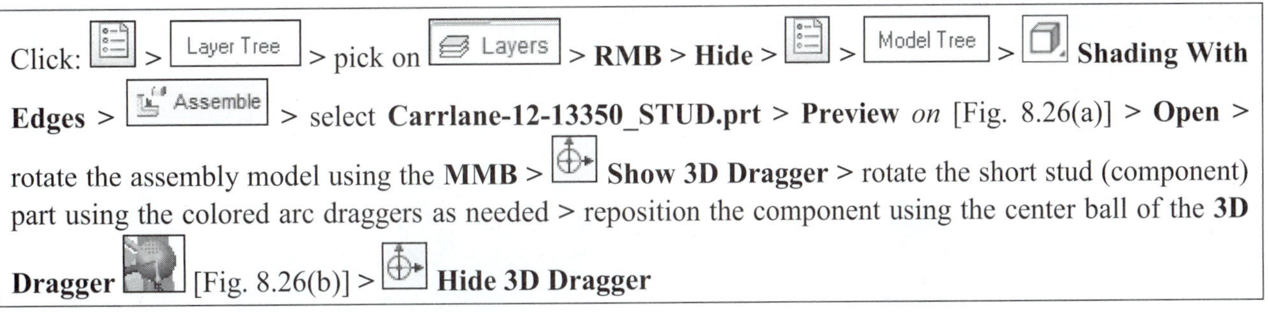

Click: Layer Tree > pick on Layers > **RMB** > **Hide** > Model Tree > **Shading With Edges** > Assemble > select **Carrlane-12-13350_STUD.prt** > **Preview** *on* [Fig. 8.26(a)] > **Open** > rotate the assembly model using the **MMB** > **Show 3D Dragger** > rotate the short stud (component) part using the colored arc draggers as needed > reposition the component using the center ball of the **3D Dragger** [Fig. 8.26(b)] > **Hide 3D Dragger**

Figure 8.26(a) Carrlane-12-13350_STUD.prt

Figure 8.26(b) Rotate the Assembly Model and Reposition the Short Stud using the 3D Dragger

Pick on the cylindrical surface of the Carrlane-12-13350_STUD [Fig. 8.26(c)] > pick on the hole surface of the Clamp_Arm_Design [Fig. 8.26(d)]

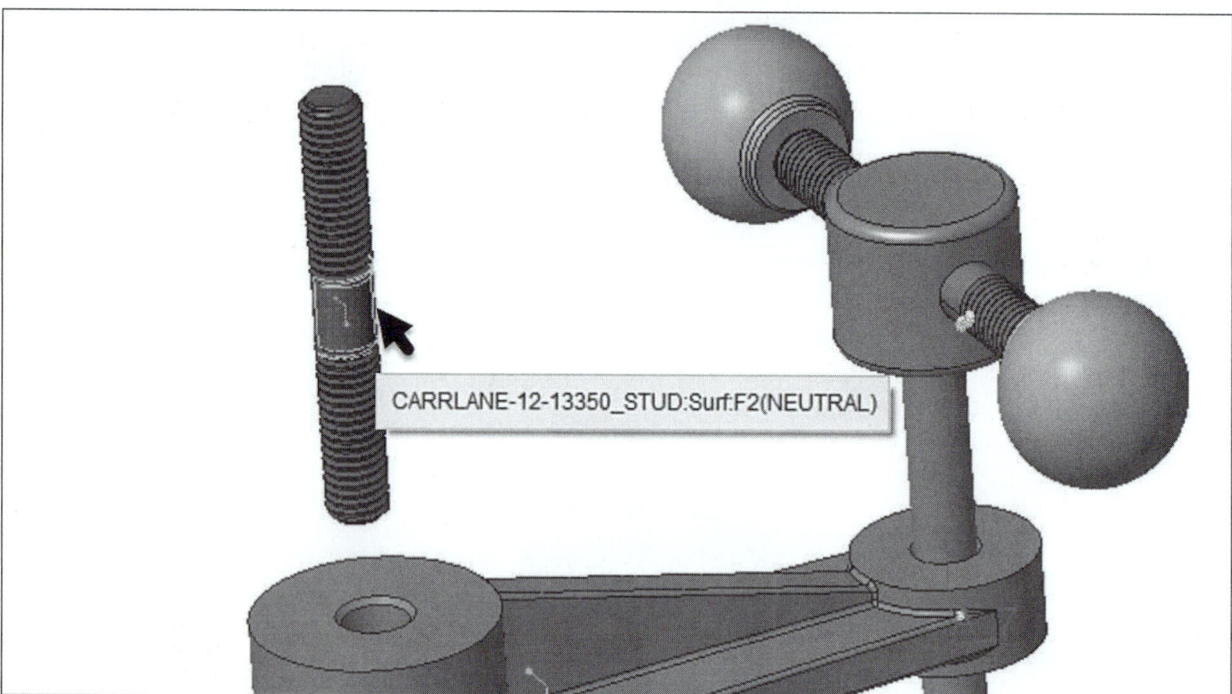

Figure 8.26(c) Select Surface of Carrlane-12-13350_STUD.prt

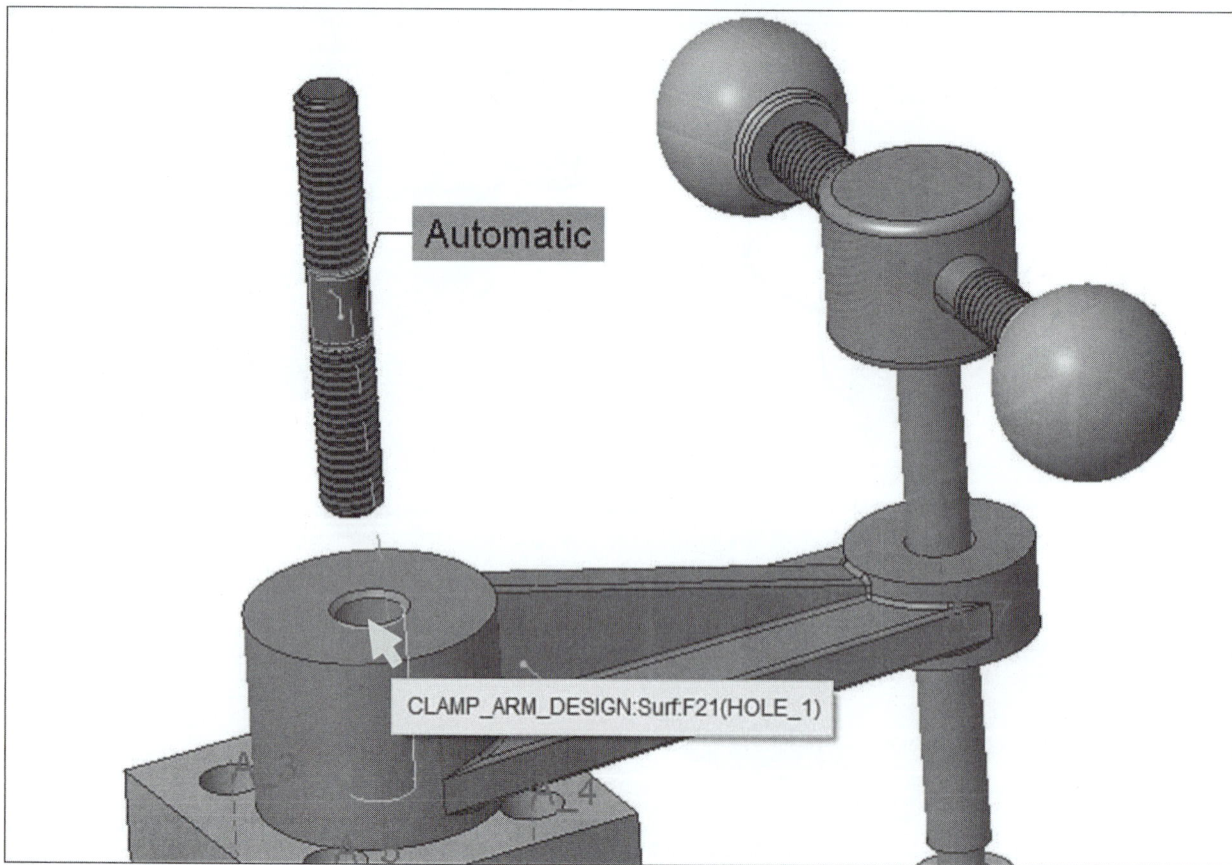

Figure 8.26(d) Select Hole Surface of the Clamp_Arm_Design

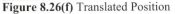

Press: **Ctrl+D** > **Move** tab > **LMB** on the Carrlane-12-13350_STUD > release the **LMB** > move stud into the hole > **LMB** to place [Fig. 8.26(e)] > **Saved Orientations** > **TOP** > > **Hidden Line** > **LMB** on the Carrlane-12-13350_STUD > translate (slide) until it is inside the Clamp_Plate [Fig. 8.26(f)] > **LMB** to place

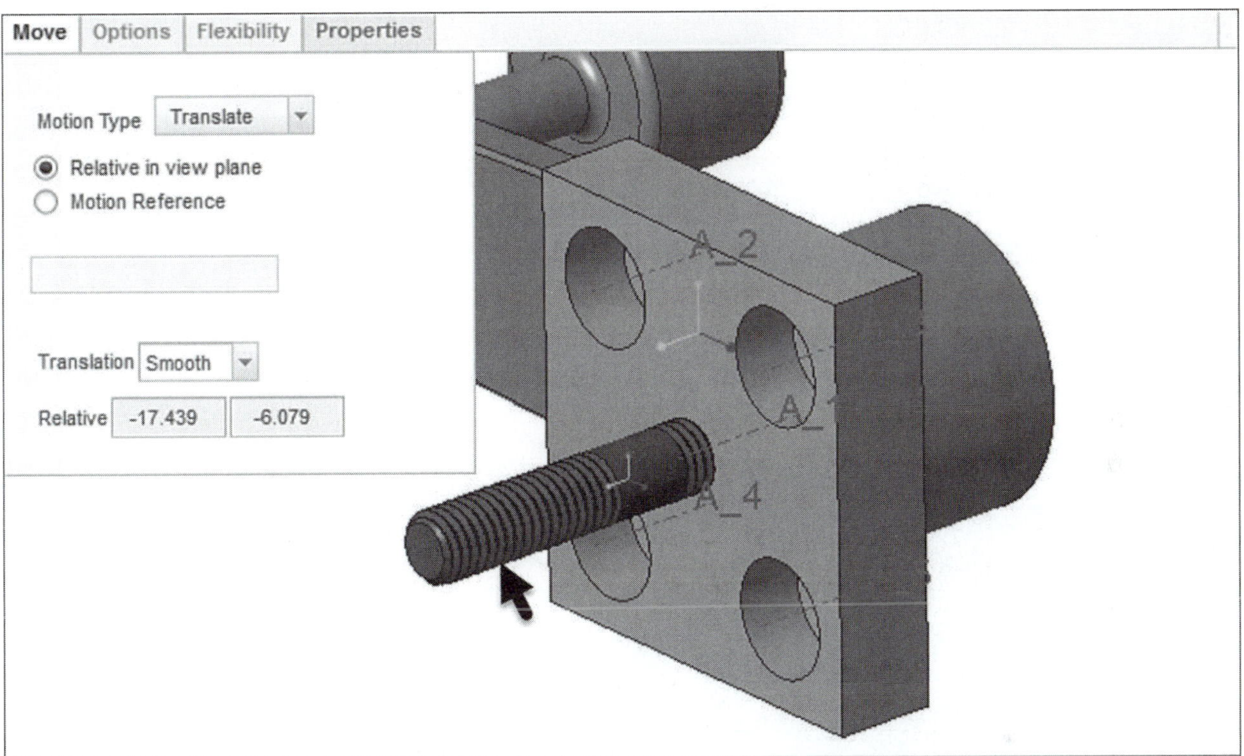

Figure 8.26(e) Move Tab, Motion Type Translate

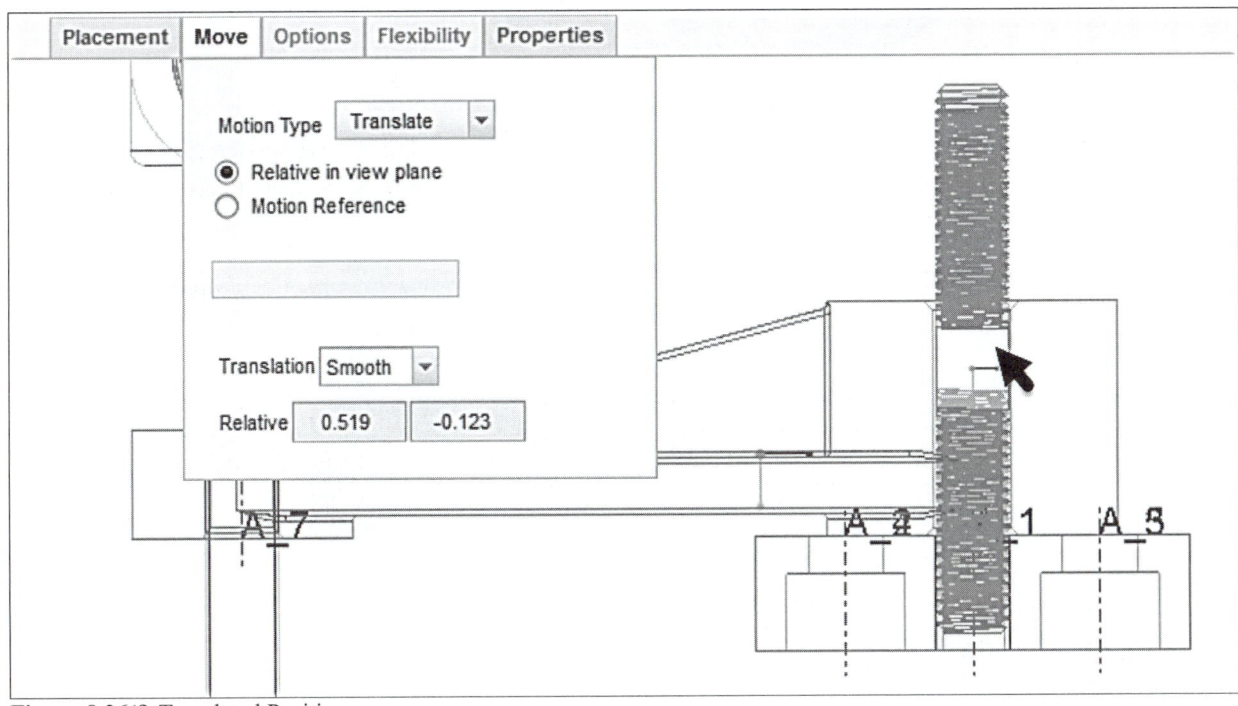

Figure 8.26(f) Translated Position

Click: > > **Ctrl+D** > **Placement** tab > **New Constraint** > Constraint Type > **Fix component to current position** [Fig. 8.26(g)] > > **Ctrl+D** > **Ctrl+S**

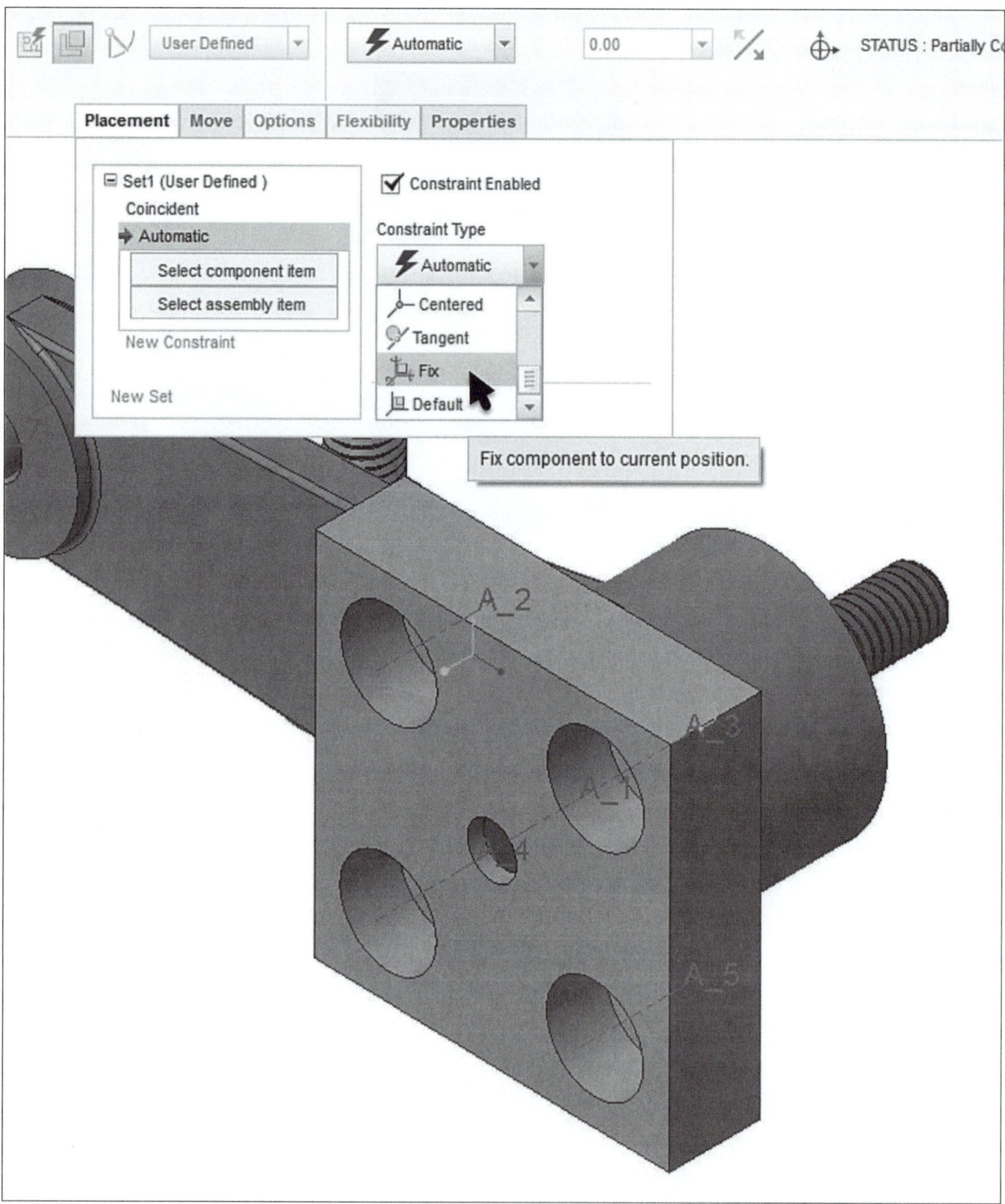

Figure 8.26(g) Fix Constraint

Click: Assemble > select **carrlane_500_fn.prt** [Fig. 8.27(a)] > **Open** > press and hold **Ctrl+Alt** > press **RMB** on the flange nut and move it closer to the assembly [Fig. 8.27(b)] > release the **RMB**

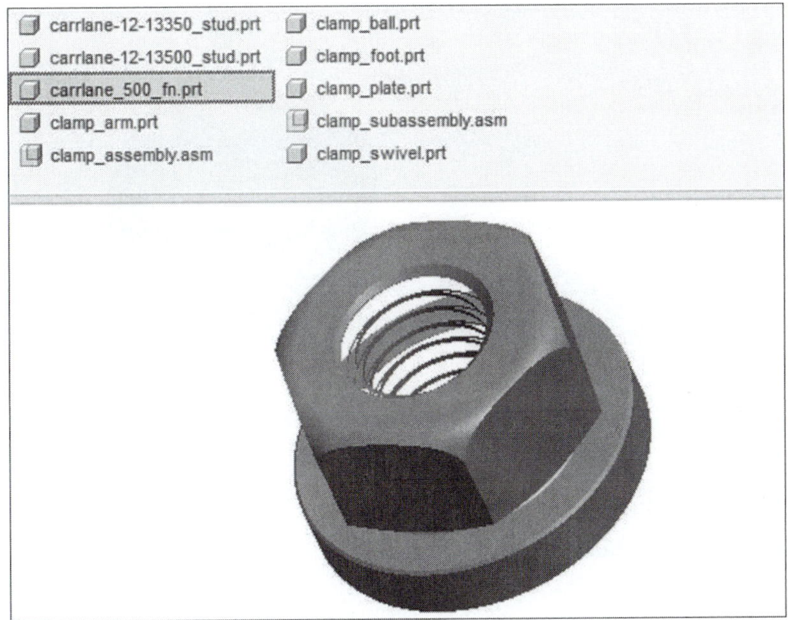

Figure 8.27(a) carrlane_500_fn.prt

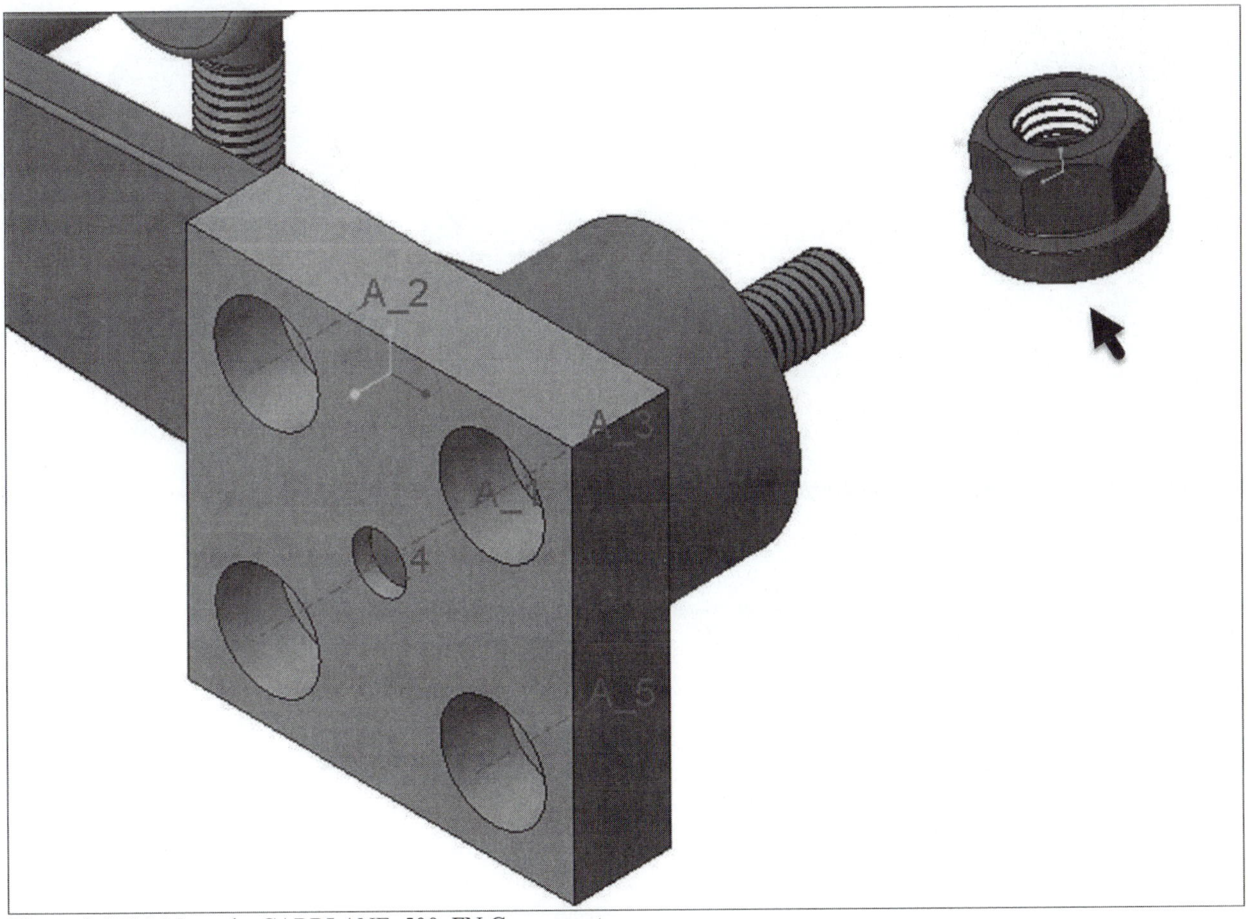

Figure 8.27(b) Move the CARRLANE_500_FN Component

Press and hold **Ctrl+Alt** > press **MMB** on the Flange_Nut and rotate [Fig. 8.27(c)] > release the **MMB** and the **Ctrl+Alt** keys > in the Graphics Window, press **RMB** > **Select component item** [Fig. 8.27(d)]

Figure 8.27(c) Rotate the Flange Nut

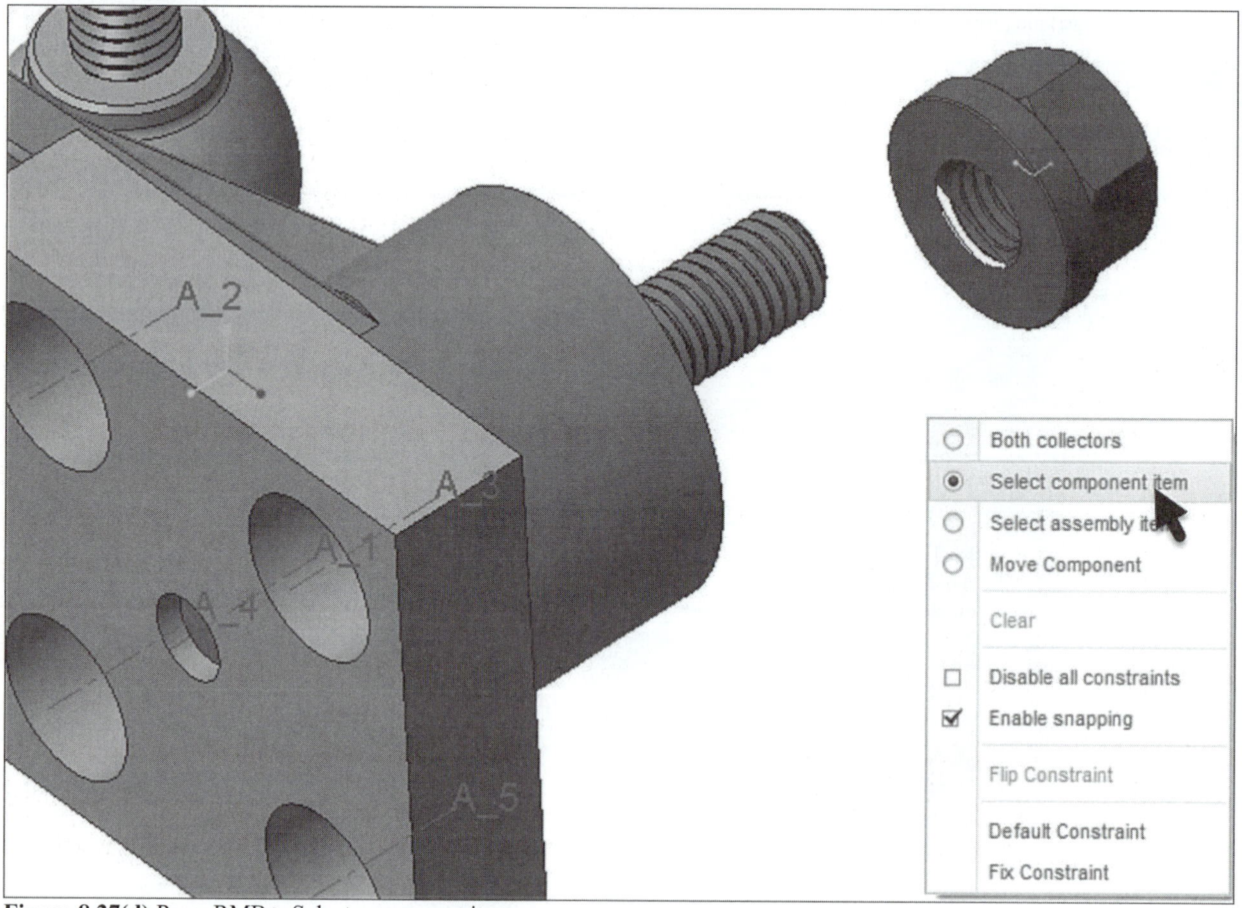

Figure 8.27(d) Press RMB > Select component item

Carefully pick the cylindrical surface of the flange nut [Fig. 8.27(e)] > carefully pick the cylindrical surface of the stud [Fig. 8.27(f)] (Coincident) *[press **RMB > Clear** if Tangent becomes the constraint]*

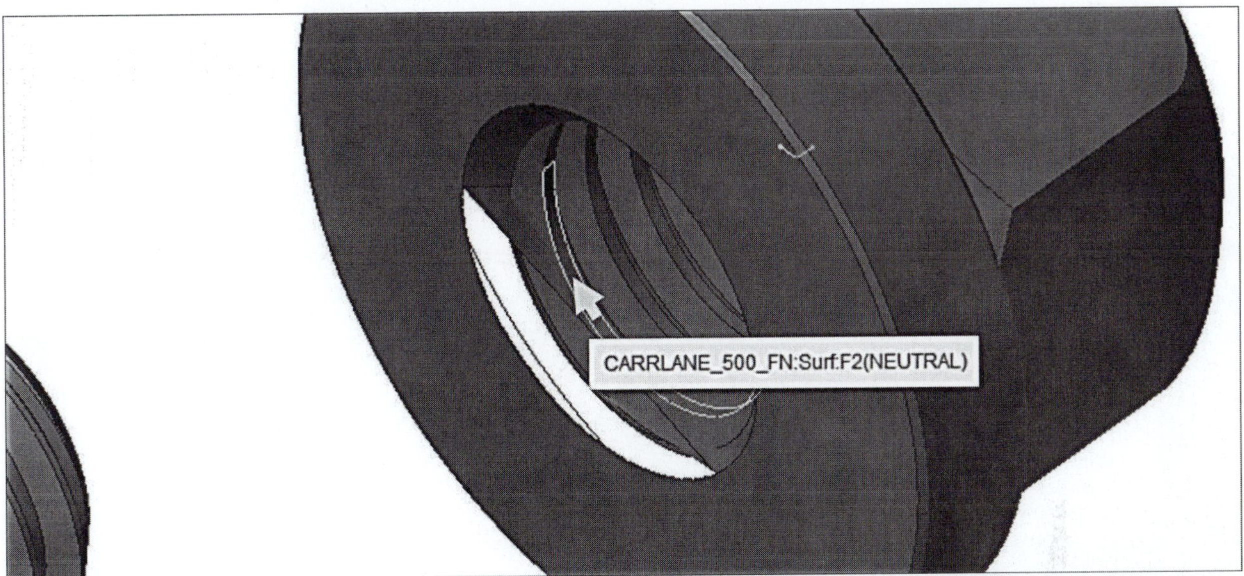

Figure 8.27(e) Select the Flange Nut Surface (zoom in if needed)

Figure 8.27(f) Select the Stud Surface (zoom in if needed)

Press: **MMB** spin the model > **Placement** tab > **New Constraint** > rotate and translate the component as required > pick on the Clamp_Arm_Design surface [Fig. 8.27(g)]

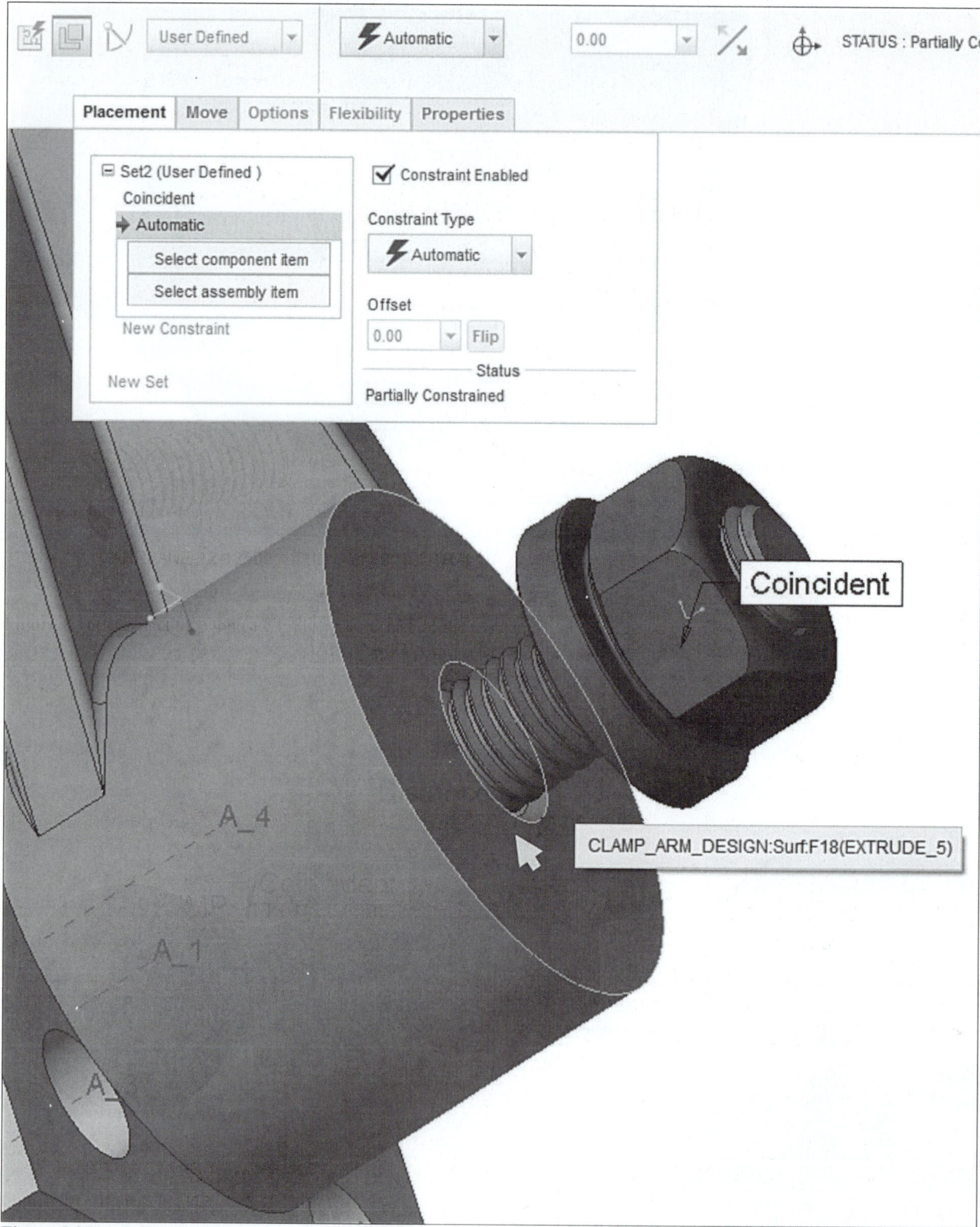

Figure 8.27(g) Select the Clamp_Arm_Design Surface

Press: **MMB** spin the model > pick on the CARRLANE_500_FN bearing surface [Figs. 8.27(h-i)]

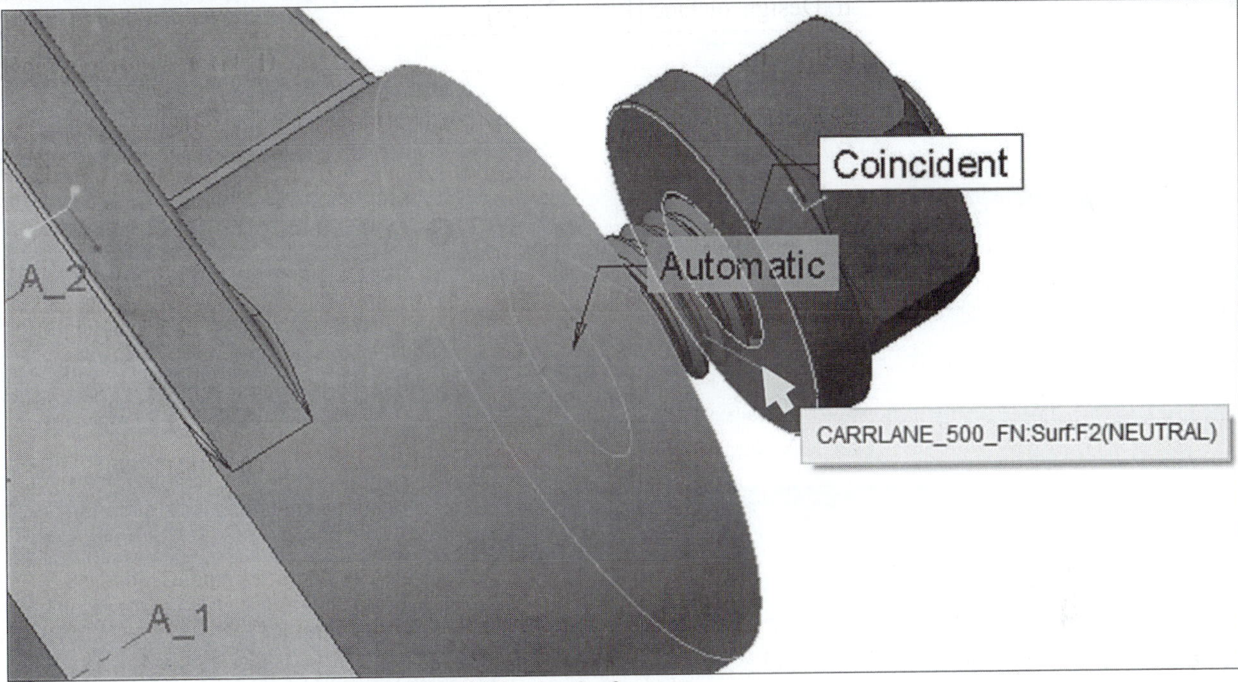

Figure 8.27(h) Select the CARRLANE_500_FN Bearing Surface

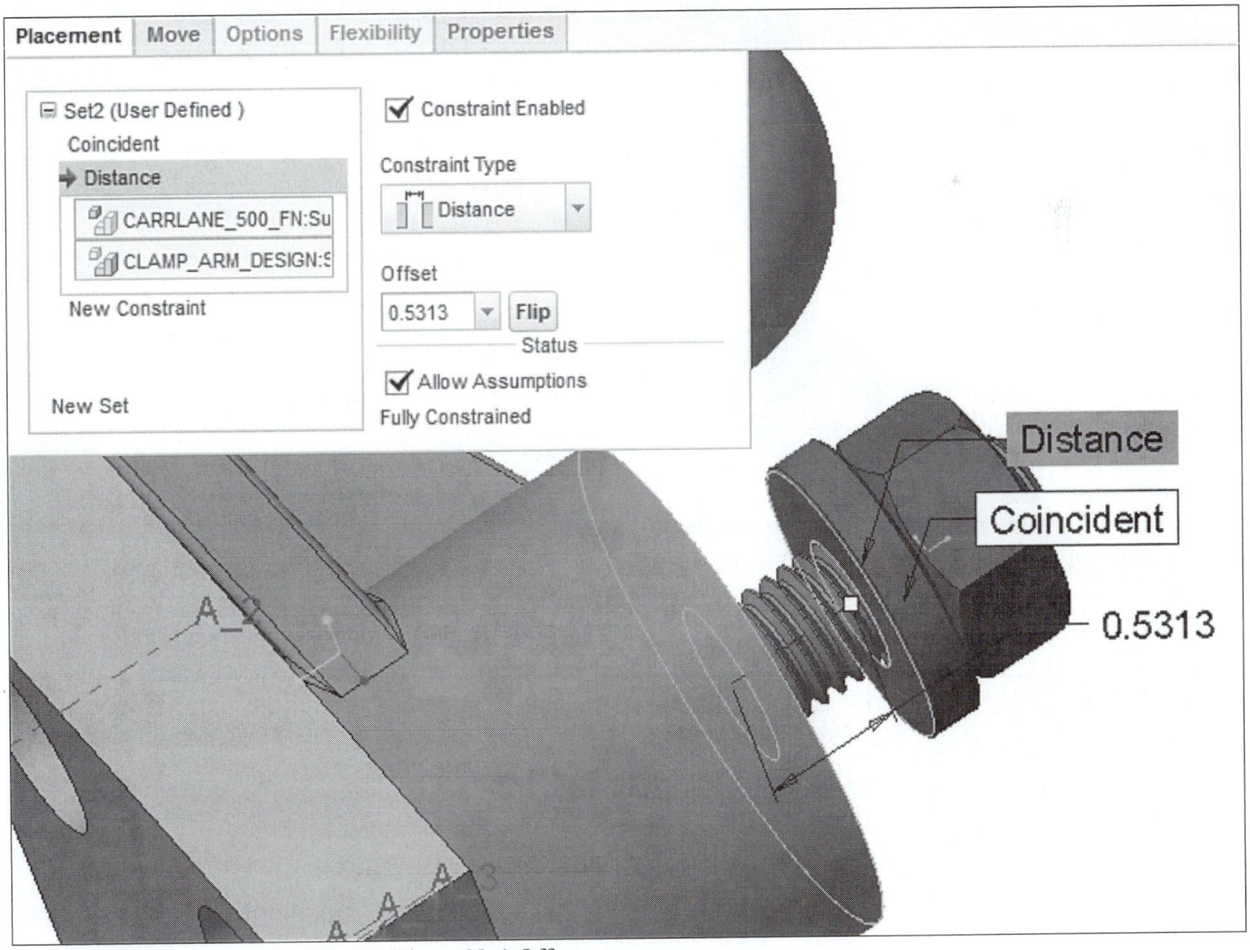

Figure 8.27(i) CARRLANE_500_FN (Flange Nut) Offset

If necessary click: Distance > Coincident [Fig. 8.27(j)] > ✓ > **Ctrl+D** > **Ctrl+S**

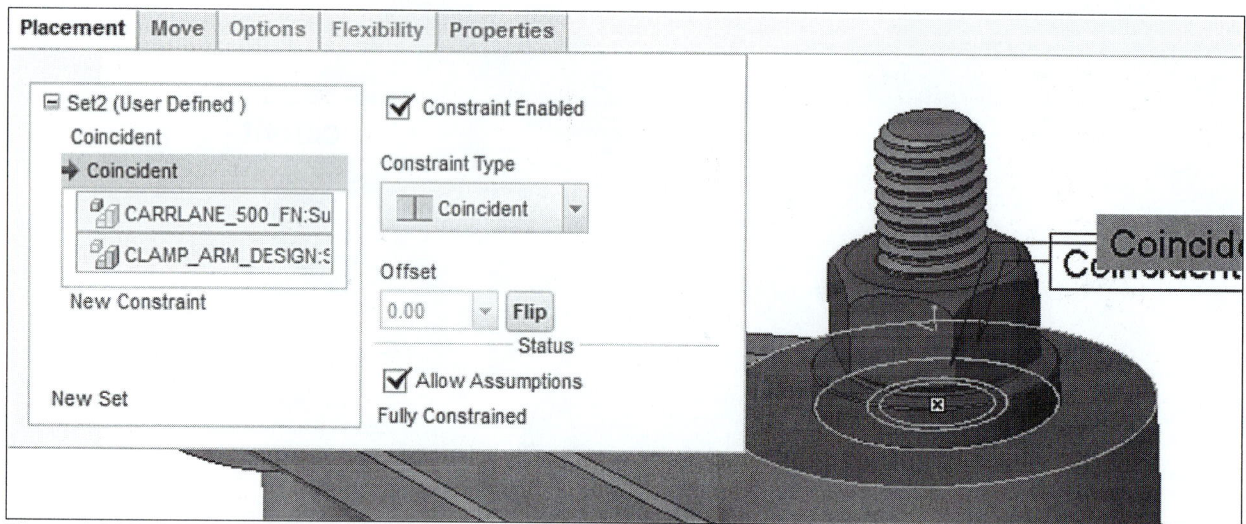

Figure 8.27(j) Constraint Type: Coincident

Perform a check on the assembly. Click: **Analysis** tab > Global Interference in the Inspect Geometry Group > Preview **Compute current analysis for preview** (Fig. 8.28) *(Select each line and note the interference between components at each threaded connection.)* > **OK** > **LMB** > **Ctrl+S**

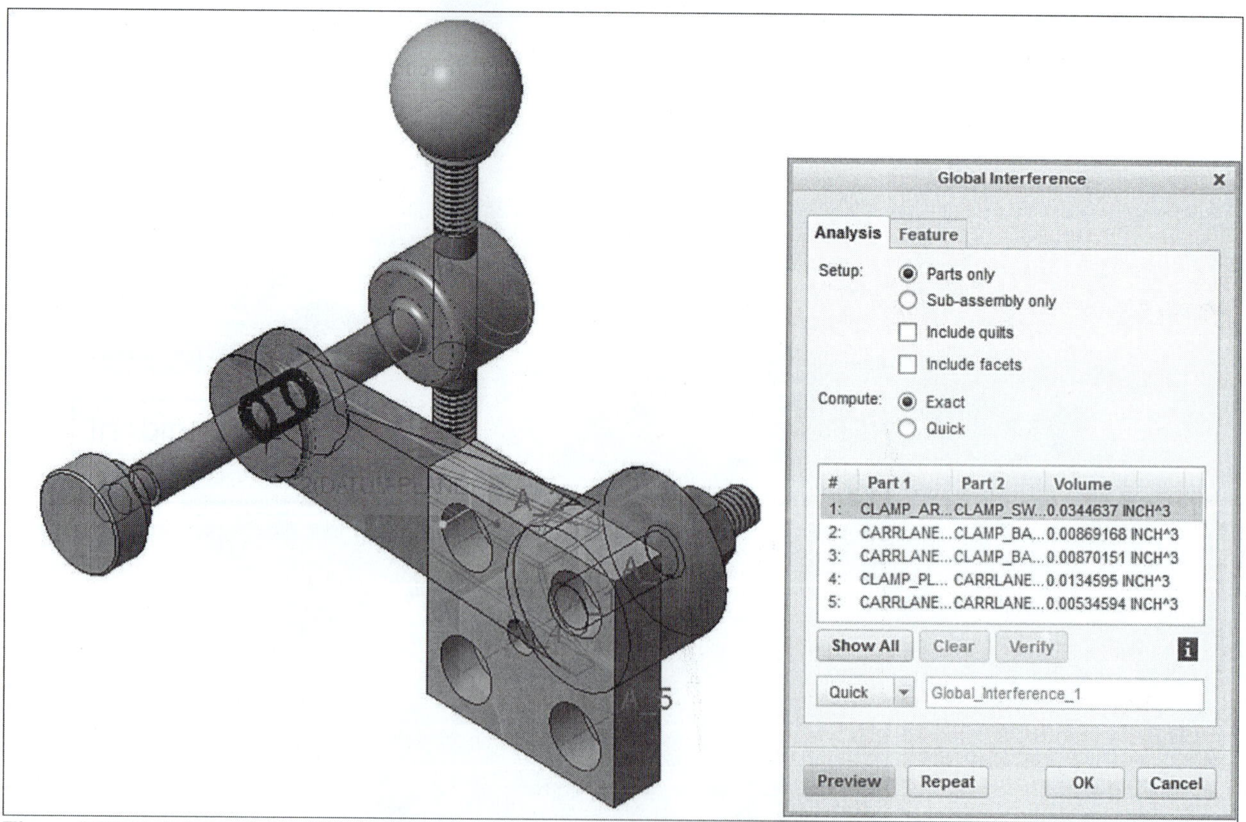

Figure 8.28 Global Interference Dialog Box

Bill of Materials

The Bill of Materials (BOM) lists all parts and part parameters in the current assembly or assembly drawing. It can be displayed in HTML or text format and is separated into two parts: breakdown and summary. The Breakdown section lists what is contained in the current assembly or part. The BOM HTML breakdown section lists quantity, type, name (hyperlink), and three actions (highlight, information and open) about each member or sub-member of your assembly:

- **Quantity** Lists the number of components or drawings
- **Type** Lists the type of the assembly component (part or sub-assembly)
- **Name** Lists the assembly component and is hyper-linked to that item. Selecting this hyperlink highlights the component in the graphics window
- **Actions** is divided into three areas:
    - **Highlight** Highlights the selected component in the assembly graphics window
    - **Information** Provides model information on the relevant component
    - **Open** Opens the component in another Creo Parametric 3.0 window

A bill of materials (BOM) can be seen by clicking: **Tools** tab > Bill of Materials > check all Include options [Fig. 8.29(a)] > **OK** [Fig. 8.29(b)] > or to close the Browser > **Model** tab

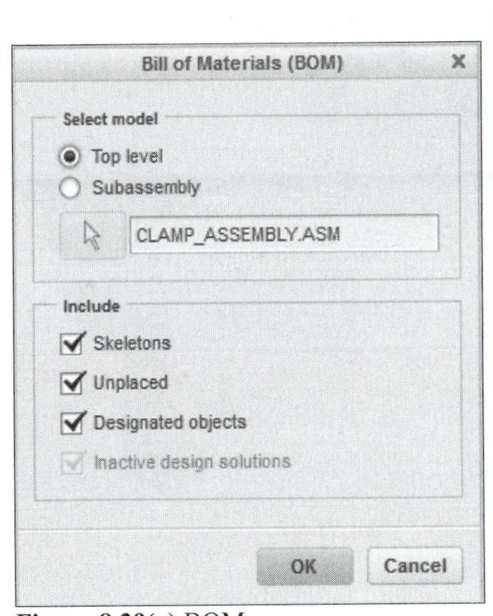

Figure 8.29(a) BOM

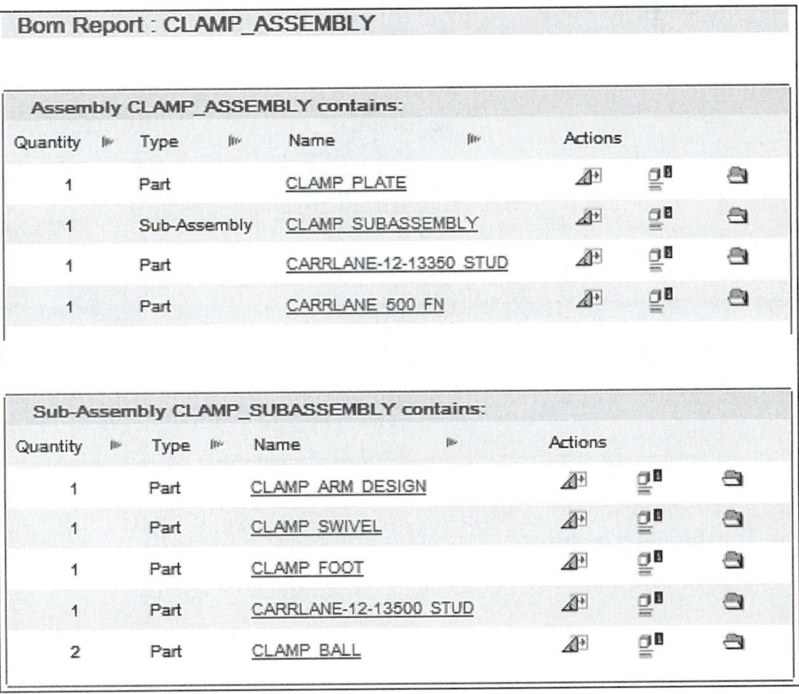
Bom Report : CLAMP_ASSEMBLY

Assembly CLAMP_ASSEMBLY contains:

Quantity	Type	Name	Actions
1	Part	CLAMP_PLATE	
1	Sub-Assembly	CLAMP_SUBASSEMBLY	
1	Part	CARRLANE-12-13350_STUD	
1	Part	CARRLANE_500_FN	

Sub-Assembly CLAMP_SUBASSEMBLY contains:

Quantity	Type	Name	Actions
1	Part	CLAMP_ARM_DESIGN	
1	Part	CLAMP_SWIVEL	
1	Part	CLAMP_FOOT	
1	Part	CARRLANE-12-13500_STUD	
2	Part	CLAMP_BALL	

Figure 8.29(b) BOM Report

Upon closer inspection, the assembly is not assembled correctly. The Clamp_Subassembly is assembled on the wrong side of the Clamp_Plate. When using Creo Parametric 3.0, you can edit items, features, components, etc., at any time. The Edit Definition command is used to change the constraint reference of assembled components.

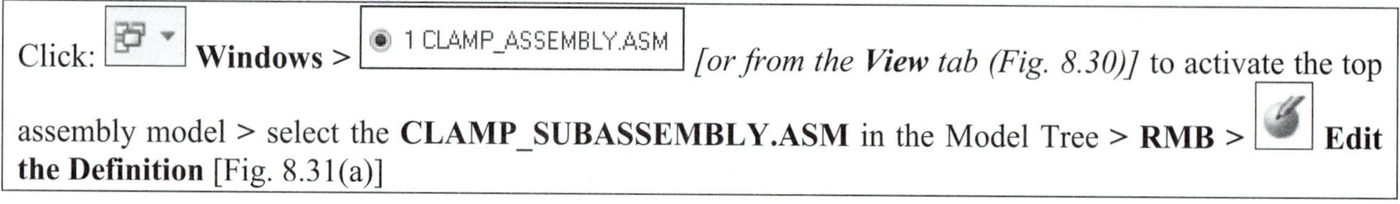

Click: Windows > 1 CLAMP_ASSEMBLY.ASM *[or from the **View** tab (Fig. 8.30)]* to activate the top assembly model > select the **CLAMP_SUBASSEMBLY.ASM** in the Model Tree > **RMB >** **Edit the Definition** [Fig. 8.31(a)]

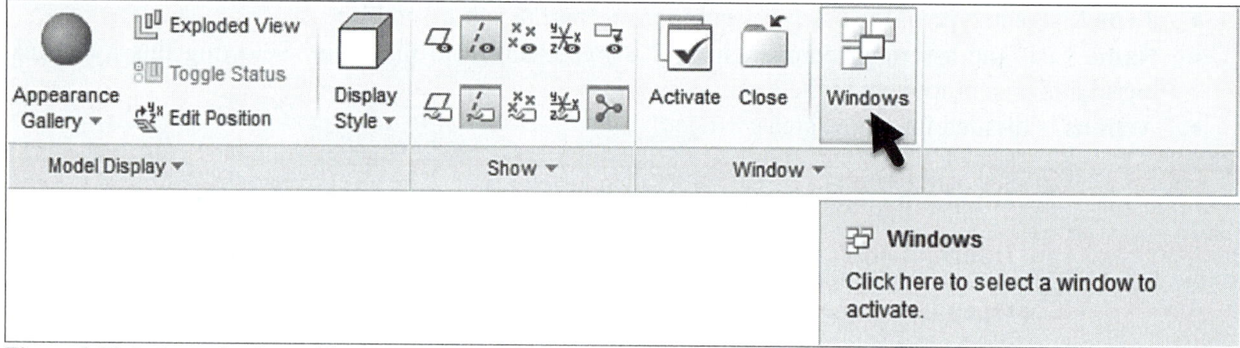

Figure 8.30 Activate the Assembly

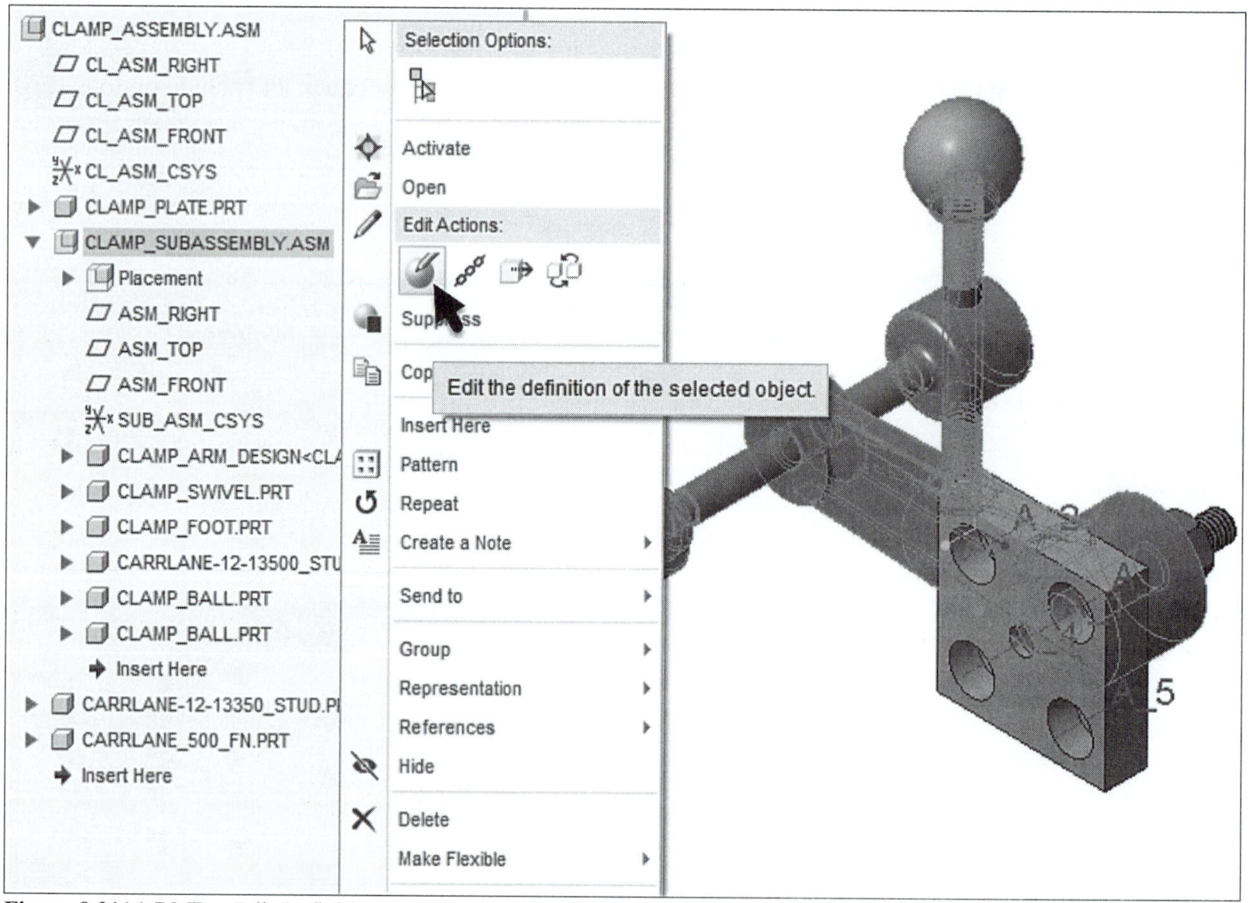

Figure 8.31(a) RMB > Edit Definition

Click: **Placement** tab (to see the existing references) > pick the second reference CLAMP_PLATE:Surf:F5 from the Placement dialog box [Fig. 8.31(b)] *(check the Tool Tip* Part Ref: CLAMP_PLATE:Surf:F5(EXTRUDE_1) *to make sure you selected the Clamp_Plate reference)*

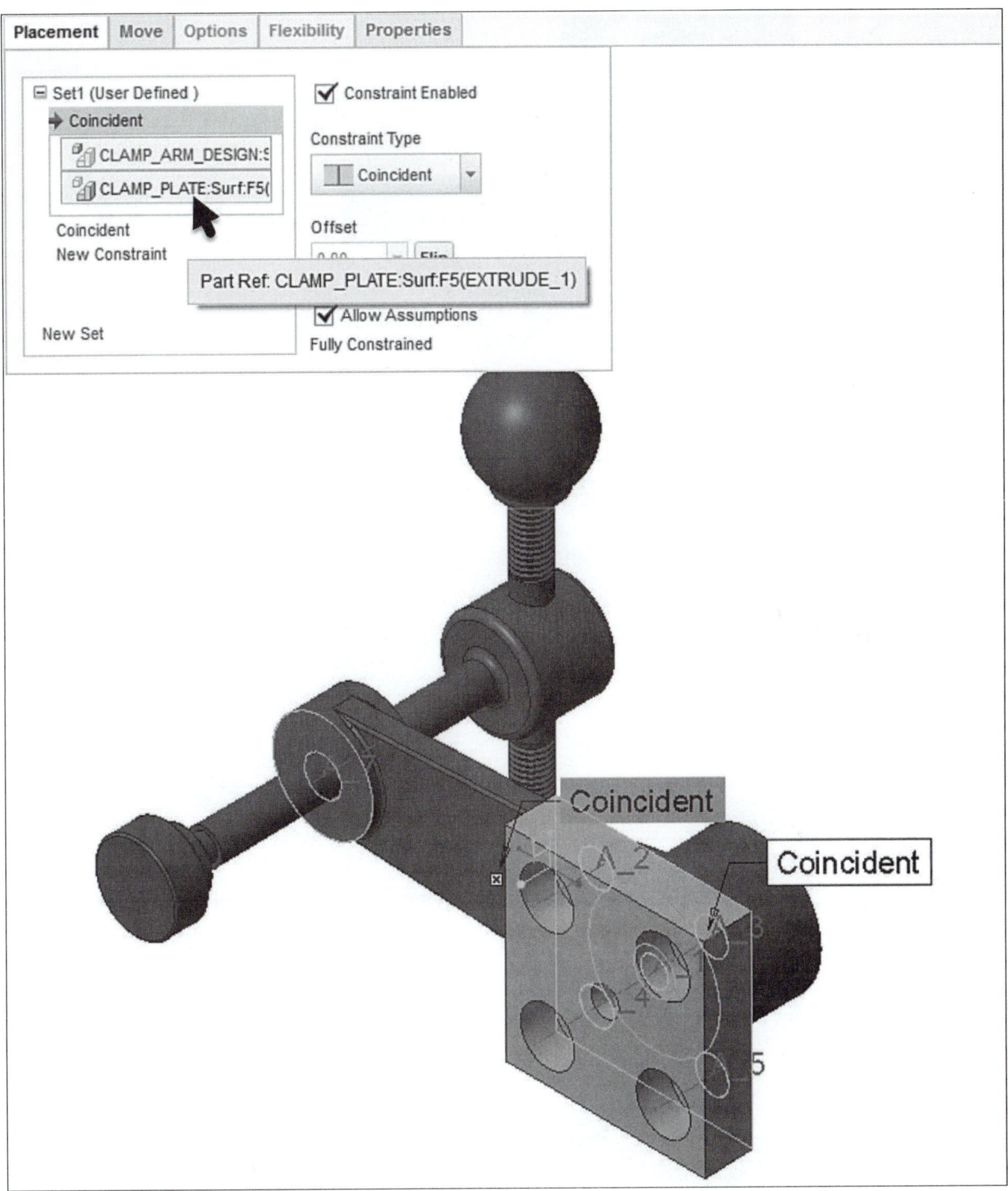

Figure 8.31(b) Select the Clamp_Plate Reference

Pick the counterbore side of the Clamp_Plate as the new assembly reference [Fig. 8.31(c)] > in the Graphics Window, press **RMB** > **Flip Constraint** [Fig. 8.31(d)] > ✓ > **Ctrl+D** [Fig. 8.31(e)]

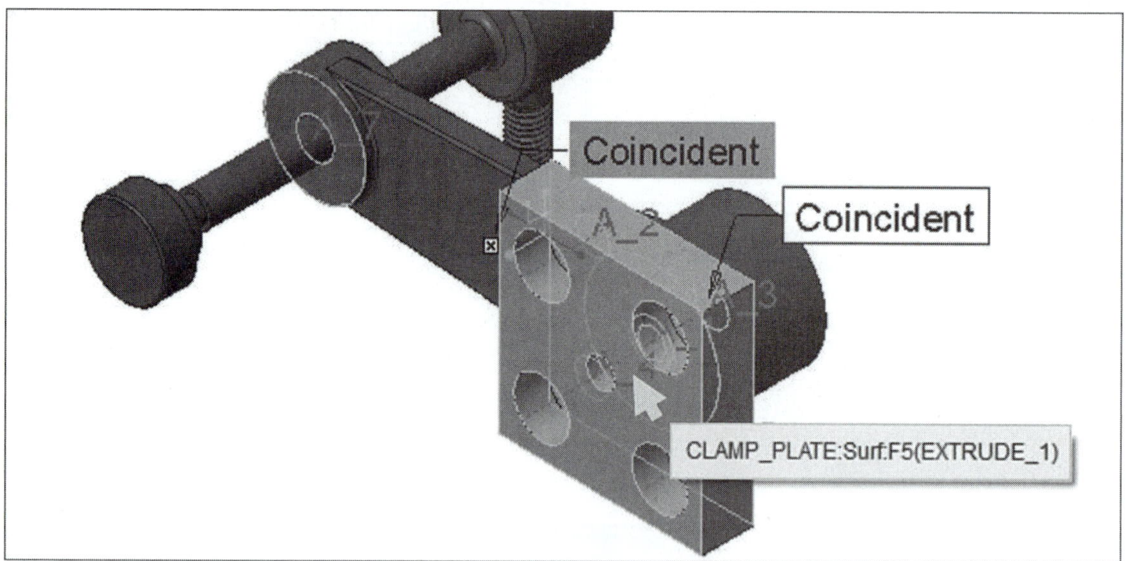

Figure 8.31(c) Select the Counterbore Side of the Plate

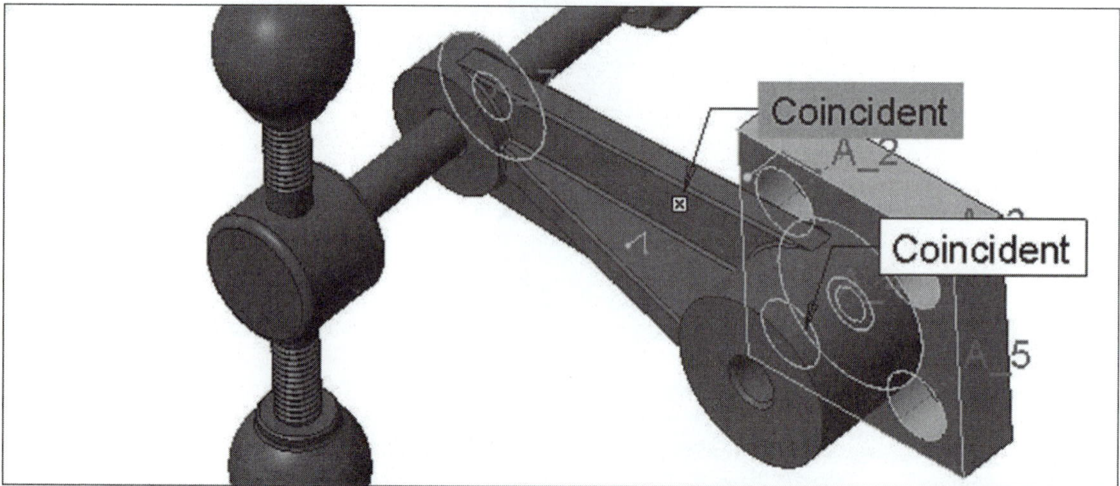

Figure 8.31(d) Clamp_Subassembly Reverses Position

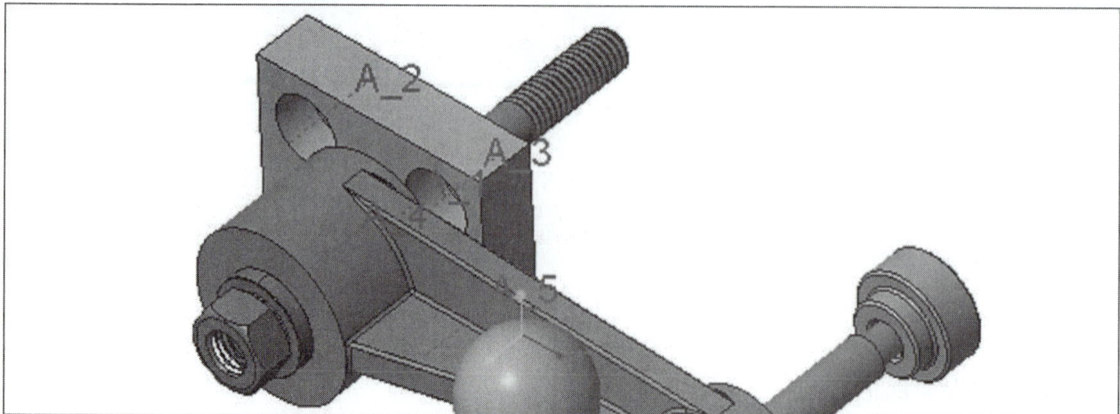

Figure 8.31(e) Correct Clamp_Subassembly Orientation

Press: **MMB** spin the model > select the **CARRLANE-12-13350_STUD.PRT** in the Model Tree > **RMB** > **Edit the Definition** > **Placement** tab > **Fix** > place the pointer on Fix, **RMB** > **Delete** [Fig. 8.32(a)] > **New Constraint** > pick the flat end surface of the stud > pick the bottom surface of the plate [Fig. 8.32(b)]

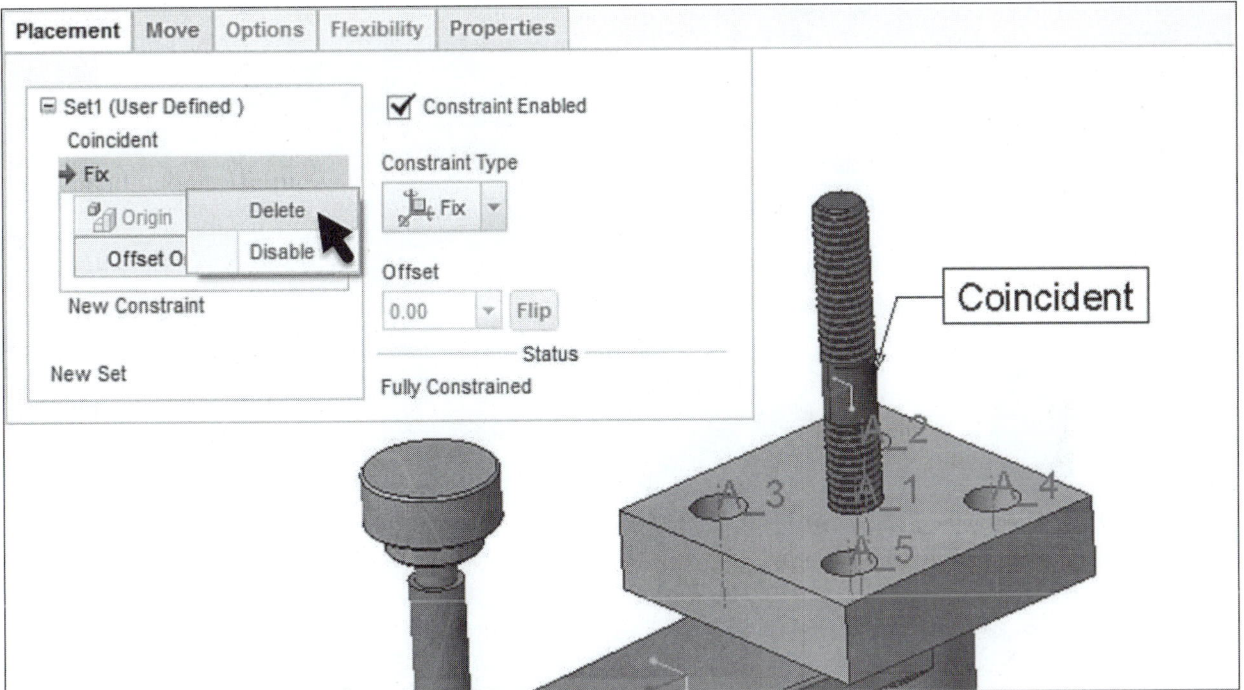

Figure 8.32(a) Delete Fix Constraint

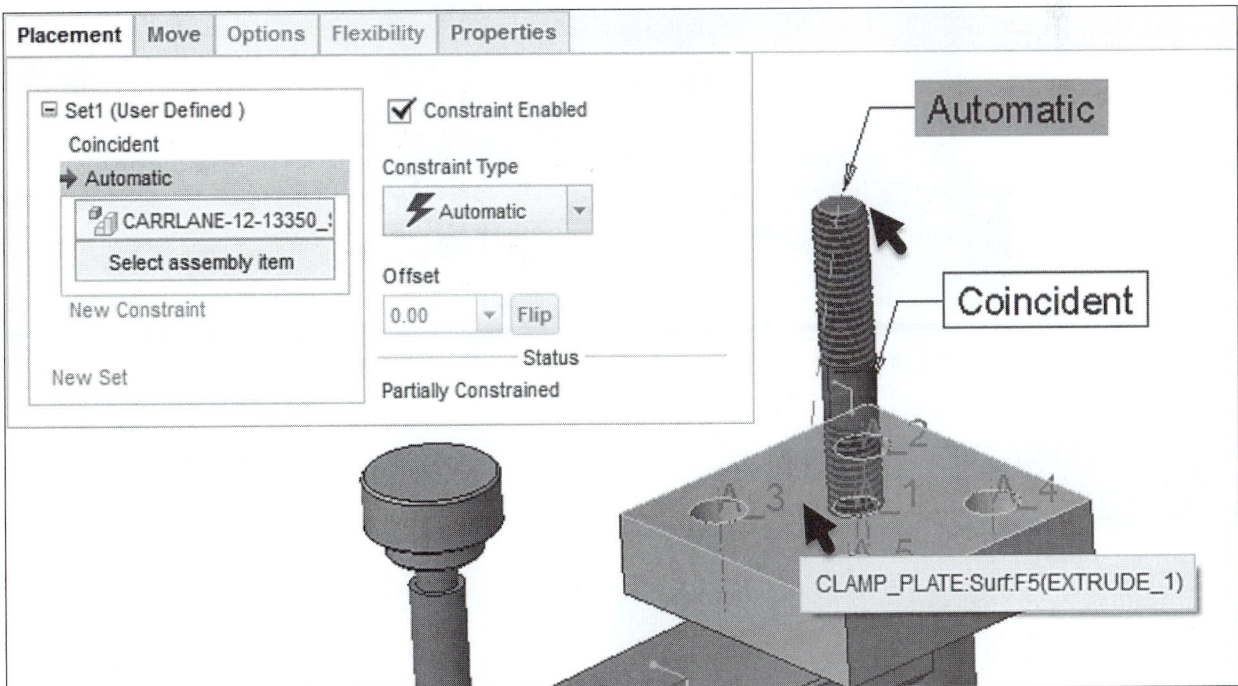

Figure 8.32(b) Select the Bottom Surface of the Clamp_Plate

Click: in the Offset field > *type*: **-.25** > **Enter** [Fig. 8.32(c)] > ✓ > **LMB** in the Graphics Window > place the pointer on the short stud to highlight [Fig. 8.32(d)]

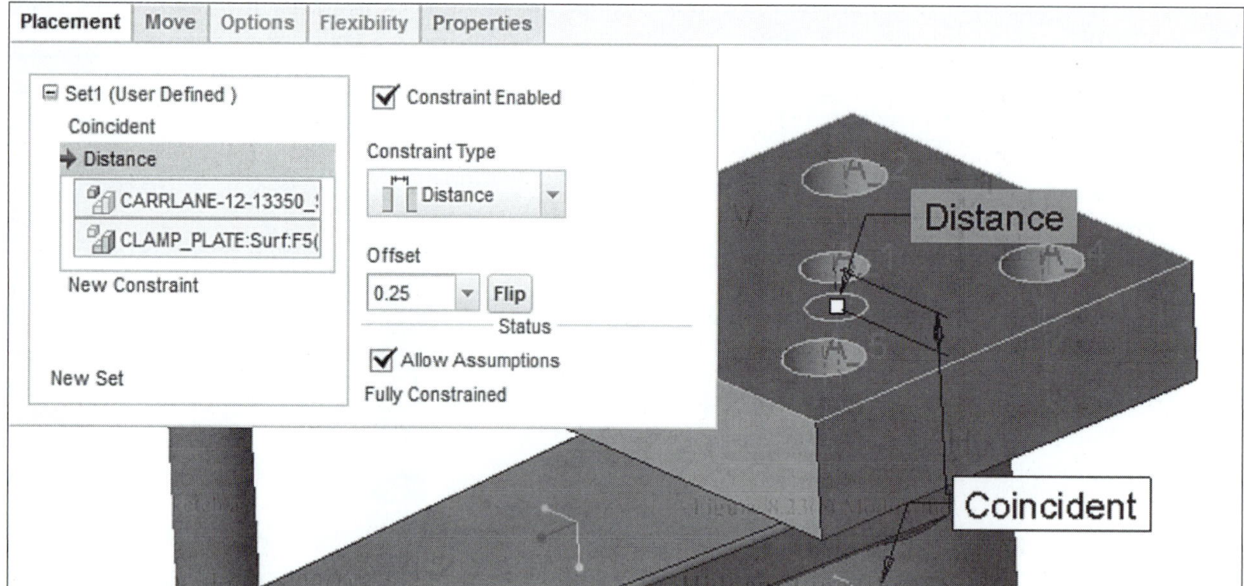

Figure 8.32(c) Offset Distance

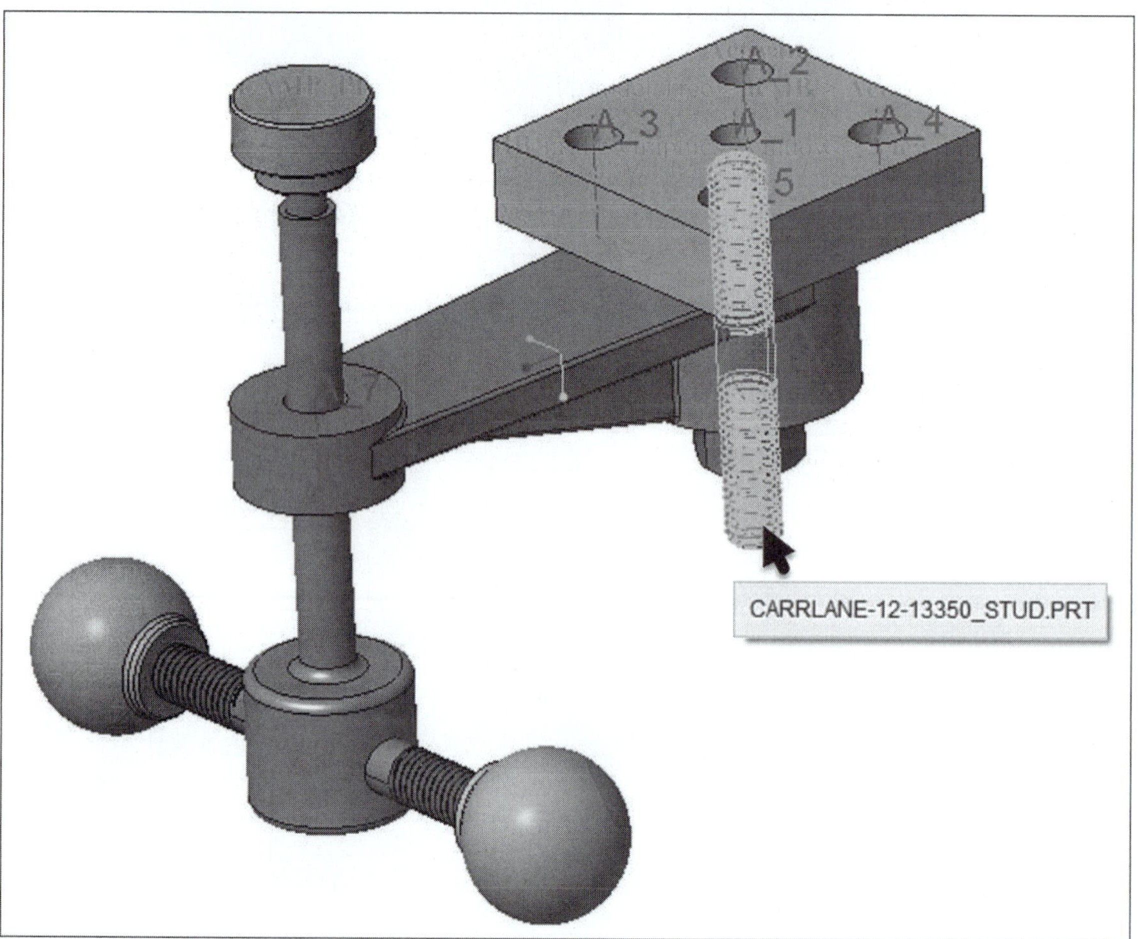

Figure 8.32(d) Highlight the Short Stud

Click: **View** tab > Standard Orientation [Fig. 8.32(e)] > **Ctrl+S**

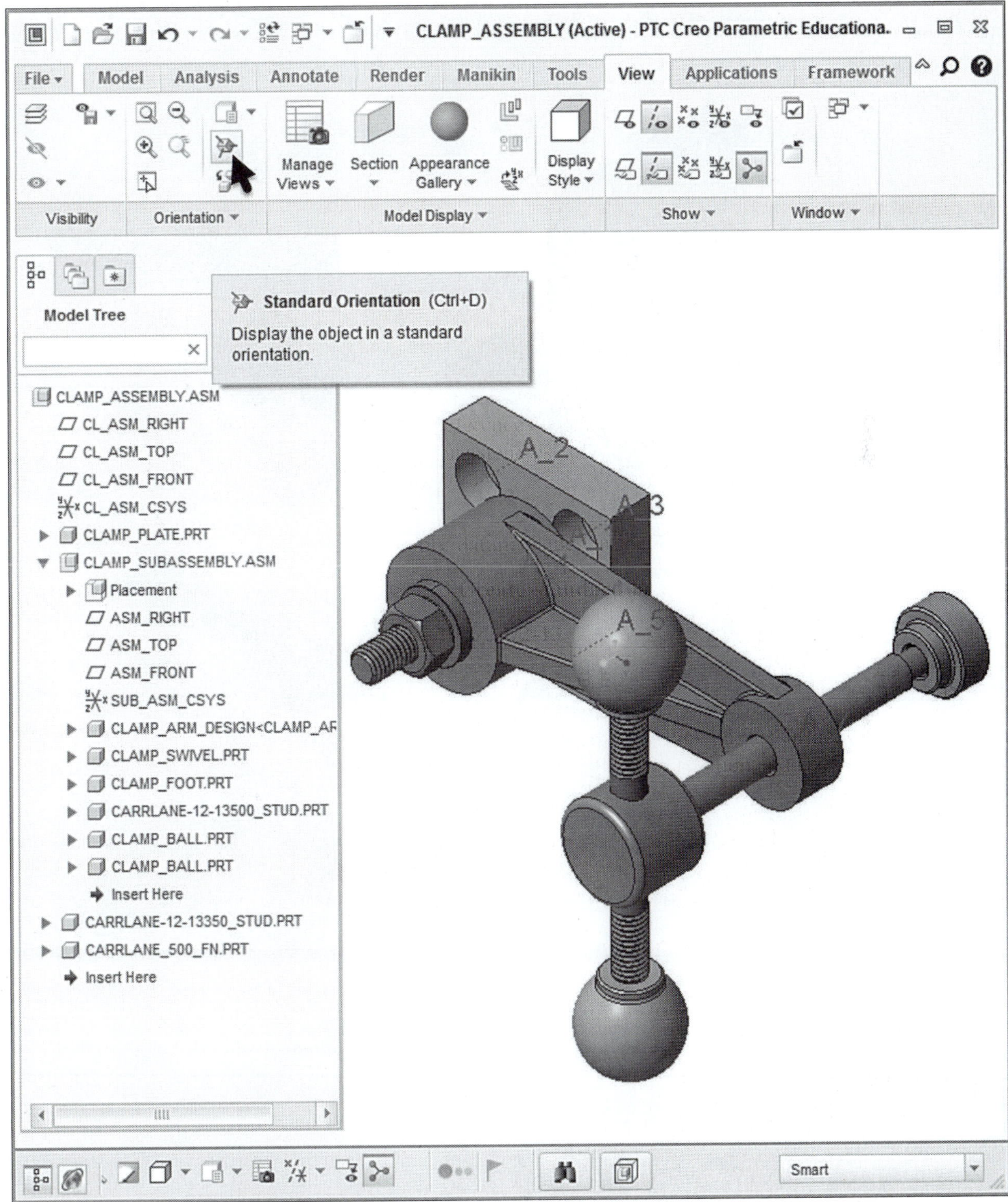

Figure 8.32(e) Standard Orientation

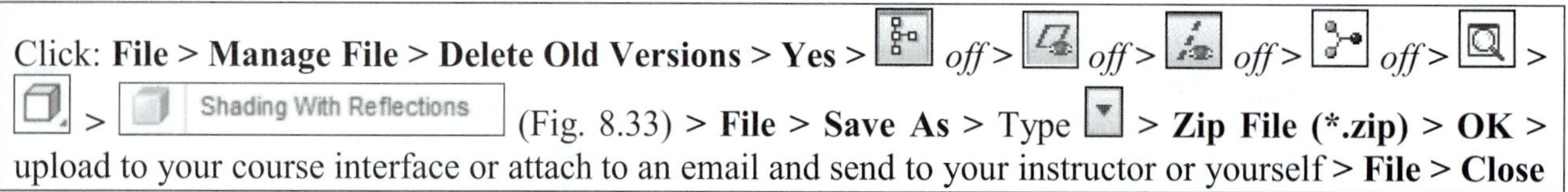
Click: **File** > **Manage File** > **Delete Old Versions** > **Yes** > *off* > *off* > *off* > *off* > > > Shading With Reflections (Fig. 8.33) > **File** > **Save As** > Type > **Zip File (*.zip)** > **OK** > upload to your course interface or attach to an email and send to your instructor or yourself > **File** > **Close**

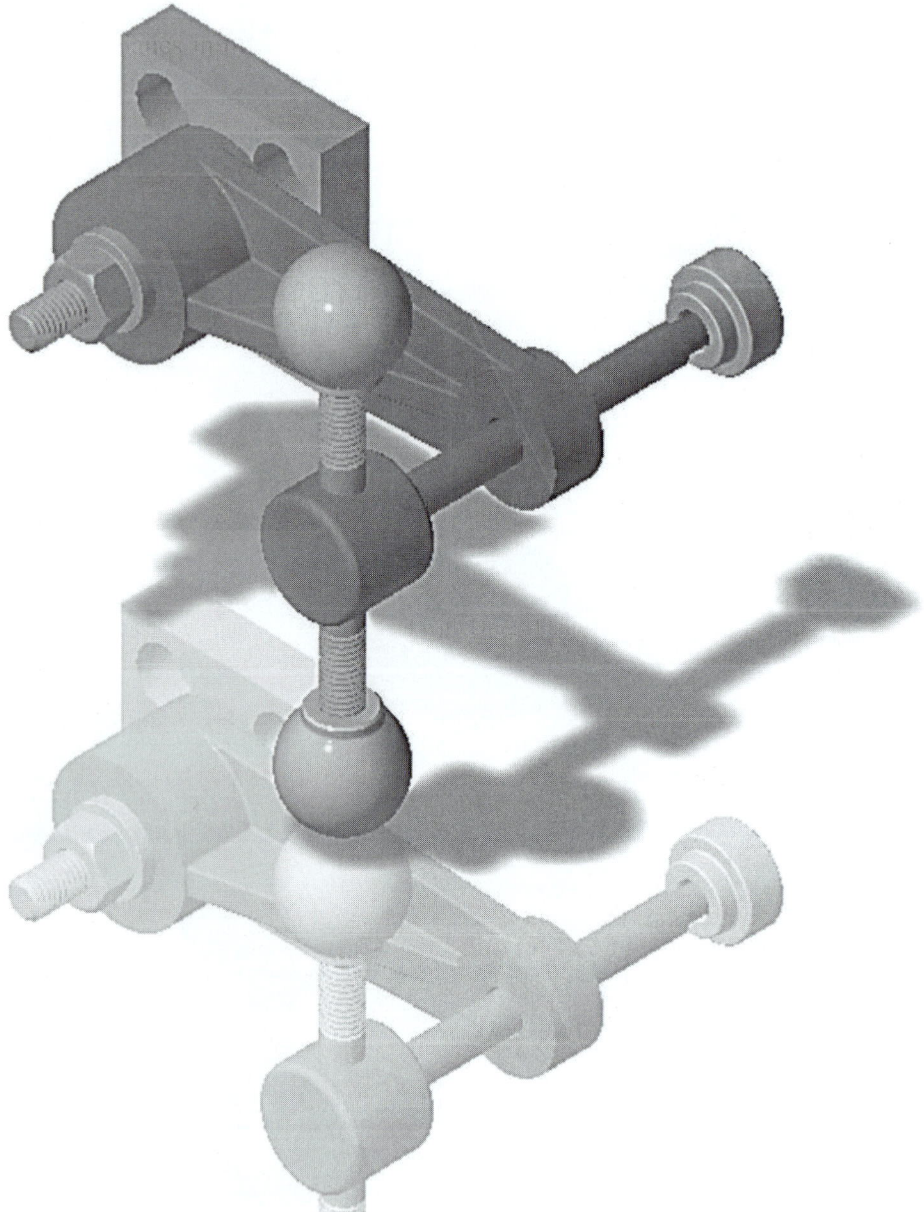

Figure 8.33 Shading With Reflections *(the quality of your graphics card and graphics settings may prevent this display)*

A complete set of projects and illustrations are available at ***www.cad-resources.com.***

Lesson 9 Exploded Assemblies and View Manager

Figure 9.1 Exploded Swing Clamp Assembly

OBJECTIVES

- Create **Exploded Views**
- Create **Explode States**
- Utilize the **View Manager** to organize and control views states
- Create unique component **visibility settings** (**Style States**)
- **Move** and **Rotate** components in an assembly
- Add a **URL** to a **3D Note**
- Create **Perspective** views of the model

REFERENCES AND RESOURCES

For **Resources** go to **www.cad-resources.com** > click on the PTC Creo Parametric 3.0 Book cover

- Lesson 9 Lecture
- Lesson 9 3D models embedded in a PDF
- Project Lectures

EXPLODED ASSEMBLIES

Pictorial illustrations, such as exploded views, are generated directly from the 3D model database (Fig. 9.1). The model can be displayed and oriented in any position. Each component in the assembly can have a different display type: wireframe, hidden line, no hidden, shading, and transparent. You can select and orient the component to provide the required view orientation to display the component from underneath or from any side or position. Perspective projections are made with selections from menus. The assembly can be spun around, reoriented, and even clipped to show the interior features. You have the choice of displaying all components and subassemblies or any combination of components in the design.

Creating Exploded Views

Using the **Explode State** option in the **View Manager**, you can automatically create an exploded view of an assembly (Fig. 9.2). Exploding an assembly affects only the display of the assembly; it does not alter true design distances between components. Explode states are created to define the exploded positions of all components. For each explode state, you can toggle the explode status of components, change the explode locations of components, and create and modify explode offset lines to show how explode components align when they are in their exploded positions. The Explode State Explode Position functionality is similar to the Package/Move functionality.

You can define multiple explode states for each assembly and then explode the assembly using any of these explode states at any time. You can also set an explode state for each drawing view of an assembly. Creo Parametric gives each component a default explode position determined by the placement constraints. By default, the reference component of the explode is the parent assembly (top-level assembly or subassembly).

To explode components, you use a drag-and-drop user interface similar to the Package/Move functionality. You select the motion reference and one or more components, and then drag the outlines to the desired positions. The component outlines drag along with the mouse cursor. You control the move options using a Preferences setting. Two types of explode instructions can be added to a set of components. The children components follow the parent component being exploded or they do not follow the parent component. Each explode instruction consists of a set of components, explode direction references, and dimensions that define the exploded position from the final (installed) position with respect to the explode direction references.

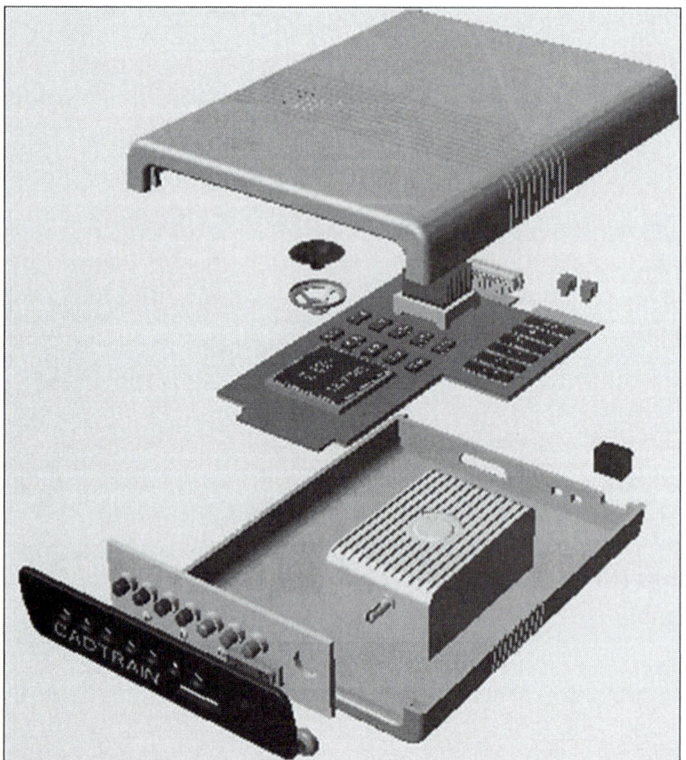

Figure 9.2 Exploded Assemblies

When using the explode functionality, keep in mind the following:

- You can select individual parts or entire subassemblies from the Model Tree or main window.
- If you explode a subassembly, in the context of a higher-level assembly, you can specify the explode state to use for each subassembly.
- You do not lose component explode information when you turn the status off. Creo Parametric retains the information so that the component has the same explode position if you turn the status back on.
- All assemblies have a default explode state called "Default Explode", which is the default explode state Creo Parametric creates from the component placement instructions.
- Multiple occurrences of the same subassembly can have different explode characteristics at a higher-level assembly.

Component Display

Style, also accessed through the View Manager, manages the display styles of an assembly. Simp Rep, Sections, Layers, Orient, Explode, and All are on separate tabs from this dialog [Figs. 9.3(a-h)]. Wireframe, hidden line, no hidden, shaded, or transparent display styles can be assigned to each component. The components will be displayed according to their assigned display styles in the current style state (that is, blanked, shaded, drawn in hidden line color, and so on). The current setting, in the View tab > Display Style, controls the display of unassigned components.

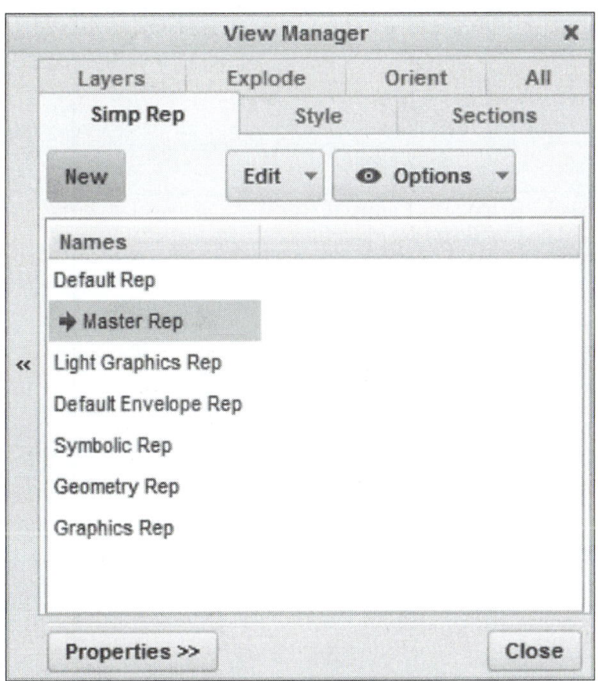

Figure 9.3(a) View Manager Dialog Box

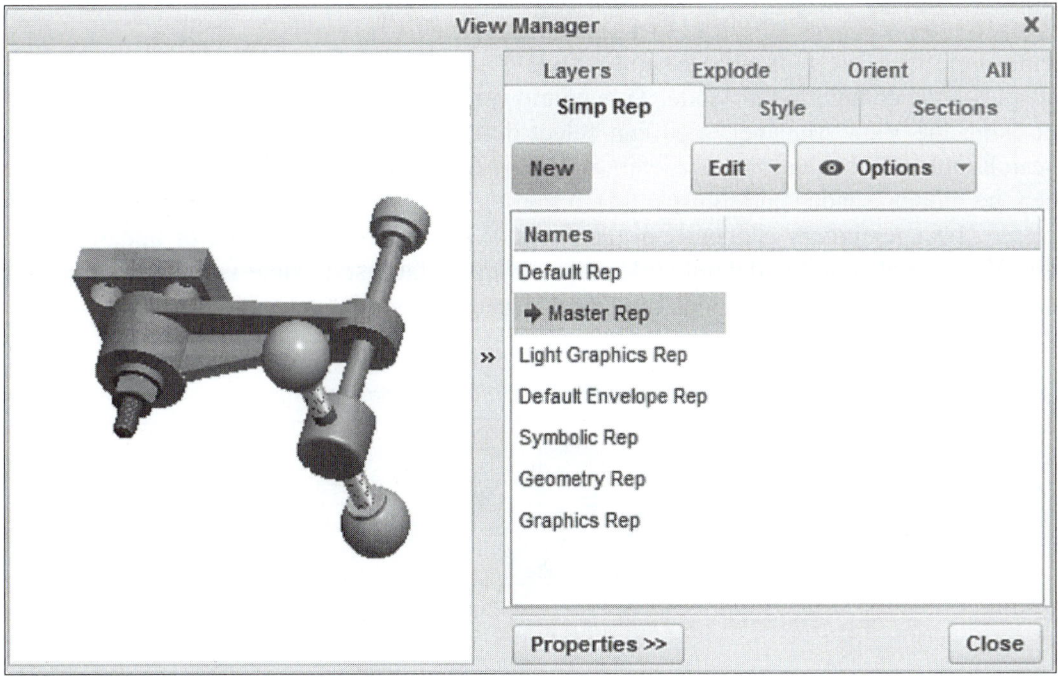

Figure 9.3(b) View Manager- Simp Rep

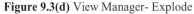

Figure 9.3(c) View Manager- Style

Figure 9.3(d) View Manager- Explode

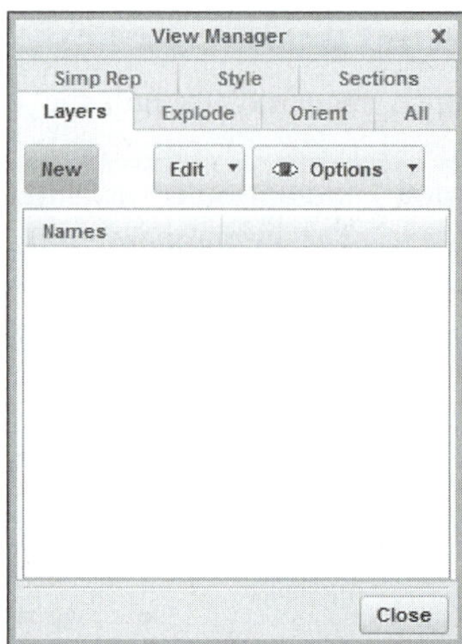

Figure 9.3(e) View Manager- Layers

Figure 9.3(f) View Manager- Sections

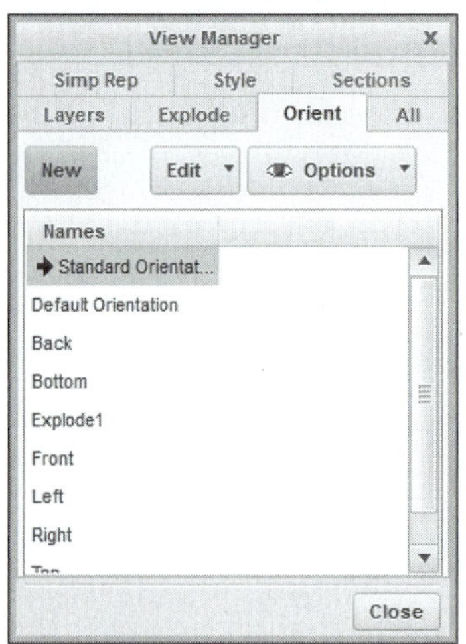

Figure 9.3(g) View Manager- Orient

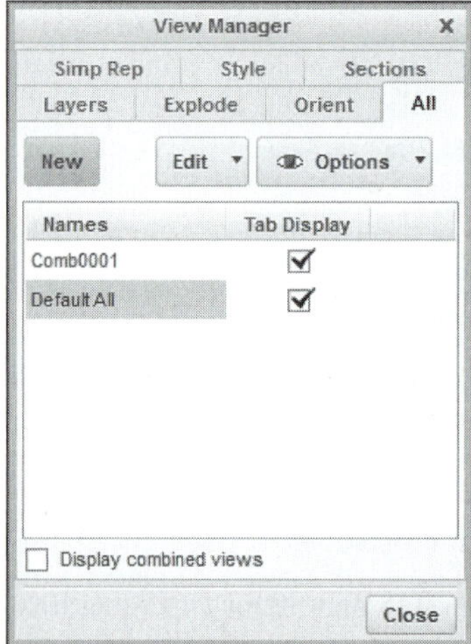

Figure 9.3(h) View Manager- All

Components appear in the currently assigned style state [Fig. 9.4(a)]. The current setting is indicated in the display style column in the Model Tree window. Component display or style states can be modified without using the View Manager. You can select desired models from the graphics window, Model Tree, or search tool, and then use the View tab > Display Style commands to assign a display style (wireframe, hidden, no hidden, shaded and transparent) to the selected models. The Style representation is temporarily changed. These temporary changes can then be stored to a new style state, or updated to an existing style state. You can also define default style states. If the default style state is updated to reflect changes different from that of the master style state, then that default style state will be reflected each time the model is retrieved.

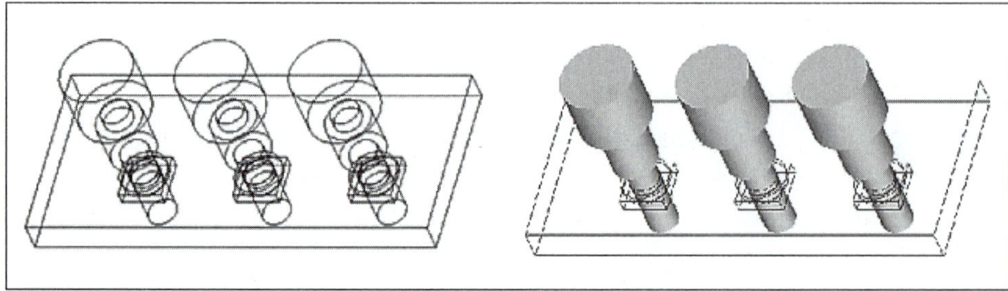

Figure 9.4(a) Style States

Types of Representations

The main types of simplified representations: **default**, **master**, **light graphics** [Fig. 9.4(b)], **default envelope**, **symbolic**, **geometry** and **graphics**; designate which representation appears, using the commands in the View Manager dialog box. Graphics and geometry representations speed up the retrieval process of large assemblies. All simplified representations provide access to components in the assembly and are based upon the Master Representation. You cannot modify a feature in a graphics representation, but you can do so in a geometry representation.

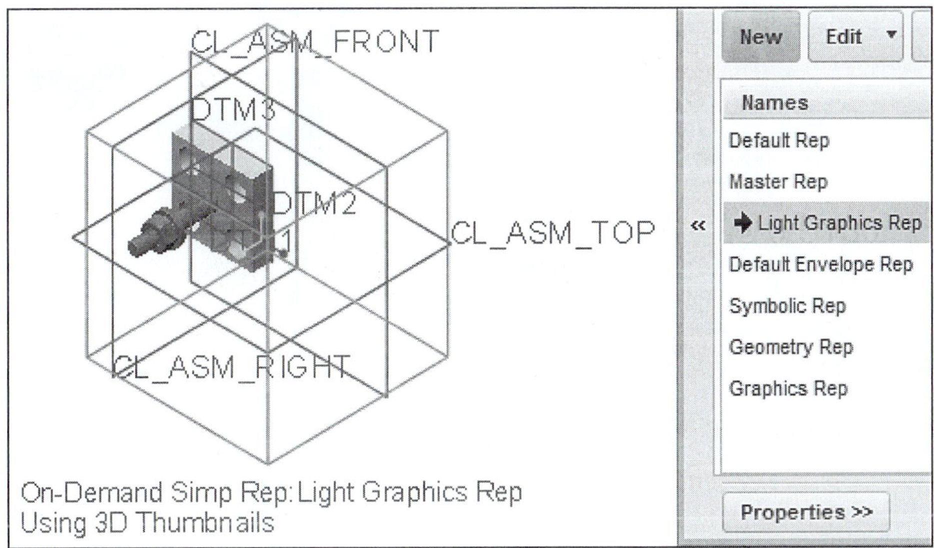

Figure 9.4(b) Light Graphics Rep

Assembly features are displayed when you retrieve a model. Subtractive assembly features such as cuts and holes are represented in graphics and geometry representations, making it possible to use these simplified representations for performance improvement while still displaying on screen a completely accurate geometric model. You can access model information for graphics and geometry representations of part models from the Information menu and from the Model Tree. Because part graphics and geometry representations do not contain feature history of the part model, information for individual features of the part is not accessible from these representations.

- The *Master Representation* always reflects the fully detailed assembly, including all of its members. The Model Tree lists all components in the Master Representation of the assembly, and indicates whether they are included, excluded, or substituted.
- The *Graphics Representation* contains information for display only and allows you to browse through a large assembly quickly. You cannot modify or reference graphics representations. The type of graphic display available depends on the setting of the *save_model_display* configuration option, the last time the assembly was saved:

    - ***wireframe*** (default) The wireframe of the components appear.
    - ***shading_low***, ***shading_med***, ***shading_high*** A shaded version of the components appears. The different levels indicate the density of the triangles used for shading.
    - ***shading_lod*** The level of detail depends on the "Shaded model display settings" dialog box setting. Click: File > Options > Model Display before you open a model to access this dialog box.

While in a simplified representation, Creo Parametric applies changes to an assembly, such as creation or assembly of new components, to the Master Representation. It reflects them in all of the simplified representations. It applies all suppressing and resuming of components to the Master Representation. However, it applies the actions of a simplified representation only to currently resumed members, which is, to members that are present in the BOM of the Master Representation.

Lesson 9 STEPS

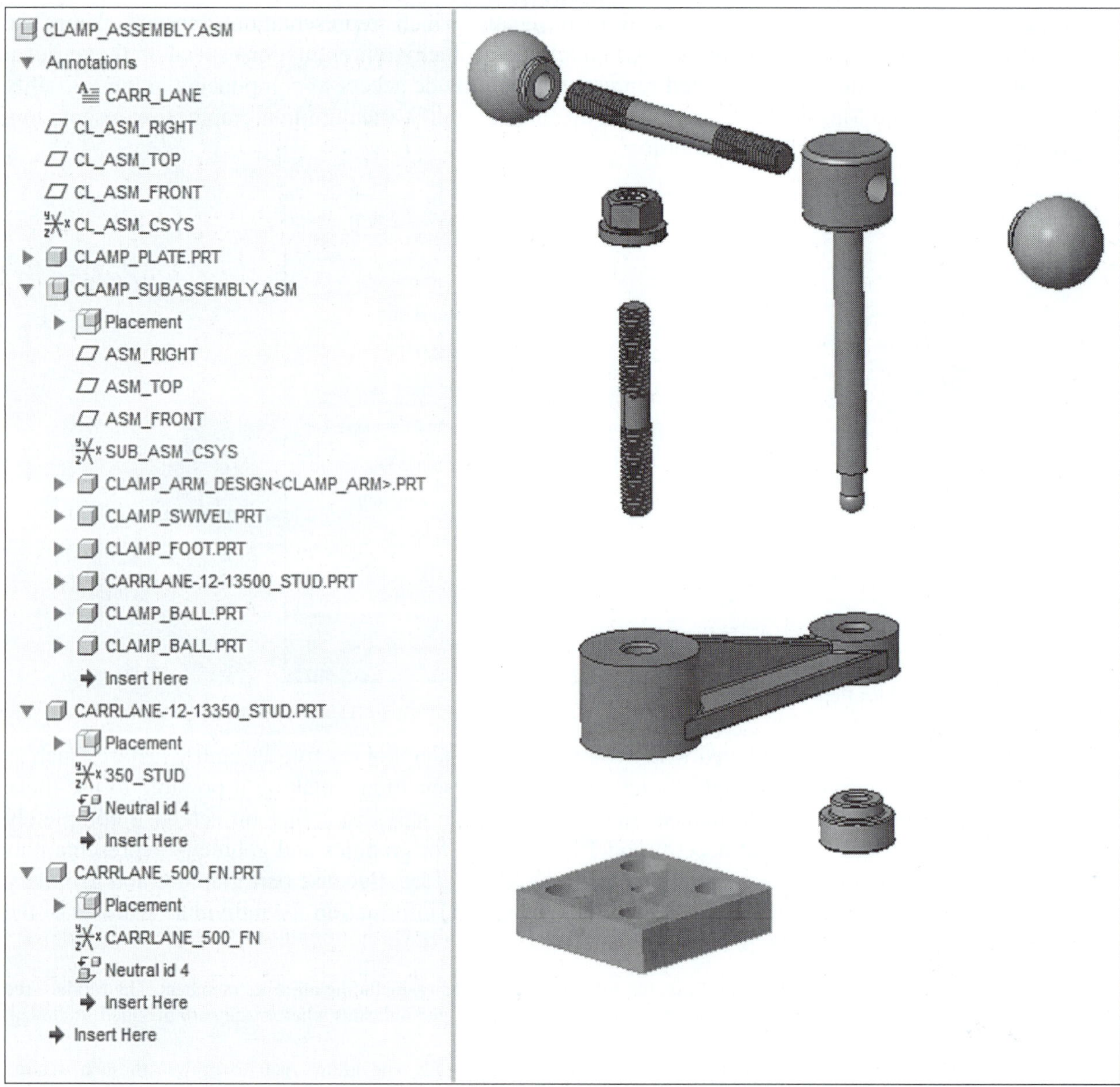

Figure 9.5 Exploded Swing Clamp Assembly

Exploded Swing Clamp Assembly

In this lesson, you will use the previously created subassembly and assembly (Fig. 9.5) to establish and save new views, exploded views, and views with component style states that differ from one another. The View Manager Dialog box will be employed to control and organize a variety of different states.

You will also be required to move and rotate components of the assembly before cosmetically displaying the assembly in an exploded state. The creation and assembly of new components will not be required. A 3D Note with a URL will be added to the model as the last information feature.

URLs and Model Notes

Since the assembly you created is a standard clamp from CARR LANE Manufacturing Co.; you could go to CARR LANE's Website and see the assembly, order it or download a 3D model. Here you will create and attach a 3D Note to the assembly, identifying the manufacturer and Website.

Launch **PTC Creo Parametric 3.0** > **File** > **Manage Session** > **Select Working Directory** > select the directory where the **clamp_assembly.asm** was saved > **OK** > **File** > **Open** > **clamp_assembly.asm** > **Open** > **File** > **Options** > **Model Display** > **Trimetric** > **Isometric** > **OK** > **No** > > **Tree Filters** > toggle all *on* > **OK** > Filters: Smart > Annotation > press **Ctrl** and select all four holes notes > **RMB** > Hide **Hide** > **View** tab > *off* > *off* > *on* > *on* > **Layers** (from the Ribbon) > pick on Layers (in the Layer Tree) > **RMB** > **Hide** > **Layers** (from the Ribbon) > > **Annotate** tab > FLAT TO SCREEN FRONT TOP RIGHT (Fig. 9.6) > **RMB** (on the FLAT TO SCREEN icon) > **Set** > Notes **Unattached Note** (in the Annotations Group) > pick a position for the note in the upper right side of the model >
type **Carr Lane Manufacturing Company** > **Enter** >
type **Component Parts** > **Enter** >
type **Jigs and Fixtures** > **MMB** > move as needed > **RMB** (Fig. 9.7) > **Add Link**

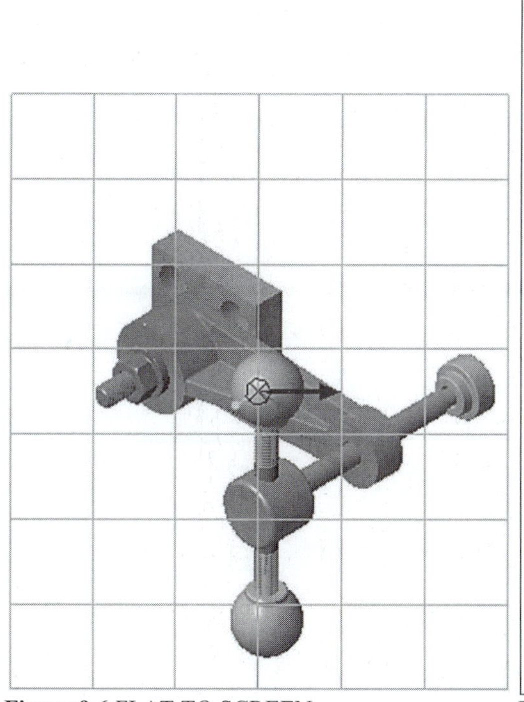

Figure 9.6 FLAT TO SCREEN

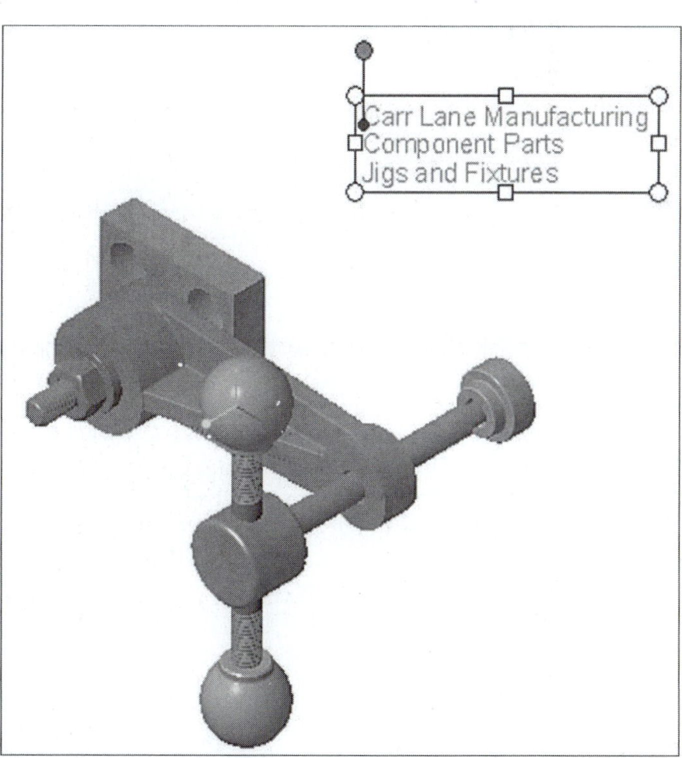

Figure 9.7 Note Dialog Box

Click: **Hyperlink** > *type* **www.carrlane.com** (Fig. 9.8) > **ScreenTip** > Screen tip text: *type* **CARR LANE Manufacturing** (Fig. 9.9) > **OK** > **OK** > **RMB** > **Text Style** > Default > Height **4** > **Enter** (Fig. 9.10) > **Apply** > **OK** > In the Model Tree ▼ Annotations > **CARR LANE** > **RMB** > **Rename** > **CARR_LANE** > **Enter** > **LMB** in the Graphics Window to deselect > **Ctrl+D** > **Ctrl+S** > place your pointer over the screen note (Fig. 9.11)

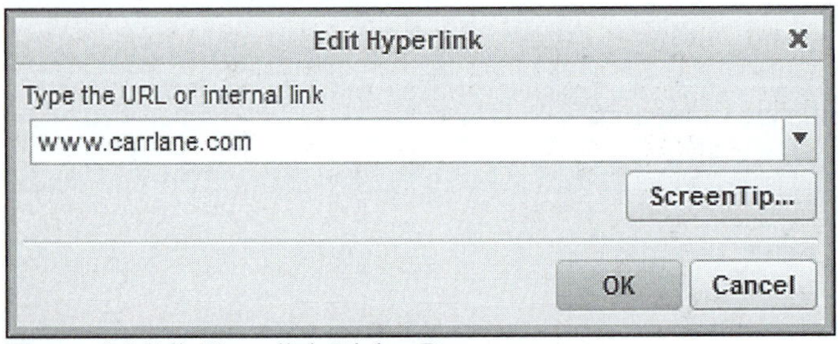

Figure 9.8 Edit Hyperlink Dialog Box

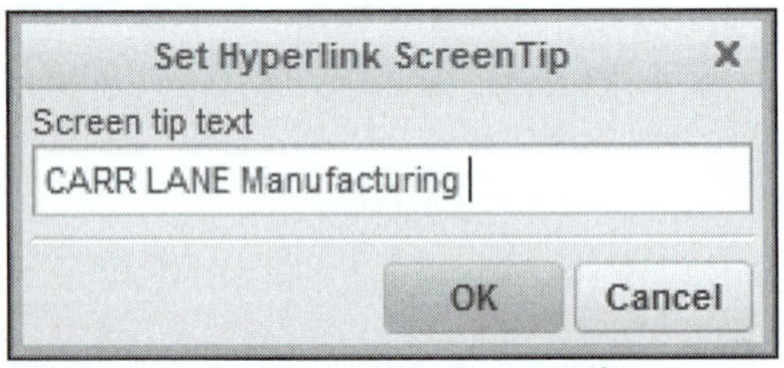

Figure 9.9 Set Hyperlink Screen Tip

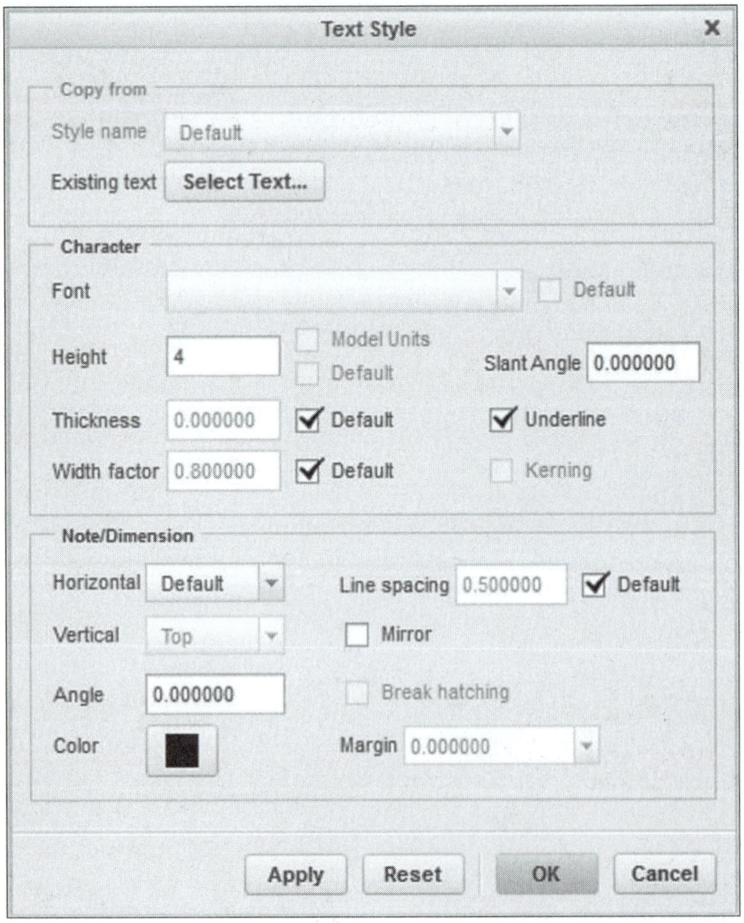

Figure 9.10 Text Style Dialog Box

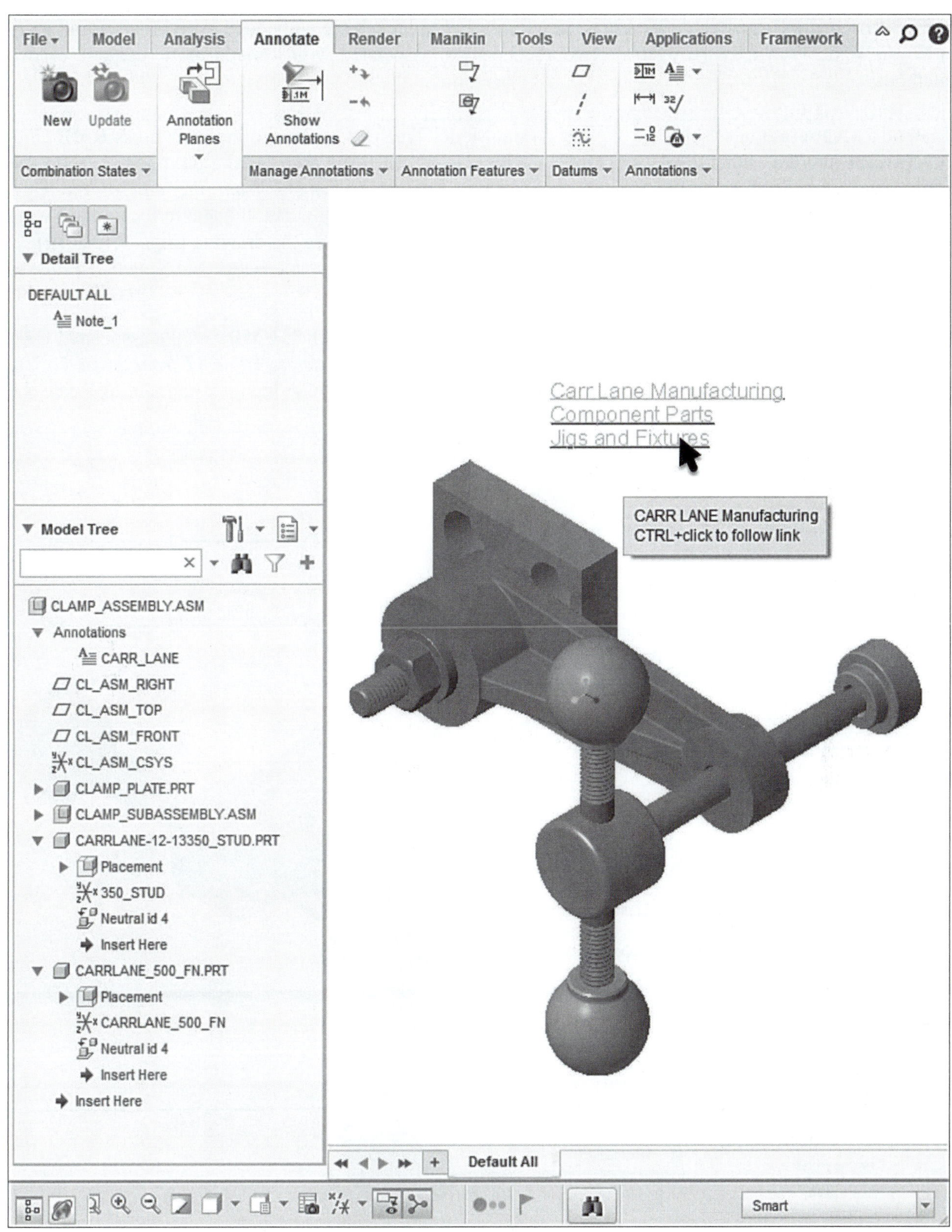

Figure 9.11 Note Displayed in Model Tree and on Model

To open a Hyperlink defined in a model note and to launch the embedded Web browser and go to a World Wide Web URL (Universal Resource Locator) associated with a model note, use one of two methods.

Method 1: **Annotate** tab > pick on the note in the Model Tree or Graphics Window > press **RMB** [Fig. 9.12(a)] the shortcut menu displays > **Open URL** [Fig. 9.12(b)] the associated URL opens in the Creo Parametric embedded browser *(if you get an error: delete **+0** or **+1** from the end of the URL **(http://www.carrlane.com/?CLAMP_ASSEMBLY.ASM** > **Enter**)* > navigate the site to locate the Swing Clamp [Figs. 9.12(c-g)] > **Dimensions** tab > use the scroll bars to the see the drawing [Figs. 9.12(h-i)]) *(Note that to pick the 3D Note in the Graphics Window; you may have to change the Selection Filter to Annotation)*

Method 2: Move the cursor over the hyperlinked note > press **Ctrl** > click **LMB** and the associated URL opens in the Creo Parametric embedded browser *(if you get an error: delete **+0** or **+1** from the end of the URL **(http://www.carrlane.com/?CLAMP_ASSEMBLY.ASM** > **Enter**)* > navigate the site to locate the Swing Clamp

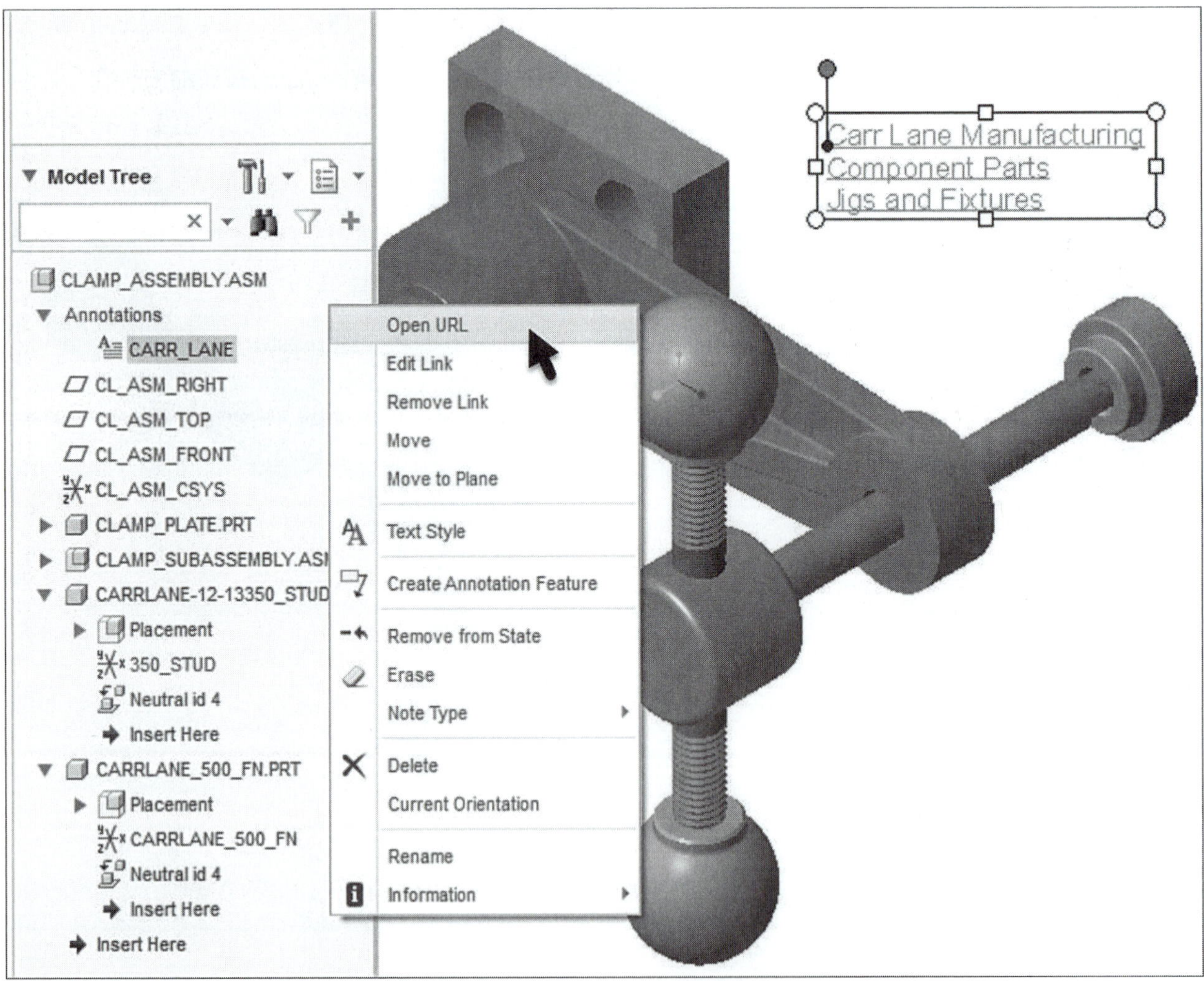

Figure 9.12(a) Open URL

CarrLane
MANUFACTURING CO.
314-647-6200
Fax: 314-647-5736
Email Carr Lane
Your Complete Source
For Component Parts of Jigs and Fixtures

CL5 QUICK CHANGE 5-AXIS WORKHOLDING
CLICK FOR MORE
CarrLane

ISO 9001-2008 CERTIFIED
FM35617
View Certificate

- Purchase Order Search
- Price & Availability
- Product Comparison
- Return Status
- Distributor Locator
- Featured Products
- Customer Survey
- CAD Drawings Search
- Trade Shows
- Articles/Press Releases
- About Carr Lane
- FAQ & Assistance
- Visitor Feedback
- Distributors Only
- Map & Directions
- Contact Carr Lane

- Online Catalog
- PDF Catalog
- Request a Catalog
- Plaques & Awards
- New Items
- Trig Book

See Our Affiliate Companies

CarrLane Castings
QUALITY Screw Machine Products, Inc.
ROEMHELD HILMA • STARK CARR LANE ROEMHELD

See Our NEW On-Line Catalog and Factory Store

Receive the Courier - our E-newsletter!
» JOIN NOW
Privacy by SafeSubscribe[SM]
For Email Marketing you can trust

Carr Lane Products

CARR LANE CO., 4200 CARR LANE CT., P.O. BOX 191970, ST. LOUIS, MO 63119-7970 USA

Carr Lane parts for engineers in all areas of manufacturing. Choose from tooling components for jigs and fixtures, toggle clamps, modular fixturing, drill jig bushings, and power workholding. Design engineers, manufacturing engineers and equipment buyers use our products, which include handles and knobs, ball plungers, hoist rings, , screw clamps, supports, rests, machine leveling feet, locators, clamp straps, fasteners, plug gages, threaded inserts, fixture bases and hydraulic clamps. View our on line catalog.

Powered by Carr Lane Advertising, LLC

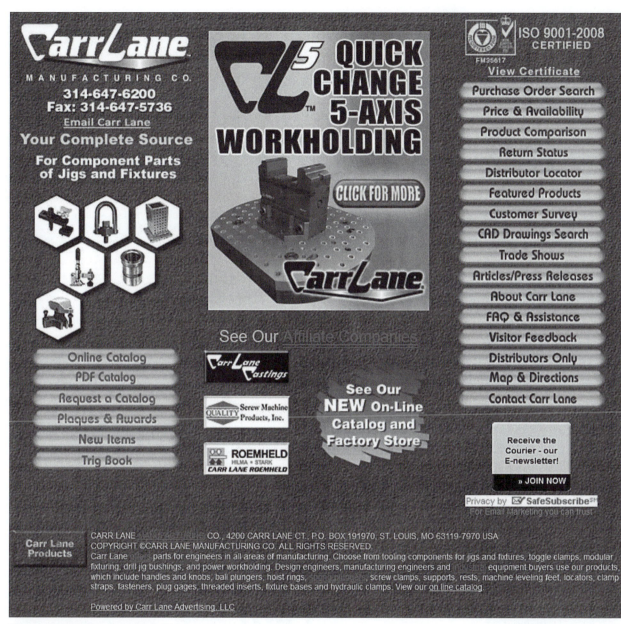

Figure 9.12(b) Carr Lane Web Site

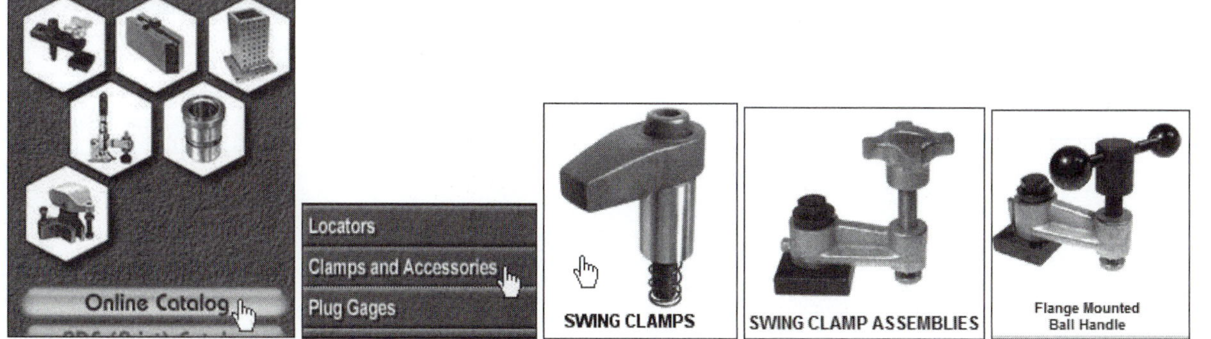

Figures 9.12(c-g) Online Catalog > Clamps and Accessories > Swing Clamps > Swing Clamp Assemblies > Flange Mounted Ball Handle

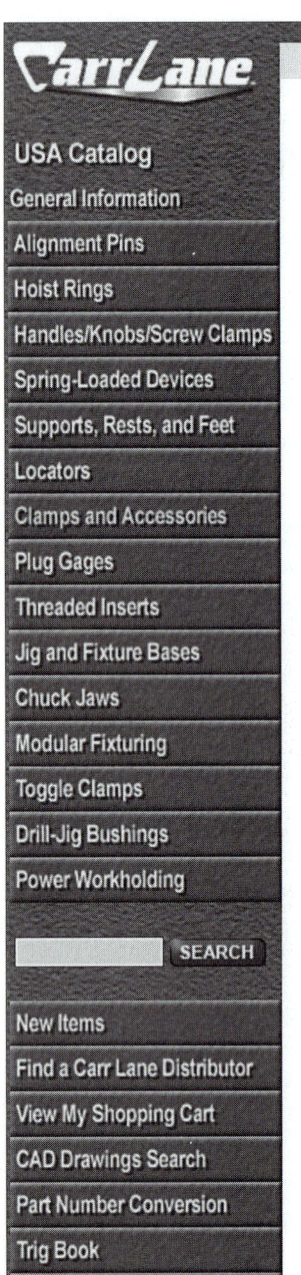

Features | Dimensions

CLAMPS AND ACCESSORIES » SWING CLAMPS » SWING CLAMP ASSEMBLIES » **Flange Mounted Ball** Handle

SWING CLAMP ASSEMBLIES

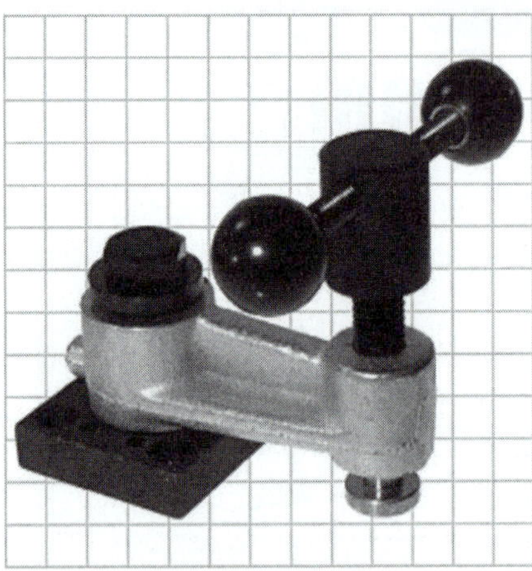

Flange Mounted Ball Handle

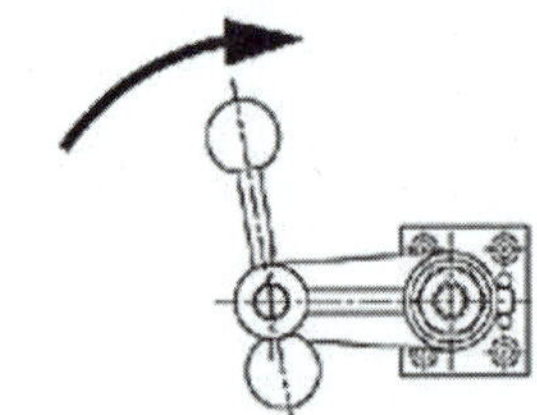

FEATURES: Long-reach swing clamp that can be hand tightened for medium clamping force. This flange-mounted version is ideal for mounting on flat surfaces. The stop pin furnished can be installed on either side for clockwise or counterclockwise swing. Made in USA.

SIZES: Available in five sizes — see Dimensions page for part numbers:

1-1/2" Reach
2" Reach
2-1/2" Reach
3" Reach
4" Reach

SWIVEL FOOT: Swing Clamp Assemblies use our patented left-hand-thread swivel-foot system. The ball end of the body screws into the foot for easy installation and removal. Once the ball threads pass beyond the foot threads, the foot can rotate and pivot freely (20° total angle). Foot threads are left-hand to positively prevent accidental black out during use. Delrin® feet are optional. To order with a Delrin® foot, add "N" to the part number, e.g. CL-1-SWA-3N. Note: Delrin® feet are not threaded – they snap on and off. For other foot types, see Feet for Swivel Screws (order separately).

MATERIAL: Swing arm > Nodular iron per ASTM A536 grade 65-45-12, zinc plated clear chromate. Post and flange base > Mild steel, black oxide finish. Other components > See catalog information for each specific item. DFARS and RoHS compliant.

en_2063

Figure 9.12(h) Ball Handle Flange Mounted Swing Clamp

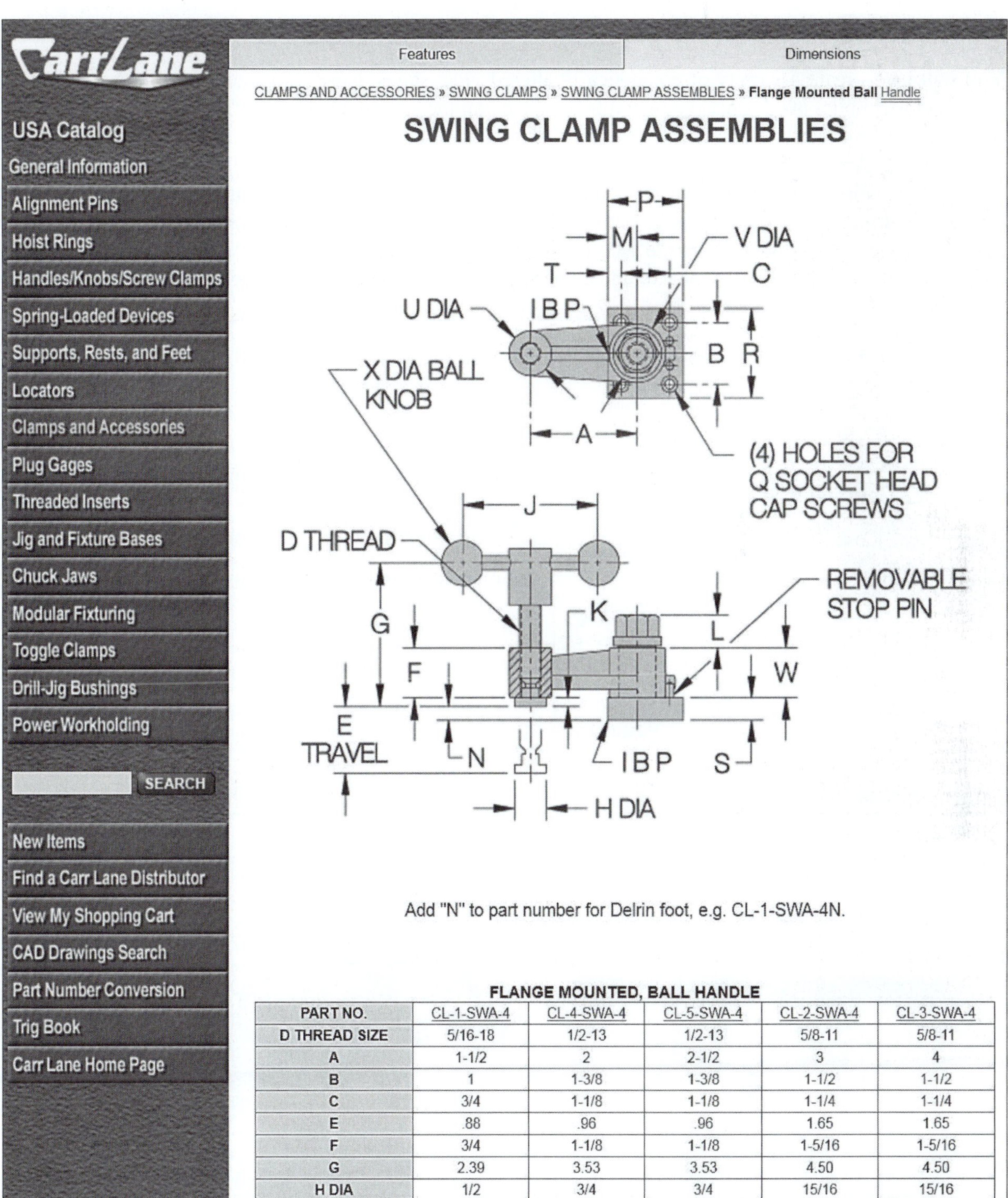

PART NO.	CL-1-SWA-4	CL-4-SWA-4	CL-5-SWA-4	CL-2-SWA-4	CL-3-SWA-4
D THREAD SIZE	5/16-18	1/2-13	1/2-13	5/8-11	5/8-11
A	1-1/2	2	2-1/2	3	4
B	1	1-3/8	1-3/8	1-1/2	1-1/2
C	3/4	1-1/8	1-1/8	1-1/4	1-1/4
E	.88	.96	.96	1.65	1.65
F	3/4	1-1/8	1-1/8	1-5/16	1-5/16
G	2.39	3.53	3.53	4.50	4.50
H DIA	1/2	3/4	3/4	15/16	15/16

Figure 9.12(i) Select Dimensions Tab

Click: [x] to close the Browser > **Annotation Display** *off* > **Ctrl+D** > **Ctrl+S** > **Enter** > in the Graphics Window, **LMB** to deselect

Views: Perspective, Saved, and Exploded

You will now create a variety of cosmetically altered view states. Cosmetic changes to the assembly do not affect the model itself, only the way it is displayed on the screen. One type of view that can be created is the perspective view.

Perspective creates a single-vanishing-point perspective view of a shaded or wireframe model. These views allow you to observe an object as the view location follows a curve, axis, cable, or edge through or around an object. To add perspective to a model view, you select a viewing path and then control the viewing position along the path in either direction. You can also rotate the perspective view in any direction, zoom the view in or out, and change the view angle at any point along the path.

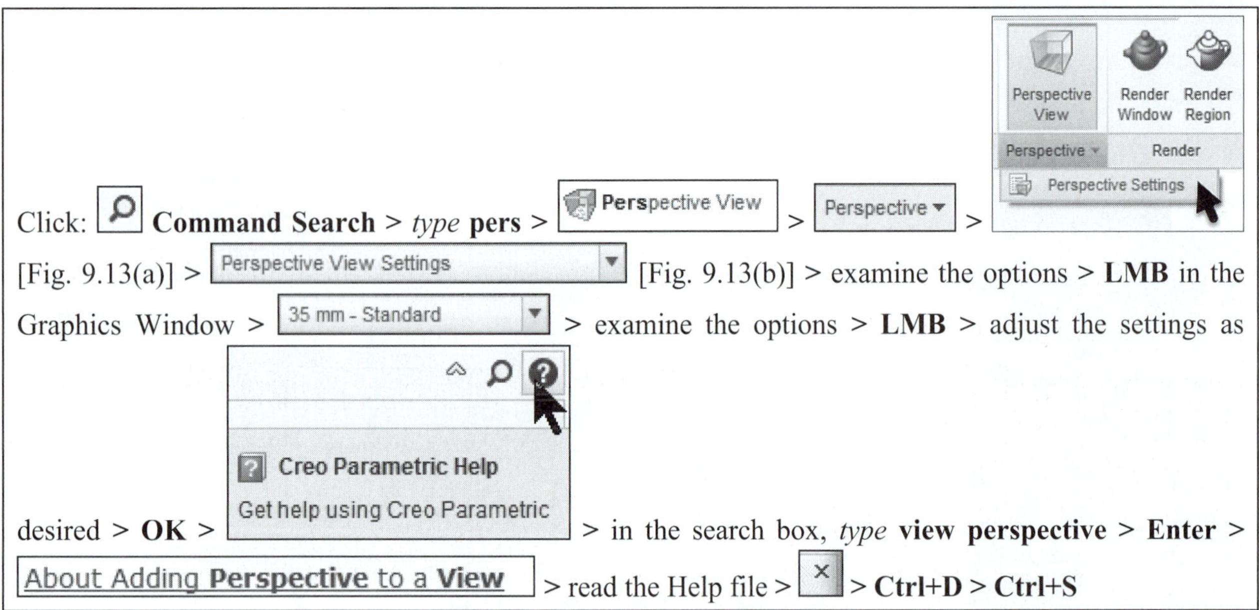

Click: **Command Search** > *type* **pers** > Perspective View > Perspective > [Fig. 9.13(a)] > Perspective View Settings [Fig. 9.13(b)] > examine the options > **LMB** in the Graphics Window > 35 mm - Standard > examine the options > **LMB** > adjust the settings as desired > **OK** > Creo Parametric Help > in the search box, *type* **view perspective** > **Enter** > About Adding **Perspective** to a **View** > read the Help file > × > **Ctrl+D** > **Ctrl+S**

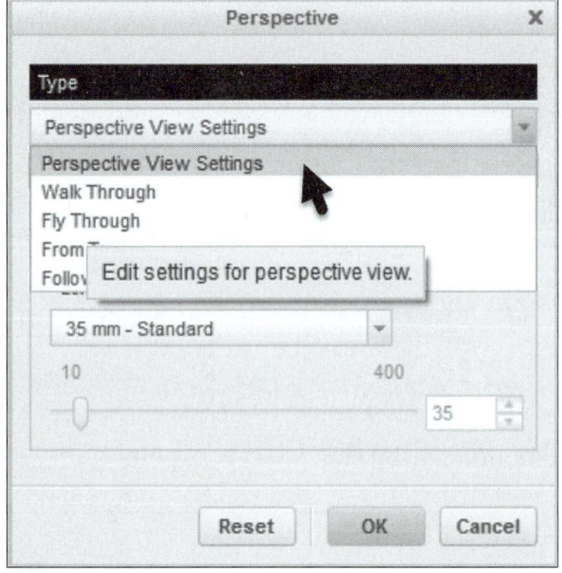

Figure 9.13(a) Perspective Dialog Box

Figure 9.13(b) Perspective View

Saved Views

Create a saved view and explode state to be used later on an assembly drawing. When using the View Manager to set display and explode states, it is a good idea to create one or more saved views to be used later for exploding. The default trimetric and isometric views do not adequately represent the assembly in its functional position.

Using the **MMB**, rotate the model from its default position [Fig. 9.14(a)] to one where the Clamp_Swivel is approximately vertical [Fig. 9.14(b)] > **Saved Orientations** from the Graphics Toolbar > [Fig. 9.14(c)] > **Saved Orientations** [Fig. 9.14(d)] > Name: *type* **EXPLODE1** > **Save** [Fig. 9.14(e)] > **OK**

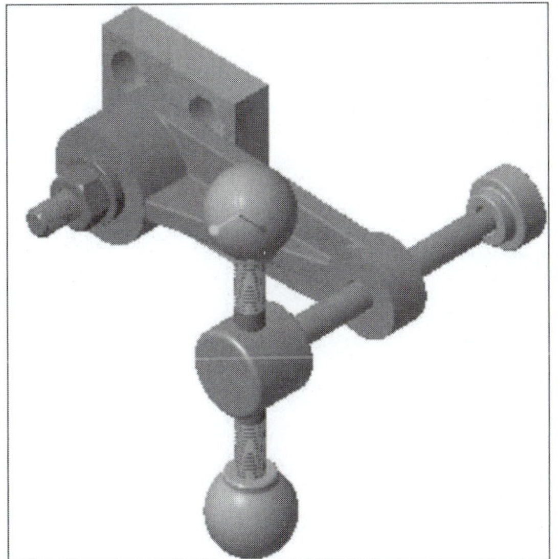

Figure 9.14(a) Isometric View

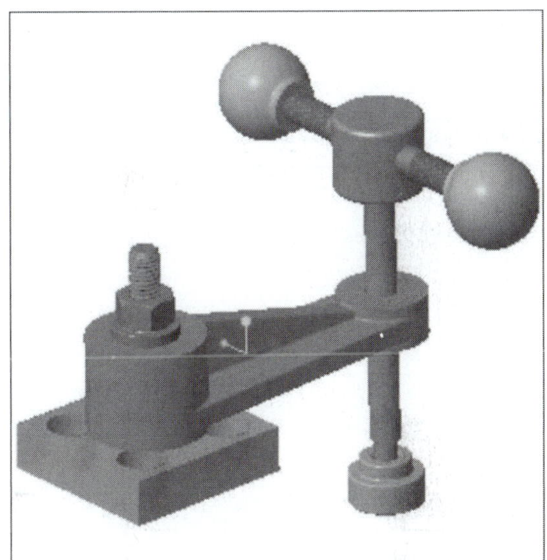

Figure 9.14(b) Reoriented View

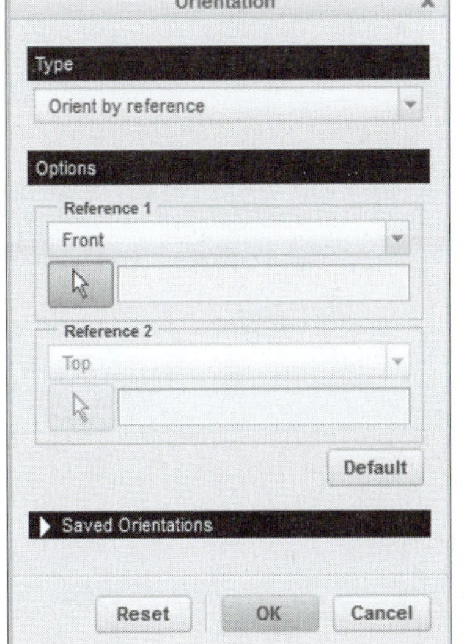

Figure 9.14(c) Orientation Dialog Box

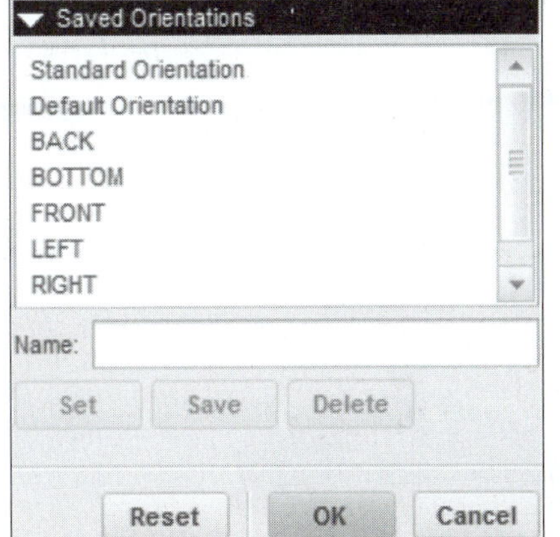

Figure 9.14(d) Saved Views

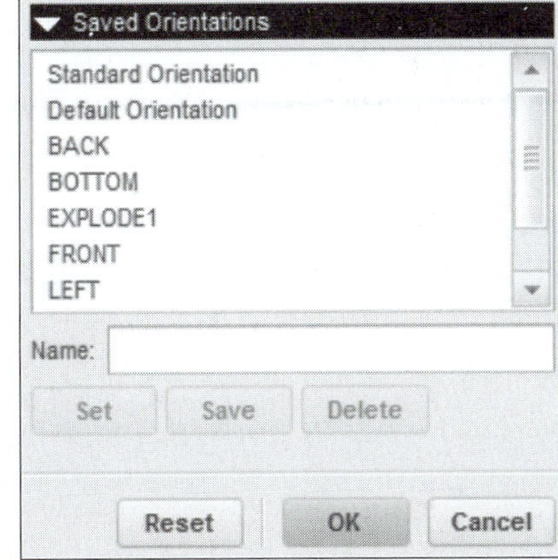

Figure 9.14(e) EXPLODE1 View

Default Exploded Views

When you create an *exploded view*, Creo Parametric moves apart the components of an assembly to a set default distance. The default position is seldom the most desirable.

Click: **View** tab > Exploded View [Figs. 9.15(a-b)] > > Exploded View >

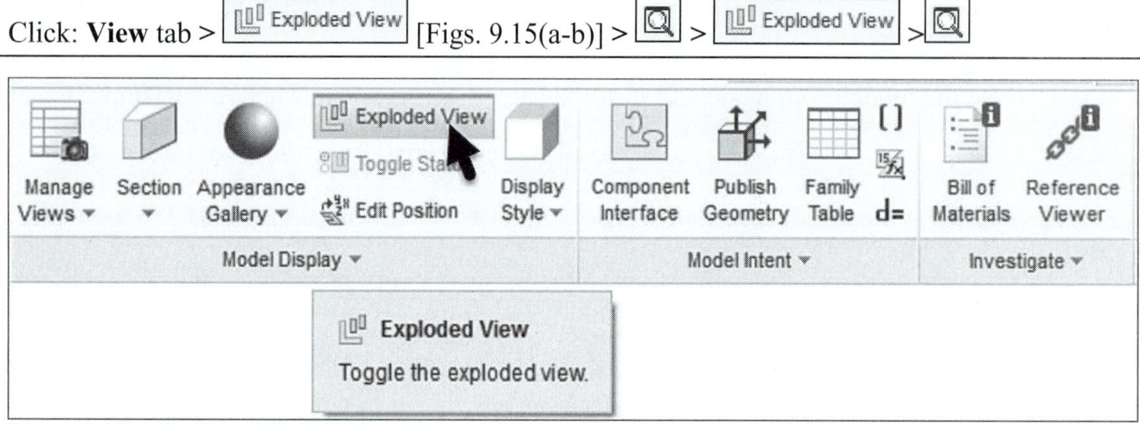

Figure 9.15(a) Exploded View

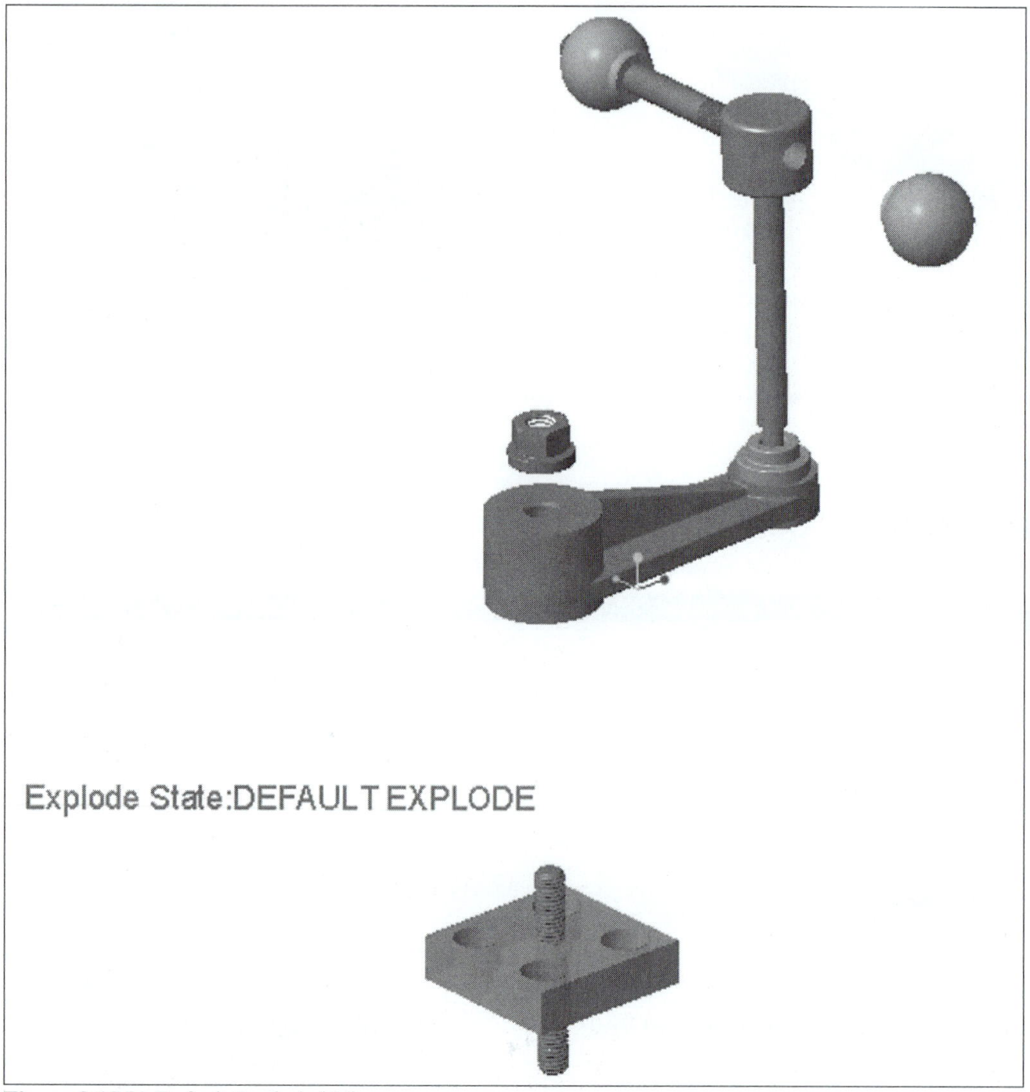

Figure 9.15(b) Default Exploded View

Click: **View** tab > Exploded View > [Refit] > Exploded View > [Refit] > place the pointer inside the Graphics Toolbar > **RMB** > **Reset to Default** > **RMB** > check all on > **Location** > **Show in Status Bar** [Fig. 9.16(a)] > press **LMB** on the bottom edge of the Graphics Window (be displaye) > move the pointer to change the height of the message area [Fig. 9.16(b)]

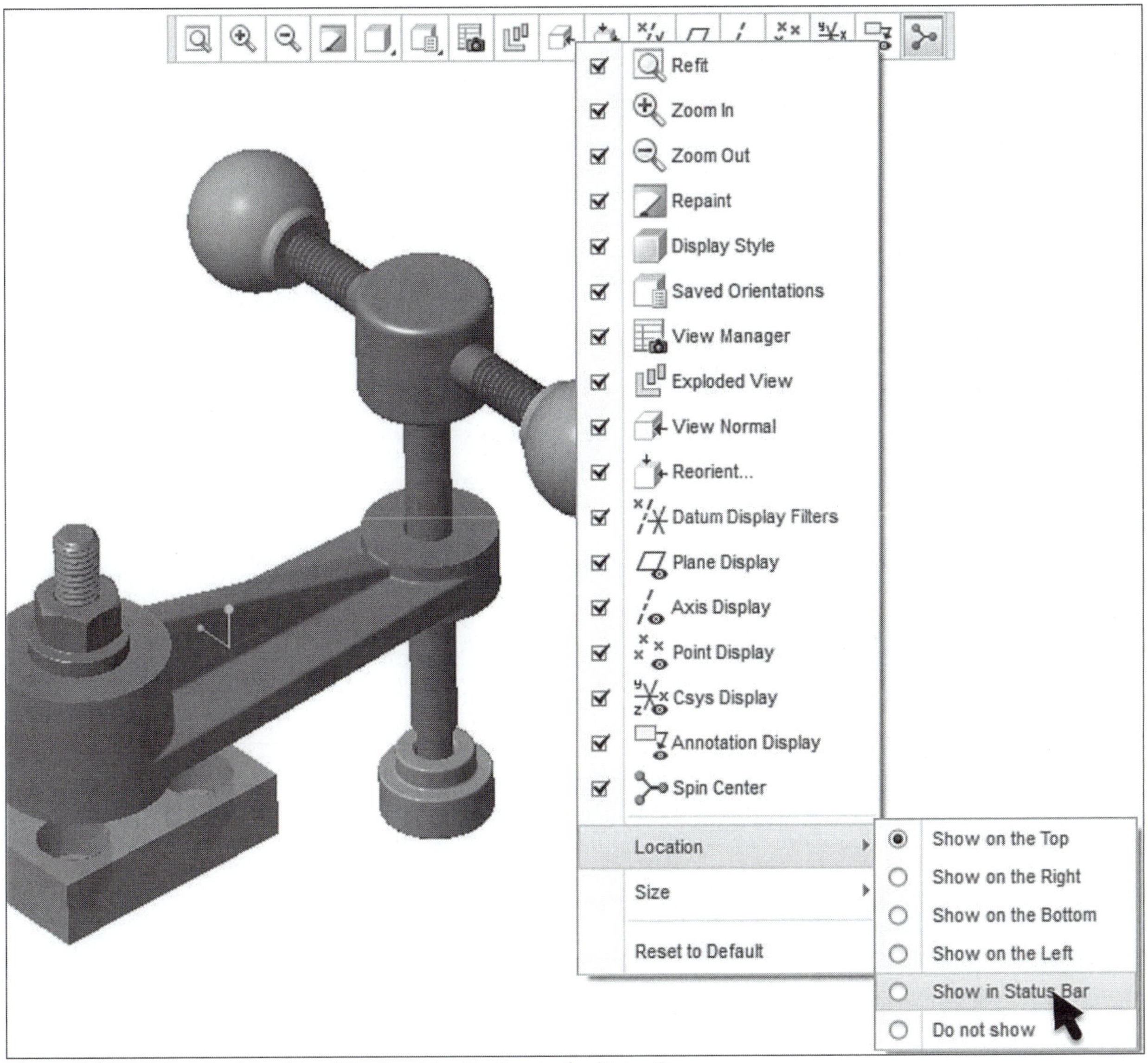

Figure 9.16(a) Move the Location of the Graphics Toolbar to the Status Bar

Figure 9.16(b) Resized Message Log and Status Bar with Integrated Graphics Toolbar *(yours may appear differently)*

Click: **Model** tab > **View Manager** from the Status Bar > **Simp Rep** tab > **Master Rep** > **Options** > **Activate** > **Orient** tab > **Explode1(+)** > **RMB** > **Save** [Fig. 9.17(a)] > **OK** [Fig. 9.17(b)] > **Explode** tab > **New** > **Enter** to accept the default name: Exp0001 [Fig. 9.17(c)] > **Properties** >> (bottom left of the View Manager dialog box) > **Edit position** [Fig. 9.17(d)] > **References** tab [Fig. 9.17(e)]

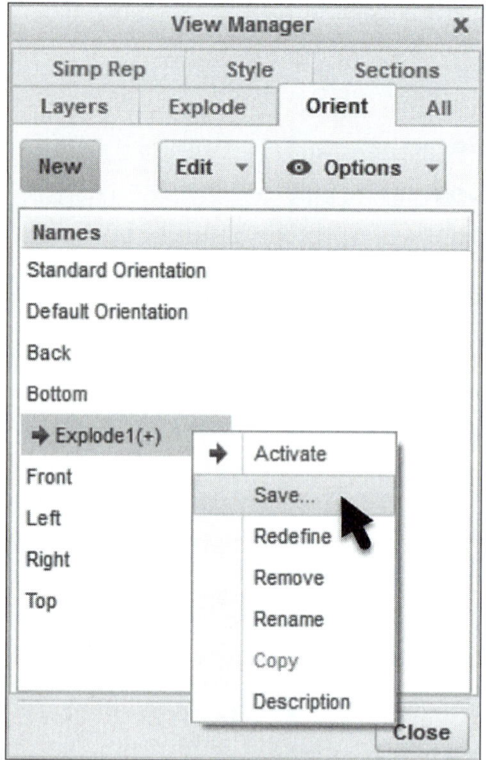

Figure 9.17(a) View Manager Orient Tab

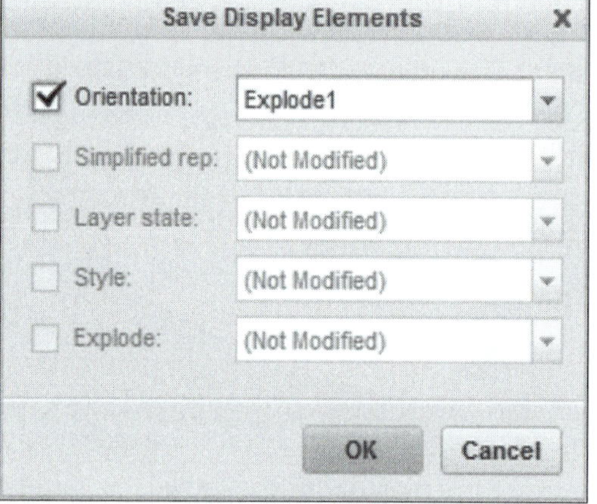

Figure 9.17(b) Save Display Elements

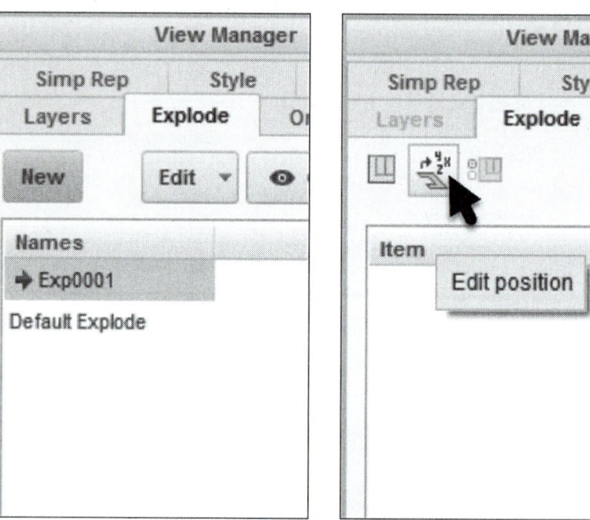

Figure 9.17(c) Exp0001

Figure 9.17(d) Edit Position

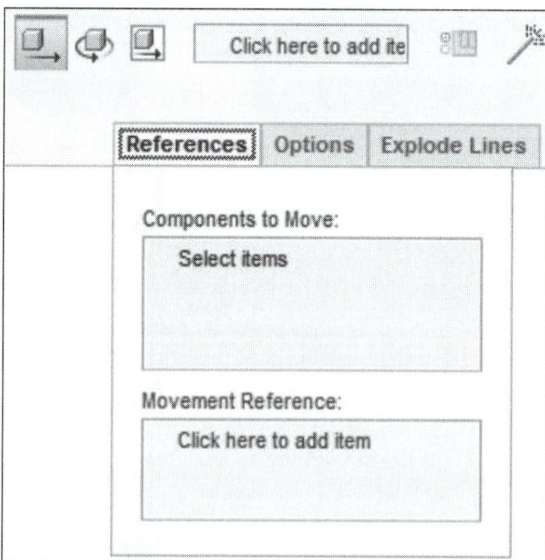

Figure 9.17(e) Explode Position Dashboard

Click: **Options** tab > ☑ Move with Children [Fig. 9.18(a)] > **References** tab > Components to Move: select the top of the **CARRLANE-12-13350_STUD** [Fig. 9.18(b)]

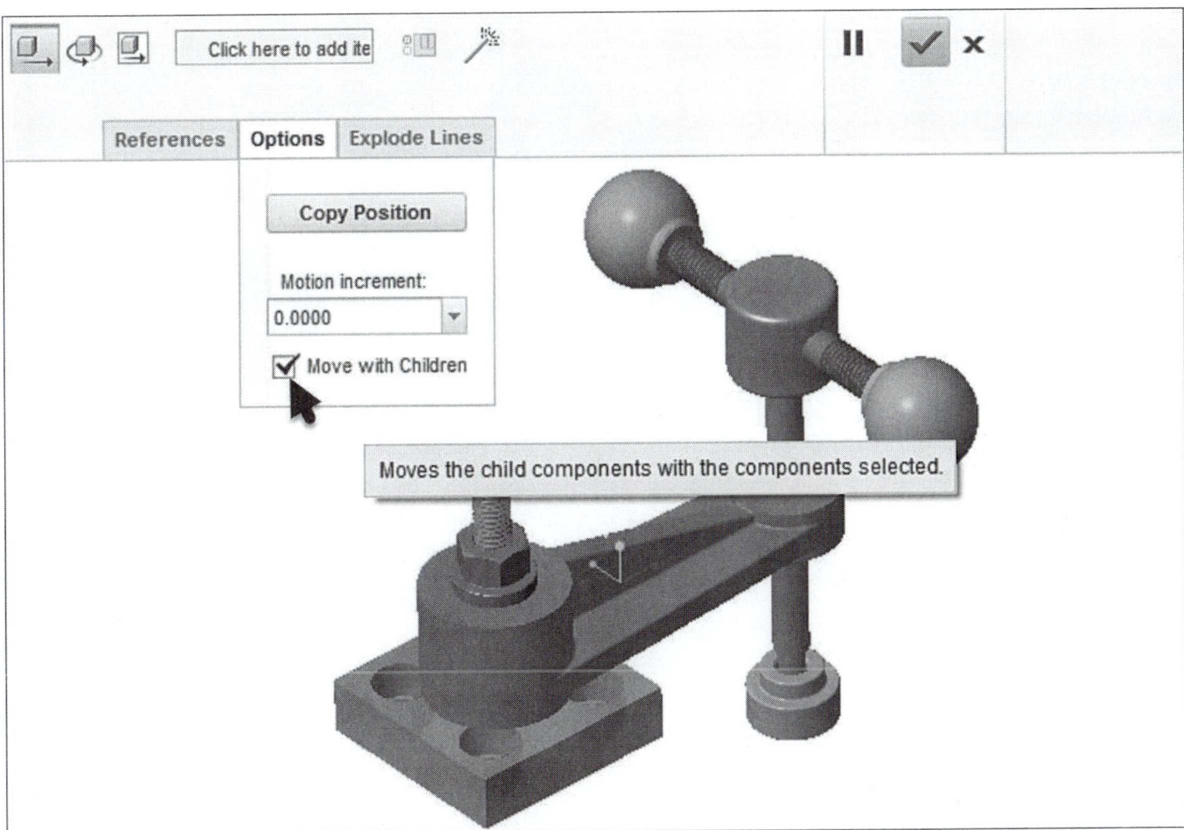

Figure 9.18(a) Options: Move with Children

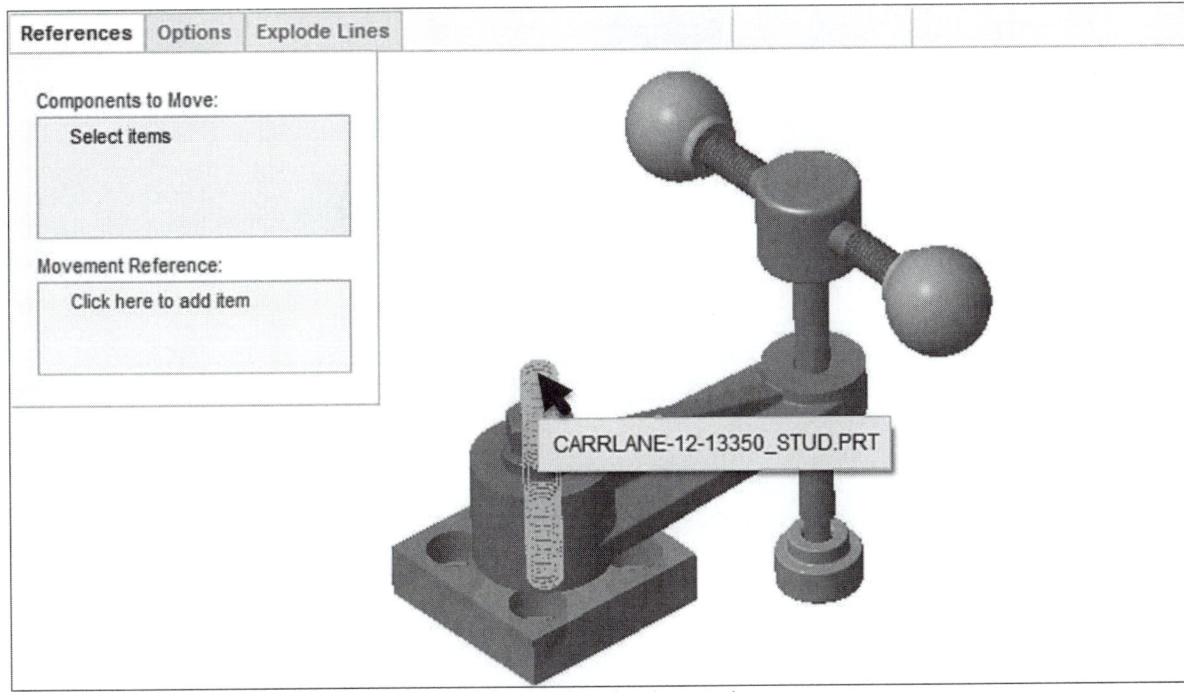

Figure 9.18(b) References tab, Components to Move: select the short stud

Using the **LMB**, select and hold the **X axis** of the temporary coordinate system [Fig. 9.18(c)] > move the pointer to drag the components to a new position > release the **LMB** [Fig. 9.18(d)]

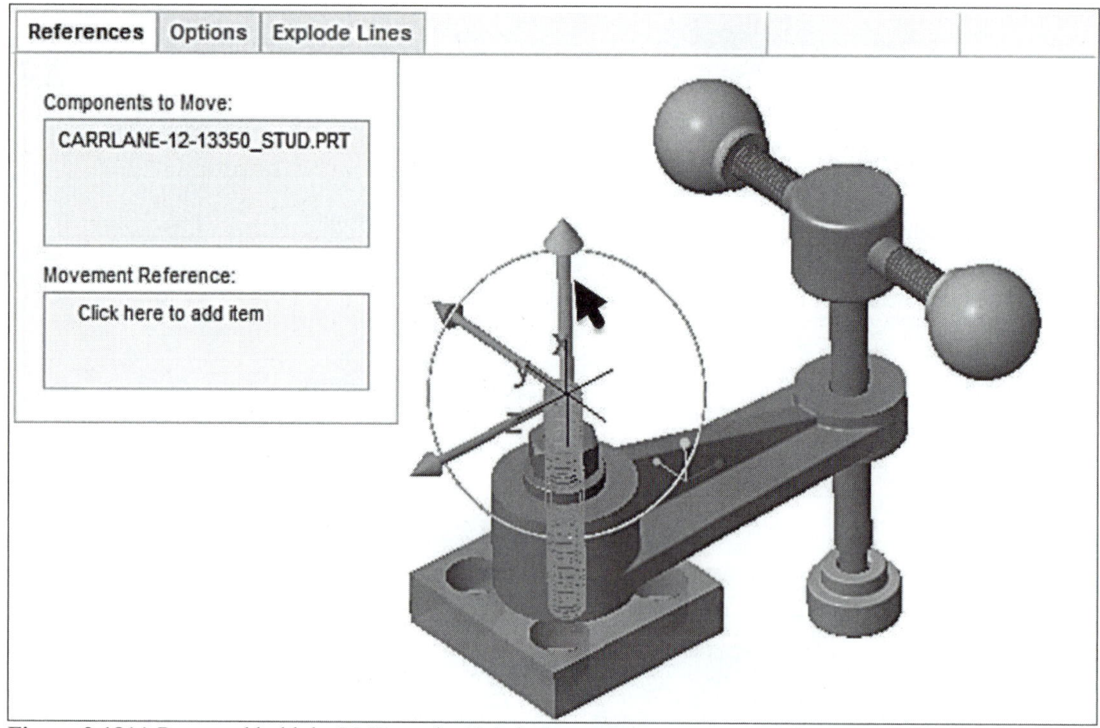

Figure 9.18(c) Press and hold the LMB on the X Axis *(your view may appear differently)*

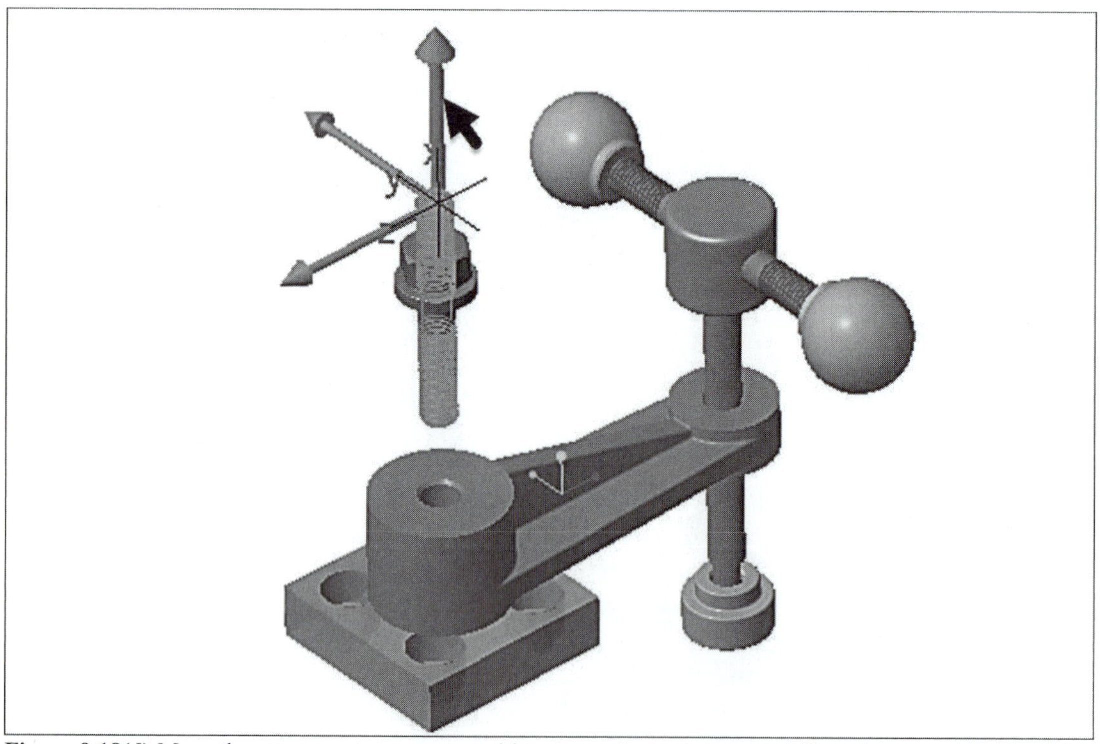

Figure 9.18(d) Move the components to a new position *(your view may appear differently)*

Click: **Options** tab > Move with Children > **References** tab > Components to Move: select the bottom corner of the **CLAMP_PLATE** [Fig. 9.18(e)] > press **LMB** on the vertical axis of the temporary coordinate system > drag the component to a new position > release the **LMB** [Figs. 9.18(f-g)]

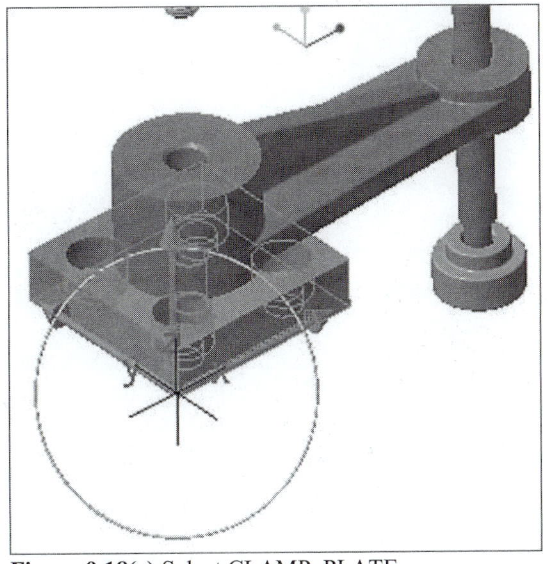

Figure 9.18(e) Select CLAMP_PLATE

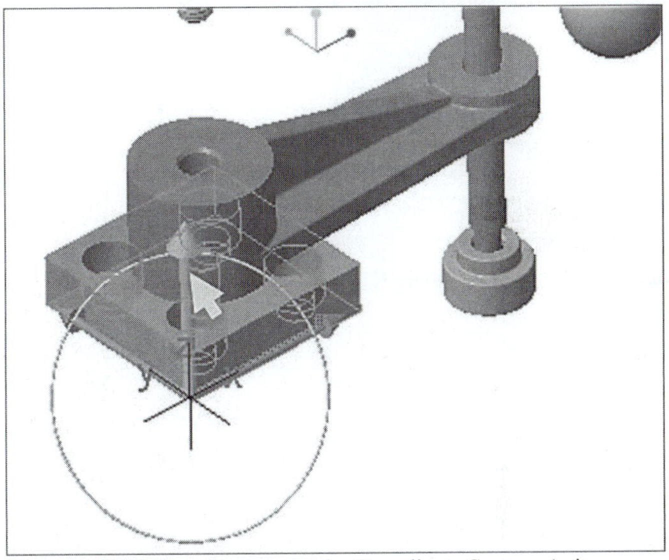

Figure 9.18(f) Drag Component by Coordinate System Axis

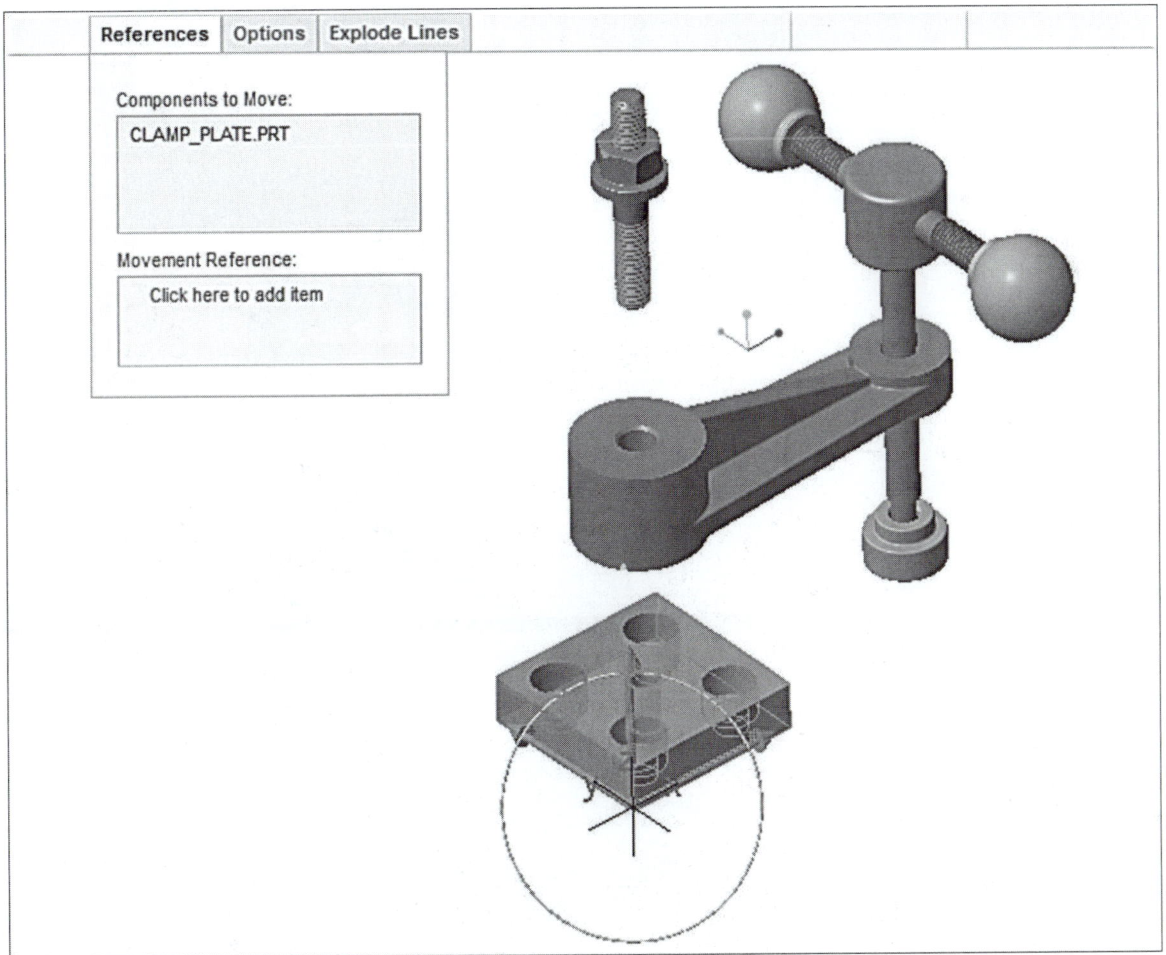

Figure 9.18(g) Moving CLAMP_PLATE > at the Desired Position, Release the LMB

Select **CARRLANE_500_FN** [Fig. 9.18(h)] > press **LMB** on the **X Axis** of the temporary coordinate system > drag to a new position [Fig. 9.18(i)] > release the **LMB** [Fig. 9.18(j)]

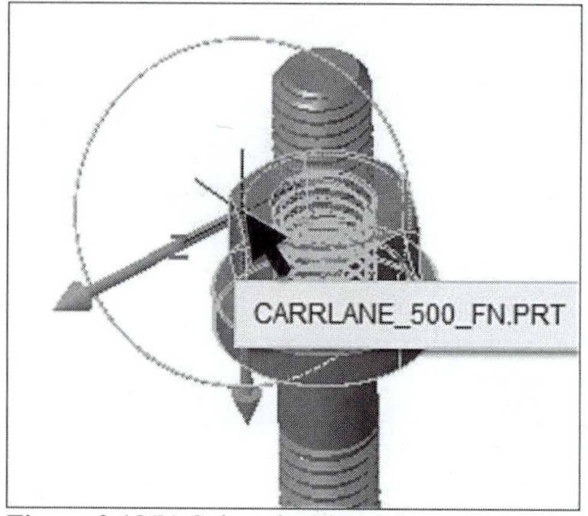

Figure 9.18(h) Select the Flange Nut

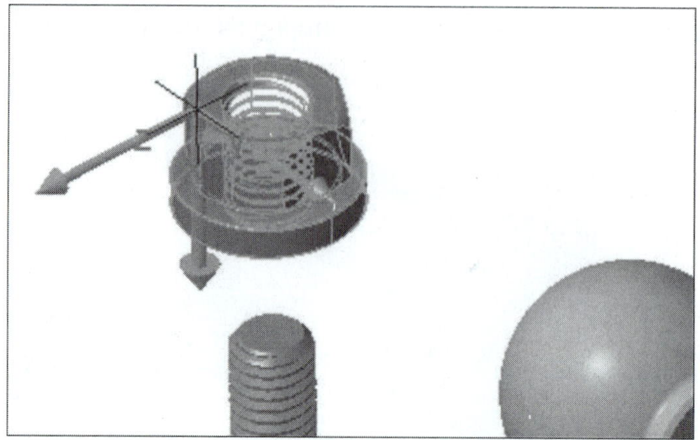

Figure 9.18(i) Drag the Flange Nut to a New Position

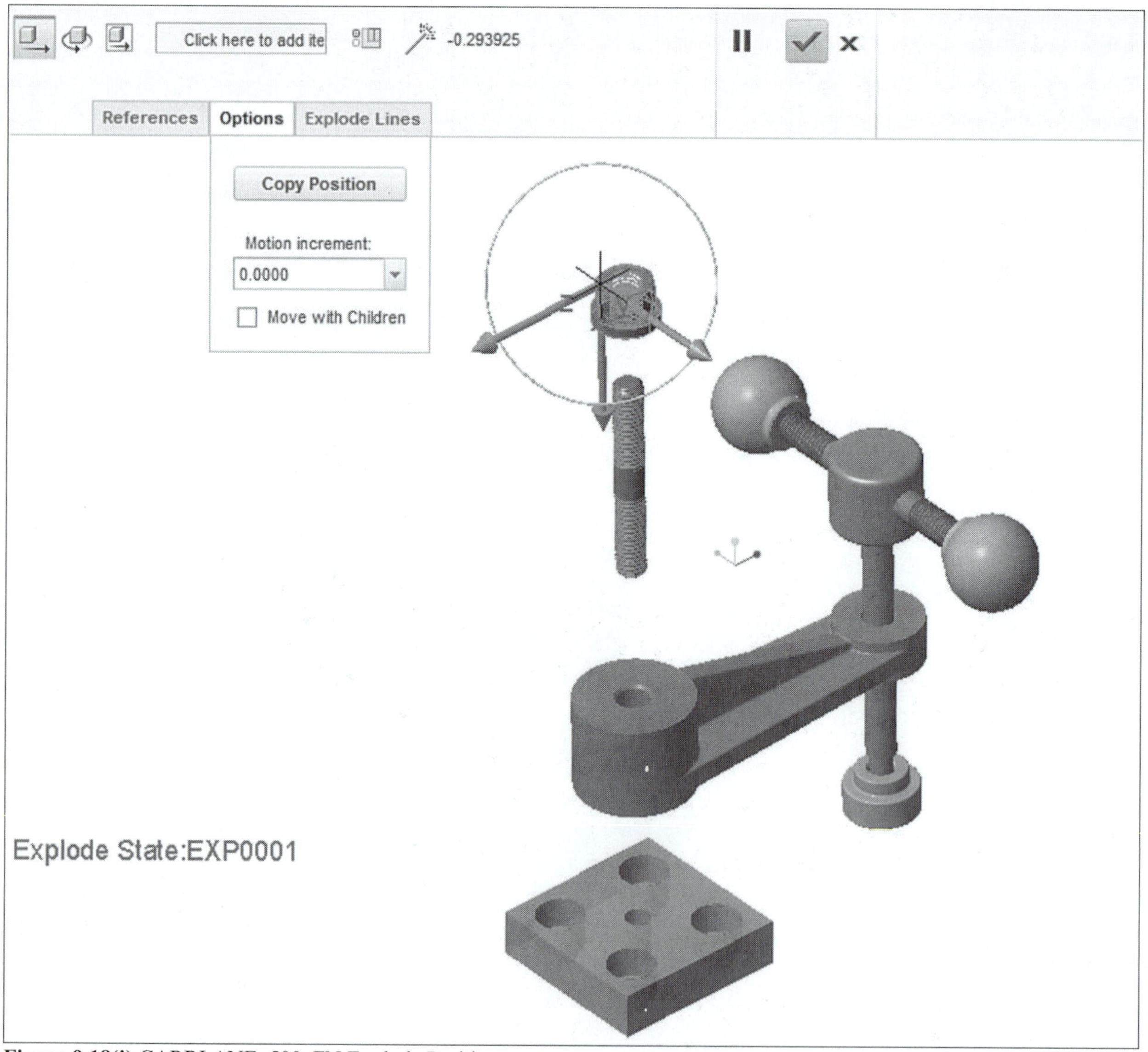

Figure 9.18(j) CARRLANE_500_FN Explode Position

Click: **Options** tab > ☑ Move with Children > **References** tab > Components to Move: select **CLAMP_SWIVEL** [Fig. 9.18(k)] > press **LMB** on the vertical axis > drag the component to a new location [Fig. 9.18(l)] > release the **LMB** > pan and zoom as required

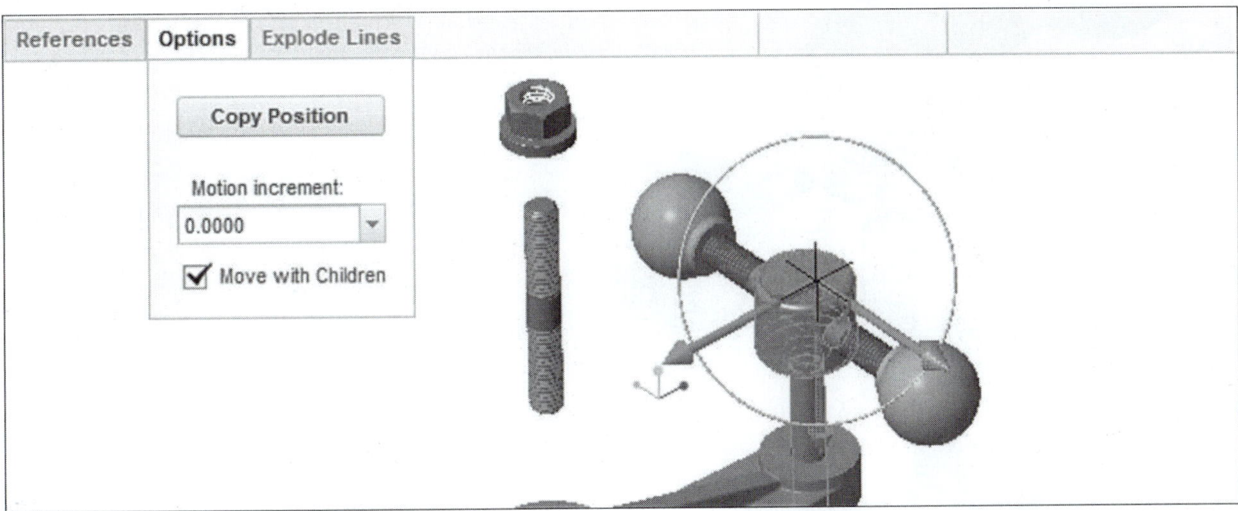

Figure 9.18(k) Select the CLAMP_SWIVEL

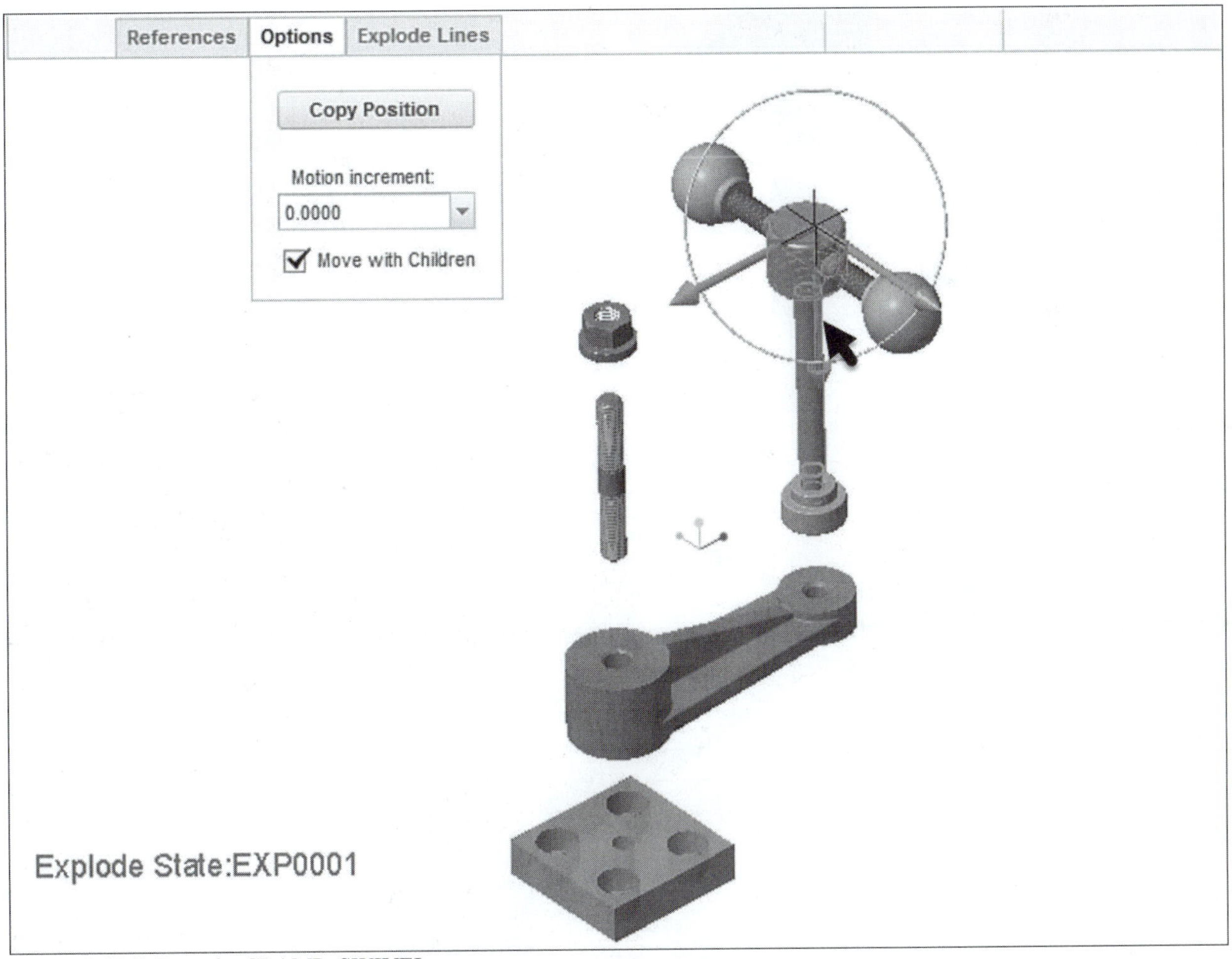

Figure 9.18(l) Drag the CLAMP_SWIVEL

Click: **Options** tab > ☐ Move with Children > **References** tab > Components to Move**:** select the **CLAMP_BALL** [Fig. 9.18(m)] > press **LMB** on the axis > move the component [Fig. 9.18(n)] > release the **LMB** > repeat the process and move the second clamp ball and the long stud [Fig. 9.18(o)]

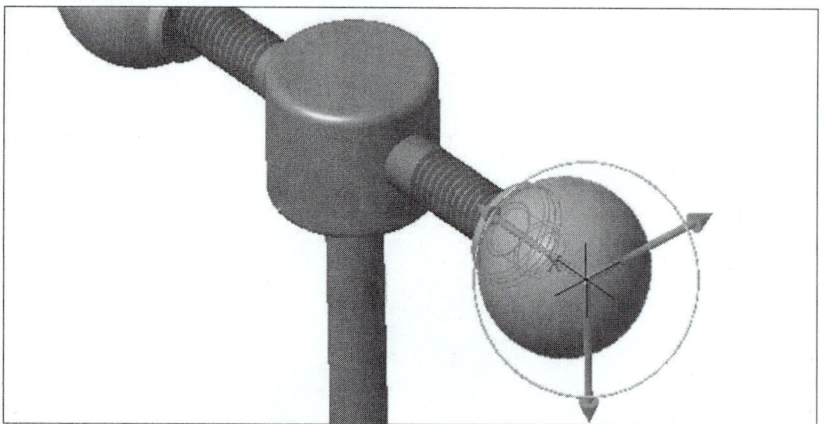

Figure 9.18(m) Select and Move the CLAMP_BALL

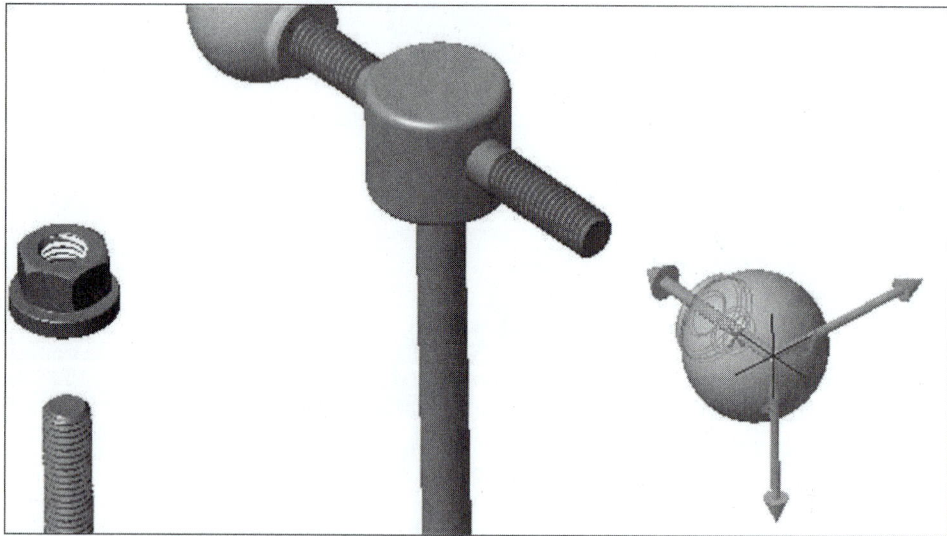

Figure 9.18(n) Moved CLAMP_BALL

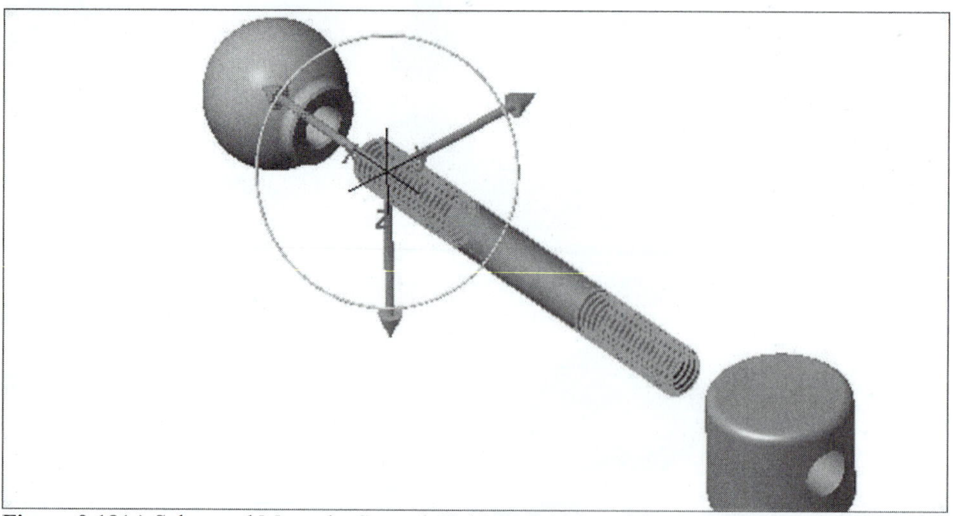

Figure 9.18(o) Select and Move the Second CLAMP_BALL and the CARRLANE-12-13500_STUD

Select the **CLAMP_FOOT** > press **LMB** on the axis of the temporary coordinate system > move the component [Fig. 9.18 (p)] > release the **LMB** > > > in the Graphics Window, **LMB** to deselect > **MMB** > adjust the view as needed [Fig. 9.18 (q)]

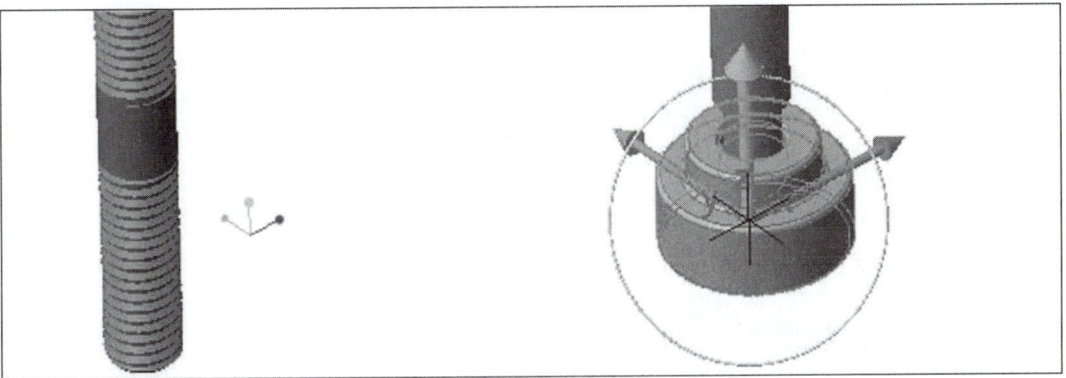

Figure 9.18(p) Move the CLAMP_FOOT

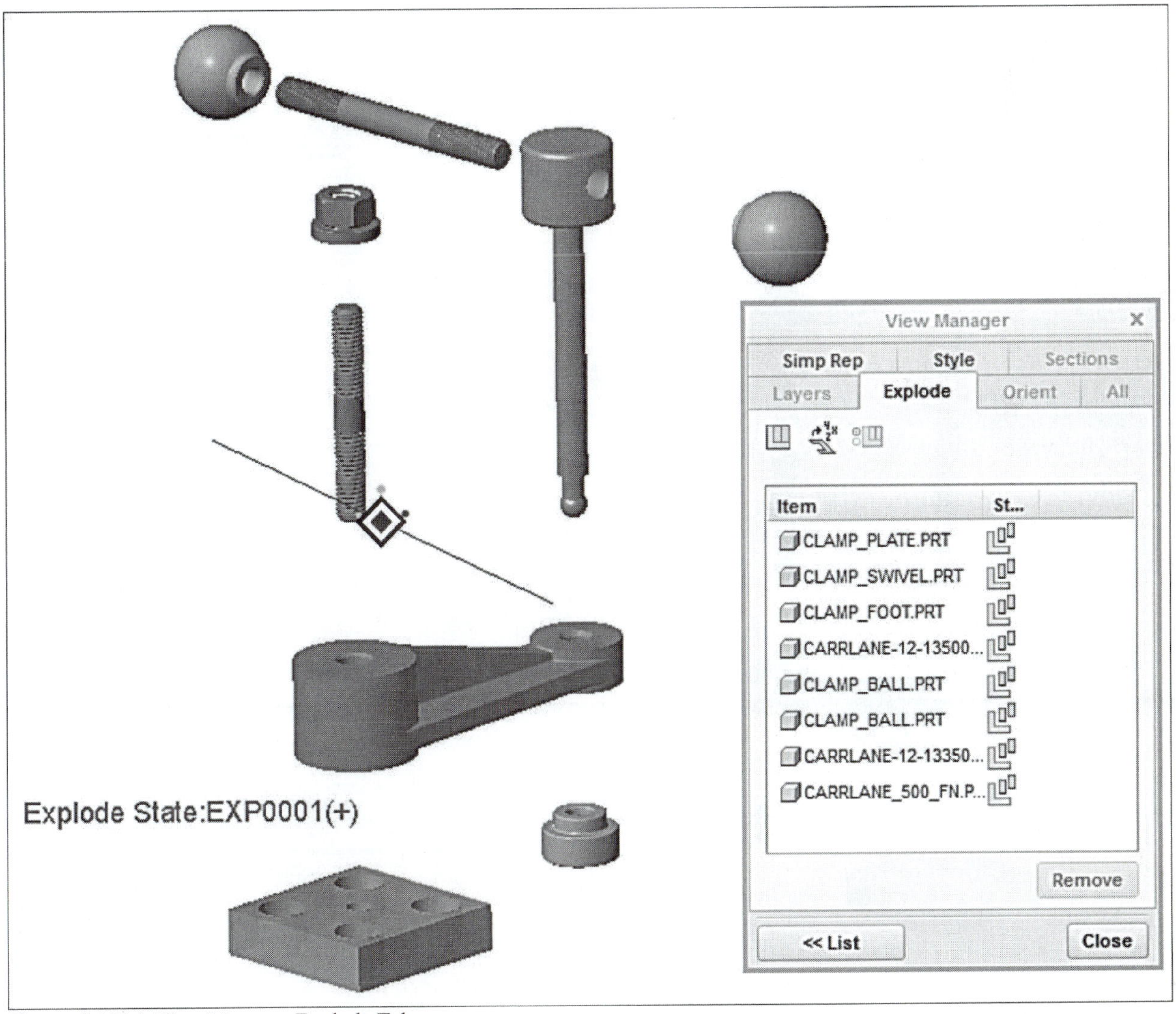

Figure 9.18(q) View Manager Explode Tab

Click: << **List** > **Exp0001(+)** > **RMB** > **Save** [Fig. 9.18(r)] > **OK** [Fig. 9.18(s)] > **Orient** tab > **Explode1(+)** > **Edit** > **Save** > **OK** > **Close**

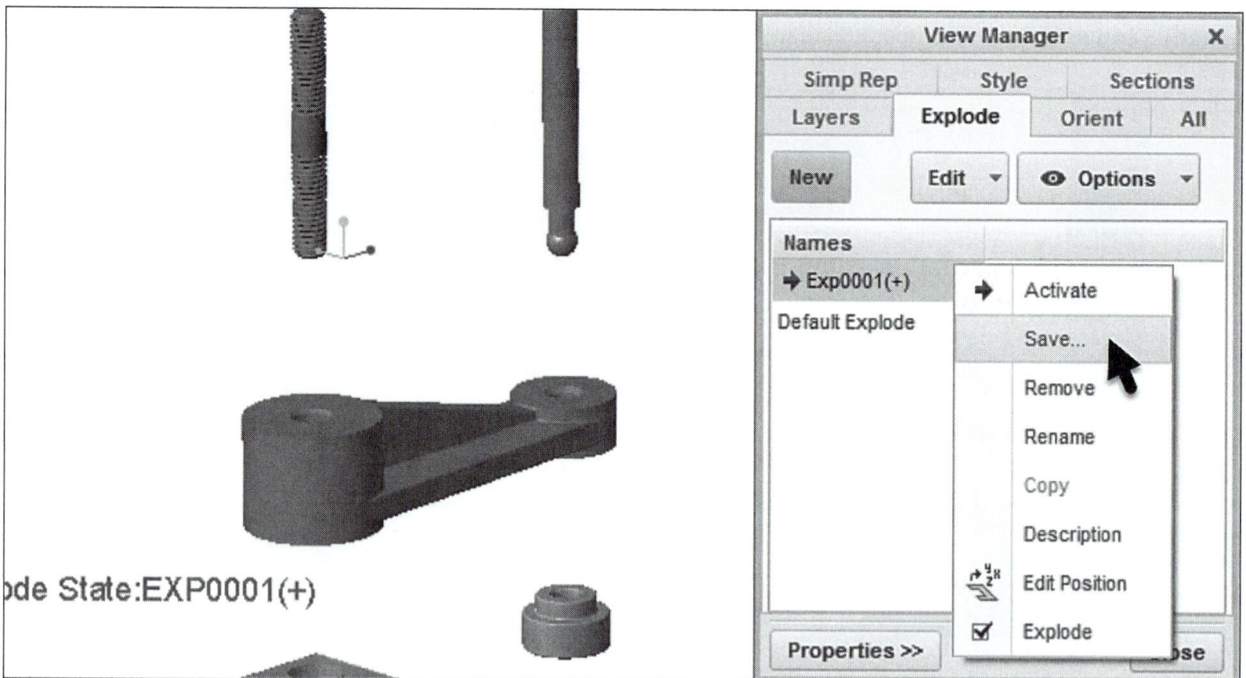

Figure 9.18(r) Save the Explode State

Figure 9.18(s) Save Display Elements Dialog Box

View Style

The components of an assembly (whether exploded or not) can be displayed individually with **Wireframe**, **Hidden Line**, **No Hidden**, **Shading**, or **Transparent**. Style is used to manage the display styles of an assembly's components. The components will be displayed according to their assigned display style in the current style state.

Click:  **View Manager** Open the View Manager > **Style** tab > **New** > **Enter** [Fig. 9.19(a)] to accept the default name: Style0001 > **Show** tab [Fig. 9.19(b)] >  Wireframe > select the **CLAMP_SWIVEL** > select the **CARRLANE-12-13350_STUD** > **Preview**

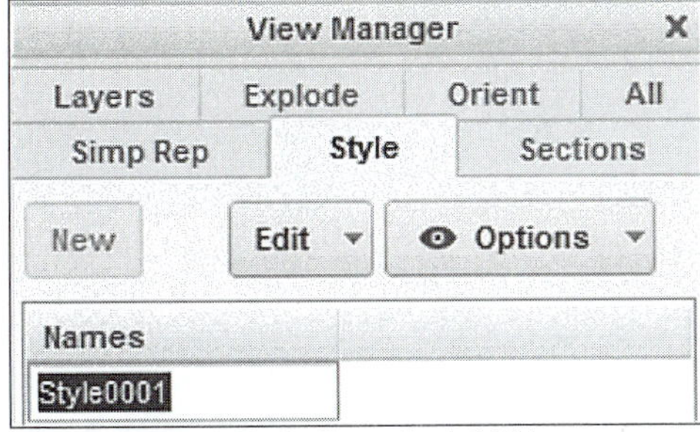

Figure 9.19(a) Style State Style0001

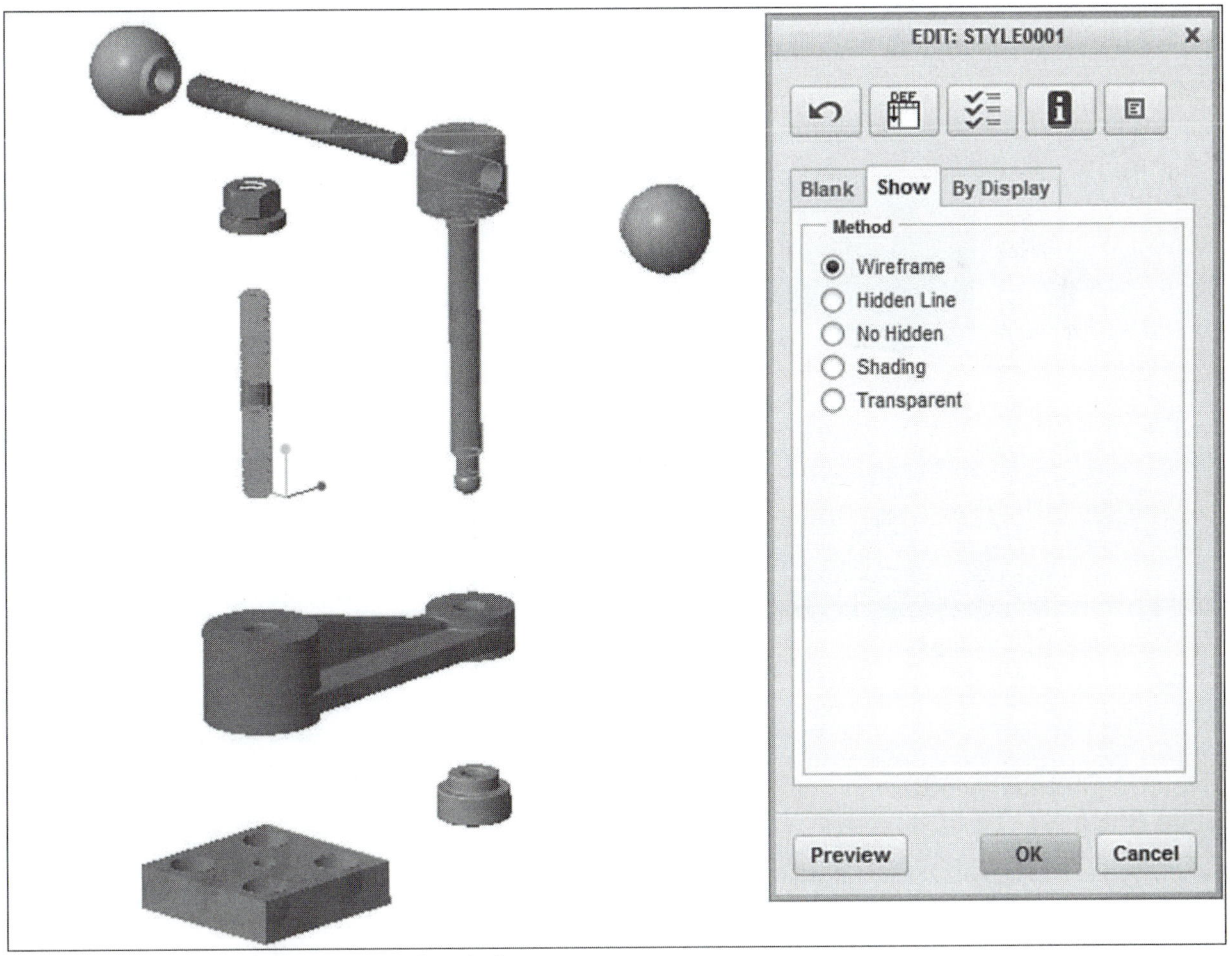

Figure 9.19(b) Edit Dialog Box, Show Tab, Wireframe

Click: [Hidden Line] > select the **CLAMP_PLATE** and the **CLAMP_FOOT** > release the **Ctrl** key > Preview [Fig. 9.19(c)]

Figure 9.19(c) Edit Dialog Box, Show Tab, Hidden Line

Click: [No Hidden] > select the two **CLAMP_BALL** components > **Preview** > [Shading] > select **CARRLANE_500_FN** and **CARRLANE-12-13500_STUD** > **Preview** [Fig. 9.19(d)]

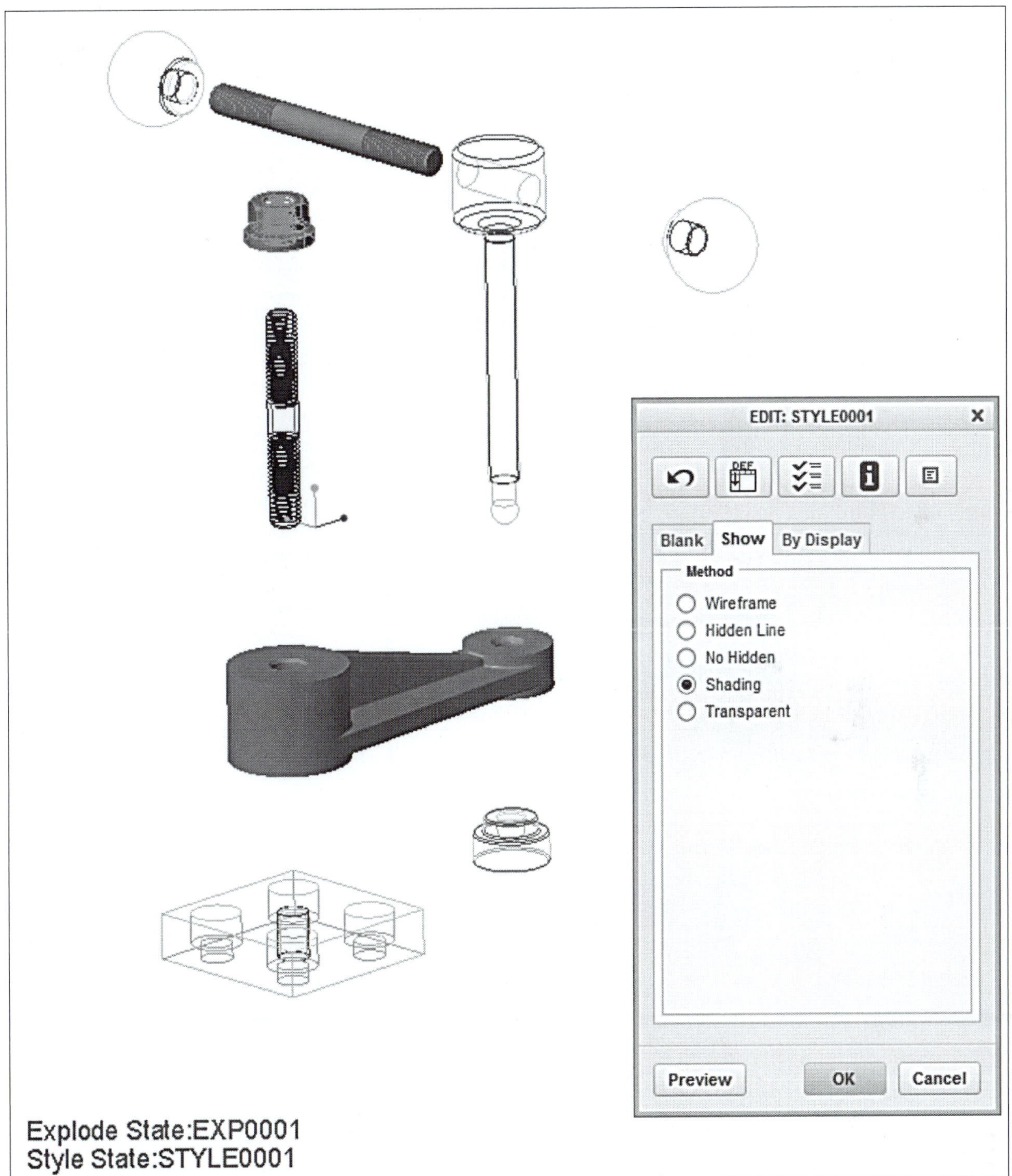

Figure 9.19(d) Edit Dialog Box, Show Tab, Shading

Click: [◉ Transparent] > select **CLAMP_ARM_DESIGN** > **Preview** [Fig. 9.19(e)] > **OK** > **OK**

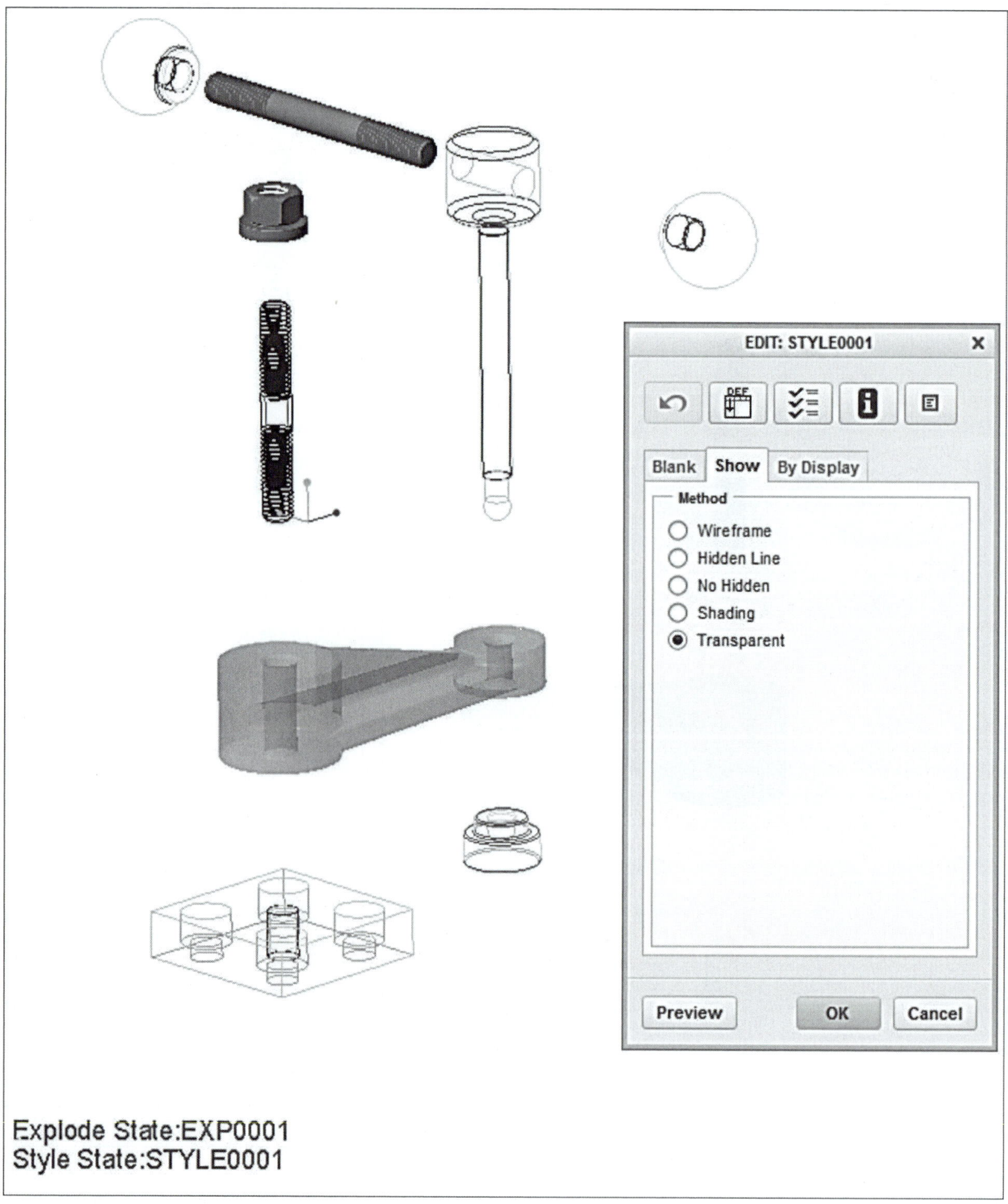

Figure 9.19(e) Completed Style State

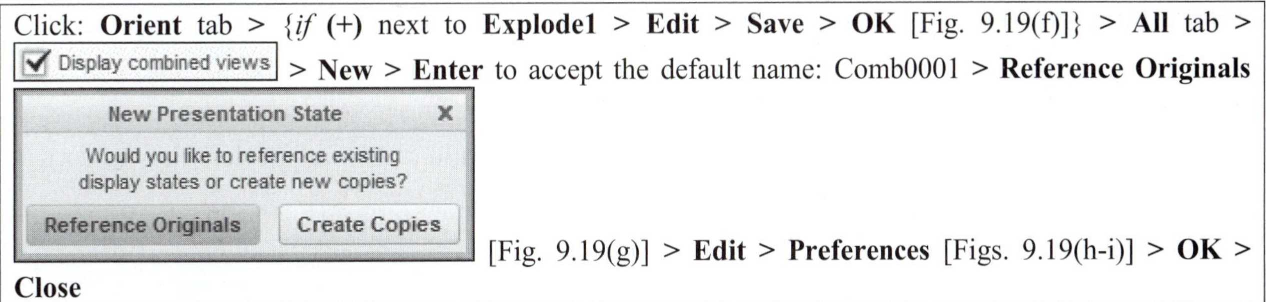
Click: **Orient** tab > {*if* (+) next to **Explode1** > **Edit** > **Save** > **OK** [Fig. 9.19(f)]} > **All** tab > Display combined views > **New** > **Enter** to accept the default name: Comb0001 > **Reference Originals**

[Fig. 9.19(g)] > **Edit** > **Preferences** [Figs. 9.19(h-i)] > **OK** > **Close**

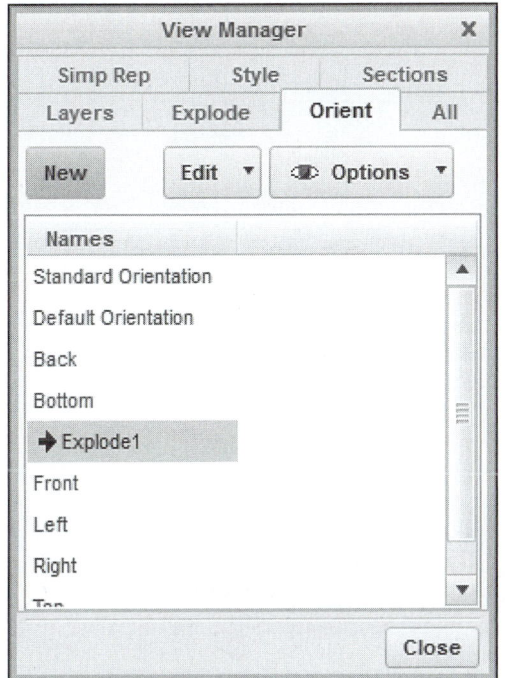

Figure 9.19(f) Orient Tab

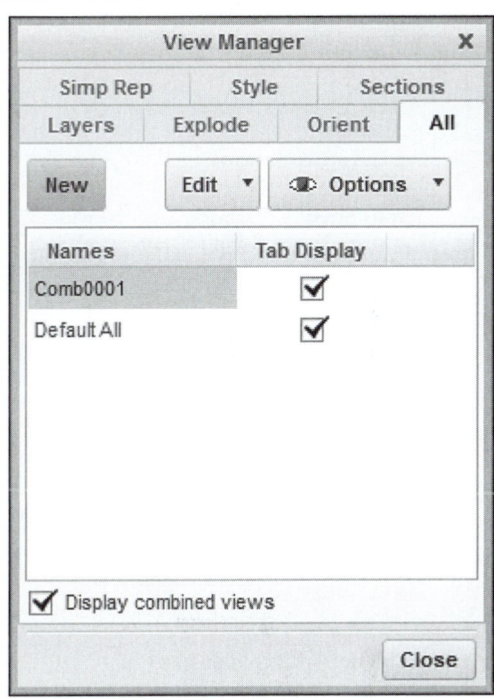

Figure 9.19(g) All Tab

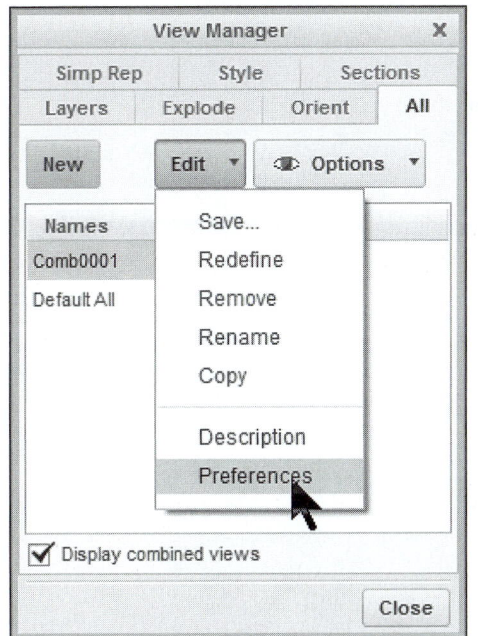

Figure 9.19(h) Preferences

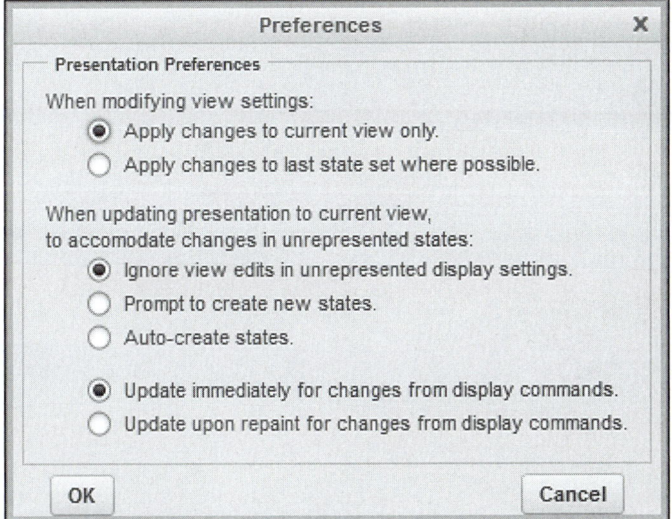

Figure 9.19(i) All Edit Preferences

Model Tree

You can display component style states, and explode states in the Model Tree.

Click: **Settings** from the Navigator > **Tree Columns** > **Info** > **Display Styles** [Fig. 9.20(a)] > **STYLE0001** > [Fig. 9.20(b)] > **Apply** > **OK** > adjust the column format [Fig. 9.20(c)]

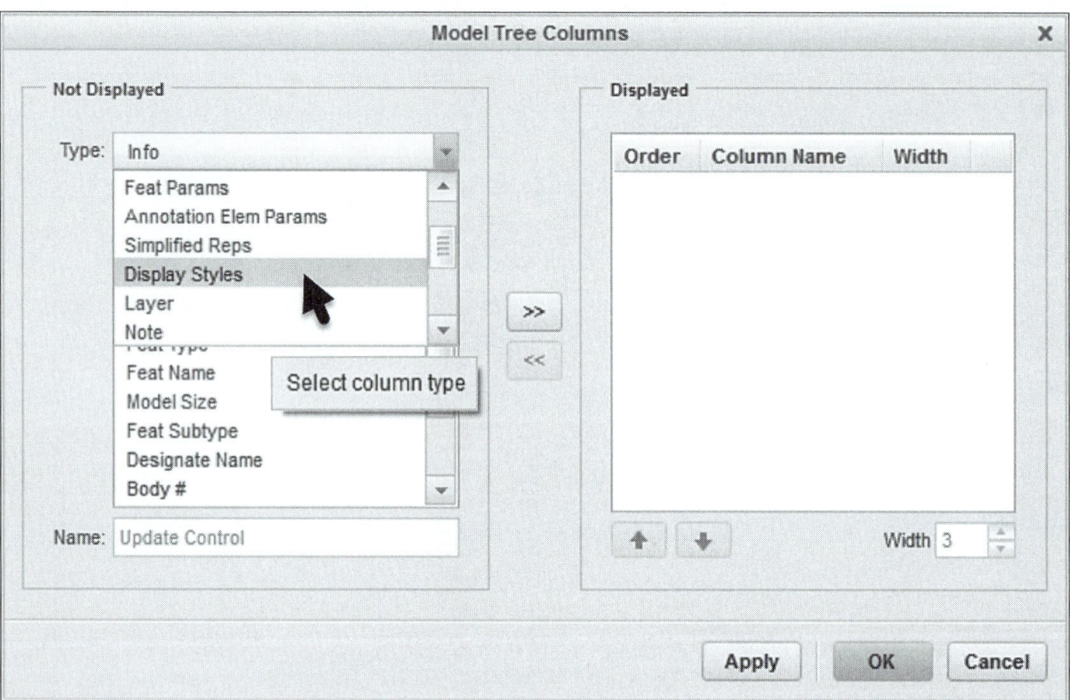

Figure 9.20(a) Model Tree Columns

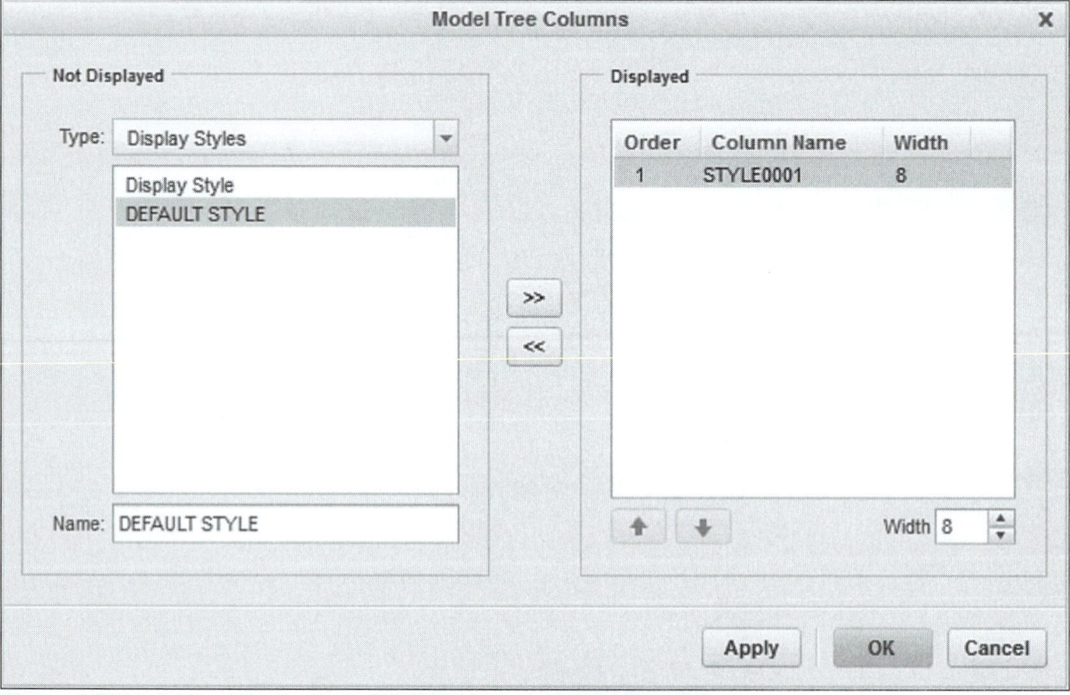

Figure 9.20(b) Adding Display Style STYLE0001 to the Model Tree Columns

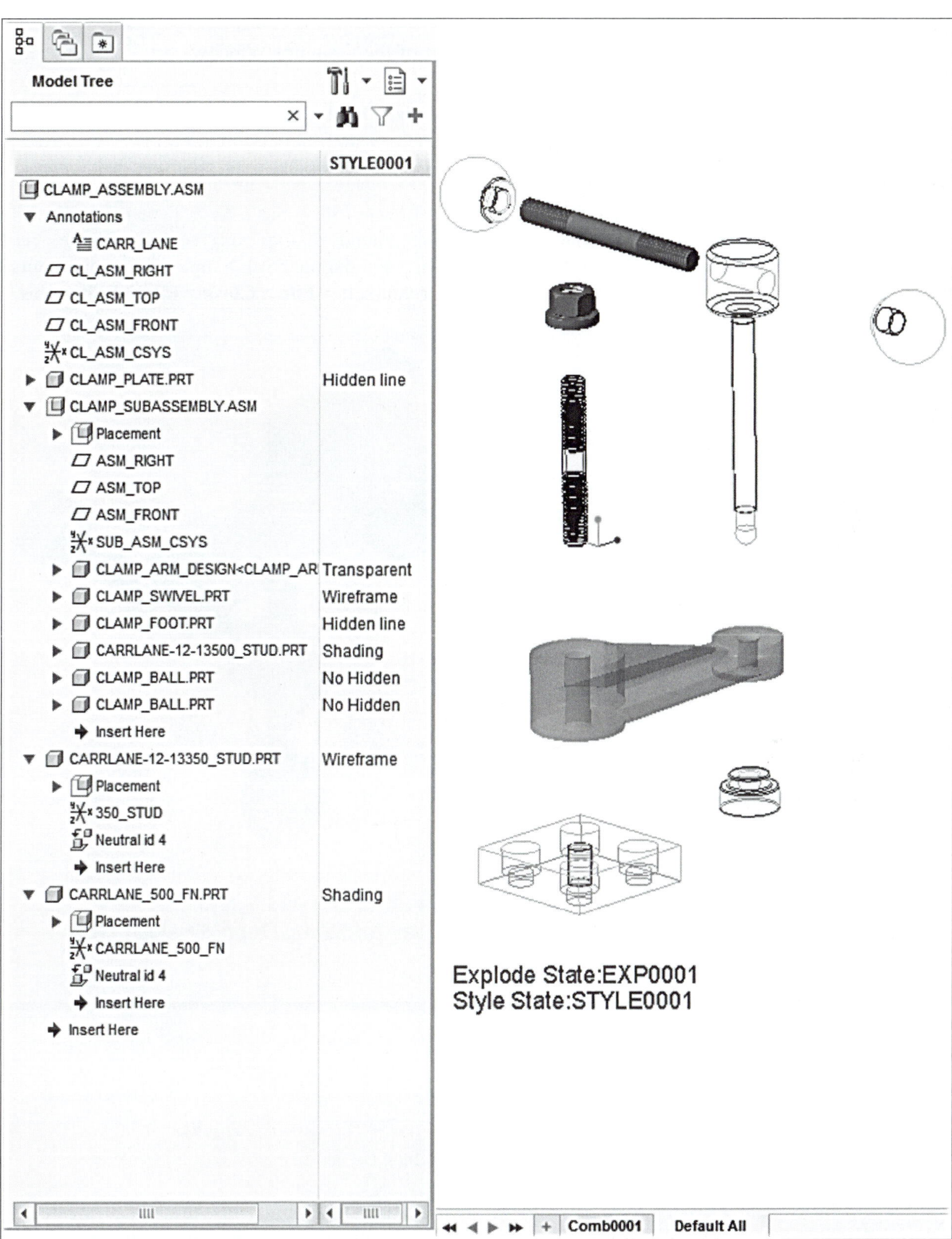

Figure 9.20(c) Adjusted Model Tree

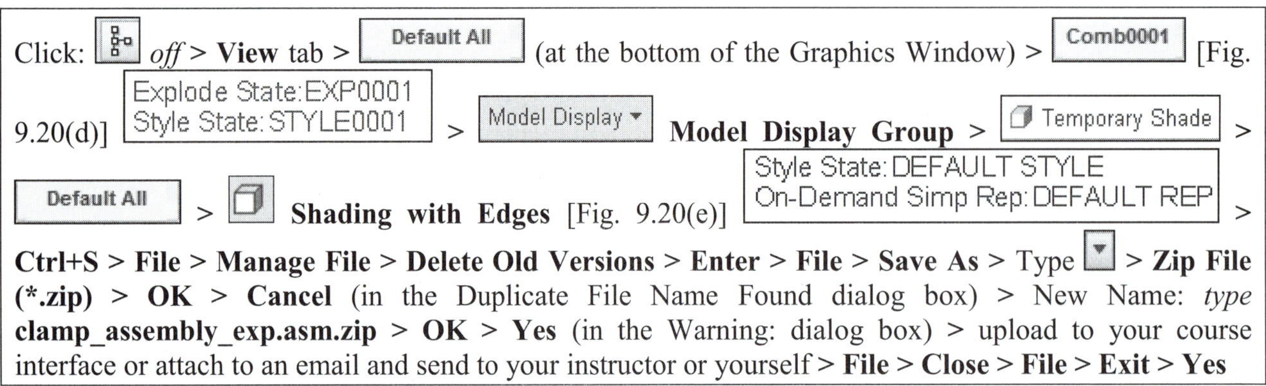
Click: off > **View** tab > Default All (at the bottom of the Graphics Window) > Comb0001 [Fig. 9.20(d)] Explode State:EXP0001 Style State:STYLE0001 > Model Display **Model Display Group** > Temporary Shade > Default All > **Shading with Edges** [Fig. 9.20(e)] Style State:DEFAULT STYLE On-Demand Simp Rep:DEFAULT REP > **Ctrl+S** > **File** > **Manage File** > **Delete Old Versions** > **Enter** > **File** > **Save As** > Type > **Zip File (*.zip)** > **OK** > **Cancel** (in the Duplicate File Name Found dialog box) > New Name: *type* **clamp_assembly_exp.asm.zip** > **OK** > **Yes** (in the Warning: dialog box) > upload to your course interface or attach to an email and send to your instructor or yourself > **File** > **Close** > **File** > **Exit** > **Yes**

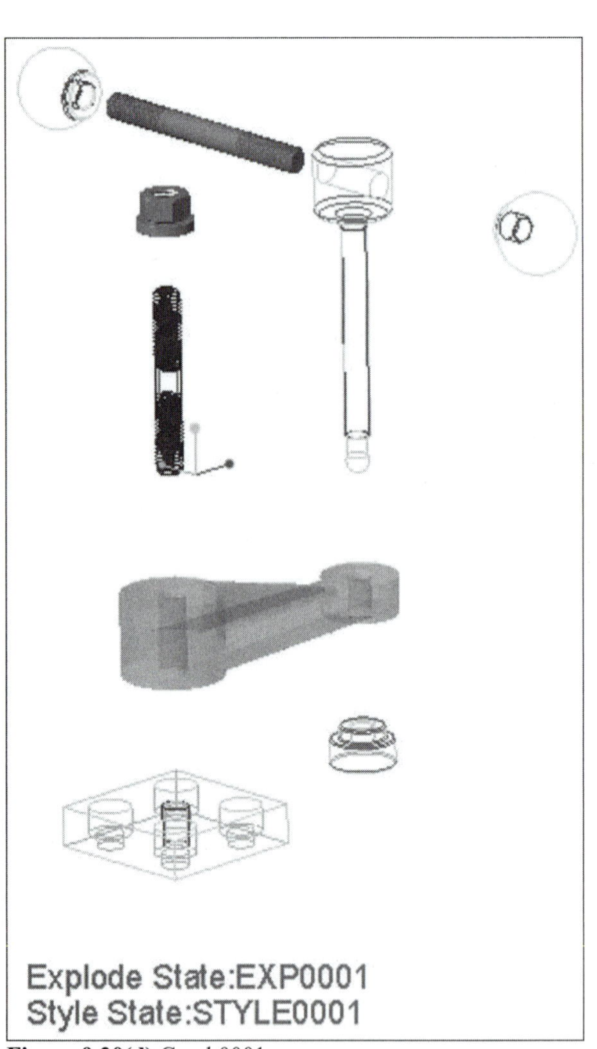

Figure 9.20(d) Comb0001

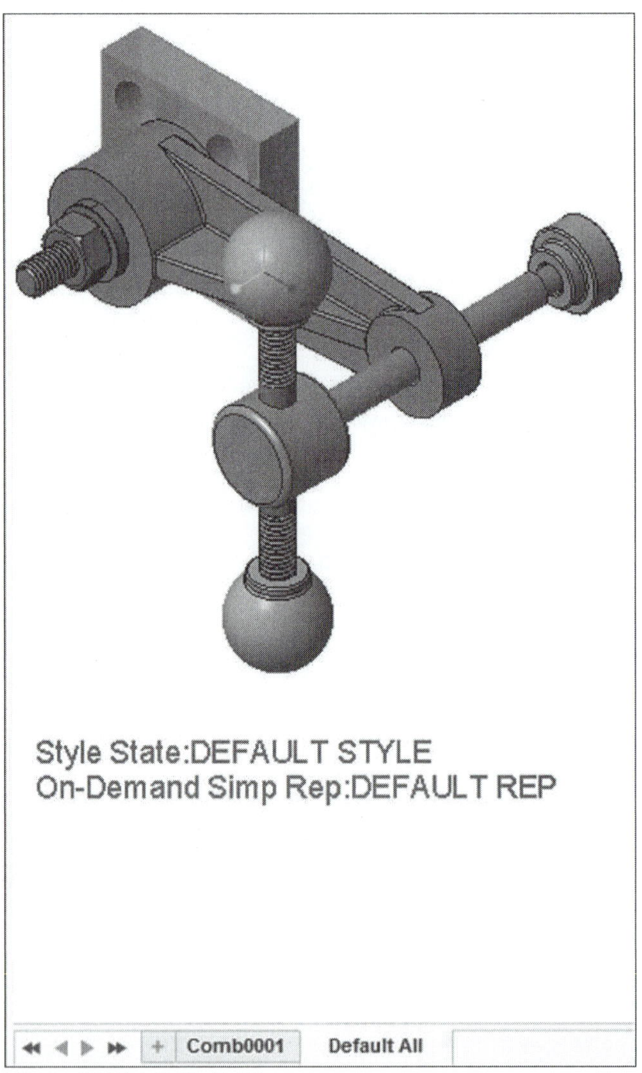

Figure 9.20(e) Default All

You have completed the Assembly Mode lessons. The assembly and view states created in Lessons 8 and 9 will be used in Lesson 12 to create assembly drawings. The component parts modeled in Lessons 6 and 7 will be used in Lesson 10 to create detail drawings.

Download additional projects from ***www.cad-resources.com.***

Lesson 10 Introduction to Drawings

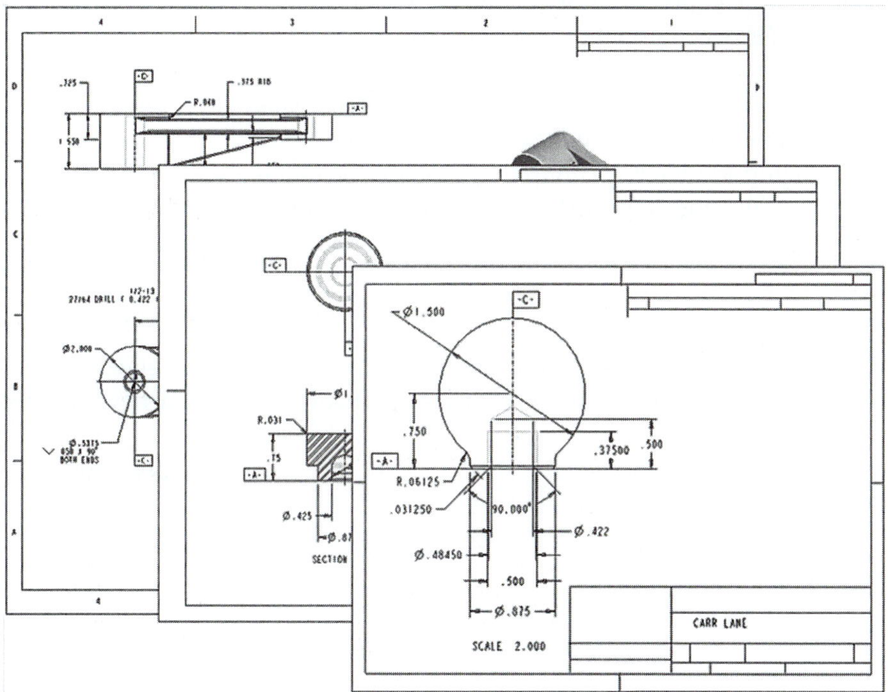

Figure 10.1 Swing Clamp Assembly Component Drawings

OBJECTIVES

- Create part drawings
- Display **standard views** using the template
- **Add, Move**, **Erase**, and **Delete** views
- Retrieve a standard **format**
- **Display** Gtol datums, centerlines, and dimensions
- **Cleanup** dimension positions
- Specify and retrieve standard **Format** paper size and units
- Change the **Scale** of a view

REFERENCES AND RESOURCES

For **Resources** go to **www.cad-resources.com** > click on the PTC Creo Parametric 3.0 Book cover

- Lesson 10 Lecture
- Lesson 10 3D models embedded in a PDF
- Book Projects PDF

Introduction to Drawings

This lesson will introduce the basic concepts and procedure for creating detail drawings (Fig. 10.1). Using parts modeled for the Swing Clamp Assembly, you will quickly create drawings, change formats, delete and add views, and display dimensions and centerlines. Since this lesson is meant to introduce drawing mode concepts and procedures, a minimum of cleanup, reformatting and repositioning of dimensions and centerlines will be required.

FORMATS, TITLE BLOCKS, AND VIEWS

Formats are user-defined drawing sheet layouts. A drawing can be created with an empty format and have a standard size format added as needed. Formats can be added to any number of drawings. The format can also be modified or replaced in a Creo Parametric drawing at any time.

Title Blocks are standard or sketched line entities that can contain parameters (object name, tolerances, scale, and so on) that will show when the format is added to the drawing.

Views created by Creo Parametric are identical to views constructed manually by a designer on paper. The same rules of projection apply; the only difference is that you choose commands in Creo Parametric to create the views as needed.

Formats

Formats consist of draft entities, not model entities. There are two types of formats: standard and sketched. You can select from a list of **Standard Formats** (**A-F** and **A0-A4**) or create a new size by entering values for length and width.

Sketched Formats created in Sketcher mode (Fig. 10.2) may be parametrically modified, enabling you to create nonstandard-size formats or families of formats.

Formats can be altered to include note text, symbols, tables, and drafting geometry, including drafting cross sections and filled areas.

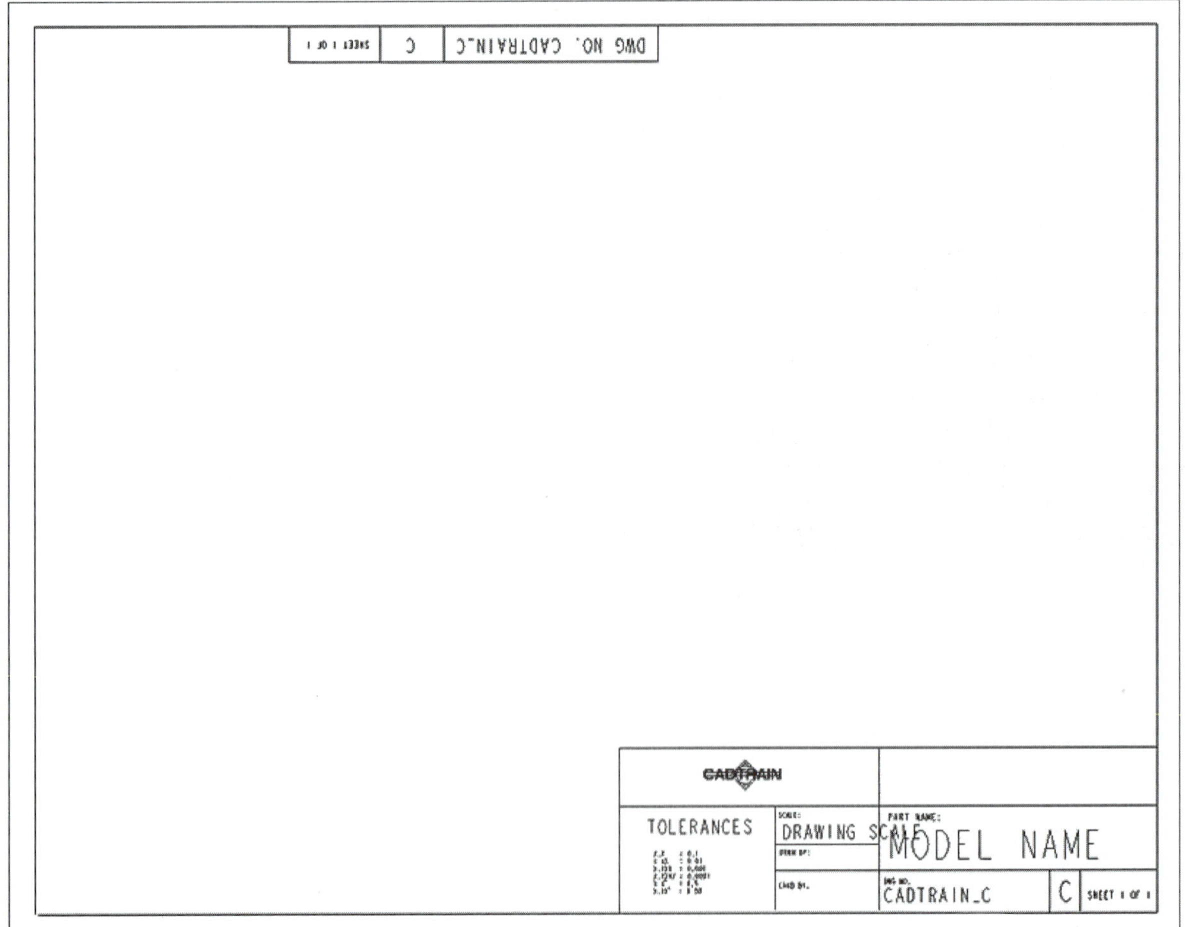

Figure 10.2 Sketched Format

With Creo Parametric, you can do the following in Format mode:

- Create draft geometry and notes (Fig. 10.3)
- Move, mirror, copy, group, translate, and intersect geometry
- Use and modify the draft grid
- Enter user attributes
- Create drawing tables
- Use interface tools to create plot, DXF, SET, and IGES files
- Import IGES, DXF, and SET files into the format
- Create user-defined line styles
- Create, use, and modify symbols
- Include drafting cross sections in a format

Whether you use a standard format or a sketched format, the format is added to a drawing that is created for a set of specified views of a parametric 3D model.

When you place a format on your drawing, the system automatically writes the appropriate notes based on information in the model you use.

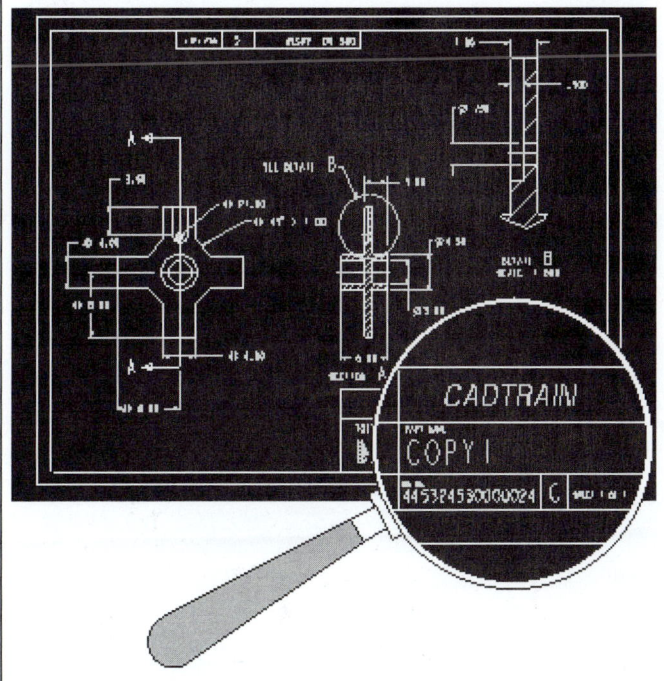

Figure 10.3 Format with Parametric Notes, Added to a Drawing

Specifying the Format Size when Creating a New Drawing

If you want to use an existing template, select a listed template [Fig. 10.4(a)], or select a template from the appropriate directory. If you want to use the existing format [Fig. 10.4(b)], using Browse will open Creo Parametric's System Formats folder [Fig. 10.4(c)]. Select from the list of standard formats or navigate to a directory with user or company created formats.

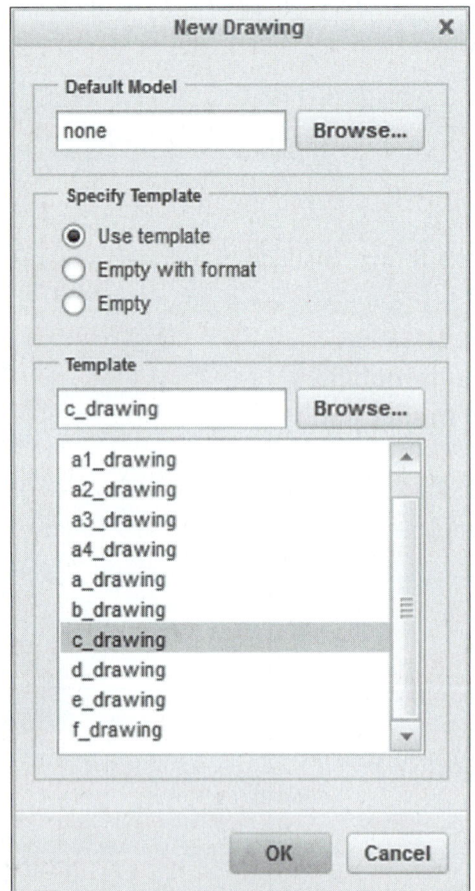

Figure 10.4(a) New Drawing Dialog Box (Use Template)

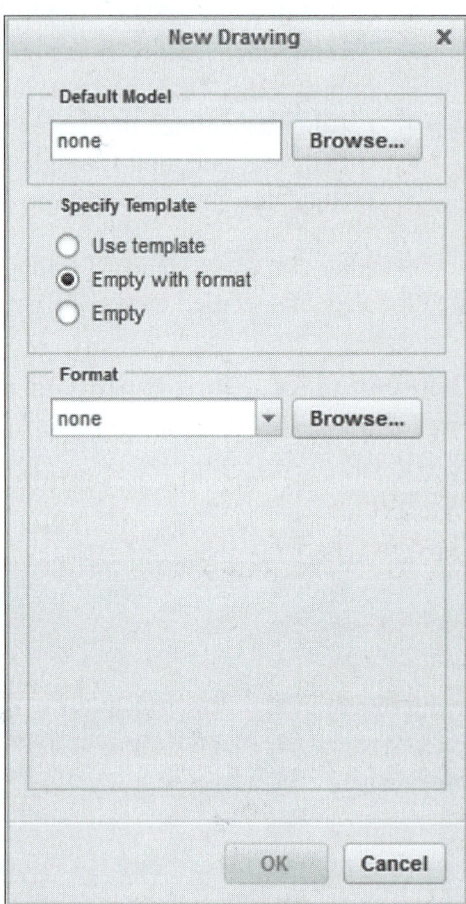

Figure 10.4(b) Empty with format

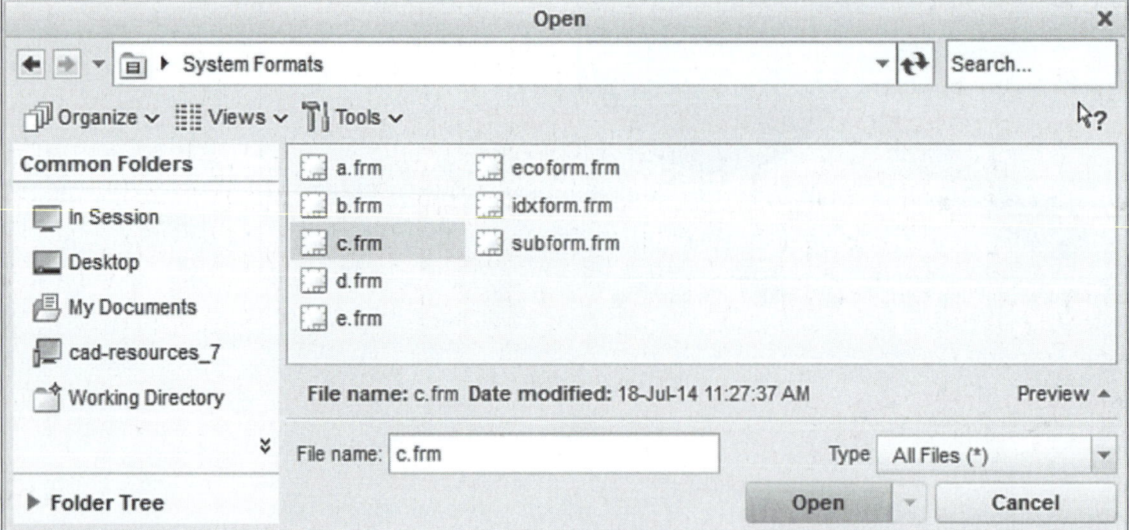

Figure 10.4(c) System Formats

If you want to create your own variable size format [Figs. 10.4(d-e)], enter values for width and length. The main grid spacing and format text units depend on the units selected for a variable size format. The New Drawing dialog box also provides options for the orientation of the format sheet:

- **Portrait** Uses the larger of the dimensions of the sheet size for the format's height; uses the smaller for the format's width
- **Landscape** Uses the larger of the dimensions of the sheet size for the format's width; uses the smaller for the format's height [Fig. 10.4(d)]
- **Variable** Select the unit type, Inches or Millimeters, and then enter specific values for the Width and Height of the format [Fig. 10.4(e)]

A0	841 X 1189	mm	**A**	8.5 X 11	in.
A1	594 X 841	mm	**B**	11 X 17	in.
A2	420 X 594	mm	**C**	17 X 22	in.
A3	297 X 420	mm	**D**	22 X 34	in.
A4	210 X 297	mm	**E**	34 X 44	in.
			F	28 X 40	in.

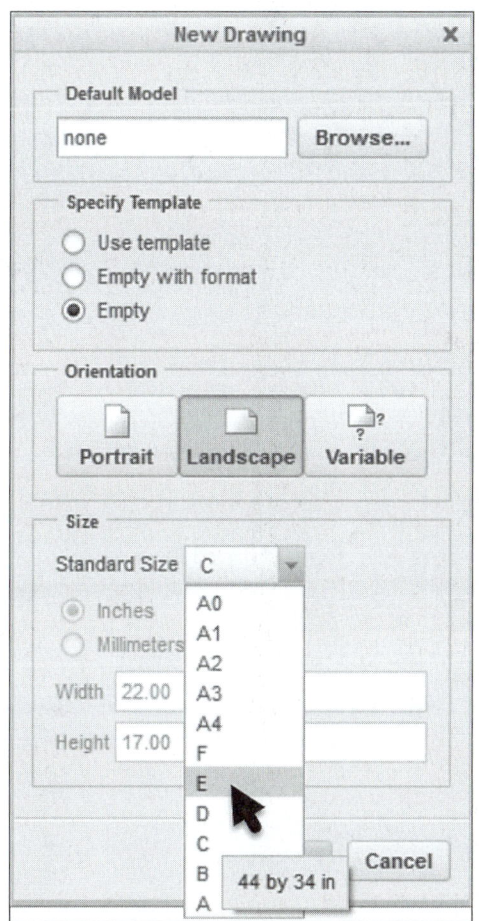

Figure 10.4(d) New Drawing Dialog Box (Empty- Landscape)

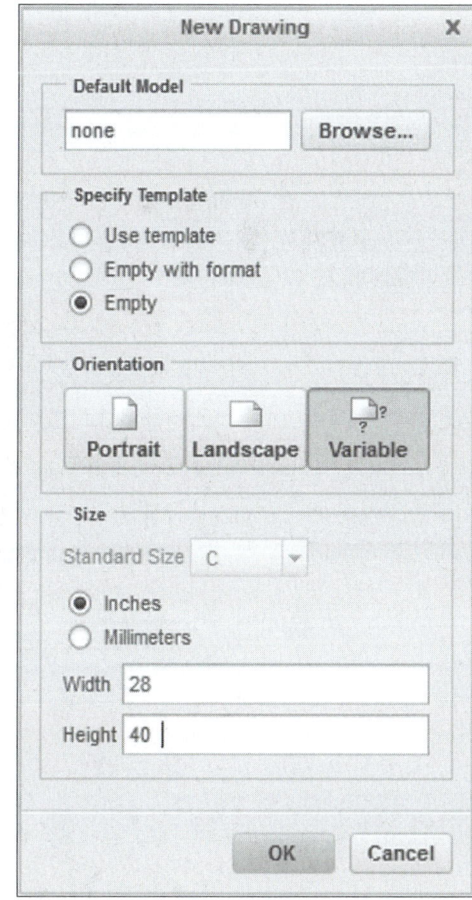

Figure 10.4(e) New Drawing (Empty- Variable)

Drawing User Interface

The Drawing user interface helps you complete your detailing tasks quickly and efficiently. The **Ribbon** optimizes the user interface according to tasks. It consists of tabs with groups of frequently-used commands organized in a logical sequence. The group overflow area contains additional commands that are not frequently used. When you click a tab, the groups on the Ribbon, the drawing object filter, the shortcut menus, and the Drawing Tree update to enable operations relevant to the task. You can minimize or maximize the Ribbon display by double-clicking on a tab or by right-clicking on the Ribbon, and clicking Minimize the Ribbon on the shortcut menu. If you have minimized the Ribbon, click on a tab to temporarily maximize its display. In this case when you try to execute a command on the Ribbon or press the ESC key, the Ribbon changes back to its minimized state.

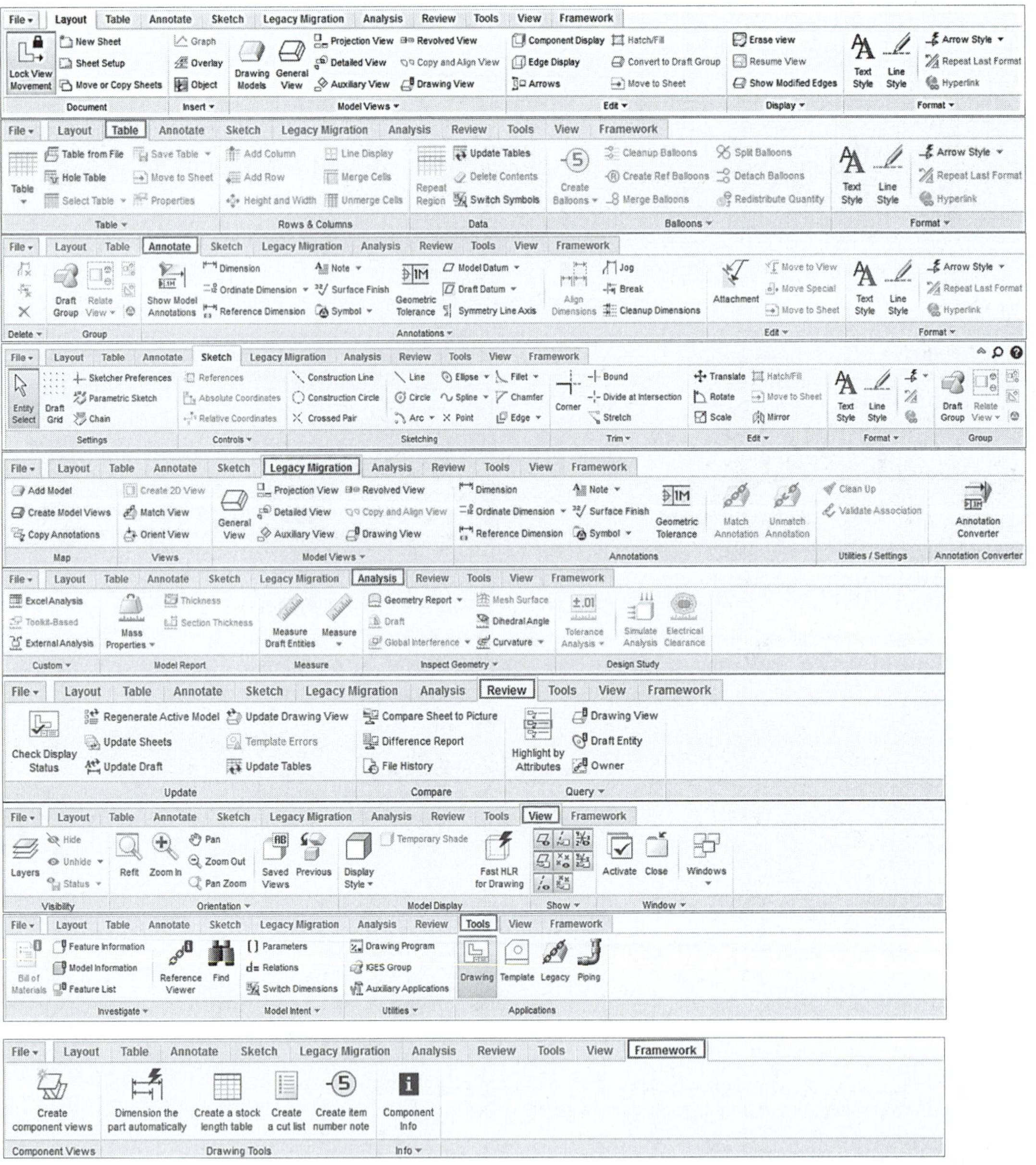

The **Navigation area** contains the Drawing Tree and the Model Tree. The Drawing Tree updates dynamically to reflect drawing objects relevant to the currently active tab. For example, when the Annotate tab is active, the annotations on the current sheet are listed in the Drawing Tree. In the navigation area, you can toggle between the Drawing Tree and the Layer Tree. Similarly you can toggle between the Model Tree and the Layer Tree. However, you cannot display the Layer Tree simultaneously in both the Drawing Tree area as well as the Model Tree area.

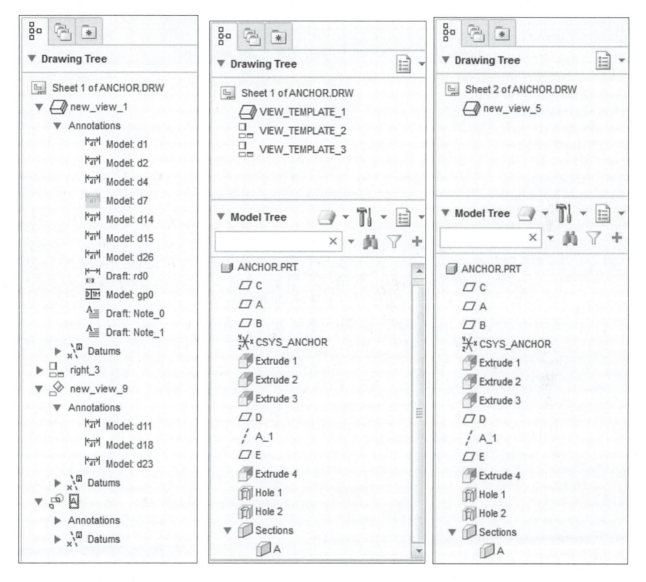

The **Drawing sheet** tabs are located below the graphics window. You can create multiple drawing sheets and move items from one sheet to another. You can also move or copy, add, rename, or delete a sheet.

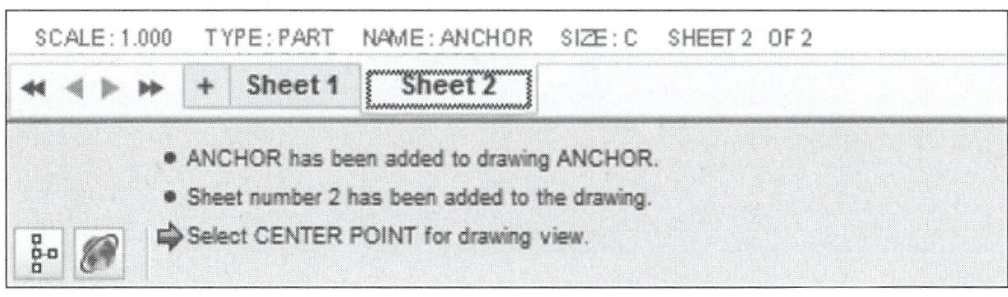

Drawing Templates

Drawing templates may be referenced when creating a new drawing. Templates can automatically create the views, set the desired view display, create snap lines, and show model dimensions. Drawing templates contain three basic types of information for creating new drawings. The first type is basic information that makes up a drawing but is not dependent on the drawing model, such as notes, symbols, and so forth. This information is copied from the template into the new drawing. The second type is instructions used to configure drawing views and the actions that are performed on their views. The instructions are used to build a new drawing object. The third type is a parametric note. Parametric notes are notes that update to new drawing model parameters and dimension values. The notes are re-parsed or updated when the template is instantiated. Use the templates to:

- Define the layout of views
- Set view display
- Place notes
- Place symbols
- Define tables
- Create snap lines
- Show dimensions

Template View

You can also create customized drawing templates for the different types of drawings that you construct. Creating a template allows you to establish portions of drawings automatically, using the customizable template. The Template View Instructions dialog box (Fig. 10.5) is accessed through Tools tab > Template > Layout tab > Template View, when in the Drawing mode.

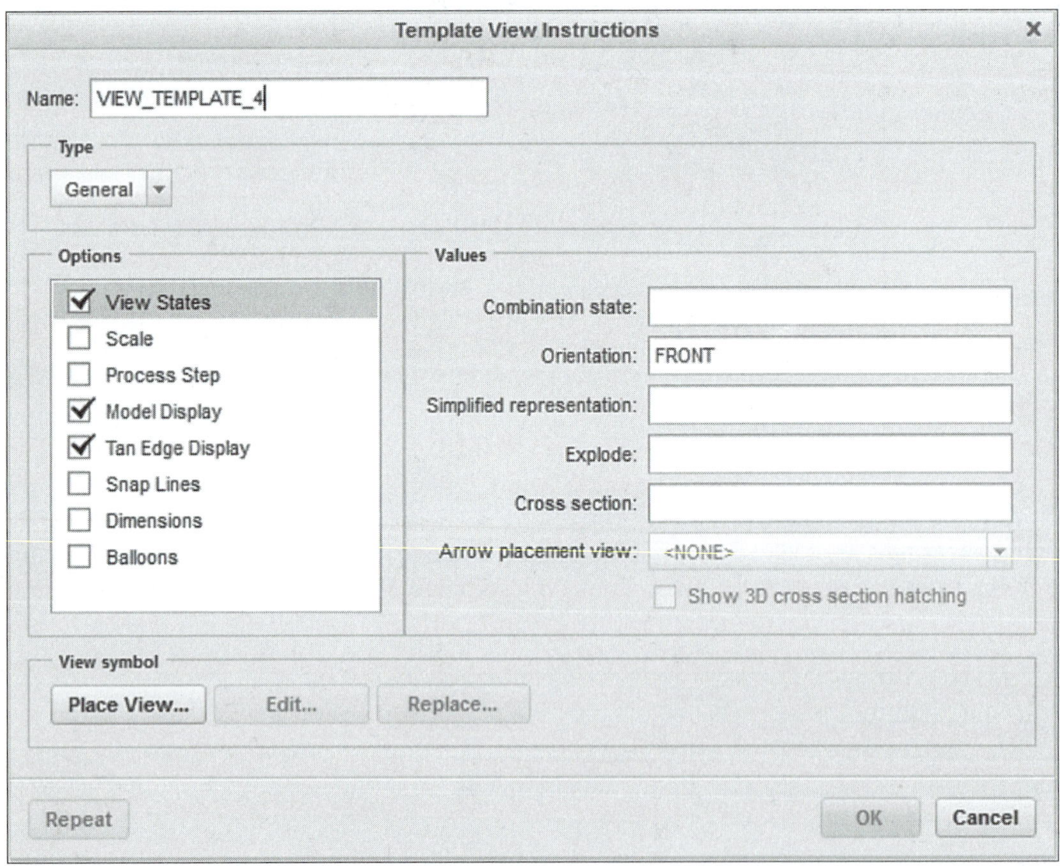

Figure 10.5 Template View Instructions Dialog Box

You can use the following options in the Template View Instructions dialog box to customize your drawing templates:

- **View Name** Set the name of the drawing view that will be used as the view symbol label
- **View Orientation** Create a General view or a Projection view
- **Model "Saved View" Name** Orient the view based on a name view in the model
- **Place View** Places the view after you have set the appropriate options and values
- **Edit View Symbol** Allows you to edit the view symbol using the Symbol Instance dialog box
- **Replace View Symbol** Allow you to replace the view symbol using the Symbol Instance dialog box

View Options	View Values
View States—Specify the type of view. By default, the **View States** check box is selected and the view state is displayed when the **Template View Instructions** dialog box opens. The value of **Orientation** is FRONT, by default. If you select **Combination State**, the **Orientation**, **Simplified Rep**, **Explode** and **Cross Section** boxes display the text Defined by Combination State. The **Arrow Placement View** and **Show X-Hatching** check boxes also become available for selection after you specify the Combination State.	Combination State, Orientation, Simplified Rep, Explode, Cross Section, Arrow Placement View
Scale—Type a new value for the scale or use the default value.	View Scale
Process Step—Set the process step for the view. You can specify the step number and set the view of the tool by checking the **Tool View** check box under **Process Step**.	Step Number, Tool View
Model Display—Set the view display for the drawing view.	Follow Environment, Wireframe, Hidden Line, No Hidden, Shading
Tan Edge Display—Set the tangent edge display.	Tan Solid, No Disp Tan, Tan Ctrln, Tan Phantom, Tan Dimmed, Tan Default
Snap Lines—Set the number, spacing, and offset of the snap lines.	Number, Incremental Spacing, Initial Offset
Dimensions—Show dimensions on the view.	Create Snap Lines, Incremental Spacing, Initial Offset
Balloons—Show balloons on the view.	

Views

A wide variety of views (Fig. 10.6) can be derived from the parametric model. Among the most common are projection views. Creo Parametric creates projection views by looking to the left of, to the right of, above, and below the picked view location to determine the orientation of a projection view. When conflicting view orientations are found, you are prompted to select the view that will be the parent view. A view will then be constructed from the selected view.

At the time when they are created, projection, auxiliary, detailed, and revolved views have the same representation and explosion offsets, if any, as their parent views. From that time onward, each view can be simplified, be restored, and have its explosion distance modified without affecting the parent view. The only exception to this is detailed views, which will always be displayed with the same explosion distances and geometry as their parent views.

Once a model has been added to the drawing, you can place views of the model on a sheet. When a view is placed, you can determine how much of the model to show in a view, whether the view is of a single surface or shows cross sections, and how the view is scaled. You can then show the associative dimensions passed from the 3D model, or add reference dimensions as necessary. Basic view types used by Creo Parametric include general, projection, auxiliary, and detailed:

General Creates a view with no particular orientation or relationship to other views in the drawing. The model must first be oriented to the desired view orientation established by you.

Projection Creates a view that is developed from another view by projecting the geometry along a horizontal or vertical direction of viewing (orthographic projection). The projection type is specified by you in the drawing setup file and can be based on third-angle (default) or first-angle rules.

Auxiliary Creates a view that is developed from another view by projecting the geometry at right angles to a selected surface or along an axis. The surface selected from the parent view must be perpendicular to the plane of the screen.

Detailed Details a portion of the model appearing in another view. Its orientation is the same as that of the view it is created from, but its scale may be different so that the portion of the model being detailed can be better visualized.

Revolved A revolved view is a cross section of an existing view, revolved 90 degrees around a cutting plane projection. You can use a cross section created in the 3D model as the cutting plane, or you can create one on the fly while placing the view. The revolved view differs from a cross-sectional view in that it includes a line noting the axis of revolution for the view.

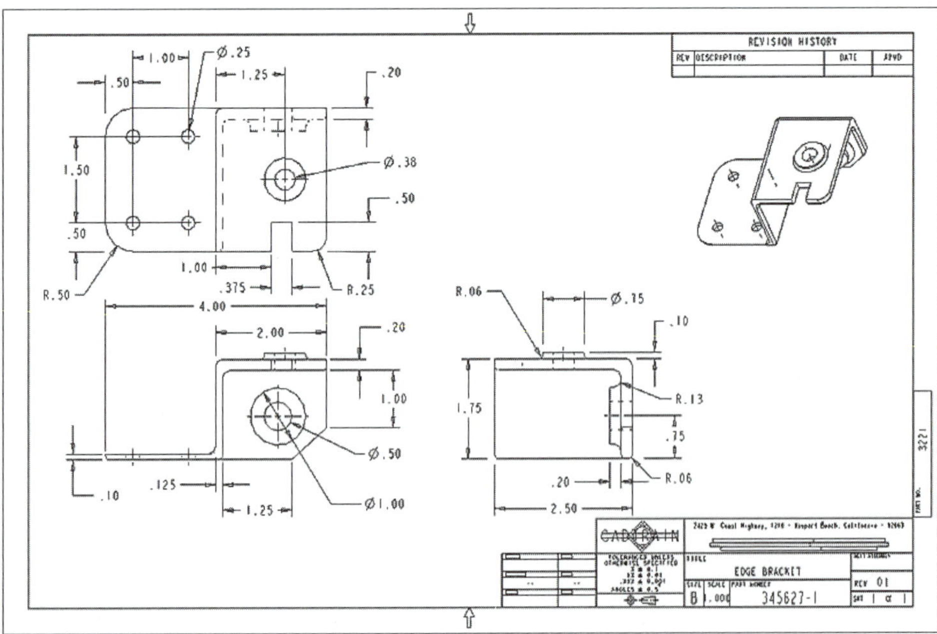

Figure 10.6 Views

The view options that determine how much of the model is visible in the view are:

- **Full View** Shows the model in its entirety.
- **Half View** Removes a portion of the model from the view on one side of a cutting plane.
- **Broken View** Removes a portion of the model from between two selected points and closes the remaining two portions together within a specified distance.
- **Partial View** Displays a portion of the model in a view within a closed boundary. The geometry appearing within the boundary is displayed; the geometry outside of it is removed.

The options that determine whether the view is of a single surface or has a cross section are:

- **Section** Displays an existing cross section of the view if the view orientation is such that the cross-sectional plane is parallel to the screen (Fig. 10.7).
- **No Xsec** Indicates that no cross section is to be displayed.
- **Of Surface** Displays a selected surface of a model in the view. The single-surface view can be of any view type except detailed.

The options that determine whether the view is scaled are:

- **Scale** Allows you to create a view with an individual scale shown under the view. When a view is being created, Creo Parametric 3.0 will prompt you for the scale value. This value can be modified later. General and detailed views can be scaled.
- **No Scale** A view will be scaled automatically using a pre-defined scale value.
- **Perspective** Creates a perspective general view.

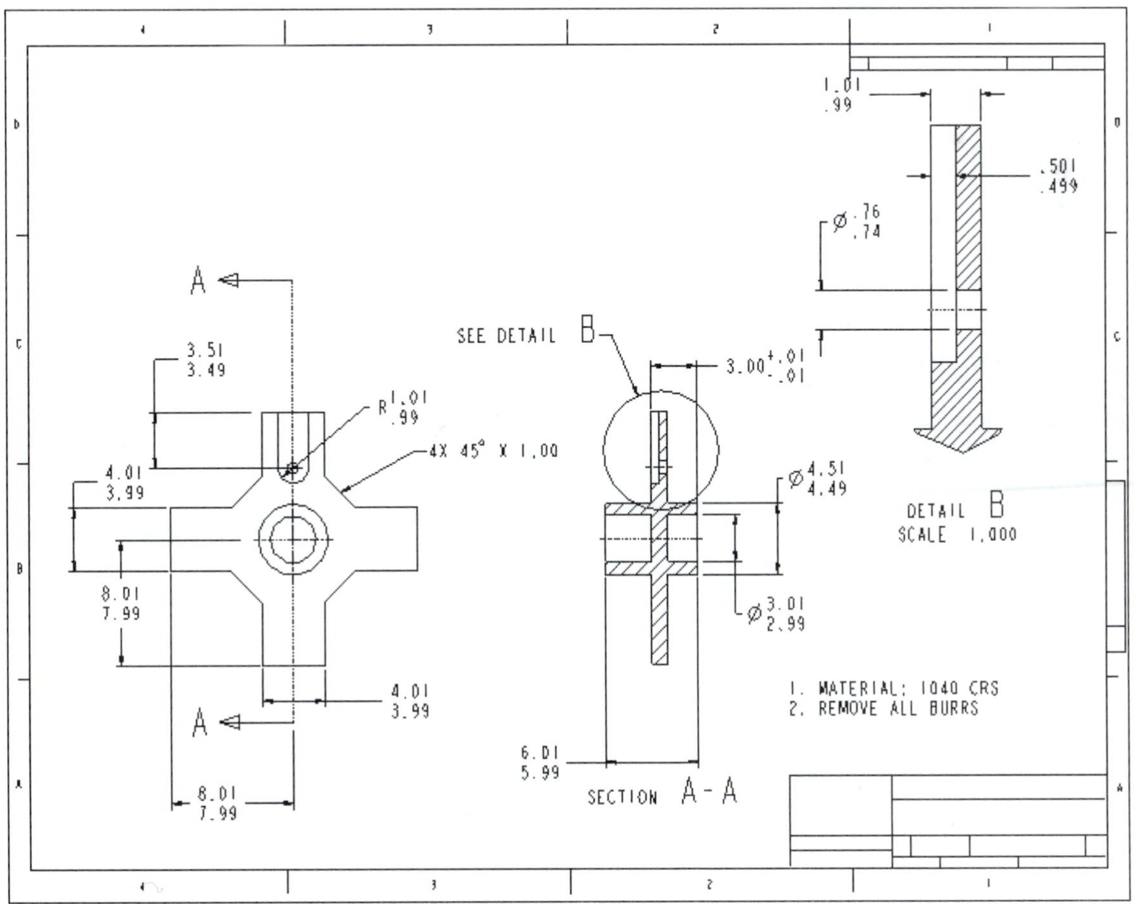

Figure 10.7 Drawing with Section Views

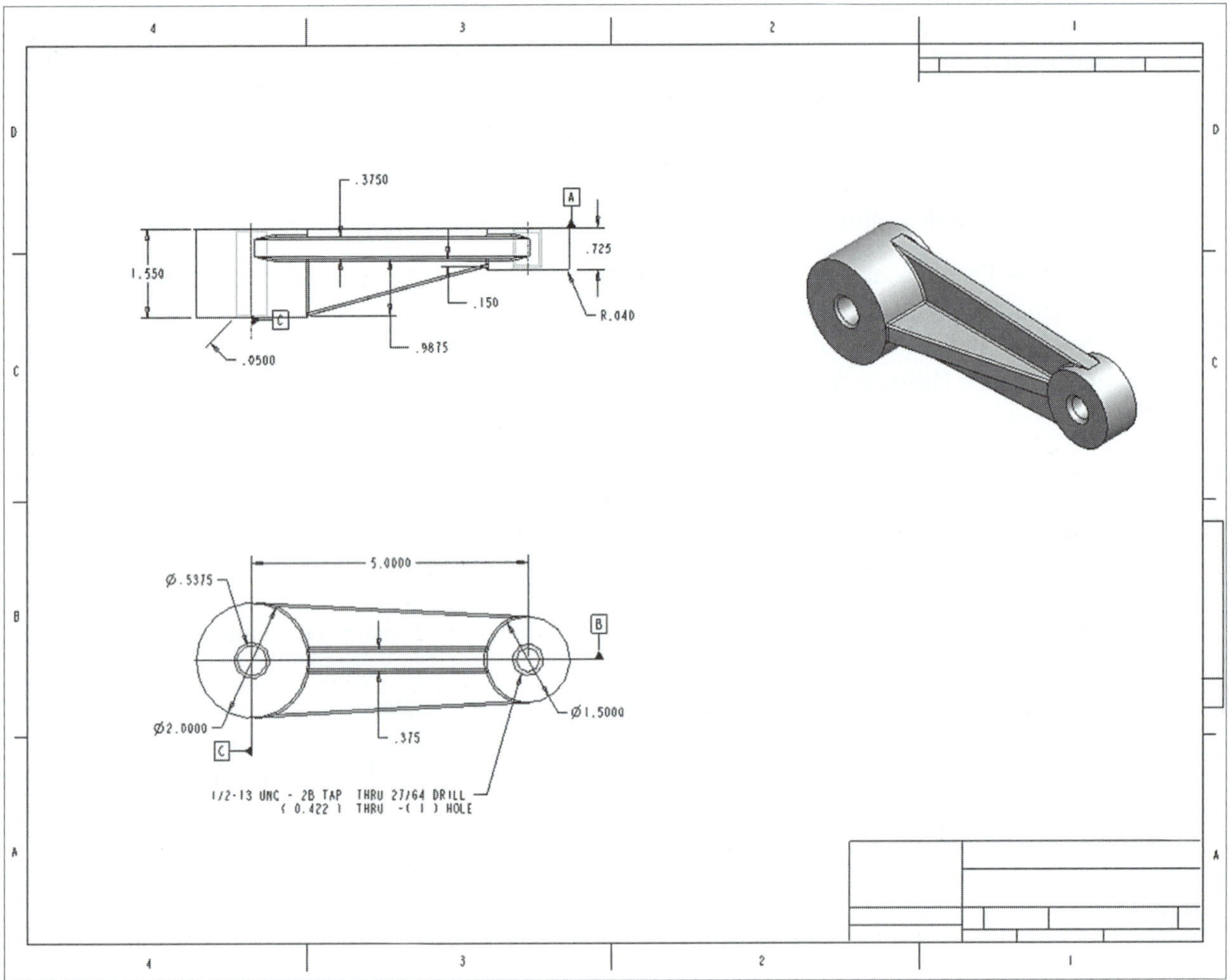

Figure 10.8 Clamp Arm Drawing

Creating a Drawing of the Clamp Arm

Before starting the drawing you will need to set your working directory to the folder that contains the Swing Clamp Assembly (and therefore all the components of the assembly). The standard three views of the Clamp_Arm (Fig. 10.8) will display automatically on the drawing, since you will use the default drawing template for this project.

Launch **PTC Creo Parametric 3.0** > **Select Working Directory** > select the directory where the **clamp_arm.prt** was saved > **OK** > [New icon] > Drawing > Name **clamp_arm** > Use default template [Fig. 10.9(a)] > **OK** [Fig. 10.9(b)] > Default Model **Browse** > pick **clamp_arm.prt** [Fig. 10.9(c)] > **Preview** *on*

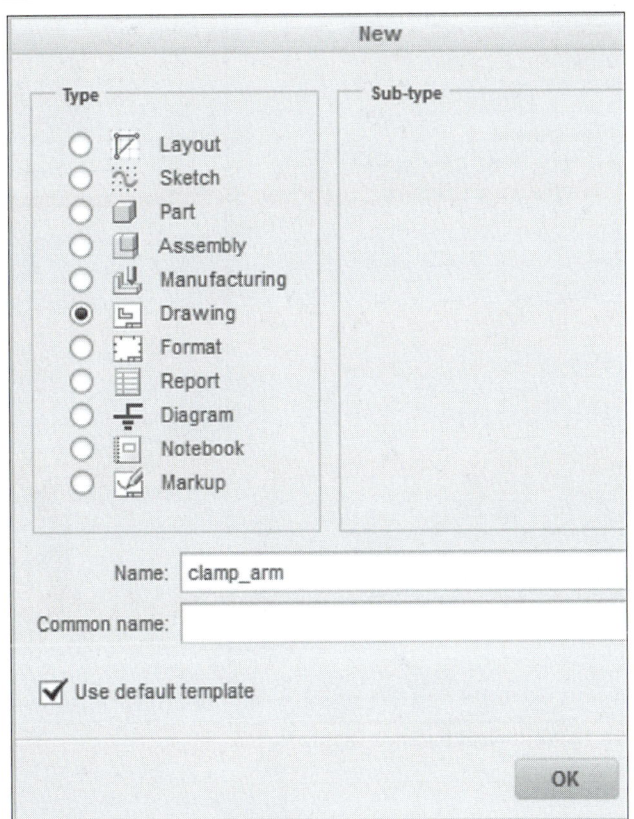

Figure 10.9(a) Type: Drawing

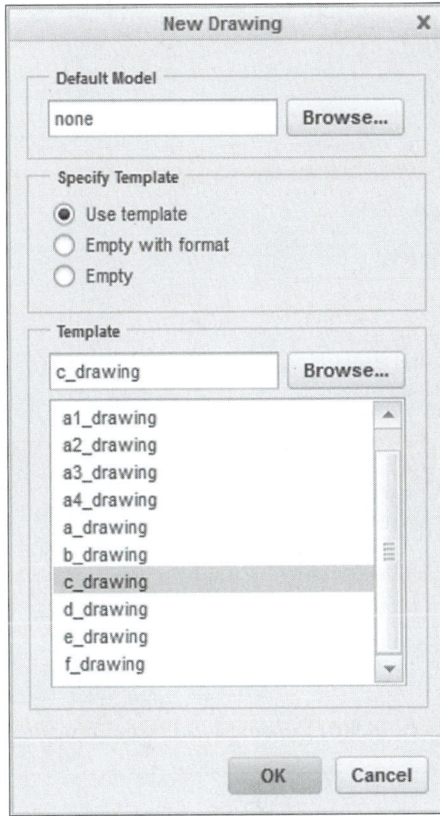

Figure 10.9(b) New Drawing Dialog

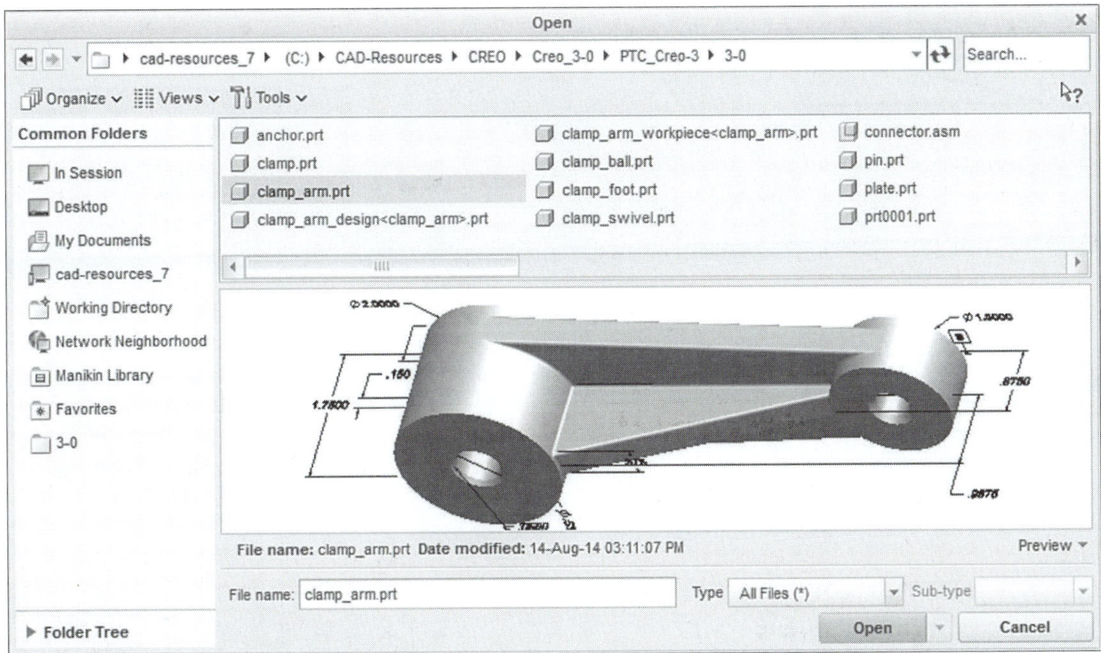

Figure 10.9(c) Preview clamp_arm.prt

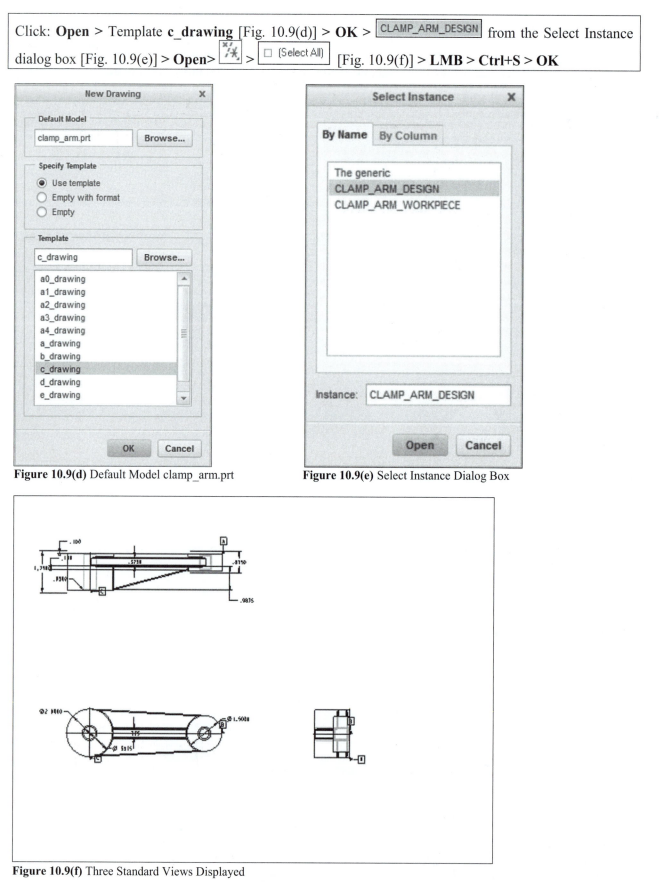

Click: **Open** > Template **c_drawing** [Fig. 10.9(d)] > **OK** > CLAMP_ARM_DESIGN from the Select Instance dialog box [Fig. 10.9(e)] > **Open**> > ☐ (Select All) [Fig. 10.9(f)] > **LMB** > **Ctrl+S** > **OK**

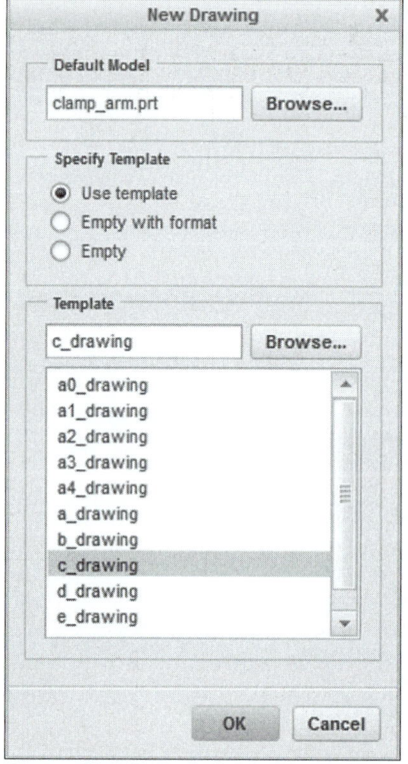

Figure 10.9(d) Default Model clamp_arm.prt

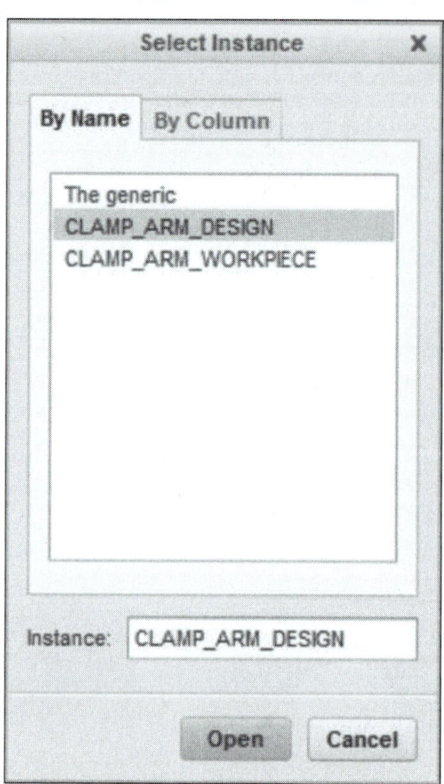

Figure 10.9(e) Select Instance Dialog Box

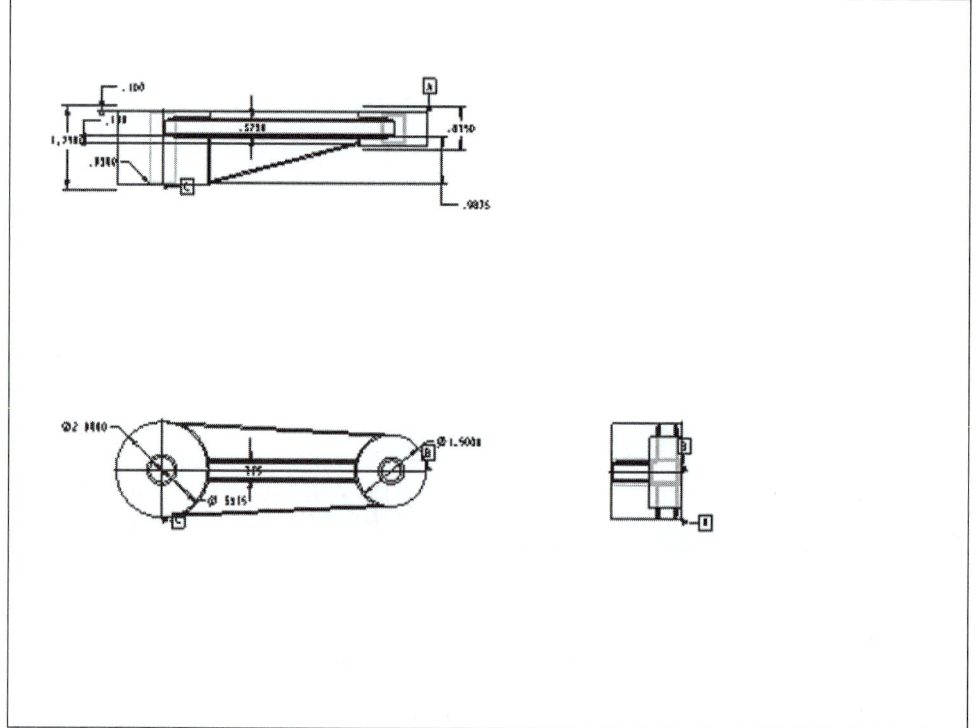

Figure 10.9(f) Three Standard Views Displayed

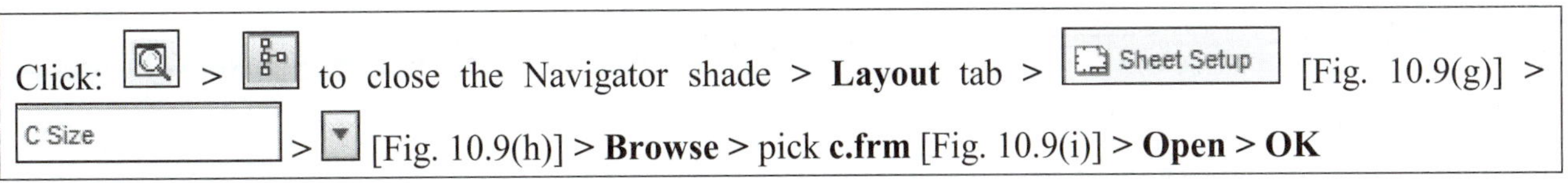

Click: > to close the Navigator shade > **Layout** tab > Sheet Setup [Fig. 10.9(g)] > C Size > [Fig. 10.9(h)] > **Browse** > pick **c.frm** [Fig. 10.9(i)] > **Open** > **OK**

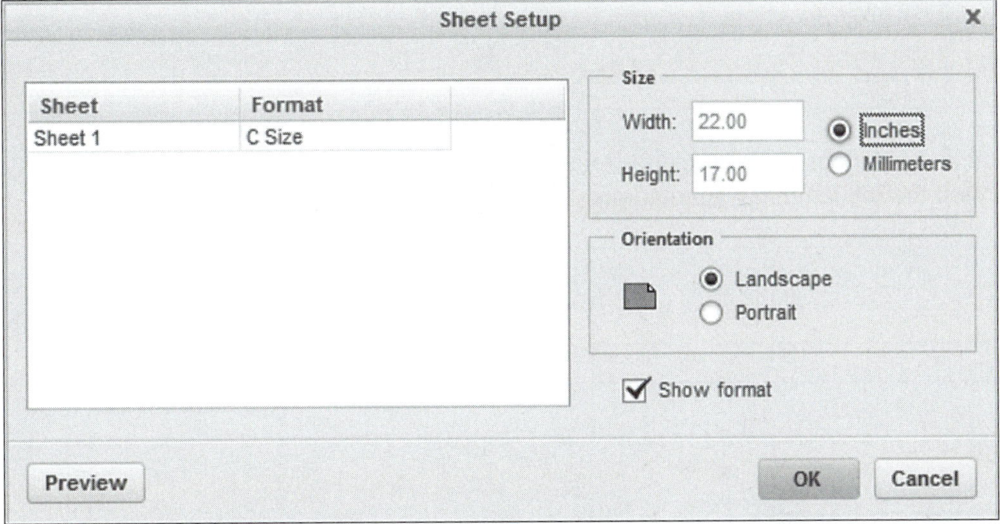

Figure 10.9(g) Sheet Setup Dialog Box

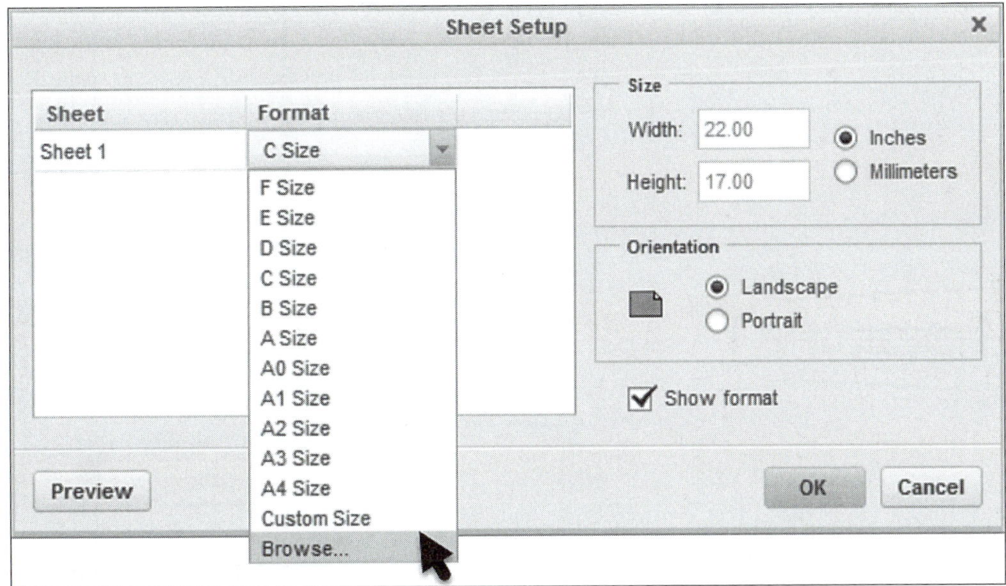

Figure 10.9(h) Sheet Setup Dialog Box > Browse

Figure 10.9(i) Open c.frm

Click: [icon] > [icon] > place the pointer inside the Graphics Toolbar > **RMB > Location > Show in Status Bar** [Fig. 10.9(j)]

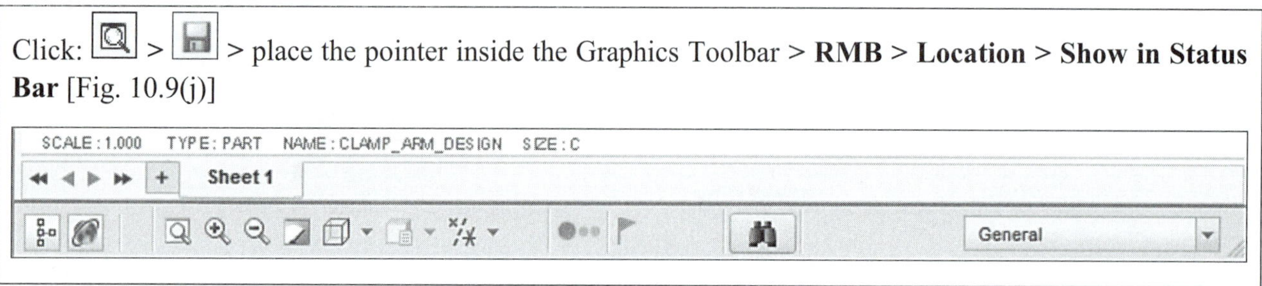

Since you created annotations of driving dimensions on the model, they will automatically display on the drawing along with the set datum tag annotations.

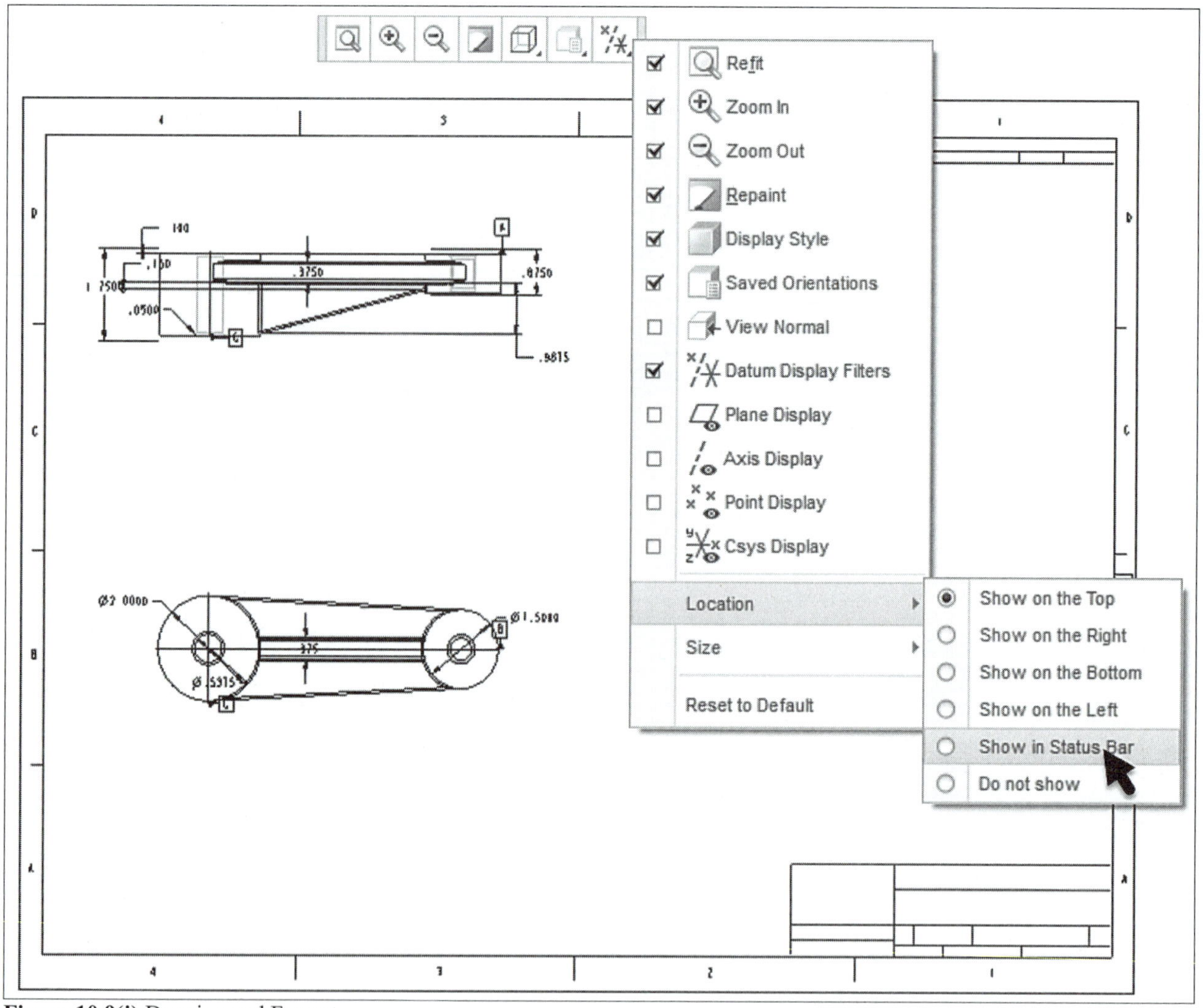

Figure 10.9(j) Drawing and Format

Press: > > **RMB** > General View **General View** [Fig. 10.9(k)] > **OK** > pick a position for the new view [Fig. 10.9(l)]

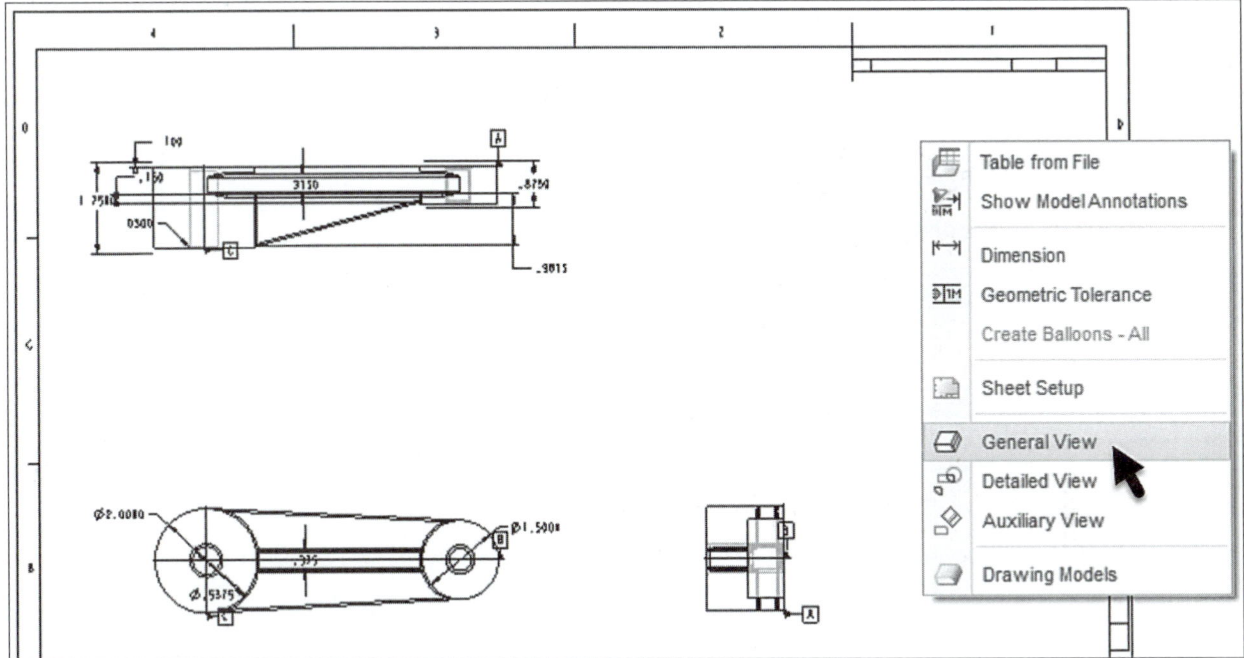

Figure 10.9(k) Press RMB > Insert General View

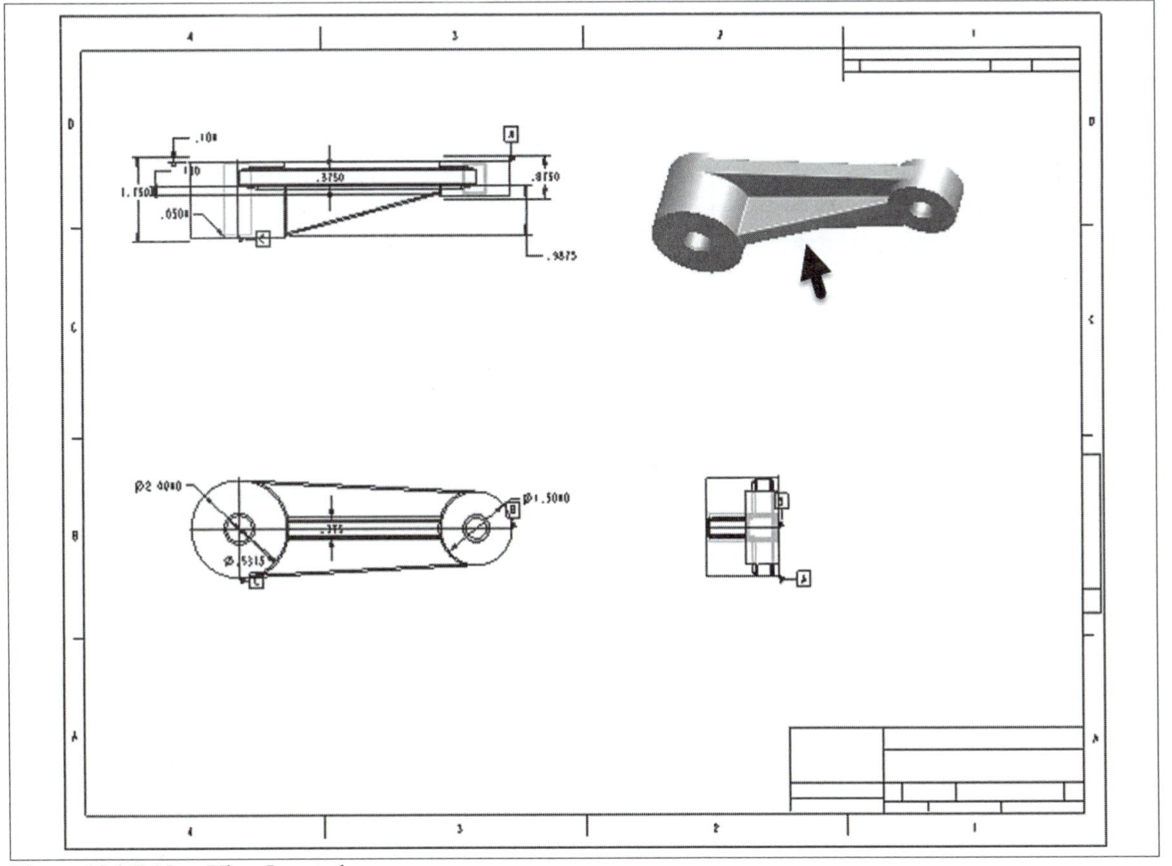

Figure 10.9(l) New View Inserted

Click: Trimetric > Default orientation: Isometric [Fig. 10.9(m)] > **OK** > **Ctrl+S** > in the Graphics Window, outside of the view boundary; **LMB** to deselect the view [Fig. 10.9(n)]

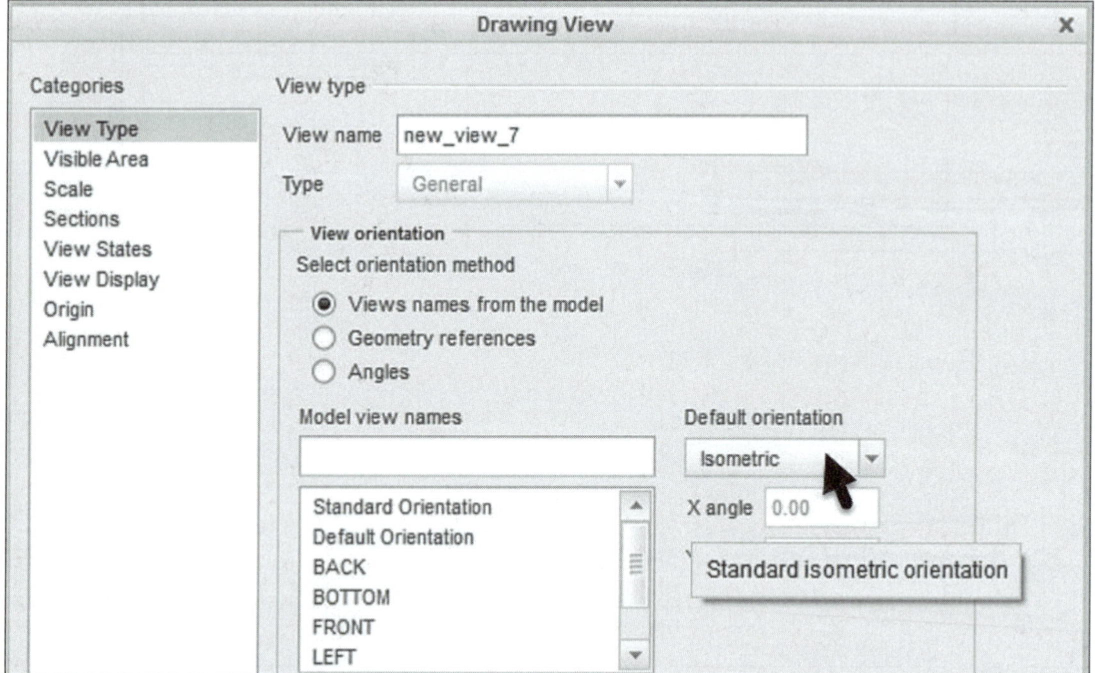

Figure 10.9(m) Default orientation: Isometric

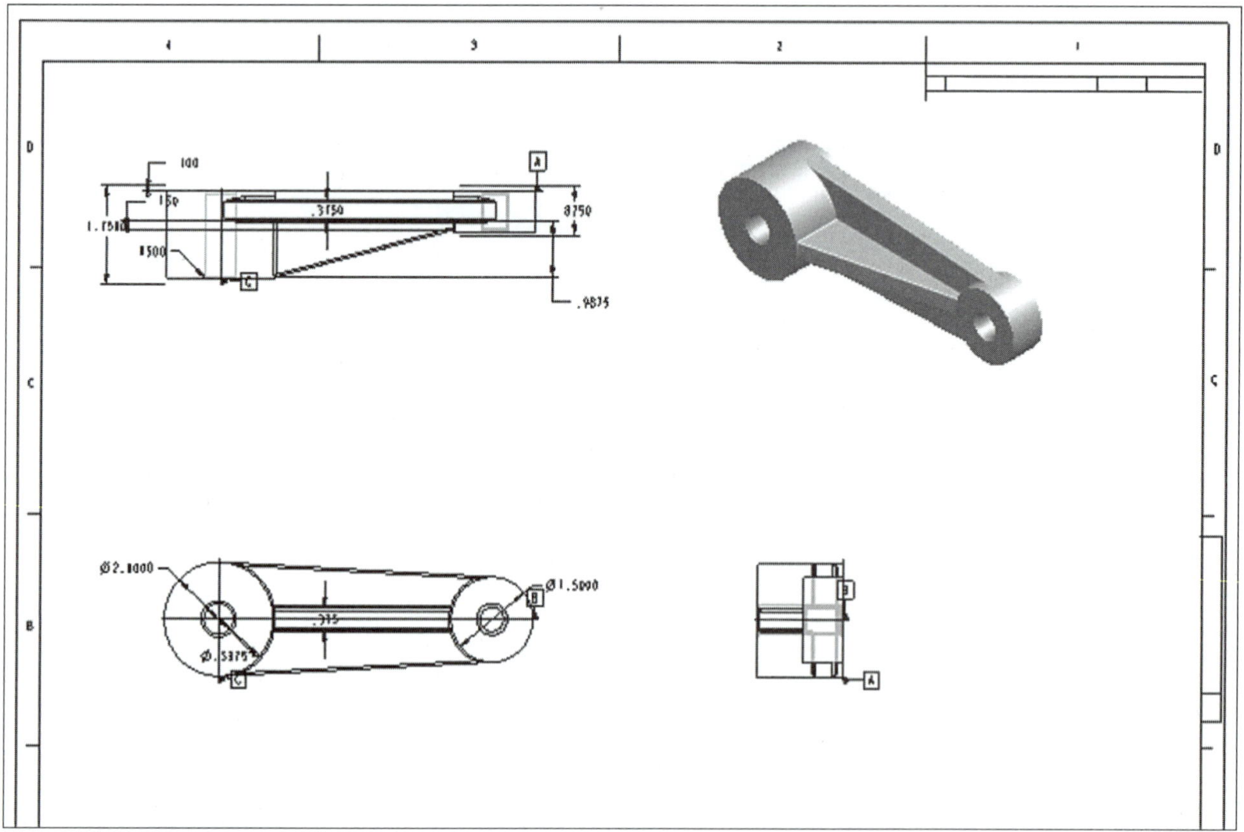

Figure 10.9(n) New Isometric View

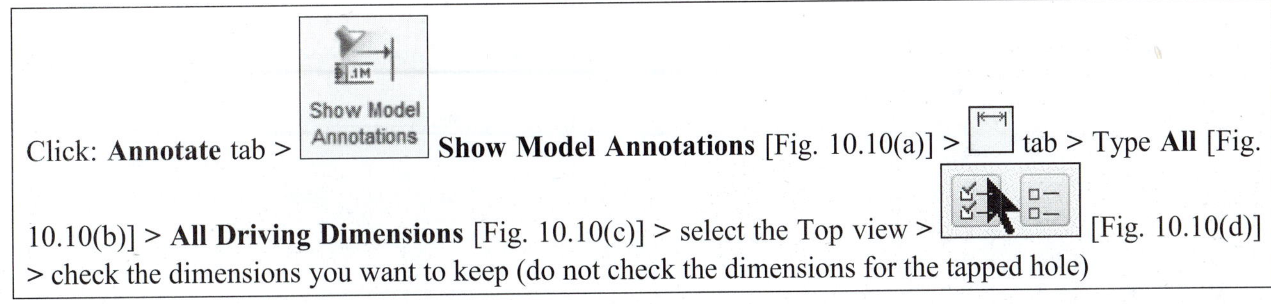

Click: **Annotate** tab > Show Model Annotations **Show Model Annotations** [Fig. 10.10(a)] > tab > Type **All** [Fig. 10.10(b)] > **All Driving Dimensions** [Fig. 10.10(c)] > select the Top view > [Fig. 10.10(d)] > check the dimensions you want to keep (do not check the dimensions for the tapped hole)

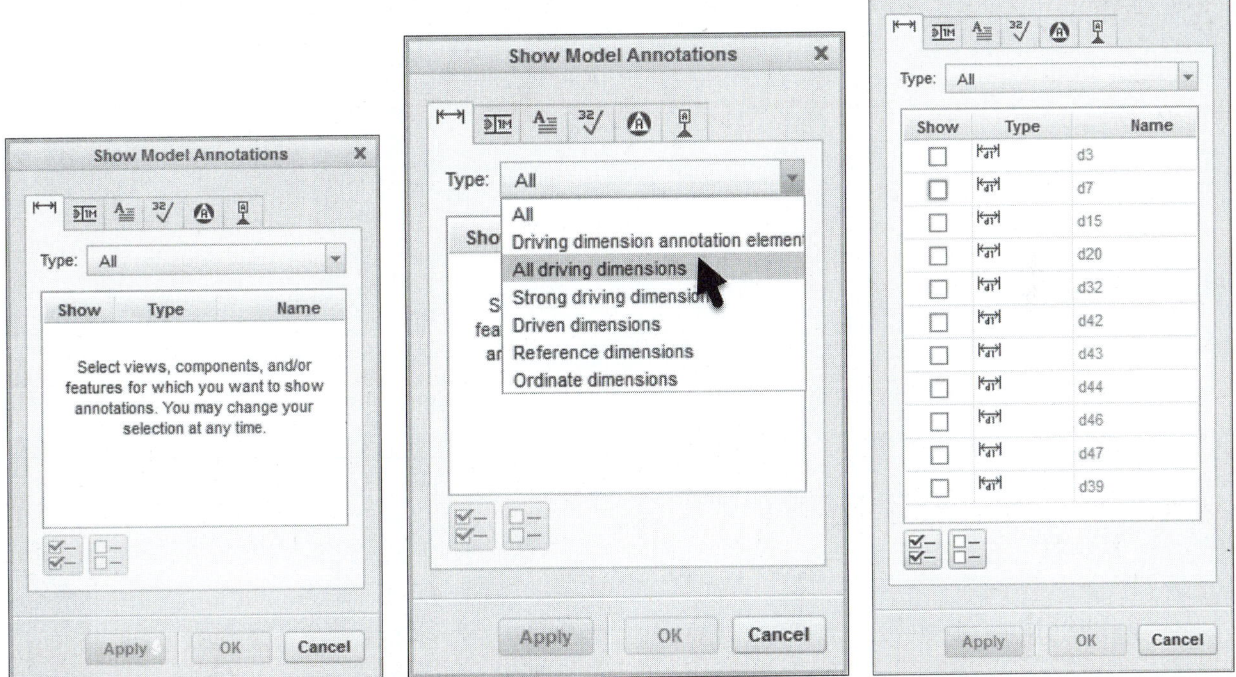

Figures 10.10(a-c) Show Model Annotations Dialog Box (Dimensions tab) *(your list may appear differently)*

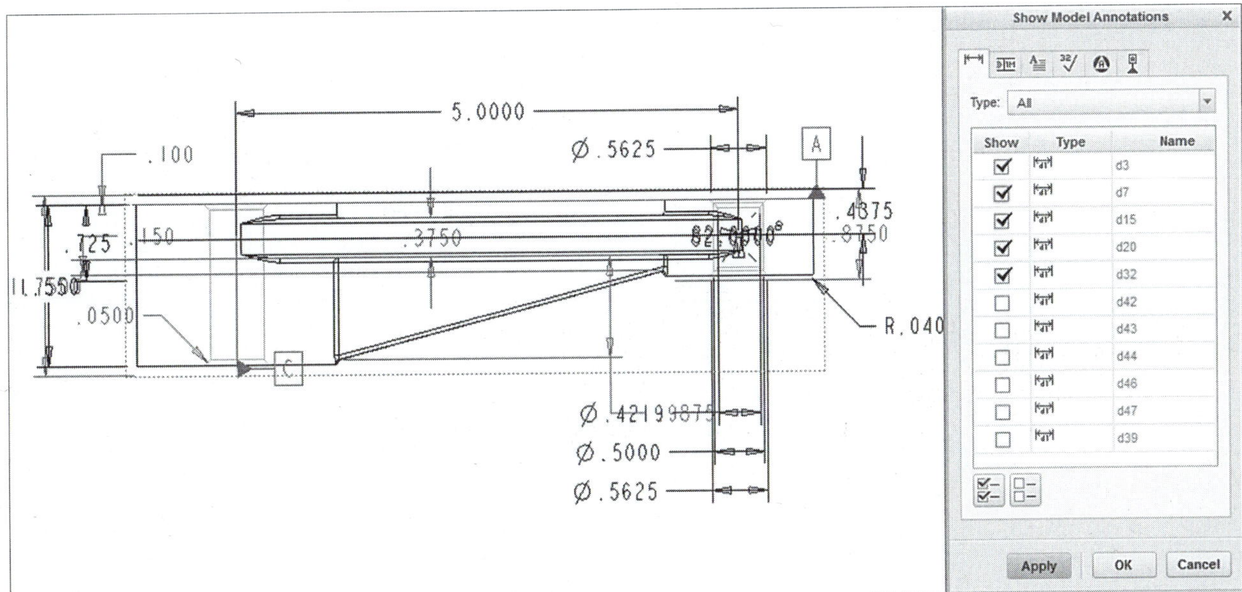

Figure 10.10(d) Select All Annotations *(your list may show more annotations)*

Click: **Apply** > [icon] tab [Fig. 10.10(e)] > [icon] [Fig. 10.10(f)] *(your list may appear differently)* > [icon] tab > [icon] [Fig. 10.10(g)] > **OK** > **LMB** to deselect [Fig. 10.10(h)] > **Ctrl+S**

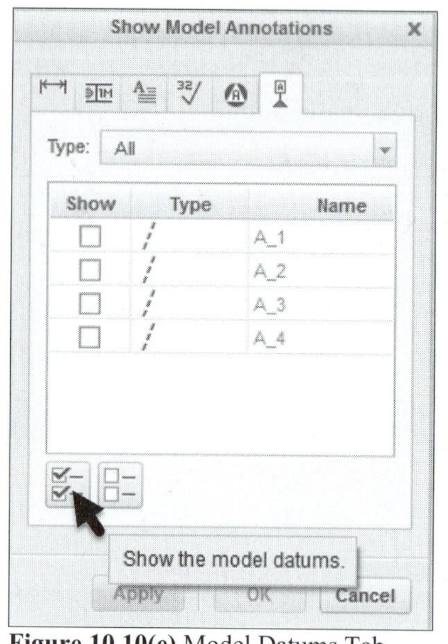

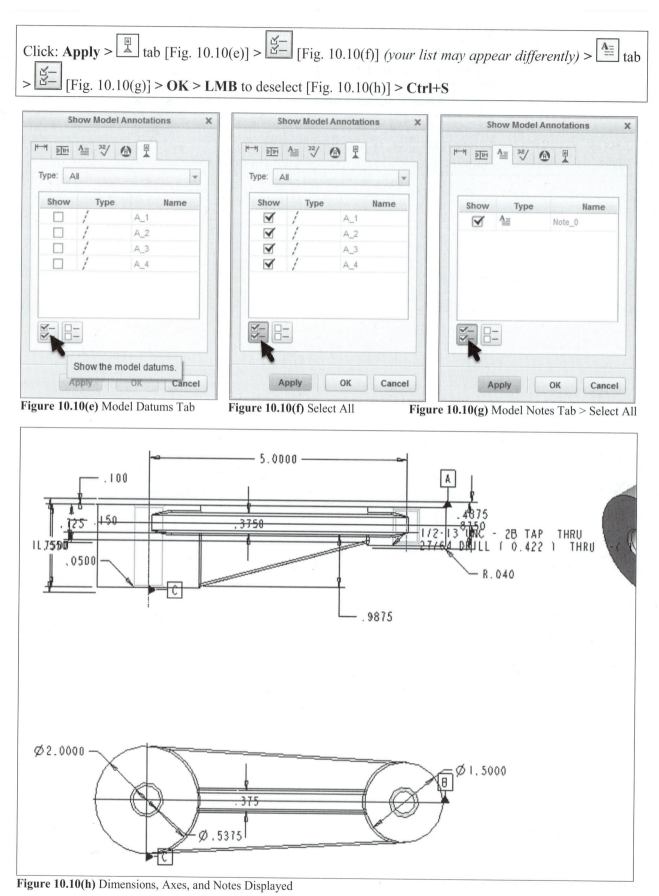

Figure 10.10(e) Model Datums Tab

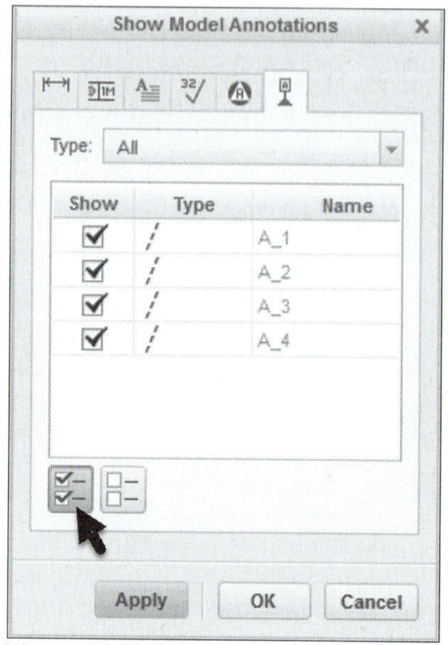

Figure 10.10(f) Select All

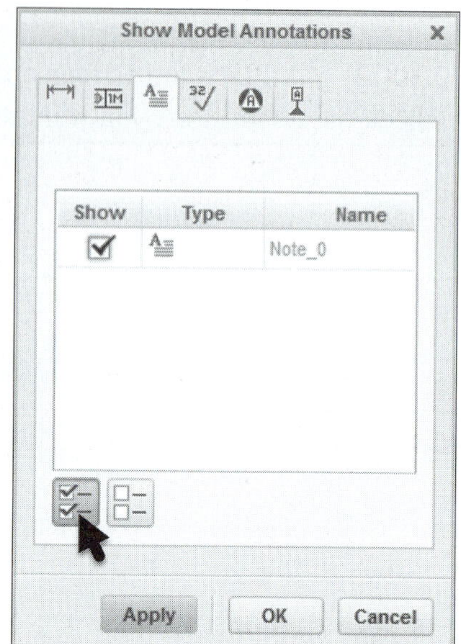

Figure 10.10(g) Model Notes Tab > Select All

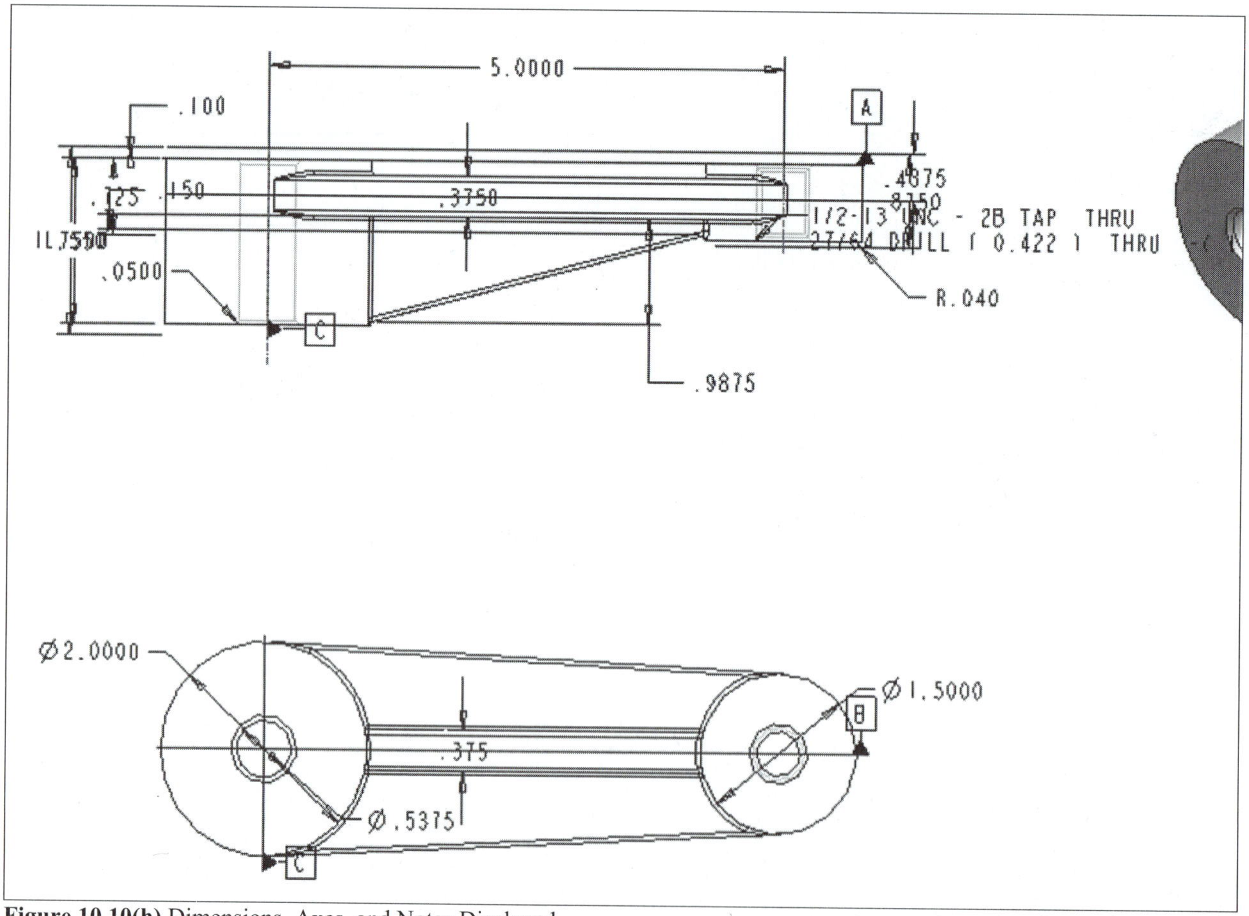

Figure 10.10(h) Dimensions, Axes, and Notes Displayed

Select the general view > press **RMB** > Lock View Movement (unlock) > hold down **LMB** and move the view [Fig. 10.11(a)] > release the **LMB** > **LMB** outside of the view to set the position [Fig. 10.11(b)]

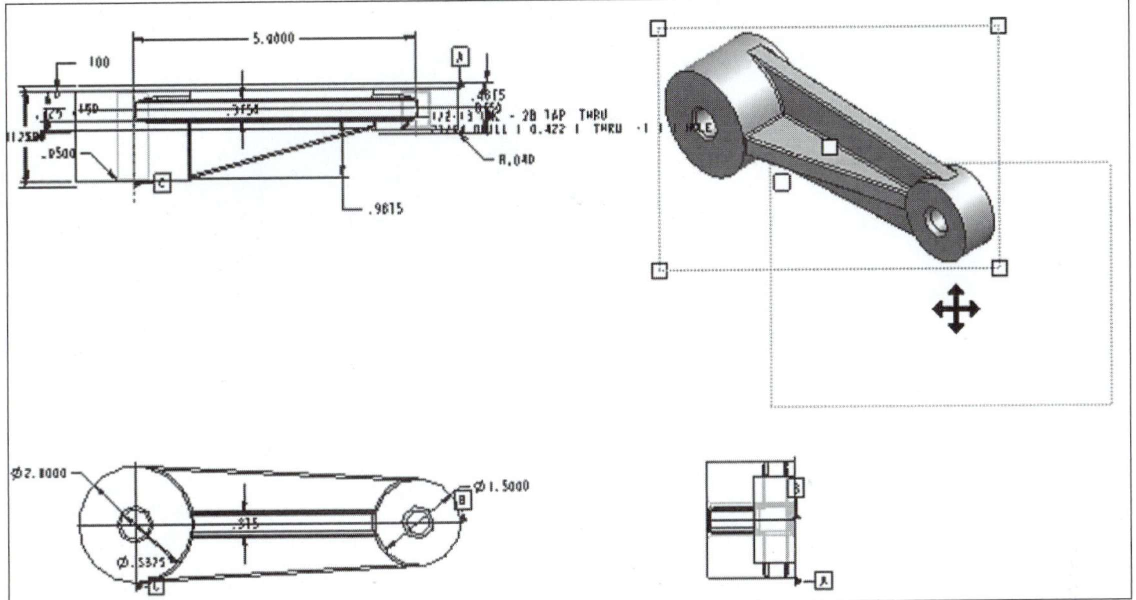

Figure 10.11(a) Move View

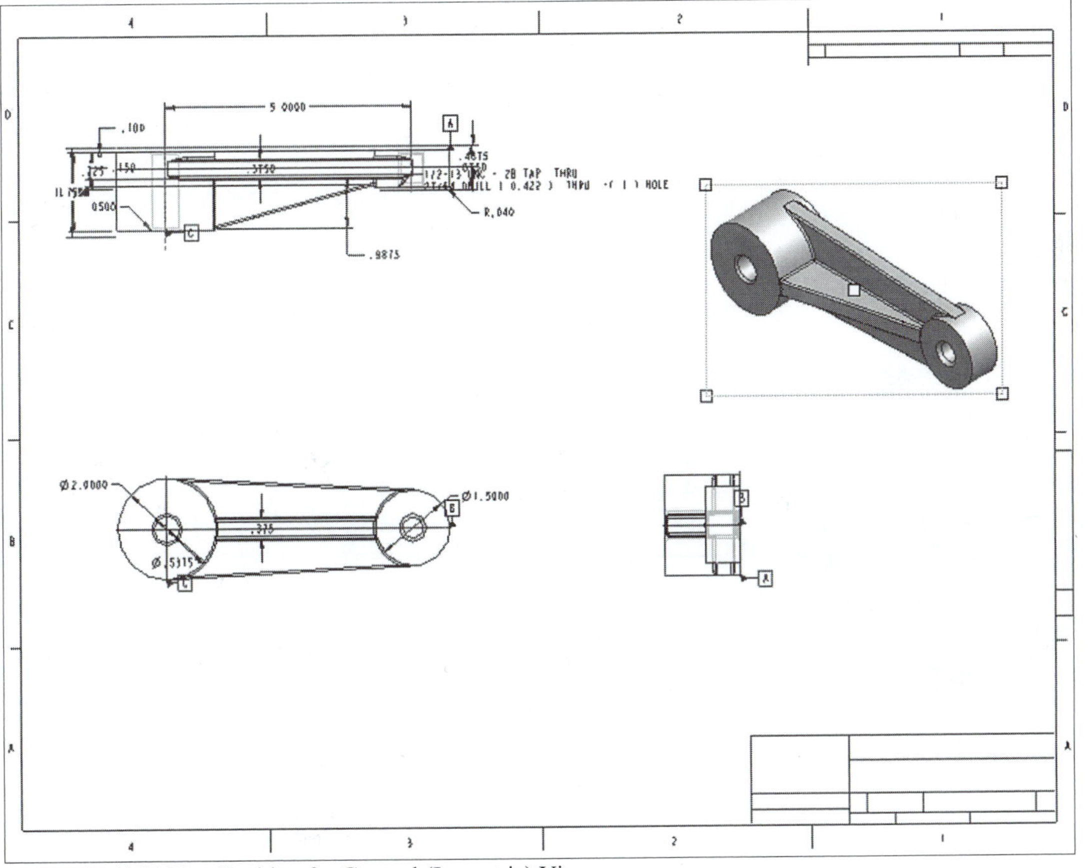

Figure 10.11(b) New Position for General (Isometric) View

Press: **Ctrl+S** > pick on the Right Side view > press **RMB** > **Delete** [Fig. 10.12(a)] > roll the (thumbwheel) **MMB** to zoom in on the Front view [Fig. 10.12(b)]

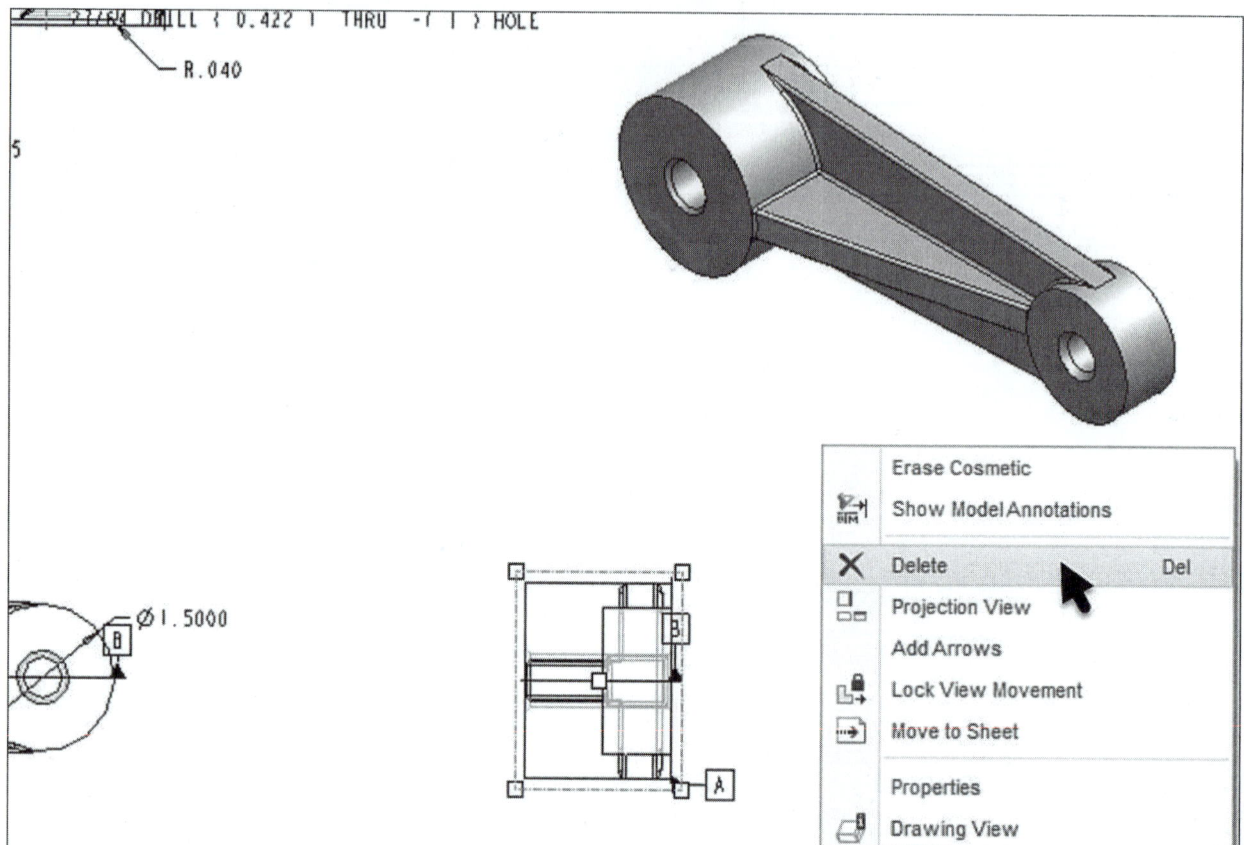

Figure 10.12(a) Press RMB > Delete the Right Side View

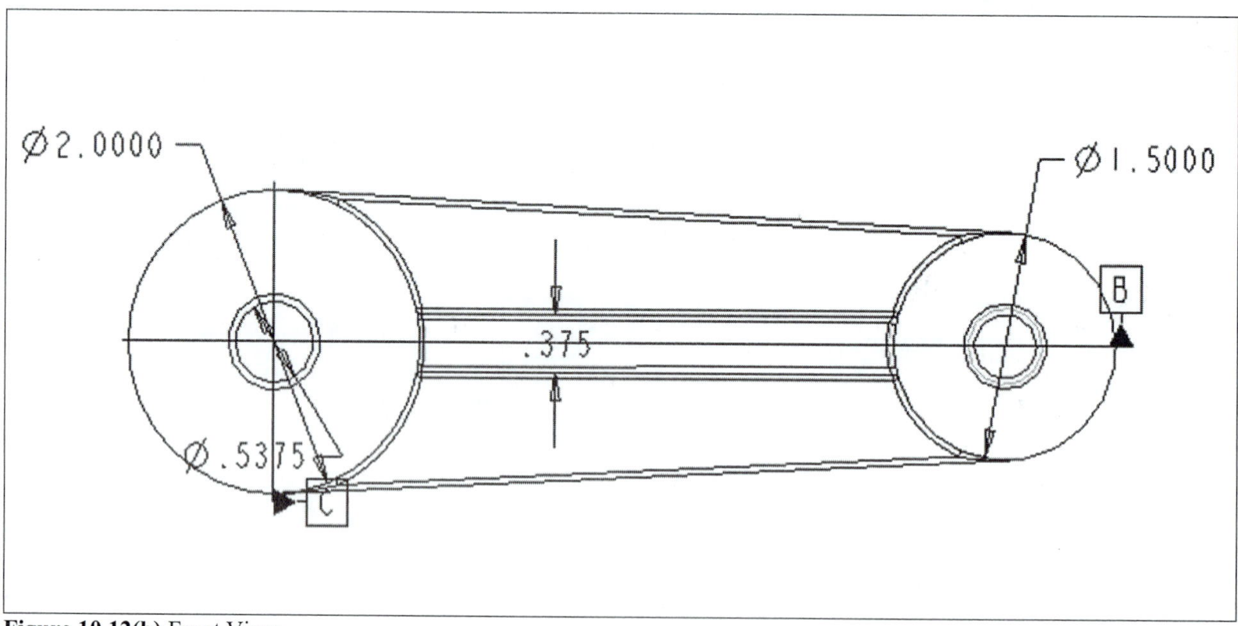

Figure 10.12(b) Front View

Click: **Annotate** tab > pick on the hole dimension > press **LMB** and move it to a new location [Fig. 10.12(c)] > release the **LMB** > reposition dimensions as needed > reposition each datum tag by picking on the item and then moving it to a new location > to erase an item, pick on it > press **RMB** > **Erase** > **LMB** to accept > pick on the **2.000** diameter dimension > press **RMB** > **Flip Arrows** [Fig. 10.12(d)]

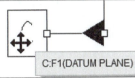

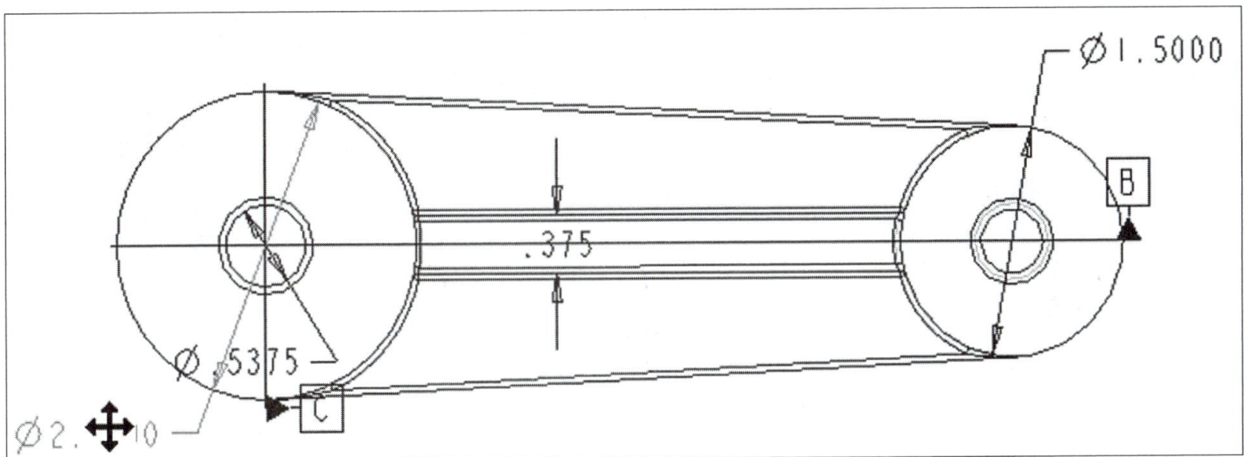

Figure 10.12(c) Move the Dimension

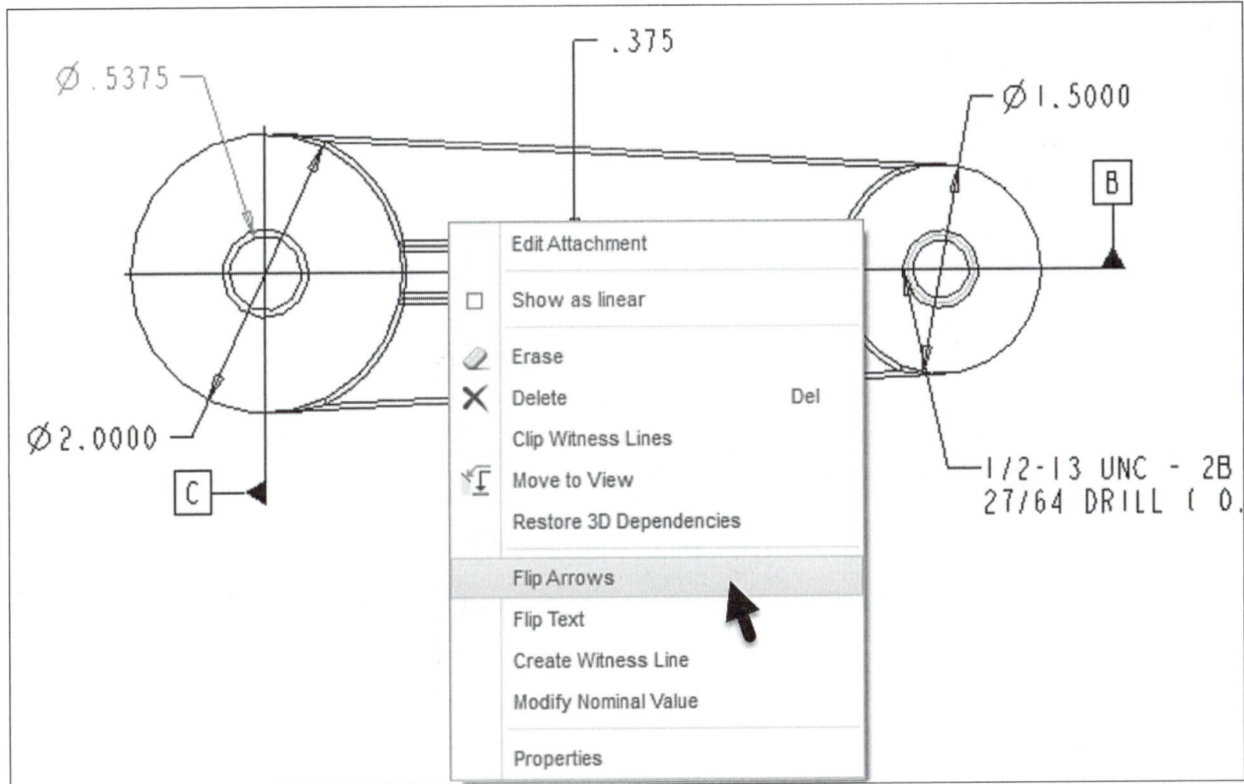

Figure 10.12(d) Flip Arrows

Pick on the note in the Top view > press **RMB** > Move to View [Fig. 10.13(a)] > select the Front view > press **LMB** on the note > move the pointer [Figs. 10.13(b-c)] > release the **LMB** > **LMB** to deselect

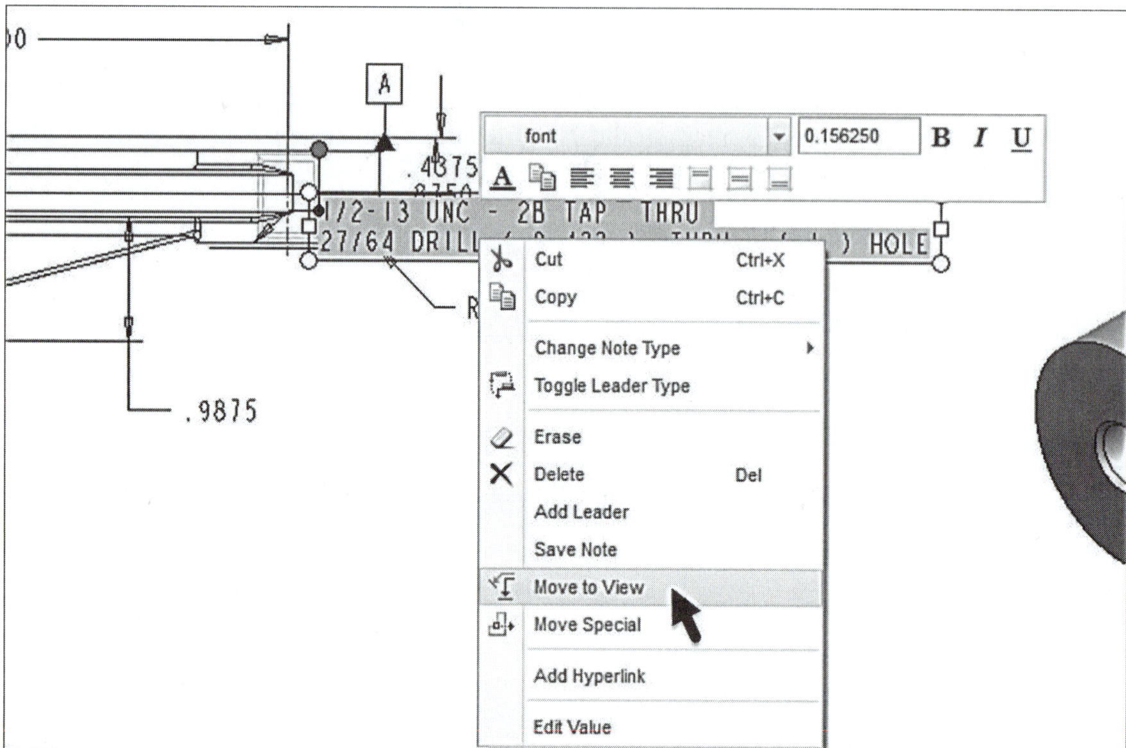

Figure 10.13(a) Press RMB > Move to View

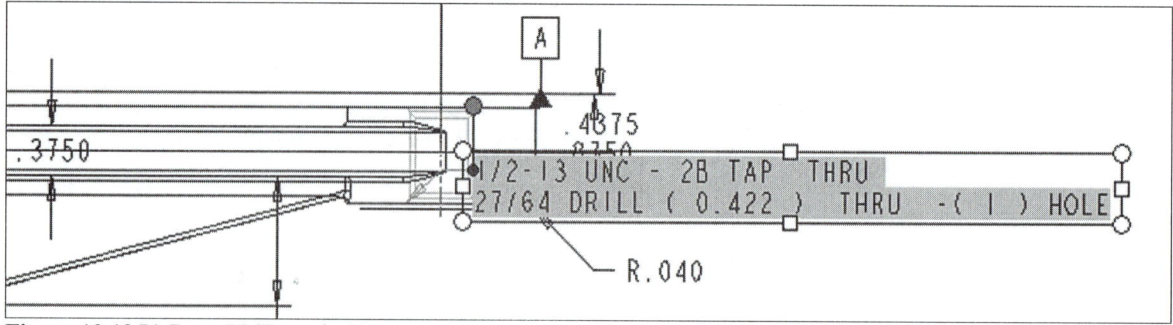

Figure 10.13(b) Press LMB on the Note

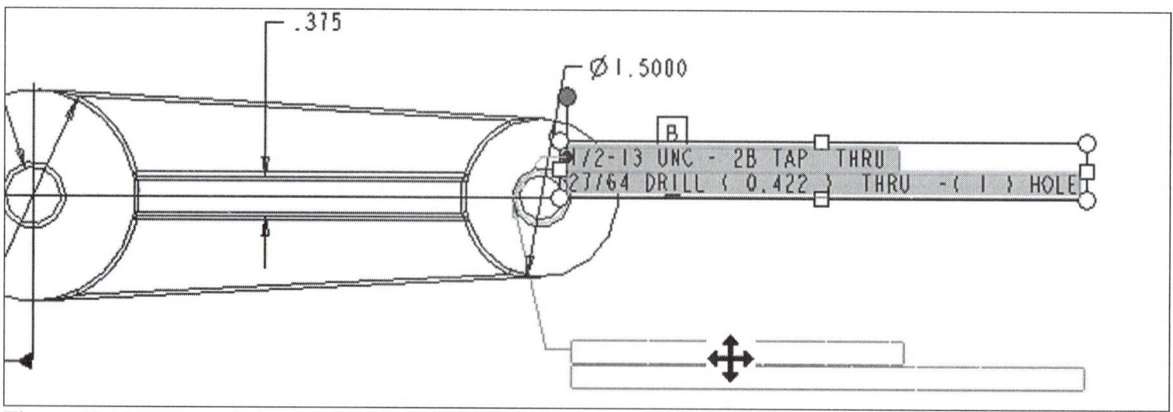

Figure 10.13(c) Move the Note

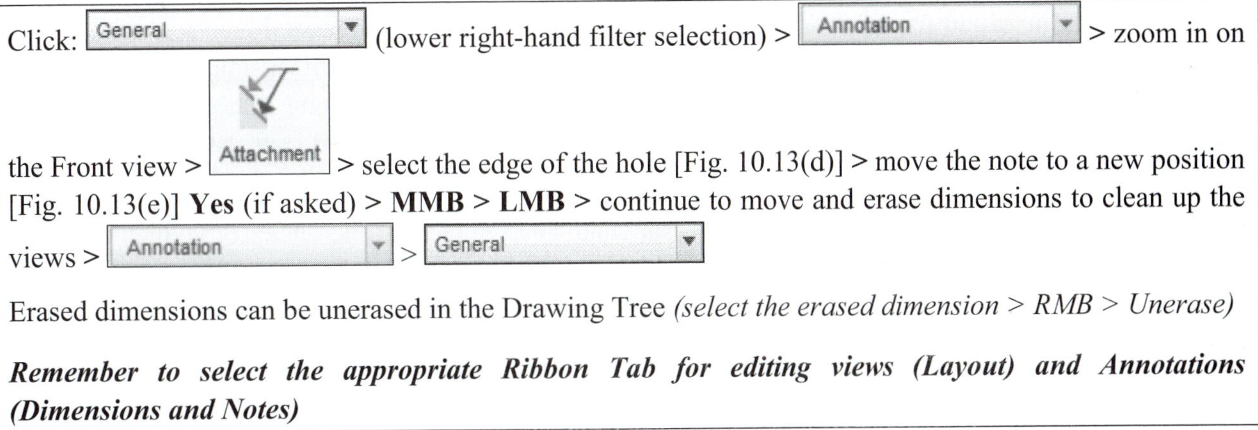

Click: General (lower right-hand filter selection) > Annotation > zoom in on the Front view > Attachment > select the edge of the hole [Fig. 10.13(d)] > move the note to a new position [Fig. 10.13(e)] **Yes** (if asked) > **MMB** > **LMB** > continue to move and erase dimensions to clean up the views > Annotation > General

Erased dimensions can be unerased in the Drawing Tree *(select the erased dimension > RMB > Unerase)*

Remember to select the appropriate Ribbon Tab for editing views (Layout) and Annotations (Dimensions and Notes)

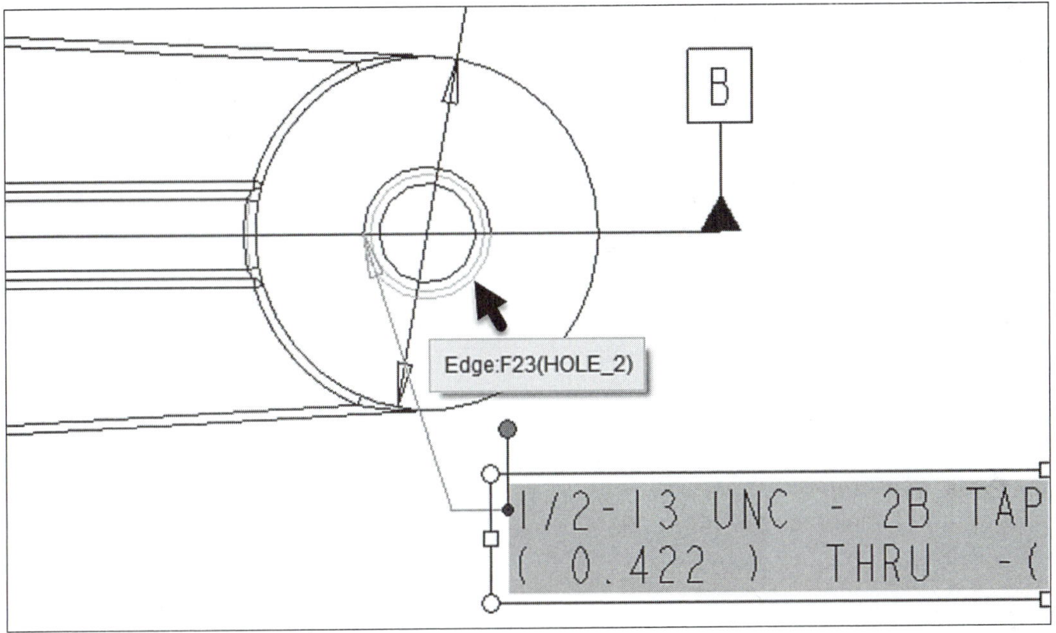

Figure 10.13(d) Erase Dimensions

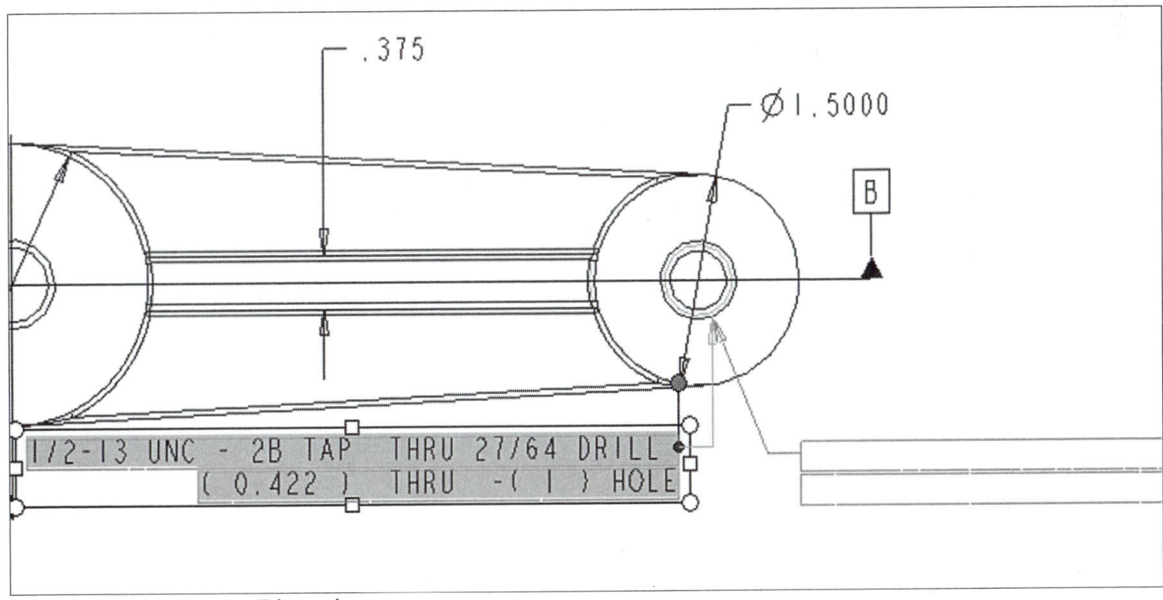

Figure 10.13(e) Unerase Dimension

Move the **5.00** dimension from the Top view to the Front view > Delete or erase dimensions that do not have a detailing purpose (dimension air- as with the 1.75 dimension). Deleted dimensions can be retrieved by using Show Model Annotations again. > Complete the drawing by erasing unneeded dimensions, flipping arrows, and repositioning dimensions, notes, and datum tags as needed [Fig. 10.13(f)]. > **Ctrl+S** > > **File** > **Manage File** > **Delete Old Versions** > **Yes** > **File** > **Save As** > Type > **Zip File (*.zip)** > **OK** > upload to your course interface or attach to an email and send to your instructor or yourself > **File** > **Close**

Though not necessarily complete or correct to ASME Y14.5M standards, we will leave this drawing as is. You may finalize this drawing after completing the next lesson.

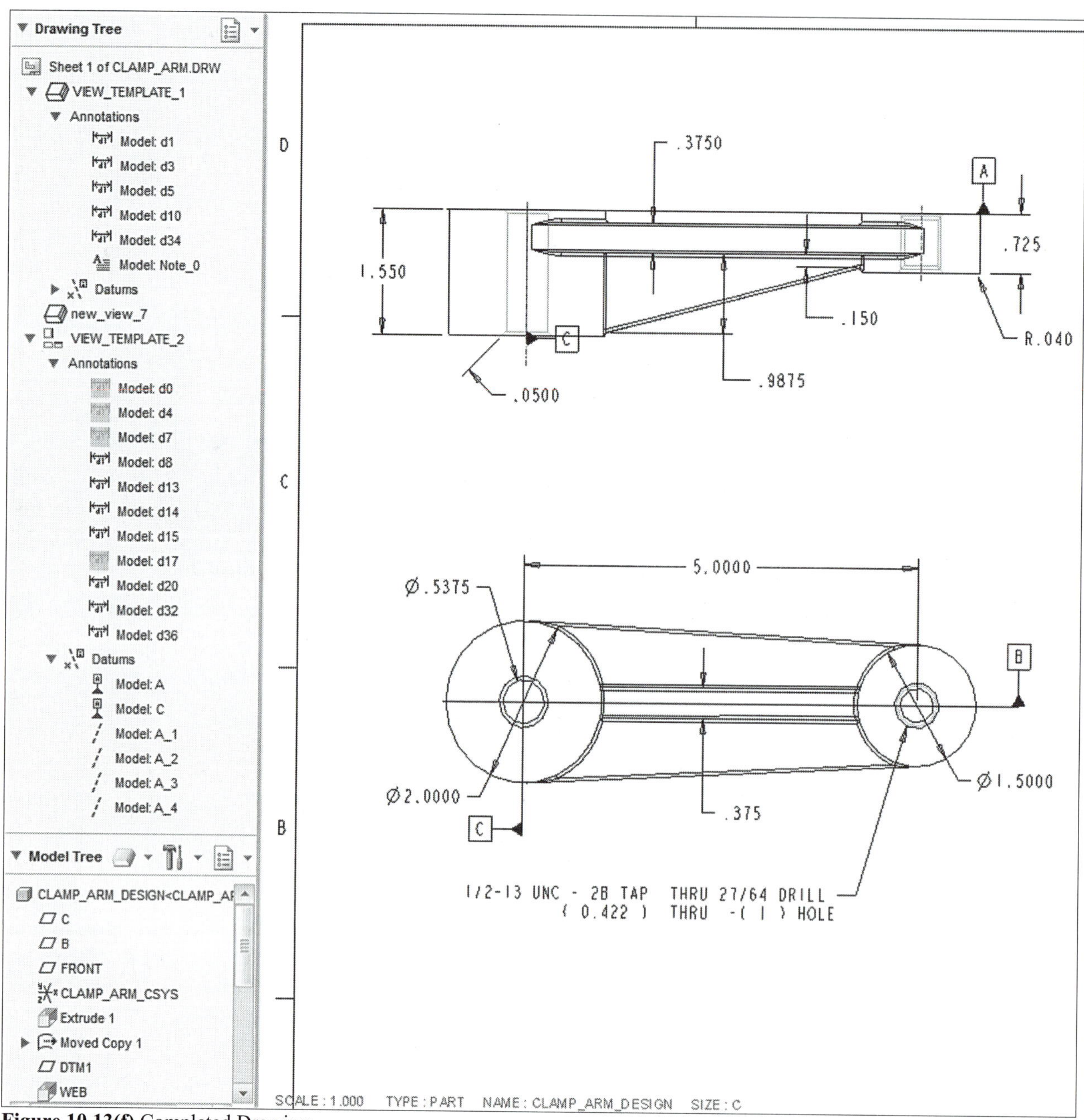

Figure 10.13(f) Completed Drawing

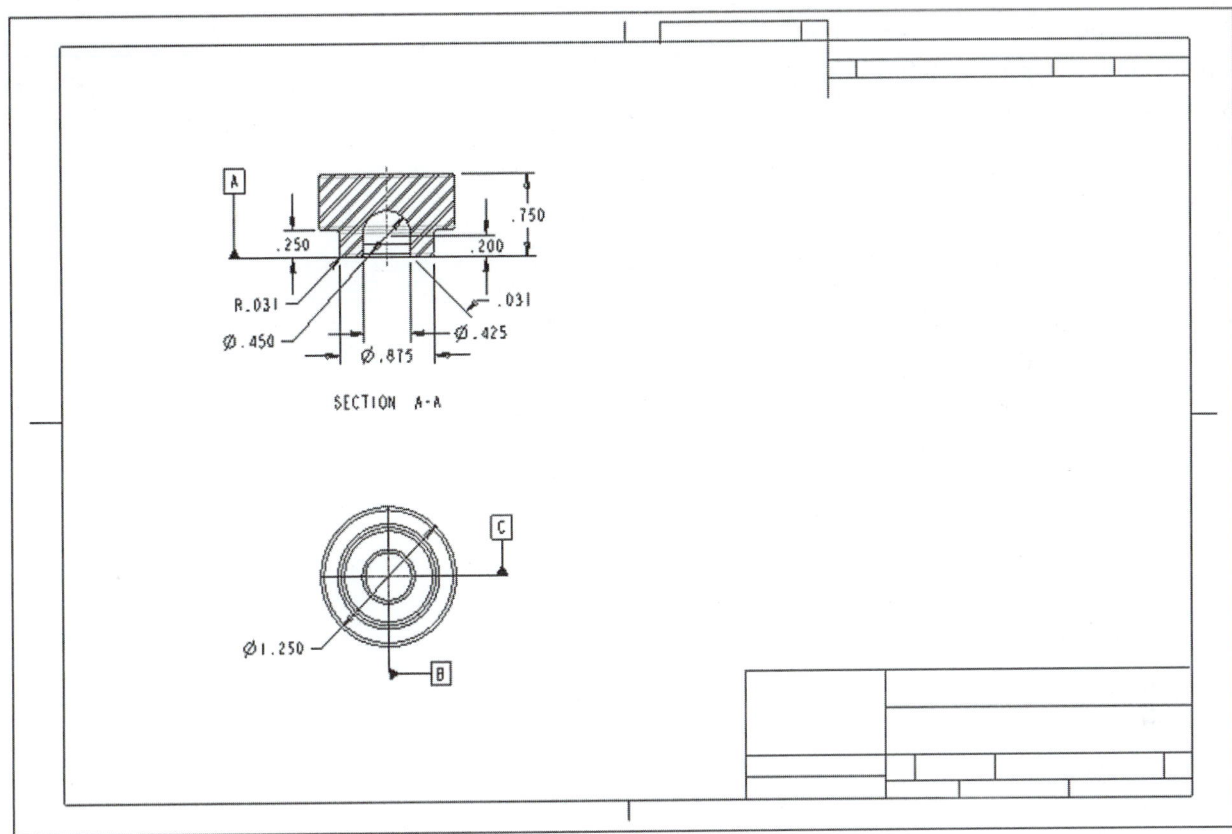

Figure 10.14 Clamp Foot Drawing

Creating a Drawing of the Clamp Foot

Before starting the drawing you will need to set your directory to the folder that contains the Swing Clamp Assembly (and therefore all the components of the assembly), and this time you will bring the Clamp_Foot (Fig. 10.14) in session by opening it. The standard three views of the model will display automatically on the drawing, since you will use the default drawing template for this project.

Click: **File** > **Manage Session** > **Select Working Directory** > select the directory where the **clamp_assembly.asm** was saved > **OK** > **File** > **Open** > **clamp_foot.prt** > **Open** (Fig. 10.15) > **File** > **Close** *(this keeps the part in session but not active or open)*

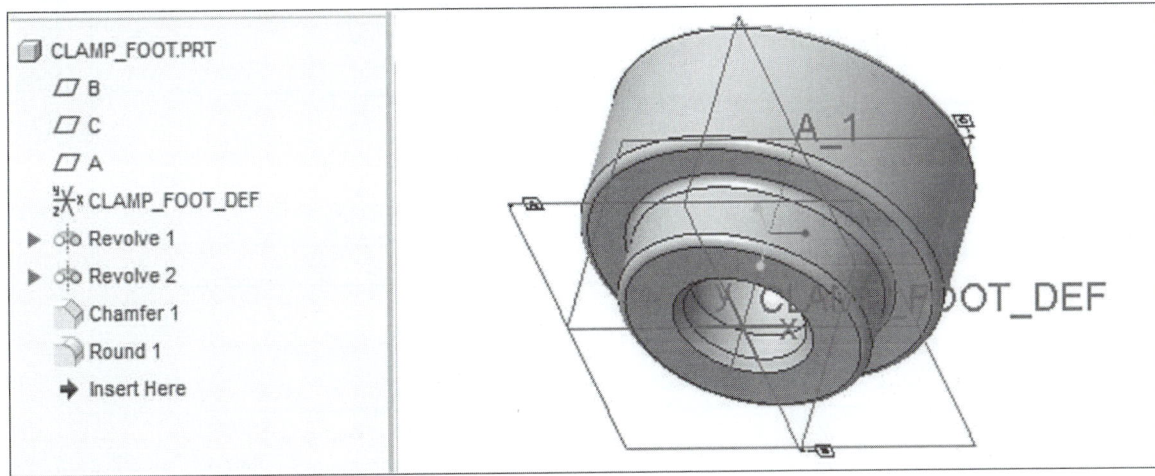

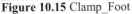

Figure 10.15 Clamp_Foot

Press: **Ctrl+N** > Drawing > Name **clamp_foot** *(keep the default settings)* > Use default template [Fig. 10.16(a)] > **OK** > Template **b_drawing** [Fig. 10.16(b)] > make sure that the Default Model is the **CLAMP_FOOT.PRT** *(if it is not, change it with **Browse**)* > **OK** [Fig. 10.16(c)] > **Ctrl+S** > **OK**

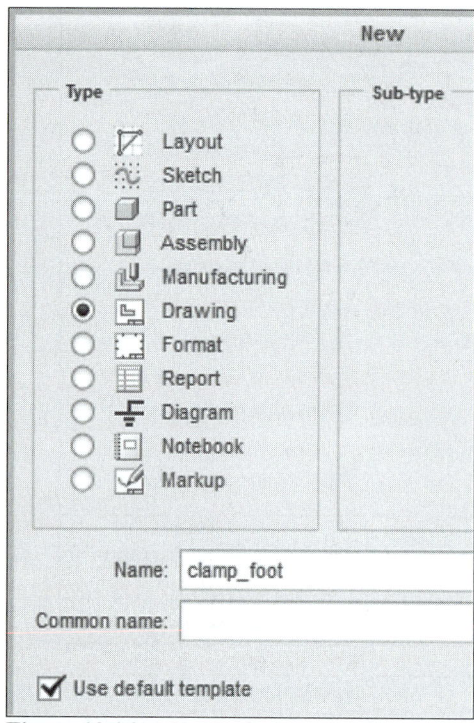

Figure 10.16(a) Use Default Template

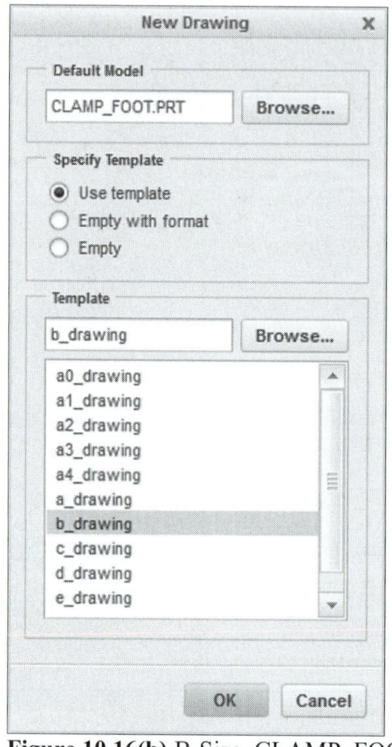

Figure 10.16(b) B-Size. CLAMP_FOOT.PRT is Default Model

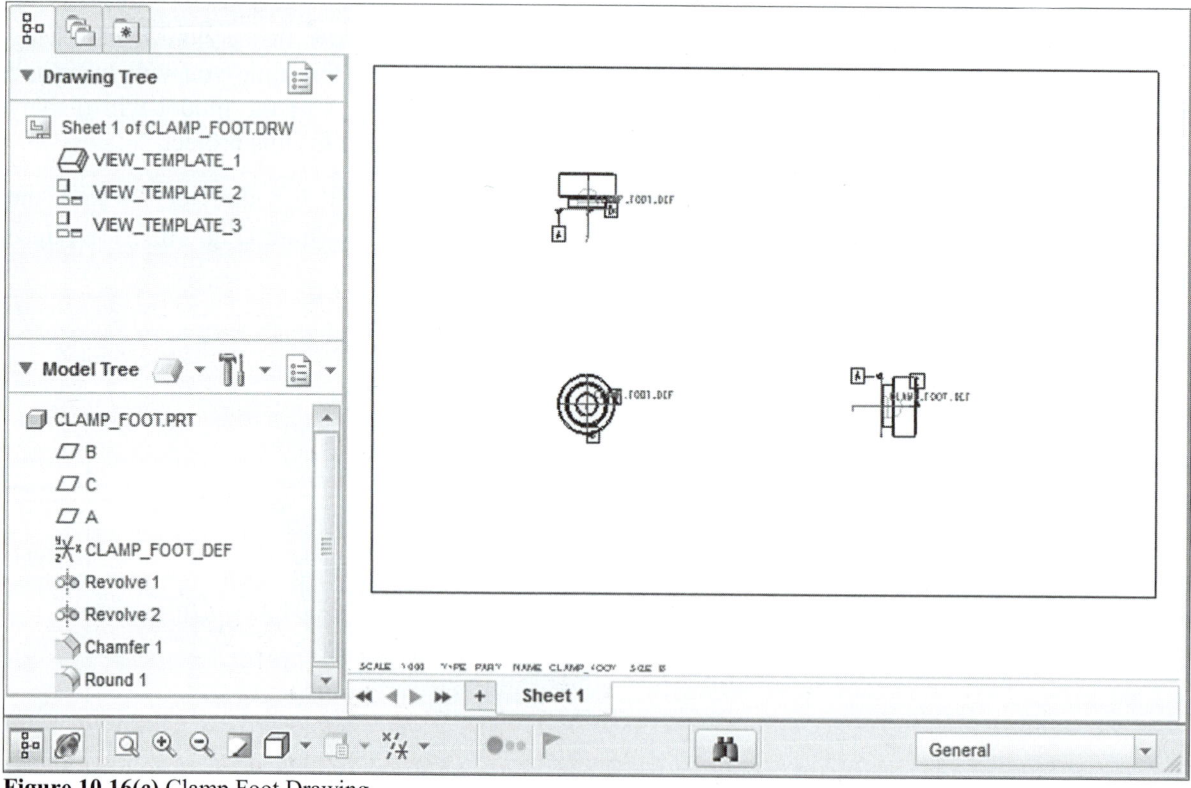

Figure 10.16(c) Clamp Foot Drawing

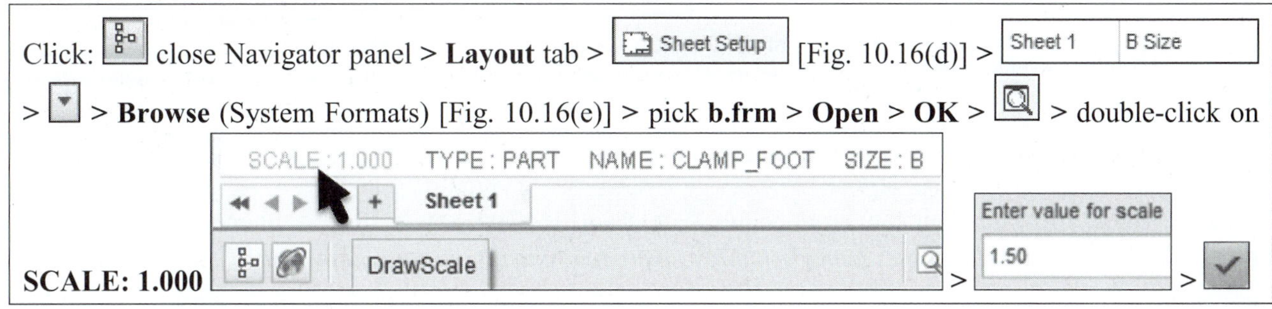

Click: close Navigator panel > **Layout** tab > Sheet Setup [Fig. 10.16(d)] > Sheet 1 B Size > > **Browse** (System Formats) [Fig. 10.16(e)] > pick **b.frm** > **Open** > **OK** > > double-click on **SCALE: 1.000** > Enter value for scale 1.50 >

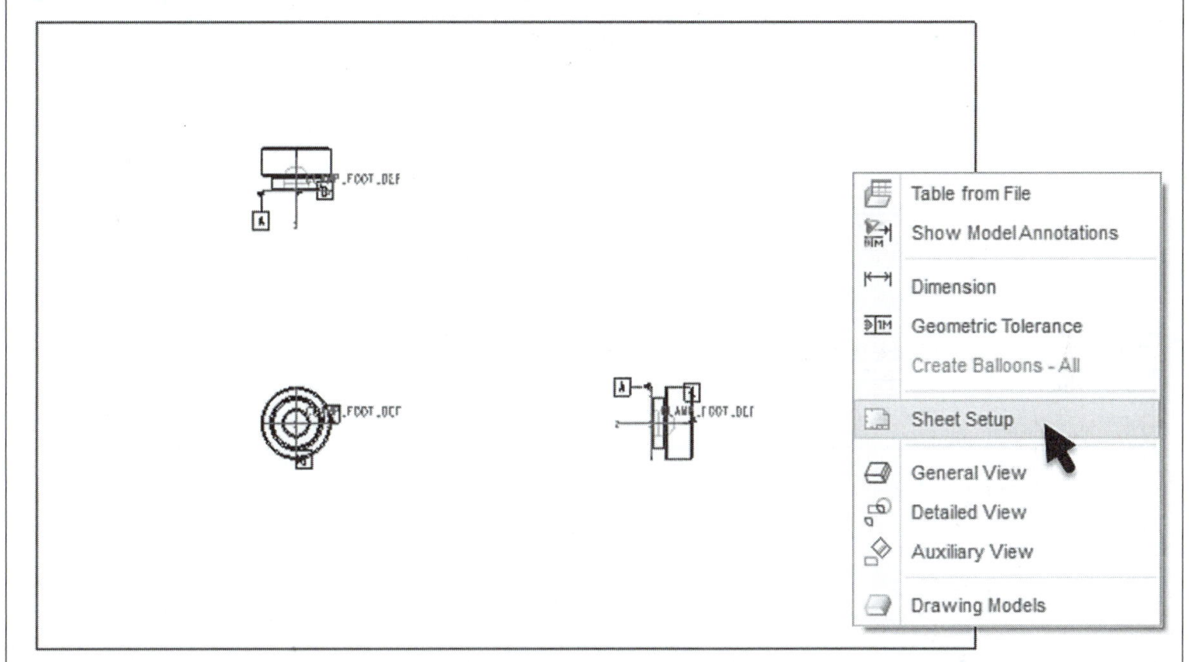

Figure 10.16(d) Clamp Foot Drawing Sheet Setup

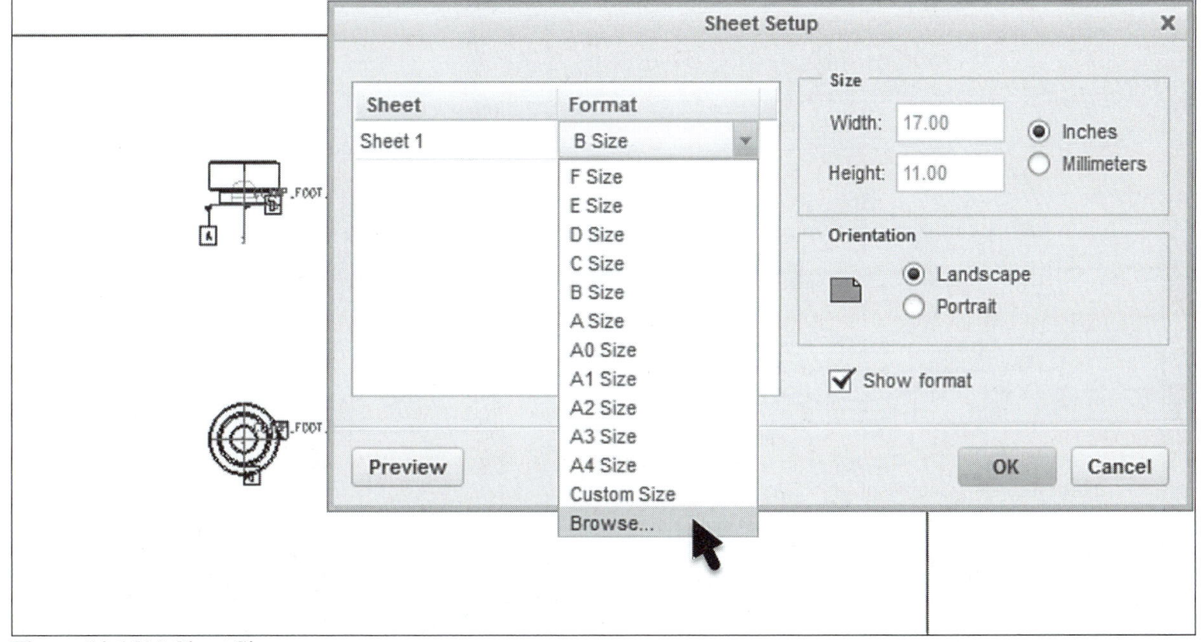

Figure 10.16(e) Sheet Size

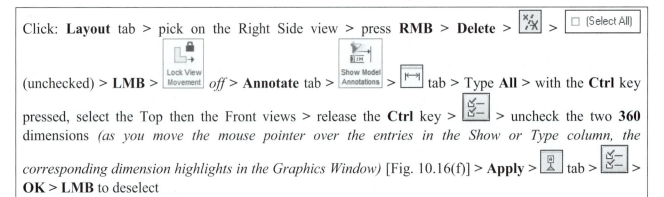

Click: **Layout** tab > pick on the Right Side view > press **RMB** > **Delete** > > (Select All) (unchecked) > **LMB** > Lock View Movement *off* > **Annotate** tab > Show Model Annotations > tab > Type **All** > with the **Ctrl** key pressed, select the Top then the Front views > release the **Ctrl** key > > uncheck the two **360** dimensions *(as you move the mouse pointer over the entries in the Show or Type column, the corresponding dimension highlights in the Graphics Window)* [Fig. 10.16(f)] > **Apply** > tab > > **OK** > **LMB** to deselect

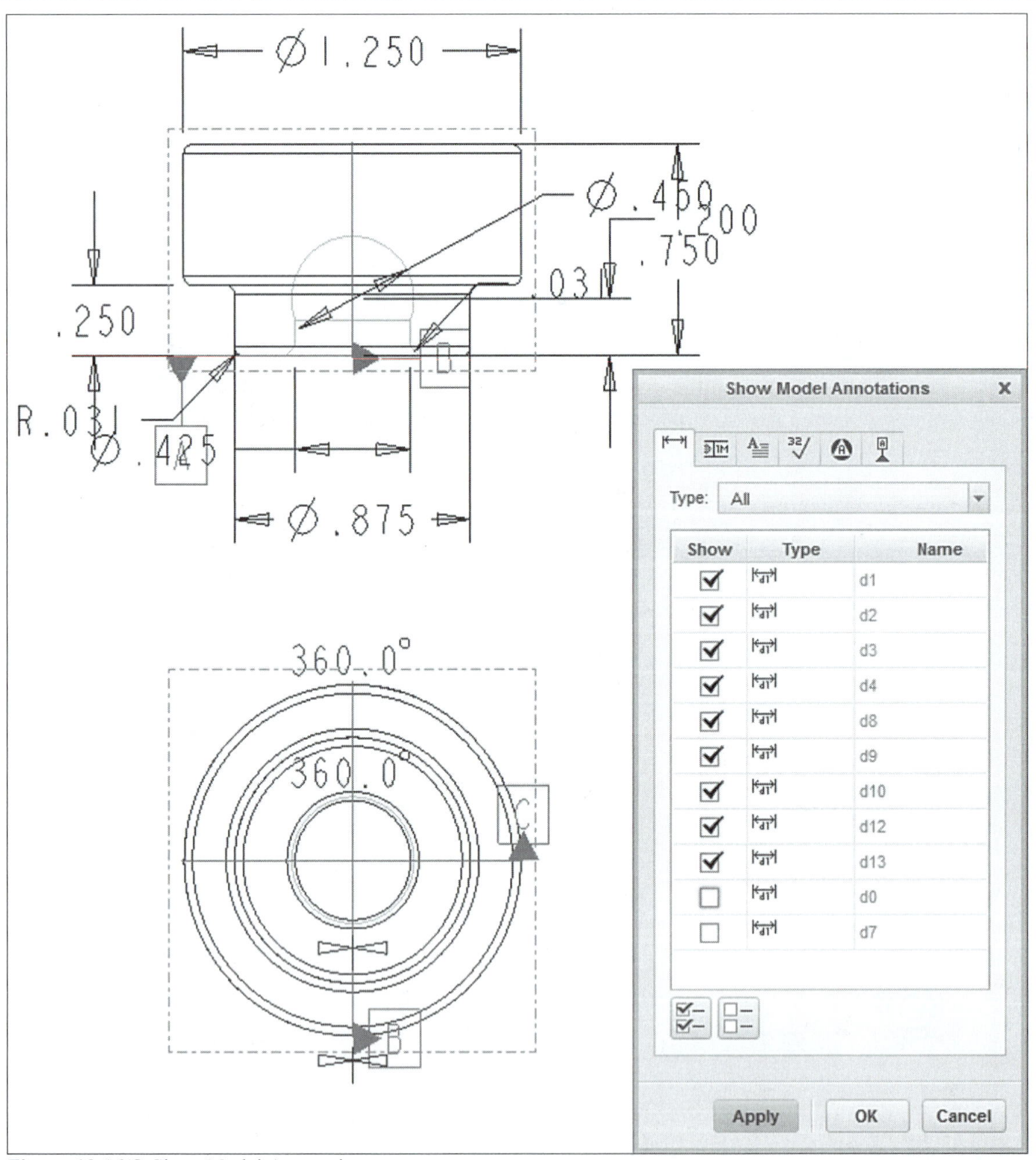

Figure 10.16(f) Show Model Annotations

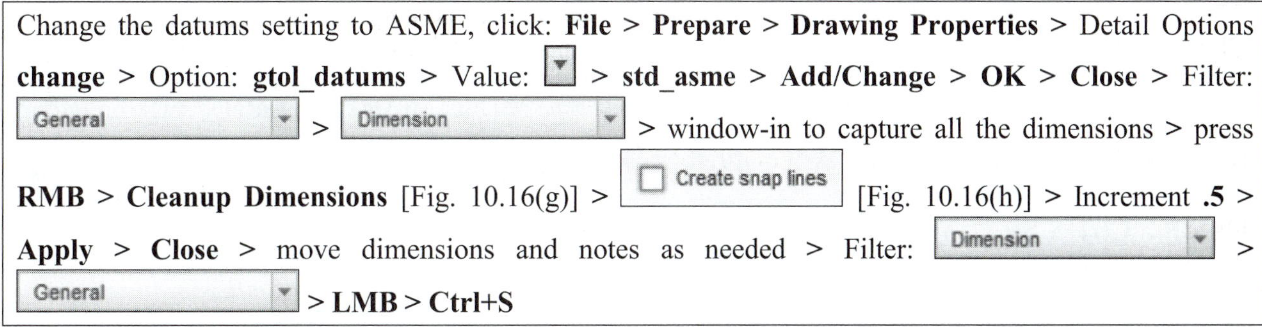

Change the datums setting to ASME, click: **File** > **Prepare** > **Drawing Properties** > Detail Options **change** > Option: **gtol_datums** > Value: ▾ > **std_asme** > **Add/Change** > **OK** > **Close** > Filter: General ▾ > Dimension ▾ > window-in to capture all the dimensions > press **RMB** > **Cleanup Dimensions** [Fig. 10.16(g)] > ☐ Create snap lines [Fig. 10.16(h)] > Increment **.5** > **Apply** > **Close** > move dimensions and notes as needed > Filter: Dimension ▾ > General ▾ > **LMB** > **Ctrl+S**

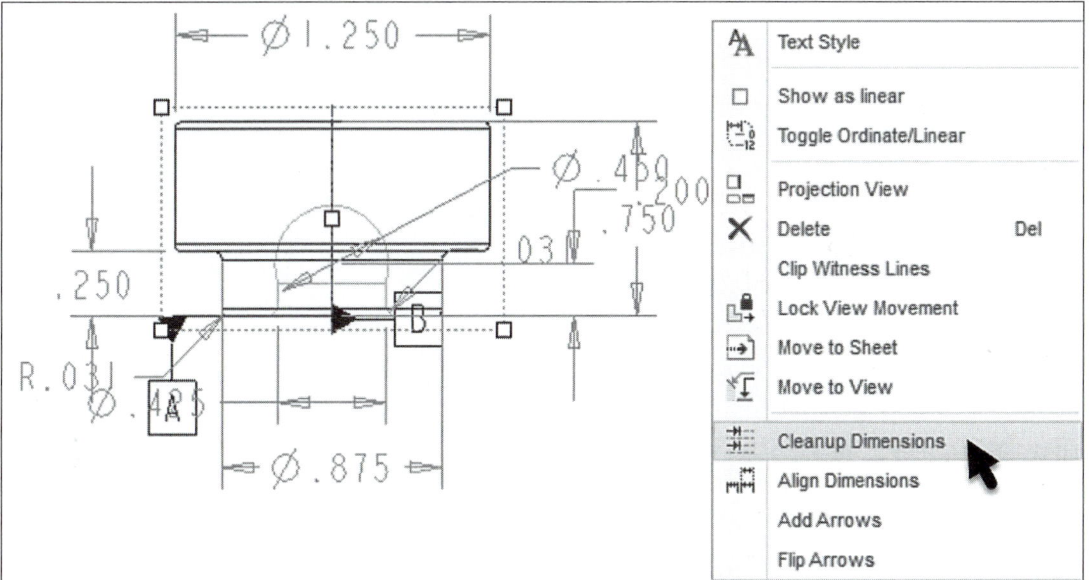

Figure 10.16(g) Press RMB > Cleanup Dimensions

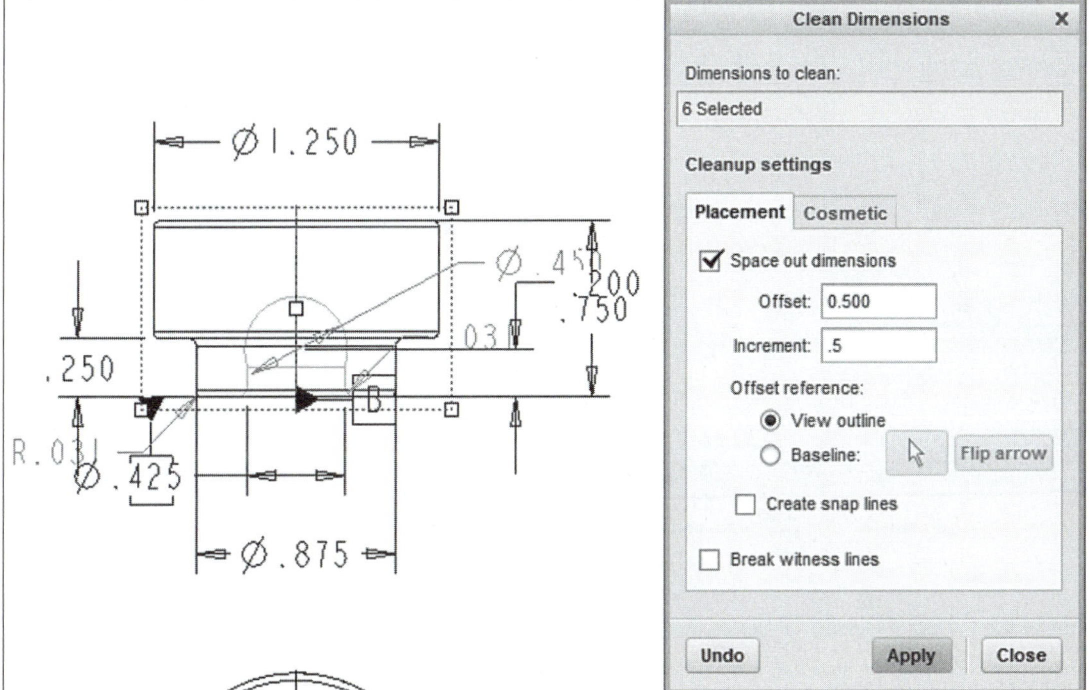

Figure 10.16(h) Clean Dimensions Dialog Box

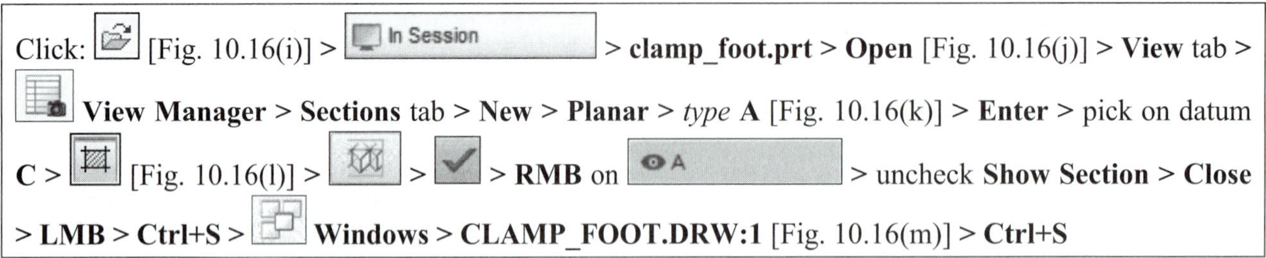

Click: [Fig. 10.16(i)] > In Session > **clamp_foot.prt** > **Open** [Fig. 10.16(j)] > **View** tab > **View Manager** > **Sections** tab > **New** > **Planar** > *type* **A** [Fig. 10.16(k)] > **Enter** > pick on datum **C** > [Fig. 10.16(l)] > > > **RMB** on A > uncheck **Show Section** > **Close** > **LMB** > **Ctrl+S** > **Windows** > **CLAMP_FOOT.DRW:1** [Fig. 10.16(m)] > **Ctrl+S**

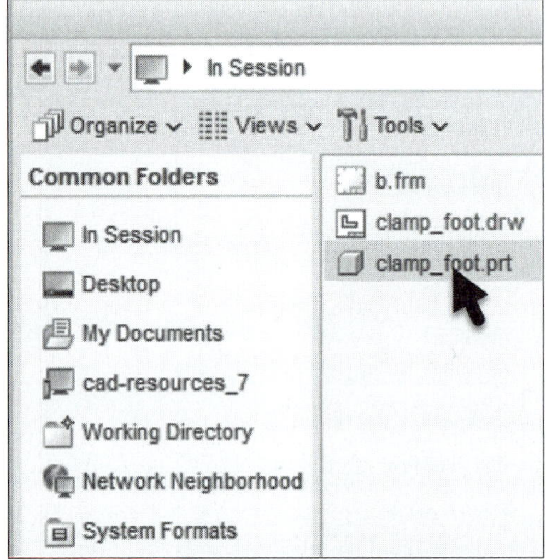

Figure 10.16(i) Open the Clamp_Foot from In Session

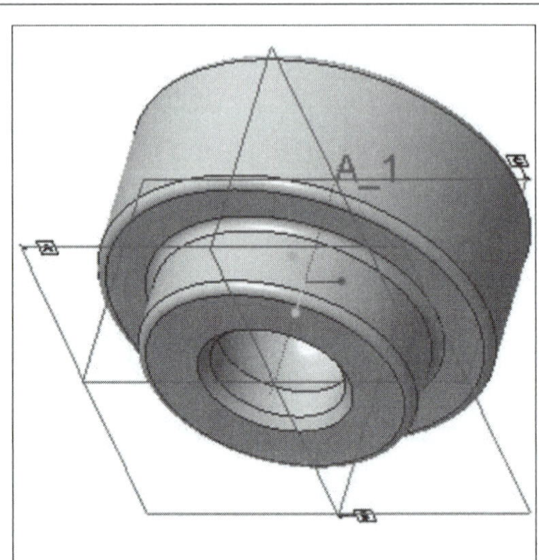

Figure 10.16(j) Part Window Activated

Figure 10.16(k) View Manager Dialog Box

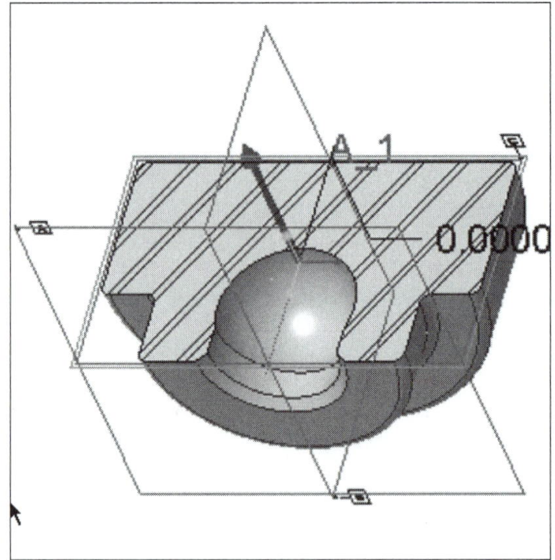

Figure 10.16(l) Part Xsection A

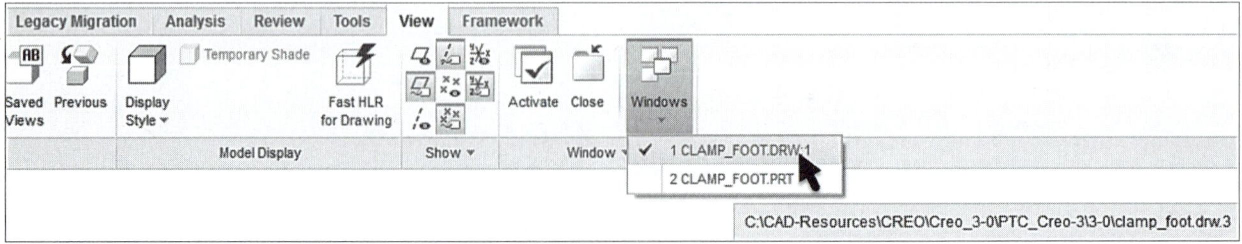

Figure 10.16(m) Drawing Active

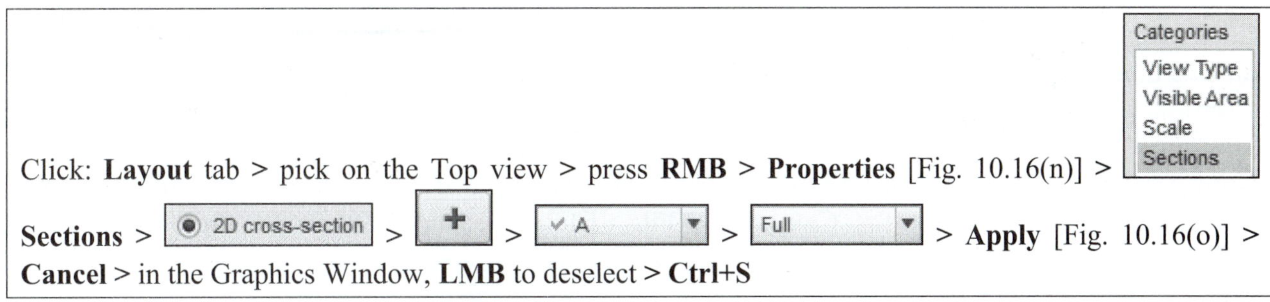

Click: **Layout** tab > pick on the Top view > press **RMB** > **Properties** [Fig. 10.16(n)] > **Sections** > 2D cross-section > + > A > Full > **Apply** [Fig. 10.16(o)] > **Cancel** > in the Graphics Window, **LMB** to deselect > **Ctrl+S**

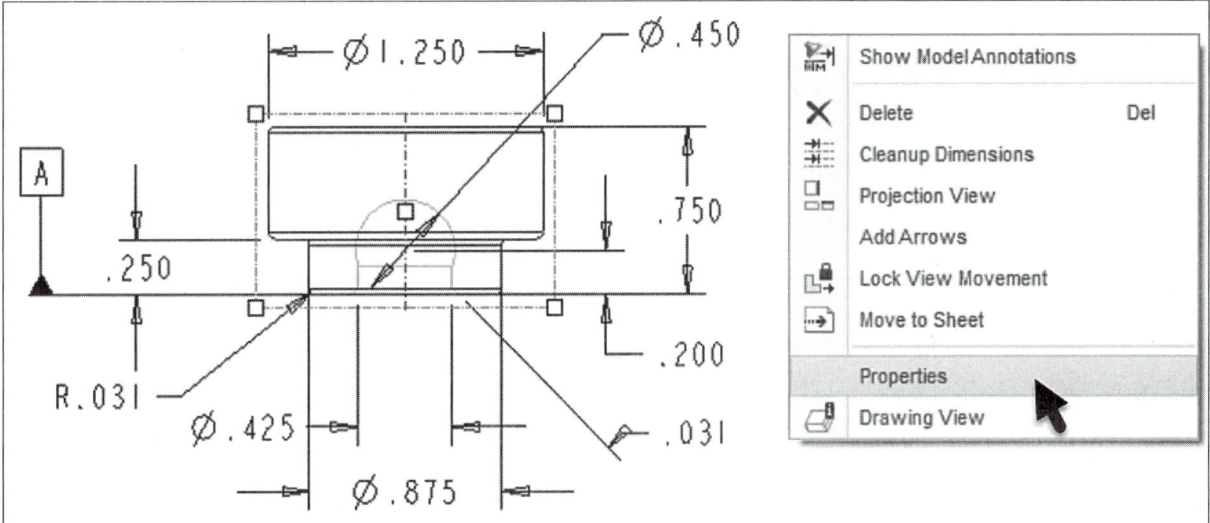

Figure 10.16(n) Press RMB > Properties

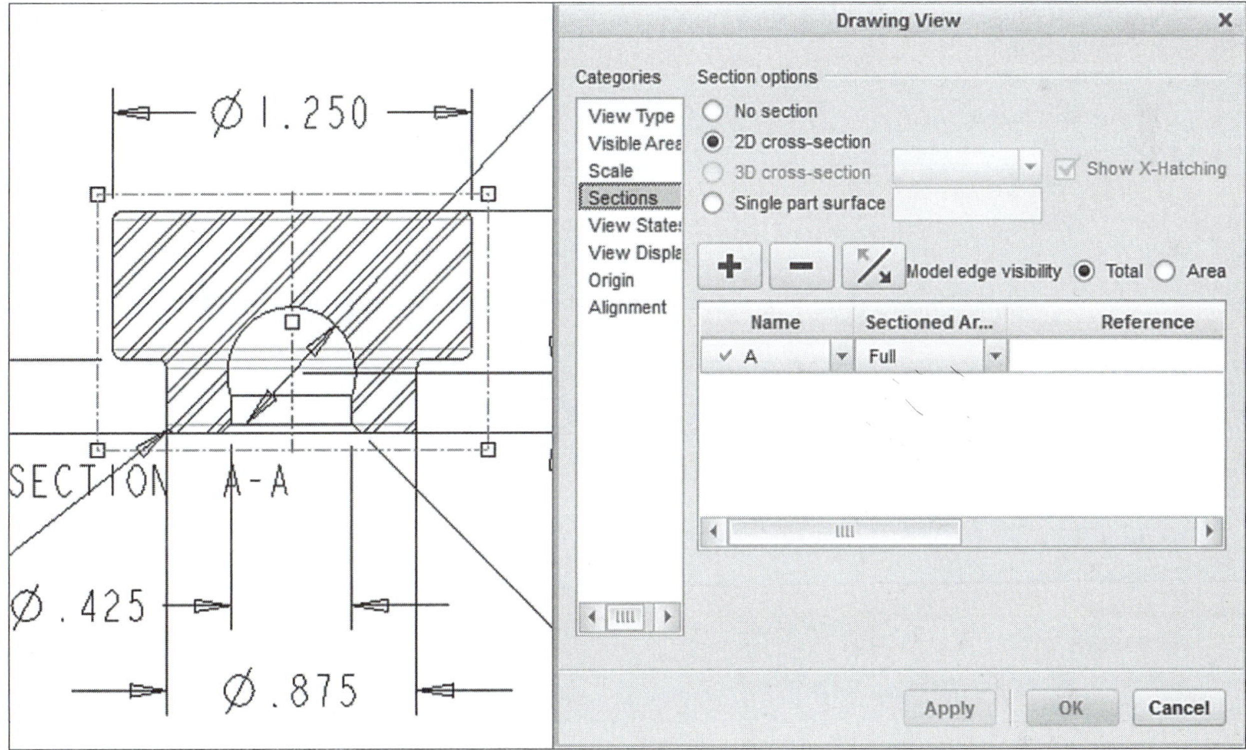

Figure 10.16(o) Section View

Click: **Annotate** tab > Cleanup the drawing views by erasing unneeded dimensions, moving datums, and moving dimensions to the appropriate view [Fig. 10.16(p)]. > in the Graphics Window, **LMB** to deselect > **Refit > Ctrl+S > File > Manage File > Delete Old Versions > Yes > File > Save As > Save a Copy** Type > **Zip File (*.zip) > OK** > upload to your course interface or attach to an email and send to your instructor or yourself > first Creo window **File > Close** > second Creo window **File > Close**

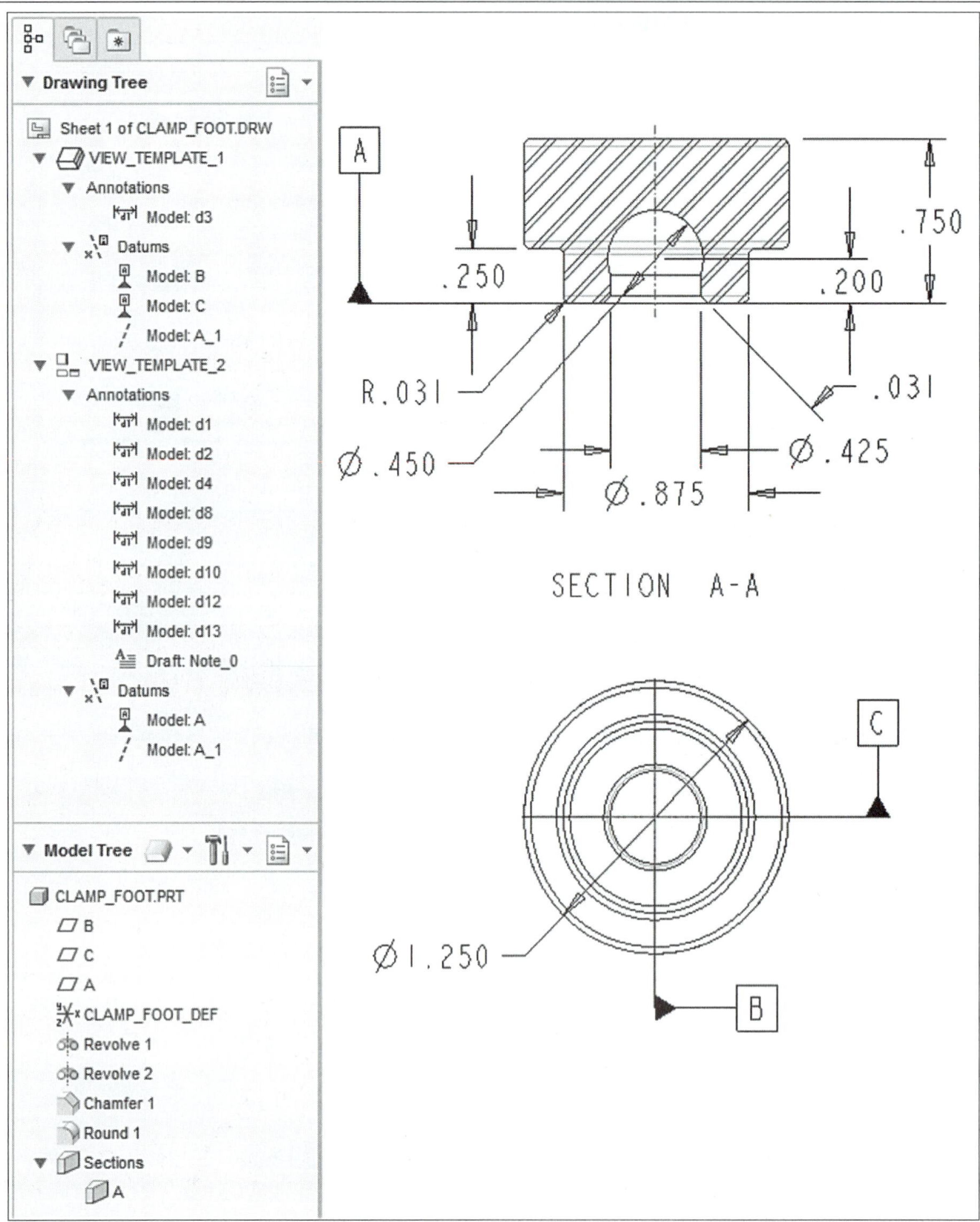

Figure 10.16(p) Section View and Front View Cleaned Up

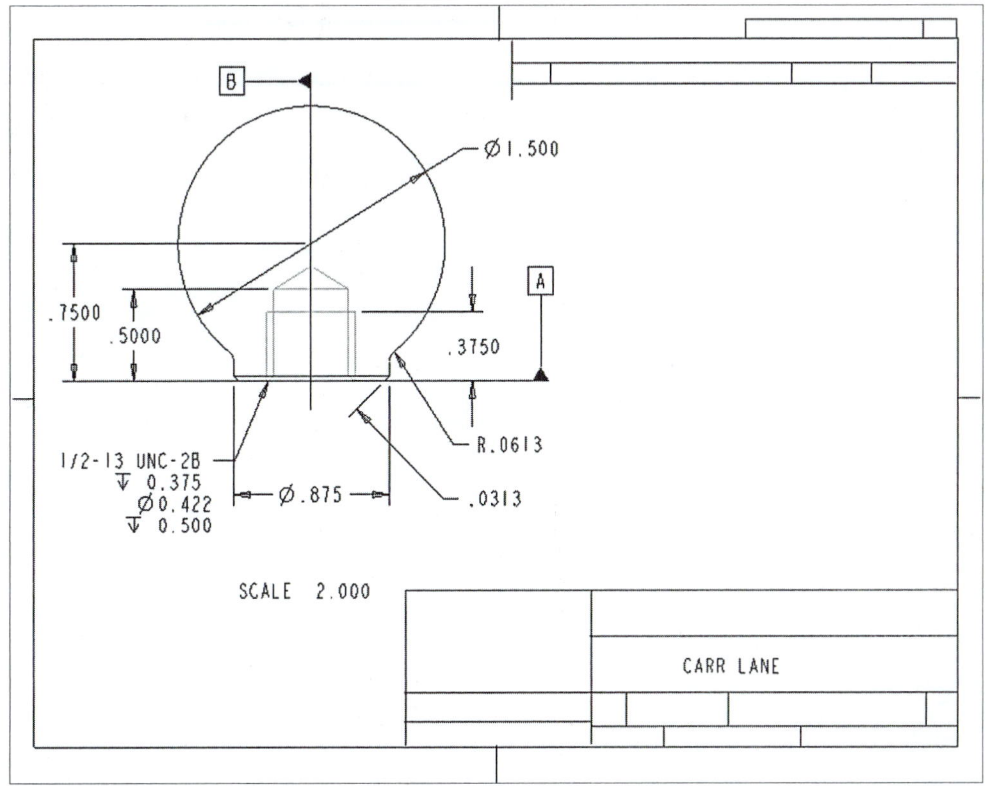

Figure 10.17 Clamp Ball Drawing

Creating a Drawing of the Clamp Ball

Before starting the Clamp Ball drawing (Fig. 10.17) for the Clamp Ball part (Fig. 10.18), you will build a template that can be used to create any drawing.

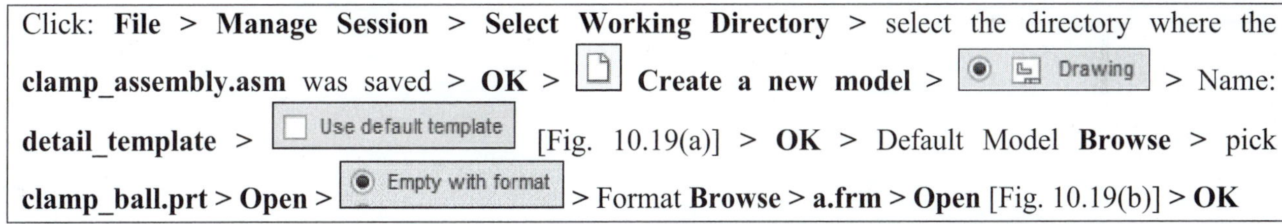

Click: **File** > **Manage Session** > **Select Working Directory** > select the directory where the **clamp_assembly.asm** was saved > **OK** > **Create a new model** > Drawing > Name: **detail_template** > Use default template [Fig. 10.19(a)] > **OK** > Default Model **Browse** > pick **clamp_ball.prt** > **Open** > Empty with format > Format **Browse** > **a.frm** > **Open** [Fig. 10.19(b)] > **OK**

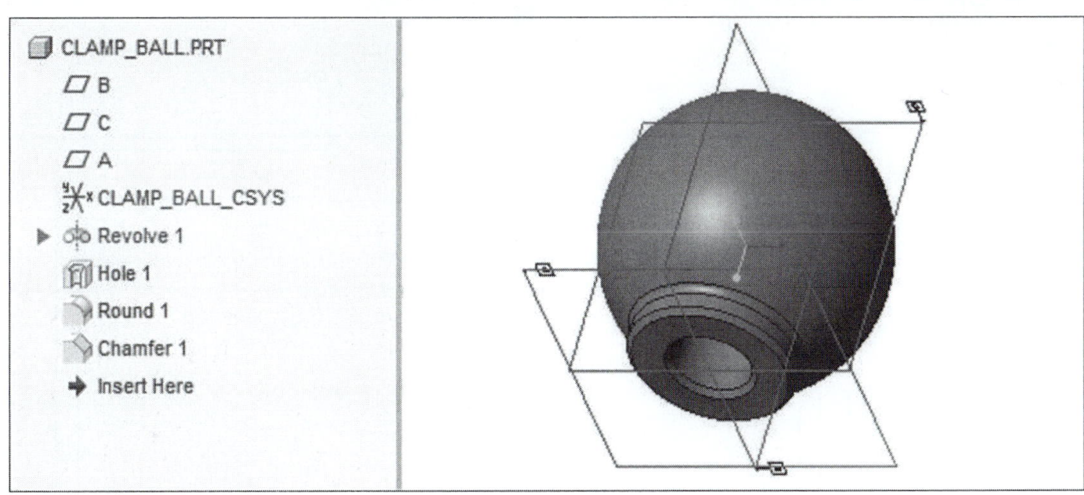

Figure 10.18 Clamp_Ball

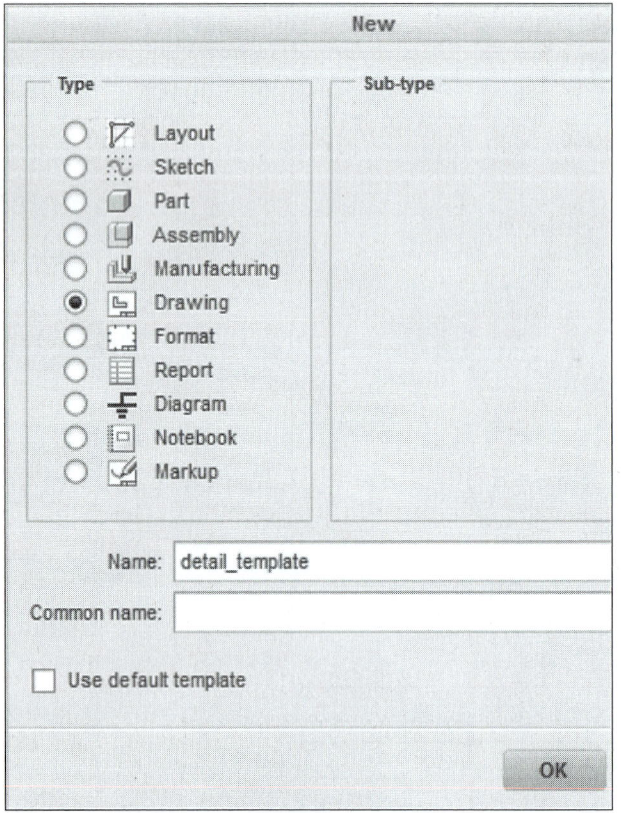

Figure 10.19(a) New Dialog Box

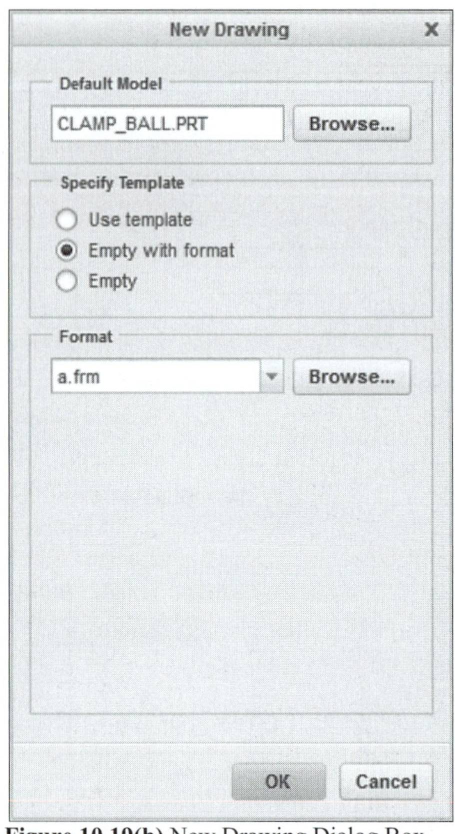

Figure 10.19(b) New Drawing Dialog Box

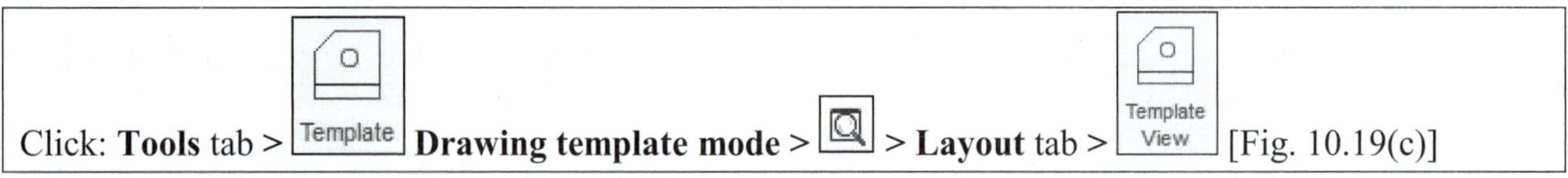

Click: **Tools** tab > Template **Drawing template mode** > > **Layout** tab > Template View [Fig. 10.19(c)]

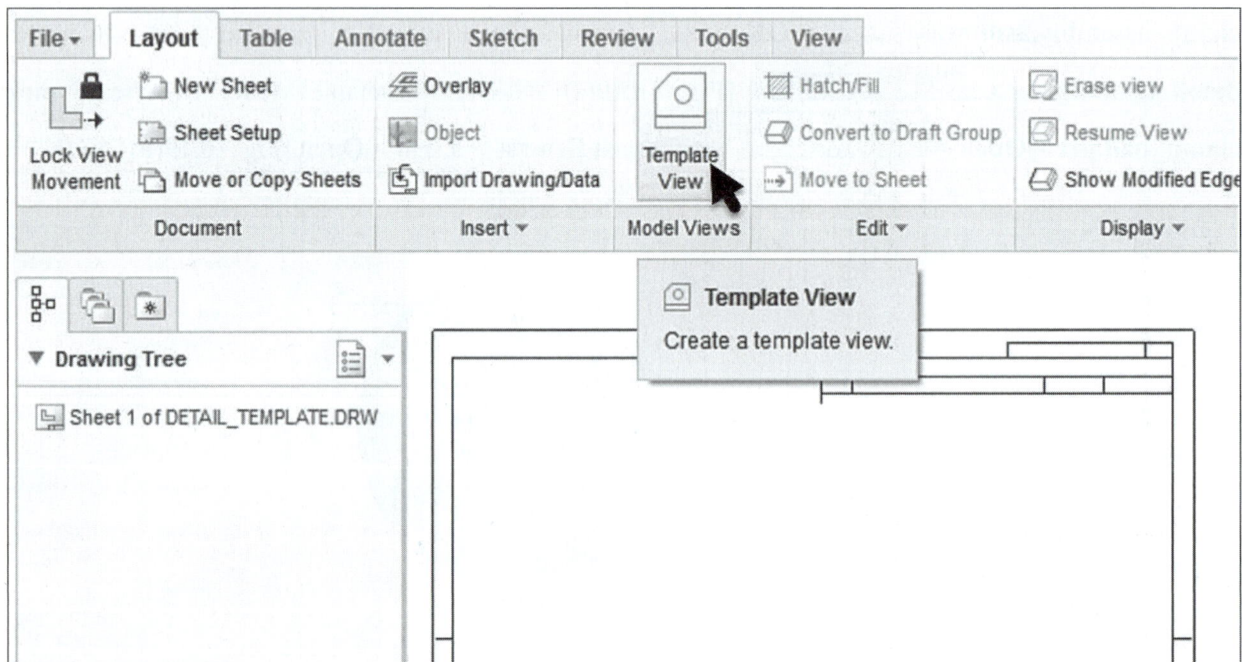

Figure 10.19(c) Tools Tab > Template > Layout Tab > Template View

The Template View Instructions dialog box opens [Fig. 10.19(d)]

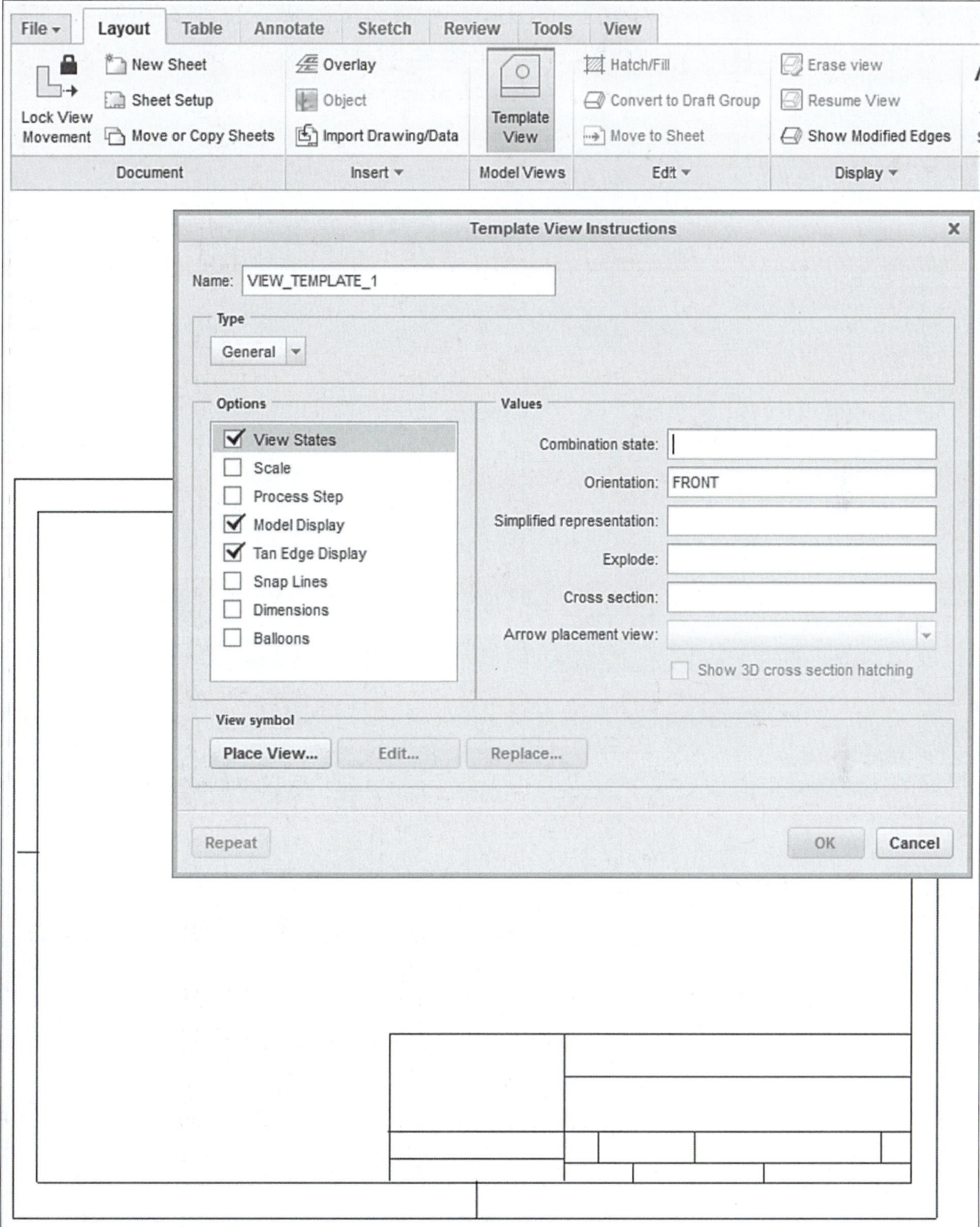

Figure 10.19(d) Repositioned Template View Instructions Dialog Box *(your template # may be different)*

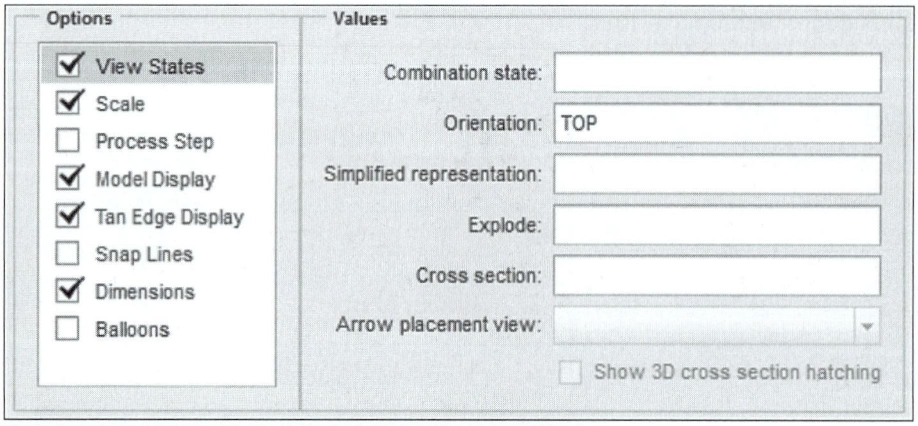

Options | Values

- [x] View States
- [x] Scale
- [ ] Process Step
- [x] Model Display
- [x] Tan Edge Display
- [ ] Snap Lines
- [x] Dimensions
- [ ] Balloons

Combination state:
Orientation: TOP
Simplified representation:
Explode:
Cross section:
Arrow placement view:
Show 3D cross section hatching

TOP [Fig. 10.19(e)] >

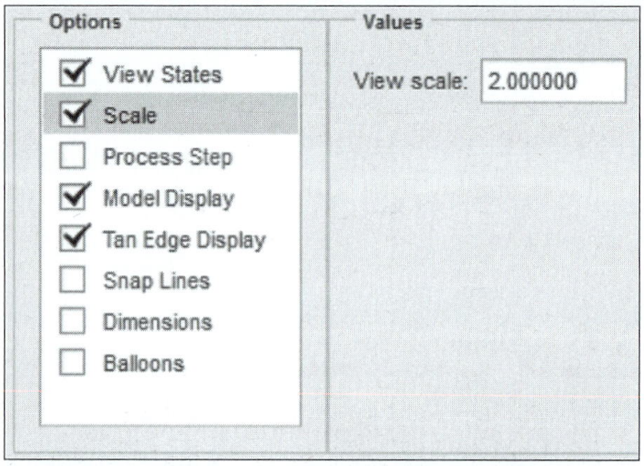

Options | Values

- [x] View States
- [x] Scale
- [ ] Process Step
- [x] Model Display
- [x] Tan Edge Display
- [ ] Snap Lines
- [ ] Dimensions
- [ ] Balloons

View scale: 2.000000

View scale **2.00** [Fig. 10.19(f)] > **Enter** >

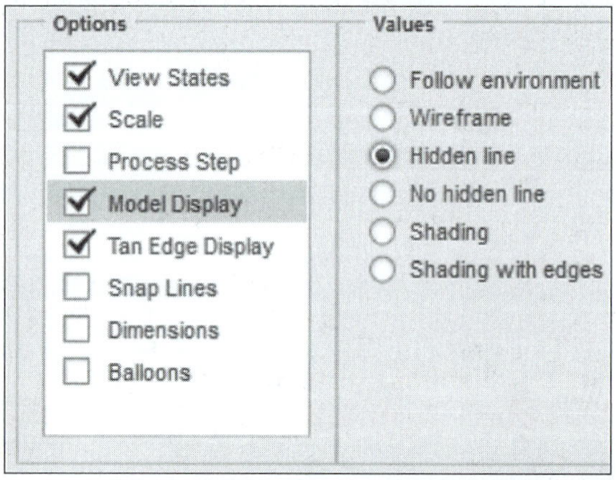

Options | Values

- [x] View States
- [x] Scale
- [ ] Process Step
- [x] Model Display
- [x] Tan Edge Display
- [ ] Snap Lines
- [ ] Dimensions
- [ ] Balloons

Follow environment
Wireframe
Hidden line
No hidden line
Shading
Shading with edges

> **Hidden line**

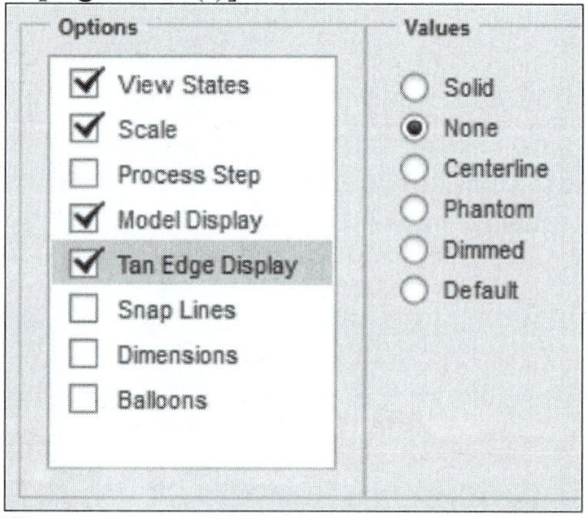

Options | Values

- [x] View States
- [x] Scale
- [ ] Process Step
- [x] Model Display
- [x] Tan Edge Display
- [ ] Snap Lines
- [ ] Dimensions
- [ ] Balloons

Solid
None
Centerline
Phantom
Dimmed
Default

None >

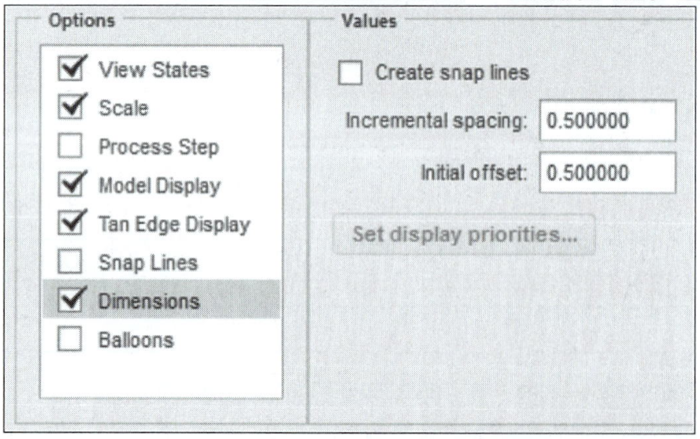

Options | Values

- [x] View States
- [x] Scale
- [ ] Process Step
- [x] Model Display
- [x] Tan Edge Display
- [ ] Snap Lines
- [x] Dimensions
- [ ] Balloons

Create snap lines
Incremental spacing: 0.500000
Initial offset: 0.500000
Set display priorities...

> **.500** > **Enter** > View States

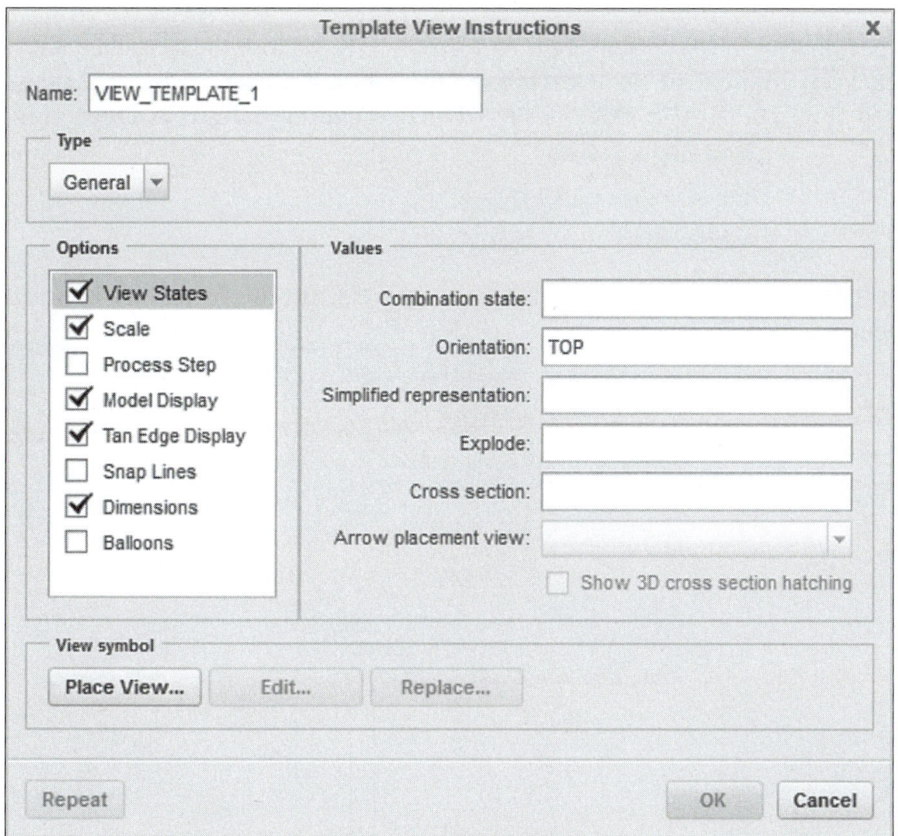

Figure 10.19(e) View States

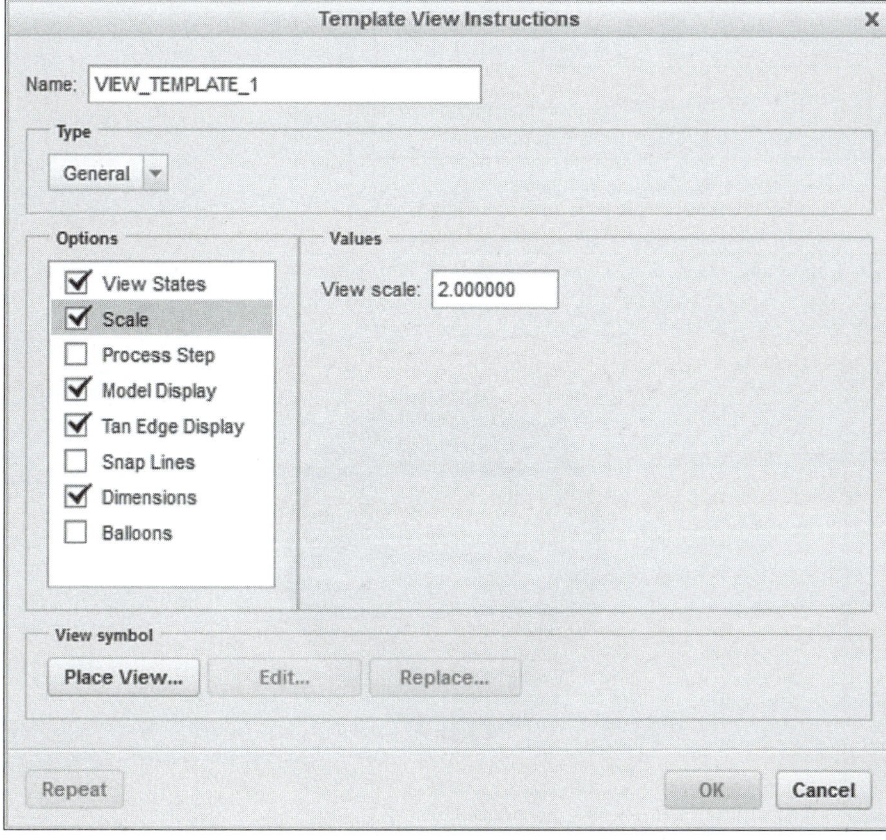

Figure 10.19(f) View Scale

Click: Place View... > pick a position for the Top view [Fig. 10.19(g)] > **OK** to close the Template View Instructions dialog box > **Lock View Movement** *on* > **Ctrl+S** > **OK** > **File** > **Close** > **File** > **Manage Session** > **Erase Not Displayed** [Fig. 10.19(h)] > **OK** > **File** > **Open** > **clamp_ball.prt** > **Open** [Fig. 10.19(i)] > **File** > **Close** > **File** > **Open** (In Session > Common Folders > In Session > clamp_ball.prt) > **Cancel** (clamp_ball is active and in session but not displayed in a window)

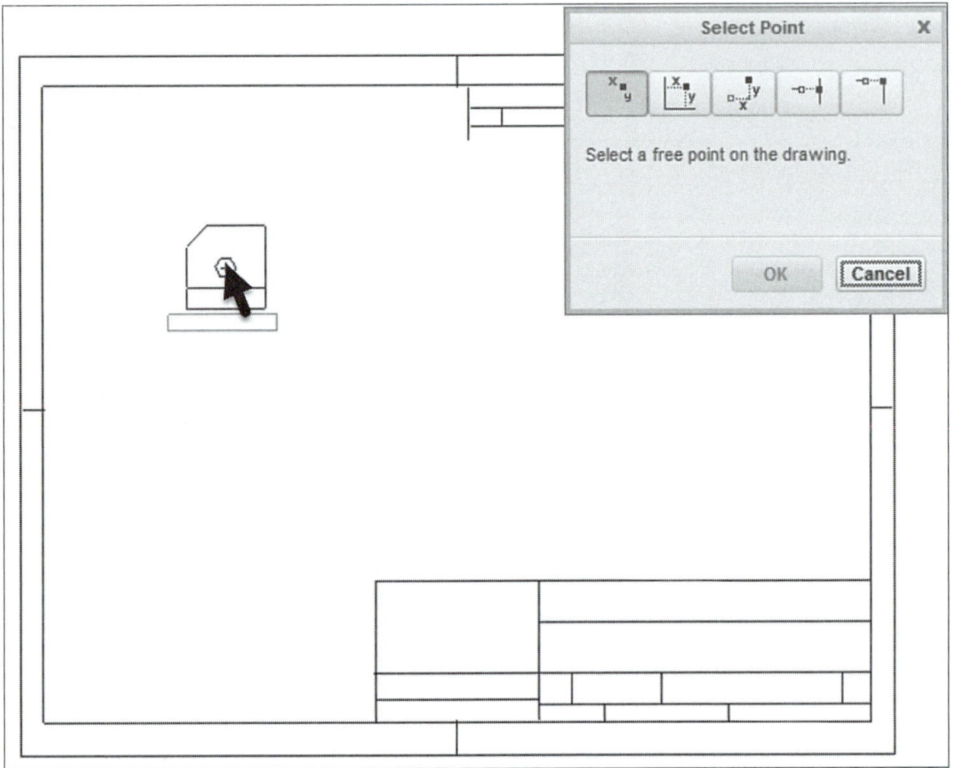

Figure 10.19(g) Place the View *(the Template View Instructions Dialog Box may be behind the Creo 3.0 Graphics Window)*

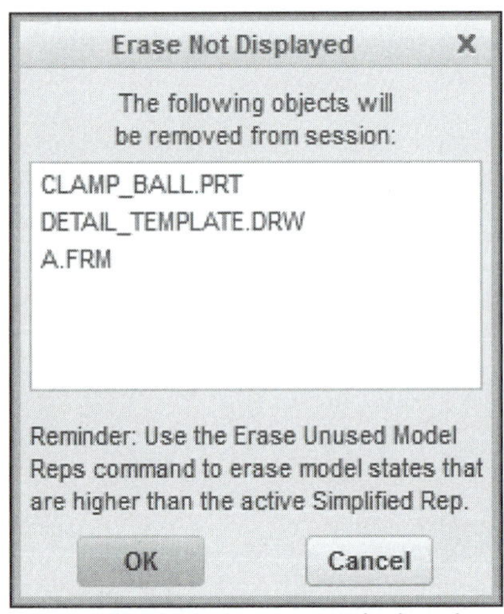

Figure 10.19(h) Erase Not Displayed

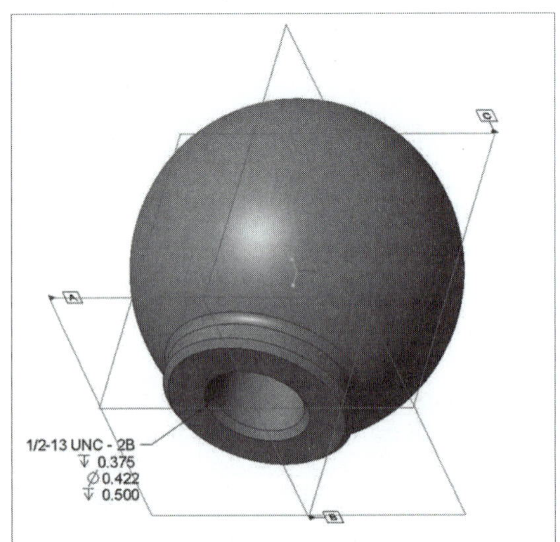

Figure 10.19(i) clamp_ball.prt

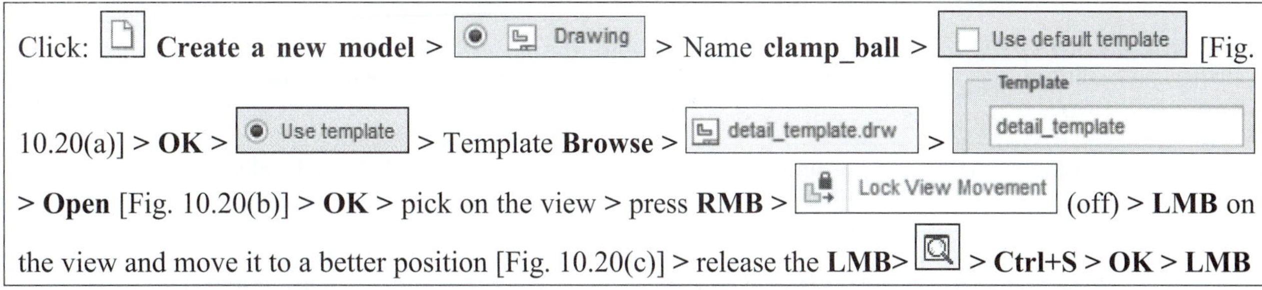

Click: Create a new model > Drawing > Name **clamp_ball** > Use default template [Fig. 10.20(a)] > **OK** > Use template > Template **Browse** > detail_template.drw > Template detail_template > **Open** [Fig. 10.20(b)] > **OK** > pick on the view > press **RMB** > Lock View Movement (off) > **LMB** on the view and move it to a better position [Fig. 10.20(c)] > release the **LMB**> > **Ctrl+S** > **OK** > **LMB**

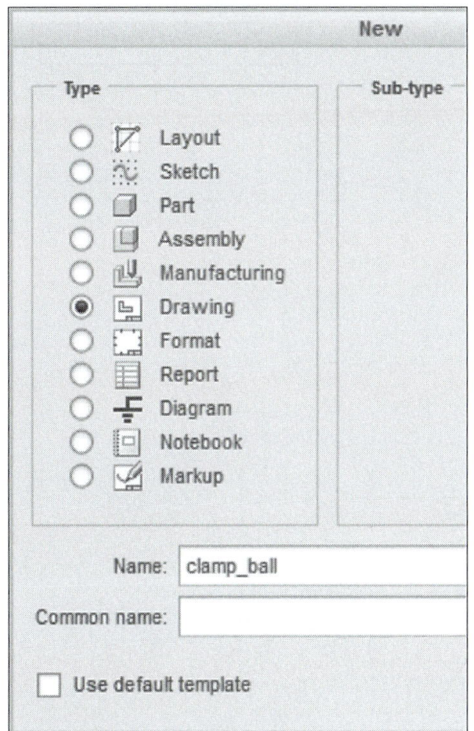

Figure 10.20(a) New Drawing

Figure 10.20(b) Template: detail_template

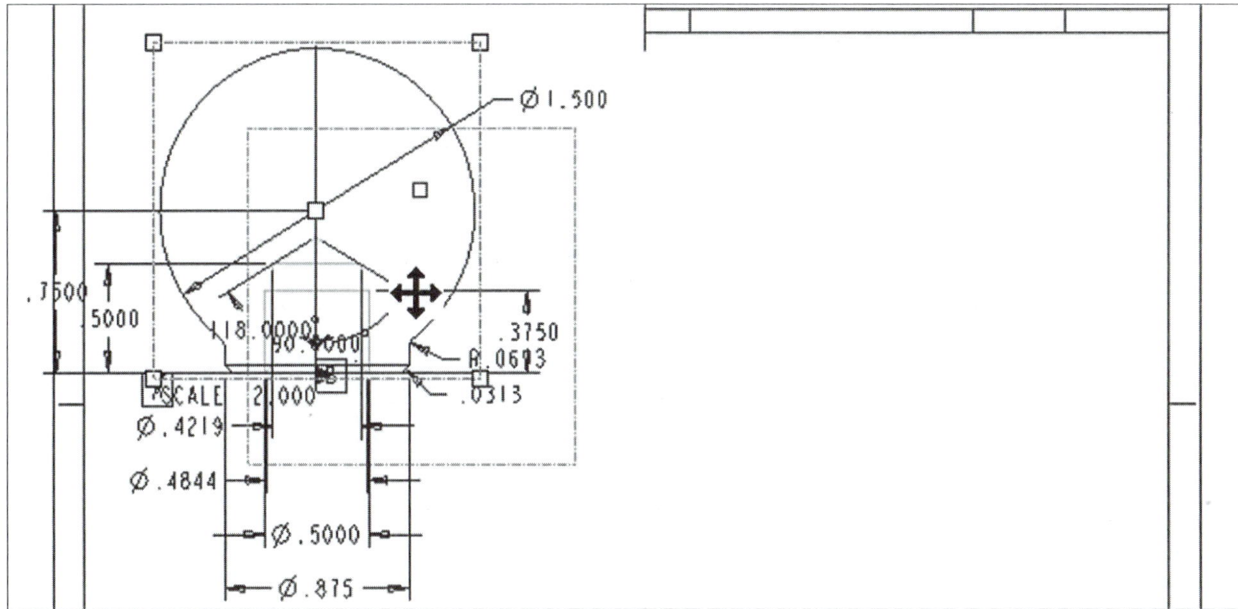

Figure 10.20(c) Repositioning View

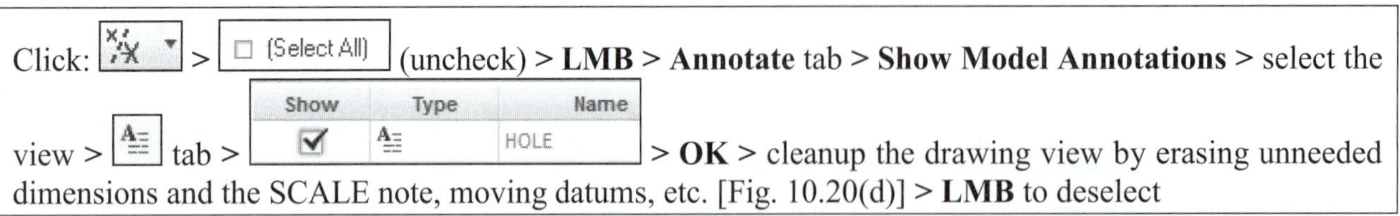

Click: [icon] > [(Select All)] (uncheck) > **LMB** > **Annotate** tab > **Show Model Annotations** > select the view > [icon] tab > [Show / Type / Name: HOLE] > **OK** > cleanup the drawing view by erasing unneeded dimensions and the SCALE note, moving datums, etc. [Fig. 10.20(d)] > **LMB** to deselect

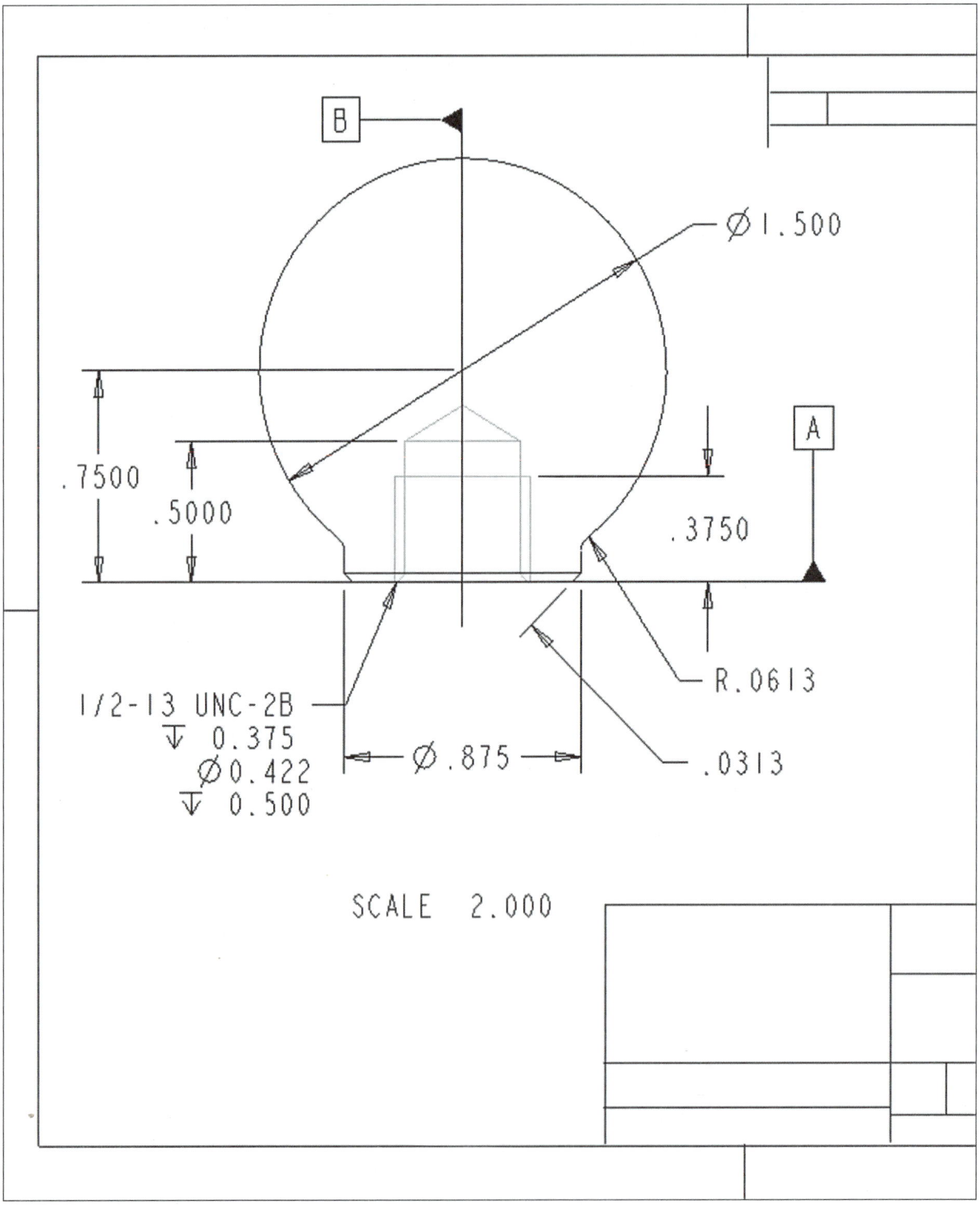

Figure 10.20(d) One-view Drawing

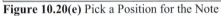

Click: Unattached Note > Select LOCATION for note [Fig. 10.20(e)] > *type* Type **CARR LANE** [Fig. 10.20(f)] > **LMB** > **LMB** [Fig. 10.20(g)]

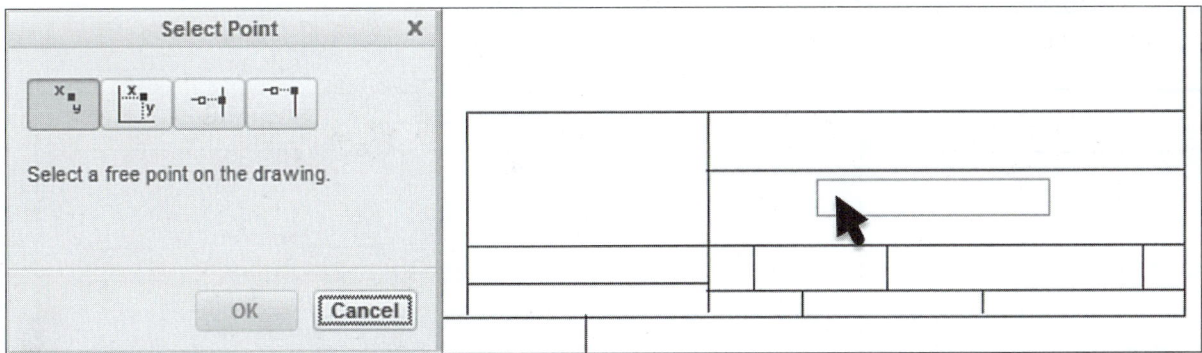

Figure 10.20(e) Pick a Position for the Note

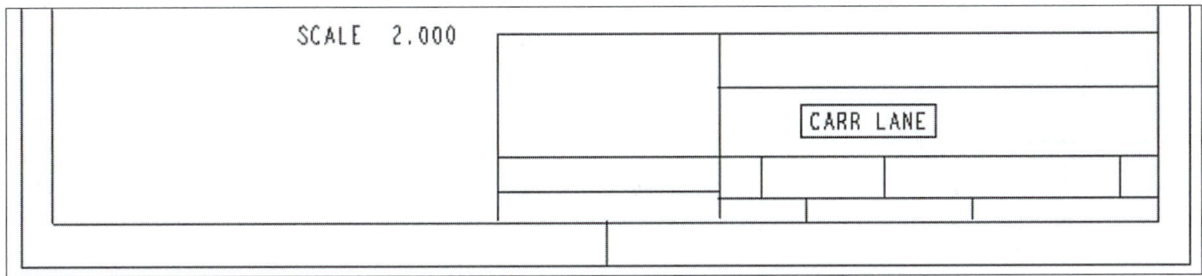

Figure 10.20(f) CARR LANE Note

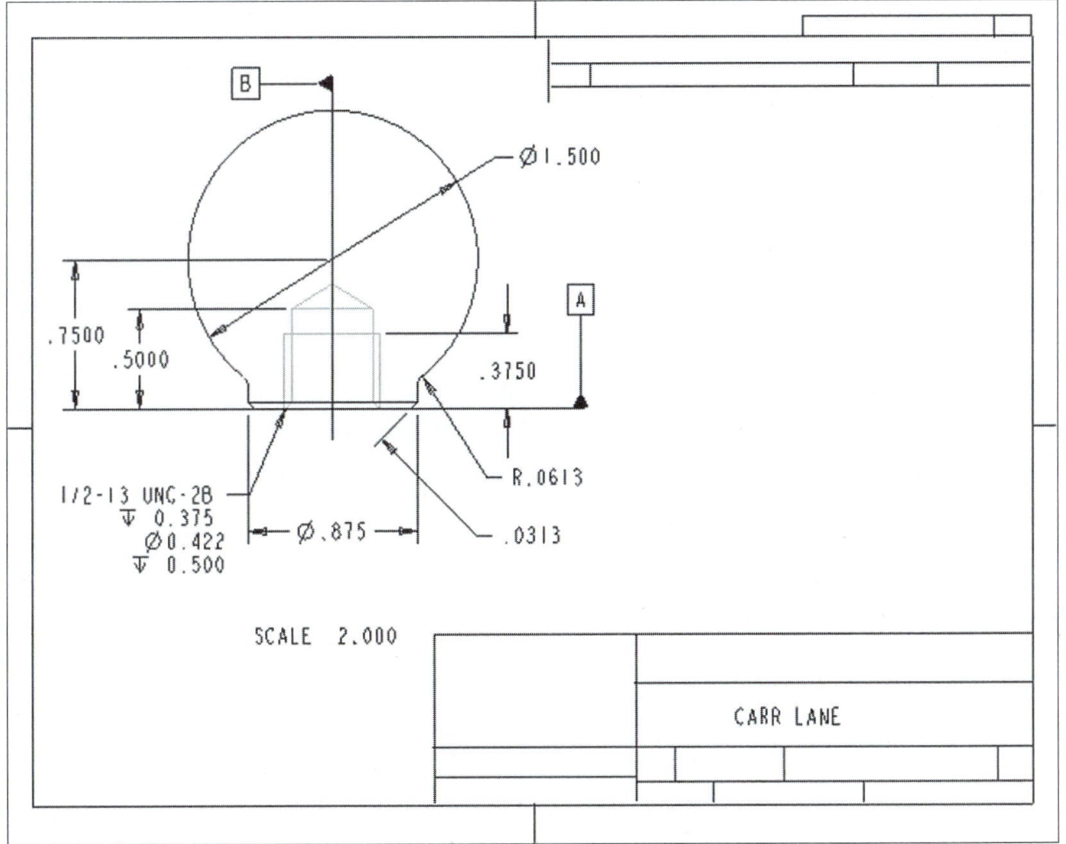

Figure 10.20(g) Completed Note

Click: [icon] [Fig. 10.20(h)] > **Ctrl+S** > **File** > **Save As** > Type [icon] > **Zip File (*.zip)** > **OK** > upload to your course interface or attach to an email and send to your instructor or yourself > **File** > **Manage File** > **Delete Old Versions** > **Enter** > **File** > **Close** > **File** > **Exit** > **Yes**

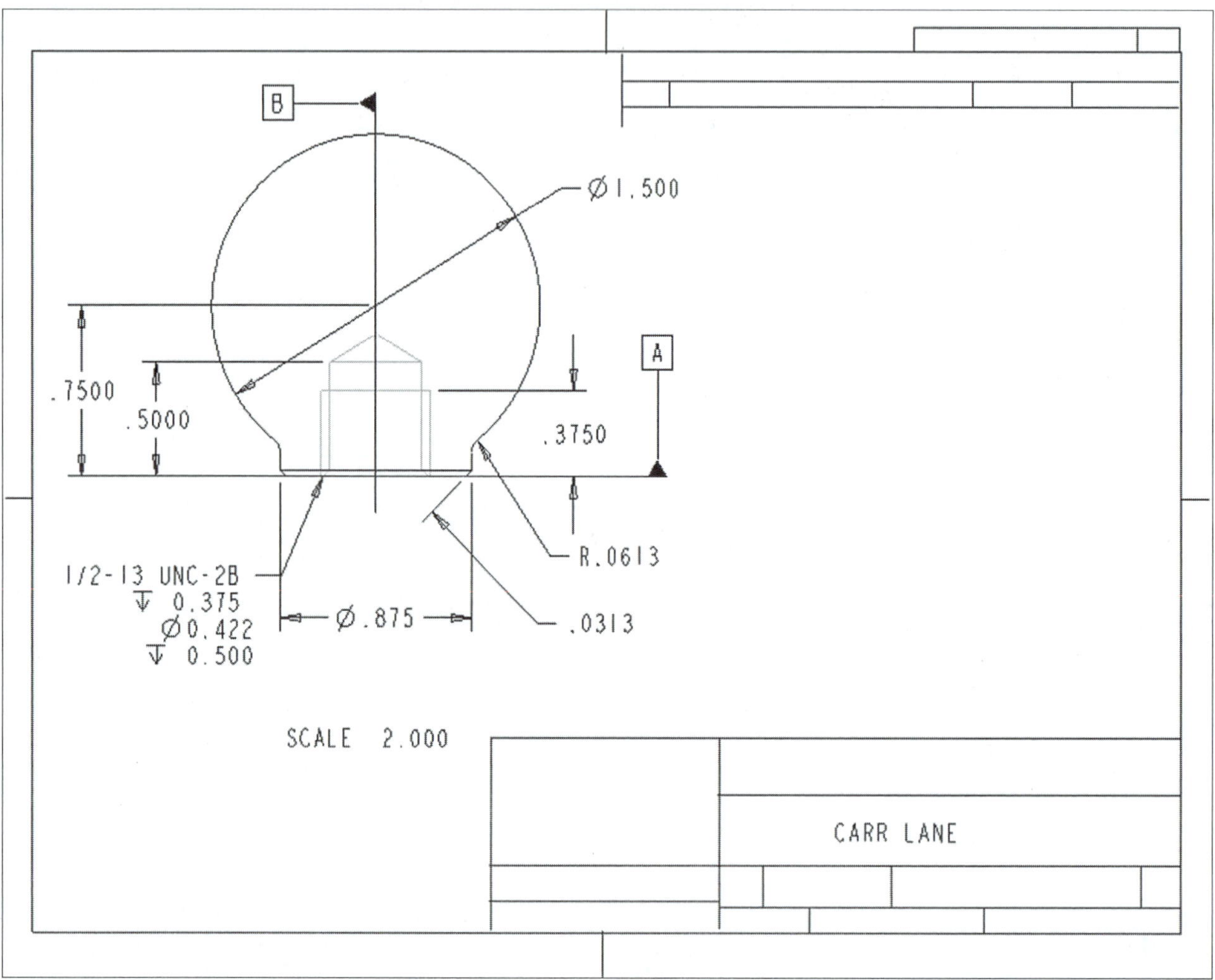

Figure 10.20(h) Completed Drawing

You can create drawings of the remaining clamp assembly parts. And, a variety of part, assembly, and drawing projects can be downloaded from the website ***www.cad-resources.com***.

Lesson 11 Part Drawings

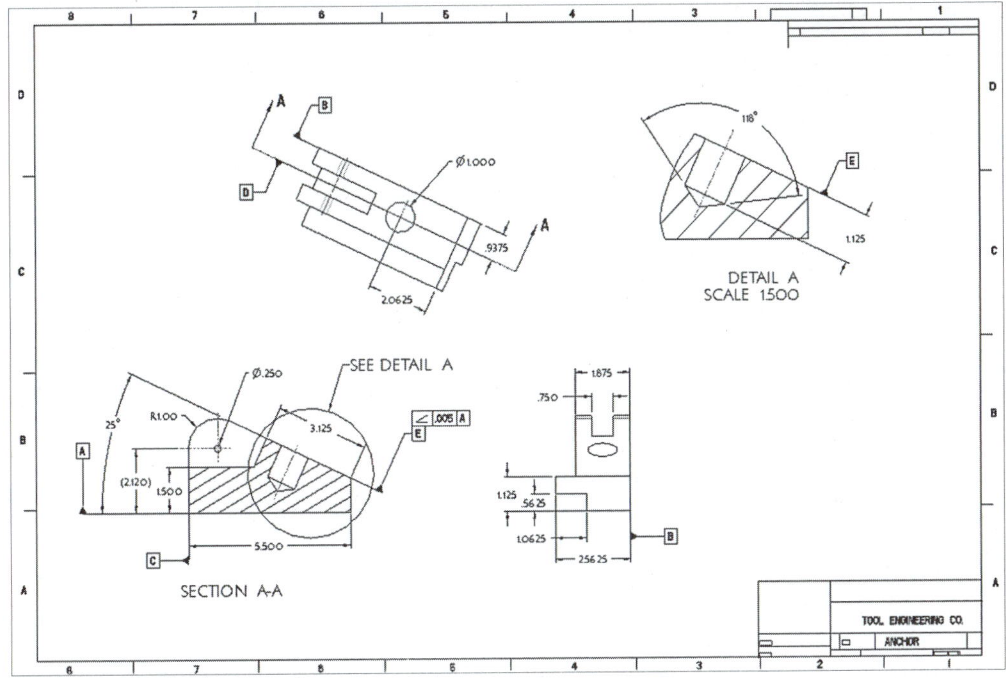

Figure 11.1(a) Anchor Drawing

OBJECTIVES

- Establish a **Drawing Options** file to use when detailing
- Identify the need for **views** to clarify interior features of a part
- Create **Cross Sections** using datum planes
- Produce **Auxiliary View** and **Detail Views**
- Use **multiple drawing sheets**

REFERENCES AND RESOURCES

For **Resources** go to **www.cad-resources.com** > click on the PTC Creo Parametric 3.0 Book cover

- Lesson 11 Lecture
- Lesson 11 3D models embedded in a PDF
- Book Projects PDF
- Project Lectures
- Quick Reference Card
- Configuration Options

PART DRAWINGS

Designers and drafters use drawings to convey design and manufacturing information. Drawings consist of a **Format** and views of a part (or assembly). Standard views, sectional views, detail views, and auxiliary views are utilized to describe the objects' features and sizes [Fig. 11.1(a)]. **Sectional views**, also called **sections**, are employed to clarify and dimension the internal construction of an object. Sections are needed for interior features that cannot be clearly described by hidden lines in conventional views. **Auxiliary views** are used to show the *true shape/size* of a feature or the relationship of features that are not parallel to any of the principal planes of projection.

Lesson 11 STEPS

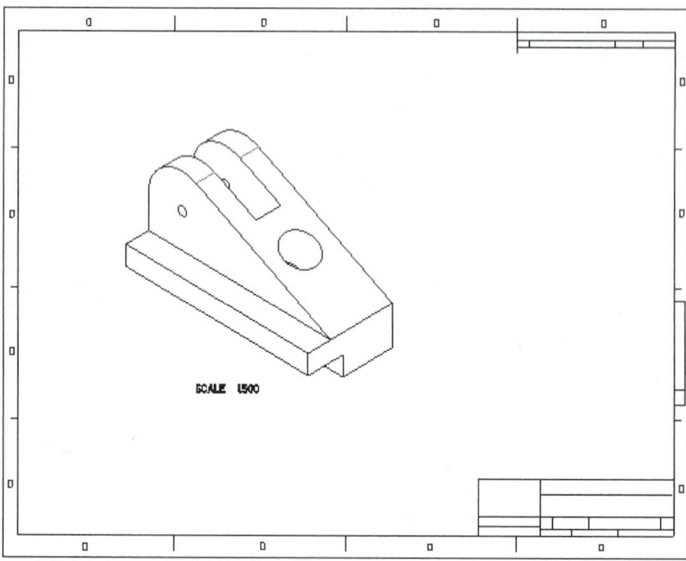

Figure 11.1(b) Anchor Drawing, Sheet 2

Anchor Drawing

You will be creating a multiple sheet detail drawing of the Anchor [Fig. 11.1(b)]. The front view will be a full section. A right side view and an auxiliary view are required to detail the part. Views will be displayed according to visibility requirements per ASME standards, such as no hidden lines in sections. The part is to be dimensioned according to ASME Y14.5M. You will add a standard format. Detailed views of other parts will be introduced to show the wide variety of view capabilities.

Launch **PTC Creo Parametric 3.0** > **Select Working Directory** > select the directory where the **anchor.prt** was saved > **OK** > > Drawing > Name **anchor** > Use default template *(if "Use default template" is checked; the Front, Top, and Right views will be automatically created)* [Fig. 11.2(a)] > **OK** > Default Model **Browse** [Fig. 11.2(b)] > pick **anchor.prt** > **Preview** *on* [Fig. 11.2(c)] > **Open**

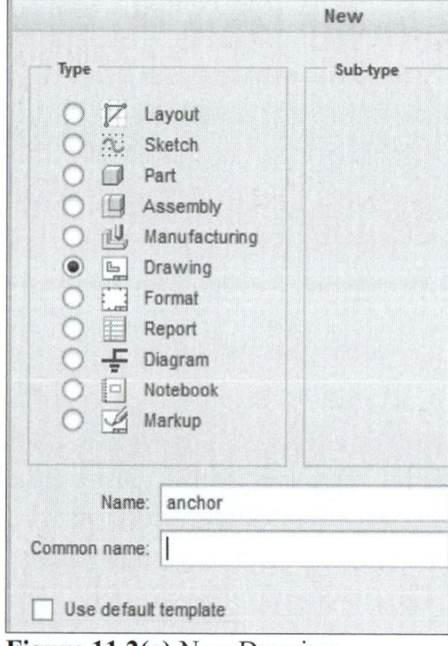

Figure 11.2(a) New Drawing

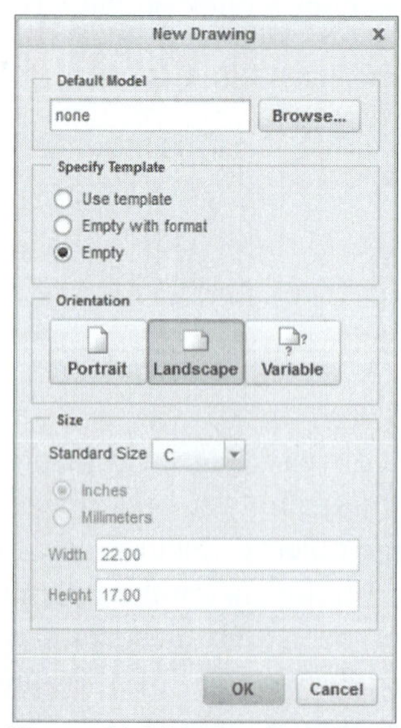

Figure 11.2(b) New Drawing Dialog Box

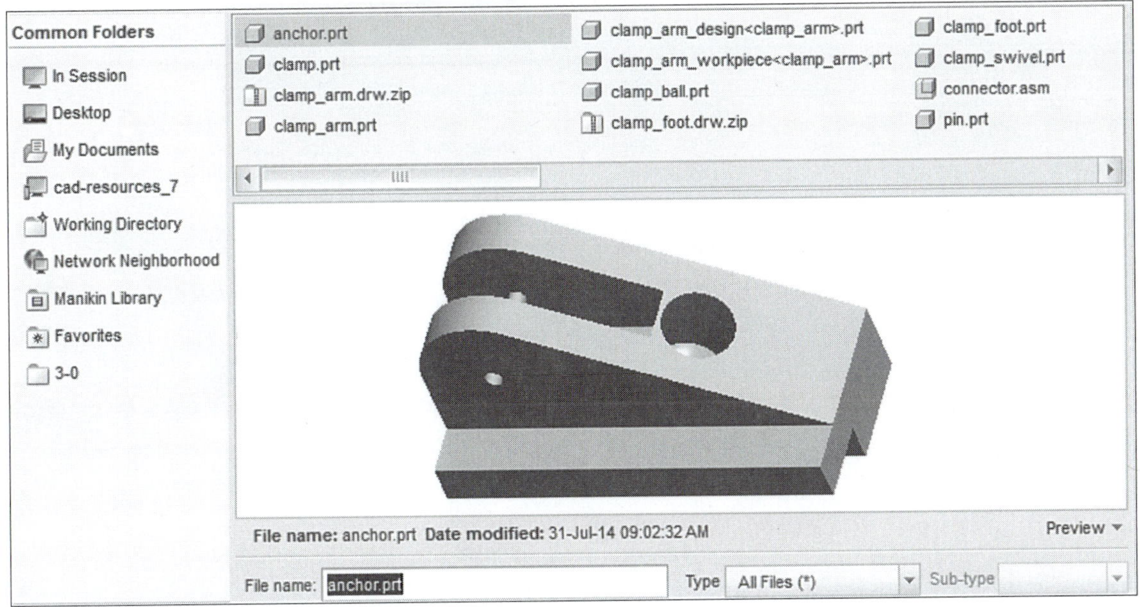

Figure 11.2(c) Anchor Part

Next to Standard Size, click: ▼ > pick **D** [Fig. 11.2(d)] > **OK** > press **RMB** > **Sheet Setup** > D Size ▼ [Fig. 11.2(e)] > ▼ > **Browse** (System Formats Open dialog box opens) > pick **d.frm** [Fig. 11.2(f)] > **Open** > **OK** [Fig. 11.2(g)] > **Ctrl+S** > **OK**

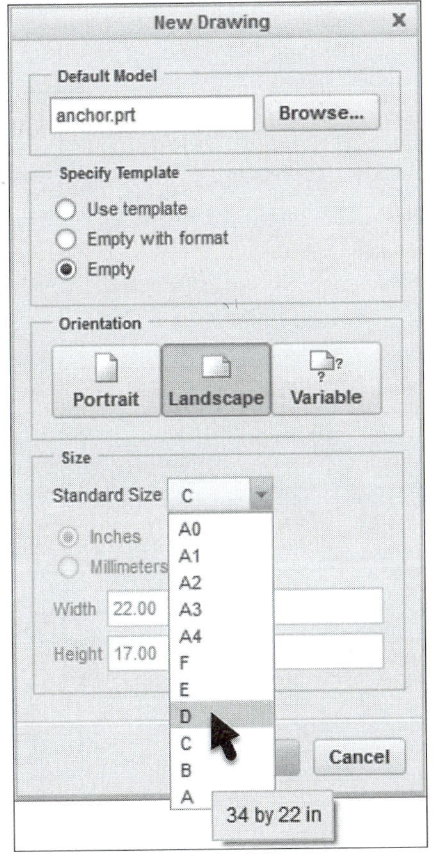

Figure 11.2(d) Standard Size D

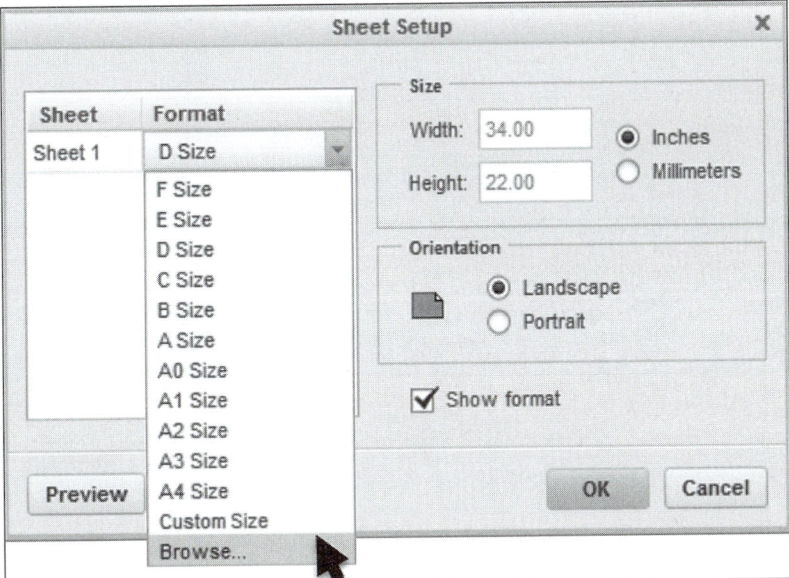

Figure 11.2(e) Sheet Setup

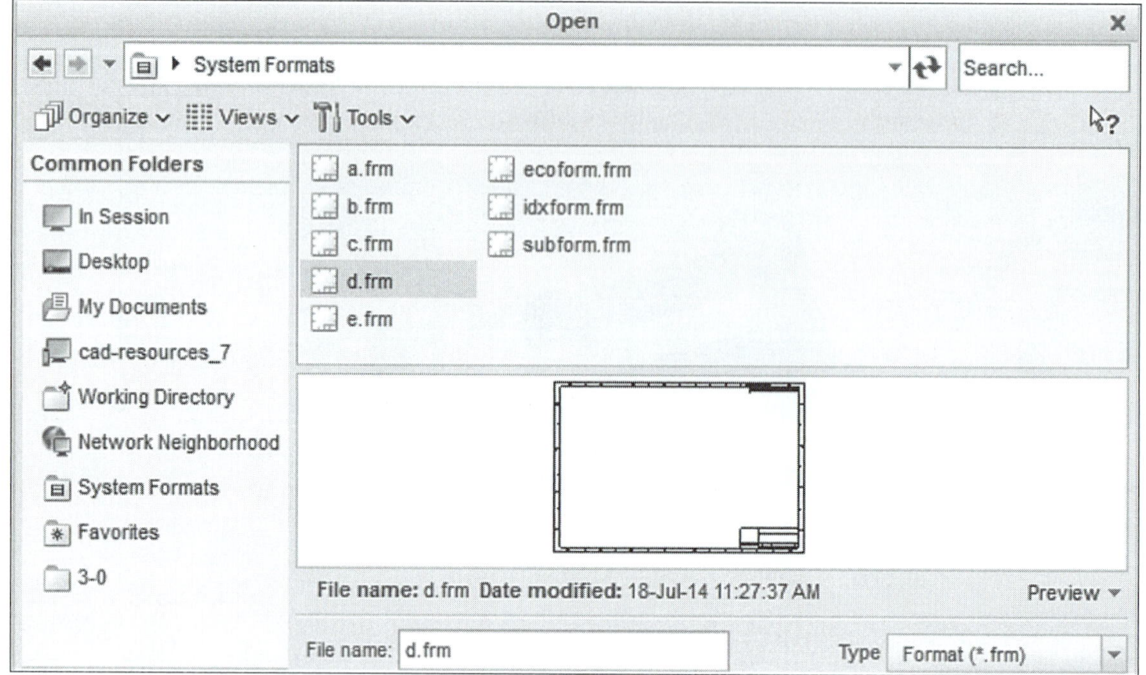

Figure 11.2(f) System Formats

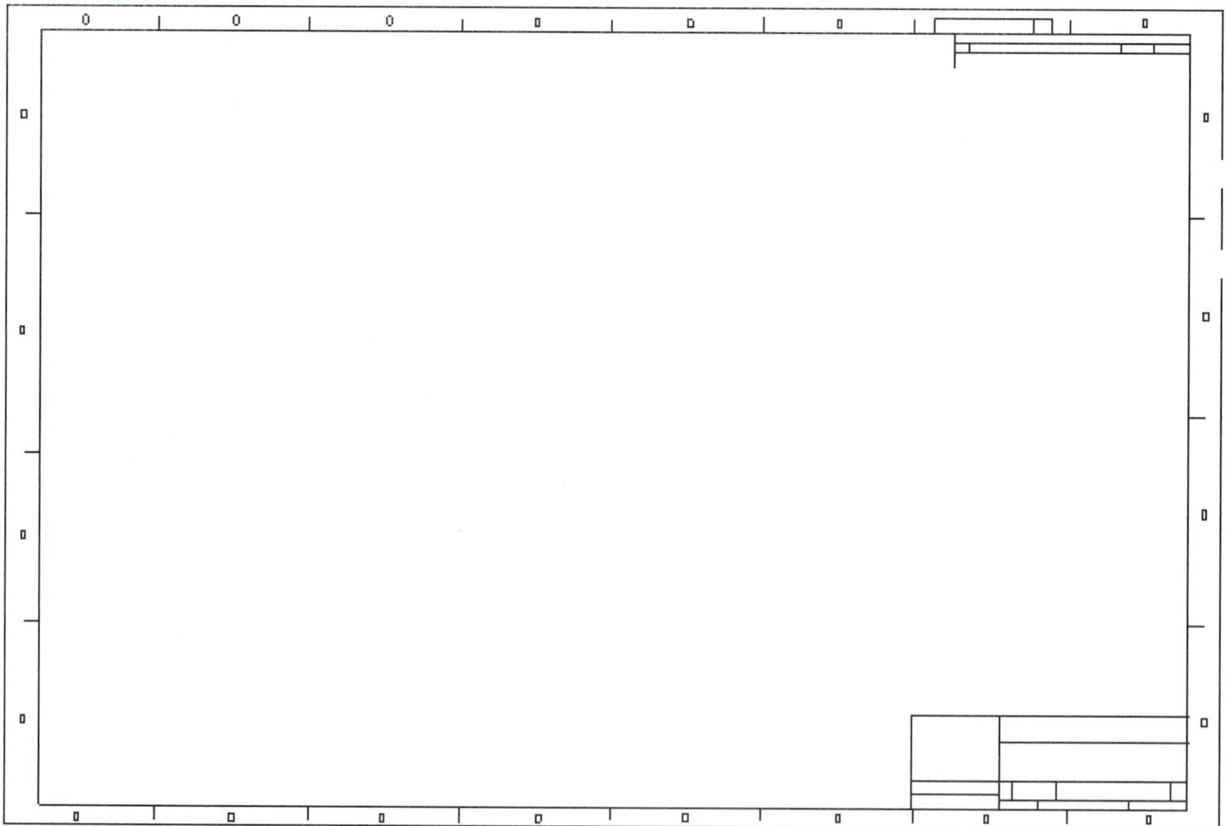
Figure 11.2(g) "D" Size Format

Click: **File** > **Prepare** > **Drawing Properties** > Detail Options **change** (the Options dialog box opens) [Fig. 11.2(h)] > make changes to create a new **.dtl** file *(or* > *pick a previously saved **.dtl** from your directory list > Open)* > Change the following options to the values listed below:

Option: ***text_height*** Value: ***.25*** > **Add/Change**
Option: ***default_font*** Value: ***filled*** > **Add/Change**
Option: ***arrow_style*** Value: ***filled*** > **Add/Change**
Option: ***allow_3d_dimensions*** Value: ***yes*** > **Add/Change**

Click: **Apply** > Sort: > **Alphabetical** [Fig. 11.2(h)] > **Save a copy of the currently displayed configuration file** > *type (a unique name for your file)* **draw_options** > **OK** > **Close**

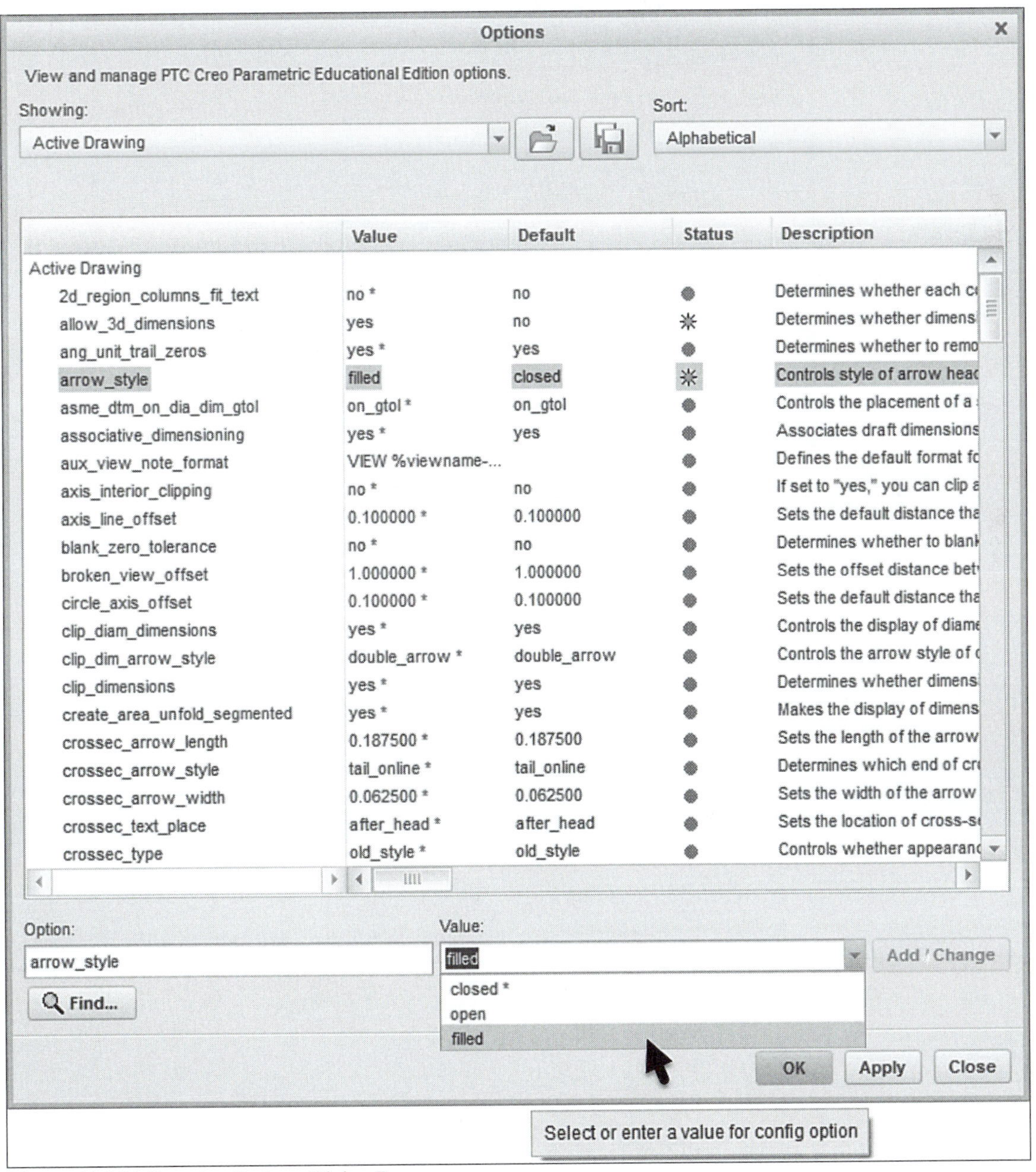

Figure 11.2(h) Drawing Options Dialog Box

Click: (to close the Navigator so that you can see a larger drawing) > > **Hidden Line >** **Layout** tab > General View **Create a general view > No Combined State** (highlighted) > **OK** > pick a position for the view [Fig. 11.3(a)] > Model view names **FRONT** > **OK** > **LMB** to deselect the view [Fig. 11.3(b)]

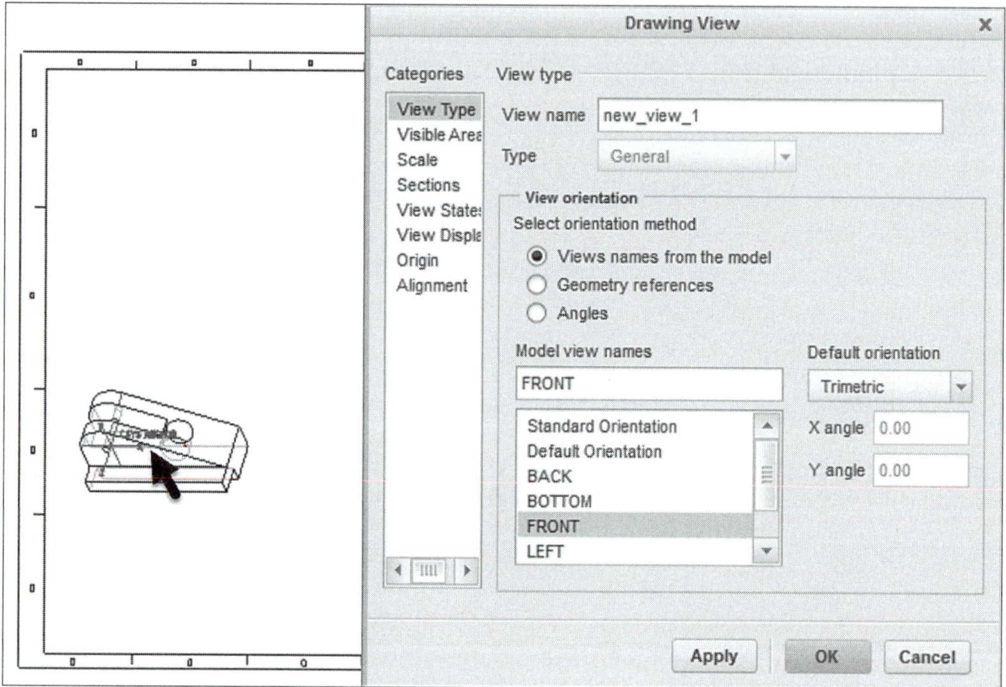

Figure 11.3(a) Select a Position for the First View

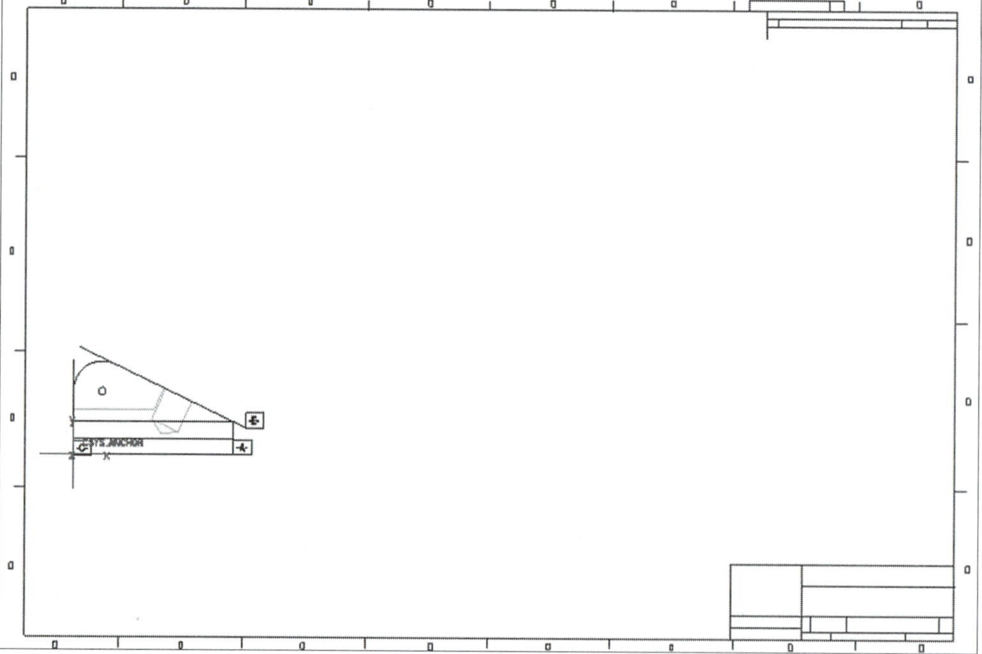
Figure 11.3(b) Front View

Click: Datum Display Filters > (Select All) *off* > LMB in the Graphics Window > Ctrl+R > add two more views; pick on the Front view (highlights) > Projection > Select CENTER POINT for drawing view. pick a position for the Right Side view [Fig. 11.3(c)] > move the mouse pointer off of the Right Side view > **LMB** to deselect > pick on the Front view > press **RMB** > **Projection View** > Select CENTER POINT for drawing view. pick a position for the Top view > move the mouse pointer off of the view > **LMB** to deselect [Fig. 11.3(d)] > from the Ribbon, **Lock View Movement** *off* > pick on a view, hold down the **LMB**, and reposition as needed > release the **LMB** > move the mouse pointer off of the view > **LMB** to deselect > **Refit** > **Save** [Fig. 11.3(e)]

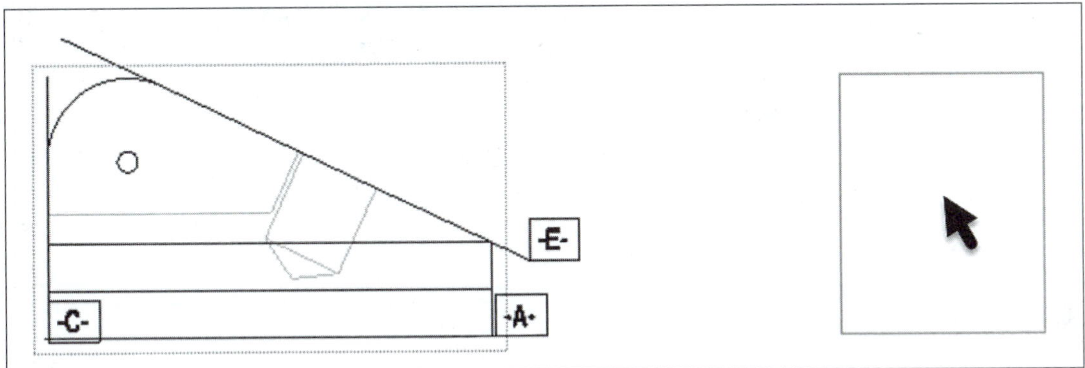

Figure 11.3(c) Add Right Side Projected View

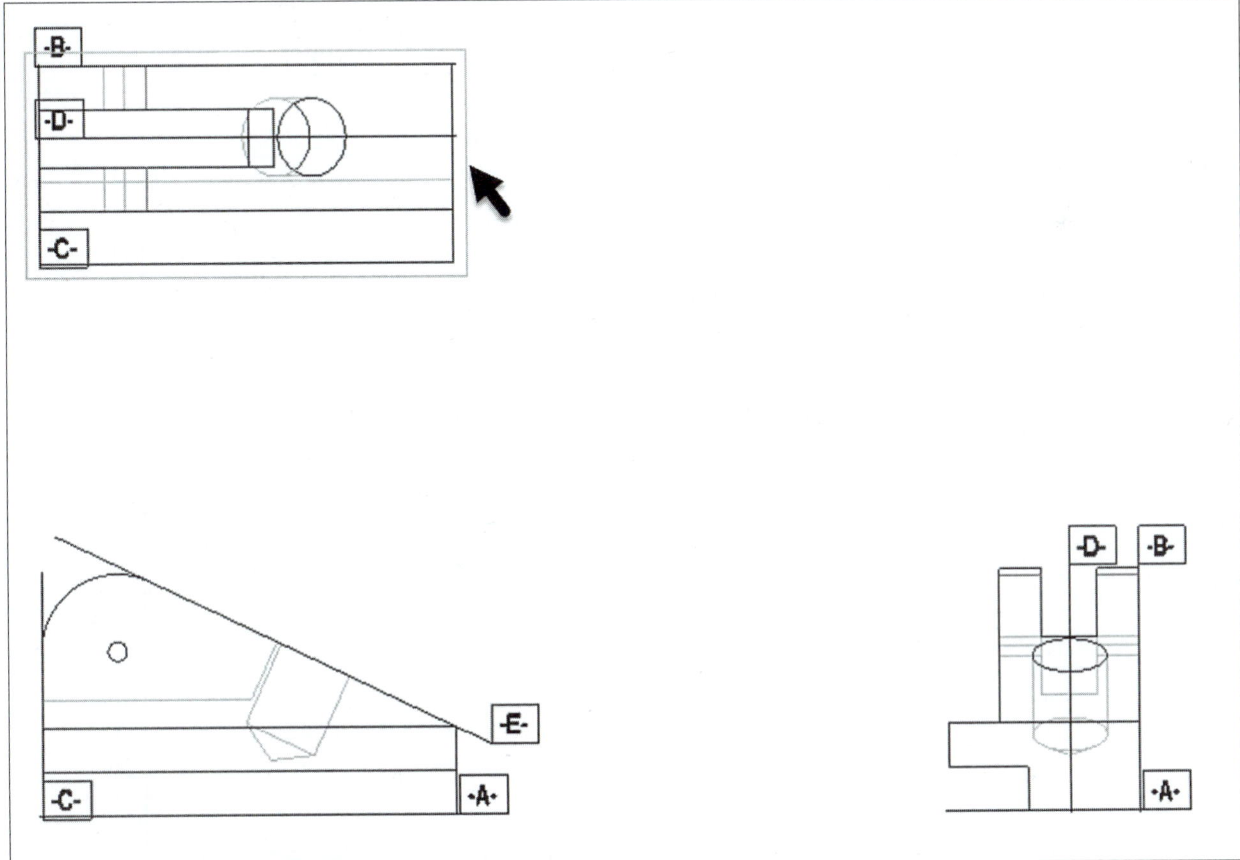

Figure 11.3(d) Add Top Projected View

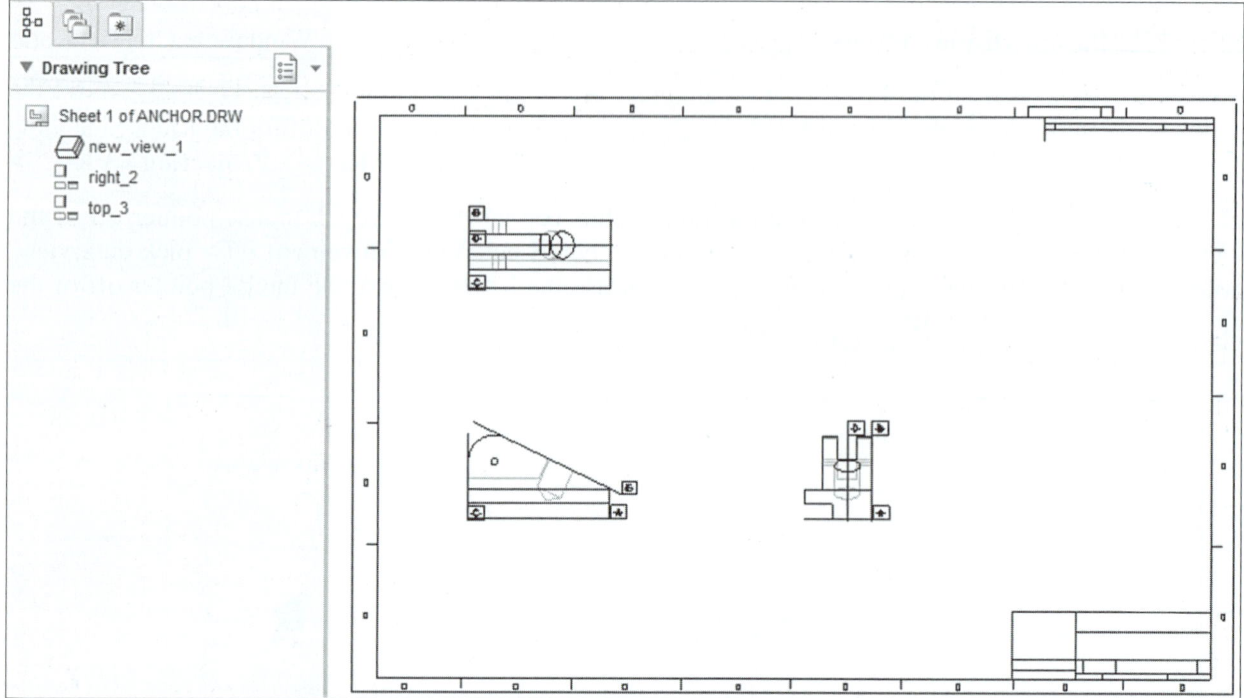

Figure 11.3(e) Repositioned Views

The top view does not help in the description of the part's geometry. Delete the top view before adding an auxiliary view that will display the true shape of the angled surface.

Pick on the Top view > press **RMB** [Fig. 11.3(f)] > **Del** key from Keyboard > [icon] > **Ctrl+S**

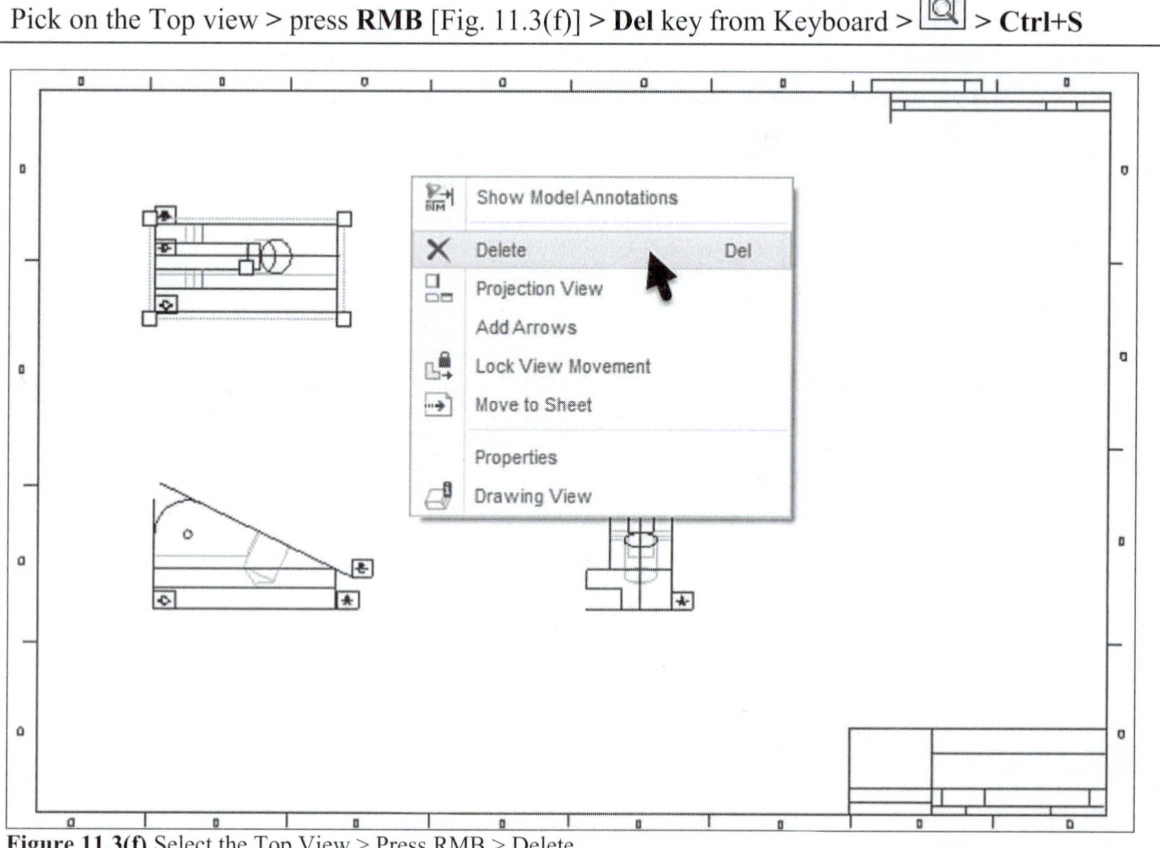

Figure 11.3(f) Select the Top View > Press RMB > Delete

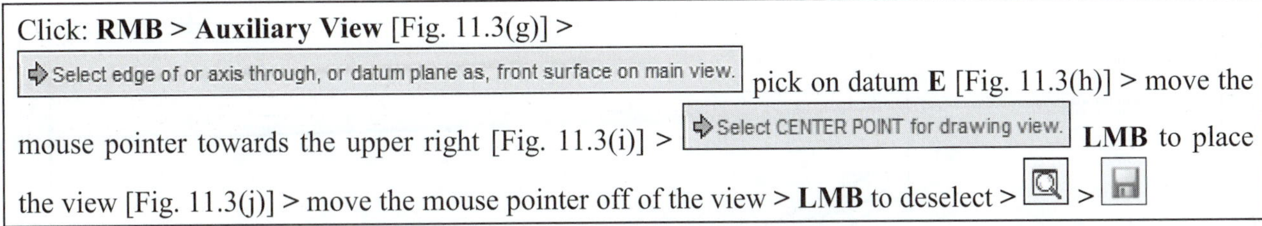
Click: **RMB > Auxiliary View** [Fig. 11.3(g)] > Select edge of or axis through, or datum plane as, front surface on main view. pick on datum **E** [Fig. 11.3(h)] > move the mouse pointer towards the upper right [Fig. 11.3(i)] > Select CENTER POINT for drawing view. **LMB** to place the view [Fig. 11.3(j)] > move the mouse pointer off of the view > **LMB** to deselect > >

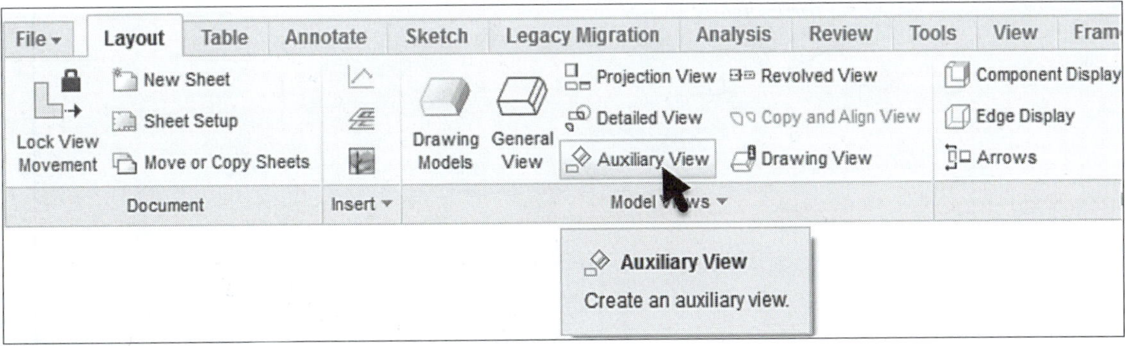

Figure 11.3(g) Create an Auxiliary View

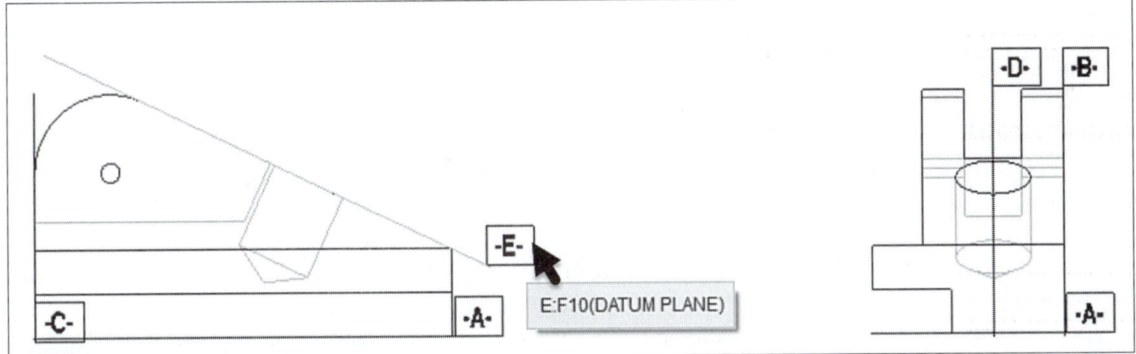

Figure 11.3(h) Select Datum Plane E or the Angled Edge

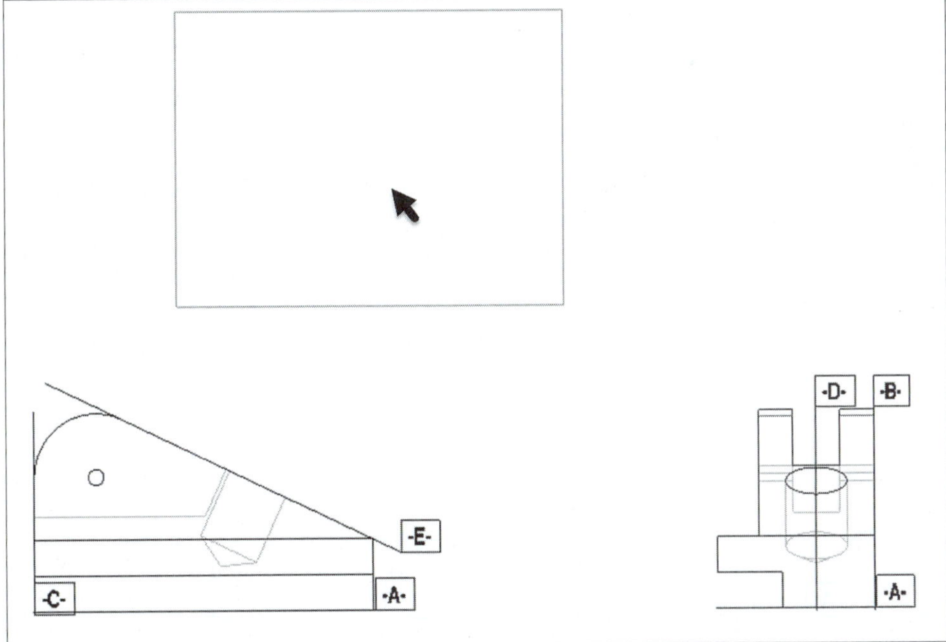

Figure 11.3(i) Move the Mouse Pointer to a new Location to Place the View

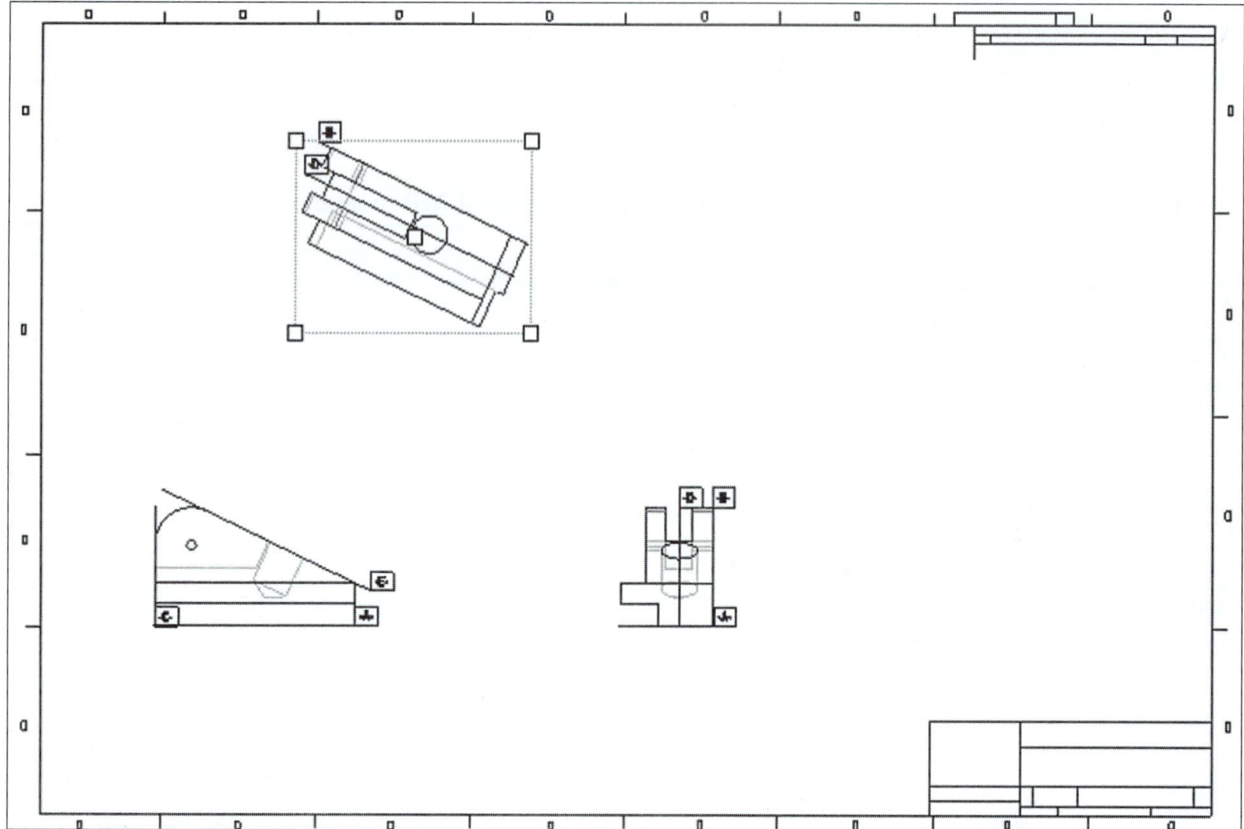

Figure 11.3(j) Auxiliary View

The newest standard was not selected when you made changes to the Drawing Options. ASME symbols for datum planes (gtol_datums) are the correct style standard used on drawings. The ANSI style [Fig. 11.3(k)] was discontinued in 1994, though retained by some companies as an "in house" standard. For outside vendors and for manufacturing internationally, the ISO-ASME standards (ASME Y14.5) should be applied to all manufacturing drawings.

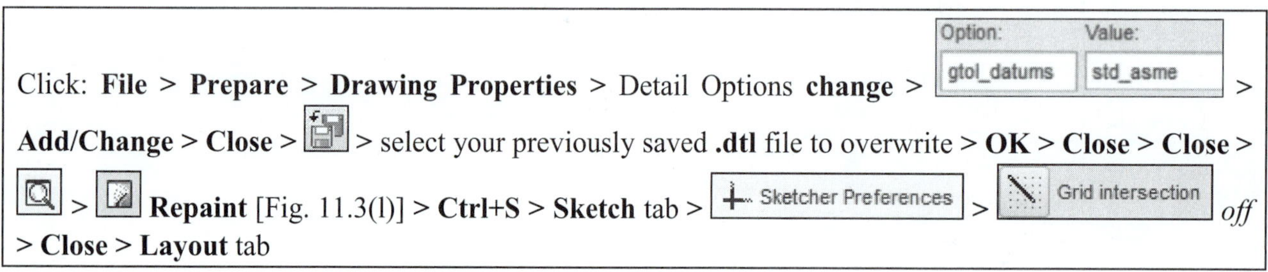

Click: **File** > **Prepare** > **Drawing Properties** > Detail Options **change** > [Option: gtol_datums, Value: std_asme] > **Add/Change** > **Close** > > select your previously saved **.dtl** file to overwrite > **OK** > **Close** > **Close** > > **Repaint** [Fig. 11.3(l)] > **Ctrl+S** > **Sketch** tab > Sketcher Preferences > Grid intersection *off* > **Close** > **Layout** tab

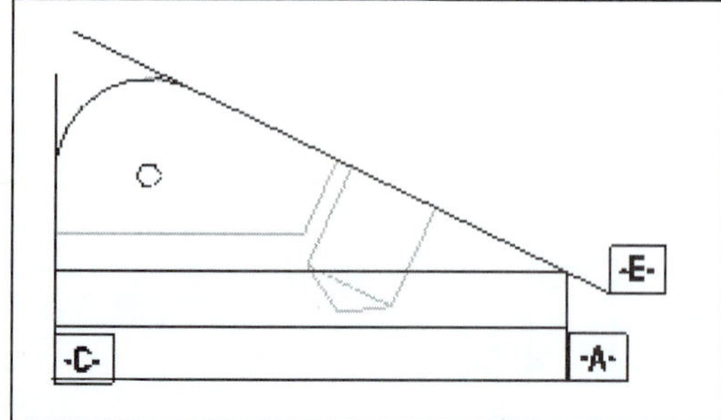

Figure 11.3(k) gtol_ansi

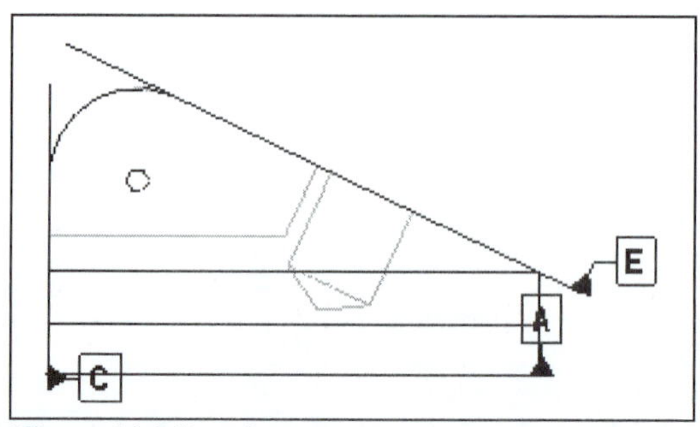

Figure 11.3(l) gtol_asme

Next, change the Front view into a sectional view. Section **A** was created in the Part mode. Pick on the Front view as the view to be modified [Fig. 11.4(a)] > press **RMB** > **Properties** [Fig. 11.4(b)] > Categories **Sections** > 2D cross-section > + **Add cross-section to view** > ✓ A from the Name list [Fig. 11.4(c)] > **Apply** > **Cancel** [Fig. 11.4(d)] > in the Graphics Window, **LMB** to deselect

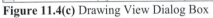

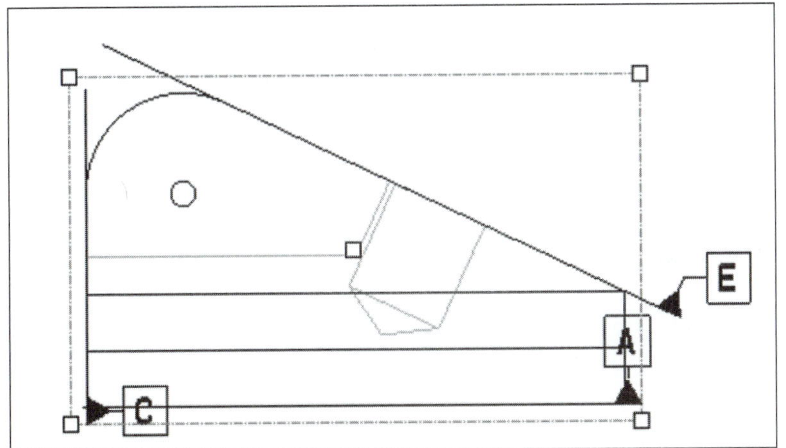

Figure 11.4(a) Select the Front View

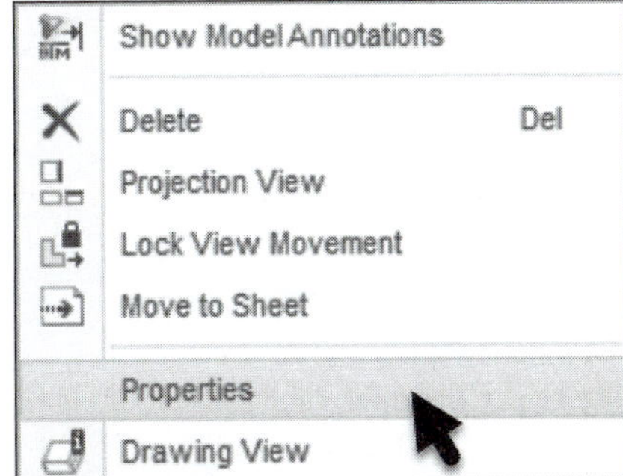

Figure 11.4(b) RMB > Properties

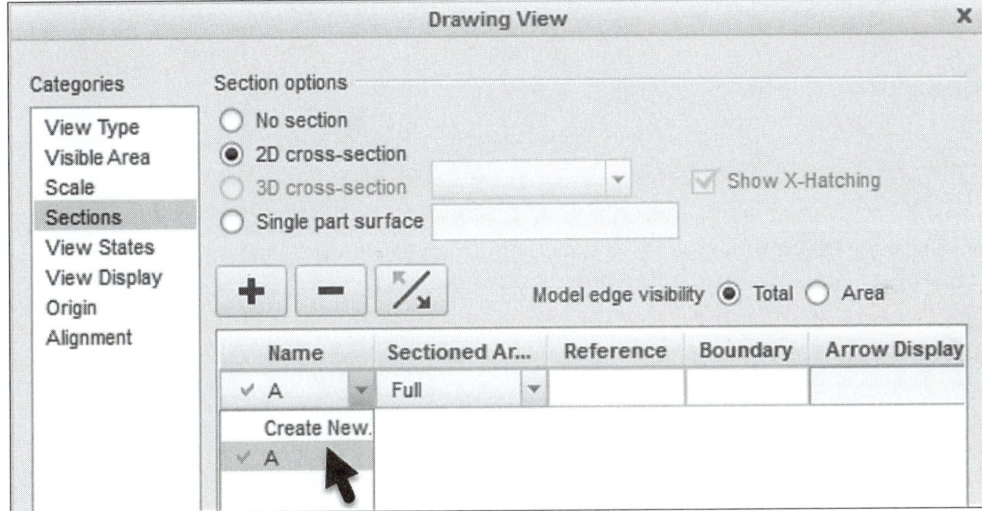

Figure 11.4(c) Drawing View Dialog Box

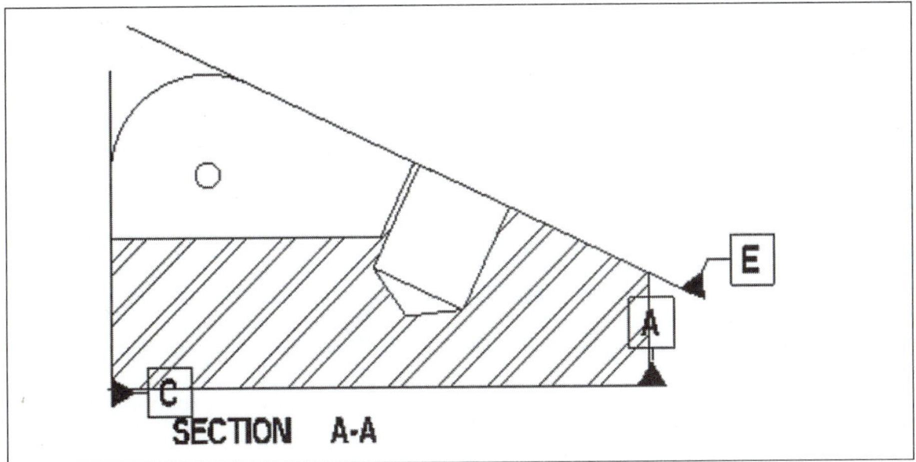

Figure 11.4(d) Section A-A

Show all dimensions and axes (centerlines), click: **Annotate** tab > [icon] > [icon] Dimensions tab > with the **Ctrl** key pressed, pick on the Front and then your Auxiliary view and Right side views [Fig. 11.5(a)] > [icon] > **Apply** > [icon] Datums tab > [icon] [Fig. 11.5(b)] > **Apply** > **Cancel**

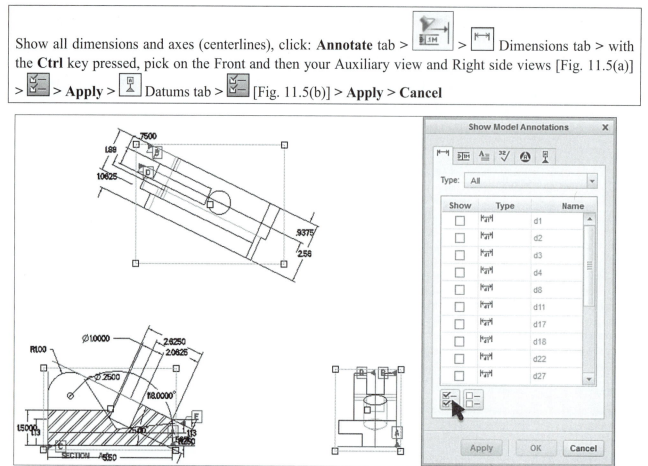

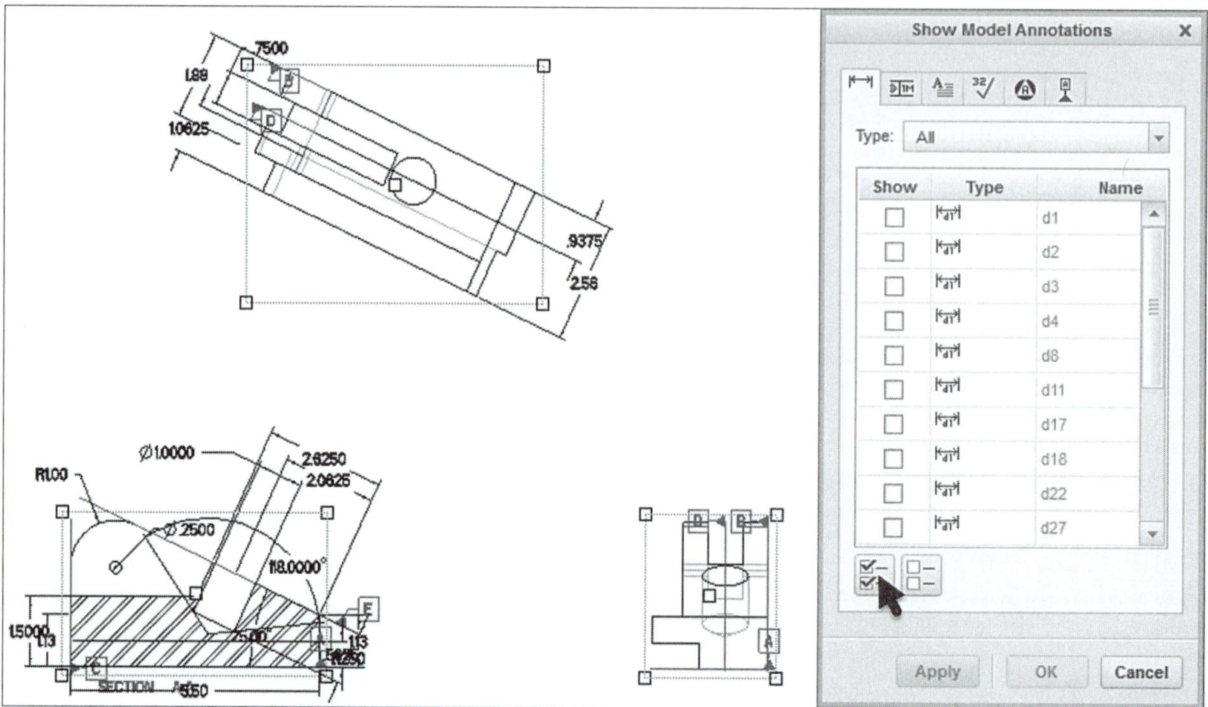

Figure 11.5(a) Show Model Annotations Dialog Box

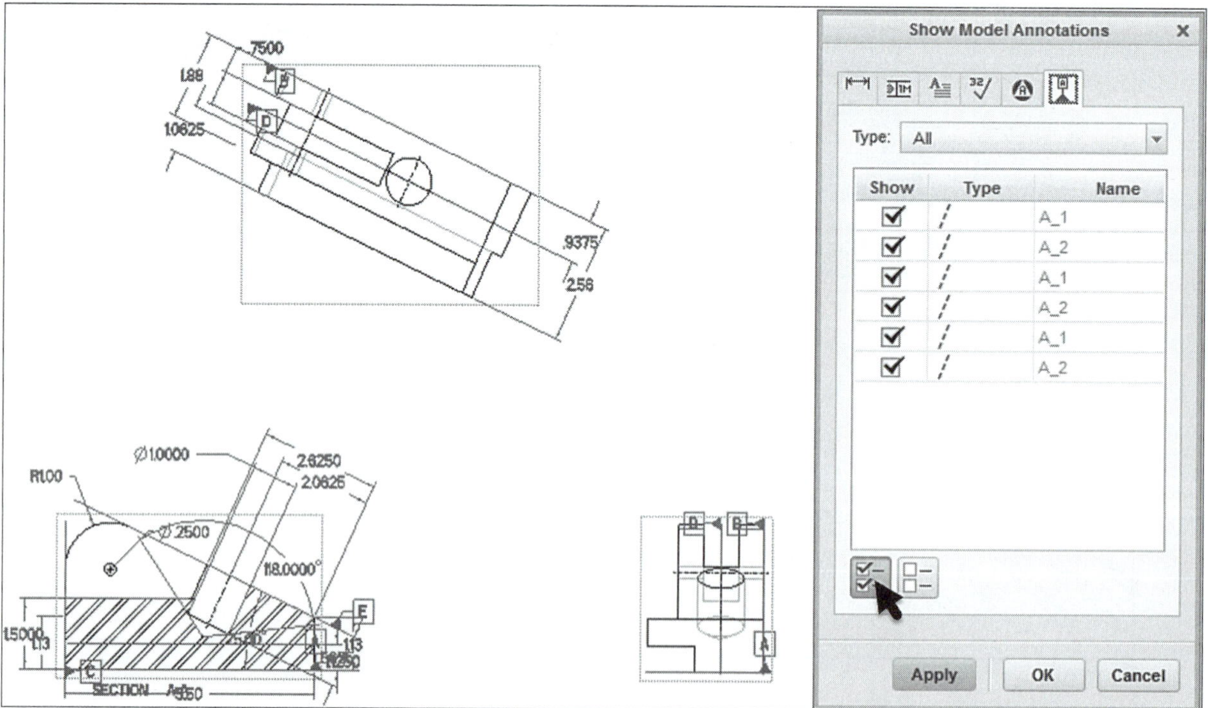

Figure 11.5(b) Dimensions and Axes displayed in the three views

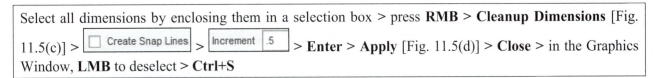

Select all dimensions by enclosing them in a selection box > press **RMB** > **Cleanup Dimensions** [Fig. 11.5(c)] > Create Snap Lines > Increment .5 > **Enter** > **Apply** [Fig. 11.5(d)] > **Close** > in the Graphics Window, **LMB** to deselect > **Ctrl+S**

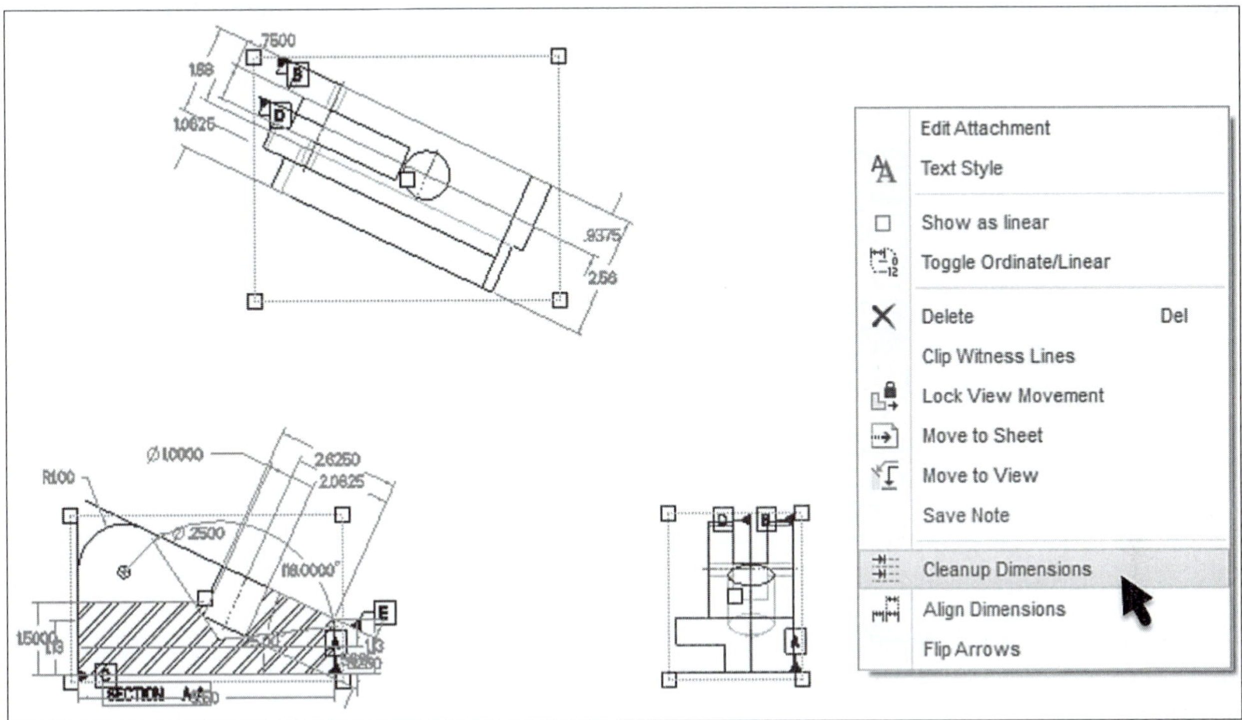

Figure 11.5(c) Press RMB > Cleanup Dimensions

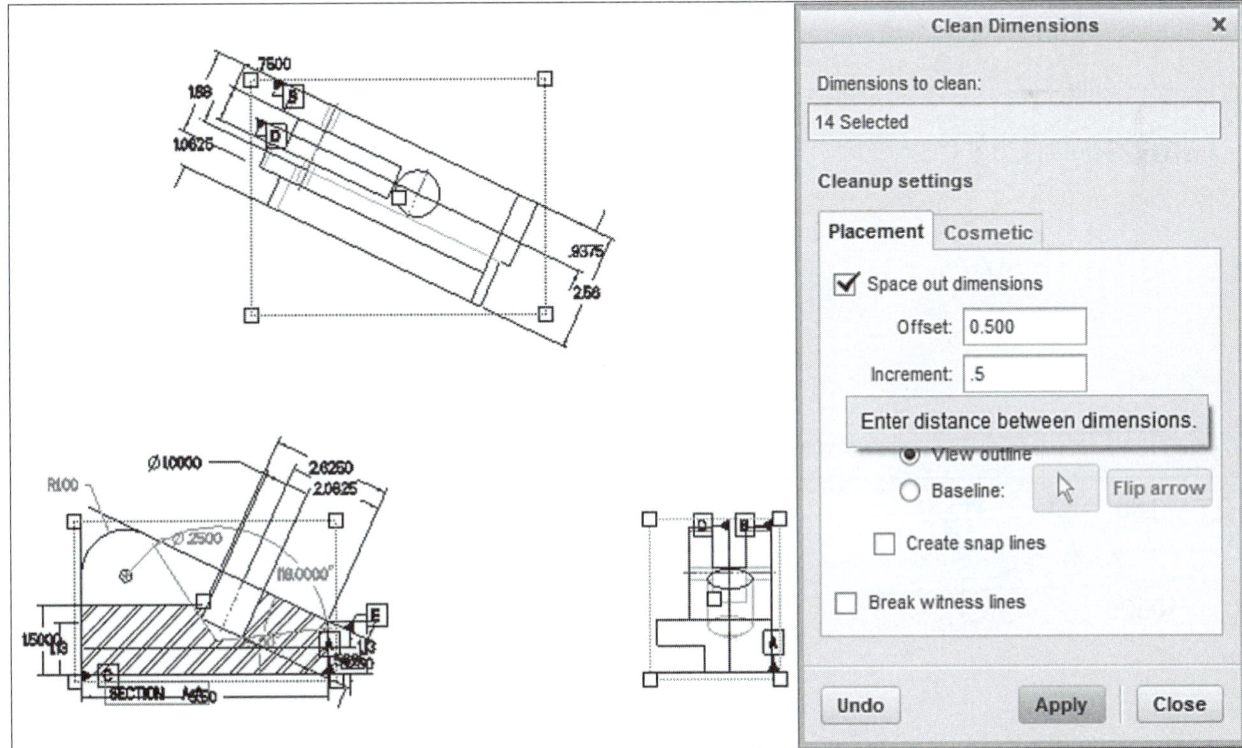

Figure 11.5(d) Cleaned Dimensions and Clean Dimensions Dialog Box *(your drawing may look different)*

Pick on the **1.00** diameter dimension > press **RMB** > **Move to View** [Fig. 11.6(a)] > pick on the Auxiliary view [Fig. 11.6(b)] > pick on and reposition the **1.00** dimension [Fig. 11.6(c)] > **LMB** to deselect > in each view, reposition dimensions, move dimension text, clip extension lines, or flip arrows where appropriate > reposition, move, clip, or erase axes and position datums > pick the Front view > press **RMB** > **Add Arrows** > pick on the Auxiliary view > move the pointer off of the view > **LMB** to deselect [Fig. 11.6(d)]

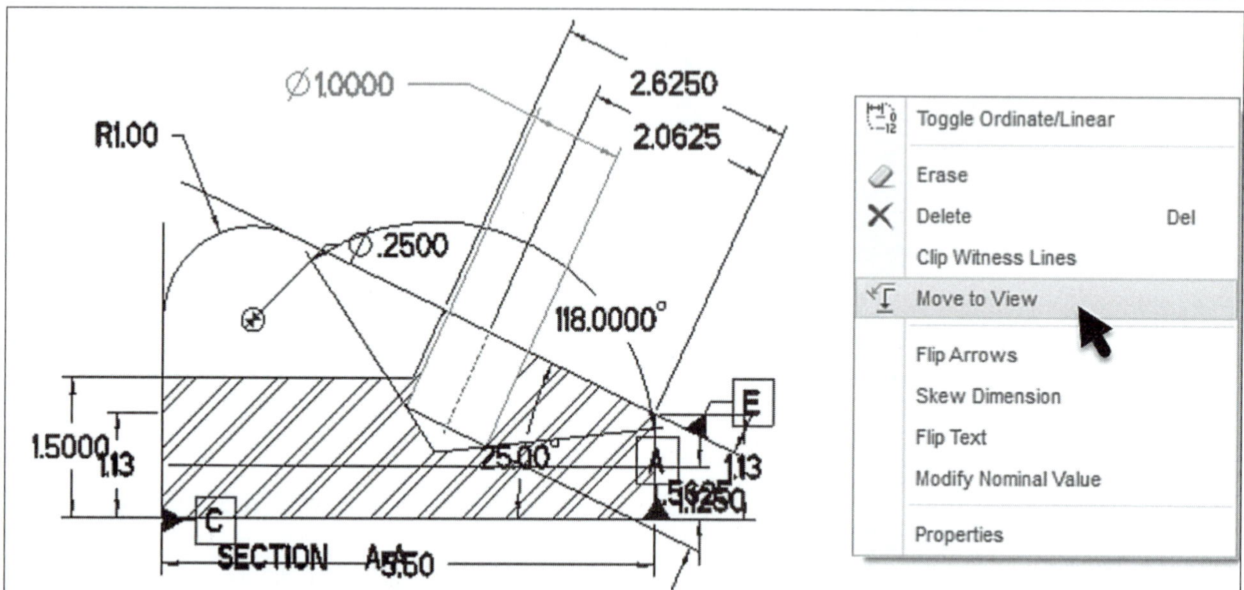

Figure 11.6(a) Select the **1.00** Diameter Dimension

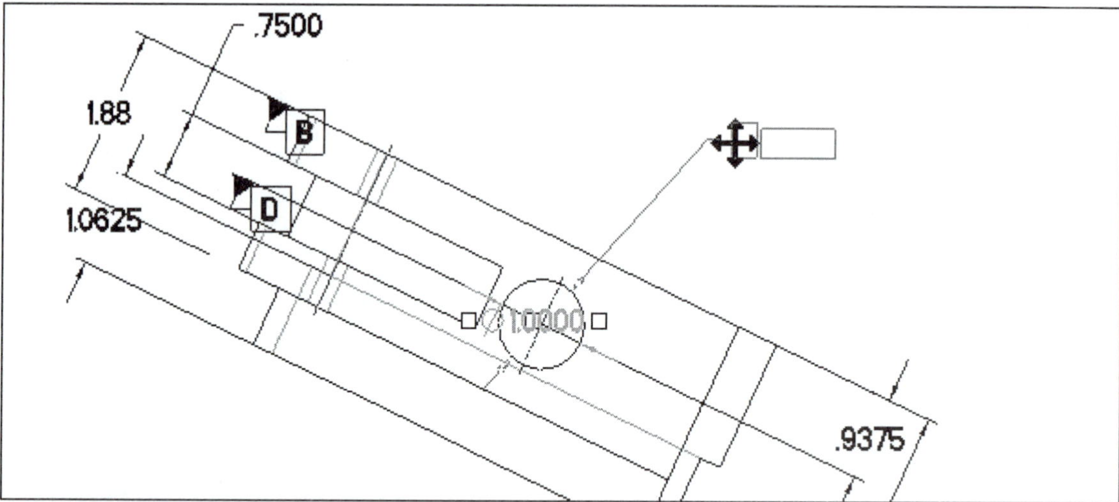

Figure 11.6(b) Diameter Dimension **1.00** Moved to Auxiliary View

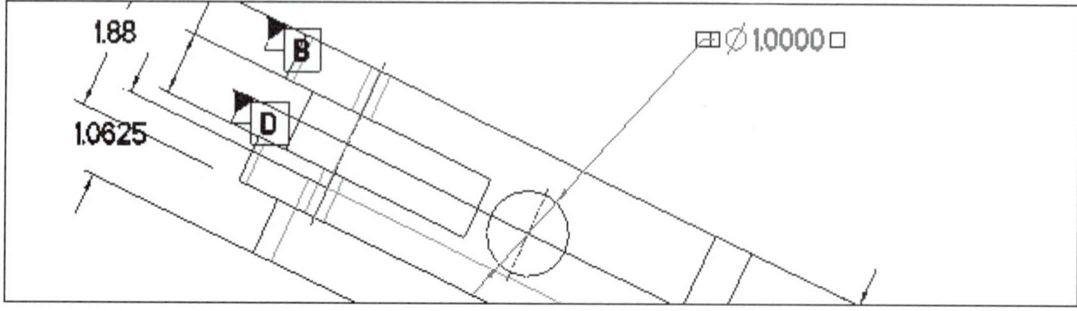

Figure 11.6(c) Repositioned **1.00** Diameter Dimension

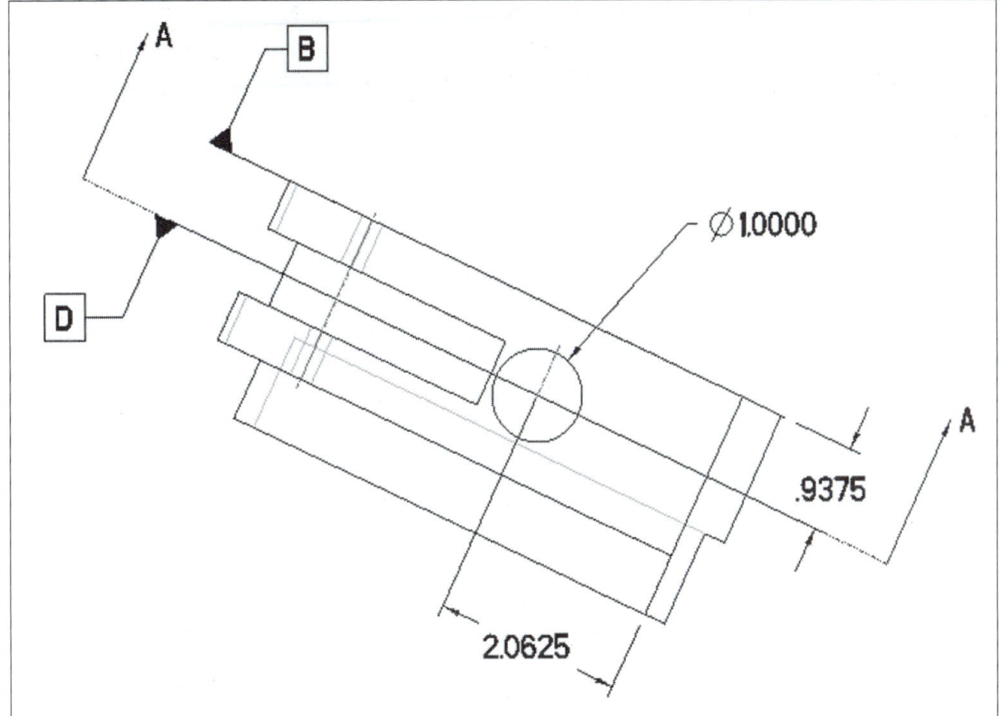

Figure 11.6(d) Edited Auxiliary View

Add a reference dimension to the small hole, click: **Reference Dimension** > press and hold the **Ctrl** key > pick the horizontal axis of the small hole > pick datum **A** [Fig. 11.7(a)] > **MMB** to place the reference dimension [Fig. 11.7(b)] > **Cancel** > **LMB** to deselect > **Ctrl+S**

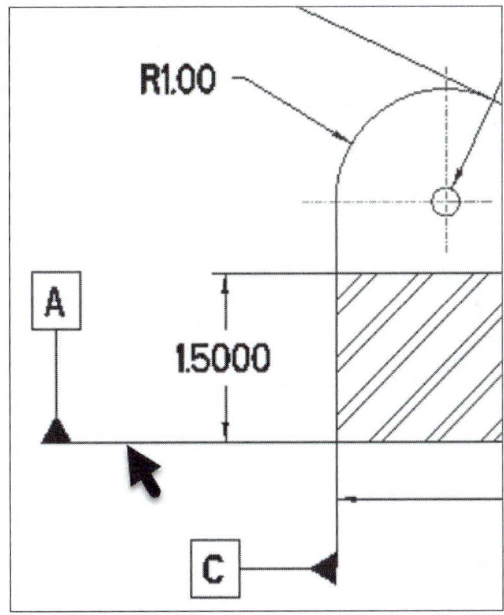

Figure 11.7(a) Select Datum A

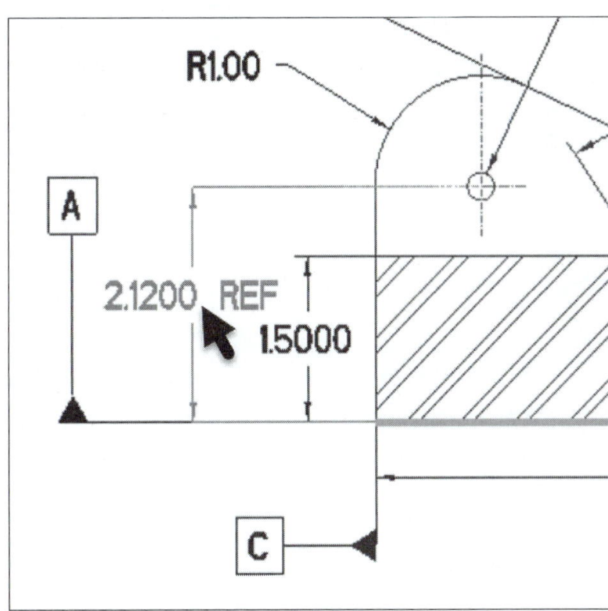

Figure 11.7(b) Placed Reference Dimension

To change the number of decimal places for a dimension > pick the dimension(s) > press **RMB** > **Properties** > Decimal Places **3** or **4** > **OK** > **LMB** to deselect > press the **Ctrl** key, select the Auxiliary and Right views > release the **Ctrl** key > press **RMB** > **Properties** > Display style > **No Hidden** > **OK** > **LMB** to deselect > **Ctrl** key > select the axes > **RMB** > **Erase** (Fig. 11.8)

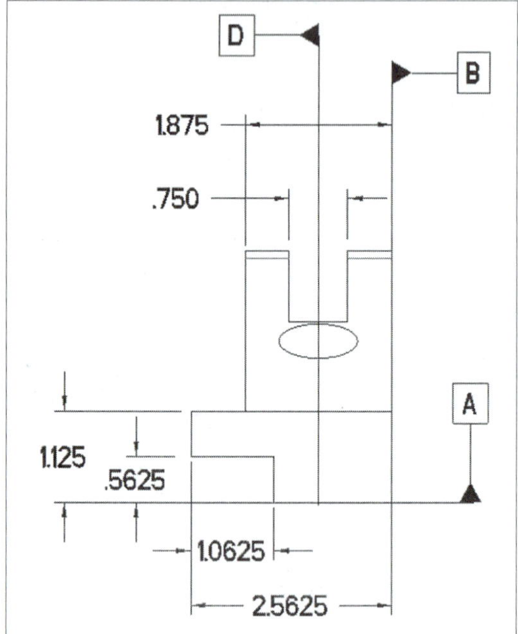

Figure 11.8 Edited Right Side View

Add the edges of the small through hole back into the auxiliary view. The edges will display in the graphics window in a lighter color, but print as dashed on the drawing plot.

Pick the Auxiliary view > **Layout** tab > Edge Display > **Hidden Line** > press and hold the **Ctrl** key and pick near where the small through hole would show as hidden in the Auxiliary view [Fig. 11.9(a)] > pick the opposite edge [Fig. 11.9(b)] > release the **Ctrl** key > **OK** > **Done** > > [Fig. 11.9(c)]

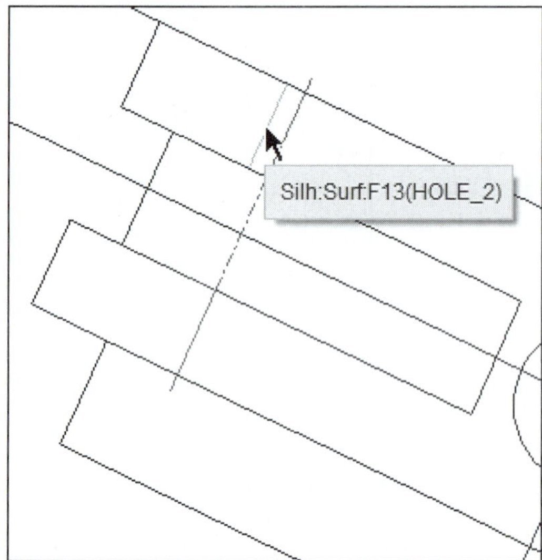

Figure 11.9(a) Select near the Hole Edge

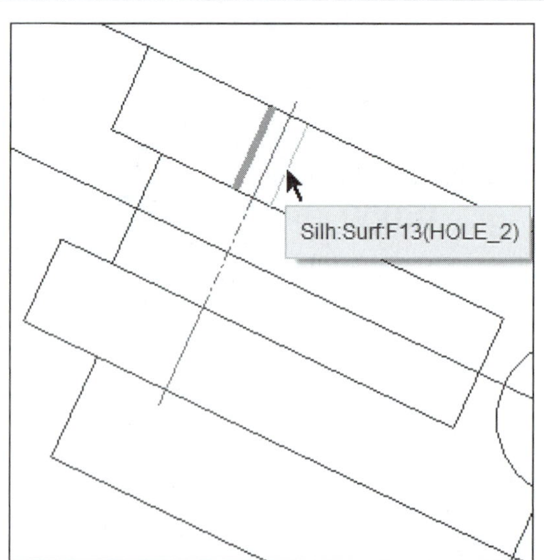

Figure 11.9(b) Select near the Opposite Edge of the Hole

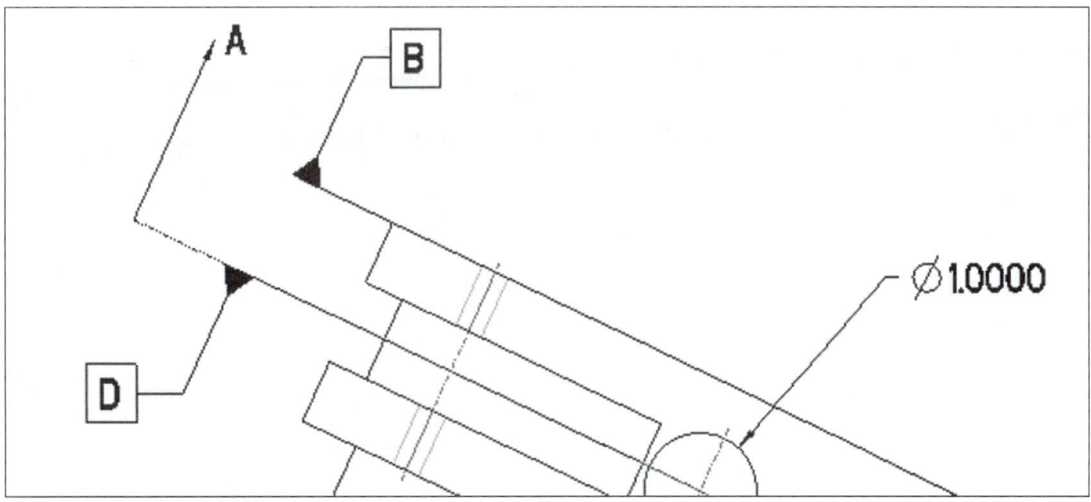

Figure 11.9(c) Small Through Hole's Edge Lines Displayed

To increase the functionality of the drawing, you will need to master a number of capabilities. Partial views, detail views, using multiple sheets, and modifying section lining (crosshatch lines) are just a few of the many options available in Drawing Mode. Create a Detail View.

Click: zoom in on the FRONT view [Detailed View] **Detailed View >** [Select center point for detail on an existing view.] pick the top edge of the hole in the Front section view [Fig. 11.10(a)] an **X** will display at the selected position > [Create points for the spline to pass through.] *each* ***LMB*** *pick adds a point to the spline* > **MMB** to end (and close) the spline [Fig. 11.10(b)] *(you do not have to "close" the spline sketch – just get near the start of the spline and clicking the MMB will close the spline for you)*

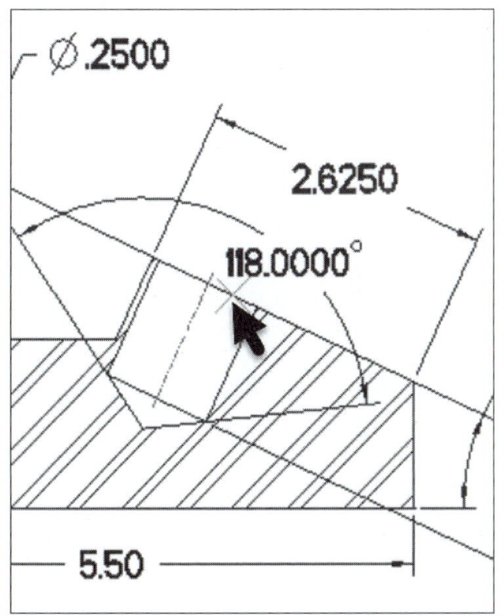

Figure 11.10(a) Select Center Point for Detail View

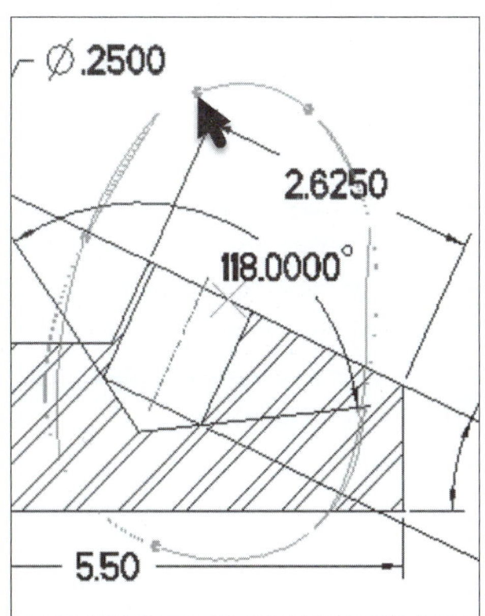

Figure 11.10(b) Sketch a Spline (MMB to close the spline)

Select CENTER POINT for drawing view. pick in the upper right corner of the drawing sheet to place the view [Fig. 11.10(c)] > with the Detailed view selected, press: **RMB** > **Properties** [Fig. 11.10(d)] > Categories **Scale** > Custom scale 2.000 > **1.500** [Fig. 11.10(e)] > **Enter** > **OK** > **LMB** > **Ctrl+S**

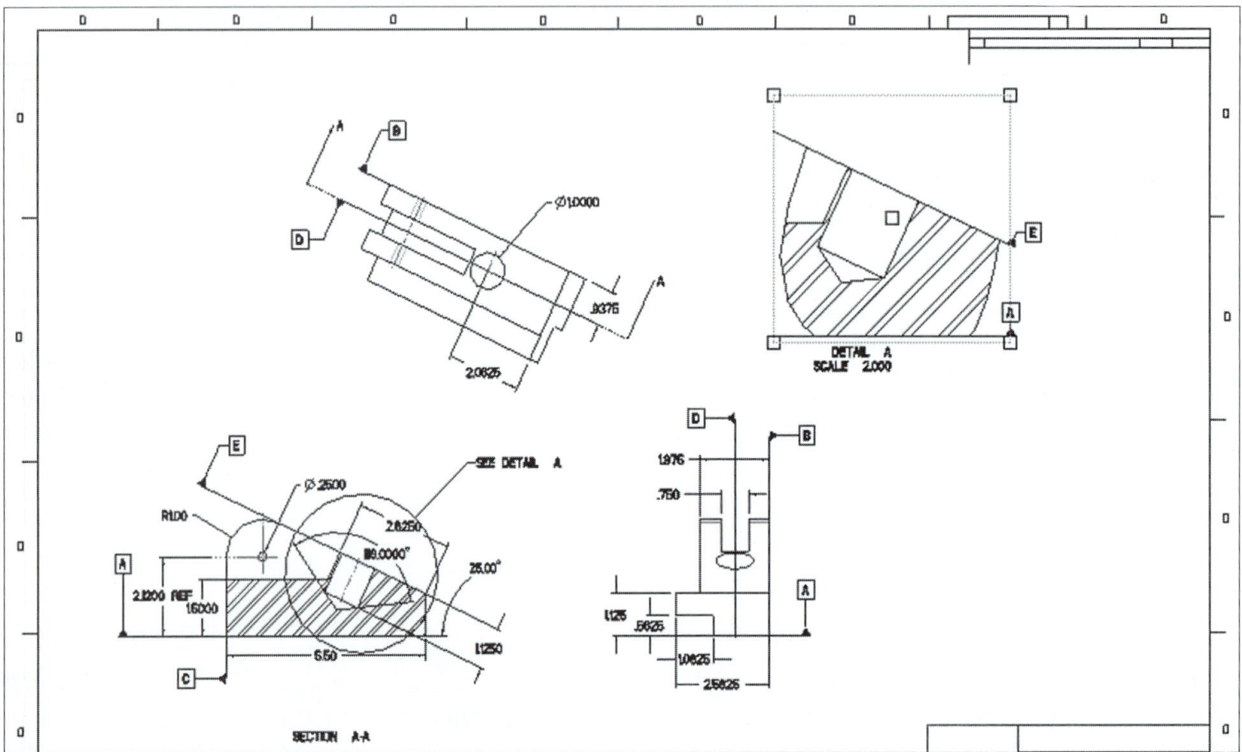

Figure 11.10(c) Detail View A

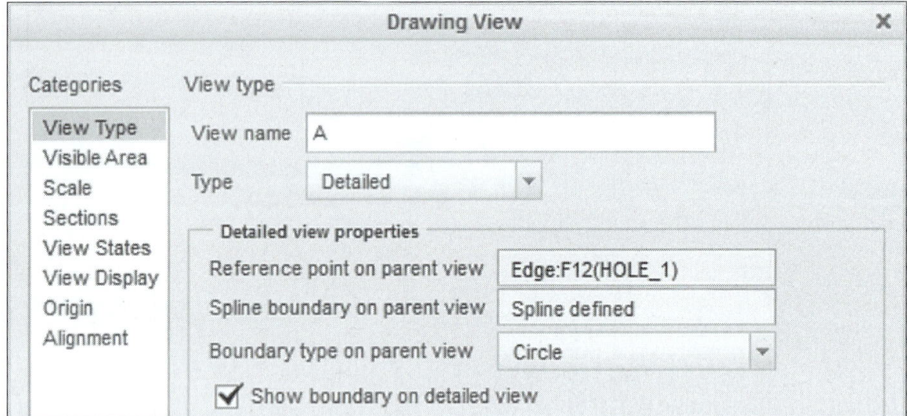

Figure 11.10(d) Drawing View Dialog Box

Drawing View
Categories
View Type
Visible Area
Scale
Sections
Scale and perspective options
Default scale for sheet (1.000)
Custom scale 1.500
Perspective

Figure 11.10(e) Custom Scale

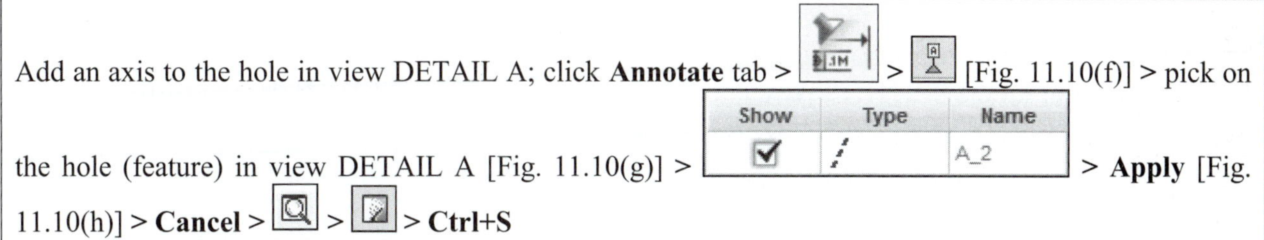

Add an axis to the hole in view DETAIL A; click **Annotate** tab > > [Fig. 11.10(f)] > pick on the hole (feature) in view DETAIL A [Fig. 11.10(g)] >

Show	Type	Name
☑	/	A_2

> **Apply** [Fig. 11.10(h)] > **Cancel** > > > **Ctrl+S**

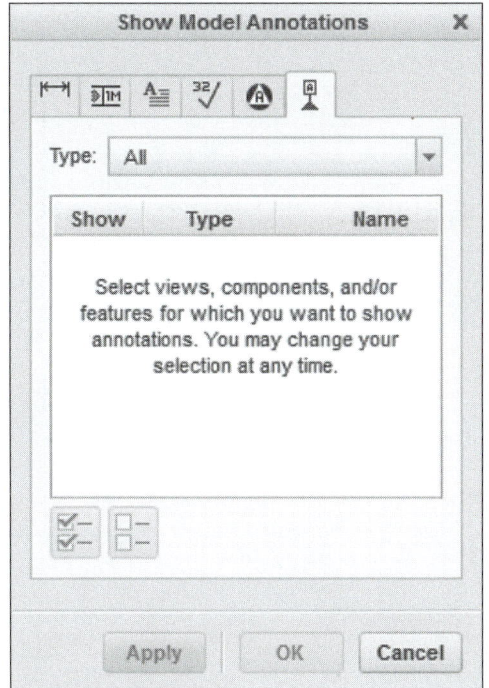

Figure 11.10(f) Datums

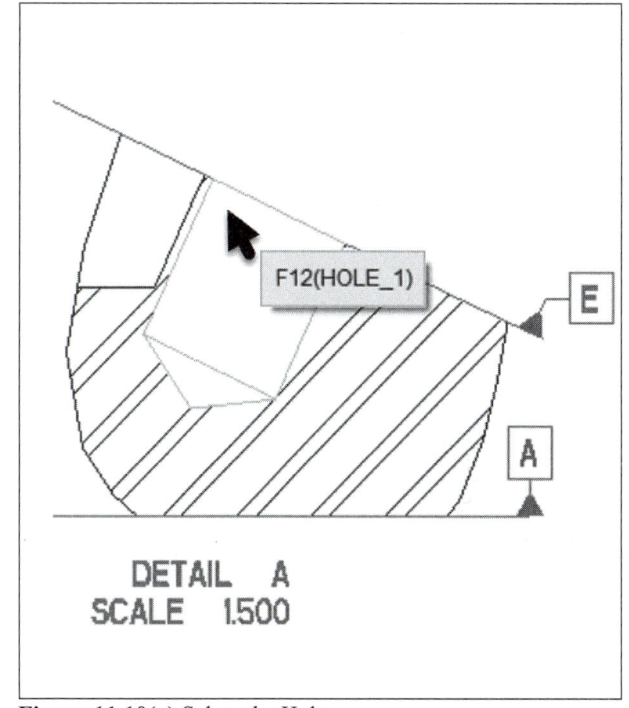

Figure 11.10(g) Select the Hole

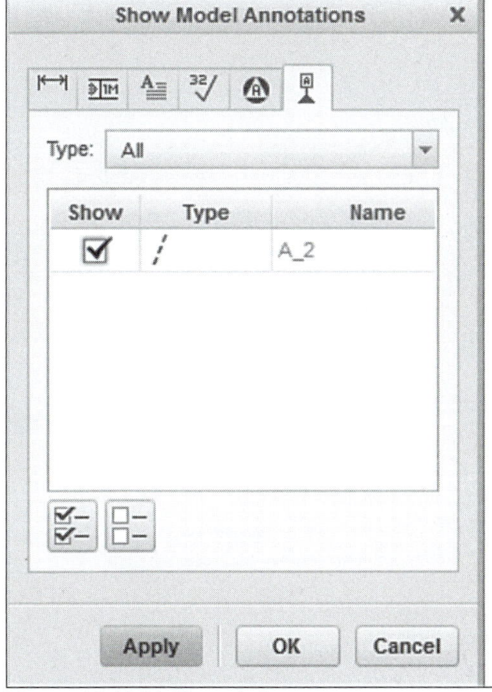

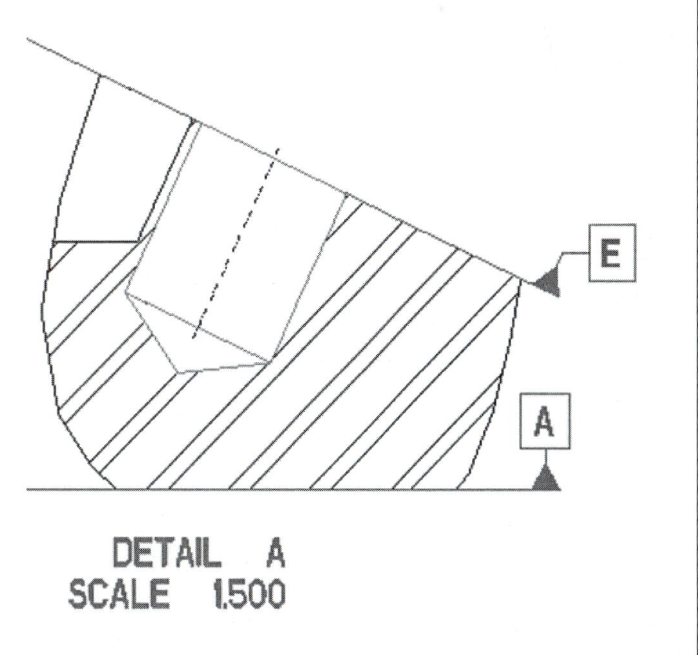

Figure 11.10(h) Show Axis

In **DETAIL A**, erase datum **A,** clip datum **E** and adjust the axis. > double-click on **DETAIL A 1.500** > **Format** tab opens > **.375** Change the text height [Fig. 11.11(a)] > **Enter** > move the note as needed [Fig. 11.11(b)] > repeat for the **SECTION A-A** and **SEE DETAIL A** notes > **Ctrl+S** [Fig. 11.11(c)]

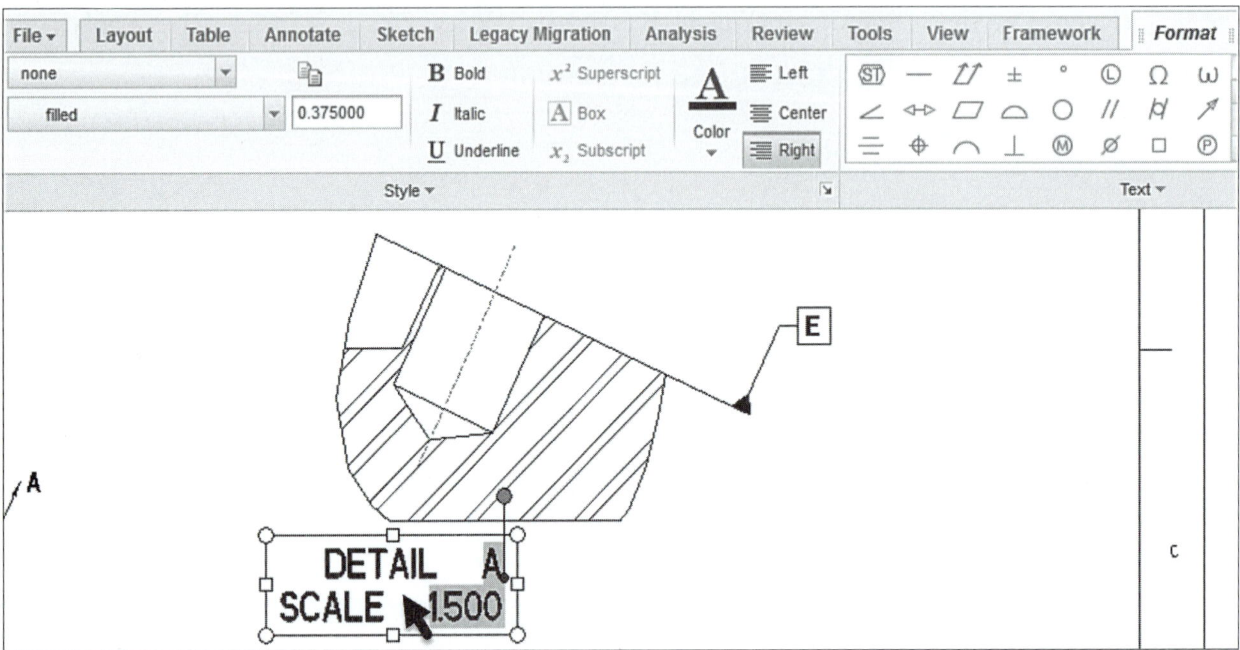

Figure 11.11(a) Select the Text Items and Change Their Height

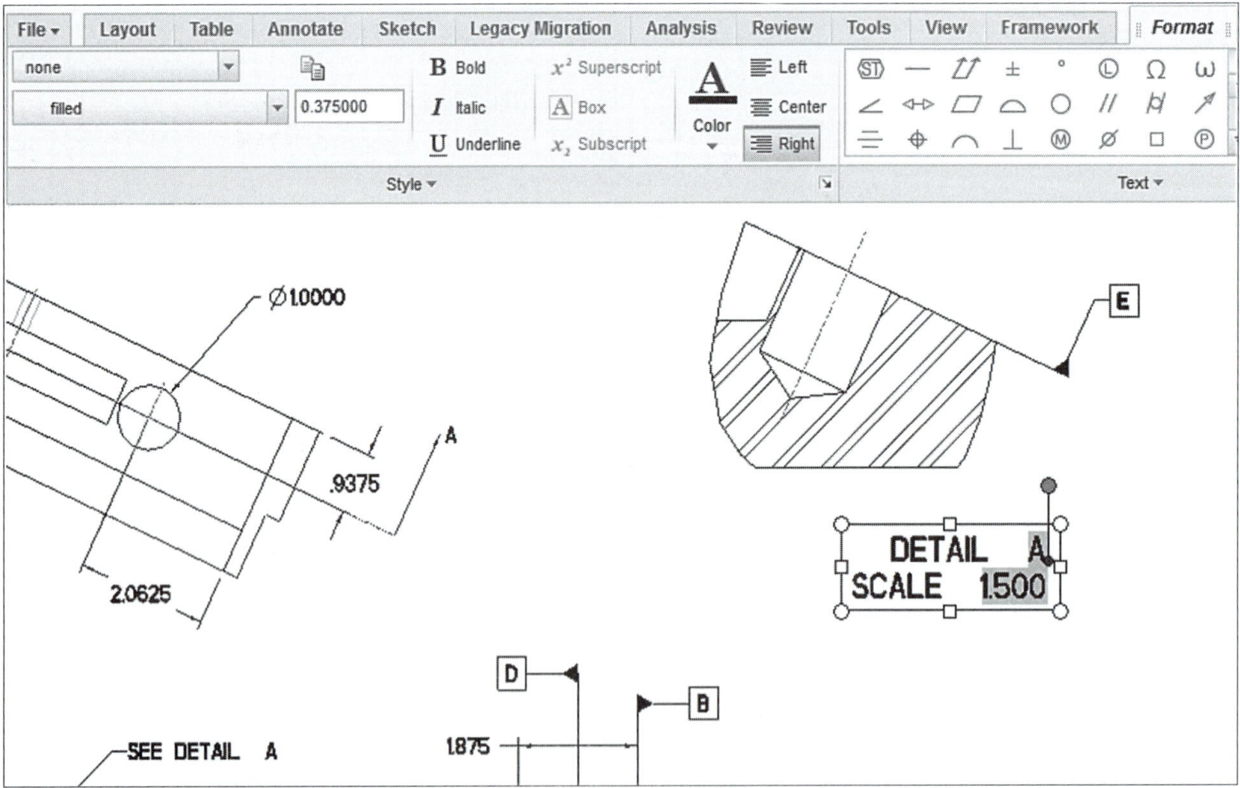

Figure 11.11(b) Text Style Dialog Box

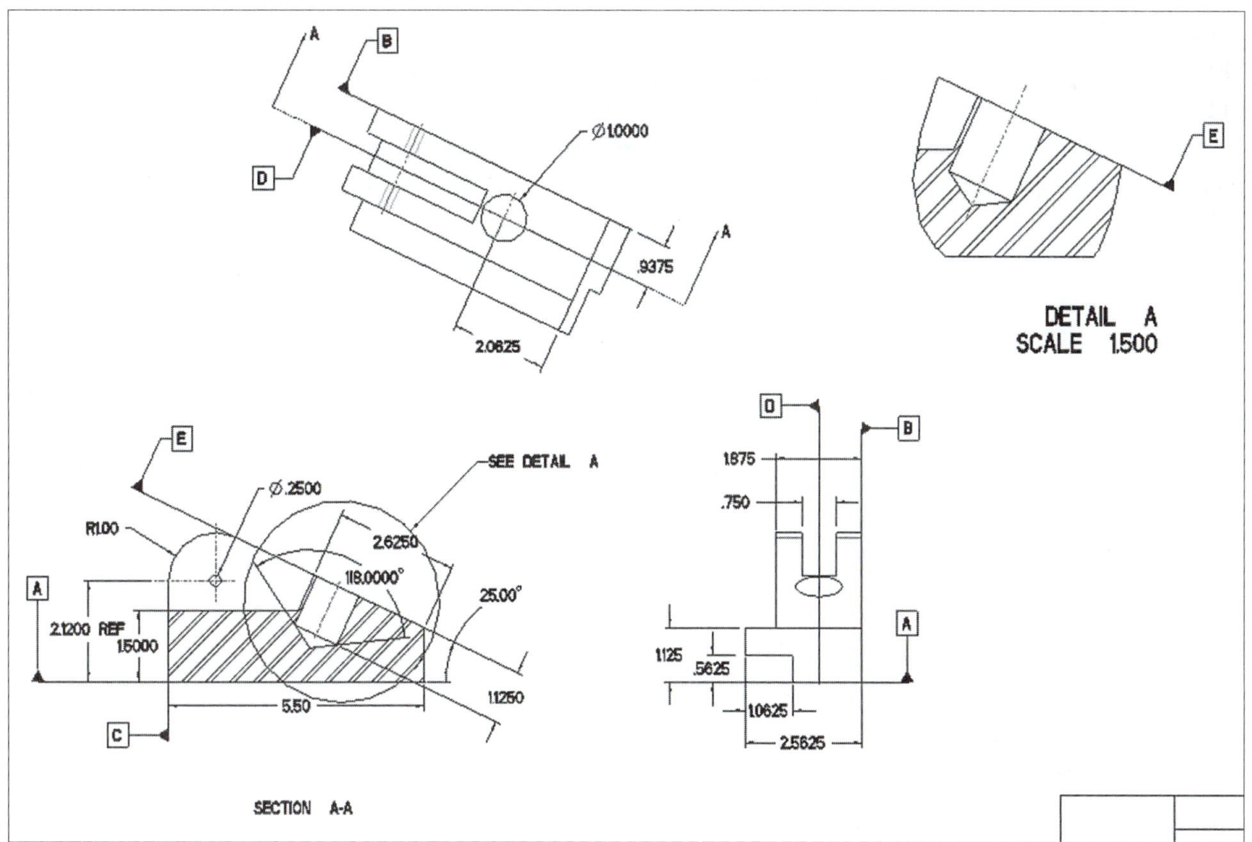

Figure 11.11(c) Changed Text Style (Height) *(your drawing may appear differently)*

Change the height of the section identification lettering to **.375**, pick on "**A**" > press **RMB** > **Text Style** [Fig. 11.11(d)] > next to Height [Default] (uncheck) > Height **.375** > **Enter** > **Apply** [Fig. 11.11(e)] > **OK** > **LMB** to deselect > window-in all drawing text and dimensions > press **RMB** > **Text Style** > next to Font [Default] (uncheck) > [▾] > [Font | font] > **Apply** > **OK** > **LMB** to deselect > **Ctrl+S**

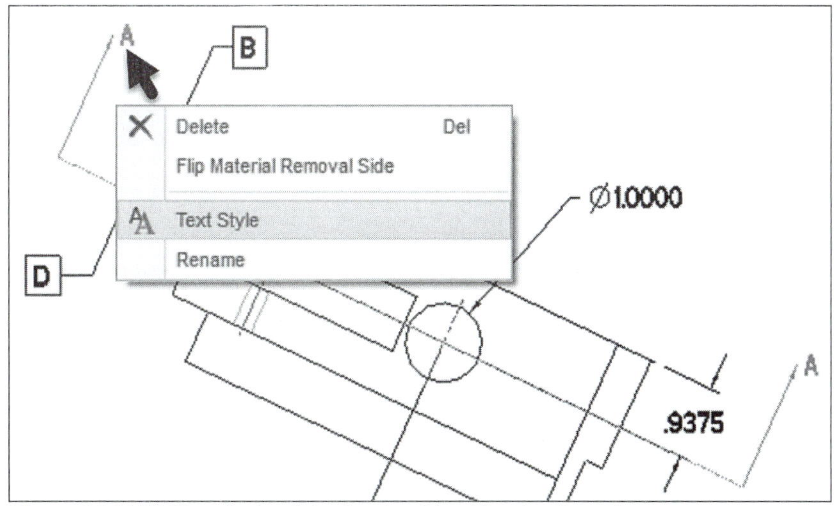

Figure 11.11(d) Text Style of the Section Identification Lettering

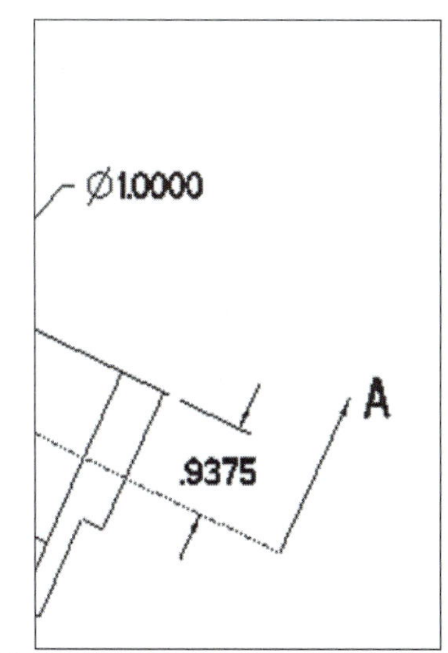

Figure 11.11(e) New Text Height for **A**

The shape and size of the Arrows for the cutting plane and dimensions are too small. These are controlled by the Drawing Options, click: **File** > **Prepare** > **Drawing Properties** > Detail Options **change** > ***draw_arrow_length .25*** > **Enter** > ***draw_arrow_width .10*** > **Enter** > ***crossec_arrow_length .375*** > **Enter** > ***crossec_arrow_width .125*** > **Enter** > ***default_font*** > **default** [Fig. 11.12(a)] > **Add / Change** > ***arrow_style*** > ***filled*** > **Enter** > **Apply** > **Close** > **Close** > scroll the **MMB** to zoom in [Fig. 11.12(b)] >

▼ These options control cross sections and their arrows				
crossec_arrow_length	.375	0.187500	●	Sets the length of the
crossec_arrow_style	tail_online *	tail_online	●	Determines which en
crossec_arrow_width	.125	0.062500	●	Sets the width of the
crossec_text_place	after_head *	after_head	●	Sets the location of c

Figure 11.12(a) Cross Section Drawing Options

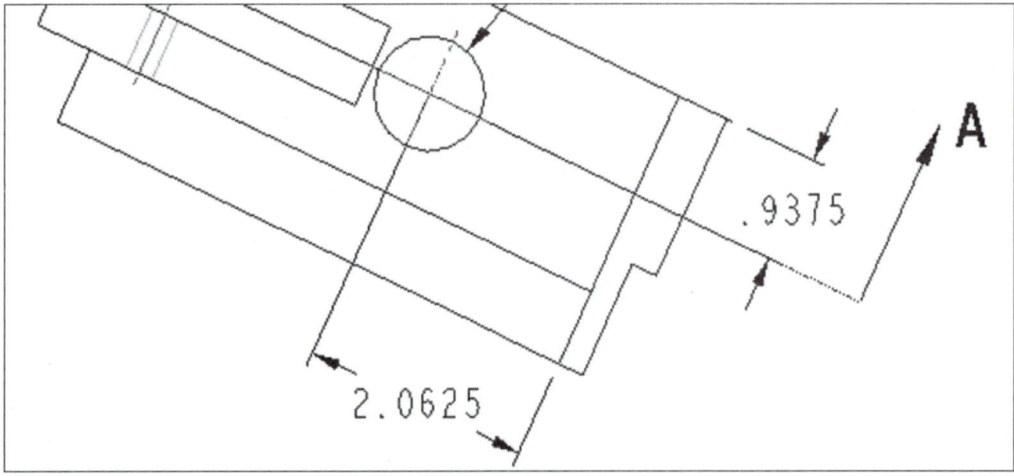

Figure 11.12(b) Section Arrows Length and Width Changed

Erase the **118.0** angle dimension from the Front view. > Since the model and the drawing are associative, make a modification to the slot dimension. Pick on the slot dimension [Fig. 11.12(c)] > press **RMB** > **Modify Nominal Value** > *type* **3.125** [Fig. 11.12(d)] > **Enter** > **LMB** to deselect (dimension changes color) > **Review** tab > Regenerate Active Model > reposition dimensions, axes, datums, and notes [Fig. 11.12(e)]

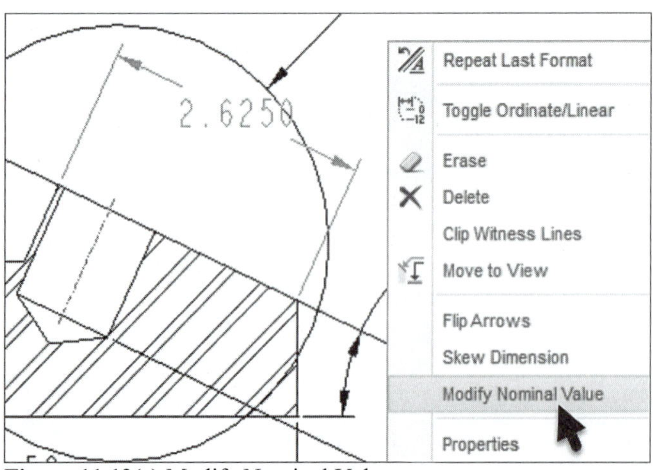

Figure 11.12(c) Modify Nominal Value

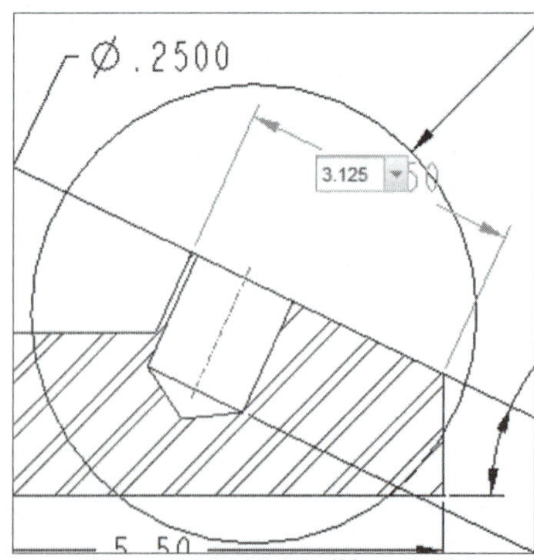

Figure 11.12(d) Modify to **3.125**

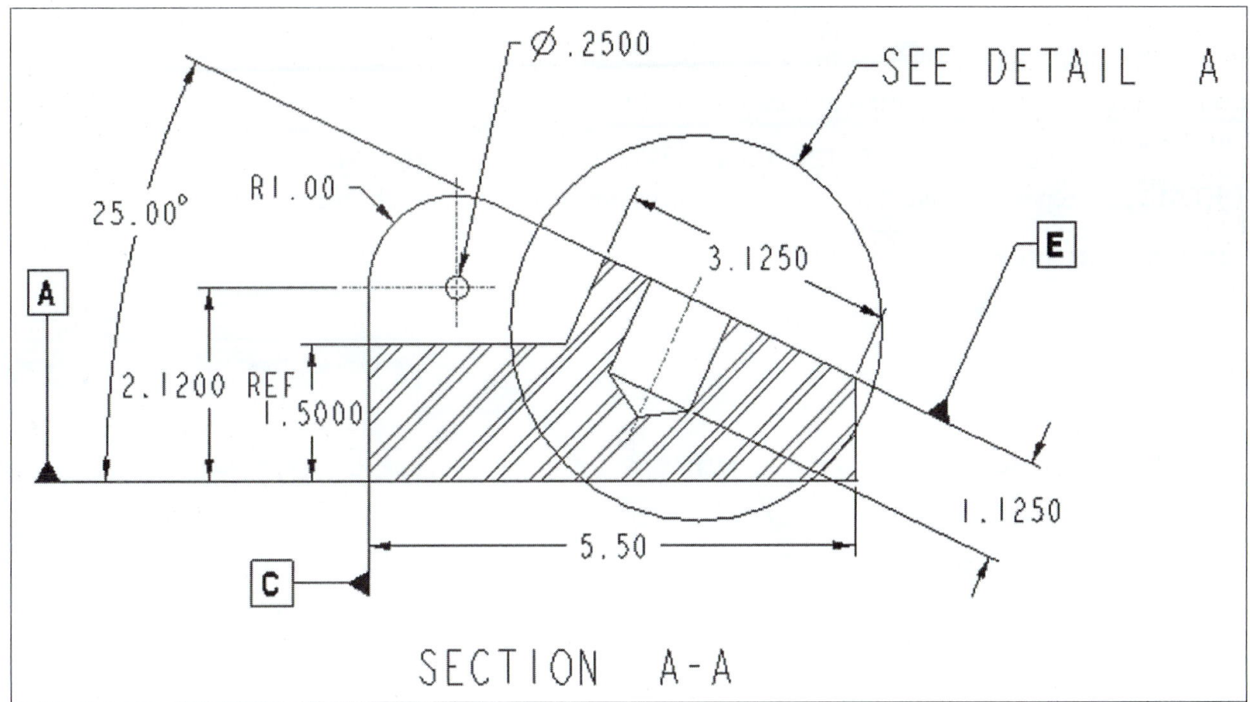

Figure 11.12(e) Regenerated Part and Drawing

Pick on the **1.125** hole location dimension in the Front section view [Fig. 11.13(a)] > press **RMB > Move to View** > pick on view DETAIL A [Fig. 11.13(b)]

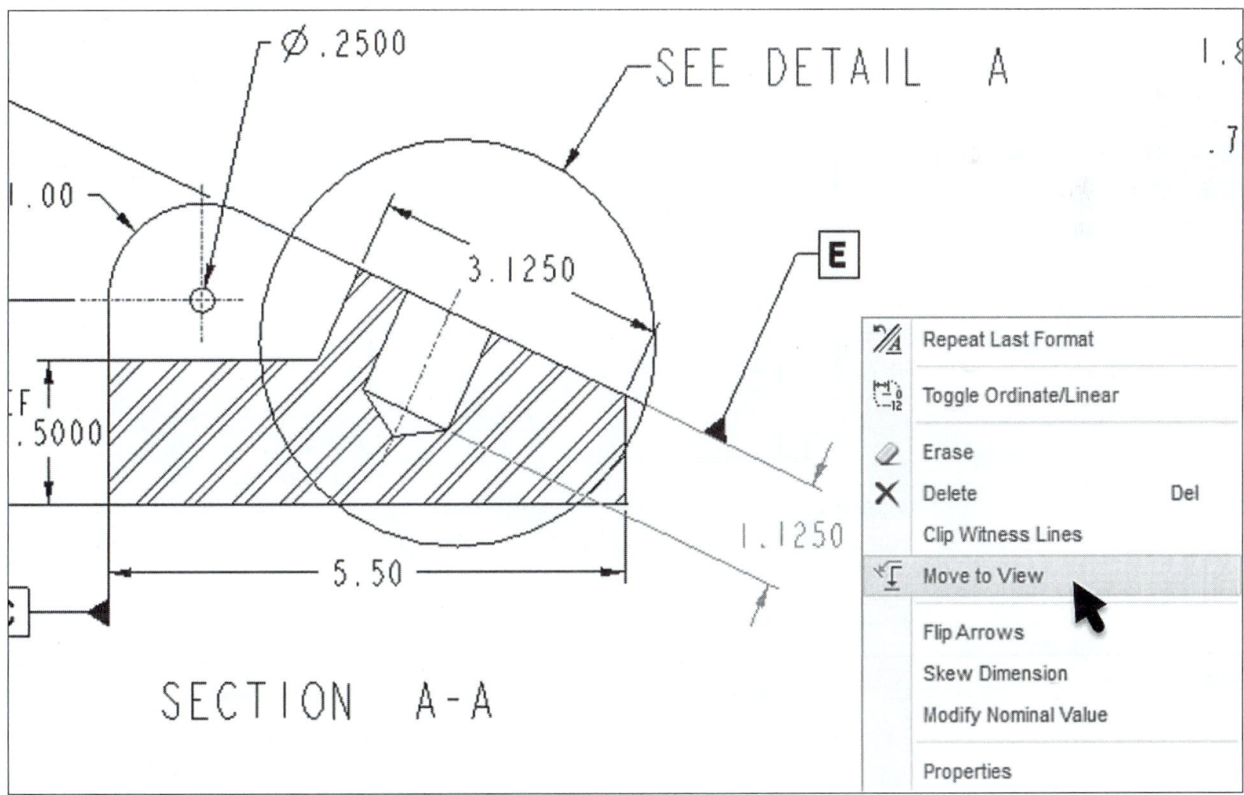

Figure 11.13(a) Select the **1.125** Hole Depth Dimension

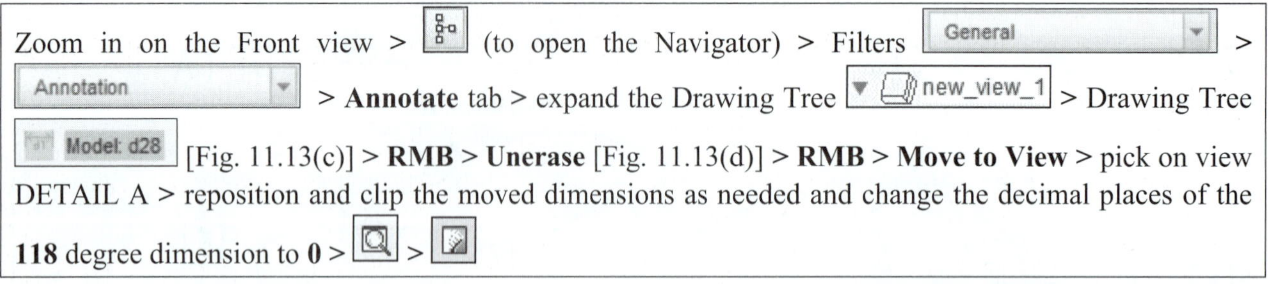

Zoom in on the Front view > (to open the Navigator) > Filters General > Annotation > **Annotate** tab > expand the Drawing Tree new_view_1 > Drawing Tree Model: d28 [Fig. 11.13(c)] > **RMB** > **Unerase** [Fig. 11.13(d)] > **RMB** > **Move to View** > pick on view DETAIL A > reposition and clip the moved dimensions as needed and change the decimal places of the **118** degree dimension to **0** > >

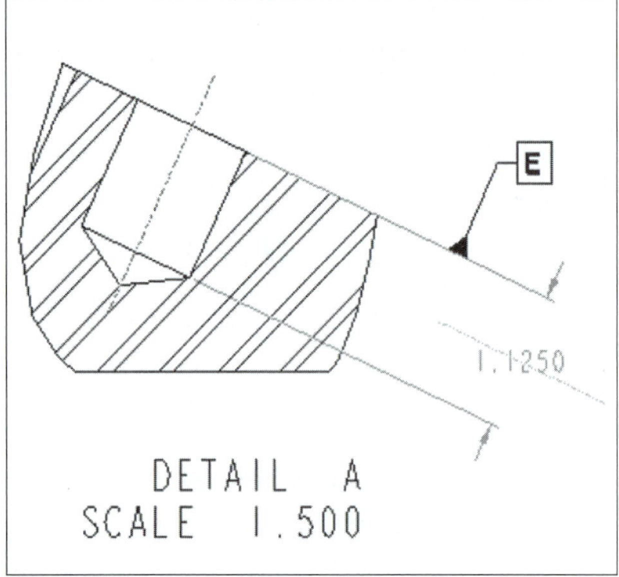

Figure 11.13(b) Move Dimension to DETAIL A

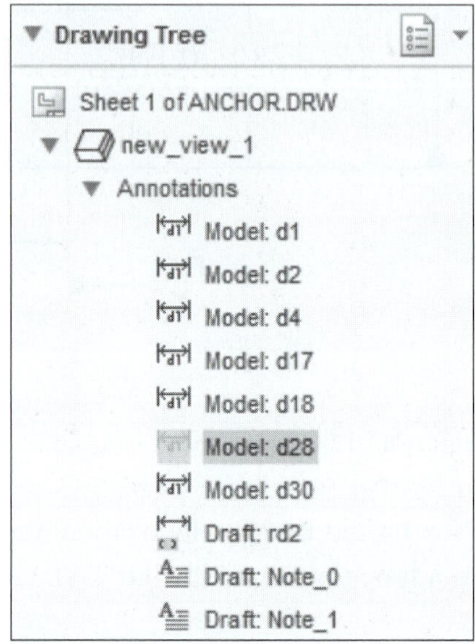

Figure 11.13(c) Drawing Tree

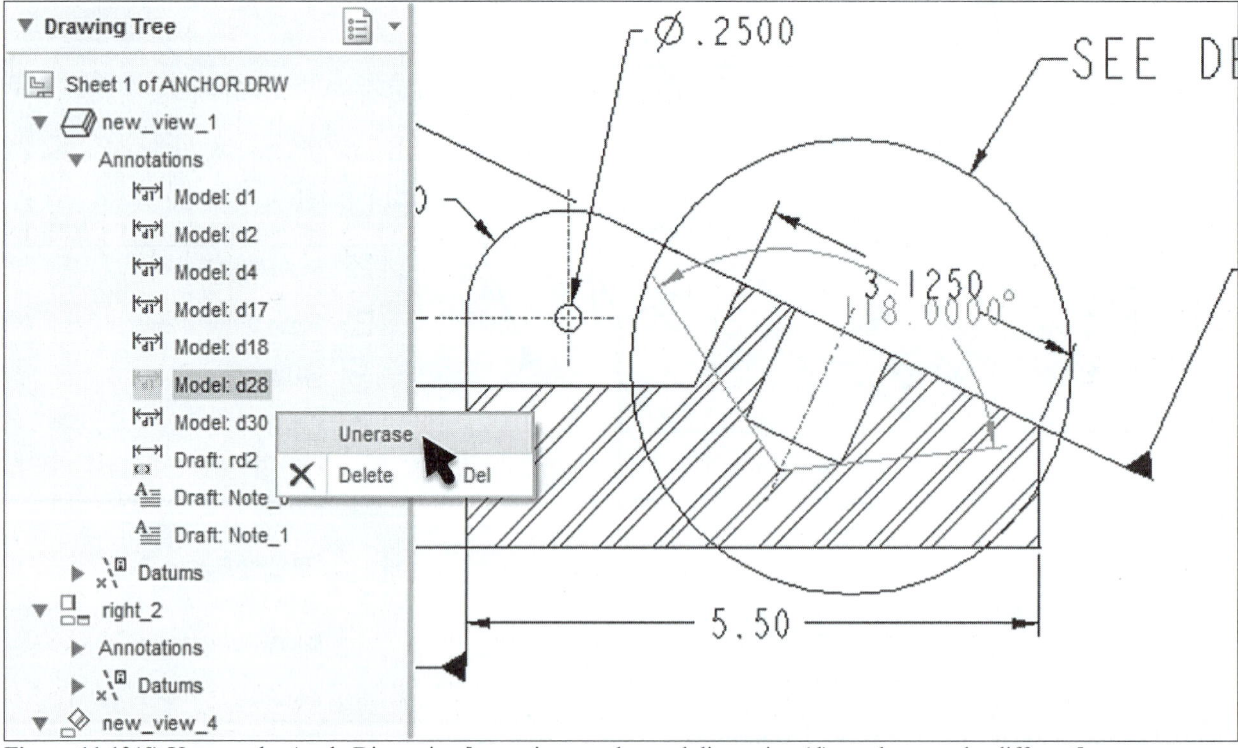

Figure 11.13(d) Unerase the Angle Dimension [your view number and dimension (d) number may be different]

Clean up your drawing views as necessary [Figs. 11.14(a-c)] > > > **File** > **Manage File** > **Delete Old Versions** > **Enter** > **LMB** to deselect > (to close the Navigator)

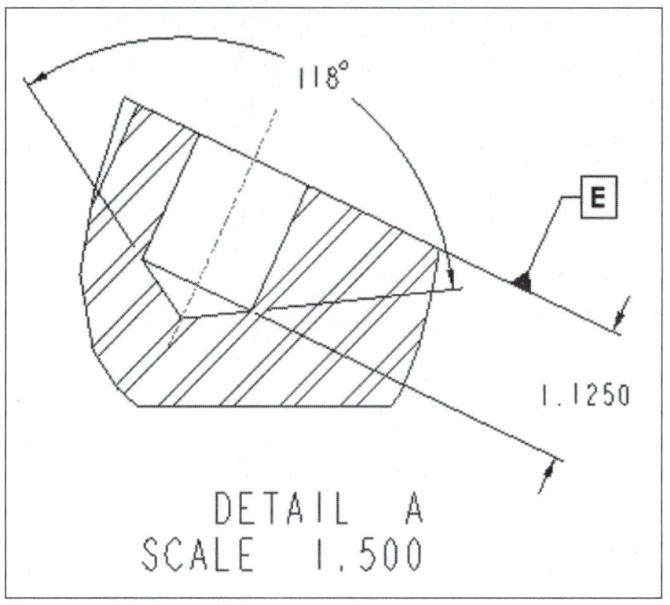

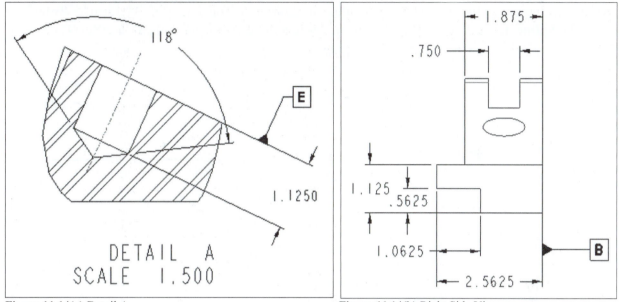

Figure 11.14(a) Detail A

Figure 11.14(b) Right Side View

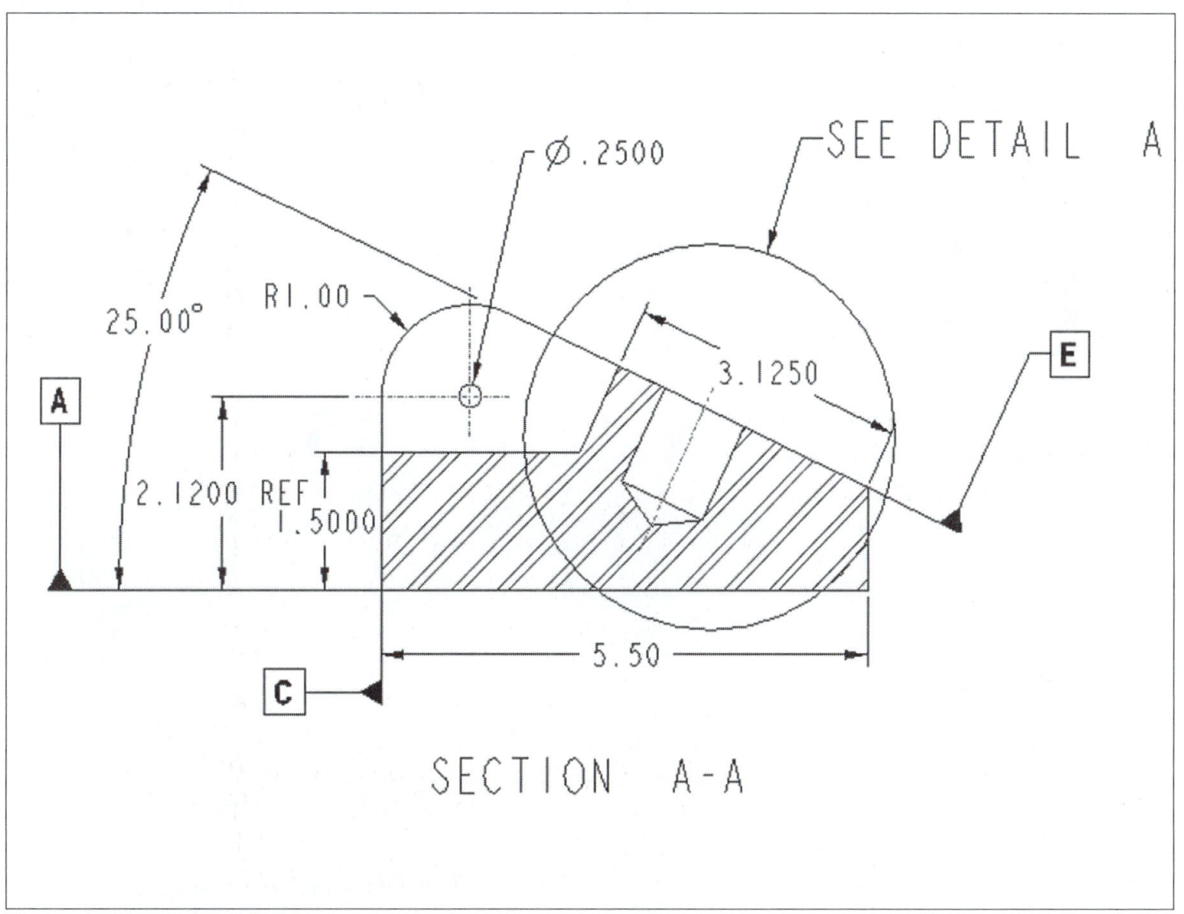

Figure 11.14(c) Section A-A

Next, you will change the boundary of view DETAIL A. > Filters Annotation General > **Layout** tab > pick on view DETAIL A > slightly move the mouse pointer > press **RMB** (if you press RMB inside the view outline, you get a pop-up list of options [Fig. 11.15(a)], whereas if you press RMB outside the view outline, there are fewer options) [Fig. 11.15(b)] > **Properties** > click the **Spline defined** button [Fig. 11.15(c)] > sketch the spline again in the Front (SECTION A-A) view [Fig. 11.15(d)] > **MMB** to end spline > **Apply**

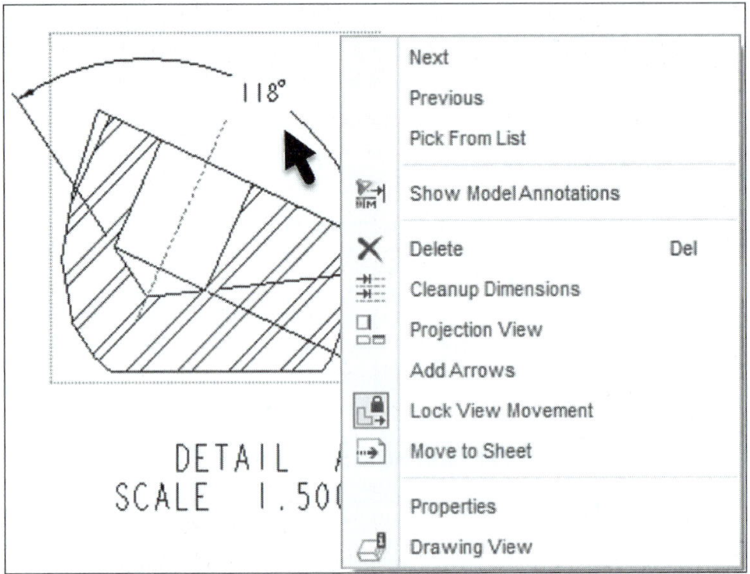

Figure 11.15(a) RMB Inside

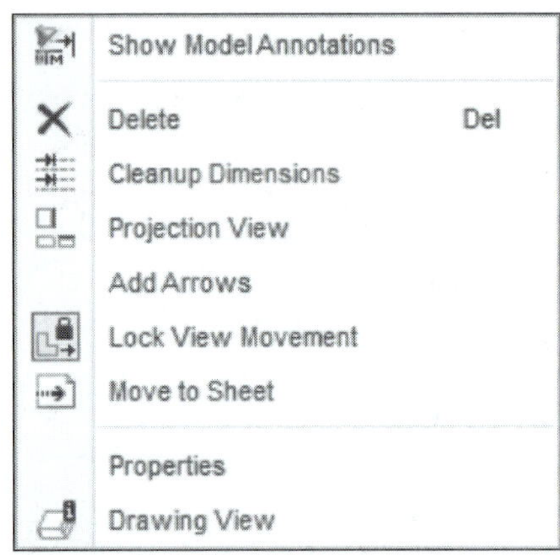

Figure 11.15(b) RMB Outside

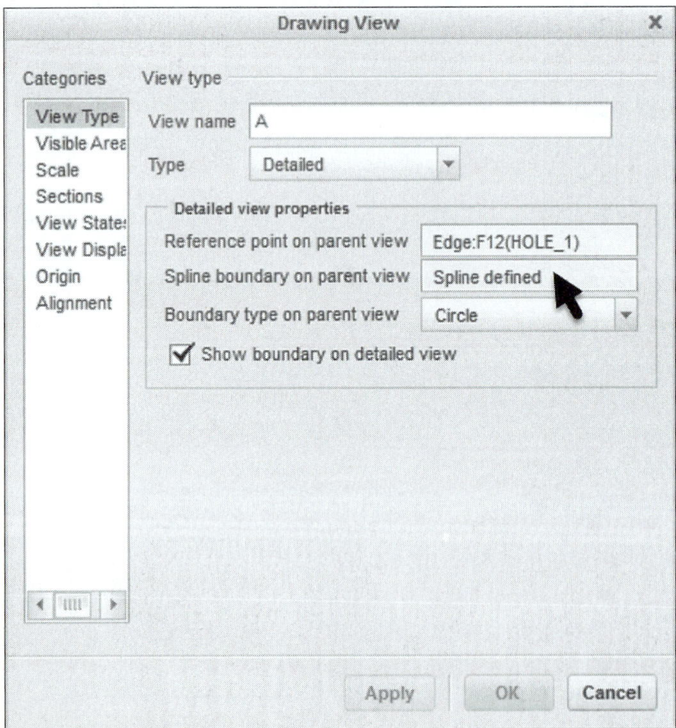

Figure 11.15(c) Select inside Spline Collector

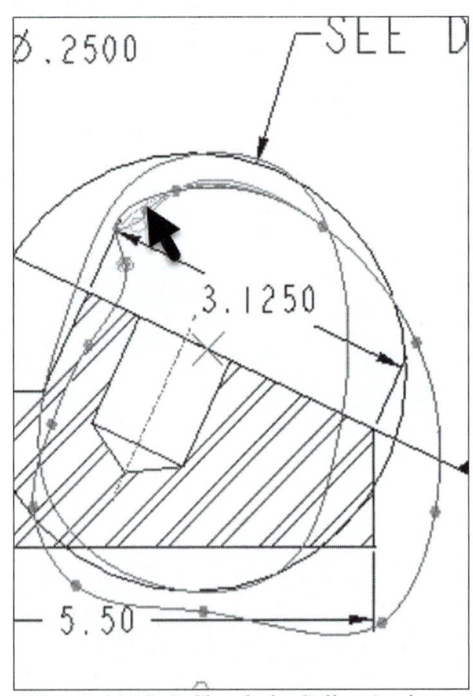

Figure 11.15(d) Sketch the Spline Again

Click: **Cancel** [Fig. 11.15(e)] > **LMB** to deselect > clean up both views [Figs. 11.15(f-g)] > **Annotate** tab > clean up the datums > **Layout** tab > [icon] > **Ctrl+S**

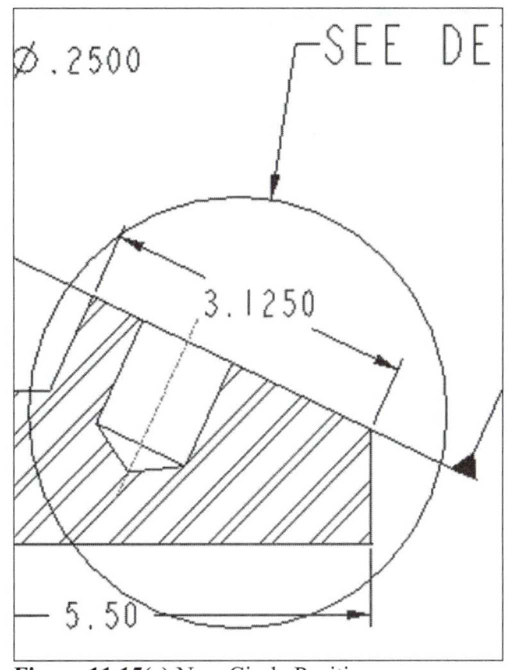

Figure 11.15(e) New Circle Position

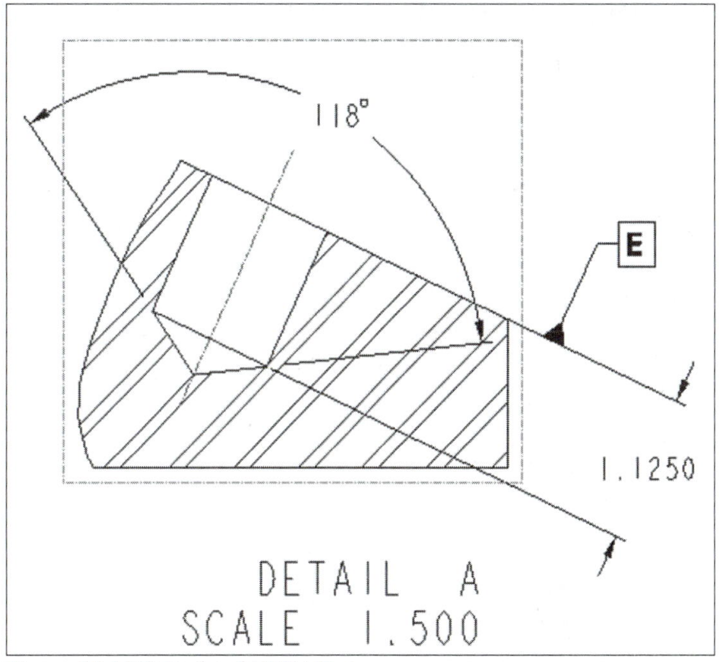

Figure 11.15(f) Updated DETAIL A

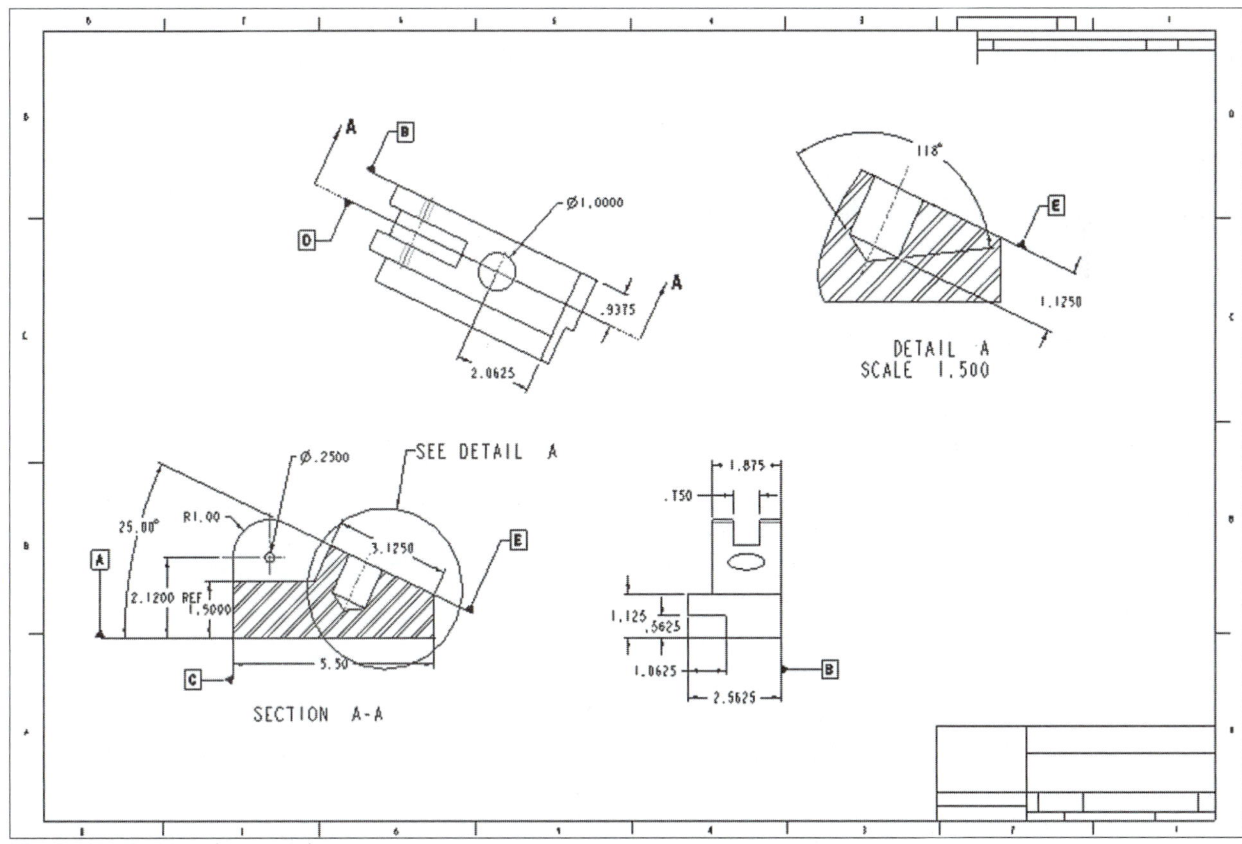

Figure 11.15(g) Anchor Drawing

Click: **Annotate** tab > pick on the **2.120 REF** dimension > press **RMB** > **Properties** > Decimal Places **3** > **Display** tab > remove the **REF** Suffix Suffix > add parentheses to the default text (@D) [Fig. 11.16(a)] > **OK** > **LMB** to deselect [Fig. 11.16(b)] > > > **Ctrl+S**

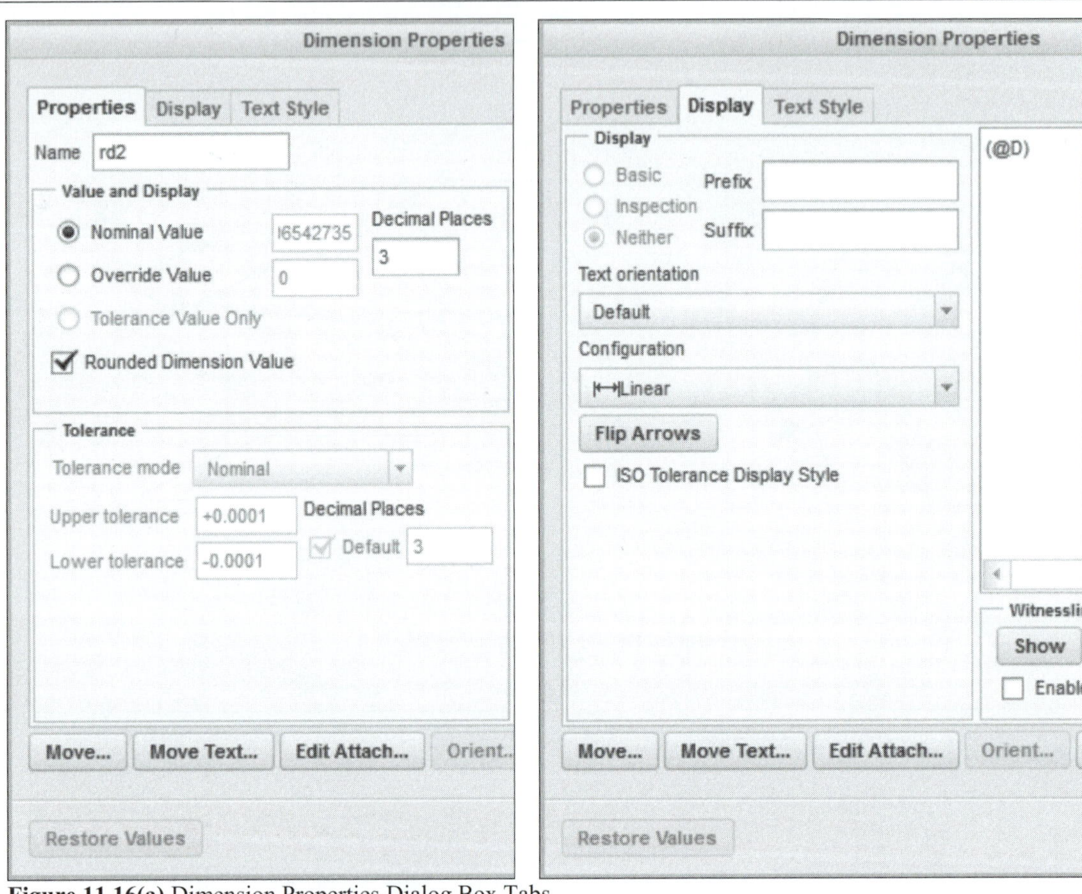

Figure 11.16(a) Dimension Properties Dialog Box Tabs

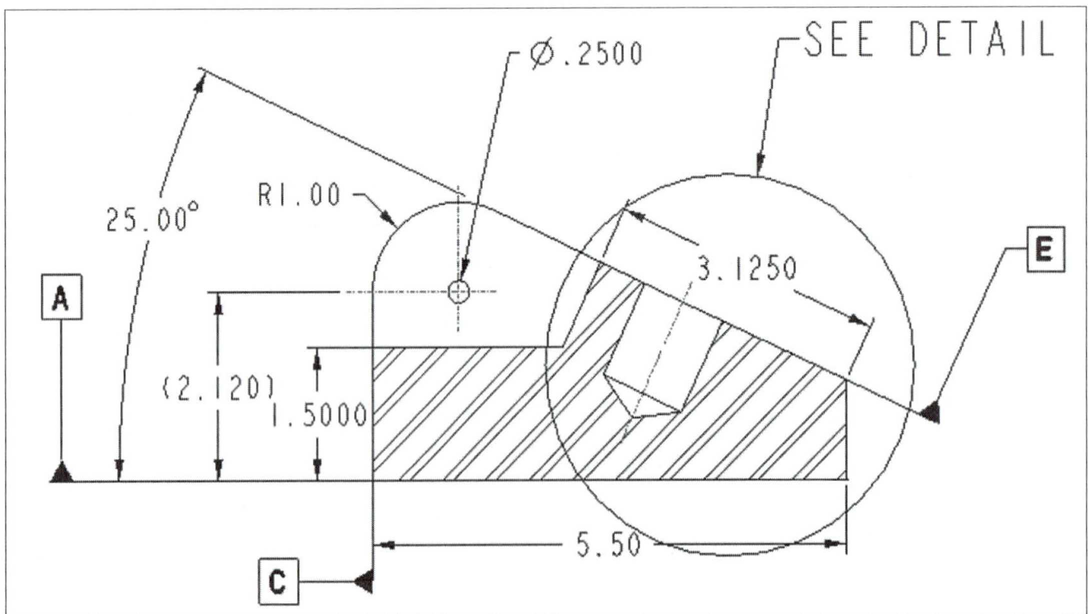

Figure 11.16(b) Correct ASME Standard Reference Dimension Style

Change the spacing and the angle of the section lining (make sure your filters are set to General). Click: **Layout** tab > double-click on the Xhatching in SECTION A-A [Fig. 11.17(a)] > **Delete line** > **Angle** > **30** > **Spacing** > **Double** > **Half** > **Done** [Fig. 11.17(b)] > **LMB** to deselect > edit the dimensions for proper decimal places and placement > [Fig. 11.17(c)] > **Ctrl+S**

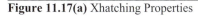

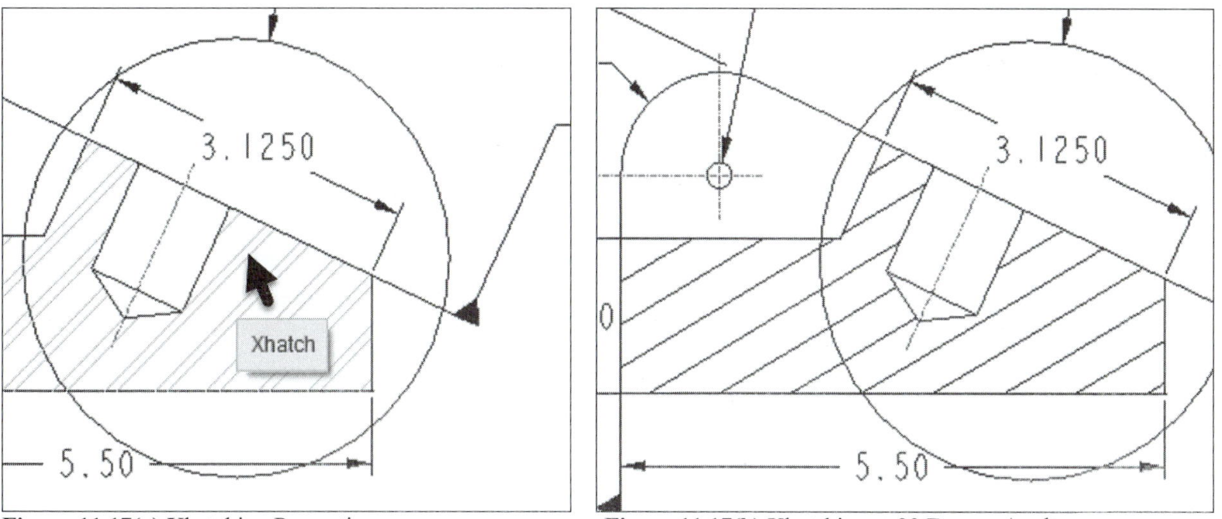

Figure 11.17(a) Xhatching Properties

Figure 11.17(b) Xhatching at **30** Degree Angle

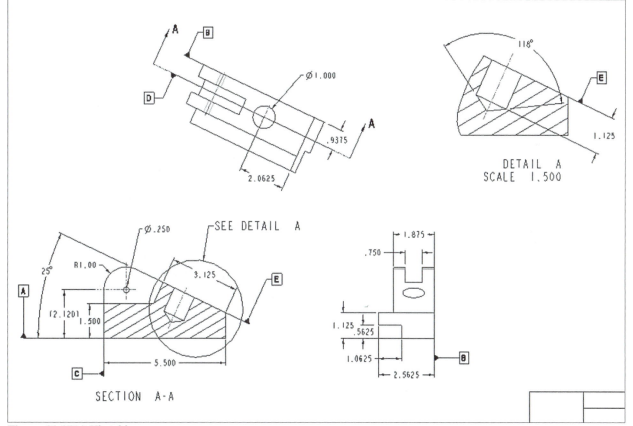

Figure 11.17(c) Xhatching

Change the spacing and the angle of the section lining in DETAIL A. Double-click on the Xhatching in view DETAIL A [Fig. 11.18(a)] > **Det Indep** > **Hatch** > **Spacing** > **Value** > **.75** > **Angle** > **45** > **Done** [Fig. 11.18(b)] > **LMB** to deselect > **Review** tab > **Update Sheets** from the Ribbon > > >

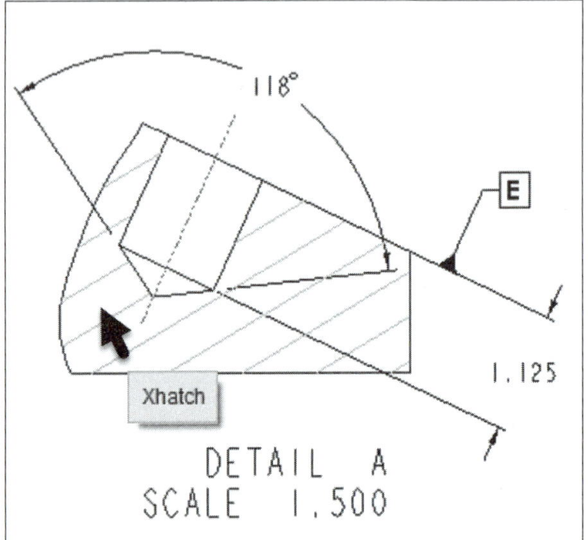

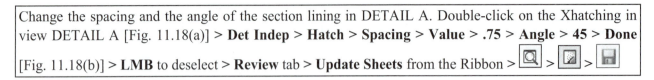

Figure 11.18(a) Detail Xhatching Properties

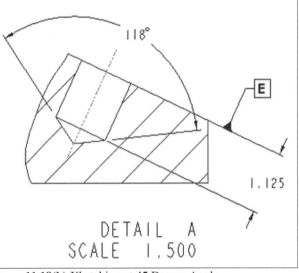

Figure 11.18(b) Xhatching at **45** Degree Angle

Change the text style used on the drawing. Click: **Annotate** tab > enclose all of the drawing text and dimensions with a selection box [Fig. 11.19(a)] > press **RMB** > **Text Style**

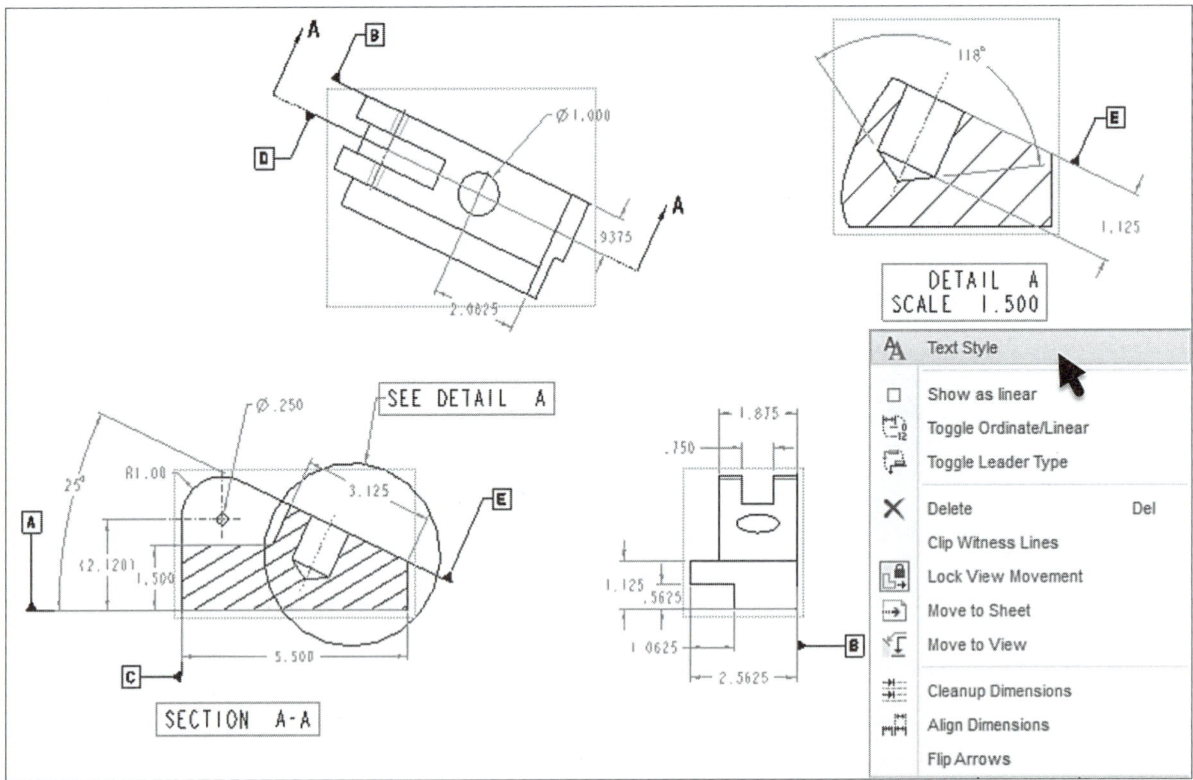

Figure 11.19(a) Select all Text and Dimensions

Click: Font ▾ **Blueprint MT** [Fig. 11.19(b)] > **Apply** > **OK** > **LMB** to deselect > change the number of decimal places for dimensions as required [Fig. 11.19(c)] > > >

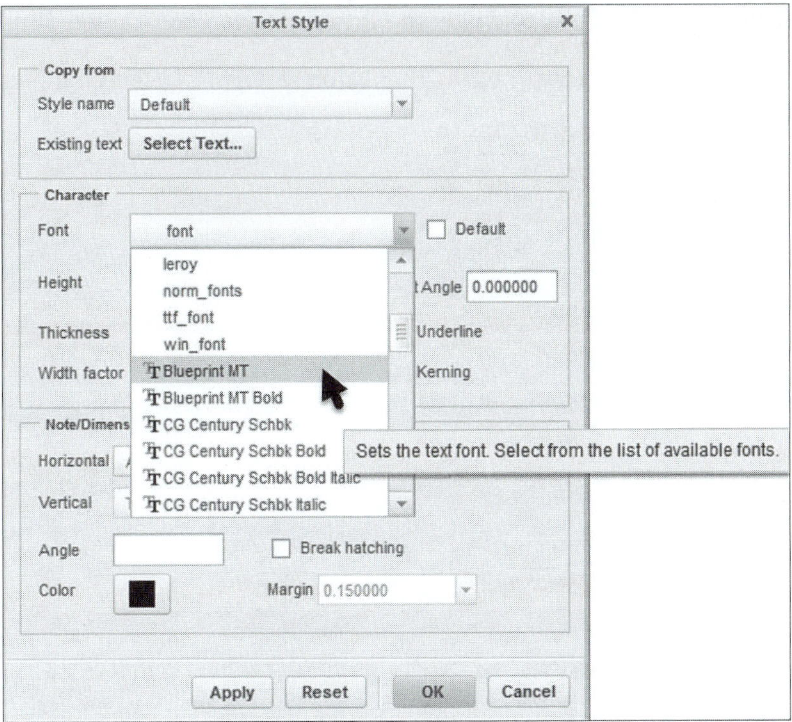

Figure 11.19(b) Character- Font- Blueprint MT

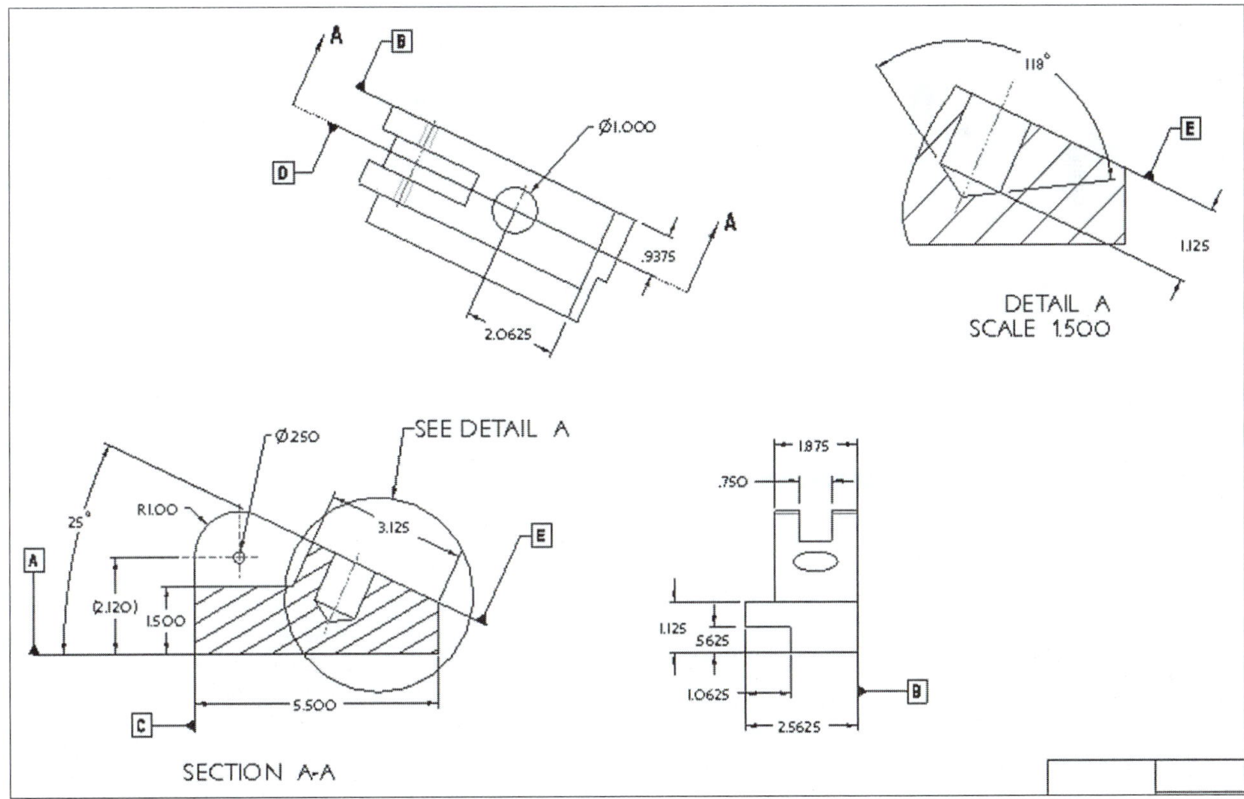

Figure 11.19(c) New Text Style

Add a geometric tolerance to the angled surface. Click: **Create geometric tolerances** > **Angularity** > Type **Datum** [Fig. 11.20(a)] > pick on datum **E** [Fig. 11.20(b)] > **Datum Refs** tab > **Primary** tab > Basic > pick on datum **A** in the Front view as the Primary Reference Datum [Fig. 11.20(c)] > **Tol Value** tab > Overall Tolerance 0.005 [Fig. 11.20(d)] > **Enter** > **OK** > in the Graphics Window, **LMB** to deselect [Fig. 11.20(e)] > **Ctrl+S**

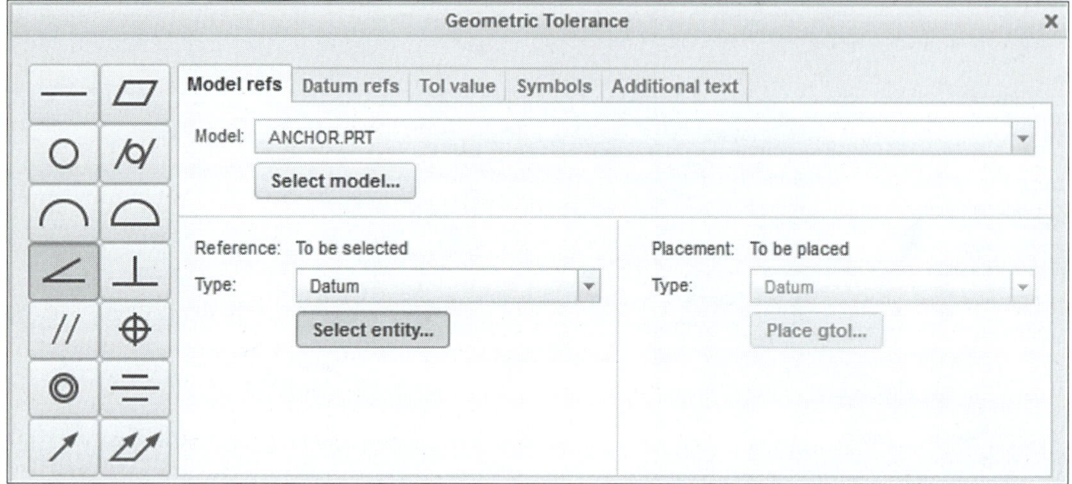

Figure 11.20(a) Angularity, Type Datum

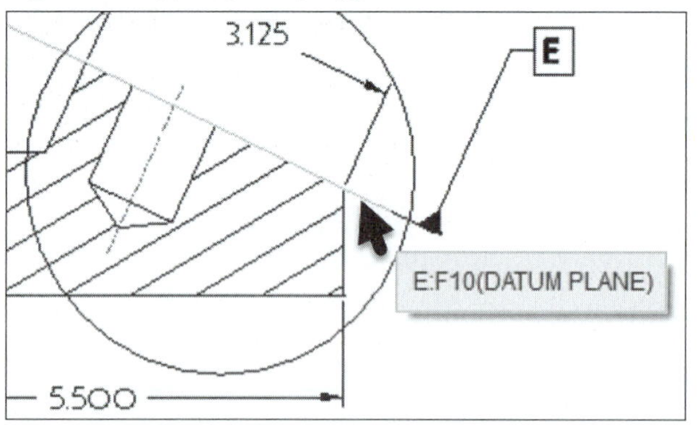

Figure 11.20(b) Pick Datum E

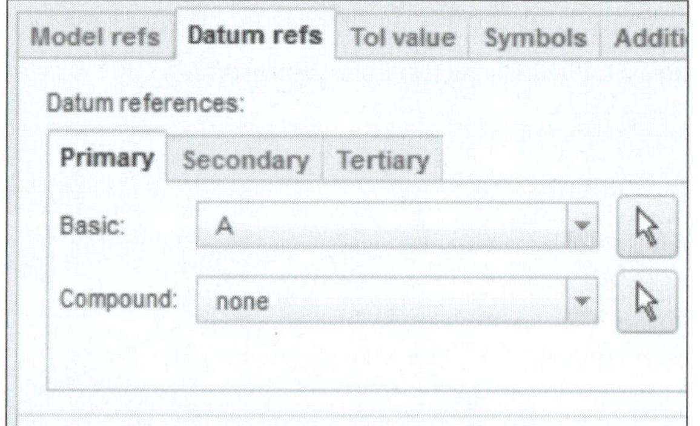

Figure 11.20(c) Primary Reference Datum A

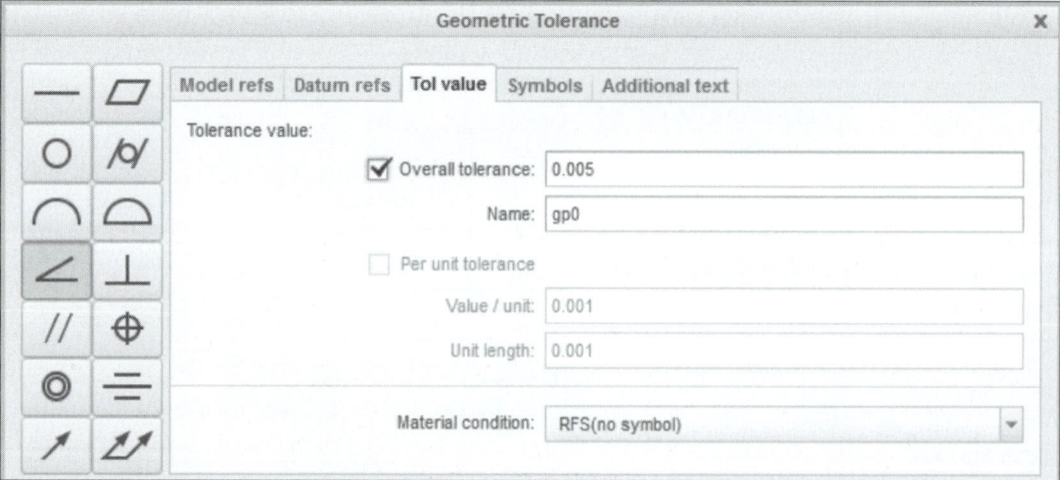

Figure 11.20(d) Overall Tolerance Value **.005**

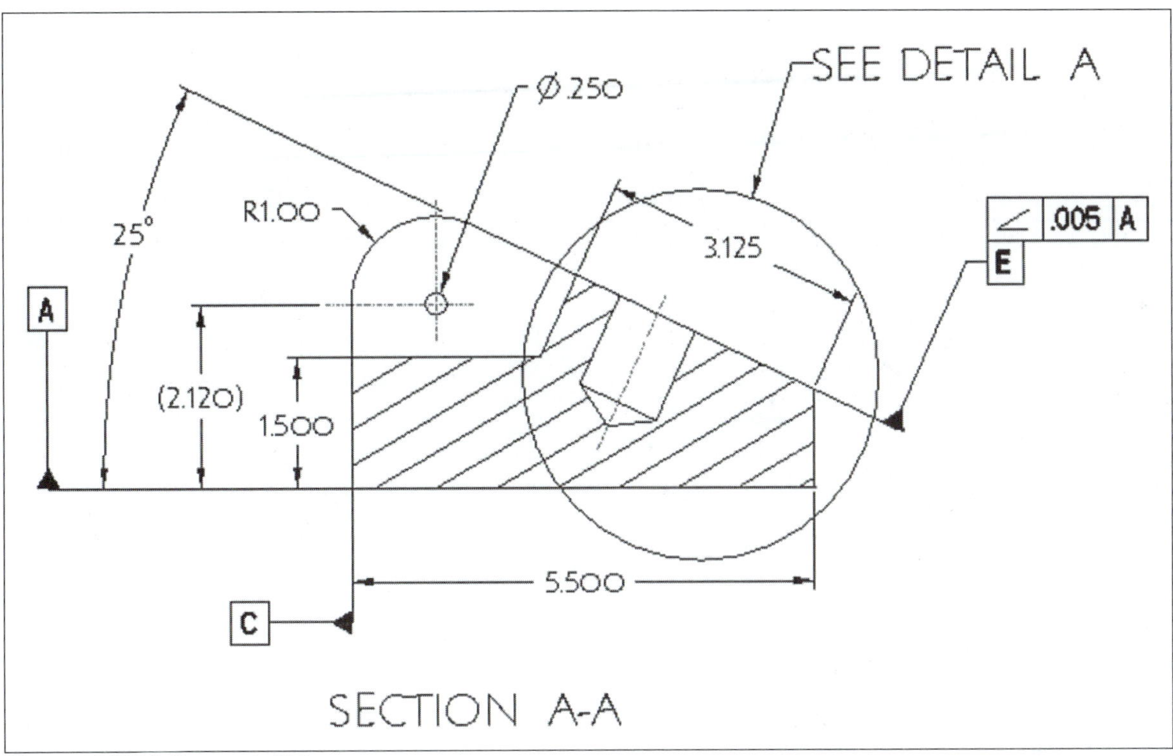

Figure 11.20(e) Geometric Tolerance

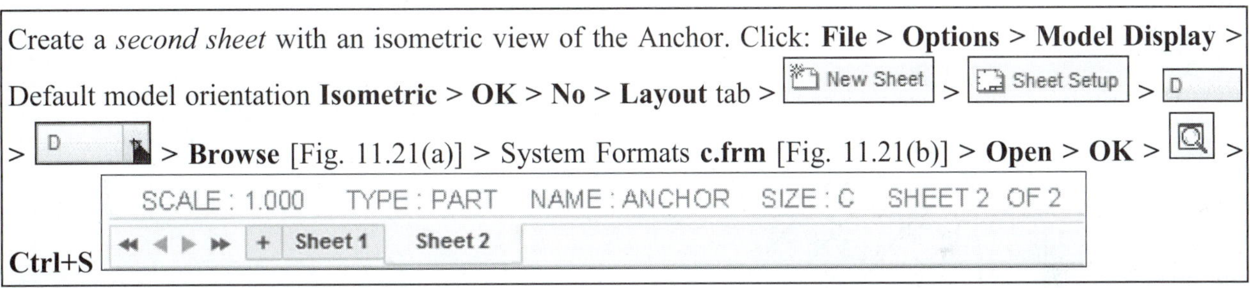

Create a *second sheet* with an isometric view of the Anchor. Click: **File** > **Options** > **Model Display** > Default model orientation **Isometric** > **OK** > **No** > **Layout** tab > New Sheet > Sheet Setup > D > D > **Browse** [Fig. 11.21(a)] > System Formats **c.frm** [Fig. 11.21(b)] > **Open** > **OK** > > **Ctrl+S**

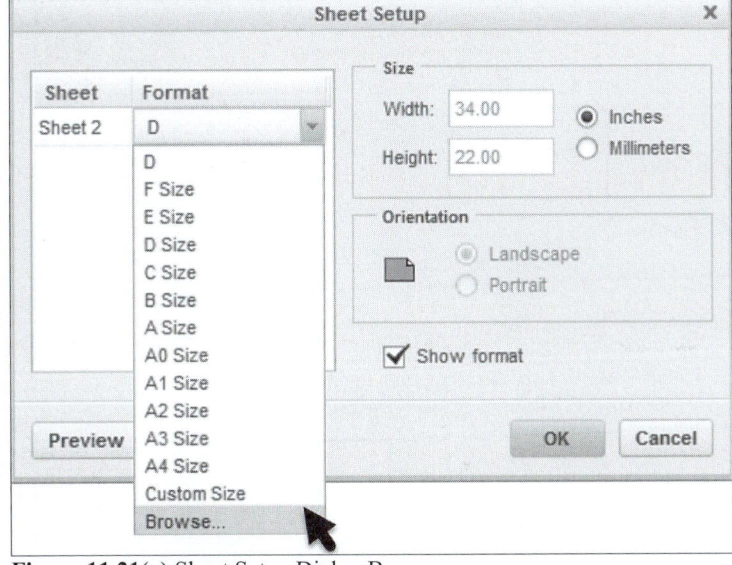

Figure 11.21(a) Sheet Setup Dialog Box

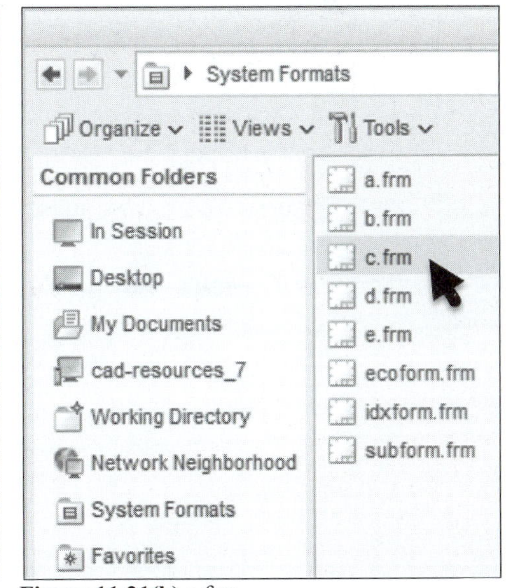

Figure 11.21(b) c.frm

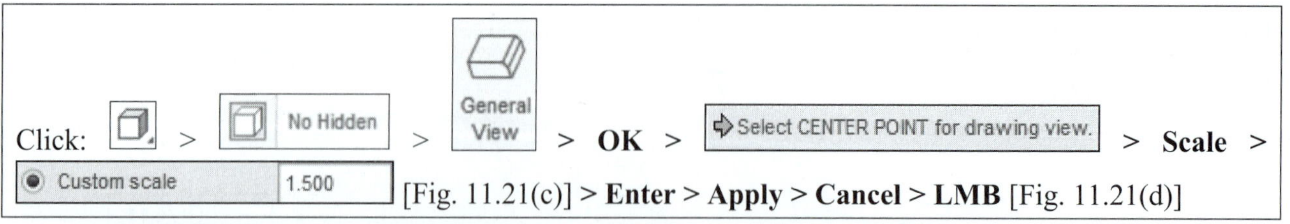

Click: > No Hidden > General View > OK > Select CENTER POINT for drawing view. > Scale >
Custom scale 1.500 [Fig. 11.21(c)] > Enter > Apply > Cancel > LMB [Fig. 11.21(d)]

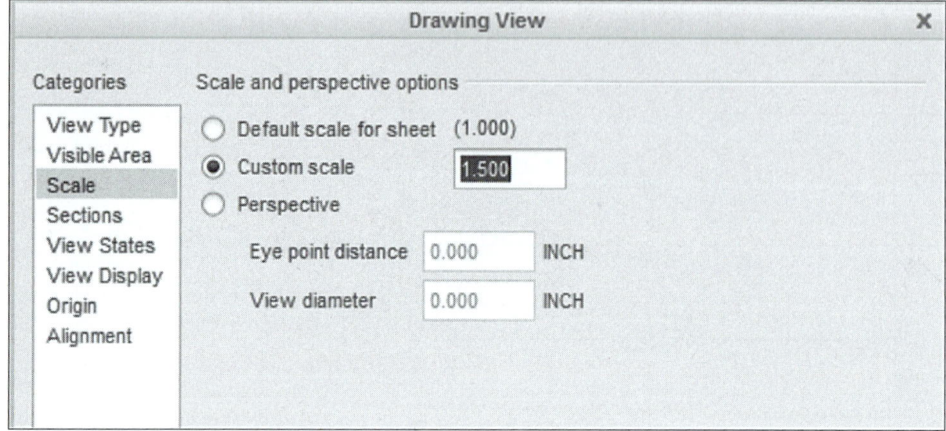

Figure 11.21(c) Drawing View Custom Scale

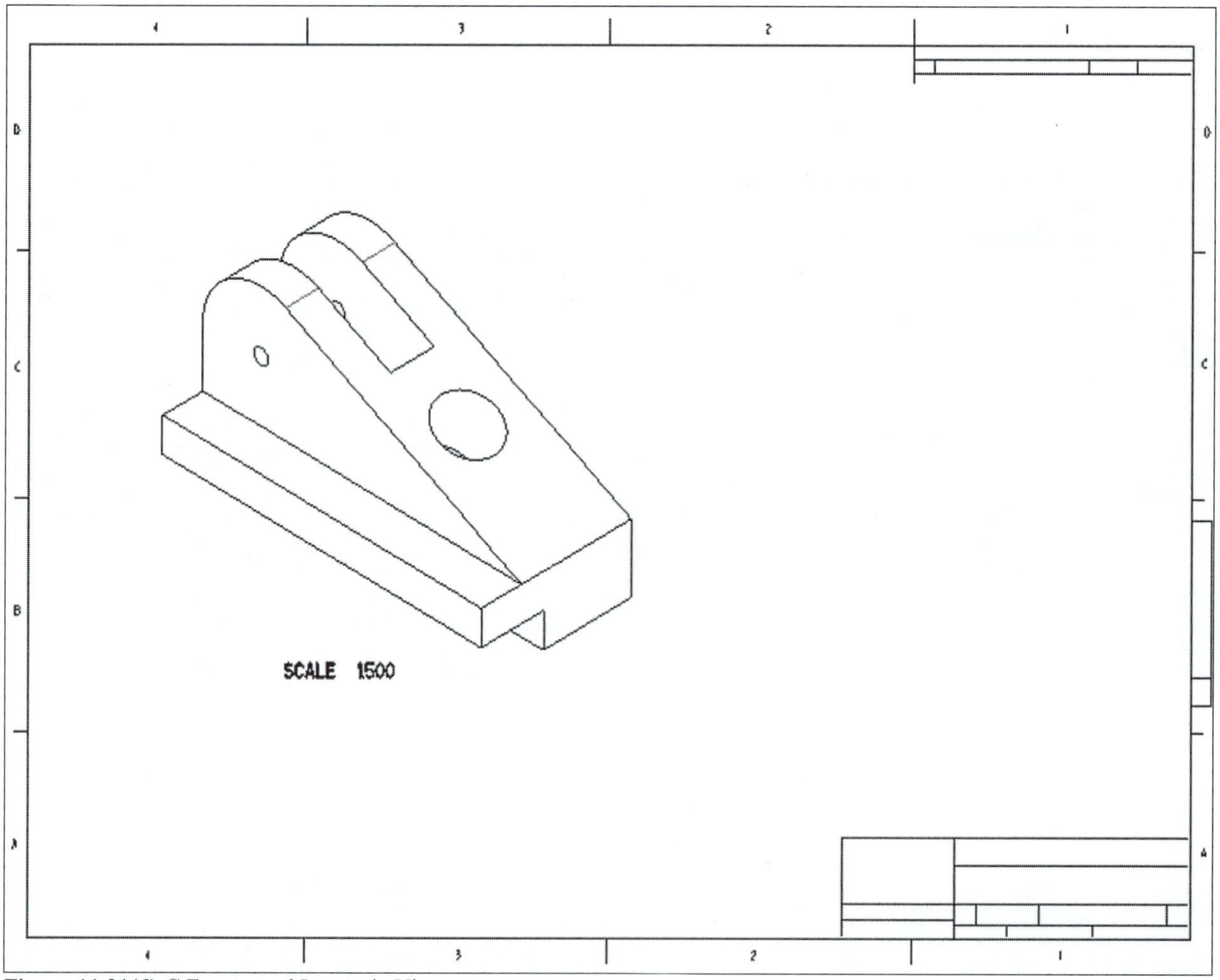

Figure 11.21(d) C Format and Isometric View

Double-click on the pictorial view > **View Display** > Tangent edges display style **None** [Fig. 11.21(e)] > **OK** > **LMB** to deselect > Sheet 1 [Fig. 11.21(f)] > [icon] > [icon]

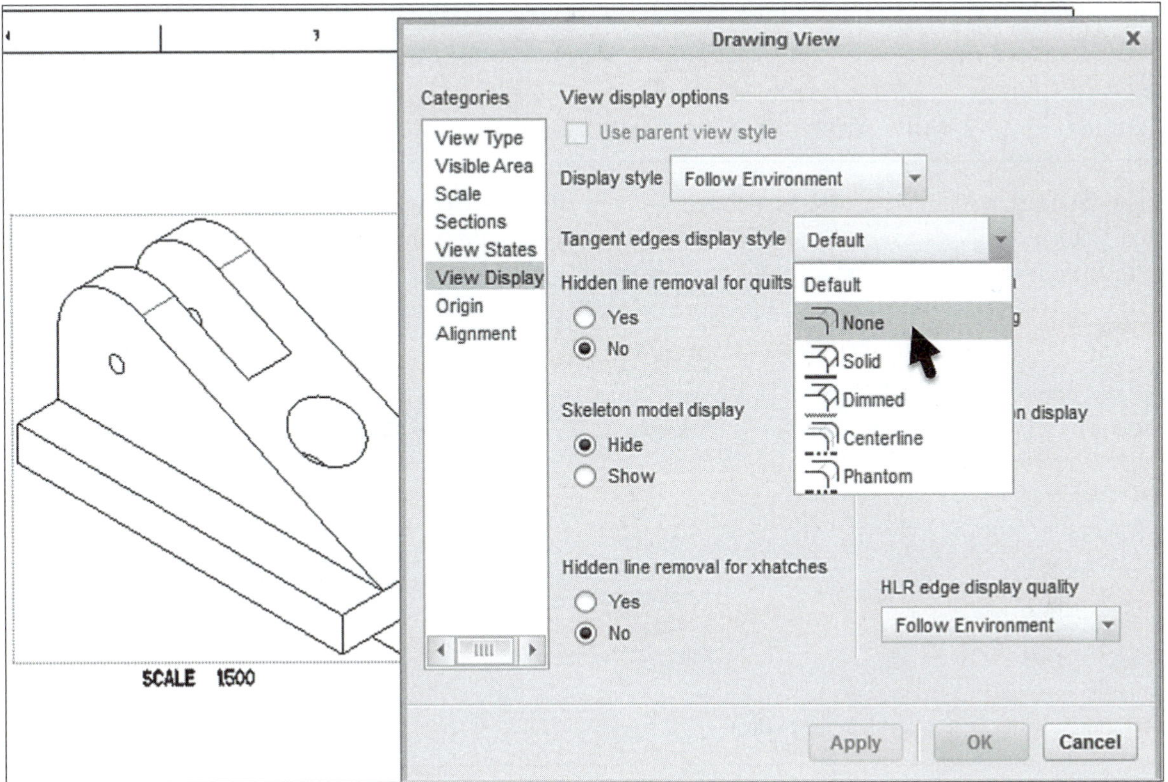

Figure 11.21(e) Drawing View Dialog Box

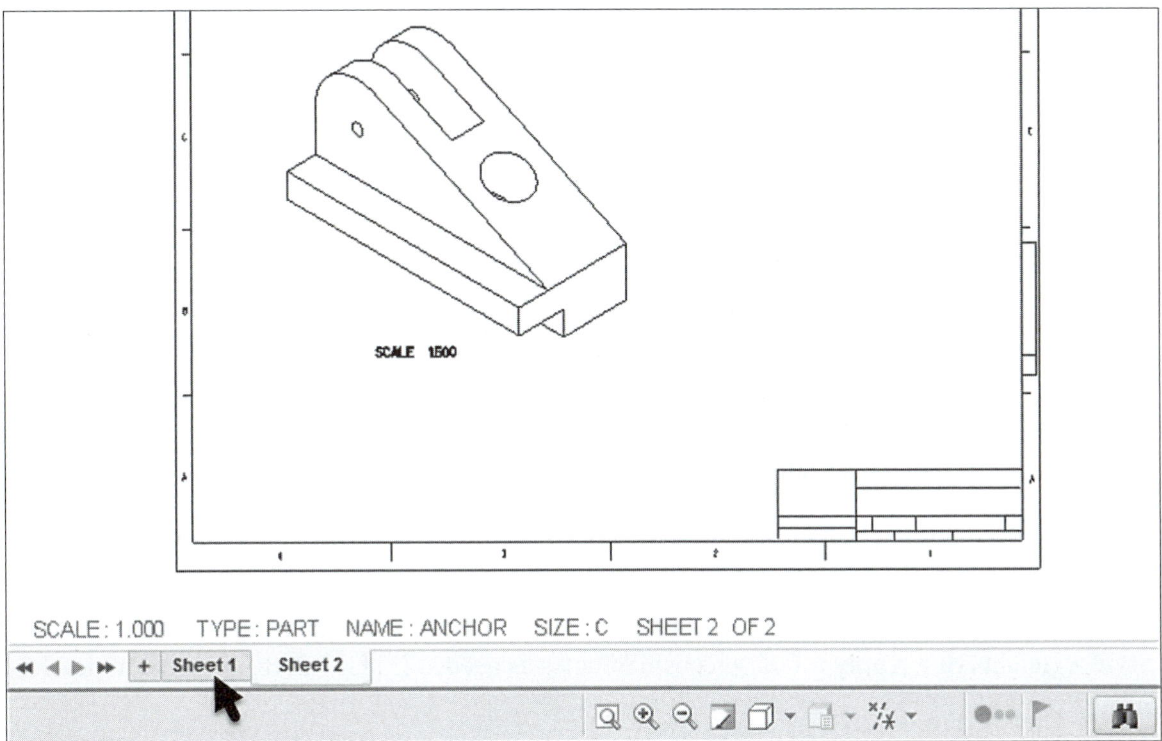

Figure 11.21(f) Anchor Drawing Sheet 1

Click: Open an existing model > System Formats > d.frm > **Open** > **File** > **Save As** > **Save a Copy** > *type a unique name for your format:* New Name **DETAIL_FORMAT_D** > **OK** > **File** > **Close** > > pick **detail_format_d.frm** > **Open** > **File** > **Prepare** > **Drawing Properties** > **change** > Sort: **Alphabetical** > ***text_height .25*** > **Add/Change** > ***default_font filled*** > **Add/Change** > ***arrow_style filled*** > **Add/Change** > ***draw_arrow_length .25*** > **Add/Change** > ***draw_arrow_width .08*** > **Add/Change** [Fig. 11.22(a)] > **Apply** > **Close** > **Close** > **Ctrl+S** > **OK**

The **format** will have a **.dtl** associated with it, and the **drawing** will have a different **.dtl** file associated with it. *They are separate **.dtl** files.* When you activate a drawing and then add a format, the **.dtl** for the format controls the font, etc. for the format only.

Active Drawing				
arrow_style	filled	closed	✳	Controls style of arrow head for all detail iter
default_bom_balloon_type	simple_circle		●	Determines the default type of BOM balloon v
default_font	filled	font.ndx	✳	Specifies the default text font. Include the ex
default_table_column_width	10		●	Specifies the default value of the column wic
default_table_columns	4		●	Specifies the default number of columns of n
default_table_row_height	1		●	Specifies the default value of the row height
default_table_rows	2		●	Specifies the default number of rows of new
draft_scale	1.000000 *	1.000000	●	Determines value of draft dimensions relative
draw_arrow_length	.25	0.187500	✳	Sets length of leader line arrows.
draw_arrow_width	.08	0.062500	✳	Sets width of leader line arrows. Drives thes
draw_attach_sym_height	DEFAULT *	default	●	Sets height of leader line slashes, integral sig

Figure 11.22(a) Drawing Option **.dtl** File for Format

Zoom into the title block region. Create notes for the title text and parameter text. Click: **File** > **Options** > **Sketcher** > ☑ Snap to grid > **OK** > **No** > **Sketch** tab > **Draft Grid** > **Show Grid** > **Grid Params** > **X&Y Spacing** > *type* **.1** > **Enter** > **Done/Return** > **Done/Return** > **Annotate** tab > (all input is case sensitive)

Note > pick a point for the note in the largest area of the title block [Fig. 11.22(b)] > *type* **TOOL ENGINEERING CO.** > **LMB** > **LMB** >

Note > pick a point > *type* **DRAWN** > **LMB** > **LMB** >

Note > pick a point > *type* **ISSUED** > **LMB** > **LMB** >

Note > pick a point > *type* **&dwg_name** > **LMB** > **LMB** >

Note > pick a point > *type* **&scale** > **LMB** > **LMB** >

Note > pick a point > *type* **SHEET ¤t_sheet OF &total_sheets** > **LMB** > **LMB** >

File > **Options** > **Sketcher** > ☐ Snap to grid > **OK** > **No** > select the four text items to be smaller > **RMB** > **Text Style** > Height **.10** > **Apply** > **OK** > reposition notes as needed [Fig. 11.22(b)] > **LMB** to deselect > > > **Ctrl+S** [Fig. 11.22(c)] > **File** > **Close**

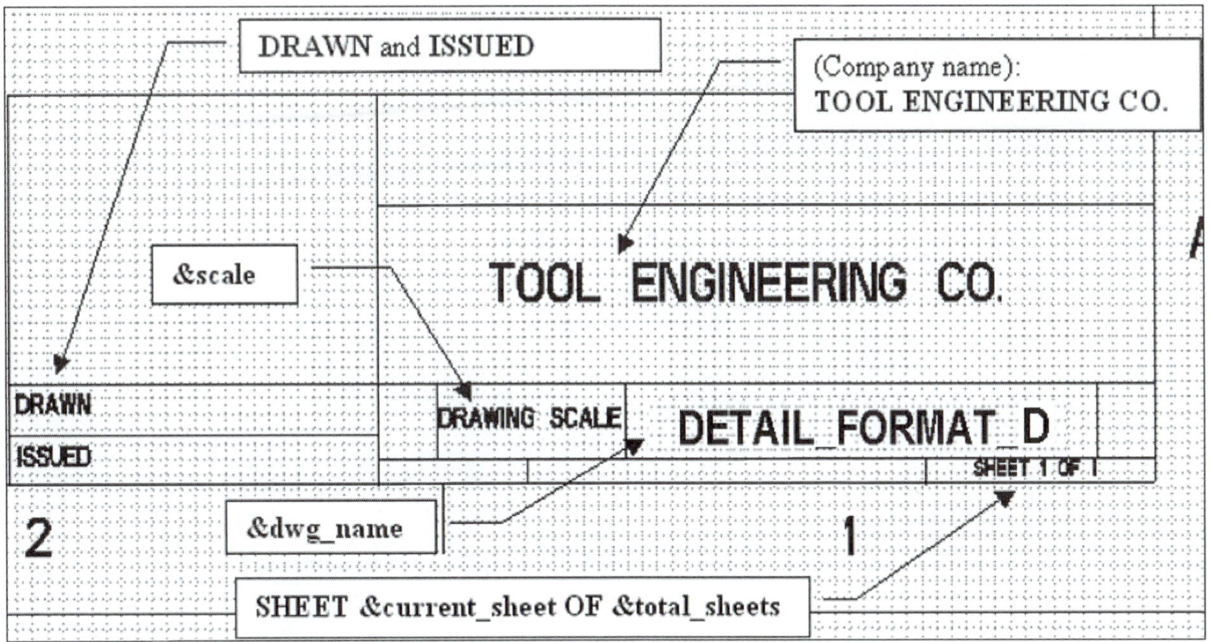

Figure 11.22(b) Parameters and Labels in the Title Block, Smaller Text is **.10** in Height

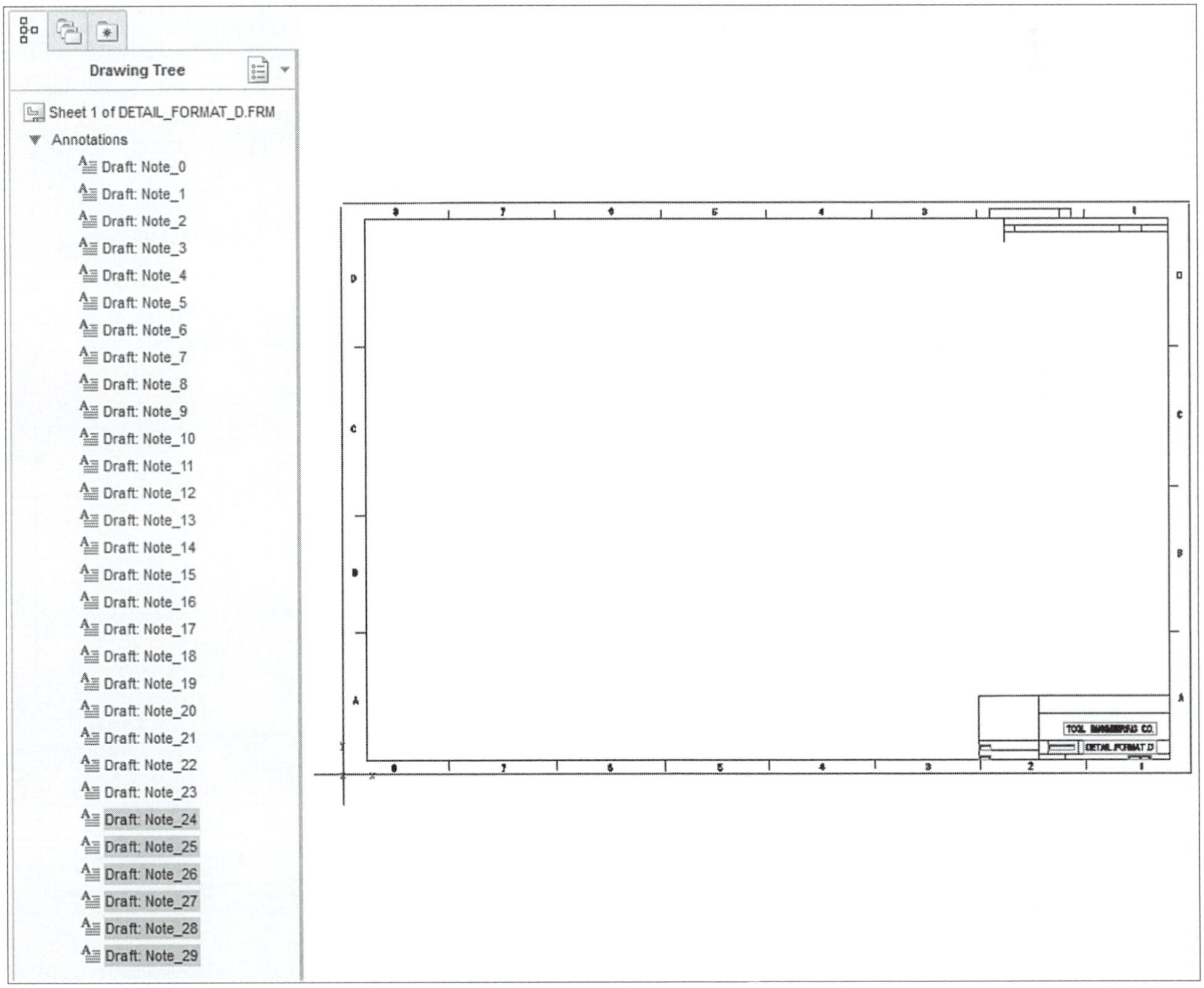

Figure 11.22(c) Format Notes

Press: **Layout** tab > **Sheet Setup** > [D] > [D ▾] > **Browse** > **Working Directory** > **detail_format_d.frm** > **Open** > **OK** [Fig. 11.22(d)] > [icon] > [icon] > [icon] [Fig. 11.22(e)] > **File** > **Save As** > Type [▾] > **Zip File (*.zip)** > **OK** > upload to your course interface or attach to an email and send to your instructor and yourself > **File** > **Close** > **File** > **Exit** > **Yes**

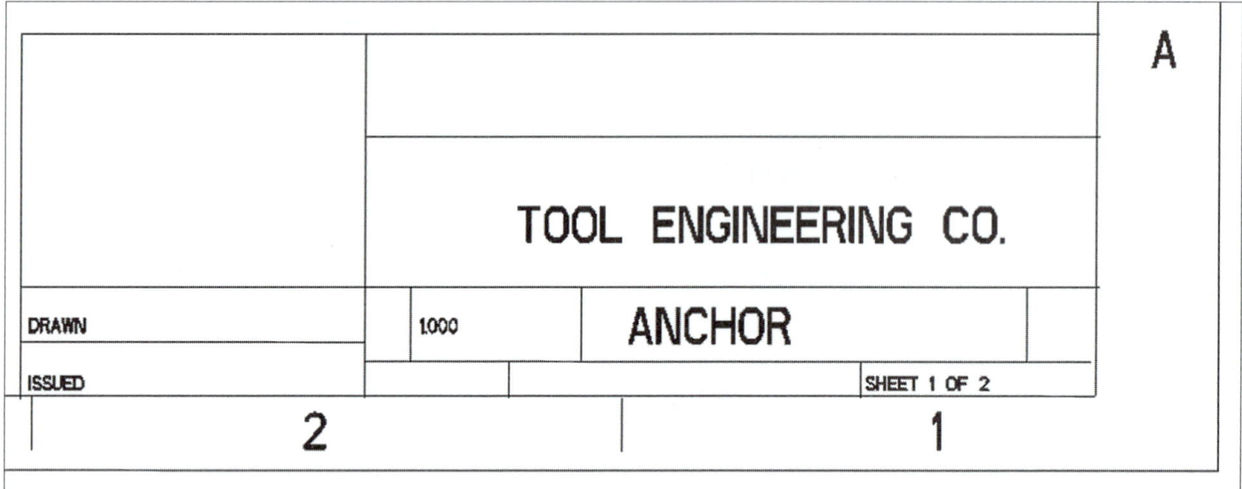

Figure 11.22(d) Title Block

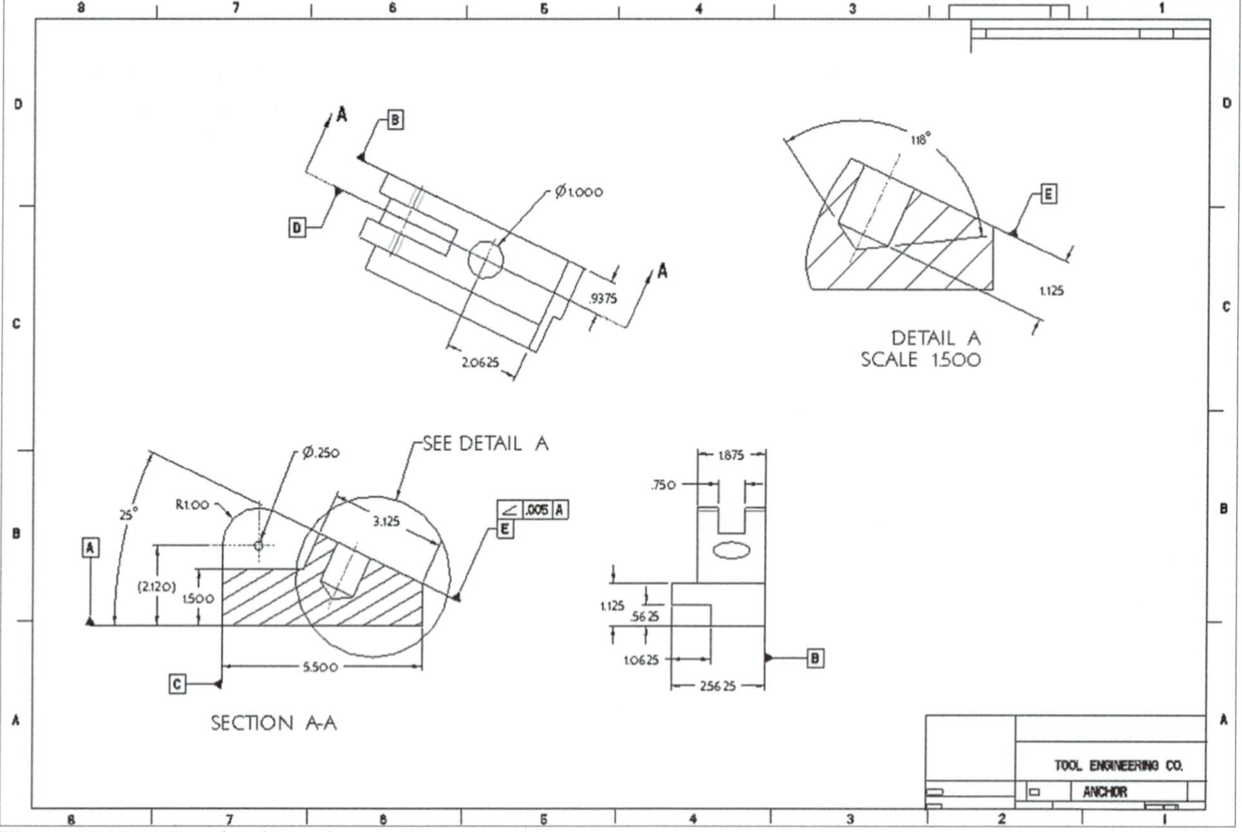

Figure 11.22(e) Completed Drawing

Projects without instructions are available at *www.cad-resources.com.*

Lesson 12 Assembly Drawings

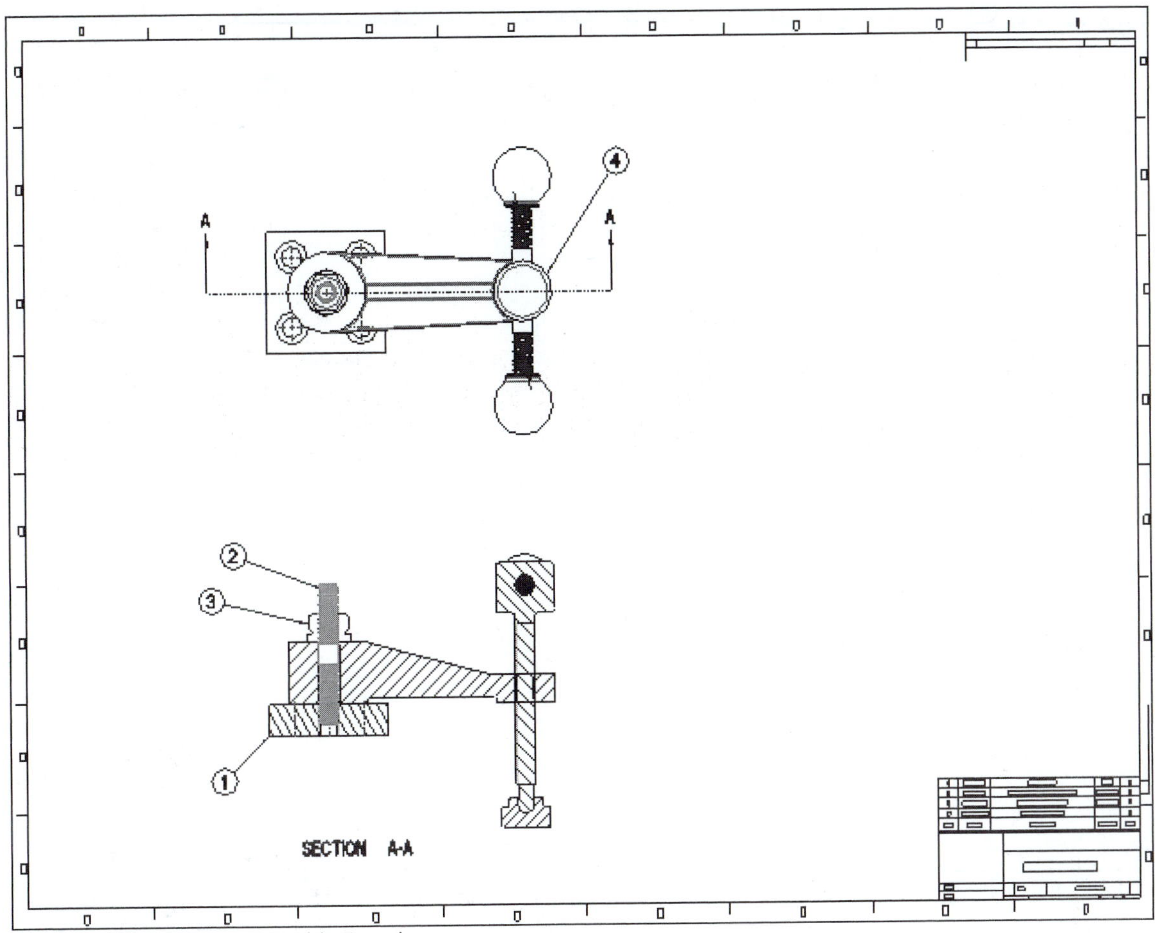

Figure 12.1(a) Clamp_Assembly Drawing

OBJECTIVES

- Create an **Assembly Drawing**
- Generate a **Parts List** from a **Bill of Materials (BOM)**
- **Balloon** an assembly drawing
- Create a **section assembly view** and change **component visibility**
- Add **Parameters** to parts
- Create a **Table** to generate a parts list automatically
- Use **Multiple Sheets**
- Make assembly **Drawing Sheets** with **multiple models**

REFERENCES AND RESOURCES

For **Resources** go to **www.cad-resources.com** > click on the PTC Creo Parametric 3.0 Book cover

- Lesson 12 Lecture
- Lesson 12 3D models embedded in a PDF
- Book Projects PDF
- Project Lectures

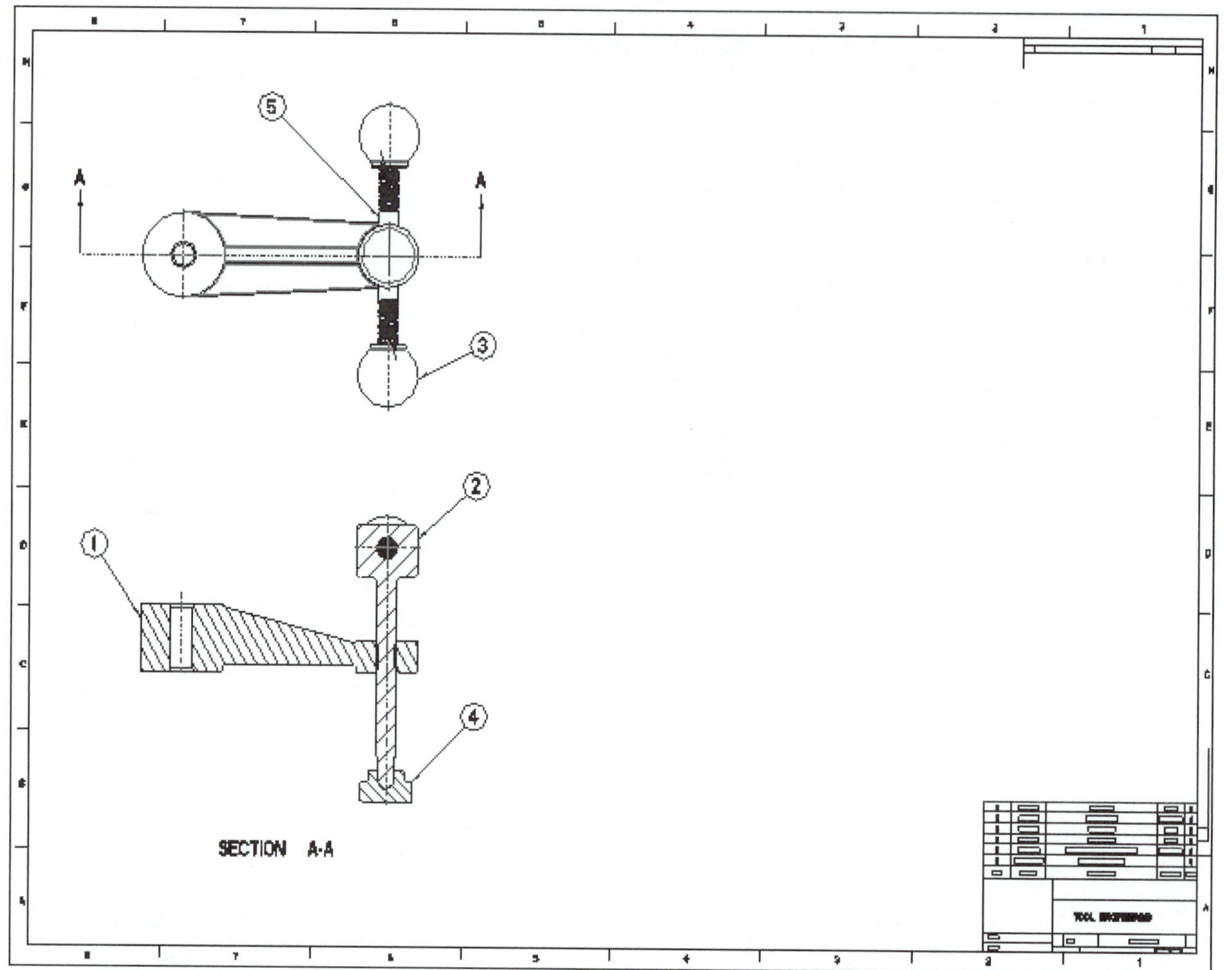

Figure 12.1(b) Clamp_Subassembly Drawing

ASSEMBLY DRAWINGS

PTC Creo Parametric 3.0 incorporates a great deal of functionality into drawings of assemblies [Figs. 12.1(a-b)]. You can assign parameters to parts in the assembly that can be displayed on a *parts list* in an assembly drawing. Creo Parametric can also generate the item balloons for each component on standard orthographic views or on an exploded view.

In addition, a variety of specialized capabilities allow you to alter the manner in which individual components are displayed in views and in sections. The format for an assembly is usually different from the format used for detail drawings. The most significant difference is the presence of a Parts List.

A parts list is actually a *Drawing Table object* that is formatted to represent a bill of materials in a drawing. By defining *parameters* in the parts in your assembly that agree with the specific format of the parts list, you make it possible for Creo Parametric to add pertinent data to the assembly drawing's parts list automatically as components are added to the assembly. After the parts list and parameters have been added, Creo Parametric can balloon the assembly drawing automatically.

In this lesson, you will create a set of assembly formats and place your standard parts list in them.

Lesson 12 STEPS

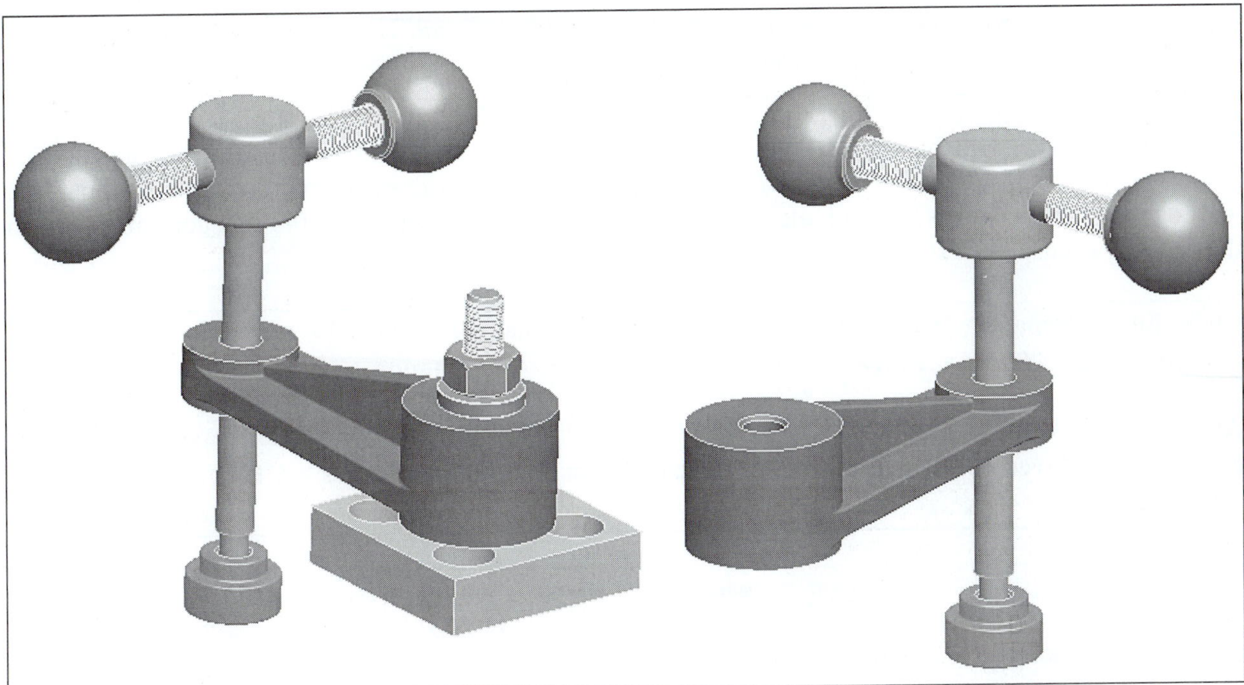

Figure 12.2 Swing Clamp Assembly and Subassembly

Swing Clamp Assembly Drawing

The format for an assembly drawing is usually different from the format used for detail drawings. The most significant difference is the presence of a parts list. You will create a standard "E" size format and place a standard parts list on it. You should create a set of assembly formats on "B", "C", "D", and "F" size sheets at your convenience.

A parts list is actually a *Drawing Table object* that is formatted to represent a bill of material (BOM) in a drawing. By defining parameters in the parts in your assembly that agree with the specific format of the parts list, you make it possible for Creo Parametric to add pertinent data to the assembly drawing parts list automatically, as components are added to the assembly.

After you create an "E" size format sheet with a parts list table, you will create two new drawings (each with two views) using your new assembly format. The Swing Clamp *subassembly* [Fig. 12.2(right)] will be used in the first drawing. The second drawing will use the Swing Clamp *assembly* [Fig. 12.2(left)]. Both drawings use the "E" size format created in the first section of this lesson. The format will have a parameter-driven title block and an integral parts list.

Launch **PTC Creo Parametric 3.0** > **Select Working Directory** > select the directory where the **clamp_assembly.asm** was saved > **OK** > **Ctrl+O** > Common Folders **System Formats** > **e.frm** > **Open** > **File** > **Save As** > **Save a Copy** > *type a unique name for your format:* New Name **asm_format_e** > **OK** > **File** > **Close** > **File** > **Open** > pick **asm_format_e.frm** > **Open**

The **format** will have a **.dtl** associated with it, and the **drawing** will have a unique **.dtl** file associated with it. *They are separate* ***.dtl*** *files.* When you activate a drawing and then add a format, the **.dtl** for the format controls the font, etc. for the format only. The drawing **.dtl** file that controls items on the drawing needs to be established separately in the Drawing mode.

Click: **File** > **Prepare** > **Detail Options** > **change** (Options dialog box opens) > ***text_height .25*** > **Enter** > ***default_font filled*** > **Enter** > ***arrow_style filled*** > **Add/Change** > ***draw_arrow_length .25*** > **Add/Change** > ***draw_arrow_width .08*** > **Add/Change** (Fig. 12.3) > **Apply** > **Save a copy of the currently displayed configuration file** > File name: **asm_format_properties** > **OK** (Fig. 12.3) > **Close** > **Close** > **Ctrl+S** > **OK**

	Value	Default	Status	Description
Active Drawing				
▼ These options control text not subject to other op				
drawing_text_color	letter_color *	letter_color	●	Controls text color in drawings. When set t
text_height	0.250000	0.156250	●	Sets the default text height for newly creat
text_thickness	0.000000 *	0.000000	●	Sets default text thickness for new text aft
text_width_factor	0.800000 *	0.800000	●	Sets default ratio between the text width a
draft_scale	1.000000 *	1.000000	●	Determines value of draft dimensions relati
▼ These options control text and line fonts				
default_font	filled.ndx	font.ndx	●	Specifies the default text font. Include the e
▼ These options control leaders				
arrow_style	closed *	closed	●	Controls style of arrow head for all detail it
draw_arrow_length	0.250000	0.187500	●	Sets length of leader line arrows.
draw_arrow_width	0.080000	0.062500	●	Sets width of leader line arrows. Drives th
draw_attach_sym_height	DEFAULT *	default	●	Sets height of leader line slashes, integral s
draw_attach_sym_width	DEFAULT *	default	●	Sets width of leader line slashes, integral s
draw_dot_diameter	DEFAULT *	default	●	Sets diameter of leader line dots. If set to "c
leader_elbow_length	0.250000 *	0.250000	●	Determines length of leader elbow (the hori
▼ These options control tables, repeat regions, and				
default_bom_balloon_type	simple_circle		●	Determines the default type of BOM balloon
default_table_column_width	10		●	Specifies the default value of the column w
default_table_columns	4		●	Specifies the default number of columns of
default_table_row_height	1		●	Specifies the default value of the row heig
default_table_rows	2		●	Specifies the default number of rows of ne
sort_method_in_region	delimited *	delimited	●	Determines repeat regions sort mechanism.
▼ Miscellaneous options				
draft_scale	1.000000 *	1.000000	●	Determines value of draft dimensions relati
drawing_units	inch		☒	Sets units for all drawing parameters.
line_style_standard	std_ansi *	std_ansi	●	Controls standard of line display in drawing
node_radius	DEFAULT *	default	●	Sets the size of the nodes displayed in syn

Figure 12.3 New Format Options

Zoom into the title block region. Create notes for the title text and parameter text required to display the proper information.

Click: **File** > **Options** > **Sketcher** > Snap to grid > **OK** > **No** > **Sketch** tab > **Draft Grid** > **Show Grid** > **Grid Params** > **X&Y Spacing** > *type* **.125** > **Enter** > **Done/Return** > **Done/Return** > **Annotate** tab > Note > **Make Note** >tool pick a point for the note in the largest area of the title block (Fig. 12.4) > *type* **TOOL ENGINEERING** > **LMB** > **LMB** >

Note > (Fig. 12.4) > *type* **DRAWN** > **LMB** > **LMB** >

Note > (Fig. 12.4) > *type* **ISSUED** > **LMB** > **LMB** >

Note > (Fig. 12.4) > *type* **&dwg_name** > **LMB** > **LMB** >

Note > (Fig. 12.4) > *type* **&scale** > **LMB** > **LMB** >

Note > (Fig. 12.4) > *type* **SHEET ¤t_sheet OF &total_sheets** > **LMB** > **LMB** > **File** > **Options** > **Sketcher** > Snap to grid > **OK** > **No** > modify some text heights to **.10** (Fig. 12.4) using one of the two methods provided here:

Method 1. Press the **Ctrl** key > select more than one note > release the **Ctrl** key > press **RMB** > **Text Style** > Default > Height 0.100000 > **OK** > in the Graphics Window, **LMB** to deselect *or*
Method 2. Double-click on a note > change the value in the height field > **Enter** > **LMB**

Reposition the notes (Fig. 12.4) select a note > > move as needed > **LMB** > **Ctrl+S**

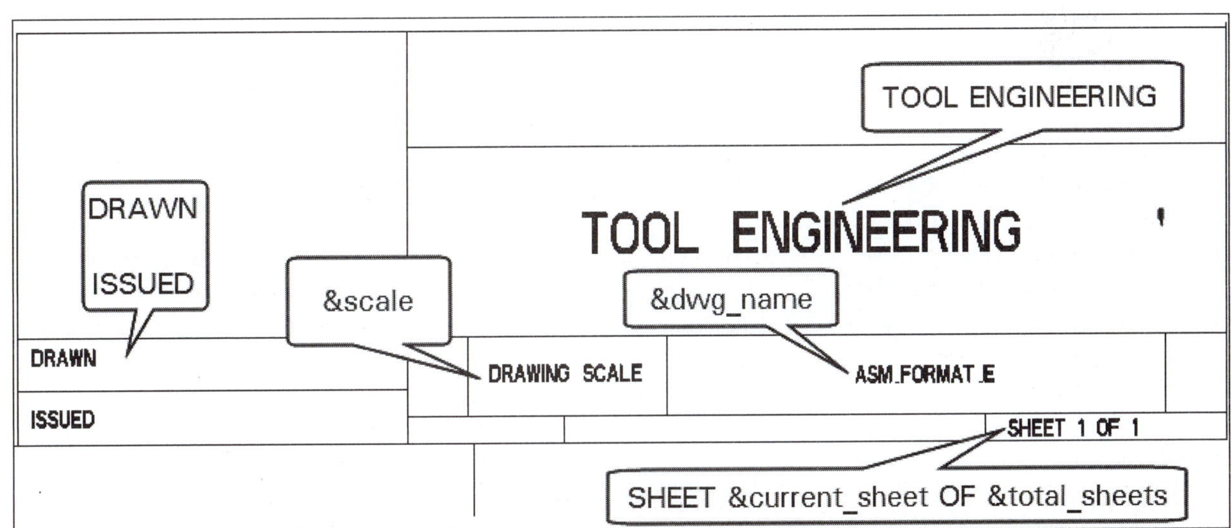

Figure 12.4 Parameters and Labels in the Title Block, Smaller Text is **.10** in Height

The parts list table can now be created and saved with this format. You can add and replace formats and still keep the table associated with the drawing. Next, start the parts list by creating a table.

Click: **Table** tab > Table > place the mouse pointer over the squares > **LMB** to select two rows and five columns [Fig. 12.5(a)] > **LMB** to place the table [Figs. 12.5(b-c)] > move the mouse pointer off of the table > **LMB** to deselect

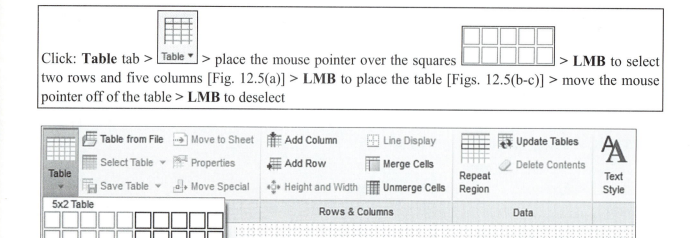

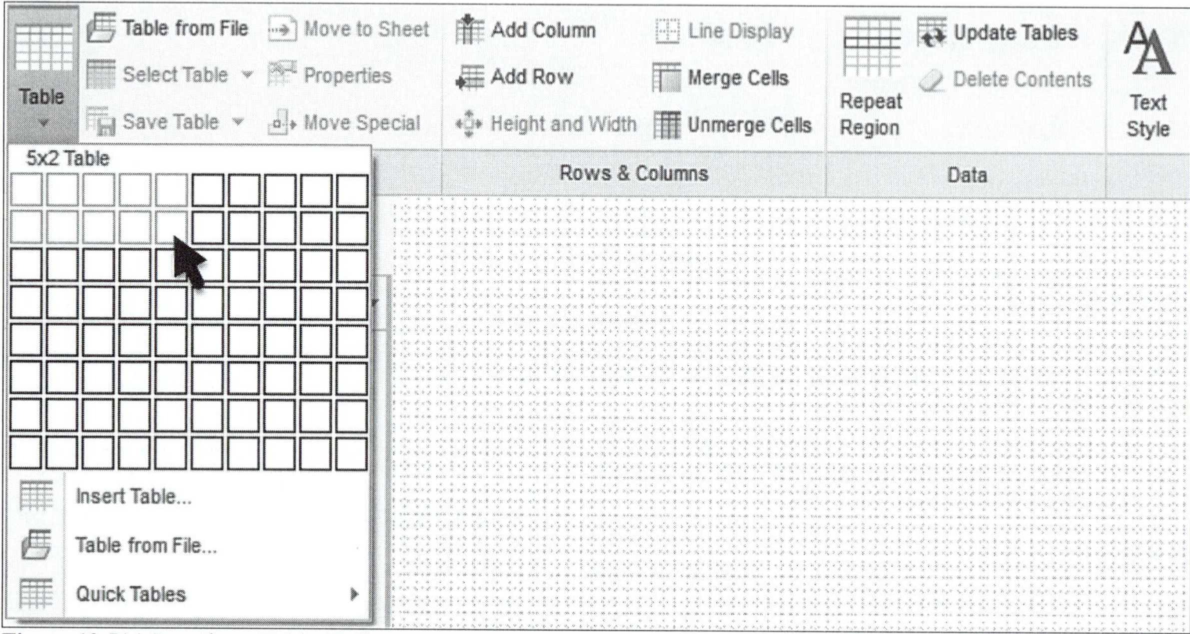

Figure 12.5(a) Inserting a Table (5x2)

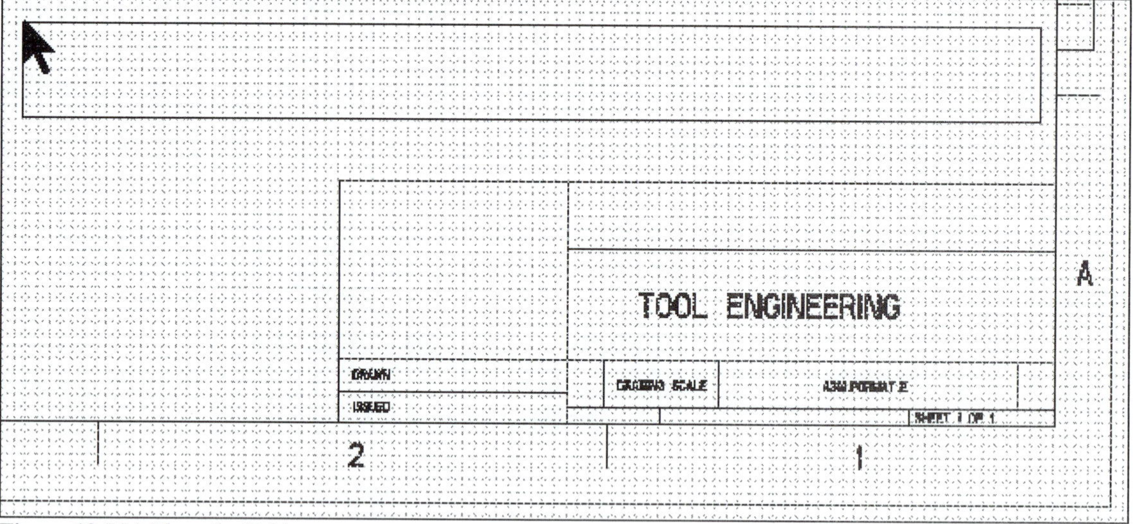

Figure 12.5(b) Place the Table

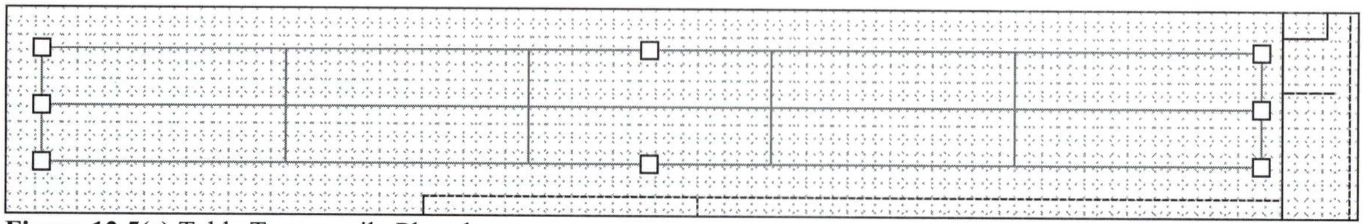
Figure 12.5(c) Table Temporarily Placed

Position the pointer over the first column until it highlights > double-click [Fig. 12.5(d)] > change the column's width to **1.000** [Fig. 12.5(e)] > **OK**

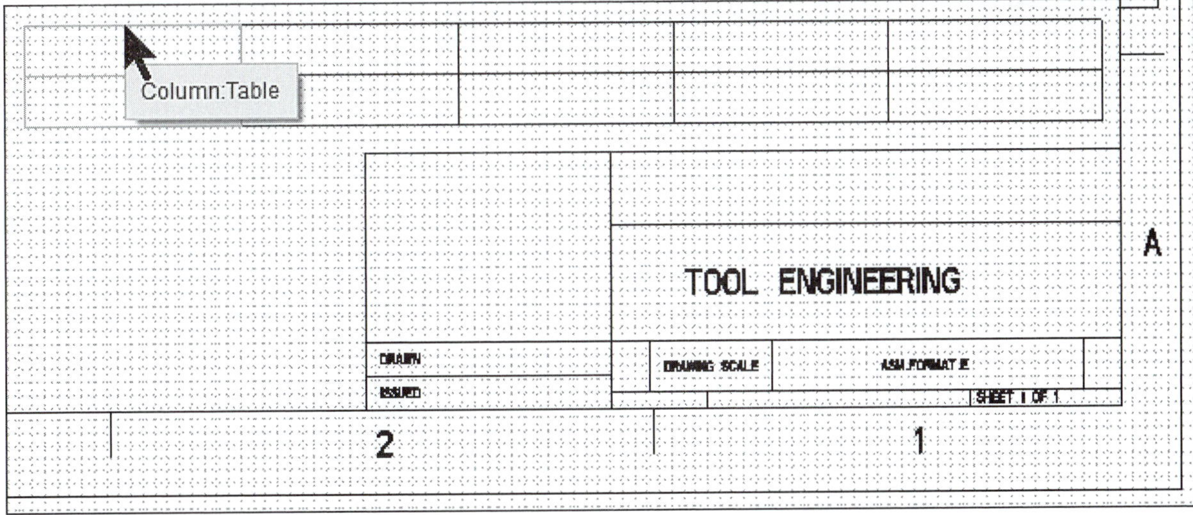

Figure 12.5(d) Highlight Column

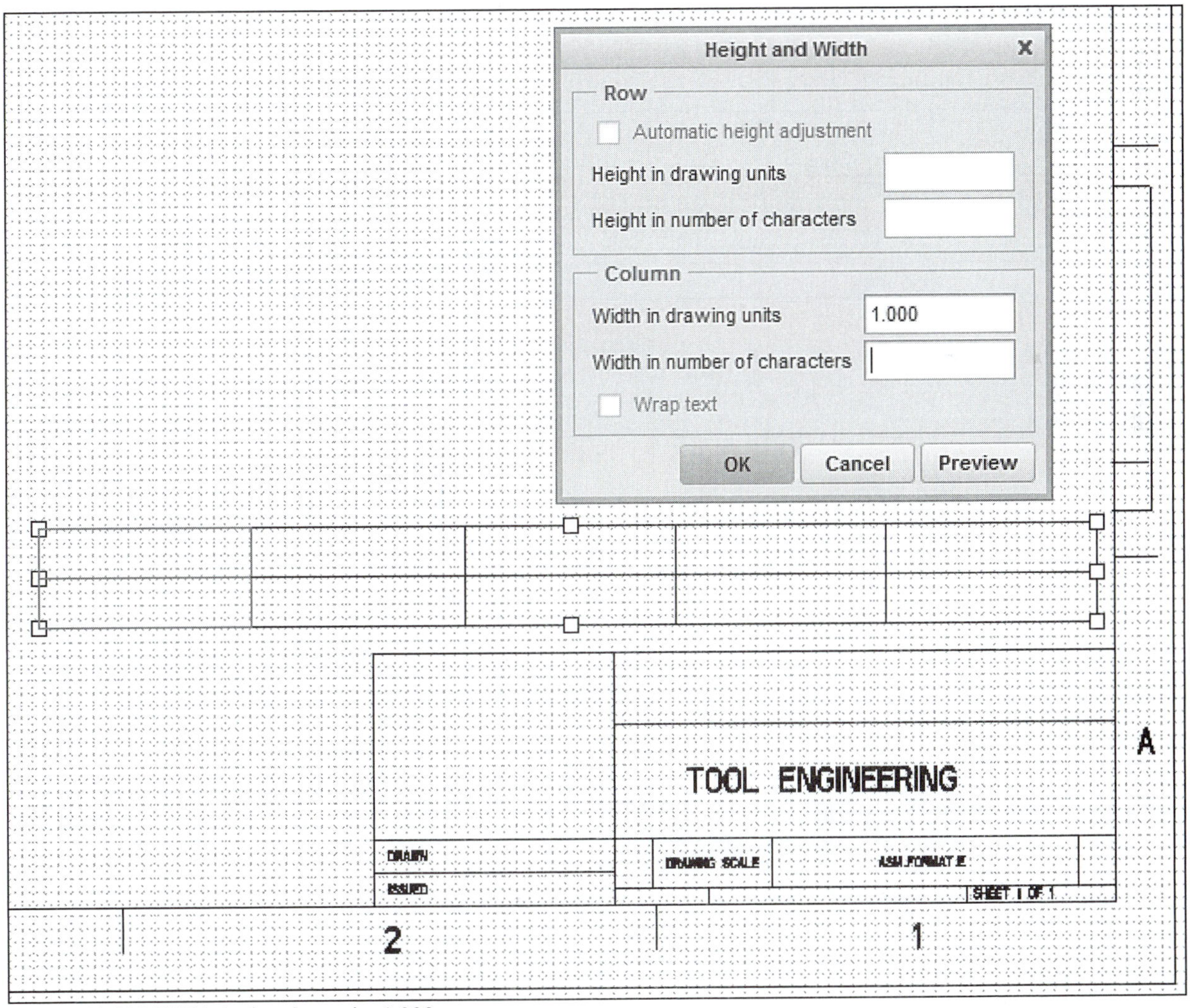

Figure 12.5(e) Width in drawing units **1.000**

Change the second column's width to **1.000** > change the third column's width to **4.000** [Fig. 12.5(f)] > change the fourth column's width to **1.000** > change the fifth column's width to **.750**

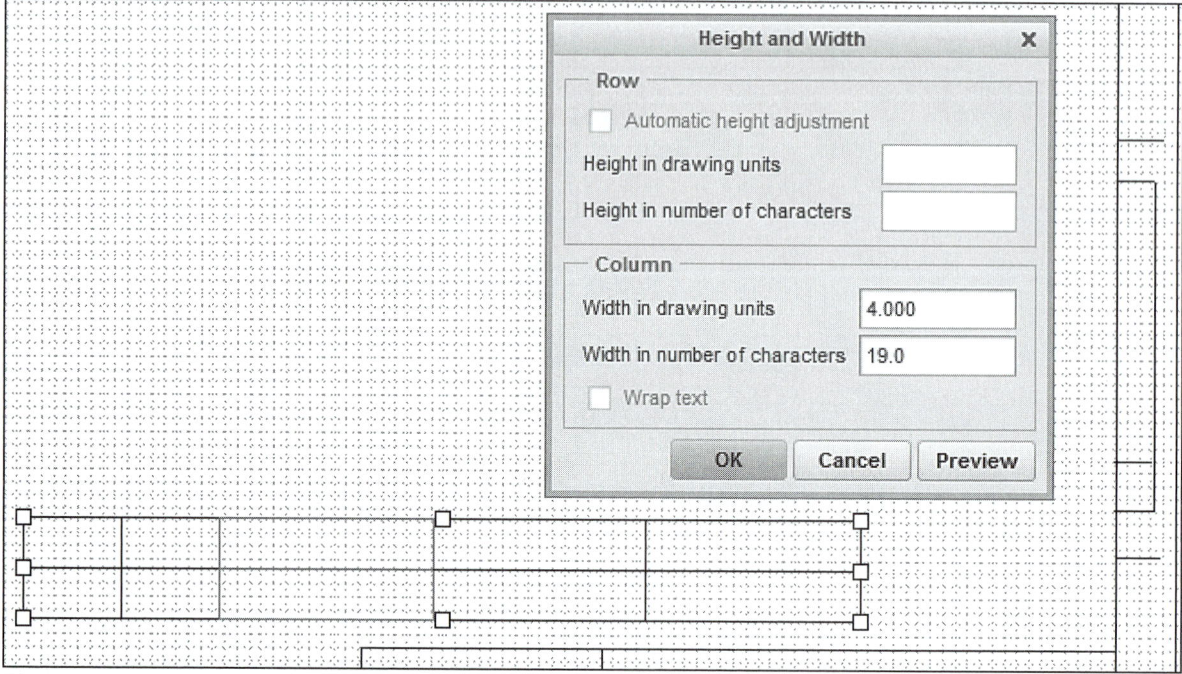

Figure 12.5(f) Width of the Third Column is **4.000**

Position the pointer over the top row (left side) until it highlights > double-click > **Automatic height adjustment** (uncheck) > change the column's height to **.375** [Fig. 12.5(g)] > **OK**

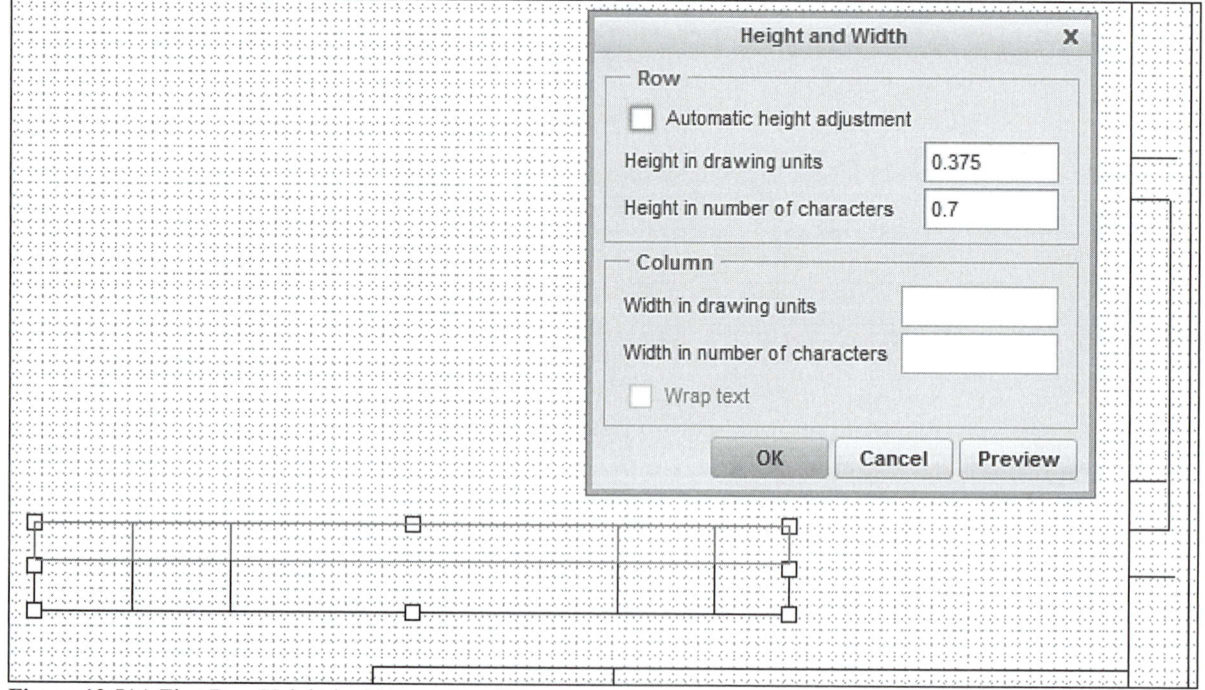

Figure 12.5(g) First Row Height is **.375**

Position the pointer over the top corner [Fig. 12.5(h)] > press and hold down the **LMB** > move the pointer to move the table [Fig. 12.5(i)] > release the **LMB** > move the pointer > **LMB** to deselect [Fig. 12.5(j)]

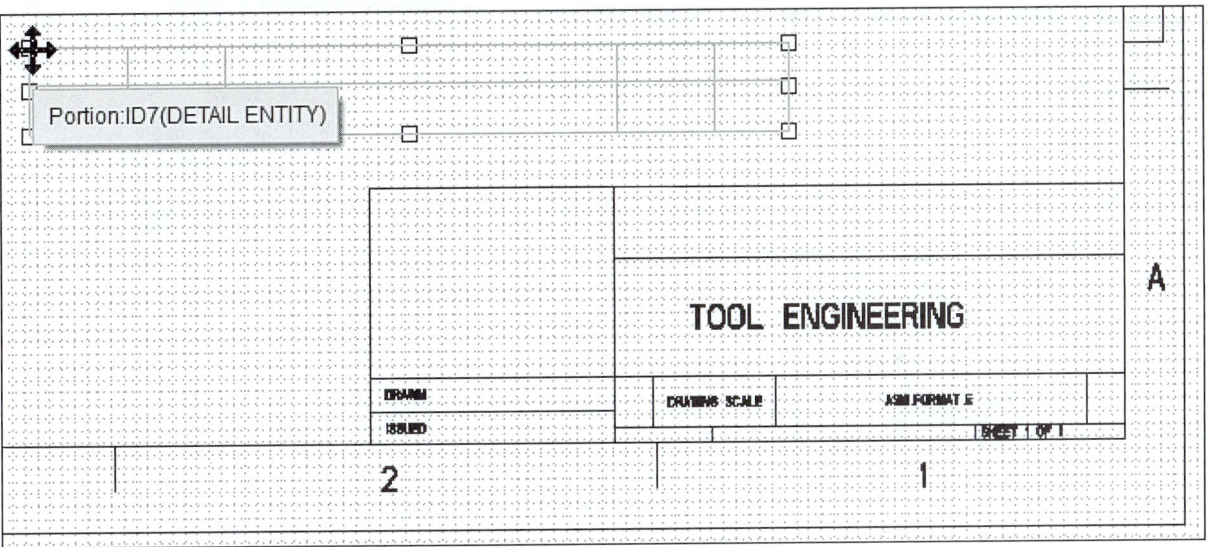

Figure 12.5(h) Place the Pointer on the Upper Left Corner of the Table

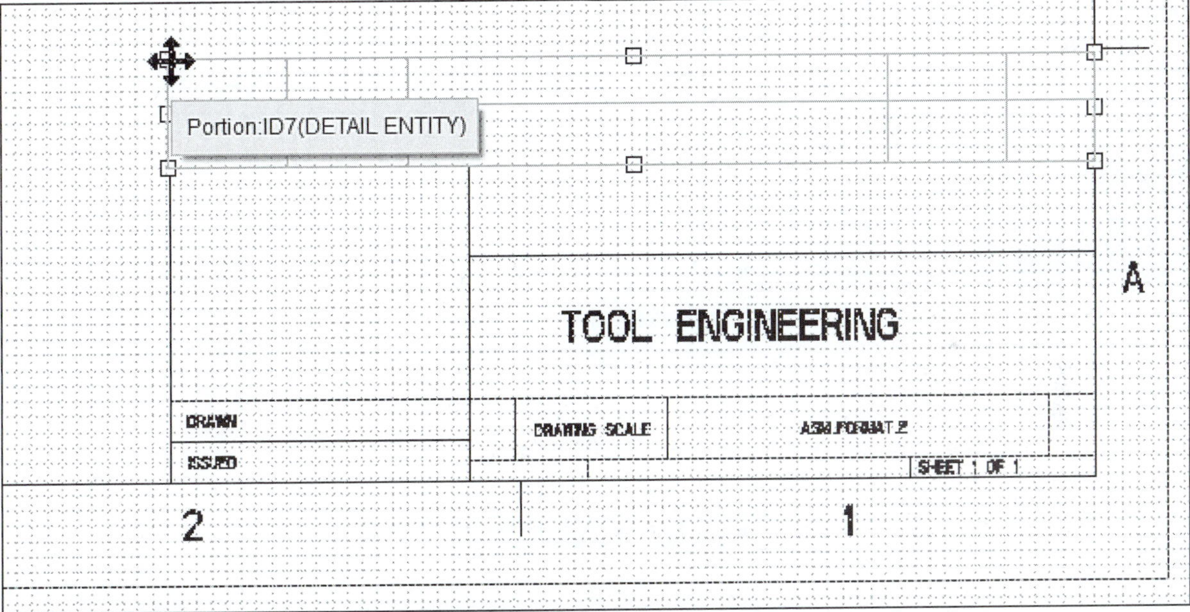

Figure 12.5(i) Move the Pointer to Position the Table

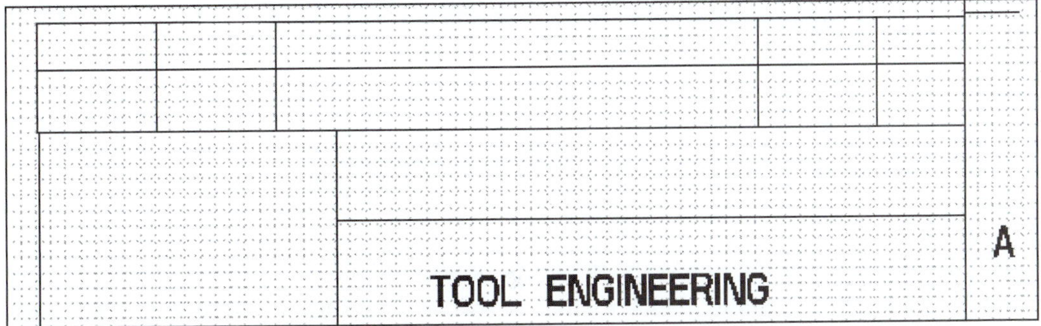

Figure 12.5(j) Positioned Table

Click: **Sketch** tab > **Draft Grid** > **Hide Grid** > **Done/Return** > **Table** tab > Repeat Region > **Add** > **Simple** > pick in the left block > pick in the right block [Fig. 12.5(k)] > **Attributes** > select the Repeat Region just created [Fig. 12.5(l)] > **No Duplicates** > **Recursive** > **Done/Return** > **Done** > **Ctrl+S**

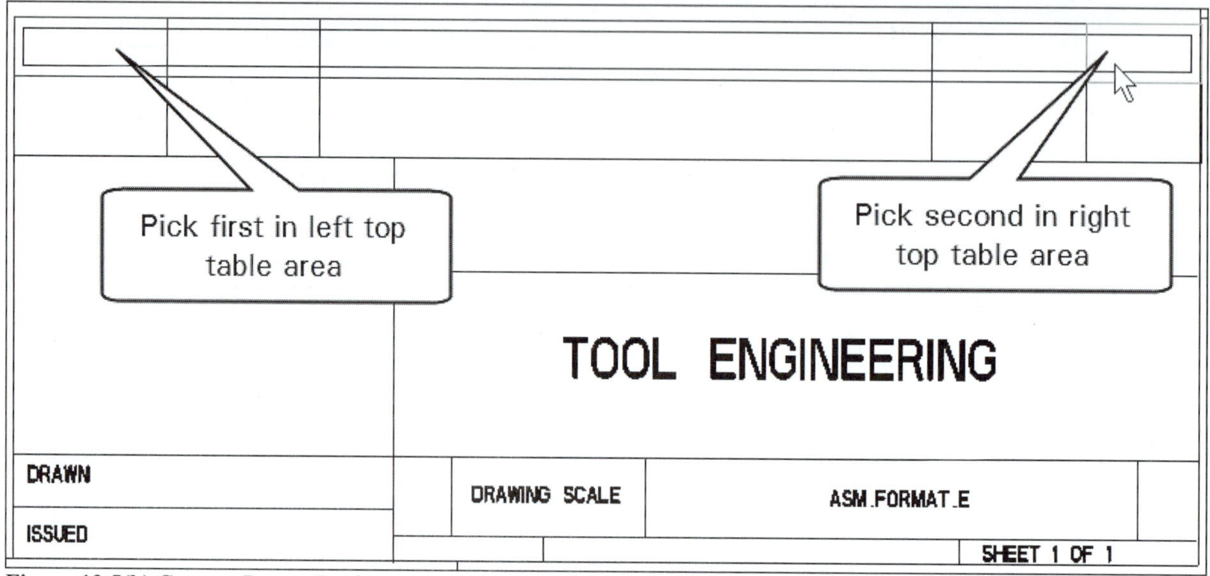

Figure 12.5(k) Create a Repeat Region

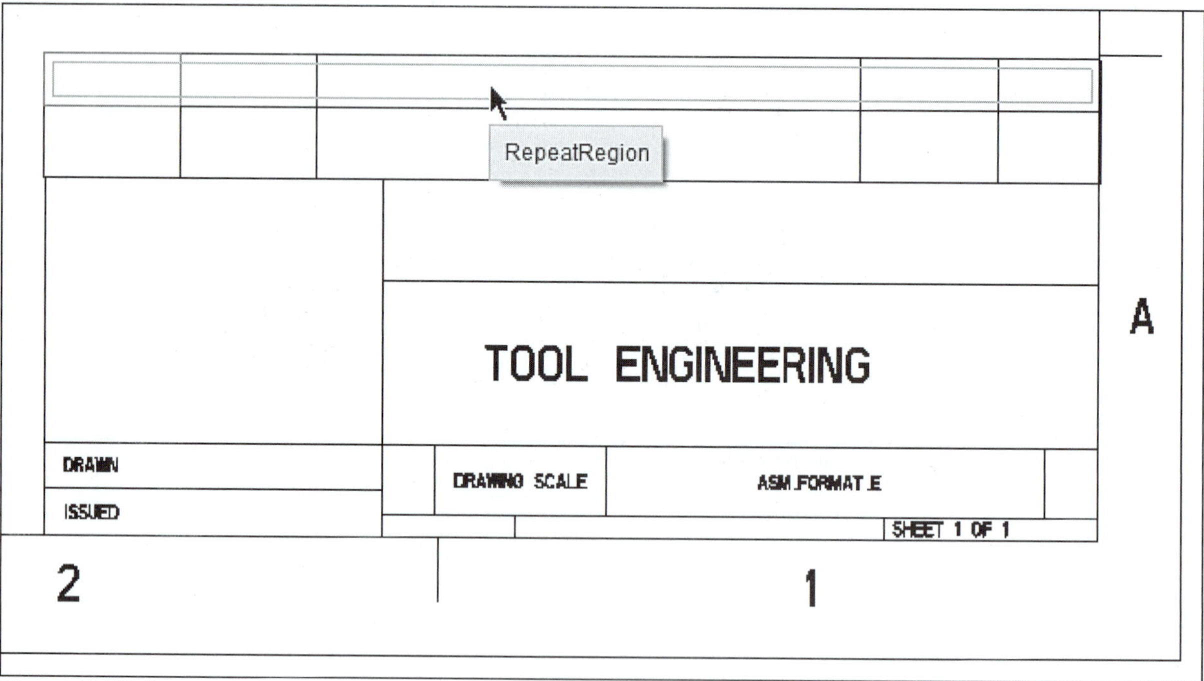

Figure 12.5(l) Repeat Region

Normally, a BOM will be ascending unless the drawing shows a subassembly. Also, in some cases a company will move the BOM to a separate sheet where the descending format is required. Here, we wanted you to move and resize the BOM table and become comfortable with the editing process.

Insert column headings using plain text, double-click on the first block of the second row [Fig. 12.6(a)] > *type* **ITEM** [Fig. 12.6(b)] (the text is case sensitive) > **LMB** to deselect

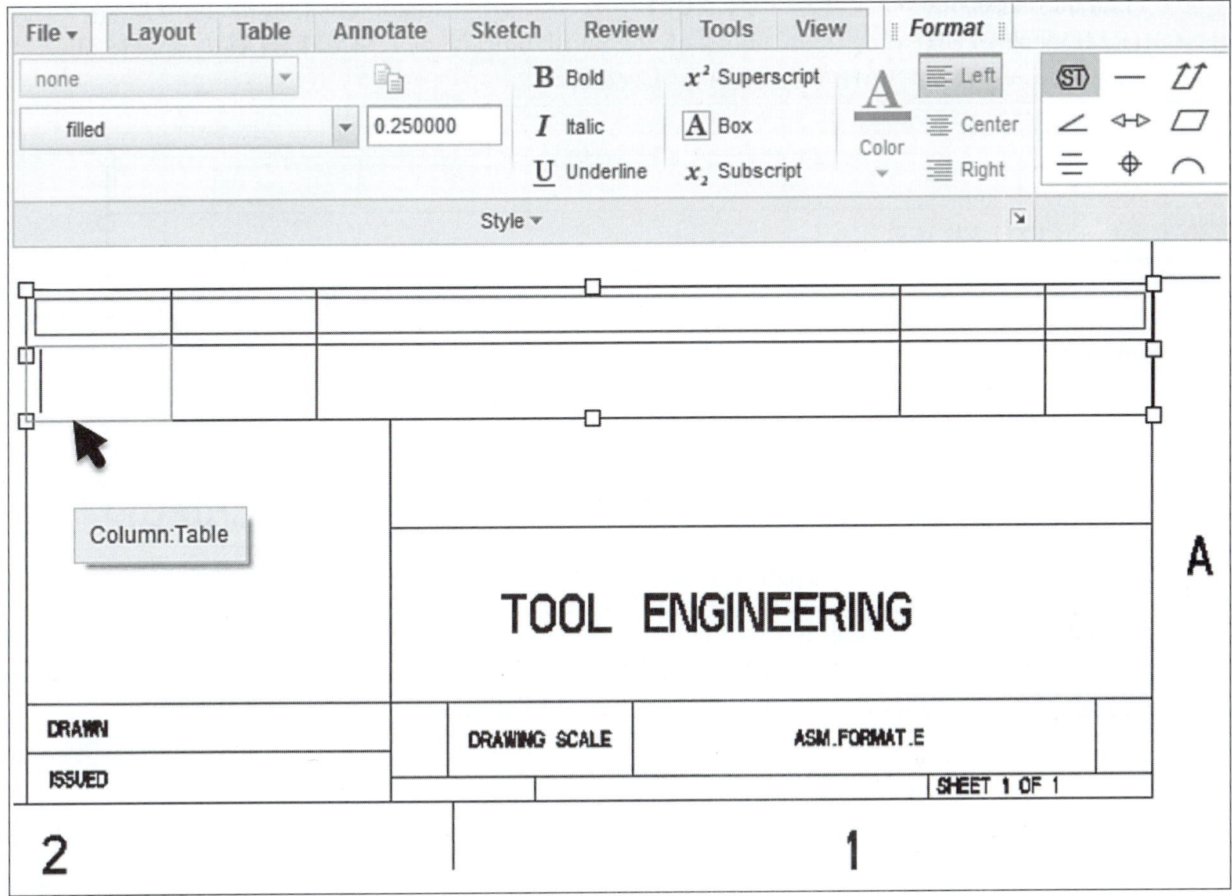

Figure 12.6(a) Select the First Block

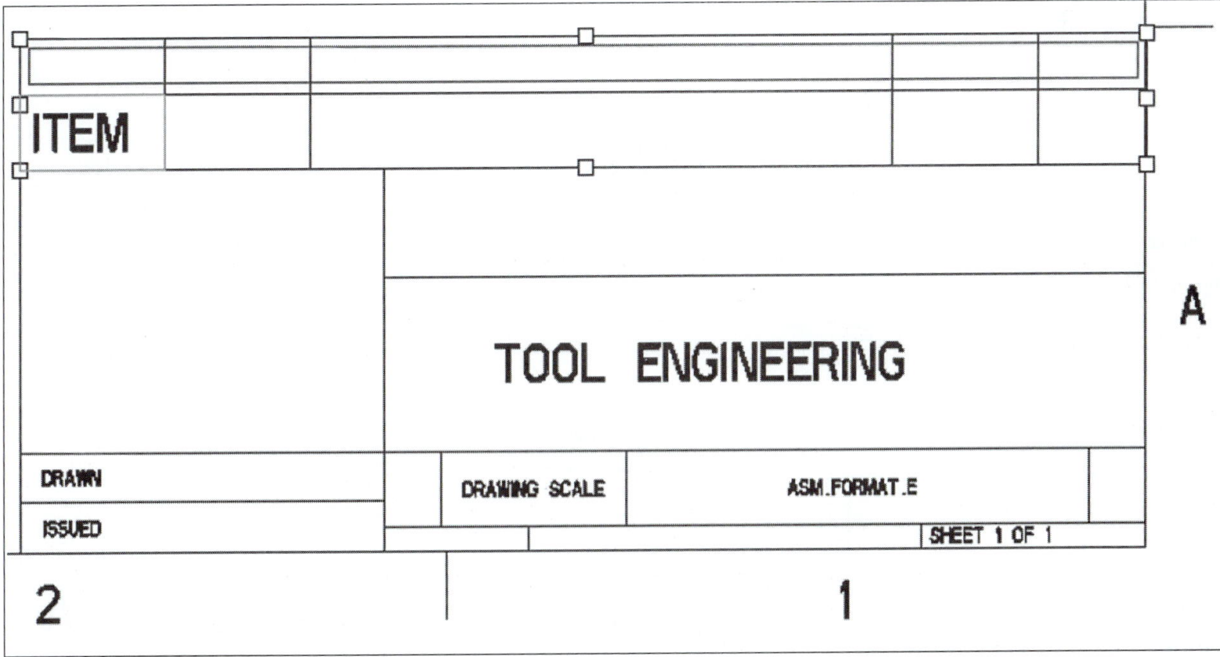

Figure 12.6(b) ITEM

The Repeat Region now needs to have some of its headings correspond to the parameters that will be created in each component model.

Double-click on the second block > *type* **PT NUM** [Fig. 12.6(c)] > double-click on the third block > *type* **DESCRIPTION MATERIAL** > double-click on the fourth block > *type* **MATERIAL** > double-click on the fifth block > *type* **QTY** > **LMB** > **LMB** [Fig. 12.6(d)]

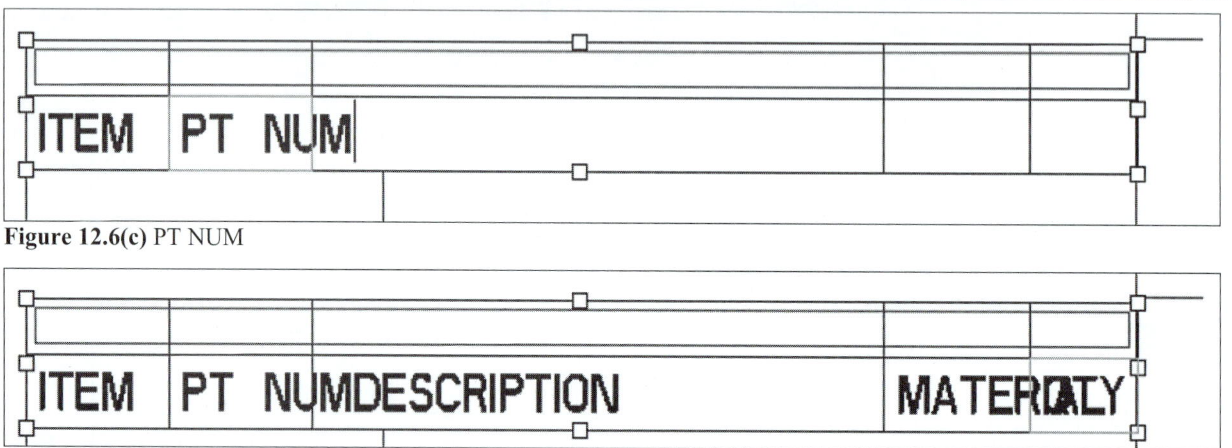

Figure 12.6(c) PT NUM

Figure 12.6(d) ITEM, PT NUM, DESCRIPTION, MATERIAL, QTY

With the **Ctrl** key pressed, select all five text cells > release the **Ctrl** key > press **RMB** > **Text Style** Text Style dialog box opens [Fig. 12.6(e)] > Character- Height **.125** > Note/Dimension- Horizontal **Center** > Vertical **Middle** > **Apply** [Figs. 12.6(f-g)] > **OK** > **LMB** > **Ctrl+S**

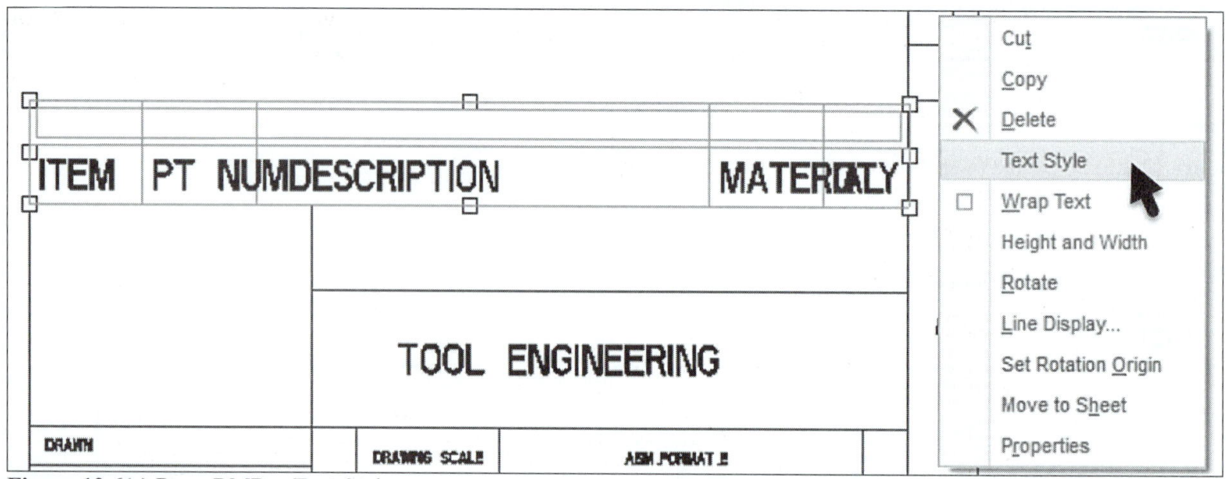

Figure 12.6(e) Press RMB > Text Style

Figure 12.6(f) Text Style Preview

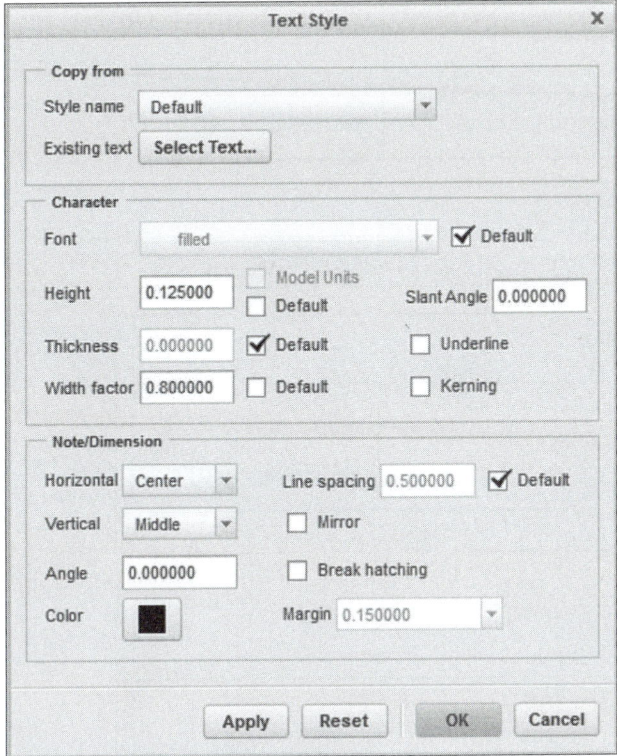

Figure 12.6(g) Text Style Dialog Box

Insert parametric text (Report Symbols) into the table cells of the Repeat Region. Double-click on the first table cell of the Repeat Region (the table cell ABOVE the text ITEM) > **rpt...** from the Report Symbol dialog box [Fig. 12.7(a)] > **index** [Figs. 12.7(b-c)]

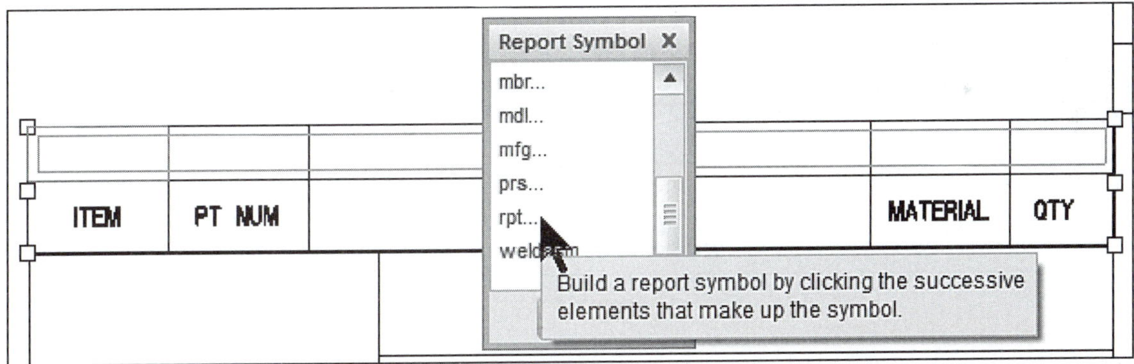

Figure 12.7(a) Report Symbol Dialog Box, Click **rpt...**

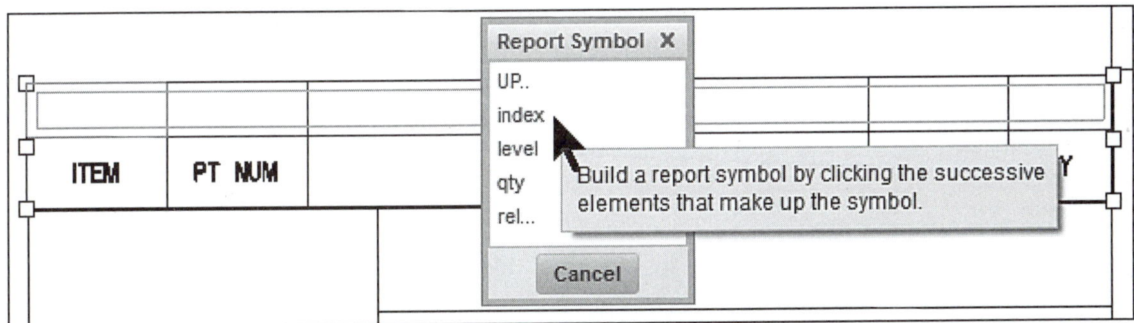

Figure 12.7(b) Click **index**

Figure 12.7(c) &rpt.index

Double-click on the fifth table cell of the Repeat Region > **rpt…** [Fig. 12.7(d)] > **qty** [Figs. 12.7(e-f)]

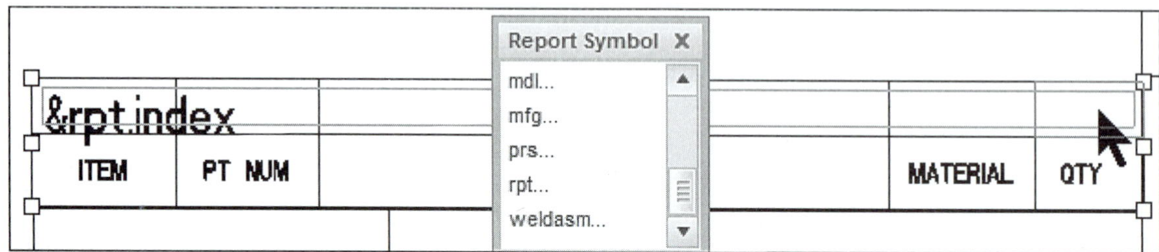

Figure 12.7(d) Click **rpt…**

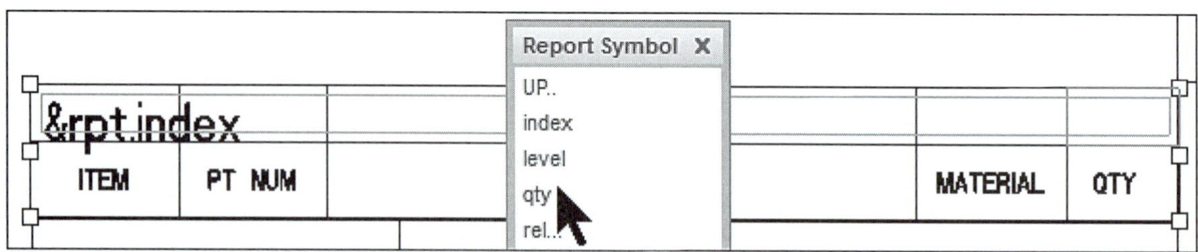

Figure 12.7(e) Click **qty**

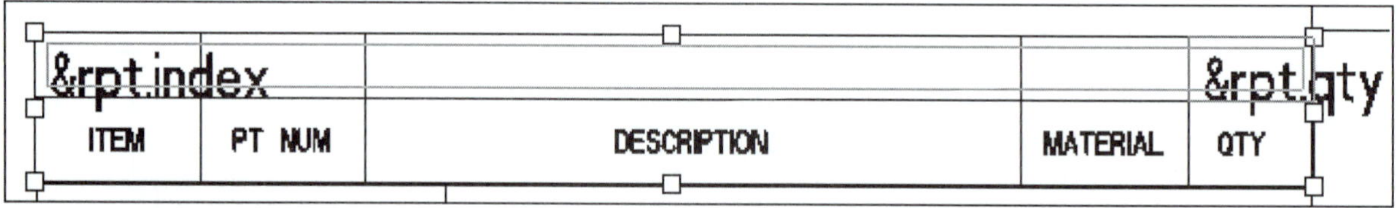

Figure 12.7(f) &rpt.qty

Double-click on the third (middle) table cell of the Repeat Region > **asm…** [Fig. 12.7(g)] > **mbr…** [Fig. 12.7(h)] > **User Defined** [Fig. 12.7(i)] > Enter symbol text: *type* **DSC** > **Enter** [Fig. 12.7(j)]

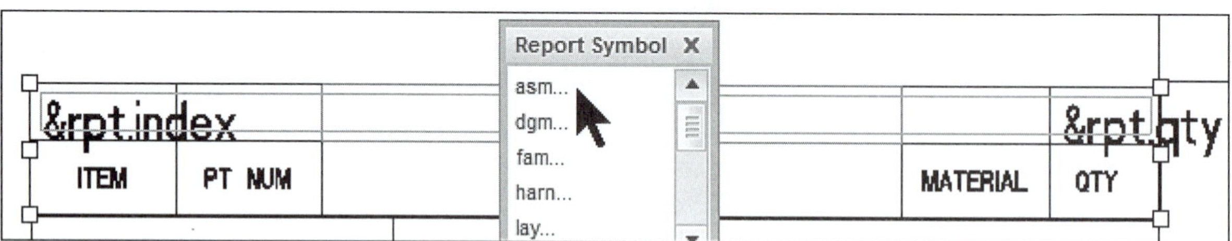

Figure 12.7(g) Click **asm…**

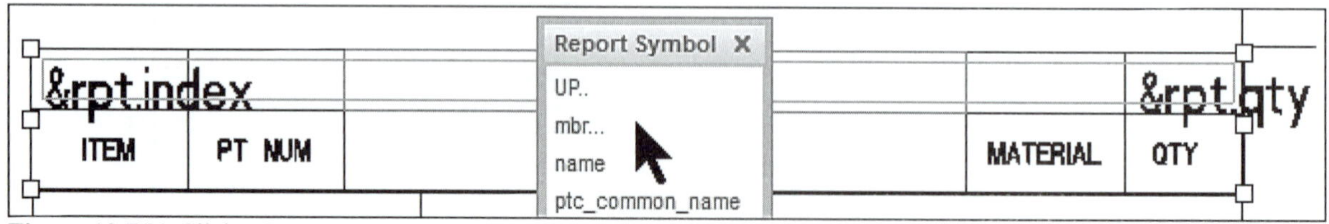

Figure 12.7(h) Click **mbr…**

Figure 12.7(i) Click **User Defined**

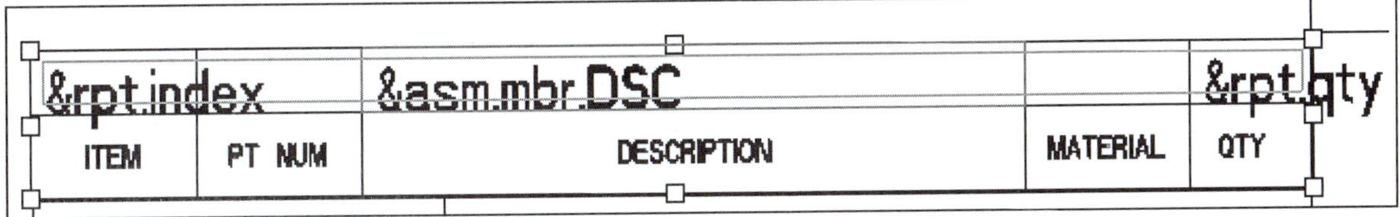

Figure 12.7(j) &asm.mbr.DSC

Double-click on the fourth table cell of the Repeat Region > **asm…** > **mbr…** > **ptc_material…** > **PTC_MATERIAL_NAME** [Figs. 12.7(k-l)] > double-click on the second table cell of the Repeat Region > **asm…** > **mbr…** > **User Defined** > Enter symbol text: *type* **PRTNO** > **Enter** [Fig. 12.7(m)] > in the Graphics Window, **LMB** to deselect > **Ctrl+S**

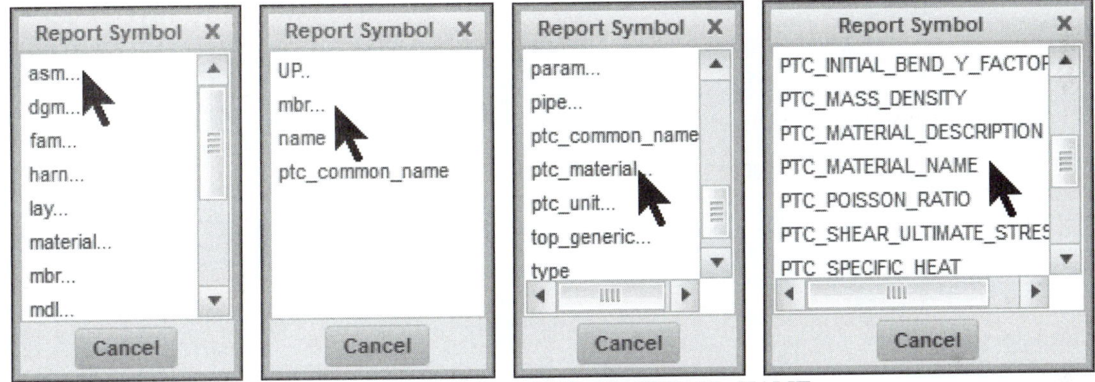

Figure 12.7(k) asm… > mbr… > ptc_material… > PTC_MATERIAL_NAME

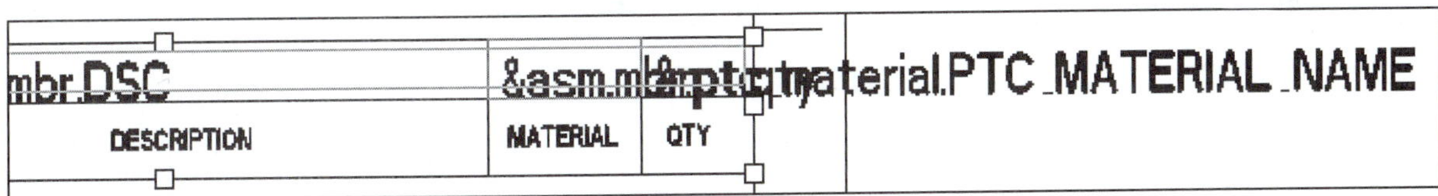

Figure 12.7(l) &asm.mbr.ptc_material.PTC_MATERIAL_NAME

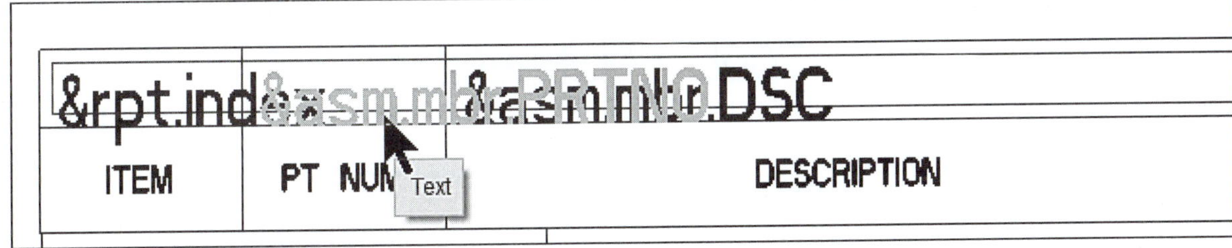

Figure 12.7(m) &asm.mbr.PRTNO

Change the text height of the Report Symbols in the table cells of the Repeat Region to **.125**.

Window in the table > press **RMB** > **Text Style** Text Style dialog box opens > Character- Height **.125** > Note/Dimension- Horzontal **Center** > Vertical **Middle** > **Apply** [Fig. 12.7(n)] > **OK** > **LMB** to deselect > Update Tables [Fig. 12.7(o)] > > **Ctrl+R** > **Ctrl+S** > **File** > **Close**

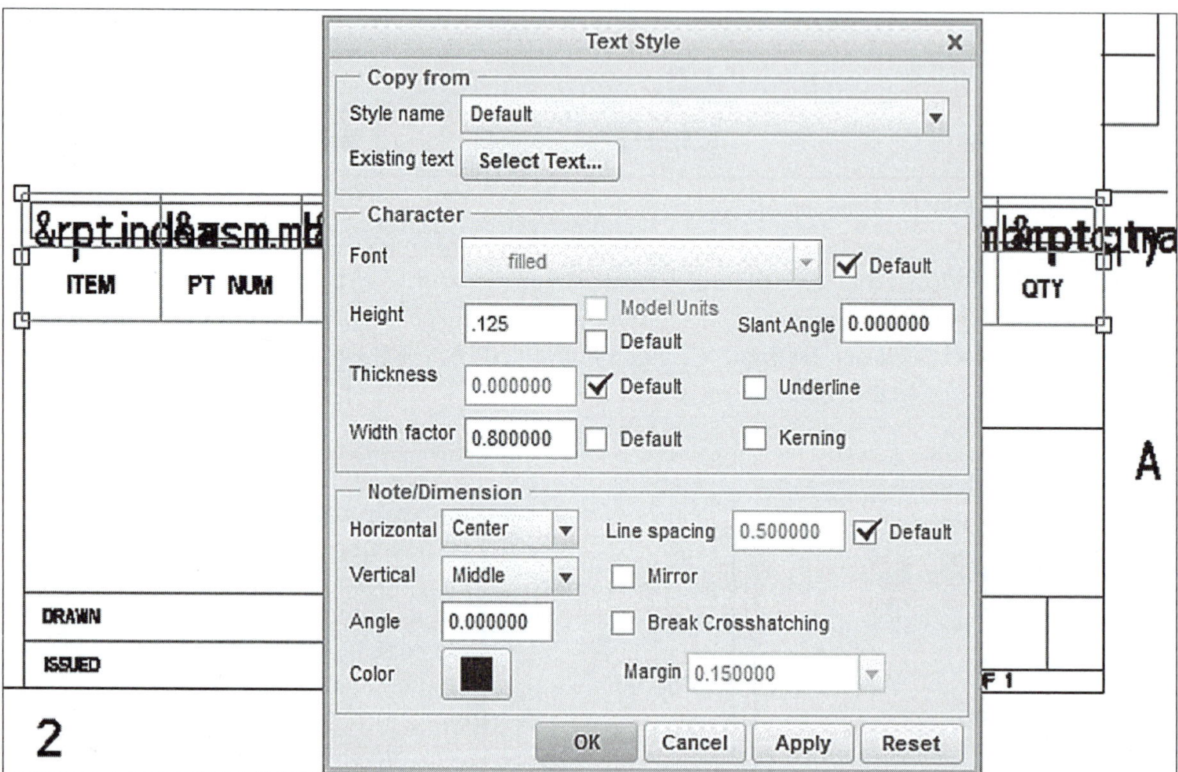

Figure 12.7(n) Text Style Dialog Box

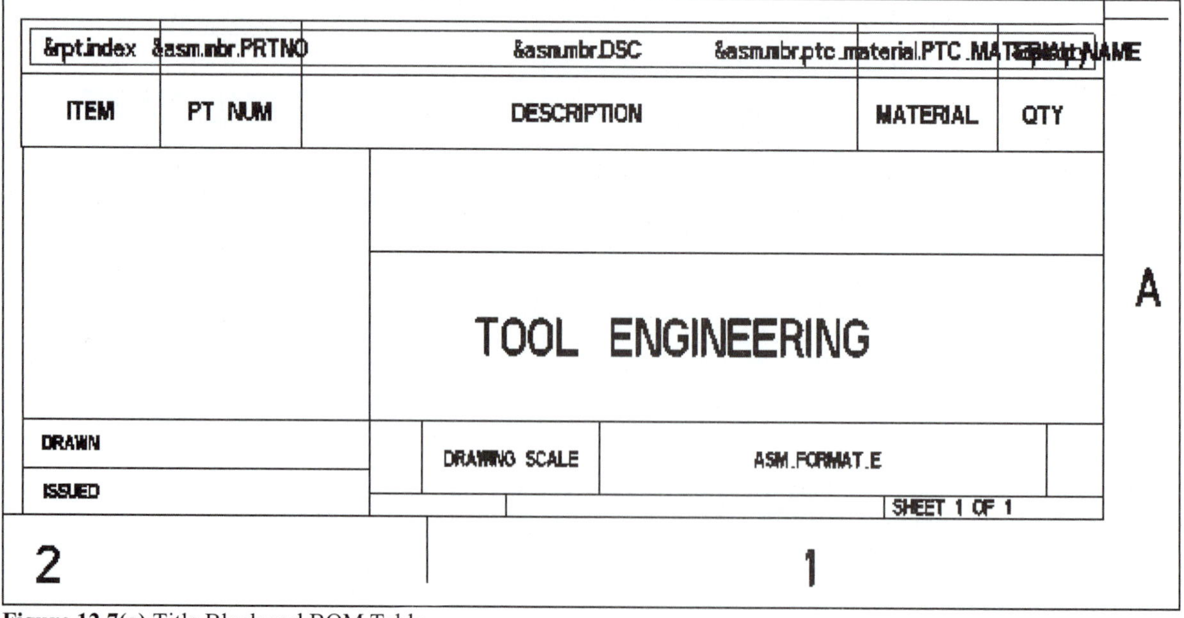

Figure 12.7(o) Title Block and BOM Table

Adding Parts List (BOM) Data

When you save your standard assembly format, the Drawing Table that represents your standard parts list is now included. You must be aware of the titles of the parameters under which the data is stored, so that you can add them properly to your components. As you add components to an assembly, Creo Parametric 3.0 reads the parameters from them and updates the parts list. You can also see the same effect by adding these parameters after the drawing has been created. Creo Parametric 3.0 also provides the capability of displaying Item Balloons on the first view that was placed on the drawing. To improve their appearance, you can move these balloons to other views and alter the locations where they attach.

Retrieve the clamp arm, click: **File** > **Open** > **clamp_arm.prt** > **Open** > **The generic** > **Open** > **File** > **Options** > **Model Display** > Default model orientation: **Isometric** > **OK** > **No** > **Ctrl+D** > [icon] > ☐ (Select All) *off* > **LMB** > [icon] *off* [Fig. 12.8(a)] > **Tools** tab > **Parameters** [Fig. 12.8(b)]

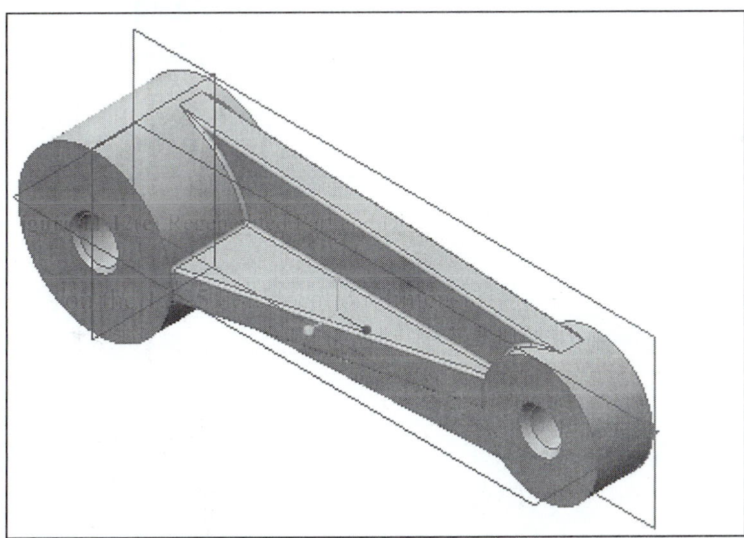

Figure 12.8(a) Clamp_Arm

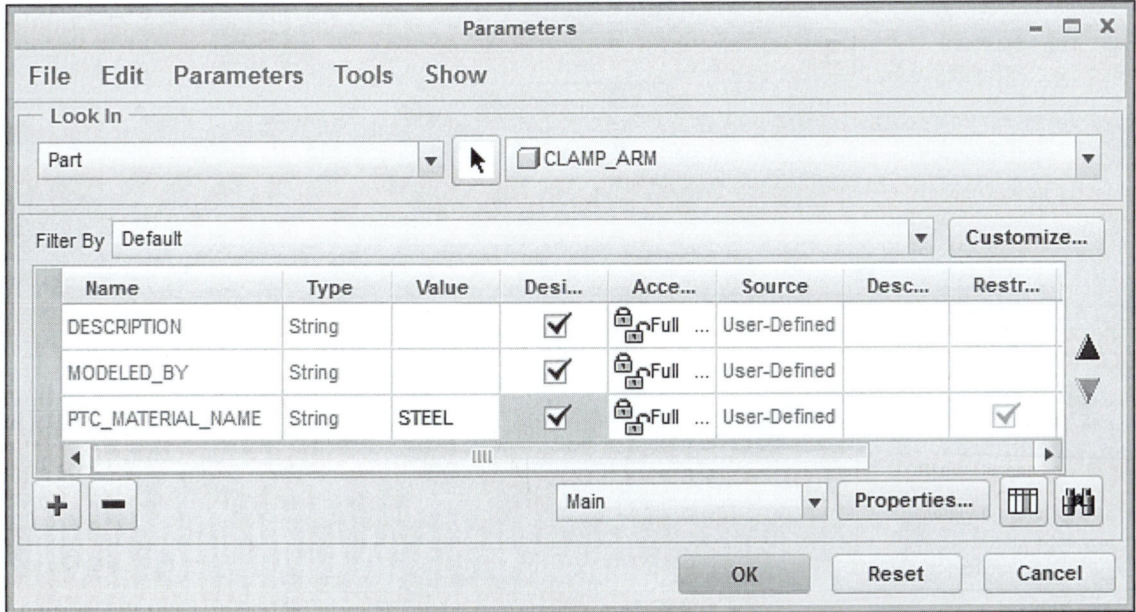

Figure 12.8(b) Parameters Dialog Box

Click: **Parameters** from Parameters dialog box > **Add Parameter** [Fig. 12.8(c)] > Name field should be highlighted (if not, click in field and highlight) [Fig. 12.8(d)], *type* **PRTNO** [Fig. 12.8(e)] > Designate > click in Type field [Fig. 12.8(f)] > [Fig. 12.8(g)] > **String**

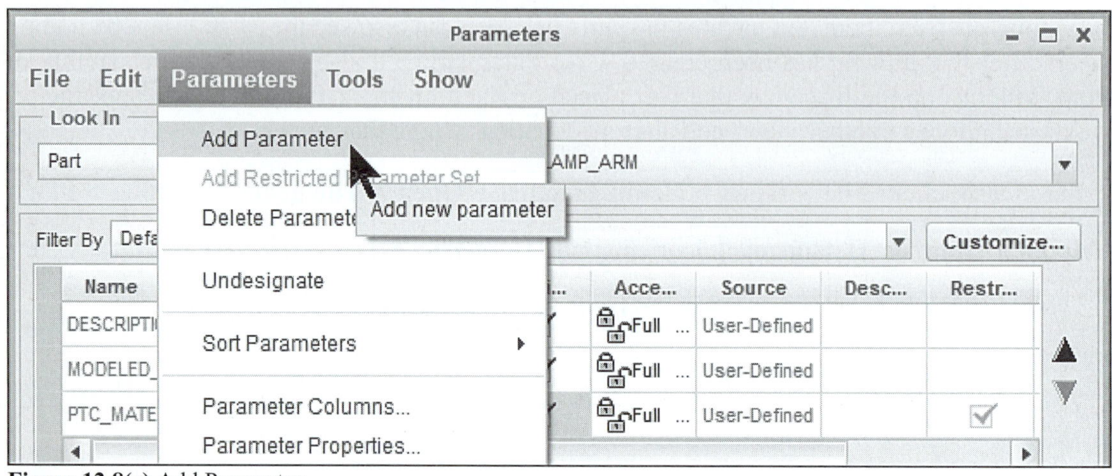

Figure 12.8(c) Add Parameter

DESCRIPTION	String		✓	Full ...	User-Defined			
MODELED_BY	String		✓	Full ...	User-Defined			
PTC_MATERIAL_NAME	String	STEEL	✓	Full ...	User-Defined		✓	
PARAMETER_1	Real Num...	0.000000	☐	Full ...	User-Defined			

Figure 12.8(d) Adding Parameters

DESCRIPTION	String		✓	Full ...	User-Defined			
MODELED_BY	String		✓	Full ...	User-Defined			
PTC_MATERIAL_NAME	String	STEEL	✓	Full ...	User-Defined		✓	
PRTNO	Real Num...	0.000000	✓	Full ...	User-Defined			

Figure 12.8(e) Name PRTNO

DESCRIPTION	String		✓	Full ...	User-Defined			
MODELED_BY	String		✓	Full ...	User-Defined			
PTC_MATERIAL_NAME	String	STEEL	✓	Full ...	User-Defined		✓	
PRTNO	Real Number	0.000000	✓	Full ...	User-Defined			

Figure 12.8(f) Click in Type Field- Real Number

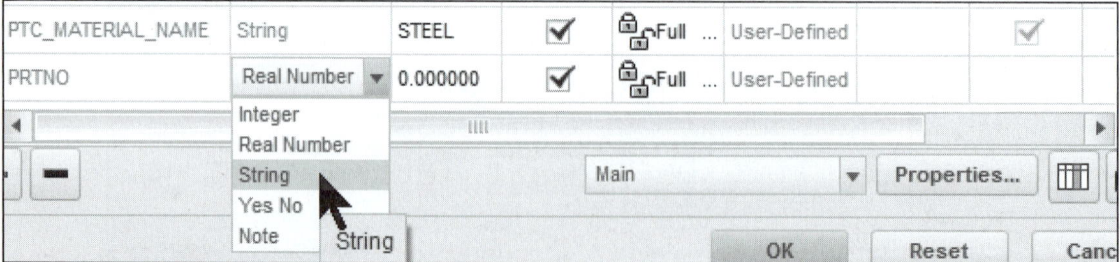

Figure 12.8(g) String

Click in Value field > *type* **SW101-5AR** [Fig. 12.8(h)] > click in Description field > *type* **part number** [Fig. 12.8(i)] > for PTC_MATERIAL_NAME, click: Designate

DESCRIPTION	String		✓	Full ...	User-Defined		
MODELED_BY	String		✓	Full ...	User-Defined		
PTC_MATERIAL_NAME	String	STEEL	✓	Full ...	User-Defined		✓
PRTNO	String	SW101-5AR	✓	Full ...	User-Defined		

Figure 12.8(h) Value SW101-5AR (text is case sensitive – text appears in the drawing BOM as entered here)

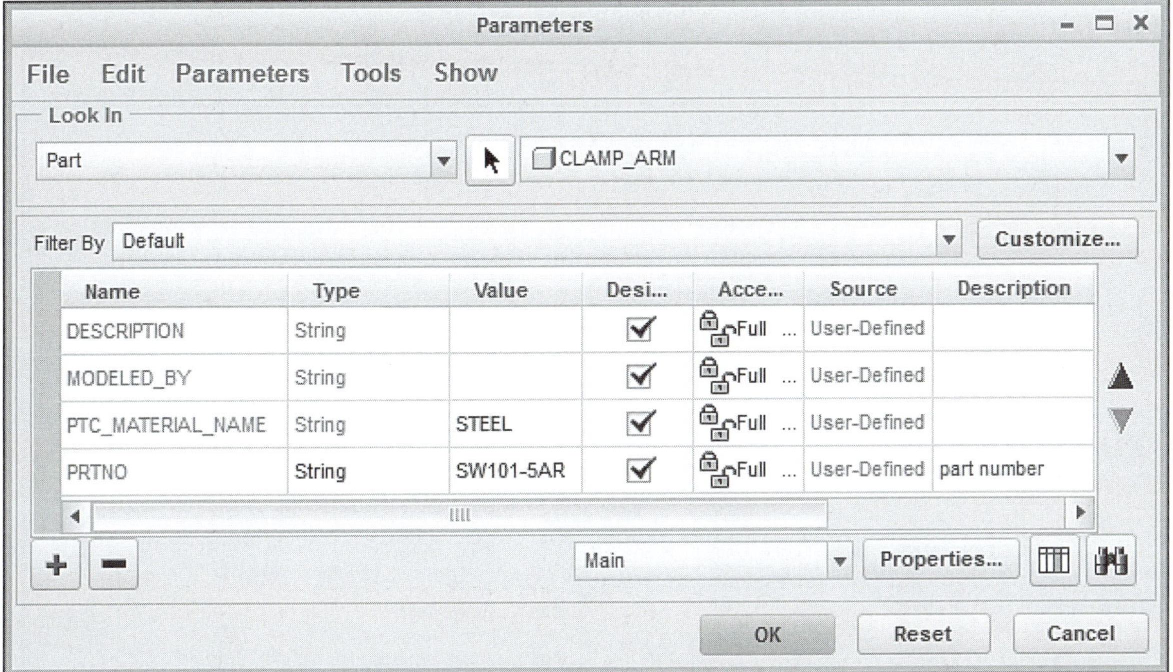

Figure 12.8(i) Add Description (describes the type of parameter – not shown in the drawing BOM)

Click: **Add new Parameter** [Fig. 12.8(j)] > in Name field, *type* **DSC** > click in Type field > > **String** > click in Value field > *type* **CLAMP ARM** > Designate > click in Description field > *type* **part description** [Fig. 12.8(k)] > **OK** > **Family Table** > **Verify** > **Verify** > **Close** > **OK** > **Ctrl+S** > **Enter**

PTC_MATERIAL_NAME	String	STEEL	✓	Full ...	User-Defined	
PRTNO	String	SW101-5AR	✓	Full ...	User-Defined	part number
PARAMETER_1	Real Number	0.000000	☐	Full ...	User-Defined	

Figure 12.8(j) Add New Parameter

PTC_MATERIAL_NAME	String	STEEL	✓	Full ...	User-Defined		✓
PRTNO	String	SW101-5AR	✓	Full ...	User-Defined	part number	
DSC	String	CLAMP ARM	✓	Full ...	User-Defined	part description	

Figure 12.8(k) Add Description

Parameters can also be displayed using the Relations dialog box. In the case of the Clamp_Arm, there was a relation created for controlling a features location. Click: **Relations >** Local Parameters [Fig. 12.8(l)] > **OK > Ctrl+S > File >** **Close**

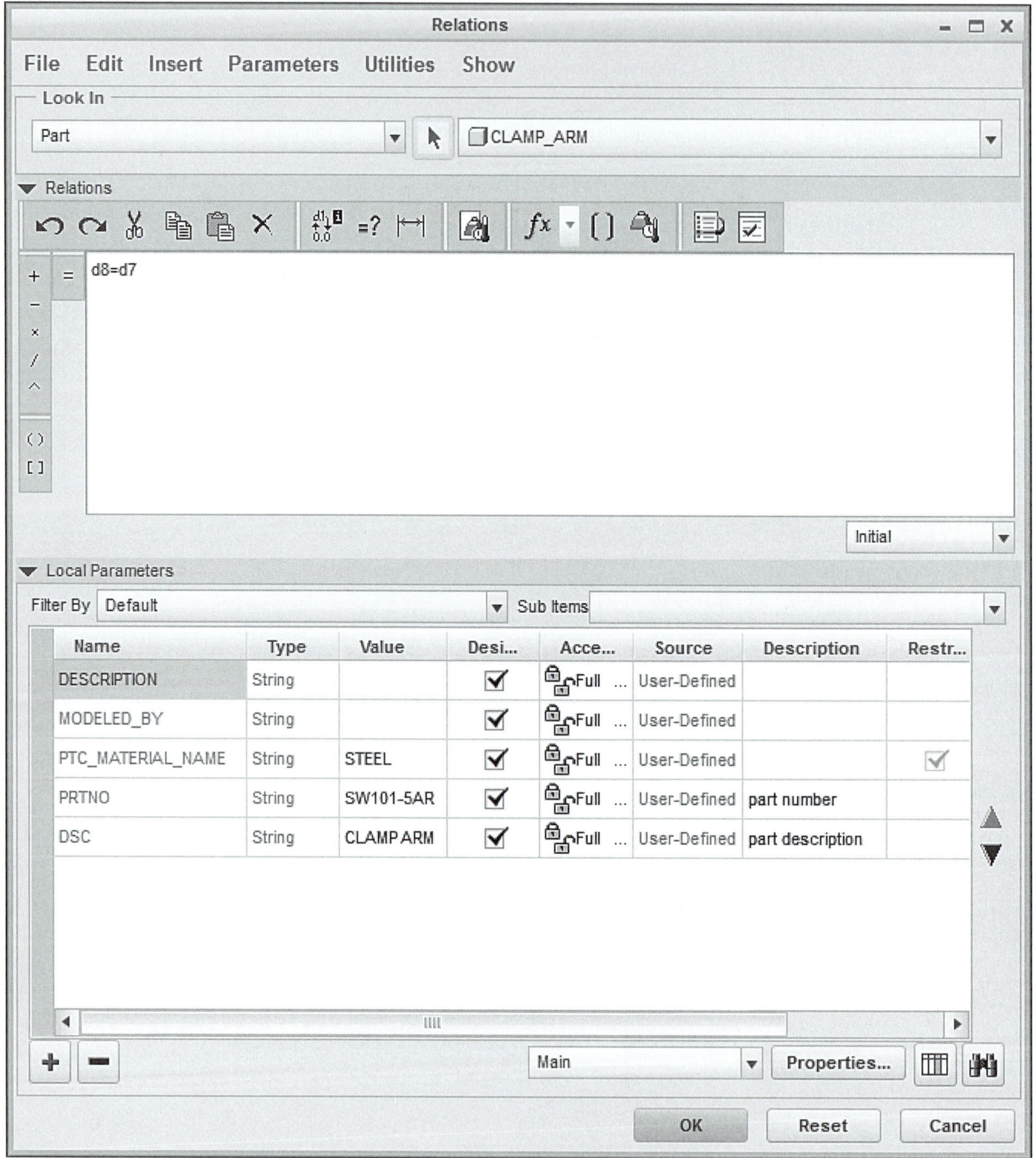

Name	Type	Value	Desi...	Acce...	Source	Description	Restr...
DESCRIPTION	String		✓	Full ...	User-Defined		
MODELED_BY	String		✓	Full ...	User-Defined		
PTC_MATERIAL_NAME	String	STEEL	✓	Full ...	User-Defined		✓
PRTNO	String	SW101-5AR	✓	Full ...	User-Defined	part number	
DSC	String	CLAMP ARM	✓	Full ...	User-Defined	part description	

Figure 12.8(l) Relations Dialog Box *(your relation values may be different)*

Retrieve the clamp swivel, click: **File > Open > clamp_swivel.prt > Open > Tools** tab **> Parameters > Parameters > Add Parameter >** complete the parameters as shown (Fig. 12.9) (remember to *type* the appropriate text in the Description fields) **> OK > File > Save > Enter > File > Close**

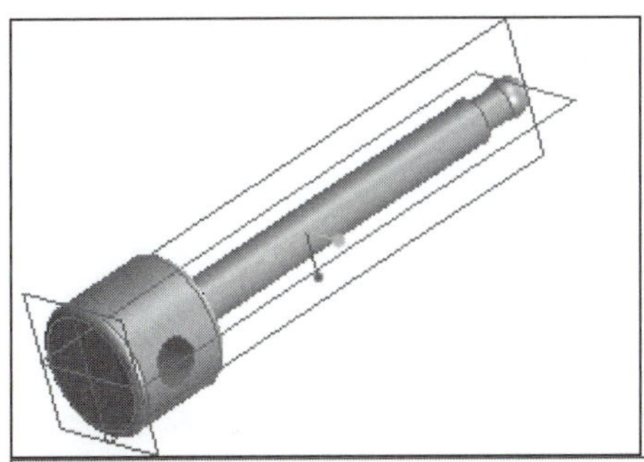

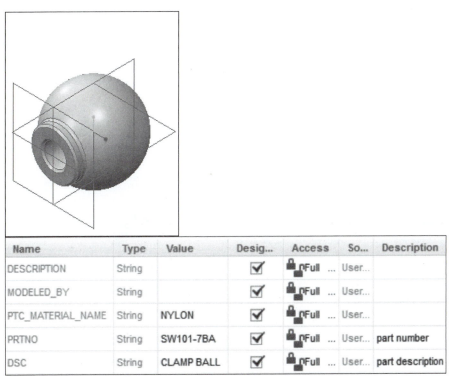

Name	Typ...	Value	Designate	Access	Source	Description
DESCRIPTION	String		☑	Full ...	User-Defined	
MODELED_BY	String		☑	Full ...	User-Defined	
PTC_MATERIAL_NAME	String	STEEL_1066	☑	Full ...	User-Defined	
PRTNO	String	SW101-6SW	☑	Full ...	User-Defined	part number
DSC	String	CLAMP SWIVEL	☑	Full ...	User-Defined	part description

Figure 12.9 Clamp_Swivel Parameters

Retrieve the clamp ball, click: **File > Open > clamp_ball.prt > Open > Tools** tab **> Parameters > Parameters > Add Parameter >** complete the parameters as shown (Fig. 12.10) (remember to *type* the appropriate text in the Description fields) **> OK > File > Save > Enter > File > Close**

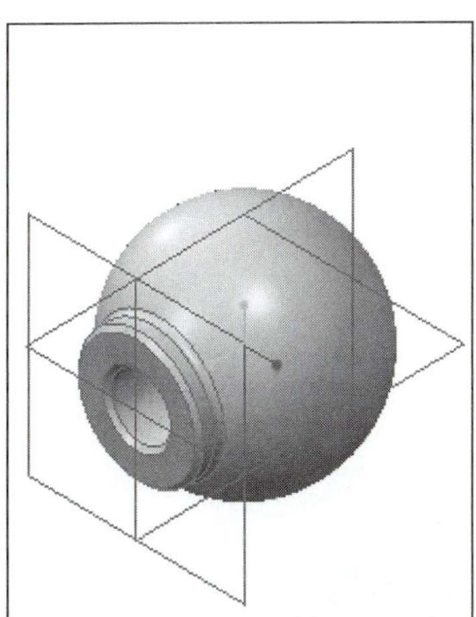

Name	Type	Value	Desig...	Access	So...	Description
DESCRIPTION	String		☑	Full ...	User...	
MODELED_BY	String		☑	Full ...	User...	
PTC_MATERIAL_NAME	String	NYLON	☑	Full ...	User...	
PRTNO	String	SW101-7BA	☑	Full ...	User...	part number
DSC	String	CLAMP BALL	☑	Full ...	User...	part description

Figure 12.10 Clamp_Ball Parameters

Retrieve the clamp foot, click: **File** > **Open** > **clamp_foot.prt** > **Open** > **Tools** tab > **Parameters** > **Parameters** > **Add Parameter** > complete the parameters as shown (Fig. 12.11) (remember to *type* the appropriate text in the Description fields) > **OK** > **File** > **Save** > **Enter** > **File** > **Close**

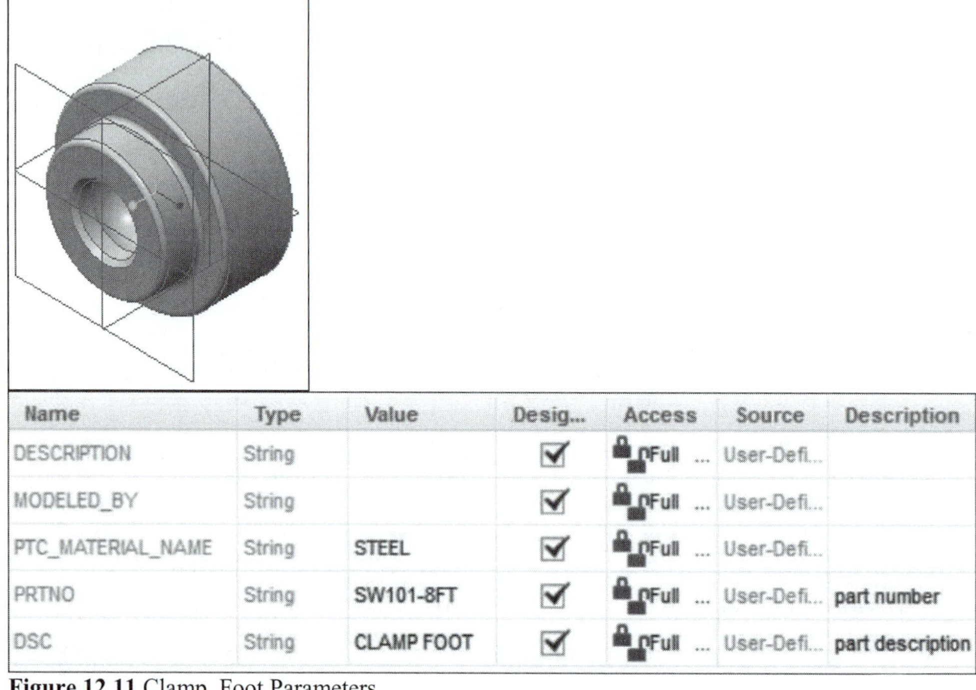

Name	Type	Value	Desig...	Access	Source	Description
DESCRIPTION	String		✓	Full ...	User-Defi...	
MODELED_BY	String		✓	Full ...	User-Defi...	
PTC_MATERIAL_NAME	String	STEEL	✓	Full ...	User-Defi...	
PRTNO	String	SW101-8FT	✓	Full ...	User-Defi...	part number
DSC	String	CLAMP FOOT	✓	Full ...	User-Defi...	part description

Figure 12.11 Clamp_Foot Parameters

Retrieve the clamp plate, click: **File** > **Open** > **clamp_plate.prt** > **Open** > **File** > **Prepare** > **Model Properties** > Material **change** > **steel.mtl** > ▶▶▶ [Fig. 12.12(a)] > **OK** > **Close** > **File** > **Save** > **Enter**

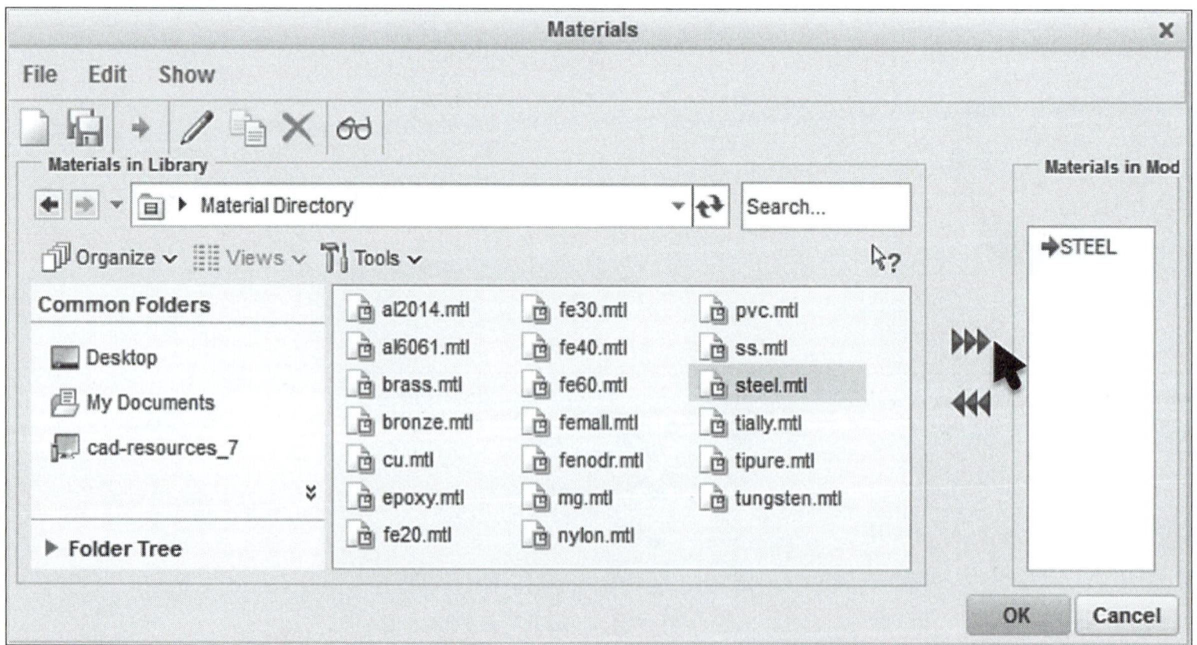

Figure 12.12(a) Materials Dialog Box

Click: [icon] > [Tree Filters...] > toggle all Display options *on* > **OK** > **View** tab > [icons] *on* > [icons] *on* > hide the hole notes in the Model Tree > pick **DTM3** from the model > press **RMB** > **Properties** > double-click in the Name field- *type* **A** > **Enter** > [icon] [Fig. 12.12(b)] > **On Datum** > **OK** [Fig. 12.12(c)] > pick **DTM2** > press **RMB** > **Properties** > double-click in the Name field- *type* **B** > **Enter** > [icon] > **On Datum** > **OK** > pick **DTM1** > press **RMB** > **Properties** > double-click in the Name field- *type* **C** > **Enter** > [icon] > **On Datum** > **OK** > **LMB** to deselect

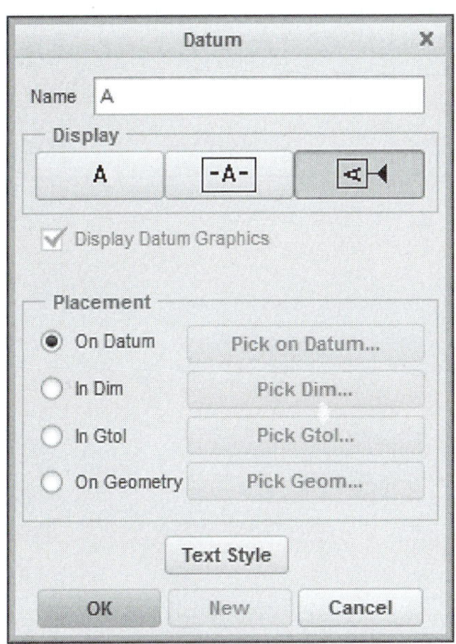

Figure 12.12(b) Datum Dialog Box

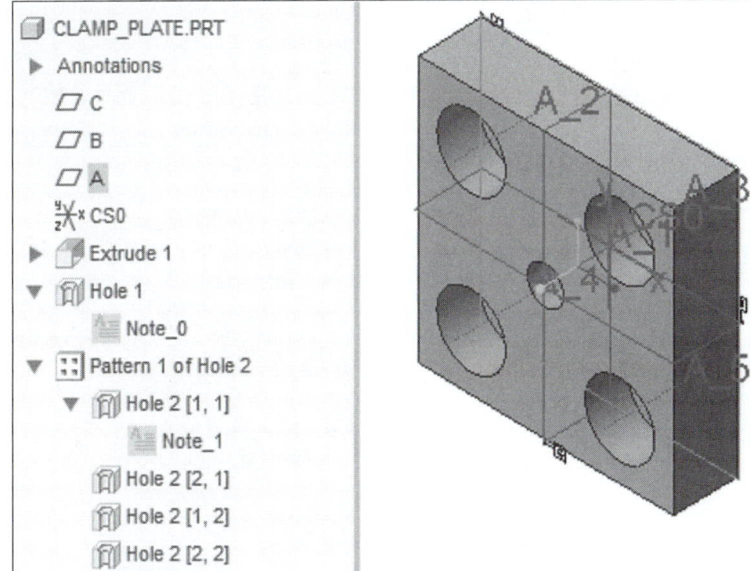

Figure 12.12(c) Set Datum Tag Annotation A

Click: **Tools** tab > **Parameters** > **Parameters** > **Add Parameter** > complete the parameters as shown [Fig. 12.12(d)] > **OK** > **File** > **Save** > **OK** > **File** > **Close**

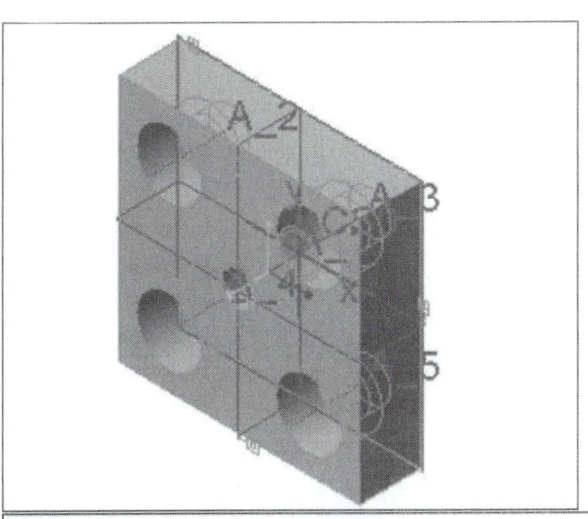

Name	Type	Value	Desig...	Access	Source	Description	Restri...	Unit Q...
PTC_MATERIAL_NAME	String	STEEL	☑	Full ...	User-Defi...		☑	
PRTNO	String	SW100-20PL	☑	Full ...	User-Defi...	part number		
DSC	String	CLAMP PLATE	☑	Full ...	User-Defi...	part description		

Figure 12.12(d) Clamp_Plate Parameters (remember to type the appropriate text in the Description fields)

Click: **File > Open > carrlane_500_fn.prt > Open > File > Prepare > Model Properties >** Material **change > File > New** [Fig. 12.13(a)] > Name field- *type* **purchased** [Fig. 12.13(b)] > **Save to Library**

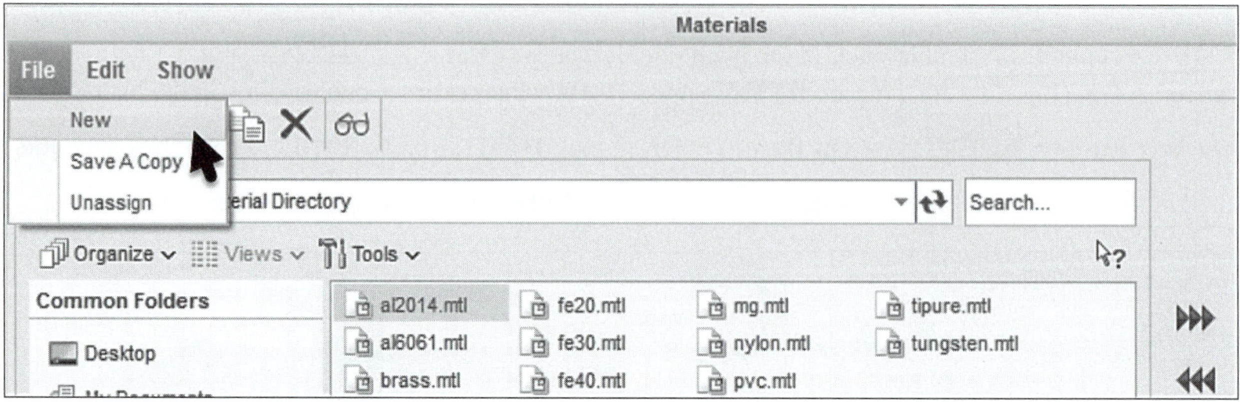

Figure 12.13(a) Materials Dialog Box

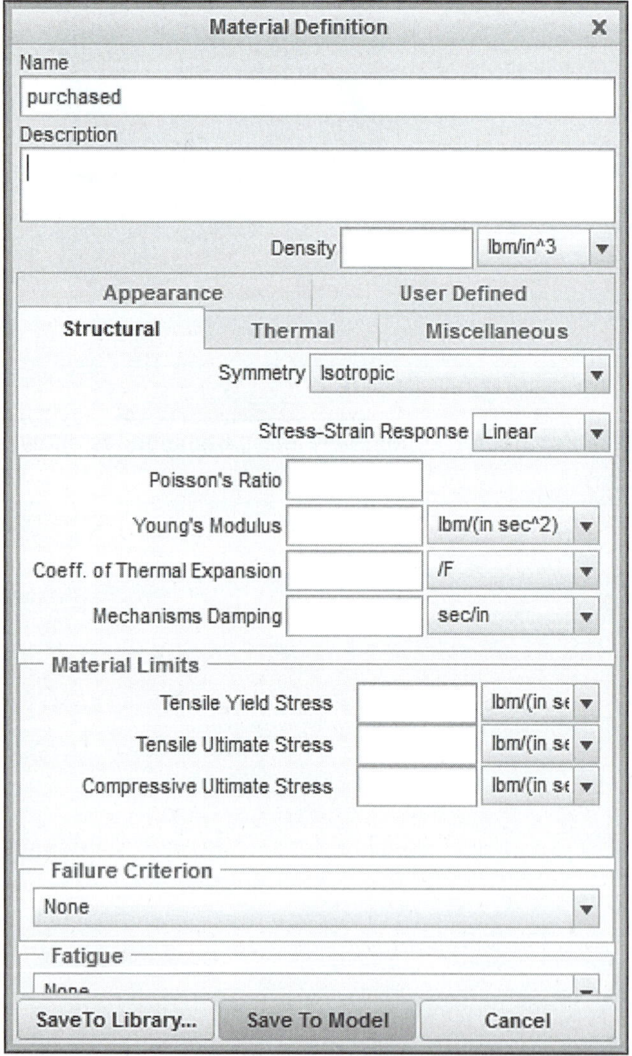

Figure 12.13(b) Material Definition Dialog Box

Navigate to the correct directory where you want the material to be saved, click: **OK** [Fig. 12.13(c)] > navigate to the correct directory (you should be able to click on **Working Directory**) > **purchased.mtl** > [Fig. 12.13(d)] > **OK** > **Close** > **Tools** tab > Model Information [Fig. 12.13(e)] > to close

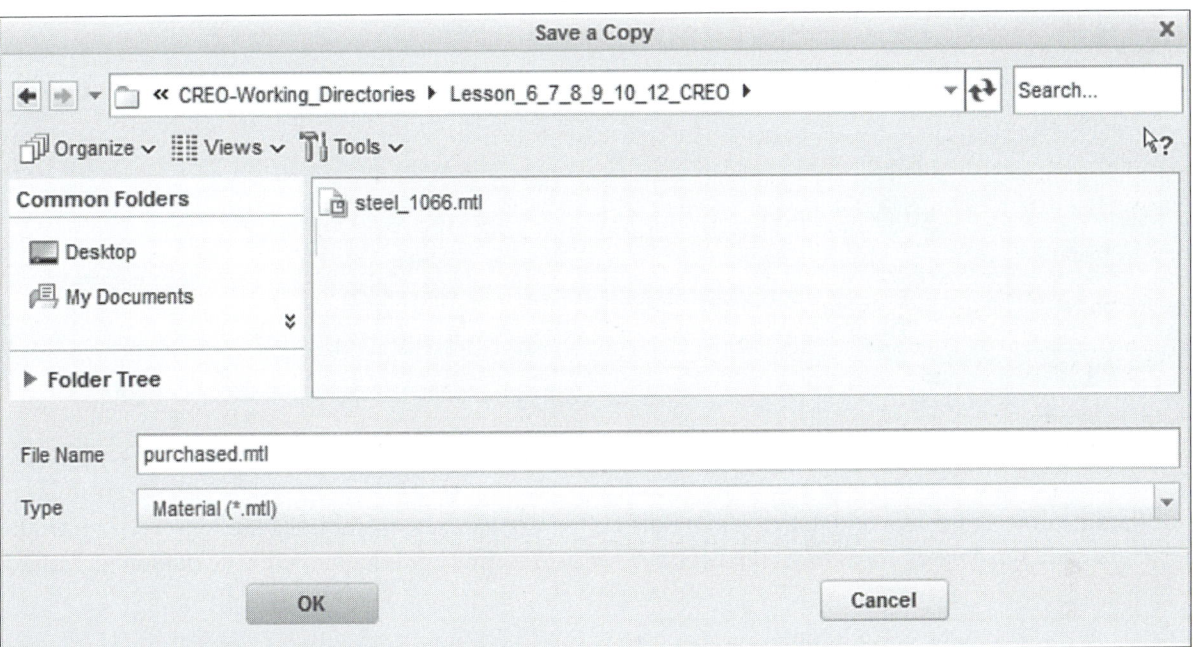

Figure 12.13(c) Save a Copy Dialog Box

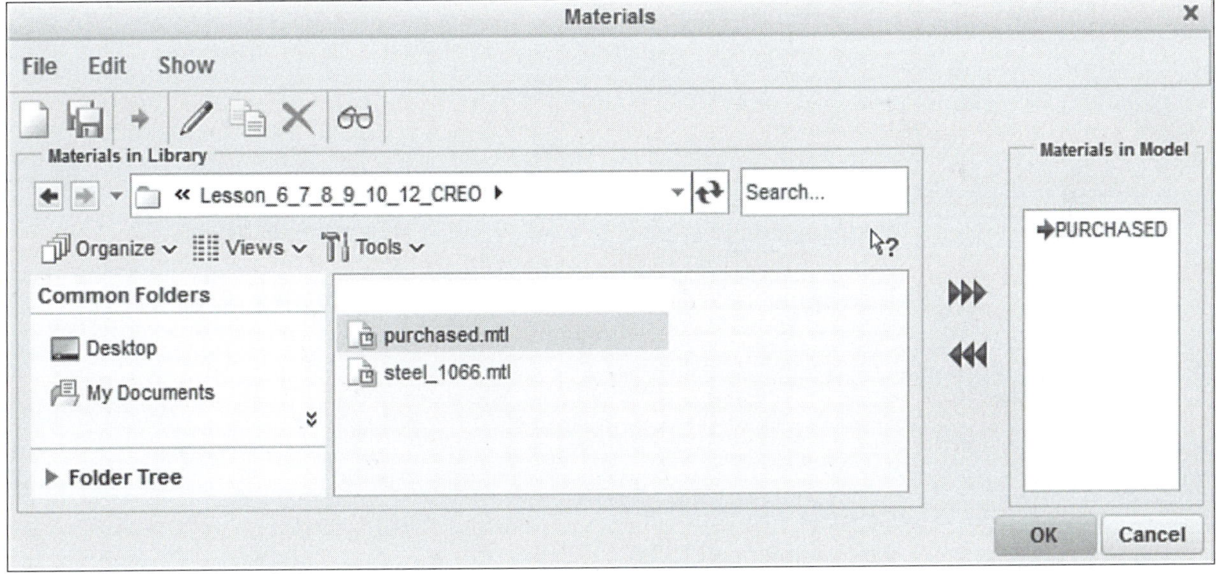

Figure 12.13(d) PURCHASED

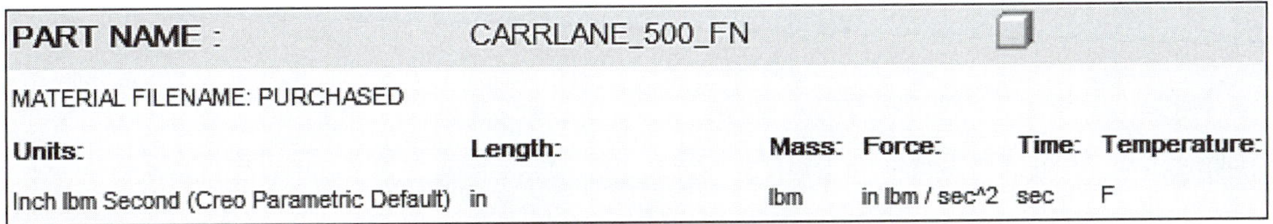

Figure 12.13(e) Model Info, MATERIAL FILENAME: PURCHASED

Click: **Tools** tab > **Parameters** > add the Parameters as shown > Designate [Fig. 12.13(f)] > **OK** > **Ctrl+S** > **OK** > **File** > **Close**

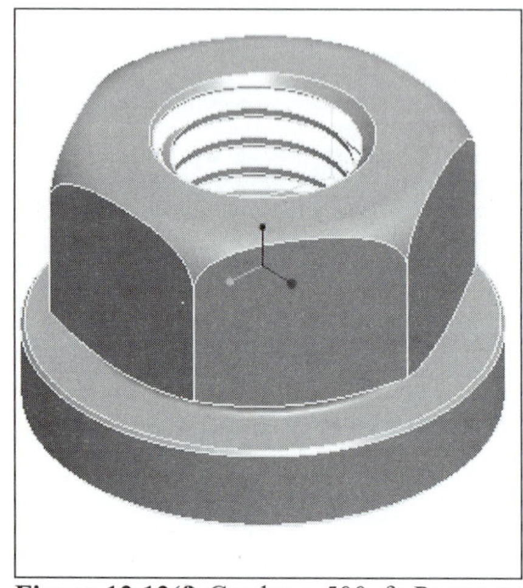

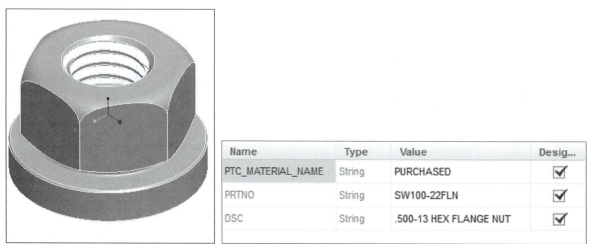

Name	Type	Value	Desig...
PTC_MATERIAL_NAME	String	PURCHASED	☑
PRTNO	String	SW100-22FLN	☑
DSC	String	.500-13 HEX FLANGE NUT	☑

Figure 12.13(f) Carrlane_500_fn Parameters

Parameters can be added, deleted, and modified in Part Mode, Assembly Mode, or Drawing Mode. You can also add *parameter columns* to the Model Tree in Assembly Mode, and edit the parameter value.

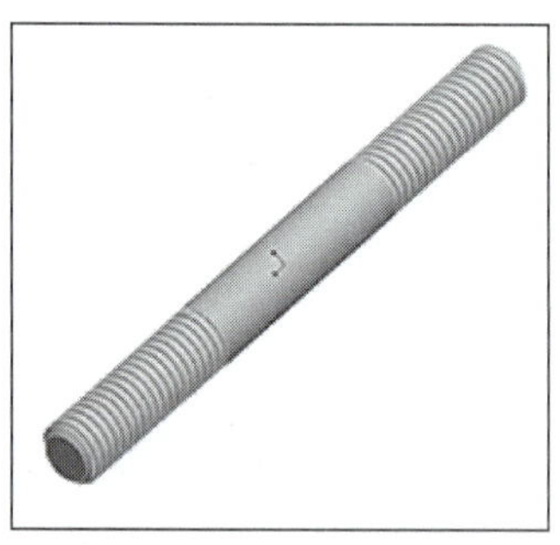

Component	**Carrlane-12-13500_STUD.prt**
Part Number (PRTNO)	**SW101-9STL**
Description (DSC)	**.500-13 X 5.00 DOUBLE END STUD**

Component	**Carrlane-12-13350_STUD.prt**
Part Number (PRTNO)	**SW100-21ST**
Description (DSC)	**.500-13 X 3.50 DOUBLE END STUD**

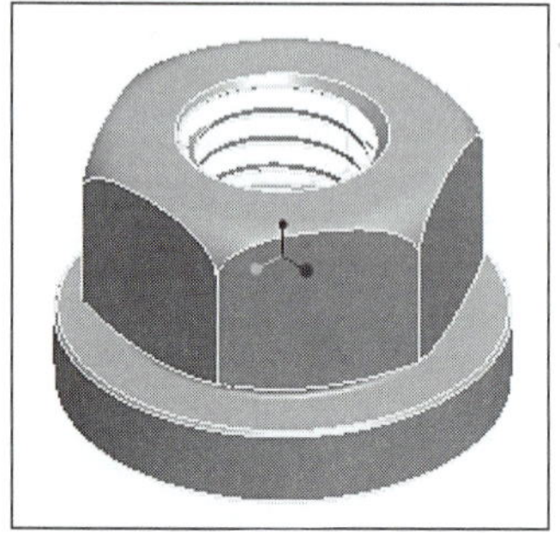

Component	**Carrlane-500_FN.prt**
Part Number (PRTNO)	**SW100-22FLN**
Description (DSC)	**.500-13 HEX FLANGE NUT**

Press: **Ctrl+O** > **Carrlane-12-13350_STUD.prt** [Fig. 12.14(a)] > **Open** > **File** > **Prepare** > **Model Properties** > Material **change** > navigate to the correct directory where you saved the *purchased.mtl* material (you should be able to click on **Working Directory**) > **purchased.mtl** > [Fig. 12.14(b)] > **OK** > **Close** > **Tools** tab > **Parameters** > Designate > **OK** > **Ctrl+S** > **Enter** > **File** > **Close**

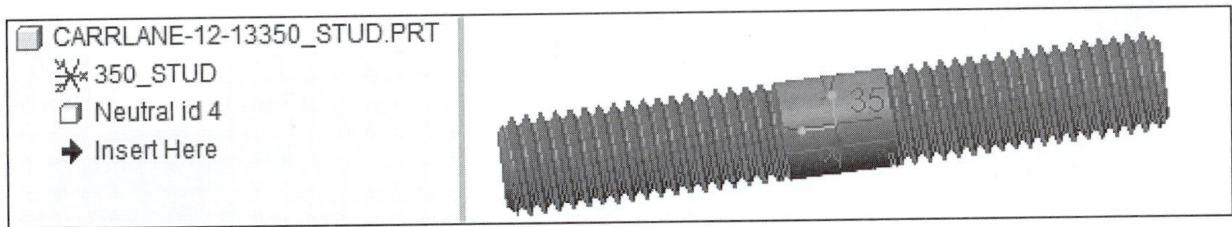

Figure 12.14(a) Carrlane-12-13350_STUD.prt

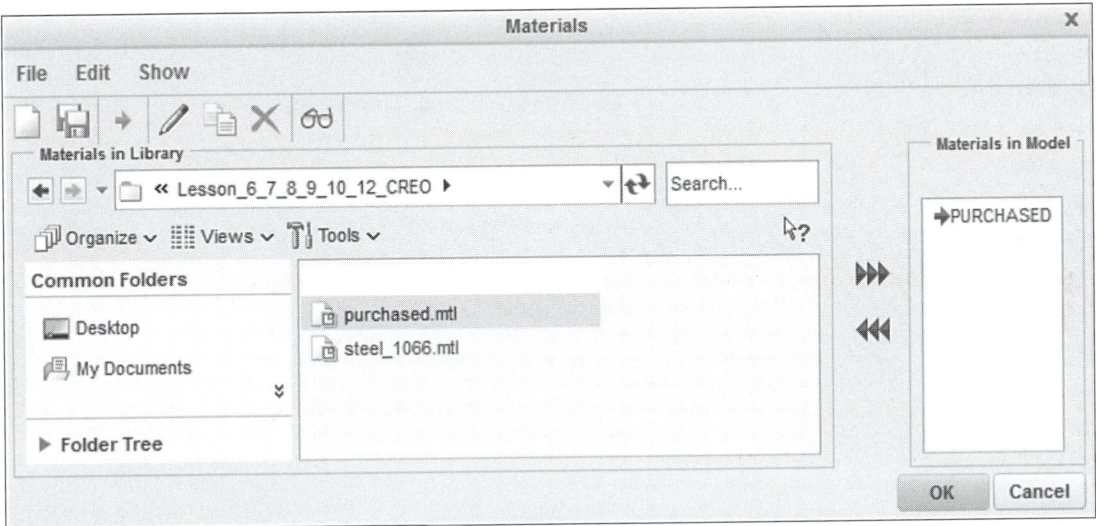

Figure 12.14(b) PURCHASED

Press: **Ctrl+O** > **Carrlane-12-13500_STUD.prt** [Fig. 12.15(a)] > **Open** > **File** > **Prepare** > **Model Properties** > Material **change** > navigate to the correct directory where you saved the *purchased.mtl* material (**Working Directory**) > **purchased.mtl** > > **OK** > **Close** > **Tools** tab > **Parameters** > Designate > **OK** > > **OK** > Model Information [Fig. 12.15(b)] > to close the browser > **File** > **Close**

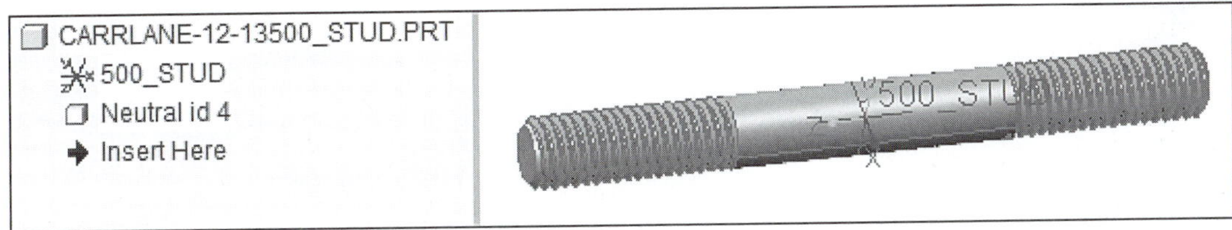

Figure 12.15(a) Carrlane-12-13500_STUD.prt

PART NAME : CARRLANE-12-13500_STUD

MATERIAL FILENAME: PURCHASED

Units:	Length:	Mass:	Force:	Time:	Temperature:
Inch lbm Second (Creo Parametric Default)	in	lbm	in lbm / sec^2	sec	F

Figure 12.15(b) Model Info

Click: **File** > **Open** > **clamp_assembly.asm** > **Open** > [Settings icon] **Settings** in the Navigator/Model Tree > [Tree Filters...] > check all Display options *on* > **Apply** > **OK** > [CLAMP_SUBASSEMBLY.ASM] expand the sub-assembly > [Settings icon] **Settings** > [Tree Columns...] [Fig. 12.16(a)] > Type: click [Info] > **Model Params** [Fig. 12.16(b)]

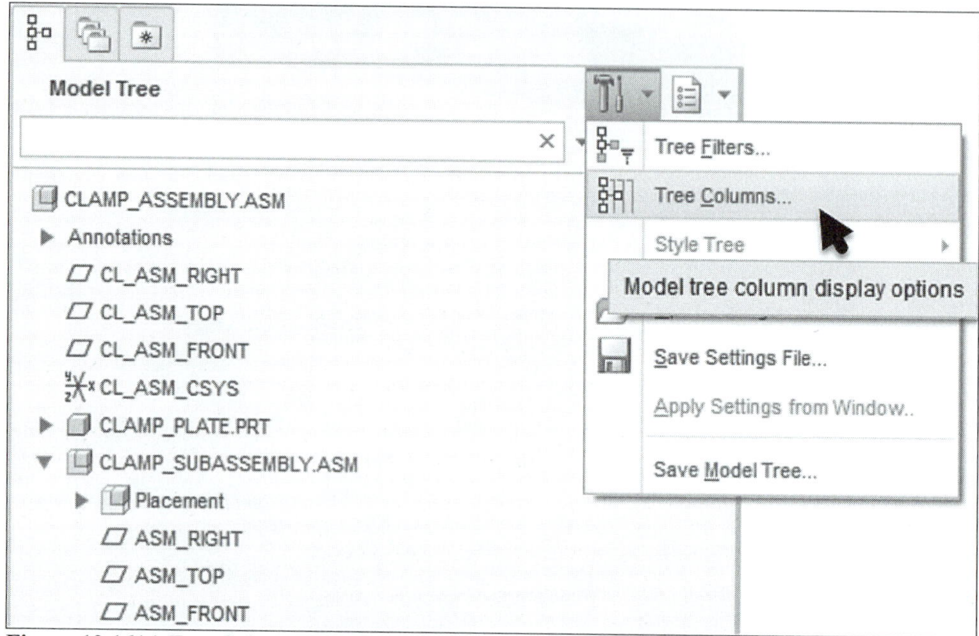

Figure 12.16(a) Tree Columns

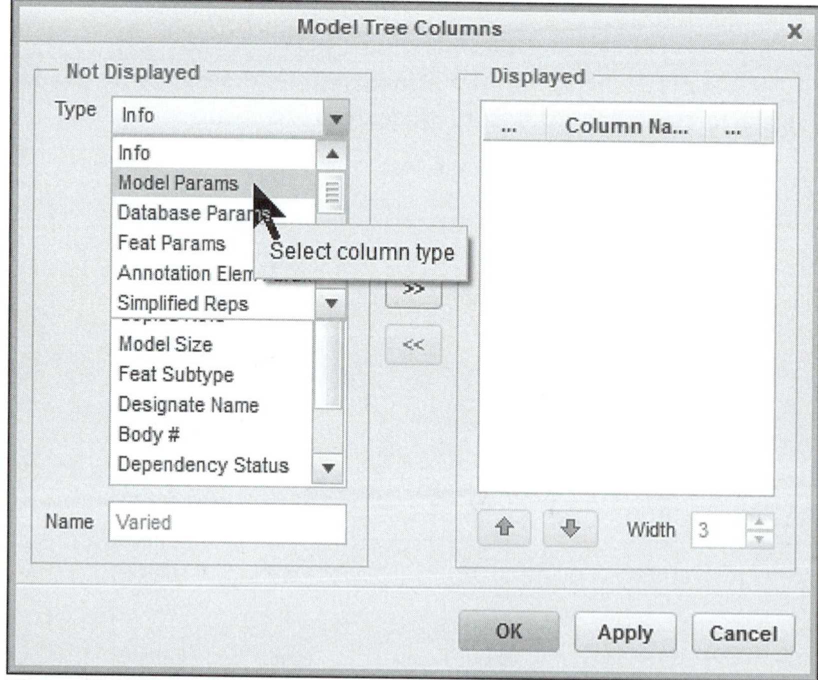

Figure 12.16(b) Model Params

Double-click in the Name field > *type* **PRTNO** [Fig. 12.16(c)] > [>>] **Add column** > double-click in the Name field > *type* **DSC** [Fig. 12.16(d)] > [>>] **Add column** [Fig. 12.16(e)]

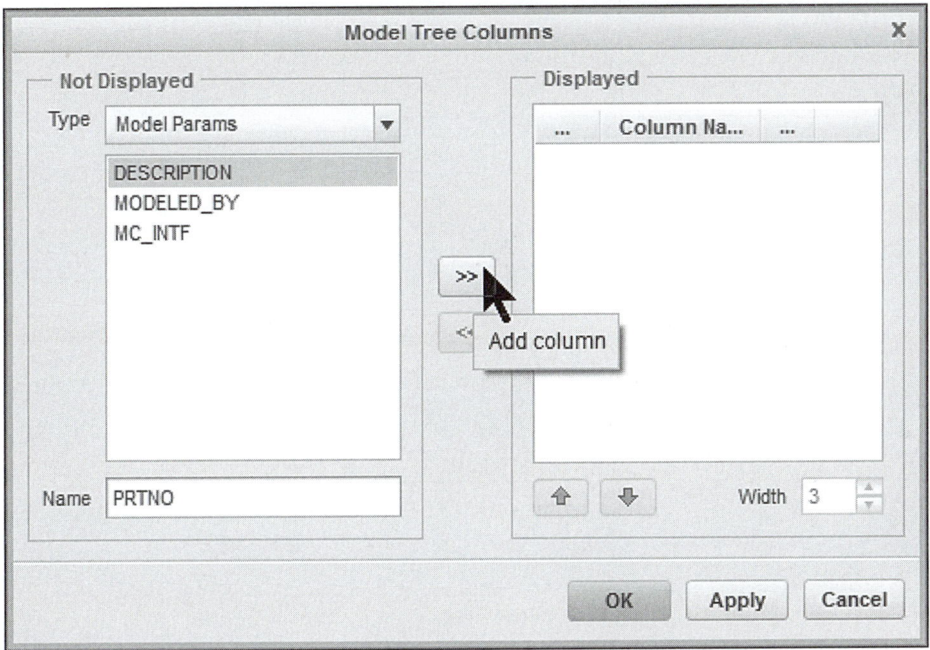

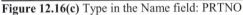

Figure 12.16(c) Type in the Name field: PRTNO

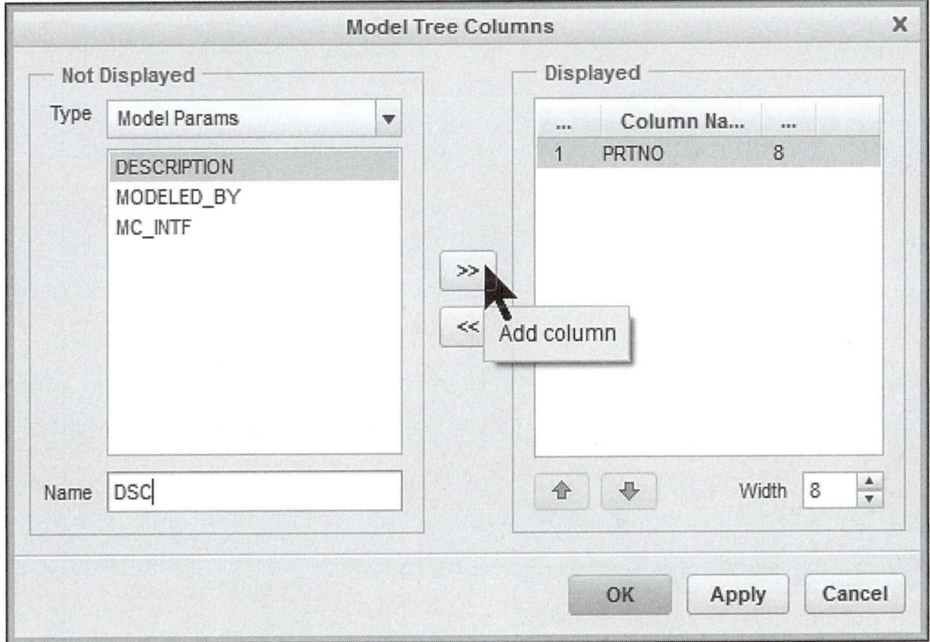

Figure 12.16(d) Type in the Name field: DSC

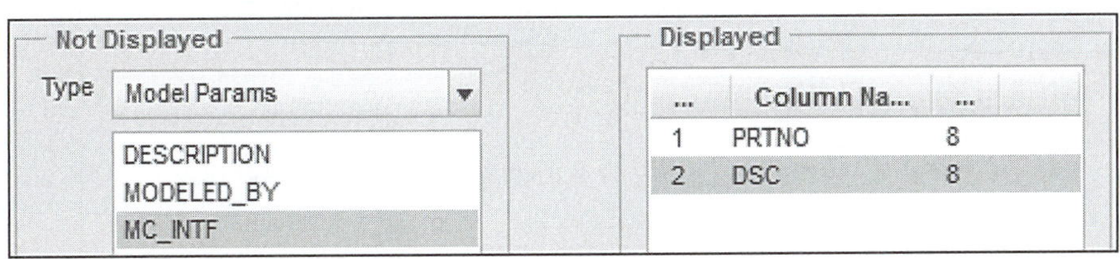

Figure 12.16(e) PRTNO and DSC

Double-click in the Name field > *type* **PTC_MATERIAL_NAME** [Fig. 12.16(f)] > >> **Add column** [Fig. 12.16(g)] > **Apply** > **OK** [Fig. 12.16(h)]

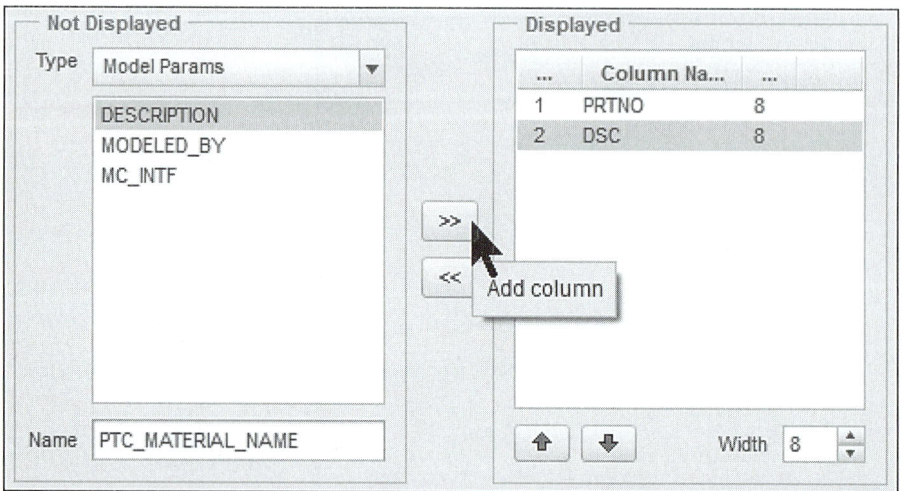

Figure 12.16(f) Type in the Name field: PTC_MATERIAL_NAME

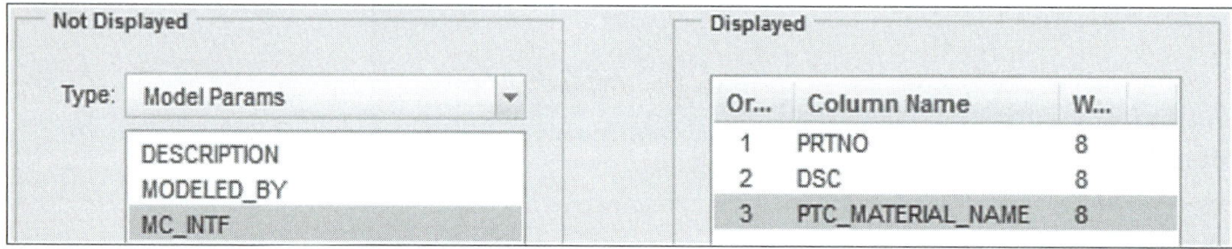

Figure 12.16(g) PTC_MATERIAL_NAME

	PRTNO	DSC	PTC_MATERIAL_NAME
CLAMP_ASSEMBLY.ASM			
Annotations			
CL_ASM_RIGHT			
CL_ASM_TOP			
CL_ASM_FRONT			
CL_ASM_CSYS			
CLAMP_PLATE.PRT	SW100-20PL	CLAMP PLATE	STEEL
CLAMP_SUBASSEMBLY.ASM			
Placement			
ASM_RIGHT			
ASM_TOP			
ASM_FRONT			
SUB_ASM_CSYS			
CLAMP_ARM_DESIGN<CL	SW101-5AR	CLAMP ARM	STEEL
CLAMP_SWIVEL.PRT	SW101-6SW	CLAMP SWIVEL	STEEL_1066
CLAMP_FOOT.PRT	SW101-8FT	CLAMP FOOT	STEEL
CARRLANE-12-13500_ST	SW101-9STL	.500-13 X 5.00 DOUBLE END STUD	PURCHASED
CLAMP_BALL.PRT	SW101-7BA	CLAMP BALL	NYLON
CLAMP_BALL.PRT	SW101-7BA	CLAMP BALL	NYLON
Insert Here			
CARRLANE-12-13350_STUD.	SW100-21ST	.500-13 X 3.50 DOUBLE END STUD	PURCHASED
CARRLANE_500_FN.PRT	SW100-22FLN	.500-13 HEX FLANGE NUT	PURCHASED

Figure 12.16(h) New Tree Columns *(your tree columns may appear differently)*

Resize the Model Tree and Model Tree columns. From the Model Tree, pick on **CLAMP_SUBASSEMBLY.ASM** > click in the **PRTNO** field [Fig. 12.17(a)] > Type: click [Real Number] > **String** > click in the Value field, *type* **SW101-SUB-ASM** > [✓ Designated] > click in the Description field, *type* **part number** [Fig. 12.17(b)] > **OK** > click in the **DSC** field > Type: [Real Number] > **String** > click in the Value field, *type* **CLAMP SUB-ASSEMBLY** > [✓ Designated] > click in the Description field, *type* **part description** > **OK** > **LMB** in the Graphics Window to deselect > **File** > **Save** > **OK**

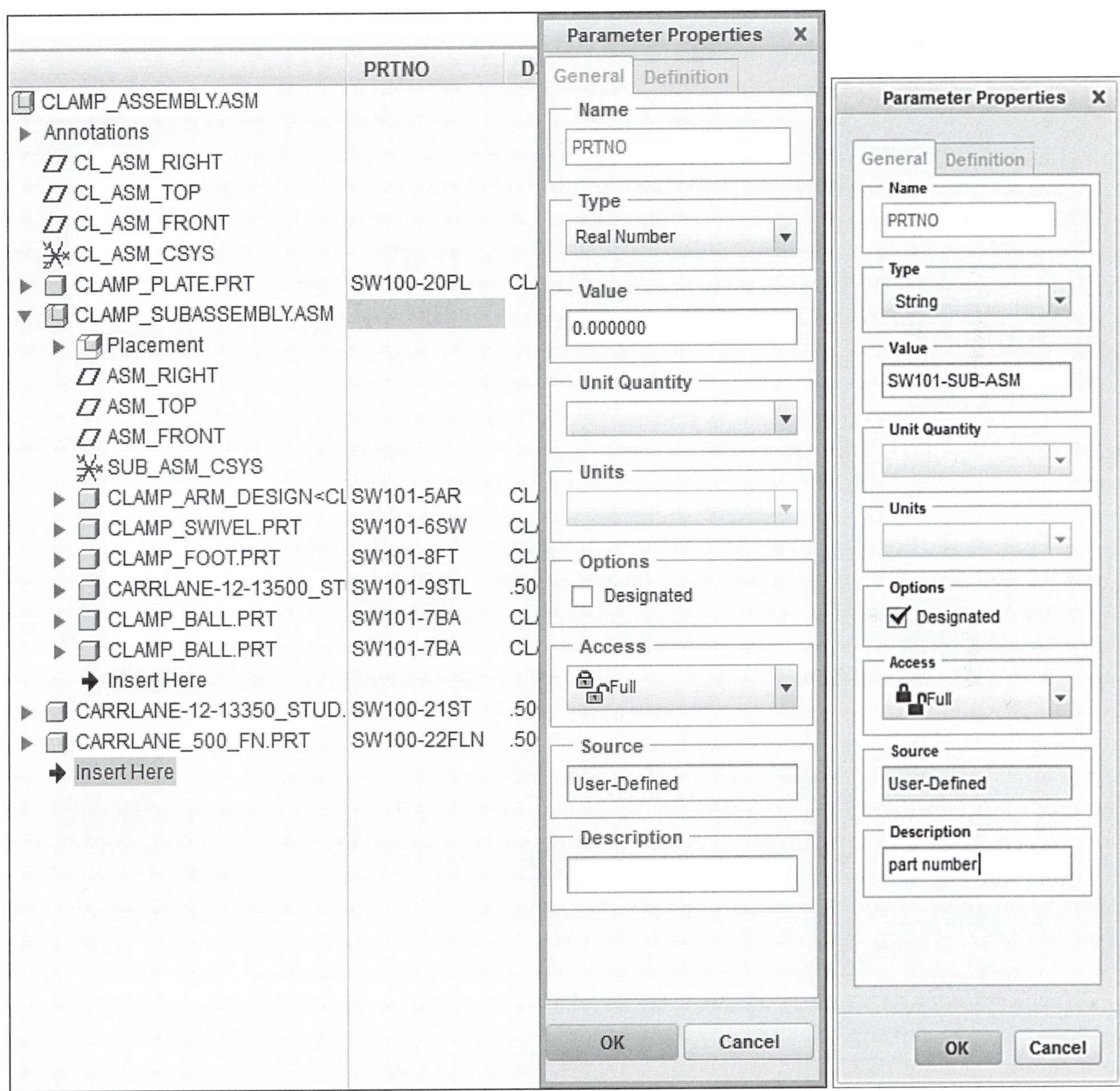

Figure 12.17(a) New Clamp Subassembly Parameter String for PRTNO

Figure 12.17(b) SW101-SUB-ASM

Complete any missing parameters for the remaining components. Where necessary modify the values to reflect what is shown here by clicking in the value field and editing the field (Fig. 12.18) (remember to click: Designated and type the appropriate text in the Description fields) > **LMB** to deselect > double-click to close the Navigator/Model Tree panel to display just the Graphics Window > **Ctrl+S** > **File** > **Close**

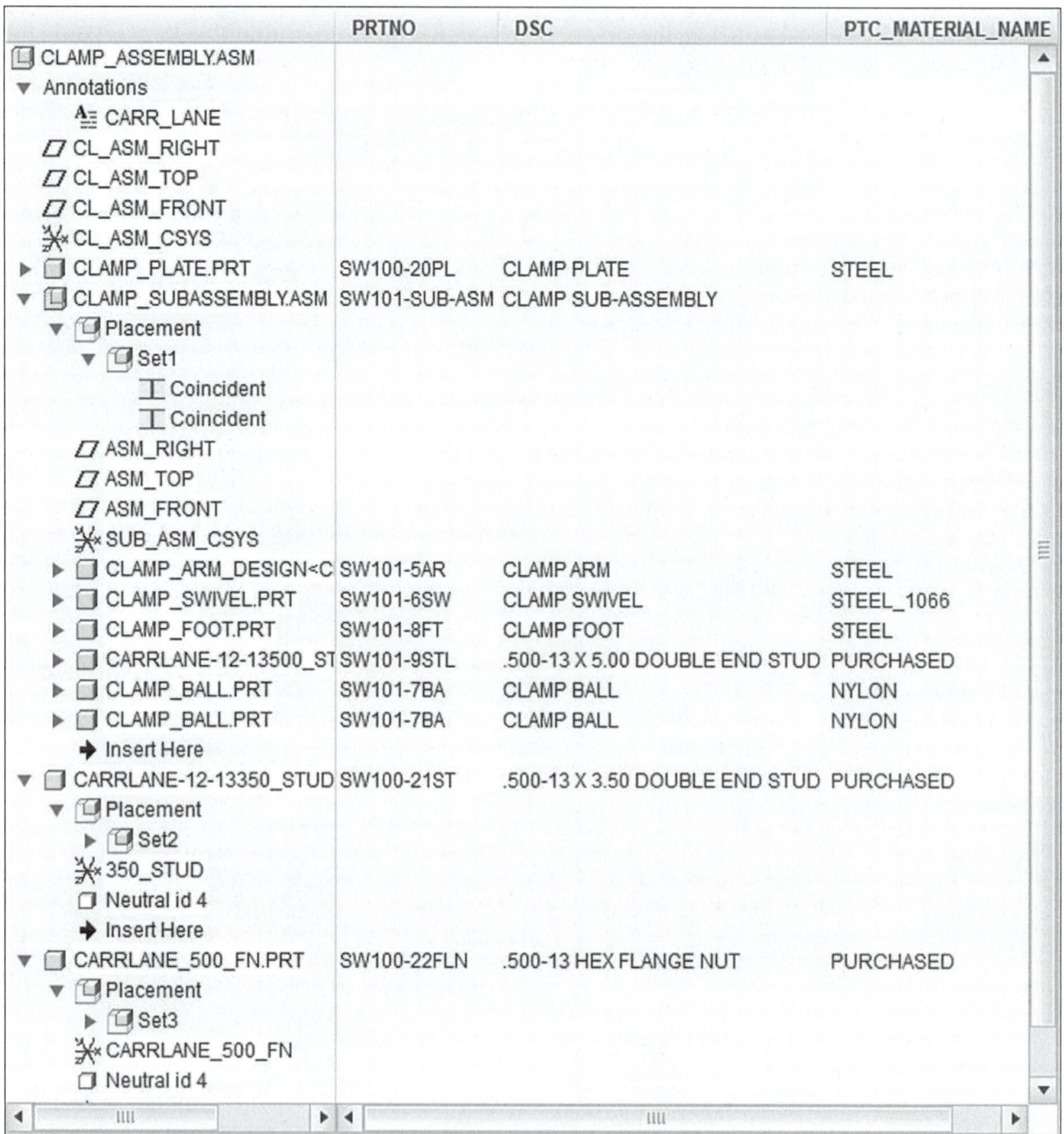

PRTNO
DSC
PTC_MATERIAL_NAME
CLAMP_ASSEMBLY.ASM
Annotations
CARR_LANE
CL_ASM_RIGHT
CL_ASM_TOP
CL_ASM_FRONT
CL_ASM_CSYS
CLAMP_PLATE.PRT SW100-20PL CLAMP PLATE STEEL
CLAMP_SUBASSEMBLY.ASM SW101-SUB-ASM CLAMP SUB-ASSEMBLY
Placement
Set1
Coincident
Coincident
ASM_RIGHT
ASM_TOP
ASM_FRONT
SUB_ASM_CSYS
CLAMP_ARM_DESIGN<C SW101-5AR CLAMP ARM STEEL
CLAMP_SWIVEL.PRT SW101-6SW CLAMP SWIVEL STEEL_1066
CLAMP_FOOT.PRT SW101-8FT CLAMP FOOT STEEL
CARRLANE-12-13500_ST SW101-9STL .500-13 X 5.00 DOUBLE END STUD PURCHASED
CLAMP_BALL.PRT SW101-7BA CLAMP BALL NYLON
CLAMP_BALL.PRT SW101-7BA CLAMP BALL NYLON
Insert Here
CARRLANE-12-13350_STUD SW100-21ST .500-13 X 3.50 DOUBLE END STUD PURCHASED
Placement
Set2
350_STUD
Neutral id 4
Insert Here
CARRLANE_500_FN.PRT SW100-22FLN .500-13 HEX FLANGE NUT PURCHASED
Placement
Set3
CARRLANE_500_FN
Neutral id 4

Figure 12.18 Assembly Model Tree and Parameters shown in the Model Tree Columns

Assembly Drawings

The parameters (and their values) have been established for each part. The assembly format with related parameters in a parts list table has been created and saved in your (format) directory. You can now create a drawing of the assembly, where the parts list will be generated automatically, and the assembly ballooned. The first assembly drawing will be of the ***Clamp_Subassembly.***

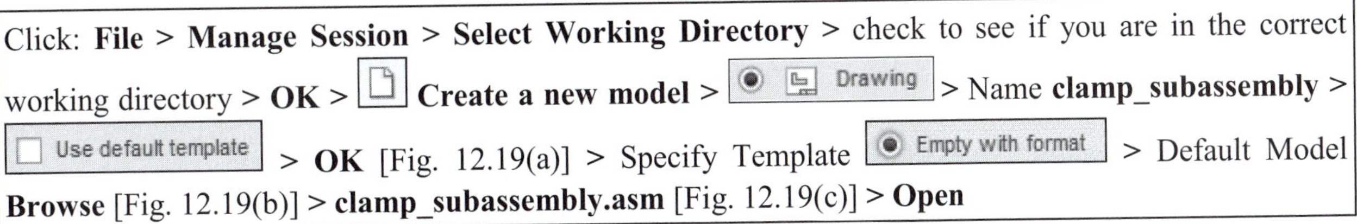

Click: **File** > **Manage Session** > **Select Working Directory** > check to see if you are in the correct working directory > **OK** > **Create a new model** > Drawing > Name **clamp_subassembly** > Use default template > **OK** [Fig. 12.19(a)] > Specify Template Empty with format > Default Model **Browse** [Fig. 12.19(b)] > **clamp_subassembly.asm** [Fig. 12.19(c)] > **Open**

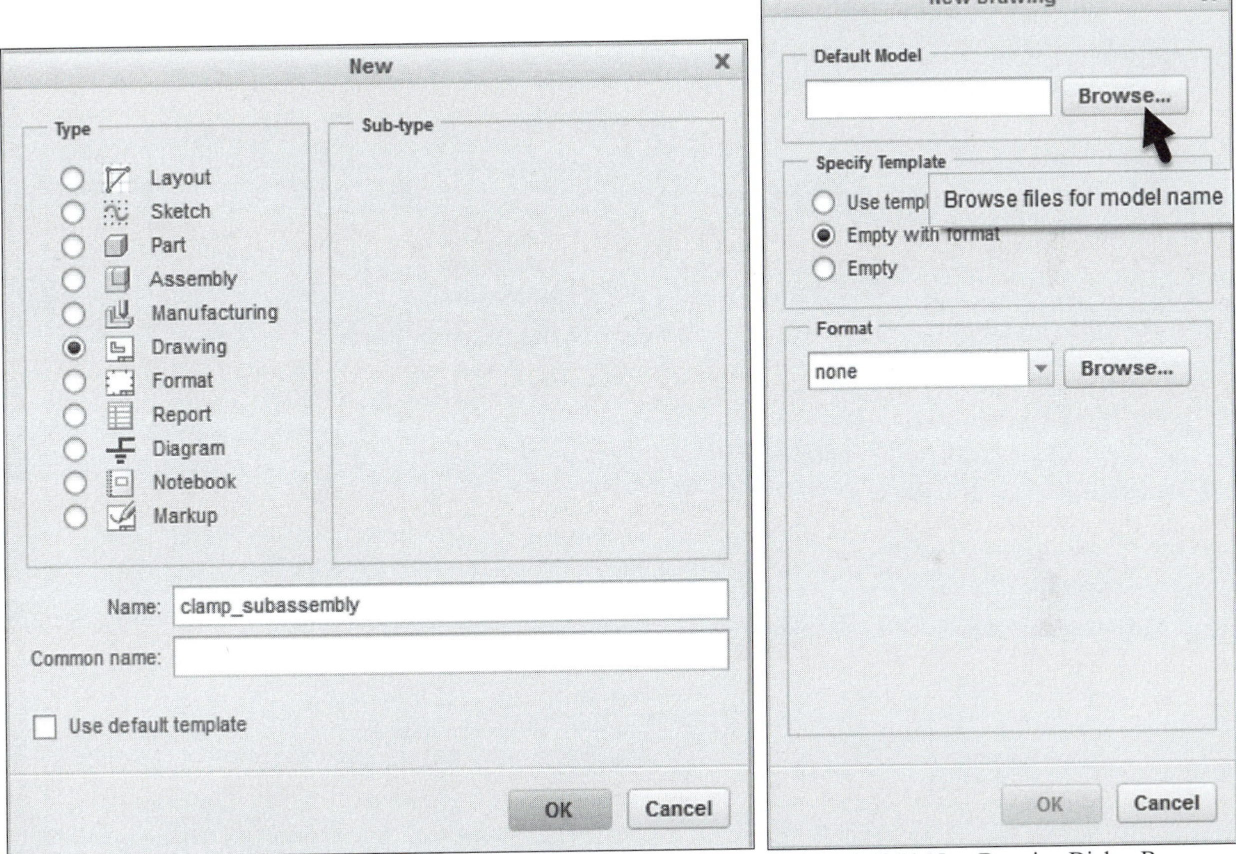

Figure 12.19(a) New Dialog Box

Figure 12.19(b) New Drawing Dialog Box

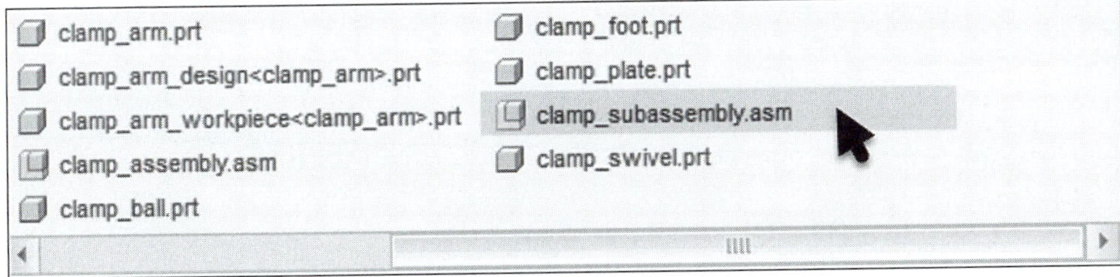

Figure 12.19(c) Open the clamp_subassembly.asm

Click: Format **Browse** [Fig. 12.19(d)] > **Working Directory** [Fig. 12.19(e)] > select **asm_format_e.frm** [Fig. 12.19(f)] > **Open** [Fig. 12.19(g)] > **OK** > **Table** tab > pan and zoom in on the title block region > pick on a table corner to highlight > hold down the **LMB** and move the table outside of the title block [Fig. 12.19(h)] > release the **LMB** > move the pointer > **LMB** to deselect

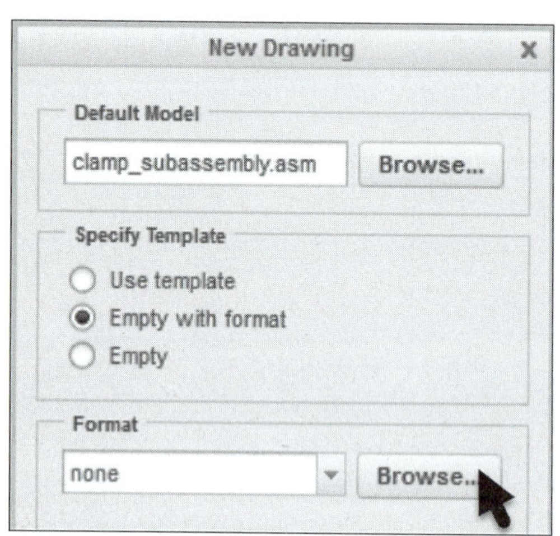

Figure 12.19(d) Format Browse

Figure 12.19(e) Working Directory

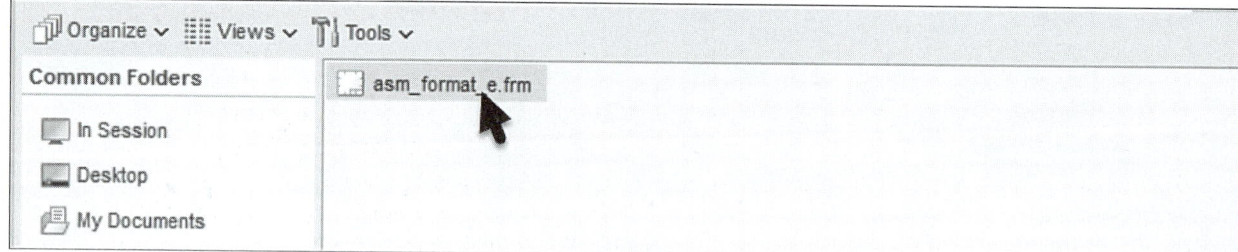

Figure 12.19(f) Select your previously created **asm_format_e.frm**

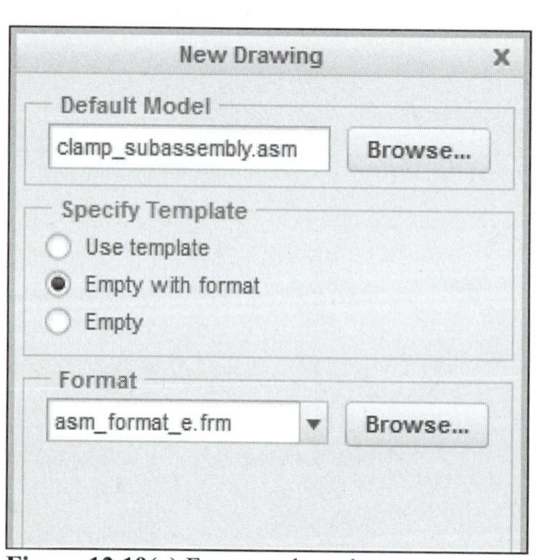

Figure 12.19(g) Format selected

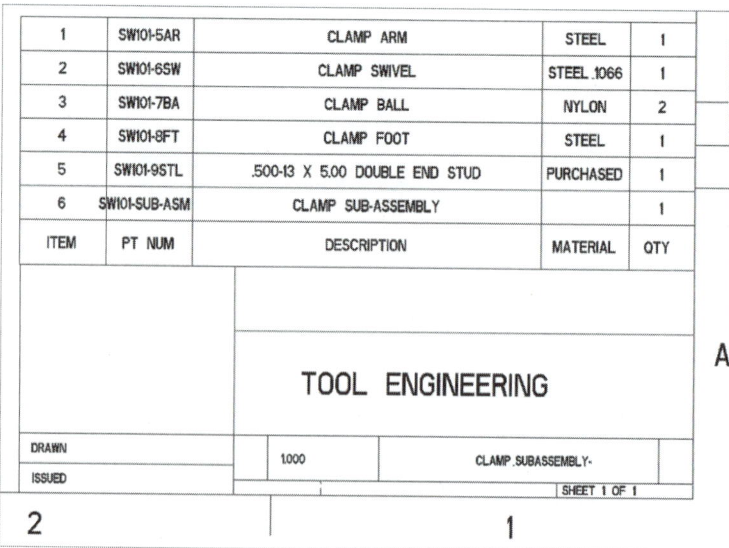

1	SW101-5AR	CLAMP ARM	STEEL	1
2	SW101-6SW	CLAMP SWIVEL	STEEL 1066	1
3	SW101-7BA	CLAMP BALL	NYLON	2
4	SW101-8FT	CLAMP FOOT	STEEL	1
5	SW101-9STL	.500-13 X 5.00 DOUBLE END STUD	PURCHASED	1
6	SW101-SUB-ASM	CLAMP SUB-ASSEMBLY		1
ITEM	PT NUM	DESCRIPTION	MATERIAL	QTY

Figure 12.19(h) Title Block and BOM

Click: **File** > **Prepare** > **Drawing Properties** > Detail Options **change** > Sort: **Alphabetical** > ***arrow_style filled*** > **Add/Change** > ***crossec_arrow_length .50*** > **Enter** > ***crossec_arrow_width .17*** > **Enter** > ***default_font filled*** > **Add/Change** > ***draw_arrow_length .375*** > **Add/Change** > ***draw_arrow_width .125*** > **Enter** > ***max_balloon_radius .50*** > **Enter** > ***min_balloon_radius .50*** > **Enter** > ***text_height .50*** > **Add/Change** > **Apply** > **Save a copy of the currently displayed configuration file** [Fig. 12.20(a)] > File Name **CLAMP_ASM** > **OK** [Fig. 12.20(b)] > **Close** > **Close** > **Ctrl+S** > **OK**

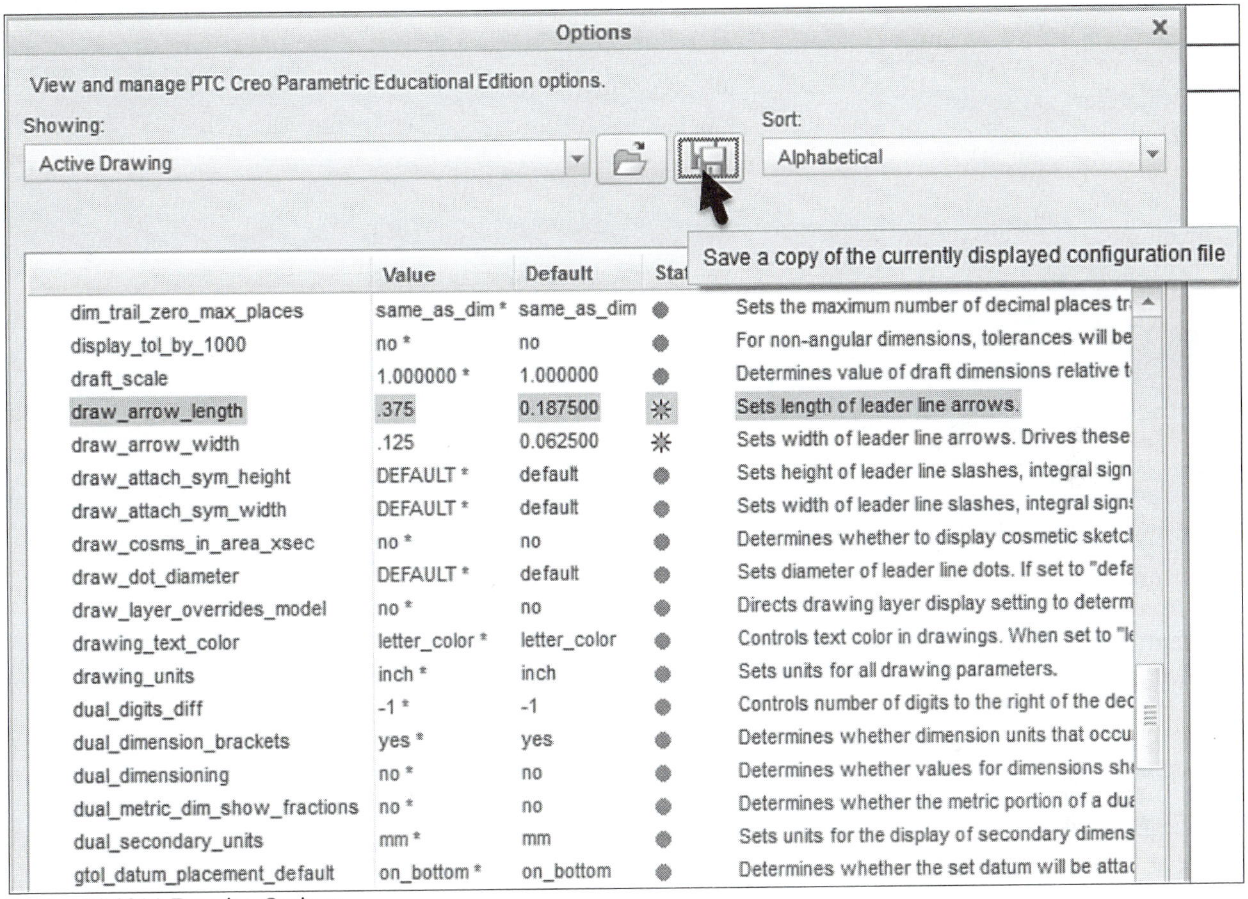

Figure 12.20(a) Drawing Options

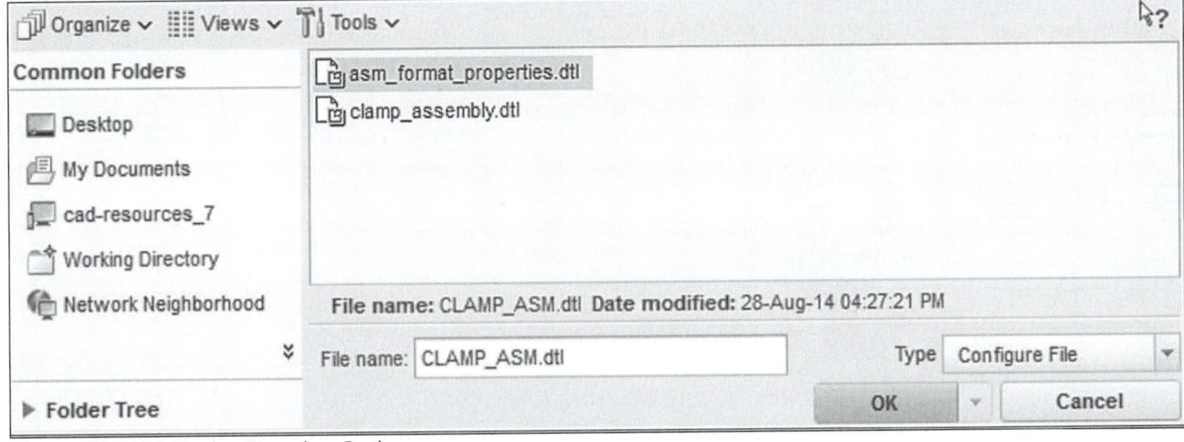

Figure 12.20(b) Saving Drawing Options

If not closed, click: to close the Navigator > > No Hidden > > (Select All) *off* > **LMB** in the Graphics Window > double-click on **SCALE: 1.000** in the lower left-hand corner of the Graphics Window > Enter value for scale, *type* **1.50** > **Enter** > **LMB** to deselect > **Layout** tab > **Lock View Movement** unlock > in the Graphics Window, press **RMB** > **General View** > **No Combined State** (highlighted) > **OK** > > Select CENTER POINT for drawing view. [Fig. 12.21(a)] > Model view names **FRONT** > **Apply** [Fig. 12.21(b)] > **Cancel** > **Ctrl+S** > **Enter**

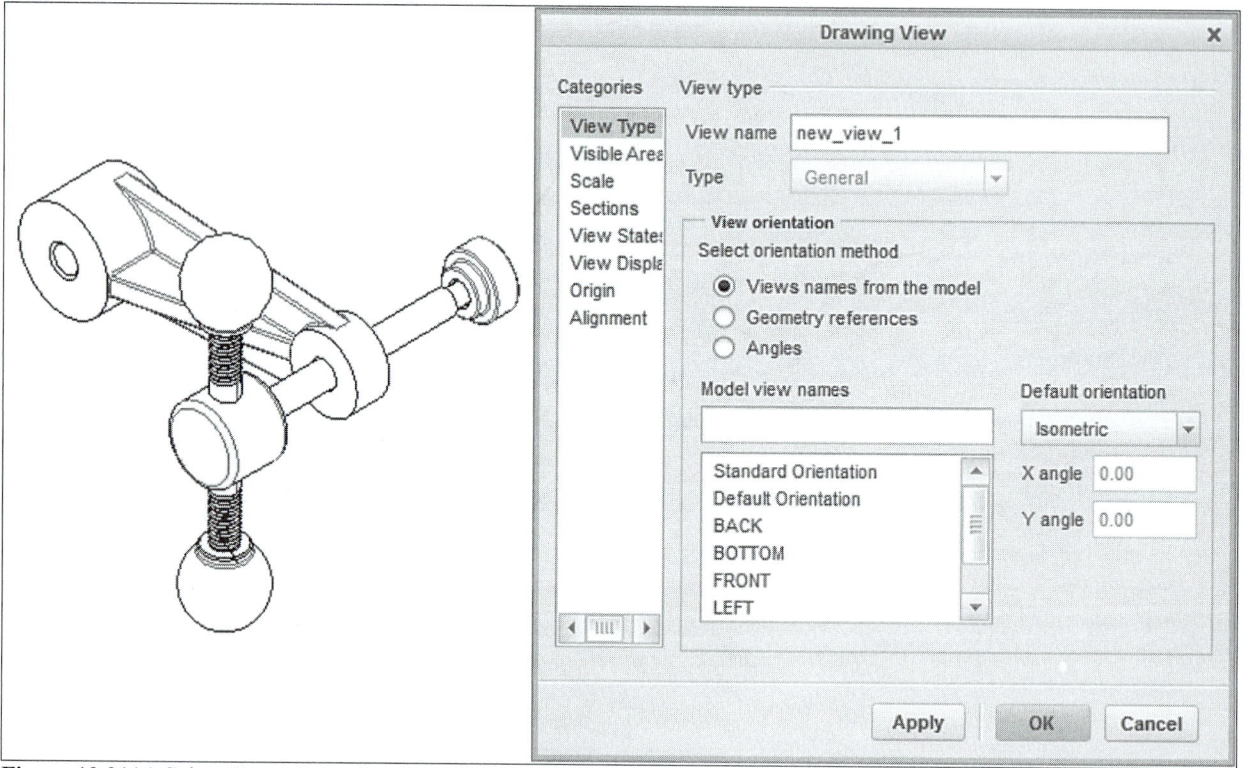

Figure 12.21(a) Select Center Point for First Drawing View

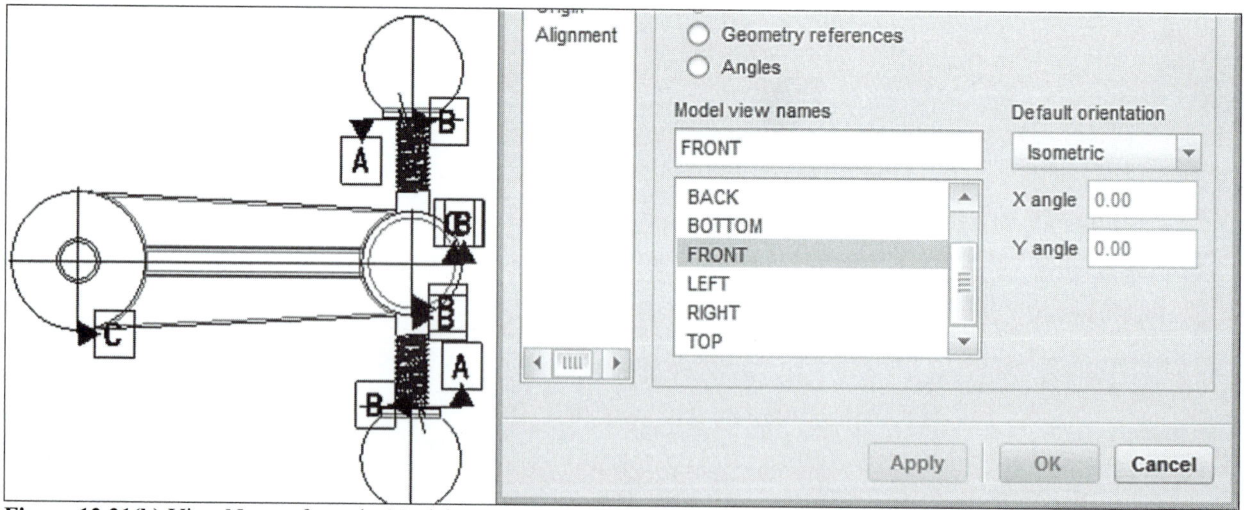

Figure 12.21(b) View Names from the Model

With the "Top" view still *selected/highlighted*; press: **RMB** > **Projection View** [Fig. 12.22(a)] > Select CENTER POINT for drawing view. (move the mouse pointer below the view previously created) [Fig. 12.22(b)] > **LMB** to select the position [Fig. 12.22(c)] > **Ctrl+S**

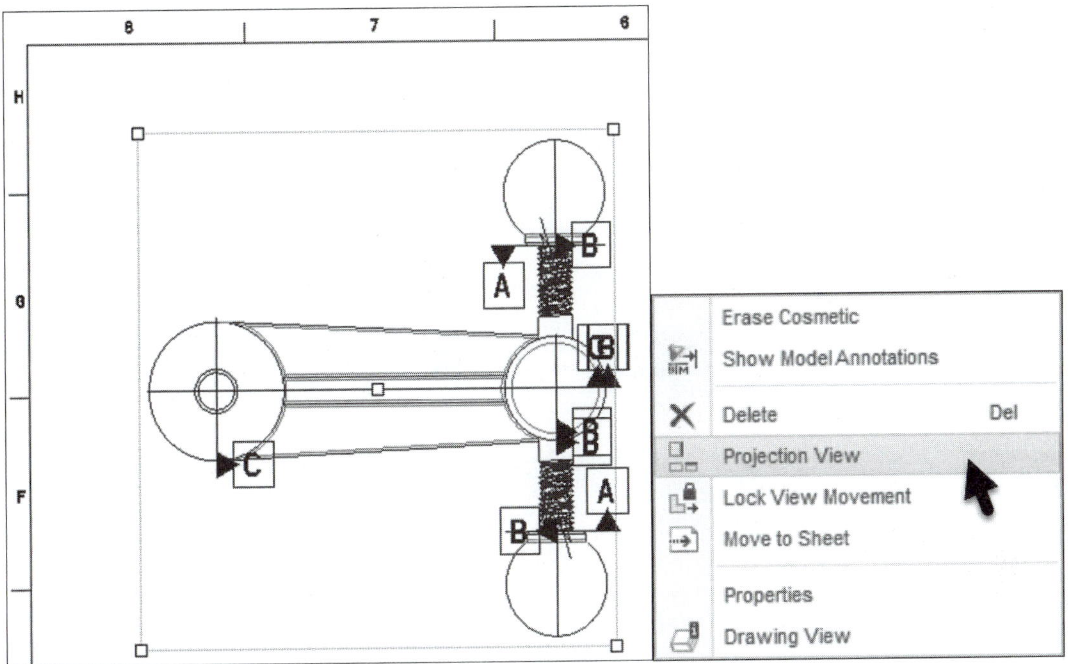

Figure 12.22(a) Insert Projection View

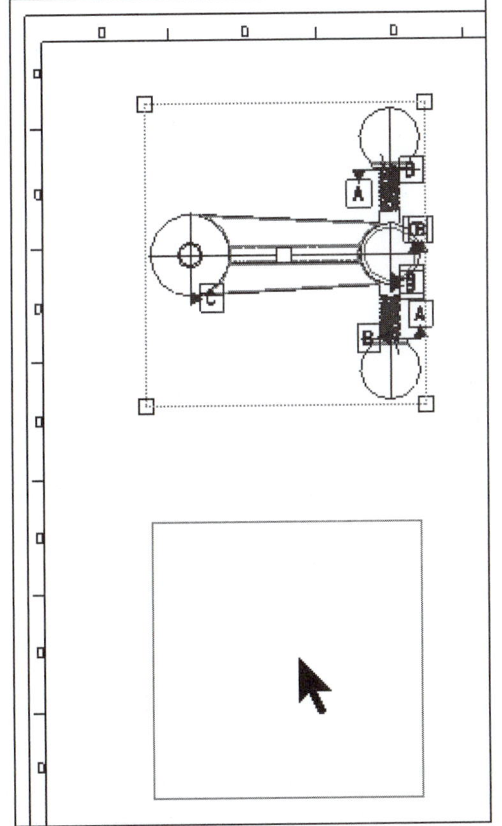

Figure 12.22(b) Location below the Previous View

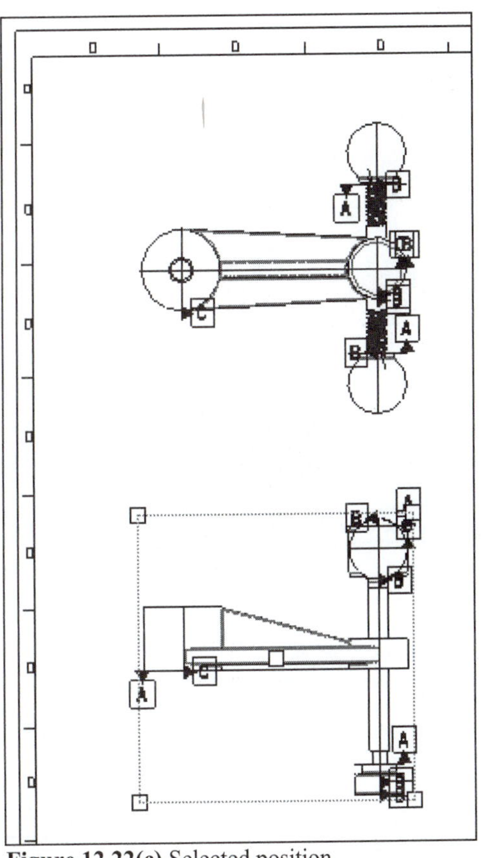

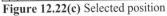

Figure 12.22(c) Selected position

With the Front view still *selected/highlighted*, press: **RMB** > **Properties** [Fig. 12.23(a)] > **Sections** > 2D cross-section > + [Fig. 12.23(b)] > ✓ A > **OK** [Fig. 12.23(c)] > **LMB** > **Ctrl+S**

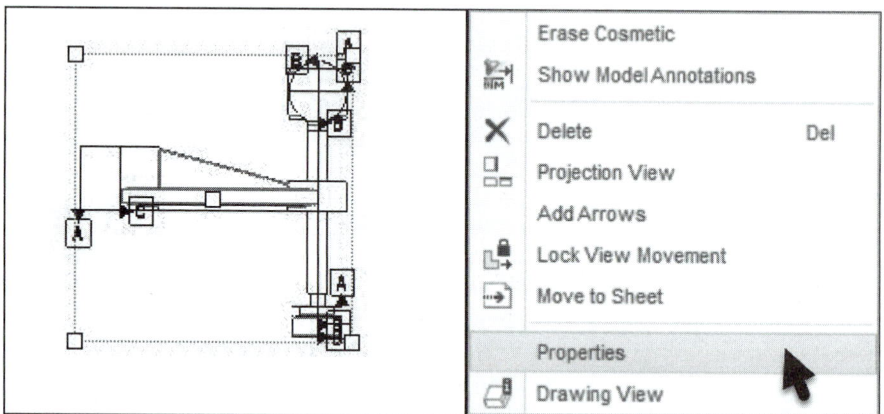

Figure 12.23(a) Properties

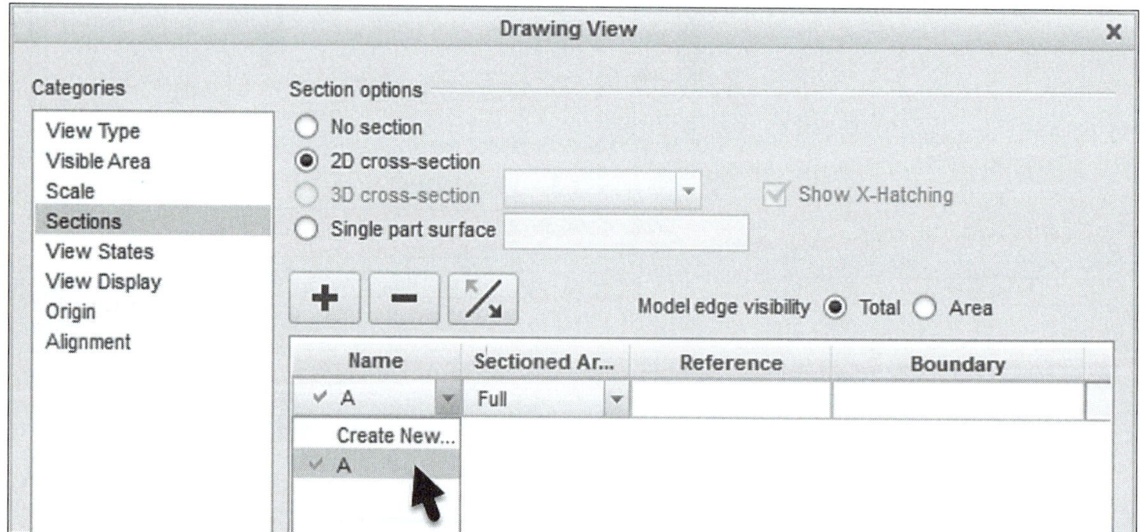

Figure 12.23(b) Drawing View 2D section (Section A should have been previously created)

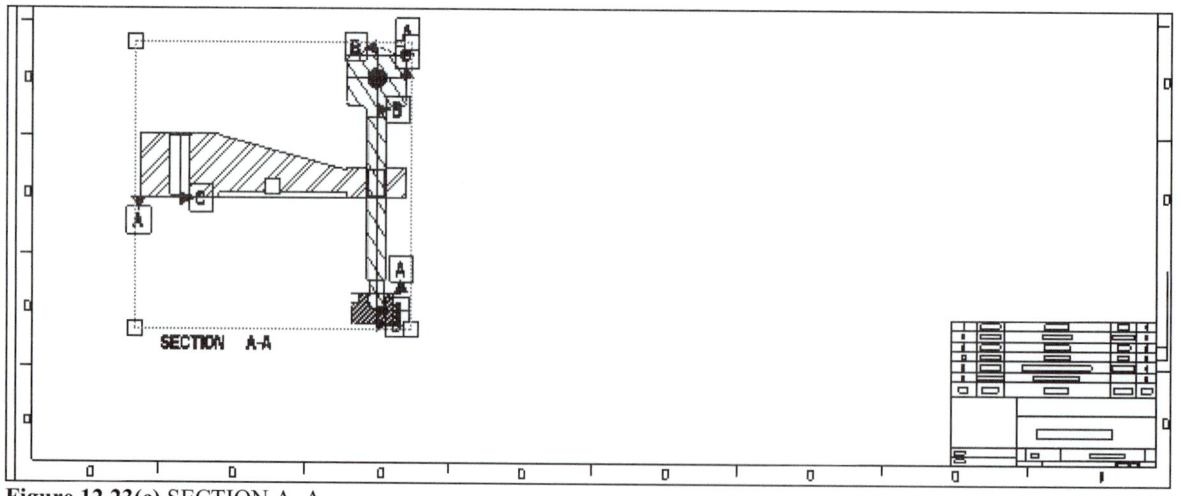

Figure 12.23(c) SECTION A_A

Click: [icon] open the Navigator > **Annotate** tab > from the Drawing Tree expand Datums [▼ Datums] > with the **Ctrl** key (or **Shift** key) pressed, select all of the Model datums > release the **Ctrl** (or **Shift**) key > **RMB** > **Erase** > repeat for the other view [Fig. 12.24(a)] > **LMB** in the Graphics Window to deselect > reposition the views as needed [Fig. 12.24(b)] > [icon] > **Ctrl+S** > pick on the Front (lower) view to select/highlight > **Layout** tab > in the Graphics Window, press **RMB** > **Add Arrows** [Fig. 12.25(a)] > pick on the Top (upper) view > move the cursor outside of the view outline > **LMB** [Fig. 12.25(b)]

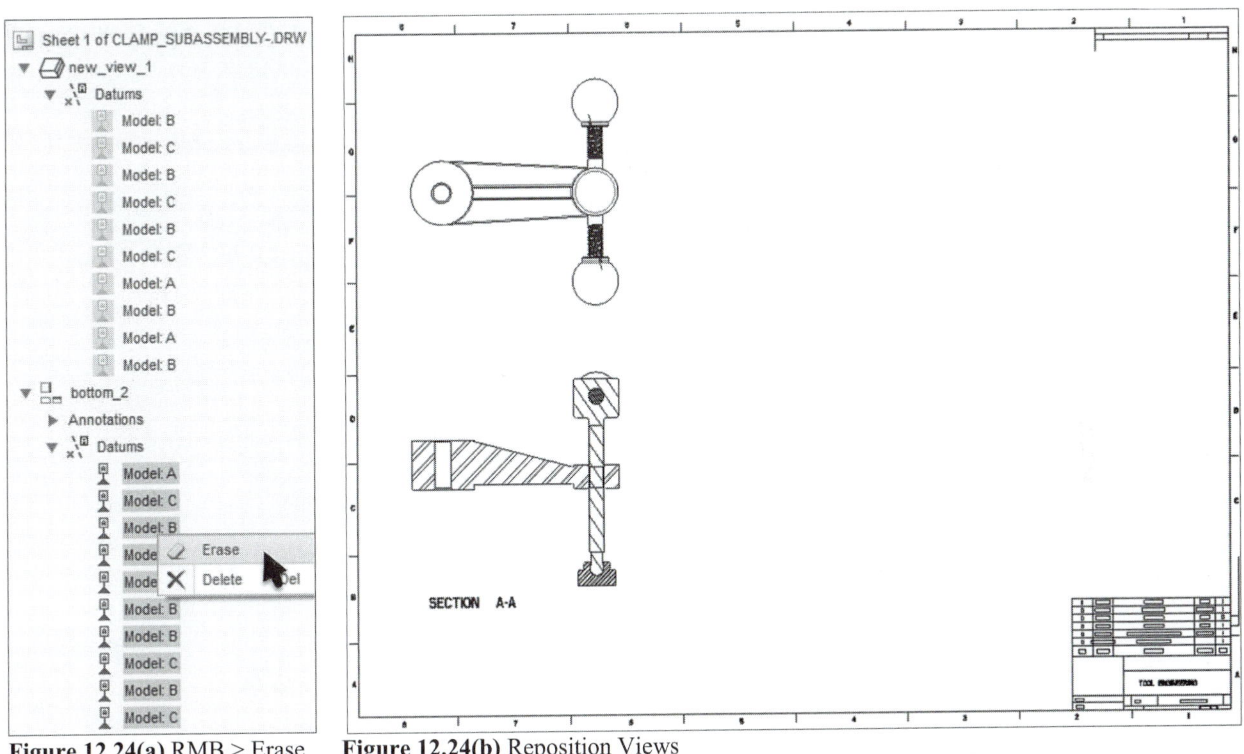

Figure 12.24(a) RMB > Erase **Figure 12.24(b)** Reposition Views

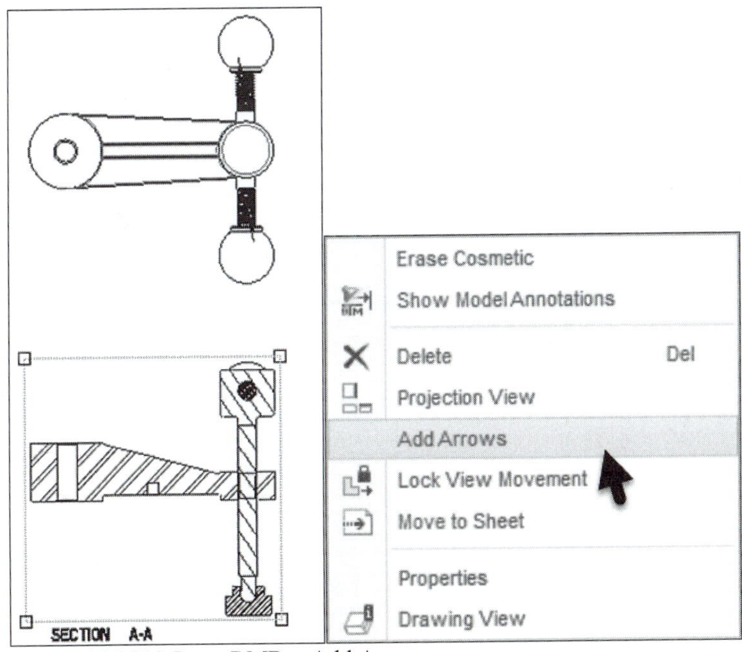

Figure 12.25(a) Press RMB > Add Arrows

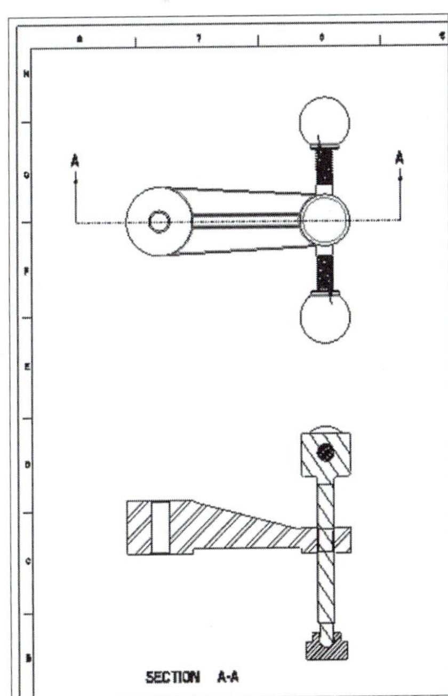

Figure 12.25(b) Section Arrows

Creo Parametric provides tools to alter the display of the section views to comply with industry ASME standard practices. Most companies require that the crosshatching on parts in section views of assemblies be "clocked" such that parts that meet do not use the same section lining (crosshatching) spacing and angle. This makes the separation between parts more distinct. First, modify the visibility of the views to remove hidden lines and make the tangent edges dimmed.

Press and hold down the **Ctrl** key > pick on both views > release the **Ctrl** key > with the cursor outside of the view outlines, press **RMB** > **Properties** > Yes Hidden line removal for xhatches > Display style No Hidden > Tangent edges display style Dimmed [Fig. 12.26(a)] > **Apply** > **Cancel** > **LMB** in the Graphics Window to deselect [Fig. 12.26(b)]

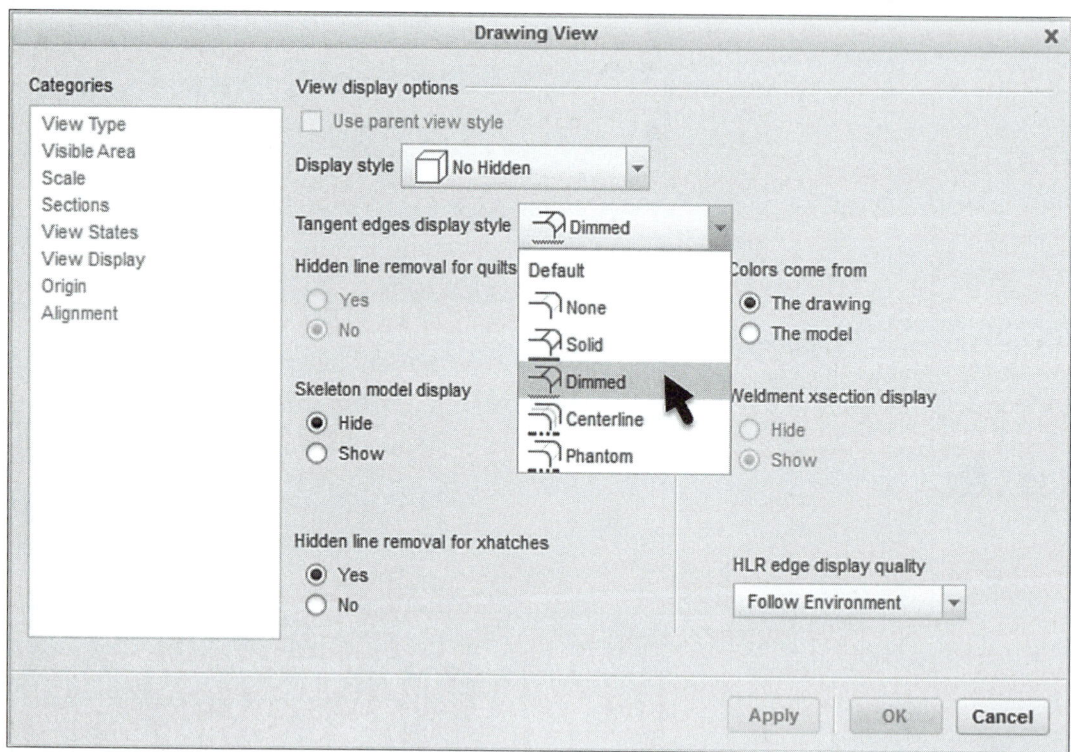

Figure 12.26(a) No Hidden

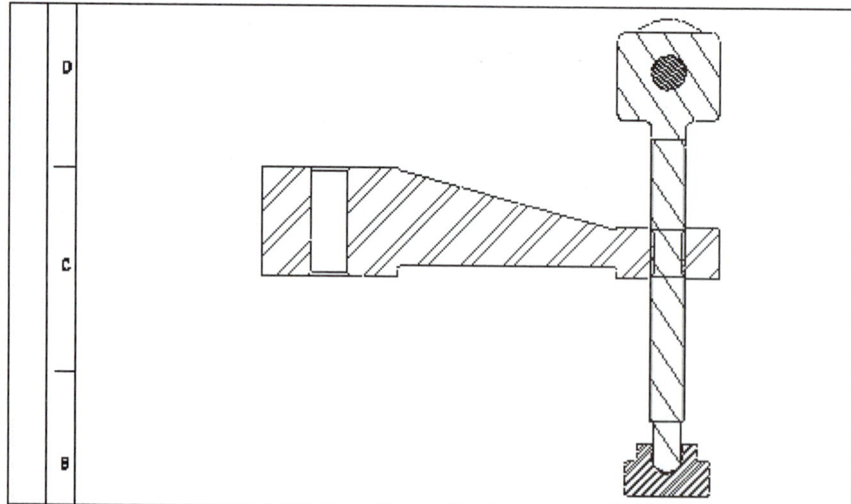

Figure 12.26(b) View Display

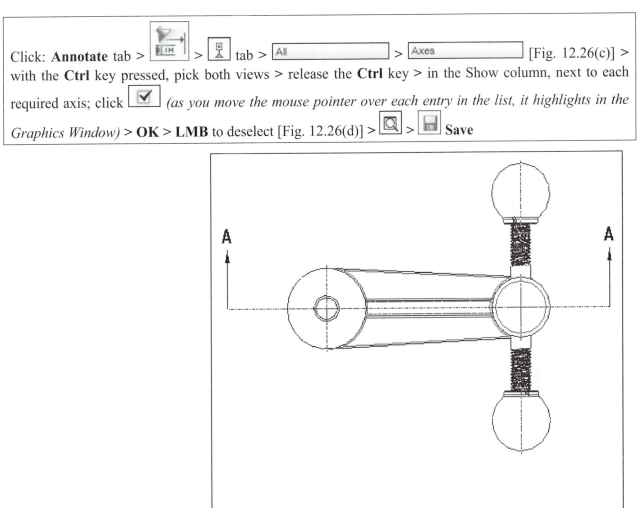

Click: **Annotate** tab > [icon] > [icon] tab > [All] > [Axes] [Fig. 12.26(c)] > with the **Ctrl** key pressed, pick both views > release the **Ctrl** key > in the Show column, next to each required axis; click [icon] *(as you move the mouse pointer over each entry in the list, it highlights in the Graphics Window)* > **OK** > **LMB** to deselect [Fig. 12.26(d)] > [icon] > [icon] **Save**

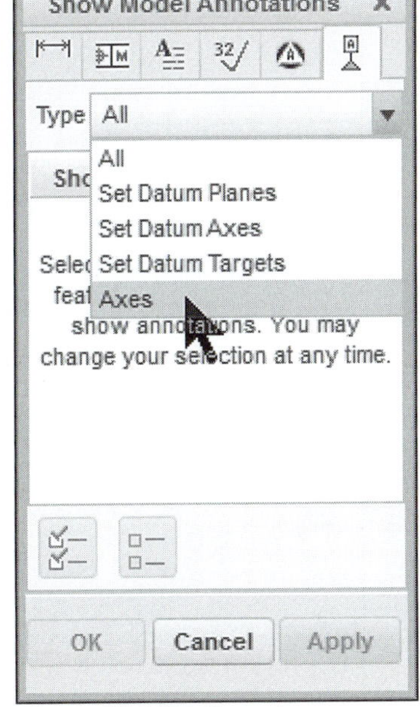

Figure 12.26(c) Type Axes

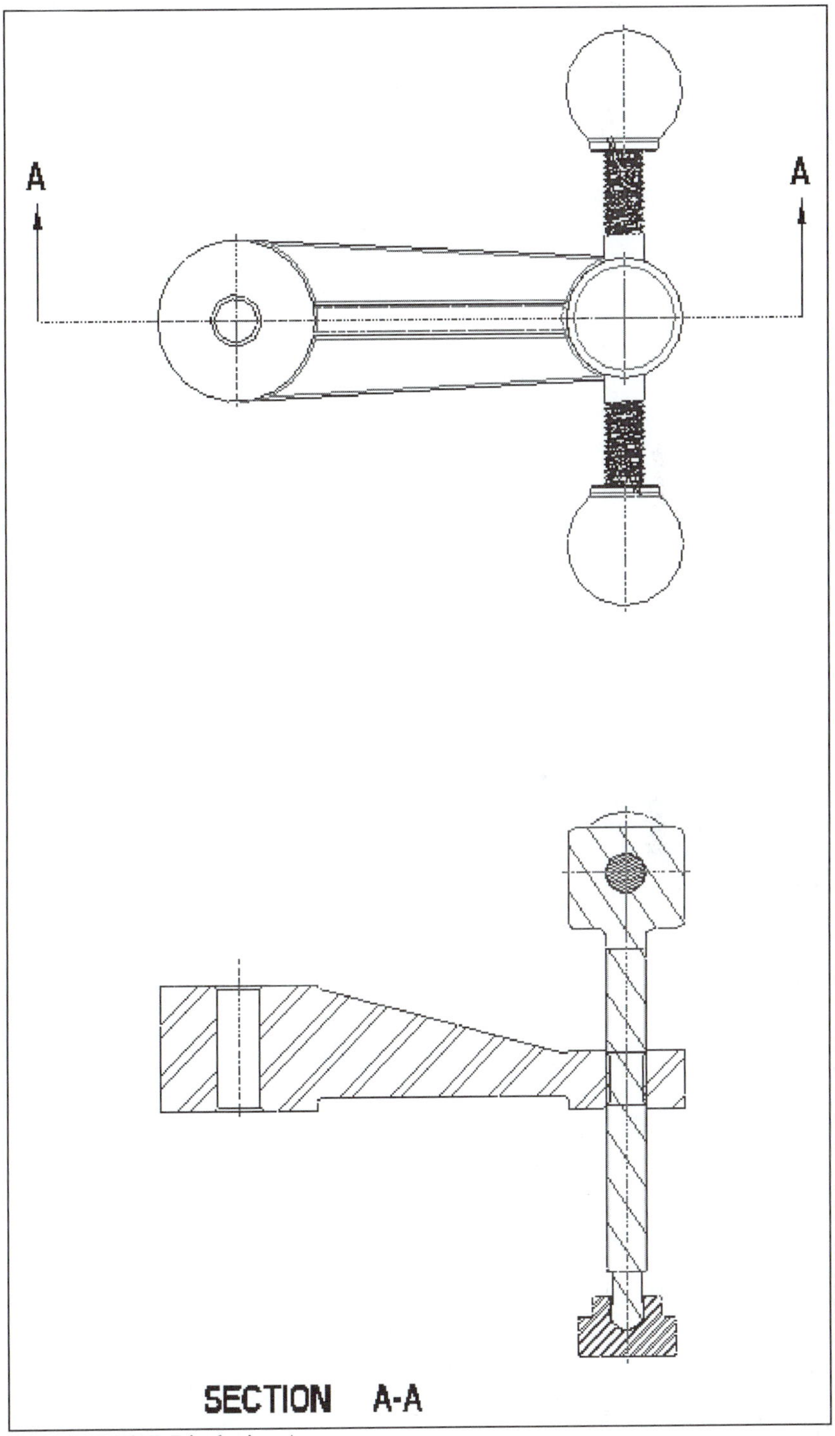

Figure 12.26(d) Displaying Axes

Click: **Layout** tab > double-click on the crosshatching in the Front view (CARRLANE-12-13500_STUD is now active) > **Fill** [Figs. 12.26(e-f)] > click **Next** until CLAMP_FOOT is active > **Delete Line** > **Hatch** > **Spacing** > **Double** > **Angle** [Fig. 12.26(g)] > **135** [Fig. 12.26(h)]

Figure 12.26(e) Modify Xhatch

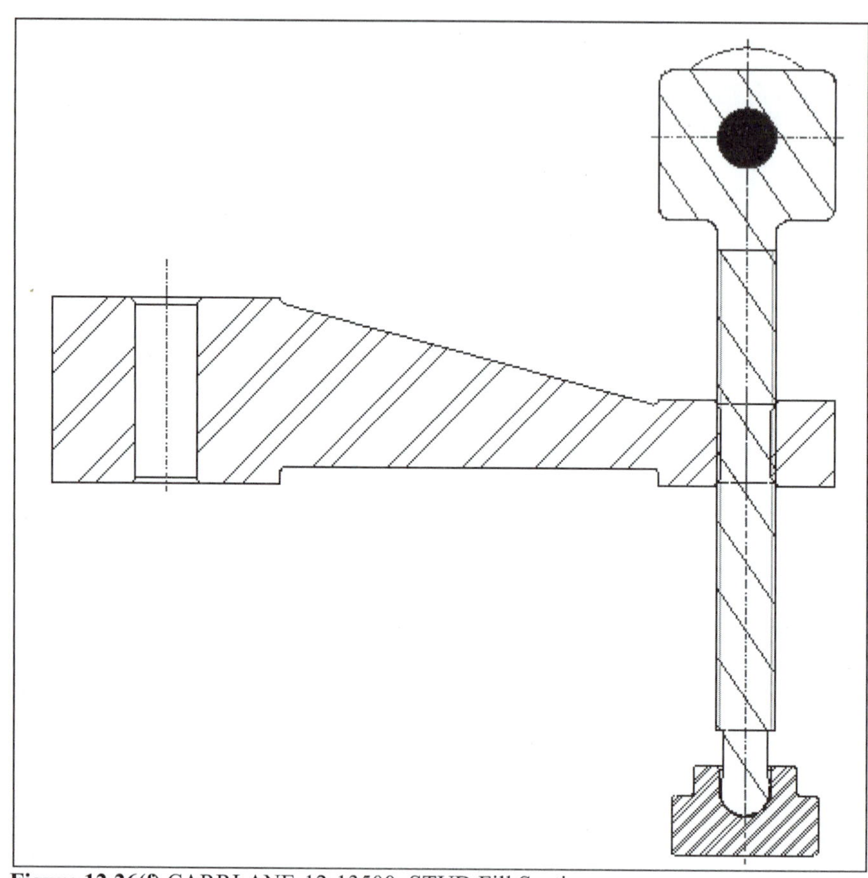

Figure 12.26(f) CARRLANE-12-13500_STUD Fill Section

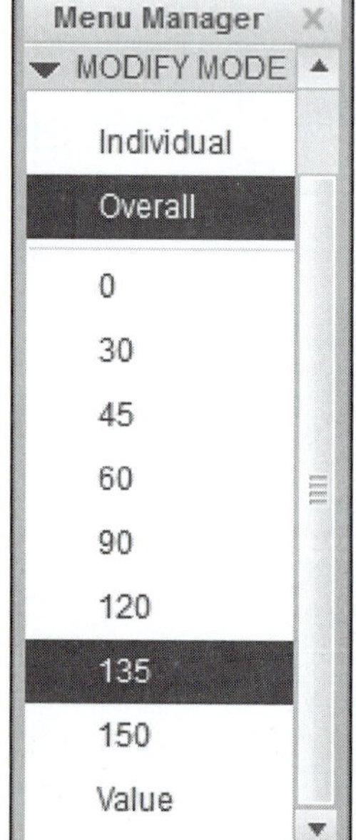

Figure 12.26(g) Angle **135**

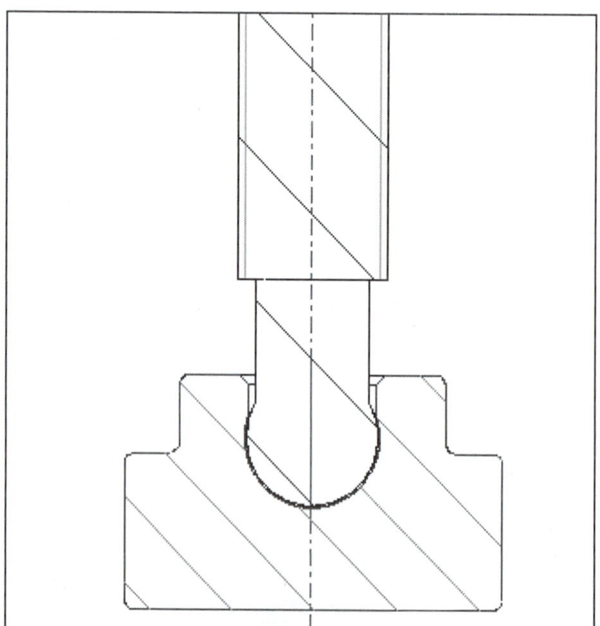

Figure 12.26(h) CLAMP_FOOT Hatching

Click: **Next** until CLAMP_SWIVEL is active > **Angle** > **45** > **Next** until CLAMP_ARM is active > **Delete Line** > **Angle** > **120** [Fig. 12.26(i)] > **Previous** until CLAMP_SWIVEL is active > **Spacing** > **Half** > **Done** > in the Graphics Window, **LMB** to deselect > > [Fig. 12.26(j)]

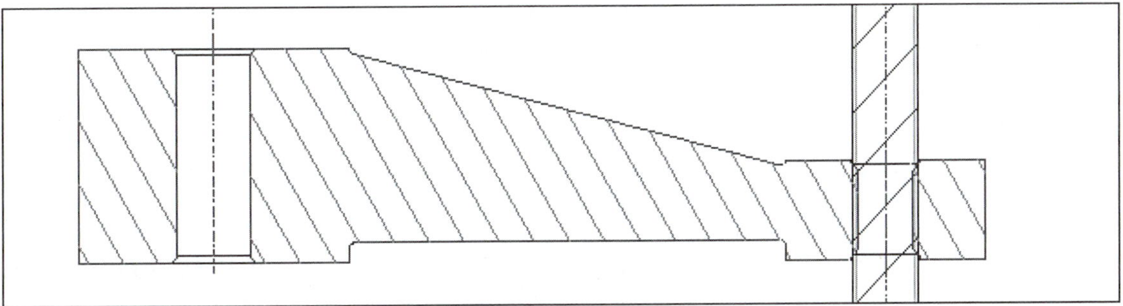

Figure 12.26(i) CLAMP_ARM, Hatch Spacing Angle **120**

SECTION A-A

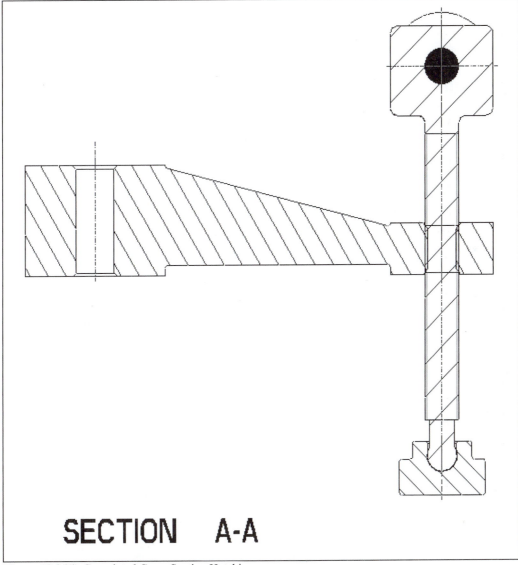

Figure 12.26(j) Completed Cross Section Hatching

Zoom in on the BOM [Fig. 12.27(a)] > **Table** tab > place the pointer to highlight the PT NUM column [Fig. 12.27(b)] > double-click > Width in drawing units **1.25** [Fig. 12.27(c)] > **OK** > place the pointer to highlight the QTY column [Fig. 12.27(d)] > double-click > Width in drawing units **.50** > **OK** [Fig. 12.27(e)] > **LMB** to deselect

1	SW101-5AR	CLAMP ARM	STEEL	1
2	SW101-6SW	CLAMP SWIVEL	STEEL 1066	1
3	SW101-7BA	CLAMP BALL	NYLON	2
4	SW101-8FT	CLAMP FOOT	STEEL	1
5	SW101-9STL	.500-13 X 5.00 DOUBLE END STUD	PURCHASED	1
6	SW101-SUB-ASM	CLAMP SUB-ASSEMBLY		1
ITEM	PT NUM	DESCRIPTION	MATERIAL	QTY

Figure 12.27(a) BOM

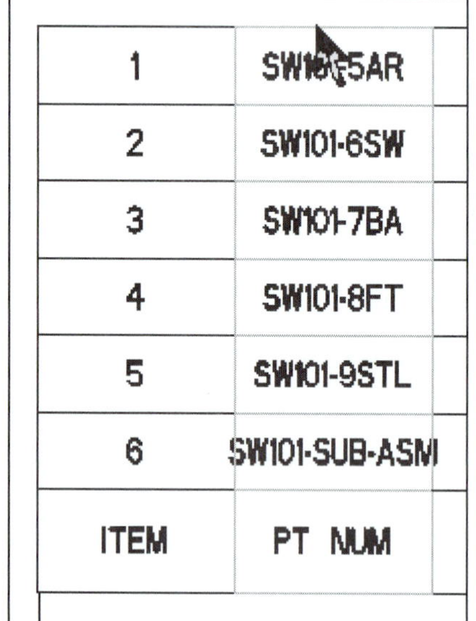

1	SW101-5AR
2	SW101-6SW
3	SW101-7BA
4	SW101-8FT
5	SW101-9STL
6	SW101-SUB-ASM
ITEM	PT NUM

Figure 12.27(b) Highlight Column

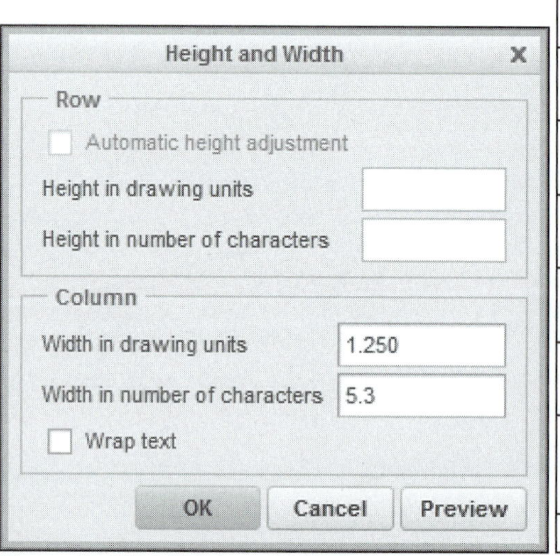

Figure 12.27(c) Height and Width Dialog Box

STEEL	1
STEEL 1066	1
NYLON	2
STEEL	1
PURCHASED	1
	1
MATERIAL	QTY

Figure 12.27(d) Qty Column

1	SW101-5AR	CLAMP ARM	STEEL	1
2	SW101-6SW	CLAMP SWIVEL	STEEL 1066	1
3	SW101-7BA	CLAMP BALL	NYLON	2
4	SW101-8FT	CLAMP FOOT	STEEL	1
5	SW101-9STL	.500-13 X 5.00 DOUBLE END STUD	PURCHASED	1
6	SW101-SUB-ASM	CLAMP SUB-ASSEMBLY		1
ITEM	PT NUM	DESCRIPTION	MATERIAL	QTY

Figure 12.27(e) Adjusted BOM Table

To complete the drawing, the balloons must be displayed for each component. Balloons are displayed in the top view as the default, because it was the first view that was created.

Click: > **Create Balloons > Create Balloons - All** [Fig. 12.28(a)] > press and hold the **Ctrl** key and pick on the balloons for the Clamp_Arm (1), Clamp_Swivel (2) and Clamp_Foot (4) > release the **Ctrl** key > press **RMB > Move Item to View** [Fig. 12.28(b)]

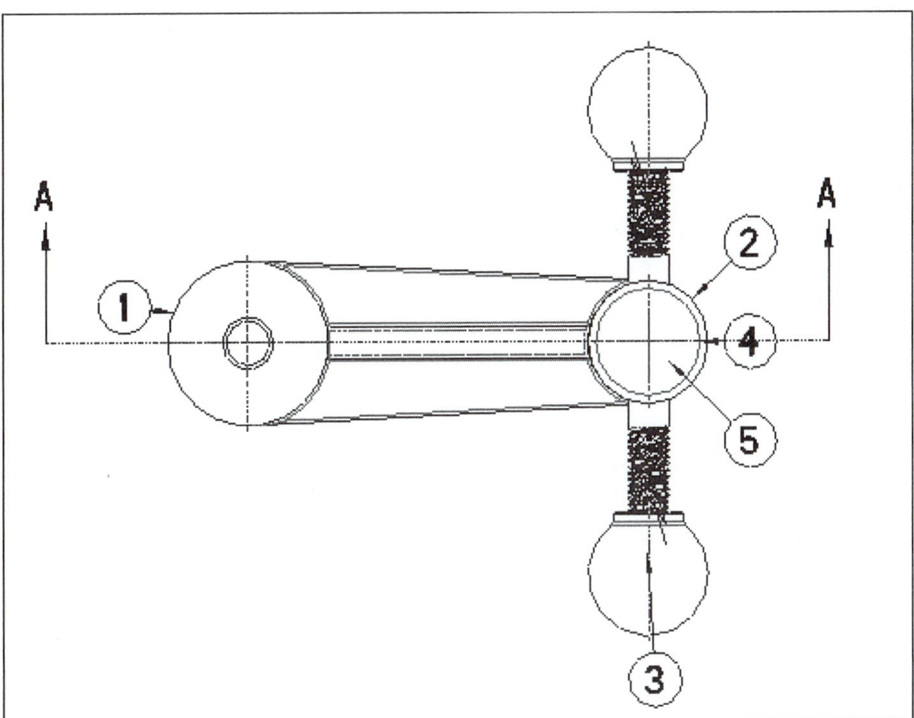

Figure 12.28(a) All Balloons Created

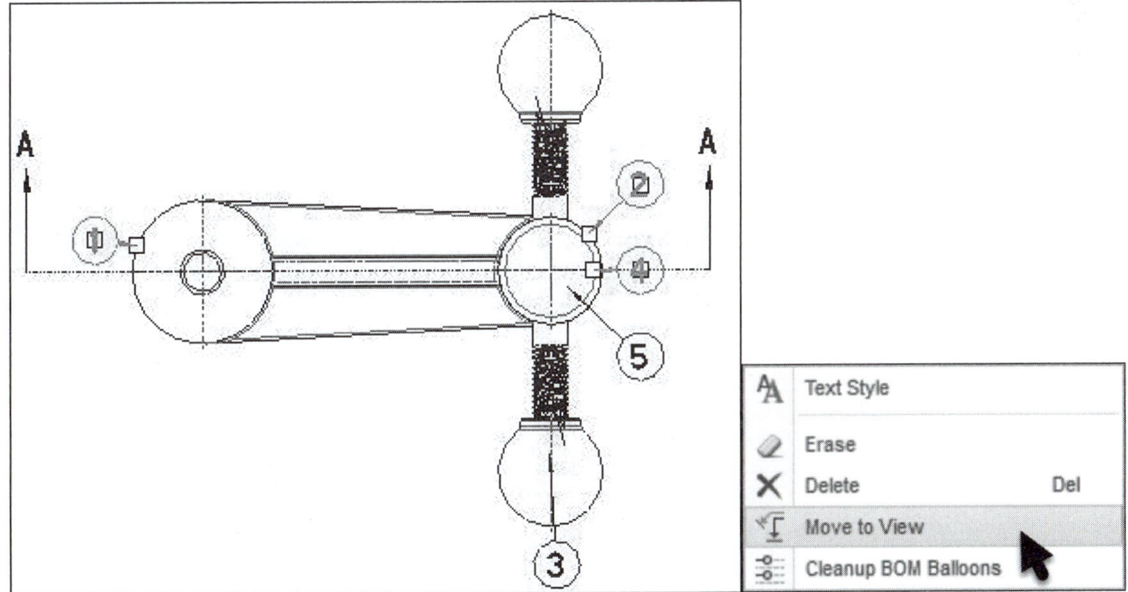

Figure 12.28(b) Move Item to View

Pick on the Front (lower) view [Fig. 12.28(c)] > move the pointer outside of the view boundary > **LMB** to deselect > pick on and reposition each balloon as needed > pick on balloon **1** > press **RMB** > **Edit Attachment** > pick a new position > **MMB** > reposition the balloon [Fig. 12.28(d)] > in the Graphics Window, **LMB** to deselect > **Ctrl+S**

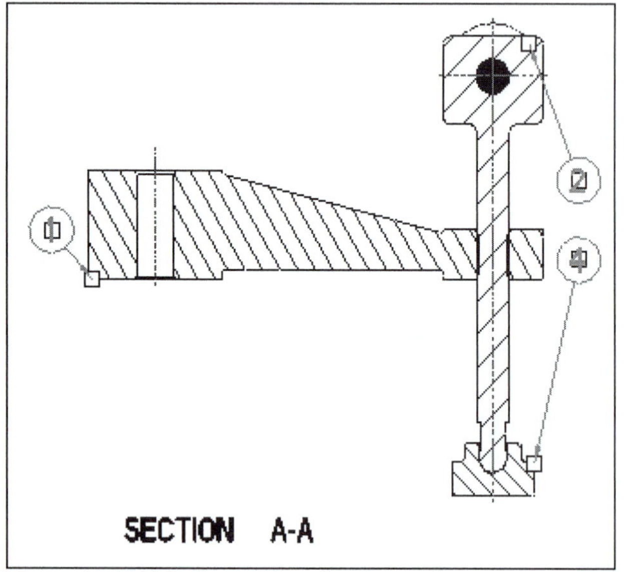

Figure 12.28(c) Balloons Moved to Front View

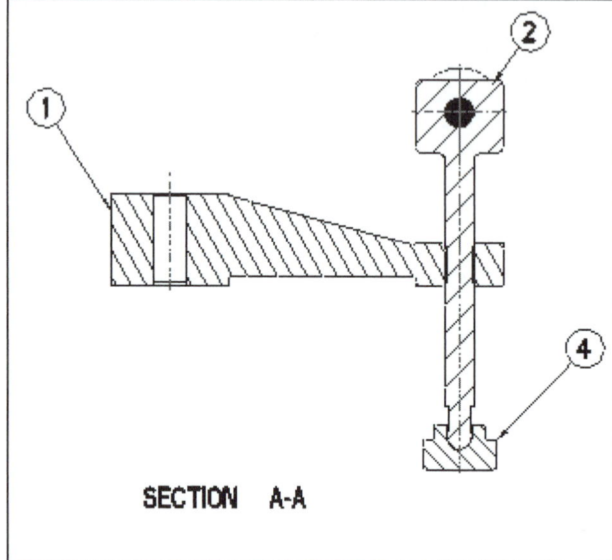

Figure 12.28(d) Repositioned Balloons in the Front View

Pick on balloon **5** (CARRLANE-12-13500_STUD) > press **RMB** [Fig. 12.28(e)] > **Edit Attachment** > pick a new position on the CARRLANE-12-13500_STUD [Fig. 12.28(f)] > **MMB** > reposition the balloon > in the Graphics Window, **LMB** to deselect

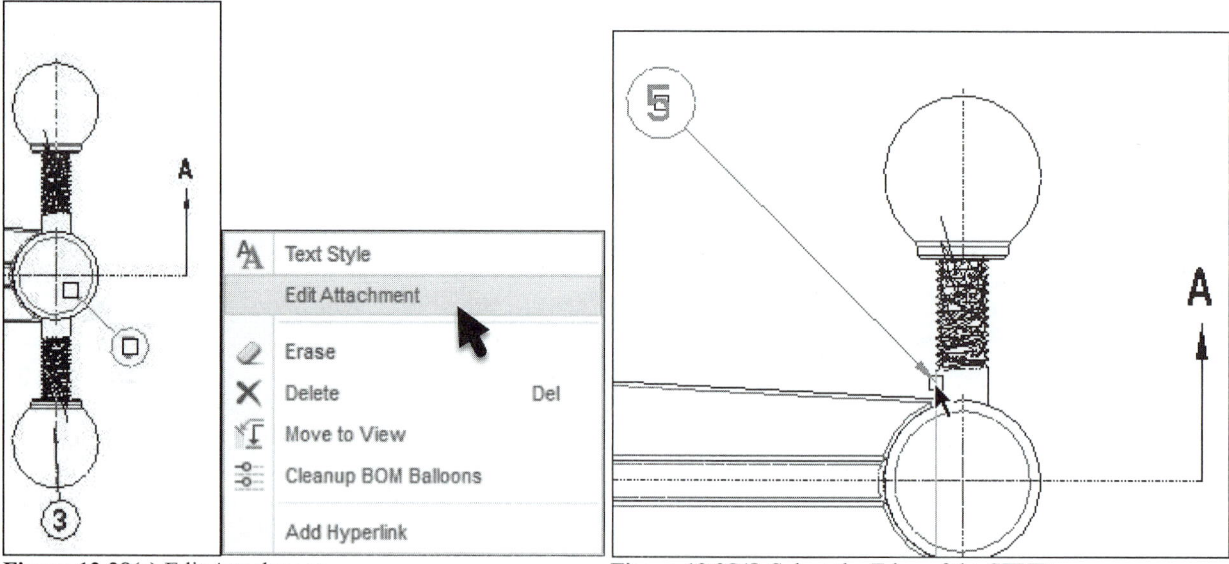

Figure 12.28(e) Edit Attachment

Figure 12.28(f) Select the Edge of the STUD

Reposition and edit the attachment of all ballons as necessary > [Figs. 12.29(a-b)] > > **File** > **Manage File** > **Delete Old Versions** > > **File** > **Close** > **File** > **Manage Session** > **Erase Not Displayed** > **OK**

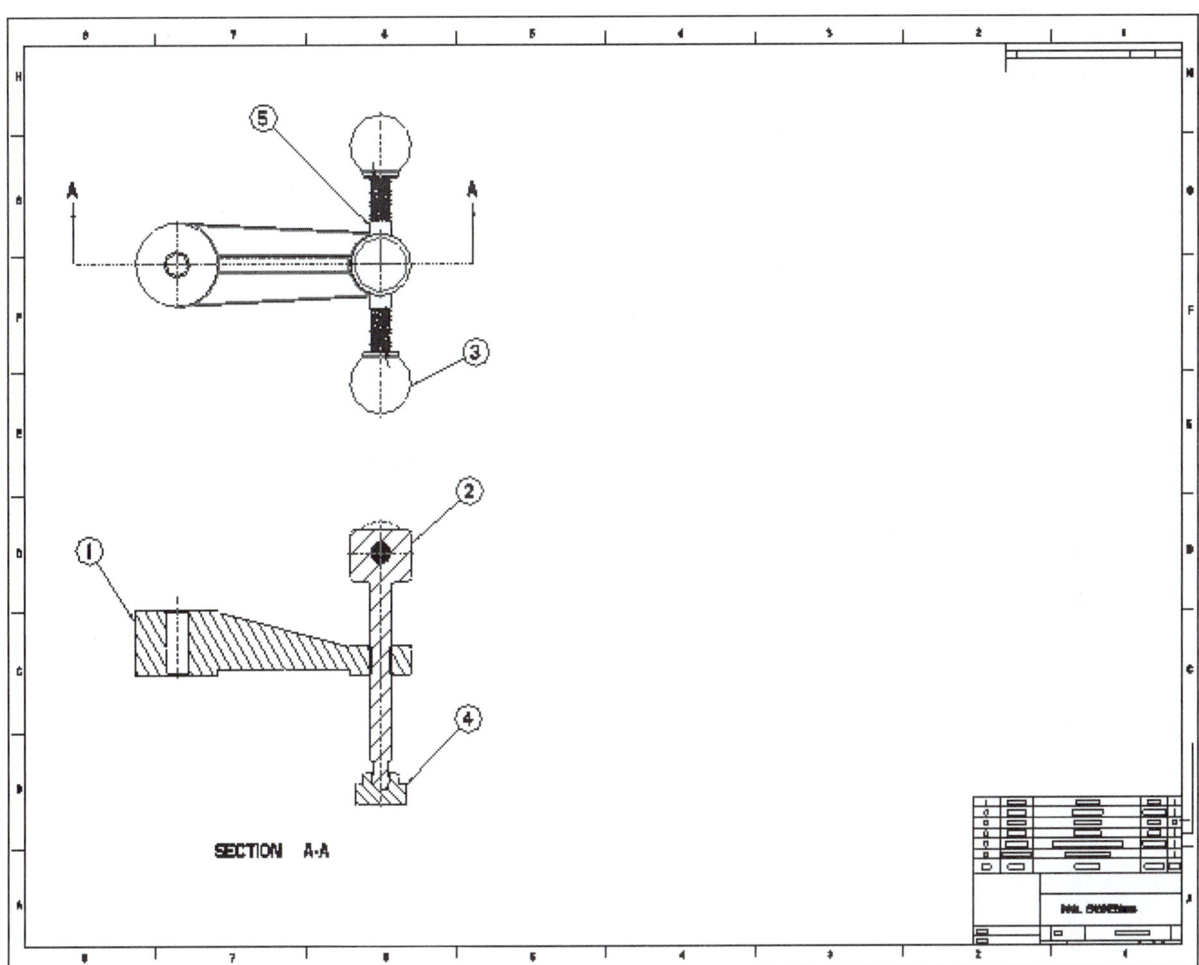

Figure 12.29(a) Completed Clamp_Subassembly Drawing

1	SW101-5AR	CLAMP ARM	STEEL	1
2	SW101-6SW	CLAMP SWIVEL	STEEL 1066	1
3	SW101-7BA	CLAMP BALL	NYLON	2
4	SW101-8FT	CLAMP FOOT	STEEL	1
5	SW101-9STL	.500-13 X 5.00 DOUBLE END STUD	PURCHASED	1
6	SW101-SUB-ASM	CLAMP SUB-ASSEMBLY		1
ITEM	PT NUM	DESCRIPTION	MATERIAL	QTY

Figure 12.29(b) Clamp_Subassembly BOM

The Swing Clamp Assembly drawing is composed of the subassembly, the plate, the short stud, and the nut. The drawing will use the same format created for the subassembly. Formats are normally read-only files that can be used as many times as needed.

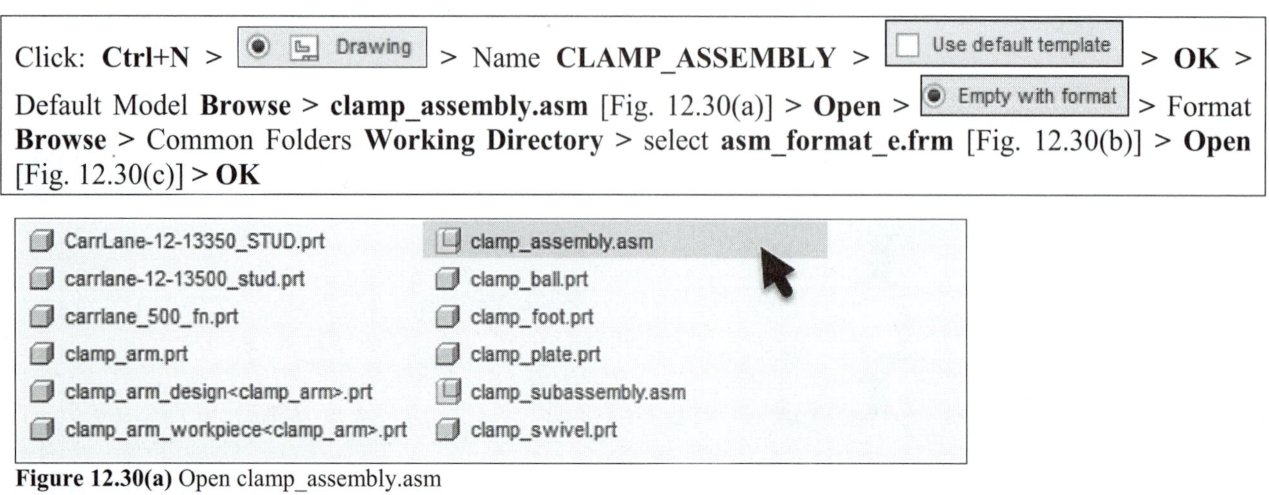

Click: **Ctrl+N** > Drawing > Name **CLAMP_ASSEMBLY** > Use default template > **OK** > Default Model **Browse** > **clamp_assembly.asm** [Fig. 12.30(a)] > **Open** > Empty with format > Format **Browse** > Common Folders **Working Directory** > select **asm_format_e.frm** [Fig. 12.30(b)] > **Open** [Fig. 12.30(c)] > **OK**

Figure 12.30(a) Open clamp_assembly.asm

asm_format_e.frm
ball_end.sec

Figure 12.30(b) asm_format_e.frm

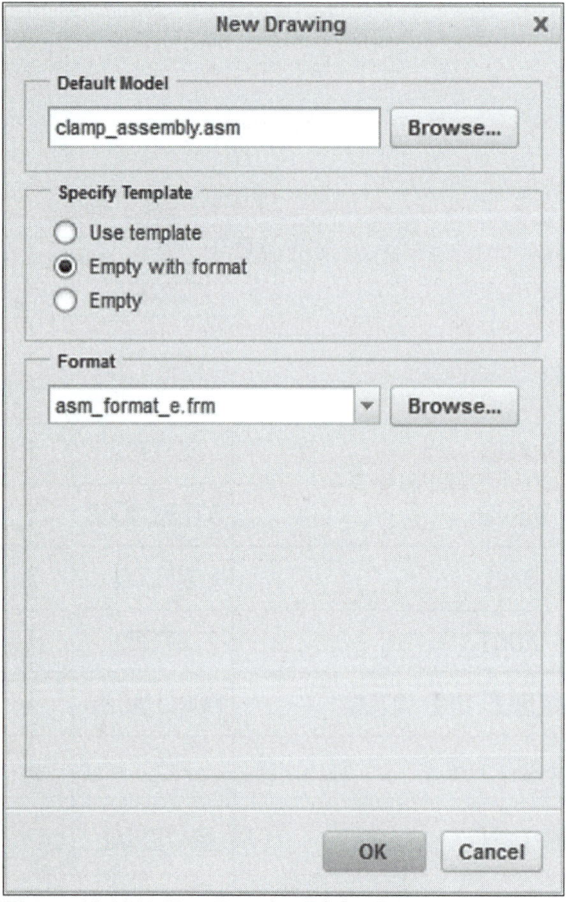

Figure 12.30(c) New Drawing Dialog Box

Zoom in on the title block > **Table** tab > window-in to select the table > move table above the title block > change the PT NUM column width to **1.25** and the QTY column width to **.50** as per the sub-assembly drawing > > > **File** > **Prepare** > **Drawing Properties** > Detail Options **change** > **Open a configuration file** > **clamp_asm.dtl** > **Open** > **Apply** > **Close** > **Close** > Double-click on **SCALE: 1.000** in the lower left-hand corner of the Graphics Window SCALE : 1.000 > Enter value for scale 1.50 [Fig. 12.31(a)] > > **LMB** to deselect > **Refit** > **Ctrl+S**

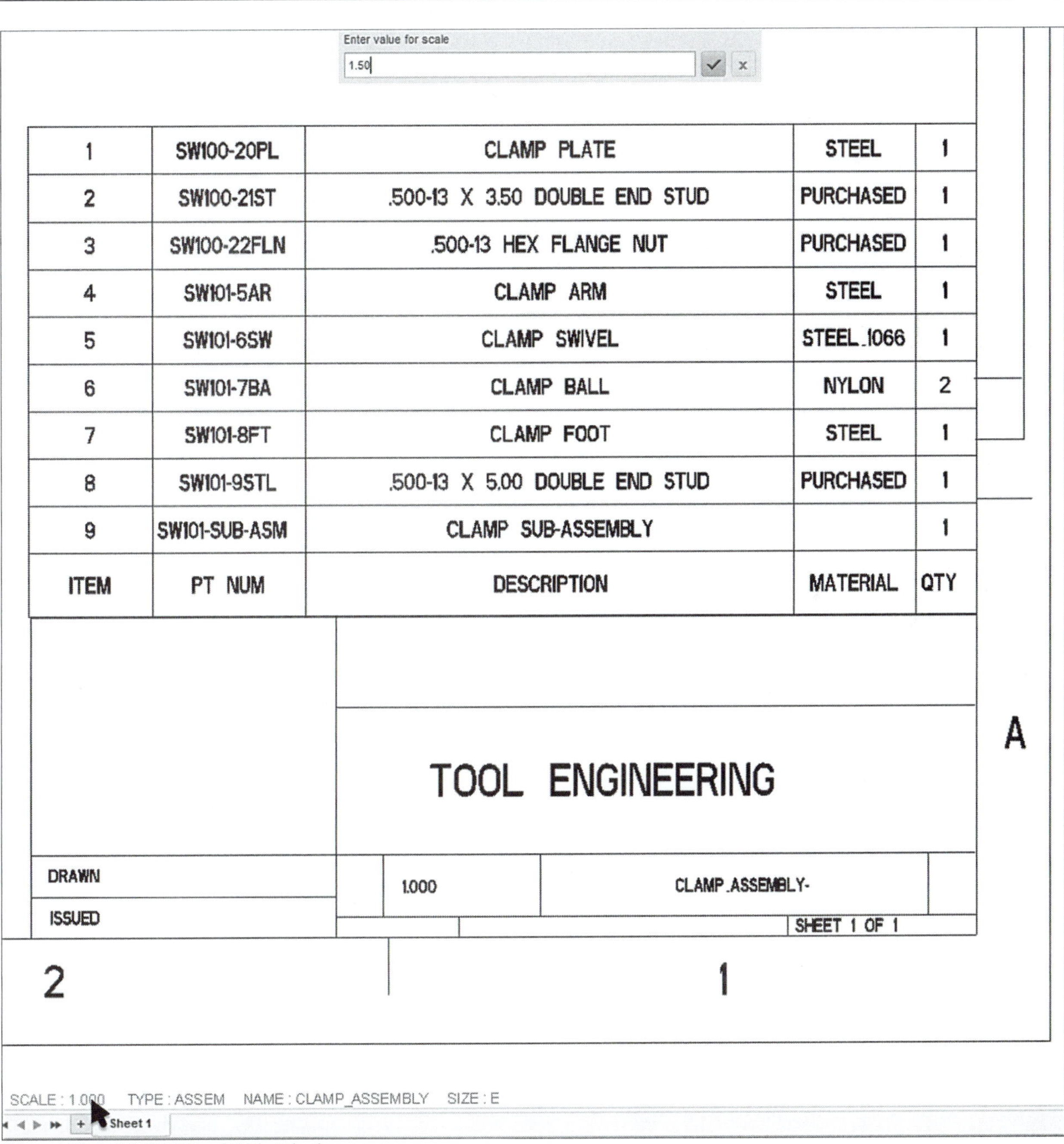

1	SW100-20PL	CLAMP PLATE	STEEL	1
2	SW100-21ST	.500-13 X 3.50 DOUBLE END STUD	PURCHASED	1
3	SW100-22FLN	.500-13 HEX FLANGE NUT	PURCHASED	1
4	SW101-5AR	CLAMP ARM	STEEL	1
5	SW101-6SW	CLAMP SWIVEL	STEEL_1066	1
6	SW101-7BA	CLAMP BALL	NYLON	2
7	SW101-8FT	CLAMP FOOT	STEEL	1
8	SW101-9STL	.500-13 X 5.00 DOUBLE END STUD	PURCHASED	1
9	SW101-SUB-ASM	CLAMP SUB-ASSEMBLY		1
ITEM	PT NUM	DESCRIPTION	MATERIAL	QTY

Figure 12.31(a) BOM and Title Block

Click: **Layout** tab > in the Graphics Window, press **RMB** > **General View** > **No Combined State** (highlighted) [Fig. 12.31(b)] > **OK** > Select CENTER POINT for drawing view. (pick where you want a Top view) > Model view names **FRONT** [Fig. 12.31(c)] > **Apply** > **View Display** > Display style **No Hidden** > Tangent edges **Dimmed** [Fig. 12.31(d)] > **Apply** > **Cancel** > in the Graphics Window, press **RMB** > **Lock View Movement** *off* > **RMB** in the Graphics Window)

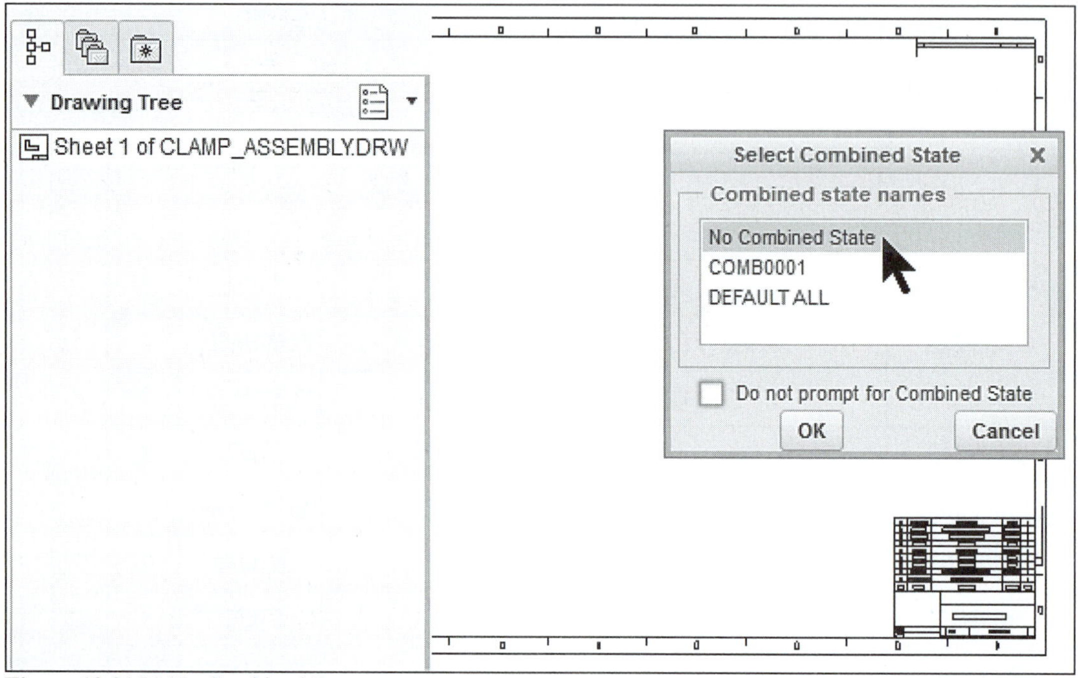

Figure 12.31(b) No Combined State

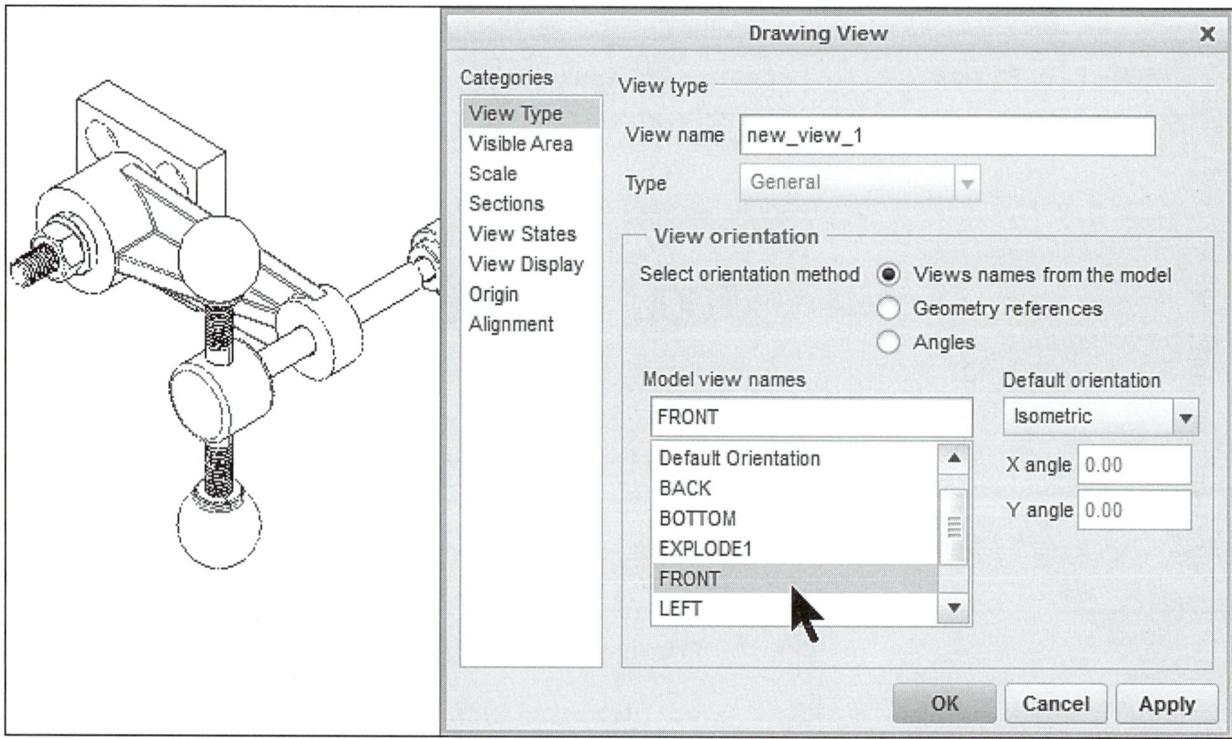

Figure 12.31(c) Drawing View Dialog Box

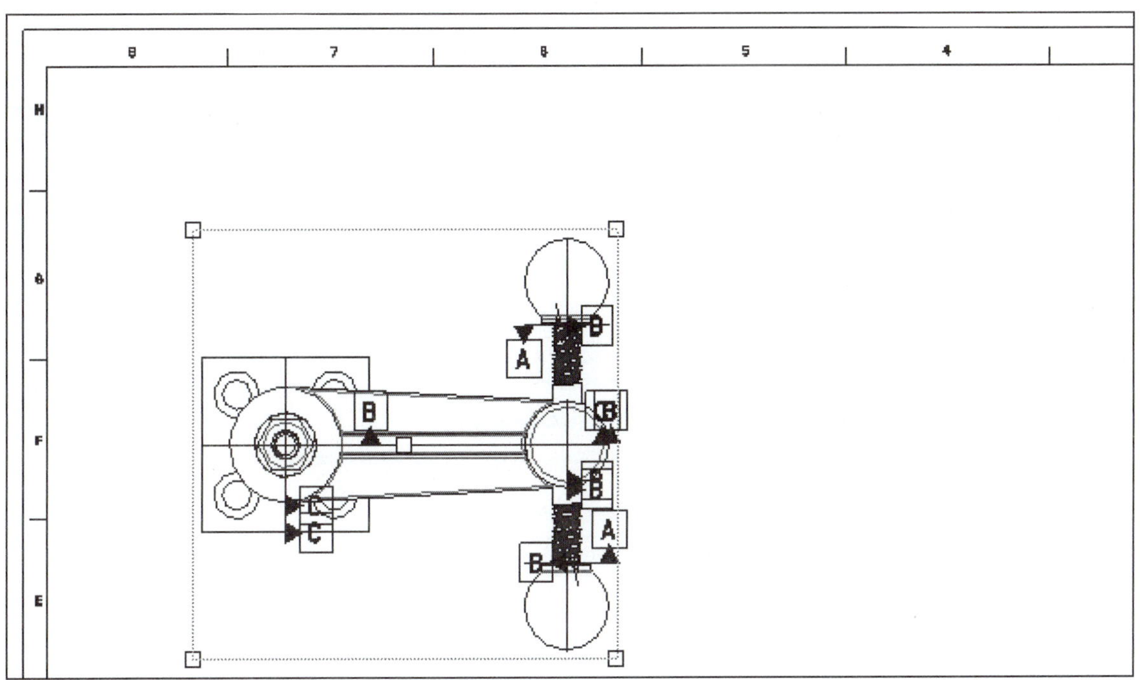

Figure 12.31(d) Top View Placed

Press: 

> **RMB** > **Projection View** [Fig. 12.32(a)] > pick a position for the Front view

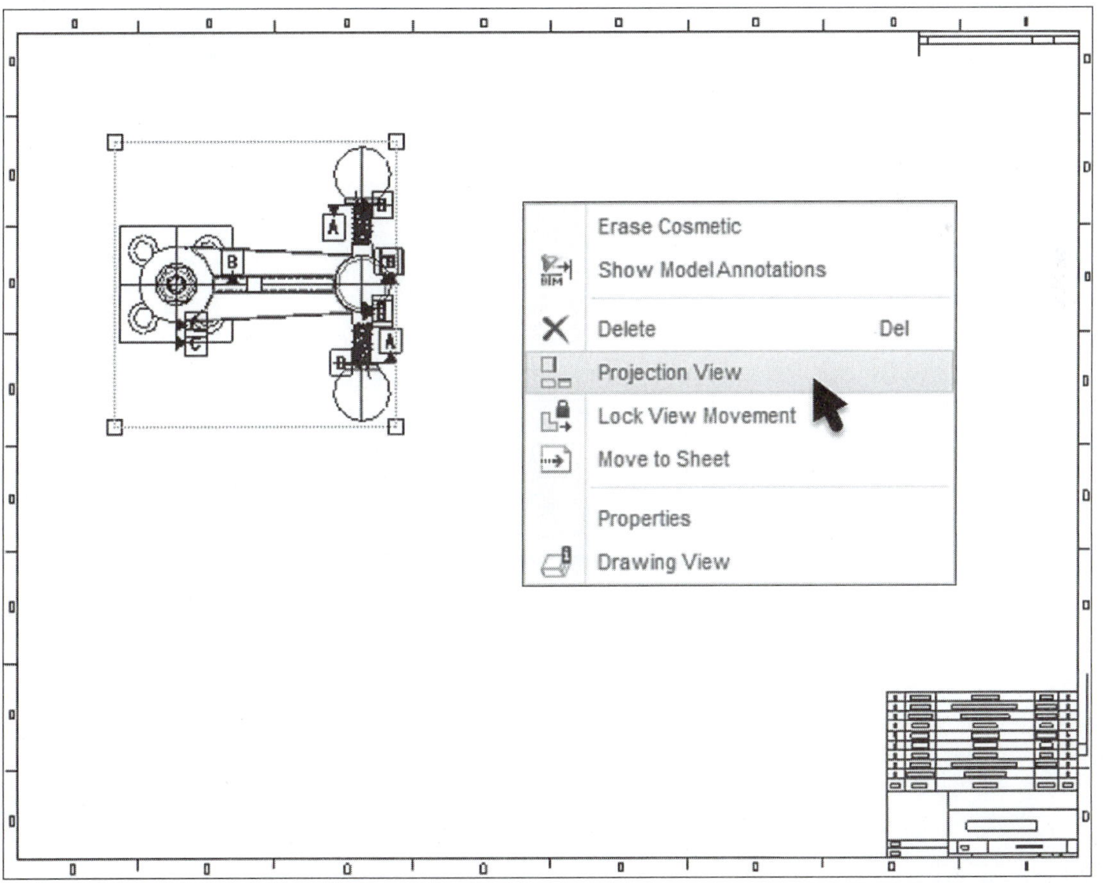

Figure 12.32(a) Press RMB > Insert Projection View

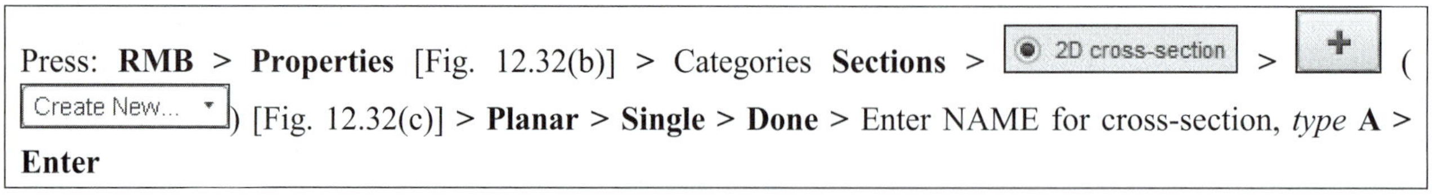

Press: **RMB** > **Properties** [Fig. 12.32(b)] > Categories **Sections** > 2D cross-section > + (Create New...) [Fig. 12.32(c)] > **Planar** > **Single** > **Done** > Enter NAME for cross-section, *type* **A** > **Enter**

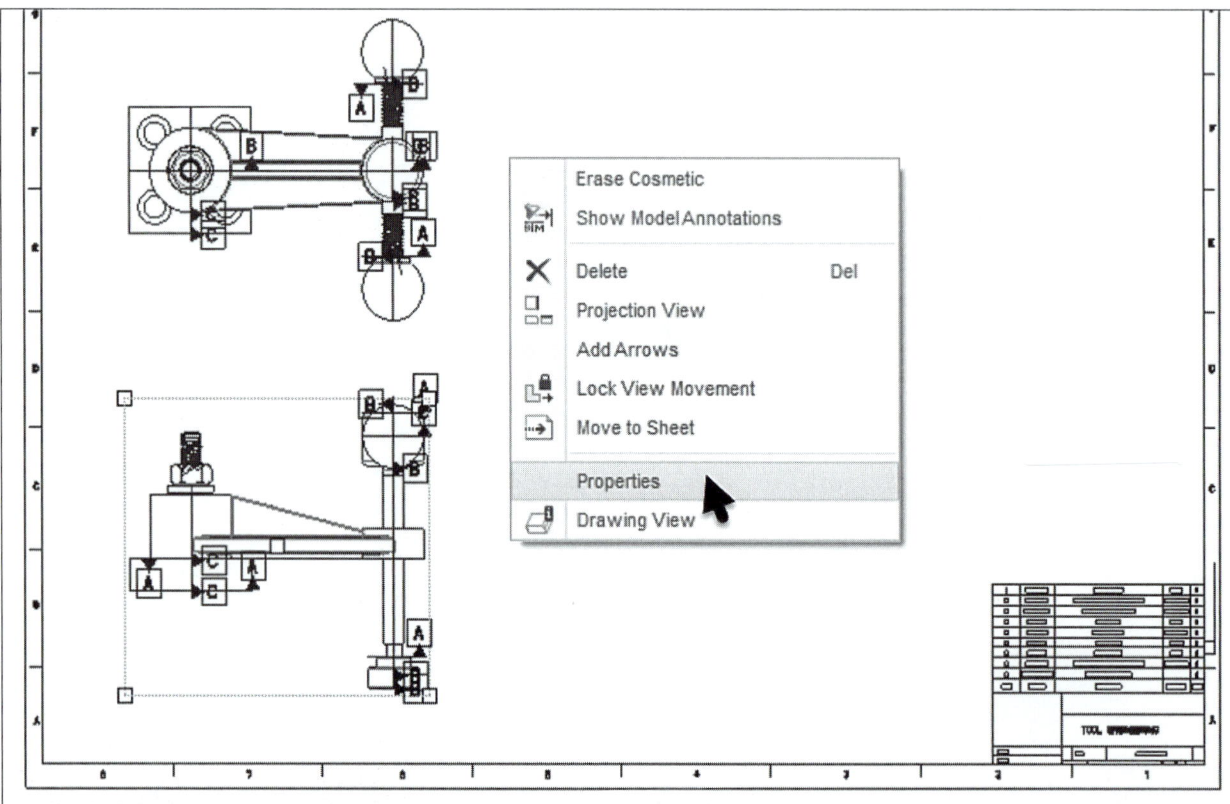

Figure 12.32(b) View Properties

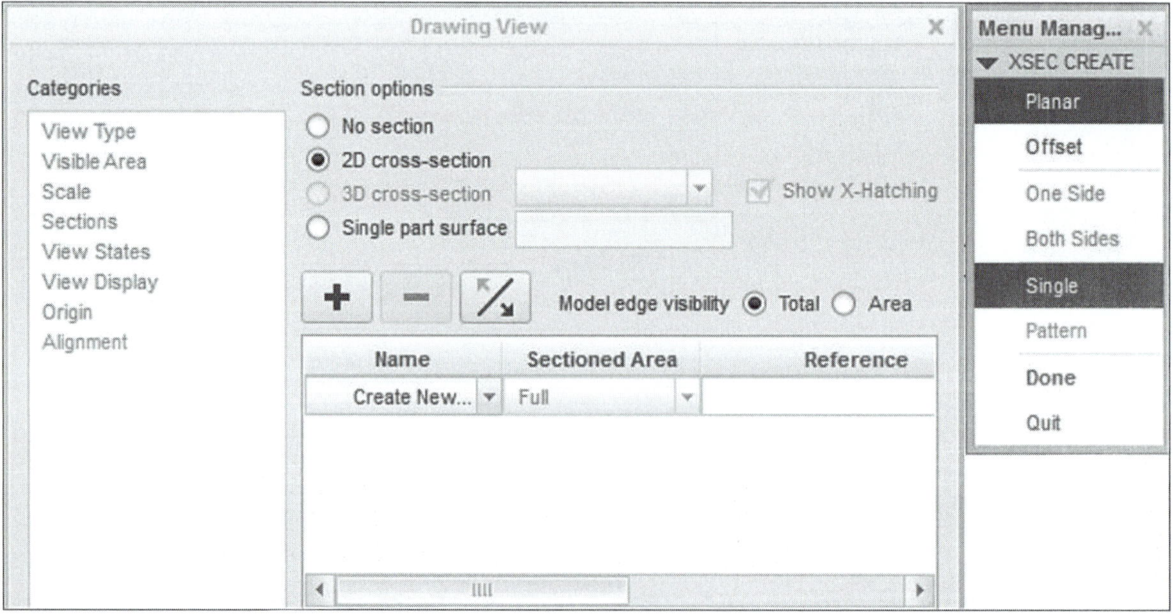

Figure 12.32(c) Add Cross-Section to View

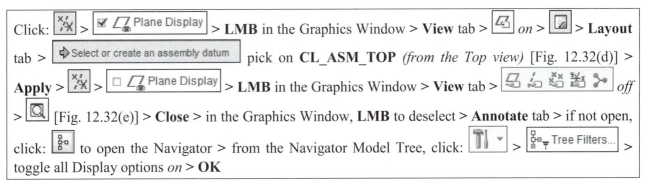

Click: [icon] > [☑ Plane Display] > **LMB** in the Graphics Window > **View** tab > [icon] *on* > [icon] > **Layout** tab > [Select or create an assembly datum] pick on **CL_ASM_TOP** *(from the Top view)* [Fig. 12.32(d)] > **Apply** > [icon] > [☐ Plane Display] > **LMB** in the Graphics Window > **View** tab > [icons] *off* > [icon] [Fig. 12.32(e)] > **Close** > in the Graphics Window, **LMB** to deselect > **Annotate** tab > if not open, click: [icon] to open the Navigator > from the Navigator Model Tree, click: [icon] > [Tree Filters...] > toggle all Display options *on* > **OK**

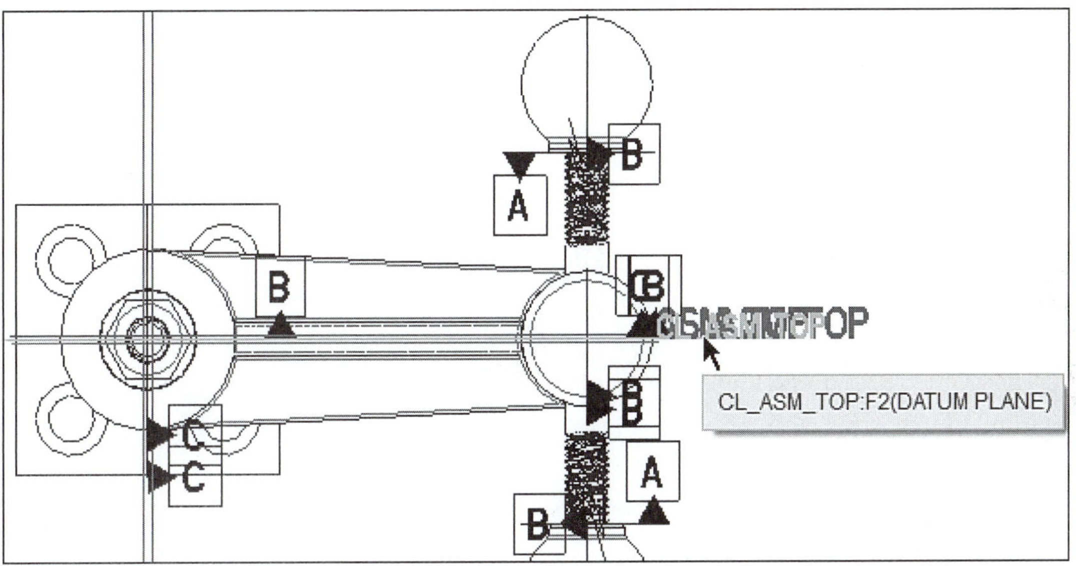

Figure 12.32(d) Select the CL_ASM_TOP Datum Plane

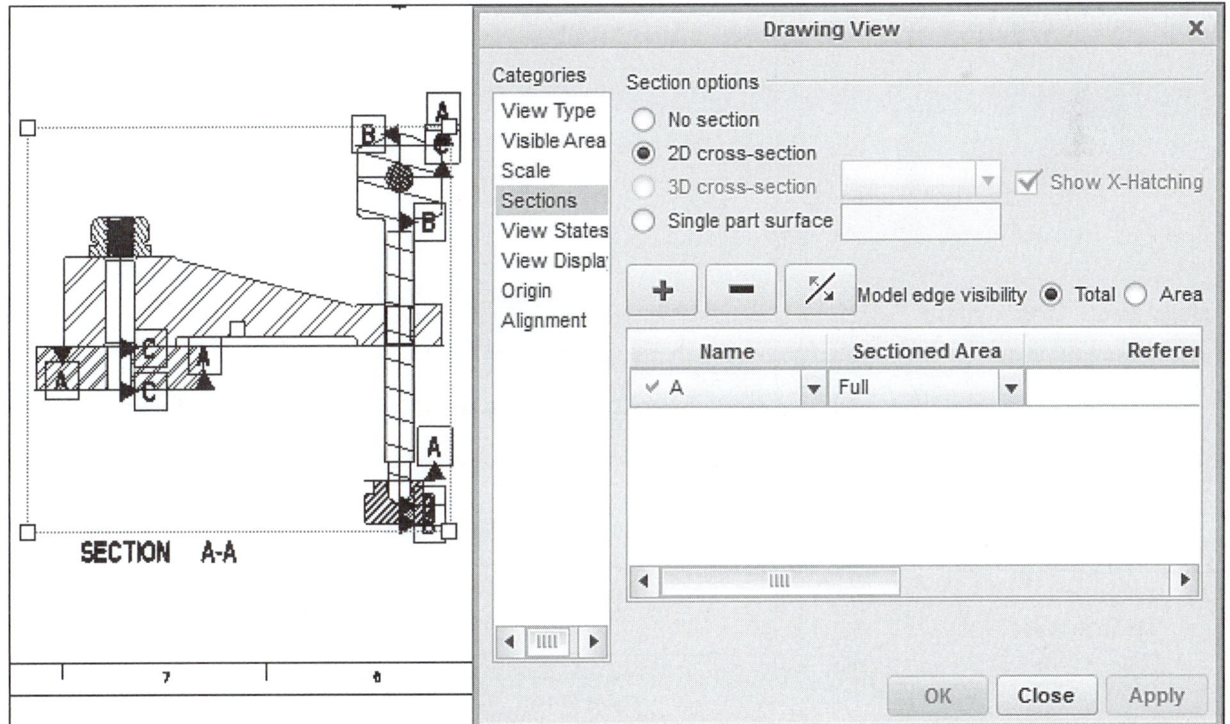

Figure 12.32(e) Categories: Sections

Click: **Annotate** tab > from the Drawing Tree, click: Datums (to expand) > with the **Ctrl** key (or **Shift** key) pressed, select all of the Model datums > release the **Ctrl** (or **Shift**) key > **RMB** > **Erase** > repeat for lower view [Fig. 12.32(f)] > in the Graphics Window, **LMB** to deselect

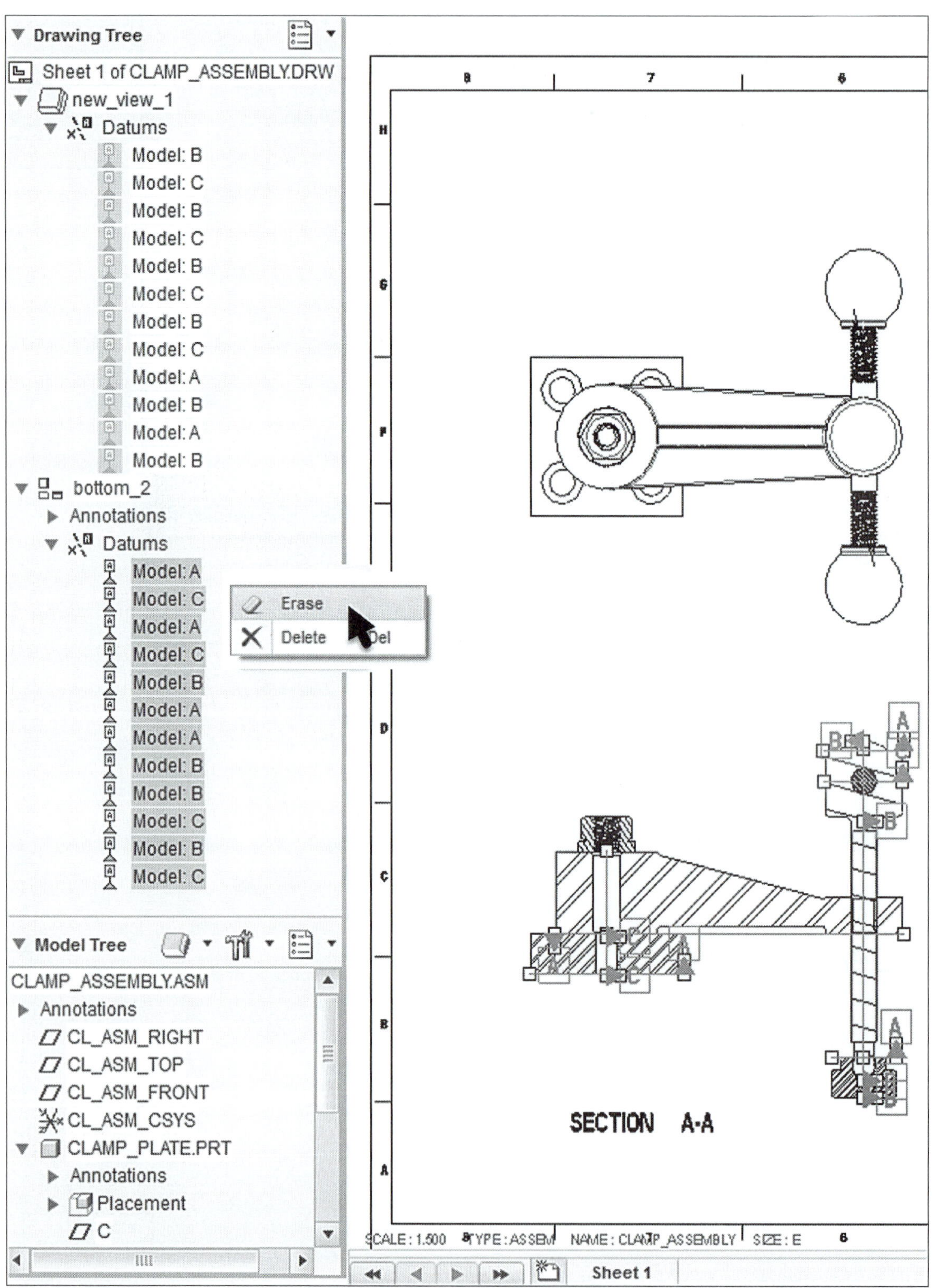

Figure 12.32(f) Drawing Tree Model Datums > RMB > Erase

Pick on the Front (lower) view > press **RMB > Add Arrows** [Fig. 12.32(g)] > Pick a view for arrows where the section is perp. MIDDLE button for none. pick on the Top (upper) view > move the cursor outside of the view outline > **LMB** [Fig. 12.32(h)] > **View** tab > **Ctrl+S**

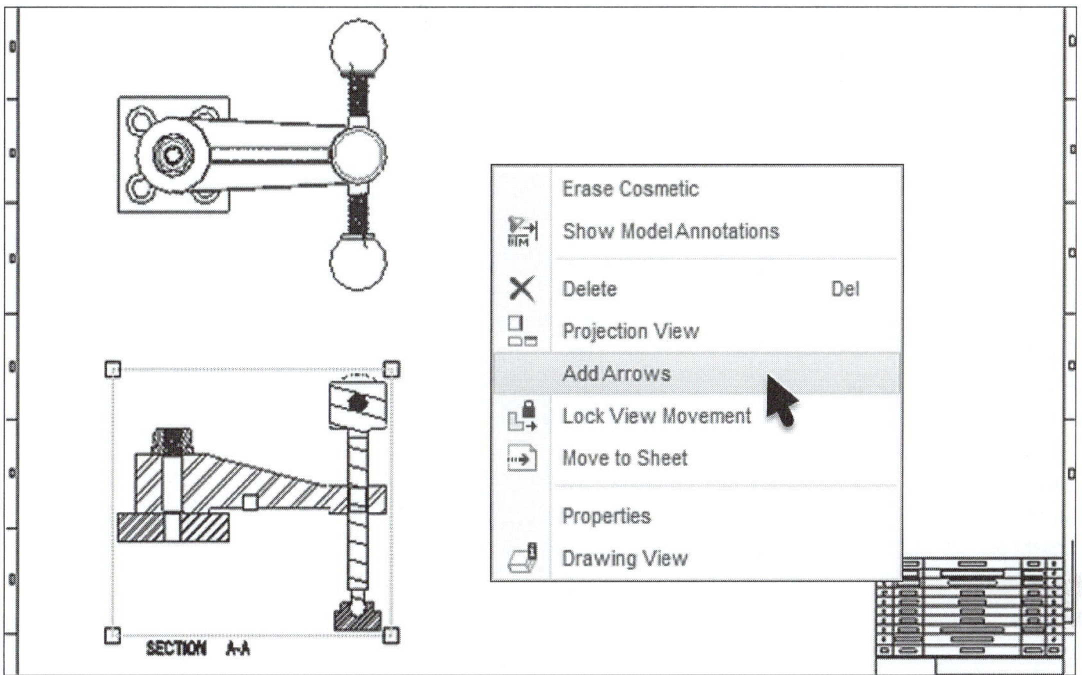

Figure 12.32(g) Press RMB > Add Arrows

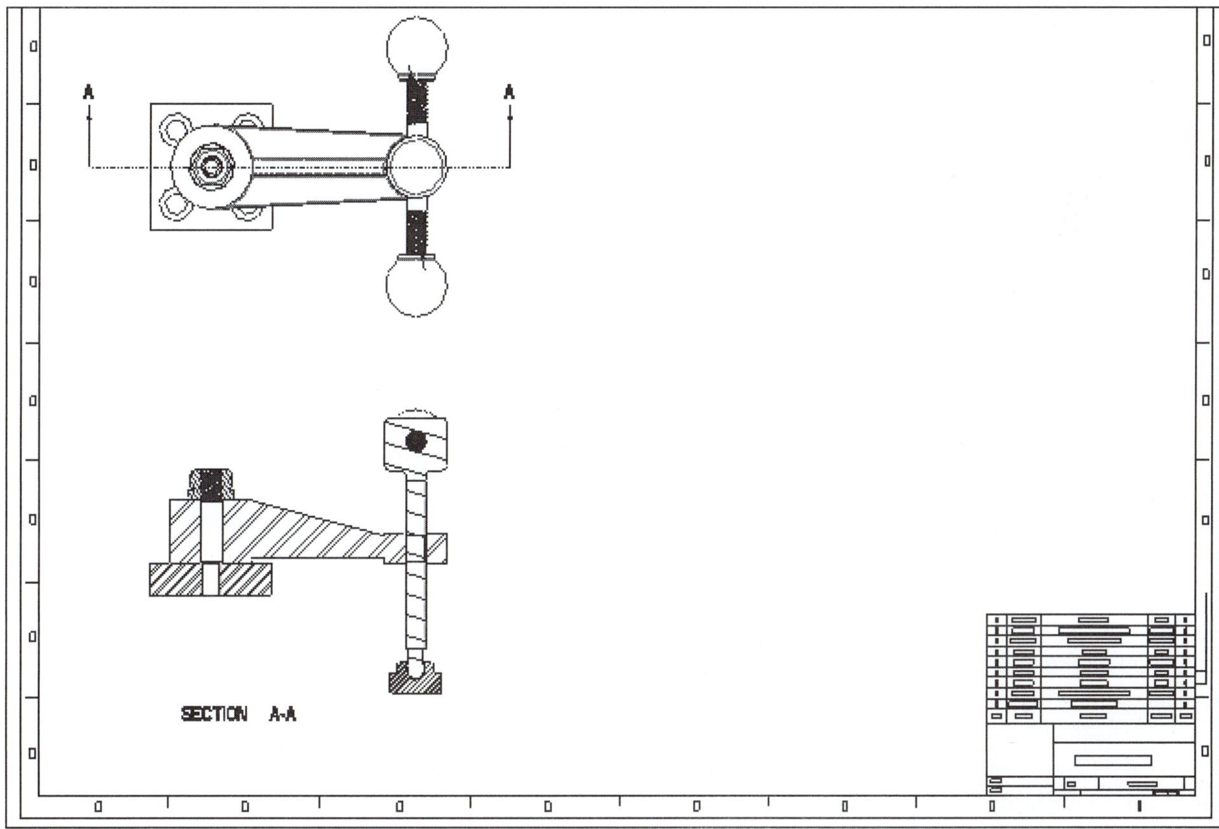

Figure 12.32(h) Cutting Plane Arrows Displayed

With the **Ctrl** key pressed; in the Model Tree, select the three components [Fig. 12.32(i)] > release the **Ctrl** key > press **RMB** > **Show Model Annotations** > [icon] > [icon] > **OK** > [icon] > **LMB** in the Graphics Window to deselect > **Ctrl+S**

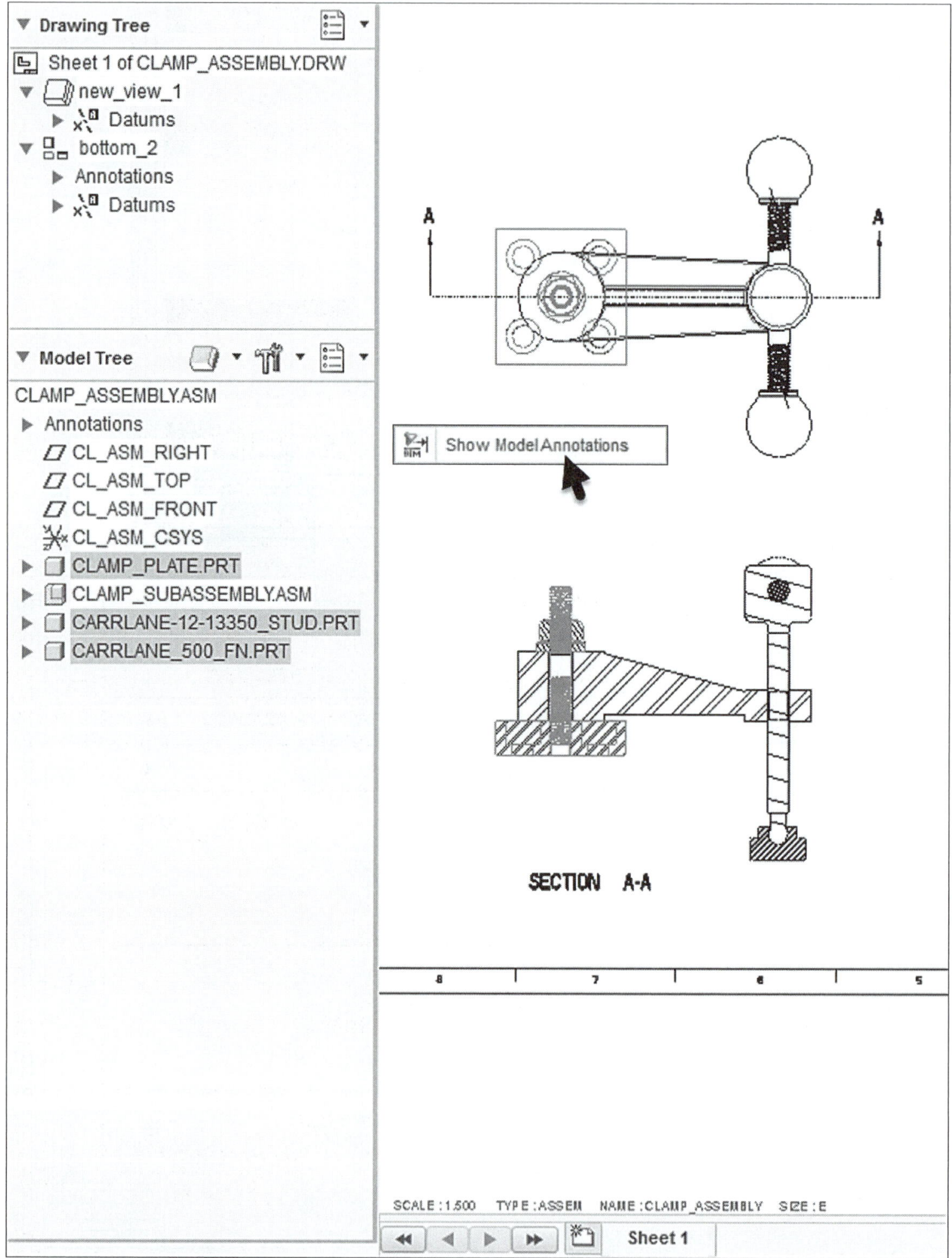

Figure 12.32(i) Assembly Drawing Views with Three Components Selected

Zoom in on the BOM, click: **Table** tab > **Repeat Region** > **Attributes** > pick in the BOM RepeatRegion field > **Flat** [Fig. 12.33(a)] > **Done/Return** > **Done** > select and move the BOM as needed [Fig. 12.33(b)]

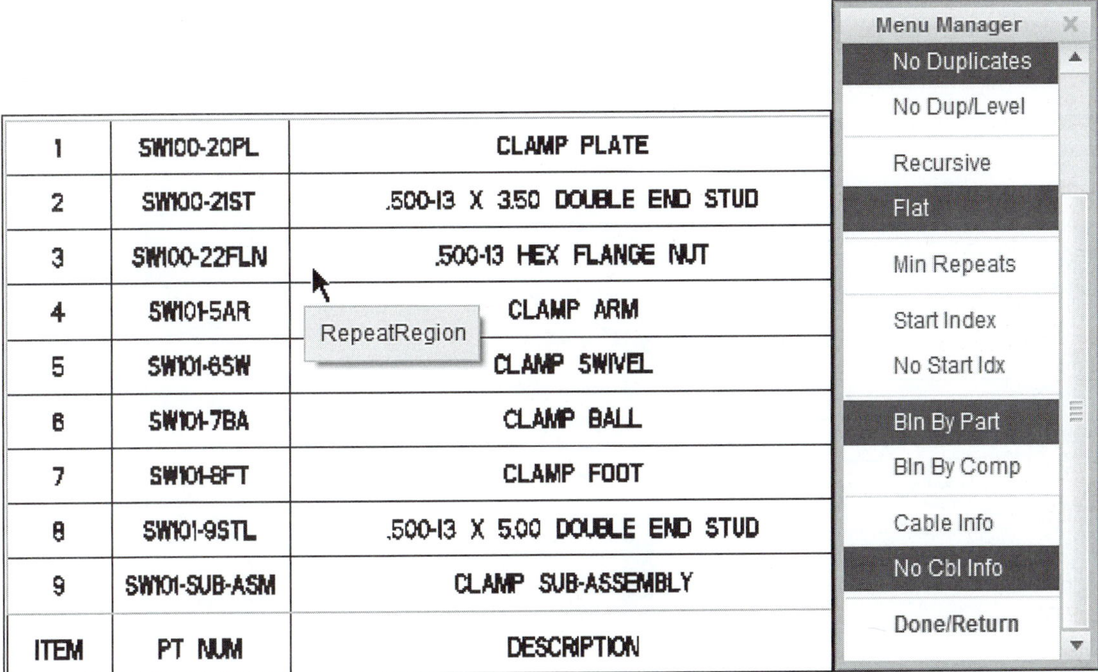

1	SW100-20PL	CLAMP PLATE
2	SW100-21ST	.500-13 X 3.50 DOUBLE END STUD
3	SW100-22FLN	.500-13 HEX FLANGE NUT
4	SW101-5AR	CLAMP ARM
5	SW101-6SW	CLAMP SWIVEL
6	SW101-7BA	CLAMP BALL
7	SW101-8FT	CLAMP FOOT
8	SW101-9STL	.500-13 X 5.00 DOUBLE END STUD
9	SW101-SUB-ASM	CLAMP SUB-ASSEMBLY
ITEM	PT NUM	DESCRIPTION

Figure 12.33(a) Table Showing with BOM Attribute Recursive, Select Flat from the Menu Manager

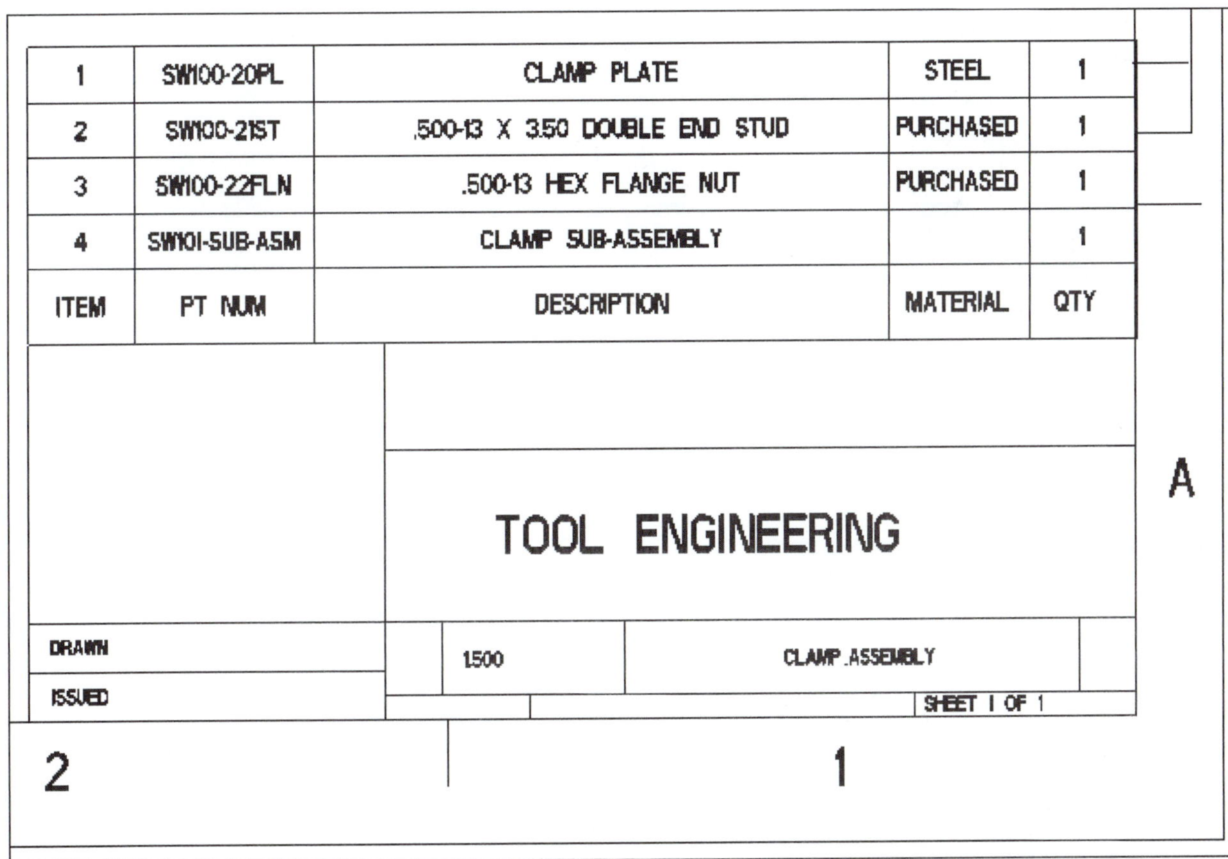

1	SW100-20PL	CLAMP PLATE	STEEL	1
2	SW100-21ST	.500-13 X 3.50 DOUBLE END STUD	PURCHASED	1
3	SW100-22FLN	.500-13 HEX FLANGE NUT	PURCHASED	1
4	SW101-SUB-ASM	CLAMP SUB-ASSEMBLY		1
ITEM	PT NUM	DESCRIPTION	MATERIAL	QTY

Figure 12.33(b) Repositioned BOM Table with Attribute Flat

Click: **LMB** to deselect > [icon] > [icon] **Create Balloons > Create Balloons - All** [Fig. 12.33(c)]

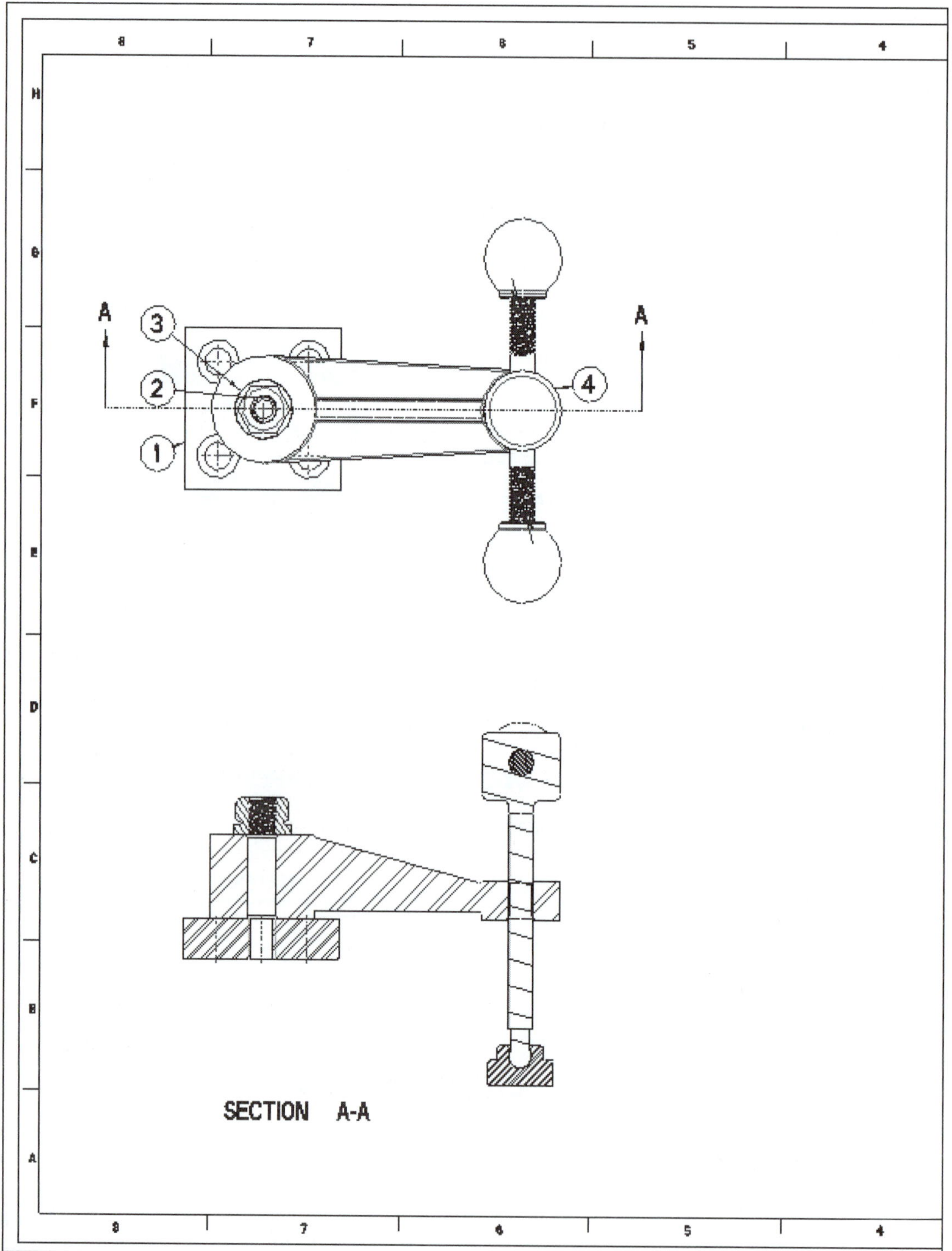

Figure 12.33(c) Balloons Displayed in Top View

While pressing the **Ctrl** key, pick on balloons 1, 2, and 3 > release the **Ctrl** key > press **RMB** > **Move Item to View** > pick on the Front view [Fig. 12.33(d)] > pick on balloon **2** > press **RMB** > **Edit Attachment** > **On Entity** > pick on an edge [Fig. 12.33(e)] > **MMB** > reposition the balloon > pick on and reposition (reattach) each balloon as needed > **LMB** to deselect

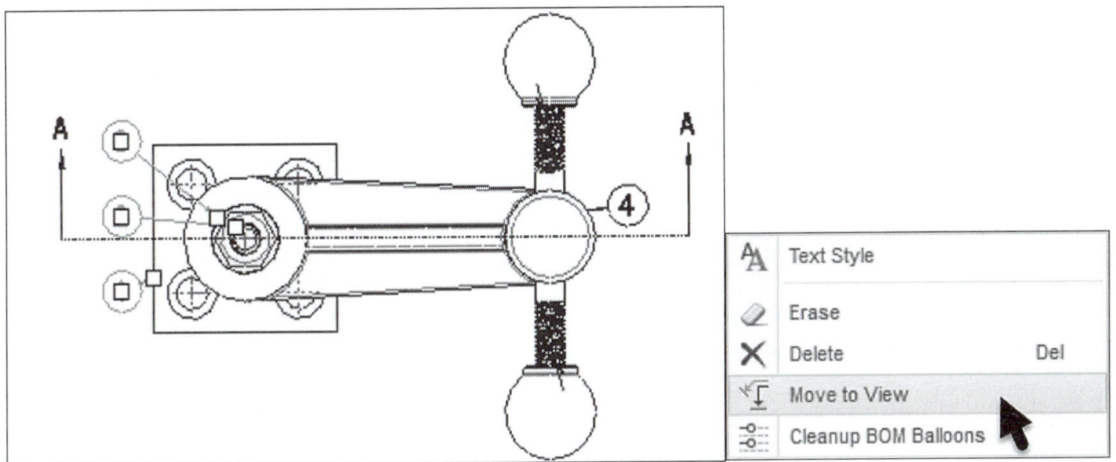

Figure 12.33(d) Press RMB > Move Item to View

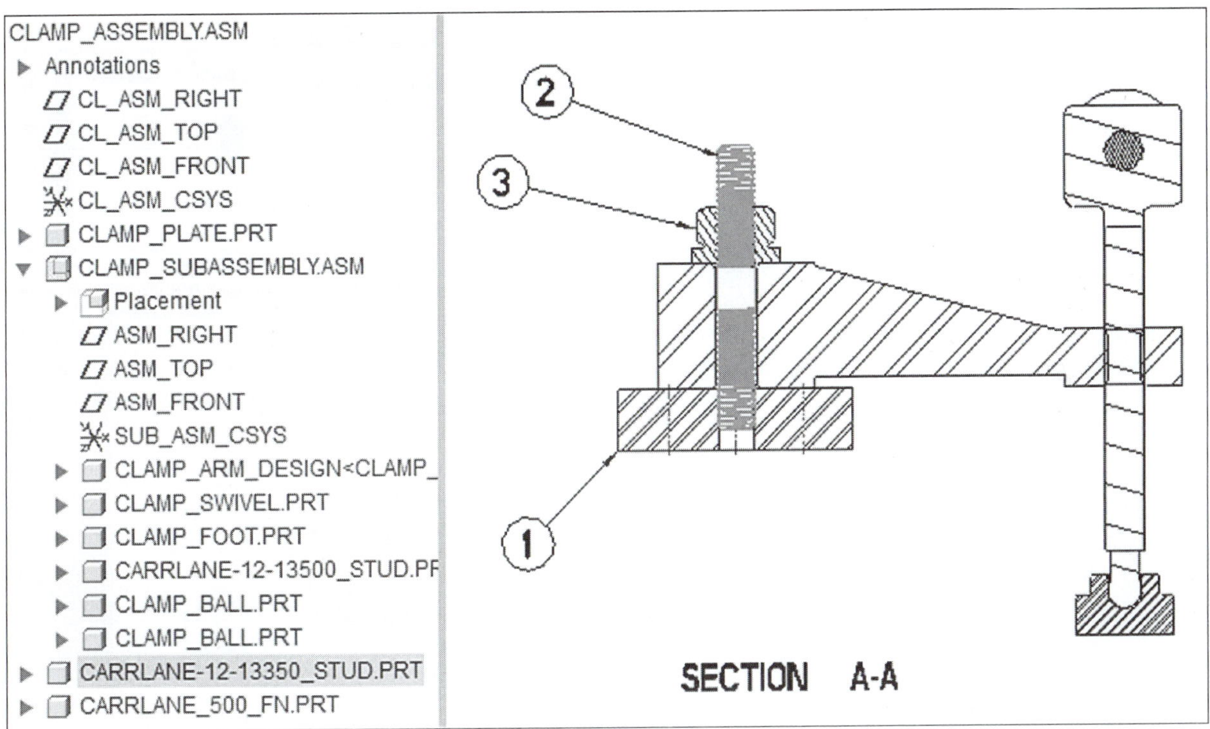

Figure 12.33(e) Repositioned Balloons

Most companies (and as per drafting standards) require that rounded purchased items, such as nuts, bolts, studs, springs, and die pins be excluded from sectioning even when the section cutting plane passes through them. *If your system does not display the short stud, highlight in the Model Tree. If it does display, then exclude the section lining from it on the drawing.*

Click: **Layout** tab > double-click on the crosshatching in the Front view [Fig. 12.33(f)] (the CARRLANE_500_FN is now active) > **Exclude** to not show the Section of CARRLANE_500_FN > **Next** (CARRLANE-12-13350_STUD is now active) > **Exclude** > **Next** (CARRLANE-12-13500_STUD is now active) > **Fill** the cross section of the long stud > **Next** (Clamp_Foot is now active) > **Spacing** > **Double** > **Delete Line** > **Next** (Clamp_Swivel is now active) > **Angle** > **135** > **Spacing** > **Half** > **Next** (Clamp_Arm is now active) > **Delete Line** > **Spacing** > **Half** > **Next** (Clamp_Plate is now active) > **Delete Line** > **Angle** > **120** > **Done** > in the Graphics Window, **LMB** to deselect > > **Ctrl+S**

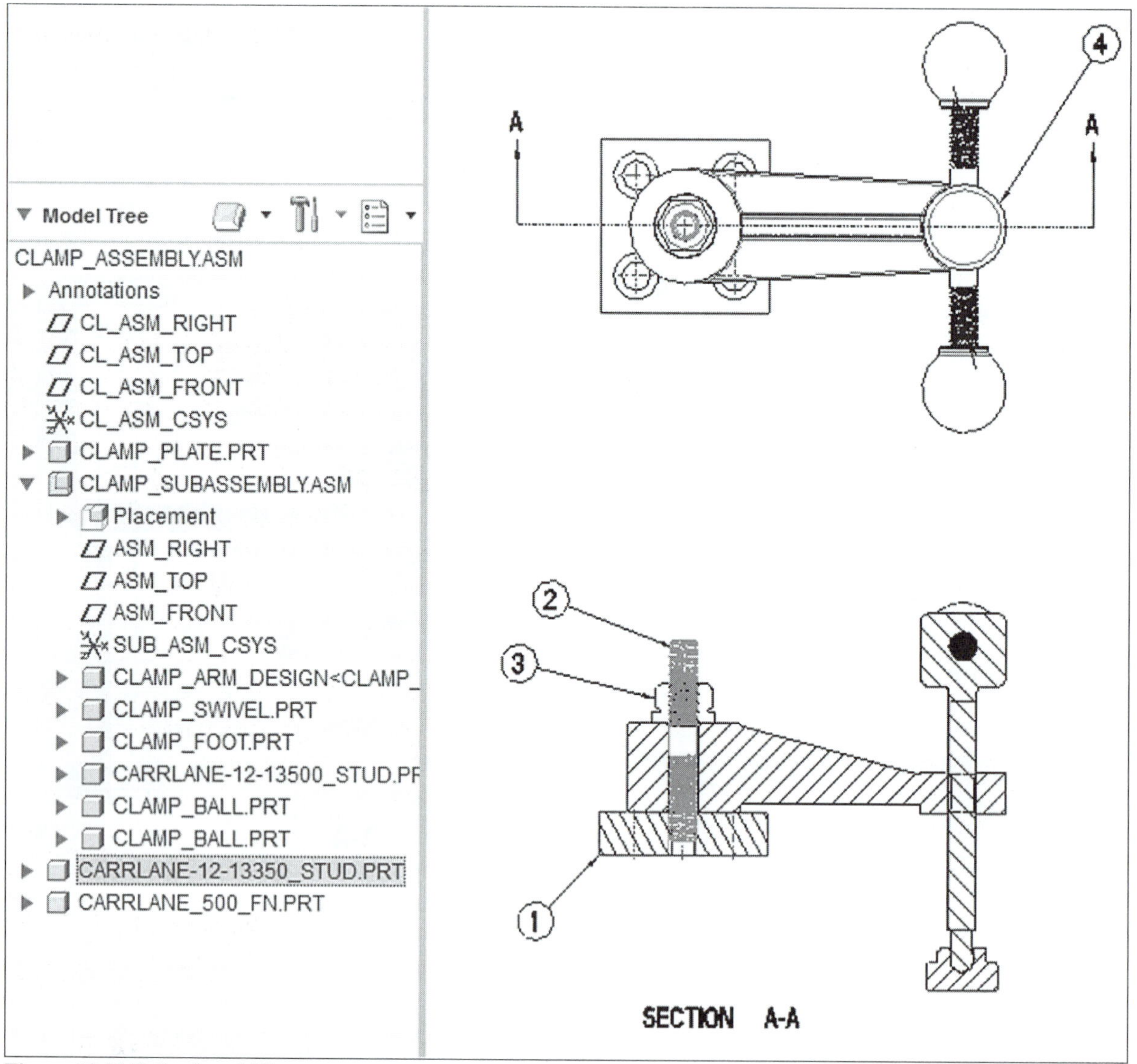

Figure 12.33(f) Sectioned View

Exploded Swing Clamp Assembly Drawings

The process required to place an exploded view on a drawing is similar to adding assembly orthographic views. The BOM will display all components on this sheet.

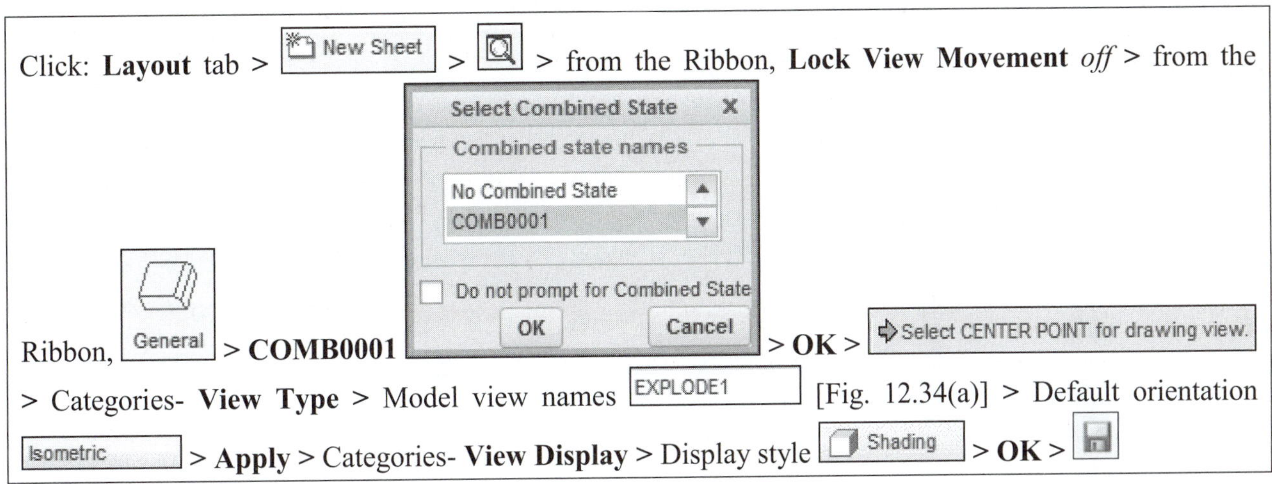

Click: **Layout** tab > New Sheet > > from the Ribbon, **Lock View Movement** *off* > from the Ribbon, General > **COMB0001** > **OK** > Select CENTER POINT for drawing view. > Categories- **View Type** > Model view names EXPLODE1 [Fig. 12.34(a)] > Default orientation Isometric > **Apply** > Categories- **View Display** > Display style Shading > **OK** >

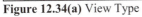

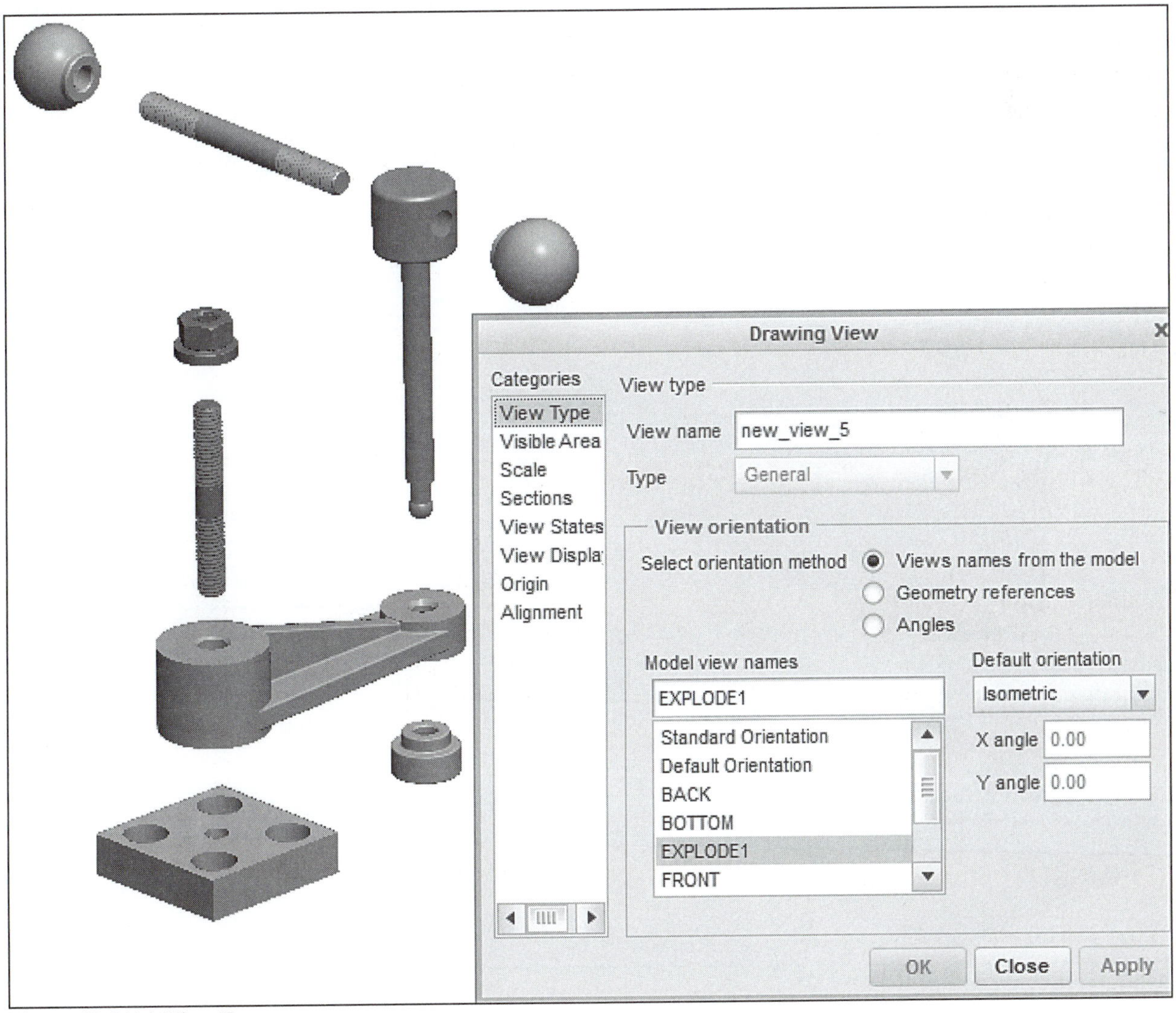

Figure 12.34(a) View Type

With the view still *selected/highlighted*, press: **RMB > Show Model Annotations >** [icon] **> All >** [icon] [Fig. 12.34(b)] > **Apply > OK > LMB > Annotate** tab > erase extra centerlines and clip each centerline (axis) to extend between components that are in line [Fig. 12.34(c)] > [icon] > [icon] **Save**

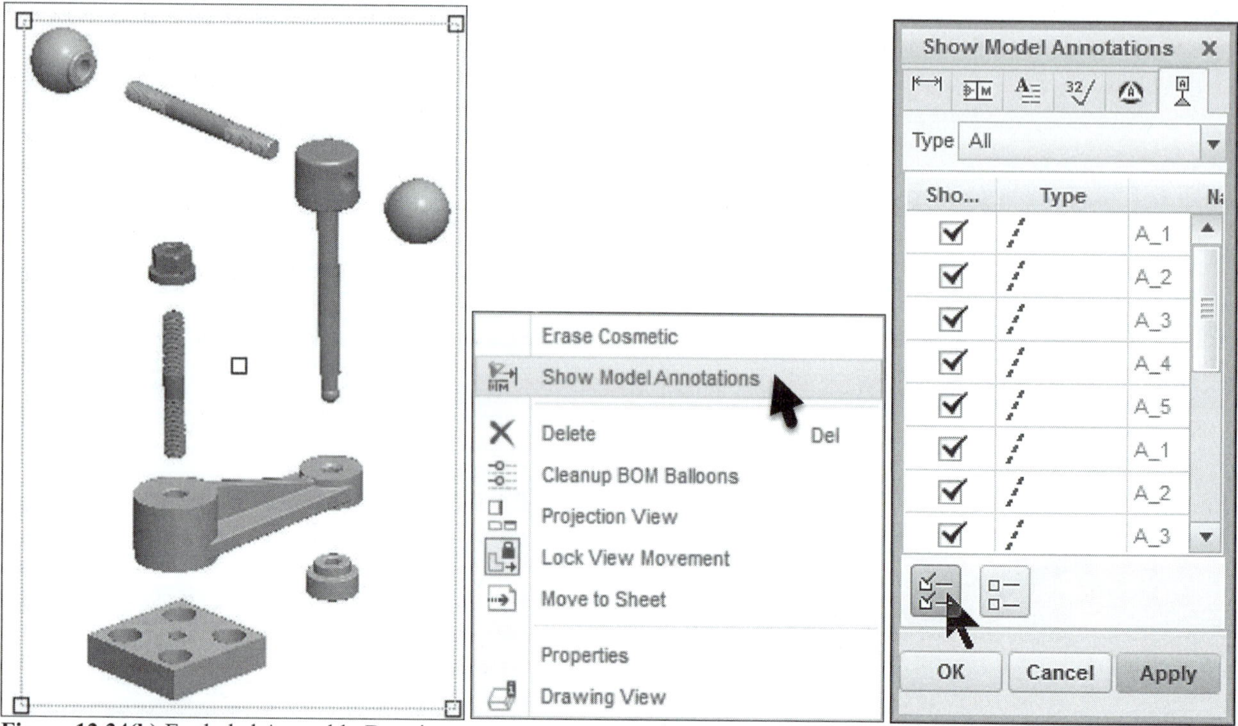

Figure 12.34(b) Exploded Assembly Drawing

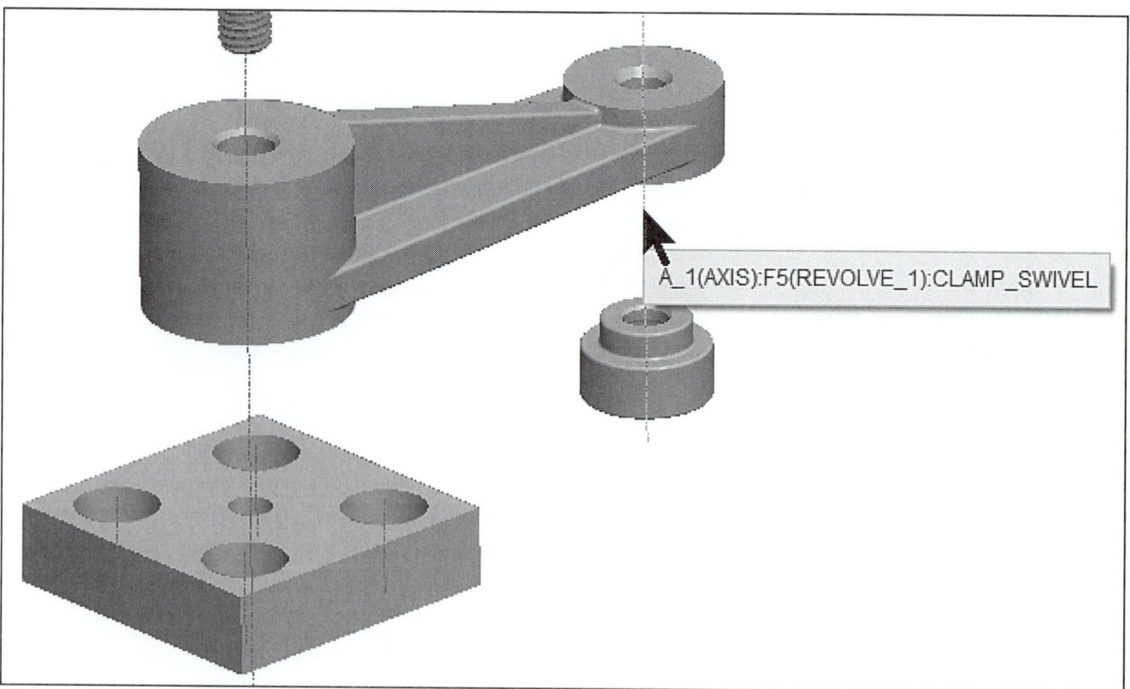

Figure 12.34(c) Clip the Centerlines Axes

Click: **Table** tab > -⑤ **Create Balloons > Create Balloons - All >** Select a region. pick in the BOM field > reposition the balloons and their attachment points as needed to clean up the drawing [Fig. 12.35(a)] > pick on pick balloon **6** > press **RMB > Edit Attachment > On Surface > Dot** > pick on the Clamp_Ball's surface > **Done/Return > LMB** > zoom in on the BOM, move table above the Title Block, and adjust the columns as necessary > > Sheet 1 > > **File > Save As** > Type > **Zip File (*.zip) > OK** > upload to your course interface or attach to an email and send to your instructor

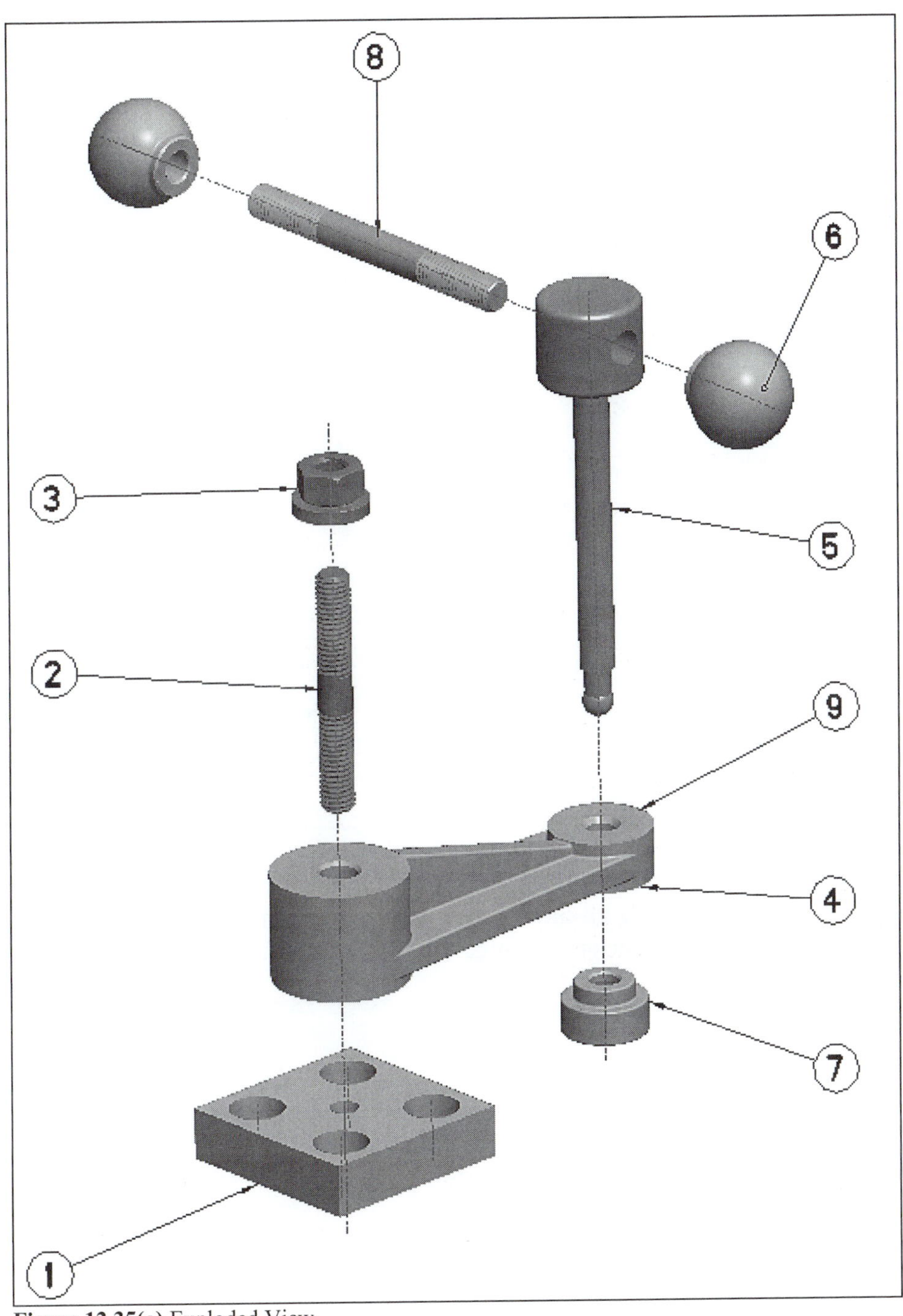

Figure 12.35(a) Exploded View

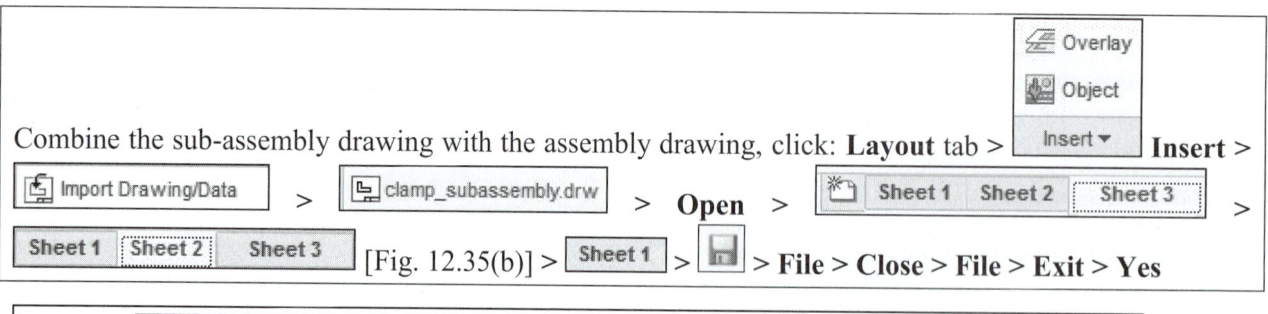

Combine the sub-assembly drawing with the assembly drawing, click: **Layout** tab > Overlay Object Insert **Insert** > Import Drawing/Data > clamp_subassembly.drw > **Open** > Sheet 1 Sheet 2 Sheet 3 > Sheet 1 Sheet 2 Sheet 3 [Fig. 12.35(b)] > Sheet 1 > > **File** > **Close** > **File** > **Exit** > **Yes**

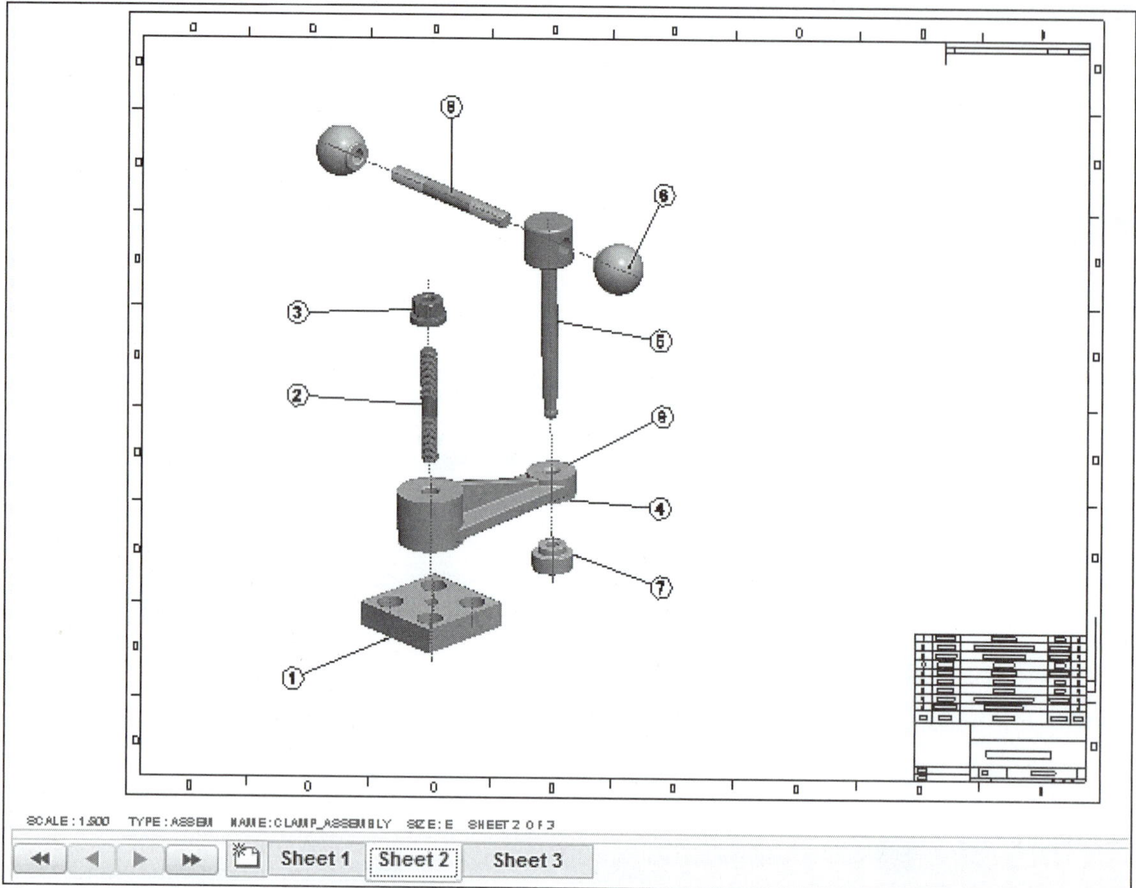

Figure 12.35(b) Completed Drawing, Sheet 2

Create a *documentation package* containing all models and drawings required to manufacture the parts and assemble the components. Your instructor may change the requirements, but in general, create and plot/print the following:

- **Part Models** for all components
- **Detail Drawings** for each nonstandard component
- **Assembly Drawings** using standard orthographic ballooned views
- **Exploded Subassembly Drawing** of the ballooned subassembly
- **Exploded Assembly Drawing** of the ballooned assembly

A different assembly Project (Coupling) is available at ***www.cad-resources.com***

In order to save space and reduce the purchase price of this book; Lessons 13 through 22 are available, at no additional cost, for download from the publisher to those purchasing this book.

Index

E

F

G